Dr. Franze,

Thank you for all you have done for me and my family. This book would have never been possible without your belief and confidence in me. I will never be able to repay you for all you have done for me.

Thank you for being a huge influence in my life and professional career. Enjoy the book!

Sincerely,

John Rayfield

Principles of Agriculture, Food, and Natural Resources

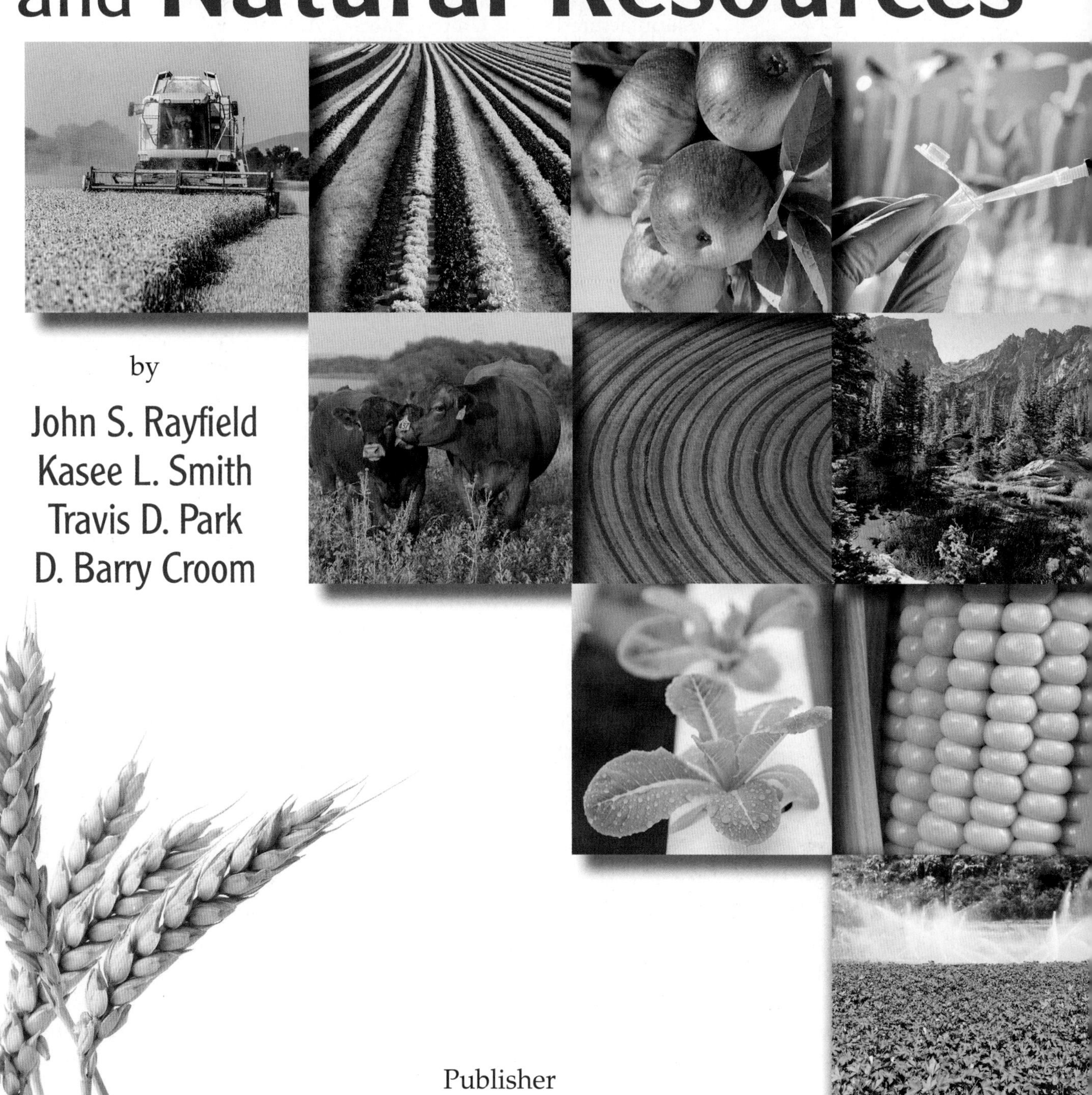

by

John S. Rayfield
Kasee L. Smith
Travis D. Park
D. Barry Croom

Publisher
The Goodheart-Willcox Company, Inc.
Tinley Park, IL
www.g-w.com

Manufactured in the United States of America.

ISBN 978-1-63126-235-7

1 2 3 4 5 6 7 8 9 – 17 – 19 18 17 16 15

The Goodheart-Willcox Company, Inc. Brand Disclaimer: Brand names, company names, and illustrations for products and services included in this text are provided for educational purposes only and do not represent or imply endorsement or recommendation by the author or the publisher.

The Goodheart-Willcox Company, Inc. Safety Notice: The reader is expressly advised to carefully read, understand, and apply all safety precautions and warnings described in this book or that might also be indicated in undertaking the activities and exercises described herein to minimize risk of personal injury or injury to others. Common sense and good judgment should also be exercised and applied to help avoid all potential hazards. The reader should always refer to the appropriate manufacturer's technical information, directions, and recommendations; then proceed with care to follow specific equipment operating instructions. The reader should understand these notices and cautions are not exhaustive.

The publisher makes no warranty or representation whatsoever, either expressed or implied, including but not limited to equipment, procedures, and applications described or referred to herein, their quality, performance, merchantability, or fitness for a particular purpose. The publisher assumes no responsibility for any changes, errors, or omissions in this book. The publisher specifically disclaims any liability whatsoever, including any direct, indirect, incidental, consequential, special, or exemplary damages resulting, in whole or in part, from the reader's use or reliance upon the information, instructions, procedures, warnings, cautions, applications, or other matter contained in this book. The publisher assumes no responsibility for the activities of the reader.

The Goodheart-Willcox Company, Inc. Internet Disclaimer: The Internet resources and listings in this Goodheart-Willcox Publisher product are provided solely as a convenience to you. These resources and listings were reviewed at the time of publication to provide you with accurate, safe, and appropriate information. Goodheart-Willcox Publisher has no control over the referenced websites and, due to the dynamic nature of the Internet, is not responsible or liable for the content, products, or performance of links to other websites or resources. Goodheart-Willcox Publisher makes no representation, either expressed or implied, regarding the content of these websites, and such references do not constitute an endorsement or recommendation of the information or content presented. It is your responsibility to take all protective measures to guard against inappropriate content, viruses, or other destructive elements.

Cover images (left to right): Stefanie Mohr Photography/Shutterstock.com; Lisa S./Shutterstock.com; Smileus/Shutterstock.com; fotohunter/Shutterstock.com; American Simmental Association, www.simmental.org; B Brown/Shutterstock.com; bjul/Shutterstock.com; hin255/Shutterstock.com; Maks Narodenko/Shutterstock.com; petefrone/Shutterstock.com; Tsekhmister/Shutterstock.com
Back cover image: Alta Oosthuizen/Shutterstock.com

Preface

Principles of Agriculture, Food, and Natural Resources was conceptualized as a modern, user-friendly, highly visual introductory text for next generation agriculture students. Today's students of agriculture need to know the basic information about food and fiber, as well as where and how that information may be used to address the grand challenges of feeding the planet, mitigating agriculture's impact on the environment, and ensuring a high quality of life for all. With a text written for the agriculture of today, complete with hundreds of high-quality images; pedagogical tools designed for learner success such as lesson outcomes, key terms, STEM connections, career connections, reviews and assessments; and extensive student and instructor supplements delivered online or in print, our integrated learning solution helps learners engage with all aspects of the field.

Why is a new introductory textbook about agriculture needed? Clearly, the world of agriculture continues to change. Sometimes, a fresh approach to organizing and presenting information is warranted. This is one of those times. Fewer and fewer of our students live on farms. More students have fewer direct experiences with agriculture. And, the agriculture that they do know oftentimes starts with *food* and not the *land*. This textbook is designed for the frame of reference for today's students. We start by presenting information where students are, and then take them into the world of opportunities in agriculture.

Secondly, our presentation covers the most relevant topics with sufficient detail for the novice agricultural education student. Our goals were twofold: (1) to allow for a solid understanding of agriculture and (2) to entice the student to study agricultural topics in greater detail with subsequent courses.

Finally, we have drawn upon modern research to create a learning design that helps students process and acquire information. We know that a logical progression of information is important for learning. Educators, science, and instructional designers also tell us that providing information in smaller chunks aids our working memory when processing that information. In addition, providing students with opportunities to process bits of information and interact with them—in reviews and assessments that foster comprehension, application, analysis, synthesis, evaluation and creation—is critical for learning success. These learning principles helped guide the design of *Principles of Agriculture, Food, and Natural Resources*.

Our text is organized and written with the National Agriculture, Food, and Natural Resources Content Standards at the forefront, ensuring necessary content is covered. Our organization was given much thought by the writing team. A panel of expert reviewers in Agricultural Education provided insight, information, debate, and ultimately validation of our content, organization, and approach. Each chapter is comprised of focused lessons: smaller, logically grouped information supported by features designed to help students acquire knowledge and demonstrate mastery. In addition, each lesson in this textbook starts with an academic literacy integration activity designed to help students engage more deeply with the agricultural content of the text. Our aim was to present information that could be easily handled in a one- or two-week time period over the course of the academic year.

Certainly, expert teachers can add additional information for content that is more expansive or relevant to a certain area. Novice teachers will find our presentation of agricultural content easy to navigate and helpful for outlining the academic year. Further, each lesson contains many ideas for critical thinking, STEM engagement, careers, FFA activities, and communicating about agriculture. These areas challenge students to engage with the content at a deeper level than most traditional texts on agriculture.

Because society is in a transition from print to digital, we know it is important to provide the content you need in the format you need it in. The student textbook and workbook is available in print or online format and is supported with professional, user-friendly online resources such as a student Online Learning Suite. These include electronic study tools, videos, and additional interactive images and activities. Quite simply, teachers have a full-service resource at their disposal to engage students in the classroom, online, and in their minds. Instructor support is provided in the form of teaching resources, class presentation, and an assessment suite in a variety of digital and online formats. Video clip libraries also enhance the text and numerous supplements by bringing concepts to life.

We hope you enjoy our presentation of the most important aspects of agriculture and welcome you to join us on our journey of enlightening young minds to the opportunities and challenges of feeding a growing planet.

About the Authors

Dr. John Rayfield is an Associate Professor in the Agricultural Leadership, Education, and Communications Department at Texas A&M University. Dr. Rayfield earned his bachelor's degree in agricultural education from Auburn University, his master's degree in agricultural extension from the University of Georgia, and his doctorate of education in agricultural education from Texas Tech University. Dr. Rayfield worked as a county extension agent for three years and as an agricultural education teacher for eight years in southwest Georgia. In 2001, his school's FFA chapter was recognized as the Outstanding Middle School Program in the nation by National FFA. After completing his doctorate in 2006, Dr. Rayfield worked at North Carolina State University as an Assistant Professor in agricultural education until 2009 when he joined the faculty at Texas A&M. Dr. Rayfield is actively involved in livestock showing and judging. He serves as superintendent for the National FFA Livestock Career Development Event and judges junior and open livestock shows across the United States.

Kasee Smith is a Graduate Teaching Assistant and PhD Candidate at Texas A&M University, where she instructs courses in agricultural education and assists in the coordination of student teaching for the Department of Agricultural Leadership, Education, and Communications. Ms. Smith earned her bachelor's and master's degrees in agricultural education from Utah State University, and taught high school agriculture in Spanish Fork, Utah, for eleven years. During her time as a secondary agricultural educator, her agriculture program was recognized as a National FFA Model of Innovation Chapter, and she had the opportunity to serve on numerous state-level curriculum and assessment development projects. Ms. Smith is anticipating using her passion for agricultural education to instruct courses on a postsecondary level after the completion of her degree in May 2016.

Dr. Travis Park is an Associate Professor at North Carolina State University, where he instructs the teaching methods course and coordinates student teaching. He also co-leads *Developing Educational Leaders and Teachers of Agriculture (DELTA)*, the alternative licensure professional development program for agriculture teachers in North Carolina. Dr. Park earned his bachelor's and master's degrees from Purdue University in agricultural education. Upon graduation, he was named the 1996 G. A. Ross Award winner, which recognizes the outstanding male student in the graduating class. After graduating from Purdue, Dr. Park taught agriculture at Tri-County High School in Wolcott, Indiana, for 5½ years. His agriculture program was recognized as one of the top six agriculture programs in the nation. Dr. Park earned his doctoral degree at the University of Florida in 2005, and then was hired as an assistant and associate professor in the College of Agriculture and Life Sciences, Cornell University. Dr. Park directed the teacher education program at Cornell and coordinated the Agricultural Outreach and Education program which provides professional development for agriculture teachers, FFA programming, and agricultural literacy through New York Agriculture in the Classroom.

Dr. Barry Croom is a professor and department head for Agricultural Education and Agricultural Sciences at Oregon State University. He is also an Alumni Distinguished Undergraduate Professor Emeritus in the Department of Agricultural and Extension Education at North Carolina State University. Prior to becoming a teacher educator, he was a high school agriculture teacher in Eastern North Carolina for nine years and later served as the State FFA Coordinator for the North Carolina FFA Association. He taught courses in teacher education at North Carolina State University for 15 years before moving to Oregon State University. Dr. Croom teaches courses in the Agricultural Science major at Oregon State University.

Dr. Croom is a former NCSU Park Faculty Scholar, a Fellow of North American Colleges and Teachers of Agriculture, and a Fellow of the American Association for Agricultural Education. He earned both the American FFA Degree and the Honorary American FFA Degree from the National FFA Organization, and the Lifetime Achievement Award by the North Carolina FFA Association. In 2015, he received the Outstanding New Professor Award from the Agricultural Executive Board at Oregon State University. His research program includes developing new teaching methods, innovative educational programming for rural and disadvantaged youth, and historical studies in career and technical education. Dr. Croom served as editor of the Journal of Agricultural Education.

Acknowledgments

The author and publisher would like to thank the following individuals and associations for their valuable input in the development of *Principles of Agriculture, Food, and Natural Resources*.

American Angus Association
American Braham Breeders Association
American Chianina Association
American Corriedale Association
American Dorper Sheep Breeders Society, Inc.
American Gelbvieh Association
American Hampshire Sheep Association
American Hereford Association
American Hereford Association
American Maine Anjou Association
American Morgan Horse Association/Jeanne Mellin
American Simmental Association, www.simmental.org
American Southdown Association
APHA/Paint Horse Journal
Appaloosa Horse Club
Arabian Horse Association, Gladys Brown Edwards painting
Bureau of Land Management (BLM)
Canadian Food Inspection Agency (CFIA)
CIMMYT (International Maize and Wheat Improvement Center)
Clark Rassi Quarter Horses
Clayton Zwilling
Critter Haven Farms/Diane Hildebrand
DeWALT Industrial Tool Co.
Diamond K Ranch, Hempstead TX
Doug Andrews Photography
Dr. Delphinium Designs in Dallas
Elgin Veterinary Hospital, Bovine Division, Elgin, Texas
Emmons Ranch Beefmasters
Forbes & Rabel Rambouillet, Kaycee, Wyoming
Hoard's Dairyman
J.P. Hancock, Texas A&M University
Jarvis Sheep Company
Jeannette Beranger/The Livestock Conservancy
Jensen Shires
La Muneca Ranch, Linn TX
Les Farms, Canada
Metzer Farms
MJ Ironwater Acres
Morgan Frederick, Three Mill Ranch
National Swine Registry
North American Limousin Foundation
Oak Haven Belfians
On Behalf of Horse Photos
P Bar T Fox Trotters, Independence, MN
Pony of the Americas Club, Inc./Impulse Photography
Poultry CRC
R Bar T Quarter Horses
Rachelsie Farm, Inc.
Red Angus Association of America
Redwood Hill Farm
Roger Hanagriff, Agricultural Experience Tracker (AET)
Santa Gertrudis Breeders International, Kingsville, Texas
Smith Family Angoras & Erbstruck Ranch
Stanley Tools
Stephanie Lastovica
TAMU Livestock Judging
Texas Longhorn Breeders Association of America and Star Creek Ranch
The Jockey Club/Susan Martin
Thomas Ranch, Harrold, South Dakota
Underwood and Underwood/Library of Congress
United Braford Breeders, Inc.
United States Department of Agriculture (USDA)
USDA Agricultural Research Service (ARS)
USDA Economic Research Service (ERS)
USDA Forest Service (USFS)
USDA Natural Resources Conservation Service (NRCS)
United States Fish and Wildlife Service (FWS)
United States Food and Drug Administration (FDA)
United States Forest Service (USFS)
United States Geological Survey (USGS)
United Suffolk Sheep Association
V8 Ranch, Wharton TX

Reviewers

The author and publisher wish to thank the following industry and teaching professionals for their valuable input into the development of *Principles of Agriculture, Food, and Natural Resources*.

John Anderson
Agriculture Science Teacher
TEA Certified Teacher
Anna High School
Anna, Texas

Tara Berescik
Agricultural Educator
Tri-Valley Central School
Grahamville, New York

Shalley K. Boles
Agricultural Science Teacher
Production Agriculture 6–12
Plano Senior High School
Plano, Texas

Carin Cason
Agricultural Science Teacher
Weimar High School
Weimar, Texas

Bryan Chisholm
Agricultural Science Teacher
Bowie High School
Bowie, Texas

Wesley Hancock
Agricultural Science Teacher
Certified Ag Science Teacher/MS Degree
Caldwell High School
Caldwell, Texas

Roxanne Herbrich
Agricultural Science Teacher
Columbus High School
Columbus, Texas

Bryce A. Hoffman
Ag Teacher/FFA Advisor
M.S. Agricultural Education, University of Illinois
Monticello High School
Monticello, Illinois

Jennifer R. Jackson
Agricultural Sciences Teacher/FFA
Centennial High CTE Center
Frisco, Texas

Kevin Jump
Area Teacher
MEd Education
Central Region Agricultural Education
Fort Valley, Georgia

Julie Kondoff
Agriculture Science Teacher
Liberty High School
Frisco, Texas

Sarah McGhee
Agriculture Educator
McCutcheon High School
Lafayette, Indiana

Tiffany Myers
Agriculture Science/CTE Teacher
Spanish Fork Junior High
Spanish Fork, Utah

Jessica Reeves
Agricultural Science Teacher
Agricultural Science Certified
Tomball High School
Tomball, Texas

Scott Robison
Agriscience Teacher
Wake County Public Schools
Raleigh, North Carolina

Loren Sell
Agriculture Science Teacher
Jacksboro ISD
Jacksboro, Texas

Alexander H. Smith
Agribusiness and Natural Resources Instructor
Dr. Phillips High School
Orlando, Florida

Camber Starling
Agriculture Teacher/CTE Department Chair
Heritage High School
Wake Forest, North Carolina

Heather Willis
Agricultural Science Teacher
Travis High School
Richmond, Texas

G-W Integrated Learning Solution

Together, We Build Careers

At Goodheart-Willcox, we take our mission seriously. Since 1921, we have been serving the career and technical education (CTE) community. Our employee-owners are driven to deliver exceptional learning solutions to CTE students to help prepare them for careers. Using educators' wisdom and our expertise, we have designed content and tools that will help students achieve success. We begin with theory and applied content based upon a strong foundation of accepted standards and curriculum. To that base, we add student-focused learning features and tools designed to help students make connections between knowledge and skills. We support our instructor with time-saving tools that help them plan, present, assess, and engage students with traditional and digital activities and assets. Because society is in a transition between print to digital, we want to provide the tools and content in any format you need. Our integrated learning solution products come in a variety of digital and online formats and we provide economical bundles so you can select the right product—print, digital, and online mix for your classroom.

Student-focused Curated Content

Goodheart-Willcox believes that student-focused content should be built from standards and/or accepted curriculum coverage. Our authors and Subject Matter Experts (SMEs) are experts in their field and give considerable thought to the best ways to present content: for some it may be a building block approach with attention devoted to a logical teaching progression; for others it may be logical content presentation in small, focused lessons. We call on industry experts and teachers from across the country to review and comment on our content, presentation, and pedagogy. Finally, in our refinement of curated content, our editors are immersed in content checking, securing, and sometimes creating figures that convey key information, and revising language and pedagogy.

Precision Exams Certification

Goodheart-Willcox is pleased to partner with Precision Exams by correlating *Principles of Agriculture, Food, and Natural Resources* to their Agricultural Science Standards. Precision Exams Standards and Career Skill Exams were created in concert with industry and subject matter experts to match real-world job skills and marketplace demands. Students that pass the exam and performance portion of the exam can earn a Career Skills Certification TM. Precision Exams provides:

- Access to over 150 Career Skills Exams™ with pre- and post-exams for all 16 Career Clusters.
- Instant reporting suite access to measure student academic growth.
- Easy-to-use, 100% online exam delivery system.

To see how *Principles of Agriculture, Food, and Natural Resources* correlates to the Precision Exams Standards, please visit www.g-w.com/principles-agriculture-food-natural-resources-2017 and click on the Correlations tab. For more information on Precision Exams, including a complete listing of their 150+ Career Skills Exams™ and Certificates, please visit https://www.precisionexams.com.

michaeljung/Shutterstock.com

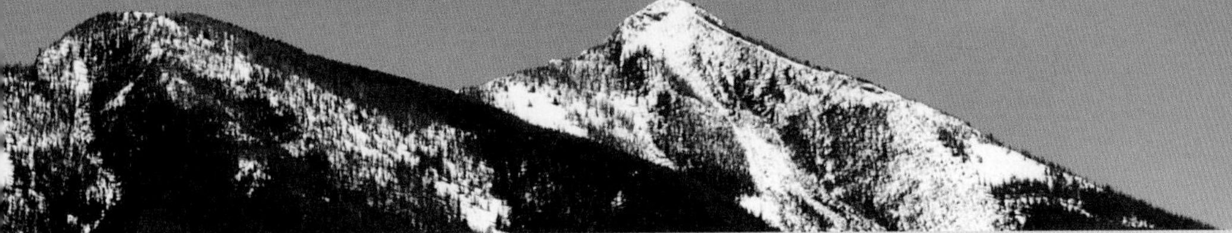

Features of the Textbook

Words to Know list the key terms to be learned in each lesson.

Learning Outcomes clearly identify the knowledge and skills to be obtained when the lesson is completed.

Before You Read literacy integration activities at the beginning of each lesson encourage development of confidence and skill in literacy and learning.

Lesson 3.2

Your Career in Agriculture

Lesson Outcomes

By the end of this lesson, you should be able to:

- Identify career opportunities in the field of agriculture, food, and natural resources.
- Examine the characteristics employers find desirable in employees.
- Describe the skills required for employment in agriculture.
- Accurately describe the steps to obtaining employment.

Words to Know

agribusiness
agricultural communications
agricultural process
agricultural suppor
service
agricultural system
agriscience
aptitude test
cover letter
job application
natural resources management
position announce
production agricul
résumé

Before You Read

After reading each section (separated by main headings), stop and write a three- to four-sentence summary of what you just read. Be sure to paraphrase and use your own words.

On a global scale, there are more people employed in agriculture than any other industry. In the United States, more than 22 million people are employed in agriculture careers. What might surprise you is that most of the people employed by agriculture are *not* traditional farmers and ranchers. In fact, production agriculturalists only make up a small percentage of the overall agricultural workforce, **Figure 3-16**.

In this lesson, you will learn about some of the many careers available in the different sectors of the agriculture industry. We will also discuss the traits employers are looking for in employees, and what you can do to stand out from the competition when applying for a job.

Career Sectors in Agriculture

According to the Council for Agricultural Education, there are eight different pathways of careers in agriculture. Each of these pathways is specific to the type of agriculture in which you are involved. Agriculture has a diverse range of careers. Much like choosing an SAE, there are agricultural careers certain to meet your interests.

Visions/Shutterstock.com

Figure 3-16. Even though a farmer is the career that most people think of when they think about careers in agriculture, less than 2% of Americans are production agriculturalists. There are many other careers to consider.

Copyright Goodheart-Willcox Co., Inc. Chapter 3 Agriculture as a Career 117

Lesson 6.1

Agricultural Structures

Words to Know

aquaculture
broiler production
confinement operation
equine
farrowing facility
finishing facility
grain bin
greenhouse
milking parlor
silage
silo

Lesson Outcomes

By the end of this lesson, you should be able to:

- Identify the importance of materials used in agricultural structures.
- Describe functional and efficient facilities for agricultural use.
- Identify criteria in selecting materials in agricultural construction/fabrication.

Before You Read

Before reading the lesson, make a list of as many types of agricultural buildings you can identify. Think about the different types of agricultural buildings and facilities you have seen in your community. How many different facilities can you identify? Make a list of the types of buildings. Compare your list to the types of buildings covered in the lesson. Share your list with a peer or during class discussion.

The landscape across rural America is decorated with agricultural structures of all shapes, sizes, and purposes, **Figure 6-1**. Barns, animal housing, greenhouses, grain storage and drying facilities, and waste handling facilities are essential to agricultural production. It would be impossible to carry out any agricultural activity without them. Uses for agricultural structures include:

- Plant growth
- Animal housing
- Animal birthing and nursing
- Milking and storage
- Crop storage and processing
- Animal feed storage
- Mechanical shops for maintenance and repair
- Equipment storage
- Waste storage and processing

Elena Elisseeva/Shutterstock.com

Figure 6-1. Many of the barns in use across the American countryside date back over 100 years. Farmers maintain these buildings because they are used to house and store the animals, crops, and equipment from which a farmer makes a living. *Are there any historic barns in your area?*

258 Principles of Agriculture, Food, and Natural Resources Copyright Goodheart-Willcox Co., Inc.

STEM Connections integrate all four components of STEM education as well the social sciences and language arts.

Career Connection features introduce students to careers in every field of agriculture.

112 Principles of Agriculture, Food, and Natural Resources

Career Connection What Can Your SAE Do for Your Future?

Your SAE is an essential part of your agricultural education experience, but what can it do for you after high school? Many agricultural education students have taken their SAE enterprises and turned them into businesses that support them financially. Exploratory, Research, and Placement SAEs can give you valuable experience that will make you more marketable as an employee in fields related to your SAE category. Entrepreneurship SAEs can also give you employment skills, which you could use to grow your SAE into a productive and successful full-time business venture.

Some of the most common examples of SAEs that easily translate into viable companies include landscaping businesses and livestock operations.

Landscaping Business: One of the most easily transferrable businesses that many FFA members use as their adult career are landscaping businesses. These business owners have taken their high school business and built it into a thriving, full-time employment for themselves. Can you think of some of the benefits of proper SAE setup for those who plan to continue their venture after high school?

Livestock Operations: Many students begin with small numbers of animals during their SAE, and are able to build their enterprise to be large enough to raise animals full-time after high school. These students can then pursue livestock management as a suitable means of income to fund their education, or as a full-time source of employment income.

There are many other ways that your SAE can grow into something much more than just a high school project. In your SAE planning, think not only about how your SAE can play a role in your life today, but in your future.

Pixsooz/Shutterstock.com

Providing professional service to your customers will ensure a solid customer base. Written estimates will help you keep track of accurate and profitable job quotes.

Nate Allred/Shutterstock.com

Raising animals during your SAE will provide you with valuable experience you can apply to your advanced education and future agricultural employment.

Copyright Goodheart-Willcox Co., Inc.

Chapter 5 Agriculture Science 237

STEM Connection DNA and Genetic Engineering

Simply stated, genetic engineering is the process of taking an isolated section of DNA from one organism and fusing it to the DNA of another organism. To understand how genetic engineering works, we need to first understand the structure of DNA. The DNA for all organisms is made of only six components: a phosphate group, deoxyribose sugar (which make the sides of the ladder-shaped molecule), and four different nitrogenous bases: adenine, thymine, cytosine, and guanine. The order in which the bases combine determines the way the genes are read, proteins are created, and organisms are formed.

To genetically modify an organism, scientists perform four basic steps:

Step 1: Identify the desired gene in the host organism.

Step 2: Using enzymes, cut the desired gene and a corresponding section of the DNA in the organism you are genetically engineering.

Step 3: Insert the host gene into the DNA of the genetically engineered organism.

Step 4: Insert the genetically engineered DNA into a cell of the organism and replicate to create a genetically modified organism (GMO).

The new DNA sequence is called ...combinant DNA, because it is a recombination ...the genetic sequence. Recombinant DNA ...used for a variety of purposes, including ...armaceuticals and cheese production.

...**armaceuticals.** By modifying the DNA ...uence of bacteria, we can alter them so ...that they produce specific chemicals. ...is the process used to create almost all ...he insulin given to diabetic humans. The ...e process is used to create vaccines, ...an growth hormone supplements, and ...y other chemicals used in the health care ...stry.

...**ese production.** One of the chemicals ...ved in making cheese is a substance called ...et. Rennet contains a chemical found in ...tomach of calves. Today, about 60% of the ...se produced worldwide uses recombinant ...technology to produce this chemical, rather ...harvesting it from calves stomachs.

...Goodheart-Willcox Co., Inc.

SAE Connections help students make real-life connections to a variety of new and interesting SAE opportunities.

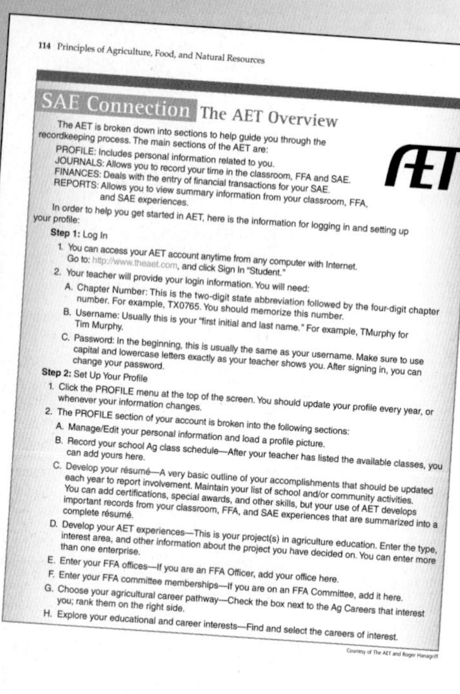

114 Principles of Agriculture, Food, and Natural Resources

SAE Connection The AET Overview

The AET is broken down into sections to help guide you through the recordkeeping process. The main sections of the AET are:

PROFILE: Includes personal information related to you.
JOURNALS: Allows you to record your time in the classroom, FFA and SAE.
FINANCES: Deals with the entry of financial transactions for your SAE.
REPORTS: Allows you to view summary information from your classroom, FFA, and SAE experiences.

In order to help you get started in AET, here is the information for logging in and setting up your profile:

Step 1: Log In

1. You can access your AET account anytime from any computer with Internet. Go to: http://www.theaet.com, and click Sign In "Student."
2. Your teacher will provide your login information. You will need:
   A. Chapter Number: This is the two-digit state abbreviation followed by the four-digit chapter number. For example, TX0765. You should memorize this number.
   B. Username: Usually this is your "first initial and last name." For example, TMurphy for Tim Murphy.
   C. Password: In the beginning, this is usually the same as your username. Make sure to use capital and lowercase letters exactly as your teacher shows you. After signing in, you can change your password.

Step 2: Set Up Your Profile

1. Click the PROFILE menu at the top of the screen. You should update your profile every year, or whenever your information changes.
2. The PROFILE section of your account is broken into the following sections:
   A. Manage/Edit your personal information and load a profile picture.
   B. Record your school Ag class schedule—After your teacher has listed the available classes, you can add yours here.
   C. Develop your résumé—A very basic outline of your accomplishments that should be updated each year to report involvement. Maintain your list of school and/or community activities. You can add certifications, special awards, and other skills, but your use of AET develops important records from your classroom, FFA, and SAE experiences that are summarized into a complete résumé.
   D. Develop your AET experiences—This is your project(s) in agriculture education. Enter the type, interest area, and other information about the project you have decided on. You can enter more than one enterprise.
   E. Enter your FFA offices—If you are an FFA Officer, add your office here.
   F. Enter your FFA committee memberships—If you are on an FFA Committee, add it here.
   G. Choose your agricultural career pathway—Check the box next to the Ag Careers that interest you; rank them on the right side.
   H. Explore your educational and career interests—Find and select the careers of interest.

Courtesy of The AET and Roger Hanagriff

Copyright Goodheart-Willcox Co., Inc.

Chapter 8 Agricultural Mathematics 405

or hunting lease business. Almost all of your income and expenses would occur during the hunting season. Having a solid business plan could help you show potential investors or loan officers that the initial spending would have a return in the future.

Market

The business description should also include information on the market for your products. A ***market*** is the group of consumers that are likely to purchase a specific product. Considering who your market will be is an integral part of the planning. Think about items related to your consumer including:

- Is there a certain age that this product will appeal to?
- What level of income will these consumers likely have?
- How do consumers typically buy this type of product?

Part of understanding your market may include conducting an analysis of the comparative advantage of the specific business, **Figure 8-4.** A ***comparative advantage*** is a statement looking at potential competitors for a certain business, and how the business fits into the same market.

If you wanted to create a new business selling animal feed, you would likely want to do a comparative advantage that included how

Minerva Studio/Shutterstock.com

Figure 8-4. Understanding what your product can offer over your competitors is an important part of developing a business plan. Knowing how your product stacks up to others that consumers can choose from will allow you to better market your product.

FFA Connection Marketing Plan CDE

The Marketing Plan CDE gives a team of FFA members the opportunity to create a marketing plan for a real agricultural business in their area. In this CDE, students have the chance to analyze products available from a company and think of creative ways to increase the profitability of the company through marketing goods and services. Do you think you might like a career in advertising, business management, or owning your own company? If so, the Marketing Plan CDE might be a great fit for you.

Rawpixel/Shutterstock.com

Copyright Goodheart-Willcox Co., Inc.

FFA Connection features introduce students to the exciting world of FFA opportunities including Career Development Events in a variety of agricultural areas.

Illustrations have been designed to clearly and simply communicate the specific topic.

Caption questions help the students connect the text to the art, while encouraging active reading and critical thinking.

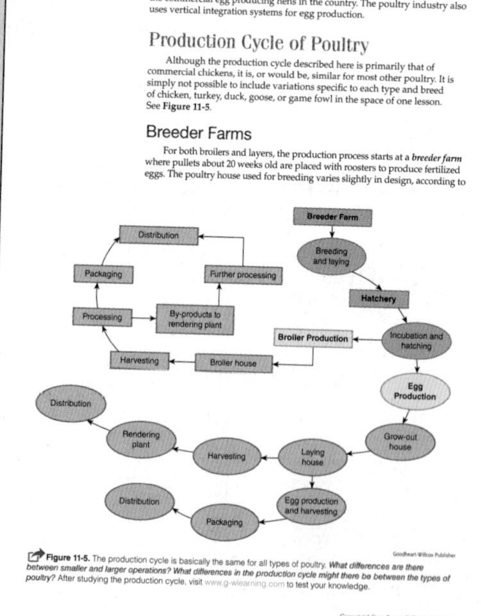

578 Principles of Agriculture, Food, and Natural Resources

the commercial egg producing hens in the country. The poultry industry also uses vertical integration systems for egg production.

Production Cycle of Poultry

Although the production cycle described here is primarily that of commercial chickens, it is, or would be, similar for most other poultry. It is simply not possible to include variations specific to each type and breed of chicken, turkey, duck, goose, or game fowl in the space of one lesson. See **Figure 11-5.**

Breeder Farms

For both broilers and layers, the production process starts at a ***breeder farm*** where pullets about 20 weeks old are placed with roosters to produce fertilized eggs. The poultry house used for breeding varies slightly in design, according to

Goodheart-Willcox Publisher

Figure 11-5. The production cycle is basically the same for all types of poultry. *What differences are there between smaller and larger operations? What differences in the production cycle might there be between the types of poultry?* After studying the production cycle, visit www.g-wlearning.com to test your knowledge.

Copyright Goodheart-Willcox Co., Inc.

Chapter 4 Agricultural Safety 185

prevent electrical accidents, as well

Safety Note

Do not try to grab a person who is being shocked by electricity. You will become a part of the "circuit" and also be shocked. If it is safe to do so, disconnect the power source by turning off the circuit breaker. Do not try to unplug the cord or move the energized line with any object. You will also be shocked.

Did You Know?

Fifteen amps, which is typically the smallest size breaker found in a normal household, is 250 times greater than what is required to cause cardiac arrest in an individual.

Ethics in AG

You and your partner are working outside on a very hot day. He suddenly begins to complain of stomach cramps. What will you do?

Frostbite and Hypothermia

Not all farm and ranch work is done during the warmer part of the year. This is especially true of livestock operations. Overexposure to low temperatures, cold wind, and cold moisture may result in cold-related illnesses such as frostbite and hypothermia.

- ***Frostbite***—body tissue injury caused by overexposure to extreme cold. It typically affects exposed skin and extremities like the nose, fingers,

Copyright Goodheart-Willcox Co., Inc.

Safety Notes alert students of potentially dangerous materials and practices.

Did You Know? features point out interesting and helpful facts about the agricultural industry.

Ethics in AG present students with everyday situations in which they must make vital decisions.

Words to Know matching activities reinforce vocabulary development and retention. All key terms are included in the text glossary and are connected to numerous online review activities.

Know and Understand review questions allow students to demonstrate knowledge, identification, and comprehension of lesson material.

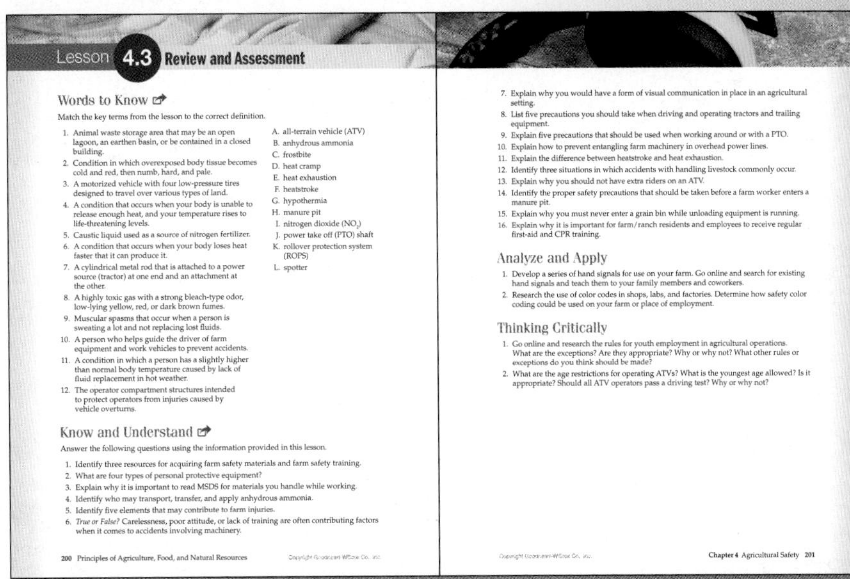

Lesson 4.3 Review and Assessment

Words to Know

Match the key terms from the lesson to the correct definition.

1. Animal waste storage area that may be an open lagoon, an earthen basin, or be contained in a closed building.
2. Condition in which overexposed body tissue becomes cold and red, then numb, hard, and pale.
3. A motorized vehicle with four low-pressure tires designed to travel over various types of land.
4. A condition that occurs when your body is unable to release enough heat, and your temperature rises to life-threatening levels.
5. Caustic liquid used as a source of nitrogen fertilizer.
6. A condition that occurs when your body loses heat faster that it can produce it.
7. A cylindrical metal rod that is attached to a power source (tractor) at one end and an attachment at the other.
8. A highly toxic gas with a strong bleach-type odor, low-lying yellow, red, or dark brown fumes.
9. Muscular spasms that occur when a person is sweating a lot and not replacing lost fluids.
10. A person who helps guide the driver of farm equipment and work vehicles to prevent accidents.
11. A condition in which a person has a slightly higher than normal body temperature caused by lack of fluid replacement in hot weather.
12. The operator compartment structures intended to protect operators from injuries caused by vehicle overturns.

A. all-terrain vehicle (ATV)
B. anhydrous ammonia
C. frostbite
D. heat cramp
E. heat exhaustion
F. heatstroke
G. hypothermia
H. manure pit
I. nitrogen dioxide (NO_2)
J. power take off (PTO) shaft
K. rollover protection system (ROPS)
L. spotter

Know and Understand

Answer the following questions using the information provided in this lesson.

1. Identify three resources for acquiring farm safety materials and farm safety training.
2. What are four types of personal protective equipment?
3. Explain why it is important to read MSDS for materials you handle while working.
4. Identify who may transport, transfer, and apply anhydrous ammonia.
5. Identify five elements that may contribute to farm injuries.
6. *True or False?* Carelessness, poor attitude, or lack of training are often contributing factors when it comes to accidents involving machinery.

200 Principles of Agriculture, Food, and Natural Resources Copyright Goodheart-Willcox Co., Inc.

7. Explain why you would have a form of visual communication in place in an agricultural setting.
8. List five precautions you should take when driving and operating tractors and trailing equipment.
9. Explain five precautions that should be used when working around or with a PTO.
10. Explain how to prevent entangling farm machinery in overhead power lines.
11. Explain the difference between heatstroke and heat exhaustion.
12. Identify three situations in which accidents with handling livestock commonly occur.
13. Explain why you should not have extra riders on an ATV.
14. Identify the proper safety precautions that should be taken before a farm worker enters a manure pit.
15. Explain why you must never enter a grain bin while unloading equipment is running.
16. Explain why it is important for farm/ranch residents and employees to receive regular first-aid and CPR training.

Analyze and Apply

1. Develop a series of hand signals for use on your farm. Go online and search for existing hand signals and teach them to your family members and coworkers.
2. Research the use of color codes in shops, labs, and factories. Determine how safety color coding could be used on your farm or place of employment.

Thinking Critically

1. Go online and research the rules for youth employment in agricultural operations. What are the exceptions? Are they appropriate? Why or why not? What other rules or exceptions do you think should be made?
2. What are the age restrictions for operating ATVs? What is the youngest age allowed? Is it appropriate? Should all ATV operators pass a driving test? Why or why not?

Copyright Goodheart-Willcox Co., Inc. Chapter 4 Agricultural Safety 201

Analyze and Apply features allow students to transfer their understanding of concepts to new applications both in and out of the agricultural field.

Thinking Critically questions develop higher-order thinking, problem solving, personal, and workplace skills.

Key Points feature provides an additional review tool for students and reinforces key learning outcomes.

Words to Know are repeated for each lesson as a means of vocabulary review.

Check Your Understanding review questions extend students learning and help them analyze and apply knowledge from each lesson in the chapter.

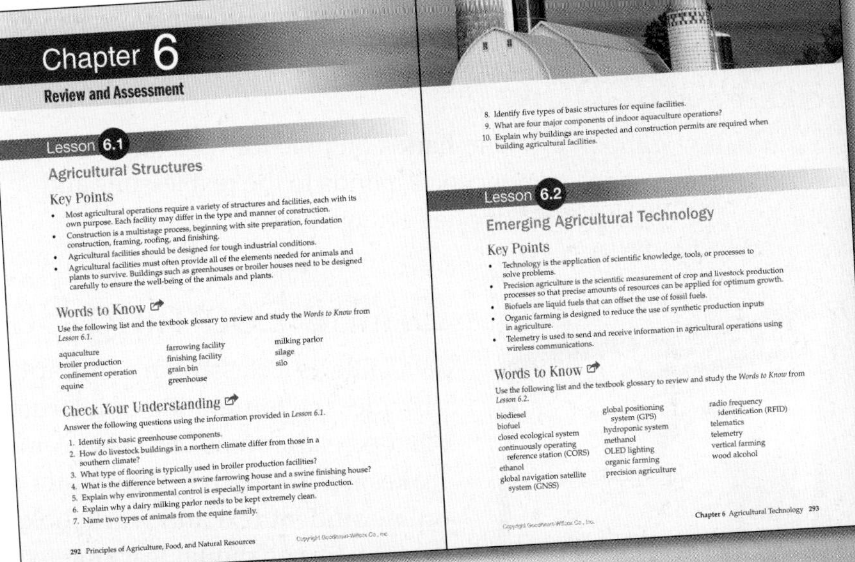

Chapter 6

Review and Assessment

Lesson 6.1

Agricultural Structures

Key Points

- Most agricultural operations require a variety of structures and facilities, each with its own purpose. Each facility may differ in the type and manner of construction.
- Construction is a multistage process, beginning with site preparation, foundation construction, framing, roofing, and finishing.
- Agricultural facilities should be designed for tough industrial conditions.
- Agricultural facilities must often provide all of the elements needed for animals and plants to survive. Buildings such as greenhouses or broiler houses need to be designed carefully to ensure the well-being of the animals and plants.

Words to Know

Use the following list and the textbook glossary to review and study the *Words to Know* from *Lesson 6.1.*

aquaculture
broiler production
confinement operation
equine
farrowing facility
finishing facility
grain bin
greenhouse
milking parlor
silage
silo

Check Your Understanding

Answer the following questions using the information provided in *Lesson 6.1.*

1. Identify six basic greenhouse components.
2. How do livestock buildings in a northern climate differ from those in a southern climate?
3. What type of flooring is typically used in broiler production facilities?
4. What is the difference between a swine farrowing house and a swine finishing house?
5. Explain why environmental control is especially important in swine production.
6. Explain why a dairy milking parlor needs to be kept extremely clean.
7. Name two types of animals from the equine family.

292 Principles of Agriculture, Food, and Natural Resources Copyright Goodheart-Willcox Co., Inc.

8. Identify five types of basic structures for equine facilities.
9. What are four major components of indoor aquaculture operations?
10. Explain why buildings are inspected and construction permits are required when building agricultural facilities.

Lesson 6.2

Emerging Agricultural Technology

Key Points

- Technology is the application of scientific knowledge, tools, or processes to solve problems.
- Precision agriculture is the scientific measurement of crop and livestock production processes so that precise amounts of resources can be applied for optimum growth.
- Biofuels are liquid fuels that can offset the use of fossil fuels.
- Organic farming is designed to reduce the use of synthetic production inputs in agriculture.
- Telemetry is used to send and receive information in agricultural operations using wireless communications.

Words to Know

Use the following list and the textbook glossary to review and study the *Words to Know* from *Lesson 6.2.*

biodiesel
biofuel
closed ecological system
continuously operating reference station (CORS)
ethanol
global navigation satellite system (GNSS)
global positioning system (GPS)
hydroponic system
methanol
OLED lighting
organic farming
precision agriculture
radio frequency identification (RFID)
telematics
telemetry
vertical farming
wood alcohol

Copyright Goodheart-Willcox Co., Inc. Chapter 6 Agricultural Technology 293

STEM and Academic Activities are provided in the areas of science, technology, engineering, math, social science, and language arts.

8. How often should a horse's hooves be trimmed or reshod?
9. Explain why it is important to establish a pest management plan for use with and around equines.
10. What is the most common cause of death in equines and what are its symptoms?

Chapter 10 Skill Development

STEM and Academic Activities

1. **Science.** Cheese is made from a chemical process that alters the milk protein using enzymes. What makes different varieties of cheese different? Choose three types of cheese to research and write a paragraph on each, describing the specific process for making that type of cheese. Then, compare and contrast your three cheese types so that you can explain how they are different. Be prepared to share your findings with the class.
2. **Technology.** Milking machines are incredibly fascinating machinery. To find out how the milking machine mirrors the actual suckling of a calf, conduct some research or ask a local dairyman. Draw a diagram that shows the parts of the milking machine that are required to ensure that dairy cows are milked efficiently.
3. **Engineering.** Cattle handling facilities are incredibly diverse. Imagine that you are the owner of a cow-calf operation and need to design a cattle handling area. In groups of two or three, research the common components of a cattle handling area, and design your layout. Make sure you include a squeeze chute or calf table, holding pen, and any other pens or structures that you think would be important.
4. **Math.** Working with a partner, measure the stalls in an equine stable. Calculate the number of square feet for each stall. Obtain prices for various types of bedding used for equines. Using the total square footage of the stalls, calculate how much bedding will be needed to provide adequate bedding for each stall. Calculate the cost of each type of bedding. Determine the cost for one month's bedding for all the stalls. If possible, include additional costs such as shipping and delivery.
5. **Social Science.** Why are businesses, homes, and livestock facilities usually in separate areas of a community? Are there instances where livestock may be kept on property that is not zoned as agricultural land? Interview the person in your community who is in charge of planning or zoning. Find out why zoning is considered important for the community and how a property owner can seek a zoning change. Obtain a copy of the community's zoning map and use it for a visual aid as you report your findings to the class.
6. **Language Arts.** Many stories, books, and poems have been written about horses. Choose one of the following titles, or one of your own favorites, to read and write a detailed book report. Use a standard format from your agriculture or English teacher. Create a diorama of your favorite or most memorable scene. Books to consider: *Black Beauty* by Anna Sewell, *The Black Stallion* by Walter Farley, *The Horse Whisperer* by Nicholas Evans, *A Horse Called Wonder* by Joanna Campbell.

570 Principles of Agriculture, Food, and Natural Resources Copyright Goodheart-Willcox Co., Inc.

Communicating about Agriculture

1. **Reading and Speaking.** Research the products and services available for raising either cattle or equines. Collect promotional materials for a variety of products and services from product manufacturers. Analyze the data in these materials based on the knowledge gained from this chapter. With your group, review the list for words that can be used in the subject area of raising livestock. Practice pronouncing the word, and discuss its meaning. As a fun challenge, work together to compose a creative narrative using as many words as you can from your new list.
2. **Speaking and Listening.** Divide into groups of four or five students. Each group should choose one of the following types of beef cattle operations: backgrounding operation, cow-calf operation, and stocker operation. Using your textbook as a starting point, research your topic and prepare a report on the operation. Include topics such as costs, land needs, structural needs, and employees, as well as the types of challenges the operation faces. As a group, deliver your presentation to the rest of the class. Take notes while other students give their reports. Ask questions about any details that you would like clarified.
3. **Reading and Writing.** The ability to read and interpret information is an important workplace skill. Presume you work for a local, well-known dairy operation. Your employer is considering upgrading the milking parlor with new milking equipment. Your supervisor wants you to evaluate and interpret some research on several new milking systems. Locate at least three reliable resources for the most current information on new milking systems. If possible, contact representatives from the manufacturers, and (after explaining your project) ask them about their products and additional costs, such as delivery and installation and employee training on the new equipment. Write a report summarizing your findings in an organized manner.

Extending Your Knowledge

1. Buying a horse is a huge financial responsibility. Even if you already have the horse and the equipment needed, there are additional costs such as entry fees for shows or rodeos, and medical fees. Figure the operating costs to keep a horse for one year. Research the cost for horse-related expenses in your area, and find the total. You will need to provide:
   A. Housing—contact a local boarding facility to determine the cost per month for stall, paddock, or pasture rent.
   B. Horse shoeing—calculate the number of trimmings or shoeings your horse will require per year (once every 6–8 weeks).
   C. Feed—light breed horses will eat on average one ton of hay every three months. Find this cost plus the cost of any feed additives.
   D. Vaccinations—most horses will receive annual and semiannual vaccinations. Contact an animal health provider to calculate this cost. Remember that this cost does *not* include other common costs like additional veterinary bills or paperwork for out-of-state travel.

Copyright Goodheart-Willcox Co., Inc. Chapter 10 Large-Animal Production 571

Communicating about Agriculture questions and activities help integrate reading, writing, listening, and speaking skills while extending their knowledge on the chapter topics.

Extending Your Knowledge provides additional activities for students to explore and deepen their understanding of key concepts from each chapter.

Student Resources

Textbook

The *Principles of Agriculture, Food, and Natural Resources* textbook provides an exciting, full-color, and highly-illustrated learning resource. The textbook is available in a print or online version.

G-W Learning Companion Website/ Student Textbook

The G-W Learning companion website is a study reference that contains photo identification and matching activities, animations and videos, review questions, vocabulary exercises, and more! Accessible from any digital device, the G-W Learning companion website complements the textbook and is available to the student at no charge.

Lab Workbook

The student workbook provides hands-on practice with questions and lab activities. Each chapter corresponds to the text lessons and reinforces key concepts and applied knowledge.

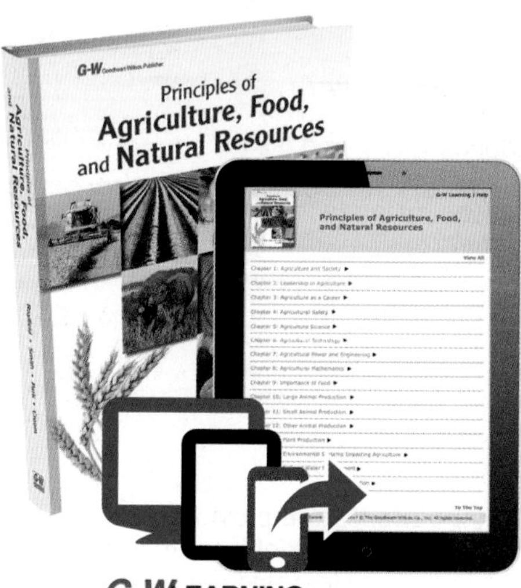

Online Learning Suite

Available as a classroom subscription, the Online Learning Suite provides the foundation of instruction and learning for digital and blended classrooms. An easy-to-manage shared classroom subscription makes it a hassle-free solution for both students and instructors. An online student text and workbook, along with rich supplemental content, brings digital learning to the classroom. All instructional materials are found on a convenient online bookshelf and are accessible at home, at school, or on the go.

Online Learning Suite/ Student Textbook Bundle

Looking for a blended solution? Goodheart-Willcox offers the Online Learning Suite bundled with the printed text in one easy-to-access package. Students have the flexibility to use the print version, the Online Learning Suite, or a combination of both components to meet their individual learning style. The convenient packaging makes managing and accessing content easy and efficient.

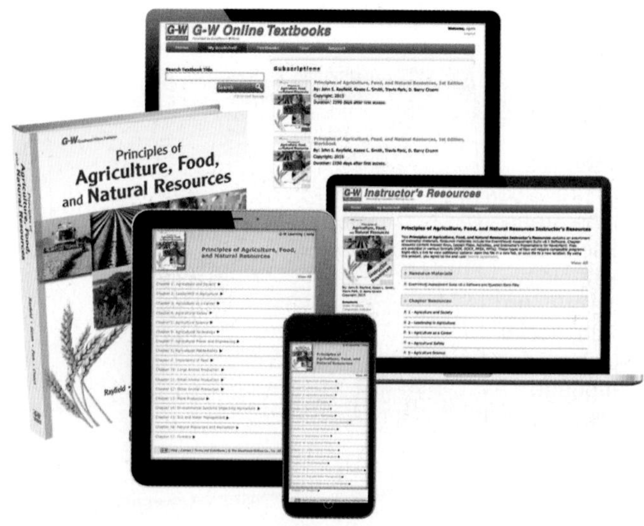

Instructor Resources

Instructor's Presentations for PowerPoint®

Help teach and visually reinforce key concepts with prepared lectures. These presentations are designed to allow for customization to meet daily teaching needs. They include objectives, outlines, and images from the textbook.

ExamView® Assessment Suite

Quickly and easily prepare, print, and administer tests with the ExamView® Assessment Suite. With hundreds of questions in the test bank corresponding to each lesson, you can choose which questions to include in each test, create multiple versions of a single test, and automatically generate answer keys. Existing questions may be modified and new questions may be added. You can prepare pretests, formative, and summative tests easily with the ExamView® Assessment Suite.

Instructor's Resource CD

One resource provides instructors with time-saving preparation tools such as answer key, rubrics, lesson plans, correlation charts to standards, and other teaching aids.

Online Instructor Resources

Online Instructor Resources provide all the support needed to make preparation and classroom instruction easier than ever. Available in one accessible location, support materials include Answer Keys, Lesson Plans, Instructor Presentations for PowerPoint®, ExamView® Assessment Suite, and more! Online Instructor Resources are available as a subscription and can be accessed at school or at home.

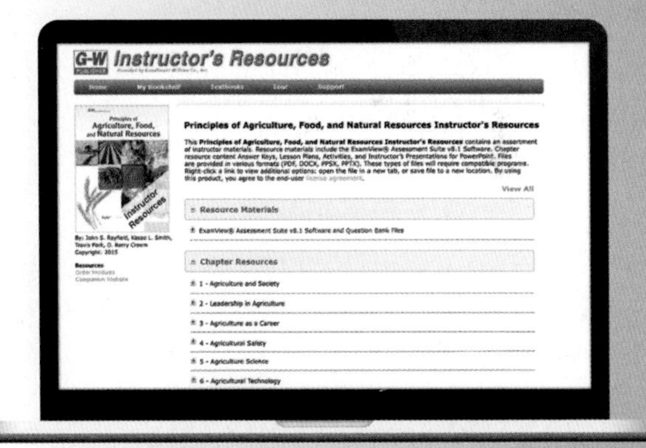

Video Clip Library

Live-action videos provide students with solid information on 30 agricultural topics. High-quality video footage and crisp narrations enable visual learners to clearly understand the content.

Contents in Brief

Contents

Chapter 7
Agricultural Power and Engineering. 296

Chapter 8
Agricultural Mathematics. 400

Chapter 9
Importance of Food 440

Lesson 9.2

Lesson 9.3

Lesson 9.4

Chapter 10

Lesson 10.1

Lesson 10.2

Lesson 10.3

Chapter 11
Small-Animal Production 572

Chapter 1

Agriculture and Society

Adisa/Shutterstock.com

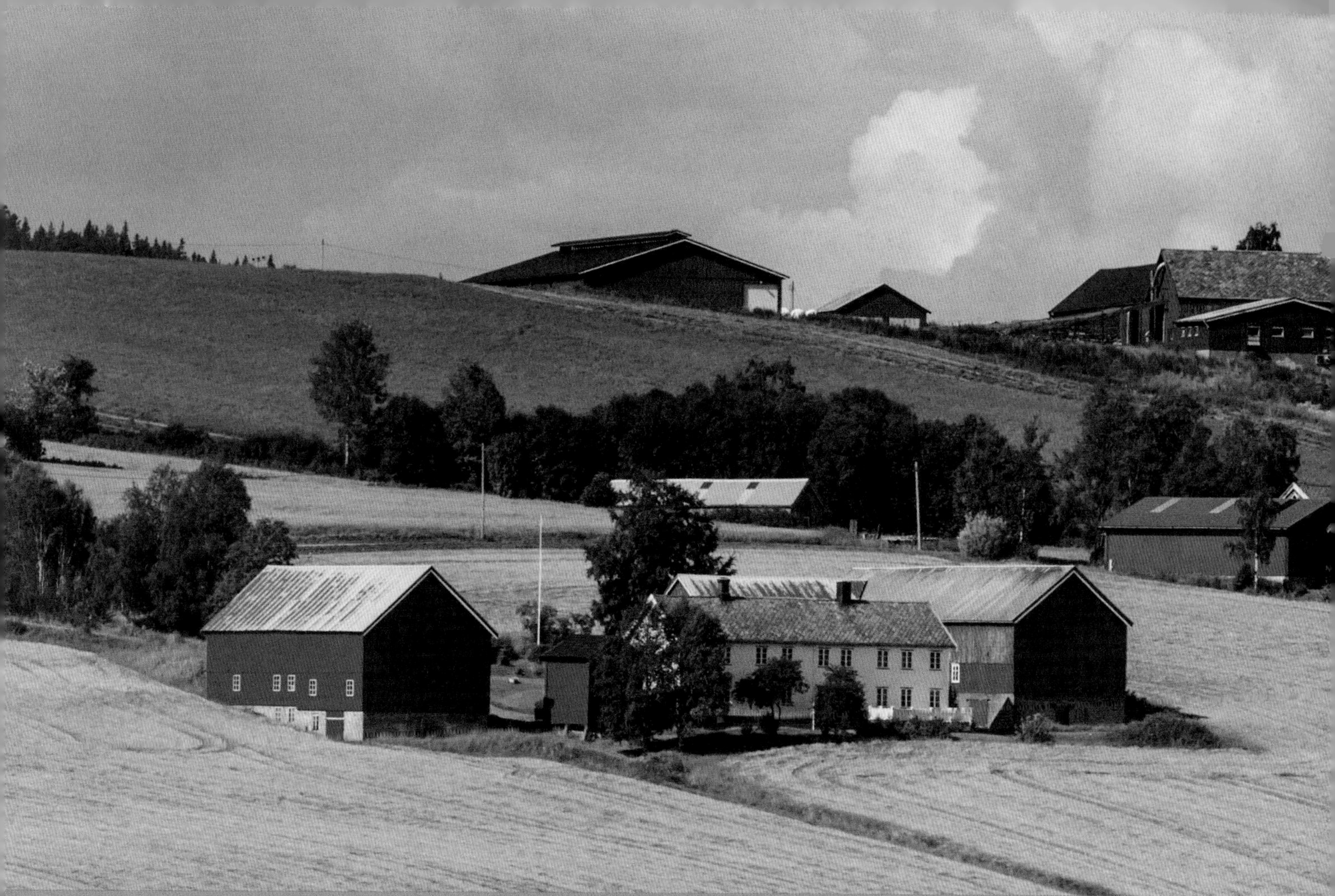

Galyna Andrushko/Shutterstock.com

Everett Historical

KYTan/Shutterstock.com

While studying, look for the activity icon  **to:**

- **Practice** vocabulary terms with e-flash cards and matching activities.
- **Expand** learning with video clips, animations, and interactive activities.
- **Reinforce** what you learn by completing the end-of-lesson activities.
- **Test your knowledge** by completing the end-of-chapter questions.

Lesson 1.1

Agriculture's Impact on Society

Words to Know

agriculture
arable land
commodity
consumer
fiber
food
natural resource
natural resources management
producer

Lesson Outcomes

By the end of this lesson, you should be able to:

- Define agriculture and the components of the agricultural industry.
- Explain the scope of agriculture as it relates to land use, economic impact, and employment.
- Give the characteristics of the three main areas of agricultural commodities.
- Analyze the importance of the human-agriculture connection.

Before You Read

Find a partner and read the lesson title together. Tell your partner what you have experienced or already know about this topic. Based on your prior knowledge, write a paragraph describing what additional information you would like to learn about the topic. After reading the lesson, share two things you have learned with the classmate.

What comes to your mind when you hear the word agriculture? Do you think of something similar to Old MacDonald's farm, like the image in **Figure 1-1**?

MaxyM/Shutterstock.com

Figure 1-1. When people think about agriculture, they picture a single farmer working with many different commodities. ***Does your view of agriculture look like this traditional farm?***

Or, do you picture something like **Figure 1-2**, a modern industry, complete with integrated business models, cutting-edge technology, and billions of dollars in revenue each year? Over the course of this text, we will examine principles of agriculture as they relate to your life, the lives of those around you, and the lives of those around the world.

Syaochka/Shutterstock.com; Kalinovsky/Shutterstock.com; Marcin Balcerzak/Shutterstock.com

Figure 1-2. Modern agriculture is different from the traditional image that many think of when they hear the word agriculture. Today's agriculture includes highly technical and specialized operations. ***Does your view of agriculture look like this?***

What Is Agriculture?

In its most basic definition, ***agriculture*** is the art and science of cultivating plants, animals, and other life forms for use by humans to sustain life. If a human is controlling the patterns of another life form for human benefit, it is agriculture.

Sectors of Agriculture

Because agriculture is such a broad industry, it can be divided into subcategories or sectors. These sectors are classified by the type of work they include, **Figure 1-3**. Have you ever wondered how food from around the world makes its way to your grocery store year-round? The relationships between these specialized sectors allow the agricultural industry to continually produce and distribute goods around the world. By specializing, each sector performs tasks with maximum efficiency, getting products to consumers on a predicted schedule.

Scope of Agriculture

Before we can fully understand how the principles of agriculture, food, and natural resources impact our lives, we need to examine the broad scope of agriculture. In addition to cultivating other life forms, agriculture includes the industries that process, market, and distribute everything from raw products to processed agricultural goods.

The scope of agriculture is based on three broad categories:

- The amount of land agriculture uses.
- The economic impact of agricultural industries.
- The number of people employed in agriculture worldwide.

Did You Know?

According to the U.S. Department of Agriculture (USDA), when all women involved with farming are added up, they are nearly one million strong and account for 30% of U.S. farmers.

Sectors of Agriculture

Sector	Description	Jobs
Agribusiness	Managing the profitability of agricultural goods	Commodity trader Farm and/or ranch manager Agricultural loan officer
Agriscience	Research and development of emerging agricultural technologies	Geneticist Biotechnology engineer Biologist Food chemist
Agricultural Communications	Informing those in the agricultural industry as well as consumers regarding topics concerning agricultural products	Journalist Public relations manager Sales representative
Agricultural Processing	Transforming raw agricultural goods into modified products for consumers	Grain mill operator Food technician Quality control manager
Agricultural Support Services	Industries and careers that provide support for the equipment and technology associated with agricultural production	Veterinarian Distribution coordinator Mechanic
Agricultural Systems	Design and build agricultural equipment and machinery	Agricultural engineer Irrigation systems specialist
Natural Resources Management	Focus on the ecology and conservation of cultivated and uncultivated lands	Range manager Fish hatchery technician Wildlife biologist
Production Agriculture	Production of raw animal and plant goods for human use	Cattle rancher Grain producer Catfish farm owner Cotton grower

Goodheart-Willcox Publisher

Figure 1-3. The agricultural industry is separated into many sectors based on the type of work conducted. ***Which sector do you think is most fitting for your personality?***

Svend77/Shutterstock.com

Figure 1-4. The amount of arable land for cultivating crops is decreasing daily.

Agricultural Land Use

More than three-quarters of Earth's surface is covered by water. It is estimated that half of the remaining quarter is used for some type of agriculture. Agricultural land use includes the land used for crops, forests, rangeland, pastures, national parks, and agricultural businesses. ***Arable land*** is the term used to describe land that is suitable for growing crops, **Figure 1-4**. It is estimated that only one-thirty second (1/32) of Earth's surface is arable land.

The United States covers nearly 2.3 billion acres. According to the U.S. Census Bureau, land use is broken down as shown in **Figure 1-5**. Agricultural land is in danger. Some reports state that over 3,000 acres of cropland are lost each day to development. More than two-thirds of the economic income from agriculture is generated from land adjacent to urban areas, the type of agricultural land most threatened by urban growth, **Figure 1-6**.

Proper management of agricultural land is critical to the success of the agricultural industry. Learning skills that enable agriculturalists to increase production capabilities will be a large part of agriculture as we move into the future.

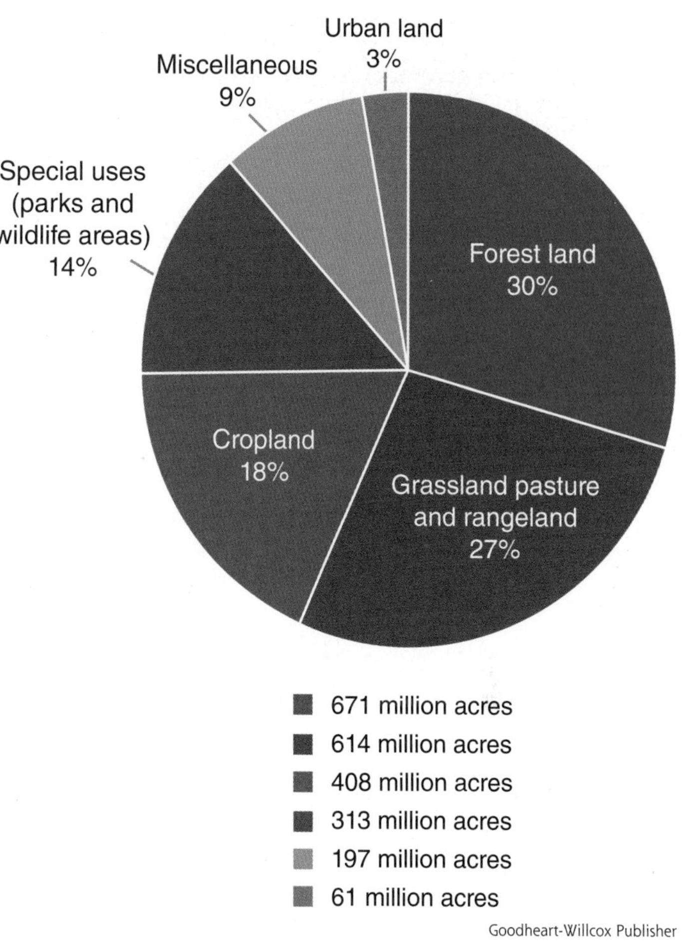

Goodheart-Willcox Publisher

Figure 1-5. Breakdown of land use in the United States. (Percentages are expressed in rounded numbers.)

Economic Impact of Agriculture

Agriculture as an industry is estimated to account for approximately 10% of all the money spent on trade worldwide. That amounts to more than $7.2 trillion in agricultural goods sold each year. The influence of agriculture varies by country. In some countries, agriculture accounts for more than half of all product sales. As a country's level of industry increases, the influence of

Richard Thornton/Shutterstock.com

Figure 1-6. Protecting agricultural land from development and increasing crop yields are important considerations for agriculture. *What types of action can you take to help protect agricultural land?*

Did You Know?

According to the USDA, more than $6 million of U.S. agricultural products are contracted for export each hour.

Agriculture Commodities in the U.S.	
Imports	**Exports**
Vegetables and preparations	Soybeans
Fruit and preparations (i.e. juice)	Feed grains and products
Feed grains and products	Corn
Wine	Meat and meat products
Oilseeds and products	Wheat

U.S. Census Bureau, 2013

Figure 1-7. Top agricultural imports and exports in the United States.

agricultural products on overall sales diminishes.

According to the USDA, American agriculture accounts for more than $775 billion in sales annually. Most of what is produced through agriculture in the United States is used by American consumers, amounting to about $635 billion in goods each year. American agriculture also exports more than $140 billion of products each year and imports around $90 billion in agricultural goods annually. The most commonly imported and exported agricultural goods in the United States are shown in **Figure 1-7**.

Employment in Agriculture

Did You Know?

More than 97% of the farms in the United States are family-owned and operated.

Understanding the scope of agriculture from a land use and financial standpoint is important, but how does that relate to people? When we examine the scope of agriculture as it relates to human interaction, we should consider the number of people the industry employs.

On a global level, agriculture is the world's largest supplier of jobs, with some experts placing the global agricultural workforce at well over a billion people. In around 50 countries, more than half of the population is employed in agriculture, and in many countries, that number exceeds more than three-quarters of the workforce. The percentage of agricultural workers in a country is largely driven by the level of development in the country, **Figure 1-8**; the more industrialized a country is, the more likely a smaller segment of the population is working in agriculture. While there are still large numbers of people employed in agriculture worldwide, the number of farmers has been in a steady global decline since the 1950s.

nofilm2011/Shutterstock.com

Figure 1-8. In developing countries, more than half of the workforce may be employed by agriculture. *Why do you think so many people are employed by agriculture in these countries?*

In the United States, nearly one out of six workers is employed directly or indirectly in the agricultural industry. Most of these jobs are related to the processing, marketing, and distribution of agricultural products. The number of Americans involved in production agriculture has been decreasing since the early 1900s. Less than 2% of U.S. workers are currently classified as farmers or ranchers. This small portion of the population is still responsible for producing the food and fiber that sustains life for the rest of the nation and for export around the world. This means that fewer production agriculturalists are now

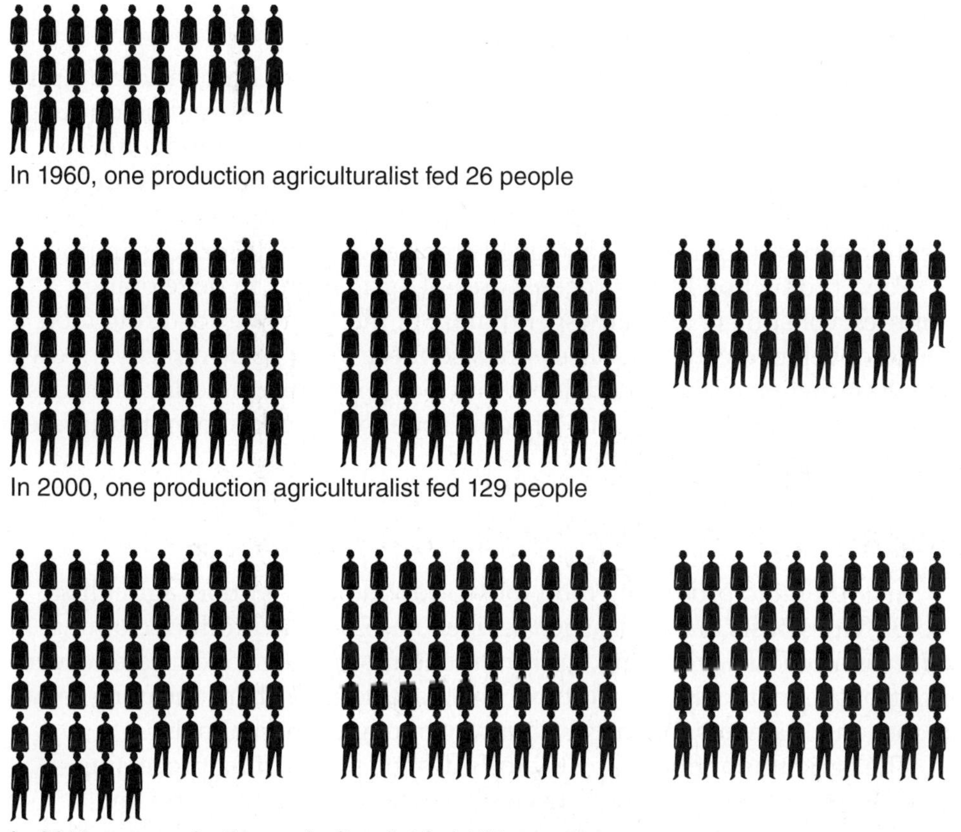

Goodheart-Willcox Publisher

Figure 1-9. There is a lot of pressure on individual production agriculturalists as they each become responsible for feeding more and more people. *Is your family in production agriculture? Can you calculate how much food your farm produces and how many people it would feed?*

responsible for feeding more of the world than ever before, **Figure 1-9**.

The agricultural industry offers many career opportunities, many of which will be examined in *Lesson 3.2*. A broad overview of the percentage of people employed in each type of agricultural career is shown in **Figure 1-10**.

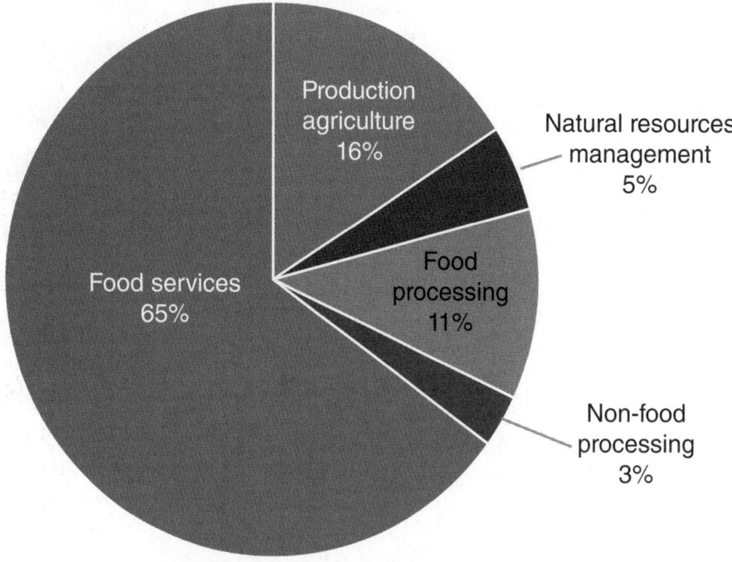

USDA, Economic Research Service using data from U.S. Department of Commerce, Bureau of Economic Analysis

Figure 1-10. The breakdown of employment in agriculture in the United States.

Areas of Agriculture

Agriculture is driven by three main commodity areas: food, fiber, and natural resources. A ***commodity*** is a raw material that can be bought and sold. Corn, market hogs, oranges, and cotton are all examples of agricultural commodities. Understanding these commodity areas will give you more insight into how humans interact with agriculture.

Food

Food includes all materials and substances used for human consumption. Food production includes the growing, harvesting, processing, distributing, marketing, and sale of all products used for human consumption. It also includes the industries related to producing food for humans, including the production of feed for animals that will enter the food chain, and the scientific support for raising improved crops and animals for consumption.

Food production is the broadest area of agriculture. In developing countries, consumers typically spend more of their income acquiring food than in developed countries. They are also more likely to be directly involved in the production of their food. How much of your household income goes to purchase food? The percent of income spent on food is shown for different countries in **Figure 1-11**.

Fiber

Did You Know?

Rayon is not a synthetic fiber. It is a manufactured fiber (ca. 1905) made mainly from wood pulp.

In addition to producing food for worldwide distribution, agriculture also provides basic materials, such as fiber, necessary to produce clothing. A ***fiber*** is a product with long, thin components, often used to create woven or composite materials. Agricultural fibers are natural fibers, meaning they are the long filaments that occur naturally in plants and animals.

The agricultural fiber industry includes the production, processing, and manufacturing of products from fibers, along with the marketing, distribution, and sales of fiber products. Agricultural fibers include cotton, wool, and pulpwood.

Figure 1-11. The percent of total income spent on food by country. *Do you come from a large or a small family? How much is your family's annual food budget?*

STEM Connection Measurement Conversion

How many is a billion? It is estimated that the United States produces more than $775 billion in agricultural products annually. A billion is a word we use often, but do we really understand how many a billion is? Let's use our math skills to figure out how many granules of sugar it would take to equal the number of dollars for annual U.S. agricultural production.

Ranglen/Shutterstock.com

If each granule of sugar were $1, it would take approximately $23,625 to fill a teaspoon.

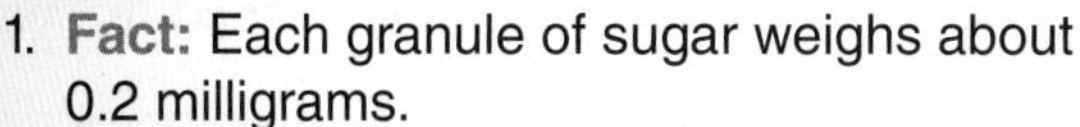

1. **Fact:** Each granule of sugar weighs about 0.2 milligrams.
   Calculation: Find out how many granules of sugar per milligram:
   1_____ ÷ _____ = _____
   (number of milligrams ÷ weight per granule of sugar)

2. **Fact:** There are 28,350 milligrams in an ounce.
   Calculation: Find out how many granules of sugar in an ounce:
   _____ × _____ = _____
   (number of granules in a milligram × number of milligrams in an ounce)
3. **Fact:** There are 16 ounces in a pound.
   Calculation: Find out how many granules of sugar are in a pound:
   _____ × _____ = _____
   (number of granules of sugar in an ounce × number of ounces in a pound)
4. **Fact:** There are 1000 million in a billion.
   Calculation: How many pounds of sugar would it take to make a billion granules?
   1,000,000,000 ÷ _____ = _____
   (one billion ÷ number of sugar granules in a pound)
5. **Fact:** The United States produces more than $775 billion in agricultural goods each year.
   Calculation: How many pounds of sugar would it take to make the number of sugar granules equal the annual value of U.S. agricultural products?
   _____ × _____ = _____
   (number of pounds of sugar to make a billion granules × number of billions of dollars)

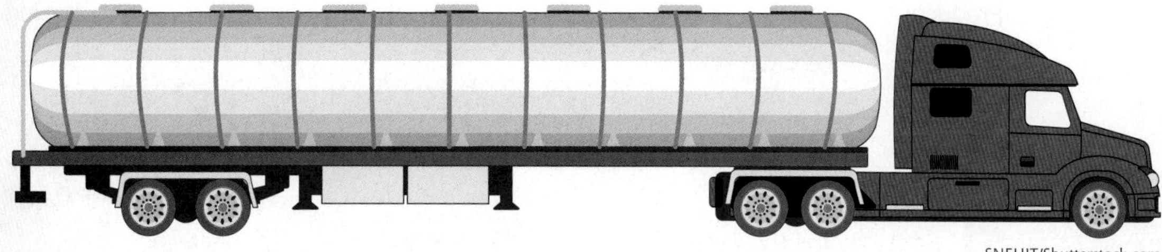

SNEHIT/Shutterstock.com

How many trailers like this full of sugar do you think it would take to equal 775 billion grains?

ANSWERS: 1. 1 ÷ 0.2 = 5 2. 28,350 × 5 = 141,750 3. 16 × 141,750 = 2,268,000 4. 1,000,000,000 ÷ 2,268,000 = 440.92 lb 5. 440.92 × 775 = 341,713 lb

Natural Resources

Natural resources are materials and substances that occur in nature, such as minerals, wood, water, and wildlife. Some natural resources are essential for our survival, while others are used to satisfy human needs and wants. Most natural resources are sold for economic gain. This area of agriculture includes ***natural resources management***, and the processing, marketing, and distribution of natural resources to consumers.

Natural resources management includes individual resource management and managing the ecosystems in which the natural resources are found. In recent years, there has been a renewed commitment to responsible natural resources management in the United States. This commitment has developed as consumers have shown an increased interest in environmental preservation.

The Human-Agriculture Connection

How we interact with agriculture can be categorized in one of two roles: *producer* or *consumer*. ***Producers*** are those actively involved in raising raw agricultural products. In the United States, very little of the population falls in this category. ***Consumers*** are those who purchase and use agricultural products. You act in the role of a consumer every time you check out at the grocery store.

The production of some products involves more people than others before they reach the consumer. Consider a fresh apple and a frozen apple pie. Which item do you think would have more human interaction before reaching the consumer? Those items that require less processing typically reach the consumer more quickly, transfer a greater share of the profit to the producer, and have a less complicated journey prior to reaching the consumer, **Figure 1-12**.

kwest/Shutterstock.com; boscorelli/Shutterstock.com; Richard Thornton/Shutterstock.com; Pavel L. Photo and Video/Shutterstock.com; Alastair Wallace/Shutterstock.com; Kondor83/Shutterstock.com; wavebreakmedia/Shutterstock.com

Figure 1-12. Think about what you had for breakfast or lunch today. *How many steps did your food go through before it reached your table?*

When a group of high school freshmen were asked to describe a world without agriculture, they described humans much like prehistoric cavemen. Were they correct? Not quite. You see, even in the earliest record of mankind, humans had already begun hunting specific types of animals and using wild plants for specific purposes. So, using items from the natural environment for our benefit is standard. It is the concept that has allowed humans to grow and evolve. Neither agriculture nor human beings would exist without the other.

A World without Agriculture?

Now that you know more about the agricultural industry, imagine a world without agriculture. Take away all the domesticated crops and animals, all the renewable sources of energy that come from natural resources, and all the by-products that come from processing agricultural products. It sounds like a pretty bleak existence. Do you have a different view of what that world would be like without agriculture after learning what agriculture is and how broad of a scope it has?

Career Connection Production Agriculturalist

Job description: Production agriculturalist is the term used to describe the profession of farming and/or ranching. Production agriculturalists work to grow plants and raise animals for food and fiber. Being a production agriculturalist requires a wide variety of skills, as this career often encompasses many other careers in the agricultural industry.

Education required: Varies; many production agriculturalists have two- or four-year degrees related to agricultural business management or specific to the commodity they produce.

Job fit: This job may be a fit for you if you enjoy working for yourself, you have a strong connection to a specific type of agricultural commodity, or you like to work outdoors.

Goodluz/Shutterstock.com Iakov Filimonov/Shutterstock.com

Production agriculturalists grow and raise raw agricultural commodities.

Lesson 1.1 Review and Assessment

Words to Know

Match the key terms from the lesson to the correct definition.

1. Those who purchase and use agricultural products.
2. Those actively involved in raising raw, agricultural products.
3. A product with long, thin components often used to create woven or composite materials.
4. All materials and substances used for human consumption.
5. Land that is suitable for growing crops.
6. Material or substance that occurs in nature, such as minerals, wood, water, and wildlife.
7. A raw material that can be bought and sold.
8. The processing, marketing, and distribution of natural resources to consumers.
9. The art and science of cultivating plants, animals, and other life forms for use by humans to sustain life.

A. agriculture
B. arable land
C. commodity
D. consumers
E. fiber
F. food
G. natural resource
H. natural resources management
I. producers

Know and Understand

Answer the following questions using the information provided in this lesson.

1. What is agriculture?
2. On what three broad categories is the scope of agriculture based?
3. Briefly explain how the relationships between specialized sectors of agriculture function.
4. Land that is suitable for growing crops is referred to as _____ land.
5. List five types of agricultural land use.
6. Agriculture as an industry is estimated to account for approximately _____ of all the money spent on trade worldwide.
   A. 5%
   B. 10%
   C. 23%
   D. 48%
7. American agriculture accounts for more than _____ dollars in sales annually.
   A. 98.5 million
   B. 775 billion
   C. 7.2 trillion
   D. 20 trillion

8. American agriculture also exports more than _____ dollars of products each year.
   A. 10.5 billion
   B. 80 billion
   C. 140 billion
   D. 15 trillion
9. *True or False?* On a global level, agriculture is the world's largest supplier of jobs.
10. *True or False?* The number of Americans involved in production agriculture has been on a steady incline since the early 1900s.
11. List the three main commodity areas that drive agriculture.
12. Identify the six areas included in food production.
13. What are natural resources? Explain why natural resources management is important.
14. Explain the difference between the agricultural roles of *producers* and *consumers*.

Analyze and Apply

1. Think about the scope of agriculture that was discussed in this lesson. List ten things that you have used in the last 24 hours that are tied to agriculture. Then identify which of the sectors of agriculture these items came from.
2. Which sector of agriculture do you think is most important to human survival? Please explain your answer with at least three reasons you think this is the most important sector.

Thinking Critically

1. Agricultural land is decreasing every day. With a partner, come up with a step-by-step plan that you feel should be implemented to stop the loss of agricultural land to development. Make sure you include to whom you will address your plan and at least three specific things you feel should be done to prevent land loss.
2. **Figure 1-7** shows the top imports and exports of agricultural goods in the United States. Carefully examine the top imports and think of specific products that would fall in each category. Develop your ideas of why the United States imports these types of products.

Lesson 1.2

History of Agriculture, Food Systems, and Natural Resources

Words to Know

agrarian civilization
artificial selection
biotechnology
domestication
Dust Bowl
Fertile Crescent
feudal system
Green Revolution
indentured servitude
land grant institutions
nomadic tribe
Norman Borlaug
precision agriculture
sharecropping

Lesson Outcomes

By the end of this lesson, you should be able to:

- Describe advancements made in agricultural production systems since 10,000 BCE.
- Understand the role of increasing global population on the agricultural industry.
- Analyze the impact of inventions and new technology on agriculture throughout history.
- Compare and contrast agriculture systems of the past to those of the present.
- Speculate on new advances that will be needed by agriculture to meet global demands.

Before You Read

Review the chapter headings and use them to create an outline for taking notes during reading and class discussion. List the main headings and sub-headings for each section. After looking over the outline, write two questions that you expect the chapter to answer.

chippix/Shutterstock.com

Figure 1-13. Agriculture has changed drastically since the time of draft animals for labor. *Can you imagine being a farmer that relied solely on draft animals to pull equipment?*

Have you ever heard the saying, "The only constant is change"? This saying certainly applies to the nature of human society. Think about things like cars, television, telephones, and music media. Are these things the same as they were a few years ago when you were in fifth or sixth grade? How have they changed in the past two or three years? How different are they from the way they were when you were born? How have changes in these technologies affected society since you were born or were in grade school? Now, think about how different these things were for your parents and grandparents when they were in high school.

Just as society changes and adapts to new ideas and technologies, the agricultural field has changed and adapted through the course of history. Advancements in agricultural management, production systems, and equipment make today's agriculture drastically different than agriculture 1000, 100, 50, or even 20 years ago, **Figure 1-13**. This lesson will examine the evolution of agriculture and allow us to gain a small glimpse into the ways agriculture has adapted to meet society's needs.

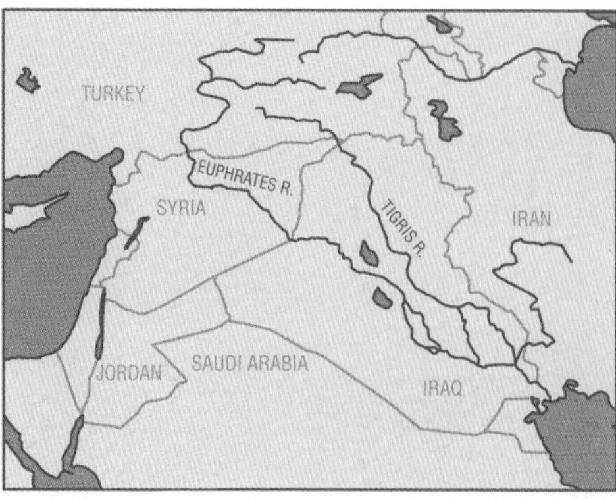

Robert Simmon/NASA

Figure 1-14. The Fertile Crescent has been termed the "birthplace of agriculture."

The Birthplace of Agriculture

Most anthropologists believe an area called the *Fertile Crescent*, shown in **Figure 1-14**, to be the "birthplace of agriculture." The ***Fertile Crescent*** is an area that extends from the eastern part of the Mediterranean to the lower Zagros Mountains in Iraq and Iran. As the climate changed around 10,000 BCE, the land in this area became more suitable for growing crops. ***Nomadic tribes*** (groups who traveled from place to place, hunting available food sources) began settling in this region, causing the shift to agrarian civilization. ***Agrarian civilizations*** are those that are based in agricultural production. Many anthropologists believe the movement of these tribes from a nomadic to agrarian society was the largest single factor in developing the modern world.

Did You Know?

The Egyptians grew their crops along the banks of the Nile River in the rich black soil that was left behind after the yearly floods.

STEM Connection Using Satellites to Protect the Fertile Crescent

Over the course of the last 30 years, major dam construction on the Tigris and Euphrates Rivers has caused a fundamental shift in the environment within the Fertile Crescent. It is estimated that the marshlands in the Fertile Crescent have decreased by 90% over this period of time. Scientists have been aided in their study of this phenomenon through the use of satellite imaging.

From 1992 to 2000, more than 16,000 pictures were taken by U.S. satellites that showed a change in the climate of the region. Over this time, the satellites showed a decrease in rainfall and a reduction in the amount of land that was covered with freshwater (marshes). These images are a more accurate way to depict the change in such a large area of land over time. Continued use of satellite monitoring has allowed researchers to more completely understand the severity of the marshland reduction.

Because of information gained through this technology, the United Nations Environment Program has allocated funds aimed at protecting this important ecological and historical region.

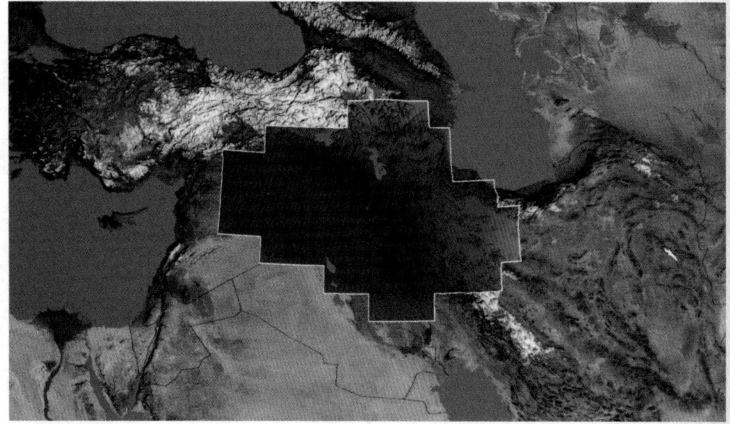

NASA's Goddard Space Flight Center

Satellite images show a sharp decline in underground freshwater reserves over the last decade. After a drought in 2007, hundreds of wells were drilled in the Tigris and Euphrates river basin to obtain water for drinking and agriculture. NASA scientists say water extraction through these wells is happening much faster than rainfall is restoring the groundwater.

Bzzuspajk/Shutterstock.com

Figure 1-15. A move to an agrarian society led to the development of permanent villages and cities, like this one in Egypt.

The Shift to an Agrarian Society

While there is much debate on the reasons nomadic tribes decided to settle and begin cultivating crops and animals, **Figure 1-15**, the shift to an agrarian society had definitive impact on human civilization. The shift to an agrarian lifestyle transformed life for early societies. The development of agrarian societies allowed people to:

- Develop more complex social organizations.
- Formulate methods for governing people and resources.
- Develop concepts of property ownership.
- Build permanent villages and cities.
- Build larger communities through an increase in population supported by an agrarian lifestyle.
- Establish business concepts through bartering, markets, and commodity exchange.
- Analyze and develop methods of controlling water flow through irrigation.
- Develop metal-working technologies.

Plant and Animal Domestication

A large part of the shift to an agrarian society is the concept of *domestication*. ***Domestication*** is the process of humans changing a plant or animal from its wild form to a domesticated form for the benefit of humans, **Figure 1-16**. The earliest domestication of plants and animals from the wild are believed to have occurred in the Fertile Crescent during the shift to agrarian society. In order

Career Connection Archaeologist

Examining ancient agriculture and society

Job description: Archaeologists uncover evidence left behind by earlier civilizations to gather information about human history. Their work can occur locally or in remote locations around the world, as they work to unearth ancient villages and artifacts.

Education required: Most archaeologists have a minimum of a four-year degree in archaeology.

Job fit: This job may be for you if you are fascinated by ancient cultures, like to solve mysteries, want to travel, or like to work outdoors.

Patricia Hofmeester/Shutterstock.com

Most archaeologists study ancient civilizations.

for an animal or plant to be considered domesticated, it must be different from its wild variety in five ways:

- First, the organism must have a structural difference.
- Second, the organism must behave differently.
  For plants, the growth pattern or structure is changed.
  For animals, their fear or aggressiveness to humans is diminished or eliminated.
- Third, the organism must rely on humans for sustenance.
- Fourth, the organism's reproduction must be subject to human control.
- Finally, there must be a clear purpose for humans to use and cultivate the organism.

cycnoclub/Shutterstock.com

Figure 1-16. Trees have been cultivated by man for hundreds of years. Fig trees are one of the earliest plants domesticated by humans. ***Have you ever eaten fresh figs?***

STEM Connection Artificial Selection

The process of domesticating plants and animals is rooted in the scientific principle of artificial selection. ***Artificial selection*** is the process of selecting which organisms to breed to create offspring with more desirable characteristics. Look at the picture of the animal in the figure below. Which of its characteristics could be helpful to humans? Design a plan for selectively breeding this wild animal to create a domesticated version.

1. First, decide what the purpose of domestication would be. For which physical traits would you selectively breed?
2. Next, determine ways that you would want the animal's behaviors to change. Which behavioral traits would you artificially select?

By determining what the animal's purpose for humans would be, then determining which traits to select to meet that purpose, you are employing the scientific concepts of artificial selection. That is exactly what early agriculturalists did to domesticate ancestors of the livestock we use for food and fiber today.

Oleg Znamenskiy/Shutterstock.com

How would you use artificial selection to domesticate these gazelles?

Did You Know?

The undomesticated ancestor of the potato was poisonous to humans. Ancient people in Peru would eat them with a mud made from clay and water to prevent the potatoes from killing them.

Most experts believe that fig trees, wheat, and millet were the first domesticated crops. Domestication of animals occurred at roughly the same time as the earliest crop domestication, with one exception, the dog. Most anthropologists agree that the dog was domesticated as early as 30,000 BCE. Evidence suggests nomadic tribes kept dogs with them for protection, hunting, and companionship. Livestock animals were domesticated as humans moved into a more civilized agrarian society. Timetables of plant and animal domestication are shown in **Figure 1-17**.

Early Plant Domestication

Plant	Region	Date
Fig trees	Fertile Crescent	9000 BCE
Wheat	Fertile Crescent	9000 BCE
Foxtail millet	East Asia	9000 BCE
Chickpeas	Turkey	8500 BCE
Gourds	Central America	8000 BCE
Rice	Asia	8000 BCE
Potatoes	Andes Mountains	8000 BCE
Beans	South America	8000 BCE
Squash	Central America	7500 BCE
Maize	Central America	7000 BCE

Early Animal Domestication

Animal	Region	Date
Dogs	Unknown	14,000 – 30,000 BCE
Sheep	Fertile Crescent	8500 BCE
Cats	Fertile Crescent	8500 BCE
Goats	Fertile Crescent	8000 BCE
Pigs	Fertile Crescent	7000 BCE
Cattle	Eastern Sahara	7000 BCE
Chickens	Asia	6000 BCE
Llamas	Andes Mountains	4500 BCE
Alpacas	Andes Mountains	4500 BCE
Donkeys	Africa	4000 BCE

Goodheart-Willcox Publisher

Figure 1-17. Through research and excavation, archeologists have determined the time periods in which many plants and animals became domesticated. *Use the information above to create a timeline and to determine periods of overlap.*

Early Agriculture on a Global Scale

Since its beginning in the Fertile Crescent, agriculture has been a driving factor in the evolution of civilization. Major agricultural developments have spurred changes in society, and developments in society have provided the stimulus for advancements in agriculture. While the impact of adaptation and developments may be limited to local areas, it often has, or has had, a ripple effect that travels on a global scale.

Fertile Crescent

Agriculture in the Fertile Crescent was largely managed by individual farmers. Each farmer owned the land he farmed and managed the sale of his commodities. Successful farming in the Fertile Crescent gave rise to some of the largest civilizations recorded in ancient history, including that of the Ancient Egyptians.

East Asia

Agriculture was driven largely by desire for products produced in Asia by countries in the Middle East and Europe. The government would often employ large numbers of people to work in rice fields, like those shown in **Figure 1-18**, or collect and produce products from silkworms. Even today, a large number of people in Asia are employed in the production of agricultural goods for export to western markets.

Did You Know?

Silk is considered an animal fiber because it has a protein structure. When unwound, a single filament can be as long as 1,600 yards!

John Bill/Shutterstock.com

Figure 1-18. Rice production was a force that led to the expansion of early agriculture in Asia.

Europe

As society developed between 800 ADE and 1400 ADE, many European nations managed agriculture with a feudal system. In the ***feudal system*** of agricultural management, land was granted to those who completed military service. These landowners would lease the land, along with supplies, to a farmer in exchange for the payment of taxes and goods as rent. While the feudal system worked well in some cases, many historians cite abuse of the system by the landowners as a key factor leading to its end.

Central and South America

Did You Know?

The Incan farmers did not have beasts of burden or iron tools but were still able to grow crops in all sorts of terrain from the deserts to the high mountains.

In Central America, advances in agriculture were the driving forces behind the success of ancient civilizations like the Mayans, Aztecs, and Incas. Well-developed agricultural systems allowed these groups to produce enough food and fiber to support large populations. In South America, specialized production of crops such as sugar cane and coffee became important when settlers from Europe first visited these regions.

Agriculture in North America

Although new technologies have brought into question exactly when agriculture was introduced in North America, it has long been accepted that Native Americans were the first to widely develop agriculture systems in the United States.

While many tribes were hunters and gatherers relying on the abundance of wild plants and animals for their diet, other tribes were agrarian, using vast tracts of fertile land to cultivate crops like maize, beans, sweet potatoes, peanuts, and tomatoes. Unlike other early agriculturalists, Native Americans did not raise many domesticated animals. An ample supply of fish and wild game provided sufficient protein to their diets.

Peter Dennis/Thinkstock

Figure 1-19. Native American knowledge of agriculture helped save hundreds of early European settlers from starvation.

Settlers in the New World

Many of the earliest European settlers coming to North America had little or no agricultural experience. Many of the settlers came to the New World with hopes of finding natural resources they could sell on the European market. These early settlers were ill-prepared to obtain their own food, and without assistance and teaching from Native Americans, would have likely perished due to starvation, **Figure 1-19**.

As time passed, settlers brought additional agricultural skills and commodities to the colonies. These settlers introduced crops such as rice, wheat, barley, and oats to the New World and also brought livestock such as cattle, sheep, and pigs. As the settlers became more adept at farming in the New World, they began exporting crops such as corn, chocolate, cotton, and tobacco to Europe. The growing European desire for New World agricultural products was a large driving factor for the settlement of North America.

The American Revolution

Following the American Revolution, the newly formed United States was even more reliant on agricultural export to European markets than the colonies had been. No longer supported by Great Britain, the American citizens needed to create more income to support the burgeoning nation. The strong demand for crops like tobacco, cotton, and rice, led to the widespread use of slavery on large row crop farms called plantations. The reliance on income from agricultural crops necessitated the increase of farm laborers, hence the importation of more slaves and the beginning of a dark period in American agricultural and societal history.

Did You Know?

George Washington first suggested that Congress form an agriculture department in 1799. The U.S. Department of Agriculture was not officially created until 1862, when Abraham Lincoln authorized the formation of the "people's department."

Post-Civil War

After the Civil War and the abolishment of slavery, the agricultural industry needed to once again establish a labor force. This increase in labor force was accomplished through two types of agricultural labor systems: sharecropping and indentured servitude.

Sharecropping

Sharecropping was a method of agricultural labor in which the landowner provided the use of arable land to a farmer in exchange for a share of the crop produced. Sharecropping agreements differed across the country and over time, but almost always favored the landlord. At the end of a season, workers were often paid with only a third of the crop produced and, in many cases, were indebted to the landowner or other supplier for necessities (seed, fertilizer, clothing, shoes, etc.) purchased on credit from the "plantation store."

Sharecropping came to an end as people moved to cities and new technologies, such as tractors and cotton pickers, allowed landowners to work their land with less labor and more efficiency.

Indentured Servitude

Indentured servitude is a system of agricultural labor in which a person is indebted to someone and required to work for that person until their debt is paid. After the Civil War, indentured servants were generally European immigrants who came to America on tickets purchased by a landowner with the agreement that they would work for the landowner until their debt, and often a large amount of interest on the debt, was paid off. As with sharecropping, indentured servitude agreements almost always favored the landowner.

Academic Connection The Life of a Sharecropper

There are many opportunities to examine sharecropping in the time after the Civil War. Consider these facts about sharecropping:

- At the end of the Civil War, there were many more black Americans with agricultural skills in the South than white Americans with agricultural skills.
- Many freed blacks returned to the plantations where they had been slaves to work as sharecroppers.
- Both white and black Americans worked as sharecroppers for wealthy landowners.
- Often, sharecropping contracts were written in ways that left little chance for the sharecropper to make a profit.

With a group of 2–3 students, use these facts to answer the question: "Was sharecropping good for the culture of the South in the late 1800s?"

If you are interested in knowing more about the life of a sharecropper, books you may want to read include *Tobacco Road* by Erskine Caldwell, or *A Childhood: The Biography of a Place* by Harry Crews.

Arthur Rothstein, Library of Congress

Figure 1-20. Poor practices led to the Dust Bowl of the 1930s. Much of the fertile Midwest soil appeared as barren desert. *Could there be another Dust Bowl in the United States?*

The Dust Bowl

American agriculture was extremely prosperous in the early 1900s. This prosperity led farmers to plow hundreds of thousands of acres of range and pasture to produce wheat in the years from 1925 to 1930. This poor management of agricultural land, combined with years of drought, led to the most devastating human-created ecological disaster in American history: the Dust Bowl. The ***Dust Bowl*** is the name used to describe the time period from 1930–1936 when the mismanagement of American cropland led, in part, to massive dust storms spreading throughout the Midwest region of the United States, **Figure 1-20**. The Dust Bowl had devastating effects on the land, its farmers, and society as a whole. The Dust Bowl:

- Resulted in the loss of topsoil from almost one million acres of farmland.
- Was a contributing factor to the Great Depression.
- Forced thousands of Americans to leave their homes in search of more inhabitable areas.
- Changed the social climate of the country by breeding animosity and resentment amongst and toward fellow Americans searching for homes and employment.
- Drastically changed the methods of farming and regulations for soil conservation in the United States.

Did You Know?

The dust storms during the Dust Bowl could generate enough static electricity to knock people to the ground if they tried to shake hands during a storm.

Green Revolution

Between 1940 and the late 1960s, many scientists focused on developing new technology that would increase agricultural production on a worldwide scale. Their goals included:

- The development of higher-yield crops.
- The development and implementation of irrigation systems.
- Improved pest and disease control.
- The development of synthetic fertilizers.

Their work led to huge technological advances in agricultural production and increased agricultural production worldwide. The time period became known as the ***Green Revolution*, Figure 1-21**. ***Norman Borlaug*** (1914–2009) is credited as being the father of the Green Revolution. His scientific exploration of the genetics and production of wheat is credited with saving more than a billion people from starvation, and earned him a Nobel Peace prize. Lessons learned from the Green Revolution are used today as scientists continue to use new technology to help secure a more stable world food supply.

luchschen/Shutterstock.com

Figure 1-21. Widespread scientific research in the mid-twentieth century led to great advances in agriculture called the Green Revolution

Legislation That Influenced Modern Agricultural Education

There are many important pieces of governmental regulation that helped shape agriculture and agricultural education in the United States. Some of the most influential legislation includes:

- Morrill Act (1862)—this act set aside land in every state for college-level education in agriculture. These colleges were called ***land grant institutions***.
- Homestead Act (1862)—this legislation allowed settlers to claim 160 acres of land in a designated area and earn ownership rights by fencing it, digging a well, plowing ten acres for agricultural production, building a house, and living there. This act had a large impact on western expansion in the United States.
- Hatch Act (1887)—this act designed the setup of agricultural experiment stations. These stations were developed for regional research in agricultural commodities.
- Smith-Lever Act (1914)—provided federal funding for outreach education for land grant institutions. These outreach programs are still in place today in the form of cooperative extension programs.
- Smith-Hughes Act (1917)—set up federal funding for agricultural education for students prior to graduation from high school. This act essentially started high school agricultural education.

Did You Know?

The first high school to offer agricultural education courses under the Smith-Hughes Act legislation was Woodlawn High School in Woodlawn, Virginia.

Inventions That Changed Agriculture

The agricultural industry relies heavily on machinery and equipment, and many of the biggest changes in agriculture have come about because of mechanical inventions that changed the way crops were produced, harvested, and processed. Each invention or new technology affects every aspect of agricultural production—from the seeds planted to the amount of labor required to harvest and process commodities. For example, the amount of time required to produce 100 bushels of wheat or corn has significantly decreased since 1830, **Figure 1-22**. This drastic reduction can be directly attributed to mechanical inventions that have made harvesting these crops much less labor-intensive.

Although it is not possible to list all the inventions that have contributed to advances in the agricultural industry, we can review a few key inventions that revolutionized American agriculture.

Did You Know?

The plow has been cited as one of the most important inventions in the advancement of society.

Cotton Gin

One of the most important advances in American agriculture was Eli Whitney's 1793 invention of the cotton gin. The cotton gin replaced hand labor with a machine that could effectively remove seeds from the cotton fibers. Using Whitney's revolutionary cotton gin, one worker could clean

Hours of Labor to Produce Crops

Year	100 Bushels of Wheat	100 Bushels of Corn	Methods
1830	250–300	75–90	Walking plow, harrow, hand planting
1890	40–50	35–40	Gang plow, seeder, thresher, wagons, horses
1930	15–20	15–20	2-bottom gang plow, disk, cultivators, pickers, tractor, combine
1945	10–12	10–14	Tractor, 3-bottom plow, tandem disk, 4-section harrow, 4-row planters and cultivators, and 2-row picker
1975	3.75	3	Tractor, 30-foot sweep disk, 27-foot drill, 22-foot self-propelled combine, 5-bottom plow, 20-foot tandem disk, planter, 20-foot herbicide applicator, 12-foot self-propelled combine, and trucks
Today	Less than 3 hours	Less than 2.75 hours	Machinery, GPS-enhanced equipment using precision pesticide and herbicide applications

Goodheart-Willcox Publisher

Figure 1-22. The use of agricultural machines has greatly reduced the number of hours needed to produce crops.

50 pounds of cotton in one day! This was 50 times more cotton (per day) that a worker could clean than when removing seeds by hand, **Figure 1-23**.

Mechanical Reaper

In the earliest parts of the nineteenth century, most grains were reaped, or cut down, by hand using blades on long handles. In 1837, Cyrus McCormick developed the first horse-drawn reaper that could be used to reap grains mechanically, **Figure 1-24**. This machine dramatically reduced the amount of time required to harvest cereal grains and led to a massive increase in the amount of grains planted each year in the United States.

Steel Plow

A broken sawmill blade inspired a blacksmith to create one of the most important agricultural inventions in history. In 1837, John Deere, a blacksmith from Illinois, saw the steel saw blade as a possible solution to tilling the sticky clay soil of the Midwest. Prior to the steel plow, farmers would have to stop and clean the dense soil off their cast iron plows

Jerry Hobert/Shutterstock.com; Steven Wynn/iStock/Thinkstock

Figure 1-23. Before Eli Whitney's cotton gin revolutionized cotton processing in the United States, plantation workers would spend hours removing the small cotton seeds from the cotton fiber by hand. ***How did the cotton gin remove the seeds?***

overcrew/Shutterstock.com; Hein Nouwens/Shutterstock.com

Figure 1-24. Imagine the difference in harvest time that farmers experienced when moving from hand reaping (left) to the use of the McCormick horse-drawn reaper (right).

smereka/Shutterstock.com

Figure 1-25. Before the invention of the steel plow, clay-based soils would stick to the plow and required the farmer to carry a large wooden paddle for scraping excess soil, which could prevent the plow from working.

with a wooden paddle. The steel plow helped to decrease downtime and increase the speed with which the tough Midwestern soil could be turned. See **Figure 1-25**.

Steam Tractor

The booming railroad industry helped lead to the invention of the first steam-powered tractor in 1868. Although these machines were very heavy and not incredibly well suited to farm work, their invention was the first engine driven labor source in agriculture and led to the decreased use of draft animals for agricultural power. See **Figure 1-26**.

pwrmc/Shutterstock.com

Figure 1-26. While not perfect by any means, the first steam tractors symbolized the advancement of agricultural machinery in the twentieth century. ***What types of problems were there with steam-powered tractors?***

Barbed Wire

The patent for what is considered the first barbed wire was awarded to Joseph Glidden in 1874. Barbed wire became an important tool in managing the large tracts of land where cattle were being housed in the Great Plains region. In contrast to open range and long cattle drives moving animals between grazing lands and market, barbed wire allowed cattle ranchers to keep their animals in managed pastures. Having more control over the area where cattle could roam led to an increase in selective breeding and improvements in cattle genetics.

General Purpose Tractor

Although tractors had been used since the 1860s, a widely available general purpose tractor was not available until the mid-1920s. In order for the tractor to be considered general purpose, it needed to be cost effective for the average farmer and replace the majority of dependence on draft animals for farm labor. While some debate exists, it is widely accepted that the Farmall tractor, produced by the International Harvester company in 1924, was the first general purpose tractor. The widespread use of tractors made a huge impact on modern agriculture.

Did You Know?

The only way to get the first tractors delivered from the rail station to the fields was to pull them with draft animals.

Satellite Technology

In 1994, the first global positioning systems (GPS) for agriculturalists were used. GPS technology allows farmers to use precision agriculture techniques. ***Precision agriculture*** is the method of managing agricultural land with the assistance of computer or satellite information, **Figure 1-27**. Agriculturalists can use GPS technology to control fertilizer and pesticide application, monitor erosion, collect information related to yields in specific areas within larger fields, monitor pest infestation, monitor crops for disease, and evaluate irrigation issues. GPS technology may also be used to monitor and maintain machinery as well as monitor the location of employees working alone in distant areas.

Did You Know?

There are three main groups of satellites: fixed satellites that handle voice, data, and video; mobile satellites that are used for navigation; and scientific satellites that handle meteorological data, land survey images, and other scientific research functions.

Biotechnology

Biotechnology is the use of scientific modification to the genetic material of living cells in order to produce new substances or functions. Biotechnology also includes the process of genetic engineering, where a gene from one living organism is inserted into the genetic makeup of another organism. The first biotechnology crop approved for human consumption was the Flavrsavr™ tomato, which was first available to consumers in 1994. In 1997, the first crops with genetically engineered pest resistance and the first crops with herbicide resistance became commercially available. Since that time, biotechnology has become a valuable tool for agricultural production. Biotechnology in agriculture is explored more in *Lesson 5.3*.

Denton Rumsey/Shutterstock.com

Figure 1-27. It is amazing to see just how far agriculture has progressed in recent years. These potatoes are being planted in Idaho with a GPS guided tractor. ***In what other ways are farmers using modern technology to produce higher yields?***

Lesson 1.2 Review and Assessment

Words to Know

Match the key terms from the lesson to the correct definition.

1. Time period between 1940 and the late 1960s in which scientists focused on developing new technology that would increase agricultural production on a worldwide scale.
2. The use of scientific modification to the genetic material of living cells to produce new substances or functions.
3. Groups of people who traveled from place to place, hunting available food sources.
4. The process of humans changing a plant or animal from its wild form to a domesticated form for the benefit of humans.
5. A system of agricultural labor in which a person is indebted to someone and required to work for that person until their debt is paid.
6. The process of selecting which organisms to breed to create offspring with more desirable characteristics.
7. The time period from 1930–1936 when the mismanagement of American cropland and climate changes led to the loss of topsoil throughout the Midwest.
8. Scientist whose exploration of the genetics and production of wheat is credited with saving more than a billion people from starvation.
9. The area referred to as the birthplace of agriculture.
10. Organized groups whose society is based in agricultural production.
11. An agricultural management system in which the land was owned by those who completed military service.
12. A method of agricultural labor in which the landowner provided the use of arable land to a farmer in exchange for a share of the crop produced.
13. The method of managing agricultural land with the assistance of computer or satellite information.

A. agrarian civilization
B. artificial selection
C. biotechnology
D. domestication
E. Dust Bowl
F. Fertile Crescent
G. feudal system
H. Green Revolution
I. indentured servitude
J. nomadic tribe
K. Norman Borlaug
L. precision agriculture
M. sharecropping

Know and Understand

Answer the following questions using the information provided in this lesson.

1. What is the Fertile Crescent and why is it important to the history of agriculture?
2. Briefly explain the difference between a nomadic tribe and an agrarian civilization.
3. *True or False?* Many anthropologists believe the movement from wandering tribes to agrarian society was the largest single factor in developing the modern world.

4. List five significant social changes that came about because of the development of agrarian societies.
5. In what ways must an animal or plant be different from its wild variety in order to be considered domesticated?
6. *True or False?* Agriculture in the Fertile Crescent was largely maintained by sharecroppers.
7. *True or False?* Agriculture was driven largely by the desire for products produced in Asia by countries in the Middle East and Europe.
8. In feudal systems of agricultural management, land ownership was _____.
   A. restricted to those who belonged to royal families
   B. exchanged for harvested crops
   C. granted to those who completed military service
   D. None of the above.
9. *True or False?* Well-developed agricultural systems allowed ancient civilizations like the Aztecs to produce enough food and fiber to support large populations.
10. List four crops cultivated by early Native Americans.
11. What was a large driving factor for agricultural products in the early settlement days of North America?
12. Briefly explain the difference between *sharecropping* and *indentured servitude.*
13. How did agricultural practices contribute to the Dust Bowl in the 1930s?
14. What was the *Green Revolution* and what were some of its goals?
15. List five key inventions that revolutionized American agriculture.

Analyze and Apply

1. Use the Internet and your local library to look for period articles on land conservation. Focus on a particular conservationist and the area(s) with which they were involved. Find a historic illustration or famous painting of the same area. Find current images and articles of the same area. Is it the same? How different is it? Make a list of differences or changes. Create a compare/contrast chart with the information.
2. There are many reports that the Fertile Crescent is "drying up." Explain how humans have affected this area and if it is possible to restore it to its once bountiful splendor.

Thinking Critically

1. Imagine that you were alive and owned a greenhouse and florist retail shop during and just after World War II. What effects would the war and its aftermath have on you as the owner?
2. You are thinking about taking a vacation to a historically accurate 1920s farm. This farm uses only equipment available before 1929. What will you expect to see when observing the workers plowing, cultivating, and harvesting crops? Please explain which equipment they might be using and which equipment you would not see in this operation.

Lesson 1.3

Future of Agriculture, Food Systems, and Natural Resources Management

Words to Know

engineering
information system
informed consumer
STEM
sustainability
sustainable
sustainable agriculture
sustainable energy
technology
trend

Lesson Outcomes

By the end of this lesson, you should be able to:

- Examine factors that determine the future of agriculture, food systems, and natural resources.
- Explain the areas of technology that will affect the future of agriculture, food, and natural resources.
- Analyze your role in the future of agriculture, food systems, and natural resources.

Before You Read

Be prepared to read this lesson with paper for recording questions. As you read the lesson, record any questions that come to mind. Indicate where the answer to each question can be found: within the text, by asking your teacher, in another book, on the Internet, or by reflecting on your own knowledge and experiences. Pursue the answers to your questions.

denisgo/Shutterstock.com

Figure 1-28. Experts in the 1960s had a very different idea of how society would look today. *What do you think the world will look like in 50 years?*

In the mid-twentieth century, it was hypothesized by a group of scientists that by the year 2010, most Americans would have an at-home "computation terminal," drive a flying car, and eat mainly freeze-dried foods, **Figure 1-28**. While it is true that most homes today do have computers, these well-qualified experts missed the mark on many of their predictions. We may not be able to predict the future, but careful study of past changes and the factors that drove those changes, will give us a better idea of what the future holds for agriculture, food systems, and natural resources management. Taking a close look at evolving technologies and their involvement in the agricultural industry will give us a better idea of how the future of agriculture will affect us as individuals and society as a whole.

Driving Factors

In order to determine what will drive the future of agriculture, food systems, and natural resources, we need to use our deductive thinking skills to examine what has happened to create the agriculture of today. When we look at the things that have stimulated change in the past, we can see that many changes have been driven by three main factors: population growth, consumer demands, and long-term sustainability.

Population Growth

The first factor in determining the future of agriculture, food systems, and natural resources management, is population growth. As stated in *Lesson 1.1*, agriculture is the art and science of cultivating plants, animals, and other life forms for use by humans to sustain life. Because agriculture is production for human use, it would make sense to conclude that the future of agriculture is directly dependent on the number of people that will be on the planet.

According to the Population Division of the United Nations, the world population is projected to grow to 9.6 billion by the year 2050, **Figure 1-29**. More than half of the world population lives in urban areas, away from the land required to produce their food. This fact requires the timely shipping of fresh goods and adds more pressure to those involved in production agriculture.

Think about how the rate of global population increase will affect agriculture. How does a rapid population increase affect the overall need for change in agricultural systems?

In the early twentieth century, agriculturalists would simply have put more arable land into production in order to provide food and fiber for a growing population. The issue facing agriculture now is that arable land is decreasing at an alarming rate. The only remaining option for the agricultural industry is to adapt for the future by developing more efficient ways to produce agricultural commodities with the land currently in production.

Did You Know?

At current population growth rates, the global population increases by one person every 0.42 seconds. That is an increase of around 142 people per minute.

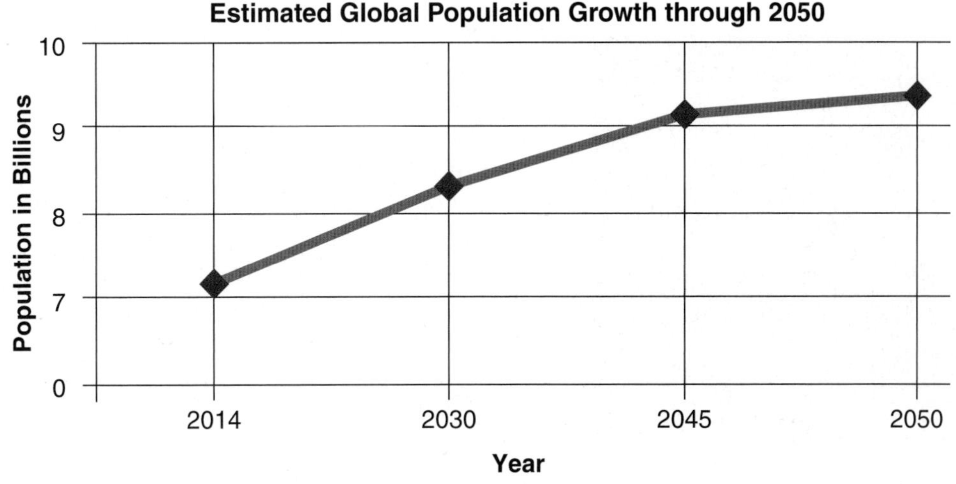

Figure 1-29. With such a quickly growing population, today's agriculturists are under increased pressure to produce enough food to meet the growing demand.

Career Connection Market Research Analyst

Determining what consumers want

Job description: Market research analysts use a variety of tools in order to determine consumer demographics, sales trends, and buying habits, along with developing projections for future consumer demands.

Education required: Most market research analysts have a four-year degree in a field related to communication, statistics, or business administration.

Job fit: This job may be for you if you like working with both people and numbers, if you have a natural curiosity about why people buy certain items, or if you enjoy interacting with all types of people.

Andresr/Shutterstock.com

Market research analysts collect and interpret data from consumers to project future trends.

Consumer Demands

Like any industry, agriculture relies on consumers to purchase and use the commodities produced. Production agriculturalists are constantly working to develop and promote products that consumers will find desirable. Oftentimes, trends will emerge in consumer demands that become the driving factor in agricultural production. A ***trend*** is the general direction that something is developing or changing. Understanding exactly what the consumer will desire is important in projecting what future agricultural products might be.

There are several examples of consumers driving change in the agricultural industry. When the desire for leaner pork products emerged in the 1970s, the industry modified to breed swine that carried less body fat. The swine we raise today look nothing like the swine we raised prior to the shift. A more modern example of consumer demands driving change in the agricultural industry is the increase in desire for gluten-free products. As more and more consumers desire products that are made without gluten, those involved in food processing and production are continually exploring methods for creating new products without this ingredient. These goods are marketed to a growing group of gluten-free consumers. The gluten-free market, which was once a very small component of the total food production industry, gained in popularity and production in a short amount of time due to consumer demands, **Figure 1-30**.

ChameleonsEye/Shutterstock.com

Figure 1-30. The increased market for gluten-free products is an example of how consumer demands will play a role in the future of agriculture, food systems, and natural resources. *Can you name food products that were brought about by consumer demands?*

Sustainability

Another important factor that will determine the future of agriculture, food, and natural resources management, is the concept of sustainability. When something is ***sustainable***, it is able to be used without being used up or permanently damaged. Therefore, in terms of our natural environment, ***sustainability*** is based on the fact that everything we need to survive depends on our natural environment and how we take care of it.

Sustainability related to agriculture, food, and natural resources can be looked at in terms of energy sustainability or agricultural sustainability. ***Sustainable energy*** is energy that can be used to meet current needs without compromising the ability to meet future energy needs. For example, solar energy, wind energy, and geothermal energy are renewable energy sources that can be harnessed indefinitely. Agriculture and natural resources play an important role in energy sustainability through the use of renewable energy sources. The future of renewable energy is definitely a factor that will affect the future of agriculture, food systems, and natural resources management.

Did You Know?

Careful stewardship by farmers has spurred a nearly 50% decline in erosion of cropland by wind and water since 1982. (*American Farm Bureau Federation*)

Sustainable Agriculture

Sustainable agriculture, **Figure 1-31**, is the process of producing agricultural products using techniques that protect the environment and all living beings while allowing agricultural land to maintain production for many years.

Because agricultural land is lost daily to development, using sustainable agricultural practices allows producers to ensure that the land currently in production is being managed in a way that will not reduce its ability to produce food and fiber for future generations. Sustainable agricultural practices are discussed more completely in Chapters 13 and 15 and include practices such as rotating crops, integrated pest management plans, and managing erosion.

romrf/Shutterstock.com

Figure 1-31. Sustainable agriculture involves making management decisions that will protect the usefulness of the environment for future generations. *What actions can you take in your daily life to help protect our environment?*

Technology and Agriculture

Technology is the application of scientific knowledge, tools, or processes for practical purposes. As the global population, consumer demands, and sustainability factors lead to changes in the industry, scientific principles will be used to make changes and bring about new technology. Advances in technology that will affect the agricultural industry include technological advances in engineering, biotechnology, and information systems.

MOLPIX/Shutterstock.com

Figure 1-32. Complex agricultural machinery, like this rice mill, will rely on engineering technology to make them more efficient in the future. ***What are some of the benefits of more efficient agricultural machinery?***

Photosebia/Shutterstock.com

Figure 1-33. Small drone aircrafts are being developed for monitoring and herding livestock. ***For what other agricultural purposes are drones being used?***

Engineering

Engineering is the branch of science concerned with designing and building new machines, power systems, and structures. This section of technology is incredibly important to agriculture since it relies heavily on equipment to cultivate, harvest, and process raw products. Many of the advances in agriculture are made through the use of more efficiently engineered equipment, as shown in **Figure 1-32**.

The future of the agricultural industry will depend on engineering advances that allow more commodities to be grown on the available land and transferred to consumers. While we cannot predict the future of engineering technology in agriculture, there are already some advanced machines being developed. Testing is already underway for remote-controlled tractors and harvesting equipment, planting equipment that can test the soil and plant different varieties in the same field, and small drone aircrafts designed for checking and moving livestock, **Figure 1-33**.

Biotechnology

The next area of technology that will be important to the future of agriculture, food, and natural resources is biotechnology. Simply stated, *biotechnology* is a collection of scientific techniques used to modify plants, animals, and microorganisms to improve their:

- Longevity
- Disease resistance
- Drought resistance
- Flavor
- Adaptability

Scientists are working on creating organisms that will be able to withstand conditions outside of arable environments, including land that has too much salt or too little water for traditional crops, **Figure 1-34**. The application of biotechnology may be a way to solve many production issues facing agriculture. Biotechnology allows changes to be made more quickly than traditional animal and plant breeding.

A more thorough discussion of biotechnology, including the processes used and social concerns, is available in *Lesson 5.3*. Currently, many agricultural products are being used that have been developed using biotechnology. What changes do you think the future holds?

Did You Know?

Advances in biotechnology have allowed a gene from the Arctic flounder to be transferred to plants like tomatoes and strawberries to ease their susceptibility to cold.

Charles Brutlag/Shutterstock.com

Figure 1-34. Biotechnology is working to develop new varieties of crops that can grow in areas that have not historically been arable land. ***Do you foresee drawbacks with this technology?***

Information Systems

Information systems include the equipment and software that are used to collect, filter, process, and distribute information. In other words, the systems used to process data. Vast networks of computers and mobile devices are used in the agricultural industry, making this section of technology one of the most rapidly growing contributors to the future of agriculture.

One of the biggest advantages to increasing the use of information systems technology in the agricultural industry is increasing the amount of precise data collected about production and commodity yields. Having precise information of how much of a commodity is being produced under certain conditions allows producers to make better management decisions and customize production to maximize yields, **Figure 1-35**. Some examples of information systems technologies that are being developed and used include items such as:

- A smartphone accessory that can test cattle blood for diseases.
- A combine that collects harvest yields accurate to the square foot of a field.
- A dairy system that monitors production of each individual cow through a computerized ankle band.

swissmacky/Shutterstock.com

Figure 1-35. Advances in information technology is allowing production agriculturalists around the world greater access to collecting data and making informed management decisions.

You and the Future of Agriculture

As a young adult, the future of agriculture, food systems, and natural resources will definitely include you. Whether you choose to pursue a career in agriculture that helps shape the future or are simply a consumer of agricultural products, you will play a role in the path that the future will take.

Production

Have you ever thought about a career in agriculture or natural resources? If so, there are jobs in every facet of the agricultural industry that would allow you to have a hand in developing the future. *Lesson 3.2* outlines opportunities for you to use your passion, skills, and talents to help the future of agriculture and natural resources.

STEM Connection

One of the best ways to be involved in the production of agriculture and natural resources is by continuing your education and gaining more knowledge in the science, technology, engineering, and mathematics fields. ***STEM*** is the acronym for science, technology, engineering, and mathematics. It is a term coined by the National Science Foundation in the early 2000s to describe the areas of education that are vital to the development of students and new technologies.

There are many careers in STEM-related fields. By choosing a career in any of these areas, you can play a vital role in developing the new technologies needed to adapt and meet the needs of population growth, consumer demands, and sustainability in the future.

Because agriculture is an industry that combines so many areas, think about all the ways that STEM is applied! For example, agriculture combines science and technology by using the scientific principle of methane production to develop technology that harnesses methane from animal waste.

Can you think of a place where agriculture, food, and natural resources combine the following concepts of STEM?

- Science and engineering
- Engineering and mathematics
- Science, engineering, and technology

STEM Careers

Agricultural Architects
Agricultural Engineers
Agricultural Scientists
Airplane Pilots
Animal Scientists
Aquatic Technicians
Biologists
Broadcast Technicians
Chemical Engineers
Chemists
Civil Engineers
Communications Equipment Mechanics
Computer Engineers
Computer Operators
Computer Programmers
Computer Support Specialists
Computer Systems Administrators
Computer Systems Analysts
Conservation Scientists
Construction and Building Inspectors
Construction Managers
Data Communications Analysts
Dietitians
Drafters
Electrical and Electronics Engineers
Electricians
Engineering Managers
Environmental Engineers
Environmental Scientists
Food Scientists
Foresters
Forestry Technicians
Geologists and Geophysicists
High School Teachers
Marine Biologists
Mechanical Engineers
Meteorologists
Occupational Health and Safety Specialists
Park Rangers
Quality Control Inspectors
Science Technicians
Statisticians
Surveying and Mapping Technicians
Surveyors
Transportation Inspectors
University and College Teachers
Urban and Regional Planners
Veterinarians
Veterinary Assistants
Water Treatment Plant Operators
Zoologists

Is production agriculture for you? If you enjoy working with a specific agricultural commodity, you might want to consider being involved in helping to produce the crops or livestock needed to help feed the global population of the future. You can be involved in production agriculture even if it is not your main employment. In fact, according to the USDA, 91% of families that own a farm have at least one household member with an off-farm job. That means that much of our agricultural production is done by those who enjoy producing agricultural commodities even though it is not their main source of employment.

Consumption

Even if production agriculture is not something you want to pursue as a career, you will still have a place in the future of agriculture, food, and natural resources. By eating, purchasing clothing, and using any of the thousands of products that are developed from agricultural commodities, you will be a lifelong consumer. Becoming an ***informed consumer*** is an important step to fulfilling your role in the future of agriculture, food, and natural resources. The two main steps of becoming an informed consumer are:

- Knowing the origin and understanding the processes used to produce agricultural products.
- Making decisions about trends in products.

Informed Consumers

In the beginning of the lesson, you read how consumer demands will help determine the types of products produced in the future. The products you purchase are included in these consumer demands, **Figure 1-36**. Understanding where agricultural products come from and the production processes used to produce them will prepare you to make smart consumer decisions. Reading this textbook and enrolling in high school agricultural classes will give you a good foundation in your understanding of the agricultural industry and also help you become a well-informed consumer.

Once you understand how commodities are produced, the next step is to develop knowledge about the wants and needs of consumers or your own desires as a consumer. As in any industry, trends exist in the agricultural industry. The move toward gluten-free products is a good example of an agricultural trend. Being able to follow and understand, and maybe even predict, trends in the agricultural industry will help you become a well-informed consumer. This ability may also help those involved in agricultural production to become more efficient producers.

Lisa S./Shutterstock.com

Figure 1-36. Your choices will help determine the future of agriculture, food, and natural resources. ***What choices will you make as a consumer?***

Lesson 1.3 Review and Assessment

Words to Know

Match the key terms from the lesson to the correct definition.

1. Used to describe the areas of education that are vital to the development of students and new technologies.
2. The application of scientific knowledge, tools, or processes for practical purposes.
3. The equipment and software used to collect, filter, process, and distribute information, or to process data.
4. The concept that those in the agricultural fields need to use natural resources wisely and practice sustainable agriculture.
5. The branch of science and technology concerned with the design, building, and use of engines, machines, and structures.
6. The process of producing agricultural products using techniques that protect the environment and all living beings while allowing agricultural land to maintain production for many years.
7. A person who is knowledgeable about the agricultural processes and origins of agricultural commodities who can make good decisions based on this knowledge.
8. When something is able to be used without being used up or permanently damaged.
9. The general direction that something is developing or changing.
10. Energy that can be used to meet current needs without compromising the ability to meet future energy needs.

A. engineering
B. information system
C. informed consumer
D. STEM
E. sustainability
F. sustainable
G. sustainable agriculture
H. sustainable energy
I. technology
J. trend

Know and Understand

Answer the following questions using the information provided in this lesson.

1. Explain how studying past changes and the factors that drove those changes will help determine the future of the agricultural industry.
2. List the three main factors that stimulate changes in the agricultural industry. Briefly explain how each of these factors affects the agricultural industry.
3. Why does the distance between urban dwellers and the land where food is produced add pressure to those involved in production agriculture?
4. Explain why agriculturalists cannot simply put more arable land into production to provide more food and fiber for a growing population.
5. How do trends in food consumption become driving factors in agricultural production?
6. Something that is able to be used without being used up or permanently damaged is considered to be _____.

7. Explain sustainability as it relates to agriculture, food, and natural resources.
8. Solar energy, wind energy, and geothermal energy are all examples of _____ energy sources.
9. Briefly explain why sustainable agricultural practices should be used.
10. Advances in technology that will affect the agricultural industry include advances in _____.
    A. information systems
    B. engineering
    C. biotechnology
    D. All of the above.
11. The future of the agricultural industry will depend on _____ advances that allow more commodities to be grown on the available land and transferred to consumers.
12. List five characteristics modified by the collection of scientific techniques known as biotechnology.
13. How does having precise information of how much of a commodity is being produced under certain conditions help producers?
14. Give an agricultural example for each of the four STEM fields.
15. What are the two main steps of becoming an informed consumer?

Analyze and Apply

1. We know that new technologies will impact the future. Design a new technology in agriculture, food systems, or natural resources management and explain how you think this new technology will impact the future.
2. The future will depend on the changes that are made in the agricultural industry today. Find a current issue related to a new trend in agriculture, food systems, or natural resources management. An easy way to find them is to search for "new technology in agriculture" under the news section of a search engine. For your issue, play the role of the market analyst and answer the following questions: What is the trend? What types of consumers do you think are most likely to like the new trend? What impact will this trend have on agricultural production? What do you project will happen with this trend one, five, and ten years in the future?

Thinking Critically

1. Research shipping methods used by growers in the United States. Create a table that shows the most common transportation methods. List situations where retailers request things such as daily deliveries or non-refrigerated storage for certain fruits or vegetables and the ways in which they could be resolved.
2. Research the topic of food preservatives. What would happen if we did not use food preservatives? Are there natural alternatives? Are they as effective as synthetic ones? Has biotechnology made advances in making harvested foods last longer? What are the pros and cons of using or not using food preservatives? Do additional research to find expert opinions, associated costs, and other relevant information.

Chapter 1

Review and Assessment

Lesson 1.1

Agriculture's Impact on Society

Key Points

- Agriculture is essential to human survival on Earth.
- Understanding how humans interact with agriculture and how it impacts the world is essential to managing our food, fiber, and natural resources.
- By examining the main sectors and the scope of agriculture, we can see how each person fits into the human-agriculture connection.
- Much of the available land on Earth is used for agriculture, although that amount is diminishing daily.
- Agriculture is an important part of the economy in all countries and generates approximately 10% of the revenue from sales of goods worldwide.
- Less than 2% of the U.S. population is considered production agriculturalists.
- Agriculture has three broad areas: food, fiber, and natural resources.
- Human interaction with agriculture occurs when a producer grows or cultivates a raw agricultural product or when a consumer purchases a product.
- We interact with agriculture every day as we eat, dress, and use products made through agriculture or from natural resources.

Words to Know

Use the following list and the textbook glossary to review and study the *Words to Know* from *Lesson 1.1*.

agriculture	consumer	natural resource
arable land	fiber	natural resources management
commodity	food	producer

Check Your Understanding

Answer the following questions using the information provided in *Lesson 1.1.*

1. Briefly explain why agriculture is essential to human life on Earth.
2. In addition to cultivating other life forms, what types of industries does agriculture include?
3. *True or False?* It is estimated that only one-thirty second (1/32) of Earth's surface is arable land.
4. Explain why proper management of agricultural land is critical to the success of the agricultural industry.
5. *True or False?* The percentage of agricultural workers in a country is largely driven by the level of development in the country.
6. *True or False?* Less than 2% of U.S. workers are currently classified as farmers or ranchers.
7. Food production includes the _____ of all products used for human consumption.
   A. growing and harvesting
   B. distributing and marketing
   C. processing and selling
   D. All of the above.
8. *True or False?* In developing countries, consumers typically spend more of their income acquiring food than in developed countries.
9. The agricultural fiber industry includes the _____ of fibers and fiber products.
   A. marketing, distribution, and sales
   B. processing and manufacturing
   C. production and harvesting
   D. All of the above.
10. What areas of agriculture are covered under natural resources management?

Lesson 1.2

History of Agriculture, Food Systems, and Natural Resources

Key Points

- The changes in agriculture from ancient times to modern day are staggering.
- Studying the origins of agriculture allows us to see the impact agriculture has had on the building and development of society.
- Examining the process of domestication allows us to see changes that have been made to wild plants and animals to suit human needs.
- History shows that agriculture plays a large role in the ways that society functions.
- Studying change in global agriculture production provides insight into useful methods for agricultural management.
- Understanding the history of agriculture in the United States allows us opportunity to see how agriculture has shaped our nation.
- Knowing the history of agriculture shows the progress we have made, along with the dangers of mismanagement.
- Examining major agricultural inventions shows us how important it is to continually improve agricultural practices and equipment.
- Biotechnology has become an important tool in making agriculture more effective.

Words to Know

Use the following list and the textbook glossary to review and study the *Words to Know* from *Lesson 1.2*.

agrarian civilization
artificial selection
biotechnology
domestication
Dust Bowl
Fertile Crescent
feudal system
Green Revolution
indentured servitude
land grant institutions
nomadic tribe
Norman Borlaug
precision agriculture
sharecropping

Check Your Understanding

Answer the following questions using the information provided in *Lesson 1.2*.

1. List three significant social changes that came about because of the development of agrarian societies.

2. *True or False?* The earliest domestication of plants and animals from the wild are believed to have occurred in East Asia.
3. What was the key factor leading to the end of the feudal system?
4. *True or False?* The early Native Americans did not raise domesticated animals.
5. *True or False?* The early settlers taught the Native Americans how to raise native crops and store food for the winter.
6. What led to the widespread use of slavery after the American Revolution?
7. Explain why sharecropping came to an end in the United States.
8. *True or False?* The Dust Bowl resulted in the loss of topsoil from almost one million acres of farmland.
9. Explain why Eli Whitney's cotton gin is considered one of the most important advances in American agriculture.
10. The use of scientific modification to the genetic material of living cells in order to produce new substances or functions is referred to as _____.

Lesson 1.3

Future of Agriculture, Food Systems, and Natural Resources Management

Key Points

- The future of agriculture, food systems, and natural resources will be driven by three main factors: population growth, consumer demands, and sustainability.
- In order to fulfill the needs and desires of people in the future, the agricultural industry will need to use new technologies to bring about change.
- Many technological advances are expected to be made in engineering, biotechnology, and information systems that will lead to increased production of agricultural commodities.

Words to Know

Use the following list and the textbook glossary to review and study the *Words to Know* from *Lesson 1.3*.

engineering
information system
informed consumer
STEM
sustainability
sustainable
sustainable agriculture
sustainable energy
technology
trend

Check Your Understanding

Answer the following questions using the information provided in *Lesson 1.3*.

1. Briefly explain how population growth affects the agricultural industry.
2. In terms of our natural environment, _____ is based on the fact that everything we need to survive depends on our natural environment and how we take care of it.
3. Sustainable agricultural practices include _____.
   A. erosion management
   B. crop rotation
   C. integrated pest management
   D. All of the above.
4. Identify three technological areas in which advances will affect the agricultural industry.
5. List five types of production agriculture. What types of challenges do you think these areas will face in the near future?
6. Explain why it is important to be an informed consumer.

Chapter 1 Skill Development

STEM and Academic Activities

1. **Science.** Investigate current research programs in biotechnology. Choose a research program with a particular focus that interests you. Prepare a report on the scientific methods used in this program and how the results affect the agricultural industry.
2. **Technology.** Research environmental issues related to horticulture and find out what role technology has played in its advancement. Choose two specific topics and write a report explaining how technology has helped (or hurt) efforts in these specific areas to become more environmentally friendly.
3. **Engineering.** Choose a piece of agricultural machinery and research the changes in the machinery since it originated. Make note of any of the mechanical advancements and how they have made the machinery more efficient. What changes do you foresee in the future of this equipment?
4. **Social Science.** Choose a civilization from the past and research land ownership customs of that civilization. Write a report comparing and contrasting those customs with customs generally followed by people today.
5. **Language Arts.** Many famous writers have written poems about farming. Conduct research to find and read one or more poems written by famous poets. Note the rhythm of the words and the speech patterns. You may want to read the poems aloud or have

someone read it aloud while you listen. Using the poems you have read as inspiration, write your own poem about agriculture. Share your poem with the class.

Communicating about Agriculture

1. **Reading and Speaking.** With a partner, make flash cards of the *Words to Know* listed at the beginning of each lesson. On the front of the card, write the term. On the back, write the phonetic spelling as found in a dictionary. Practice reading the terms aloud, clarifying pronunciations where needed.
2. **Speaking and Listening.** In small groups, review the key terms listed at the beginning of each lesson. For each term, discuss the meaning of the term and describe the term in simple, everyday language. Record your group's initial description, and then make suggestions to improve your description. Compare your descriptions with those of the other groups in a classroom discussion.
3. **Reading and Speaking.** Complete a timeline listing the major agricultural advancements over the past 100 years. Research agricultural history in the past 100 years. Determine the ten most important advancements in agriculture during that time period. Explain to a partner the timeline sequence and the reasons why each event/invention was important.
4. **Speaking and Listening.** Make a collage. Using pictures from magazines or free online resources, create a collage of agricultural advances over the past 100 years. Include the dates and names of the people involved. Show and discuss your collage in a group of four to five classmates. Are the other members of your group able to determine the succession of advancements that you tried to represent?
5. **Reading and Speaking.** Select a historical era that interests you. Using at least three resources, research the agricultural history of that era. You may narrow the topic to a particular area of interest (mechanical advances, technological advances, biotechnology, etc.). Using the information gathered through your research, write a report. Present your report to the class using visuals such as PowerPoint®.
6. **Reading and Listening.** In small groups, discuss the main topics in the chapter. Ask questions of other group members to clarify concepts or terms as needed.

Extending Your Knowledge

1. What is agriculture? Conduct research to gather people's perceptions about agriculture. Ask at least one person born in every decade (go back as far as you can find participants) to draw a picture of what they think about when you say the word agriculture. Compare and contrast the differences between the drawings collected by yourself and your classmates.

Chapter 2

Leadership in Agriculture

©iStock/Rawpixel Ltd

SNEHIT/Shutterstock.com

©iStock/Chagin

USDAgov

While studying, look for the activity icon  **to:**

- **Practice** vocabulary terms with e-flash cards and matching activities.
- **Expand** learning with video clips, animations, and interactive activities.
- **Reinforce** what you learn by completing the end-of-lesson activities.
- **Test your knowledge** by completing the end-of-chapter questions.

Lesson 2.1

Building Leadership Skills through Agriculture

Words to Know

autocratic leadership
character
delegation
delegative leadership
democratic leadership
etiquette
followership
4-H
laissez-faire leadership
leader
leadership
mentor
mentorship
mission statement
National FFA Organization
participative leadership
personal leadership
personal leadership plan
servant leadership
SMART goals
team

Lesson Outcomes

By the end of this lesson, you should be able to:

- Distinguish between types of leadership.
- Explain the importance of developing personal leadership skills.
- Analyze the factors of team dynamics.
- Discuss the characteristics of a good leader.
- Identify opportunities for personal leadership.

Before You Read

Before you read the lesson, interview someone that you consider to be a leader. Ask the person why they feel it is important to know about leadership and how leadership affects their life. Take notes during the interview. As you read the lesson, highlight the items from your notes that are discussed in the lesson.

Think about the word *leader*. Who comes to mind? Do you picture the founding fathers of our country? Do you think about great leaders related to specific movements or actions, **Figure 2-1**? Do you see yourself as a leader, now, or in the future?

There are many views on what makes someone a leader. By definition, a ***leader*** is someone who guides or directs a group or organization, and ***leadership*** is a personal quality related to being able to guide or direct others. Understanding leadership requires knowing:

- The types of leadership.
- Characteristics of a good leader.
- How to develop your own leadership.
- Opportunities available to help you develop your leadership skills.

Types of Leadership

Although there are numerous ways that leadership can be classified, most experts agree that all leadership styles fall into one of three basic types: *autocratic*, *participative*, and *laissez-faire*. These three types differ in the way that the leader approaches those they serve.

stockelements/Shutterstock.com

Critterbiz/Shutterstock.com

eric Broder Van Dyke/Shutterstock.com

Figure 2-1. Monuments, memorials, and statues are often erected in honor of great leaders such as Frederick Douglas, George Washington, Thomas Jefferson, Theodore Roosevelt, Abraham Lincoln, and Mahatma Ghandi. ***Are these the type of people you think of when you think of someone who is a leader?***

Autocratic Leadership

In ***autocratic leadership***, the leader makes most of the decisions and sets the expectations for the followers, **Figure 2-2**. There is a clear division between the leader and the followers, and the followers generally have little input related to the decisions being made. One of the benefits of autocratic leadership is that decisions can be made quickly. One of the drawbacks is that followers may see the leader as "bossy." Can you think of times in your life when you were the follower of an autocratic leader? Can you identify a country currently under autocratic leadership?

Goodheart-Willcox Publisher

Figure 2-2. In autocratic leadership, a strong leader will take charge, assign tasks, and establish solid deadlines. ***Are there disadvantages to this type of leadership?***

Participative Leadership

In ***participative leadership*** (also referred to as ***democratic leadership***), the leader provides guidance while encouraging group members to share their ideas and opinions, **Figure 2-3**. There is less of a division between the leader and followers. One advantage of this leadership type is the greater contribution of ideas and creativity from the group. One disadvantage is that while everyone gets input, the leader may have to make decisions that are not in line with the thinking of some of the followers. Can you think of other benefits of participative leadership?

Goodheart-Willcox Publisher

Figure 2-3. Participative leadership encourages the creativity of the group's members, which may lead to unique ideas and innovative solutions. ***Are there disadvantages to participative leadership?***

Laissez-Faire Leadership

■ = Decision Maker

Leader

Group members

Goodheart-Willcox Publisher

Figure 2-4. Although laissez-faire leadership is a type of hands-off leadership, it is best if the leader remains open and available to group members for feedback and advice. *Would this type of leadership work well in a classroom?*

Laissez-Faire Leadership

In ***laissez-faire leadership***, also called ***delegative leadership***, the leader provides little or no guidance and allows the group members to make the decisions, **Figure 2-4**. The leader is present merely to provide tools and resources. This leadership style may work well when followers are highly skilled and knowledgeable, but in general, leads to low productivity among group members. Can you think of situations in which laissez-faire leadership would work well?

Characteristics of a Good Leader

Being a good leader can be hard work. Think of someone in your life who you think is a great leader. What are some of the challenges they have faced? What are some of the personal traits they possess that allowed them to overcome those challenges? Let's analyze some of the common skills and traits associated with being a good leader.

To be a good leader requires strong skills and distinctive traits in four main areas: *character*, *resourcefulness*, *mentorship*, and *focusing on results*. A person with a strong combination of skills and traits from these four areas may develop into a leader who is equipped to handle the unique challenges that come with any leadership role. As we examine the skills and traits of good leaders, think about your areas of strength, and areas you may need to improve to become a well-rounded leader, **Figure 2-5**.

Ulrich Willmunder/Shutterstock.com

Figure 2-5. *Have you ever thought about what leadership characteristics you possess?*

Character

Character is a group of personal traits that define moral or ethical quality. For a leader to have good character, he or she must be a moral and ethical person. Can you think of someone who you consider to have good character? When seen as a moral and ethical person, a leader will build trust and rapport with his or her followers. The same perception will also allow a leader to maintain a reputation of being fair and honest. Some of the traits related to good character are shown in **Figure 2-6**.

Resourcefulness

In addition to having good character, effective leaders are resourceful. They have the ability to find and use resources to help their group reach goals. These "resources" are not only physical resources such as funding and supplies, but human resources also.

Have you ever been in charge of a group and felt like you were doing most

Which of the following traits related to good character do you have?	
Adaptability	Humility
Compassion	Innovation
Conviction	Integrity
Coping	Intuition
Courage	Listening
Creativity	Responsibility
Desire for lifelong improvement	Self-discipline
Enthusiasm	Service
Ethical	Strong values
Globally aware	Trustworthy
Good citizenship	Understanding and appreciating others
Honesty	

Goodheart-Willcox Publisher

Figure 2-6. Traits related to good character. *Can you think of particular actions that represent each of the character traits?*

of the work? The best leaders realize that, in order for a group to be successful, they need to distribute the workload and make group members responsible for various parts of a project. They need to *delegate*. ***Delegation*** is the process of giving another person the control and responsibility for a given task. By delegating, leaders lighten their workload, and increase the followers' stake in the project. Good leaders also use their problem-solving skills to look for places within the organization that need improvement, and then assess possible resources that can be used to improve those areas.

Mentorship

Strong leaders keep an eye out for followers with potential, and develop those individuals' skills through ***mentorship***. Mentorship helps the individual, as well as the group, become stronger and more successful. So, another character trait of a good leader is the ability to *mentor*. A ***mentor*** is someone who advises and gives guidance to another person to help develop their potential, **Figure 2-7**.

Leaders who are good mentors invest time in getting to know members as individuals and by learning their strengths and weaknesses. Leaders must also have respect for individual differences in order to accurately assess the areas in which members can improve. By working as a mentor, leaders can help empower followers to be future leaders of the organization and allow everyone to feel as though they have a stake in the organization's goals.

SergeBertasiusPhotography/Shutterstock.com

Figure 2-7. Being a mentor involves helping someone else develop their leadership potential. ***Who are the mentors in your life?***

Focus

The fourth area of characteristics of good leaders is the ability to focus on the results. Leaders are often given the challenge of developing a mission or goals for the group they are leading. After this vision is created, the leaders are most effective if they provide support and guidance toward achieving the goals. To do this, strong leaders need to be able to make decisions for the benefit of the group, adapt to change, communicate the mission and any potential obstacles to the followers, and maintain their perseverance. Think of a time when you were in a group setting and given a complicated task. Did you have a leader who helped you remain focused on the end goal? Without this ability, even the best leaders often find themselves frustrated with the direction the group is heading, **Figure 2-8**.

Zurijeta/Shutterstock.com

Figure 2-8. Without the ability to remain focused and on task, even the best leaders can become frustrated. ***What types of actions or events cause you to lose focus? Is it easy for you to regain focus and get back on task?***

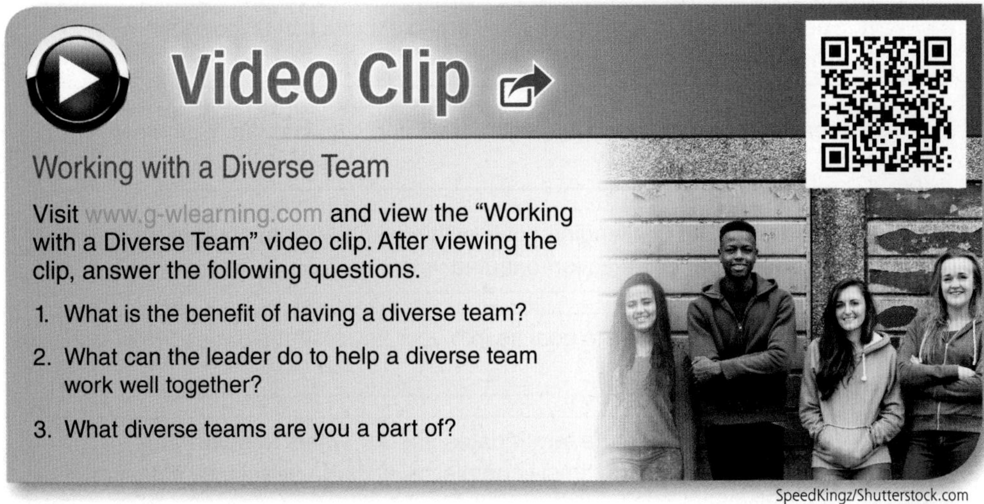

SpeedKingz/Shutterstock.com

Personal Leadership

Do you think you are a leader? Have you ever been in charge of coordinating a project, team, or event? Did you enjoy the experience? When you look back at the experience, can you determine what your leadership strengths and weaknesses were? Everyone has leadership potential. ***Personal leadership*** is the ability of a person to embody the characteristics of a good leader, and work toward becoming a better leader. Determining and improving your personal strengths and weaknesses can help you develop skills that will prepare you for any leadership opportunity that comes your way. To develop your personal leadership, you should:

- Develop a personal leadership plan.
- Set personal goals.
- Maintain good character.
- Develop and use proper etiquette.

Personal Leadership Plan

A ***personal leadership plan*** is a strategy for how you will accomplish your goals. When designing your plan, first set your goal, take your strengths and weaknesses into account, and then set a path for reaching that goal.

Goals

Setting personal goals is an important step toward personal leadership development and personal success. Setting personal goals will encourage you to persist and keep you focused. Keep the acronym *SMART* in mind when establishing your list of goals. ***SMART goals*** are goals that are *specific*, *measurable*, *attainable*, *realistic*, and *timely*, **Figure 2-9**. By taking the first step and setting your goals, you have already begun sharpening your leadership skills.

Character

Having good character will also impact your personal leadership skills. Understanding that your actions affect others is an important step in building

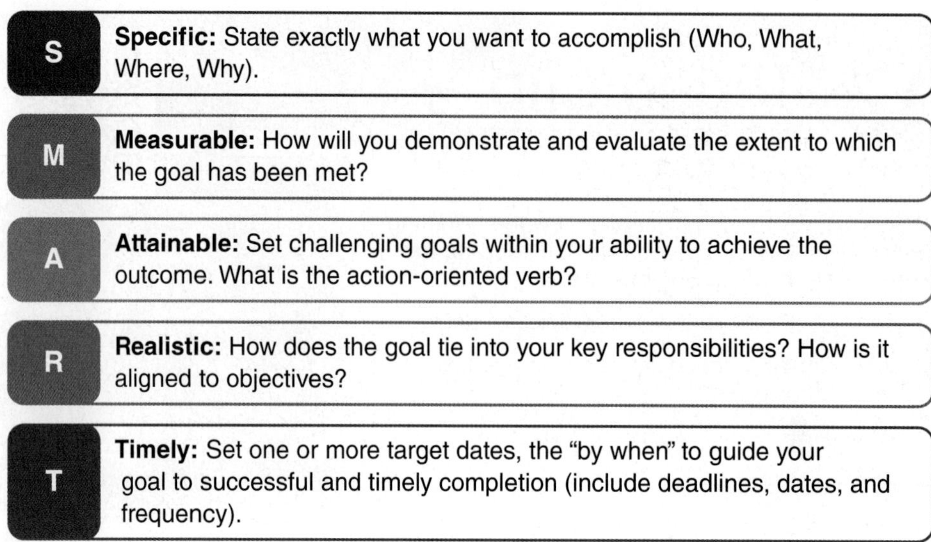

Goodheart-Willcox Publisher

Figure 2-9. Setting up SMART goals is a best practice framework used by athletes, students, instructors, and business people alike. *Have you taken the time to develop your own SMART goals?*

your character. When we looked at the traits related to having good character earlier in the chapter, did you make note of the things you were good at doing? Did you recognize traits that you have not mastered? These are areas you may want to work on improving.

Etiquette

Etiquette is behavior that is considered polite by society. Because leaders are seen as representatives of their group, it is incredibly important they exhibit proper etiquette in all situations. It is also important to remember that etiquette changes based on the group of people or culture with which you are

Hands-On Agriculture

Developing a Personal Leadership Plan

In order to fulfill your goals and missions, it is helpful to develop your own personal leadership plan. Grab a piece of paper or index card and think about the following:

- Think of a goal you want to accomplish; it can be as simple as earning a specific grade in a class, or as complicated as becoming the president. The most important thing is that it is important to you.
- Write the goal down in a format that ensures that you follow the conventions of a SMART goal.
- Beneath the goal, list the characteristics you already have that will allow you to reach the goal.
- Next, list some skills or knowledge you will need to gain in order to accomplish the goal.
- Now that you have assessed your skills, think about some potential barriers or roadblocks that might come your way. Write those down.
- Next to each of the roadblocks, write down something you can do to avoid or overcome it.
- Keep this paper within easy reach so you can refer to it often and stay focused.

interacting. For example, most people in the United States would not be offended by someone giving a hand signal with two fingers, **Figure 2-10**. In other countries, the same gesture may mean something incredibly offensive. Using proper etiquette can strengthen your personal leadership skills, set a good example for your followers, and allow people to view you in a more dignified manner.

Rido/Shutterstock.com

Figure 2-10. Although this hand symbol might be proper etiquette in the United States, the etiquette in other countries may find it very offensive. When working with new and unfamiliar cultures, take time to research or consult with someone in the community who can advise you. ***Have you ever unknowingly offended someone with improper etiquette?***

Team Leadership

A ***team*** is defined as a group of people who come together to achieve a common goal. The common goal could be winning a competition, having a successful fundraiser, or even building a community garden. Almost everyone has been on a team of some kind. Even families can be considered a type of team. Research shows that a successful team can accomplish more by working as a group than if each person worked on their own and combined their individual efforts. To help facilitate good team dynamics, the team should have:

- A shared mission or vision.
- The ability to work with the strengths of everyone on the team.
- Ground rules and mutual respect.
- The ability to resolve conflict.

Tom Reichner/Shutterstock.com

Figure 2-11. Much like a team of dogs in sled races, all members of a team should be moving in the same direction. ***Do sled dogs automatically move in the same direction or do they require extensive training to learn to work together? How important is the lead dog?***

Shared Mission

Having a shared mission is important to the overall success of the team. Picture a dog sled moving across the snow, **Figure 2-11**. How successful would the team be if all the dogs pulled in different directions? To be successful, team members need to work interdependently and cooperatively. The leader in team settings is much like the dog sled driver—he or she is there to ensure progress toward the end goal. A ***mission statement*** is a summary of the aims and values of the group. All team members should have input when creating the mission statement, and should consider the purpose, actions, and outcomes that the group expects to complete during their time working together. Can you think of some mission statements for companies or groups in which you are involved?

STILLFX/Shutterstock.com

Figure 2-12. Delegation can be approached in the same way you would select the best tool for a job. ***Are you comfortable delegating assignments?***

Strengths and Weaknesses

Teams work best when they use the strengths of each member for the benefit of the team. This requires the leader to know the strengths and weaknesses of each member and to delegate tasks to those who are most likely to perform the task well. It is common sense to allow team members to do what they do best. If you had a toolbox filled with tools, **Figure 2-12**, you would always select the tool that best fits the task, like using a screwdriver to turn a screw or a wrench to tighten a bolt. The difference with people is that, while each has their own specialty, they can also be mentored to learn new tasks. The best leaders use team members' current strengths while helping them develop new skills.

Ground Rules

Setting ground rules and having mutual respect is important to teams as well. By making sure that everyone understands the expectations for attending meetings, performing team-related tasks, reporting problems, and sharing credit, many team disagreements can be avoided. Use the first team meeting to determine jointly what ground rules your team will function under. Having ground rules can also be a great resource when conflict occurs, **Figure 2-13**. To resolve conflict, the leader and team members can refer to the ground rules to determine which behaviors need to be modified to get the team back on track.

Pressmaster/Shutterstock.com

Figure 2-13. Having ground rules can help resolve conflict in groups quickly and efficiently. ***Do you clearly communicate your sentiments or feelings with your family members? How about with your friends?***

Conflict Resolution

It would be nice if every team had the ability to work without disagreements all the time, but that is not really possible. Anytime a group of people with differing backgrounds, thoughts, and opinions work together, there are bound to be conflicts. Knowing how to resolve the conflict is an important part of working as a team. A few tips on how to resolve conflict include:

- Wait until everyone involved in the conflict can speak rationally before talking. Conflict is best resolved when people's emotions are in check.
- Take responsibility for your own actions and role in the conflict; apologize if needed.
- Be open and flexible; try to come to a solution from which everyone can benefit.
- Keep the mission of the group in mind. Seek a solution that allows the mission to remain unchanged.

Servant Leadership

Servant leadership is leadership in which the leaders approach their role as a way to serve those who follow them. These leaders are concerned with the broad mission of those who follow them, and with helping their followers grow as individuals. Think back to those in your life who you consider to be great leaders. Did they help you grow as an individual? Are you a better person because of their leadership influence in your life? If so, they were completing the mission of servant leadership. Being a servant leader requires conscious thought about how your actions will help others fulfill their potential.

Robert Greenleaf, who was the modern-day advocate for servant leadership, outlined ten additional leadership characteristics that leaders should have to follow the principle of servant leadership. These characteristics are:

- Listening—being able to hear the suggestions, concerns, and desires of those you are leading.
- Empathy—the ability to see things from someone else's point of view.
- Healing—desire to help people overcome setbacks.
- Awareness—understanding the social situation surrounding the group and the group members' feelings.
- Persuasion—the ability to sway the opinions and actions of others.
- Conceptualization—being able to develop a plan of action that is best for the group.
- Foresight—understanding the implications of the group decisions.
- Stewardship—maintaining care for the well-being of group members.
- Commitment to the growth of people—desire to help everyone reach their full potential.
- Building community—understanding the importance of the group working collaboratively.

Followership

Some people like to be the leader of any group they are in; others would rather not take a leadership role. Regardless of which group you fall in, the reality is that you will not be the leader all the time. ***Followership*** is the ability to *actively* follow a leader. Because you will be a follower at some point in your life, it is necessary to know the qualities of being an effective follower, **Figure 2-14**. To be a good follower, you should be able to:

- Work independently.
- Be committed to the shared vision or goal.
- Have skills or talents that allow you to contribute to the group.

sculpies/Shutterstock.com

Figure 2-14. Being a good follower can be just as important as being a good leader. ***Do you see yourself as a good follower?***

STEM Connection Leadership in STEM

With all of the changes needed in science, technology, engineering, and mathematics in the upcoming years, strong leadership will be important to the future of STEM. Which leadership traits do you think will play the most important role for the following STEM careers?

- Scientific Research Team Leader
- Genetic Engineer
- Computer Technology Coordinator
- Statistician

- Be willing to hold true to your values even when they disagree with the guidance of the leader.

Not everyone approaches followership the same. Some people are more inclined to want to make decisions on their own, rather than listening to the leader; others are more accepting of decisions made by the leaders. Related to supervision, some followers require constant support and supervision from the leader, while others are more self-directed. Finding your own style of followership will allow you to build your skills and strengths for the times when you are the leader.

Opportunities to Lead

Hopefully, by this point, you can see that you already possess many of the characteristics of an effective leader. To further develop these traits, look for opportunities to lead. Even if you are already a leader in your school, home, or community, be on the lookout for additional opportunities to practice and develop your leadership skills.

Student Organizations

School is a great place to find opportunities to lead. By helping to lead a student organization, you can have a broad impact on the school, community, and those who follow you.

The ***National FFA Organization*** is the middle school and high school organization most closely tied to agriculture, food systems, and natural resources management, **Figure 2-15**. In fact, agricultural education students just like you set up the organization in 1928 in order to develop leadership skills and encourage collaboration between agriculture students across the country. This intracurricular program works to make a positive difference in the lives of students by developing their potential for premier leadership, personal growth, and career success.

Have you ever thought about being a leader in FFA? FFA provides numerous opportunities for you to develop your skills in agriculture and as a leader. This organization has more than half a million members and is set up to be run by student members. Leadership opportunities in the FFA include

Did You Know?

The National FFA Organization awards members more than $2.2 million in scholarships each year, and state associations contribute more than an additional $2 million to members, all based in part on the leadership skills of applicants.

University of Kentucky College of Agriculture

Figure 2-15. A newly elected FFA National Vice President speaking to the crowd at an event celebrating her election. ***Which Career Development Events focus on developing leadership skills?***

local, regional, and national leadership positions. Participation in FFA also includes Career Development Events (CDEs) that are designed to develop your leadership skills. In addition, FFA also presents opportunities for you to help coordinate FFA community service. There are many opportunities waiting for you in the National FFA Organization. Your agriculture science teacher can direct you to the specific opportunities available in your local chapter.

Many other opportunities to serve in a leadership role exist in your school. They include:

- Student government
- Content specific clubs (FBLA, FCCLA, DECA)
- Specialized clubs (languages, activities, arts)
- Leading team sports

Community Organizations

Many leadership opportunities exist for you outside of school as well. The ***4-H*** youth development organization is a community-based youth development program that is coordinated by the USDA. This program allows youth from ages 8–18 to participate in many agriculture- and leadership-oriented activities, **Figure 2-16**. Talk to your county extension agent or agriculture science teacher to find the best resources for you to get involved in 4-H in your area.

Julija Sapic/Shutterstock.com

Figure 2-16. The 4-H is a national youth organization coordinated by the USDA to provide leadership and career opportunities.

Lesson 2.1 Review and Assessment

Words to Know

Match the key terms from the lesson to the correct definition.

1. Behavior that is considered polite by society.
2. The act of advising and giving guidance to another person to help develop their potential.
3. The ability to actively follow a leader.
4. Leadership style in which the leader provides guidance while encouraging group members to share their ideas and opinions.
5. The process of giving another person the control and responsibility for a given task.
6. A community-based youth development organization coordinated by the USDA.
7. Leadership style in which the leader provides little or no guidance and allows the group members to make the decisions.
8. Leadership style in which the leader makes most of the decisions and sets the expectations for the followers.
9. A personal quality related to being able to guide or direct others.
10. A leadership method in which the leader approaches their role as a way to serve those who follow them.
11. A group of personal traits that define moral or ethical quality.
12. A group of people who come together to achieve a common goal.
13. The middle school and high school organization most closely tied to agriculture, food systems, and natural resources management.
14. A strategy for how you will accomplish your goals.
15. Personal goals that are specific, measurable, attainable, realistic, and timely.

A. autocratic leadership
B. character
C. delegation
D. etiquette
E. followership
F. 4-H
G. laissez-faire leadership
H. leadership
I. mentorship
J. National FFA Organization
K. participative leadership
L. personal leadership plan
M. servant leadership
N. SMART goals
O. team

Know and Understand

Answer the following questions using the information provided in this lesson.

1. *True or False?* In autocratic leadership, the leader makes most of the decisions and sets the expectations for the followers.

2. *True or False?* In democratic leadership, the leader is present merely to provide tools and resources.
3. *True or False?* In laissez-faire leadership, the leader provides guidance while encouraging group members to share their ideas and opinions.
4. *True or False?* Followership is not a leadership skill.
5. In what four areas does a person require strong skills and distinctive traits to become a good leader?
6. Briefly explain why good character is important to leadership.
7. Explain the importance of delegation.
8. Explain the benefit of mentorship to a group.
9. List four steps you should take to help develop your personal leadership.
10. Briefly explain why it is important for leaders to exhibit proper etiquette in all situations.
11. Identify three "things" a group should have to help facilitate good team dynamics.
12. What is servant leadership? What are ten additional leadership characteristics that leaders should have to follow the principle of servant leadership?
13. What is followership? Why do you think it is important to be a good follower?

Analyze and Apply

1. What are five characteristics of a good leader that you feel you already have?
2. Think about the characteristics of a good leader. Select three characteristics and analyze the following for each of them: Why is this characteristic important to a leader? If someone wanted to develop this skill, what could they do?
3. Followership is an important characteristic to have. Make a list of ten things you think a good follower does.
4. Do you think good leaders are born or made? Please explain your answer.
5. Which characteristic of a leader do you feel you should focus on developing? How will developing this skill help you reach your goals?

Thinking Critically

1. Using what you know about leadership, pick a historical leader to research. For the leader you select, analyze the following: What leadership type do you think they worked under? List five leadership characteristics they had, along with evidence from their life to justify that leadership quality. Prepare a short summary about what this leader did and how you would rank their leadership skills.
2. Think about a movie you watched recently where someone showed a change in their leadership skills. Explain the person at the beginning of the movie, and discuss what events led them to develop their leadership skills. Describe the character at the end of the movie. How did they change?
3. Using at least 150 words, write your own obituary. Think about all of the things you would like to accomplish in your life and the things that you hope people will say about you after you pass away. What leadership traits do you hope people will accredit to you?

Lesson 2.2

Communication Skills in the Agricultural Industry

Words to Know

active listening
communication
communication systems
conversation
correspondence
creed speaking
dyadic communication
extemporaneous speaking
group communication
impromptu speaking
informational written correspondence
intrapersonal communication
listening
mass media
nonverbal communication
oral communication
passive listening
prepared public speaking
presentation
social media
speech
verbal communication
written communication

Lesson Outcomes

By the end of this lesson, you should be able to:

- Analyze the components of verbal, written, and visual communication.
- Describe the differences in modes of communication including intrapersonal, small group, mass media, and social media.
- Understand the components of written communication.
- Examine the skills required to make an effective verbal presentation.

Before You Read

Write down the list of words to know. For each term, write down what you think it means; make your best guess for terms that you do not know. Compare your list with a partner prior to reading. As you read, change the definitions of terms to match the definitions given in the lesson. After completing the lesson and updating your definitions, discuss the changes you made with your partner.

Have you ever taken a class and not understood what the teacher was talking about, even though you were paying attention? Have you ever done something based on someone else's instructions, only to find out that what you did was not what the person wanted you to do? ***Communication*** is the process of sending and receiving information using verbal and nonverbal cues. Can you think of a time when something you said was misinterpreted? *Miscommunication* can cause frustration, confusion, and even embarrassment, **Figure 2-17**. Proper communication can make you a more effective leader, employee, and member of society.

In this lesson, we will look at the different types of communication, along with the modes through which information is transferred. We will also examine the principles associated with written and verbal communication, and discuss ways you can hone your communication skills.

Communication Basics

Communication occurs in two basic ways: verbal and nonverbal. ***Nonverbal communication*** is the exchange of knowledge through the senses. ***Verbal communication*** is the transfer of information using words and language. The verbal communication category includes ***oral communication*** (communicating using spoken words and language) and ***written communication*** (which is information transferred through writing).

Communicating clearly relies on the same principles regardless of which type of communication you are using. To ensure accurate information transfer, consider the following:

- Target audience—the people receiving the information.
- Clarity—information should be easy to understand.
- Tone—the underlying message behind the information.

If you are able to tailor your message to the target audience and transfer information in a manner that is easy to understand and sends the correct message, you will be well on your way to effectively communicating, regardless of the mode of communication you are using.

Vladimir Gjorgiev/Shutterstock.com

Figure 2-17. Being an effective communicator can save you from the embarrassment of miscommunication. *Can you think of an instance when miscommunication could lead to a dangerous or damaging situation?*

Nonverbal Communication

When was the last time you knew how someone felt about a topic without them saying a word? Have you ever looked at a picture and knew what the message was from just the image? Nonverbal communication is communication that occurs without the use of spoken or written words, **Figure 2-18**. This type of communication includes:

- Gestures
- Eye contact
- Touch
- Space
- Voice
- Visual images

Robert Hoetink/Shutterstock.com

Figure 2-18. Your brain instinctively picks up on and interprets nonverbal communication. *What do you think is going on between the people in this picture?*

By consciously using nonverbal cues, you are accessing the primitive instincts of your audience. Human beings naturally use and understand nonverbal communication. So, there is some truth to the concept that *how* you send a message may often be more important than the message itself.

Did You Know?

Most researchers believe that somewhere between 50% and 70% of all communication occurs nonverbally.

Verbal Communication

We are involved in verbal communication every day. Whether it is listening to your teacher give instructions for an assignment, talking with

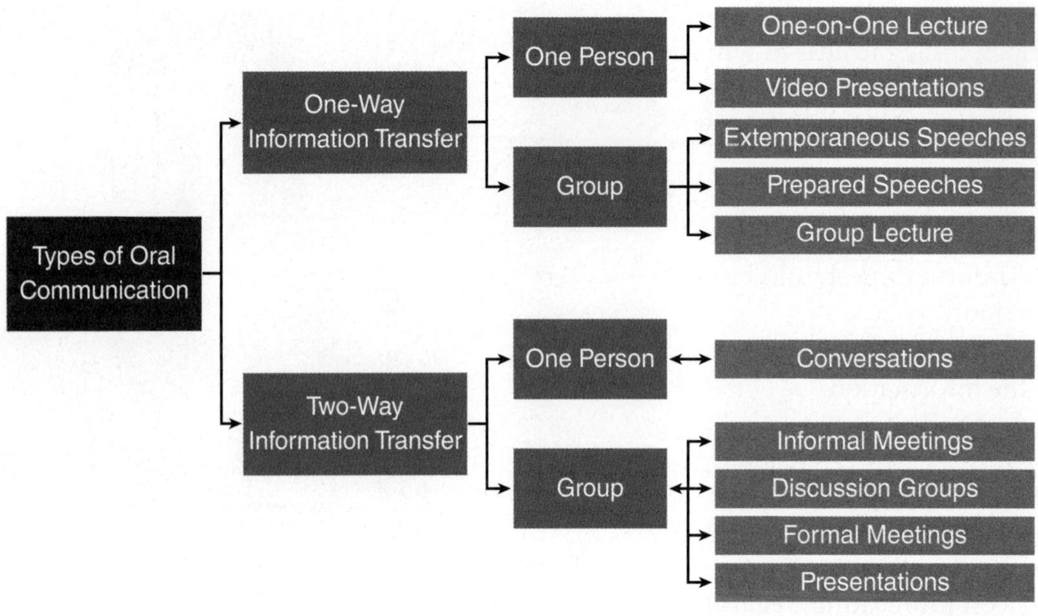

Figure 2-19. Conversation is the most common type of verbal communication. ***How many conversations do you have on a typical school day? Do you usually have multiple conversations with the same people on a daily basis?***

your friends about your plans for the weekend, or watching your favorite TV show, verbal communication is a core component of information exchange. The types of verbal communication are shown in **Figure 2-19**. Verbal communication can occur with one person transferring the information and another receiving the information, with two-way communication in which individuals exchange information, and it may also be communication to a single person or a group of people.

Conversation

A ***conversation*** is two-way oral communication that occurs between two people. This is the most common type of communication, **Figure 2-20**. Think about the last conversation you had with your friends. Did it come naturally to you? Can you recall a conversation with someone that did not flow smoothly? Effectively communicating might seem like something that should come to us naturally because we have conversations all the time. In reality, being a truly effective communicator takes work. Have you ever had to start a conversation with someone that you did not know? Was it easy for you, or did you feel awkward? Review the following guidelines and use them to develop your ability to start conversations in any situation:

- Break the ice—introduce yourself with a friendly demeanor and a question.
- Find common ground—look for something you have in common.
- Practice active listening—repeat what is said, and give nonverbal cues that you are interested in the conversation.
- Ask open-ended questions—avoid questions that can be answered with yes or no.

Public Speaking

A ***speech*** is a type of formal, oral communication that typically involves one-way transfer of information. The purpose of a speech may be to inform, instruct, persuade, or entertain. Speeches may be classified by the amount of time a speaker is given to prepare the speech.

Prepared public speaking is the category of speeches in which the speaker has had ample time to prepare the content and delivery of the information. This type of public speaking is the most common type of speech. It allows the speaker to develop, practice, and revise the speech before delivering it to the target audience.

pio3/Shutterstock.com

Figure 2-20. Two-way oral communication is the most common type of communication. ***How many different people do you speak with on a typical school day?***

In contrast, ***extemporaneous speaking*** is a speech format in which the speaker has a very short period of time between receiving the topic and presenting the speech. This type of speaking requires the speaker to be able to "think on his or her feet." ***Impromptu speaking*** is extemporaneous speaking in which the speaker is expected to give a speech immediately following the receipt of the topic.

FFA Connection Speaking Events

If you love to speak in public, or if you want the opportunity to practice your speaking skills, there are several opportunities available for you within your FFA experience.

The three speaking events that you can compete in at the local, regional, state, and national level include:

Creed Speaking—First-year FFA members memorize the FFA Creed and present it to judges who score them on their presentation skills, along with their answers to questions based on the creed.

Extemporaneous Speaking—Members draw three agricultural topics and select one to speak on. The member then has 30 minutes, using only 5 resources, to write a 4–6 minute speech to present to the judges.

USDA

Prepared Public Speaking—Members plan, research and write a 6–8 minute speech on an agricultural topic of their choice. Judges score the member based on their manuscript, presentation, and answers to questions related to their topic.

Many other speaking opportunities exist for you on a local level. Talk to your FFA advisors to determine which of these events might be right for you.

Writing a Speech

Have you ever written an essay in English class? If you have, you are already familiar with the basics of speech writing. Good speeches follow a format very similar to a five-paragraph essay format. Before writing your speech, you should:

- Define your target audience.
- Research your topic.
- Organize the information.
- Develop an outline.

Once you are ready to write your speech, make sure you use a well-defined format to ensure that your speech has good flow.

Two important definitions for the term ***presentation*** are:

- A type of oral communication where the audience has the opportunity to ask for clarification after the information has been presented.
- The specific set of skills required to effectively transfer information through oral communication.

So, to become a good presenter, you must master the art of both verbal and nonverbal communication. You need to develop what you are going to say, how you are going to say it, and how you are going to present it. Tips for being an effective presenter include:

- Make a connection with the audience through stories or examples.
- Speak with a clear and concise voice.
- Remember nonverbal cues including eye contact, gestures, and body language.
- Practice! You will be much more at ease if you are comfortable with the information.

Hands-On Agriculture

Writing a Speech

Step 1: Develop an introduction.
An introduction should include something to catch the attention of your audience, along with giving the audience an outline of the information that will be included in the main portion of the speech. This outline of information is similar to the thesis statement of a five-paragraph essay.

Step 2: Write the body.
The body of the speech should be written following the outline that you developed from your research. Talk about each point separately, making sure that they are presented to the audience in a sequence that is logical.

Step 3: Tie it together with a conclusion.
The conclusion of the speech should tie together all of the main points that you discussed in your speech. It should also include a final statement that will help your audience remember the information that you presented.

Step 4: Practice your presentation.

FFA Connection Presenting

Good presentation skills are key components to your success in competitive FFA events. In fact, at the national level, 21 Career Development Events have a presentation component. Honing your presentation skills will help you compete in many FFA events!

Listening

Do you think you are a good listener? Do your parents think you are a good listener? It is important to hone your verbal communication skills, even when you are not the one doing the talking. ***Listening*** is the act of bringing in information through hearing verbal communication. There is a definite difference between hearing something, and actually listening to it. Our brains are designed to filter out distracting sounds and only listen to the things that have our attention. Take a minute and just listen to the sounds around you—what can you hear? Can you hear other students moving around in their seats or talking? Can you hear the hum of the lights in the classroom? Developing your listening skills will help you become a better communicator.

Passive and Active Listening

Passive listening is the act of hearing information, but not interacting with it. If you are listening to a teacher present a lesson, you are participating in active listening. You are also a passive listener whenever you are listening to the radio or television. Sometimes, we are passive listeners when we are talking with other people because we are not giving them our full attention or actively engaging with them. ***Active listening*** is listening and engaging with the conversation. Giving verbal and nonverbal cues is essential to developing your active listening skills.

Becoming an Active Listener

To become an active listener, you should focus on giving your complete attention to the other person or people in the conversation. Body language like making eye contact, shaking your head, or making facial expressions, are indicators that you are receiving the message. Asking questions also shows that you are actively listening to the speaker, as does making appropriate comments. The best way to be an active listener is to give the speaker your full and complete attention, which means that any distractions (like cell phones) should be avoided or ignored during the interaction. Speaking to the person next to you and wearing earbuds or headphones are also indicators that you are not actively listening to the speaker.

Written Communication

Written communication is vital to your success. Imagine you have graduated and are applying for your dream job. How will you go about getting hired? First, you will need to write a résumé and a cover letter, then you may need to fill out an application. After your interview, perhaps write a thank-you note or follow-up email. All of these items require you to have written communication skills. Written communication skills, good *and* bad, will make you stand out from other applicants.

Written communication has varying levels of formality. For example, a letter or email to a college admissions officer requires a much different tone than a quick text or email to your friend. Knowing how to write professionally is one of the top five skills desired in the workplace. For the purpose of this lesson, we will discuss written communication as it relates to correspondence and one-way communication.

Correspondence

Correspondence is two-way written communication between two or more people. It may be as simple as a series of text messages sent back and forth, or may be as complicated as formal, printed professional letters mailed back and forth between businesses. Informal, nonprofessional correspondence sent between friends requires little formatting. Professional correspondence has more structured guidelines.

Professional correspondence includes three main parts: a greeting, the body, and a closing, **Figure 2-21**. The *greeting* is the opening information and welcome line. In business letters, the greeting includes the contact information for the sender and the date, along with the recipient's address and a salutation, or line of welcome such as "Dear Mr. Jenkins." It is not usually necessary to insert the sender's address or the date in the opening of a professional email because it is automatically generated through the sender's email program. Therefore, professional emails typically begin with only the salutation.

The *body* of the letter includes the letter's purpose along with information that the sender is requesting from the recipient.

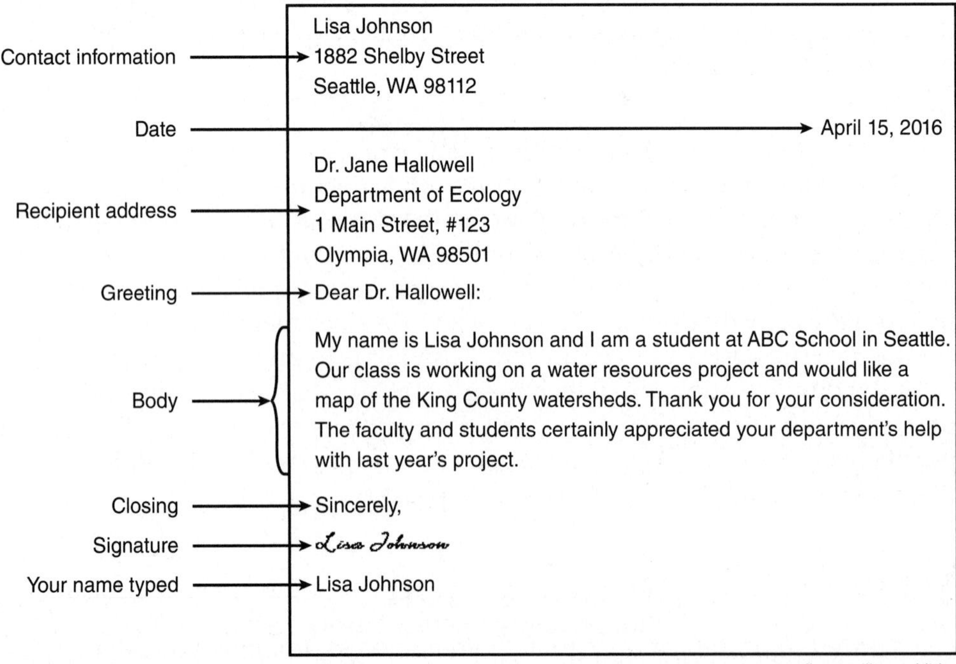

Lisa Johnson
1882 Shelby Street
Seattle, WA 98112

April 15, 2016

Dr. Jane Hallowell
Department of Ecology
1 Main Street, #123
Olympia, WA 98501

Dear Dr. Hallowell:

My name is Lisa Johnson and I am a student at ABC School in Seattle. Our class is working on a water resources project and would like a map of the King County watersheds. Thank you for your consideration. The faculty and students certainly appreciated your department's help with last year's project.

Sincerely,

Lisa Johnson

Lisa Johnson

Goodheart-Willcox Publisher

Figure 2-21. It is always best to err on the side of caution and follow formal standards when writing professional correspondence. No one is likely to be offended if you are *too* formal or *too* polite.

The final part of the letter is the *closing*, which includes a parting statement such as "Sincerely," and the sender's name. Although the sender's name is automatically included in the email, it is still considered good etiquette to include the name in the closing of a professional email.

Ensuring Professional Correspondence

Have you sent or received a text message in the last hour? Chances are, you send a lot of correspondence throughout the day, most of which is likely casual, informal communication to friends and family. There are some important tips that can add to your professionalism when sending professional correspondence:

- Use proper grammar, punctuation, and spelling. While sending your "bff" an "lol" or saying that "i'll b there b4 u" is perfectly acceptable for informal correspondence, it is not professional. When sending information to those you have a professional relationship with, even through text messages, conventional rules of English still apply.
- Beware of auto-correct. Prior to hitting the send button on mobile communications, make sure that the message on the screen is exactly what you wanted to say. This can save you the embarrassment of an awkwardly worded auto-correct statement.

One-Way Communication

Informational written correspondence is written communication in which information is being transferred one way. There are thousands of applications for informational written correspondence. How many times have you read something today that did not require a formal response? In fact, by reading this textbook right now, you are experiencing informational written correspondence. This type of communication is used for many different reasons, including everything from selling a product, to career seeking, to simply transferring information to the public, **Figure 2-22**.

Marketing

Informational written correspondence related to marketing a product includes written advertisements, promotional materials like pamphlets or brochures, along with technical manuals, instructions, and other items that are included with new products.

Career Seeking

Examples of career-seeking written correspondence would be résumés, cover letters, job applications, and follow-up letters. These types of communication inform potential employers about an applicant's qualities related to a specific job description.

Information Transfer

Can you think of written correspondence that exists solely as a means to inform the public? Although newspapers were once the main source of written information for the public, most people would now list the Internet as the main source of information written for public consumption. Other examples include press releases, magazine articles, novels, and millions of websites with information in written form.

NAN728/Shutterstock.com

kraphix/Shutterstock.com; racorn/Shutterstock.com

Figure 2-22. Written communication can be created for career seeking, marketing, or informational purposes. *Have you considered working on the school newspaper or writing the newsletter for an organization to which you belong?*

Communication Systems

Types of communication are how information is transferred, while ***communication systems*** describe the interaction of people during the act of communication. There are five basic communication systems:

- Intrapersonal
- Dyadic
- Group
- Mass media
- Social media

Each system involves a different level of interaction between people.

Intrapersonal Communication

Intrapersonal communication is often called self-reflection. This is communication in which the thought or language use is internal to the communicator. It is used as a person evaluates a situation, processes information, and clarifies his or her thoughts, **Figure 2-23**. It may also take on written form through a personal journal or blog. All communication that occurs between people is considered intrapersonal communication because it happens between people.

Dyadic Communication

Dyadic communication is communication that occurs between two people. A one-on-one conversation would be an example of dyadic communication. Dyadic communication includes phone calls, emails, and text messaging between two people.

d13/Shutterstock.com

Figure 2-23. Intrapersonal communication is communication with yourself; keeping a journal is one type of intrapersonal communication. *Have you ever kept a journal?*

Group Communication

Group communication involves a selected group of people. This may be as few as three people discussing a topic, but could also be as large as a speaker giving a prepared speech at a convention.

Mass Media

Mass media is a communication system in which information is transferred to a broad and diverse public audience, mediated by technology. Examples of mass media are television or radio broadcasts, newspapers and magazines, and informational websites.

Social Media

The final communication system is social media. Using ***social media*** to communicate involves communication through publically available websites and the use of applications that enable users to share content and engage in social networking. While social media is a large part of communicating in today's world, it is important to be cautious when posting, commenting, and sharing content.

Social Media Safety

Social media technology allows us to connect in real-time to our friends, family, and colleagues around the world. With these great benefits come some unlikely challenges. While social media is a great place to share your thoughts and opinions, there are several things that you should keep in mind:

- Do not assume that everyone on social media is who they say they are. If you get a message or a friend request from someone you don't know in real life, you may not want to interact with them until you can verify their identity.
- Remember that what you post can, and will, come back to haunt you. A recent study showed that up to 90% of employers and college admissions coordinators check candidates' social media profiles. Make sure what you post represents the image you want to portray.
- *Always* make sure you involve your parents or guardians in any decision you make related to bringing your virtual friends into your real life.
- *Never* give out personal information through social media. Sharing your home address, schedule, or other personal information on your social media page can tell the world more about you then you may want them to know. It could even lead someone right to you.
- *Always* think twice about how much someone could figure out about you from your profile or the things you post.

Lesson 2.2 Review and Assessment

Words to Know

Match the key terms from the lesson to the correct definition.

1. When first-year FFA members memorize the FFA Creed and present it to judges who score them on their presentation skills, along with their answers to questions based on the creed.
2. The process of sending and receiving information using verbal and nonverbal cues.
3. Communication that involves a selected group of people. It may be as few as three people or a large crowd.
4. The transfer of information using words and language.
5. Communication through publically available websites and the use of applications that enable users to share content and engage in social networking.
6. Written communication in which information is being transferred one way.
7. The transfer of information through writing.
8. A type of oral communication where the audience has the opportunity to ask for clarification after the information has been presented. The specific set of skills required to effectively transfer information through oral communication.
9. Communication in which the thought or language use is internal to the communicator. It is often called self-reflection.
10. Communication that occurs between two people. Includes phone calls, emails, and text messaging between two people.
11. A communication system in which information is transferred to a broad and diverse public audience, mediated by technology.
12. The interaction of people during the act of communication.
13. The act of communicating using spoken words and language.
14. The exchange of knowledge through the senses.
15. A type of formal, oral communication that typically involves one-way transfer of information.

A. communication
B. communication systems
C. creed speaking
D. dyadic communication
E. group communication
F. informational written correspondence
G. intrapersonal communication
H. mass media
I. nonverbal communication
J. oral communication
K. presentation
L. social media
M. speech
N. verbal communication
O. written communication

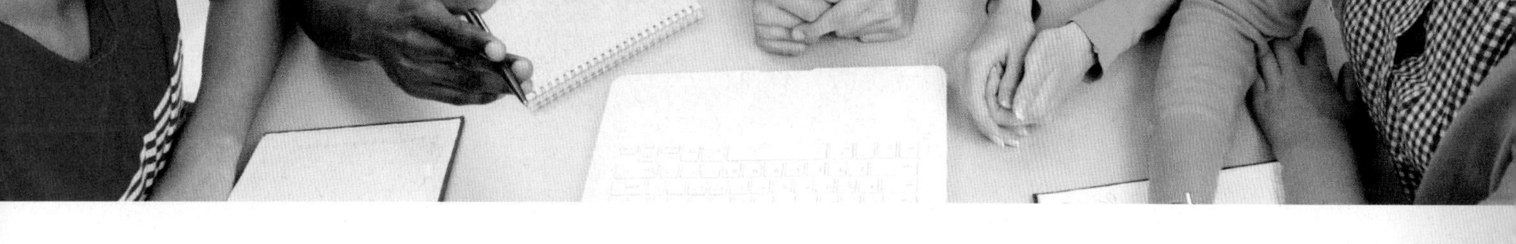

Know and Understand

Answer the following questions using the information provided in this lesson.

1. What are the main types of communication? Give an example of each.
2. To ensure accurate information transfer, _____ should be considered.
   A. tone
   B. the target audience
   C. clarity
   D. All of the above.
3. In addition to physical gestures, list five types of nonverbal communication.
4. Look at the types of oral communication in **Figure 2-19**. Give an example of when you might use each of the types of oral communication.
5. In what three ways does verbal communication occur?
6. List four ways that will help you develop your ability to start conversations in any situation.
7. _____ speaking is when the speaker has had ample time to prepare the content and delivery of the information.
   A. Extemporaneous
   B. Prepared public
   C. Impromptu
   D. Indirect delivery
8. _____ speaking is a speech format in which the speaker has a very short period of time between receiving the topic and presenting the speech.
   A. Extemporaneous
   B. Prepared public
   C. Impromptu
   D. Indirect delivery
9. _____ speaking is speech in which the speaker is expected to give a speech immediately following the receipt of the topic.
   A. Extemporaneous
   B. Prepared public
   C. Impromptu
   D. Indirect delivery
10. What are the two definitions of *presentation*?
11. Explain the difference between passive and active listening.
12. *True or False?* Written communication skills, good *and* bad, will make you stand out from other applicants.

13. List the three main parts of professional correspondence.
14. Identify three reasons for which informational written communication may be used.
15. What are the five basic communication systems?
16. In your own words, explain why it is important to exercise caution when posting, commenting, and sharing content on social media.

Analyze and Apply

1. Social media is publically available written communication. It is also fairly new in the world of communication (considering that the first written communication was started by cavemen, and the printing press was not invented until around 1450 CE). Create a "social media users guide" with tips someone over the age of 55 would need to know in order to begin using social media.
2. Prepare a speech on an agricultural topic suitable for presentation in the National FFA Organization Prepared Public Speaking Career Development Event.
3. Write a formal letter to someone in the community asking them to sponsor an award for an agricultural science student at your high school. Let your advisor review the letter, then send it to the potential sponsor.

Thinking Critically

1. Communication occurs in one-way and two-way applications. Write two lists: one with a list of skills that a person would need to be an effective one-way communicator, and one with a list of skills that a person would need to be an effective two-way communicator. Compare the lists to determine which characteristics are different.
2. Select a news article that interests you and that you feel deserves some commentary (positive or negative). Make a list of reasons you agree or disagree with the author. Use your list to compose a response to the article. Mail (or email) your commentary to the publication and watch for it to be published or see if you receive a response. You may work with a peer on this project.

USDA

U.S. Department of Agriculture Secretary Tom Vilsack was joined by members of the Iowa state delegation for a press event at the National FFA Convention. *How far will your leadership skills take you?*

Lesson 2.3

Conducting Meetings in Agricultural Organizations

Words to Know

amendable motion
chair
chairman
debatable motion
gavel
germane
incidental motion
main motion
minutes
motion
order of business
parliamentary procedure
precedence
privileged motion
quorum
second
subsidiary motion

Lesson Outcomes

By the end of this lesson, you should be able to:

- Explain the purpose for conducting meetings using parliamentary law.
- Demonstrate the ability to make, discuss, and vote on topics in a formal meeting setting.
- Analyze the effectiveness of a chairman in conducting a meeting.
- Classify parliamentary motions based on their intent.

Before You Read

As you read the lesson, be aware of the questions you have about the content. Write your questions on sticky notes and put them next to the section that prompted the question. After you are done reading the lesson, discuss the questions with your classmates or teacher.

It can be difficult to get a group of people to make a joint decision. Becoming a good leader requires you to analyze input from all members of a group and then help the group reach a consensus. Think about the last time you were with a group of people trying to reach a decision. Was the decision made quickly? Did you feel like you were able to share your opinion? Did the outcome please all of the people in the group? If you answered yes to all of these questions, it is likely your group was using some of the concepts of *parliamentary procedure*.

Parliamentary procedure is a set of rules and regulations for properly conducting meetings. This lesson will focus on the purpose of parliamentary procedure and will help you understand some of the basic motions used. Various aspects of the proper way to conduct a meeting using parliamentary procedure will also be discussed.

Purpose of Parliamentary Procedure

The main purpose of parliamentary procedure is to conduct meetings effectively, **Figure 2-24**. To accomplish this, there are four main goals:

- Handle one item at a time.
- Extend courtesy to everyone.
- Ensure that the rights of the minority are protected.
- Ensure that decisions are made with the majority.

Many organizations use parliamentary procedure in order to keep their meetings running effectively. The United States and nearly 200 other

GarethPriceGFX/iStock/Thinkstock

Figure 2-24. This is the House of Parliament in England. Parliamentary procedure is used to conduct all meetings. The U.S. Congress, state governments, local governments, and even your local FFA chapter can hold meetings using the rules of parliamentary procedure.

countries use the principles of parliamentary procedure to guide their governmental decisions. Other groups that rely on parliamentary procedure for conducting their meetings include: state and local governments, civic groups and committees, school organizations, nonprofit organizations, and many student clubs. Knowing parliamentary procedure can make you a valuable asset to any of these organizations.

To correctly implement parliamentary rules and keep meetings on track, most organizations use a set of guidelines called *Robert's Rules of Order*. A condensed version of parliamentary rules, tailored to meet the specific needs of FFA members, is *Gray's Parliamentary Guide.*

Conducting a Meeting

Parliamentary procedure sets a specific course of action for conducting meetings. Can you recall a time that you have been to a formal meeting? Did you find that it took a while to get to new business or information? That is because the order of business must follow parliamentary rules.

Before a meeting can officially begin, it must be determined whether or not a *quorum* is present. A ***quorum*** is the number of members that must be in attendance at a meeting for business to be conducted. It is typically set at one more than half of the total members in the group. The main reason a quorum must be present is to prevent a small number of the group from making decisions that may not be in the best interest or a true representation of the group's intentions.

Chip Somodevilla/Getty Images News/Thinkstock

Figure 2-25. It is the responsibility of the chairman to help keep the meeting on track and make sure that parliamentary rules are being followed. *How does the gavel help the chairman accomplish the task?*

Chairman

The ***chairman***, or ***chair***, is the person responsible for conducting the meeting, **Figure 2-25**. Once it is determined a quorum is present, the chairman can continue with the meeting. Much of the responsibility for the business being conducted properly falls on the chairman. To be effective, the chairman must:

- Have a good knowledge of parliamentary procedure.
- Keep track of the pending business.
- Be unbiased.
- Speak clearly.

Typically, the chairman is not allowed to enter into discussion on a topic, and the chairman is only allowed to vote on a motion when his or her vote is the deciding vote.

StudioSmart/Shutterstock.com

Figure 2-26. The gavel is the most important tool used in parliamentary procedure.

Gavel

Have you ever been in a courtroom or watched a courtroom show on TV? If so, then you probably know what a gavel is. A ***gavel*** is the hammer-shaped tool that the chairman uses to send specific cues to the assembly, **Figure 2-26**. What you might not be aware of is the technical purpose and rules behind the gavel. Different gavel taps signify different messages to the members of the assembly:

- One tap of the gavel signals the result of a vote, lets the members know to be seated, signifies the end of the meeting, or is used to gain attention.
- Two gavel taps are used to begin the meeting.
- Three gavel taps signify to the members to rise.
- When the gavel is tapped in a series, it signifies to the members to come to order and refocus on the business the group is discussing.

Minutes

The ***minutes*** are a complete, written record of the proceedings of a formal meeting. Information in the minutes include:

- The type of meeting.
- The date and time the meeting began.
- A list of those in attendance, including the chairman.
- A record of all main motions and secondary motions discussed during the meeting.

- The results of voting on all motions.
- The time of adjournment.

The minutes of the meeting are prepared and shown to the members at the next meeting for review and approval. Can you think of a reason why the members would need to vote and accept the minutes?

Order of Business

The ***order of business*** is the order in which items should be presented in a meeting. The proper order of business should be:

1. Calling the meeting to order/opening ceremonies.
2. Reading and approval of the minutes of previous meetings.
3. Reports from standing committees.
4. Reports from special committees.
5. Special orders.
6. Unfinished business.
7. New business.

This flow allows the meeting to run smoothly. Which of the purposes of parliamentary procedure do you think guides the rule that new business should only come up once unfinished business is handled?

Types of Motions

A ***motion*** in parliamentary procedure is a formal proposal by a member which, if agreed upon by the members of the assembly, will result in a certain action. Making a motion brings something before the group. Because one of the goals of parliamentary procedure is to handle only one item at a time, there is an order of *precedence* for motions. ***Precedence*** in parliamentary procedure is the list of motions in the order that they must be handled. A motion with lower precedence cannot be brought before the assembly if a motion with higher precedence is already before the group for discussion or vote. Motions listed in order of precedence are shown in **Figure 2-27**.There are five classifications of motions: privileged, incidental, subsidiary, main motions, and motions that bring a question again before the assembly.

Privileged Motions

Privileged motions have the highest precedence of all motions, and deal with the rights or needs of the organization. These motions include those concerned with starting, stopping, pausing, or focusing the purpose of the meeting.

Incidental Motions

Incidental motions are related to parliamentary rules and procedures, not to other motions. They include motions that ask questions, correct

Motions in Order of Precedence	
Classification	**Motion**
Privileged	Fix Time to Adjourn
	Adjourn
	Recess
	Question of Privilege
	Orders of the Day
Incidental	Appeal
	Point of Order
	Parliamentary Inquiry
	Suspend the Rules
	Withdraw
	Divide the Question
	Division of Assembly
Subsidiary	Lay on Table
	Previous Question
	Limit or Extend Debate
	Postpone Definitely
	Refer to a Committee
	Amend
	Postpone Indefinitely
Main Motion	Main Motion
Motions that bring a question again before the assembly	Take from the Table
	Reconsider
	Rescind

Goodheart-Willcox Publisher

Figure 2-27. Motions in order of precedence.

mistakes, ensure the accuracy of a vote, or change or modify rules. This group of motions is lower in precedence than the privileged motions.

Subsidiary Motions

Subsidiary motions are motions that deal with managing other motions. These motions work to set aside a motion, modify the amount of debate, or change the motion in some way. They have an order of precedence among themselves. The main motion is sometimes classified as the lowest precedence of the subsidiary classification, when not given its own classification.

Main Motions

A ***main motion*** is a motion that brings up a new topic of discussion before the assembly. Because it introduces a new item of business for the group to discuss, this motion cannot be made when any other motions are pending. This means that the main motion has the lowest precedence of all motions.

Motions That Bring a Question Again before the Assembly

The final classification of motions includes motions that bring a question before the assembly again. These motions allow the group to reconsider, rescind, or discuss a motion that has been previously set aside or voted on.

Basics of Handling Motions

In order to get information introduced, discussed, modified, and decided, there are set parliamentary rules that must be followed. The basic steps involved in processing a motion include:

- Bringing the motion before the assembly.
- Debating the motion.
- Amending the motion.
- Voting on the motion.

Understanding how the motions are managed will get you on your way to becoming a parliamentary procedure professional.

Bringing Motions before the Assembly

In order for a motion to be discussed, or decided upon, it must first be brought formally before the group. The first thing to consider when making a motion is how it should be presented to the chairman. Some motions require the chair to call on the floor member and get recognition before making the motion. To obtain the chairman's recognition, the member should stand and say "Mr. or Madam Chairman," then wait for the chairman to call on them. Once recognized, all motions begin with the words "I move." Some privileged or incidental motions can be made without recognition from the chair. These motions are then dealt with by the chairman before returning to other business.

Some motions require a ***second*** to signify that more than one person wants to discuss or cast a vote on the motion. The person who seconds a motion does not necessarily need to be in favor of the motion.

Once the motion has been made and seconded, the chair restates the motion, which allows floor members to consider the motion through debate or vote.

Debating Motions

After the motion has been made and restated, the next consideration is whether it is a ***debatable motion***. All main motions and amendments to main motions are debatable. For debatable motions, standard rules of parliamentary procedure state that:

- The person who made the motion has the right to debate the motion first.
- Each member has the right to debate twice per debatable motion.
- Debates are limited to ten minutes.

When debating motions, the standard leadership concepts of good communication apply. The member should use his or her speaking skills to present the debate using logical arguments and clearly stated reasons for his or her position on the motion.

Amending Motions

While the motion is being considered by the group members, some motions can also be changed. If parliamentary rules allow for a motion to be modified, it is said to be an ***amendable motion***. Amendable motions may be amended up to two times. There are three ways to amend a main motion:

- Inserting—information is added to the motion.
- Striking out—sections of the motion are deleted.
- Striking out and inserting words (substitution)—a portion of the motion is substituted for different information.

A key factor about amendments is that they must be *germane* to the main motion. ***Germane*** means directly related to the subject being considered. In short, this rule means that the amendment made to a main motion must have something to do with the main motion. For example, if the main motion was "I move that our group attend a leadership workshop," a germane amendment might be to add "in the fall semester." In contrast, if you tried to amend the motion by adding "...and that we have steak at the year-end banquet," the amendment would not be in order, because it would not be germane to the original motion.

In addition, amendments to a motion cannot be hostile to the intentions of the original motion. Using the same example as above, if the main motion were "I move that our group attend a leadership workshop," it would be out

Hands-On Agriculture

Practice Writing and Amending Motions

Now that you know how to write and amend motions, it is time to practice. To complete this activity, you will need a partner and a blank sheet of paper.

To begin, each person should write ten main motions. Make sure that each of your motions is written exactly how you would state them to the chairman. An example might be: "I move that we award all of the seniors in our agriculture program $1,000 scholarships."

After you and your partner have both completed your motions, trade papers. Think about how you could amend each of your partner's motions. Using the proper wording, write an amendment for each of the motions that your partner listed. For example, you could amend the motion about scholarships by writing "I move to amend the motion by substituting $2,000 for $1,000."

Once you have amended each other's motions, trade papers and discuss the reasons you made the amendments. This discussion is a good glimpse into what you might want to debate on the amendments for those topics.

of order to amend the motion by inserting the word "not" to make the motion read "I move that our group *not* attend a leadership workshop," because it goes against the intent of the main motion. If the member truly felt that attending the leadership workshop was not a good idea, what should they do instead of proposing a hostile amendment?

Voting on a Motion

After the debate and amendments are made on a motion, and the group is ready to make a decision, it is time to vote on the motion. Voting can take place through several different methods:

- Voice—the chairman asks for a response from all members in favor of the motion, and then from all opposed to the motion.
- Rising—the chairman asks for all in favor to stand or raise hands, and the responses are counted. The chairman then asks for all those opposed to stand or raise hands, and those responses are counted. The chairman announces the number in favor and the number opposed, along with the decision.
- Roll call—each member is asked for their vote and their vote is recorded as such.
- Ballots—anonymous votes that are written, tallied, and recorded.

Some motions do not require a vote while others need a certain number of votes to pass. Motions that do not require a vote are motions on which the chair makes the decision. Motions that need to be voted on typically require a majority vote or a two-thirds vote.

The number of votes needed for a motion to pass depends on the type of motion. A majority vote requires one more than half of the members to vote in favor in order for the motion to pass. The two-thirds vote requires at least 2/3 of the members to vote in favor in order to pass. Motions requiring a two-thirds vote are motions that restrict the rights of members in some way. For example, the previous question motion works to end debate on a motion and moves it directly to a vote. Because a previous question motion limits the rights of members to have the opportunity to debate the topic, it requires a two-thirds vote.

STEM Connection Calculating Votes

To properly calculate the number of votes required for a motion to pass requires you to apply several math concepts. Motions can pass with either a majority or a two-thirds portion of the members.

A majority vote requires more than half of the members to vote in favor of the motion for it to pass. To calculate this number, you can divide the total number of members in half. If you get a whole number, simply add one to make it more than half. If there is an odd number of members, you will not get a whole number when you divide. In this case, round the number to the next number higher to find the number of members required to pass.

A two-thirds vote means that at least 2/3 of the members must vote in favor for the motion to pass. To calculate this number, take the total number of members, divide it by three (which gives you 1/3 of the membership) and multiply it by two (to get 2/3 of the members). If the number you get for an answer is not a whole number, round it to the next highest number.

Motions

There are parliamentary motions designed to properly handle any number of situations. In fact, there are 25 commonly used motions in parliamentary procedure, each with a different function. Can you think of any circumstances regarding motions that were not covered in the preceding section? Did any of the following questions come to mind?

- What if someone makes a main motion incorrectly?
- What if the group decides they need more information before they vote?
- What if someone wants to debate more than three times?

These are all valid questions. Refer to **Figure 2-28** for a list of the common motions and their individual purposes.

FFA Connection Parliamentary Procedure CDE

Do you enjoy debating, speaking, or conducting meetings? If so, ask your FFA advisor about participating in the Parliamentary Procedure Career Development Event.

Number of members per team: 6 (one chairman, one secretary, four floor members)

Event components include a knowledge test, presentation of parliamentary skills, oral questions about presentation and parliamentary laws, and preparation of minutes.

Scoring is based on one's knowledge of parliamentary law and presentation skills.

Nata-Lia/Shutterstock.com

In addition to the National FFA Parliamentary Procedure CDE, many states have additional FFA events related to conducting meetings. For example, in Texas, members can compete in the Chapter Conducting event, which relies on knowledge of parliamentary procedure and FFA opening ceremonies. Other states have events particularly related to presenting opening and closing ceremonies. Check with your advisor to see what is available in your state.

Parliamentary Procedure Motions Guide

Motion	Debatable?	Amendable?	Vote	What the Motion Does
Privileged Motions				
Fix Time to Adjourn	N	Y	Majority	Sets the time you will start the next meeting after you adjourn
Adjourn	N	N	Majority	Ends the meeting
Recess	N	Y	Majority	Allows for a break in the meeting
Question of Privilege	N	N	Chair Decision	Allows a member to ask privilege to perform a personal need
Call for the Orders of the Day	N	N	None*	Guides the members to the items that the meeting was intended to discuss

*There are additional rules that apply to these statements; refer to the most recent parliamentary resource for additional information.

Goodheart-Willcox Publisher

Figure 2-28. Common motions and their uses.

Motion	Debatable?	Amendable?	Vote	What the Motion Does
Incidental Motions				
Appeal	N	Sometimes	Sometimes	Allows the floor to vote to overturn the chairman's decision
Point of Order	N	N	None	Corrects a parliamentary mistake
Parliamentary Inquiry	N	N	None	Allows members to ask a question about parliamentary law (answered by the chairman)
Suspend the Rules	N	N	2/3*	Allows organizational rules to be changed
Withdraw a Motion	N	N	Majority*	
Object to the Consideration	N	N	2/3*	Allows a member to not put any motion before the floor
Division of the Assembly	N	N	None	Gives a counted vote after the results of a voice vote are unclear
Subsidiary Motions				
Lay on the Table	N	N	Majority	Sets aside a motion (does not give a specific time)
Previous Question	N	N	2/3	Takes a vote to close debate and proceed immediately to voting
Limit or Extend Debate	N	Y	2/3	Changes the number or length of allowed debates
Postpone Definitely	Y	N	Majority	Postpones a motion to a specific time
Refer to Committee	Y	Y	Majority	Sends a motion to a committee to gather more information
Amend	Y	Y	Majority	Modifies a motion
Postpone Indefinitely	Y	N	Majority	Sends the motion away, kills the motion
Main Motion				
Main Motion	Y	Y	Majority	Introduces a new item of business to the group
Motions That Bring a Question Again before the Assembly				
Reconsider	Y*	N	Majority	Allows the group to bring back up a motion that was previously voted on to discuss and vote on again
Rescind	Y	Y	Majority or 2/3*	Allows the group to completely remove record of a motion that was made
Take from the Table	N	N	Majority	Brings back up a motion that was postponed with lay on the table
*There are additional rules that apply to these statements; refer to the most recent parliamentary resource for additional information.				

Goodheart-Willcox Publisher

Figure 2-28. *(Continued)*

Lesson 2.3 Review and Assessment

Words to Know

Match the key terms from the lesson to the correct definition.

1. A set of rules and regulations for properly conducting meetings.
2. A motion that deals with managing other motions.
3. The person responsible for conducting the meeting.
4. It means that the amendment made to a main motion must have something to do with the main motion.
5. A motion related to parliamentary rules and procedures, not to other motions.
6. A motion that may be modified per parliamentary rules.
7. The order in which items should be presented in a meeting.
8. The list of motions in the order that they must be handled.
9. A motion that may be put up for debate. All main motions and amendments to main motions are debatable.
10. A motion with the highest precedence of all motions that deals with the rights or needs of the organization. Privileged motions include those concerned with starting, stopping, pausing, or focusing the purpose of the meeting.
11. The number of members that must be in attendance at a meeting for business to be conducted. It is typically set at one more than half of the total members in the group.

A. amendable motion
B. chairman
C. debatable motion
D. germane
E. incidental motion
F. order of business
G. parliamentary procedure
H. precedence
I. privileged motion
J. quorum
K. subsidiary motion

Know and Understand

Answer the following questions using the information provided in this lesson.

1. *True or False?* Parliamentary procedure is a set of rules and regulations for properly conducting meetings.
2. List the four main goals of parliamentary procedure.
3. The number of members that must be in attendance at a meeting for business to be conducted is called a(n) _____.
4. The person responsible for conducting the meeting is the _____.
5. The hammer-shaped tool that the person conducting the meeting uses to send specific cues to the assembly is a(n) _____.
6. A complete, written record of the proceedings of a formal meeting are called the meeting's _____.
7. List the *order of business* in proper order.
8. A formal proposal by a member which, if agreed upon by the members of the assembly, will result in a certain action is a(n) _____.
9. *True or False?* Precedence is the list of members in the order in which they are allowed to speak.
10. What is the purpose of privileged motions?

11. If a new motion were to come about that allowed members to take a break from the meeting to check their cell phones and post social media about the meeting, what classification of motions do you think it should belong to?
12. Motions that ask questions, correct mistakes, ensure the accuracy of a vote, or change or modify rules are called _____ motions.
13. Motions that work to set aside a motion, modify the amount of debate, or change the motion in some way are called _____ motions.
14. A motion that brings up a new topic of discussion before the assembly is a(n) _____ motion.
15. An indication that there is at least one person besides the maker of the motion that is interested in seeing a motion come before a meeting is called a(n) _____.
16. *True or False?* For debatable motions, standard rules state that each member has the right to debate only once per debatable motion.
17. *True or False?* Amendable motions may be amended up to four times.
18. Briefly explain what the term *germane* means in regard to the main motion.
19. What are the four different methods through which voting on motions may take place?
20. Why do some motions require a two-thirds vote?

Analyze and Apply

Use your knowledge of parliamentary procedure and math to answer the following. Based on the number of members, find the number required for passing a majority and two-thirds vote.

1. 30 members
2. 31 members
3. 100 members
4. 1500 members
5. 10 members

Thinking Critically

For each of the following meeting scenarios, please find the motion you could use to manage the situation and achieve your goal.

1. You think that the topic the group is discussing will require more information. You want a group of people to research the topic and bring suggestions back to the group.
2. The meeting has been going for a while, and you feel like there should be a break so that members can get a drink, use the restroom, or quickly attend to other personal matters.
3. You would like to find out what vote a motion requires.
4. You think that the chairman announced the results of a voice vote incorrectly.

Give an example of a situation in which you would use each of the following motions.

5. Question of Privilege
6. Point of Order
7. Postpone Definitely
8. Previous Question
9. Reconsider

Chapter 2

Review and Assessment

Lesson 2.1

Building Leadership Skills through Agriculture

Key Points

- Developing your leadership skills is important to your ability to effectively guide others in the future.
- Understanding the types of leadership, along with the characteristics of a good leader, will allow you to develop and work toward a personal leadership plan.
- Taking advantage of opportunities available to you now will be beneficial to your future as a leader.

Words to Know

Use the following list and the textbook glossary to review and study the *Words to Know* from *Lesson 2.1.*

autocratic leadership
character
delegation
delegative leadership
democratic leadership
etiquette
followership
4-H
laissez-faire leadership
leader
leadership
mentor
mentorship
mission statement
National FFA Organization
participative leadership
personal leadership
personal leadership plan
servant leadership
SMART goals
team

Check Your Understanding

Answer the following questions using the information provided in *Lesson 2.1.*

1. Understanding leadership requires knowing _____.
   A. the characteristics of a good leader
   B. how to develop your own leadership
   C. the types of leadership
   D. All of the above.
2. List the three basic types of leadership. Give an example of each type.

3. Briefly explain why resourcefulness is important to leadership.
4. Explain how good character may impact your personal leadership skills.
5. Explain how organizations like FFA and 4-H provide opportunities for students to develop leadership skills.

Lesson 2.2

Communication Skills in the Agricultural Industry

Key Points

- The components of verbal, written, and visual communication make up the three basic types of communication.
- The differences in communication allow information to be passed from person to person through different channels.
- Written communication can be conducted formally or informally and should follow grammar and structure rules.
- Having the skills to make an effective verbal presentation are essential to becoming a good communicator.

Words to Know

Use the following list and the textbook glossary to review and study the *Words to Know* from *Lesson 2.2*.

active listening
communication
communication systems
conversation
correspondence
creed speaking
dyadic communication
extemporaneous speaking
group communication
impromptu speaking
informational written correspondence
intrapersonal communication
listening
mass media
nonverbal communication
oral communication
passive listening
prepared public speaking
presentation
social media
speech
verbal communication
written communication

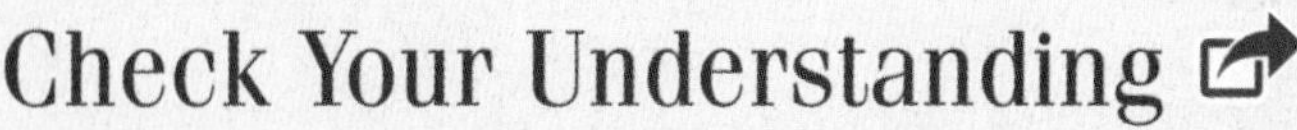

Check Your Understanding

Answer the following questions using the information provided in *Lesson 2.2*.

1. List the main types of communication.
2. *True or False?* Communicating clearly relies on the same principles regardless of which type of communication you are using.
3. Identify five types of nonverbal communication. Give an example of each type.
4. List three types of public speaking.
5. Identify four things you should do before writing a speech.
6. *True or False?* Giving verbal and nonverbal clues is essential to developing your active listening skills.
7. List three types of body language that indicate you are receiving the message. List three types of behavior that indicate you are not actively listening to the speaker.
8. Explain the importance of safe social media practices.

Lesson 2.3

Conducting Meetings in Agricultural Organizations

Key Points

- There are set rules and procedures to follow when conducting meetings.
- Understanding the role of the chairman, gavel use, purpose of minutes, and order of business, will allow you to help meetings run smoothly.
- Main motions are presented, then might be debated or amended before they are voted on.
- There are motions used to handle all situations as well as modifications that may be desired over the course of a meeting.
- Understanding how parliamentary procedure works allows you to be a valuable member of a formal group as you develop your leadership skills.

Words to Know

Use the following list and the textbook glossary to review and study the *Words to Know* from *Lesson 2.3*.

amendable motion
chair
chairman
debatable motion
gavel
germane
incidental motion
main motion
minutes
motion
order of business
parliamentary procedure
precedence
privileged motion
quorum
second
subsidiary motion

Check Your Understanding

Answer the following questions using the information provided in *Lesson 2.3*.

1. *True or False?* The main purpose of parliamentary procedure is to conduct meetings as quickly as possible.
2. Explain how to determine if there are enough members in attendance for a meeting to be held.
3. The chairman must _____ to effectively run a meeting.
   A. speak clearly and be unbiased
   B. have good knowledge of parliamentary procedure
   C. keep track of the pending business
   D. All of the above.
4. One tap of the gavel _____.
   A. is used to begin the meeting
   B. signifies to the members to rise
   C. signals the results of a vote
   D. signals to the members to come to order and refocus
5. Two taps of the gavel _____.
   A. are used to begin the meeting
   B. signify to the members to rise
   C. signal the results of a vote
   D. signal to the members to come to order and refocus
6. Three taps of the gavel _____.
   A. are used to begin the meeting
   B. signify to the members to rise
   C. signal the results of a vote
   D. signal to the members to come to order and refocus

Chapter 2 Skill Development

STEM and Academic Activities

1. **Social Science—Research.** Find a popular personal leadership test (some examples include the Myers-Briggs Type Indicator, Strengthsfinder, or the Kolb Learning Styles Inventory). Have the members of your class take the test to see what it says about their leadership skills. Create a graph or chart that depicts your findings and make predictions about how these findings will influence the dynamics of your class.
2. **Technology.** There is increasing information that human emotions and expressions are able to be read by computers. Do an Internet search for computers detecting emotion and find an article about this new technology. Create a short half-page summary of your article and share it with the rest of the class, along with your thoughts on how this technology could be used to help communication in the future.
3. **Math.** Calculating the number of members required for a vote is an important part of conducting parliamentary procedure. For each of the following groups, calculate the number required for a majority vote and a two-thirds vote.
   A. 50 members
   B. 100 members
   C. 250 members
   D. 500 members
   E. 1000 members
4. **Social Science—New Words.** In the first edition of the Oxford English Dictionary, there were 252,200 entries. As of the year 2010, there were almost half a million words in the English language. Do you think the addition of new words makes it difficult to communicate? Please list and define ten terms that you think are "new" and commonly used by your peers that some adults may not understand.
5. **Language Arts.** Using the guidelines in *Lesson 2.2*, write a speech that follows the prepared public speaking guidelines for the National FFA Organization Prepared Public Speaking CDE. Choose a topic in which you are interested. Make sure your speech has a well-defined outline.

Communicating about Agriculture

1. **Speaking and Listening.** Divide into groups of four to five students. Shoot a short video with two people having a discussion on a situation or topic on which they strongly disagree or strongly agree. Present the video to fellow students, first without sound. Ask them to devise a short script based on their observations. (Or ask the following questions: Did the two people already know each other? What is their relationship? What are their moods? Are they in agreement? How did you know this?)

   Replay the video with sound. Have your fellow students compare what they heard in the second viewing to what they observed in the first viewing. How accurate were their observations? Ask your fellow students to identify nonverbal cues (facial expressions, gestures, posture, body movement, body contact) that helped them interpret the situation. Ask for examples from the video.

2. **Listening.** Record a half-hour news broadcast. Before viewing the broadcast, turn on your radio, and have your phone or laptop on. View the broadcast while using your phone or laptop. Once the broadcast is over, write down everything you remember seeing or hearing. Set aside the phone or laptop and turn off the radio. Watch the broadcast again. How much did you remember? How accurate were your observations? How were your observations affected by the distractions? Perform the experiment again (with a different broadcast), but eliminate all distractions. Compare your findings.
3. **Writing.** Make a two-column chart and list the names and occupation or relationship of ten people with whom you communicate on a regular basis in the first column. How is the way you communicate with these people influenced by your relationship? Use the second column to identify and write the different ways you speak and behave when in the presence of these individuals.
4. **Speaking and Listening.** Working with a peer, make two lists of five people you both consider to be leaders. In the first list, write the names of historical figures who are no longer alive or in office. In the second list, write the names of people who are currently active in a leadership position. Beside each name, list at least three characteristics you both agree makes the person a good leader. Compare the characteristics you listed. Are they the same? Does the current era make a difference in leadership qualities needed to be successful?
5. **Speaking and Writing.** Working with a peer and using the lists from the previous activity, write at least three character flaws for each person. How do they compare? Were these flaws detrimental to their careers or success? How? Write a short report on your findings and present it to the class.

Extending Your Knowledge

1. Think about the following scenario: you are the team leader for a science project; team members cannot agree on anything and nothing is being accomplished. Choose a leadership style and describe the method you would use to overcome the problem.
2. Watch a speaking contest, or find videos of people giving speeches on YouTube®. Make notes about the speaker's voice, volume, tone, gestures, and expressions. Analyze their speaking style and write down two things the speaker did well, and two things the speaker could improve on.
3. In groups of five or six, ask your agricultural science instructor to give you practice problems from the local area or state parliamentary procedure or chapter conducting CDEs. Work as a team to select a chairman and practice your skills in parliamentary procedure. Make sure that you can make motions, debate, and come to a resolution of the problem. Have a class contest to determine which team is the best.

Chapter 3

Agriculture as a Career

SpeedKingz/Shutterstock.com

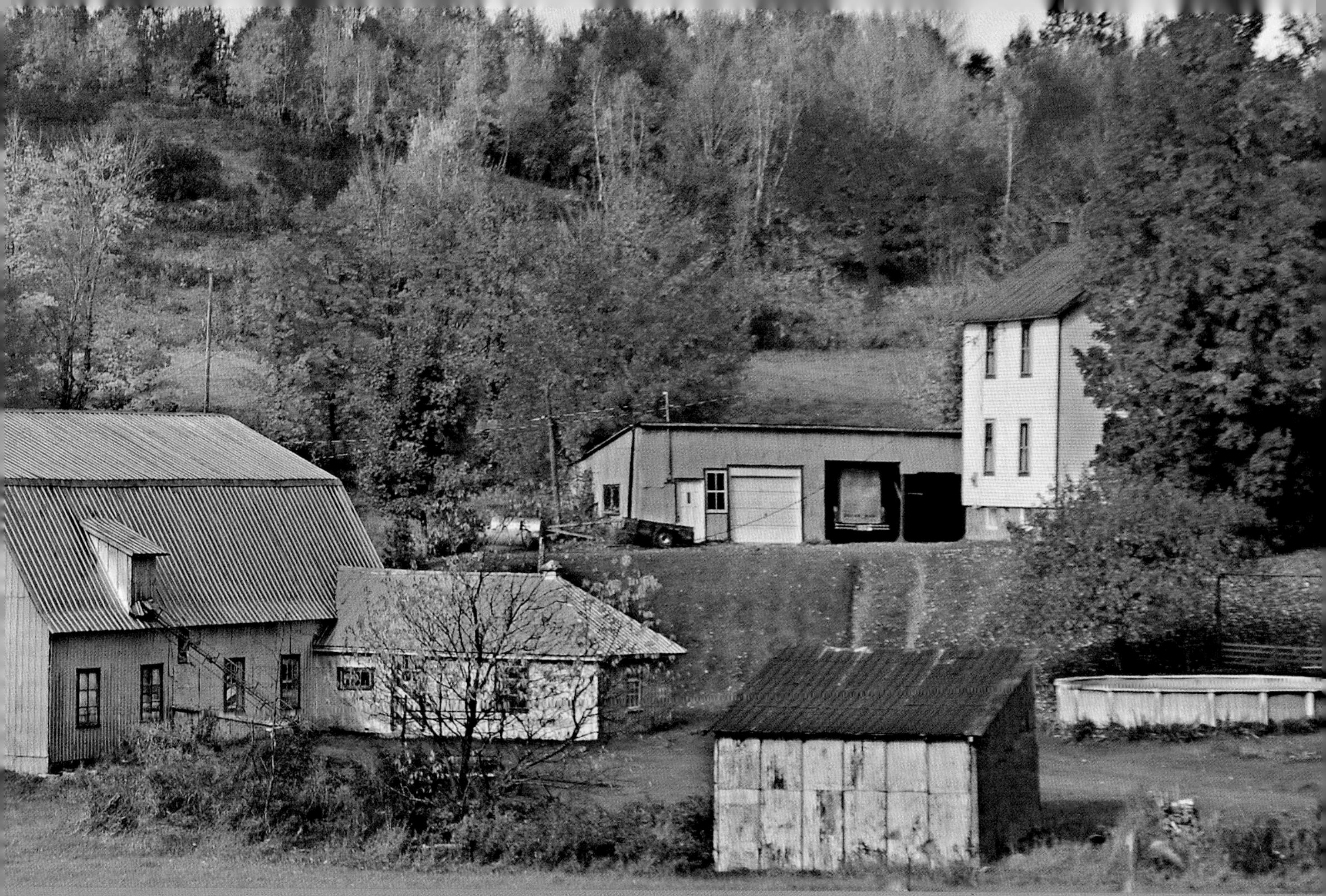

meunierd/Shutterstock.com

sezer66/Shutterstock.com

mythja/Shutterstock.com

While studying, look for the activity icon  **to:**

- **Practice** vocabulary terms with e-flash cards and matching activities.
- **Expand** learning with video clips, animations, and interactive activities.
- **Reinforce** what you learn by completing the end-of-lesson activities.
- **Test your knowledge** by completing the end-of-chapter questions.

Lesson 3.1

Experiential Learning through Agriculture (SAE)

Words to Know

enterprise
entrepreneurship SAE
experiential learning
exploratory SAE
hands-on learning
long-term goal
placement SAE
proficiency awards
research SAE
SAE improvement project
short-term goal
Supervised Agricultural Experience (SAE)

Lesson Outcomes

By the end of this lesson, you should be able to:

- Explain the importance of learning through experience.
- Compare and contrast the four main types of Supervised Agricultural Experiences (SAE).
- Develop a plan for a personal SAE, including evaluation of interests and resources.
- Set both short-term and long-term goals for your SAE.
- Apply proper recordkeeping skills related to an SAE program.

Before You Read

Before you read the lesson, read all of the table and photo captions. Share what you know about the material covered in this lesson just from reading the captions with a partner.

Learning to Do, Doing to Learn; these are the last two lines of the National FFA motto. Agricultural education has a long tradition of basing instruction on actual experiences. To facilitate learning through doing, agricultural classes are designed to provide you with the opportunity to participate in experiential learning. ***Experiential learning*** is the process of creating understanding from doing something. For example, you might be able to give someone all the instructions for driving a car, but they cannot really *learn* to drive until they have the experience of being behind the wheel.

This lesson will give you an overview of why experiential learning is important and allow you to develop your plan for learning through experience.

Did You Know?

Studies have shown that most people are more than four times as likely to remember something if they have a hands-on learning experience.

Experiential Learning

The concept of experiential learning is not new. Confucius (551 BCE–479 BCE) said, "I hear and I forget, I see and I remember, I do and I understand," outlining the basic concept of experiential learning. Think about something that you know how to do really well; something that you feel qualified to teach someone to do. Think about the skills that are needed to perform this task. What mental images come to mind? Do you see words and a list of instructions for performing the task? Do you see a mental image of yourself going through the activity? It is a safe guess that you see images of the activity being performed. Our brains are hardwired to learn through experience, and that's exactly what experiential learning is all about.

Agricultural Education

Agricultural education classes provide many opportunities for you to learn through experience. **Figure 3-1** shows the agricultural education model, including the places where experiential learning is likely to occur. In agricultural classes, there are many ways that your teacher can incorporate hands-on learning. ***Hands-on learning*** is instruction that includes activities provided for you to better understand and apply knowledge in a way that allows you to interact with material and objects. Agricultural education is set up to provide experiential learning in three components:

- Classroom and laboratory instruction.
- Leadership development through FFA activities.
- Hands-on application and learning through Supervised Agricultural Experiences (SAEs).

Classroom and Laboratory

The classroom and laboratory part of agricultural education classes allows for experiential learning through teaching methods that let you participate in the learning process. There are many different agriculture classes from which to choose. Many classes are specific to the agriculture in your particular region. Sometimes, an agriculture program is made up of only one instructor, who teaches all of the different agriculture classes. Quite often, agricultural education programs will have multiple teachers, who each specialize in a particular area of agricultural science. Who are the agricultural educators at your school?

One of the unique things about agriculture classes is their predisposition to laboratory activities. The structure and content of agricultural education courses makes them great places to explore new information through hands-on learning. Think about the different lab areas you have in your agriculture program. Greenhouses, gardens, animal labs, agricultural mechanics facilities, school farm plots, and even woodworking and food science areas are all used as lab spaces by agriculture programs around the country, **Figure 3-2**.

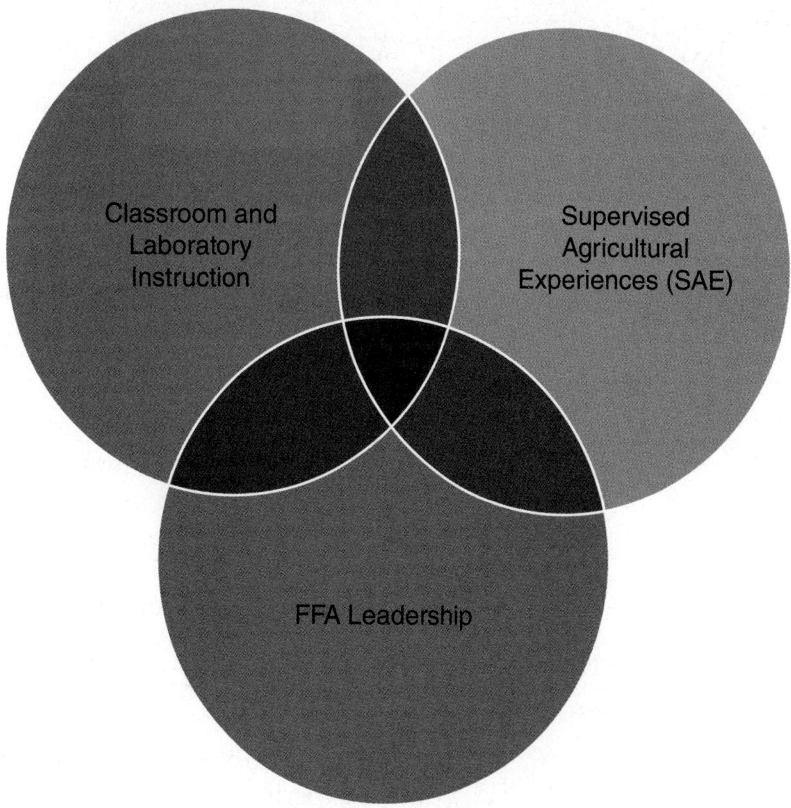

Goodheart-Willcox Publisher

Figure 3-1. The agricultural education model shows the three main components of a total program. *How balanced do you find your current agricultural education experience?*

Courtesy of Jason McKibben

Figure 3-2. There are many laboratory spaces available in a local agricultural education program. These students are working on developing their skills with electrical systems. ***What skills can you develop in the lab of your agricultural education program?***

Your program does not need a special lab space to be hands-on; classrooms are also great places to perform smaller scale agricultural science laboratory activities.

FFA Activities

FFA activities allow you to learn concepts in class and apply them through leadership development and career development events. The National FFA Organization is an intracurricular part of an agricultural education program. Through FFA, agriculture students are provided the opportunity to not only gain skills related to agriculture, but also skills that will make them better communicators, leaders, and citizens.

Have you been to an FFA activity? Your local chapter has the opportunity to plan and conduct a wide variety of activities to help members develop leadership while doing service and having fun. Ask your agriculture teacher when the next FFA activity in your chapter is scheduled and how you can be a part of it. Consider getting involved and helping plan activities for the FFA members at your school. There are also activities at the area and state levels that will allow you to interact with other FFA members from around the country.

Supervised Agricultural Experiences (SAEs)

Every student in an agricultural education program or class is required to have an SAE project or program. ***Supervised Agricultural Experiences (SAEs)*** are personalized experiential learning programs made up of projects tailored to meet your needs and interests. SAEs are designed and conducted by students and supervised by agriculture teachers. Students choose an SAE based on their likes, interests, and talents. There are four guiding questions used to determine if something can be used as an SAE:

- Has the project been planned and developed based on student interests and resources?
- Does the student plan on documenting and recording the activities of the experience?
- Will the project be supervised and overseen by the agriculture teacher or FFA Advisor?
- Is the project based in agriculture, food, or natural resources?

SAE Type	Description	Investment of Time	Investment of Money
Exploratory	Learning something new about the agricultural industry	Yes	Typically none
Research	Using the scientific method to analyze a research question	Yes	Typically none
Placement	Working for someone else in the agricultural industry	Yes	No input, may be paid
Entrepreneurship	Owning and managing an agricultural enterprise	Yes	Yes, responsible for all cost input and profit

Goodheart-Willcox Publisher

Figure 3-3. The four types of SAEs have been developed to help you make the most of your current interests and resources. ***Which type is best for you?***

Did You Know?

The blue corduroy jacket was adopted as official dress in 1933, when FFA members saw the band from Fredricktown, Ohio, wearing matching jackets to the National FFA Convention. Gus Lintner, the director of the band, is credited with its creation. Do you have an FFA jacket yet?

If all four of these questions can be answered yes, then the project can be considered a viable option for an SAE. Your agricultural science teacher is the expert on this topic. If you have an idea and are not sure it fits the SAE requirements, talk to your teacher and let him or her help you. Because no two students are alike, and SAE programs should be individualized, there are many options for what you can do for your SAE. Each specific project you include in your SAE is called an ***enterprise***.

Types of SAEs

There are four types of SAEs (**Figure 3-3**):

- Exploratory
- Research
- Placement
- Entrepreneurship

The main differences between SAE types is the nature of resources invested and the outcome of the project. All SAE types require records of the experiences and resources to be kept.

Exploratory

Exploratory SAEs allow you to invest time in learning something new about the agricultural industry, **Figure 3-4**. Exploratory SAEs are a great way to learn about an agricultural area before you undertake a full-scale investment of money and resources. The primary investment in an exploratory SAE is time. The records kept on

wavebreakmedia/Shutterstock.com

Figure 3-4. Exploratory SAEs allow you the opportunity to learn more about a career or special topic in agriculture.

an exploratory SAE will include the new information or skills learned, along with the hours you spend learning about your chosen topic.

Research

Research SAEs are designed for you to use the scientific method to analyze a research question or test a hypothesis. This type of SAE, as shown in **Figure 3-5**, is a great way to learn more about something you are interested in, and develop your scientific research skills. This type of SAE typically involves only an investment in time. Records are kept related to the hours spent on the research, along with the initial research, procedures, data collected from research, and conclusions that can be made from the project.

Placement

The third type of SAE is a ***placement SAE***. In placement SAEs, you work for someone else, **Figure 3-6**. Placement SAEs can occur when you work for an agricultural business, for an individual for wages, or unpaid internships. There is no investment of money, only an investment of time spent working. Records are kept on the hours spent, skills learned, and of any money that has been made through the hours worked.

skyfish/Shutterstock.com

Figure 3-5. Completing a Research SAE allows you to conduct an experiment based on agricultural products. They are also perfect for competing in the Agriscience Fair program.

MJTH/Shutterstock.com

Figure 3-6. Placement SAEs involve you working for someone else. These SAEs can be paid or unpaid, and keep track of your investment in time.

Entrepreneurship

The final type of SAE is an ***entrepreneurship SAE***. In an entrepreneurship SAE, **Figure 3-7**, you have the chance to own your own SAE resources and make management decisions related to the entire enterprise. There is typically an investment of both time and money in this type of SAE. Records are kept on the time spent with the project, the skills learned, the money spent, and the income received from the project.

CREATISTA/Shutterstock.com

Figure 3-7. Owning a stockshow animal is an example of an entrepreneurship project. You are the sole owner of the enterprise and make all management decisions.

Developing an SAE Plan

Now that you know the types of SAEs, it is time to start thinking about what SAE enterprises you might like to have. Some people would assume that there is a *best* type of SAE, or that having a stockshow animal is a more valuable experience than having a placement experience. That could not be further from the truth. One of the most important aspects of an SAE program is that it is custom tailored to you.

It is important to remember that you can have more than one SAE enterprise, and that your SAEs might be different types. For example, if you work at a local flower shop, and also have a market hog for a junior livestock show, you have two SAE enterprises: one placement and one entrepreneurship.

There are several considerations that you need to think about when selecting an SAE. Most of the considerations you will need to make are related to developing a personal SAE inventory. Your personal inventory should be conducted related to your interests and your resources.

SAE Interests

One of the coolest things about an SAE is that it allows you to completely control what you want to learn more about, and your personal interests come into play, **Figure 3-8**. There are many different paths for specialization that you can take. Generally, SAEs will fall into one of the following description categories:

- Animal—includes all animals kept for entrepreneurship SAEs, along with all employment, research, and exploration with animals of all types.
- Agribusiness—deals with the financial end of agriculture including lending, accounting, and commodity markets.
- Leadership, education, and communication—involves working with producers and consumers to help develop understanding about agriculture.

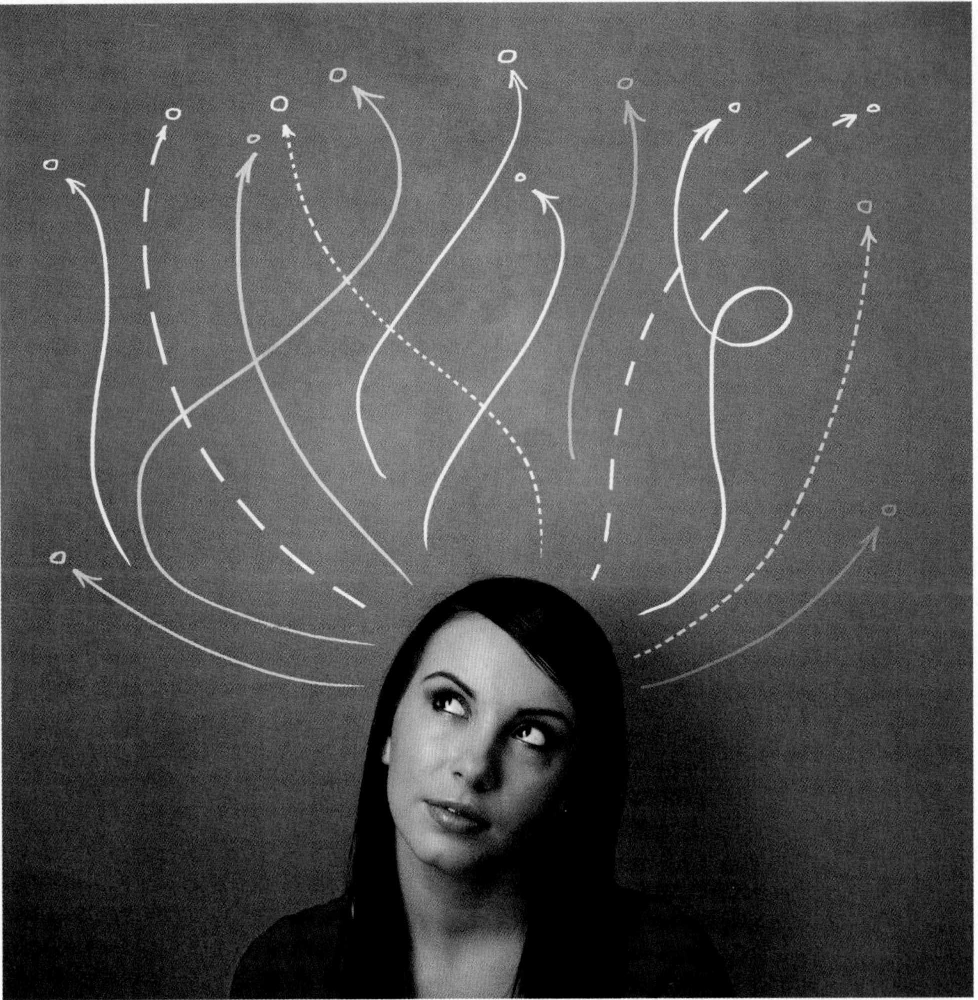

ra2studio/Shutterstock.com

Figure 3-8. Your interests can take your SAE path in many different directions.

Did You Know?

There are several good SAE inventories available online, all of which will take into account your interests and available resources to help you determine which SAE is right for you. One of the most commonly used online SAE inventories is Explore SAE, a website that allows you to learn about the different SAE areas and explore which SAE might be right for you. Ask your teacher about using an online inventory in your SAE selection process.

- Environmental service—related to preserving and reclaiming our natural environment.
- Food products and processing—includes everything required to process and regulate agricultural goods entering the food supply.
- Power, structural, and technical—deals with the design, fabrication, and maintenance for agricultural equipment and structures.
- Natural resources—involved with the management and conservation of natural resources including minerals, water, air, soil, and wildlife.
- Plants—deals with the cultivation and growth of plants used in the food system or in ornamental applications.
- Biotechnology—related to using technology to enhance the natural world through genetic engineering or other technologies.

Understanding which of these categories you are interested in is an important step in finding the right SAE for you. The table in **Figure 3-9** gives you some examples of each type of SAE in each interest category.

Examples of SAEs by Type and Category				
Description Categories	**Exploratory**	**Research**	**Placement**	**Entrepreneurship**
Animal	Finding the requirements to become a veterinarian	Testing which feed makes laying hens produce the most eggs	An internship at the local dairy farm	Owning and exhibiting a market steer
Agribusiness	Examining the trends in price for a certain agricultural commodity	Conducting market research to decide which product consumers prefer	Working for a local agricultural lender	Owning a service that completes payroll services for local farmers
Leadership, Education, and Communication	Looking at which universities offer Agricultural Leadership Degree Programs	Experimenting to see if students in agriculture classes with required SAE grades complete more SAE hours	Interning at a local radio station	Designing and maintaining websites for local agricultural businesses
Environmental Service	Exploring the number of water pollution complaints in your county	Testing air quality at various places in your community	Volunteering to help with a community recycling project	Running a company that collects scrap metals and brings them to recycling centers
Food Products and Processing	Determining the requirements for safe food handling in the United States	Testing the differences between organic and nonorganic foods	Working at a local farmers market	Selling homemade jams and jellies
Power, Structural, and Technical	Finding a two-year certificate program in diesel mechanics	Conducting load tests for different shapes of metal roofing	Working in a local farm implement dealer service shop	Owning a mobile equipment repair company that provides infield service to producers
Natural Resources	Compiling a list of the most utilized natural resources in your state	Conducting soil tests for farms in different locations in your county	Interning with the local division of wildlife services	Owning a firewood and fence post company
Plant	Looking at the media for information about how drought affects crops in your area	Testing soil amendments to see their impact on corn plants	Working for a local farmer to harvest forage crops	Owning a landscaping and yard-mowing company
Biotechnology	Examining the job requirements for becoming a genetic engineer	Conducting an experiment to compare the growth of GMO and non-GMO plants	Serving as a research assistant in a biotechnology lab	Producing GMO crops

Goodheart-Willcox Publisher

Figure 3-9. Looking at different examples of SAEs can help you understand where your SAE might fit. *Do you see an SAE in this table that you might find interesting?*

SAE Resources

Once you understand your interest areas, you can begin to determine what type of SAE will fit you best. There are three main considerations that should be taken into account regarding your resources: time, money, and equipment. Examining your expectations in these three categories can provide great clarity to your SAE selection process.

Time

The amount of time you have to spend on your SAE will be a big factor in determining what type of SAE you should take on, **Figure 3-10**. Although every SAE is a little bit different, entrepreneurship SAEs typically take more time than exploratory or research projects. Placement SAEs require varying amounts of time, depending on how many hours you are scheduled to work at your SAE job. Determine how much free time you have to dedicate to an SAE by making a schedule of all of your responsibilities. Once you can determine how much available time you have, you can see where your SAE can fit into your time. Remember that the amount of time you can spend on your SAE will be a large determinant of how successful the project will be.

Money

Not all SAEs require an investment of money. Exploratory and research SAEs are typically conducted with no monetary input at all, and placement SAEs can even be conducted to help you earn money, **Figure 3-11**.

Syda Productions/Shutterstock.com

Figure 3-10. Time is one of the biggest considerations to keep in mind when selecting a potential SAE. Making sure you have enough time to work on your SAE will help you strengthen your experience.

Entrepreneurship SAEs are the type most likely to require a financial investment up front. This is because you are essentially starting your own business enterprise. Thinking through how much money you have available can guide your selection of an SAE. Will you need to have money to purchase an animal for your show animal project? Will you need to purchase advertising materials for your landscaping business? Many students rely on short-term loans from their parents to get the funds needed to start their SAE. In addition, there are several organizations that will provide grants and loans to students for initial SAE costs.

OLJ Studio/Shutterstock.com

Figure 3-11. The amount of money you have to spend on SAE resources, and the amount you hope to profit from your SAE, are factors to use in making your selection.

Equipment/Facilities

The last category of resources you should consider is the special equipment or facilities you have available. For example, if you want to conduct a research SAE on growth patterns of greenhouse plants, you would need to have access to a greenhouse. Look around your agriculture program for available resources. By making a list of the equipment and facilities you have at home, school, or in other locations, you might be able to come up with a great idea for an SAE.

Your equipment or facilities may change over the course of time. An ***SAE improvement project*** is an enterprise undertaken to make changes to your SAE for the better. Some examples of improvement projects are designing a new website for your business, restoring an old piece of equipment, or building a new facility for use with your existing SAE enterprises, **Figure 3-12**.

rawmn/Shutterstock.com

Figure 3-12. Restoring an old piece of equipment for use with one of your SAE enterprises is an example of an improvement project.

CLS Design/Shutterstock.com

Figure 3-13. Selecting an SAE is very much like choosing a new sports team or after school activity. Pick one that fits both your interests and your schedule.

Selecting Your SAE

Once you have taken stock of your interests and available resources, you are ready to select your SAE. Deciding on an SAE is a little bit like trying out a new sport for the first time, **Figure 3-13**. You typically choose to try a new sport because it interests you, the schedule for practices and games fits into your current schedule, and you can commit to learning the skills required to be good at the sport. The same is true of an SAE. Select an SAE you are interested in and that fits into your schedule and financial situation, and commit yourself to being willing to learn the new skills so your SAE will be successful. Also like a new sport, if you decide the SAE you picked is not the best fit for you, you can reassess and move to another SAE that will be a better match.

The decision tree shown in **Figure 3-14** allows you to look at general types of SAEs you may want to pursue. Once you determine a general area, you can choose an interest area and finalize the specific details of your project.

Setting SAE Goals

The best way to make sure your SAE is successful is to develop SAE goals. Just like the personal leadership goals we discussed in *Lesson 2.1*, SAE goals should be SMART (specific, measurable, attainable, realistic, and timely). Setting goals for both short-term and long-term time frames will allow your SAE to have a guided path for the rest of your high school career, and in many cases, even beyond high school. When you are developing SAE goals, you can think about them in two broad categories: short-term and long-term.

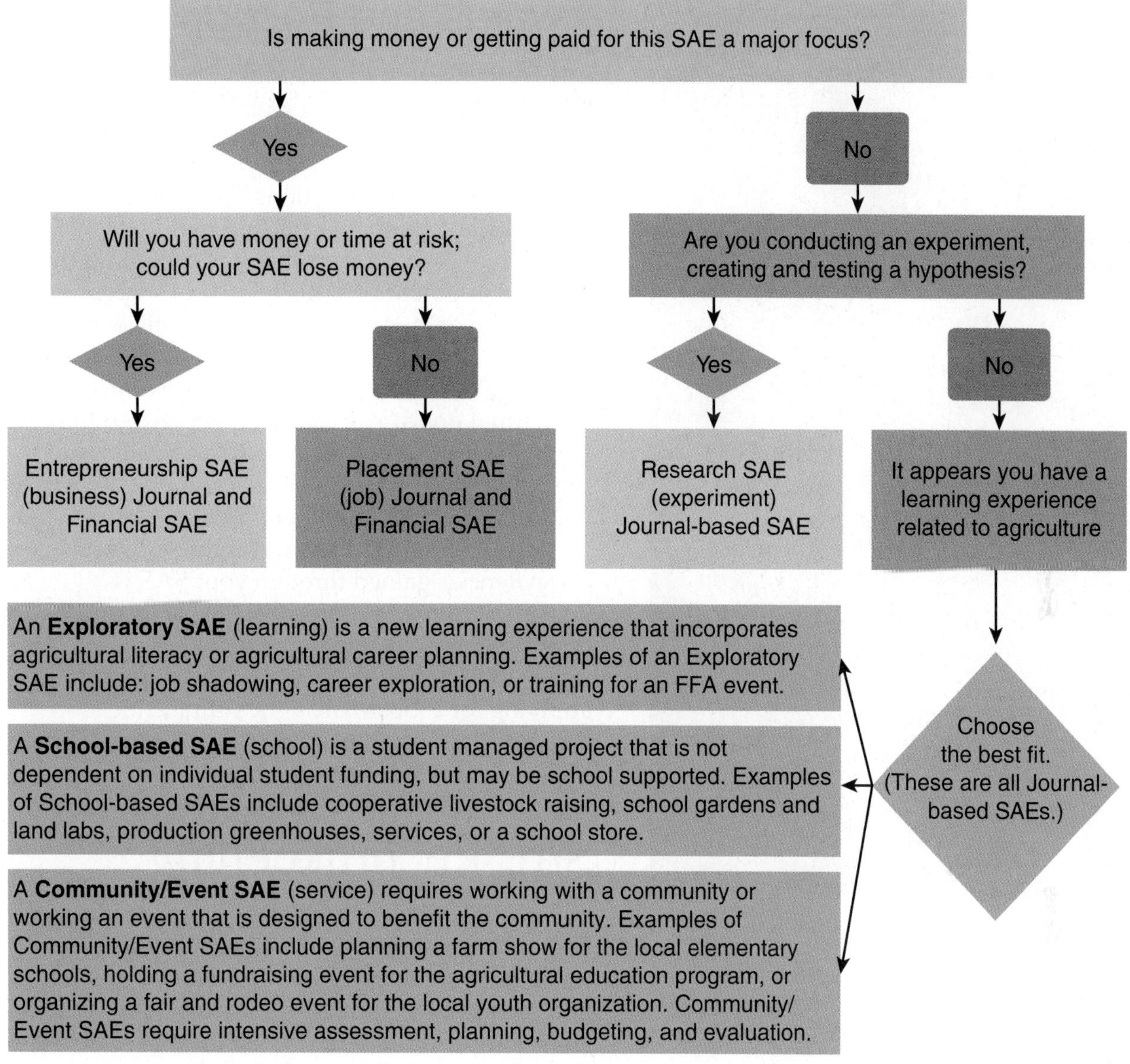

Roger Hanagriff, The AET Record Book

Figure 3-14. If you are having difficulty deciding on an SAE, this decision tree can help you. To use the decision tree, begin answering the questions from the top down and follow the arrows. Complete this exercise for each of your SAE project ideas.

Short-term goals are those goals that you plan on accomplishing in the immediate future. A general guideline is that these goals can be accomplished in a time frame of less than one year. ***Long-term goals*** are those that may take several years to accomplish. When you determine which type of goal you are setting, you should consider at what point you are in your SAE development. For example, the goal of exhibiting the grand champion heifer at the county fair may be a short-term goal for someone who has shown for many years, and has already put in the time and resources to select and exhibit quality animals, **Figure 3-15**. The same goal might be a long-term goal for someone who is new to showing cattle.

As you develop goals, it is a good practice to write them down and refer to them often. To take the goal setting process one step further, you can make a list of the smaller steps required to accomplish your goals and begin working toward them.

CREATISTA/Shutterstock.com

Figure 3-15. The same goal may be a short-term goal for one person and a long-term goal for another. For example, someone who has been showing livestock for many years may have a short-term goal of exhibiting a grand champion at the county fair. For someone who is new to showing livestock, the same goal would be a long-term goal.

Awards and Recognition

Having a quality personal SAE can be reward enough for all of your hard work and dedication, but there are also other ways to be recognized for meeting your SAE goals. ***Proficiency awards*** are awards given by the National FFA Organization on the local, regional, state, and national levels for student SAE projects that have shown growth and development of skills related to success in the SAE area. Applications for proficiency awards take into account all the resources used, competencies gained, and progress toward individual goals. You may also get recognition for your SAE through your school science fair and the National Agriscience Fair Program run by the National FFA Organization.

Keeping detailed records of the skills and competencies gained through your SAE is important for various reasons. One of the most important reasons is using them to get accepted into college or university programs, or for applying for a job. Many agricultural education students have used their SAEs as a springboard to a successful career.

SAE Recordkeeping

An essential component of having a quality SAE is keeping records. Keeping accurate records will allow you to measure the scope of your SAE and see progress toward your goals. In order to have complete records, you should be recording both the experiences and financial transactions that occur in your program.

How to Know What to Record

One of the biggest decisions you will be faced with over the course of your SAE is what to record in your recordbook. In a nutshell, the answer is "everything." The specific types of records you will keep depend largely on the type of each of your SAE enterprises.

- Exploratory SAEs will typically require you only to journal your experiences of activities and skills learned. In general, you will have no financial records associated with this type of SAE.
- Research SAEs will require you to journal the research methods, results, and conclusions. In addition, you should account for your time while conducting the project, and record any expenses you incurred while conducting research.
- Placement SAEs require you to record the time, either paid or unpaid, that you spent working. You will also need to journal exactly what you did on

FFA Connection Proficiency Awards

One of the most common goals for SAE projects is to use them to compete and win a proficiency award. Proficiency awards are given to members that excel in their SAE programs. The list of official proficiency award areas changes on an annual basis, but the following can be used as a guideline for determining which proficiency category might be right for you.

- Agricultural Communications
- Agricultural Education
- Agricultural Mechanics Design and Fabrication
- Agricultural Mechanics Energy Systems
- Agricultural Mechanics Repair and Maintenance—Entrepreneurship
- Agricultural Mechanics Repair and Maintenance—Placement
- Agricultural Processing
- Agricultural Sales—Entrepreneurship
- Agricultural Sales—Placement
- Agricultural Services
- Agriscience Research—Animal Systems
- Agriscience Research—Integrated Systems
- Agriscience Research—Plant Systems
- Beef Production—Entrepreneurship
- Beef Production—Placement
- Dairy Production—Entrepreneurship
- Dairy Production—Placement
- Diversified Agricultural Production
- Diversified Crop Production—Entrepreneurship
- Diversified Crop Production—Placement
- Diversified Horticulture
- Diversified Livestock Production
- Emerging Agricultural Technology
- Environmental Science and Natural Resources
- Equine Science—Entrepreneurship
- Equine Science—Placement
- Fiber and/or Oil Crop Production
- Food Science and Technology
- Forage Production
- Forest Management and Products
- Fruit Production
- Goat Production
- Grain Production—Entrepreneurship
- Grain Production—Placement
- Home and/or Community Development
- Landscape Management
- Nursery Operations
- Outdoor Recreation
- Poultry Production
- Sheep Production
- Small Animal Production and Care
- Specialty Animal Production
- Specialty Crop Production
- Swine Production—Entrepreneurship
- Swine Production—Placement
- Turf Grass Management
- Vegetable Production
- Veterinary Science
- Wildlife Production and Management

the job each day and describe any new skills that you develop during your employment. If you are working in a paid placement area, you will also need to record the financial income related to each paycheck.

- Entrepreneurship SAEs typically require the most recordkeeping components. In addition to journaling what has been done with the enterprise, and the skills you have gained, you will need to track the inventory, income, and expenses related to the project.

Journaling Experiences

Every type of SAE requires that you keep an accurate journal of the time spent and experiences you have had with each enterprise. Keeping track of everything you are doing with your SAE is not an easy task, and is one that will likely require you to be diligent and timely in recording information. The

Career Connection What Can Your SAE Do for Your Future?

Your SAE is an essential part of your agricultural education experience, but what can it do for you after high school? Many agricultural education students have taken their SAE enterprises and turned them into businesses that support them financially. Exploratory, Research, and Placement SAEs can give you valuable experience that will make you more marketable as an employee in fields related to your SAE category. Entrepreneurship SAEs can also give you employment skills, which you could use to grow your SAE into a productive and successful full-time business venture.

Some of the most common examples of SAEs that easily translate into viable companies include landscaping businesses and livestock operations.

Landscaping Business: One of the most easily transferrable businesses that many FFA members use as their adult career are landscaping businesses. These business owners have taken their high school business and built it into a thriving, full-time employment for themselves. Can you think of some of the benefits of proper SAE setup for those who plan to continue their venture after high school?

Livestock Operations: Many students begin with small numbers of animals during their SAE, and are able to build their enterprise to be large enough to raise animals full-time after high school. These students can then pursue livestock management as a suitable means of income to fund their education, or as a full-time source of employment income.

There are many other ways that your SAE can grow into something much more than just a high school project. In your SAE planning, think not only about how your SAE can play a role in your life today, but in your future.

Pixsooz/Shutterstock.com

Providing professional service to your customers will ensure a solid customer base. Written estimates will help you keep track of accurate and profitable job quotes.

Nate Allred/Shutterstock.com

Raising animals during your SAE will provide you with valuable experience you can apply to your advanced education and future agricultural employment.

experiences that should be collected related to your SAE include the amount of time you spend with your project and any skills or competencies you have worked on or accomplished.

Detailed records will help your record book become the "story" of your SAE. Someone who picks up your record book to evaluate your project should be able to know exactly *what* you did with your project, *how* you did it, and *why* you made the management choices you made. A good rule of thumb is to record more information than you think you will need.

Even those SAEs which have many financial entries require experience entries to tell the whole picture. For example, if you had a paid placement SAE with the local veterinarian, you would enter paychecks into your record book, but those entries do not tell much about the things you learned on the job or the skills you gained while you were working. Journaling allows you to outline the specific tasks and skills that were used during each pay cycle.

Can you remember exactly what you did today? How about what you did exactly one week ago? How about one month ago? Our memories can fade over time. Failure to record information immediately can lead to holes in your SAE story. Timely recording of experiences is important to recordkeeping success. Try to record the information the same day of the experience.

Recording Finances

Depending on the type of SAE you have, you may need to record many financial transactions or none at all. Placement SAEs will require you to record any income arising from your work experience. Entrepreneurship SAEs typically have the largest number of financial entries because they have both income and expenses.

Recording finances in an SAE is a great way to gain practical experience with agribusiness principles. By recording finances for your SAE, you will become familiar with analyzing profit and loss, managing inventory, and determining net worth. These concepts are explained in detail in *Lesson 8.1*.

Methods for Recordkeeping

SAE records can be kept in several different ways. Some agricultural science programs choose to have their students keep paper records, while the majority of SAE recordkeeping has moved to computer-based systems. There are several commercially available recordkeeping systems available to help you keep track of your SAE.

Your agricultural science teacher can give you guidance on which program you will be using to keep your SAE records. The most prominent electronic recordkeeping solution for SAEs is the *Agricultural Experience Tracker*, or *AET*. This web-based program allows you to access your FFA records from anywhere with an Internet connection.

SAE Connection The AET Overview

The AET is broken down into sections to help guide you through the recordkeeping process. The main sections of the AET are:

PROFILE: Includes personal information related to you.
JOURNALS: Allows you to record your time in the classroom, FFA and SAE.
FINANCES: Deals with the entry of financial transactions for your SAE.
REPORTS: Allows you to view summary information from your classroom, FFA, and SAE experiences.

In order to help you get started in AET, here is the information for logging in and setting up your profile:

Step 1: Log In

1. You can access your AET account anytime from any computer with Internet. Go to: http://www.theaet.com, and click Sign In "Student."
2. Your teacher will provide your login information. You will need:
   A. Chapter Number: This is the two-digit state abbreviation followed by the four-digit chapter number. For example, TX0765. You should memorize this number.
   B. Username: Usually this is your "first initial and last name." For example, TMurphy for Tim Murphy.
   C. Password: In the beginning, this is usually the same as your username. Make sure to use capital and lowercase letters exactly as your teacher shows you. After signing in, you can change your password.

Step 2: Set Up Your Profile

1. Click the PROFILE menu at the top of the screen. You should update your profile every year, or whenever your information changes.
2. The PROFILE section of your account is broken into the following sections:
   A. Manage/Edit your personal information and load a profile picture.
   B. Record your school Ag class schedule—After your teacher has listed the available classes, you can add yours here.
   C. Develop your résumé—A very basic outline of your accomplishments that should be updated each year to report involvement. Maintain your list of school and/or community activities. You can add certifications, special awards, and other skills, but your use of AET develops important records from your classroom, FFA, and SAE experiences that are summarized into a complete résumé.
   D. Develop your AET experiences—This is your project(s) in agriculture education. Enter the type, interest area, and other information about the project you have decided on. You can enter more than one enterprise.
   E. Enter your FFA offices—If you are an FFA Officer, add your office here.
   F. Enter your FFA committee memberships—If you are on an FFA Committee, add it here.
   G. Choose your agricultural career pathway—Check the box next to the Ag Careers that interest you; rank them on the right side.
   H. Explore your educational and career interests—Find and select the careers of interest.

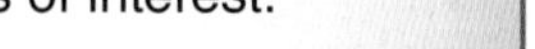

Roger Hanagriff, The AET Record Book

Lesson 3.1 Review and Assessment

Words to Know

Match the key terms from the lesson to the correct definition.

1. A Supervised Agricultural Experience where the student works for someone else in an agriculture related job.
2. An SAE in which students own and operate an agricultural business.
3. Activities provided for students to gain a better understanding of concepts by applying knowledge in a way that allows them to interact with material objects.
4. The process of creating understanding from doing something.
5. An SAE designed for students to use the scientific method to analyze a research question or test a hypothesis.
6. One specific section of a total SAE program.
7. An aim or desired result that will be accomplished over an extended period of time.
8. An enterprise undertaken to make changes and improve an existing SAE.
9. Awards given by the National FFA Organization on the local, regional, state, and national levels for student SAE projects.
10. An aim or desired result that will be accomplished in the immediate future.
11. An SAE designed to increase student agricultural career awareness through observation activities.

A. enterprise
B. entrepreneurship SAE
C. experiential learning
D. exploratory SAE
E. hands-on learning
F. long-term goal
G. placement SAE
H. proficiency awards
I. research SAE
J. SAE improvement project
K. short-term goal

Know and Understand

Answer the following questions using the information provided in this lesson.

1. What are the three components of agricultural education?
2. Briefly explain how to determine if a proposed SAE is a viable option.
3. List the four main types of SAEs.
4. *True or False?* The primary investment in an exploratory SAE is financial.
5. *True or False?* Research SAEs employ the scientific method.
6. *True or False?* When working within a placement SAE, you may or may not be paid.
7. *True or False?* There is typically an investment of both time and money in an entrepreneurship SAE.

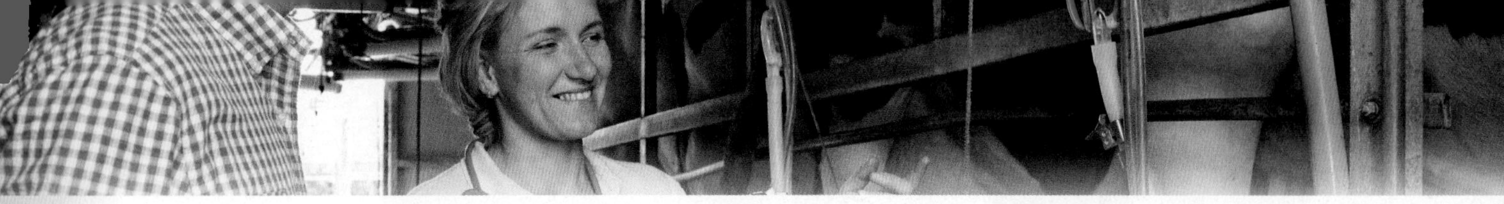

8. Briefly explain why it is important to have an SAE that is tailored just for you.
9. List five paths of specialization your SAE may take.
10. What are the three main considerations that should be taken into account regarding your resources for an SAE?
11. Explain what an improvement project is and how it works in an SAE enterprise.
12. Briefly explain SMART goals.
13. Explain the difference between a short-term and a long-term goal. Give an example of each type of goal.
14. Explain how the same goal may be a short-term goal for one person and a long-term goal for another. Give an example.
15. How could inaccurate or careless SAE recordkeeping affect the outcome of your SAE?

Analyze and Apply

1. Think about something you have learned through experiential learning (i.e. cooking a specific food, conducting maintenance on a vehicle, etc.). How do you think it would have been different if you had to learn that skill without actually doing the activity? Do you think that you can learn to perform a task without actually performing the task? Please explain.

Thinking Critically

1. Complete the setup and plan for your SAE. First, make a description of your SAE, then list the ways that you will use time and resources for the project. You should also develop some goals that you would like to accomplish with the project and define the involvement of your parents and agriculture teacher. (If you are using an AET record book, this information will directly transfer to the online system.)
2. How important do you think it is to keep records of a research project, business, or work experience? Please list ten things that could be found in records *and* how those things would be important to the enterprise. (For example, keeping track of hours worked helps an employee know how much he or she should be paid.)

Lesson 3.2 Your Career in Agriculture

Lesson Outcomes

By the end of this lesson, you should be able to:

- Identify career opportunities in the field of agriculture, food, and natural resources.
- Examine the characteristics employers find desirable in employees.
- Describe the skills required for employment in agriculture.
- Accurately describe the steps to obtaining employment.

Words to Know

agribusiness
agricultural communications
agricultural processing
agricultural support service
agricultural systems
agriscience
aptitude test
cover letter
job application
natural resources management
position announcement
production agriculture
résumé

Before You Read

After reading each section (separated by main headings), stop and write a three- to four-sentence summary of what you just read. Be sure to paraphrase and use your own words.

On a global scale, there are more people employed in agriculture than any other industry. In the United States, more than 22 million people are employed in agriculture careers. What might surprise you is that most of the people employed by agriculture are *not* traditional farmers and ranchers. In fact, production agriculturalists only make up a small percentage of the overall agricultural workforce, **Figure 3-16**.

In this lesson, you will learn about some of the many careers available in the different sectors of the agriculture industry. We will also discuss the traits employers are looking for in employees, and what you can do to stand out from the competition when applying for a job.

Visionsi/Shutterstock.com

Figure 3-16. Even though a farmer is the career that most people think of when they think about careers in agriculture, less than 2% of Americans are production agriculturalists. There are many other careers to consider.

Career Sectors in Agriculture

According to the Council for Agricultural Education, there are eight different pathways of careers in agriculture. Each of these pathways is specific to the type of agriculture in which you are involved. Agriculture has a diverse range of careers. Much like choosing an SAE, there are agricultural careers certain to meet your interests.

Sectors of Agriculture		
Sector	**Description**	**Jobs**
Agribusiness	Managing the profitability of agricultural goods	Commodity trader Farm and/or ranch manager Agricultural loan officer
Agriscience	Research and development of emerging agricultural technologies	Geneticist Biotechnology engineer Biologist Food chemist
Agricultural Communications	Informing those in the agricultural industry as well as consumers regarding topics concerning agricultural products	Journalist Public relations manager Sales representative
Agricultural Processing	Transforming raw agricultural goods into modified products for consumers	Grain mill operator Food technician Quality control manager
Agricultural Support Services	Industries and careers that provide support for the equipment and technology associated with agricultural production	Veterinarian Distribution coordinator Mechanic
Agricultural Systems	Design and build agricultural equipment and machinery	Agricultural engineer Irrigation systems specialist
Natural Resources Management	Focus on the ecology and conservation of cultivated and uncultivated lands	Range manager Fish hatchery technician Wildlife biologist
Production Agriculture	Production of raw animal and plant goods for human use	Cattle rancher Grain producer Catfish farm owner Cotton grower

Goodheart-Willcox Publisher

Figure 3-17. There are eight sectors of agriculture that you can choose a career in. *Which sector do you think is most interesting?*

If you remember from *Lesson 1.1*, there are eight main components of agriculture (**Figure 3-17**):

- Agribusiness
- Agriscience
- Agricultural Communications
- Agricultural Processing
- Agricultural Support Services
- Agricultural Systems
- Natural Resources Management
- Production Agriculture

Each of the components has a different set of specific jobs that might relate to your own unique interests.

Agribusiness

The ***agribusiness*** career area includes people who are concerned with managing the profitability of agricultural products and companies. In order to work in agribusiness, you should like to work with economics, mathematics, and finances. Careers in this area include:

- **Commodity Trader.** Much like a stock trader, commodity traders work to sell agricultural goods and commodities, sometimes months before they are ready for shipment.
- **Farm or Ranch Manager.** The manager is involved with the day-in and day-out workings of a production farm or ranch. The primary focus is to keep the operation running smoothly and making a profit. The manager is usually in charge of personnel, payroll, and buying and selling inventory.

Other careers in agribusiness are shown in **Figure 3-18.**

Agriscience

People who are interested in the scientific principles behind agriculture and the research and development of emerging agricultural technologies are often employed in ***agriscience*** careers. These careers examine the core science behind the things that happen in agriculture on a daily basis. If you have an inquisitive mind and like to explore the reasons things work the way they do, then a career in agriscience might be right for you. Nutritionists and meteorologists are examples of agriscience careers:

- **Nutritionist.** Traditionally, nutritionists plan food and nutrition programs. They understand the amount and types of food that an animal or human needs for their life functions. Depending on their training and education background, nutritionists may also be involved in the development of more nutritional varieties of plants and animals.

larry1235/Shutterstock.com

Agribusiness	
Accountant	Information systems analyst
Agribusiness manager	International marketing
Agritourism manager	Lawyer
Banker	Leasing consultant
Chemical dealer or sales representative	Loan officer
Economist	Market analyst
Farm appraiser	Marketing head
Farm labor contractor	Policy analyst
Farm manager	Political advocacy
Financer	Quality controller
Grain and livestock buyer	Resource economist
Grain merchandising	Tax consultant

Goodheart-Willcox Publisher

Figure 3-18. Many people involved in agribusiness careers are also involved in other areas of agriculture. For example, a ranch manager may manage beef cattle production as well as manage the financial aspects of the operation.

kurhan/Shutterstock.com

Agriscience	
Agricultural educator	Meteorologist
Agronomist	Microbiologist
Animal or plant pathologist	Molecular biologist
Biochemist	Nutritionist
Botanist	Physiologist
Ecologist	Soil scientist
Extension education director	Toxicologist
Field biologist	Turf scientist
Food scientist	Wildlife biologist
Geneticist	Zoologist
Marine biologist	

Goodheart-Willcox Publisher

Figure 3-19. Training in a particular area of science does not limit your career prospects. Many areas of science are intertwined and your knowledge and expertise will be applicable in many agricultural and nonagricultural situations. *Can you think of ways some of these career areas overlap? For example, how could a zoologist and a wildlife biologist help each other?*

- **Meteorologist.** A meteorologist is a scientist that examines climate and weather patterns to make predictions about the weather. Many meteorologists are employed by agricultural agencies to predict how the climate will affect agricultural commodities.

  More agriscience career options are shown in **Figure 3-19.**

Agricultural Communications

Agricultural communications is a growing industry. Careers in this field are concerned with informing consumers and industry professionals about topics related to agriculture and agricultural products. Being observant, well-spoken, and a good writer are all key traits that may help you excel in an agricultural communications career. Agricultural communications careers include:

- **Public Relations Manager.** A person who works in public relations communicates the mission or purpose of an agricultural company to the public. Public relations specialists often work to promote products or

Communications
Agribusiness manager
Agricultural educator
Agricultural journalist
Agricultural literacy coordinator
Agricultural marketing
Agricultural photographer
Broadcast journalist
Consumer counselor
Cooperative extension agent
County extension agent
Farm broadcaster
Freelance writer
Marketing representative
Political lobbyist
Public relations representative
Scientific illustrator
Technical writer
Training management

withGod/Shutterstock.com; DmitriMarula/Shutterstock.com Goodheart-Willcox Publisher

Figure 3-20. Every business requires a solid communication plan in order to succeed. In smaller businesses, employees often "wear more than one hat." For example, the office manager may also be in charge of marketing and advertising.

services or clarify any misconceptions that the media or other sources may have regarding their sector of agriculture.

- **Agricultural Journalist.** An individual who works much the same as a newspaper reporter, radio commentator, or magazine writer, only they work for a group that is dedicated to sharing information related to agricultural topics.

Additional careers in agricultural communications are listed in **Figure 3-20**.

Agricultural Processing

Many of the agricultural goods that reach consumers have been modified from their original form. Those who work in ***agricultural processing*** are involved in the transformation of raw goods to consumer-ready products. A strong work ethic, attention to detail, and desire to produce a quality product are keys to success in this field. Some of the available careers in agricultural processing include:

- **Grain Mill Operator.** A person involved with the conversion of raw grains into products like animal feeds or more finely ground products like flour, corn meal, or cottonseed meal. Their work allows the grain to be transformed into a more consumer-ready product.
- **Quality Control Manager.** A quality control manager works in processing facilities to ensure that all regulations and standards are being met, ensuring a safe food supply.

There are many other careers in the agricultural processing area, as shown in **Figure 3-21**.

Avatar_023/Shutterstock.com

Agricultural Processing	
Crop and plant graders	Machine operator
Dairy plant supervisor	Meat processing worker
Federal grain inspector	Milk processing plant supervisor
Food inspector	Product development technician
Food processing plant operator	Production supervisor
Food regulatory consultant	Quality control manager
Food safety and quality technician	Sawmill operator
Food science technician	Weights and measurements official
Grain mill operator	

Goodheart-Willcox Publisher

Figure 3-21. In the processing sector, careers range from those who pick and sort through raw food products to those who manage processing facilities. Each step of the process must be performed by someone. *How many steps do you think are required to process the sugar in the illustration above?*

Agricultural Support Services

Agriculture can only function with the guidance and support of those who work indirectly to ensure agricultural goods get from producers to consumers. People who work in ***agricultural support services*** are involved in providing the logistical, technological, and maintenance needs of the agricultural industry. This area includes careers like the following:

- **Distribution Center Manager.** Many agricultural products are sent to a distribution center before they reach consumers at retail centers. A distribution center manager works to coordinate the shipment of agricultural goods from producers to their final destination.
- **Veterinarian.** Veterinarians support animal production by working as support personnel to care for sick and injured animals. Their support for the animal industry saves producers lost time and money related to animal illness.

Goodluz/Shutterstock.com

Support Services	
Agricultural aviator	Groundskeeper
Agricultural mechanics repair technician	Pharmacologist
Animal behaviorist	Plant pathologist
Animal nutritionist	Risk management analyst
Computer support	Soil analyst
Crop consultant	Transportation dispatcher
Crop insurance salesman	Tree surgeon
Electrician	Veterinarian
Equipment salesman	Veterinary assistant
Fertilizer salesman	Waste management specialist
Firefighter or specialist	

Goodheart-Willcox Publisher

Figure 3-22. Some support services are specific to agriculture whereas others support many other industries as well. For example, a hydraulic engineer may design agricultural machines as well as machines for the oil industry.

As you can see, it takes a lot of careers in support services to keep agriculture in our nation functioning. More available careers in agricultural support services can be found in **Figure 3-22.**

Agricultural Systems

Agriculture can only function efficiently if all of the mechanical and structural systems are working correctly. Careers in ***agricultural systems*** focus on designing, manufacturing, and maintaining the mechanical equipment and structures that agriculturalists need to prepare, produce, and process agricultural commodities. Some of the careers in this area include:

- **Agricultural Engineer.** People who work as agricultural engineers design the machinery and equipment that allow agriculture to produce more products in a timely fashion.
- **Agricultural Diesel Mechanic.** These individuals troubleshoot and repair machinery in agriculture, mainly tractors and other farming implements.

A more complete list of careers in agricultural systems is shown in **Figure 3-23.**

Natural Resources Management

People who have careers in ***natural resources management*** are concerned with the conservation and use of cultivated and uncultivated lands in our country and around the world. These careers focus on the renewable and nonrenewable things in the ecosystem that can be used for human well-being. Some careers in this area include:

- **Range Manager.** Range managers work to help the delicate balance between using land for animal grazing, and ensuring the land is still able

Anton Gvozdikov/Shutterstock.com

Agricultural Systems
Agricultural electrician
Agricultural engineer
Equipment fabricators
Equipment operator
Hydraulic engineer
Irrigation engineer
Irrigation system designer
Land-leveling technician
Machine engineer
Power system mechanic
Precision farming technician
Welder
Welding/metal fabricator

Goodheart-Willcox Publisher

Figure 3-23. There are many people needed to keep agriculture systems functioning efficiently. Think of all the systems used in a dairy to gather, process, and distribute dairy products. In a large operation, maintaining the plumbing may be a full-time job.

to produce feed for years to come. They work closely with producers and government officials to ensure that joint use of the land is conducted appropriately.

- **Water Quality Tester.** This career involves those who are concerned with the contamination of surface and groundwater sources. They work to identify contaminants and potential sources of contamination in order to protect the water in an area.

More natural resources management careers are shown in **Figure 3-24.**

Production Agriculture

Production agriculture includes careers in which people are directly involved in the management and production of agricultural commodities for sale to the consumer, **Figure 3-25**. There are different categories of

Heath Johnson/Shutterstock.com

Natural Resources Management	
Cartographer	Logger
Conservationist	Miner
Environmental analyst	Outdoor recreation manager
Environmental lawyer	Park ranger
Environmental scientist	Rangeland scientist
Fish and game warden	Soil conservationist
Fisheries biologist	Soil scientist
Forester	Water quality specialist
	Wildland firefighter
	Wildlife biologist

Goodheart-Willcox Publisher

Figure 3-24. Many of the people who work in the natural resource sector do so because of their appreciation of nature and their desire to preserve and protect the environment.

Tyler Olson/Shutterstock.com

Production Agriculture	
Apiculturist (beekeeper)	Game animal producer
Aquacultural producer	Livestock producer
Arboriculturist	Nursery products grower
Christmas tree producer	Swine producer
Crop producer	Viticulturalist
Dairy producer	
Fruit, nut, and/or vegetable producer	

Goodheart-Willcox Publisher

Figure 3-25. The list of agriculture production careers is only a sampling of those available. *How many other production careers can you list?*

agricultural production, based on the commodities you grow or raise. It is interesting to note that without the small percentage of our workforce that works in production agriculture, the rest of the agricultural industry would be out of work. As a production agriculturalist, you must commit to long hours and stressful work environments. However, the rewards of knowing that you are feeding, clothing, and helping humans live better lives could make a career in production agriculture the right fit for you.

Selecting a Career

Selecting a career is not a simple task. Many of us have ideas of what we would like to be when we "grow up," but we do not always have a plan on how to get there, **Figure 3-26**. Laying out a career path, or plan, early on will allow you to focus your high school courses on your areas of interest and put

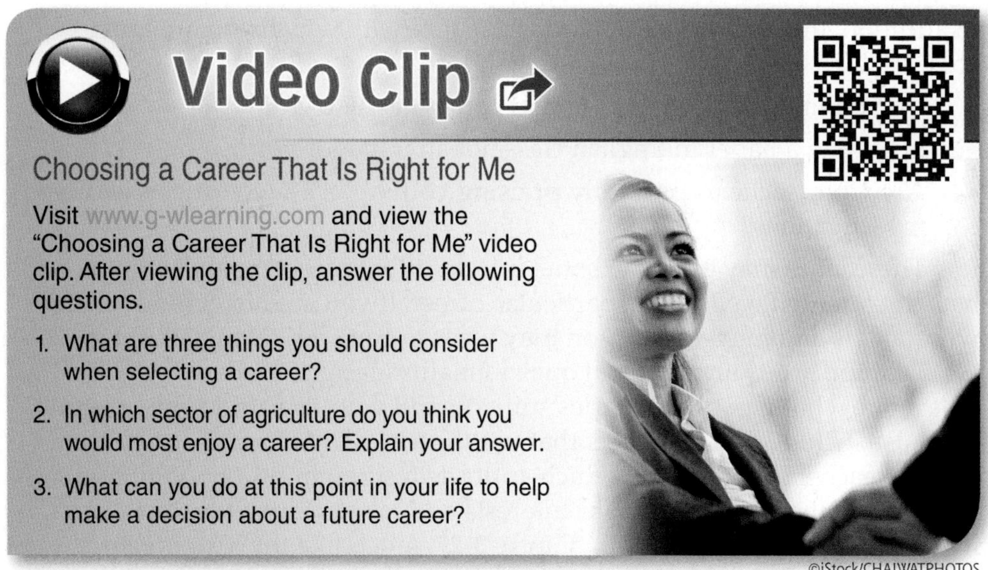

©iStock/CHAIWATPHOTOS

Hasloo Group Production Studio/Shutterstock.com

Figure 3-26. Taking the time to figure out what type of career suits you best can lead you to the career that you will enjoy. *Do you know what you want to do with your future?*

you on the fast track to a rewarding career. You can begin the process by following the same method you use to select an SAE, based on your likes, interests, and talents.

Likes and Interests

Do you know someone who really dislikes his or her job? One good reason to begin your career search by looking at your likes and interests is that no one wants to spend their adult life working in a career they do not enjoy. Make a list of the things you like to do in your spare time and activities you have enjoyed in the past. You may also want to add a list of topics that interest you. You may even want to add a column listing activities and topics you do not enjoy.

Did You Know?

In a recent study of middle school students, the potential career most selected by students was veterinarian. However, when asked to list the job duties of a veterinarian, less than one out of four understood all of the things that veterinarians do on a daily basis.

Talents

Knowing and understanding your talents will also be helpful when laying out your career path. To determine what talents you possess, first consider things at which you excel:

- Are you really good at organizing ideas?
- Do you have a natural talent for taking apart mechanical items and finding ways to improve or repair them?
- Have others commented that you are a good communicator?

In addition, think about the subjects in school with which you are most comfortable.

- Do you find math something that comes easily to you?
- Is writing a paper in English class not an issue for you?
- Are your teachers constantly praising your ability to conduct scientific investigations?

These clues may give you some insight into underlying skills that you have that may fit well with a particular career. If you are not certain where your talents and interests lie, you may want to consider taking an ***aptitude test***. Aptitude tests are personal questionnaires designed to help you determine what types of activities suit you and your talents. Many aptitude tests also include lists of careers that use those skills. Ask your agricultural science teacher for advice on which aptitude test might be a good fit for you. Taking the time to put your interests and talents into a personal career inventory like the one shown in **Figure 3-27**, much like the SAE inventory, can also be a helpful step in selecting a career.

Instructions: On a separate piece of paper, list the attributes in order of importance to you below. Those you find most important should be at the top of the list, and those you find least important should be at the bottom. You can use your top five attributes to help select a job that is right for you.

Ranking	Job Attribute
	Ability to make a difference in the world and lives of others
	Ability to make your own work decisions, be your own boss
	Being able to work and perfect a specific task
	Freedom to make your own work schedule
	Getting to work as quickly as possible after graduating high school
	Having a flexible workload and working at your own pace
	Having a set work schedule each week
	Having varied work experiences, doing something new every day
	High demand, meaning that you will be able to find a position easily
	Interacting with people
	Large income potential
	Meeting deadlines and having set job responsibilities and demands
	Opportunities to advance in leadership roles
	Steady income, set by the hours you put in or a flat rate
	Utilizing your physical strength to perform job tasks
	Variable income, set by the effort you put in or the accomplishments you make
	Working in an environment where you can be on your own
	Working inside, out of the elements
	Working outdoors
	Working with your hands

Goodheart-Willcox Publisher

Figure 3-27. Looking at your job priorities is a great way to take the first step in choosing a career. This tool will help you select the attributes of a job that you might enjoy.

Job Descriptions

Knowing what is included in a career is a key factor in knowing whether or not it is a career you want to pursue. Not fully understanding the requirements for a job could lead to your spending years getting training in a career that you do not really enjoy. You can find job descriptions online, or

Maxx-Studio/Shutterstock.com

Figure 3-28. Sample position announcements used to occur only through the classified ads. Today, online job boards also list the job requirements for career openings. *What kind of job will suit you best?*

you can find someone who currently works in that career to give you a complete overview of what the job entails. It may also be helpful to look at ***position announcements***, which are full job descriptions published by companies when they are looking to hire for a specific position, **Figure 3-28**. They include items like job duties, work environment, skills and training required, educational requirements, and potential salary. Once you find the job description for a career that you are interested in, crosswalk it to the list of your interests and talents.

Some questions you might want to ask yourself as you analyze the job description include:

- Where do the things in the job description align with my talents and interests?
- Are there points in the job description that I might not enjoy doing? Are they things I can deal with, or will they really bother me?
- Can I make a plan that will allow me to gain the required education?
- Will the salary cover my expenses?
- What opportunities are there for growth in this career?

On-Site Experience

The next step in selecting a career is to gain some real-life experience. You can conduct interviews with someone who already has this career, or you can gain an in-depth view through on-site experiences. On-site career experiences vary from job shadowing for a short period of time, to unpaid internships as an assistant, to full-time paid employment in jobs closely related to your career of choice, **Figure 3-29**. Remember that these types of experiences qualify as part of your overall SAE program.

auremar/Shutterstock.com

Juergen Faelchle/Shutterstock.com

Figure 3-29. On-site career experience is sometimes called an internship. Having the opportunity to learn the job through unpaid experiences is a great way to prepare yourself for a future career. *Can you think of an internship opportunity in a field in which you are interested?*

Preparing to Be an Employee

Once you have selected a particular career that interests you, take some time to think about how you will prepare to be successful in that job. Studies have shown that the skills employers want most in an employee are not related to the actual job description. To make a

SAE Connection Exploratory SAEs

By keeping track of the hours you spend and the research you conduct in a career search, you are fulfilling the requirements for an Exploratory SAE. Entering the information into your record book can not only make it readily accessible, but can also allow you to make progress toward SAE hours for FFA degree applications.

long story short, before you can be good at a specific job, you have to be good at being an employee.

Important Skills

Can you look at a situation and make a decision about how to make it better? Are you able to take verbal instructions and use them to complete a task? These are examples of the types of skills employers are looking for. The skills that will make you a successful employee, regardless of your position, include:

- Critical thinking—being able to logically reason to reach a solution.
- Problem solving—identifying and evaluating different parts of a situation to reach a desired outcome.
- Decision making—being able to weigh the benefits and risks of an action and make the best choice for a given situation.
- Active listening—giving someone undivided attention and taking the time to understand instructions and guidance.
- Computer use—ability to use computers and electronic equipment, including basic computer applications.

Ethics in AG

Think about the following scenario and decide how you would handle it. You are responsible for hiring the new cashier at the local feed store. You have interviewed two candidates. Notes from your interviews are as follows:

Potential Employee #1: Was 15 minutes late to the interview. Came in dressed in dirty jeans and a stained T-shirt. Has a lot of knowledge about animal feeding and nutrition and worked for a local feedlot feeding cattle. Mumbled at the ground during the interview; never made eye contact. When asked why he was leaving the feedlot job, he said "because I hate the guy who owns it."

Potential Employee #2: Showed up five minutes early for the interview. Dressed in nice jeans and a short-sleeve, button-down shirt. Said that he had little experience with animal feeds but is anxious to learn about animal nutrition and feeding. Was warm and pleasant to speak with. He also made eye contact throughout the interview.

Which potential employee would you hire? In groups, discuss the pros and cons of each individual and decide why you would make the hiring decision you agreed on.

- Mathematics—practical understanding of how to figure out real-world problems using mathematical equations and formulas.
- Operational analysis—the ability to project changes to the outcome of a situation based on changes to the initial stages of the situation.
- Evaluation—monitoring your own progress and the progress of others and changing behavior to ensure that deadlines and goals are met.
- Marketing—knowing how to promote the company or products to potential consumers.

By looking at these traits, you can determine your own strengths and weaknesses in order to develop your potential to be the kind of employee that employers want to hire.

In addition to the skills employers want, there are also personality traits that will help you get (and keep) the job of your dreams. Overall, employers are looking for a person who is:

- Professional
- High-energy
- Confident
- Self-monitoring
- Curious

As you think about applying for a job, think about how you can share those attributes about yourself with those who are making the hiring decisions.

Career Connection Human Resources Director

Helping find the right people for agricultural jobs

Job description: Those who work in human resources are responsible for managing the employees of the organization. For smaller agricultural businesses, these are tasks that are usually performed by someone in addition to their regular job. In larger corporations, there are often entire human resources, or HR, departments. Some of the tasks involved in working in human resources are hiring and termination decisions, managing payroll and benefits, and resolving employee concerns or disputes.

kurhan/Shutterstock.com

Depending on the size of the organization, the human resources director may be involved in the hiring of employees from every department of the company.

Education required: Typically a bachelor's degree or higher in human resources or business administration.

Job fit: This job may be a fit for you if you enjoy working with people, think you are a good judge of character, and are detail oriented.

Getting Hired

Selecting the career or job that you are interested in can be difficult. Getting hired for a specific job opening can be just as challenging. After looking at yourself and committing to be the type of employee that employers want to hire, you need to market yourself to those who make the hiring decisions. Typically, there are four main stages to getting hired:

- Initial contact
- Application and paperwork
- Interview
- Follow-up

At each step in this process, there are things you can do to make sure that you are visible and memorable to the employer.

Odua Images/Shutterstock.com

Figure 3-30. When making your initial contact by phone, use good phone etiquette: stand or sit up straight, turn off electronics, do not eat or chew gum, speak clearly, and smile. You should also make sure no one will interrupt you. ***Why do you think it is a good idea to smile when you answer or speak on the phone?***

Initial Contact

The first step in getting a job is making contact with the employer. You may hear about the job from a friend or relative, through an online or newspaper posting, or from a job board at your school or a local business. It is advisable that you make your initial contact either in person or through a phone call, **Figure 3-30**. However, some businesses prefer you communicate through email. Conduct research about the position, the company, and its hiring policies. Figure out how you would fit this job before you make the call. The initial contact will likely set the impression the employer has of you through the entire hiring process.

If possible, set up an interview during your initial contact. Write down the date, time, and location of the interview, along with the pertinent contact information. Be prepared to answer questions about your qualifications and interest in the job. The most important thing about this contact is to portray yourself as a pleasant, professional individual who can communicate well and as someone the potential employer would consider hiring.

Applying for the Job

After you have made initial contact, you may be asked to provide a résumé and complete an application form. Many businesses now require you fill out an online application, but some still use paper forms. The most common documents to accompany a job application are a résumé and cover letter. Understanding the key factors to include in this paperwork can give you an advantage over other applicants.

AndreyPopov/iStock/Thinkstock

Figure 3-31. Completing job applications truthfully and legibly is an important step in obtaining a job. ***What other tips should you follow when completing an application?***

Job Applications

Job applications are formal requests for employment, and are generally filled out at the initial contact or directly prior to the interview, **Figure 3-31**. The application typically requires contact information and general questions about work history and skills. To make sure that you are impressing potential employers, your application should be complete, consistent with your résumé, and show that you have good written communication skills (including punctuation and spelling).

Some of the information you will need to have on hand when filling out an application includes:

- Your social security number (SSN).
- Your personal contact information.
- Emergency contact information.
- Names, addresses, and phone numbers of current and past employers.
- Names, titles, and contact information for people who have agreed to act as personal or professional references.
- Names, addresses, and dates of attendance for schools you have attended or are attending.

With the exception of your social security number, write the information in a small notebook or fill out a "practice" application and keep it on hand. It is recommended you memorize your social security number and disclose it only when absolutely necessary.

Résumés

A ***résumé*** is a brief overview of your education, qualifications, and skills. Having a complete résumé is a key to being able to show the employer that you are a good fit for the job. The information in your résumé allows the employer to see your skills and experience at a glance. Most résumés have six sections: contact information, objective, education, experience/skills, awards/honors, and references, **Figure 3-32**. Understanding how these parts

John Doe | **johndoe@email.com**

123-456-7890 | **www.johndoe.com**
1234 Main Street, Anytown, ST 12345

Objective	To obtain a job that will allow me to use my skills to build or do something for your company
Education	2006–2010 Some State University, Collegetown, ST Bachelor's of Science, This Field Emphasis areas including: science of this field, philosophy of this field, communication, computer skills 2002–2006 Anytown High School, Anytown, ST Graduated with High Honors Class Valedictorian Completed advanced courses in: this area, English, Mathematics
Work History	2008–Present Intern in This Field Some State University, Collegetown, ST • In this job I completed these tasks related to the job I am applying for: • Some job achievement while working here • Some job achievement while working here 2007–2008 Customer Service Manager Some Company, Collegetown, ST • In this job I completed these tasks related to the job I am applying for: • Some job achievement while working here • Some job achievement while working here
Awards/Honors	Awesome Award Recipient Some Organization 2010 Another Really Cool Award Another Organization 2009
References	Available upon request

Goodheart-Willcox Publisher

Figure 3-32. Typical résumés are laid out so that the employer can find the information related to employment easily. *Can you easily locate each of the six sections of this résumé?*

Section	Tips
Contact Information	• Make sure the information is up-to-date and current • Be sure that your email address is straightforward and includes your name
Objective	• Look at the job description for the job you are applying for; customize your objective to the job description • Make sure that you include proactive words in your objective so the employer knows you are motivated
Education	• If you have not yet graduated, it is appropriate to list the "anticipated graduation" date in the education section • Make sure you highlight the things in your education that will be helpful to the specific job you are applying for
Work History	• You can list positions even if they weren't paid; internships are work experience, too • List the work experience beginning with your most recent experience • Think about the skills that you acquired at each position and include relevant skills with the work history
Awards and Honors	• Think about any school or community awards that you have received; these can all be helpful to obtaining a job • Leadership positions in student, community, or church groups can also be included in this section
References	• Rather than writing "available upon request," try to identify 2–3 trusted adults who would give a good report of your skills • Make sure that you ask someone before listing them as a reference • References can make or break your chances; be certain that the people you list have confidence in your abilities

Goodheart-Willcox Publisher

Figure 3-33. *Have you written a résumé before?* Consider these tips to make sure that your résumé will stand out from the others.

work together is the first step in developing a quality résumé. Some tips for writing each section are included in **Figure 3-33**.

Cover Letter

A ***cover letter*** is a document that is sent with a résumé to give the employer information about your interest in the position and additional information about your qualifications. The same rules for writing a formal letter, as discussed in *Lesson 2.2*, apply to writing a cover letter. A good cover letter should include information related to the position, how you learned about the job, how your qualifications fit the job description, and contact information for the employer to reach you. An example cover letter is shown in **Figure 3-34**.

Interview

Doing a good job on your paperwork will lead you to the next step in the hiring process, the interview. Although the application, cover letter, and résumé will give the employer an overview of your skills and abilities, the

John Doe
1234 Main Street
Anytown, ST, 12345

123-456-7890
johndoe@email.com

January _, 20__

Mr. John Employer, CEO
Important Job Industries
1000 Moneymaker Lane
Big City, ST, 54321

Dear Mr. John Employer:

I am writing in response to your advertisement online for a temporary worker. Based on the requirements listed in the ad, I feel that my skills and experience are a perfect match for this position.

I have worked in this area and have experience including three things that I can use to help your company. I am available to start in a new position immediately.

I have enclosed my resume for your review. I look forward to further discussing opportunities with Important Job Industries. If you have any questions or would like to schedule an interview, please call me at 123-456-7890.

Sincerely,

John Doe
Enclosure

Goodheart-Willcox Publisher

Figure 3-34. A cover letter introduces you to the potential employer and expresses your interest. ***Why do you think it is a good idea to make sure this document is professional and well-written?***

interview is where most of the hiring decisions will be made. To make sure that you are the candidate selected for the job, you should be mindful of your appearance, your body language, responses to questions, communication skills, and how you wrap up the interview.

For your interview, dress one step above the clothing required for the position, **Figure 3-35**. For example, if employees wear T-shirts and jeans on the job, wear nice jeans and a collared shirt to the interview. If you are unsure as to what may be appropriate, err on the side of caution and wear formal business attire. It is always better to be overdressed for the interview than to wear inappropriate clothing. In addition to choosing appropriate clothing, make sure your overall look, including hair and accessories, portray professionalism.

baranq/Shutterstock.com

Figure 3-35. Making sure that you are dressed appropriately can be key to getting the job. A good rule of thumb is to dress one step nicer than the clothes you would wear to the job each day.

Do	Do Not
Prepare ahead of time, know who you will meet with, and find out as much as you can about them; research the company and its goals	Go into the interview process blind
Know yourself, be honest about your strengths and weaknesses	Focus on your weaknesses
Analyze reasons why you are the best person for this position	Interview or apply for a position that you think you will dislike
Practice your route to the interview, know exactly how to get there	Risk getting lost on your way to the interview
Be early to the interview	Arrive late (even on time is considered late by interview terms)
Be polite and friendly to everyone you meet at the location of the interview, you never know who you'll bump into	Be so preoccupied with the interview that you are not able to be friendly and personable
Wear something that is suited to the position and you feel great in	Wear uncomfortable clothing, no matter how professional you look; fidgeting is a negative
Be confident in your abilities	Shortchange yourself or try to oversell your abilities
Practice your answers in front of a mirror	Go into an interview without some trial interviews

Goodheart-Willcox Publisher

Figure 3-36. Understanding some basic tips about interviewing could help you better prepare to get the job.

First impressions in an interview matter. Greet the interviewer and make sure you use open and confident body language. Smile, be pleasant, and try your best to calm your nerves and relax during the interview. Answer questions completely and honestly, tying information you share back to the information included on your résumé. Give the interviewer examples and share relevant experiences to help communicate who you are and how your experiences have prepared you for the job. Be honest! If the employer asks you a question you cannot answer, do not try to guess your way through it. Additional interviewing *do's* and *don'ts* are included in **Figure 3-36**.

Follow-Up

The completion of an interview is not the end of the hiring process. Completing follow-up correspondence will allow you to make a final contact with those making hiring decisions. Your follow-up should be addressed to the person with whom you interviewed, and should express appreciation for the interview, along with a recap of your qualifications and interest in the job. Writing a follow-up can be a way to make yourself memorable to the employer, **Figure 3-37**.

Syda Productions/Shutterstock.com

Figure 3-37. A follow-up letter shows the potential employer how serious you were about the job, shows gratitude for their time, and leaves one last impression in the minds of the interviewers.

FFA Connection Job Interview CDE

Do you want to hone your skills related to getting hired? The National FFA Organization has a Career Development Event directly related to the employment process. In the Job Interview CDE, participants are scored on their ability to make initial contact for a job, fill out an application, submit a complete résumé and cover letter, present themselves in a face-to-face interview, and write a follow-up letter.

Ask you agricultural science teacher about the chance to participate in this CDE on the local, regional, or state level. State winners are invited to compete at the National FFA Convention.

Andrey_Popov/Shutterstock.com

Lesson 3.2 Review and Assessment

Words to Know

Match the key terms from the lesson to the correct definition.

1. A document sent to give the employer information about your interest in the position and additional information about your qualifications.
2. The agricultural industry area concerned with the profitability of agricultural products and companies.
3. A personal questionnaire designed to help you determine what types of activities suit you and your talents.
4. The agricultural industry area concerned with providing the logistical, technological, and maintenance needs of the agricultural industry.
5. A formal request for employment; may be electronic or paper format.
6. The agricultural industry area concerned with the conservation and use of cultivated and uncultivated lands in our country and around the world.
7. A brief overview of your education, qualifications, and skills.
8. The agricultural industry area concerned with informing consumers and industry professionals about topics related to agriculture and agricultural products.
9. A full job description published by companies when they are looking to hire for a specific position.
10. The agricultural industry area concerned with designing, manufacturing, and maintaining the mechanical equipment and structures that agriculturalists need to prepare, produce, and process agricultural commodities.
11. The agricultural industry area involved directly in the management and production of agricultural commodities.
12. The agricultural industry area concerned with the scientific principles behind agriculture.
13. The agricultural industry area concerned with the transformation of raw goods to consumer-ready products.

A. agribusiness
B. agricultural communications
C. agricultural processing
D. agricultural support service
E. agricultural systems
F. agriscience
G. aptitude test
H. cover letter
I. job application
J. natural resources management
K. position announcement
L. production agriculture
M. résumé

Know and Understand

Answer the following questions using the information provided in this lesson.

1. If you like working with finances and economics, you might enjoy a career in the _____ sector of the agricultural industry.

2. List three types of agriscience careers and give a brief description of each job.
3. What types of traits are essential if you want to work in agricultural communications?
4. How does veterinary work qualify as an agricultural support service?
5. Explain why the work of people in agricultural systems careers is vital to the agricultural industry.
6. *True or False?* Careers in natural resources management are concerned with the conservation of land and water.
7. *True or False?* Production agriculturists make up the largest portion of the agricultural workforce.
8. Explain why it is important to begin laying out a career plan early in your education.
9. How can you use your likes and interests to give you direction when looking at careers?
10. Explain why it is a good idea to evaluate your talents when laying out a career plan.
11. *True or False?* Aptitude tests measure IQ.
12. What are the different types of on-site career experiences available?
13. *True or False?* Employers are looking for employee skills other than those related directly to the job description.
14. List three types of skills that will help make you a successful employee.
15. What are the four main stages to getting hired?
16. *True or False?* The initial contact will likely set the impression the employer has of you.
17. Explain why it is better to overdress for an interview if you are not sure what to wear.

Analyze and Apply

1. Think about a company that you might want to own someday. Now write a job description for an employee who you might want to work for you. Make sure you include the job responsibilities and personal characteristics that you would look for in someone you were going to hire.
2. Write a list of potential interview questions that you feel could be used in any interview. Pair up with a partner and practice your answers to each other's questions.

Thinking Critically

1. You have applied to a local grower for a job working with field plants. The company has invited you to come in for an interview. What would you consider appropriate dress for this interview?
2. You are interviewing for a position as a delivery person for a local retail florist. The interviewer asks, "How old are you?" How should you respond?
3. How will you benefit from researching about a company before you schedule an interview?
4. You are the president of your local FFA chapter. While conducting a chapter meeting, what would you do when a fellow member constantly whispers to others or starts side conversations during discussions?

Chapter 3

Review and Assessment

Lesson 3.1

Experiential Learning through Agriculture (SAE)

Key Points

- Having an SAE that is custom fit to your interests and needs can be one of the most rewarding parts of an agricultural education program.
- To be considered an SAE, a project should be supervised by your agricultural science teacher, related to agriculture, well-planned, and have records.
- There are four main types of SAEs: exploratory, research, placement, and entrepreneurship.
- By looking at your interests and available resources, you will be able to determine the SAE enterprise that is the best fit for you.
- Once you have selected an SAE, you can work toward goals and measure your progress by keeping records of your experiences, time invested, and related finances.
- Having a quality SAE program can lead to the development of career skills and recognition for a program of which you are proud.

Words to Know

Use the following list and the textbook glossary to review and study the *Words to Know* from *Lesson 3.1*.

enterprise
entrepreneurship SAE
experiential learning
exploratory SAE
hands-on learning
long-term goal
placement SAE
proficiency awards
research SAE
SAE improvement project
short-term goal
Supervised Agricultural Experience (SAE)

Check Your Understanding

Answer the following questions using the information provided in *Lesson 3.1*.

1. Identify four lab areas in an agriculture education program.
2. *True or False?* Through FFA, agriculture students are provided the opportunity to gain skills that will make them better communicators, leaders, and citizens.

3. Students choose an SAE based on _____.
   A. their instructor's decision
   B. likes, interests, and talents
   C. their employment status
   D. All of the above.
4. What are the main differences between SAE types?
5. *True or False?* Exploratory SAEs are the most costly and time-consuming types of SAEs.
6. What type of information is recorded for a research SAE?
7. What type of information is recorded for a placement SAE?
8. *True or False?* In an entrepreneurship SAE, you have the chance to own your own resources and make management decisions related to the entire enterprise.
9. If someone were to pick up your SAE record book to evaluate your project, what types of information should they be able to gain?
10. Explain why it is best to record information the same day of your experience.

Lesson 3.2

Your Career in Agriculture

Key Points

- Choosing a career can be a daunting task. By examining careers that are available in the different categories of agriculture, you might be able to find a career that you think would be a perfect fit for you.
- Looking closely at your interests, hobbies, and skills can help you narrow down your career search to a specific area or job.
- Developing your employability skills will allow you to look at the characteristics that employers want, and work on those things that will make you more likely to get hired.
- Once you are ready for a job, you can complete the four steps of making initial contact, completing paperwork, interviewing, and following up with the employer to land the job you desire.

Words to Know

Use the following list and the textbook glossary to review and study the *Words to Know* from *Lesson 3.2*.

agribusiness
agricultural communications
agricultural processing
agricultural support service
agricultural systems
agriscience
aptitude test
cover letter
job application
natural resources management
position announcement
production agriculture
résumé

Check Your Understanding

Answer the following questions using the information provided in *Lesson 3.2*.

1. What benefit is there to listing the types of activities and topics you do not enjoy when examining your likes and interests?
2. Explain why it is important to fully understand the requirements for a job before you lay out your career plan.
3. What benefit is there to on-site experiences when you are laying out your career plan?
4. List three ways in which a person can be a good employee.
5. Identify five skills that will help make you a successful employee.
6. Explain why the initial contact may be a decisive factor in whether or not you get hired.
7. List four types of information you should have on hand when filling out a job application.
8. List the six sections of a standard résumé. Why is it important for this information to be consistent with the information you include in your application?
9. Explain what type of information should be included in your cover letter.
10. Identify the types of information that should be included in your follow-up.

Chapter 3 Skill Development

STEM and Academic Activities

1. **Science.** Understanding how agriculture is intertwined with science is a key factor in becoming knowledgeable about the industry of agriculture. List ten agricultural careers which relate to science and list how science plays a role in each of those careers.
2. **Technology.** Create a chart listing ten aspects of job searching and gaining employment. Include aspects such as looking for jobs in a particular field. In one column, identify

traditional (or dated) methods used to perform a task. In the second column, identify new technological methods of performing the task. Use your chart to write 3–4 paragraphs on how technology has made job searching and employment easier.

3. **Math.** Find information on the average hourly pay rate in your area for three farm- or ranch-related occupations. Calculate what a person in each of these occupations would earn (before taxes) in one week, one year, and over 20 years. Also assume a rate of inflation of about 3% per year and a pay increase of the same. Develop a table to show your results. Can you draw any conclusions from the results?
4. **Language Arts.** Create an informational pamphlet on the agricultural career of your choice. Research the education requirements, job responsibilities, advancement opportunities, etc. Include images in your pamphlet. Present your pamphlet to the class.

Communicating about Agriculture

1. **Reading and Speaking.** Working with two partners, research the type of interview questions you may be asked when applying for a job in the agricultural industry. Look for examples of the best way to reply to interview questions. Create a script with one partner applying for the position and the other two partners performing the interview. Perform the skit for your class.
2. **Reading and Speaking.** Create an informational pamphlet on how to apply for a job in the agricultural industry. Narrow your focus to a particular area such as greenhouse production or small animal production. Research résumé strategies and portfolio organization and download a sample job application. Present your pamphlet to the class. After your project has been graded and returned to you, review the instructor's comments. List the types of changes you could make to improve your project.
3. **Writing and Speaking.** Interview someone local who works in an agricultural field. For example, you could interview a local veterinarian, nursery manager/owner, park ranger, or swine producer. Choose an area you are interested in and/or with which you are not familiar. Ask the person to describe a typical day at work. Prepare a list of questions similar to the following: *How long have you been in the _____ industry? Did you go to school? Did you work as an intern? What is the work environment like? What are your job duties? What other types of professionals do you work with?* Report your findings to the class, giving reasons why you would or would not want to pursue a career similar to that of the person you interviewed. (Do not forget to send a note thanking the person for their time and help.)

Chapter 4

Agricultural Safety

©iStock/steve everts

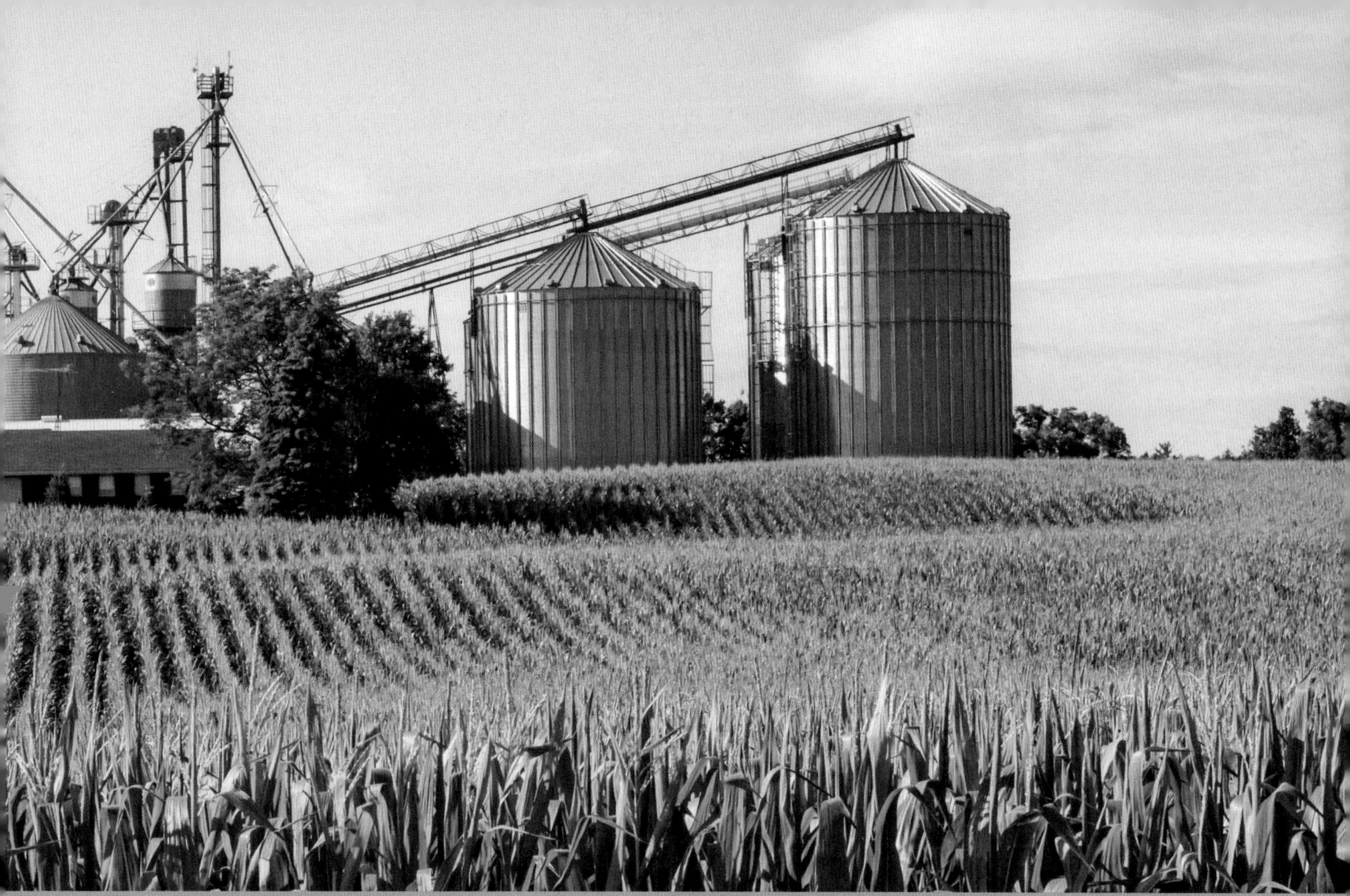

Labrynthe/Shutterstock.com

Halfpoint/Shutterstock.com

©iStock/Igor Sokolov

While studying, look for the activity icon to:

- **Practice** vocabulary terms with e-flash cards and matching activities.
- **Expand** learning with video clips, animations, and interactive activities.
- **Reinforce** what you learn by completing the end-of-lesson activities.
- **Test your knowledge** by completing the end-of-chapter questions.

Lesson 4.1

Occupational Safety and Health

Words to Know

apprentice
Fair Labor Standards Act (FLSA)
horsepower
indentured servant
Occupational Safety and Health Administration (OSHA)
YouthRules!

Lesson Outcomes

By the end of this lesson, you should be able to:

- Know the history of child labor laws and how they apply to you.
- Explain when and where you can work as a youth.
- Describe the type of job tasks you can do at your age.
- Understand your rights as an employee.
- Demonstrate knowledge of rules for agricultural jobs and nonagricultural jobs.
- Know which agricultural jobs are considered hazardous.
- Understand how to help maintain a safe work environment.
- Develop an SAE training plan that includes a description of job hazards and safety precautions to follow.

Before You Read

Write down everything you know about the topic of this lesson. It is okay if you do not know much about the topic, just write down what you are certain you know. Look at the "Words to Know" above. Write down your definition of the words you know. Put the words you cannot define into a separate list. Finally, skim through the lesson and write down the topics that you think you might know well, and those that you do not know much about. Once you are finished, pair up with another student and share your lists.

Production agriculture employs almost two million people in the United States. These individuals work in jobs that create the food, fiber, and natural resources on which we depend. Almost one million youth live on farms in the United States, and approximately half of them work on these farms. There is a very good chance that you will find yourself working in an agricultural occupation once you graduate from high school or college. You may even find yourself working in an agricultural job before you graduate.

More than 15,000 workers under the age of 20 are injured in accidents on the farm each year. On average, about 113 youth are killed in on-farm accidents annually, and most of these deaths occur in youth aged 16 to 19. Not only do agricultural accidents result in pain and suffering for those injured and their families, but they also cause damage to equipment, which results in lost profits and efficiency Accidents happen, and many are preventable by understanding how to perform your work safely, **Figure 4-1**.

Why So Many Rules?

Although it is up to you to know and follow the safety rules of your job, it is up to your employer to provide the proper training and follow employment regulations. The government has created laws and regulations for the health and safety of working youth. To understand how and why there are so many rules and restrictions, we must review life in the early days of our country.

Scott Bauer, USDA

Figure 4-1. Many youth live on farms where they spend a lot of their free time working. *Do you live or work on a farm? What types of chores do you have to do on a daily basis?*

Indentured Servants

During colonial times, American youth worked as indentured servants and apprentices to skilled craftsmen. An ***indentured servant*** was a person who was indebted to someone and required to work for that person until their debt was paid. During colonial times, many people became indentured servants in order to travel to America. Their debt was often to a person who paid their lodging and transportation to the colonies.

Apprentices

Youth were also employed as ***apprentices*** to skilled craftsmen for the purpose of learning a specific trade, **Figure 4-2**. Youth were taken on as apprentices to silversmiths, blacksmiths, clothing makers, printers, and other trades. At the end of the apprentice period, the youth would be trained for a skilled trade and his debt would be paid for by the labor he had performed during the apprentice period. Because colonial schooling was all but nonexistent for lower-class children, an apprenticeship was valuable training that would enable these young adults to work in a particular trade.

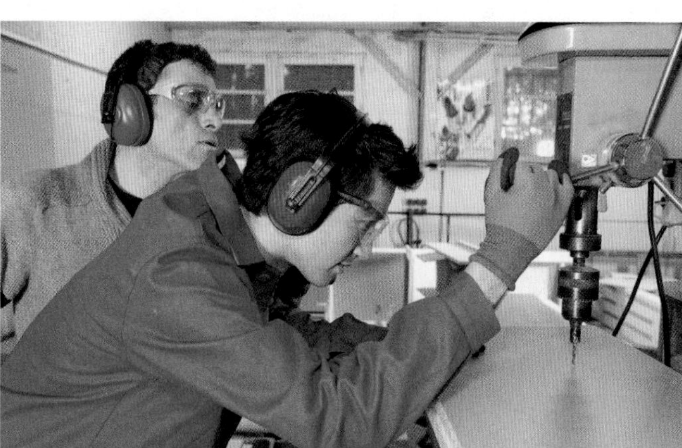

auremar/Shutterstock.com

Goodluz/Shutterstock.com

Figure 4-2. An apprenticeship is a good way to become familiar with a profession. The compensation you receive may be wages, or it could be the knowledge and experience you gain from the work. Working for free may not seem appealing, but you gain valuable experience and have a better idea if you want to pursue a career in a particular field. *What type of apprenticeship interests you?*

Lewis Kline/U.S. Archives

Figure 4-3. The Keating-Owen Child Labor Act of 1916 limited the working hours of children and made it illegal to sell goods produced through child labor. *Why was the act ruled unconstitutional?*

There were no official rules about how youth were treated in indentured servitude and apprenticeships, and they were often mistreated and taken advantage of by their employers. There were no workplace rules or safety initiatives, and young servants or apprentices were often in danger of serious injury.

Child Labor

As businesses became more industrialized, youth were hired to work in factories because they were a source of cheap, manageable labor. According to the U.S. Census of 1900, more than two million children were employed in coal mines, textile mills, and manufacturing facilities. The work was very hard and dangerous for adults, and even more so for children. Once again, there were no employment restrictions or safety programs in place to protect the young workers.

Keating-Owen Child Labor Act of 1916

In 1908, the National Child Labor Committee hired a staff photographer to travel across the country to photograph and report on child labor, **Figure 4-3**. The photographs were enough to incite the public to demand reform. In 1916, the Keating-Owen Act was passed by Congress and signed into law by President Woodrow Wilson. The law was based on the government's ability to regulate interstate commerce and ruled unconstitutional by the Supreme Court in 1918.

Did You Know?

In 1900, almost 20% of all American workers were under the age of 16.

The Child Labor Tax Law

A second child labor bill was passed in December of 1918 as part of the Revenue Act of 1919, also called the Child Labor Tax Law. This law was based on the government's power to levy taxes. In 1922, the Supreme Court found it to be unconstitutional.

Although the nation wanted federal laws regulating child labor, the Supreme Court's rulings left little room for federal legislation. Campaigns for a Child Labor Amendment met resistance in the 1920s, and it was not until 1938 that federal protection became a reality.

Did You Know?

More than 168 million children worldwide perform child labor, and 86 million of them perform hazardous jobs.

The Fair Labor Standards Act (FLSA)

The maiming and deaths of child workers spurred Congress to act on behalf of the country's workers and enact the ***Fair Labor Standards Act (FLSA)*** in 1938. The FLSA set rules establishing a minimum wage, granted overtime pay, and prohibited the employment of most minors in industrial jobs. In 1941, the FLSA was challenged before the Supreme Court, who upheld its constitutionality. The FLSA is still in force today.

The Department of Labor (DOL)

Today, the U.S. Department of Labor (DOL) protects American youth by restricting the number of hours young people may work, and prohibiting youth from working in hazardous jobs. The ***Occupational Safety and Health Administration (OSHA)*** is part of the U.S. Department of Labor. OSHA enforces the laws regarding worker safety, and provides training and education that promotes safe working conditions for all workers. Some of the rules enforced by the U.S. Department of Labor go beyond just safety standards. For instance, the law now requires that young people be paid a minimum wage of not less than $4.25 per hour.

Ethics in AG

You are working in an agricultural business and notice that another employee is not following the approved safety procedures for operating equipment. What will you do?

YouthRules!

The Department of Labor launched a "***YouthRules!***" initiative to promote positive and safe work experiences for young workers. Their website contains information regarding what your employer can and cannot require of you, teen work stories, access to state and federal offices, and many other helpful resources. Know your rights; visit the site to become an informed and safer worker.

Agricultural and Nonagricultural Rules

In the United States, there are different rules for youth working in agricultural jobs than those for youth working in nonagricultural jobs. Youth in agricultural jobs are allowed to do more tasks that may be hazardous. This allows youth who live and work on a farm to contribute to the overall success of the family farming operation.

Agricultural Jobs

For young people working in agricultural jobs, the rules are different. According to the U.S. Department of Labor, young people can work in any agricultural job as long as it is on a farm where the youth is supervised by a parent, **Figure 4-4**. Parents may designate another adult to stand in their place.

Young people 14 and 15 years of age can work in certain hazardous agricultural occupations. However, the youth must work under a written agreement that contains the following statements:

- The work performed by the youth must be required as part of the learning activities on the job.
- Youth may only work for short periods of time and under the direct and close supervision of a qualified adult.

Corepics VOF/Shutterstock.com

Figure 4-4. Many young people work on farms to help maintain a family business. This young man needs to know how to operate this tractor safely and efficiently. *What does he need to know in order to operate this tractor safely?*

Safety Note

It is important to remember that you should never perform any task or job unless you have been trained or have knowledge of how the task should be safely performed.

- The student's school must provide safety instruction that coincides with on-the-job training by the employer.
- A work schedule must be developed that explains the type of hazardous work to be performed.

Fourteen- and fifteen-year-olds may also operate tractors as long as they have completed the 4-H Federal Extension Training Program for tractor operation. Youth must also receive training in the safe operation of the equipment from the employer. Every tractor operates differently, and it is the employer's duty to train you to safely operate the tractor you will be using. Federal law also requires your employer to closely supervise you when you operate or perform hazardous tasks. Your employer must perform a safety check at least three times per day, once in the morning, at noon, and at midafternoon. Your employer must keep records of training agreements on file.

Did You Know?

All states have rules regarding the employment of young workers. In addition, some states have separate minimum wage requirements. When federal and state rules are different, the rules that provide the most protection will apply. Be sure to find out about the rules in your state.

Nonagricultural Jobs

At the age of 14, youth may work in grocery stores, offices, retail stores, restaurants, movie theaters, and amusement parks and camps. Federal law does not allow people aged 15 and younger to work in public utilities, construction, warehouses, or in jobs that require them to operate motor vehicles. Youth aged 15 or younger cannot operate power equipment or machinery. Once you reach the age of 16, youth may work in any job, except for:

- Jobs requiring the handling, manufacturing, or storing of explosives.
- Driving a motor vehicle, or being an outside helper for a motor vehicle operator.
- Mining, logging, and sawmilling, **Figure 4-5**.
- Jobs using power-driven woodworking and metalworking machines.
- Positions using cranes and hoisting equipment.
- Jobs using radioactive substances.
- Meat packing and using meat slicing machines.
- Wrecking and demolition.
- Any type of work on the roof of a building.
- Driving earthmoving and excavation machines.

U.S. Archives

Figure 4-5. These boys worked in a coal mine in the early 1900s, sorting coal and rock. Many young boys were injured or killed in mining operations. *Can you imagine working in a mine when you were only ten years old?*

Hazardous Jobs

The Fair Labor Standards Act (FLSA) identifies the following as hazardous agricultural occupations:

- Operating a tractor larger than 20 ***horsepower***.
- Connecting tools and implements to tractors larger than 20 horsepower.
- Working from ladders and scaffolds above a height of 20′.

Alaettin YILDIRIM/Shutterstock.com

Tish1/Shutterstock.com

Figure 4-6. Machines like the cotton picker and combine illustrated above have modernized farming, but also introduced new hazards that should not be taken lightly. *Can you list five immediate dangers you might encounter when working around this type of equipment?*

- Operating any of the following machines:
  - Augers and conveyors
  - Cotton picker (**Figure 4-6**)
  - Crop drying equipment
  - Cutting timber and transporting logs
  - Earthmoving equipment
  - Feed grinders and mixers
  - Forklifts (**Figure 4-7**)
  - Grain combine (**Figure 4-6**)
  - Hay and forage harvesting equipment
  - Portable or stationary power saws
  - Potato diggers
  - Rotary tillers

Hazardous agricultural occupations also include working around livestock, in dairy operations, and in poultry facilities. Working in livestock pens occupied by aggressive animals can be hazardous. In addition, machinery used to convey feed and move animals from one pen to another and to manage waste may also pose hazards. With so many possible hazards, you can see why farm workers need to exercise care and good judgment while working on a farm.

Did You Know?

Horsepower is a standard unit of measure of power output for many types of farm machinery. One horsepower is equal to the ability of a machine to move 1,000 lb a distance of 33′ in one minute.

Working Hours

If you are age 14 or 15, you can work outside of school hours, provided that you do not work before 7 am or after 7 pm. You can only work three hours on a school day and eight hours on a non-school day. The maximum number of hours you can work in a school week is 18, and the most you can work on a nonschool week is 40 hours.

Dmitry Kalinovsky/Shutterstock.com

Figure 4-7. Youth are restricted from using certain types of equipment, such as this forklift, until they reach a certain age. Make sure that you and your employer comply with federal and state rules for equipment operation. *Why do you think youth are restricted from using certain types of equipment?*

Safety Note

Get enough rest! Aside from health issues that arise from lack of sleep, excessive sleepiness also contributes to a greater than twofold higher risk of sustaining an occupational injury!

If you are 16 or older, federal law allows you to work any day, any time of day, and for any number of hours. Although there are no federal restrictions on the work hours of youth age 16 or older, individual states may have stricter rules. Once you reach 18 years of age, there are no special youth rules regarding how or when you can be employed.

Questions for Your Employer

Assume you have been hired to work in a part-time job at a local agricultural business. You are ready and excited to get to work on your first day. Typically, your new employer will give you an orientation to the job, **Figure 4-8**. During this orientation, make sure you ask the following questions (and any others you may have):

- How do I go about doing the job safely?
- What type of hazards are there, and where are they?
- What personal protective equipment do I need to use or wear?
- What do I do if there is an emergency or someone is injured?

You should ask these questions the first day you report to work on the job, every time you are given a new task, or whenever a new, potential hazard is introduced to the workplace. If you are unsure how

Lester Balajadia/Shutterstock.com

T.Dallas/Shutterstock.com

Figure 4-8. Employers must make certain that appropriate warning signs are in place and that both employees and visitors to a job site obey these warning signs. *Look around your school lab or workshop or your place of employment. Do you see any unsafe situations? What can you do to make the area safer?*

J.D.S./Shutterstock.com

to complete a task, it is okay to ask questions. It is better to be safe now than sorry later.

The first time you are asked to complete a task that may be hazardous, make sure you are physically able, and you have been trained properly. Your supervisor is responsible for teaching you how to use equipment safely, and for teaching you the safety procedures to follow. Do not be nervous if your supervisor observes your work habits; he or she is making sure that you are doing the job safely.

Maintaining a Safe Environment

Although your employer is responsible for providing a safe work environment, employees should help maintain that environment. Employees may contribute to workplace safety by observing safety rules, reporting unsafe situations, reporting malfunctioning machinery, and wearing the proper protective equipment. Employers and supervisors should regularly monitor the work environment to identify potential hazards, and investigate the cause of accidents when they occur. Employers should also have emergency response plans in place, and periodically review theses plans with employees. Reacting promptly and properly to an emergency situation may save someone's life.

Safety Note

Periodically check the U.S. Department of Labor rules and regulations to ensure you are following the most current rules.

Working Outdoors

Agricultural jobs involve a lot of outdoor work. Your supervisor or employer is responsible for providing you, and any other workers, with the things you need in order to keep outdoor work safe. When working outdoors, an employer must provide:

- Safe drinking water and toilet and hand-washing facilities.
- Access to first aid and a protected area in which employees may rest, **Figure 4-9**.
- Access to shelter in the event of severe weather.

You also need to take precautions of your own, such as:

- Appropriate clothing for the season.
- Adequate protection from sunlight (sunscreen, sunglasses, hat).
- Proper equipment for working with pesticides and fertilizers.

When you work outside, there are more environmental factors that may affect your job and safety. Know your surroundings and be aware of others, especially children, who may not be paying attention to you and the work you are performing.

SNEHIT/Shutterstock.com

Figure 4-9. It is important that workers have adequate shelter available when working outside. Your employer should have a safe location for you to go to and a plan to get everyone to safety if the weather turns bad. ***Do you work outdoors? If so, can you identify safe locations you could get to if the weather changed abruptly?***

Chris Geszvain/Shutterstock.com

Figure 4-10. Eye wash stations are designed to flush chemicals and objects from your eyes and face. Review the instructions for using an eye wash station or shower.

Praphan Jampala/Shutterstock.com

Figure 4-11. Employers are responsible for making sure that every employee has the appropriate personal protective equipment necessary for the job. This welder has the appropriate helmet, gloves, and overalls for this welding job.

Best Practices in Workplace Safety

Draft horses and mules were used in early 1900s coal mining operations. If a mule was killed, the mine owners had to purchase another one. In that same era, if a youth worker was killed in a mining accident, mine owners would hire another boy to take his place. In essence, the young workers were valued no more than the mules.

Today, the standards for workplace safety are much tougher and protect you from dangerous conditions on the job. Employers must develop plans to improve health, safety, and environmental performance. These plans should include:

- Fire, earthquake, and tornado evacuation routes and shelter zones.
- Information on hazardous materials and processes.
- Training on the use of emergency equipment such as fire extinguishers, fire blankets, eye wash stations, and emergency shower stations, **Figure 4-10.**
- Procedure for selecting and storing personal protective equipment, and directions on how to use this equipment correctly, **Figure 4-11.**
- Training on safety procedures for all hazardous equipment.
- Procedures for storing flammable materials, waste disposal, and recycling.
- Procedures for inspecting shop and lab equipment and testing to make sure that all safety shields and guards work as intended, **Figure 4-12.**

Supervised Agricultural Experience (SAE)

Students in agricultural education complete *Supervised Agricultural Experience (SAE)* programs that allow them to gain knowledge about agricultural occupations. As part of your SAE program, you will develop a training plan with the help of your parents, your agriculture teacher, and your employer, **Figure 4-13**. This training plan will describe the types of work you will do, the hours you will work, and any other conditions of employment that need to be addressed. Make sure that your training agreement includes a description of job hazards and safety precautions to follow. You should also keep a copy of your training agreement with your Supervised Agricultural Experience (SAE) records.

Nagy-Bagoly Arpad/Shutterstock.com

Figure 4-12. Only use the equipment that has the required shields and guards in place. *Can you identify the safety guard on this power miter saw?*

Iakov Filimonov/Shutterstock.com

Figure 4-13. Supervised Agricultural Experience projects help you develop the work skills you will need in the future. It should also include safe work practices. *Have you created a training plan for your SAE? Did you include safe work practices?*

Career Connection Agricultural Engineer

Job description: Agricultural engineers (also referred to as biological and agricultural engineers) work in a variety of agricultural areas. They may specialize in anything from agricultural machinery, equipment, and structure design to post harvest processing, storage, and transport. Agricultural engineers must be knowledgeable and skillful in a variety of fields to work effectively in the diverse agricultural and agribusiness industries.

Agricultural engineers work indoors and outdoors. They visit worksites, travel to farms, ranches, dairies, and biogas plants. Some travel across the nation and even around the world.

Dejan Dundjerski/Shutterstock.com

Education required: Minimally, a bachelor's degree in your area of expertise, preferably in agricultural or biological engineering.

Job fit: This job may be a fit for you if you enjoy variety in your work and are interested in many agricultural areas.

Lesson 4.1 Review and Assessment

Words to Know

Match the key terms from the lesson to the correct definition.

1. A person paid to work for an employer for a certain number of years in return for paying a debt.
2. A unit for measuring the work of a motor.
3. A safety initiative launched to promote positive and safe work experiences for young workers.
4. A law that defines the standards for paying wages.
5. A division of the government that enforces the laws regarding worker safety.
6. A worker bound by legal agreement to work for an employer for the purpose of learning a craft.

A. apprentice
B. Fair Labor Standards Act (FLSA)
C. horsepower
D. indentured servant
E. Occupational Safety and Health Administration (OSHA)
F. YouthRules!

Know and Understand

Answer the following questions using the information provided in this lesson.

1. *True or False?* It is your employer's responsibility to provide proper training.
2. Explain the difference between an indentured servant and an apprentice.
3. What brought about the Fair Labor Standards Act?
4. Which government department enforces the laws regarding worker safety?
5. Explain what types of information you can find from the Department of Labor's YouthRules! initiative.
6. Explain why the rules for young people working in agricultural jobs are different than for those working in nonagricultural jobs.
7. List five jobs identified as hazardous agricultural jobs by the Fair Labor Standards Act.
8. What are the working hour restrictions if you are 14 or 15 years of age?
9. Why is it important to ask your employer questions about new tasks or potentially hazardous jobs you are assigned?
10. Identify ways employees can contribute to workplace safety.
11. Identify five things an employer must provide for employees working outdoors.
12. Why must you develop a training plan/agreement as part of your SAE?
13. Briefly describe the importance of safety, health, and environmental regulations and procedures in the workplace.

Analyze and Apply

1. Create a safety poster for a job or activity that you perform. You may use your SAE projects as the basis for the poster. You may also create the poster using a shop or lab activity that you perform in your agriculture class. The purpose of this poster is to teach others about the basic safety rules associated with a task or activity.
2. Walk through the area where you work as part of a job that you have or as part of your SAE projects. Identify any potential safety hazards and bring them to the attention of your supervisor.

Thinking Critically

1. You are in disagreement with a company safety policy. How do you handle the situation?
2. As the supervisor for your work team, how would you rectify a situation in which a certain team member refuses to work with another team member?
3. While working for a grocer, you walk into the back room and find that water containing bleach and detergent has spilled from a walk-in cooler onto the workroom floor and seeped under pallets containing boxes of melons. Determine the most appropriate solution to the problem while addressing the following points: using personal safety precautions, using proper safety procedures in cleaning up the situation, properly disposing of waste materials, and using proper follow-up procedures. Write your solutions down and submit them to your instructor.

Lesson 4.2

Shop and Lab Safety

Words to Know

air-purifying respirator
atmosphere-supplying respirator
circuit breaker
filtering facepiece (dust mask)
gas mask
ground
ground fault circuit interrupter (GFCI)
material safety data sheet (MSDS)
personal protective equipment (PPE)
powered air-purifying respirator (PAPR)
safety data sheet (SDS)
self-contained breathing apparatus (SCBA)
short circuit
shutoff switch
supplied-air respirator (SAR)

Lesson Outcomes

By the end of this lesson, you should be able to:

- Identify and use the proper personal protective equipment for a given task.
- Distinguish between types of fires and use proper fire extinguishers.
- Identify, understand, and use material safety data sheets.
- Understand electrical safety in the shop or lab.
- Understand ladder and scaffolding restrictions and safe practices.
- Explain the importance of first-aid training.

Before You Read

The next time you enter your school's greenhouse, shop, or lab facility, look for safety equipment and posted safety rules. Why do you think the equipment is necessary? Which rules make sense to you, and which rules do not? Pair with a classmate and share your answers.

Working in an agricultural shop or lab can be very enjoyable. Working in a shop offers many opportunities for learning agricultural mechanics, metal fabrication, and carpentry skills. Some agricultural labs also provide opportunities for scientific experiments. As enjoyable and educational as the shop is, it also contains many potential dangers. To keep the shop or lab a safe place to learn and work, safety rules and procedures must always be followed.

What Makes Agricultural Shops and Labs Dangerous?

Agricultural shops and labs contain industrial equipment, **Figure 4-14**. This equipment is designed to work efficiently under tough conditions. A portable grinder operates at about the same revolutions per minute as a jet engine. If improperly used or handled, this type of grinder could cause a person serious harm. If the operator is not wearing the proper eye protection, materials could fly into his or her eyes and cause irreparable damage. If carelessly handled, the grinder could cut off fingers or cause deep, serious cuts to arms and legs.

Many other machines used in an agricultural shop or lab also have potential to cause serious harm to the operator or anyone standing nearby.

ProfStocker/Shutterstock.com

auremar/Shutterstock.com

Figure 4-14. Shop equipment is designed to work under tough conditions. Injuries occur when safety equipment is not used properly. These workers are wearing the appropriate personal protective equipment.

Welding torches and plasma cutting tools operate at extremely high temperatures, and failing to wear the proper eye protection while welding can damage the operator's eyes. Table saws and band saws use very sharp, fast-moving blades. Pesticides and other chemicals used in agricultural applications may also cause injury if they are used or stored improperly. As you can see, the potential for harm in an agricultural shop or lab is boundless.

Safety Procedures

How can you stay safe and keep your coworkers safe while working in the shop? Primarily by learning and following safety procedures and insisting your peers do the same. Each shop or lab varies in the types of equipment it contains and the types of jobs that may be performed. Your agricultural shop or lab instructor will review the safety procedures used in his or her lab. These will include such items as using the proper personal protective equipment and not operating machinery or tools you have not been trained to use. You will most likely have to sign a form stating you know the safety procedures and will adhere to them while working in the shop or lab.

Do not take safety lightly. Many of the safety procedures you learn in school will carry over into your working life, and will help keep you safe and healthy. One of the first lessons for safety in the shop is the proper use of personal protective equipment.

Personal Protective Equipment (PPE)

Your first task upon entering an agricultural shop or lab is to wear the appropriate ***personal protective equipment (PPE)***. Wearing the proper PPE is essential for personal safety and productivity. Personal protective equipment includes:

- Safety eyewear
- Hearing protection

- Dust masks and respirators
- Protective clothing and shoes

Safety Eyewear

In a typical agricultural shop or lab, there are innumerable hazards to your eyes, including chemical splashes, splinters, wood chips, dirt particles, radiation, sparks, and even hand tools. Eye injury can be as minor as slight irritation to serious abrasions, loss of vision, or even loss of an eye. Fortunately, there are many types of eye protection available. If you are unsure as to which type of eye protection you need, ask your teacher or supervisor for advice.

Eye protection should be worn whenever there is the slightest possibility of eye injury. Make it a habit to wear safety glasses at all times in shop and laboratory settings. Eye protection devices include safety eyeglasses, goggles, face shields, welding helmets, and full-face respirators, **Figure 4-15**. When choosing protective eyewear, keep the following points in mind:

- Ordinary eyeglasses and sunglasses will not provide the protection you need; however, there are special safety glasses and shields designed to be worn over eyeglasses. Make sure your safety glasses have side shields.
- If you wear prescription eyewear, you can purchase safety glasses in your prescription.
- If you wear contact lenses, be extra cautious around dust, gases, and fumes. Some facilities prohibit the use of contact lenses because of added dangers.
- Sunglasses must not be used in a shop or lab setting because they do not provide adequate protection and can impair your vision and limit your ability to safely operate equipment.

Ethics in AG

While working on a lab project, you notice two students who are not wearing safety glasses. What will you do?

Andrey_Popov/Shutterstock.com

Eremin Sergey/Shutterstock.com

Figure 4-15. Safety glasses and face shields are two types of eye protection devices. The type of job you are performing will determine which type of protective eyewear you need to use.

The Occupational Safety and Health Administration (OSHA) sets the rules for safety in shops and labs and on the farm. OSHA rules specify that safety glasses should be durable, easy to clean, and above all, protect you from the hazards for which they are designed.

Hearing Protection

It is just as important to protect your hearing as it is to protect your eyesight. Hearing loss accumulates over time and may be prevented if you always wear the proper hearing protection. Hearing protection must be worn in the lab, shop, or field when operating loud equipment. How do you know if you need hearing protection? If you have to raise your voice to be heard over the equipment noise, you need to use hearing protection. See **Figure 4-16**.

Hearing protection comes in many styles. Over-the-ear hearing protectors, such as the ones shown in **Figure 4-17A**, are easy and comfortable to wear. In some situations, you may find it more comfortable to use ear plugs, **Figure 4-17B**. Make certain the method of hearing protection you choose provides adequate protection from the noise hazards you will encounter. Your agriculture teacher can help you select the right hearing protectors for use at school, home, and on the farm.

How Loud Is Too Loud?

Decibels	Sound Source
150	Firecracker or gunshot
120	Ambulance siren
110	Chain saw, rock concert
100	Wood shop
95	Motorcycle
90	Lawn mower
85	City traffic
60	Normal conversation
If you have to raise your voice to be heard in a noisy environment, you need ear protection.	

Goodheart Willcox Publisher

Figure 4-16. Use the noise intensity/decibels listed here to help you identify situations requiring hearing protection. *How loud is the volume on your headphones or ear buds when you listen to music? Is it loud enough to affect your hearing?*

Safety Note

Keep in mind that when wearing hearing protection, you may not be able to hear coworkers. Pay attention to your work but stay alert to your surroundings.

Respiratory Protection

To protect your lungs against airborne dangers, it is important to wear the right type of respiratory equipment for the job at hand. There are many types of equipment including simple dust masks, cartridge respirators, and gas masks. Air purifying respirators use either mechanical filters or a

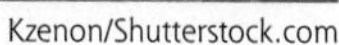
Kzenon/Shutterstock.com

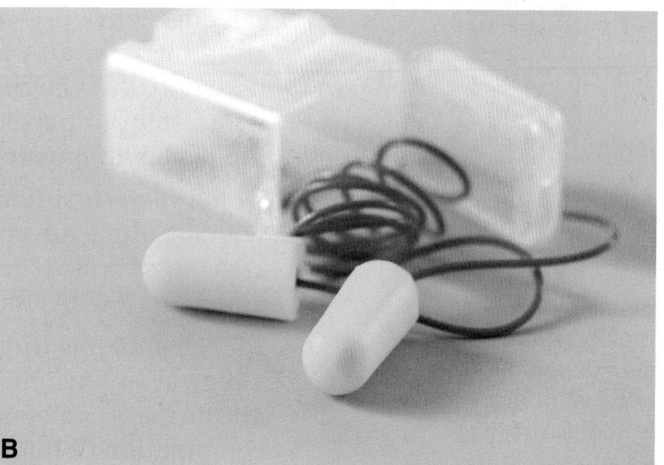

florin oprea/shutterstock.com

Figure 4-17. Hearing protection should not be taken lightly. Choose the appropriate type of protection for the job you are performing. A—Over-the-ear are available with different levels of protection. B—In many situations, ear plugs will be sufficient and also more convenient.

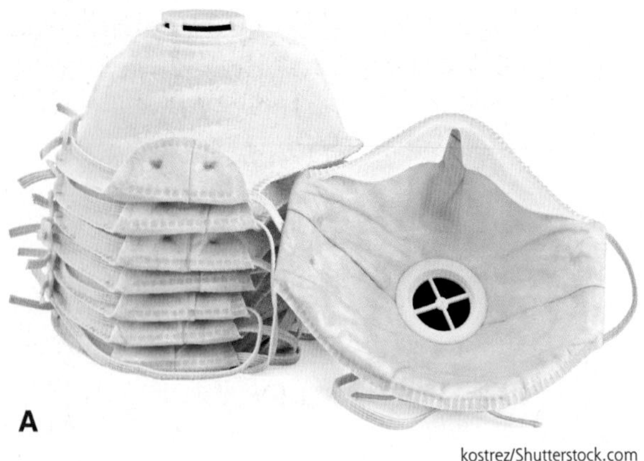
A

kostrez/Shutterstock.com

B

Kletr/Shutterstock.com

Figure 4-18. As with many types of personal protective equipment, the type you use will depend on the task at hand. A—Dust masks are the simplest types of respirators. Although there are various types available, their use is limited. B—Gas masks are not effective in situations where breathable air is limited.

Flashon Studio/Shutterstock.com

Figure 4-19. Each firefighter wears a self-contained breathing apparatus to protect himself/herself from dangerous gases and fumes, and from the lack of oxygen in burning structures. *Can you identify three situations in which the use of a self-contained breathing apparatus is warranted?*

chemical cartridge to clean the air. Mechanical filter purifiers are not for use around toxic gases. Chemical cartridge respirators protect against certain gases and all but the most toxic organic vapors.

Typical agricultural settings that require protection against airborne dangers include when working in dusty or moldy conditions, chemical application, spray painting, and when working in silos, tanks, bins, and manure storage places. Your teacher or supervisor can explain which type of apparatus you will need for the specific work you will be doing. The following is a brief list of different types of available respiratory equipment.

- *Filtering facepiece (dust mask)*—a respirator with a filter as an integral part of the facepiece or with the entire facepiece composed of the filtering medium, **Figure 4-18A**.
- *Gas mask*—a respirator that is more effective than chemical cartridge respirators against high concentrations of toxic gases, **Figure 4-18B**. Gas masks should not be used in oxygen-limited environments.
- *Air-purifying respirator*—a respirator with an air-purifying filter, cartridge, or canister that removes specific air contaminants by passing air through the air-purifying element. Air-purifying respirators can be used only in an environment that has enough oxygen to sustain life. Do not use air purifiers to provide protection from the dangers of oxygen-limited environments such as silos or manure storage areas.
- *Atmosphere-supplying respirator*—a respirator that supplies the user with breathing air from a source other than the air in the surrounding area (i.e. supplied-air respirators (SARs) and self-contained breathing apparatus (SCBA) units).
- *Powered air-purifying respirator (PAPR)*—a respirator that uses a blower to force the air through air-purifying elements. Powered air purifiers can be mechanical filter, chemical cartridge, or a combination of both. They are not designed for use in oxygen-limited environments.
- *Self-contained breathing apparatus (SCBA)*—a respirator for which the breathing air source is carried by the user (i.e. a tank), **Figure 4-19**.

- *Supplied-air respirator (SAR)*—a respirator for which the source of breathing air is not designed to be carried by the user. An SAR will have a hose mask with a blower and an emergency air supply. SARs are for use in oxygen-deficient areas such as manure pits, silos containing silo gas, airtight silos, or bins containing high-moisture grain.

Sedlacek/Shutterstock.com

Figure 4-20. This worker is dressed appropriately for dusty working conditions. The thick coveralls protect against flying wood or metal chips. A long sleeve shirt would provide more protection against sunburn and flying debris.

Protective Clothing and Shoes

Just as you should always use the right tool for the job, you should always wear the proper clothing and shoes, **Figure 4-20**. For example, loose fitting clothing should not be worn in an agricultural shop or lab because it can get caught in the moving parts of machinery. The same goes for hanging jewelry or long hair. If you have long hair, it should be tucked inside your collar or underneath a cap to prevent it from being caught in the rotating parts of shop machinery.

Clothing that is highly flammable should also be avoided. Sparks from welding equipment and metal grinders can ignite clothing, causing serious burns. Do not wear sandals or open-toed shoes in an agricultural shop or lab. Shop equipment and materials can cut or crush toes. Wear sturdy shoes or boots with non-skid soles to protect your feet while working. In a welding shop, closed-toed shoes can prevent burns from sparks and molten metal. Good boots will also protect your feet when working with livestock.

Additional protective clothing includes hard hats, gloves (welding, lined rubber), aprons, rubber suits, and chaps. These items protect you from various hazards, including:

- Falling objects
- Burns
- Frostbite
- Sparks
- Chemicals
- Cuts and abrasions

Wear the proper safety equipment necessary to perform shop and lab tasks.

Ethics in AG

While entering the laboratory, you notice a puddle of liquid on the floor. What will you do?

Shop Safety

As stated earlier, the first task of any shop or lab activity is to follow the safety rules. While working in an agricultural shop can be fun and interesting, it can also be a hazardous place to work if safety is not a consideration.

Personal Protective Equipment

The first thing you must do when entering the shop or lab is to put on safety glasses or goggles. If you do this often enough, it becomes a habit and you will quickly realize when you are *not* wearing your safety glasses. Safety glasses, hearing protection, respiration protection, and protective clothing are all "tools of the trade" in agriculture. Remember, none of these things will protect you if you do not wear them properly in the shop, lab, or on the farm. It only takes a moment to put on the safety equipment you need for the job at hand.

LesPalenik/Shutterstock.com

Figure 4-21. Keep all power saw shields and guards in place at all times. ***Can you identify the safety guard in the illustration?***

Michael Warwick/Shutterstock.com

Figure 4-22. Emergency shutoff switches on power tools should be clearly marked and easy to reach. ***Do you know where the emergency shutoff switch is for the power tools in your school shop?***

Work Area

Before beginning any project, look over your work area to make sure no hazards exist. For instance, you would not grind metal next to a waste can filled with sawdust and paper. The sparks from the grinding process could ignite the waste and cause a fire. Your work area should also be clean, well lit, and have enough room to work. If any of these are a problem, notify your agriculture teacher or shop supervisor, so the problem may be corrected. If you are welding, or working in an area with fumes and smoke, make sure you have adequate ventilation.

Hand and Power Tool Use

Make sure that you are fully aware of how to use the shop tools before beginning any task. If you do not know how to properly use a hand or power tool, ask your agriculture teacher or shop supervisor for help. Most shop accidents with hand tools occur when the user fails to use the tool in the manner for which it was designed. Always select and use the correct tool for the job. Screwdrivers are not made to be used as chisels, and wrenches do not make good hammers.

When operating power tools, keep all shields and guards in place. Never remove a safety guard, **Figure 4-21**. Make sure that hand tools are in proper working order before using them. If a tool or piece of equipment appears damaged, report this to your agriculture teacher or shop supervisor. Additional safety guidelines for using specific power tools and hand tools are included in *Lesson 7.1, Agricultural Tools and Equipment*.

Emergency Procedures

As with safety procedures, emergency procedures will vary according to the shop or lab setup. Emergency procedures are in place to prevent panic and get everyone to safety in case of emergency. Your agriculture shop or lab teacher will review emergency procedures before any lab work begins. Pay close attention to his or her instruction and do not hesitate to ask questions. Know where the exits, fire extinguishers, first-aid kit, and eyewash or shower stations are located. You should also note the location of the shop equipment shutoff switch, **Figure 4-22**. The emergency ***shutoff switch*** is designed to immediately turn off all power to the machines in the shop. Know who to contact in case the emergency involves your instructor and he or she is unable to take charge of the situation.

Fire Extinguishers

Fire extinguishers should be easy to locate and operate, and appropriate for the type of fire hazard. Use the correct fire extinguisher for the fire hazard, **Figure 4-23**. Fire extinguishers in agricultural shops are designated as Class A, B, C, or D, according to the type of fire hazard.

- Class A fires are those involving wood, paper, plastic, or cloth.
- Class B fires are those involving flammable liquids.
- Class C fires are fires resulting from a fault in an electrical system.
- Class D fires are those involving flammable metals such as magnesium.

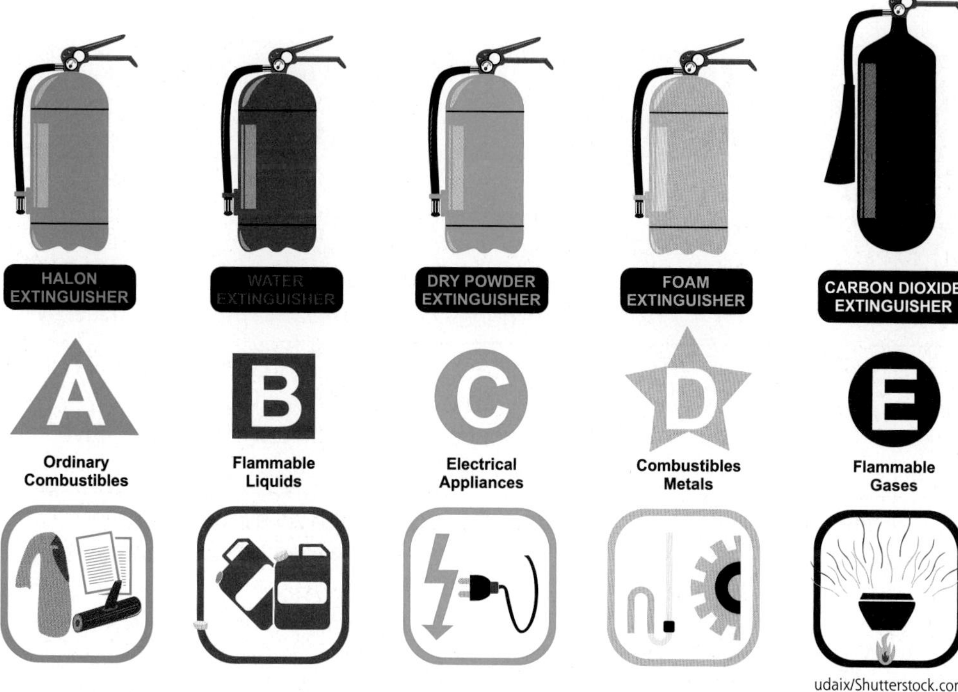

udaix/Shutterstock.com

Using a Fire Extinguisher	
Step 1	Identify the class of fire.
Step 2	Locate an extinguisher labeled for the class of fire you are extinguishing.
Step 3	Pull the pin on the extinguisher.
Step 4	Aim the hose nozzle low toward the base of the fire.
Step 5	Squeeze the handle to release the extinguishing agent.
Step 6	Sweep the nozzle from side to side at the base of the flames until extinguished.

Goodheart-Willcox Publisher

Figure 4-23. Special color coding and letters indicate which type of fire extinguisher is needed for particular types of fires. Every worker should understand where to find and how to use the appropriate fire extinguisher to extinguish a fire in the workplace. Remember these simple steps when using a fire extinguisher.

In many cases, a fire extinguisher can be used to extinguish multiple hazards. A fire in an agriculture shop will likely involve Class A, B, and C hazards. In this situation, a Class ABC extinguisher should be available. The label should clearly indicate on which type of fires the extinguisher can be used. Always be aware of the location of fire extinguishers in the shop or lab, *before* an emergency arises.

Color Codes

In order to make specific types of hazards and safety equipment easy to identify, various organizations combined forces and established a safety color coding system, **Figure 4-24**. A poster may be displayed in the shop and lab to ensure each worker or student understands the color coding system.

Safety Color Coding		
Color	**General Meaning**	**Uses**
Red (or predominately **red** with letters or symbols in contrasting color)	Danger Stop	Safety containers for flammable materials Fire protection equipment (extinguishers, hoses, axes) Emergency devices (shutoff switches, stop bar, emergency buttons)
Orange (or predominately **orange** with letters or symbols in contrasting color)	Warning	Machinery or equipment that can crush, shock, cut, or cause other serious injury Electrical controls, switches, and levers have orange backgrounds
Yellow (or predominately **yellow** with letters or symbols in contrasting color)	Caution	Physical dangers such as slipping, tripping, falling, caught-in-between hazards, and striking-against hazards Levers and adjustment knobs may also be yellow
Green	Safety	To identify first-aid and other safety equipment
Blue	Information	Informs the operator of something that requires extra attention or that equipment is under repair
Black and **yellow** stripes or **Magenta** and **yellow** stripes	Danger	Radiation hazards
Black or **white**, or a combination of the two	Caution	Traffic movement cautions (used in aisles, housekeeping areas, forklift areas)
Gray		On work area floors as well as tabletops and the bodies of machines

Goodheart-Willcox Publisher

Figure 4-24. Safety color coding may be applied to machinery, walkways, work areas, and both large and small equipment.

Hazardous Materials

Many agricultural applications require the use of potentially hazardous materials. The key to safe use of these materials depends on knowledge, proper storage, and safe handling.

Safety Data Sheets (SDS)

All places of employment, schools, and training facilities that have any type of hazardous materials are required to have ***safety data sheets (SDS)*** available for each hazardous material on the premises. (The SDSs were formally referred to as ***MSDSs***, or ***material safety data sheets***.) An SDS contains information on the potential hazards (fire, health, reactivity, and environmental) and how to work safely with the product. It also contains essential information for emergency crews in case of an emergency. If you are working with any type of hazardous material, take the time to read the SDS. You will have a better understanding of how to safely handle the material and what to do in case of an accident or spill, **Figure 4-25**. Some of the topics included on SDS include the melting point, boiling point, flash point, toxicity, health effects, first-aid treatment, storage requirements, and disposal requirements of each chemical.

Travis Klein/Shutterstock.com

Figure 4-25. Safety data sheets (SDS) should be available to every worker. These sheets describe the hazardous materials in the workplace, and their handling and storage. The SDS also provide information about spill cleanup, first aid, and the appropriate emergency response. ***Do you know where the SDS are located in your school lab? How about at your place of employment?***

Hazardous Material Storage

Petroleum fuels, paints, pesticides, and lab chemicals are frequently used in agriculture shops and labs. Caustic chemicals can burn your skin, irritate or injure your eyes and lungs, and can be a fire hazard. These chemicals need to be stored in a safe and secure manner, with plenty of ventilation to prevent the buildup of dangerous fumes. Poisonous chemicals such as pesticides and petroleum products need to be stored in a locked cabinet designed for chemical storage, **Figure 4-26**.

Baloncici/Shutterstock.com

Figure 4-26. Storage cabinets are designed to securely store flammable liquids. ***Do you work with flammable liquids? What type of storage space is used to house flammable liquids at your place of employment?***

Electrical Safety

Electrical service to an agricultural lab or shop enables the flow of electricity from its source to the lights, air conditioning, heating, and outlets and machinery being used in the shop. The shop's electrical service operates on the most basic principle of electric theory: electricity, or electric current, is the flow of electrons from one atom to another in any material. Materials that conduct electricity better than others are used for wiring, and materials that do not allow electricity to flow through them are used for insulating the wiring. Unfortunately, the human body is not naturally insulated from electricity and will easily conduct electrical current, causing anything from a mild shock to cardiac arrest. For this reason, it is essential to follow all safety rules and regulations regarding the installation and use of electric current.

The following list of safety procedures is not all-inclusive, and has been provided as a basic guideline to help ensure a safe working environment. Follow your teacher's instructions and use common sense when working around any electrical equipment.

- First and foremost, *all electrical circuits or receptacles should only be repaired or installed by a qualified and licensed electrician. There are national and local electrical codes and restrictions that must be followed for the safe installation of electrical systems.*
- Read the operators manual before plugging a power tool into a receptacle outlet and before operating the tool. Some tools require higher amperage than others and the use of special outlets designed to handle the higher output.
- Examine electrical power cords for fraying, loose connections, and damage to insulation. Replace or repair them if they are damaged.
- Do not remove the ground from any electrical cord. The ***ground*** on a plug is the third prong on the plug, **Figure 4-27**. It serves as a return path for electrical current to return safely to "ground" without danger to anyone.
- Electrical extension cords should not be placed across walkways or doorways where they can become trip hazards. They should not be placed under carpet or rugs where they can overheat and cause a fire. *Always* unroll the extension cord to prevent heat buildup.
- Do not overload circuits. In most instances, circuit breakers will protect electrical circuits from overload and prevent electrical shock and fire hazards, **Figure 4-28**. A ***circuit breaker*** is an automatic device for stopping the flow of current in an electric circuit. However, in older

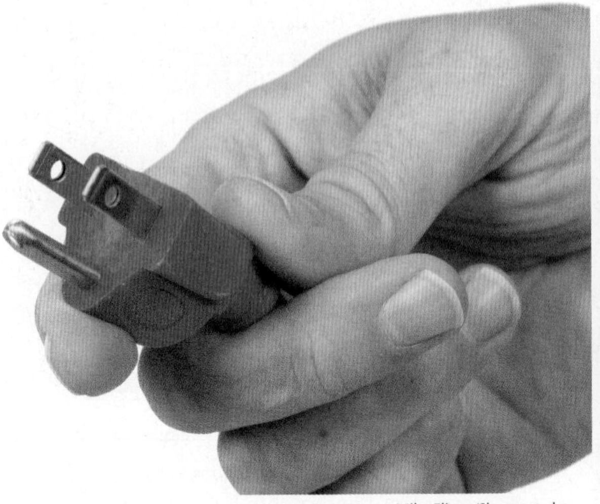

Mike Flippo/Shutterstock.com

Mike Flippo/Shutterstock.com

Figure 4-27. Grounded tools (three-pronged cords) must be plugged into a properly grounded installed outlet. Do *not* use adapter plugs, and *never* remove or cut off the grounding prong or modify the plug in any way.

Lisa F. Young/Shutterstock.com

Figure 4-28. Circuit breakers protect electrical circuits from overload, preventing electrical shock and fire hazards. *Do you know where the breaker panel is in your home? Have you been told how to check and reset the circuit breakers?*

homes and structures, the electrical system may be inadequate and may easily overload.

- Do not plug more tools into a circuit than it is designed to carry. Most labs or shops are designed with specific areas for specific tool use and this should not be a problem. Too many power tools plugged into the same circuit can cause the circuit breaker to trip, disconnecting power to the whole circuit. If this happens, unplug some of the power tools and move them to another circuit before attempting to reset the circuit breaker. An additional circuit may need to be installed if this is a recurring problem.
- If you notice that an electrical outlet or switch feels warm to the touch, immediately notify the shop supervisor or your agriculture teacher. This could be an indication of a short circuit. A ***short circuit*** occurs when something interrupts the proper flow of electricity along its intended path.
- A ***ground fault circuit interrupter (GFCI)*** is a fast-acting circuit breaker designed to cut off the current within as little as 1/40 of a second if the outlet becomes ungrounded, **Figure 4-29**. Ground fault circuit interrupters (GFCIs) must be installed in wet or moist areas where there is a heightened risk of a shock hazard (water easily conducts electricity).

Brandon Blinkenberg/Shutterstock.com

Figure 4-29. Ground fault circuit interrupters (GFCIs) are designed to prevent shock hazards in wet areas and if the circuit becomes "ungrounded."

Safety Note

By law, if you are under the age of 16 and employed in an agricultural occupation, you cannot work from a ladder at heights over 20′. If you are under the age of 16 and employed in a nonagricultural occupation, you cannot work from a ladder at *any* height.

When properly installed and used, electricity infinitely improves our lives at home, at school, on the job, and for most forms of leisure.

Ladder and Scaffolding Safety

According to the CDC, falls from ladders are one of the leading causes of injury nationwide. Falls from ladders cause head injuries, dislocated limbs, and broken bones. The following list is not all-inclusive, but it contains safety practices that will help you use ladders safely. As you are allowed to work from a ladder, follow these safety rules, as well as any given to you by your teacher or supervisor, to prevent injury.

- Follow all safety directions and labels on the ladder.
- Do not use the ladder for any task for which it is not intended, such as using it for scaffolding or to support a structure.
- Do not exceed the load weight of the ladder. The load rating should be printed on a sign or placard somewhere on the ladder.
- Be cautious when using ladders around electrical power lines. If a metal ladder touches a power line, an electrical shock hazard is likely.
- Always inspect a ladder for damage before you use it. If a ladder is damaged, it should be repaired or discarded. Make sure that all components of the ladder are working properly, including locks and latches.
- Use the three-point method when climbing a ladder. This means that of your four limbs (hands and feet), at least three must be touching the ladder at all times when climbing.
- Always face the ladder and keep your body close to the ladder when climbing.
- When using a step ladder, always use it in its unfolded position. Do *not* climb a step ladder that is propped against a wall or other object **Figure 4-30**.
- Never use the top rung of any ladder as a step. Using the top rung as a step may cause the ladder to slide out from underneath you or topple over.
- Use ladders only on firm, level surfaces. The exception to this rule would be when using a tripod orchard ladder. These ladders are designed for use on softer, uneven ground.

kryzhov/Shutterstock.com

Figure 4-30. It is important to never overreach on a ladder because it could cause the ladder to tip over or cause you to lose your balance.

- Do not attempt to move or shift a ladder while you or someone else is on the ladder.
- Only one person should be on a ladder at one time.
- Extension ladders must extend at least three feet above the top of the surface you are climbing.
- Do not use ladders in strong winds or place them in front of doors.

There are many important aspects to choosing and using ladders in a safe manner. Follow your supervisor's or instructor's directions for safely using and handling ladders. Do additional research on the types of ladders that should be used in various situations, as well as on how to maintain ladders to keep them safe and in usable condition.

Did You Know?

Many falls occur because of slipping or tripping and could be avoided by wearing shoes with nonslip soles and following safe work procedures.

First Aid

It is always a good idea to know basic first aid regardless of where you work or live. However, when your home or job is far from medical facilities, it is especially important to know how to respond in case of an emergency. Some schools offer CPR (cardiopulmonary resuscitation) training as part of the curriculum. If this is the case, enroll in the class and get certified. Other places you may receive basic first-aid training, often for a nominal (if any) fee, include:

- Local hospitals
- Fire departments
- The local YMCA
- Community organizations
- Student organizations
- The American Red Cross

If you cannot find these types of classes, contact your local fire department or hospital public relations department and ask if they will work with you to organize training for high school students in your area. Prepare a list of topics you would like them to cover including:

- Why it is important to learn first aid.
- What to do and say when making an emergency call.
- How to treat a sprain.
- How to assess if a person is unconscious.
- How to help someone who is unconscious and breathing.
- How to help someone who is choking.
- How to treat a burn.
- What to do if someone is bleeding heavily from a wound.
- How to perform CPR.
- What items should be in every first-aid kit.

Encourage your family members or coworkers to take these classes also. When possible, set up emergency drills or simulations to practice

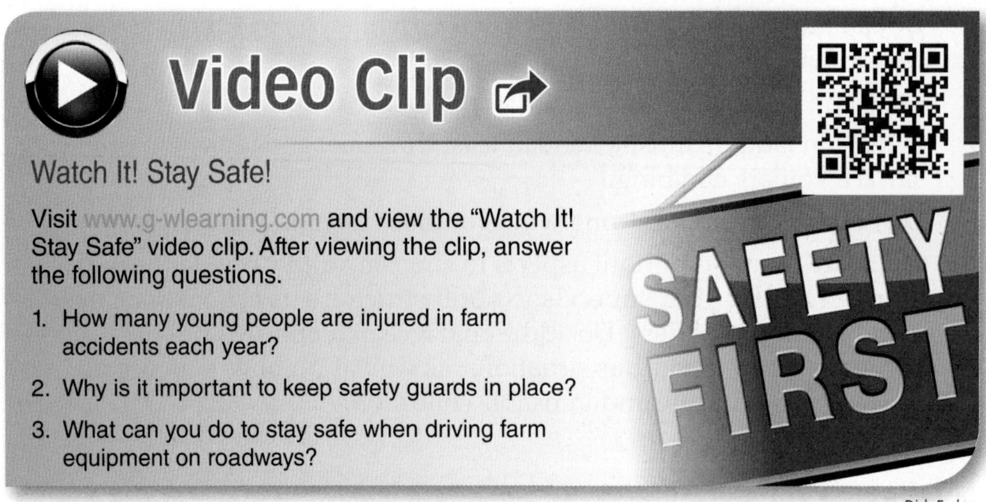

Dirk Ercken

and maintain your skills. If it is not possible to perform drills, create questionnaires or download material for everyone to review proper responses to various situations.

First-Aid Kits

It is important to keep fully stocked and up-to-date first-aid kits where they are easily accessible, **Figure 4-31**. First-aid kits may be kept in work vehicles, mounted near fire extinguishers and emergency plans, in shop buildings, with tool boxes, in tack rooms, and in the home. Family members and workers must be aware of locations for quick access in case of emergency. Schedule regular reviews and restock kits as necessary.

jennyt/Shutterstock.com

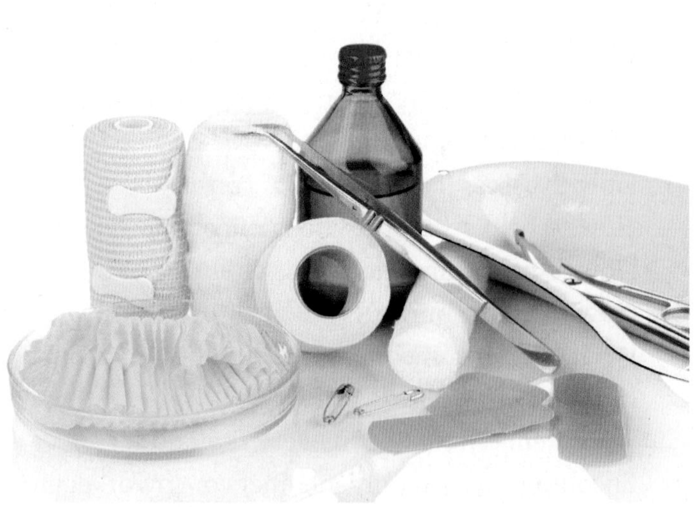

Africa Studio/Shutterstock.com

Figure 4-31. First-aid kits must be kept clean and up-to-date. It is best to assign one person to maintain and refill kits. *Do you have a first-aid kit in your home or vehicle?*

Lesson 4.2 Review and Assessment

Words to Know

Match the key terms from the lesson to the correct definition.

1. A respirator that uses a blower to force the air through air-purifying elements.
2. A respirator that supplies the user with breathing air from a source other than the air in the surrounding area.
3. A respirator with an air-purifying filter, cartridge, or canister that removes specific air contaminants by passing air through the air-purifying element.
4. A respirator with a filter as an integral part of the facepiece or with the entire facepiece composed of the filtering medium.
5. A respirator that is more effective than chemical cartridge respirators against high concentrations of toxic gases.
6. Protective clothing and gear designed to be used on the job to prevent injury.
7. On an electrical plug, the third prong that serves as a return path for electrical current to return safely to "ground" without danger to anyone.
8. An emergency switch designed to immediately cut off all power to the machinery in the shop.
9. Printed data sheet that contains information on the potential hazards and how to work safely with a potentially hazardous product.
10. An automatic device for stopping the flow of current in an electric circuit.
11. A respirator for which the source of breathing air is not designed to be carried by the user.
12. When something interrupts the proper flow of electricity along its intended path.
13. A fast-acting circuit breaker designed to cut off the current within as little as 1/40 of a second if the outlet becomes ungrounded.
14. A respirator for which the breathing air source is carried by the user.

A. air-purifying respirator
B. atmosphere-supplying respirator
C. circuit breaker
D. filtering facepiece (dust mask)
E. gas mask
F. ground
G. ground fault circuit interrupter (GFCI)
H. material safety data sheet (MSDS)
I. personal protective equipment (PPE)
J. powered air-purifying respirator (PAPR)
K. self-contained breathing apparatus (SCBA)
L. short circuit
M. shutoff switch
N. supplied-air respirator (SAR)

Know and Understand

Answer the following questions using the information provided in this lesson.

1. Identify four types of personal protective equipment.
2. What is the role of OSHA in shop and lab safety?
3. How do you know if you need hearing protection in a shop setting?
4. What personal protective equipment should you always wear in a shop or lab?
5. Explain the different types of fire extinguishers by their A, B, C, D designations.
6. What is an MSDS? What is the purpose of an MSDS?
7. What are the important safety rules to consider when using extension power cords?
8. Explain the purpose of a ground fault circuit interrupter (GFCI).
9. What are the rules about youth and ladder use?
10. Where should a first-aid kit for a shop or lab be located?

Analyze and Apply

1. Acquire a copy of an MSDS and answer these questions:
   A. What is the name of the chemical?
   B. What is its melting point or boiling point?
   C. At what temperature will the chemical burst into flames?
   D. What is its level of toxicity?
   E. How could exposure to this chemical affect your health?
   F. What is the recommended first-aid treatment in case of injury?
   G. Where should this chemical be stored?
   H. What is the safest method of disposing of this chemical?
   I. What should you do if this chemical is spilled accidently?

Thinking Critically

1. What should you do in the following situations to protect yourself and other students?
   A. During the lab activity, you notice a group of students throwing small blocks of wood at each other and pushing each other.
   B. You are finished using the portable electric drill and are ready to move on to the next step of the lab activity. What will you do with the drill?
   C. While walking to your lab station in the school shop, you notice that another student's shoes are not tied and the laces are dangling.
   D. You are assembling a wooden bookcase in the school shop when you accidentally smash your thumb with a hammer. The thumb is starting to bleed.

E. Your lab partner is getting ready to turn on the portable power drill and has not noticed that her hair is dangling and could get caught in the tool.

F. In the agricultural lab, your lab partner dares you to taste a solution he concocted.

G. While out collecting plant samples/specimens, you notice a shrub with berries on it. You have never seen this shrub before, but some of the other students want to taste them to see if they are good enough to eat.

2. Your teacher will give an overview of basic first aid and emergency procedures to follow in the event of an accident. You should be able to answer the following:

   A. If an accident happens, who should I tell and when?

   B. Where are the emergency exits located in the lab or classroom?

   C. What is the evacuation route?

Lesson 4.3 Farm and Work Safety

Words to Know

all-terrain vehicle (ATV)
ammonia (NH_3)
anhydrous ammonia
carbon dioxide (CO_2)
frostbite
heat cramp
heat exhaustion
heatstroke
hydrogen sulfide (H_2S)
hypothermia
manure pit
methane (CH_4)
nitrogen dioxide (NO_2)
power take off (PTO) shaft
rollover protection system (ROPS)
silo gas
spotter

Lesson Outcomes

By the end of this lesson, you should be able to:

- Identify, understand, and apply safe practices when using farm machinery.
- Apply safe practices when working with and around a power take off shaft.
- Identify heat-related illnesses and understand the means to prevent them.
- Identify cold-related illnesses and understand the means to prevent them.
- Understand and apply precautions when using ATVs.
- Understand dangers of manure gas and safety precautions that should be taken when working around manure storage areas.
- Identify hazards associated with silos and silo gas.
- Explain the importance of first-aid training and maintenance of first-aid kits.

Before You Read

Think about a time when you or a friend had an accident of some type. How did the accident happen? What was the result of the accident? After reading the lesson, determine what you or your friend could have done to avoid the accident. Pair with a classmate and share your answers.

Working in agriculture is interesting and challenging. It can be a dangerous occupation as well. In an average year, thousands of agricultural workers are injured or killed in accidents while on the job.

What Makes Agriculture Jobs So Dangerous?

Most agricultural accidents occur to people under the age of 15 and over the age of 65. Accidents involving young people may occur because they do not understand or follow safety rules when using tools and equipment, or they lack the physical strength to operate the equipment properly. Accidents involving older workers may also be caused by the physical inability to properly operate equipment.

Careless Behavior

Agricultural accidents also occur due to careless behavior or improper use of powerful equipment, tools, and chemicals, **Figure 4-32**. Farm tractors, power tools, and pesticides are designed to work in very tough environmental conditions. Powerful farm implements are designed to cut grain stalks, grind feed, and overturn soil during cultivation. Injuries occur when people are careless or inattentive and they get in the way of these powerful tools.

Dale A Stork/Shutterstock.com

Figure 4-32. All too often, on-the-road accidents occur with agricultural machinery. Aside from the dangers of personal injury, hazardous materials used in agricultural applications may pose additional dangers that must be attended to by trained professionals. ***What hazards does this overturned propane tank present?***

Improper Tool Usage

Smaller, hand-operated tools may also be dangerous when used improperly. In an agriculture shop or on the farm, things can happen quickly. A portable grinder operates at about the same revolutions per minute as a jet engine. Welding equipment can liquefy metals to temperatures up to 2800°F (1538°C). Some plasma cutting tools operate at temperatures up to 4500°F (2482°C).

Farms are often located a long way from hospitals and emergency medical care. If you are injured on the farm, there may be a delay in getting you the medical assistance you need, **Figure 4-33**. To avoid serious injury or death, always follow the safety rules associated with agricultural jobs and do not fool around when working with, or around, agricultural equipment.

Training Programs

In addition to the materials and safety training available through OSHA, many university agricultural programs have created safety training programs and written materials to educate agricultural workers on safe practices. Manufacturers have also recognized the need for safety training and work in conjunction with government and university programs to develop safe practices. Take advantage of these resources to educate yourself, your family, and coworkers to make your family farm or workplace a safer place to live and work.

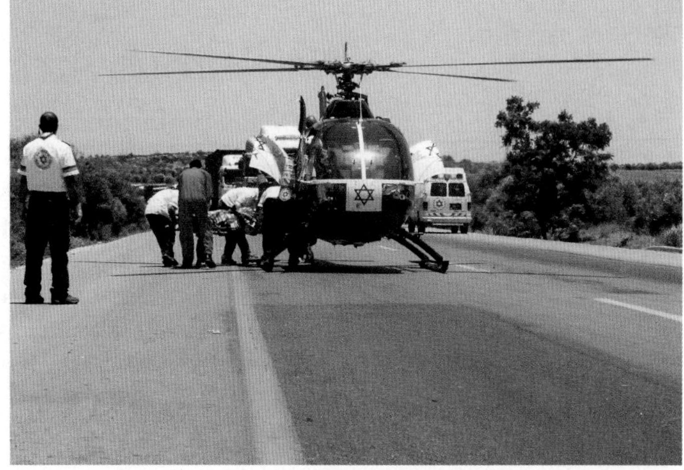

ChameleonsEye/ Shutterstock.com

Figure 4-33. For people living and working in rural areas, extreme measures may be needed to get victims lifesaving medical assistance. There is often a time delay between the accident and the arrival of help that can mean the difference between living and dying from injuries.

Personal Protective Equipment (PPE)

Every agricultural worker must be well-informed about safety on the job. The first step in safety on the job is wearing the appropriate personal protective equipment (PPE). As discussed in *Lesson 4.2*, personal protective equipment includes: safety eyewear, hearing protection, dust masks and respirators, and protective clothing and shoes.

Alex Kosev/Shutterstock.com

Figure 4-34. Using the proper type of personal protective equipment will help keep you safe on the job. ***What types of PPE do you use in school or at your job?***

Wearing the proper PPE may reduce the number and severity of agricultural-related injuries and illnesses, **Figure 4-34**. Maintaining a safe work environment and using personal protective equipment also allows workers to be more productive. *Lesson 4.2, Shop and Lab Safety*, has detailed information on the types and uses of personal protective equipment.

Hazardous Materials

The use of potentially hazardous materials is often essential to many agricultural applications. It is important to learn how to safely and properly handle these materials. Any facility that has any type of hazardous materials is required to have safety data sheets (SDS) available for each hazardous material on the premises. If you are working with any type of hazardous material, take the time to read the SDS. You will have a better understanding of how to safely handle the material and what to do in case of an accident or spill. The SDS also contains essential information for emergency crews in case of an emergency.

Hazardous Material Storage

As explained in the preceding lesson, caustic chemicals can burn your skin, irritate or injure your eyes and lungs, and can be a fire hazard. These chemicals need to be stored in a safe and secure manner, with plenty of ventilation to prevent the buildup of dangerous fumes. Poisonous chemicals such as pesticides and petroleum products need to be stored in a locked cabinet designed for chemical storage.

Anhydrous Ammonia

Many production agriculturists use ***anhydrous ammonia*** as a source of nitrogen fertilizer. It is extremely caustic and special care must be taken to prevent injuries to eyes, skin, and the respiratory tract. Personal protective equipment such as tight-fitting, chemical-proof goggles, rubber gloves with an extended cuff, a rubber suit (if possible), long-sleeve heavy-duty work shirt, and coveralls or rubber apron. Only trained and qualified personnel should transport, transfer, or apply anhydrous ammonia, **Figure 4-35**. The distributor, manufacturer, or dealer can help provide proper training and advice.

Safety Note

It is against the law for anyone under the age of 16 to transport, transfer, or apply anhydrous ammonia.

Farm Safety

Almost 1.25 million youth live and work on farms in the United States. Tragically, almost 8000 young people are injured in a farm accident each year, and approximately 120 young people die each year as the result of a farm accident. If you work on a farm or ranch, you can make a difference by developing safe work practices and making sure your family members and coworkers do the same. Many elements may contribute to farm injuries, including: machinery, weather conditions, livestock handling, gas hazards, chemical hazards, water hazards, electrical hazards, and road hazards.

Charles Brutlag/Shutterstock.com

Figure 4-35. Anhydrous ammonia is a necessary but dangerous chemical used on farms.

Machinery Safety

Accidents involving farm machinery are the leading cause of death and injury among young people working in the agricultural industry. Carelessness, poor attitude, or lack of training are often contributing factors when it comes to accidents involving machinery. It is important to know and understand the safety rules to follow when working around farm machinery. Keep the following points in mind when working with or around farm machinery:

- Until children are old enough, mature enough, and large enough to be trained to operate machinery, they should not be allowed on or near equipment.
- Lack of training is often to blame for injuries sustained on farm equipment. Do not perform work with any machine until you are properly trained and capable of using the equipment.
- Do not rush through jobs. Taking adequate time to perform a job allows you to think and plan ahead. You can reduce accidents because you will have more time to identify hazardous situations.
- Do not permit any person (children or coworkers) to walk closer than six feet beside operating harvesting machinery.
- Have some form of visual communication (hand) signals in place. There should be some way to communicate with a machine operator who is wearing ear protection or riding in a tractor with a noise reduction cab.
- It is important to get adequate rest. Your reaction time increases with increased fatigue, and so does the risk of personal injury.
- Machine failure can cause serious injury. As the operator, it is important to check the conditions of the machine on a regular basis. Proper maintenance schedules should also be in place.
- Forcing a machine beyond its designed limits creates a hazardous situation. The machine will not function efficiently and may even malfunction, causing the operator and bystanders serious injury.
- Keeping work areas free of debris will make the areas safer. Knowing you can move machinery around freely makes it easier and safer to do your job.
- When using any self-propelled machine, stop the engine and remain in the operator's seat until all machine elements have ceased movement

Ethics in AG

You are operating a tractor, and a young child signals to you that he wants to ride along with you. What will you do?

Pablo Debat/Shutterstock.com

Figure 4-36. A simple job can turn deadly when a tractor rolls over. *What safety precautions would be taken when loading hay rolls on a flat bed trailer?*

before attempting any adjustments, maintenance, repair, or unclogging operations.

- Keep all safety guards or shields in good repair and in their place. Removing and failing to replace safety guards is careless. If safety guards must be removed for machine maintenance, make sure they are properly replaced before anyone uses the machinery.

Tractors

Tractor *rollovers* and *runovers* are the cause of many farm injuries and deaths, **Figure 4-36**. Side overturns occur when the tractors are operated on slopes and near ditches. Turning the tractor uphill on a slope increases the risk of overturn, as does traveling too close to the edge of a ditch. The wheels of the tractor could slip down into the ditch, causing an overturn. Tractors can also overturn when implements are improperly attached to the tractor, or when implements are too large for the size and horsepower of the tractor. Speed is also a factor in overturns. Sharp turns at high speeds can cause a tractor to overturn.

Although no one is immune from accidents, employing safe practices and using safety equipment can help reduce these incidences. Keep the following points in mind when working on or around farm tractors and trailing equipment:

- Inspect the tractor regularly and perform scheduled maintenance.
- Use the proper safety equipment, including ***rollover protection systems (ROPS)***, seat belts, handholds, and footholds.
- Keep the steps, platforms, and foot-plates clear of debris, including mud, snow, and manure.
- Before starting or moving a tractor, check to make sure no one has climbed aboard the tractor or trailing equipment.
- Do not carry small children in the cab or on your lap. They can easy slip and fall beneath the tractor. The rule to remember is, one seat, one rider. Two people should not occupy the same seat. There is also no seat belt for a second rider.
- Use the appropriate speed on roads, hills, and uneven ground, **Figure 4-37**.
- Do not jump off a moving tractor! Wait for the tractor to come to a complete stop, set the brakes, and use the handrails and footholds to climb down.
- Make sure trailing equipment is properly attached.

Tractors may have hydraulic systems for operating implements. The hydraulic fluid in the hoses is under very high pressure. A tiny leak can allow fluid to escape at high velocity and with enough force to puncture your skin. When working around tractor hydraulic systems, wear appropriate safety

Know how to safely operate the tractor.
Make sure you are familiar enough with the tractor to drive it on a public road. If you have questions about how to operate a tractor, ask a knowledgeable person to assist you.
Obey the traffic laws.
Make sure you follow all traffic laws, including speed limits and traffic signs and signals. Tractor drivers must follow the rules of the road at all times. Display the slow moving vehicle (SMV) symbol in its proper location on the tractor, usually at the rear, facing traffic coming up behind you. Make sure your SMV sign is clean and visible for an appropriate distance behind your tractor.  Mark Herreid/Shutterstock.com
Drive with your lights on.
If your tractor is equipped with headlights and brake lights, make sure they are in proper working order. Use your lights to improve your visibility on the road. Some states may require that you operate yellow or red flashing hazard lights while operating a tractor on a public road. Check the local laws in your state to determine the right course of action to follow. Make certain that your lights are visible from the rear when hauling implements behind the tractor. Most collisions occur when the tractor is rear-ended or when making a left turn. Make sure your brake lights, turn signals, and SMV sign are clearly visible to traffic approaching from the rear.
Pick the right time to drive on a public road.
Choose a time to move farm equipment when traffic is light. Plan the route so you avoid major highways where traffic may be heavy with cars traveling at high speeds. Drive when highway visibility is good. Avoid driving when visibility on the road is 500′ or less.  David Scheuber/Shutterstock.com
Using tractor brakes safely on the road.
Before taking your tractor onto a public road, make sure you have locked the brake pedals together. If you have to hit the brakes in a hurry, this will assure that both sets of rear wheel brakes will engage simultaneously.
Use an escort vehicle.
Use an escort vehicle with a flashing warning light if the equipment you are hauling extends into the opposite lane of travel.
Wear the seat belt.
If the tractor is equipped with rollover protection, wear the seat belt. Without the seat belt holding you in place, it would be easy to be thrown off the tractor in the event of a crash.

Figure 4-37. About one-third of all fatal accidents involving tractors occur on public roads. You can avoid injury by following these practical safety tips.

Dmitry Kalinovsky/Shutterstock.com

Figure 4-38. Tractor hydraulic systems operate under extremely high pressures. Regular maintenance of these systems is essential to keep them in safe, working order.

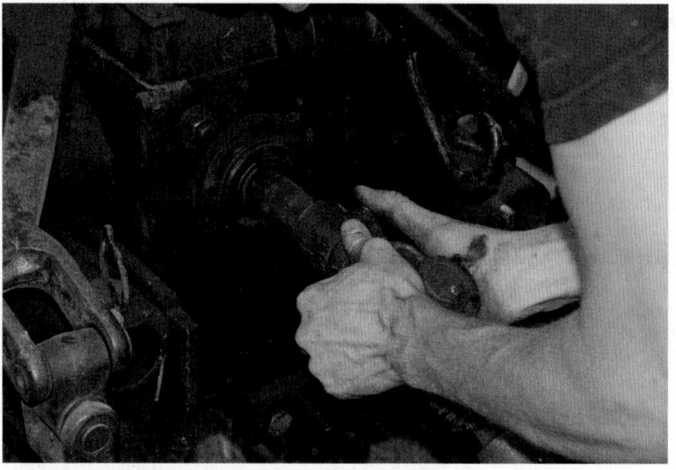

Tumarkin Igor - ITPS/Shutterstock.com

Figure 4-39. Power take off (PTO) shafts rotate at 740 to 1000 revolutions per minute. Use extreme caution when working around these rotating shafts to avoid getting tangled. Serious injury will result. ***Never work on a PTO shaft while the tractor engine is running.***

glasses and check for leaks using a broom or long wooden stick. Some leaks are so small that you may not detect the fluid escaping under high pressure, **Figure 4-38**.

Power Take Off (PTO) Shaft

The ***power take off (PTO) shaft*** feature on many tractors converts power from the tractor's engine into rotary power that can operate grain augers, feed grinders, mowers, and irrigation pumps. The PTO is a shaft typically located at the rear of the tractor, **Figure 4-39**. When the engine is running, power flows along the shaft and transfers energy from the engine to the attachment. When engaged, the PTO spins at either 540 revolutions per minute or 1000 revolutions per minute. Tractor manufacturers make shields that cover the PTO shaft when it is not being used.

The PTO's introduction helped revolutionize agriculture in the 1930s. It is also the source of many farm machinery injuries. According to the National Agricultural Safety Database, most PTO accidents occur when clothing and/or limbs are entangled in the rotating PTO shaft.

- Keep all shields and guards in place, and wear the personal protective equipment necessary for the job.
- PTO shafts are very dangerous. They rotate at very high speeds with a high torque.
- A PTO shield should be in place, and it should cover all of the operating parts of the PTO.
- When looking at a PTO shaft in operation, you should not see any part of the moving shaft.
- Use extra caution when adjusting running equipment attached to a PTO shaft.
- Inspect the PTO shaft regularly for defects and proper attachment.
- Never step over a PTO shaft when it is running, and avoid wearing loose clothing or jewelry that can become tangled in the running PTO shaft.
- Avoid going near PTO shafts when they are operating.
- Disconnect the PTO shaft when it is not in use.

Electrical Safety

As discussed in the preceding lesson, the human body is not naturally insulated from electricity and will easily conduct electrical current, causing anything from a mild shock to cardiac arrest. For this reason, it is essential to follow all safety rules and regulations regarding the installation and use of electric current. In agricultural settings, electrical accidents commonly occur because of:

- Tall equipment, such as grain augers, combines, and raised dump truck beds becoming entangled in overhead power lines. Most overhead power lines have no protective insulation and any contact with them may be dangerous.
- Tractors or other large machinery snagging the guy wires on electrical utility poles and causing the overhead lines to sag or break.
- Irrigation or well pipe coming in contact with overhead power lines when being moved.
- Electrical wire insulation becoming damaged due to rodents, weathering, improper wiring, improper wire size or type, and corrosion of electrical connections.
- The dusty, moist, and corrosive environments of livestock houses and other farm structures.

Safety Procedures and Preventive Measures

Fortunately, electrical accidents can be avoided through the proper installation of electrical service and safety devices designed for environments such as those found in agricultural applications. The following list of safety procedures and preventive measures is not all-inclusive, and has been provided as a basic guideline to help ensure a safe working environment. Follow your teacher's instructions and use common sense when working with or around electricity.

- All electrical work must be done by a qualified electrician who meets state, local, and national certification requirements, **Figure 4-40**. All work done must follow local and national codes. This includes new electrical service and wiring in new structures, updating or replacing older wiring and fuse boxes, and the installation of exhaust and ventilation systems.
- When possible, have all farmyard overhead electrical wire buried underground. This eliminates the possibility of entanglement and the dangers associated with fallen power lines.
- To prevent entanglements in overhead wires, ensure all augers, sprayer booms, bale elevators, dump truck beds, and other tall equipment are lowered to their lowest possible level before moving them. Use a ***spotter*** (someone to help guide the driver) to make certain contact is not made with power lines.

131pixfoto/Shutterstock.com

Figure 4-40. Functioning switches, like the makeshift one illustrated here, that have been poorly installed should be replaced by a licensed electrician. *What types of problems or accidents could this type of installation cause?*

Safety Note

Local utility services must be called before any digging is performed to avoid contact with buried power or gas lines.

- Install waterproof, dust proof, and explosion proof electrical boxes, outlets, and motors in farm structures as required by the National Electric Code (NEC). This includes livestock facilities, silos and adjoining buildings, manure storage areas, barns, and any other farm structures where moisture, gases, and dust may create hazards.
- Treat every electrical wire as though it is a "hot," or "live," wire and make sure power is disconnected before working on any equipment. Use lock-out devices and tags to ensure no one accidentally turns on the power.
- Install a grounding electrode system as required by the electrical code in addition to any lightning protection grounding system.
- Check grounding rods and wires around buildings and power poles. Grounding rods and wires provide a means for electricity to follow the easiest path to the ground and prevent electrical shock. If damaged or broken, the systems will not provide adequate protection.
- Ensure all heated livestock waterers and metal guard rails animals can touch are properly grounded. Livestock are as susceptible to electrical hazards as are people.
- Do not operate under the assumption that nonmetallic materials such as lumber, tree limbs, tires, or ropes will not conduct electricity. These materials will conduct electricity, depending on their moisture content and surface contamination.
- Do not clear storm-damaged trees, limbs, or other debris that are touching power lines, **Figure 4-41**. Call the utility service and keep all animals and people away from downed power lines.
- Do not trim or cut down trees that may fall onto power lines. Call the utility service if these trees or limbs need to be removed.
- Do not approach downed power lines. Everything around them, including the soil, could be energized. Immediately call the utility service and keep all animals and people away from the area.
- Establish a routine maintenance schedule of electrical equipment and devices that includes regular cleaning of service areas. Routine maintenance can reduce the possibility of fire and accidents.

Guy J. Sagi/Shutterstock.com

Figure 4-41. Trees and other nonmetallic materials may conduct electricity. Do not attempt to move trees or other materials that are touching power lines. Always call the local utilities or emergency personnel to take care of the situation.

As you can see, there are many ways electricity can be hazardous on the farm. Many universities and government agencies offer advice and programs to help people learn how to keep their home and agricultural business safe from electrical hazards. Take advantage of these resources to learn more on how to prevent electrical accidents, as well as other types of accidents, on the farm.

Weather Safety

Weather conditions will create or amplify hazards when you are working with farm equipment. Consider weather and ground conditions before operating any large machinery. Mud, ice, snow, and frozen ground will affect the way the machinery handles. Strong winds will blow dirt or snow that may obscure your vision. You should always have a plan in case of extreme weather, like a tornado, when you are working outside.

Extreme weather conditions will also affect your body. Take the proper precautions to avoid weather-related injuries.

Sun and Heat Exposure

The human body is designed to maintain a temperature average of 98.6°F. When exposed to extreme heat conditions, your body must release excess heat to maintain its optimum temperature, primarily through perspiration. If you consistently replace fluids while working in extreme conditions, you can more easily avoid heat-related illnesses such as heatstroke, heat exhaustion, and heat cramps.

- ***Heatstroke***—condition that occurs when your body is unable to release enough heat, and your temperature rises to life-threatening levels. Symptoms include skin that feels feverish, confusion or disorientation, seizures, and ultimately, the victim may fall into a coma.
- ***Heat exhaustion***—a condition that occurs when you do not drink enough fluids to replace what is being lost through heavy perspiration and you have a slightly higher than normal body temperature. Symptoms include severe thirst, headache, nausea, diarrhea, profuse sweating, dizziness, clammy skin, and a rapid pulse. Heat exhaustion may lead to heatstroke if not treated.
- ***Heat cramps***—muscular spasms that occur when a person is sweating a lot and not replacing lost fluids.

To prevent these heat-related illnesses, keep yourself hydrated and make sure your coworkers do the same. Keep clean drinking water available at all times and take regular breaks to cool off. It is also advisable to keep some type of sports drink available when working in extreme conditions. These drinks contain sodium and electrolytes to replace those lost through extreme perspiration.

You should also protect your skin from overexposure to the sun. Wear a broad-brimmed hat and apply sunscreen regularly throughout the day.

Frostbite and Hypothermia

Not all farm and ranch work is done during the warmer part of the year. This is especially true of livestock operations. Overexposure to low temperatures, cold wind, and cold moisture may result in cold-related illnesses such as frostbite and hypothermia.

- ***Frostbite***—body tissue injury caused by overexposure to extreme cold. It typically affects exposed skin and extremities like the nose, fingers,

Safety Note

Do not try to grab a person who is being shocked by electricity. You will become a part of the "circuit" and also be shocked. If it is safe to do so, disconnect the power source by turning off the circuit breaker. Do not try to unplug the cord or move the energized line with any object. You will also be shocked.

Did You Know?

Fifteen amps, which is typically the smallest size breaker found in a normal household, is 250 times greater than what is required to cause cardiac arrest in an individual.

Ethics in AG

You and your partner are working outside on a very hot day. He suddenly begins to complain of stomach cramps. What will you do?

and toes. The skin first becomes cold and red, then numb, hard, and pale. Blisters may appear. In extreme cases, it may result in gangrene.

- *Hypothermia*—a condition that occurs when your body loses heat faster that it can produce it, causing a dangerously low body temperature. It can occur if you are underdressed for cold weather and exposed for an extended period of time, you are wet in extreme cold conditions, or you become immersed in extremely cold water.

The best recourse for cold-related illnesses is to prevent them from occurring. Keep the following points in mind when you are working outdoors in cold weather:

- Wear the proper protective clothing and wear multiple layers.
- Make sure someone knows where you working and what time you expect to return.
- Keep extra, dry clothes in your vehicle, along with blankets, high-energy snacks, and water.
- Make sure your cell phone or radio is charged and accessible.

Working with Livestock and Other Animals

Many farm or ranch injuries occur when workers, family members, or even visitors are handling livestock. Animals are unpredictable and not every action can be foreseen, but there are precautions that can be taken to reduce hazards and injuries to both handlers and animals. In addition to understanding certain animal behavior, you can create safe and secure enclosures, maintain handling and restraining equipment, and properly train anyone working around livestock, **Figure 4-42**.

Animal behavior is very different from that of humans. By understanding the characteristics of animal behavior, you have a better chance of avoiding injury when interacting with them. Farm animals can only see in black and white and have difficulty judging distances. By the same token, healthy farm animals have excellent hearing and are sensitive to loud noises.

operator1975/Shutterstock.com

Figure 4-42. Many people are injured when working around or with livestock. ***How does knowledge about animal behavior help those working with livestock?***

Many farm animals are much larger and more powerful than humans. Beef cattle, swine, and horses can kick, crush, trip, or trample humans by accident. The key to safely working around farm animals is to avoid situations that startle or frighten them. Frightened animals can bolt, causing serious injury to you and perhaps even the animals.

The most common accidents occur with livestock when:

- Animals are prodded to go in a direction that makes them nervous.

- Young animals are lifted improperly.
- Wrapping ropes or animal leads around your hand or body.
- Not wearing safety equipment such as gloves or helmets.
- Trying to move livestock without enough helpers.

A farm is an industrial workplace. Running and playing around livestock can cause them to become nervous and bolt. Never approach large farm animals from the rear as they might kick you. If you need to work around large animals, stand close to the side or shoulder of the animal. This will protect you from the full force of the animal should it decide to kick you. Never stand directly behind large farm animals. Animals can be aggressive at feeding times.

Animals can harbor diseases that can be transferred to humans. Make certain that you use good hygiene after working with livestock, including:

- Washing hands after working around or with animals.
- Changing out of soiled clothing before entering your home.
- Checking for ticks and removing them before they become attached to your body.

Additional livestock handling and safety procedures are covered in *Chapters 10* and *11*.

All-Terrain Vehicle (ATV) Safety

All-terrain vehicles (ATVs) are used for work and recreation on many farms and ranches. ***All-terrain vehicles (ATVs)*** are motorized vehicles that usually have four low-pressure tires designed to travel over various types of rough terrain, **Figure 4-43**. The tires have deep grooves and treads designed to handle muddy and rocky land and do not perform well on paved surfaces. They are not designed for use on paved surfaces and, in most areas, cannot be legally driven on roads.

Each year, many people are severely injured or killed on ATVs. Many of these injuries and deaths could be prevented if the operators were properly trained and employed certain safety precautions. The following list is not all-inclusive, but it does include key points you should keep in mind when purchasing and using an ATV—for fun or for work.

- Anyone who will be using an ATV should receive training through an accredited program. There are programs available through 4-H, the All-Terrain Vehicle Safety Institute, and state-run programs.
- Use the right size ATV for your age. Children should never drive an adult-sized ATV. ATVs with an engine size of 70cc to 90cc should only be operated by people 12 years or older. ATVs with an engine size greater than 90cc should only be operated by people 16 years or older.

Supertrooper/Shutterstock.com

Figure 4-43. ATVs are useful vehicles on a farm or ranch when used safely.

- Appropriate riding gear should always be worn. Approved riding gear is designed to help protect you from head and spinal injuries, as well as cuts and abrasions. Appropriate riding gear includes a DOT- or Snell ANSI-approved helmet, goggles, gloves, over-the-ankle boots, long-sleeve shirt, and long pants.
- Do not allow additional riders. Most ATVs are designed for single riders. There are no handholds or footholds for additional riders, and they are easily thrown from the vehicle.
- Do not drive on paved roads. ATVs are not designed for on-road use and do not handle well on smooth pavement. When you must cross a paved road, drive slowly and watch for vehicles.
- Use only manufacturer-approved attachments. Use extra caution when using added attachments because they will affect the operation, braking, and stability of the ATV.

As with any machinery, you should regularly inspect and perform standard maintenance procedures to keep ATVs in safe operating condition. Use the Internet to research ATV safety and locate training programs available in your area.

Manure Pit Gas Hazards

The traditional method of shoveling and removing solid animal waste is not efficient for larger livestock and dairy operations. These larger operations typically use ***manure pits*** to store and process animal waste, **Figure 4-44**. Manure pits vary in design and may be open lagoons, earthen basins, below grade, above ground, and contained in closed buildings. Many waste removal systems feature grated floors through which waste can fall and be transferred to the manure pit, **Figure 4-45**. Waste may be flushed to the manure pit with water or removed through a pumping system.

The dangers of manure pits lie mainly in the manure gas produced through decomposition. The accumulation of manure gas creates an

Matthijs Wetterauw/Shutterstock.com

pinthong nakon/Shutterstock.com

Figure 4-44. Manure storage facilities may be open lagoons or elaborate processing systems. Always use caution when working around manure storage to ensure you are not overcome by toxic gases.

Corepics VOF/Shutterstock.com

Corepics VOF/Shutterstock.com

Figure 4-45. Grates on floor allowing waste to be washed down and flushed to the manure storage and/or processing facilities

oxygen-deficient, toxic, and potentially explosive atmosphere. The four gases in manure pits that are of primary concern are ***methane (CH_4), hydrogen sulfide (H_2S), carbon dioxide (CO_2)***, and ***ammonia (NH_3)***. These gases are highly toxic, may be odorless, and will quickly overcome anyone exposed to the concentrations present in manure pits, especially those in confined spaces. Many of the deaths that occur in and around manure pits are due to asphyxiation by one or more of the toxic gases. The following list of precautions is not all-inclusive but contains important key points you should keep in mind if you are working around manure pits.

- Hazard signs must be posted at all entrances to the storage areas. Signs in multiple languages may be necessary in some areas.
- Prepare and post a written confined space entry procedure and make all workers and family members aware of the proper entry procedure.
- The need for entry into the manure pit may be eliminated or reduced if access to all serviceable parts is provided outside the facility.
- Install gas detection monitors or use portable gas detectors to determine oxygen level and the levels of toxic gases. These should be accessible from the outside of the facility.
- All enclosed manure pits should have fully operational, continually operating, fresh-air ventilation systems. Ventilation systems may be portable or permanent, but must have an explosion proof design due to the explosive properties of methane and hydrogen sulfide.
- No one should enter a manure pit or transfer structure unless the proper safety precautions have been taken. These precautions include wearing a safety harness attached to a mechanical retrieval device, the proper breathing apparatus, a standby person in place with the ability to use the retrieval device, and the implementation of an emergency plan.

- Gas traps should be installed in transfer pipes to prevent gases from flowing back into the animal housing area.
- No one should enter the storage area a when the manure is being agitated or while the manure is being removed. The manure gas will also be agitated and possibly more hazardous.

In addition, anyone who is under the age of 16, who are not the children of the farm operator, are completely prohibited working inside confined manure storages.

Silo Safety

A *silo* is a tall metal or concrete storage container that holds silage, **Figure 4-46**. *Silage* is grass, fermented feed, or other fodder that is harvested green and used to feed cattle. Silos may also be used to store grain. Silos present several hazards to farmers and farm workers, including toxic silo gas and entrapment by flowing grain.

Silo Gas

Silo gas is formed by the natural fermentation of the silage shortly after it is placed in the silo. The type and quantity of gas depends on the type of silo. For example, in sealed silos, both nitrogen and carbon dioxide gases are created, but carbon dioxide is produced in greater amounts. (High carbon dioxide levels help maintain high silage quantity.) In conventional or open-top silos, nitrogen dioxide is abundant.

digitalreflections/Shutterstock.com

Figure 4-46. Silos are used for the bulk storage of grain and silage. Silo gas formed by the fermentation of chopped silage can irritate the nose, mouth, and lungs and cause difficulty breathing. Use caution around these structures.

Carbon Dioxide

Carbon dioxide (CO_2) is an odorless and colorless and highly dangerous gas. High concentrations of carbon dioxide can cause asphyxiation in a very short time. It replaces the silo's oxygen in high concentrations, and gives little warning to the victim. Carbon dioxide is a heavier gas so higher concentrations will settle at the top of the silage and/or flow down the silo chute and collect in an adjoining room or other areas near the base of the silo.

Nitrogen Dioxide

Nitrogen dioxide (NO_2) has a strong bleach-type odor, low-lying yellow, red, or dark brown fumes, and is also highly toxic. Nitrogen dioxide causes severe irritation to the nose and throat and may lead to inflammation of the lungs. Death may occur quickly with higher concentrations, but low-level exposure may also be deadly. Low-level exposure may cause minimal immediate pain or discomfort, but the victim may die just hours later from fluid collecting in his or her lungs. Victims may also suffer relapses two to six weeks after the initial exposure. Anyone who is exposed to this gas should seek immediate medical attention.

Nitrogen dioxide is also a heavier gas and will settle near the top of the silage and/or flow down the silo chute and collect in adjoining areas, including the barn itself. Livestock is also susceptible to asphyxiation by silo gases and care should be taken to prevent gases from seeping into livestock areas.

Although silo gases will be present as long as silage is being used, there are safety measures to help prevent these gases from harming farm workers and livestock. The following list is not all-inclusive, but does contain key safety points you should keep in mind during harvesting and feeding.

- Post silo gas warning signs around silos. Fencing installed around the silo base will help keep people and animals a safe distance from the silo and toxic gases.
- Block access to silos with door locks and gate locks where applicable.
- Use monitoring equipment to determine oxygen and silo gas levels.
- Do not enter the silo for four to six weeks after filling stops. Nitrogen dioxide levels peak about three days after filling, but hazards remain, especially if the gas has not been able to escape the silo.
- Ventilate the silo room for at least two weeks after filling. Keep all doors closed and the roof panel open for more effective ventilation. These actions do not guarantee that all dangerous gases have been removed.
- Never enter a silo alone. There should be *at least* one person, preferably two, outside to help in case of emergency.
- Acquire, use, and maintain the proper type of self-contained breathing apparatus (SCBA). Make sure all workers and family members receive training on its use.
- If someone must enter the silo, they must wear a self-contained breathing apparatus (SCBA) and a safety harness. There should be at least one person outside the silo to help in case of emergency. Run the silo blower for 45 minutes before entering the silo, and leave the blower running while inside.

- Watch for yellowish or reddish fumes in or near the silo and be alert for bleach-like gas odors.
- Provide adequate ventilation for adjacent livestock quarters to keep animals safe from silo gases.
- Lock out and tag all loading and unloading mechanisms when not in use. Tags should also be used to indicate someone is working inside the silo.
- Do not attempt to help somebody who has succumbed to gas hazards unless you are adequately protected or you will suffer the same fate.
- Do not enter the silo when loading and unloading equipment is in use.
- Prevent excess nitrates through proper fertilization and proper weed and insect and disease control.
- Anyone exhibiting exposure symptoms such as respiratory congestion, watery eyes, cough, difficulty breathing, fatigue, and nausea should be removed to fresh air and receive immediate medical care.

Ethics in AG

The top surface of grain in a grain bin is crusted over. Your coworker climbs into the bin to "walk down" the grain so that it flows to the unloading auger. What will you do?

Flowing Grain

In addition to the dangers presented by silo gases, workers may be suffocated by flowing grain when working inside silos/grain bins or while loading and unloading grain. Workers are especially in danger of entrapment when entering the structure to loosen crusted, spoiled, or frozen silage or grain while the equipment is running. Entrapment also occurs when workers fall into grain transport vehicles while they are being loaded or unloaded. Safe practices to prevent entrapment are listed in the following section on grain bin safety.

Grain Bin Safety

Many farms have grain bins. The tall structures receive, handle, and store grain until it is ready to be transported to market, **Figure 4-47**. Grain bins are very hazardous for several reasons. Hazards in grain bins include oxygen

Charles Brutlag/Shutterstock.com

Denton Rumsey/Shutterstock.com

Figure 4-47. Grain bins use augers to move grain from one location to another. Use extreme caution when working around augers and rotating shafts.

deficiency, accumulation of toxic gases, grain dust explosions and fires, and molds. One of the biggest dangers is entrapment in the grain itself. The grain in grain bins acts like quicksand and can pull a worker underneath the grain where they suffocate from the lack of air.

The most dangerous time for entrapment is when the auger is running and grain is flowing out of the bin. It creates suction in a cone of flowing grain. A human body can be sucked down into the flow in only seconds, and covered with the flowing grain in less than 20 seconds. The suction created by the flowing grain is powerful enough to immobilize a grown man standing only knee-deep in the grain. Once the grain reaches waist-high, it will take over 600 pounds of force to pull the person free. Using this amount of force to free someone may also break or dislocate limbs. The best practice is to prohibit anyone from entering the bin while the equipment is running and grain is flowing.

The augers that move grain from one location to another can cut or crush workers, and falls from heights can cause serious injury. Never enter grain bins, grain trucks, or wagons when the grain is being unloaded. Shut off power to the grain bin before entering it to avoid injury if the auger system were to restart. Sometimes grain can become crusted on the inside of a grain bin. The safest method for breaking loose crushed grain is to use a long pole while standing outside of the bin. If you must enter a grain bin, wear a safety harness attached to a rope so that you can be pulled out of the grain if you get trapped. You should also have two people outside the bin who can help you if you get trapped.

Safety Note

If you are under the age of 16, you are prohibited under federal law from entering confined spaces or environments, including grain storage structures.

Most grain bin accidents can be prevented by using safe practices and the proper safety equipment. Keep the following in mind when working in and around grain bins.

- All workers must be properly trained for the hazardous work involved when entering and working inside of grain bins. Training should include using a lifeline to rescue coworkers.
- Grain dust can cause respiratory problems. Wear the appropriate dust mask when working in areas where grain dust is high.
- Anyone entering a grain bin or silo must wear a body harness with a lifeline connected to a proper anchor point. The lifeline must be long enough to prevent a worker from sinking farther than waist-deep. A boatswains chair may also be used, **Figure 4-48**.
- Before entering a grain bin, use a meter to test the air for the oxygen level and presence of combustible and toxic gases. If high concentrations of combustible and toxic gas levels are detected, do not enter a bin or silo until they are reduced to nonhazardous levels and an appropriate oxygen level is maintained. It is recommended the proper respirator be worn when working in grain bins or silos to guard against toxic gases as well as mold and dust.
- Ladders should be installed inside and outside of all bins. Ladders installed on the outside must end at least seven feet off the ground to prevent unauthorized entry to the bin. Do not rely on a rope, chain, or pipe ladder hanging from the roof.
- Regardless if grain is or is not flowing, never enter a bin when unloading equipment is running. If someone must enter a grain bin, make sure all

Sailom/Shutterstock.com

Konstantin Sutyagin/Shutterstock.com

Figure 4-48. Follow federal rules and guidelines when working above ground level, and wear the appropriate safety equipment. ***Have you ever used a safety harness when working?***

powered equipment, including augers used to move the grain, is turned off and locked out. This will prevent anyone from turning the equipment on while someone is working inside.

- When someone must enter a grain bin or silo, station a second person outside to track the person in the bin and to provide assistance if necessary. A third person should be available should an emergency situation arise, as it may take two people to lift the person inside to safety. If possible, equip workers with radios to aid communication.
- Exercise extra caution when working with grain that has begun to spoil. Toxic gases, molds, and surface crusts over hollow cavities are more common with spoiled grain.
- Do not stand on moving grain or allow anyone to "walk down grain" to make it flow.
- Use a long pole to check and dislodge surface crusts, bridges, steep piles of grain, and grain build-up on the sides of the bin. The surface crust or bridge could collapse and steep piles or grain buildup may avalanche when dislodged.
- Inhibit mold growth by preventing water from entering the bin and by storing only fully dried grain at the correct moisture content.
- Fire and explosion hazards may be reduced through equipment maintenance and ventilation (if appropriate). Prohibit smoking or the use of fire, for any reason, near flammable dusts, gases, and fumes.

Grain Machinery and Transport

According to the National Institute for Occupational Safety and Health (NIOSH), grain handling machinery is the second largest cause of farm machinery related deaths and injuries. Many of these accidents occur when someone's limb or body becomes entangled in the machinery, or loose-fitting clothing, loose hair, or jewelry get caught in rotating machine parts. Accidents also occur when loading or unloading grain from truck grain beds

Denton Rumsey/Shutterstock.com

Bryan Eastham/Shutterstock.com

Figure 4-49. Make sure that you can be easily seen and heard when working around trucks loading and unloading grain. Avoid getting into "blind spots" where the driver cannot see you. ***What types of safety gear could workers wear to make themselves more visible to drivers?***

and dump wagons, **Figure 4-49**. Using common sense and safe practices can help prevent these types of tragedies. Search the Internet to locate training programs and materials that will help make your workplace safer. Keep the following safety tips in mind when working around grain machinery and trucks used to haul grain.

- Make sure all safety shields and devices are in place when unloading or loading grain.
- When using any form of PTO shaft, make sure it is fastened securely to the tractor drawbar. Vibration may cause the attachment to move, allowing the PTO shaft to come apart and move.
- Unload grain wagons on a level surface. Do not elevate them to speed the unloading process.
- Tractor brakes must be in a locked position, especially when tractor power is used to run the auger.
- Do not allow anyone to ride on trucks using grain beds or gravity dump wagons. Just as in the grain bin, there is danger of entrapment in the grain bed or dump wagon.
- Do not climb into the hopper or wagon to help move the grain.
- Minimize electrocution and fire hazards by maintaining electrical equipment and using the correct fuses, wiring, switches, and insulation.
- Use lock out and tag procedures to ensure no one will activate machinery while it is being serviced or someone is inside the bin or silo.
- Always be aware of overhead power lines and avoid equipment or personal contact.
- Use rollover protective structures and seat belts on tractors when packing silage in bunker silos.

This list is by no means all-inclusive. Take the time to understand and implement safety precautions to prevent accidents. Use the resources available from safety councils, government, manufacturers, and university programs to educate yourself and your coworkers.

Ethics in AG

You and a coworker are working in an on-farm storage unit when a pesticide spill occurs. Your coworker gets pesticide on her clothing and skin. What will you do?

Federico Rostagno/Shutterstock.com

Tyler Olson/Shutterstock.com

Figure 4-50. The safe use of pesticides begins before mixing and applying pesticides. Wear the appropriate protective equipment, then read and follow the manufacturer's directions on the safe use and application of pesticides. *Do weather conditions affect when pesticide spraying should or should not be done?*

Thinking Green

Container Disposal

Safe disposal of pesticide containers is a must. Clean and recycle containers according to the regulations in your state.

Pesticide Safety

Pesticides should only be applied by those who are trained and certified in safe pesticide application. However, you should be familiar enough with safe pesticide handling to know what to do in the event of an accident, **Figure 4-50**.

Pesticides are a useful method for controlling pests in crops and livestock. Because they are poisonous, they should be used carefully and in the correct amounts for the pest being controlled. Overapplication of pesticides does not mean that more pests will be controlled, but it does often mean that the risk of pesticide poisoning for humans and animals is increased.

- Those individuals applying pesticides should wear appropriate personal safety equipment. This also includes those individuals working near a location where pesticides are being applied.
- Pesticides for field crops should not be applied on windy days or very warm days. Wind may cause the pesticide droplets to drift from the target field to nearby houses and lawns. On very warm days, the liquid

Safety Note

- If a pesticide has been swallowed, induce vomiting only if instructed to do so by qualified emergency personnel.
- If a pesticide splashes into the eye, use clean running water to flush the eye.
- If a pesticide splashes onto the skin, use water to quickly remove as much of the poison as possible, then use soap and water to thoroughly wash the affected area. Remove clothing contaminated with pesticides. Keep a gallon jug of soapy water and a jug of rinse water handy.
- If a pesticide is inhaled, move the victim to fresh air, and start artificial respiration if the victim is not breathing and you are trained to do so.

pesticides can evaporate from crop plants and the vapor can travel to homes and lawns.

- Pesticides should be stored out of reach of children, preferably in a locked cabinet or storage room.
- Pesticides should always be stored in their original containers.
- In the event that you or someone else is exposed to pesticides and is having trouble breathing or convulsions, give first aid immediately. Call emergency services or 911 immediately.

First Aid

It is always a good idea to know basic first aid, regardless of where you work or live. However, when your home or job is far from medical facilities, it is especially important to know how to respond in case of an emergency. Oftentimes, certain conditions on a farm may turn a minor injury into a life-threatening emergency before professional help is available. This is especially true if the injury occurs in a remote location. When farm workers know how to deal with a medical emergency, it may save the life of someone who is seriously injured.

Training

First and foremost, get training in first aid and CPR, and encourage family members and fellow employees to do the same. Contact the American Red Cross, National Safety Council, or even the local fire department to find training programs in your area. It is also important to take refresher courses on a regular basis to refresh your training and learn new techniques and treatments. Remember, if you do not know how to use it, even the best prepared first-aid kit will be useless.

First-Aid Kits

Since workplaces vary in their location, facilities, size, and the types of possible accidents that may occur, it is necessary to customize first-aid kits. Use the following list to establish the number of first-aid kits you will need, where they should be located, and what they should contain.

Location and Maintenance

Strategic locations for first-aid kits include areas such as on the most frequently used tractors and harvesting machines, in the workshop, livestock facilities, tack room, home, and work vehicles. Small kits may also be kept in ATVs, truck toolboxes, snowmobile storage compartments, and saddle bags. You will have to evaluate areas of your farm or ranch to determine where first-aid kits should be placed and what each one should contain. First-aid kits should be readily available but also kept where the contents will stay clean and dry.

First-aid kits may be mounted on walls, stored in designated cabinets, or kept in portable bags. To make them highly visible, clearly label each kit with bright labels or tape. Store portable kits in large, brightly colored travel or sports bags with a visible label.

To keep first-aid kits up-to-date and fully stocked, assign the task of inspecting and restocking first-aid kits to one person and establish a schedule to replace expired items on a regular basis.

General Contents

The rural aspect of most agricultural facilities require first-aid kits to contain additional information to help administer proper care by coworkers and emergency personnel alike. Use the following list to ensure your first-aid kits contain all necessary items and information.

- Each first-aid kit should include a list of emergency numbers as well as instructions on how to contact an ambulance, the closest hospitals, police departments, and fire departments. Keep in mind that 911 numbers may not be a standard service in some locations.
- Include simple, clearly written directions on how to get to the farm, ranch, field, and work areas. It will make it much easier to convey this information to emergency personnel.
- Compile personal information for everyone who is regularly on the premises. This includes people who live and work on the farm or ranch. Each kit should contain personal information such as existing medical conditions, prescription medications that may affect treatment, allergies (i.e. bee stings, medicines, latex), name and phone number of personal physician, and emergency contact information. Use index cards to record the information and place the cards in a sealable plastic bag.
- Flares, flashlights, and waterproof matches will be useful after dark and in isolated areas. Flashlight batteries must be checked and replaced on a regular basis unless the flashlight is manually rechargeable.
- Include a current first-aid manual or first-aid chart in each first-aid kit. It is important to have this information on hand during an emergency when it may be difficult to think clearly. First-aid manuals and charts list steps for administering first aid to victims of drowning, shock, fractures, and burns. They also include guidelines for preventing additional injury.

First-Aid Supplies

Some of the most common farm-related injuries occur because of accidents with machinery or livestock, and due to slips and falls. These accidents may result in small wounds, fractures or sprains, concussions or other head injuries, or even severed limbs. For this reason, first-aid kits must be tailored to cover specific types of injuries. The chart in **Figure 4-51** lists supplies that would be adequate to cover most farm-related injuries. However, it is advisable to ask your first-aid training professional to review your first-aid kits to ensure they contain all necessary supplies.

First-Aid Supplies Required	
For major and minor trauma including: Entanglement Fractures Minor bleeding Major bleeding Severed limbs Small wounds Sprains	First-aid manual or charts Antiseptic treatment (spray, liquid, swabs, wipes, or towelettes, not in pressurized can) to disinfect contaminated wounds Scissors strong enough to cut through clothing such as denim Stainless steel scissors to cut bandages Tweezers Tongue depressors (individually wrapped) Rubber or latex gloves Safety glasses Chemical ice packs Emergency blanket Saline or sterile water Soap or sanitizer (61% ethyl alcohol) to clean hands Wooden or plastic splints (1⁄4″ × 3″ × 12″ – 15″) to immobilize injured limbs, or air inflatable roll of elastic wrap (to attach splint) **Dressings** 2 - 36″ triangular bandages to make slings, control bleeding, and splint fractures 12 - large adhesive bandages for small cuts, puncture wounds, abrasions 4 - safety pins or clips to anchor triangular bandages Sterile compress bandages to dress wounds and/or control bleeding (four each: 2″ × 2″, 3″ × 3″, 4″ × 4″; and two 4″ × 8″) 1 - 2″ × 9″ roll of adhesive tape to anchor dressing 6 - pressure bandages (8″ × 10″) to control bleeding, splint fracture 2 - rolls of elastic wrap to anchor dressings 1 each - 1″, 2″, 6″ wide rolls of gauze bandages 5 - clean plastic bags (one garbage, 2 kitchen, 2 bread-sized) to transport amputated tissue
For poisonings	Emergency and poison control center number Syrup of Ipecac (use only if advised by medical professional) Tongue depressors Blanket for treating shock Rubber or latex gloves 2 - quart containers of clean, potable water 2 - small, clean plastic jars with tight-fitting lids 1 - can of evaporated milk with attached can opener or pull top lid
Additional supplies	Burn dressings in various sizes Burn treatment for use on minor burns only Antibiotic treatment Analgesic (should not contain ingredients known to cause drowsiness) Sterile eye wash solution Eye coverings Several small packages of sugar Mouth protection for mouth-to-mouth resuscitation Nonperishable food items in pest-resistant packaging
Larger first-aid kits should be located at main farm or ranch buildings or in the home. Smaller first-aid kits should be kept on major pieces of farm equipment and in vehicles.	

Goodheart-Willcox Publisher

Figure 4-51. Use this list to check your first-aid kits to ensure they are properly supplied.

Lesson 4.3 Review and Assessment

Words to Know

Match the key terms from the lesson to the correct definition.

1. Animal waste storage area that may be an open lagoon, an earthen basin, or be contained in a closed building.
2. Condition in which overexposed body tissue becomes cold and red, then numb, hard, and pale.
3. A motorized vehicle with four low-pressure tires designed to travel over various types of land.
4. A condition that occurs when your body is unable to release enough heat, and your temperature rises to life-threatening levels.
5. Caustic liquid used as a source of nitrogen fertilizer.
6. A condition that occurs when your body loses heat faster that it can produce it.
7. A cylindrical metal rod that is attached to a power source (tractor) at one end and an attachment at the other.
8. A highly toxic gas with a strong bleach-type odor, low-lying yellow, red, or dark brown fumes.
9. Muscular spasms that occur when a person is sweating a lot and not replacing lost fluids.
10. A person who helps guide the driver of farm equipment and work vehicles to prevent accidents.
11. A condition in which a person has a slightly higher than normal body temperature caused by lack of fluid replacement in hot weather.
12. The operator compartment structures intended to protect operators from injuries caused by vehicle overturns.

A. all-terrain vehicle (ATV)
B. anhydrous ammonia
C. frostbite
D. heat cramp
E. heat exhaustion
F. heatstroke
G. hypothermia
H. manure pit
I. nitrogen dioxide (NO_2)
J. power take off (PTO) shaft
K. rollover protection system (ROPS)
L. spotter

Know and Understand

Answer the following questions using the information provided in this lesson.

1. Identify three resources for acquiring farm safety materials and farm safety training.
2. What are four types of personal protective equipment?
3. Explain why it is important to read MSDS for materials you handle while working.
4. Identify who may transport, transfer, and apply anhydrous ammonia.
5. Identify five elements that may contribute to farm injuries.
6. *True or False?* Carelessness, poor attitude, or lack of training are often contributing factors when it comes to accidents involving machinery.

7. Explain why you would have a form of visual communication in place in an agricultural setting.
8. List five precautions you should take when driving and operating tractors and trailing equipment.
9. Explain five precautions that should be used when working around or with a PTO.
10. Explain how to prevent entangling farm machinery in overhead power lines.
11. Explain the difference between heatstroke and heat exhaustion.
12. Identify three situations in which accidents with handling livestock commonly occur.
13. Explain why you should not have extra riders on an ATV.
14. Identify the proper safety precautions that should be taken before a farm worker enters a manure pit.
15. Explain why you must never enter a grain bin while unloading equipment is running.
16. Explain why it is important for farm/ranch residents and employees to receive regular first aid and CPR training.

Analyze and Apply

1. Develop a series of hand signals for use on your farm. Go online and search for existing hand signals and teach them to your family members and coworkers.
2. Research the use of color codes in shops, labs, and factories. Determine how safety color coding could be used on your farm or place of employment.

Thinking Critically

1. Go online and research the rules for youth employment in agricultural operations. What are the exceptions? Are they appropriate? Why or why not? What other rules or exceptions do you think should be made?
2. What are the age restrictions for operating ATVs? What is the youngest age allowed? Is it appropriate? Should all ATV operators pass a driving test? Why or why not?

Chapter 4

Review and Assessment

Lesson 4.1

Occupational Safety and Health

Key Points

- Workplace safety is no accident. You need to know the rules about when and where you can work, and the types of work tasks you can legally and safely do.
- If you have any questions about how to do your job safely and effectively, ask your employer, agriculture teacher, or parents.
- Make sure that you have an SAE training plan that describes the type of work you are to perform in the job.
- Report unsafe working conditions to your employer.

Words to Know

Use the following list and the textbook glossary to review and study the *Words to Know* from *Lesson 4.1*.

apprentice
Fair Labor Standards Act (FLSA)
horsepower
indentured servant
Occupational Safety and Health Administration (OSHA)
YouthRules!

Check Your Understanding

Answer the following questions using the information provided in *Lesson 4.1*.

1. Production agriculture employs almost _____ people in the United States.
2. Identify three child labor laws. Which of the three is still in force today?
3. Explain the purpose of the Occupational Safety and Health Administration.
4. *True or False?* Young people can work in any agricultural job as long as it is on a farm where the youth is supervised by a parent or designated adult.
5. As a 14- or 15-year-old, you may operate a tractor as long as you have completed which training program?
6. *True or False?* Once you reach the age of 16, you may work in any job.

7. List five agricultural occupations identified as hazardous by the FLSA.
8. If you are 14 or 15, you can work _____.
   A. any day, any time of day, and for any number of hours
   B. outside school hours, provided you do not work before 7 am
   C. five hours on a school day, and 10 hours on a nonschool day
   D. whenever you choose between 7 am and 7 pm
9. Explain why it is important to ask your employer questions about new assignments or tasks.
10. Explain what you can do to help maintain a safe work environment.

Lesson 4.2

Shop and Lab Safety

Key Points

- Agricultural labs and shops provide opportunities for learning agricultural mechanics, metal fabrication, and carpentry skills.
- Agricultural labs and shops contain many potential dangers, and procedures must always be followed. Learning and applying the safety rules in the shop and lab is the professional approach to any job.
- Personal protective equipment (PPE) such as safety glasses, hearing protection, dust masks, and protective clothing are all "tools of the trade" in agriculture. Always wear the appropriate personal protective equipment.
- It is important to follow all safety procedures given by your shop teacher or lab instructor in order to keep the lab or shop a safe place to work. Know the safety rules, as well as the procedures to follow in the event of an accident.
- Know the location and proper use of fire extinguishers.
- Review the MSDS for hazardous products you may be using in the shop or lab.
- Practice proper and safe use of all electrical equipment.
- Legal regulations for ladder and scaffolding use are in place to protect users.
- Learning first aid and keeping properly supplied first-aid kits are important means to keeping the lab and shop safe places to learn and work. Keep a well-stocked first-aid kit available and in clear view of all shop and lab workers.

Words to Know

Use the following list and the textbook glossary to review and study the *Words to Know* from *Lesson 4.2*.

air-purifying respirator
atmosphere-supplying respirator
circuit breaker
filtering facepiece (dust mask)
gas mask
ground
ground fault circuit interrupter (GFCI)
material safety data sheet (MSDS)
personal protective equipment (PPE)
powered air-purifying respirator (PAPR)
safety data sheet (SDS)
self-contained breathing apparatus (SCBA)
short circuit
shutoff switch
supplied-air respirator (SAR)

Check Your Understanding

Answer the following questions using the information provided in *Lesson 4.2*.

1. *True or False?* Personal protective equipment includes equipment to protect your eyes, hearing, and lungs.
2. Explain why sunglasses must not be used in a shop or lab setting for eye protection.
3. What is the purpose of the Occupational Safety and Health Administration with regard to agricultural shops and labs?
4. List four types of protective respiratory equipment.
5. What is the main cause of accidents with regard to hand tools?
6. Explain the purpose of an emergency shutoff switch.
7. Fire extinguishers classified as "ABC" are effective against which types of fires?
8. What is the purpose of the material safety data sheets (MSDS)?
9. Explain the purpose of the ground wire on an electrical plug.
10. What is the purpose of a ground fault circuit interrupter (GFCI)?
11. If you are under the age of 16 and working in a nonagricultural job, at what height can you work from a ladder?
12. What is the three-point method used in climbing ladders?
13. Explain the proper method for climbing a ladder.
14. Why should you never stand on the top rung of a ladder?
15. List five places first-aid kits may be kept.

Lesson 4.3

Farm and Work Safety

Key Points

- Take the time to do the job safely. Trying to do things in a hurry is a recipe for disaster.
- The proper personal protective equipment must be worn for all work performed on a farm or other agricultural operation.
- It is vital to worker safety to know and understand the safety rules to follow when working around farm machinery.
- PTO shafts are very dangerous and all workers must follow safe practices to prevent entanglement in rotating PTO shafts.
- All workers must be aware of potential electrical accidents in agricultural settings and how to avoid commonly occurring accidents.
- Ensuring safety precautions are taken when working around livestock will prevent injuries to both handlers and animals.
- Proper training and the employment of safety precautions will help ensure ATV operator safety.
- Waste storage and management facilities can be very dangerous to the health of humans and animals alike. Extreme safety measures must be taken to ensure the health and safety of workers and animals.
- Silos and grain bins also present many hazards to workers and animals, and proper safety measures must be taken to ensure worker and animal health and safety.
- All workers should be familiar with safe pesticide handling in order to be properly prepared in the event of an accident.
- First-aid and CPR training is recommended for all family members and workers employed in agricultural operations.

Words to Know

Use the following list and the textbook glossary to review and study the *Words to Know* from *Lesson 4.3*.

all-terrain vehicle (ATV)
ammonia (NH_3)
anhydrous ammonia
carbon dioxide (CO_2)
frostbite
heat cramp
heat exhaustion
heatstroke
hydrogen sulfide (H_2S)
hypothermia
manure pit
methane (CH_4)
nitrogen dioxide (NO_2)
power take off (PTO) shaft
rollover protection system (ROPS)
silo gas
spotter

Check Your Understanding

Answer the following questions using the information provided in *Lesson 4.3*.

1. Explain why you should read and understand the MSDS for any hazardous material being used where you work or live.
2. *True or False?* It is against the law for anyone under the age of 16 to transport, transfer, or apply anhydrous ammonia.
3. Elements that contribute to farm injuries include _____.
   A. gas and chemical hazards
   B. electrical, water, and road hazards
   C. livestock handling and machinery
   D. All of the above.
4. The lack of _____ is often to blame for injuries sustained on farm equipment.
5. *True or False?* Your reaction time decreases with increased fatigue.
6. List and explain three driving actions that will increase the likelihood of a tractor overturn.
7. *True or False?* Most PTO accidents occur when clothing or limbs are entangled in the rotating PTO shaft.
8. *True or False?* You should not see any part of the moving shaft when looking at a PTO in operation.
9. List three common reasons electrical accidents occur in agricultural settings.
10. Explain why it is important for all electrical work to be done by a qualified electrician.
11. List five types of injuries or illnesses associated with hot and cold weather conditions.

Chapter 4 Skill Development

STEM and Academic Activities

1. **Social Science.** Colors have certain associations, such as red for danger and yellow for caution. Research the emotional associations of color, such as blue conveying calm skies or water. Make a list of 10 colors and take a survey of at least 15 people. Ask each person to respond with one word that describes their association to each of the colors. Collate your results to see if you can find any patterns. For example, did people in different age groups respond differently to certain colors? Write a short report describing your results and any conclusions you were able to make.
2. **Language Arts.** The Federal Occupational Safety and Health Act requires the use of fall protection systems on many types of construction projects where people will be working above the ground. Contact the nearest OSHA office or go to the OSHA website and find information on the circumstances where fall protection systems must be used, what those systems consist of, and how they work. Present an oral report to the class.
3. **Language Arts.** List three potentially hazardous products commonly used on a crop production operation. Obtain safety data sheets (SDS) for these products. SDS can often

be downloaded on the Internet or obtained from retailers that sell the products. Some SDS might be available in your school. Write a report on one of those products, explaining the information you found on the SDS and recommendations for working with that product.

Communicating about Agriculture

1. **Reading and Speaking.** With a partner, make flash cards for the key terms in Chapter 4. On the front of the card, write a term. On the back of the card, write the pronunciation and a brief definition. (You may consult a dictionary.) Then take turns quizzing one another on the pronunciation and definitions of key terms. Practice reading aloud the terms, clarifying pronunciations where needed.
2. **Writing and Speaking.** Make a series of safety posters for working in various areas on the farm. Posters may cover working with machinery, driving a tractor, loading and unloading grain, working with livestock, using an ATV, working around manure storage facilities, or other areas of your choosing. Explain your posters to the class.
3. **Reading and Speaking.** As the construction foreman in charge of a work crew, it is your responsibility to ensure the safety of your workers in the event of a fire. You must post signs in the work area to educate your workers on the different types of fires and how to extinguish them. Research the types of fires and categorize them as Class A, Class B, or Class C. Create your signs in the form of a presentation. Share the presentation with the class, as though the class were your crew. Ask for *and* answer any questions your crew may have.
4. **Writing and Speaking.** Search online for "farm safety walkabout" worksheets or a workbook. Perform the walkabout on your family farm, the school facilities, or your place of employment. After completing the evaluation, come up with a reasonable plan to correct any hazards or unsafe practices you have found. Present your plan to your family, your instructor, or your supervisor. If possible, implement your plan and evaluate its success.

Extending Your Knowledge

1. Map it out. Use a sheet of paper to draw an overhead diagram of the agricultural shop or similar facility at home, on the farm, or the place where you work. Label the location of the fire exits, fire alarms, walkways, and fire extinguishers. Label the location of the eye wash station if the facility has one. Share and compare your drawing with that of another student. Did you identify all of the safety equipment and exits? What safety features are missing?
2. Obtain a copy of a safety data sheet (SDS) from the Internet or your instructor. Answer the following questions:

   What is the name of the chemical?

   What is its melting point or boiling point?

   At what temperature will the chemical burst into flames?

   What is its level of toxicity?

   How could exposure to this chemical affect your health?

   What is the recommended first-aid treatment in case of injury?

   Where should this chemical be stored?

   What is the safest method of disposing of this chemical?

   What should you do if this chemical is spilled accidently?

Chapter 5

Agriculture Science

Lesson 5.1
Agriscience and the Scientific Method

Lesson 5.2
Practical Science in Agriculture

Lesson 5.3
Biotechnology in Agriculture

©iStock/monkeybusinessimages

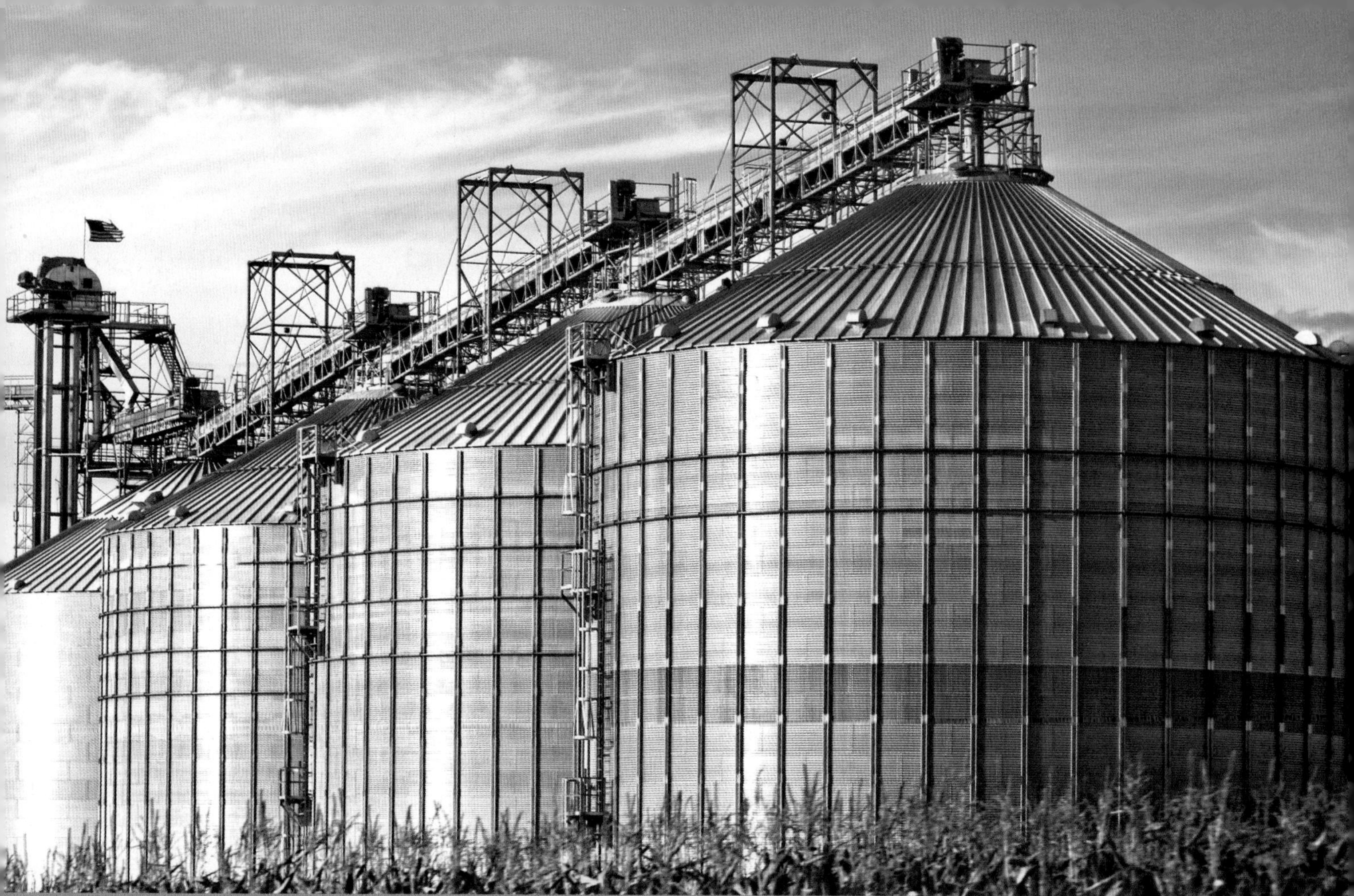

Stuart Monk/Shutterstock.com

©iStock/praisaeng

©iStock/chemicalbilly

While studying, look for the activity icon to:

- **Practice** vocabulary terms with e-flash cards and matching activities.
- **Expand** learning with video clips, animations, and interactive activities.
- **Reinforce** what you learn by completing the end-of-lesson activities.
- **Test your knowledge** by completing the end-of-chapter questions.

Lesson 5.1

Agriscience and the Scientific Method

Words to Know

agriscience
chart
constant
control
dependent variable (DV)
experiment
graph
hypothesis
independent variable (IV)
logbook
scientific method
treatment
trial
variable

Lesson Outcomes

By the end of this lesson, you should be able to:

- Describe scientific methods of research.
- Identify and apply research in agriculture, food, and natural resources.
- Differentiate between facts and inferences.
- Design a testable experiment related to an agricultural field of study.

Before You Read

As you read the lesson, record any questions that come to mind. Indicate where the answer to each question can be found: within the text, by asking your teacher, in another book, on the Internet, or by reflecting on your own knowledge and experiences. Pursue the answers to your questions.

Did you know that a gate made only of duct tape is strong enough to stop a car moving at 50 miles per hour? Did you know that if you interweave the pages of two phonebooks, they cannot be pulled apart, even by two trucks? Did you know that you cannot make your right foot move in a clockwise circle while drawing a number six with your right hand? It is human nature to be curious and want to test theories to reach conclusions.

This lesson deals with agriscience and the ways in which you can go about designing and testing research questions related to agriculture. We will first look at exactly what agriscience is and why it is important to the future of agriculture. Then you will have the chance to look at all the parts of an experiment in order to develop your own skills related to becoming a good researcher.

Agriculture as a Science

In *Lesson 3.2*, we discussed the term ***agriscience*** as the research and development of emerging agricultural technologies. It may also be described as the application of science to agriculture. Agriculture is an industry that is deeply rooted in scientific principles. In *Lesson 5.2*, we will examine the interaction between the different fields of science and agriculture. Science helps agriculture by providing the facts and information that lead to new technology, products, and practices that will increase the effectiveness of the agricultural industry, **Figure 5-1**.

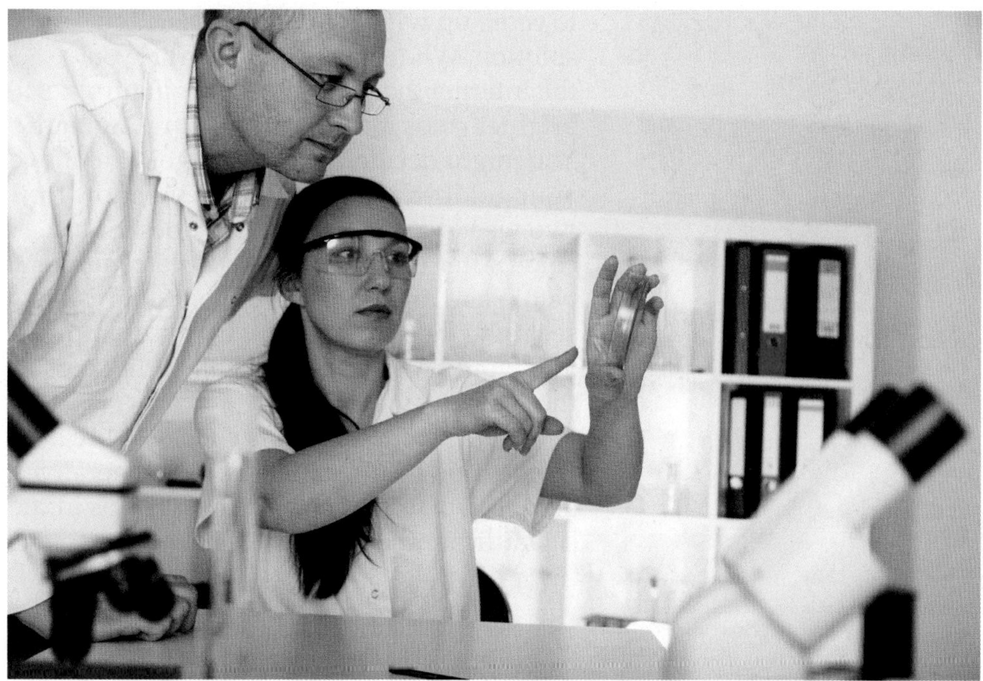

Alex011973/Shutterstock.com

Figure 5-1. Agriscience involves the scientific development of research questions related to agriculture. *Can you think of an agricultural topic you might like to investigate?*

Before we can discuss the application of scientific principles to agriculture, we should first examine the basics of research using scientific methods.

The Scientific Method

The ***scientific method*** is a means of research that uses systematic rules and procedures to investigate a problem. It includes the entire research process, including:

- Locating facts.
- Making observations.
- Measurement.
- Experimentation.
- Formulating and testing a hypothesis.

Using the scientific method ensures that research leads to reliable and true results, **Figure 5-2**.

Air Images/Shutterstock.com

Figure 5-2. By paying careful attention to the steps in the scientific method, you will have the opportunity to create a science experiment that gives you the most accurate answer to your research question.

You probably use the scientific method to make decisions on a daily basis without even thinking about it. For example, let us say that you notice your goldfish does not look as bright orange as he has in the past, **Figure 5-3**. You could rely on the information you know about your goldfish, or you could research goldfish

Richard Lyons/Shutterstock.com

Figure 5-3. What could be the possible causes of your goldfish losing his brilliant color? *How do you think the scientific method could help you solve this problem?*

to come up with a possible reason and/or solution. What if you read an article on the Internet that says goldfish will turn a brighter color if exposed to natural sunlight? You might decide that you are going to move your goldfish tank closer to the window for two weeks, and see what that does for his color. After two weeks, you could compare before and after pictures of your goldfish to see if the theory was right. In this example, you would have completed all of the steps in the scientific method.

While some scientists choose to break the method into more steps, it is generally accepted that there are five basic steps in the scientific method, **Figure 5-4**:

1. Identify the Problem
2. Formulate the Hypothesis
3. Conduct an Experiment to Test Hypothesis
4. Collect and Analyze Data
5. Reach a Conclusion

Step 1: Identify the Problem

Make Observations. All scientific tests begin with an observation of something that the researcher cannot explain. For example, for someone raising lambs, an identified problem could be that his or her lambs are

Career Connection Agricultural Research Experiment Station Director

Conducting experiments to advance agriculture

Job description: Through the Hatch Act in 1887, each state received funding for agricultural experiment stations. The directors of these stations work to manage the field research and data collection for the experiments being conducted at their sites.

Education required: Master's degree or higher in Agricultural Education or Extension.

Job fit: This job might be for you if you love agriculture and conducting science experiments, and are good at managing people.

State Library and Archives of Florida

These boys were all members of the 4-H Corn Club in Gainesville, Florida, circa 1911. The corn on display in the experiment station building was grown by the club members.

not growing quickly and may not reach market weight in time for the show.

Refine Observations. The lamb owner would take the information he or she is unsure about and make additional observations about the topic. For example, the owner must determine why the lambs are not growing quickly to amplify their growth and weight gain. The owner may decide he or she needs to try a feed supplement to help get the lambs to market weight before the show, **Figure 5-5**.

Research the Topic. Before determining the research question, the lamb owner should draw on the information he or she knows about the topic, and/or research new information to get a better idea of things that may be influencing the problem. A good example of this would be the lamb owner researching nutritional factors that lead to faster lamb growth rates.

Research Question. The final research question should be specific enough that it can be used to design a testable experiment. For this example, the lamb owner's research question might be:

- Which feed supplement will lead to the greatest rate of gain in market lambs?

Some research questions can be answered with a single experiment, like the lamb problem. More complex questions might require multiple rounds of testing. Think about the number of research questions that scientists must work with trying to cure a major disease, **Figure 5-6**.

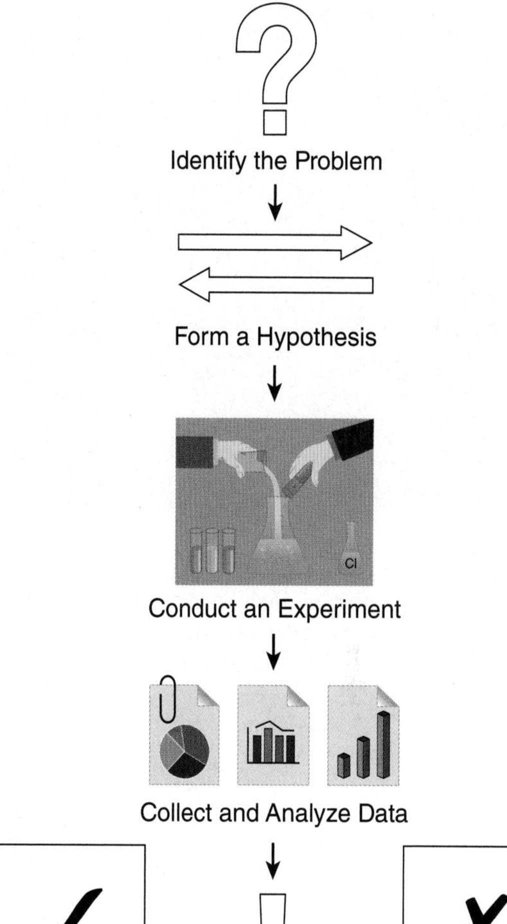

Goodheart-Willcox Publisher

Figure 5-4. Using the scientific method ensures that you are making conclusions based on factual data. *Why would making this type of decision be helpful?*

Step 2: Formulate the Hypothesis

A ***hypothesis*** is an educated prediction of the outcome of a scientific experiment. To ensure the credibility of your hypothesis, make sure you have documented background information and viable reasons your hypothesis may explain the unknown information in your research question. For example, the lamb owner (from step 1) might use the claims made by a feed supplement manufacturer that says lambs fed its brand will gain more per day than those being fed another manufacturer's supplement.

Step 3: Conduct an Experiment to Test the Hypothesis

Once you have identified the problem and formulated a hypothesis, you can design an experiment to test your hypothesis. An ***experiment*** is a controlled scientific test to determine the validity of a hypothesis.

nomadphoto/Shutterstock.com

Figure 5-5. Depending on the manufacturer, livestock feed contains most nutrients needed for animals to thrive. *How could using the scientific method help you determine if a feed supplement would increase the daily rate of gain in your market lambs?*

There are many considerations to take into account when designing an experiment, including:

- Variables
- Treatments
- The control
- Constants
- Trials

The best place to begin designing an experiment is by determining how to isolate the problem so that you are testing only the hypothesis. In the lamb feeding experiment, it may be best to split the lambs into smaller groups so they could be fed different supplements. This would allow the comparison of the different groups and how the supplements affected them.

Variables

Variables are the things that will change over the course of the experiment. There are two basic types of variables in any scientific experiment, independent and dependent. ***Independent variables***, sometimes called ***IVs***, are the things the researcher changes in order to test the hypothesis. For example, in the lamb experiment, the independent variable is the type of supplement. The different levels of the independent variables are called the ***treatments***. If you decided to test *Super Amazing*, *Co-op*, and *Local Producer* brand supplements in determining your lamb growth rates, these three things would be your treatments.

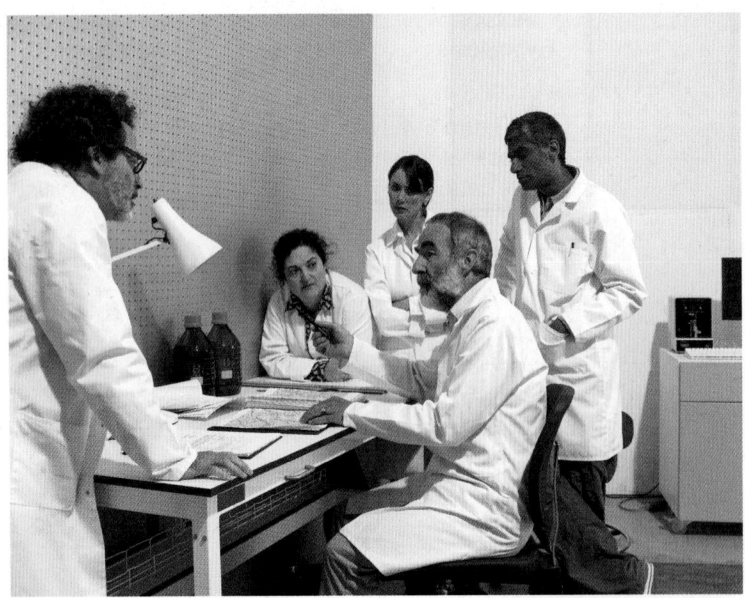

Micahel Blann/Digital Vision/Thinkstock

Figure 5-6. Complicated problems often require teams of scientists working together for years in order to research all of the different components. *Can you think of some scientific endeavors that would require this level of scientific investigation?*

The other type of variable is a dependent variable. ***Dependent variables***, or ***DVs***, are the variables that change because of the outcome of the experiment. Essentially, the DV is your measurement, or how you will know which portion of the experiment performed best, **Figure 5-7**. In the lamb feeding experiment, the dependent variable would be the weight gained by the lambs per day in pounds. Understanding the difference between IVs and DVs is one of the most important parts of designing an experiment. An easy way to remember the difference is:

- *"I"* change the IV.
- The DV *"depends"* on what treatment has the most effect.

Control

The ***control*** in an experiment is a section of the experiment that does not receive treatment. For the lamb feed experiment, the control would be a group of lambs that would not be fed a supplement. The control group is used to compare results to ensure that it was actually the treatment that caused the change in the DV. Having a control allows the researcher to decide if the results are based on the IV rather than an external factor.

Goodheart-Willcox Publisher

Figure 5-7. If your research involves the best feed to make lambs grow, weight gain in the lambs would be your dependent variable. These scales are the most common way to weigh lambs. ***Have you ever had the chance to use these?***

Constants

Constants are things that are kept the same between the experimental groups to ensure that only one thing is being changed at a time. What sort of things would you want to make sure are the same between the groups of lambs in the feed experiment? Did you consider the following:

- Did you think about having lambs that were the same age, gender, and breed?
- How about making sure that the lambs were being fed the same amount of feed and feed supplement?
- Should they all have the same environment including exercise, pen structure, and access to water?

There are many things that would need to be held constant in this experiment to make sure that you could determine the outcome was based on the feed supplement, and not on other factors.

Ethics in AG

Ethical considerations exist in the conducting of information. Making sure you collect all data from the experiment and do not alter it is important to maintaining both the integrity of the experiment and your own character. It is never okay to make up data or omit results that do not support your hypothesis.

Trials

Because researchers cannot always control all the external factors that affect their research, research is often conducted with multiple trials. The term ***trials*** refers to the number of organisms the treatments are applied to, or the number of times the experiment is repeated. Trials are always counted as the *number per treatment*. In our lamb feed example, if there were 5 lambs fed each of the supplements, the number of trials would be 5 trials per treatment. By collecting data multiple times on the same experimental topic, researchers have a larger amount of evidence from which to draw their conclusions.

Did You Know?

Falsifying results of scientific experiments is not only unethical, but also illegal. One researcher from Iowa State University was sentenced to more than 20 years in prison for falsifying the results of a project he was working on relating to the development of an HIV/AIDS vaccine. Many other researchers have lost their credentials and faced lawsuits due to falsifying data.

Step 4: Collect and Analyze Data

During the experiment, it is important to collect all the data on the outcomes, constants, control, and any other information that may have affected the outcome of the experiment. Collecting organized data is critical to being able to answer the research question, and to determine if the hypothesis was right or wrong. One of the best ways to collect information related to your experiment

Lisa F. Young/Shutterstock.com

Figure 5-8. Keeping a detailed logbook of your scientific experiment, including all your results and observations, will allow you to analyze your results in more detail. *What are some of the things that would need to be kept in the logbook for the lamb feeding experiment?*

is with a logbook. A ***logbook*** is a notebook containing detailed notes of your project, **Figure 5-8**. This would include all information related to the methods, management, and outcomes of the experiment. Keeping an accurate and up-to-date logbook is essential to the success of any scientific experiment.

How data is analyzed depends greatly on the type of data you collect. For numerical data, the analysis often involves comparing the averages of different groups to the control or to the other treatments. You might also want to analyze your information related to some of the factors that you could not control for. For example, in our lamb feeding experiment, it might be worthwhile to analyze the information for the supplement performance related to the gender of the lamb, to see if one supplement worked better than the others when fed specifically to ewes rather than wethers, or vice versa.

Facts and Inferences

When collecting data, be certain that you are making the distinction between a fact and an inference. Facts are those items which are actually observable in the data, while inferences are guesses about a situation based on interpretation of facts. True scientific research should always be based in fact, rather than inference.

FFA Connection The National FFA Agriscience Fair

In order to encourage agriculture students to pursue scientific research in agriscience areas, the National FFA Organization has a competition specifically tailored to student agriscience projects.

To compete, FFA members develop and complete an agriscience project, keep a logbook of their experience, and complete a written paper on their methods and results along with creating a display for judging. Judges evaluate the paper, the display, and student responses to questions about their project.

The agriscience fair allows students to compete in one of six categories: animal systems; environmental services/natural resource systems; food products and processing systems; plant systems; power, structural, and processing systems; and social systems.

There are four divisions:

I. Individuals in grades 7–9

II. Individuals in grades 10–12

III. Teams in grades 7–9

IV. Teams in grades 10–12

Awards are given in each division and category (24 in total) at the local, regional, state, and national levels. If you are interested in the National FFA Agriscience Fair, ask your agriculture science teacher how you can start the process.

Step 5: Reach a Conclusion

After analyzing the data from the experiment, you can make a determination of the outcome based on your hypothesis. This conclusion should include the information you found, along with the reasoning behind what you found. A good conclusion includes all the results and relates them to the hypothesis. It also includes information about how these results relate to the original problem being investigated.

Communicating Results

One of the major considerations in developing agriscience experiments is the method for communicating and presenting information gathered through the research process. The most common methods for communicating results are through written reports, graphs, and charts.

Written Reports

For some research studies, the most practical method for sharing results is through a written report. Written research reports, including those for the National FFA Agriscience Fair, generally include the following sections:

- Abstract—a brief summary of the research and findings.
- Introduction—gives the reason that this study was conducted.
- Literature review—includes information on the topic and related studies that have already been completed in this area.
- Methods—describes the experimental process in detail.
- Results—gives all of the findings of the study.
- Discussion/conclusion—shares the implications of the findings.
- References—a complete list of the sources that were used to develop the background and methods information.

Graphs and Charts

A ***chart*** is information shown in a diagram form rather than in a written narrative. A ***graph*** is a type of chart that shows the relation between two or more variables. There are many different types of charts that are important in showing the results of scientific research. Selection of a specific type of chart often depends on the type of data and the audience to which the information is being presented. Some common types of charts and their applications are shown in **Figure 5-9**.

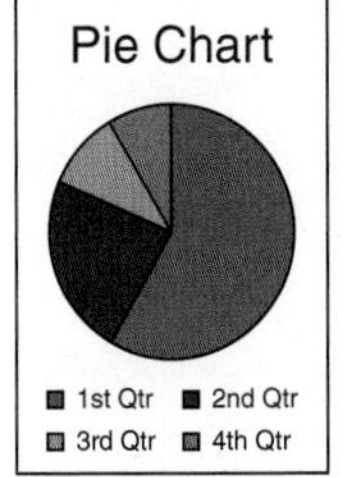

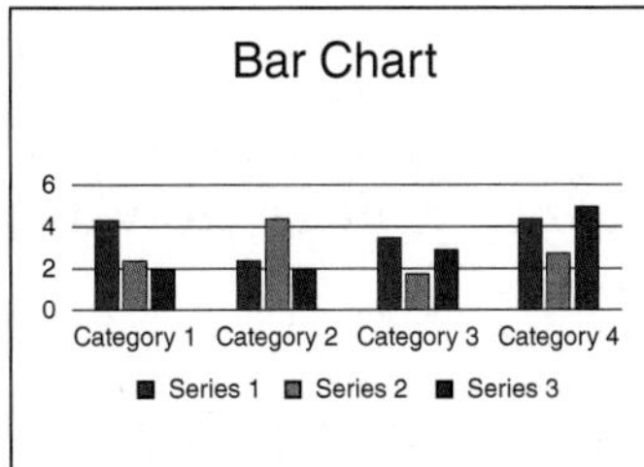

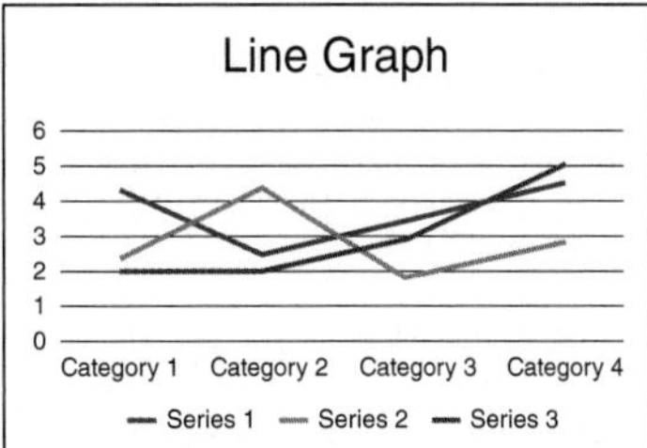

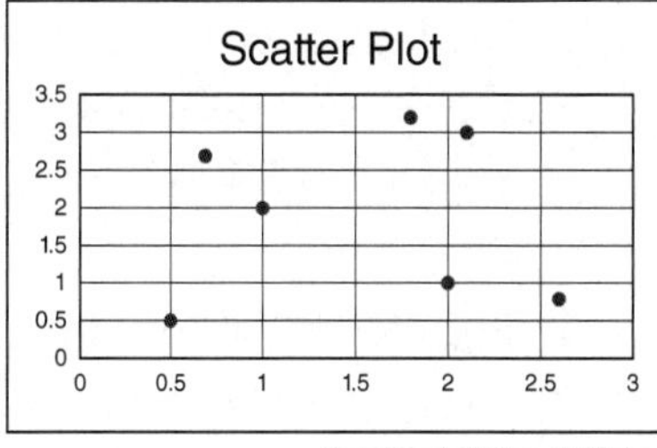

Goodheart-Willcox Publisher

Figure 5-9. Communicating your findings is a necessary step in scientific experimentation. *How do you think charts and graphs help aid the communication process?*

Lesson 5.1 Review and Assessment

Words to Know

Match the key terms from the lesson to the correct definition.

1. Something that will change over the course of the experiment.
2. Information shown in a diagram form rather than in a written narrative.
3. A notebook containing detailed notes of a scientific experiment.
4. An educated prediction of the outcome of a scientific experiment.
5. A section of a scientific experiment that does not receive treatment.
6. The different levels of the independent variables in a scientific experiment.
7. The number of times the experiment is repeated.
8. A type of chart that shows the relation between two or more variables.
9. Something that the researcher changes in order to test the hypothesis.
10. Something that changes because of the outcome of a scientific experiment.
11. Something that is kept the same between experimental groups to ensure that only one thing is being changed at a time.
12. A means of research that uses systematic rules and procedures to investigate a problem.
13. A controlled scientific test to determine the validity of a hypothesis.

A. chart
B. constant
C. control
D. dependent variable (DV)
E. experiment
F. graph
G. hypothesis
H. independent variable (IV)
I. logbook
J. scientific method
K. treatment
L. trial
M. variable

Know and Understand

Answer the following questions using the information provided in this lesson.

1. Explain how science helps agriculture.
2. What is at the beginning of all scientific tests?
3. Give an example of how you might use the scientific method to make an everyday decision.
4. Identify the five basic *steps* used in the scientific method.
5. Give an example of a possible problem (that could be used to create an experiment) for each of the following agricultural areas: animal production, water/irrigation, and crop production.

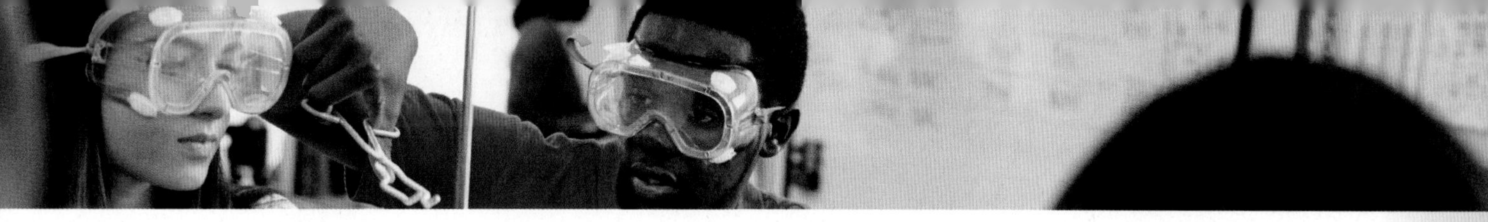

6. Formulate a hypothesis for the lamb feed experiment used as an example in this lesson.
7. Identify four considerations that must be taken into account when designing an experiment.
8. Explain the difference between independent variables and dependent variables.
9. *True or False?* Omitting minor data from an experiment is okay when it does not support your hypothesis.
10. Explain what is included in a "good conclusion" to a scientific experiment.

Analyze and Apply

1. You have two healthy houseplants that are at the same height. Please design an experiment that will determine whether or not Miracle-Gro™ will help them to grow faster. *Do not write in the textbook.* On a separate sheet of paper, use the following format to write your experiment. Explain exactly what you would do in this experiment.
   1) Problem: ______________________
   2) Hypothesis: ______________________
   3) Experiment Design
      A) Control: ______________________
      B) Test: ______________________
   4) Explain the experiment: ______________________
2. Frankie owns a trout farming operation. Frankie has heard that global warming is making the world's oceans and natural waters become warmer. He wants to know how water temperature affects the growth of fish. Please design an experiment to explore Frankie's observation. Use the following outline format to write out your experiment. Explain exactly what you would do in this experiment.
   1) Problem: ______________________
   2) Hypothesis: ______________________
   3) Experiment Design
      A) Control: ______________________
      B) Test: ______________________
   4) Explain the experiment: ______________________

Thinking Critically

For each of the scenarios, please find the requested information.

1. **Scenario One.** The effect of fertilizer on plants.

   Description. Edwards's agriculture class was studying various ways to recycle materials, including the use of compost as fertilizer. Edward decided to compare the effect of cafeteria compost and commercial compost on plant growth.

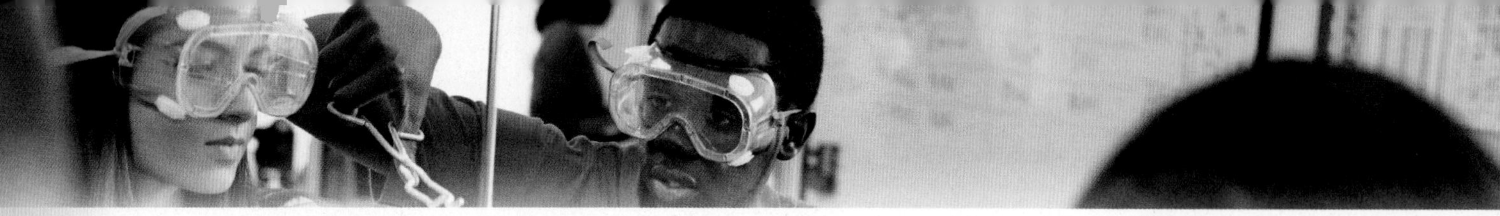

Three flats of pea plants (25 plants per flat) were grown for five days. The plants were then fertilized as follows:

Flat A received 10g of cafeteria compost.

Flat B received 10g of commercial compost.

Flat C received no compost.

The plants received the same amount of sunlight and water each day. At the end of 20 days, he recorded the height of the plants in centimeters.

A. Problem

B. Hypothesis

C. IV

D. Treatments

E. Trials

F. DV

G. Constants

2. **Scenario Two.** The death of catfish.

Description. You are concerned about the rate of death in catfish in your aquaculture operation tanks that have a high amount of algae growth. You collect some of the algae and send it to the university to have it identified. You find that it is a blue-green algae called *Anabaena*. *Anabaena* is known to be to toxic to fish. You design an experiment to test how much of the algae the fish can stand before they die.

You obtain six large aquariums (all the same size) and fill them with water taken from a healthy catfish pond at your operation. You put each aquarium on the same bench in the laboratory, where the light and temperature values are identical. You let the water stand for one day before starting the experiment.

Aquarium 1: 2 grams of algae and no fish

Aquarium 2: 20 catfish and no algae

Aquarium 3: 20 catfish and 2 grams of algae

Aquarium 4: 20 catfish and 4 grams of algae

Aquarium 5: 20 catfish and 8 grams of algae

Aquarium 6: 20 catfish and 16 grams of algae

The aeration rate of each aquarium is identical and the fish are the same size and age. You make two observations at the same time each day for two weeks. You keep track of the numbers of fish in each tank that die.

A. Problem

B. Hypothesis

C. IV

D. Treatments

E. Trials

F. DV

G. Constants

Lesson 5.2

Practical Science in Agriculture

Lesson Outcomes

By the end of this lesson, you should be able to:

- Describe the fields of science as they relate to agriculture.
- Explore the places where agriculture and science interact with each other.
- Explain the importance of using scientific principles in agriculture.
- Give examples of agricultural applications for scientific principles.

Words to Know 

biology
biotechnology
botany
chemistry
Earth science
economics
formal science
mathematics
natural science
physical science
physics
political science
social science
statistics
theoretical computer science
zoology

Before You Read

Think about the title of this lesson, then write a list of five to ten topics that you think should be covered related to the title. After reading the lesson, revisit the list and examine the differences between what you thought would be covered and the actual lesson content.

What comes to mind when someone says the word *science*? Do you imagine a scientist in a white lab coat working in a lab? Do you think of microscopes and test tubes? The typical view of science is something like the image in **Figure 5-10**. Now, take a moment to think how science applies to agriculture. Do you think about genetically modified organisms? How about cloned animals? The various ways that science and agriculture interact may surprise you. This lesson covers practical uses for science in agriculture.

Fields of Science

Science is typically broken down into three broad areas, or fields: natural, formal, and social. These fields of science have subcategories that relate directly to agriculture.

- Natural Sciences:
  Physical Sciences—Physics, Astronomy, Chemistry, Earth Science
  Biology—Zoology, Botany, Biotechnology, Microbiology
- Formal Sciences—Mathematics, Theoretical Computer Science, Statistics
- Social Sciences—Education, Communication, Political Science, Economics

Natural Science

Natural science is the field of science that deals with the study of the physical world, **Figure 5-11**. It can be broken down into two main categories, physical science and biology. One of the defining characteristics of natural

anyaivanova/Shutterstock.com

Figure 5-10. *Do you think about a scientist conducting experiments in a sterile laboratory when you hear the word* science? Science can take place in many settings outside a laboratory.

Photodiem/Shutterstock.com

Figure 5-11. Natural science includes the investigation of everything in the natural world, including both the living and nonliving components. *What types of organisms would you encounter in the setting illustrated above?*

science is that it is based on observable and measureable data. Agriculture is deeply rooted in the principles of natural science.

Physical Science

Physical science is related to the study of inanimate objects. This field includes areas like physics, chemistry, and Earth science. The importance of physical science to agriculture is that it provides the means by which humans can interact with the natural world.

Physics

Physics involves the study of the properties of matter and energy and the forces and interactions they exert on one another. Physics is regarded as a fundamental science because all other natural sciences use and obey its principles and laws. Anytime energy is required or matter needs to be moved, physics is the science that provides the principles. The subject matters that fall under physics include mechanics, heat, sound, magnetism, electricity, and light and other radiation.

Think about the moving of agricultural materials and goods. Consider agriculture's involvement in energy creation and consumption. **Figure 5-12** highlights some ways that agriculture uses physics.

Chemistry

Chemistry is a branch of science that studies the composition, structure, properties, and change of matter. Simply put, it deals with the interaction of molecules to form different substances. Agriculture interacts with chemistry in a number of ways, including:

- The development of animal nutrition.
- How animals digest formulated foods.
- Maintenance and improvement of animal health.
- The development and use of chemicals for crop growth and pest management, **Figure 5-13**.

Physics and Agriculture
Designing New Equipment
Moving Animal Feeds
Creating Alternative Energy Sources
Power Systems Design
Product Packaging for Safe Delivery
Water Systems
Soil Structure
Hydraulic Systems

Goodheart-Willcox Publisher

Phillip Lange/Hemera/Thinkstock

Figure 5-12. Although physics apply to everyday life, special installations like this irrigation system in the desert in Dubai require a thorough understanding. *What principles apply to such an irrigation system?*

Chemistry and Agriculture
Enzyme Interaction for Food Taste
Food Preservatives
Food Quality
Pesticides
Herbicides
Biotechnology
Fertilizers
Microorganism Control
Animal Health

Goodheart-Willcox Publisher

Brian Brown/iStock/Thinkstock

Figure 5-13. Without chemistry, we would not have the benefits of insecticides and fertilizers.

Career Connection Food Scientist

Testing the food we consume on a molecular level

Job description: Food scientists examine human food supply products on a molecular level. These specialists not only make sure that the food we are consuming is safe and free of potentially harmful ingredients, but they also work on a molecular level to improve the taste and quality of foods. They may work in food processing facilities, at universities, or for privately owned laboratories.

Education required: A bachelor's degree or higher in food science, microbiology, chemistry, or a related field.

Job fit: This job may be for you if you enjoy doing scientific experiments, like the process of testing food, or are fascinated by the way that raw agricultural products are changed into consumer-ready goods.

anyaivanova/Shutterstock.com

Food scientists examine food quality and food safety.

Earth Science and Agriculture
Weather Broadcasts
Meteorology
Soil Testing Services
Mining Equipment
Irrigation Systems
Water Conservation
Soil Conservation
Commercial Ocean Fishing
Groundwater Testing
Erosion Control

Goodheart-Willcox Publisher

ra2studio/Shutterstock.com

Figure 5-14. With today's technology, we can access detailed weather broadcasts from our phone. *Can you think of at least five ways in which meteorology and weather are important to the agricultural industry?*

Did You Know?

Chemical Regulation. The use of chemicals is highly regulated by the USDA, Environmental Protection Agency (EPA), and other governmental agencies to ensure that they are safe.

If you have a good understanding of chemistry, you will develop a better understanding of how important chemistry is to growing crops and raising animals. You will learn that chemical reactions happen both naturally and through human manipulation. You will also understand how the proper use and application of chemicals can be beneficial to people and the environment, and how abuse of the same chemicals can be detrimental.

Earth Science

Another area of physical science that agriculture is very involved with is Earth science. ***Earth science*** involves the examination of the physical Earth and its atmosphere. This area of science includes topics like geology, oceanography, meteorology, and ecology, **Figure 5-14**. The most obvious place where interaction between Earth science and agriculture occurs is in natural resources management. Agriculturalists working with natural resources need to have an extensive understanding of how Earth works, and what geology could lead to deposits of certain types of natural resources.

Do you think that production agriculturalists need to be more concerned with the weather than the average person? Understanding climate is also an Earth science area that is important to agriculture. Tracking precipitation, temperature, and global climate change are all areas where agriculture and Earth science interact.

Biology

Aside from physical science, which is the area of natural science related to inanimate objects, the other main category of natural science is biology. ***Biology*** is literally translated to "the study of life." This field of science is the area that has perhaps the most overlap with agriculture because, as we learned in *Lesson 1.1*, agriculture is the human cultivation of plants, animals, and other life forms.

Multiple biological principles apply to agriculture, and biology can be broken down into two main areas, zoology and botany.

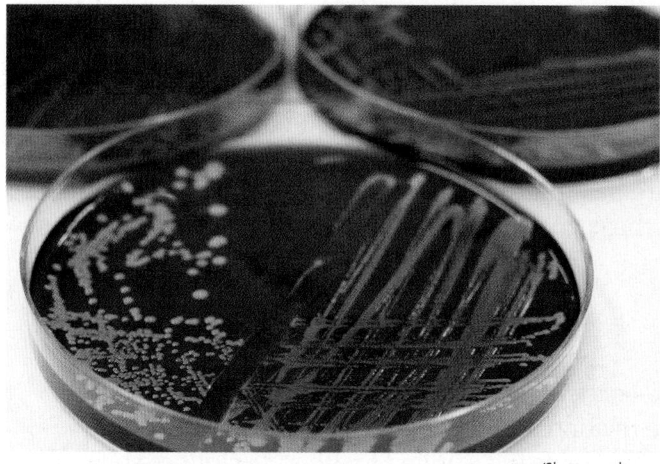
attem/Shutterstock.com

Figure 5-15. *When you think about animals, do you also think about bacteria?* These tiny animals are included in the field of zoology, and studying them is incredibly important to agriculture.

YanLev/Shutterstock.com

Figure 5-16. The study of all plants is botany. Agriculture involves the use of many different kinds of plants, each with their own special field of botany.

Zoology

Zoology is the scientific study of the behavior, structure, and physiology of animals. Animal agriculture has obvious interaction with this field of science; everything that happens from conception to death for animals in agricultural settings is rooted in zoological principles. The less obvious interaction between agriculture and zoology occurs with regard to microscopic animals like bacteria and protozoa, **Figure 5-15**.

Botany

Botany is the plant life section of biology and includes the study of plant structure, physiology, genetics, and classification, **Figure 5-16**. This area of science relates to agriculture because it is the science behind the growth and reproduction of all agricultural crops. If the agricultural industry is to grow the crops required to sustain human life on the planet, it will need to employ botanists or other people with extensive botanical knowledge. It is also important to understand that plant life, beyond what is being cultivated, is also vital to maintaining natural resources.

DVARG/Shutterstock.com

Figure 5-17. Biotechnology involves using science to alter organisms for the benefit of society. Agriculture has many different applications for biotechnology. *Can you think of all the ways that the DNA of strawberries could be changed for human benefit?*

Biotechnology

One of the areas where biology and agriculture intersect on an important level is through biotechnology. ***Biotechnology*** is the use of living systems and organisms to develop or make products useful to society. Agriculture is an industry that makes use of the principles of biotechnology in order to increase effectiveness and efficiency of raw agricultural products, **Figure 5-17**. Biotechnology and its relation to agriculture is discussed in *Lesson 5.3*.

Formal Science

Formal science includes disciplines concerned with abstract scientific concepts. Examples of formal systems are logic, mathematics, theoretical computer science, statistics, and guiding theories behind the systems we use on a daily basis. Agriculture has interactions with formal science as well. Can you think of places where these abstract science concepts might be found in the agricultural industry?

Mathematics

Mathematics is defined as the abstract science of numbers, quantities, and space or shapes and the relationships between them. Mathematics may be studied as pure mathematics, or as it is applied to other disciplines such as physics, chemistry, and engineering.

One area where the use of mathematics is very apparent is in the field of agricultural business. Agricultural businesses use math to help them determine profit share, market analysis, and general business operations like accounts payable and accounts receivable.

In agricultural engineering, engineers work with the principles of mathematics, specifically geometry, all the time. In fact, without a strong foundation in mathematics, the engineering for modern agricultural machinery and equipment would not be possible. Agricultural power systems rely on mathematics for information related to design as well as torque, horsepower, power distribution, and many other applications.

Production agriculture and agricultural processing are also deeply rooted in mathematics. For example, a feedlot worker must be able to calculate the amount of feed to put in a proper livestock ration, and a wheat farmer needs to be able to calculate his seed purchase for the year.

As you can see, agriculture is an industry that is very related to mathematics, and understanding math will be helpful to you as you pursue a career in an agricultural field, **Figure 5-18**. More specific mathematics in agriculture will be covered in *Chapter 8*.

Math and Agriculture
Calculating Planting Dates
Geometry in Agricultural Design and Fabrication
Cost of Production Calculations
Profit/Loss Margins
Balancing Animal Rations
Calculating Pesticide and Herbicide Concentrations
Interest Rates on Agricultural Loans
Soil Analysis
Calculating Area of Agricultural Land
Materials List for Agricultural Mechanics Projects

Goodheart-Willcox Publisher

Dan Comaniciu/iStock/Thinkstock

Figure 5-18. Yes, you will use the math you learn in school in your everyday life. This is just a small list of agricultural areas where math is used.

Theoretical Computer Science

Theoretical computer science is, in essence, the science behind creating computer programs. As agriculture becomes more and more technologically advanced, there are more and more ties to computer science. This area of formal science is where mathematics and logic come together to enable computers to process information. In agriculture, the way we plant, harvest, cultivate, process, and market agricultural products are all being advanced due to the hard work of computer scientists and programmers, **Figure 5-19**.

Fuse/Thinkstock

Figure 5-19. Theoretical computer science is a growing field in agriculture as new computer applications are developed to help produce food and fiber. *How do you think this producer can use his cell phone to help grow better crops?*

Statistics

Agriculture is an industry that relies on data to make decisions. ***Statistics*** is the formal science related to collecting and analyzing large quantities of numerical data and to relate findings to unknown data. Agricultural business uses statistics to determine the future markets for agricultural products. Agricultural companies and farms may use statistical information to help them decide the best time to harvest or sell their product. They may also use statistical data to determine how much of a commodity to plant, or how to market a new product to consumers.

Social Science

Social science is the study of human society and social relationships. Agriculture is unique in the fact that it is an industry that is based in both natural and social sciences. Natural science provides the principles for cultivating and raising agricultural goods. Social science in agriculture is related to educating, communicating, selling, and regulating agricultural products and information in society.

Did You Know?

It is estimated that one of the largest areas for occupational growth in agriculture in the next 30 years will be in the area of computer programming.

Education

You might not consider your agricultural science teacher a scientist, but agricultural education is a very specific area of social science. This area of social science studies the methods used to present the topic of agriculture

FFA Connection Agriscience Fair

The agriscience fair held through the National FFA Organization has an entire category devoted to student projects in the social sciences. If you are interested in how people and society interact with agriculture, ask your agricultural science teacher how you can develop an agriscience fair project in this category.

Courtesy of Lesleigh Bagley

Figure 5-20. Agricultural educators have been trained to use the social science skills related to education. ***Do you consider your agricultural science teacher a "scientist"?***

to students in agricultural classes and to the public. Agricultural science teachers learn about the scientific principles behind teaching and learning in order to effectively transfer knowledge to their students, **Figure 5-20.**

Communication

Agricultural communication is an important aspect of social science in agriculture. By understanding the scientific principles related to communication, agriculturalists can best select the platform for sharing the message of agriculture to the public. (Refer to *Lesson 2.2.*) To effectively communicate about agriculture, we must share information with agriculturalists *and* the general public.

Political Science

Agriculture is an industry that can be highly affected by public policy and legislation. ***Political science*** is the study of how public policy is regulated and developed. By understanding the ways that political science principles interact with agriculture, the industry can be better prepared to develop regulations that allow agriculture to feed the world in a safe, conscientious manner. Can you think of some areas of politics that would have implications for agriculture? One of the most important pieces of agricultural policy is the Farm Bill, which regulates many of the policies for agriculture in the United States. It is also important to understand how political science affects international trade and relations, because many of the agricultural products produced in the United States are distributed globally.

Economics

Economics is the study of the production, distribution, and consumption of commodities. This area of science includes the business aspect of determining what products consumers are likely to purchase, and how those purchases affect the agricultural industry. Basing decisions on sound economic principles is vital to the success of any agricultural business.

Putting It All Together

Agriscience is a very unique area of study because it is truly a multidisciplinary science. From what we now know about agriculture and practical science, can you guess what the term "multidisciplinary" means? It means that agriculture does not exist in just one area. Agriculture is a broad field that encompasses natural, formal, and social sciences. Think about all of the other things you can learn while studying agriculture!

STEM Connection Where Do Agriculture and STEM Connect?

STEM is the acronym for Science, Technology, Engineering, and Mathematics. Throughout this book, you will find features designed to share STEM principles that are being shown in the text. But where are the places that STEM principles are shown in agriculture?

Lesson 5.2 Review and Assessment

Words to Know

Match the key terms from the lesson to the correct definition.

1. The formal science related to collecting and analyzing large quantities of numerical data and to relating findings to unknown data.
2. Includes the study of plant structure, physiology, genetics, and classification.
3. Disciplines concerned with abstract scientific concepts; includes logic, mathematics, theoretical computer science, and statistics.
4. Science field that includes geology, oceanography, meteorology, and ecology.
5. The scientific study of the behavior, structure, and physiology of animals.
6. The use of living systems and organisms to develop or make products useful to society.
7. The study of human society and social relationships.
8. The study of the production, distribution, and consumption of commodities.
9. The study of how public policy is regulated and developed.
10. The branch of science that studies the composition, structure, properties, and change of matter.
11. The abstract science of numbers, quantities, and space or shapes and the relationships between them.
12. The study of life.
13. The study of the properties of matter and energy.

A. biology
B. biotechnology
C. botany
D. chemistry
E. Earth science
F. economics
G. formal science
H. mathematics
I. physics
J. political science
K. social science
L. statistics
M. zoology

Know and Understand

Answer the following questions using the information provided in this lesson.

1. Identify the three broad areas of science.
2. What subject matters are included in physics?
3. Explain why a good understanding of chemistry is useful to someone working in agriculture.
4. Explain why it is important for someone working in any natural resources area to have a strong background in Earth science.
5. Identify the two main areas of biology.
6. List three agricultural applications for mathematics.
7. *True or False?* Political science in agriculture is related to educating, communicating, selling, and regulating agricultural products and information in society.

8. The study of the production, distribution, and consumption of commodities is _____.
9. *True or False?* Economics includes the business aspect of determining what products consumers are likely to purchase.
10. Explain why agriscience is considered a multidisciplinary science.

Analyze and Apply

1. Which area of science do you think is most important to the future of agriculture? Please list three reasons that you think this area is critical to agriculture's future success.
2. Annually, there are many jobs as an agricultural educator that go unfilled. What do you think we can do to solve the shortage of agricultural educators?

Thinking Critically

1. Imagine that you are confronted by someone who says that agriculture has no relationship to science. What would you say to this person? Formulate a 2–3 paragraph argument to their claim.
2. Most people think that science can only take place in a laboratory. Write a list of ten places that science happens outside of a lab, and then think of an experiment that could be conducted in that setting.

Lesson 5.3

Biotechnology in Agriculture

Words to Know

allergen
artificial insemination (AI)
artificial selection
biodiversity
cell tissue culture
cloning
embryo transfer (ET)
genetic engineering
genetically modified organism (GMO)
micropropagation
natural selection
phytoremediation
recipient
sexed semen
transgenic animal
transgenic organism

Lesson Outcomes

By the end of this lesson, you should be able to:

- Define biotechnology and explore the impact it has on agriculture.
- Analyze the scope and impact of agricultural biotechnology in the global society.
- Compare and contrast the potential benefits and risks of genetically modified organisms (GMOs).
- Discuss issues related to the current and future use of GMOs.

Before You Read

As you read the lesson, put sticky notes next to any section where you have questions. Write your questions on the sticky notes. Discuss the questions with your classmates or teacher.

What would you think if you watched crop farmers intentionally spray thousands of acres of crops with a substance designed to kill all plant life, **Figure 5-21**? Do you think it is odd that some farmers have decided to stop spraying their crops with pesticides, leaving their crops on their own against insect invaders? What if you encountered a rancher who told you his best cow mothered 24 calves this year? Are these just examples of really crazy farmers, or could these actions be a result of production agriculture using biotechnology?

This lesson deals with the science of biotechnology and the implications this field of science has for agriculture in our nation and the rest of the world. We will examine the methods through which biotechnology exists, the areas of agriculture using biotechnology, and the potential benefits and risks associated with the use of biotechnology in agricultural applications.

oticki/Shutterstock.com

Figure 5-21. Some farmers have planted crops in which the plants create their own pesticides and do not require spraying. *How do you think the crops produce their own pesticides?*

What Is Biotechnology?

Biotechnology is the use of living systems and organisms to develop or make products that are useful to society. Anytime that biological

Olga Markova/Shutterstock.com

Yellowj/Shutterstock.com

Figure 5-22. Biotechnology can help in the production of yeast bacteria and the enzymes used in cheesemaking. *Did you know that biotechnology was a very "cheesy" topic?*

processes are influenced by humans, biotechnology is taking place. This manipulation of organisms is not new. It has actually been going on for thousands of years. Common examples of using living organisms to alter, improve, or create agricultural products include:

- The use of yeast to make bread rise.
- Promoting or enabling bacteria growth to produce cheeses, **Figure 5-22**.
- The use of bacteria to produce yogurts, sour cream, and buttermilk.
- Creating favorable environments for microorganisms to ferment fruits and grains into vinegars, wines, and alcohol.
- Creating airtight environments in which bacteria work to produce silage through fermentation of grains and green grasses stored in silos.

Humans have also long used biotechnology to manipulate the development of domesticated animals and plants.

Natural and Artificial Selection

Animals and plants have been genetically altered by natural selection since the origins of life. ***Natural selection*** is the process of genetic selection in which only the strongest organisms will live to reproduce. Through this process, plant and animal species retain or develop traits that are more favorable for survival. The concept of natural selection has driven animal and plant evolution on Earth since the beginning of life on our planet. ***Artificial selection*** is the process of humans selectively breeding plants and animals to choose traits which will be more beneficial to humans. Through this process, as seen in **Figure 5-23**, domesticated animals and plants have become drastically different than their wild counterparts. Biotechnology is essentially using science to perform artificial selection.

G and L Showpigs

KOO/Shutterstock.com

Figure 5-23. The process of artificial selection has led to drastic changes in the structure and behavior of wild animals compared to their domestic counterparts. ***Can you believe that the domestic pig (left) came from wild hogs (right)?***

The processes of biotechnology have been in use for a very long time, and have allowed humans to make tremendous advances in the medical, agricultural, and industrial industries.

Traditional and Modern Biotechnology

The traditional uses of biotechnology in agriculture, including traditional breeding techniques, are well-accepted and do not incite public debate. Artificial insemination is commonplace as a means of breeding livestock, pets, and even endangered animals. Crossbreeding heirloom fruits and vegetables and grafting fruit trees as a means of creating new varieties are also accepted agricultural practices, **Figure 5-24**. It is the new modern biotechnology, which includes the tools of genetic engineering, that stirs debate.

STEM Connection Genetics in Agriculture

Agriculture relies on the principles of genetics, especially when using artificial selection to improve the quality of plants and animals. The principles of genetics were first examined by Gregor Mendel, an Austrian monk, who is considered the father of genetics.

Several principles of genetics are important to understand when looking at how traits are passed on to offspring.

- Almost all organisms receive two copies of every gene, one from the female parent and one from the male parent.
- **Genotype.** The actual genes an organism has for a specific trait are called the organism's *genotype*.
    Homozygous means the genes an organism has from both parents are the same for a specific trait.

(continued)

Heterozygous means the genes an organism has from both parents are different for a specific trait.

Some genes are expressed, while others are not:

- **Phenotype.** The outward expression of the genes an organism has for a specific trait are called the organism's ***phenotype***.

  Dominant genes are those that will be expressed if they are in the organism's genetic makeup; they are expressed by a capital letter. For example, in pea plants, green color is dominant over yellow color. Plants with the genotype GG and Gg will both appear to be green.

  Recessive genes are those that are expressed only when no dominant gene is found in the genetic makeup. These genes are generally shown with a lowercase letter. Pea plants that have a yellow phenotype would have the genotype gg.

- Some traits show as a mixture when the genotype is heterozygous. In some flowering plants, the homozygous genotypes RR and WW produce flowers that are red (RR) and white (WW). If the plant is heterozygous (RW), the phenotype for the flowers is pink. These situations are called *codominance* or *incomplete dominance*, depending on the nature of the specific combination.

The odds of an organism receiving different genotypes and phenotypes from their parents can be calculated by using a *Punnett Square*. To use a Punnett Square, you place the two genes from the male parent above the columns of the square and the two genes from the female parent next to the rows of the square. By combining the genes from each column and row, you can see the chances that each offspring has for receiving a specific genotype or phenotype.

	G	g
G	GG	Gg
g	Gg	gg

The example shown is the cross between two heterozygous green pea plants, both of which have a heterozygous genotype for color (Gg × Gg). What phenotype would each of the parents exhibit?

The chance for each offspring to have a specific genotype or phenotype is typically expressed as a fraction. For example, there is a 1/4 chance that the offspring in the cross above will be homozygous and green (GG). What is the chance the offspring will be heterozygous? What is the chance the phenotype of the offspring will be green?

Understanding how Punnett Squares work can give you great insight into the role that genetics plays in the development of improved animals through artificial selection.

Al-xVadinska/Shutterstock.com

Figure 5-24. Grafting is a widely accepted form of plant propagation in which different varieties of trees are grafted together so that the resulting plant can be grown with the traits of both trees. Most commonly, fruit trees are grafted onto varieties of trees that have superior rooting traits.

Genetic Engineering

The USDA defines ***genetic engineering*** as the technique of removing, modifying, or adding genes to a DNA molecule (of an organism) in order to change the information it contains. By changing this information, genetic engineering changes the type or amount of proteins an organism is capable of producing, thus enabling it to make new substances or perform new functions. The intended outcome of genetic engineering is to produce desired characteristics and to eliminate undesirable ones.

The introduction of any technology can be controversial, and genetic engineering is no exception. Genetic engineering in agriculture is not new, but its increased use has brought the issue to public attention. As with every controversy, there are at least two sides: those for and those against. Advocates point out the limitless benefits, whereas the opposition enumerates potential risks and fears of the unknown, as well as numerous ethical issues. Let us examine some examples of genetic engineering applications in both the agricultural (plants and animals) and medical fields, and the discussion regarding the potential benefits and risks of GMOs.

Genetically Modified Organisms (GMOs)

Plants and animals that have been created using genetic engineering are called ***genetically modified organisms***, or ***GMOs***. A GMO may also be referred to as a ***transgenic organism***. Currently, GMOs are widely used in medical applications, food production, feed production, and in animal production. In fact, genetic engineering is used to mass produce insulin, and more than three-fourths of the soybeans and cotton, and more than one-third of the corn grown in the United States is genetically modified.

The use of genetically modified organisms is a topic that has been met with controversy. Some groups see the use of GMOs as the best way to feed the growing global population. Other groups are concerned about the implications or effects of GMOs on public health and the environment. Let us first address the benefits of the genetic engineering of plants.

Did You Know?

Scientists have recently developed a genetically modified potato that glows when it needs water.

STEM Connection DNA and Genetic Engineering

Simply stated, genetic engineering is the process of taking an isolated section of DNA from one organism and fusing it to the DNA of another organism. To understand how genetic engineering works, we need to first understand the structure of DNA. The DNA for all organisms is made of only six components: a phosphate group, deoxyribose sugar (which make the sides of the ladder-shaped molecule), and four different nitrogenous bases: adenine, thymine, cytosine, and guanine. The order in which the bases combine determines the way the genes are read, proteins are created, and organisms are formed.

To genetically modify an organism, scientists perform four basic steps:

Step 1: Identify the desired gene in the host organism.

Step 2: Using enzymes, cut the desired gene and a corresponding section of the DNA in the organism you are genetically engineering.

Step 3: Insert the host gene into the DNA of the genetically engineered organism.

Step 4: Insert the genetically engineered DNA into a cell of the organism and replicate to create a genetically modified organism (GMO).

The new DNA sequence is called recombinant DNA, because it is a recombination of the genetic sequence. Recombinant DNA is used for a variety of purposes, including pharmaceuticals and cheese production.

Pharmaceuticals. By modifying the DNA sequence of bacteria, we can alter them so that that they produce specific chemicals. This is the process used to create almost all of the insulin given to diabetic humans. The same process is used to create vaccines, human growth hormone supplements, and many other chemicals used in the health care industry.

Cheese production. One of the chemicals involved in making cheese is a substance called rennet. Rennet contains a chemical found in the stomach of calves. Today, about 60% of the cheese produced worldwide uses recombinant DNA technology to produce this chemical, rather than harvesting it from calves stomachs.

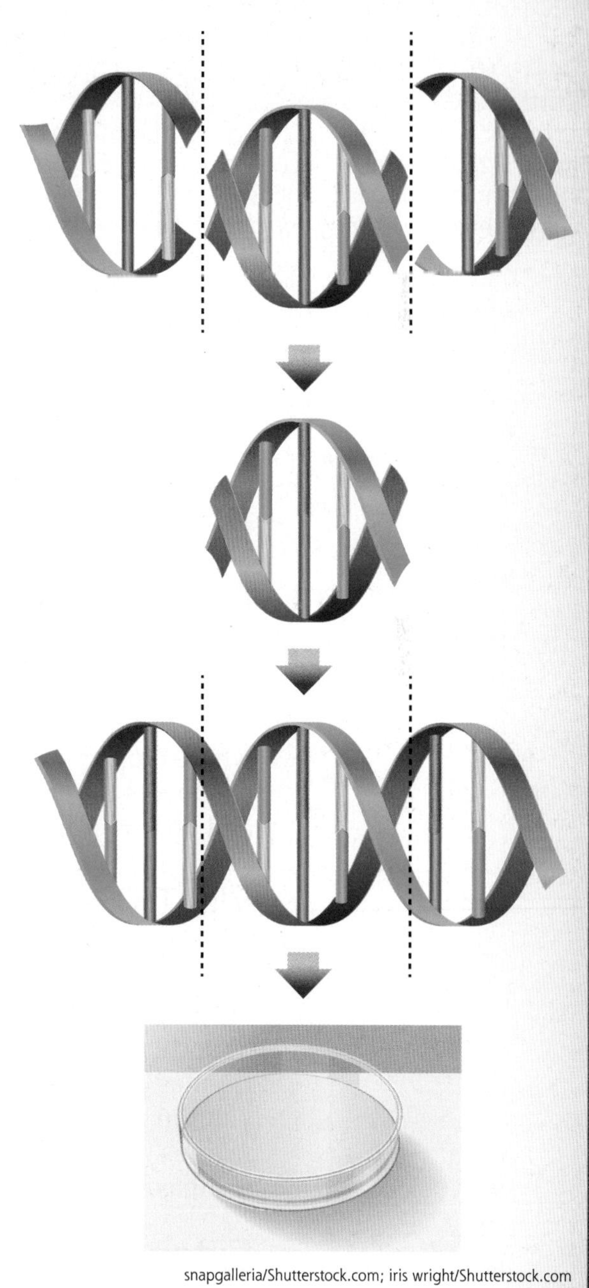

snapgalleria/Shutterstock.com; iris wright/Shutterstock.com

EdgeOfReason/Shutterstock.com

Figure 5-25. The use of biotechnology can help producers use land that has traditionally not been useful for growing crops. *Do you think genetic engineering could help the crops in this climate?*

Benefits of GMO Crops

By targeting specific genes and inserting them into crops, scientists have been able to create plants with specific characteristics that lead to:

- An improvement in quality and an increase in crop yields.
- Growing crops in areas where crops have been grown, but conditions are not optimal.
- Growing crops in areas that have not traditionally been used for agriculture, **Figure 5-25**.
- A reduction in the application of pesticides and fertilizers.

Genetically modified organisms currently used in agriculture include crops and plants that:

- Are resistant to drought and frost.
- Are infused with valuable vitamins and nutrients.
- Have longer shelf life.
- Are resistant to herbicides.
- Make their own pesticides.

Drought and Frost Resistance

Plants designed to withstand drought conditions will continue to grow at their normal rate during times of minimal rainfall. Certain varieties of drought-resistant crops show promise for being cultivated in arid areas where crops would previously never have survived. Frost resistance allows producers to plant without concern for losses in cold weather situations, and allows crops to be sown in nontraditional, colder locations.

Vitamin and Nutrient Infusion

Perhaps one of the most promising potential benefits to using GMOs is the ability to create crops that include essential vitamins and nutrients, like the golden rice shown in **Figure 5-26**. Swiss scientists inserted genes from a daffodil and bacterium into rice plants to produce this "golden rice." Golden rice has all of the nutritional value of traditional rice, but also includes vitamin A. Vitamin A deficiencies are a huge concern in underdeveloped regions. These types of crops could play an important role in feeding people in areas where nutrient deficiencies are common.

Extended Shelf Life

Genetic engineering of some produce has produced characteristics that allow for easier transport. For example, genetically modified tomatoes with delayed softening may be vine-ripened and shipped without bruising. Other examples include sweet corn that has been altered to taste sweeter, and peppers that have

STEM Connection Reproductive Biotechnology in Plants

Reproductive biotechnology in plants involves the process of cell tissue culture. ***Cell tissue culture*** is the growth of new plants or animals from a cell or small group of cells taken from the parent tissue. Through this process, thousands of identical plants can be created from a parent plant with desirable characteristics. The process of using tissue culture to create new plants is called ***micropropagation***.

Do the tiny plants illustrated to the right look like they are identical? Micropropagation involves the growing of thousands of plants from a single parent plant.

KYTan/Shutterstock.com

been modified to be spicier. Through genetic engineering, consumers are given extended access to fresh, nutritional foods that resist decay and damage as well as a loss of nutrients. The shelf life of some processed foods has also been extended through the use of GMO ingredients.

Herbicide Resistance

Herbicides cannot differentiate between crop plants and weeds, and their application may harm or kill crop plants. The most effective herbicides are nonselective and will kill a wide range of weeds and valuable crop plants. Selective herbicides are used with traditional crops. This type of herbicide does not harm the crop, but will not rid the fields of all weed types. Scientists have developed herbicide-resistant crops that are essentially immune to nonselective herbicides. That means that the entire field can be sprayed with nonselective herbicide once, killing all of the weeds, but leaving the crop plants unharmed.

JIANG HONGYAN/Shutterstock.com

Figure 5-26. *What if a food could help cure one of the most critical nutrient deficiencies in the world?* Thousands of people in developing countries suffer the effects of vitamin A deficiency, which can lead to blindness, weakness, and even death. This golden rice has been modified to have vitamin A and could be a cure for vitamin A-related deficiencies.

Did You Know?

Scientists are currently working on creating plants that can detoxify soil. After absorbing the pollutants, the plants would be harvested and disposed of safely. The process is called ***phytoremediation***.

Insect and Pest Resistance

Plants have been genetically modified to make their own pesticide by inserting a gene from a bacterium that is toxic to insects. When this bacterium is inserted into the plant DNA, it creates a crop that does not need to be sprayed with pesticide to deter the insects. The insects will die if they eat the plants, **Figure 5-27**.

AFNR/Shutterstock.com

Figure 5-27. Thousands of dollars are lost each year to corn pests. Fields of genetically modified corn produce a chemical that is toxic to insects so they do not need to be sprayed with a pesticide. *Have you ever encountered a worm in your corn?*

Risks of GMO Crops

While many scientists and experts are in favor of using GMOs, others are concerned about the potential risks to using this type of organism in the food supply. The potential risks include:

- Allergen production.
- Creating resistant pests (insects, bacteria, viruses, and weeds).
- A change in natural biodiversity.
- Impact on non-target species.

Allergen Production

Do you know someone who has food allergies? People with food allergies have varying immune reactions when they are exposed to specific proteins, called ***allergens***, in food. Fortunately, most people do not have food allergies.

Career Connection Cell Culture Technician

Developing the practice of genetic engineering

Job description: Cell culture technicians work to grow organisms from single cells, like those that are used for micropropagation of plants. Their responsibilities include monitoring the growth and maintaining the growing media and nutrients to ensure optimal conditions for the cells.

Education required: Associate's degree or higher in biomanufacturing or biology.

Job fit: This job may be a fit for you if you want to work in a scientific lab, are detail oriented, but do not want to spend six or more years at the university level to earn a biochemical degree.

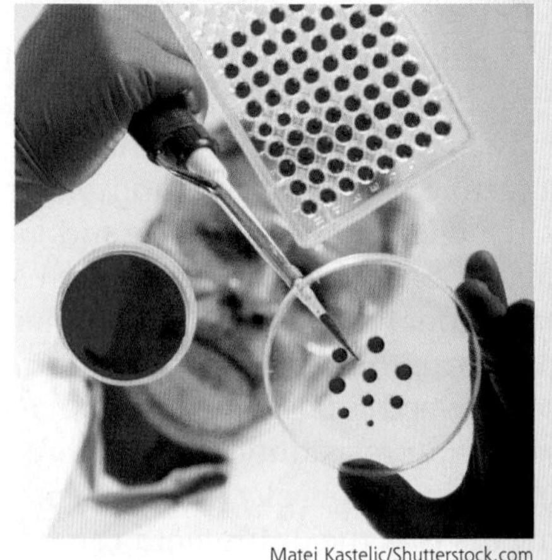

Matej Kastelic/Shutterstock.com

Figure 5-28. This image shows some of the most common allergens in human food. Some people worry that GMOs could cause unknown allergen risks. *Do you know someone with food allergies?*

However, those who do may have life-threatening or fatal reactions when exposed to specific allergens.

There is some concern that because GMOs include genes from other organisms, there is an increased chance that they could create allergens that have not been identified, **Figure 5-28**. Because of this, all genetically modified organisms that will enter the human food supply are regulated by government agencies such as the Food and Drug Administration (FDA). Currently, the FDA checks for increased levels of naturally occurring allergens in transgenic foods. Although it may pose a potential threat to those with allergies, through the process of genetic engineering, researchers are also investigating methods to use genetic engineering to remove allergens from foods such as peanuts.

Did You Know?

To help reduce the risks of allergen introduction in GMOs, none of the eight most common allergens are used for any transgenic organisms.

Pest and Weed Resistance

Another potential risk of using GMOs is the potential to create pests and weeds that are resistant to the control methods we are currently using. In every pest population, there are a small percentage of individuals that are immune to the control methods. For example, if the pesticide in a GMO corn plant kills 99.9% of all the pest insects that eat it, the remaining 0.1% have an immunity to the toxin in the plant. If the insects with immunity reproduce, they have the potential to create an entire subspecies of insect that will not be affected by the transgenic corn. Likewise, there is a risk that weeds can develop the same immunity, leading to "superweeds," **Figure 5-29**, that cannot be controlled with current herbicides. There is also the fear that transgenic crops will cross pollinate with related weeds and create superweeds through the crossbreeding.

sauletas/istock/Thinkstock

Figure 5-29. *Have you heard about weed resistance?* A potential risk of using GMO crops could lead to the evolution of pesticide- and herbicide-resistant organisms.

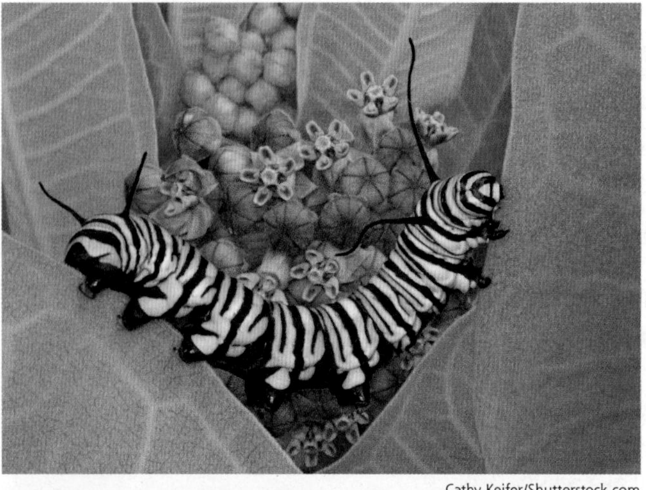

Cathy Keifer/Shutterstock.com

Paul Reeves Photography/Shutterstock.com

Figure 5-30. One of the concerns about using GMOs is that they could harm non-target species. Monarch butterflies often lay their eggs on milkweeds at the edge of the corn field. *What impacts would there be if the corn was engineered to be toxic to insects like Monarch caterpillars?*

Biodiversity Issues

In addition to allergen and resistance concerns, there are opponents of GMO use who are concerned that the use of transgenic organisms will result in a loss of the planet's natural biodiversity. ***Biodiversity*** is the biological diversity in an environment as indicated by the numbers of different plant and animal species present. Since the beginning of time, animals and plants have changed based on natural selection in order to create a delicate balance of biodiversity. There are concerns that by using genetic engineering to accelerate genetic diversity, we may upset the biodiversity of the planet.

Impact on Non-Target Species

There are significant concerns that transgenic crops may have unforeseen effects on other species in the environment. For example, a GMO corn plant that produces a specific pesticide to kill only pests that feed on the corn was found to be toxic to the caterpillars of the Monarch butterfly, **Figure 5-30**. Although the caterpillars do not feed on corn, pollen from the transgenic corn may drift onto the milkweed leaves that they do eat.

Other organisms that may be affected by GMOs include birds, mammals, insects, fish, other aquatic life, and worms.

Genetic Engineering and Animal Production

Genetic engineering is not new to animal production. For years, breeders have steered animal reproduction to produce desirable traits. Initially, they chose two animals to breed and then used the most desirable offspring to use in breeding another generation. Traditional breeding is time consuming, often taking years to produce stock with the most desirable traits. The introduction of reproductive biotechnologies such as artificial insemination, cloning, and embryo transfer has led to more efficient and precise animal production, **Figure 5-31**. Current technology in genetic engineering has taken reproductive technologies a step further, and enabled scientists and breeders to produce animals with traits beyond

Dmitry Kalinovsky/Shutterstock.com

Figure 5-31. Reproductive technologies have allowed producers to increase breeding efficiency and quality in their animals. This reproductive scientist is looking for ovulation through ultrasound in a sow.

those created with traditional breeding methods. While not in everyday production at this time, some examples of genetic modification in animal science include technologies such as:

- Cows that produce milk with a higher protein content to facilitate cheese production.
- Animals genetically modified to produce higher levels of growth hormones to increase their size.
- Goats engineered to express drugs and other proteins in their milk.
- Genetic engineering to create enhanced muscle mass in animals.

Reproductive Biotechnology

As stated earlier, the introduction of reproductive biotechnologies such as artificial insemination, cloning, and embryonic transfer has led to more efficient and precise animal production. These technologies have been around for some time, are widely used, and allow producers to develop higher quality animals and scientists to research more advanced biotechnologies.

Artificial Insemination

Artificial insemination, or ***AI,*** is the process of placing semen in the female reproductive tract through artificial means. Through this process, animal breeding can take place more efficiently. Using artificial insemination allows for more females to be inseminated by a single male. In addition, using AI results in higher conception rates, a decrease in the spread of disease between breeding animals, and there is less chance for harm to either animal. It is also more efficient because only the

Did You Know?

More than 85% of the dairy cattle in the United States are produced using artificial insemination.

Jason Benz Bennee/Shutterstock.com

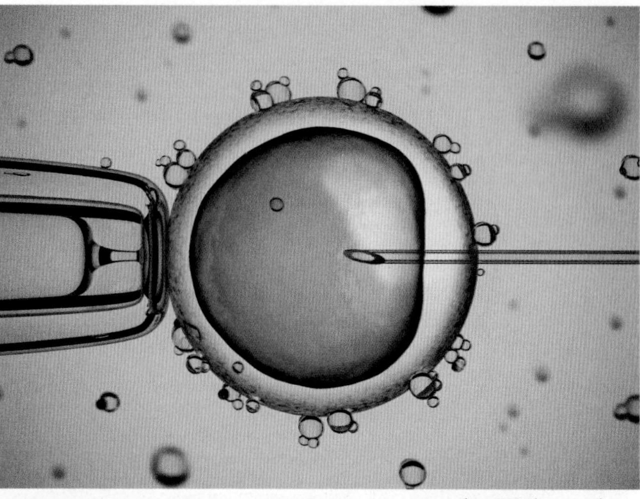

koya979/Shutterstock.com

Figure 5-32. Cloning is a biotechnological procedure in which the DNA in an egg is replaced with DNA from a donor cell so that it grows into a genetically identical copy of the donor animal.

semen needs to be transported, allowing for the cost-effective breeding of animals who live thousands of miles apart. Artificial insemination is widely used in the livestock industry, most prominently in the breeding of dairy cattle and swine.

Did You Know?

Dolly, a Finn-Dorset ewe, was the first mammal to have been successfully cloned. It took more than 400 attempts before an embryo was successful. Dolly was the only lamb to live after 277 eggs were used to create 29 embryos, which only produced three lambs.

Cloning

Cloning is the process of creating a genetically identical copy of an organism, **Figure 5-32**. Cloning is a complicated and time-consuming process. In the process of cloning, a fertilized egg is collected and the DNA is removed. The DNA from an adult cell of the animal you are trying to clone is then inserted in the egg through specialized scientific procedures. Then the egg is placed into another female animal for the rest of gestation. The resulting offspring is genetically identical to the animal you have cloned.

Through the use of cloning, agriculturalists have the ability to produce animals who are genetically identical to superior animals. There are also important research implications in creating animals that have the same genetics. These animals are perfect candidates for research into

STEM Connection Sexed Semen

Scientists have developed many different reproductive technologies to help livestock producers. One of the most recent is the widespread use of ***sexed semen*** to produce offspring of a desired gender. Because sperm cells carry either an X or a Y chromosome, the gender of the offspring depends on which chromosome the sperm cell contains. Can you think of some livestock scenarios where you would prefer one gender over another? If you wanted to produce show steers, you may want to use male sexed semen to artificially inseminate your cows. On the other hand, if you were a dairy cattle producer, you would want to have the highest possible percentage of female offspring, and could used female sexed semen. Through several different methods, genetics companies can sex semen with more than 80% assurance of the desired gender.

things like animal nutrition and feeding because they do not have genetic differences as a variable.

Embryo Transfer

To perform ***embryo transfer***, or ***ET***, fertilized embryos are collected from a female in order to store them or place them into other females for growth and development. The process of embryo transfer is most widely used to allow female animals with desirable traits to produce more offspring. A ***recipient*** is a female animal who carries the embryo collected from the donor animal.

To perform an embryo transfer, the donor female is given hormones to stimulate the ovulation of multiple unfertilized eggs. The female is then artificially inseminated and several days later, the uterus is flushed to collect the embryos, **Figure 5-33**. The embryos are then evaluated under a microscope and embryos that are likely to produce offspring are placed in the recipient females, who have been given hormones to ensure they are in the

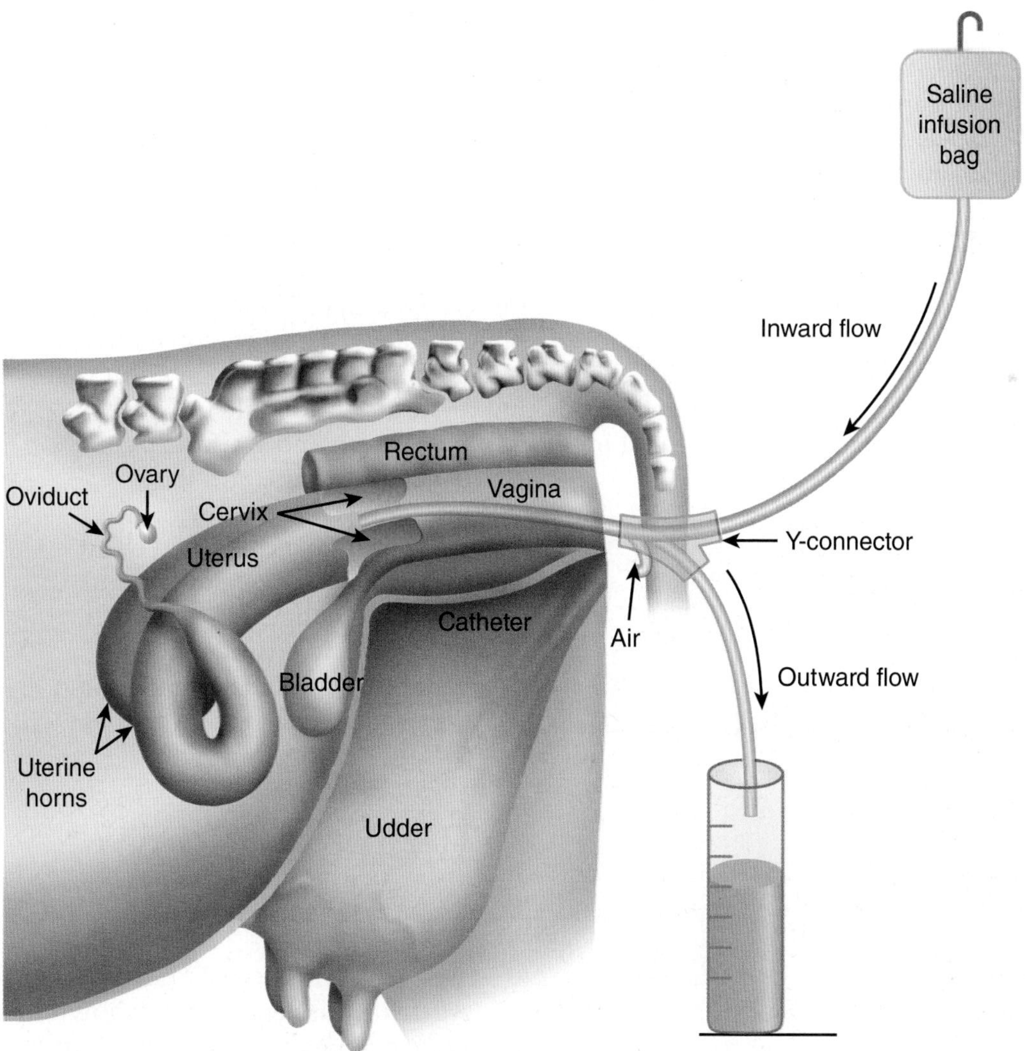

Peter Hermes Furian/Shutterstock.com

Figure 5-33. The process of flushing a donor animal involves using a liquid solution to fill the uterus and oviducts and then retrieving the fluid along with any fertilized embryos.

same reproductive stage as the donor. In addition, embryos can be frozen in liquid nitrogen and used at a later time.

This technology is becoming increasingly common with superior female animals and allows for greater efficiency. For example, if you have a very high-performing racehorse mare, she can only have one offspring per year through conventional technologies. Using ET, the same mare can produce many more foals in the same period of time and still be able to compete because she is not carrying a foal herself.

Did You Know?

Just like in humans, animal embryos can be flushed and then fertilized through in vitro fertilization. This process is used in special cases with animals and can even be used to collect unfertilized eggs from female animals right after they have died.

Benefits of Transgenic Animals

A ***transgenic animal*** is an animal whose genome has been changed to carry genes from other species. Transgenic animals in agricultural applications include benefits to breeding, quality, and possibly disease resistance. Farmers and ranchers may be able to use transgenic animals to breed larger herds with specific traits that will ultimately deliver higher yields. The quality and the amount of food from animal sources may also be raised. This includes animals with more meat, higher milk production, and milk with added benefits.

According to the American Institute of Biological Sciences, the creation of transgenic animals has resulted in a change in the use of higher order species as laboratory animals, **Figure 5-34**. It has also decreased the number of animals used in experimentation.

Risks of Transgenic Animals

Comparatively speaking, little research has been done on the risks of transgenic animals released into the marketplace. Farmers, consumers, and scientists all want safe food; therefore, research of transgenic animals must continue before freely producing them for wide-scale use in the marketplace. Research must include animal health and welfare, public concerns, ethics, and long-term effects.

Through a combined effort of researchers, developers, and government agencies, research is ongoing to ensure that any genetically modified animals and the products derived from them will be safe.

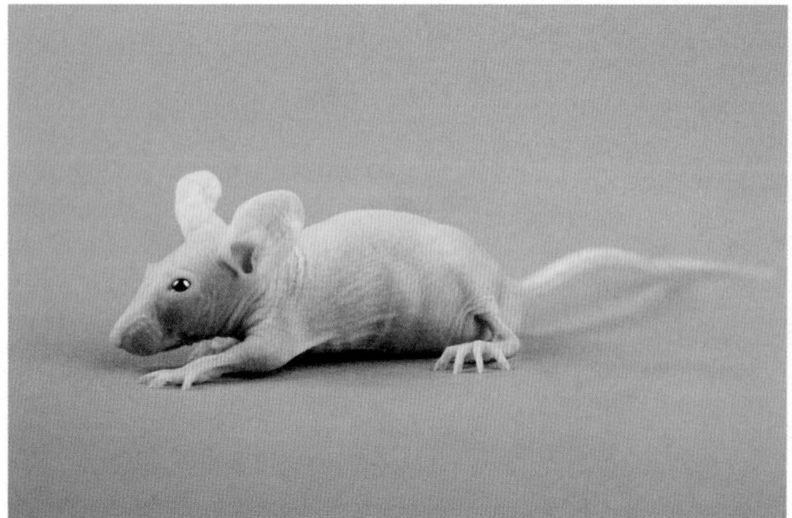

Vasiliy Koval/Shutterstock.com

Figure 5-34. The use of transgenic animals has widespread implications for research animals. If you had hundreds of genetically identical animals, you could more accurately research a specific condition or disease.

Biotechnology Regulation

According to the USDA Foreign Agricultural Service, and shown in **Figure 5-35**, there are more than 70 countries that currently approve production, import, or field trials of genetically engineered crops. In the United States, agricultural

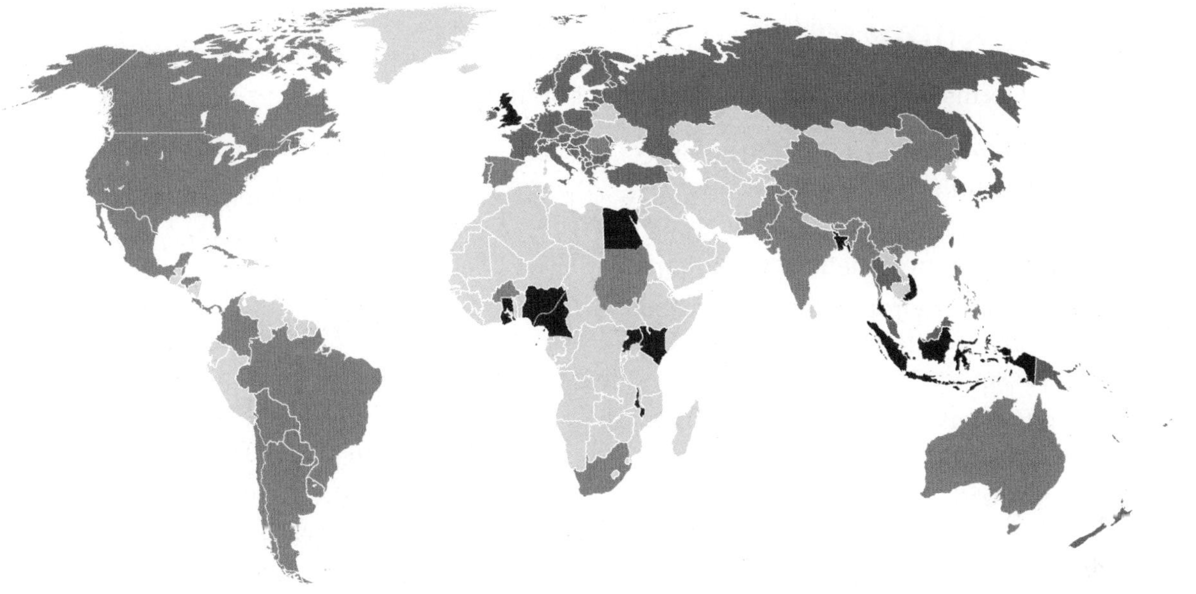

ekler/Shutterstock.com

Figure 5-35. Although there is controversy about the use of GMOs, many countries have accepted their use and allow import of products containing genetically modified ingredients.

genetically modified organisms must go through testing at both the USDA and Food and Drug Administration (FDA). Depending on the nature of the genetic modification, they may also undergo regulation at the Environmental Protection Agency (EPA).

Labeling

Labeling of GMO food products is something that is an ongoing discussion. Those trying to market transgenic crops do not want labeling because they feel it sends a message that GMOs are unsafe. There are consumer groups who are in favor of requiring all foods which have GMO ingredients to have a label stating that they include GMOs, citing that the public should know if they are included. Currently, there is no requirement in the United States to label transgenic products, but those *without* GMO ingredients sometimes choose to add a GMO-free label. What do you think the regulation should be on labeling transgenic foods?

Lesson 5.3 Review and Assessment

Words to Know

Match the key terms from the lesson to the correct definition.

1. A female animal who carries the embryo collected from the donor animal.
2. Selectively breeding plants and animals with traits deemed more beneficial to humans.
3. The state of an environment as indicated by the numbers of different plant and animal species present.
4. Altering the genetic code of an organism in order to create desired characteristics.
5. Sperm collection which has been separated to have a higher concentration of either X or Y chromosome sperm cells.
6. The process of genetic selection in which only the strongest organisms will live to reproduce.
7. An animal whose genome has been changed to carry genes from other species.
8. Specific protein in food that causes some people to have an immune reaction.
9. The process of creating a genetically identical copy of an organism.
10. The process of placing semen in the female reproductive tract through artificial means.
11. The use of living plants to remove organic and inorganic contaminants from soil.
12. Fertilization process in which embryos are removed from one female and placed in others.
13. The growth of new plants or animals from cells taken from the parent tissue.
14. A plant or animal that has been created using genetic engineering.
15. The process of using tissue culture to create new plants.

A. allergen
B. artificial insemination (AI)
C. artificial selection
D. biodiversity
E. cell tissue culture
F. cloning
G. embryo transfer (ET)
H. genetic engineering
I. genetically modified organism (GMO)
J. micropropagation
K. natural selection
L. phytoremediation
M. recipient
N. sexed semen
O. transgenic animal

Know and Understand

Answer the following questions using the information provided in this lesson.

1. List three common examples of using living organisms to alter food products.
2. *True or False?* Artificial selection is responsible for the drastic differences in domesticated animals and their wild counterparts.

3. Explain why you think genetic engineering in agriculture stirs so much debate.
4. List and explain three benefits of using genetically engineered crops.
5. List and explain three potential risks of using genetically engineered crops.
6. How would a "superweed" develop?
7. *True or False?* The use of GMOs is highly regulated.
8. *True or False?* Genetic engineering was introduced to animal production in the twenty-first century.
9. List three means of biotechnological animal reproduction.
10. Identify three potential benefits to the use of transgenic animals in agriculture.

Analyze and Apply

1. Why would someone clone an animal rather than perform an ET?
2. Explain how the results of natural selection and artificial selection differ.

Thinking Critically

1. Much controversy exists over the use of GMOs in the food supply. Some countries do not even allow the import of products which contain GMOs. Which side are you on? Do you think GMOs should be widely used or not used in the food supply? Think about sound facts for your argument and write a letter to those who oppose your view to try and convince them to see your side of the argument. Be prepared to share your letter with the class.
2. Genetic engineering is the transfer of small sections of DNA for a single trait from one organism into another. Think of an organism that you would like to genetically engineer. Then choose a *single* trait from another organism that you think would be beneficial to engineer into your organism. Explain how the trait you insert would be beneficial and draw a picture of your new organism. (Remember, this is *one* trait only; you should not be creating 1/2 and 1/2 mutants.)

Chapter 5

Review and Assessment

Lesson 5.1

Agriscience and the Scientific Method

Key Points

- All scientists begin with an observation of something they cannot explain.
- Using the scientific method allows researchers to conduct experiments in a way that ensures they measure the hypothesis.
- The variables are the items in an experiment that change—the independent variable is changed by the researcher; the dependent variable changes because of the treatment.
- Experiments should be set up so that careful attention is paid to the control, constant, treatments, and trials.
- Conclusions should be based on the outcome of the experiment and communicated through a narrative or charts.

Words to Know

Use the following list and the textbook glossary to review and study the *Words to Know* from *Lesson 5.1*.

agriscience
chart
constant
control
dependent variable (DV)
experiment
graph
hypothesis
independent variable (IV)
logbook
scientific method
treatment
trial
variable

Check Your Understanding

Answer the following questions using the information provided in *Lesson 5.1*.

1. *True or False?* Using the scientific method can help ensure that research leads to reliable results.
2. Explain why some research questions require multiple rounds of testing.
3. Explain how you can ensure the credibility of your hypothesis.
4. What is the best way to begin designing an experiment?

5. *True or False?* Having a control allows the researcher to decide if the results are based on the independent variable rather than an external factor.
6. Explain the importance of having a constant in a scientific experiment.
7. *True or False?* Researchers collect data multiple times on the same experimental topic so they will have a larger amount of data from which to choose to apply to their conclusions.
8. What are the most common methods for communicating the results of a scientific experiment?
9. List five sections generally included in a written research report.
10. What is the main difference between a graph and a chart?

Lesson 5.2

Practical Science in Agriculture

Key Points

- There are three broad fields of science: natural sciences, formal science, and social science.
- Natural science is based on observable and measureable data.
- Formal science includes abstract scientific concepts like mathematics, statistics, and theoretical computer science.
- Social science includes the study of humans and social relationships. This area includes communication and education.
- Agriculture is multidisciplinary because every field of science has agricultural implications. Understanding the ties between agriculture and science helps producers and researchers know which core science concepts can be applied to make the agricultural industry successful.

Words to Know

Use the following list and the textbook glossary to review and study the *Words to Know* from *Lesson 5.2*.

biology
biotechnology
botany
chemistry
Earth science
economics

formal science
mathematics
natural science
physical science
physics
political science
social science
statistics
theoretical computer science
zoology

Check Your Understanding

Answer the following questions using the information provided in *Lesson 5.2*.

1. _____ science can be broken down into two main categories, physical sciences and biology.
2. _____ science includes areas like physics, chemistry, and Earth science.
3. _____ is regarded as a fundamental science because all other natural sciences use and obey its principles and laws.
4. _____ deals with the interaction of molecules to form different substances.
5. _____ science includes topics like geology, oceanography, meteorology, and ecology.
6. _____ is the field of science that has the most overlap with agriculture because agriculture is the human cultivation of plants, animals, and other life forms.
7. *True or False?* Everything that happens from conception to death for animals in agricultural settings is rooted in zoological principles.
8. _____ is the field of science that relates to agriculture because it is the science behind the growth and reproduction of all agricultural crops.
9. Agricultural business uses the formal science of _____ to determine the future markets for agricultural products.
10. _____ science in agriculture is related to educating, communicating, selling, and regulating agricultural products and information in society.

Lesson 5.3

Biotechnology in Agriculture

Key Points

- The genetic engineering of agricultural crops may increase growth rates, resistance to pests and diseases, and reduce the use of fertilizers and pesticides.
- Biotechnology has made pest and weed control safer and easier.
- Through biotechnological applications, less pesticides and herbicides that would contaminate groundwater and the environment could be applied.

- Genetically engineered plants can be used to protect crops from devastating diseases and infestations.
- The risks of GMO crops transferring traits to wild relatives or neighboring non-GMO crops must be thoroughly researched.

Words to Know

Use the following list and the textbook glossary to review and study the *Words to Know* from *Lesson 5.3*.

allergen
artificial insemination (AI)
artificial selection
biodiversity
cell tissue culture
cloning
embryo transfer (ET)
genetic engineering
genetically modified organism (GMO)
micropropagation
natural selection
phytoremediation
recipient
sexed semen
transgenic animal
transgenic organism

Check Your Understanding

Answer the following questions using the information provided in *Lesson 5.3*.

1. Explain how genetic engineers create genetically modified organisms.
2. *True or False?* Biotechnology may be used to perform artificial selection.
3. What is the main difference between traditional and modern biotechnology?
4. According to the USDA, what is the intended outcome of genetic engineering?
5. How much of the corn grown in the United States is genetically modified?
6. Explain how the use of GMO crops may increase crop yields.
7. The process of using tissue culture to create new plants is called _____.
8. Explain why allergens are a concern with the use of transgenic crops.
9. *True or False?* Less than 30 countries currently approve production, import, or field trials of genetically engineered crops.
10. *True or False?* It is currently required for all produce that has been genetically modified to be labeled as such.

Chapter 5 Skill Development

STEM and Academic Activities

1. **Science.** Design an agricultural science experiment. Write down the steps of the scientific method and list what you will do in each step. (Tip: this is a great way to get the setup done for an Agriscience Fair project.)
2. **Technology.** A key factor in genetic engineering is understanding where which genes are located on the DNA strand. Over the last few decades, scientists have worked to develop the "genome" of many different species. In what area of science do you think these scientists are working?

   A genome is a map of the entire DNA sequence of the organism with the location of specific traits identified. Choose a livestock species and conduct an Internet search to see what you can find out about the genome of that species. Most specifically, identify the number of base pairs included in the DNA strand. That number will tell you just how many "rungs" there are on the DNA ladder of that species. (Hint: it will be in the *billions*.)
3. **Engineering.** One of the coolest tools developed by engineers to help biotechnology is called the "gene gun." Use a search engine to find out what the gene gun is and how this tool is important in genetic engineering. Then create an advertisement for a gene gun that includes the key features and a description of the product.
4. **Math.** One reason to use artificial insemination is to increase conception rates. What does it really cost to have cows "open" (not pregnant)? Let us assume you have the following scenario:

   You have 100 cows. You have calculated the management costs (including artificial insemination) at $500 each year. For the upcoming year, you have a contract to sell calves when they average 700 lb for $1.05 per pound. How much profit can you expect if 90% of your cows calve? What happens to your profit if you only have 80% of your cows calve?
5. **Social Science.** There are many opinions of genetically modified organism use in the United States. Conduct an Internet search to find one organization that supports GMO use and one that opposes it. Compare and contrast the information from the organization websites. Explain which organization you feel makes the most compelling argument on the topic.
6. **Language Arts.** Which government agencies are responsible for researching and reviewing transgenic crops/food used in the human food supply? What roles does each agency play? Research the subject and write a brief report.

Communicating about Agriculture

1. **Reading and Speaking.** Make a timeline of biotechnological advancements in the last 20–30 years. Focus on those relating the most to agriculture. Research advancements and the scientists who brought about these changes. Create a 10- to 15-point timeline showing significant milestones that have led to new discoveries and theories. Describe your findings to the class.
2. **Speaking.** Debate the topic of the use of GMOs in food for human consumption. Divide into two groups. Each group should gather information in support of either the "pro" argument (GMOs should be used) or the "con" argument (GMOs should not be used). Use definitions and descriptions from this chapter to support your side of the debate and to clarify word meanings as necessary. You will want to do further research to find expert opinions, costs associated with GMO research and creation, and other relevant information.
3. **Listening.** As classmates deliver their presentations on the use of GMOs in food for human consumption, listen carefully to their arguments. Take notes on important points and write down any questions that occur to you. Later, ask questions to obtain additional information or clarification from your classmates as necessary.
4. **Listening.** In a group, create a presentation sharing specific examples of how agriculture interacts with one of the following fields of science: natural, formal, or social. Present two to three examples of how this field of science is used in a specific sector of agriculture. Be prepared to share your examples with the class.

Extending Your Knowledge

1. Conduct an Agriscience Fair project using the guidelines set forth by the National FFA Organization. In order to make sure that your experiment is complete, keep a detailed logbook of your work on the project. Remember that you can also include this information as a portion of your Supervised Agricultural Experience (SAE) program.
2. Think about your response to the following statement: "Agriculture courses should not be counted for science credit in high schools." After reading through this chapter, could you explain to someone why agriculture courses include a science component? Make sure you think about the fields of science and the role they each play in agriculture.
3. You have been hired to create a children's storybook about genetically modified organisms. Please create a simplified, illustrated story that would allow children in elementary school to gain an understanding of what happens in genetic engineering.

Chapter 6

Agricultural Technology

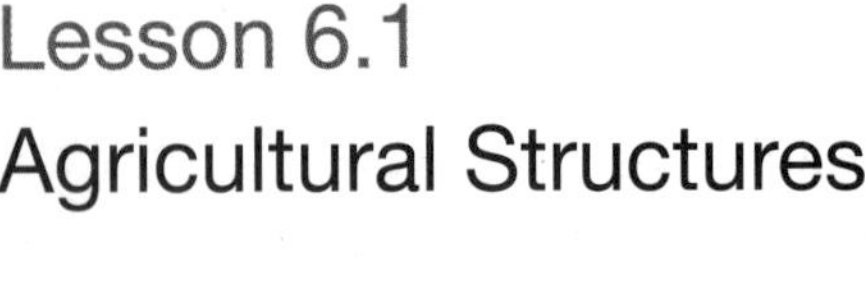

Lesson 6.1
Agricultural Structures

Lesson 6.2
Emerging Agricultural Technology

©iStock/36clicks

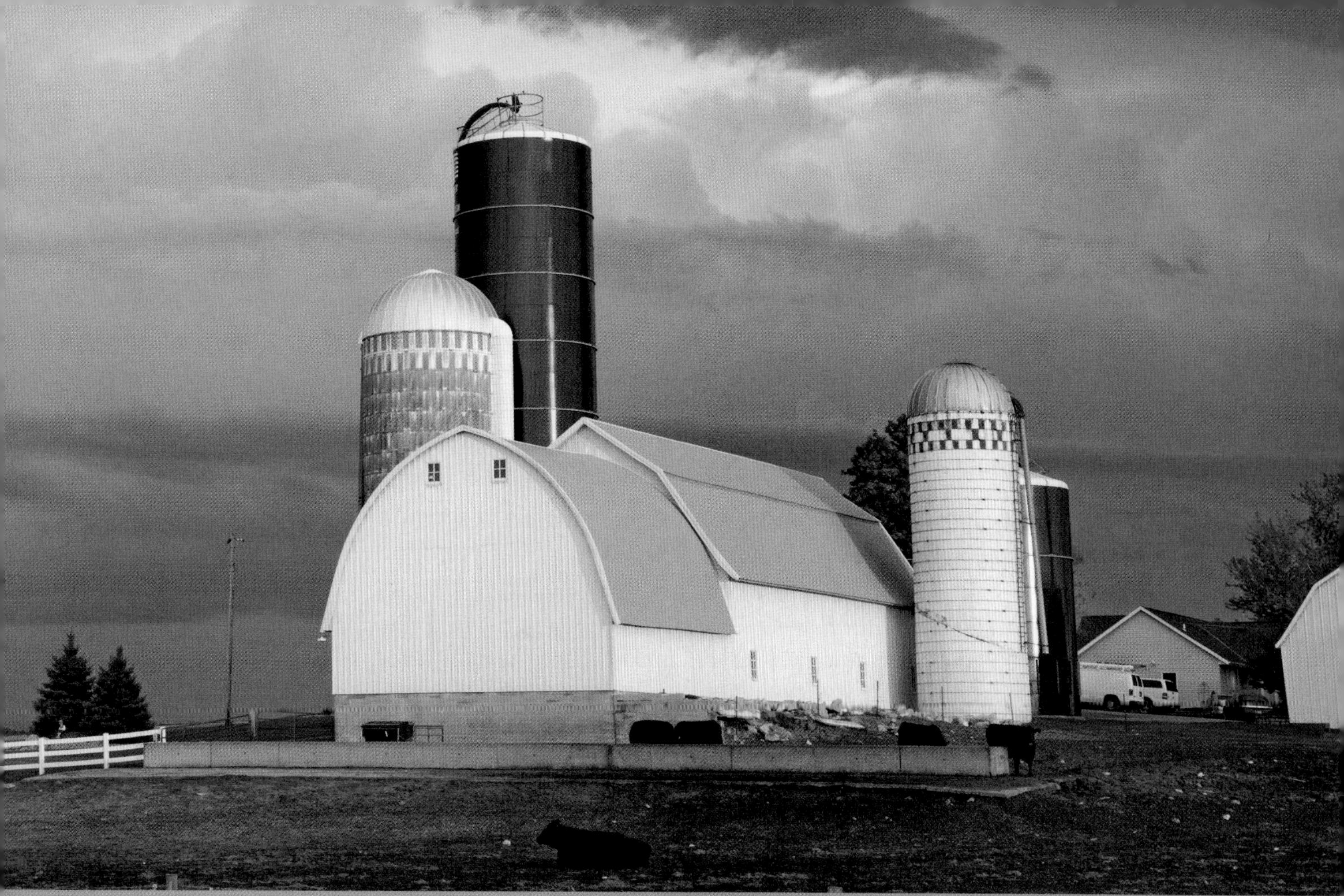

Dan Thornberg/Shutterstock.com

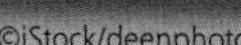
©iStock/deenphoto

©iStock/defun

While studying, look for the activity icon  **to:**

- **Practice** vocabulary terms with e-flash cards and matching activities.
- **Expand** learning with video clips, animations, and interactive activities.
- **Reinforce** what you learn by completing the end-of-lesson activities.
- **Test your knowledge** by completing the end-of-chapter questions.

Lesson 6.1

Agricultural Structures

Words to Know

aquaculture
broiler production
confinement operation
equine
farrowing facility
finishing facility
grain bin
greenhouse
milking parlor
silage
silo

Lesson Outcomes

By the end of this lesson, you should be able to:

- Identify the importance of materials used in agricultural structures.
- Describe functional and efficient facilities for agricultural use.
- Identify criteria in selecting materials in agricultural construction/fabrication.

Before You Read

Before reading the lesson, make a list of as many types of agricultural buildings you can identify. Think about the different types of agricultural buildings and facilities you have seen in your community. How many different facilities can you identify? Make a list of the types of buildings. Compare your list to the types of buildings covered in the lesson. Share your list with a peer or during class discussion.

The landscape across rural America is decorated with agricultural structures of all shapes, sizes, and purposes, **Figure 6-1**. Barns, animal housing, greenhouses, grain storage and drying facilities, and waste handling facilities are essential to agricultural production. It would be impossible to carry out any agricultural activity without them. Uses for agricultural structures include:

- Plant growth
- Animal housing
- Animal birthing and nursing
- Milking and storage
- Crop storage and processing
- Animal feed storage
- Mechanical shops for maintenance and repair
- Equipment storage
- Waste storage and processing

Elena Elisseeva/Shutterstock.com

Figure 6-1. Many of the barns in use across the American countryside date back over 100 years. Farmers maintain these buildings because they are used to house and store the animals, crops, and equipment from which a farmer makes a living. ***Are there any historic barns in your area?***

Structure Applications

Agricultural structures are different than structures designed for human use, such as the home in which you live. However, there are some similarities:

- You may have an air conditioning and heating system in your home. Animal housing facilities have climate control systems to keep livestock and poultry at a comfortable temperature, **Figure 6-2**.
- Your home probably has a kitchen where food is prepared. Livestock facilities often have feed storage and handling systems.
- You may have a television and computer in your home. Animal facilities may have these as well. No, livestock and poultry do not watch television or answer email, but many modern facilities use closed-circuit cameras to monitor animal activity and health, or to provide security around mechanical shops and storage facilities.

Gualberto Becerra/Shutterstock.com

Figure 6-2. Animal housing must have adequate ventilation and temperature control for the benefit of the animals. *How much energy do you think is needed to run this type of poultry-house ventilation system?*

Computers are used in many agricultural facilities to operate and monitor feed delivery systems, environmental controls, and manage the movement of harvested crops or livestock in the facility, **Figure 6-3**.

auremar/Shutterstock.com

Figure 6-3. Today, computers are used in every aspect of farming. There are agricultural software programs designed to do everything from monitor milking schedules, remind farmers to perform routine machinery maintenance, and to help distribute feed accurately. *Are you using a computer to record data for your SAE?*

- Most homes have multiple bathrooms connected to public sewer systems or septic tanks to handle human waste. Livestock facilities also have some type of waste removal and management system.
- Houses provide space for growing families as well as protection from the elements. Agricultural buildings are much larger than those structures designed for human use, but are also designed to provide space for growing as well as protection from the elements.

Agricultural structures are designed to last under tough environmental conditions. Imagine if beef cattle or dairy animals lived in a home like yours. These animals would quickly wreck a house designed for human inhabitants.

Each type of agricultural structure has its own requirements regarding the building materials best suited for the application, the need for electrical service, the need for plumbing, and the type of environmental control system that should be installed. This lesson will give a brief overview of various agricultural structures and some of their specific requirements.

Plant Growth Facilities

Greenhouses are the most common form of agricultural structure for growing plants, **Figure 6-4**. Greenhouses are used to recreate ideal growing conditions for plants underneath a transparent glass or plastic cover for the purpose of extending the growing range and season. Modern commercial greenhouses have precise environmental controls to control the air flow, temperature, light levels, and moisture available to plants. Greenhouse operations must also be carefully monitored to ensure the health of the crops being grown.

Some of the basic components of greenhouses include:

- Structural materials—the framework may be made of metal or plastic tubing and the covering may be fabric, plastic, or glass. Permanent structures will require footings, drainage, and flooring. Structural materials must be designed to endure moist, warm conditions. For example, a wooden frame would be more susceptible to mold and rot.
- Floor materials—depending on the growing system (in-ground planting, container planting, or hydroponics), the floor may be soil with gravel or dirt pathways, gravel, or concrete with drains. Other types of mulch may be used to line pathways as needed.
- Growing system—in-ground planting and hydroponics may be used in greenhouses with unfinished floors. If

pixinoo/Shutterstock.com

Figure 6-4. Greenhouses must provide all of the inputs a plant needs in order to grow—water, nutrients, warmth, and sunlight. Some structures are designed so the roofing material can be rolled down or up to take advantage of sunlight or to protect the plants from the elements.

dailin/Shutterstock.com

V_ace/Shutterstock.com

Sukpaiboonwat/Shutterstock.com

Figure 6-5. There are many types of greenhouse installations varying in size and the planting and irrigation system used. A—In-ground planting. B—Container planting. C—Hydroponic planting. ***Can you think of advantages and disadvantages for each type of growing system?***

the flooring is finished, container planting or hydroponics may be used. See **Figure 6-5**.

- Irrigation system—a steady source of water is required for all greenhouse structures. In rural areas, this will most likely be a dedicated well. The most commonly used greenhouse watering systems are drip irrigation or micro sprinklers. Irrigation systems are also used to distribute nutrients to greenhouse plants.
- Environmental control system—adequate ventilation systems are required regardless of the greenhouse location. In warmer climates, the ventilation system will often be used to prevent the temperature from becoming too warm due to excess heat buildup from the sun. In cooler climates, the ventilation system will often include a heat source to maintain warmer temperatures.
- Lighting system—greenhouses are used to grow plants out of season and may require supplemental lighting, **Figure 6-6**. Lighting systems must be designed to withstand the warm, damp conditions in growing facilities and to operate safely in wet conditions.

Commercial greenhouses may hold thousands of plants, and all of them need the same resources for growth. Once established, a greenhouse operation may be a very lucrative business. However, there are many factors to consider, including the financial investment a facility would

A.Krotov/Shutterstock.com

Figure 6-6. Many greenhouse structures require the installation of lighting systems. Artificial lighting can be used to extend the growing time of fruits and vegetables and to force certain flowering plants into bloom for specific holiday seasons. ***Can you name any varieties of flowering plants that are forced into bloom in time for holiday sales?***

polat/Shutterstock.com

Figure 6-7. Confining animals enables producers to move livestock more efficiently, to manage animal health, and to feed the animals more efficiently.

entail. A greenhouse operation requires careful planning and design in order to be successful.

Animal Facilities

In the same manner as plant growing structures, livestock, poultry, and dairy facilities provide all the resources needed for the animals to grow, plus a few others. Animal facilities also protect livestock and poultry from predators, and provide some resistance to the spread of disease and parasites. Animal housing also keeps livestock and poultry in one location for ease of management, **Figure 6-7**.

Livestock and poultry facilities are designed according to the weather of a specific geographic region. For instance, buildings in northern climates must be able to withstand the weight of snow and ice during winter months. In warmer climates, buildings may need supplemental cooling systems to control heat buildup.

The location of livestock and poultry facilities is also important. Buildings that house livestock should be housed a suitable distance from housing developments. Facilities should be located near good roads that provide access to emergency and fire vehicles in the event they are needed, **Figure 6-8**. Agricultural structures use a lot of water and electrical power, so careful planning is necessary to ensure that these utilities are available.

Agricultural buildings also require special construction techniques so that there is adequate space for equipment, animals, or plants. Large, open floor plans are necessary, and this means that the roof and walls must be sturdy enough to support the weight of the building. External siding and roofing must be able to withstand the weather. In some cases, insulation is needed to protect plants and animals from the extremes of weather.

Poultry Houses

Poultry houses must provide a safe and efficient environment for birds. Modern poultry houses use automated feeding and watering systems. They

American Spirit/Shutterstock.com

American Spirit/Shutterstock.com

Figure 6-8. Farm facilities should be located near adequate transportation resources and utilities.

also provide supplemental heat for growing chicks. In ***broiler production*** (poultry raised as a meat source), sufficient floor space is needed for the growing birds. In egg production facilities, special equipment is used to move eggs from the laying cages to a central collection area for grading and packaging.

Most poultry houses are large buildings with open floor plans, **Figure 6-9**. Major components of poultry houses are the electrical service, the environmental control system, and the water supply system.

Waste Management

Large operations must also have a means of collecting, storing, and processing waste. Poultry waste may be collected, stored in large sheds, and used as a crop fertilizer. The type of facility will determine the type of waste collection used. In broiler production facilities, absorbent material is

Kharkhan Oleg/Shutterstock.com

sittitap/Shutterstock.com

Figure 6-9. Modern poultry houses provide adequate protection from predators and the weather. Modern egg production facilities use sophisticated equipment to move eggs from the laying cages to a central collection area. ***What are some of the daily expenses of a poultry house?***

laid on the ground (dirt or clay pad) to absorb the poultry manure. This may be manually removed. Poultry house manufacturers recommend pressure-washing all walls before removing the bedding materials so the bedding can absorb the waste water. Many growers recommend an acid shock treatment to the dirt floor, or pad, once all bedding and manure is removed. The acid treatment is used to kill viruses and bacteria still present in the flooring. Some broiler operations may use a slatted or grid-type flooring that is raised off the dirt pad to allow more air circulation.

In egg production facilities, waste may be collected in trays and rinsed into a drainage system to flow into a storage and processing area.

Swine Facilities

Swine facilities include farrowing facilities and various types of finishing facilities. ***Farrowing facilities*** provide an area for the sow to birth and nurse pigs until weaning. ***Finishing facilities*** are designed for pigs to grow from weaning weight to market weight. Swine operations may use completely enclosed, indoor facilities, outdoor facilities, or a combination of indoor and outdoor facilities. The majority of swine operations in the United States are indoor operations referred to as ***confinement operations***.

Structures designed for confinement operations are usually wide-open structures divided by individual pens. In farrowing operations, the pens are designed to keep the sow and her piglets apart, yet allow the piglets to safely nurse, **Figure 6-10**. In finishing operations, the pigs are usually divided into pens according to age and weight for more effective feeding and weight gain.

Environmental Control

Indoor swine operations are designed to allow the producers to control the climate within their facilities to the desired temperature needed for any stage of production. This is especially important since swine are prone to overheating. Swine do not perspire and may easily overheat if they do not have the means to cool off as they would naturally in mud or water.

Aumsama/Shutterstock.com

Figure 6-10. Modern swine facilities provide everything that hogs need to grow to market weight. The farrowing pen is designed to prevent the sow from laying on and smothering the nursing piglets.

Waste Management

Manure from swine facilities must be removed before it accumulates. Most operations feature slatted flooring through which waste matter is flushed into storage and processing facilities.

Cattle Chutes and Gates

Beef cattle are not commonly housed in winter because their thick hair coats protect them from the elements. However, cattle farming and ranching does require holding pens, feedlots, and loading and unloading facilities. Permanent structures are built with wood or steel fencing, **Figure 6-11**.

Wendy Townrow/Shutterstock.com

Richard Thornton/Shutterstock.com

Figure 6-11. Cattle chutes are often constructed of steel or wood. They may be permanent or semi-permanent structures ranchers can relocate and reassemble as needed. Fencing in feed lots may be fairly simple and also constructed of metal or wood.

Facilities for calving can range from open pasture in the Southern United States to permanent barns in Northern climates where freezing temperatures can be a problem.

Dairy Facilities

Dairy facilities include separate structures for dairy cattle housing, storage for grain and forage, milking, and milk processing and storage. Important structure components include electrical service and plumbing.

Housing

Dairy cattle housing structures are usually open-plan buildings with steel pens for separating livestock as needed. Dairy cattle housing should provide adequate shelter from the weather, while at the same time reducing stress on the animals, **Figure 6-12**. Housing facilities may vary in design and needs due to the local climate.

MaxyM/Shutterstock.com

Laila Kazakevica/Shutterstock.com

Figure 6-12. Dairy housing barns may be partially open structures to allow natural ventilation. Some facility designs may allow the cattle to enter and exit the housing area/pasture at will.

Charodeyka/Shutterstock.com

Figure 6-13. Milking parlors must be kept clean to prevent milk contamination. Most dairy farms have strict protocol for the milking process in order to maintain high-quality products. *Have you ever visited a large dairy production operation? Were visitors restricted to certain areas of the farm?*

Milking Parlor

The ***milking parlor*** is the section of a dairy where the cow is moved to in order for milking to occur. Milking parlors are elaborate facilities with complicated plumbing designs, **Figure 6-13**. They must be kept extremely clean to prevent milk contamination and to help prevent udder infections. The temperature is also usually controlled to help maintain milk at a specific temperature before it is processed. Careful planning and design will allow efficient movement of cows into and out of the parlor.

Andrew F. Kazmierski/Shutterstock.com

Figure 6-14. Sheep farms provide adequate grazing resources and housing. *How much grazing and housing space should be allotted per animal?*

Sheep Facilities

The housing need for sheep is determined by the climate and weather where they are raised. Sheep are hardy, grazing animals with natural protection against cold weather, **Figure 6-14**. Many sheep producers have standard barn facilities for use during the lambing season. This is especially important if lambing occurs during extreme weather conditions.

In addition to feed and equipment storage structures, sheep facilities may require a shearing facility with storage space for fleece. Facilities may also require fencing for feedlots and runs.

Goat Facilities

The housing need for goats is determined by the climate where they are grown. Goats are usually content with simple housing facilities and, depending on the climate, may not require more than simple structures to keep them dry, **Figure 6-15**. They are also adept at escaping their confinements and require sturdy, goat-proof fencing.

Equine Facilities

The term ***equine*** refers to those members of the horse family, which also includes donkeys, mules, and wild species such as the African Zebra. Domesticated species of the equine family are housed in equine facilities. Riding stables that provide boarding and riding or training lessons have much different needs than an operation designed mainly for breeding or training racehorses. Facility needs will also vary by climate.

Judy Marie Stepanian

Figure 6-15. Goat facilities may be simple structures designed to keep the animals dry.

Iriana Shiyan/Shutterstock.com

forestpath/Shutterstock.com

Figure 6-16. Horse barns provide clean and safe housing for horses. Separate washing areas come in handy when preparing animals for showing. ***What other types of special amenities might come in handy in a horse barn?***

Basic structures for equine facilities include:

- Stables—to provide safe and efficient space for feeding and housing horses, **Figure 6-16**. Most facilities use separate stalls for each horse. Concrete or dirt flooring may be used along with bedding materials. Rubber mats designed for horse stalls should be used with concrete floors. Many stables are built of wood but metal structures are also used. When metal structures are used, wooden stalls should be installed with wood covering exposed metal walls to prevent damage to the wall as well as to the animals.
- Arenas—indoor arenas are used for riding as well as horse training. In colder climates, indoor arenas are especially useful during winter months. These facilities may or may not be heated, **Figure 6-17**. The flooring is typically bare ground with some type of mulching material, such as

Eduard Kyslynksyy/Shutterstock.com

Iriana Shiyan/Shutterstock.com

Figure 6-17. Indoor arenas are used for riding, riding instruction, and training. In addition to protection from the elements, indoor arenas may help the rider or trainer limit distractions from other animals, people, and vehicles.

pecan shells, sand, or shredded asphalt shingles. As with most materials, each has its advantages and disadvantages.

- Feed and forage storage—in areas where grazing is limited or during cold winter months, horse owners must have adequate storage space for bales of hay and other forage. Hay used for horses must be kept free of mold as they are more susceptible to health problems caused by molds than most other ruminants. Inadequate storage space will require the purchase of forage at higher, off-season prices.
- Tack room—to properly store expensive bridles and saddles and any other equipment needed for equine care. This area may include a small refrigerator to store perishable medical supplies.
- Large equipment storage—to properly store equipment ranging from racing sulkies to carriages and hayride wagons.

Other important components of equine facilities include a water supply system and drainage as well as electrical service. Horses and other equines require large amounts of clean drinking water. Some facilities use a well and buckets whereas others may have more elaborate plumbing to provide fresh water to each stall. A specific area for bathing horses may also be desired. Storage space for bedding materials may also be required, depending on the size of the operation.

Waste Management

Most equine facilities use manual labor to remove waste from individual stalls. To keep stables clean, manure must be removed on a regular basis. Many facilities use open areas to keep manure until it is hauled away for processing into fertilizer.

Aquaculture

Although ***aquaculture***, or fish farming, is often done outdoors in natural lakes, ponds, or even the sea, some farmers have smaller scale operations using tanks inside barns, **Figure 6-18**. As with traditional livestock

Vladislav Gajic/Shutterstock.com

Pan Xunbin/Shutterstock.com

Figure 6-18. Aquaculture operations may be set up in open waters or in indoor facilities with an elaborate system of tanks. Indoor operations must provide an adequate supply of clean oxygenated water for fish in addition to nutrients and adequate temperature control. ***Can the water from the tanks be recycled into the system or used for another purpose?***

production, fish farming requires equipment and facilities for feeding and housing fish. Depending on the climate, the structure may require insulation and a heating or cooling system. The flooring may be concrete or gravel to provide drainage and reduce dust.

Major components of indoor aquaculture operations include:

- Electrical service—the system must be designed to withstand moist environments. It must also be designed to provide adequate power for pumps, heating and cooling (water and air), lighting, and other specialty equipment.
- Water supply and filtration system—neither tap water nor well water is usually suited to house fish without treatment. An indoor aquaculture operation will require a steady supply of clean water to maintain water levels.
- Environmental control system—the type of system needed will depend on the local climate and the efficiency of the structure. Most fish require cooler temperatures to thrive so both the ambient temperature and water temperature must be controlled.
- Waste management—all animal production generates waste, and fish are no different. Facility design and management must include a plan for fish waste disposal.

Storage Facilities

Each agricultural operation varies in its structural needs. However, most operations are not limited to one production area and require multiple types of structures. For example, many dairy farms grow and harvest hay and grains to feed the dairy cattle. In addition to the animal housing and milking facilities, the dairy farm would need structures to store the hay and grains. The dairy farm would also need structures to store and maintain machinery such as tractors and harvesting equipment. A large operation would also need structures to store and process waste. Typical storage structures found on many farms and ranches include:

- Barns
- Grain bins
- Silos
- Waste storage (manure pits/lagoons)

Barns

Did You Know?

Why are barns red? Some people say barns are red because farmers once added ferrous oxide (rust) to the oil mixture used to coat the barn wood. The rust deterred the growth of fungi, including mold and moss. These fungi would trap moisture in the wood, increasing its rate of decay.

Barns come in all shapes and sizes, and are usually built for a specific purpose. Barns for hay storage are larger than those designed for tobacco curing, and horse stables are typically arranged differently than barns designed to house dairy cattle. Older barns were built entirely of wood whereas many newer barns may be framed with wood, but are finished with metal or vinyl siding. The intended use may determine the type of structural and finishing materials that should be used, or the owner's desired appearance may be the deciding factor, **Figure 6-19**. Another factor to consider is the amount of maintenance each type of material will require.

Grain Bins

Grain bins are highly specialized structures designed to store grain until it is transported to market, **Figure 6-20A**. Most grain bins are made from metal and may include machinery used to move the grain from trucks into the bin. Grain bins are designed to withstand and contain the immense pressure and weight of their contents. The main components of grain bins include:

- Electrical service—to run augers and conveyors when unloading or loading grain. It is also used to run ventilation systems.

Brian Goodman/Shutterstock.com

Modfos/Shutterstock.com

Figure 6-19. Barns vary greatly by size, layout, and building materials. The barn on the left provides hay storage on the second level (or loft), and may be used to house animals in stalls or equipment on the ground floor. The prefabricated barn on the right is designed to house livestock and most likely does not include a storage area.

marchello74/Shutterstock.com

V. J. Matthew/Shutterstock.com

Figure 6-20. Grain facilities store and dry grain and have a system of augers and conveyors to move grain into and out of grain bins (A) and silos (B).

- Ventilation systems—to remove gases and maintain proper moisture levels to prevent spoilage.
- Built-in ladders—to provide safe access when loading and unloading grain.

Grain bins must also provide the content with protection from the elements, especially precipitation.

Silos

Silos are tall metal or concrete storage containers used to hold silage, **Figure 6-20B**. ***Silage*** is grass, fermented feed, or other fodder that is harvested green and used to feed cattle. Silos may also be used to store grain. Silos are often designed so the height may be increased by adding another "ring" of building material. Like grain bins, silos require electrical service to run conveyance equipment and run ventilation systems. When attached to animal housing, silos require special construction to prevent silo gases from entering and settling in the housing areas.

Waste Storage Structures

Livestock and dairy operations generate animal waste. Animal waste has value. It can be used to fertilize growing crops. Efficient systems for managing animal waste make good use of the nutrients in manure while at the same time preventing contamination of water and other natural resources, **Figure 6-21**.

Waste storage and processing systems will vary in size and design according to the size of the production operation. There are many hazards

rtem/Shutterstock.com

CreativeNature R.Zwerver/Shutterstock.com

Figure 6-21. Slatted floors allow hog waste to drop into a holding area where it is flushed out of the house and into a holding lagoon. Many larger livestock production facilities have installed waste processing systems to create biofuels for use on the farm and for sale.

involved with waste storage and management, and facilities should be carefully planned and constructed with safe operation in mind.

The Construction Process

Once the building or structure has been designed and the blueprints created, the construction process may begin. The next step is to select materials for the project. As stated earlier, the intended use of the structure and the maintenance needs may determine which materials would best suit the project. Many modern agricultural facilities use prefabricated components for ease of assembly and lower installation costs.

Permits and Inspection

One of the next steps to building new structures is permit acquisition. Construction projects usually require building permits and periodic inspections during the building process. The permitting process ensures that builders use the correct building methods and materials, and that they construct buildings on appropriate sites.

Building inspectors conduct inspection throughout the construction process to verify that approved practices are being used for electrical and plumbing installation, and that the building structure is sound.

Site Preparation

Construction begins with site preparation. To prevent accidents and service interruption, local utility services must first be called to mark the location of any buried lines, cables, or pipes on the property. Once these areas have been marked, a surveyor is hired to survey the site and mark appropriate boundaries. At this point, the area may safely be cleared and leveled in preparation for the foundation, **Figure 6-22**. Improper or

Darryl Brooks/Shutterstock.com

Figure 6-22. Good site preparation is needed for sturdy foundations. *What type of equipment is used to clear building sites?*

inadequate site leveling and preparation can result in cracked foundations and damage to the building.

Footings

Almost all structures require the installation of footings before any other part of the building is constructed or installed. Footings are made of concrete and designed to distribute the weight of a structure over a greater area to minimize settling or movement of the structure. Local building codes often determine the size, spacing, and depth of footings.

Foundation

Once the footings are poured and cured, the building's foundation may be installed. The foundation is typically constructed of concrete, masonry block, or brick. Many agricultural structures are constructed on concrete slab foundations. This type of foundation is poured over a gravel base and usually has reinforcement bars placed throughout. All drains must be in place before the concrete is poured.

Framing

Once the foundation is in place, the framing process begins. The external walls and roof are constructed on-site, or prefabricated sections are transported to the job site and assembled. The frame of the building supports the exterior walls, interior walls, and the roof, **Figure 6-23**. All framing must meet approved building standards. Once the wood

Christian Delbert/Shutterstock.com

Figure 6-23. Framing is the construction of the interior frame of the building, the roof, and siding.

Christian Richards/Shutterstock.com

Figure 6-24. The finishing stage involves the installation of the electrical and plumbing fixtures.

framing and roofing is completed, the electrical wiring, plumbing, and ventilation ducts may be installed.

Roofing

Roofing materials vary according to the purpose and use of the building as well as the climate of the area. Sheet metal, asphalt shingles, or fiberglass panels may be used to cover most agricultural buildings. Greenhouses may use shade fabric, plastic sheeting, polycarbonate panels, or glass panes and panels.

Finishing

The finishing process includes the outdoor siding as well as various interior components. Many agricultural buildings are not "finished" on the interior. For these types of interiors, the finishing process usually involves the installation of electrical and plumbing fixtures as well as the construction of pens, stalls, and work areas, **Figure 6-24**. Structures that are completely finished on the interior also require the installation of insulation, wall coverings, ventilation systems, and appliances such as furnaces, water heaters, and air conditioners. Inspectors will examine the building to ensure that all systems meet safety standards and approved practices for construction.

Career Connection General Contractor

Job description: The general contractor, who is often a tradesman, is responsible for the overall coordination of a project. The general contractor is employed by the client and may be recommended to the client by the architect or engineer designing the project.

As a general contractor, you must hire (contract) and organize the scheduling of work crews for each step of the project. The crews include the surveyors, site preparation crew, foundation/concrete crew, framers, roofers, electricians, plumbers, drywall crew, painters, and landscapers. You may be responsible for accounting and payroll for your own crew, as well as paying the contracted workers on behalf of the employer. The general contractor is also responsible for providing the materials and equipment necessary for the construction of the project.

Bikeriderlondon/Shutterstock.com

As a general contractor, you must be able to work with multiple crews at the same time and have good problem-solving skills.

Education required: High school diploma or equivalent at minimum. You must also acquire a license through the state licensing exam. An associate's or bachelor's in business management is also useful. Experience in the trades industry is also necessary.

Job fit: This job may be a fit for you if you have good oral and written communication skills, as well as excellent organizational skills. You must also be willing to work in all types of weather conditions and have good leadership skills. Good problem-solving skills are essential in this type of position.

Lesson 6.1 Review and Assessment

Words to Know

Match the key terms from the lesson to the correct definition.

1. An agricultural structure designed to provide space and feed for pigs to grow from weaning weight to market weight.
2. The section of a dairy operation where the cow is moved to in order for milking to occur.
3. The animals in the family *Equidae*, which includes single-toed hooved animals like horses and donkeys.
4. A tall metal or concrete storage container used to hold silage.
5. An agricultural structure designed to provide an area for the sow to birth and nurse pigs until weaning.
6. A swine production facility in which pigs are not allowed outdoors.
7. Highly specialized structures designed to store grain until it is transported to market.
8. A production operation in which poultry are raised as a meat source.
9. A structure used to replicate ideal growing conditions for plants to extend the growing range and season.
10. Grasses such as corn and sorghum that are harvested green and fermented to create feed for cattle.
11. Fish production or farming that may be done outdoors in open waters or indoors in tanks.

A. aquaculture
B. broiler production
C. confinement operation
D. equine
E. farrowing facility
F. finishing facility
G. grain bin
H. greenhouse
I. milking parlor
J. silage
K. silo

Know and Understand

Answer the following questions using the information provided in this lesson.

1. List five uses for agricultural structures.
2. How are livestock and plant facilities similar to those built for humans?
3. List five important greenhouse components.
4. Explain why agricultural facilities are best located near roads and other modes of transport.
5. List three basic structures used in equine operations.
6. How can animal waste be used on a farm?

7. What are the steps to the construction project?
8. Why must buildings under construction be inspected?
9. Explain why site preparation is important in building construction.
10. List the types of roofing materials used in agricultural construction.

Analyze and Apply

1. Examine farming operations in your community and choose two that produce livestock, dairy, poultry, or some type of specialty animal. Compare and contrast the different facilities and resources needed by each specific farm.
2. How are livestock facilities similar to the homes in which we live?
3. Describe the process you would use to build a greenhouse.

Thinking Critically

1. A friend of yours is planning to start a small business raising quail. He plans to sell quail eggs to the local community. He has done the research about all areas of the business, but has come to you for advice on what structures he needs for the birds. What advice would you give him?
2. In some areas of the Northern United States, it may be necessary to provide a calving barn for a beef herd. Other than this structure, beef animals tend to do well in colder climates as long as they have a windbreak to prevent windchill. Why not build barns for beef animals to get them out of the cold weather?
3. Horse stables are different from most other types of barns on the farm. Why do you suppose this is so?

Lesson 6.2

Emerging Agricultural Technology

Words to Know

biodiesel
biofuel
closed ecological system
continuously operating reference station (CORS)
ethanol
global navigation satellite system (GNSS)
global positioning system (GPS)
hydroponic system
methanol
OLED lighting
organic farming
precision agriculture
radio frequency identification (RFID)
telematics
telemetry
vertical farming
wood alcohol

Lesson Outcomes

By the end of this lesson, you should be able to:

- Identify emerging technologies in agriculture.
- Explain the intended benefits of these emerging technologies.
- Identify current technologies at work in the agricultural industry.
- Explain the benefits of current agricultural technologies.

Before You Read

Before you read the lesson, interview someone in the workforce (your supervisor, a parent, relative, or friend). Ask the person why it is important to know about the lesson topic and how this topic affects the workplace. Take notes during the interview. As you read the lesson, highlight the items from your notes that are discussed in the lesson.

The population of planet Earth now exceeds more than seven billion people, and more than 800 million people go to bed each night without enough food to eat. According to the Population Division of the United Nations, the world population is projected to grow to 9.6 billion by the year 2050. As Earth's population continues to grow, producing enough food and fiber to feed and clothe so many people continues to be a challenge for the agriculture industry. To meet these needs, the agriculture industry is continuously researching and developing safer and more economical methods to improve food and fiber production. Many of these methods include new and exciting technologies, **Figure 6-25**.

Technology in Agriculture

Technology is the application of scientific knowledge, tools, or processes for practical purposes and problem solving. Technology in agriculture includes the application of these tools and processes to:

- Develop more economical methods to increase food and fiber production.
- Develop more responsible methods of water use and reduce contamination of water resources.
- Responsibly reduce the effect of pests on crops and livestock, and reduce the amount of pesticides required for pest control.
- Reduce soil erosion due to water and wind as well as tilling practices.

alik/Shutterstock.com

Figure 6-25. Newer technologies like drones and GPS have been adapted for use in the agriculture industry. *Can you think of other technologies that have been adapted for agricultural applications?*

- Build more efficient and environmentally friendly machinery.
- More accurately apply fertilizers and reduce excessive runoff.
- Produce alternative forms of fuel.
- Develop alternative farming methods.

As you can see, technology is applied to every aspect of agriculture.

Technological Advances

It is a common sight during the spring months to see a tractor pulling a large plow through a field, turning the soil and preparing it for planting. The tractor makes the job look easy, but it has taken hundreds of years of technological advances to get us to where we are now with regard to agricultural mechanization, **Figure 6-26.**

Consider the moldboard plow whose simple design evolved over more than five decades before being mass produced.

Moldboard Plow

In 1788, while serving as minister to France, Thomas Jefferson observed the difficulties European farmers had in tilling the soil. The plows of the 1700s did not adequately turn the soil. Inspired by this problem, Jefferson, upon his return to the United States, developed the *moldboard plow* (1794). Early plows barely scratched the surface of the soil. Jefferson's moldboard plow lifted and turned the soil, making a suitable seedbed for crops. First being made of wood, Jefferson's plow was prone to break. Although Jefferson

Boo Heisey/Shutterstock.com Gerard Koudenburg/Shutterstock.com

Figure 6-26. This grain harvester (left) was a state-of-the-art machine in the early 1900s. Modern grain combines (right) can do the work of several of their early predecessors in a fraction of the time. ***How were the early machines powered?***

began having his plows cast in iron in 1814, he never patented the design and made no further alterations.

Around 1833, Illinois blacksmith John Lane expanded on Jefferson's design by creating a plow surface made from a steel saw blade. The steel face of the plow allowed the soil to slide easily across the plow and be lifted and turned. John Deere perfected the design and began mass production of the plow in the 1850s. Deere's plow made the task of seedbed preparation and cultivation much easier for us than it was for our ancestors. The emergence of the steel plow is symbolic of the continual improvement of agriculture, **Figure 6-27**.

Precision Agriculture

Precision agriculture is a set of technologies that have helped agriculture advance into the digital information-based world. Together, these technologies help agriculturalists, such as farmers and ranchers, gain greater control over the management of their production and harvesting operations.

Precision agriculture is based on specific measurements of crop and livestock production factors. Once collected, these measurements are used to supply resources in precise amounts for optimum growth of both plants and animals. For example, the soil in a farm field is not uniform in its characteristics. Some areas of a field may have the right amount of nutrients for the crop plants, while other areas may be deficient. In the early days, farmers would apply a uniform amount of fertilizer across the field regardless of the soil's varying nutrient characteristics. Some areas would receive more nutrients than were needed, while others would still be deficient. This

Weldon Schloneger/Shutterstock.com

Figure 6-27. John Deere's moldboard plow made the task of plowing the prairie easier and faster. *What other mechanical inventions revolutionized agriculture?*

was a waste of fertilizer, much of which would end up in runoff water. Today, this waste can be avoided through a combination of technologies including global navigation and positioning systems.

Global Navigation Satellite Systems (GNSS)

Farmers and ranchers across the world can now use global navigation satellite systems (GNSS) with their computer-operated machinery to precisely apply fertilizers, spray pesticides, and irrigate their fields. ***Global navigation satellite systems (GNSS)*** are satellite systems that use receivers and specialized computer programs to pinpoint the geographic location of a user's receiver. In the United States, agriculturalists may use a GNSS referred to as ***GPS***, or ***global positioning system***, to coordinate their livestock feeding and application of fertilizers, pesticides, and water.

Global Positioning System (GPS) Application

When using GPS to optimize fertilizer application, farmers first go through their fields, taking soil samples and marking the locations of the samples by inputting the GPS coordinates, **Figure 6-28**. Once the soil samples are analyzed, this data is also put into the computer program. With the use of precision maps (developed with GPS technology) and in-cab guidance systems, computer-controlled implements vary the amount of fertilizer application as the tractor moves across the fields. Through this GPS technology, the exact nutrient needs of the crop are met and fertilizer waste and pollution is reduced, **Figure 6-29**. The same principles are used to apply pesticides and control irrigation. Farmers using precision agriculture techniques will reduce labor costs, improve fuel efficiency, conserve irrigation water, and improve soil health.

Sander van der Werf/Shutterstock.com

Figure 6-28. GPS systems allow the precise application of fertilizers and pesticides in a field.

auremar/Shutterstock.com

Figure 6-29. Data collected from the field is analyzed and used to manage resources with a minimum of waste. *What type of electronic devices are needed to use these types of GPS systems?*

Telematics and Mobile Applications

Agricultural equipment is expensive to purchase and costly to maintain. To help maintain agricultural equipment and prevent costly damage, newer machinery uses telematics to communicate with the

nulinukas/Shutterstock.com

Figure 6-30. These beef animals are wearing devices around their necks that help the farmer determine the amount of feed they are consuming. *What type of transmission technology do these types of devices use? How is the data collected?*

WilleeCole Photography/Shutterstock.com

Figure 6-31. RFID chips are used in small animal care for tracking purposes. These chips can also be implanted in cattle to deter rustling.

operator and the implement dealer when a mechanical problem arises. ***Telematics*** is the sending and receiving of data using wireless telecommunications. This is typically used in the transportation industry to track vehicles and is used in agriculture for the same purpose.

Telemetry is the automated communication system that collects data from a remote location and transmits it to receiving equipment in another location. A self-diagnostic computer sends a message or code to the user describing the problem with the equipment. These computers are usually also programmed to notify the operator when routine maintenance, like an oil change, is needed.

Livestock Telemetry

Telematics are not just reserved for use in agricultural equipment. The technology is being used in the livestock industry as well. Beef cattle wear electronic collars or have microchips implanted that pinpoint their location and physical condition, **Figure 6-30**. By examining the data received from these tracking devices, the farmer or rancher can make well-informed management decisions regarding areas such as breeding and feeding, and assessing the overall health of animals.

Livestock producers may also use ***radio frequency identification (RFID)*** tags on animals to automatically identify and track their movement, **Figure 6-31**. Much like the ID chips used in companion animals, RFID tags are a useful deterrent to the practice of cattle rustling.

Crop Telemetry

Telemetry is also in use in crop production. Hay growers plant microchips in hay bales to monitor moisture content. This reduces hay storage loss due to excessive moisture. Telemetric devices called ***continuously operating reference stations (CORS)*** are also used to locate pest infestations, measure soil moisture and temperature, and monitor weather conditions,

Figure 6-32. Telemetry systems may also be used to control the amount of irrigation a field will receive based on measurements of soil moisture.

claudio zaccherini/Shutterstock.com

Figure 6-32. CORS monitor weather, and provide information to help determine the amount of irrigation needed.

Drones

Farmers use drones to acquire information about their crops. Drones are an inexpensive method for taking aerial photographs and video of farm fields. These photographs and videos reveal patterns in the foliage of a crop. These patterns may indicate differences in soil moisture and the presence of pest or fungal infestations. The cameras mounted on a drone can also take infrared photographs which further highlight the differences between healthy and stressed plants. See **Figure 6-33**.

Stockr/Shutterstock.com

Figure 6-33. Using drones to scout crops can save a grower time and effort. *How far can a typical drone travel? What types of things might interfere with a drone's controls?*

Video Cameras

Video cameras strategically mounted throughout multiple facilities allow workers to observe and monitor several farm operations at once. Observation may take place in a central location using multiple monitors, or on tablets, laptops, and cell phones with mobile applications. This technology may allow some flexibility in the owner/operator's day-to-day schedule and added efficiency for his or her operation.

Video cameras may be used to monitor activities in and around livestock feeding facilities, to check grain elevator and feed mill production, and to provide security in and around farm buildings and equipment. Cameras may also be used to monitor hazardous areas, such as the top of grain handling equipment and inside grain bins.

Agricultural Robotics

The agriculture industry is developing agricultural robotics, or Agbots, to handle some of the repetitive, labor-intensive tasks on the farm. In the near future, Agbots may be manually removing weeds, scouting for crop pests, taking soil samples, and harvesting crops. In citrus groves, Agbots may be used to locate and harvest ripe fruit. In time, these robots may be able to

KYTan/Shutterstock.com

Figure 6-34. Biofuels use energy derived from plants. ***Have you ever been to a biofuel plant?***

measure the moisture content and ripeness of the fruit, and detect the presence of disease or pesticide residues.

Biofuels

Biofuels are liquid fuels made from organic materials and are intended to replace fossil fuels, **Figure 6-34**. There are three types of biofuels:

- ***Ethanol***—the product of the fermentation of sugars derived from plants.
- ***Methanol*** or ***wood alcohol***—derived from natural gas, coal, plant vegetation, and agricultural crop biomass.
- ***Biodiesel***—a fuel derived from vegetable oils and animal fats.

In the United States, ethanol is the most common biofuel. It is primarily made from corn starch. These fuels from biological materials are an emerging technology. The production of corn for biofuel increases the demand for corn and drives up the price per bushel. At the same time, this higher price for corn pushes up the cost for corn-based livestock feeds. There is also the concern that biofuels may be more harmful to the environment than traditional fossil fuels.

Organic Farming

Organic farming is a system of agricultural production that attempts to reduce or eliminate the use of synthetic agricultural inputs such as synthetic antibiotics, fertilizers, and pesticides. The goal of organic agriculture is to produce healthy sustainable agricultural ecosystems. The organic foods movement in the United States continues to gain momentum.

Organic foods are now available in more than 20,000 retail stores and more than 75% of all grocery stores, **Figure 6-35**. The sale of organic foods is more than $35 billion annually in the United States. Fruits and vegetables make up the largest segment of the organic foods market. However, consumers can also buy organic beef, poultry, eggs, pork, lamb, and dairy products.

Farmers can apply to receive recognition for growing organic crops and livestock on their farm through the U.S. Department of Agriculture organic certification program. This helps to ensure that products labeled as organically grown actually meet standards for organic production. General requirements for organic certification through the USDA include:

- Agricultural practices must preserve natural resources.
- Livestock must be confined in a space that allows animals to have natural movement and exercise.
- Farmers must only use production inputs that meet organic standards.
- No genetically modified organisms or products may be used as inputs.

Mike McDonald/Shutterstock.com

Figure 6-35. Organic farming is emerging as a major economic force.

In Vitro Meat Production

Scientists are experimenting with the possibility of creating meat products through tissue engineering. This involves taking stem cells from livestock animals and growing the meat in a culture medium in a lab, **Figure 6-36**. The cells divide and grow. The goal is to produce meat products without having to slaughter livestock. The term *in vitro* means that the meat is produced in a glass container such as a petri dish. The in vitro production of meats would be relatively easy for products such as sausage and hamburger. Creating highly structured meats such as beefsteak and pork chops would be more difficult because they are comprised of muscle and tissue. Muscle cell tissue is more difficult to create in vitro.

Advantages and Disadvantages

The advantage of in vitro meat production is that it does not require the slaughter of animals. The stem cells can be removed painlessly from the animal, which reduces the stress on the livestock. The disadvantage of in vitro meat production is that it requires artificial structures and processes to

anyaivanova/Shutterstock.com

Figure 6-36. In vitro meat production is an emerging technology. *Can you imagine eating a steak grown in your refrigerator?*

create something that nature manages to do on its own. Stem cells are cells that have not decided what they are going to be. In nature, stem cells might become muscle tissue or organ tissue. Stem cells may require coaxing to become the type of cells that form marketable meat products.

Growth Medium

Stem cells need a growth medium to provide nutrients while they grow. Developing a suitable growth medium for the in vitro process is a challenge. It must provide nutrients without contaminating the meat cells with other animal cells. Therefore, the growth medium, or serum, cannot be derived from animal cells. The growth medium must also serve as a structure for growing into the type of cells needed in meat production. Muscle fibers, the essential element in most meat products, needs to be stretched or exercised in order to develop. Typically, a livestock animal stretches muscle fibers by moving around. This is not possible in the in vitro process.

Once the stem cells, growth medium, and structure are provided, they must be combined at the proper temperature and environment. Research into the optimal development for stem cell production is ongoing.

Vertical Farming

Vertical farming has been in existence in rural areas since the 1950s. Multiple-story farm structures, like two-story poultry and swine facilities, have been used in small farming operations to house livestock and crops. In urban areas, ***vertical farming*** is an emerging technology in which buildings

Kingarion/Shutterstock.com

Figure 6-37. Tomatoes are well suited for vertical farming. *What attributes make a tomato plant well suited to vertical farming?*

taller than one story may be used to cultivate raw agricultural products, **Figure 6-37**. Vertical farming involves the conversion of existing structures into spaces for growing plants or raising small farm animals for consumption by the local population. Most of the transportation needs and costs associated with food production are reduced or eliminated since the food is being cultivated close to the population that is consuming it.

Making vertical farming in urban areas a reality requires solving problems different from those experienced in traditional agricultural practices, including:

- Lighting—artificial lighting needed to supplement natural sunlight.
- Irrigation—water must be precisely distributed throughout the facility.
- Fertilization—nutrients must be applied in precise amounts.
- Pest control—the use of pesticides in a controlled atmosphere.
- Waste control—using wastewater from commercial fish tanks to fertilize plants.

Did You Know?

More than 80% of the U.S. population lives in an urban area.

Lighting Systems

Lighting for a vertical farm presents some unique challenges, but proponents of vertical farming are quickly finding solutions such as OLED lighting (organic light-emitting diode) and the use of parabolic mirrors or reflectors, such as those used in solar energy systems. ***OLED lighting*** uses light emitted by organic material to which voltage has been applied. OLED lighting is a surface light source, it does not heat up, it is slim and lightweight, it can be curved, and it does not contain harmful substances or consume much power. See **Figure 6-38**.

A.Krotov/Shutterstock.com

Figure 6-38. Lighting systems provide extra "sunlight" for plants to encourage growth during short winter day periods.

Irrigation and Fertilization

Most vertical farms use, or will use, hydroponic systems to provide water and nutrients. ***Hydroponic systems*** grow plants in nutrient-rich water, without the use of soil, **Figure 6-39**. This allows the precise application of the correct amount of fertilizer for plants. Water used in the growing process is recycled repeatedly.

udeyismail/Shutterstock.com

Figure 6-39. Hydroponic systems eliminate the need for soil, and yet provide the nutrients plants need to thrive. *Have you eaten food grown in a hydroponic system? Did it taste different from the same food grown in soil?*

Pest and Waste Control

Because of the highly controlled environment in which plants are grown, pests are not as much of a problem as in conventional agriculture. The use of pesticides is kept to a minimum. Very little waste is produced and unused plant materials may be composted.

Cost

Another challenge for implementing vertical farming is the cost and effort of renovating buildings to house agricultural crops. Skyscrapers are built as office and living space for humans, and converting buildings for agricultural may be expensive. New construction will require extensive research and the development of new designs.

Rooftop Gardening

One important variation of vertical farming is the development of rooftop gardening. Rooftop gardens are emerging as a solution to the problem of providing locally grown produce in urban areas, **Figure 6-40**. Rooftop gardens provide insulation for the building and help reduce energy costs. This form of vertical farming allows the inhabitants of a city to enjoy gardening as well.

Alison Hancock/Shutterstock.com

Figure 6-40. Rooftop gardening is not only a hobby, but can be a business opportunity for urban farm enthusiasts. *What would be the biggest challenge to establishing a rooftop garden?*

Closed Ecological Systems

An ecosystem is a community of living organisms. These living communities often share resources with other ecosystems. For example, water in a mountain forest ecosystem is used by the plants in that ecosystem. This same water travels downstream in a watershed to lowlands where it can also be used in crop production. ***Closed ecological systems*** are a specific type of community that do not allow for the exchange of any material inputs outside of the system. They are an attempt to create an ecosystem that is self-sustaining. Organisms in the ecosystem create resources for use by the other organisms in the ecosystem. Waste products are converted into nutrients and energy. An example of a closed ecological system is the Biosphere 2 project in Arizona, **Figure 6-41**. Ongoing research into these systems may eventually develop technology that could be used to reduce the cost of agricultural production. Closed ecological systems are also of significant importance to long-duration space flights.

Spirit of America/Shutterstock.com

luchschen/Shutterstock.com

Figure 6-41. Experiments at the Biosphere 2 in Arizona measure the feasibility of closed environmental systems. *Could you live in this type of isolated environment? What would be the most difficult part of adapting to life in a biosphere?*

Lesson 6.2 Review and Assessment

Words to Know

Match the key terms from the lesson to the correct definition.

1. A biofuel derived from natural gas, coal, plant vegetation, and agricultural crop biomass.
2. A biofuel that is the product of the fermentation of sugars derived from plants.
3. An automated communication system that collects data from a remote location and transmits it to receiving equipment in another location.
4. A biofuel derived from vegetable oils and animal fats.
5. A crop growing system in which plants are grown in nutrient-rich water, without the use of soil.
6. Telemetric system used to locate pest infestations, measure soil moisture and temperature, and monitor weather conditions.
7. A lightweight, cool, lighting system that uses organic light-emitting diodes.
8. A set of technologies that helps agriculturalists gain greater control over the management of their production and harvesting operations.
9. An enclosed community that does not allow for the exchange of any material inputs outside of the system.
10. The sending and receiving of data using wireless telecommunications.
11. A system of agricultural production that attempts to reduce or eliminate the use of synthetic agricultural inputs such as synthetic antibiotics, fertilizers, and pesticides.
12. An emerging technology in urban areas in which buildings taller than one story may be used to cultivate raw agricultural products.
13. A small electronic tag that is attached to livestock in some way and used to identify and track animal movement.
14. A system of navigation that uses space satellites to provide information on the location of places and objects.

A. biodiesel
B. closed ecological system
C. CORS (continuously operating reference station)
D. ethanol
E. global positioning system (GPS)
F. hydroponic system
G. methanol
H. OLED lighting
I. organic farming
J. precision agriculture
K. radio frequency identification (RFID)
L. telematics
M. telemetry
N. vertical farming

Know and Understand

Answer the following questions using the information provided in this lesson.

1. Explain how the agriculture industry is preparing to meet the needs of Earth's growing population.
2. List three ways in which technology is applied in agriculture.
3. *True or False?* Most technological advances occur quickly.
4. Explain how precision agriculture helps farmers and ranchers gain greater control over the management of their production and harvesting operations.
5. Agriculturalists may use _____ technology to coordinate their livestock feeding and application of fertilizers, pesticides, and water.
6. List three types of biofuels.
7. What are four general requirements for organic certification through the USDA?
8. *True or False?* Creating highly structured meats in vitro would be easier than producing meats like sausage and hamburger.
9. List five areas that will require problem solving in order to make vertical farming in urban areas a reality.
10. Do you think pest control would be a major issue in a closed ecological system? Explain your answer.

Analyze and Apply

1. How has technology improved something in your own life experience?
2. How could agricultural robots be used to control pests in a greenhouse?
3. What are the key factors to consider when installing a vertical farming operation?

Thinking Critically

1. In your opinion, what is the technology that has made the greatest positive impact on agriculture?
2. If you are growing a small garden in your backyard, how could you apply precision agriculture methods to save resources?
3. Are there drawbacks to genetic engineering? Explain your answer.

Chapter 6

Review and Assessment

Lesson 6.1

Agricultural Structures

Key Points

- Most agricultural operations require a variety of structures and facilities, each with its own purpose. Each facility may differ in the type and manner of construction.
- Construction is a multistage process, beginning with site preparation, foundation construction, framing, roofing, and finishing.
- Agricultural facilities should be designed for tough industrial conditions.
- Agricultural facilities must often provide all of the elements needed for animals and plants to survive. Buildings such as greenhouses or broiler houses need to be designed carefully to ensure the well-being of the animals and plants.

Words to Know

Use the following list and the textbook glossary to review and study the *Words to Know* from *Lesson 6.1*.

aquaculture
broiler production
confinement operation
equine
farrowing facility
finishing facility
grain bin
greenhouse
milking parlor
silage
silo

Check Your Understanding

Answer the following questions using the information provided in *Lesson 6.1*.

1. Identify six basic greenhouse components.
2. How do livestock buildings in a northern climate differ from those in a southern climate?
3. What type of flooring is typically used in broiler production facilities?
4. What is the difference between a swine farrowing house and a swine finishing house?
5. Explain why environmental control is especially important in swine production.
6. Explain why a dairy milking parlor needs to be kept extremely clean.
7. Name two types of animals from the equine family.

8. Identify five types of basic structures for equine facilities.
9. What are four major components of indoor aquaculture operations?
10. Explain why buildings are inspected and construction permits are required when building agricultural facilities.

Lesson 6.2

Emerging Agricultural Technology

Key Points

- Technology is the application of scientific knowledge, tools, or processes to solve problems.
- Precision agriculture is the scientific measurement of crop and livestock production processes so that precise amounts of resources can be applied for optimum growth.
- Biofuels are liquid fuels that can offset the use of fossil fuels.
- Organic farming is designed to reduce the use of synthetic production inputs in agriculture.
- Telemetry is used to send and receive information in agricultural operations using wireless communications.

Words to Know

Use the following list and the textbook glossary to review and study the *Words to Know* from *Lesson 6.2*.

biodiesel
biofuel
closed ecological system
continuously operating reference station (CORS)
ethanol
global navigation satellite system (GNSS)
global positioning system (GPS)
hydroponic system
methanol
OLED lighting
organic farming
precision agriculture
radio frequency identification (RFID)
telematics
telemetry
vertical farming
wood alcohol

Check Your Understanding

Answer the following questions using the information provided in *Lesson 6.2.*

1. List four ways in which technology benefits agriculture.
2. How is telemetry used to improve livestock operations?
3. List three ways in which drones can be used in a farming operation.
4. Explain how video cameras may be used to improve farming operations.
5. What are the key agricultural ingredients of biofuels?
6. What is vertical farming in urban areas? What other types of vertical farming are used in livestock and poultry production?
7. What are four USDA requirements for having an agricultural product classified as organic?
8. Describe how in vitro meat production might be useful, and describe its disadvantages as well.
9. What is a hydroponic system in a greenhouse? How does it differ from a traditional greenhouse operation?
10. What is the usefulness of a closed ecological system?

Chapter 6 Skill Development

STEM and Academic Activities

1. **Science.** Do research to compare the properties of several types of siding used in barn construction (wood, metal, prefabricated panels, etc.). Compare impact resistance, fire resistance, useful life, maintenance needed, environmental impact, and cost per square foot. Create a table showing the results of your research.
2. **Math.** Using reference books and manufacturers' catalogs, prepare a list of hand tools you believe a carpenter will need for rough framing an exterior finish of a typical agricultural structure. Determine the cost of equipping a carpenter with a complete hand tool kit based on your list. Using catalogs, the Internet, and visiting a hardware store, find the cost of each tool. Add the costs and report to the class.
3. **Social Science.** Why are businesses, homes, livestock/poultry production facilities usually in separate areas of a community? Interview the person in your community who is in charge of planning or zoning. Find out why zoning is considered important for the community and how a property owner can seek a zoning change. Obtain a copy of the community's zoning map and use it for a visual aid as you report your findings to the class.

Communicating about Agriculture

1. **Reading and Speaking.** With a partner, make flash cards for the key terms in Chapter 6. On the front of the card, write a term. On the back of the card, write the pronunciation and a brief definition. (You may consult a dictionary.) Then take turns quizzing one another on the pronunciation and definitions of key terms. Practice reading aloud the terms, clarifying pronunciations where needed.
2. **Reading and Speaking.** Go to your town or county's website, or visit your town or county building office to view the requirements for acquiring building permits. Collect material showing the permits needed for various installations in agricultural buildings. Analyze the data in these materials based on the knowledge gained from this lesson. Recommend to the class the items that are needed for various buildings.
3. **Writing.** Visit a historical farm. Make notes of the unusual details used in the construction of the farm's structures. Write an essay describing the historical farm's structures. Be as specific as possible while trying to identify the different construction materials and techniques used. Compare the historical materials with those used today to construct most farm structures. Point out the differences. Include your opinion on the pros and cons of each type of construction.

Extending Your Knowledge

1. Lay out a diagram of a modern diversified farming operation that raises hogs and beef cattle. Draw and position all of the facilities needed for successful operation of this farm.

Chapter 7

Agricultural Power and Engineering

xuanhuongho/Shutterstock.com

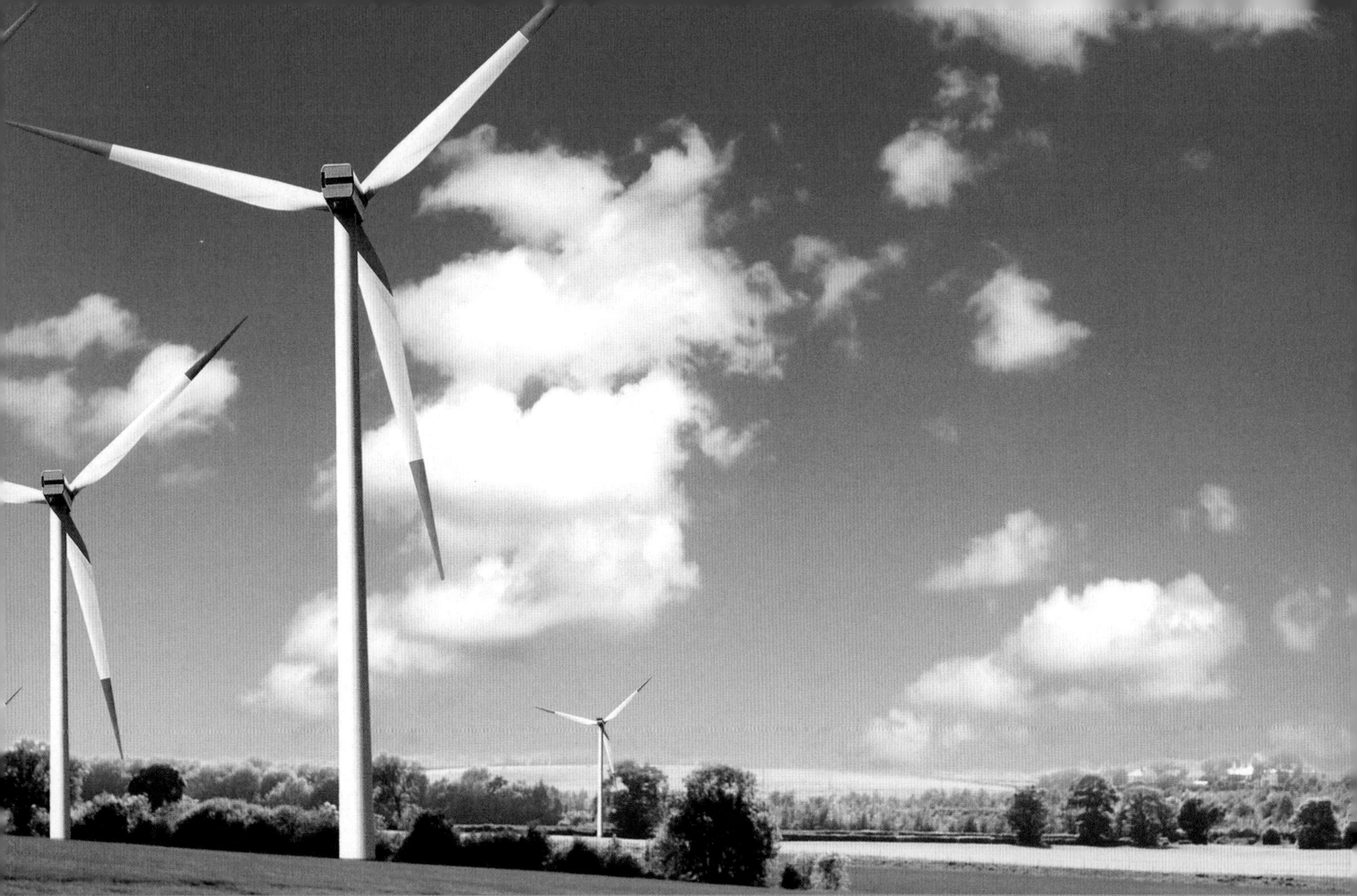

WDG Photo/Shutterstock.com

AVANGARD Photography/Shutterstock.com

maggee/Shutterstock.com

While studying, look for the activity icon  **to:**

- **Practice** vocabulary terms with e-flash cards and matching activities.
- **Expand** learning with video clips, animations, and interactive activities.
- **Reinforce** what you learn by completing the end-of-lesson activities.
- **Test your knowledge** by completing the end-of-chapter questions.

Lesson 7.1

Energy Systems

Words to Know

air turbine
anaerobic digestion
biomass
biomass energy
coal
crude oil
derrick
diode
drag
drilling fluid
drilling mud
fossil fuel
geothermal energy
geothermal heat pump
ground-source heat pump
hydroelectric power plant
hydropower
in-situ leach (ISL) mining
in-situ recover (ISR) mining
kinetic energy
lift
nacelle
natural gas
nonrenewable energy
open-pit mining
photovoltaic cell (PV cell)
renewable energy
solar energy
solar power

(Continued)

Lesson Outcomes

By the end of this lesson, you should be able to:

- Describe renewable energy sources.
- Describe nonrenewable energy sources.
- Identify energy resources at work in an agricultural business or farm.
- Explain the advantages and disadvantages of energy sources.

Before You Read

You are probably reading this at your school or at home with the help of an overhead light or lamp. Have you ever thought about where the electricity comes from that illuminates those lights? Where was it generated? What energy source or method is used to generate it—wind, solar, coal, oil, nuclear, or geothermal? See if you can determine the source of the electrical power that you use on a daily basis.

The lights we absently turn on each day, the cell phones that have become our constant companions, the hot water we use to shower, and the cars we drive to work all require energy. Do you ever stop and consider the sources of the energy you use each day? Are the sources in their raw form? How much processing is involved? How about their methods of transmission? The energy sources we use each day come from a variety of resources, some renewable and some nonrenewable. Each of these energy sources must be converted from their raw form into a form that we can use safely and efficiently. This lesson will identify and briefly explain the energy systems in use today.

Renewable Energy Sources

Renewable energy is energy that comes from sources continually replenished by nature. Solar energy, wind energy, hydropower energy, geothermal energy, and biomass energy are the most common sources of renewable energy. About 20% of the energy used in the United States comes from renewable resources. ***Nonrenewable energy*** is energy that comes from natural sources that took millions of years to develop and cannot be replenished. Nonrenewable resources include ***fossil fuels*** such as coal, natural gas, and crude oil, and raw metals like uranium.

Solar Energy

Solar energy is radiant energy emitted by the sun. ***Solar power*** is power obtained by harnessing this energy. Solar energy is the cleanest form of energy we use. It produces no air or water pollution. Solar panels and solar thermal technology are the two most common systems used to harness the sun's energy.

Solar Panels

Most solar energy systems use large, flat solar panels containing photovoltaic cells that convert sunlight into electricity, **Figure 7-1**. ***Photovoltaic cells (PV cells)*** are specialized semiconductor diodes that convert visible light into direct current (DC) electricity. (A ***diode*** is a semiconductor device that allows the flow of electrical current in one direction only.) This electricity is usually stored in batteries. The electricity may be drawn directly from the batteries to immediately power machinery, or it may be converted to AC power and used through a building's wiring system.

Solar Thermal Energy

Solar thermal energy systems also use solar panels to collect the sun's energy. However, the heat from the sun is used to heat water or create

Words to Know

(Continued)

strip mining
syngas
uranium
waste rock
wind energy
wind farm
wind power
wind turbine

Elena Elisseeva/Shutterstock.com

Figure 7-1. Stationary solar panels are positioned to receive the most sunlight. Some solar energy systems use movable panels to take full advantage of the sunlight throughout the day. *In which direction should solar panels face to receive the most sunlight? Does it vary by region or climate?*

aurielaki/Shutterstock.com

Roy Pedersen/Shutterstock.com

Figure 7-2. Small systems consisting of one or two panels may be used to heat water for household use. *How much energy do you think your family could save if you used a solar energy system to provide hot water for your home?*

steam. Depending on the system, this heat can be used to warm swimming pools, heat water or air for residential and commercial use, and for power generation on a commercial scale. See **Figure 7-2**.

Did You Know?

The world's highest-situated wind turbine is located at the base of the Patoruri Glacier in Peru. It is 16,001′ (1,877 m) above sea level.

Agricultural Applications

Solar energy systems can be used to provide power to a home and agricultural structures on a farm or ranch. They may also be used in remote locations, such as fields and pastures, not equipped with power or gas lines. A solar energy system can be used to prevent livestock drinking water from freezing or to provide electricity to run a water pump. It may even be used to power electric fences. Simple, portable solar stoves may be used to boil water, cook food, or dry fruits and vegetables.

Wind Energy

Wind energy, or ***wind power***, is the kinetic energy of air in motion. (***Kinetic energy*** is the energy possessed by an object or system as a result of its motion.) Wind energy is considered a form of solar power. As Earth's surface is heated by the sun and cooled, wind is created as air moves from areas of high pressure to low pressure.

Wind energy has been harnessed and used by humans for thousands of years. Sails and the wind were used to move boats on the Nile as far back as 5000 BCE. Windmills were used to power grain mills as early as 200 BCE. Windmills have also long been used to pump water and generate electricity in many rural areas across Europe and the United States. Today, modern versions of the windmill are used to convert kinetic wind energy into mechanical power or electricity for single homes, small communities, and even large cities. These modern-day windmills are called ***wind*** or ***air turbines***.

Wind Turbines

Although there are a variety of wind turbine designs in use, the most common is the horizontal axis wind turbine (HAWT). Much like the traditional farm windmill, today's horizontal axis turbines use blades, a shaft, and a tower. Together, two or three blades form the rotor that turns the main shaft as the blowing wind spins the blades. The shaft is connected to a transmission that is used to increase the turning speed high enough for the generator to make the electricity. The transmission is connected to the generator with a high-speed shaft. A controller prevents the rotor from exceeding its maximum safe operating speed by operating a brake, installed between the transmission and the generator. The components are housed in a fixture called a ***nacelle***, which is located at the top of the tower, **Figure 7-3**.

Teun van den Dries/Shutterstock.com

Figure 7-3. The axle and generator are housed inside the nacelle of this wind turbine.

The turbine's blades are located 100′ or more aboveground to take advantage of the faster and less turbulent wind. Wind speeds usually must be above 12 to 14 miles per hour in order for a wind turbine to work efficiently. Electricity generated by the turbines is used directly or fed into photovoltaic cells or batteries for storage and transmission. The electricity may also be fed into the utility power grid.

HAWTs must be pointed into the wind. Large turbines use a wind sensor and a computer-controlled motor to keep them pointed to the wind. Small turbines use a simple wind vane.

Did You Know?

Three large wind farms in California generate enough electricity to supply a city the size of San Francisco.

Wind Farms

Wind farms are being built in areas around the world where wind is a steady and reliable resource. (A ***wind farm*** is a group of wind turbines in the same location used for production of electricity.) A wind farm may consist of a few turbines or have hundreds of individual turbines over an extended area. A wind farm may also be located on open water to harness offshore wind resources, **Figure 7-4**. On most wind farms, turbines are spaced about 6–10 times the rotor diameter apart.

v.schlichting/Shutterstock.com

Figure 7-4. Offshore wind farms can take advantage of consistent winds.

Single Turbines

Smaller, individual wind turbines are located close to where the generated electricity will be used. For example, near homes, telecommunication towers or dishes, or water pumping

STEM Connection Lift and Drag

The blades on air turbines behave much like airplane wings. A pocket of low-pressure air forms on the downwind side of the blade when the wind blows. The low-pressure air pocket pulls the blade toward it, causing the rotor to turn. This action is called ***lift***. The force of the lift is much stronger than the wind's force against the front side of the blade. The force against the front side of the blade is called ***drag***. The combination of lift and drag causes the rotor to spin like a propeller.

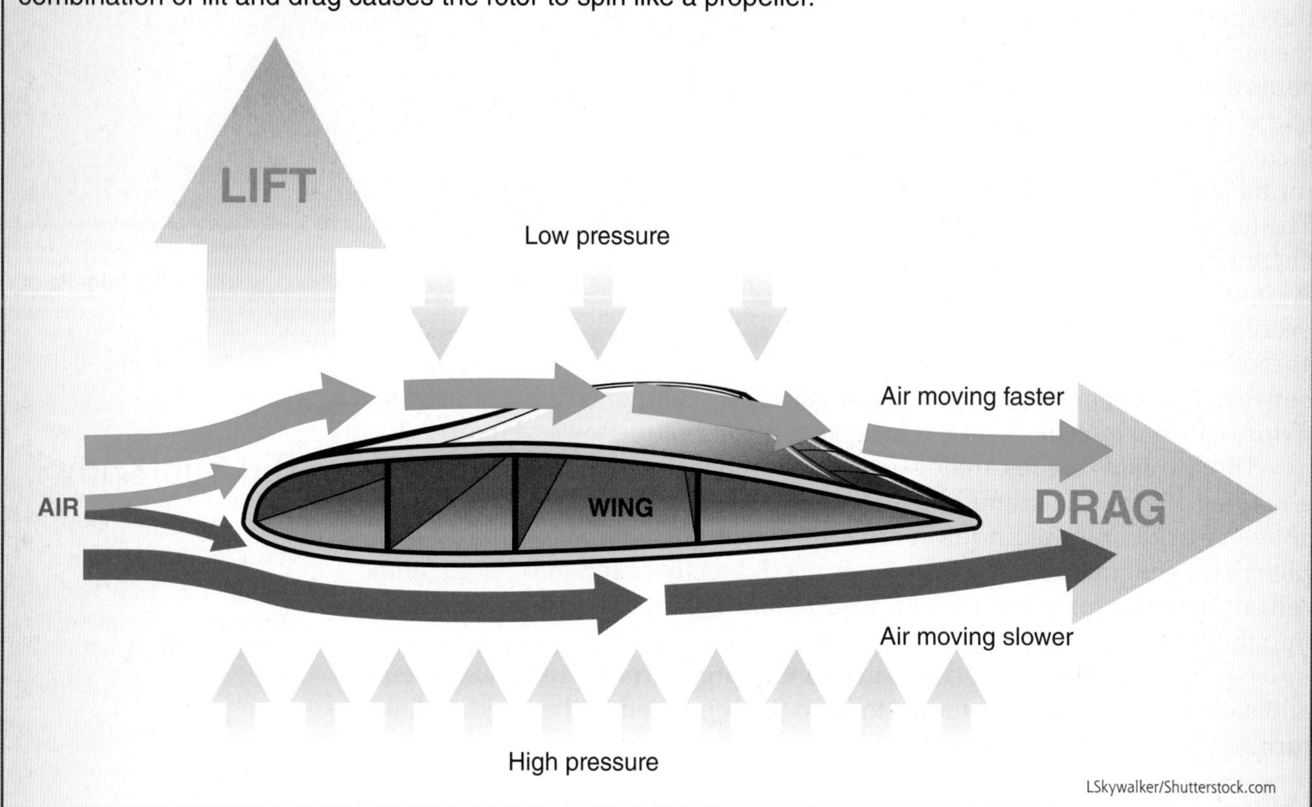

LSkywalker/Shutterstock.com

stations. Turbines connected directly to diesel generators, storage batteries, and photovoltaic systems are commonly used in remote locations where a connection to the utility grid is not available. Smaller turbines may also be used on rooftops, on boats to charge auxiliary power units, and even for traffic signs.

Cost

The technology used to harness and use wind energy has decreased immensely in cost. However, it still requires a higher initial investment than fossil-fueled generators. The U.S. Department of Energy is working with industry partners to improve the efficiency and reliability of wind turbine technology and reduce its costs. Wind energy is a source of clean, non-polluting electricity, and continued growth of the industry will reduce carbon pollution and water use connected with conventional energy production and use.

Did You Know?

Washington, Oregon, California, New York, and Alabama are the top hydropower producing states.

Hydropower

Hydropower is another form of renewable energy. ***Hydropower***, is power derived from the kinetic energy of moving water. In ***hydroelectric***

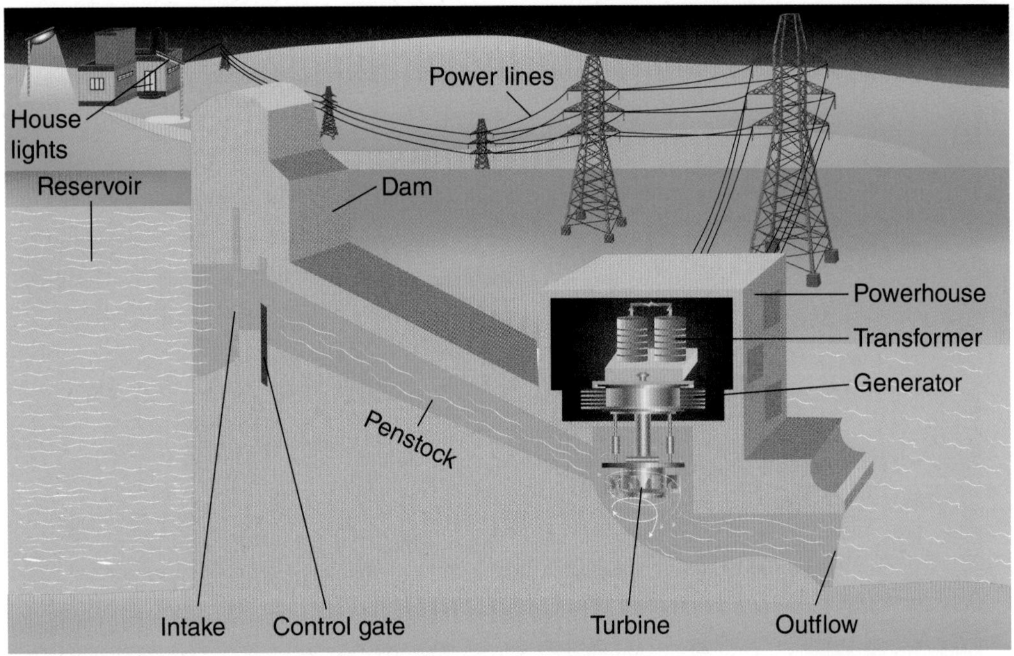

snapgalleria/Shutterstock.com

Figure 7-5. The water behind the dam flows through an intake (penstock) and pushes against blades in a turbine, causing them to turn. The turbine spins the generator to produce electricity.

power plants, water passing through gates in a dam is directed through turbine generators that create electricity on a commercial scale, **Figure 7-5**. Harnessing water to create power is not a new concept. Humans have used moving water to generate power for centuries. Since ancient times, water wheels have been used to convert the kinetic energy of moving water into mechanical energy to run grinding mills.

Hydropower in the United States

In the United States, hydropower became a source for generating electricity in the late 1800s. Today, hydropower accounts for about 6% of total U.S. electricity generation. It also accounts for more than 50% of power generation from renewable resources in the United States. Hydroelectric power generates a significant source of reliable electricity for homes and businesses. Once dams are constructed and the turbine generators installed, the water source is essentially free, making hydropower a cost-efficient energy source in the long term.

Today's hydropower plants have an efficiency of around 90%, meaning that only a small amount of energy is wasted in the process of generating electricity. Modern hydropower plants have no waste disposal issues, have low operating and maintenance costs, and allow great flexibility in controlling power generation. The water flow can easily be reduced when power consumption is low and increased when consumption is higher.

Hydropower Systems

Hydropower may be produced using a storage system (dams and reservoirs) or run-of-the-river system (diversion of running water). The biggest hydropower plant in the United States is located at the Grand Coulee

Benedictus/Shutterstock.com

Figure 7-6. The Grand Coulee Dam is 550′ tall and 5,223′ long, and contains almost 12 million cubic yards of concrete. Power from the dam is supplied to Canada and eleven western states.

Dam on the Columbia River in northern Washington, **Figure 7-6**. It generates more than 21 billion kilowatt-hours of electricity each year. The reservoir that serves the dam is called the Franklin D. Roosevelt Lake. In addition to power generation, the dam provides flood control, river regulation, irrigation, and water storage. It also serves as a recreational area allowing boating, water skiing, swimming, and fishing.

Dam Reservoirs

The reservoirs created by dams are often used for recreational purposes such as swimming, fishing, and boating. For example, the Hoover Dam, on the Colorado River between Arizona and Nevada, created Lake Mead, a popular, 110-mile-long national recreational area. Water from the reservoir may also be used for irrigation purposes.

Environmental Impact

Hydropower generators are one of the cleanest forms of power generation. They do not directly produce emissions of air pollutants, and the water source is renewable yearly by snow and rainfall. However, they do have substantial impact on the environment. Dams that create reservoirs and run-of-river systems that divert water flow may obstruct fish migration, change natural water temperatures, water chemistry, river flow characteristics, and silt loads. These changes can affect the ecology and physical characteristics of the river and surrounding areas, **Figure 7-7**. The reservoirs may also cover natural areas or agricultural lands and displace people as well as fauna. In some instances, flooding caused by reservoirs may displace even more people from their homes and animals from their natural habitats.

Hydropower Limitations

One of the most substantial limitations to expanding the amount of hydroelectric power produced is the limited number of suitable locations available for building reservoirs and dams. In addition to potential environmental impact, potential power production must also be considered. How far the water drops and how much water moves through the system determines how much electricity can be generated. If the potential for power production is limited, it may not be cost-effective to harness the water power in a particular location.

Rigucci Bonneville Dam, OR

Figure 7-7. Fish ladders provide a way for spawning fish species to navigate around hydroelectric dams.

Geothermal Energy

Geothermal energy is energy harnessed from the natural supply of heat beneath Earth's surface. Geothermal energy may be heat found in shallow ground, hot water, and hot rock found miles beneath Earth's surface, or even deeper in molten rock called magma. Geothermal energy can be used to generate electricity on a commercial scale, heat and cool individual homes or businesses, or power individual systems in remote locations.

Geothermal Power Plants

Geothermal power plants use the hot water or the steam from geothermal reservoirs to generate electricity. The largest concentration of geothermal reservoirs of hot water occur along the Pacific Rim, which is located along the coasts of the Pacific Ocean. Therefore, the largest potential for geothermal power plants in the United States is in states along the west coast, in Hawaii, and in Alaska. There are also a number of "hot spots" in Nevada. Geothermal power plants use one of three systems to generate electricity:

- The steam in a *dry steam system* goes directly through the turbine and then into a condenser where it is condensed into water.
- In a *flash steam system*, hot water is depressurized, or flashed, into steam, which then drives the turbine, **Figure 7-8**.
- In a *binary cycle system*, the hot water is passed through a heat exchanger where it heats a second liquid in a closed loop. The second liquid boils at a lower temperature than water so it is more easily converted into steam to run the turbine. All of the water is returned underground.

In most geothermal systems, the warm water is returned to the geothermal reservoir. However, some systems are open and much of the moisture escapes into the atmosphere after it passes through the turbines. To compensate for this loss, one large plant in California is recycling clean wastewater by injecting it into the ground to replenish the reservoir and

Did You Know?

More than 40 geothermal power plants are in operation in California. They provide nearly 7% of the state's electricity.

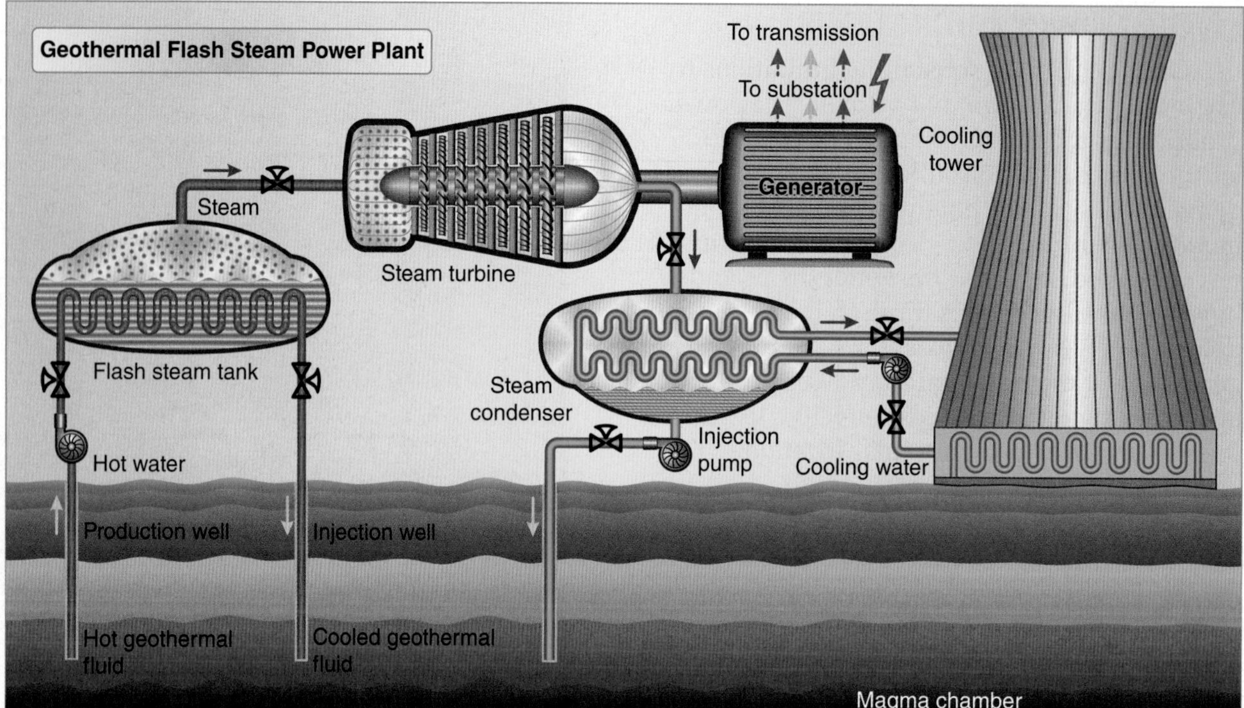

Fouad A. Saad/Shutterstock.com

Figure 7-8. A geothermal flash steam power plant pumps the hot water from below Earth's surface and flashes it into steam to drive the turbines.

prolong the life of the heat source. The layers of rock below Earth's surface are still hot and will heat the recycled water.

Geothermal Heat Pumps

Although the most efficient areas to locate larger geothermal energy systems are in regions with the highest underground temperatures, ***geothermal heat pumps*** may be used almost anywhere in the world. Depending on the location, the temperature beneath the upper 10′ to 20′ of Earth's surface remains nearly constant at 50°F–60°F (10°C–16°C). In the winter, geothermal heat pump systems can remove this heat from the ground, transfer it to a heat exchanger, and pump it into an indoor air delivery system. In summer, the heat pump moves heat from the indoor air into the heat exchanger. The heat removed from the indoor air may also be used to heat water. Geothermal heat pumps may also be referred to as ***ground-source heat pumps***. See **Figure 7-9**.

Environmental Impact

Open geothermal plants release some air pollutants in the steam, including hydrogen sulfide, trace amounts of arsenic, and minerals. Depending on the location, salt buildup on the pipes may also be a problem. Geothermal plants using a binary system have no emissions, and properly planned and installed geothermal pump systems are some of the most energy-efficient and environmentally clean heating and cooling systems available.

Biomass Energy

Nonrenewable fossil fuels are not the only organic source of energy. Biomass is another useful energy source. ***Biomass*** is any organic matter, besides fossil fuels, that can be used as an energy source. This includes wood (firewood, charcoal, chips, sheets, pellets, sawdust, pulp sludge), animal waste, and selected agricultural crops. Biomass is a renewable energy source because it can easily and quickly be replaced by nature at a rate that meets or exceeds consumption of it as fuel. Using biomass as an energy source also keeps large amounts of organic waste from being sent to landfills. ***Biomass energy*** is the energy harnessed from organic matter as it decomposes, when it is burned, or when it is treated with chemicals.

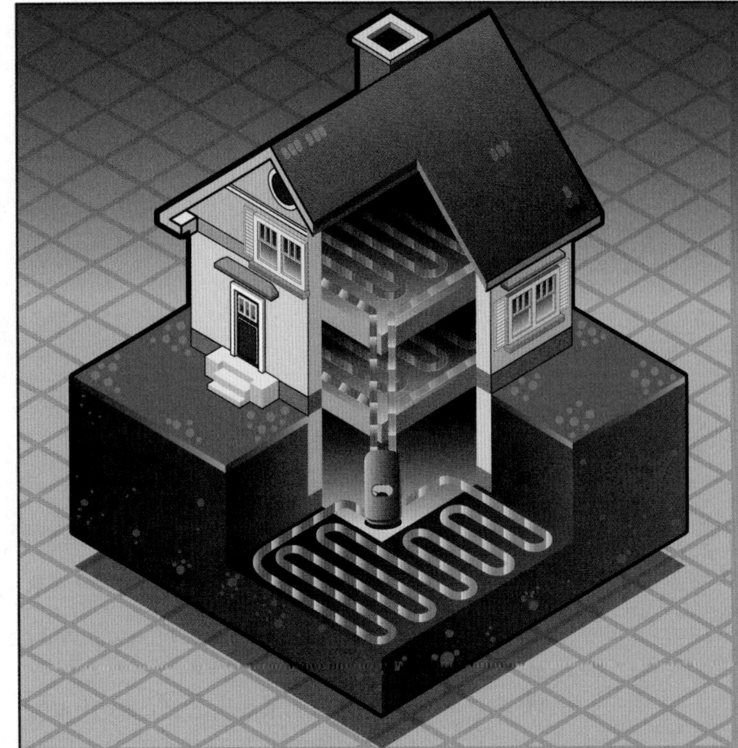

aurielaki/Shutterstock.com

Figure 7-9. Air or antifreeze liquid is pumped through the buried pipes and circulated into the building. In some designs, piping runs under the floors and the heat radiates upward. The most effective systems use compressors and pumps to maximize the heat transfer. Geothermal energy systems are especially useful in rural areas where there is no utility service.

Converting Biomass to Energy

Biomass is converted into energy through a thermal or biochemical process.

Thermal processes include direct combustion to produce heat or steam, and gasification (using heat, oxygen, carbon monoxide, and pressure) to produce syngas. ***Syngas*** (synthesis gas) is a mixture of hydrogen and carbon monoxide produced from biomass. After contaminants and carbon dioxide are removed, the syngas may be run directly through a gas turbine or burned to produce steam for steam turbines. Syngas may also be processed to make liquid biofuels or other useful chemicals.

Biochemical processes such as anaerobic digestion (decomposition) are also used to convert biomass to energy. ***Anaerobic digestion*** is a process in which microorganisms break down organic materials such as food scraps, manure, and plant residue. The process creates biogas, such as methane that can be harnessed and used to create energy, and solids that may be used to amend soil.

Biomass Power Plants

Biomass power plants may be dedicated to one type of biomass, able to process various types of biomass, or designed to co-fire biomass with coal or another fossil fuel. For example, coal plants may be converted to run entirely on biomass or modified to co-fire coal with biomass. Anaerobic digesters are used to control the decomposition process of animal manure and raw sewage and capture the resulting biogas.

For practical purposes, many biomass processing plants are situated close to their material source. For example, ethanol production facilities are located near farms and large livestock production to reduce transportation

LianeM/Shutterstock.com

Figure 7-10. Biomass plants located near their fuel source incur less transportation costs. *Can you think of additional advantages?*

costs of shipping agricultural feedstocks to the plant, and shipping the resulting nutrition supplements back to the livestock facilities. Manure processing facilities are also usually located near larger livestock production operations to reduce processing and shipping costs, **Figure 7-10**.

Environmental Impact

Using biomass to produce energy has substantial environmental impact, most of which has been very positive.

- Using biomass to produce energy greatly reduces the amount of organic material that would otherwise end up in landfills or polluting other areas of the environment.
- Using the biogas produced by decomposing organic materials as an energy source reduces the amount of methane and carbon dioxide released into the atmosphere.
- Using biomass instead of fossil fuels reduces harmful emissions of pollutants such as carbon, sulfur, and mercury.
- Using biomass to create products usually made from petroleum products reduces our dependence on crude oil and creates more biodegradable products.

Agricultural Impact

Using biomass for energy production creates many opportunities for agriculturists. With proper planning, farmers and ranchers can effectively sell or process by-products that might otherwise go unused.

Nonrenewable Energy Sources

Nonrenewable energy sources are organic compounds requiring thousands or even millions of years to form. The organic compounds, referred to as fossil fuels, we use to generate energy include coal, natural gas, and crude oil, and metals such as uranium. Coal has been used for many years to produce heat and electricity. Crude oil is used to make gasoline and diesel fuel. Natural gas and crude oil are both used to make propane gas and other gases used for energy. Uranium is mined from the ground and converted into fuel for nuclear power plants.

Fossil fuels are a major source of energy in the United States. Our dependence on fossil fuels, the pollution they generate, and the depleting supplies, are the main reasons alternative fuel sources are being researched and developed.

abutyrin/Shutterstock.com

Figure 7-11. The more carbon in coal, the more efficiently it burns.

Coal

Coal is used in the United States primarily to generate electricity. ***Coal*** is a combustible, dark rock consisting mainly of carbonized or fossilized plant matter, **Figure 7-11**. Coal is mined from both surface (open-pit) and underground deposits. ***Open-pit mining***, or ***strip mining***, requires the removal of surface soil and rock to reach the coal below.

Once mined, coal must be processed and shipped to power generation plants by railroad. At the power plants, the coal is burned to heat water and create steam under high pressure. When the steam is released, it operates the turbines which, in turn, operate the generator to create electricity.

Did You Know?

Coal-generated power accounts for almost 40% of the electricity production in the United States.

Environmental Impact

Over the years, as mining procedures and exhaust and filtering systems have been improved, mines and coal-burning plants have greatly reduced their pollution emissions. However, the use of coal as an energy source still impacts the environment.

- Mining processes often strip the land of both its natural resources and beauty, and often cause pollution of nearby water resources.
- Underground mining is dangerous to the miners and often disrupts ecological cycles of the surrounding wildlife.
- Shipping coal across the country requires fuel and often results in some particulates being released into the atmosphere.
- If improperly stored, dust from stockpiles may pollute the local air. Rain may also cause runoff to nearby land or water.
- Water intake and warm water discharge may greatly affect local aquatic life.
- Waste from the power plant includes ash and sludge that may contaminate water supplies if not treated properly.

Did You Know?

Most of the oil and natural gas used in the United States is pumped through over 190,000 miles of liquid pipelines and over 300,000 miles of natural gas pipelines.

Natural Gas

Natural gas is an odorless, colorless, flammable gas consisting mainly of methane and other hydrocarbons. Natural gas occurs underground and is often found with coal or crude oil deposits, but may also be found in natural gas fields. Natural gas is formed when layers of buried plants, gases, and animals are exposed to intense heat and pressure over millions of years. Like coal, it is considered a nonrenewable energy source because it cannot be replenished.

Natural gas is acquired by drilling directly into natural gas pockets beneath Earth's surface or as a by-product when drilling for oil or mining for coal. Depending on its source, natural gas may contain impurities such as ethane, propane, butane, carbon dioxide, nitrogen, and hydrogen sulfide. These impurities must be removed before the gas can be transported by pipeline to storage facilities and distributed to the end users.

Environmental Impact

Natural gas is a good source of energy for cooking, space and central heating, heating water, and drying clothes. It is very popular in the United States, as more than half of all American homes use some form of natural gas for heating and cooking. The more than one million miles of natural gas pipelines used to transport the fuel to American homes and businesses also attests to its popularity, **Figure 7-12**.

Natural gas has very low emissions and can be efficiently and safely stored. Its carbon dioxide emissions are 30% less than that of oil and 45% less than that of coal. It burns without leaving any smell, ash, or smoke, and it does not release sulfur dioxide. It is also used to produce some paints and plastics as well as hydrogen and ammonia for fertilizers. Natural gas can be used as compressed natural gas (CNG) or as liquefied petroleum gas (LPG).

Not unlike most fuels, natural gas does have some disadvantages, including:

- Natural gas is highly volatile and very toxic. Leaks are extremely hazardous. Gas companies add a sulfuric-like odor to the fuel for easier leak detection.
- It is less expensive than diesel or gasoline, but provides less efficient vehicle mileage.
- Its combustion releases carbon dioxide, carbon monoxide, and other carbon compounds into the atmosphere.
- Processing to eliminate impurities results in many by-products (hydrocarbons, sulfur, water, helium, nitrogen, and carbon dioxide).
- Creating and managing pipelines can be costly.
- Certain drilling processes may contaminate nearby water sources.

Zorandim/Shutterstock.com

Figure 7-12. Natural gas is moved through pipelines.

Crude Oil

Crude oil is a nonrenewable fossil fuel found in underground areas called reservoirs. Crude oil is composed mainly of hydrocarbons and organic compounds. Crude oil is extracted from beneath Earth's surface and ocean floors with pumping wells, **Figure 7-13**. Like coal and natural gas, crude oil was formed when buried organic matter was exposed to intense heat and pressure over millions of years. The term *petroleum* is often used interchangeably with *crude oil*, or *oil*.

Human beings have been using crude oil for thousands of years. Initially, people collected crude oil from oil seeps bubbling up to Earth's surface or skimmed it off the tops of ponds. It was used for such things as lighting, lubrication, for making ships watertight, and for jointing masonry. As the demand for crude oil increased, oil producers began looking for additional means to extract oil. The first modern oil well was drilled in 1859, giving birth to America's oil boom and the world's oil industry. Today, crude oil is processed to make products such as gasoline, kerosene, propane, diesel fuel, and fuel oil, as well as other petroleum products such as plastics, chemicals, pharmaceuticals, wax, and even soaps, **Figure 7-14**. A small percentage of the crude oil processed in the United States is used as fuel in electricity generating plants.

huyangshu/Shutterstock.com

Figure 7-13. Oil is pumped from wells and transported to refineries.

Locating Fields

Seismologists, geologists, engineers, and other specialists work for oil companies to determine prospective drilling locations, or fields. Methods for locating oil reservoirs include measuring gravity changes, magnetic field changes, and shock waves

Products Made from Crude Oil
Liquefied petroleum gas (LPG)
Kerosene (paraffin) used for lighting oil lamps; its main use today is as jet aircraft fuel
Lubricating oils
Heavy fuel oils used in industrial boilers, power stations, and to create steam for turbines on ships
Bitumen used in road construction and for waterproofing
Tars used in roofing and road construction
Wax used in candles, electrical insulation, and waterproof covering for food cartons
Chemicals such as terephthalic acid, acetic acid, ethyl acetate (used in products from adhesives to cosmetics, clothing, curtains, bedding materials, plastics, and carpeting)

Goodheart-Willcox Publisher

Figure 7-14. Products from oil.

Did You Know?

The mass production of the internal combustion engine spurred the demand for crude oil. During the late 1800s, much of the crude oil obtained from drilling was discarded. Kerosene was the only component being used at that time.

emitted into the ground or water. Exploratory wells are dug to determine if there is a viable source of oil. Once prospective locations are determined, the drilling process begins. Drilling for oil is a complex process and hard, dirty, and dangerous work.

Drilling

Before drilling can begin, a drilling platform, or rig, is built over the drilling location to hold the drill in place. The drilling bit is attached to a drilling string suspended from the ***derrick*** (steel tower on drilling platform). Drilling is done in sections. After each section is drilled, a steel pipe (*casing*) is dropped in place to give the hole structural integrity. Concrete is usually poured around the casing to keep it in place and fill in any gaps. As the bit goes deeper, the crew continually adds new sections to the string, and the process repeats itself.

The drilling process is somewhat like drilling on a piece of wood, except the hole is filled with a mixture of fluids, solids, and chemicals that lubricate the bit. The mixture is referred to as ***drilling mud*** or ***drilling fluid***. The drilling mud is forced to bottom of the hole, through the bit, and back to the surface. The drilling mud is used to remove cuttings, cool the bit, and maintain pressure. As the drilling mud flows back to the surface, it is filtered and pumped back into the hole. To guard against dangerous blowouts while drilling, a blowout prevention system is installed to prevent pressurized oil and gas flowing up the well.

The rock cuttings and drilling mud are often disposed of in a reserve pit. The reserve pit is dug near the drilling operation and lined to prevent seepage into the surrounding soil. In some sites, the rock cuttings and mud must be removed and disposed of offsite.

Extraction

Once the hole has been drilled, it is prepared for extraction of the crude oil or gas. The casing is perforated (has holes) so the oil or gas can enter the tube. High-pressure fluids (water or acid) are pumped through to clean and fracture the rock, encouraging it to begin releasing hydrocarbons. The initial pressure inside the reservoir is usually high enough to push the oil or gas to the surface. As the pressure from the reservoir decreases, various techniques may be used to increase the pressure. These techniques include pumping in compressed air, water, gas, or steam.

When the well reaches its economic limit (the production level does not cover the cost), the tubing is removed and the hole is sealed with concrete. The pump head is excavated and removed and a cap is welded in place. The entire thing is then buried.

Did You Know?

Marine tankers may carry more than 300,000 tons of crude oil.

Transport

Because areas of supply and demand are not often in the same place, the oil must be transported to refineries and to its final destination. Depending on where it is pumped, crude oil is usually transported to refineries through pipelines or by ocean tankers, **Figure 7-15**. Marine tankers discharge the oil to pipelines connected to storage tanks in the refinery. The refined oil and fuel products are transported to their next destination by truck, train, ship, or pipelines. Pipelines may be built above

karamysh/Shutterstock.com

tcly/Shutterstock.com

Figure 7-15. Crude oil and refined petroleum products are transported by rail, ship, and pipeline in addition to highway transportation.

land, buried belowground, or underwater. The oil or gas is kept in motion by a system of pump stations built at strategic intervals along the pipeline. Devices known as pigs flow through the pipes to help clean and inspect each section of pipe.

Environmental Impact

Stricter rules and regulations have been established to lessen the environmental impact of crude oil drilling, processing, and use. Oil companies are required to perform extensive environmental studies before drilling to ensure limited environmental impact. They must review potential issues such as waste disposal, emissions, and water discharge. They must also consistently monitor environmental data. However, drilling for oil on land or offshore can still greatly impact the environment.

Did You Know?

The four quarts of oil drained in a typical oil change can contaminate one million gallons of water. If you change your own vehicle oil, please recycle it.

Land Drilling

Preparing drilling fields require land to be cleared and access roads to be built. Drilling on land requires steady access to water and may require the drilling of water wells. Underground water sources may also be contaminated, and runoff from the extraction process may affect surface waters. If the field is not located near a power supply, loud, fuel-burning generators are needed to power on-site machinery. Many drilling sites use gas flaring (burning excess gas emitted from the drill site) that pollutes the air. In addition, spilled drilling fluid contaminates the surrounding soil. All these factors impact the surrounding habitat. Native plants are disrupted and local wildlife is often displaced.

Offshore Drilling

Drilling offshore presents additional environmental issues. Setting up the rigs disrupts the seafloor habitats, and spilled drilling fluids can be toxic to marine life. Air pollution is generated from the operation of machinery on offshore oil rigs, as well as from the gas flaring. In addition, oil spills are devastating to all environments. Spilled oil spreads across the water

kajornyot/Shutterstock.com

Figure 7-16. Crude oil and refined petroleum products must be carefully managed in all phases of extraction, refinement, and transportation to reduce risk of damage to the environment.

surface and contaminates shorelines, killing birds, plants, aquatic life, and land animals, **Figure 7-16**. Even the biological and chemical methods used to clean up oil spills may be detrimental to the environment.

Refineries

According to the U.S. Energy Information Administration (EIA), there are over 130 oil refineries in the United States. Many are located along the coasts for easier access by marine tankers. Even with newer technologies and stricter laws and regulations in place, oil refining still impacts the environment, although at a much lower level than in the recent past.

- Oil refining produces wastewater sludge and other solid waste that may contain metals and toxic compounds. This waste must be treated and/or disposed of properly to prevent soil and water contamination.
- Refineries use large tanks to store crude oil, partially refined oil, finished products, and chemicals before additional refining or shipment. Some of these tanks have capacities of more than 390,000 barrels (16 million gallons). Small leaks from these tanks can contaminate both the soil and underground or surface waters. A tank rupture would be catastrophic to the surrounding environment. Fortunately, newer tanks are designed with containment areas designed to hold their capacities in the event of a tank rupture.
- Runoff water from the storage tanks, and contaminated soil surrounding them, may pollute local water sources. Some refineries use systems to divert rainwater runoff from the tanks to protect local waters.
- The refining process generates air pollutants such as carbon dioxide, carbon monoxide, sulfur dioxide, nitrogen oxides, volatile organic compounds (VOCs), and various particulates. Recovery systems can be installed to control the VOC vapors and minimize those released into the air.
- Refineries require large amounts of water consumption and generate polluted wastewater that must be treated before being released to its natural source. Oftentimes, the water is warmer than local waters and can greatly affect the aquatic life, local wildlife, and plant life of the surrounding areas.

Nuclear Energy

Most farmers do not think of having a nuclear-powered farm, but with more than 100 nuclear power plants operating in the United States, that may actually be true. These power plants generate almost 20% of the electricity used in the United States. Some of this electricity ends up at work on farms and agricultural businesses. Nuclear energy is considered nonrenewable because it requires uranium. ***Uranium*** is a naturally occurring radioactive metal used as fuel in nuclear reactors. Uranium is found in very small amounts in rocks, soil, water, plants, and animals, and contributes to low levels of natural background radiation in the environment. However, the uranium used in nuclear reactors is mined from the ground in a process similar to coal mining.

John Carnemolla/Shutterstock.com

Figure 7-17. Uranium open-pit mining affects the landscape and may increase the risk of contamination to nearby inhabitants.

Uranium Mining and Processing

Uranium may be extracted from the ground through open-pit mining, underground mining, and in-situ leach (ISL) mining.

Open-Pit Mining

Open-pit mining, or strip mining, requires the removal of surface soil and rock to reach the uranium ore below, **Figure 7-17**. Open-pit mining is used only if the uranium ore is near the surface (no more than 400′). Open-pit mining is less expensive than underground mining but generates large amounts of waste rock and may contribute to land, water, and air pollution. The ***waste rock*** is the material removed from the surface of the pit and materials from the pit that are not used. The waste materials may not be economically effective to mine, but they may become hazardous to the environment once exposed to the air. Mine workers' health, as well as that of those living nearby, may be compromised due to dust and radon exposure, noise, and water contamination.

New uranium mines must follow stricter environmental, safety, and health guidelines to minimize the environmental impact and create safer working conditions.

Underground Mining

Underground uranium mining is similar to underground coal mining. Underground mining is used when uranium deposits are too deep for open-pit mining. Both drilling and blasting techniques are used to extract uranium ore from belowground. There is less waste rock generated than with open-pit mining, but there are many potential environmental and health issues involved with underground mining. There is abundant potential for ground and water contamination, as well as worker exposure to dust, radon, and engine fumes from mining machinery.

Milling

Rocks mined through open-pit and underground mining must be crushed and pulverized to remove the uranium. Once the rocks are pulverized, water is added to create a slurry. The slurry is mixed with sulfuric acid or an alkaline solution to release the uranium. Approximately 95%–98% of the uranium can be recovered from the rocks. The uranium oxide produced from the slurry is sent on to be enriched. The remaining slurry is pumped to a tailings dam. The tailings dams expose the environment to any remaining uranium or heavy metals left in the slurry.

In-Situ Leach (ISL) Mining

In-situ leach (ISL) mining, or ***in-situ recover (ISR) mining***, involves drilling wells and pumping water from prospective uranium deposits. Oxidant and carbonates are added to the water, which is then pumped back down to the deposit. This oxygen-rich water moves through the rock, dissolving the uranium in the ground. The water is brought to the surface again, and treated and filtered to remove the uranium. The same water is treated with oxidant and carbonates (as needed), and the process is repeated.

In-situ leach mining is the least expensive and obstructive means of mining uranium. Very little waste rock is generated, there is less radiation exposure to workers, and it is less expensive to remediate. There is a stronger possibility of local water contamination with ISL mining. ISL mining cannot be used with all uranium ore deposits.

Energy Production

In order to use the uranium to generate power, it must be refined into a product known as enriched uranium. The enriched uranium is used to make fuel pellets about the size of a human finger. One of these pellets has as much energy capability as 130 gallons of oil. The pellets are stacked in metal fuel rods, and the rods are bundled into a fuel assembly. In a nuclear reactor, these fuel pellets are bombarded with neutrons, which causes fission to occur, **Figure 7-18**. In nuclear fission, the atoms are split apart to form smaller atoms. The smaller atoms do not need as much energy to hold them together as the larger atoms, so the extra energy is released as heat and radiation. The heat generated in the reactor by fission is used to boil water into steam, which is used to turn turbines. The turbines drive generators that make electricity. The steam is condensed into water and cooled in a cooling tower. The water is recycled repeatedly.

(Please note, the preceding explanation of fission is extremely simplified. It is not possible to fully explain the process in the course of one lesson.)

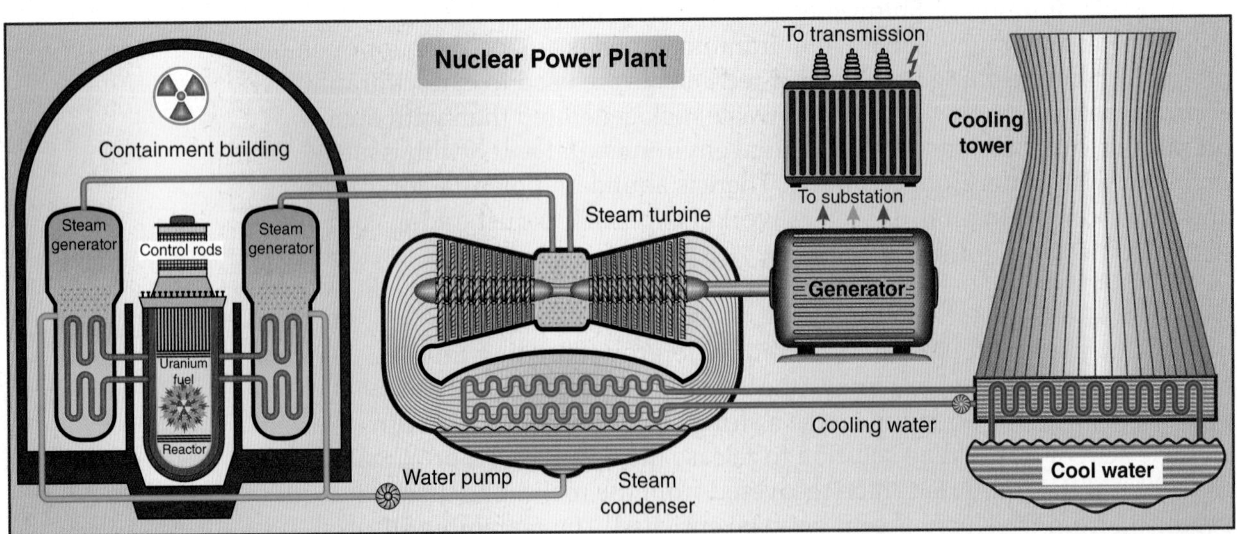

Fouad A. Saad/Shutterstock.com

Figure 7-18. The process of turning nuclear energy into electricity.

Environmental Impact

Using nuclear energy to generate electricity is a very clean way to produce energy. It does not emit carbon dioxide or other greenhouse gases. However, as with all industrial processes, nuclear power generation does have waste products. Much of the waste is low-level radioactive waste. This includes the tools, protective clothing, and disposable items used in the plant that have been contaminated with radioactive particles. These materials are disposed of under special regulations. Of greater concern are the highly radioactive spent fuel assemblies. These are stored in specially designed water pools or in specially designed dry storage containers.

Radioactive waste materials remain radioactive and dangerous to the environment and its inhabitants for thousands of years. The U.S. Nuclear Regulatory Commission has strict rules governing nuclear power plants and the cleanup and disposal of radioactive waste. Although an uncontrolled nuclear reaction could result in widespread contamination of air and water, the risk of this happening in the United States is considered to be very small. There are numerous safeguards in place to prevent such an accident.

Career Connection Agricultural Engineer

Job description: Agricultural engineers use their knowledge of science, math, and technology to solve real-world problems in the food, fuel, and fiber industries. Depending on your specific area of study, as an agricultural engineer, you may design new machinery, develop alternative energy sources, or create more efficient agricultural structures. Agricultural engineers also study environmental problems and help farmers manage soil and water resources. Some agricultural engineers work to solve problems associated with food processing. Still others create more efficient electronics and information systems.

branislavpudar/Shutterstock.com

Education required: The regular course of study requires a bachelor's degree as well as a master's degree in engineering. In addition to knowledge about physical concepts, engineers are proficient in laboratory and field work.

Job fit: This job may be a fit for you if you have good critical-thinking and problem-solving skills and enjoy working in an office setting as well as in the field.

Lesson 7.1 Review and Assessment

Words to Know

Match the key terms from the lesson to the correct definition.

1. The power derived from the kinetic energy of moving water.
2. A specialized semiconductor diode that converts visible light into direct current (DC) electricity.
3. A combustible, dark rock consisting mainly of carbonized or fossilized plant matter.
4. A naturally occurring radioactive metal used as fuel in nuclear reactors.
5. A mixture of fluids, solids, and chemicals that lubricate the bit used to drill for oil.
6. Energy that comes from sources continually replenished by nature.
7. A process in which microorganisms break down organic materials such as food scraps, manure, and plant residue.
8. Energy harnessed from the natural supply of heat beneath Earth's surface.
9. The material removed from the surface of a mining pit and materials from the pit that are not used.
10. A group of wind turbines in the same location used for production of electricity.
11. A modern version of the windmill used to convert kinetic wind energy into mechanical power or electricity.
12. The energy possessed by an object or system as a result of its motion.
13. A nonrenewable fossil fuel composed mainly of hydrocarbons and organic compounds that is found in underground reservoirs.
14. An odorless, colorless, flammable gas consisting mainly of methane and other hydrocarbons.
15. Energy that comes from natural sources that took millions of years to develop and cannot be replenished.
16. The kinetic energy of air in motion.
17. The fixture on a wind turbine that houses the operating components.

A. air turbine
B. anaerobic digestion
C. biomass energy
D. coal
E. crude oil
F. drilling mud
G. fossil fuel
H. geothermal energy
I. hydropower
J. kinetic energy
K. nacelle
L. natural gas
M. nonrenewable energy
N. open-pit mining
O. photovoltaic cell (PV cell)
P. renewable energy
Q. solar power
R. uranium
S. waste rock
T. wind energy
U. wind farm

18. The energy harnessed from organic matter as it decomposes, when it is burned, or when it is treated with chemicals.
19. An organic compound used to generate energy; includes coal, natural gas, and crude oil, and metals such as uranium.
20. Mining performed from Earth's surface.
21. Power obtained by harnessing the radiant energy emitted by the sun.

Know and Understand

Answer the following questions using the information provided in this lesson.

1. Why is wind considered a renewable resource?
2. How is wind converted to a useable energy form?
3. Explain how geothermal energy is converted to electricity.
4. What are sources of biomass energy?
5. How is biomass converted to energy?
6. Why is it better to use methane gas as an energy source as opposed to just allowing it to escape into the atmosphere?
7. How does using biomass for energy production benefit agriculturists?
8. What are three fossil fuels?
9. Which fossil fuel typically produces the least amount of pollution?
10. Why is nuclear energy considered a nonrenewable energy source?

Analyze and Apply

1. Explain how solar energy is converted to electrical energy.
2. Diagram and label the parts of a wind turbine.
3. Use materials provided by your teacher to make a solar stove.

Thinking Critically

1. Where are the largest offshore wind farms located? How much electricity does each one generate? Is it more or less expensive to install and maintain an offshore wind farm or a land-based wind farm?
2. What is the biggest challenge(s) to wind power integration in the United States?
3. Why is it more practical to ship biomass when it is reduced into pellets? What types of biomass is being reduced into pellet form?
4. Are some rocks more porous than others? Which rocks make good reservoir rocks? Research which type of rocks are most porous.

Lesson 7.2

Biofuels

Words to Know

alkyl esters
biodegradable
biodiesel
biofuel
biogas
cellulose
corn stover
crop biomass
distillation
dry mill process
ethanol
ethanol soluble
ethyl alcohol
feedstock
fermentation
glycerin
gross domestic product (GDP)
methanol
steepwater
stillage
transesterification
viscosity
wet mill process

Lesson Outcomes

By the end of this lesson, you should be able to:

- Compare the efficiency of energy production from biofuels to conventional sources.
- Explain how biofuels are manufactured.
- Identify advantages and disadvantages to biofuels.
- Evaluate the impact of biofuels on energy production.

Before you Read

Before you begin reading this lesson, consider how the author developed and presented information. Does the information provide the foundation for the next lesson? If so, how?

Did you know that the Model T automobile was originally designed to burn ethanol in addition to a petroleum-based fuel? Henry Ford expected ethanol to be the fuel of choice for most automobiles. The availability of an inexpensive petroleum fuel supply eventually caused Ford to phase out his use of ethanol as an engine fuel. It took almost a century before there was widespread use of ethanol as a fuel component in the United States. Our current dependency on this "inexpensive and abundant" petroleum fuel has inspired the development of alternative renewable fuel sources such as biofuels.

What Are Biofuels?

Biofuels are liquid fuels made from organic materials with the intention that they may one day replace nonrenewable fossil fuels. The most common biofuels are corn-based ethanol and biodiesel. These biofuels are mixed with petroleum-based fuels to create fuel blends. There are three types of biofuels:

- **Ethanol**—a natural by-product of the fermentation of sugars derived from plants.
- **Biodiesel**—a fuel derived from vegetable oils and animal fats.
- **Methanol**—a fuel derived from natural gas, coal, plant vegetation, and agricultural crop biomass. ***Crop biomass*** is the plant material or agricultural waste used as a fuel or energy source.

ArtThailand/Shutterstock.com

Luis Carlos Jimenez del rio/Shutterstock.com

Figure 7-19. A—Sugarcane is a tall perennial, or semi-perennial, grass that is typically replanted every 5–7 years. B—Sugar beets are annuals planted in the spring and harvested in the fall. Both produce sucrose, an important biofuel ingredient.

Agricultural Feedstocks

Biofuels are made from three agricultural ***feedstocks*** (raw materials supplied for processing):

- Sugar crops such as sugarcane and sugar beets, **Figure 7-19**.
- Starch crops such as corn and cereal grains.
- Oilseed crops such as canola, soybeans, and sunflowers.

Did You Know?

Sugarcane is about 7%–18% sugar by weight, while sugar beets are 8%–22% percent sugar.

Ethanol

In the United States, corn-based ethanol is the most common type of biofuel. About one bushel of corn makes about 2.5 gallons of ethanol. ***Ethanol*** is a clear, colorless, volatile liquid also known as ***ethyl alcohol***. Ethyl alcohol, or ethanol, is the intoxicating component of alcoholic beverages. Ethanol fuel uses the same type of alcohol used in alcoholic beverages, but blended with gasoline. Typical alcoholic liquors contain between 3% and 40% ethanol. Most automobiles in the United States can operate on a fuel blend that contains 10% ethanol. Ethanol may also be produced from potatoes, barley, wheat, and other common crops.

Ethanol Production

Ethanol is produced in the same way as alcoholic liquors: fermentation. ***Fermentation*** is a chemical process in which an agent, such as bacteria, yeasts, or other microorganisms, causes an organic substance to break down into simpler substances. In ethanol production, the process breaks down sugar or starch into ethyl alcohol. The two processes used to produce ethanol for commercial use are the dry mill process and the wet mill process.

Did You Know?

Each bushel of corn processed by an ethanol plant produces approximately 17 pounds of animal feed.

Dry Mill Process

In the ***dry mill process***, the grain goes through multiple steps, **Figure 7-20**:

- Milling—the washed and dried grain is ground to corn meal or flour.
- Liquefaction—liquid is added to the corn meal.

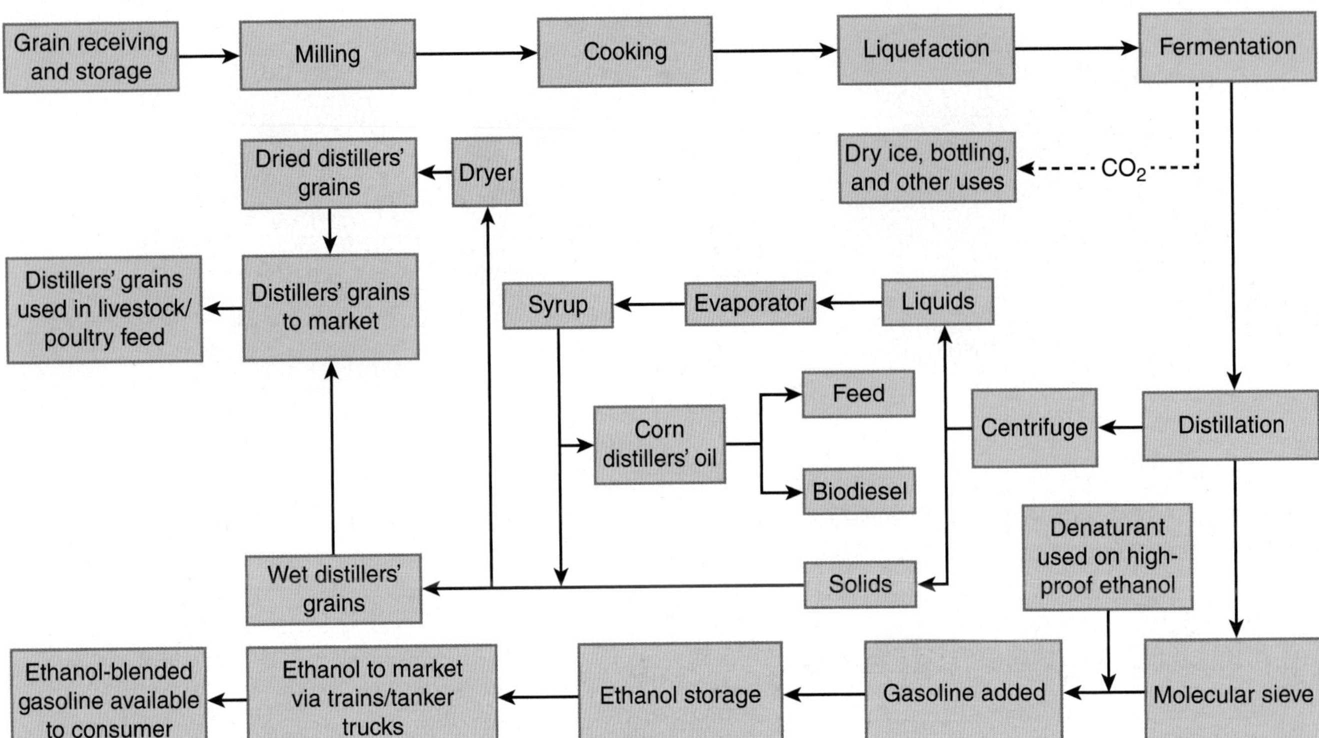

Figure 7-20. The dry mill ethanol process.

- Fermentation—the sugars are separated from the mixture and broken down into ethanol.
- ***Distillation***—the ethanol is purified through a process of heating, cooling, and recovery.

Once the ethanol is separated from the remaining solids, or ***stillage***, the stillage is processed to separate the coarse grain from the ***ethanol solubles***. These solubles are concentrated to make a syrup that is mixed and dried with the coarse grain to make nutritious livestock feed. In some ethanol production facilities (both wet and dry mill), the carbon dioxide released during fermentation is captured and sold for use in beverage carbonation and the manufacture of products such as dry ice.

Wet Mill Process

In the ***wet mill process***, the grain is soaked first and then processed through a series of grinders to separate the corn germ, **Figure 7-21**. Using various types of machinery, the corn slurry is processed to separate the oil, fiber, gluten, and starch. The oil may be processed on-site or the corn germ may be shipped for processing elsewhere. The liquid in which the grain was soaked (called ***steepwater***) is dried with the fiber and gluten and used as livestock feed. The starch and remaining liquid is processed through fermentation into ethanol, dried and sold as modified corn starch, or processed into corn syrup.

Did You Know?

The steepwater from the wet mill process is used as a component in Ice Ban™, an environmentally friendly product used to remove ice from roads.

Native and Cultivated Grasses

The use of perennial grasses may be a more environmentally sustainable solution to the development of biofuels. Grasses such as

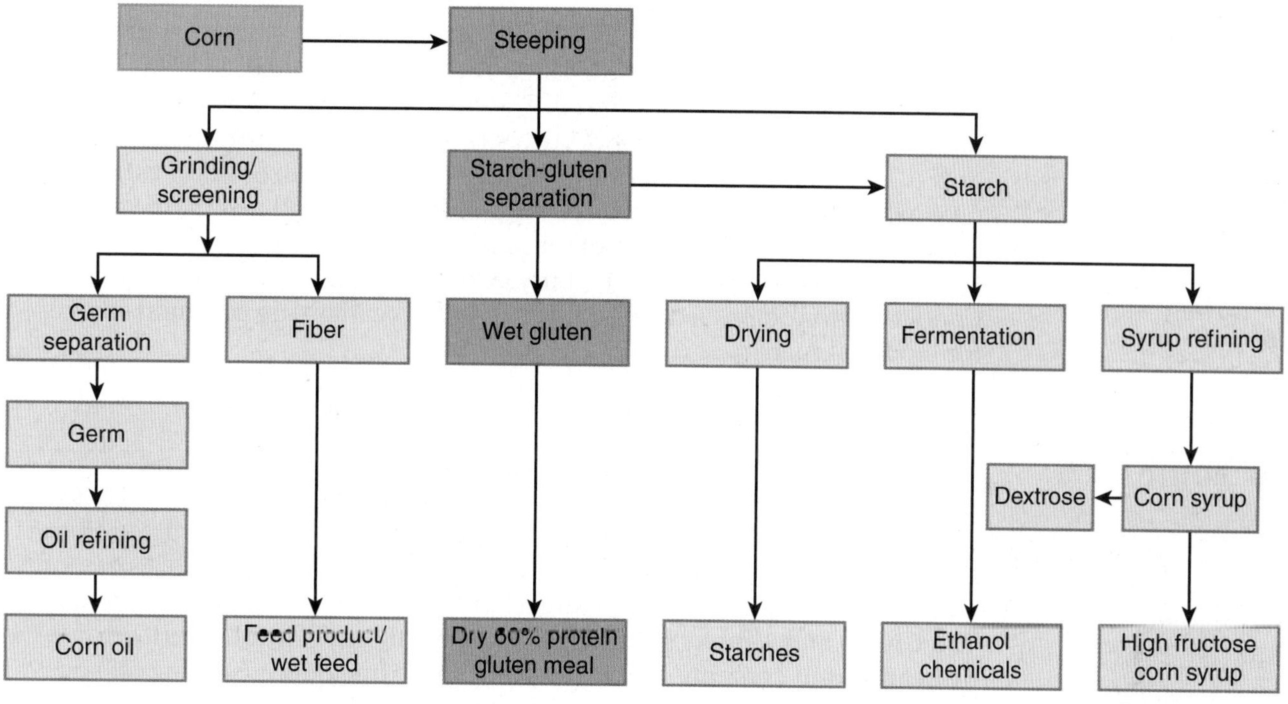

Figure 7-21. The wet mill ethanol process.

switchgrass, big bluestem, and wheatgrass contain more cellulose than grains such as corn. ***Cellulose*** is the main substance that makes up the cell walls and fibers of plants. Currently, producing ethanol from the cellulose of these types of grasses and ***corn stover*** (stalks and leaves) uses less energy than producing ethanol from the corn kernels. It can also yield about the same amount of biofuel per acre as corn. One ton of corn stover will yield about 80 gallons of ethanol, and a ton of switchgrass yields between 75–100 gallons of ethanol, **Figure 7-22**. According to the International Energy Agency, ethanol derived from cellulose may allow ethanol fuels to play a much bigger role in the future.

Safety Note

Raw vegetable oils are not the same as biodiesel and should never be used in place of biodiesel or petroleum-based diesel fuel.

hjochen/Shutterstock.com

Figure 7-22. Producing ethanol from perennial grasses like switchgrass requires less energy than producing ethanol from corn or soybeans. *How would the fact that switchgrass is a perennial affect its production costs?*

Biodiesel

Biodiesel is a fuel derived from vegetable oils, used cooking oils, and animal fats. Biodiesel may be used alone or mixed with a petroleum-based diesel fuel. Diesel fuel is heavier than gasoline and has a higher viscosity than biodiesel. ***Viscosity*** is the property of resistance to flow in a fluid or semifluid. Biodiesel's lower viscosity makes it a capable replacement of petroleum-based diesel in diesel engines. The feedstock used to make biodiesel may affect its level of viscosity as well as its ideal temperature range.

Did You Know?

All biodiesel fuel must meet strict technical fuel quality and engine performance specifications.

Biodiesel Production

A chemical reaction process called ***transesterification*** is used to convert the fat or vegetable oil to biodiesel. The chemical reaction between the fats or oils is created with an alcohol such as ethanol, along with a catalyst such as lye. The resulting products are glycerin and biodiesel, or ***alkyl esters*** (chemical name for biodiesel). The ***glycerin*** is considered a valuable co- or by-product used in soaps and other products. The glycerin originally formed the chemical bonds between the fatty acids of the oil or fats being converted. Canola oil from rapeseed, animal fats, and soybean oil are common sources for biodiesel fuel.

Did You Know?

Glycerin was originally a by-product of candle makers when candles were made with animal fats.

Canola Oil

Canola is an excellent biofuel source because it produces an oil with excellent flow characteristics in cold weather. Canola refers collectively to several varieties of rapeseed or field mustard, and to the oil extracted from the seeds of these plants. Most canola is produced in Europe and Asia, although Canada is the second largest producer of canola behind Europe. The United States ranks eighth in canola production. The oil is extracted by heating and crushing the seeds. The oil is extracted for cooking oil and biodiesel, and the rapeseed meal is used as animal feed.

Animal Fats

Approximately one-third of the fats and oils produced in the United States are derived from animal fats. This includes beef and chicken fat and pork lard. Animal fats can also be made into biodiesel fuels, provided certain conditions are met. Animal fats are highly saturated, which means that they solidify at relatively high temperatures. Beef fat has a high "cloud point," which means that it becomes cloudy when temperatures decrease below 60°F. This means that biodiesel with a high percentage of animal fat in it becomes less efficient at low temperatures, and should only be used in warm climates. Beef and pork fat is approximately 40% saturated fat and chicken is about 30% saturated fat. In comparison, soybean oil is only about 14% saturated fat.

Soybean Oil

One bushel of soybeans yields about 11 pounds of soybean oil, which makes about 1.5 gallons of biodiesel. Soybean oil is a major crop in the United States, with more than 30% of the world's soybean production occurring here. Although more than 90 million metric tons of soybeans are produced on an annual basis in the United States, soybeans as a biofuel source is not yet cost-efficient. Crops such as canola and palm yield more oil for biodiesel conversion than soybeans.

Algae

Another possible biodiesel source is algae. Studies are currently underway to develop economical and practical methods of cultivating and harvesting algae for biofuel production, **Figure 7-23**. There are currently facilities across the United States using this technology to study the process and produce viable fuel. For example, Daphne, Alabama, has an algae biofuel system that also treats wastewater, and there is a pilot plant outside Durango, Colorado, where microalgae are grown for biofuel production.

Methanol

Methanol is a biofuel derived from natural gas, coal, and organic matter such as plant vegetation and agricultural crop biomass. Methanol is a biogas. A ***biogas*** is a gaseous fuel produced by the biological breakdown or fermentation of organic matter. Methanol is primarily methane and carbon dioxide. When mixed with oxygen and combusted, methanol can create energy that can be harnessed and used for fuel. Methane is extracted from landfills, collected as a by-product of some mining operations, and collected from livestock waste in large livestock production operations, **Figure 7-24.** Since biogases can be compressed, they are useful in combustion engines used to power motor vehicles.

Steve Mann/Shutterstock.com

Figure 7-23. Advances in algae biofuel research continue to make this a promising energy source. This green energy aircraft, an EADS Diamond DA42, is powered by algae biofuel.

Did You Know?

Biogas has been in use for a long time. In the thirteenth century, Marco Polo noted in his writings that the Chinese used covered sewage tanks to generate power.

Rules and Regulations

Biofuel production in the United States is a highly debated and highly political topic. Initially, the main goals behind the development of biofuels were to reduce our dependency on fossil fuels and to reduce greenhouse gas pollutants (GHGs). Today, these are still the main goals; however,

v.schlichting/Shutterstock.com

Figure 7-24. Methane from animal waste can be captured and used as an energy source.

biofuel production has also become big business. Some will argue that the production and use of biofuels is helping us reach those goals. Others will argue the drawbacks still outweigh the benefits. Some may also argue the strict laws regarding biofuel production and use impede advancements and the development of improved processing methods. It is not a simple matter on which to take a stance, and you should do your research before you decide where to take your stand. Consider the following points:

- Biofuels, especially biodiesel, have demonstrated their potential to reduce dependence on petroleum-based fuels. This will also reduce our dependence on foreign oil production.
- Biofuels usually require subsidies and other market interventions to compete economically with fossil fuels. Many vehicles cannot use biofuels or even biofuel blends. The demand for biofuels will also vary as does the demand for petroleum-based fuels. This greatly affects the production costs and income of biofuel producers.
- The biofuel industry must ship its product by train or tanker trucks. This increases biofuel cost because it is more costly than shipping through established pipelines.
- Biofuel production includes land and water resource requirements, air and groundwater pollution, and increased food costs. These matters must be taken into account when calculating costs and benefits.
- When calculating the cost of biofuels, one must include land use, especially if land is converted to cropland or if cropland is converted for biofuel production.
- Increased biofuel production increases the demand for corn. This increased demand drives up the price of corn and increases feed prices for livestock farmers. Food prices also increase for American consumers. This may change as more cellulose is used instead of grains to produce ethanol.
- Depending on the feedstock and production process, the combustion of biofuels may emit even more greenhouse gases (GHGs) than some fossil fuels on an energy-equivalent basis.
- The process of crop production, fertilizer and pesticide production, and feedstock processing for fuel consumes a great deal of energy. Much of the energy used in the processes comes from fossil fuels. Some argue whether ethanol from crops such as corn provides more energy than is required to grow and process the corn.
- A gallon of ethanol-based fuel has approximately 30% less energy than a gallon of regular gasoline.

As you may now see, there are valid arguments for both sides and many aspects to consider regarding the benefits of biofuels. You can gain more insight by researching biofuels and the agencies and laws that guide their production.

Biofuels and the Environment

Several government agencies regularly review possible benefits and negative impacts biofuels have on the environment and the economy. For example, the Environmental Protection Agency (EPA) regularly reviews the production processes of biofuels and their effect on the environment. It also establishes fuel blend requirements and quality standards for both ethanol and biodiesel. The EPA also sets mandates in accordance with the *Renewable Fuel Standard* regarding the quantity of biofuels that should be produced, blended, and sold each year.

Ethanol

Currently, there is more controversy surrounding ethanol production than biodiesel production. As more research is performed and improved methods of ethanol processing are found, ethanol will gain more widespread use in the United States. Countries, such as Brazil, currently have large-scale ethanol production facilities and are using vehicles that run on fuel blends as high as 25% ethanol, **Figure 7-25.**

Biodiesel

EPA studies have shown that biodiesel use reduces greenhouse gas emissions from motor vehicles by more than 50% and up to 86% when compared to petroleum diesel. Biodiesel is the only EPA-designated *Advanced Biofuel* in commercial production in the United States. Because it meets such strict technical fuel quality and engine performance specifications, biodiesel can be used in existing diesel engines without modification.

According to the EPA, using biodiesel in diesel vehicles (especially older models) reduces asthma-causing soot, greenhouse gases, carbon dioxide, and sulfur dioxide in air emissions. Other benefits of increased biodiesel use

Safety Note

Before using biodiesel in any engine, carefully review engine manufacturers' warranties to ensure the use of biodiesel or biodiesel blends will not void coverage.

AFNR/Shutterstock.com

Figure 7-25. Bagasse is the fibrous matter that remains after sugarcane or sorghum stalks are crushed to extract their juice. In large-scale Brazilian production facilities, bagasse is used as a biofuel and in the manufacture of pulp and building materials.

include less need for drilling for oil and fewer opportunities for spills and the environmental disasters they cause. Biodiesel may also be made with used cooking oils and lard, reducing waste sent to landfills. Biodiesel is also non-toxic and ***biodegradable*** (capable of being broken down by bacteria or other living organisms).

Climate Change

Climate change is the change in long-term weather patterns that may be influenced by solar radiation, volcanic activity, and possibly by human activities such as fossil fuel consumption. The combustion of fossil fuels are a major source of greenhouse gases. Reduction of these greenhouse gases may have a positive effect on climate change. Replacing fossil fuels with biofuels may also help reduce harmful carbon emissions.

Biofuels and the Economy

Ethanol production in the United States currently employs about 200,000 people. Ethanol production creates more than $30 billion in household income, and adds $44 billion to the U.S. gross domestic product (GDP). A country's ***gross domestic product (GDP)*** represents the total dollar value of all goods and services produced by a country over a specific time period (usually a year). A country's GDP is also used to gauge the health of a country's economy. Farm income as a direct result of ethanol production has topped $4.5 billion. Biodiesel production is rapidly advancing, producing more than 1.8 billion gallons. The biodiesel industry currently employs more than 62,000 people.

Large, off-road industrial equipment such as that used for construction, forestry, and mining use diesel fuel. These machines can readily use biodiesel as well. Agricultural uses of biofuels are similar to those of industry. Heavy farm machinery such as tractors and combines can use biodiesel. In fact, almost any internal combustion engine in use on a farm can use biodiesel or ethanol-based fuels, **Figure 7-26**. Small gas engines, like those found on lawnmowers and string trimmers, are being designed to use ethanol blends. Biofuel use in commercial aviation is still in the development phase. However, biofuel blends are expected to increase in the next few years.

Zeljko Radojko/Shutterstock.com

Figure 7-26. Large diesel-powered farm equipment can use some biodiesel fuel blends as recommended by the equipment manufacturer.

Lesson 7.2 Review and Assessment

Words to Know

Match the key terms from the lesson to the correct definition.

1. A biofuel derived from vegetable oils, used cooking oils, and animal fats.
2. A liquid purifying process that involves heating, cooling, and recovery.
3. The state of something that can be broken down by bacteria or other living organisms.
4. The plant material or agricultural waste used as a fuel or energy source.
5. A biofuel that is a natural by-product of the fermentation of sugars derived from plants.
6. The solids remaining once the liquid ethanol is separated from the solid remnants of the feedstock.
7. The main substance that makes up the cell walls and fibers of plants.
8. An ethanol production method in which the feedstock is ground first, then mixed with liquid before further processing.
9. A chemical reaction process used to convert the fat or vegetable oil to biodiesel.
10. A biofuel derived from natural gas, coal, plant vegetation, and agricultural crop biomass.
11. The raw material supplied for processing into another product.
12. A chemical process in which an agent, such as bacteria, yeasts, or other microorganisms, causes an organic substance to break down into simpler substances.
13. The property of resistance to flow in a fluid or semifluid.
14. The liquid in which the feedstock was soaked before being processed into ethanol using the wet mill process.
15. The stalks and leaves of the corn plant.
16. A valuable by-product of biodiesel production used in soaps and other products.
17. An ethanol production method in which the dry feedstock is soaked in liquid before being ground.
18. The total dollar value of all goods and services produced by a country over a specific time period.

A. biodegradable
B. biodiesel
C. cellulose
D. corn stover
E. crop biomass
F. distillation
G. dry mill process
H. ethanol
I. feedstock
J. fermentation
K. glycerin
L. gross domestic product (GDP)
M. methanol
N. steepwater
O. stillage
P. transesterification
Q. viscosity
R. wet mill process

Know and Understand

Answer the following questions using the information provided in this lesson.

1. Explain why Henry Ford phased out the use of biofuels in his Model T cars.
2. Briefly explain the main difference between ethanol and biodiesel.
3. List at least five common crops used in ethanol production.
4. *True or False?* Fermentation is not used in the wet mill process of ethanol production.
5. What are some common agricultural sources for biodiesel production?
6. List an advantage of canola oil as a source of biodiesel.
7. Which has more energy, a gallon of regular gasoline or a gallon of ethanol-blended gasoline?
8. *True or False?* Biodiesel is nontoxic and biodegradable.
9. What is climate change? How might the increased use of biofuels influence climate change?
10. How are biofuels used in industrial and farming operations?

Analyze and Apply

1. Research the various types of crops grown in your area. Which ones would be good candidates as a biofuel source?
2. Research the concept of climate change or global warming. What may be the consequences of climate change over the next 100 years?

Thinking Critically

1. What are the advantages and disadvantages to the production and use of biofuels? What do you think is the most important argument for or against biofuels?
2. Would an algae biofuel plant work in your area? Why or why not?

Lesson 7.3

Agricultural Tools and Equipment

Words to Know

adjustable wrench
Allen wrench
aviation snips
back saw
ball-peen hammer
band saw
belt-and-disk sander
belt sander
blacksmith hammer
block plane
box end wrench
bull tongs
bush hammer
chisel
chuck
circular saw
claw hammer
club hammer
cold chisel
combination square
combination wrench
compass saw
coping saw
Crescent wrench
crimpers
crosscut saw
crowbar
drill
drill press
drywall hammer
drywall saw
flat-head screwdriver
folding rule
four-cycle engine
framing hammer
framing square
gooseneck bar
grinder
hacksaw
hammer
jack plane
jig saw
jointer
keyhole saw
lineman's pliers
lining bar
locking pliers
long tape
mason's hammer
measuring square
miter box
miter square
nail hammer
nail puller
nail set
needle-nose pliers
open end wrench
orbital sander
oscillating spindle sander
petroleum-powered tool
Phillips-head screwdriver
pincers
pinch bar
pin punch
pipe wrench
planes
pliers
pneumatic tool
pry bar
punches
push stick
radial arm saw
random orbital sander
reciprocating saw
ripping hammer
rip saw
router
rubber mallet
safety glasses
sander
screwdriver
shingler hammer
side-cutting pliers
sledge hammer
slip-joint pliers
slotted-head screwdriver
Society of Automotive Engineers (SAE)
socket wrench
soft-face hammer
speed square
strap wrench
table saw
tack hammer
tape measure
torque
try square
T-square
two-cycle engine
upholsterer's hammer
wire strippers
wood chisel
wooden mallet
wrecking bar
wrench
zigzag rule

Lesson Outcomes

By the end of this lesson, you should be able to:

- Identify standard tools and use the proper safety procedures.
- Use the appropriate procedures for the use and operation of specific tools and equipment.
- Demonstrate personal safety precautions when using tools for a specific task.

Before You Read

Before you read the lesson, read all of the table and photo captions. What do you know about the material covered in this lesson just from reading the captions?

Did You Know?

The change to a settled life, and the development of agriculture and animal domestication, increased the need for tool development.

Tools extend our ability to do things. They help us to build and create and complete tasks efficiently and effectively. We could dig holes in the ground with our hands, but using a shovel makes it a lot easier. Some jobs cannot be done at all without tools. Imagine trying to drive nails into lumber without a hammer. It would be very difficult indeed. Understanding how to safely and effectively use the tools found in agricultural occupations is the purpose of this lesson.

DJTalyor/Shutterstock.com

Figure 7-27. Wearing the proper safety gear and clothing will help prevent injury.

Tool Safety

Before using any tool or completing any work, make sure that you are using the proper personal protective equipment for the job. In most cases, ***safety glasses*** are needed. If you are using power tools or working in situations where loud noises occur, hearing protection is also needed. You should also wear clothing and footwear that protects you from injury, **Figure 7-27**.

Before using hand and power tools, make certain that you know how to use the tools properly and you are using the correct tool for the job. Read the safety manuals that come with power tools, or receive training from a qualified individual. Keep the adage "an ounce of prevention is worth a pound of cure" in mind whenever you start a project. As the tool operator, you must also make sure that bystanders and helpers are aware of safety rules and that they follow them.

General Tool Safety

The shop in which you may be working will have its own rules of operation, but there are general hand tool safety rules you should *always* follow:

- *Always* follow the safety and maintenance procedures outlined by the manufacturer.

- Do *not* use tools that are damaged or broken.
- Wear safety glasses at all times. Proper eye protection should be worn when operating any hand or power tool. Use safety glasses with side shields to fully protect your eyes. Regular eyeglasses do not provide adequate eye protection when working with hand and power tools. Use safety glasses that are designed to fit over prescription eyeglasses or purchase safety glasses made in your prescription. A full face shield may also be used over prescription eyeglasses.
- Only use tools for the job for which they are intended and for which you have been trained to operate or use.
- Properly maintain and regularly inspect hand tools to keep them in safe working order.
- Keep your work area clean and free of clutter and make sure there is adequate lighting.

Hand Tools

There are hundreds of hand tools used in the agricultural industry. Some may be specific to agricultural jobs, while many are used in a variety of professions. Tools are designed specifically for almost any job, including:

- Measuring
- Construction and demolition
- Driving or removing screws
- Turning nuts and bolts
- Prying both large and small objects
- Driving pins and setting nails
- Drilling holes
- Cutting and sawing
- Sanding and grinding

This lesson discusses the most common types found on farms and in other agricultural businesses.

Measuring Tools

Carpenters have a saying that you should always "measure twice, cut once." This means it is important to take accurate measurements when performing any job. If you make errors in measurement, the materials you have prepared for the job will not fit properly and may be unsalvageable. Remember, there is no such thing as a "board stretcher." Lumber and other materials cut to incorrect measurements cost time and money. The good news is that there are many tools available to help you take correct measurements.

Measurements are taken using either the English or metric system. *English rules* typically divide inches into 8, 16, or 32 equal sections. **Figure 7-28** shows the 1/8″ rule. The more an inch is divided, such as in the 1/32″ rule, the finer the measurement that can be taken.

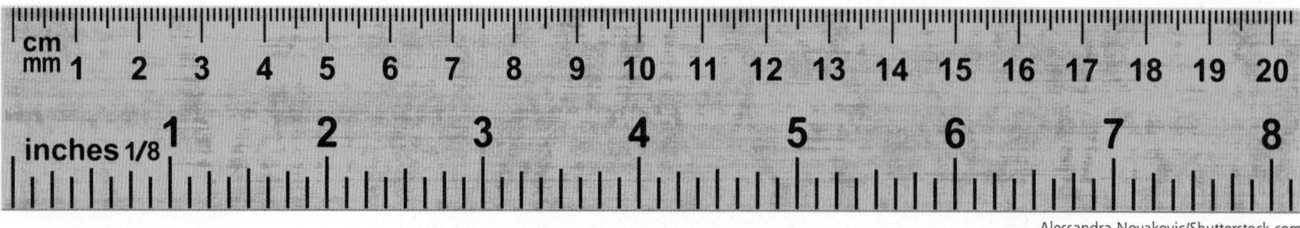

Alessandra Novakovic/Shutterstock.com

Figure 7-28. This is an example of a ruler that divides inches into eighths. The bottom set of numbers is the 1/8″ rule. The top set of numbers represents the metric equivalent in centimeters.

Tape Measures

Tape measures are useful and convenient tools for measuring a variety of spaces and materials. The tape itself may be flexible fabric, plastic, or steel, and it may be housed in a handheld case or on a reel, **Figure 7-29A**. The flexible steel tape extends and retracts from a handheld case. The steel tapes are available in different widths, with the wider tapes (1″–1 1/4″) being stiff enough for extended and overhead measurements. The tape has an L-shaped tab at the end that allows you to "hook" it to the edge of the material being measured.

Long tapes come on a reel and may or may not be fully encased, **Figure 7-29B**. Long tapes are available in lengths from 25′ to 300′ and are useful for measuring long distances for such jobs as fencing, pipes, and foundations. They also have a type of hook or clip on the end that can be looped over an object like a nail or screw to hold it in place.

Proper use and maintenance will extend a tool's life and accuracy. *Always* wipe water and mud off the steel rule before rewinding or retracting to prevent rust and damage to the rewinding mechanism. Bent or kinked steel rules will not lie flat or straight when extended and will not give an accurate reading.

B Calkins/Shutterstock.com Huntstock/Thinkstock

Figure 7-29. A—A steel tape measure is designed for portability and often includes a clip that can be used to attach the tool to your belt or apron. B—Long tapes may be encased or on an open reel.

Carmen Brunner/Shutterstock.com

Figure 7-30. Folding rules are compact enough to be carried in a pocket.

Folding Rules

Zigzag rules or ***folding rules*** have stood the test of time and are a good tool for measuring in specific situations, including: electrical work (do not conduct electricity); prints (folds to 90° angle); inside, outside, and awkward measurements; and existing odd bends. See **Figure 7-30.**

Folding rules can be maintained through careful handling, wiping off dirt and grime, and lubricating the joints regularly.

Squares

There are several common types of squares, including the framing square, the try square, and the combination square, **Figure 7-31**.

The ***framing square*** is a 90°, L-shaped tool made from one piece of metal that is used for laying out rafters and marking stair stringers. It has one wider arm called the *blade* and a narrower arm called the *tongue*. The blade is the longer of the two arms of the square, and it is usually 24″ long and 2″ wide. The tongue of the square is 1 1/2″ wide and is typically 16″ long.

The ***try square*** is an L-shaped tool used as a guide for marking 90° angles and checking the edges and ends of boards for squareness. It may also be used to determine whether a board is the same thickness for its entire length. The ***speed square,*** or ***measuring square,*** is technically also a try square. It is triangular-shaped with a flanged edge for butting against the work edge to

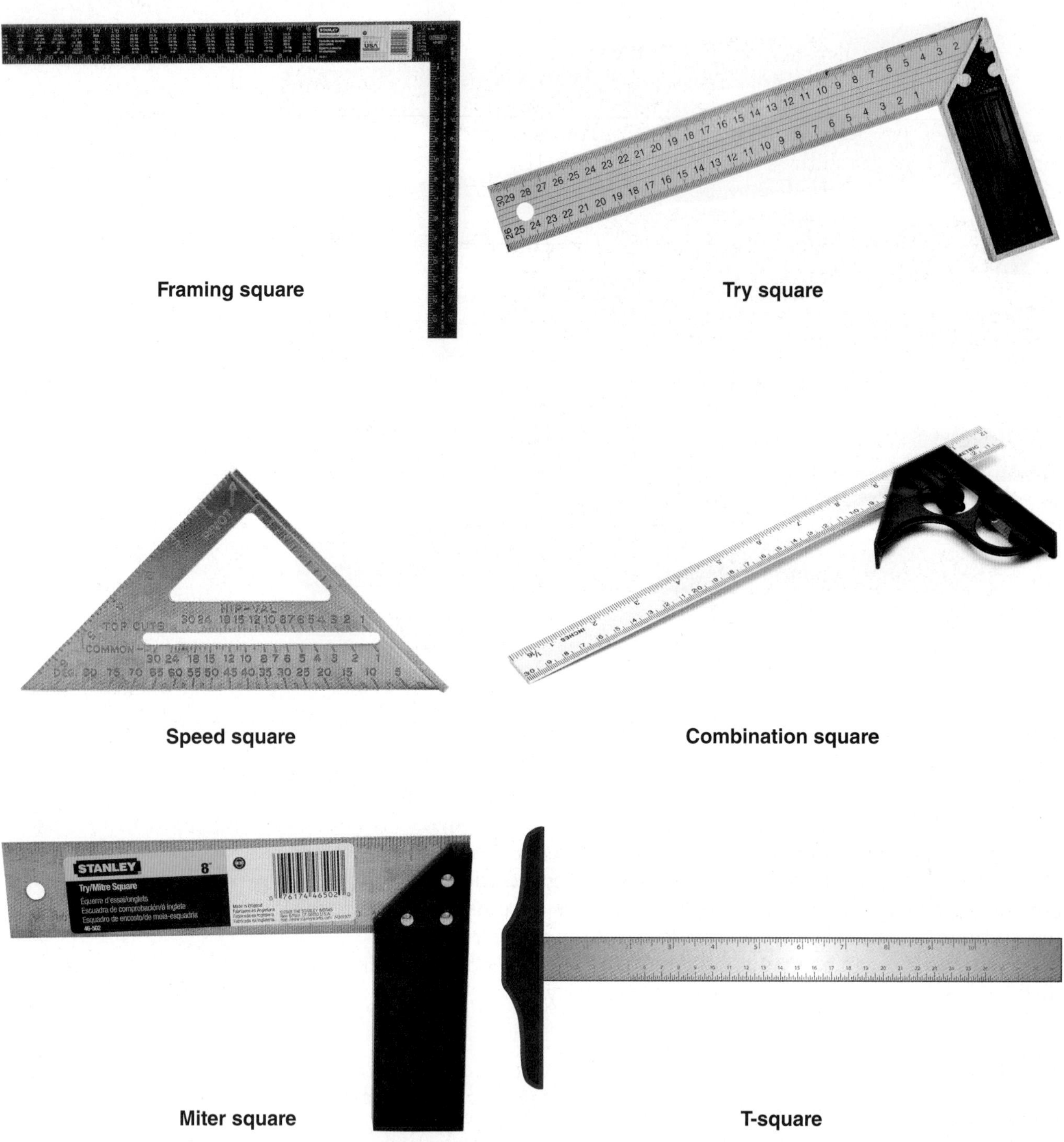

(Left to right) Provided courtesy of Stanley Tools; graja/Shutterstock.com; Scott Milless/Shutterstock.com; terekhov igor/Shutterstock.com; Provided courtesy of Stanley Tools; casejustin/Shutterstock.com

Figure 7-31. Squares come in a variety of shapes, sizes, and materials.

draw 90° or 45° angles. It may be used as a cutting fence for circular saws and for laying out rafters and stairs.

The ***combination square*** has a grooved blade and head that can be adjusted to different locations along the 12″ blade. It is useful for scribing, crosscutting, mitering, and checking for level and plumb, and it may also have a protractor for setting angles along the 180° range. The part you hold is called the *stock*, and the part at a right angle to the stock is called the *blade*.

The blade on a combination square may be graduated so it can be used as a rule.

There are also ***miter squares*** (used mostly for measuring and marking corner miter joints) and ***T-squares*** (used primarily to draw horizontal lines). T-squares come in various lengths, including 18″, 24″, 36″, and 42″. The long shaft is called the *blade,* and the short shaft is called the *stock* or *head.*

It is important to maintain a square's integrity through careful handling and proper cleaning. Do not drop them! *Always* wipe them clean and dry, and when possible, store them in a case.

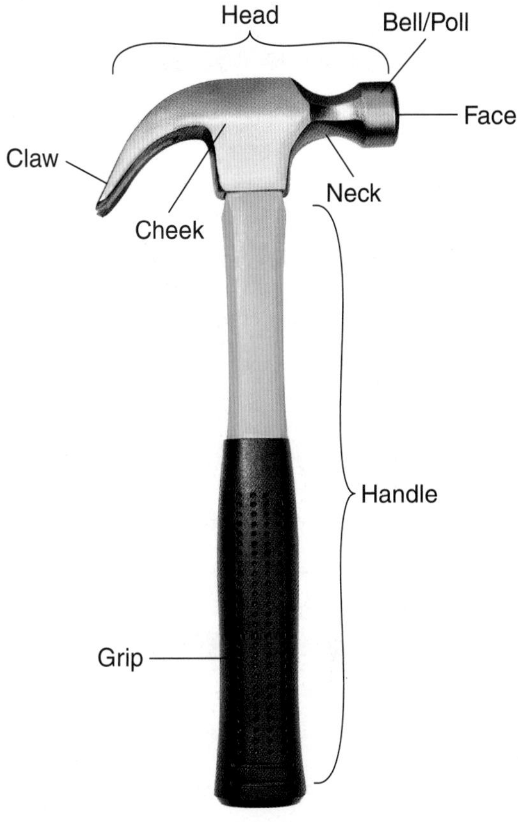

LeS/Shutterstock.com

Figure 7-32. Most hammers have the same basic components.

Hammers

The basic ***hammer*** is a tool used for driving in and pulling out nails. Most hammers have the same basic components, **Figure 7-32**. However, there are specialized types of hammers, each with a different purpose, **Figure 7-33**. Using a hammer for a task other than for which it was designed will likely result in injury and shoddy work.

Using a Hammer

Although many people would say using a hammer is simple enough, there are techniques that make hammering more efficient and more effective. A proper grip and swing is less tiring on the arm and will result in quicker, more efficient work. Keep the following points in mind:

- Make a fist when you grip the handle (wrap your thumb across the index and middle fingers or just above the index finger).
- Hold the hammer nearer the end of the handle to get maximum leverage and better tool balance. Hold the hammer closer to the head when starting a nail (for better control).
- Hold the nail near the top, not the bottom, so you do not crush your fingers between the wood and the hammer if you miss.
- Let the weight of the hammer do the work by allowing momentum and gravity help you. Using all your strength will lead to wild misses and bent nails!
- Do *not* hold the hammer too tightly! It is more tiring and stressful on your arm and shoulder.
- Drill a pilot hole when necessary and focus on the nail, not the hammer.

It is important to keep hammers clean and in good shape. Wipe them clean and dry before storing to prevent rust and deterioration of the rubber handle covering.

Did You Know?

The weight of a hammer refers to the hammer's head and not to its overall weight.

Winai Tepsuttinun/Shutterstock.com	***Ball-peen hammer***—used in metalworking and mechanics, the peen is rounded since nails are not used; available in sizes ranging from 4 oz to 2 lb; used for striking chisels and punches, and rounding the heads of metal fasteners such as rivets
Provided courtesy of Stanley Tools	***Blacksmith hammer***—used for striking unhardened metals; may also be double-headed
Vadym Zaitsev/Shutterstock.com	***Club hammer***—small sledge hammer used to drive stakes, cold chisels, or demolish masonry
Provided courtesy of Stanley Tools	***Drywall hammer***—has a serrated face for a better grip on the nail when installing drywall; the hatchet-shaped claw is used to remove old material and/or mark cut-outs for outlets and switches
Bata Zivanovic/Shutterstock.com	***Mason's hammer***—designed to cut and set bricks; the hammer head is moderately heavy with a square face for striking and a chisel edge for cutting and dressing bricks and stone
Bildagentur Zoonar GmbH/Shutterstock.com	***Nail*** or ***claw hammer***—most commonly used hammer; available in a number of sizes and weights with handles made from wood, fiberglass, steel, or composite material; it has a face for driving nails and a curved claw for pulling nails
AlexAvich/Shutterstock.com	***Ripping*** or ***framing hammer***—has a claw that is less curved to allow use as a prying tool in addition to a nail remover; typically used in construction for framing; weighs 20 to 32 oz and has a longer handle than a claw hammer for better leverage
Provided courtesy of Stanley Tools	***Rubber mallet***—for hammering on a finished metal surface to avoid marks and dents

(*Continued*)

Figure 7-33. Using the correct hammer for the job will help preserve the integrity of your hammers. Using the wrong hammer for a job may damage your materials, damage the hammer, or cause injury.

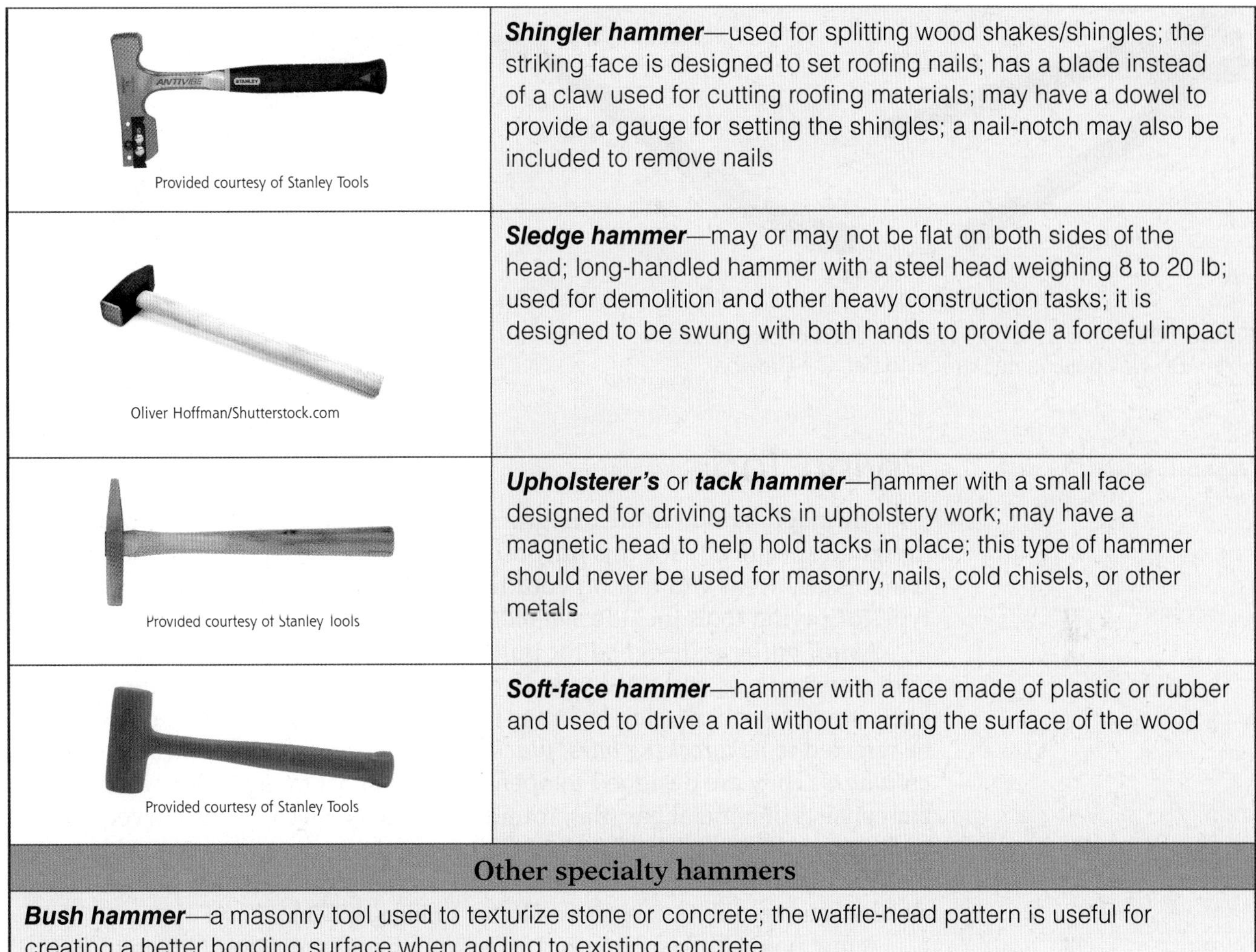

Provided courtesy of Stanley Tools	***Shingler hammer***—used for splitting wood shakes/shingles; the striking face is designed to set roofing nails; has a blade instead of a claw used for cutting roofing materials; may have a dowel to provide a gauge for setting the shingles; a nail-notch may also be included to remove nails
Oliver Hoffman/Shutterstock.com	***Sledge hammer***—may or may not be flat on both sides of the head; long-handled hammer with a steel head weighing 8 to 20 lb; used for demolition and other heavy construction tasks; it is designed to be swung with both hands to provide a forceful impact
Provided courtesy of Stanley Tools	***Upholsterer's*** or ***tack hammer***—hammer with a small face designed for driving tacks in upholstery work; may have a magnetic head to help hold tacks in place; this type of hammer should never be used for masonry, nails, cold chisels, or other metals
Provided courtesy of Stanley Tools	***Soft-face hammer***—hammer with a face made of plastic or rubber and used to drive a nail without marring the surface of the wood
Other specialty hammers	
Bush hammer—a masonry tool used to texturize stone or concrete; the waffle-head pattern is useful for creating a better bonding surface when adding to existing concrete ***Wooden mallet***—used with wood chisels	

Figure 7-33. (*Continued*)

Hammer Safety

Keep the following safety points in mind when using hammers:

- *Always* follow the safety and maintenance procedures outlined by the manufacturer.
- Proper eye protection should *always* be worn when operating any hand tool. Use safety glasses with side shields to fully protect your eyes.
- Use the correct hammer for the job to prevent personal injury and damage to the work materials. Keep the handle clean and dry to prevent it from slipping out of your hand.
- Do *not* use a hammer with a loose, cracked, or broken handle or head. The hammer head could fly off while in motion and injure you or someone else. Replace damaged handles or dispose of the hammer if it cannot be repaired.

Figure 7-34. Having multiple types or sizes of prying tools will enable you to access small or awkward areas. A—Nail puller. B—Nail puller/small wrecking bar. C—Crowbar.

Prying Tools

Although hammers are designed for use as prying tools, there are tools designed specifically for this purpose. Using these types of tools will save unnecessary wear and tear on your hammers. Some of the most common types of prying tools include *nail pullers* and *pry bars*.

A ***nail puller*** is designed for pulling out nails. It is usually a straight bar with a curved end and a V-shaped notch used to grip nail heads. It works like a small crowbar, using leverage to pull up a nail. ***Pry bars*** may also be referred to as ***wrecking bars, pinch bars, gooseneck bars, lining bars***, or ***crowbars***. They are designed to open or break apart materials. A wrecking bar consists of a metal bar (flat, round, straight, curved) with a curved end and flattened points. The flattened points are designed to be wedged between the materials being separated. The notch between the flattened points may also be used to pull out fasteners. Refer to **Figure 7-34**.

A few points to keep in mind when purchasing or using prying tools:

- Wear safety glasses when using prying tools.
- Pry bars may have handles and vary in length and thickness.
- Some pry bars are designed to be struck by hammers to spread materials apart.
- Pry bars are used for prying, moving, rolling, or lifting heavy objects.
- The size (weight and strength) of a pry bar determines its use—*always* use the correct tool for the job!

Wipe down pry bars to keep them clean and prevent rust. Keeping pry bars (and any other tool) clean of slippery substances such as oil and grease will help prevent the tool from slipping while in use.

Screwdrivers

A ***screwdriver*** is a tool for driving or removing screws. This tool uses turning force to extract or drive fasteners. The power needed for this turning motion is called ***torque***. Screwdrivers come in various sizes and types. The most common types of screwdrivers are ***flat-head,*** or ***slotted-head,*** and ***Phillips-head***. The type of screwdriver needed depends on the type of fastener requiring tightening or removal. See **Figure 7-35**. As always, it is important to use the correct tool for the task at hand. Using the wrong screwdriver will likely leave you with a stripped screw and a damaged blade.

Keep the following points in mind when using a screwdriver:

- Use a pilot hole when driving screws into wood.
- Start the screw by holding the driver tip and screw head together before tightening.
- Use a screwdriver with a shorter shank for more driving power.
- Do *not* use a screwdriver if the tip is rounded or chipped. It is likely to slip and cause injury or material damage.
- Do *not* use a screwdriver in place of a chisel.

Screwdrivers are also susceptible to corrosion and should be kept clean and dry. Use a clean cloth to wipe them down before storing. Proper storage will keep the tips sharp and help them maintain the proper shape.

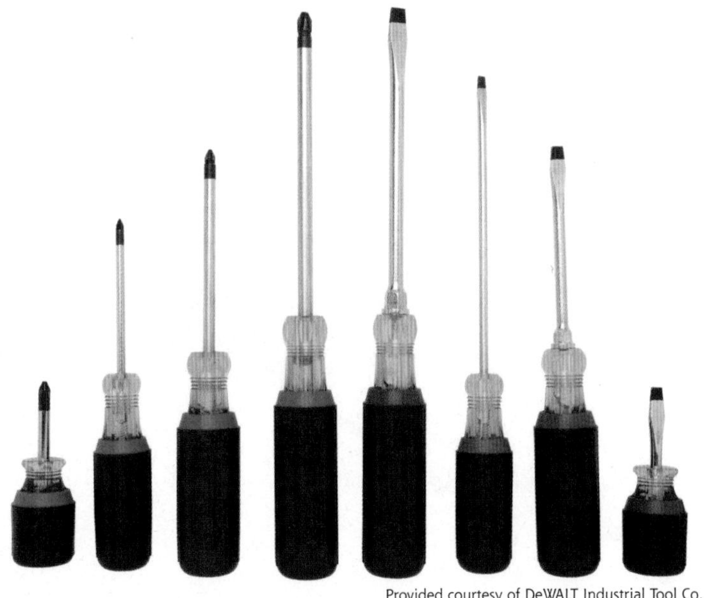

Provided courtesy of DeWALT Industrial Tool Co.

Figure 7-35. Screwdrivers come in a variety of lengths as well as types and sizes.

Wrenches

A ***wrench*** is designed to apply torque and either tighten or loosen nuts and bolts. Wrenches come in a variety of shapes and sizes, and since nuts and bolts come in metric and SAE sizes, wrenches are sized accordingly. ***SAE*** is the acronym for the ***Society of Automotive Engineers*** and, when used in conjunction with tool grade or measurement, it is a standardized unit of nonmetric measurement. SAE sizes use the inch as the standard unit of measurement. Metric wrenches may be used for SAE-sized nuts and bolts and vice versa. Conversion charts are available online and from manufacturers to determine which wrench should be used. However, most technicians prefer (and recommend) having two sets of tools, one SAE and one metric.

Some of the more commonly used wrenches include those described in the following sections.

Open End Wrench

Open end wrenches are nonadjustable wrenches that are open at each end. The open end of the tool allows it to be slid over the nut in tight spaces where other wrenches may not fit. Open end wrenches are usually double-ended with different sized openings. You will have to reposition the wrench on the fastener head after each turn.

Box End Wrench

A ***box end wrench*** is a nonadjustable wrench that is closed at each end. Each end has either six or 12 points around the inside diameter of the jaws. These points are designed to fit hexagonal heads and nuts. The 12-point design may also be used on square nuts. Box end wrenches are useful in tight spaces; however, they must be slid on over the end of the fastener and may not work in some situations. The box end helps minimize the risk of damaging the fastener. You will have to reposition the wrench on the fastener head after each turn.

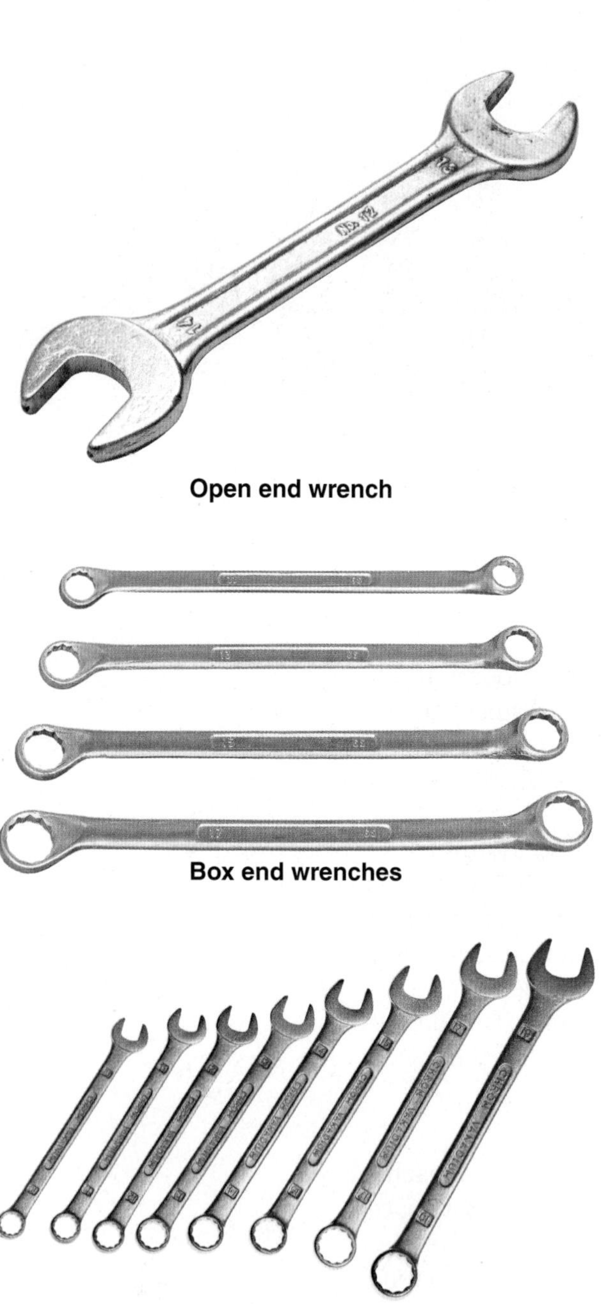

Open end wrench

Box end wrenches

Combination wrenches

aarows/Shutterstock.com; aldorado/Shutterstock.com; Andy-pix/Shutterstock.com

Figure 7-36. Open end wrenches are nonadjustable and are usually sold in sets. They are especially useful when you are tightening multiple fasteners of the same size. Box end wrenches are also nonadjustable and sold in sets, both metric and standard SAE sizes. Combination wrenches have one open end and one closed end. The ends generally fit the same size nut and bolt.

Combination Wrench

Combination wrenches have open jaws at one end and closed jaws (box end) at the other. Both ends generally fit the same size fasteners. **Figure 7-36** shows the differences between these types of wrenches.

Socket Wrench

The ***socket wrench*** is a specialized type of wrench that uses a ratcheting mechanism to turn a socket, which turns a nut or bolt. The ratcheting mechanism eliminates the need to remove the socket for repositioning after each turn. Socket wrenches allow for the rapid tightening or loosening of nuts and bolts. Socket wrenches are available with permanent sockets or interchangeable sockets, as well as a variety of attachments to suit each and every task, **Figure 7-37**. *Always* use the correct size socket or you may round off the points of the fastener.

Pipe Wrench

Pipe wrenches are adjustable, self-tightening wrenches used (especially in plumbing) to tighten and loosen threaded, round pipe. Pipe wrenches have hardened, toothed jaws and a strong grip and will leave marks on most materials. They should never be used on nuts and bolts as they will easily damage the fastener. As with most tools, pipe wrenches come in a variety of sizes (determined by the handle length), **Figure 7-38A**.

Strap Wrench

Strap wrenches are self-tightening wrenches with a strap (or chain) that can be pulled tight around cylindrical objects, **Figure 7-38B**. The friction created by the strap keeps the object from slipping. Strap wrenches are especially useful when working with wet or oily pipe. ***Bull tongs*** are a larger version of a strap wrench used with large diameter pipe.

Allen Wrench

Allen wrenches are L-shaped or T-shaped wrenches with a hexagonal-shaped shaft. On

L-shaped wrenches, the longer length is the handle and the shorter one is the head, but either end can be used for each purpose. Allen wrenches are also known as *Allen keys*, *set-screw wrenches*, and *hex wrenches*. Allen wrenches are used on set screws (recessed into material) with hexagonal-shaped sockets. See **Figure 7-38C**.

Adjustable Wrench

Adjustable wrenches are open-ended with one stationary jaw and one adjustable jaw, **Figure 7-38D**. Their versatility makes them a good purchase for beginners. When using an adjustable wrench, make sure the majority of the pressure is applied to the fixed jaw since it is stronger. This can be done simply by turning the wrench toward the movable jaw. Adjustable wrenches may

Rafa Irusta/Shutterstock.com

Figure 7-37. Socket wrenches are usually sold with a set of corresponding sockets. The benefit of the socket wrench is the ratcheting device. The ratchet device holds in place when you pull in one direction, and releases when pulled in the opposite direction.

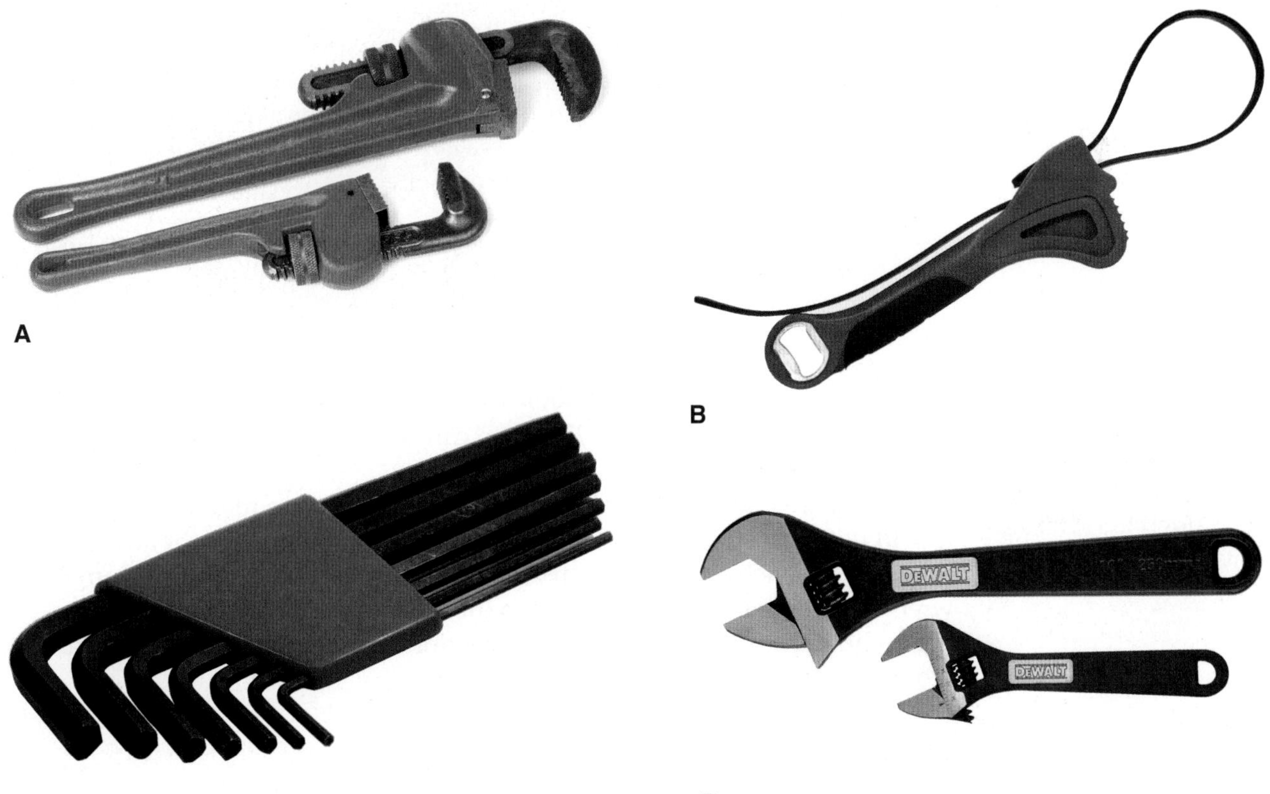

(Left to right) S1001/Shutterstock.com; dcwcreations/Shutterstock.com; Niki Crucillo/Shutterstock.com; Provided courtesy of DeWALT Industrial Tool Co.

Figure 7-38. A—Pipe wrenches are adjustable with a top jaw that moves up or down. B—Strap wrenches are versatile and may be used for tasks such as removing oil filters or installing plumbing pipes. C—Allen wrenches come in sets with various sizes. D—Crescent wrenches are adjustable and available in a range of sizes.

Wrench Safety

Keep the following safety points in mind when using wrenches:

- *Always* follow the safety and maintenance procedures outlined by the manufacturer.
- Proper eye protection should *always* be worn when operating any hand or power tool. Use safety glasses with side shields to fully protect your eyes.
- *Always* select the correct wrench type and size for the job, and *never* use a wrench on moving machinery!
- Make sure the jaw of an open end wrench is in complete contact (flat, not tilted) with the fastener before you apply pressure to prevent slippage and damage to the fastener or yourself.
- *Always* pull (not push) a wrench with slow, steady movements.
- Do *not* use a cheater bar, such as a piece of pipe over the wrench handle, to gain leverage. If you need more leverage, get a longer wrench.
- Apply penetrating oil to stubborn fasteners to reduce the risk of rounding off the fastener head and personal injury.
- Do *not* hit a wrench with a hammer to turn a fastener. You will damage the wrench and possibly injure yourself.
- Do *not* use a wrench if the handle is bent or the jaws appear to be out of shape.
- Position your body to prevent losing your balance and hurting yourself if the wrench slips or breaks.
- Use a box or socket wrench with a straight handle whenever possible.
- When using a pipe wrench, make sure the teeth are sharp and clean to prevent slippage and possible injury.
- When using a ratchet wrench, begin with a small amount of pressure to make sure the ratchet wheel is engaged in the same direction in which you are applying pressure.
- *Always* support the head of the ratchet wrench when you are using socket extensions.
- Stand to the side when you are using a wrench on overhead work.

Did You Know?

Wrench openings are slightly larger than the nuts or bolts they are designed to fit. If you use a wrench that is too loose, you will likely round off the points of the nut or bold head.

Did You Know?

Most wrenches are offset to allow two different gripping positions.

also be referred to as ***Crescent wrenches***. An adjustable wrench may easily loosen while you are working, so extra care should be taken not to round off the points on fasteners.

There are many more types of wrenches. If you are in doubt as to what tool to use for a specific task, ask your teacher or supervisor for advice.

Keep wrenches clean and dry and stored in a cool, dry place. Some wrenches may require oiling to keep their joints moving freely. Follow the manufacturer's maintenance instructions to keep wrenches in good working order.

Pliers

Pliers are hand tools designed to hold objects firmly when pressure is applied to the handles. The clamping or squeezing action brings the jaws together to hold an object. One common type of pliers found in almost every farmer's and mechanic's toolbox are ***slip-joint pliers***, **Figure 7-39**. Slip-joint pliers allow the jaws to be adjusted to allow for greater flexibility in gripping a variety of materials.

Once again, there are many types of pliers and each is designed to perform a specific task. Some of the more commonly used pliers include:

- ***Locking pliers*** are similar to slip-joint pliers, but have the added benefit of being able to be clamped down, holding materials without the operator having to maintain pressure on the handles, **Figure 7-40A**.
- ***Side-cutting pliers*** are a special type of pliers that have a set of jaws for cutting wire instead of jaws that grip and hold material. They are useful in cutting most types of wire, **Figure 7-40B**.
- ***Lineman's pliers*** combine the functions of side-cutting pliers and slip-joint pliers, **Figure 7-40C**. They can grip and hold material, but also have a built-in wire cutting edge along the jaws. These are commonly used by electricians.
- ***Needle nose pliers*** are another useful tool used by electricians. These pliers have narrow jaws that allow the user to grasp and hold small objects, **Figure 7-40D**. The jaws are used to bend and twist electrical wire.
- ***Crimpers*** are hand tools designed to crimp (connect) a connector to the end of wires, **Figure 7-40E**. They are designed to cut specific wire sizes and may also have a wire cutter near the handles.
- ***Wire strippers*** are a specialized set of pliers used to strip the insulation from various sizes of electrical wire, **Figure 7-40F**.
- ***Aviation snips*** and ***pincers*** cut sheet metals and wire, **Figure 7-40G**.

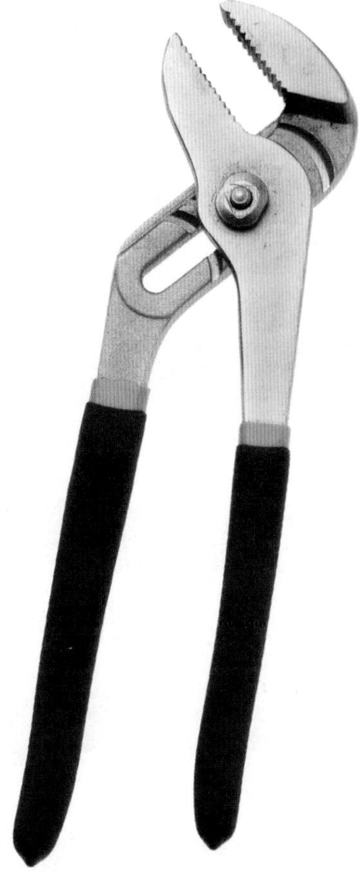

@erics/Shutterstock.com

Figure 7-39. The adjustable design of slip-joint pliers allows them to grip objects ranging in thickness from a few millimeters to more than a half inch, depending on the size of the pliers.

Pliers Safety

Keep the following safety points in mind when using pliers:

- *Always* follow the safety and maintenance procedures outlined by the manufacturer.
- *Always* use safety glasses with side shields to fully protect your eyes.
- Do *not* use pliers for anything except their intended use. For example, pliers should not be used to cut hardened wire unless manufactured for this purpose.
- Do *not* use pliers for work that is beyond their capacity. For example, do not bend stiff wire with light pliers because you will damage the tool.
- Do *not* expose pliers to excessive heat. It will damage metal properties and may cause the tool to fail.
- Preserve the sharpness of your pliers by cutting at right angles, not side to side.
- Do *not* use pliers on live electrical wire! Specially insulated tools are designed for work involving live circuits.
- Do *not* extend handle length for greater leverage! Use a larger pair of pliers.
- Pliers should *never* be used to turn nuts and bolts.
- Do *not* push items away with pliers.

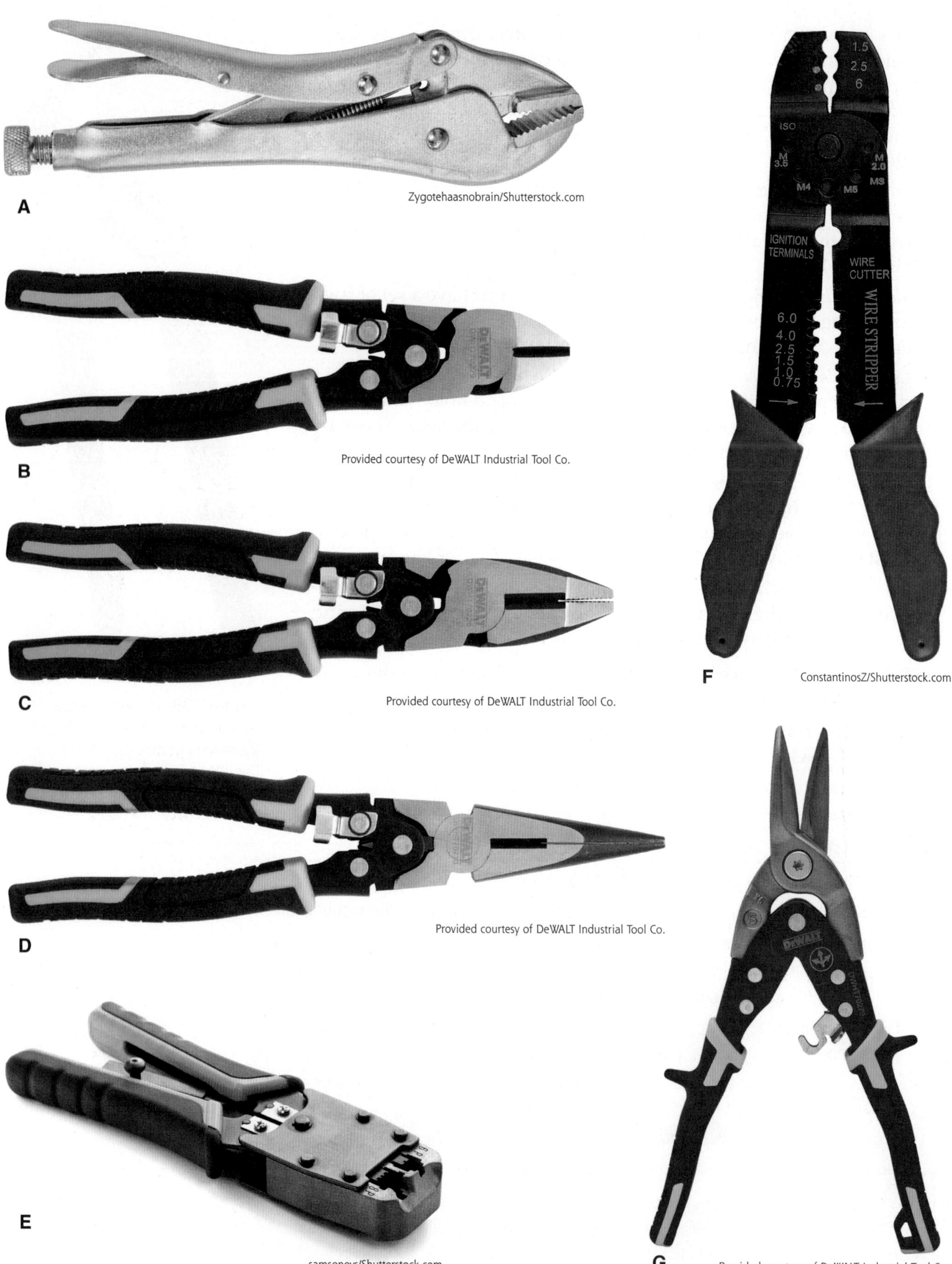

Figure 7-40. Various types of pliers will help fill out your tool collection. A—Locking pliers. B—Side-cutting pliers. C—Lineman's pliers. D—Needle-nose pliers. E—Crimpers. F—Wire strippers. G—Aviation snips.

Properly clean and maintain pliers to extend their usefulness. Oiling the joints will lengthen their life and ensure easy operation. Follow the manufacturer's maintenance instructions to keep pliers in good working order.

Punches and Chisels

Punches are made of hardened steel and are designed to be struck by a hammer or wooden mallet. ***Pin punches*** are used to drive pins out of fasteners. ***Nail sets*** are a type of punch and are used to set the heads of nails below the surface of wood. The indentation left by the nail can then be hidden with wood filler.

Chisels are used for specialized cutting. ***Wood chisels*** are designed to be struck by a mallet for the purpose of cutting wood in spaces where a saw cannot reach. Wood chisels are useful for trimming or removing small bits of wood. The chisel is placed against the wood and tapped with a mallet to drive the blade into the wood. ***Cold chisels*** are made of hardened steel and are designed to cut metal. This chisel gets its name from its ability to cut through metal that has cooled to room temperature. These chisels are made to be struck by a hammer and can cut sheet metal or cut through fasteners for removal. Refer to **Figure 7-41**.

Chisel Safety

Although cold chisels are made of hardened steel, the head of the chisel will deform after being struck by a hammer over a period of time. This may form a flattened ridge at the head of the chisel. This is a safety hazard because bits may break off when struck by the hammer and cause injury. The heads of cold chisels must be ground and shaped periodically to remove the flattened metal at the head.

J.P. Hancock, Texas A&M University

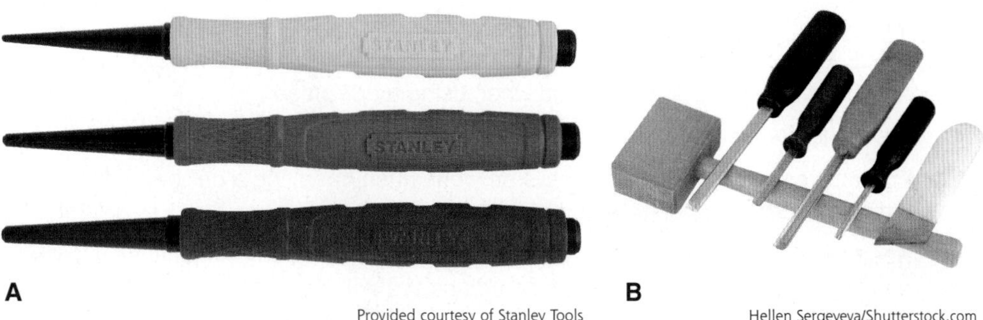

A

Provided courtesy of Stanley Tools

B

Hellen Sergeyeva/Shutterstock.com

Figure 7-41. A—A nail set is used to drive a nail so that its head is flush with or below the surface. It may also be called a nail punch. B—A wooden mallet is used with wood chisels to trim or remove small bits of wood.

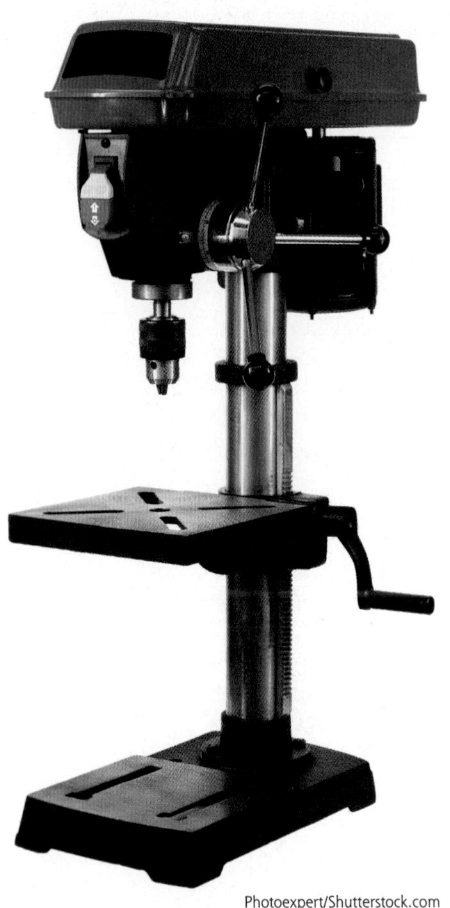
Photoexpert/Shutterstock.com

Figure 7-42. A drill press may be a countertop model, or freestanding.

Power Tools

There are power tools for most every type of task. Some are available as both stationary and portable versions. Keep the following in mind when using power tools:

- *Always* follow the safety and maintenance procedures outlined by the manufacturer.
- Wear safety glasses with side shields to fully protect your eyes. Regular eyeglasses do *not* provide adequate protection. Use safety glasses designed to fit over prescription eyeglasses or purchase safety glasses made in your prescription.
- Do *not* use tools that are damaged or broken.
- Unplug power tools when not in use or when repairing or changing tools or blades.
- Only use tools for the job for which they are intended and for which you have been trained to operate or use.
- Proper maintenance and regular inspection will help keep all tools in safe working order.
- Promptly remove any broken or damaged tools from use and report the tool damage to your teacher or supervisor.
- Maintain a clean and uncluttered work area to prevent tripping or loss of balance.
- Do *not* operate power tools near flammable materials. Sparks could ignite these materials.

Drill Press

A ***drill press*** is an electrically powered drive mounted vertically on a stand, **Figure 7-42**. The drill is manually lowered to a table where the material to be drilled is clamped. Drill presses are extremely useful and improve accuracy when making series of holes. Drill presses are similar to handheld power drills, but they are more powerful and special care should be used in their operation.

Drill Press Safety

Keep the following safety points in mind when using a drill press:

- *Always* follow the safety and maintenance procedures outlined by the manufacturer.
- Proper eye protection should *always* be worn when operating a drill press. Use safety glasses with side shields to fully protect your eyes.
- Keep clothing, long hair, and cords away from the rotating drill bit because it is very easy for material to become wrapped around the drill bit.
- Drill presses should be operated at a speed compatible with the bit being used and the type of material being drilled. The drill bit may seize in the material and cause the workpiece to rotate with the torque.
- Turn off the machine before changing the drill bit or other accessories you may be using.

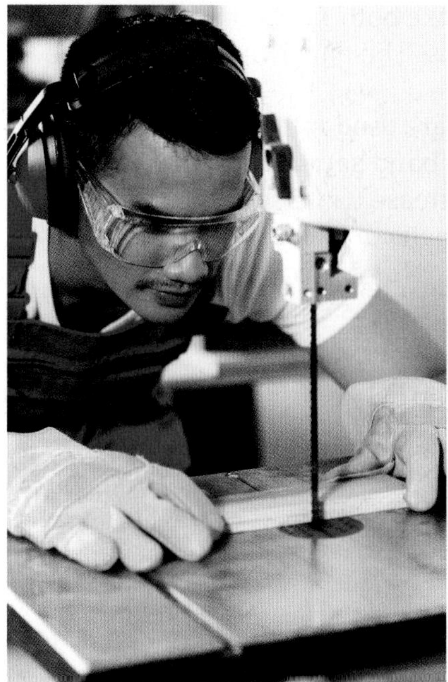

Kzenon/Shutterstock.com

dcwcreations/Shutterstock.com

Figure 7-43. Band saws are available in models ranging from small, precision models to those large enough to cut through logs.

Band Saw

A ***band saw*** is a stationary power tool that uses a circular, metal band with serrated teeth to cut through everything from bone to logs. The circular band travels at high speeds around two or more large wheels, **Figure 7-43**.

Band Saw Safety

Keep the following safety points in mind when using band saws:

- *Always* follow the safety and maintenance procedures outlined by the manufacturer.
- Proper eye protection should *always* be worn when operating a band saw. Use safety glasses with side shields to fully protect your eyes.
- Make sure the saw blade has been installed properly before using the saw. The teeth of the blade should be pointing down toward the table.
- *Never* stand on the right side of the saw while it is in operation. Do *not* allow helpers or observers to stand on the right side either. If the blade breaks, it will likely fly in that direction.
- Do *not* clean or attempt to make any changes to the saw unless the power is turned off.
- With the power off, remove sawdust and wood chips. These can interfere with moving parts and create hazardous conditions.
- Small bits of wood may become lodged in the slot through which the blade travels. *Never* try to remove these bits of wood while the saw is operating.
- If the blade breaks, shut off the power and stand clear until the machine comes to a complete stop. Saw blades occasionally break and can be ejected from the saw by the turning wheels.
- *Never* place your fingers, hands, or arms in line with the blade.
- Make sure that lumber is free of nails and other metal fasteners. Striking a metal object could cause the saw blade to break.
- Use an assistant to help you with oversized lumber. Make certain your assistant knows the safety rules and is wearing the appropriate personal protective equipment.

In woodworking, the band saw is used to make both straight and curved cuts. Band saw blades come in various widths. The narrower the blade, the easier it is to make curved cuts.

Band saws are versatile machines. They are used in the meat processing industry to prepare cuts of meat. Specialized band saws can be used to cut metals. Portable versions of the band saw increase its versatility. Lumber mills use large band saws to rip logs into dimensional lumber. Band saws have a thinner blade than most circular saws. This helps to reduce the amount of wood wasted through the cutting process.

Carpentry and Woodworking Tools

Carpentry and woodworking in the agricultural industry involve the use of tools that saw, trim edges, or bore into wood.

Hand Saws

Besides the obvious difference in shapes and sizes, hand saws vary greatly by the type of blades they use. Blades are distinguished by the number of teeth or points per inch they have, and by the shape and direction of those teeth. Saws that "rip" through wood use teeth set up like a series of chisels that grab and lift out the end grain instead of cutting it. Saws designed to cut smoother edges have teeth arranged at an angle that "score" the end grain as they cut. The type of hand saw you need for a task depends on the type and size of material you will be cutting.

The hand ***crosscut saw*** and the hand ***rip saw*** are two basic types of saws used in carpentry. The crosscut saw is used to saw across the wood grain while the hand rip saw is used to saw along the grain of the wood, **Figure 7-44**. The cuts made by crosscut and rip saws are often coarse and leave rough edges. These hand saws are very useful, but they cannot be used to cut curves. Other common hand saws include:

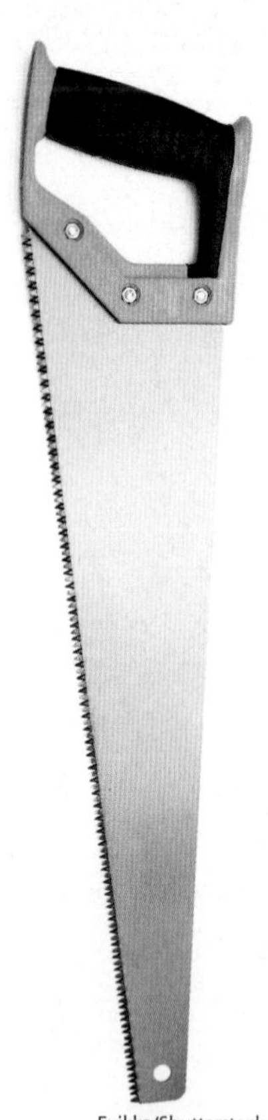

Evikka/Shutterstock.com

Figure 7-44. Although power saws are very useful, hand saws are an essential part of any tool box.

- A ***compass saw*** or ***keyhole saw*** is a push stroke hand saw with a narrow blade made for cutting holes and gentle curves in wood, **Figure 7-45A**. The keyhole saw is useful for cutting holes in walls or floors for pipes, electrical outlets, or other fixtures. It may also be referred to as a ***drywall saw*** since it is commonly used to cut holes in drywall.
- The ***back saw*** is a push stroke saw with fine teeth used to make smooth cuts. The back saw has a stiffening rib on the edge opposite the cutting edge, allowing for better control and more precise cutting, **Figure 7-45B**. Miter saws, dovetail saws, carcass saws, and tenon saws are all back saws.
- Add a ***miter box*** to the back saw and you can accurately cut wood in a variety of angles. A miter box is a three-sided box made of plastic, metal, or wood that is used to guide the saw to make precise angled cuts, **Figure 7-45C**.
- ***Hacksaws*** are push stroke saws that use disposable blades held in place with front and back pins, **Figure 7-45D**. The blades are available with different teeth configurations and are used primarily to cut metals ranging from thin metal tubing to cables and wire ropes. Hacksaw blades should be kept lightly oiled to ensure smooth cutting and to prevent overheating.

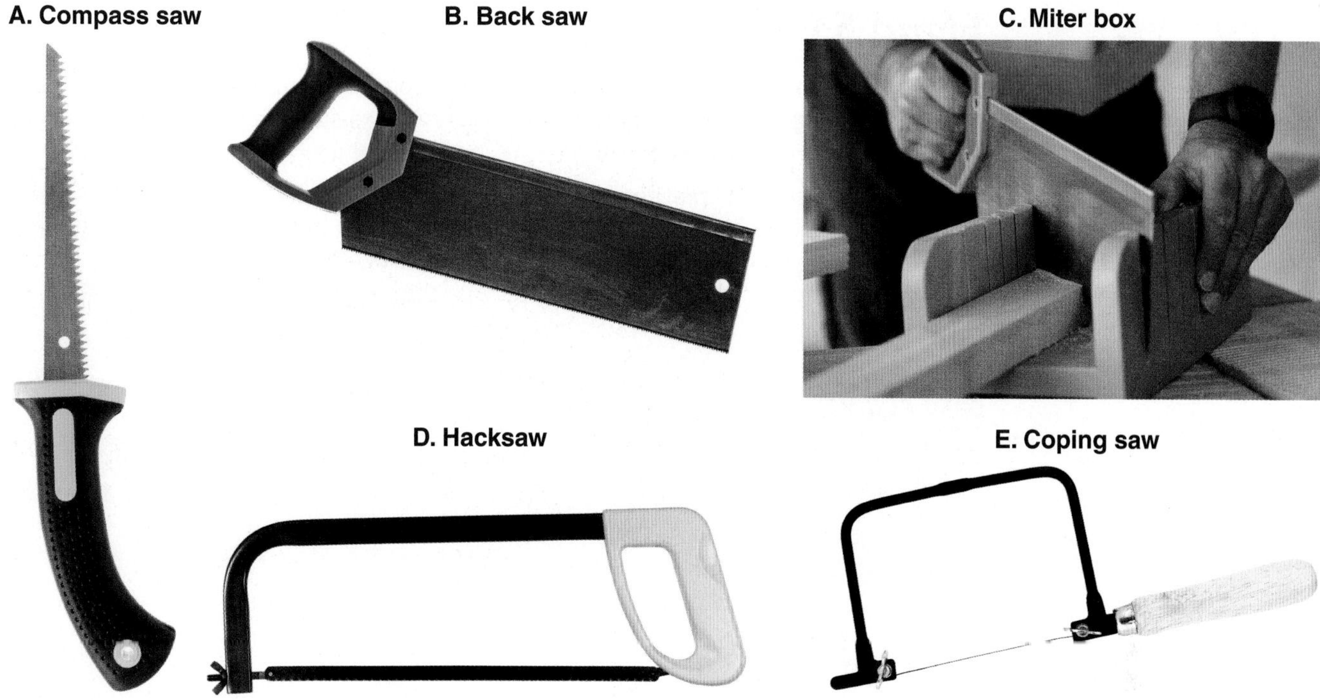

Stocksnapper/Shutterstock.com; Dario Lo Presti/Shutterstock.com; Rido/Shutterstock.com; Yanas/Shutterstock.com; Worakit Sirijinda/Shutterstock.com

Figure 7-45. It is useful to have a variety of handsaws in your tool collection.

- A ***coping saw*** is a pull stroke saw that uses a very thin blade and fine teeth to make smooth-edged, curved cuts. The steel frame of a coping saw is attached to a wooden handle that can be turned to tighten the blade. See **Figure 7-45E**. Coping saws may also be used to make cuts inside a piece of wood.

You can extend the life and usefulness of your hand saws by properly cleaning and maintaining them. The entire surface of the saw blades should be wiped down with an oily cloth to prevent rust. Use protective tooth covers to protect the teeth when saws are not in use and either hang the saw or lay it flat for storage.

Hand Saw Safety

Keep the following safety points in mind when using hand saws:

- *Always* follow the safety and maintenance procedures outlined by the manufacturer.
- Proper eye protection should *always* be worn when operating any hand or power tool. Use safety glasses with side shields to fully protect your eyes from wood dust and chips.
- *Always* inspect hand saws before use. Look for cracked handles, missing teeth, loose blades, and bent blades or frames. Remove broken or damaged saws from service, and notify your teacher or supervisor of the damage or defect.
- Do *not* test saw teeth on hands or fingers to determine if they are sharp.
- Select the proper saw shape, size, and blade type for the material being cut.
- *Always* carry hand saws by the handle with the saw end pointed down.

(*Continued*)

Hand Saw Safety (*Continued*)

- *Always* check the material being cut for nails, screws, knots, and other objects that may damage the saw or cause personal injury. Make sure the material being cut is held firmly in place before sawing.
- Start cuts *slowly* to prevent the blade from jumping, and cut harder materials more slowly than soft materials. Do *not* force a saw if it jams in a cut! Either you are using the wrong saw or the saw is not in proper condition.
- Have someone help with long stock or use some other type of support (sawhorse, vise, etc.).
- Maintain a safe, comfortable distance from the cutting area. *Always* direct the saw away from your body and use full-length strokes to get the best results.
- Do *not* cut materials over your head. Lower the work if possible, or use a ladder or scaffolding.
- *Always* store hand saws properly to prevent damage.

Planes

Planes are woodworking tools used for trimming, squaring, and smoothing the edges of lumber, **Figure 7-46**. Planes have a sharp blade and, depending on their length, one or two handles. Most planes are also adjustable. Although many other tools have been designed to perform the same tasks, planes are still useful and very portable.

- ***Jack planes*** range in length from 12″ to 18″ and are used to smooth and square rough lumber.
- ***Jointers*** are bench planes that are 12″ or longer that are used to trim, square, and straighten the edges of doors and long boards.
- ***Block planes*** are small planes used for smaller jobs such as planing the ends of wood and molding and trim.

Maintain planes by keeping them clean and dry and only using them on wood. Remove resin buildup from cutting softwoods, and keep blades sharp through proper honing. Planes should be stored on their side.

Provided courtesy of Stanley Tools

Figure 7-46. Well-maintained planes will prove useful for many years.

Portable Power Tools

Using portable power tools reduces the amount of time it takes to do a job and increases the amount of power available to do a job—provided they are used safely and as the manufacturer intended.

Maintain power tools by keeping them clean and dry and only using them for the tasks for which they are designed. If the power tool you are using has a storage/carrying case, use the case to protect the tool when it is not being used.

Portable Power Tool Safety

The same safety rules apply to portable power tools as those that apply to stationary power tools. Keep the following in mind when using power tools:

- *Always* follow the safety and maintenance procedures outlined by the manufacturer.
- Proper maintenance and regular inspection will help keep all tools in safe working order. Do *not* use tools that are damaged or broken. Promptly remove any broken or damaged tools from use and report the tool damage to your teacher or supervisor.
- Proper eye protection should *always* be worn when operating any hand or power tool. Use safety glasses with side shields to fully protect your eyes.
- Do *not* use power tools in wet conditions!
- Do *not* remove safety guards.
- Unplug power tools when not in use and when repairing or changing tools or blades.
- *Only* use tools for the job for which they are intended and for which you have been trained to operate or use.
- *Never* yank on the power cord to disconnect it.

Did You Know?

When sawing plywood with a circular saw, it is important to keep the finished side down. Otherwise, the circular saw will leave a rough or ragged edge.

Circular Saws

The portable ***circular saw*** can be used for a variety of cutting tasks, **Figure 7-47**. Saw diameters between 7″ and 8″ are the most common sizes and can be used to cut through most dimensional lumber. Circular saws often use combination blades that allow them to be used for crosscutting and ripping. Many portable circular power saws can also be used to make bevel or miter

Circular Saw Safety

The same safety rules apply to both portable power tools and stationary power tools. Keep the following safety points in mind when using a circular saw:

- *Always* follow the safety and maintenance procedures outlined by the manufacturer.
- Proper eye protection should *always* be worn when operating circular saws. Use safety glasses with side shields to fully protect your eyes.
- Power saws are loud and require the use of hearing protection.
- Do *not* remove the guards or shields from any portable power tools. Circular saws use a safety guard that retracts as the saw blade cuts through the material. It is important to keep this guard in the proper working order.
- When using a power saw, make sure the material to be cut is securely clamped down.
- Place your hands in the proper location and away from the path of the saw blade.
- Make sure the power cord will not interfere with the sawing operation and is kept safely out of the way.
- *Always* use the proper type blade for the materials you are cutting.
- Keep the saw blade sharp. It is safer to use a sharp blade than a dull one and the job will go much smoother.
- Do *not* use circular saws in wet conditions.

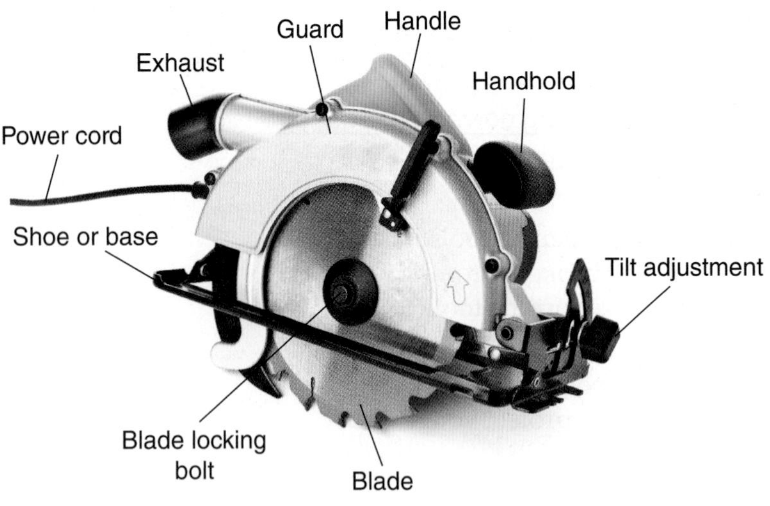

tale/Shutterstock.com

Figure 7-47. Circular saws vary by manufacturer but will have the same basic components. The saw is either right-handed or left-handed, depending on which side of the motor the blade is mounted.

cuts. That is, the base of the saw can be tilted to make angled cuts.

The blade on a circular saw spins counterclockwise, which means that the saw blade cuts from the bottom of the wood to the top. This helps hold the saw to the material and prevents the saw from "walking across" the surface of the wood.

Jig Saws

Jig saws and ***reciprocating saws*** use thin blades that move back and forth to make a cut. The thin blades allow both saws to make curved cuts. Blades can be selected and installed that are appropriate for wood or metal materials, **Figure 7-48**.

Jig Saw Safety

Keep the following safety points in mind when using a jig saw:

- *Always* follow the safety and maintenance procedures outlined by the manufacturer.
- Proper eye protection should *always* be worn when operating any hand or power tool. Use safety glasses with side shields to fully protect your eyes.
- When positioning material for cuts, make sure you will *not* damage power cords.
- Position materials so you will *not* damage the surface upon which the material is clamped. It is too easy to saw right through a sawhorse!
- Disconnect the power before changing blades.
- Use the appropriate type blade for the material being cut.
- Do *not* use jig saws in wet conditions.

Provided courtesy of DeWALT Industrial Tool Co.

Provided courtesy of DeWALT Industrial Tool Co.

Figure 7-48. As with all power tools, jig saw and reciprocating saw designs will vary by manufacturer. Review the manufacturer's guide to ensure safe and proper use of any power tool.

Grzegorz Petrykowski/Shutterstock.com

Figure 7-49. Portable power drills are essential to any tool collection. Corded drills are powerful, but quality cordless drills are less restrictive in use.

mrHanson/Shutterstock.com

Figure 7-50. A drill chuck is the device that holds the bit in place. The key used for tightening the chuck may be attached to the power cord.

Power Drills

Portable power ***drills*** are one of the most versatile power tools available, **Figure 7-49**. Depending on the type of bit used and the strength of the drill, a power drill can be used to drill holes in wood, metal, and other materials; drive screws; turn nuts and bolts; and sand or grind material. Some drills operate at a fixed speed whereas some have variable speeds. Variable-speed drills allow the operator to adjust the speed by varying the amount of finger pressure on the trigger. Power drills may be electric or battery-operated.

The ***chuck*** is a device that clamps down and holds the bit in place, **Figure 7-50**. A chuck key may be needed to tighten or loosen the chuck.

Drill Safety

Keep the following safety points in mind when using a power drill:

- *Always* follow the safety and maintenance procedures outlined by the manufacturer.
- Proper eye protection should *always* be worn when operating drills. Use safety glasses with side shields to fully protect your eyes.
- Review the manufacturer's instruction guide before using the drill.
- If corded, keep the power cord safely to the side.
- If battery-operated, keep batteries clean and properly charged.
- Position materials so you will not damage anything that may be behind it—including your fingers!
- Disconnect the power or lock the trigger before changing bits.
- Use the appropriate type bit for the task at hand.
- Do *not* use drills in wet conditions.

Router Safety

Keep the following safety points in mind when using a router:

- *Always* follow the safety and maintenance procedures outlined by the manufacturer.
- Proper eye protection should *always* be worn when operating routers. Use safety glasses with side shields to fully protect your eyes.
- *Always* unplug the router before changing bits or performing maintenance.
- Use the proper type bit for the material being cut.
- Do *not* use routers in wet conditions.

Greg Epperson/Shutterstock.com

Figure 7-51. Routers are versatile power tools. The variety of bits available enable the user to perform a variety of woodworking tasks.

Routers

Routers are typically used in cabinetmaking and millwork, and are often found in agricultural shops, **Figure 7-51**. Routers use bits similar to drill bits, but instead of drilling holes, a router bit uses blades to shape the surfaces and edges of wood stock. The portable router is moved along the stock as wood is shaped and removed.

Sanders and Grinders

Sanders and ***grinders*** both operate by grinding away the surface of materials with some type of abrasive material. Depending on the machine's design, it may use sandpaper or a grinding wheel.

Sanders

Sanders are used in woodworking to smooth rough wood surfaces and to remove small amounts of a wood surface. This could be for finishing work or for something like fixing a door that sticks. As is the case with most of the tools discussed thus far, there are specific types (and sizes) of sanders designed for specific tasks. Some of the most common types include:

- ***Belt sander***—strong power tool best used for the initial smoothing of large, flat surfaces. Belt sanders use a continuous loop or belt of sandpaper that is stretched across two pulleys. They are extremely powerful and can easily ruin a piece of stock.
- ***Orbital sander***—lightweight power tool that typically uses a square pad and 1/4 sheet of the standard 9″ × 11″ sandpaper. Orbital sanders may also be 1/3 or 1/2 sheet sanders, and some use specially shaped sanding pads. Orbital sanders are easy to control and used mainly for finish sanding and removing old finish from smaller workpieces. See **Figure 7-52**.

Norman Pogson/Shutterstock.com

VladaKela/Shutterstock.com

Figure 7-52. Small, portable belt sanders and orbital sanders may be used with a variety of grits.

- ***Random orbital sander***—lightweight power tool that uses two simultaneous motions. The circular pad rotates in small circles while simultaneously moving in a random elliptical loop; may be used for both finish sanding and slower stock removal.
- ***Belt-and-disk sander***—combination stationary sander that is useful for sanding square or curved ends on narrow boards; may also have a tilting worktable that allows precise sanding of angled workpieces.
- ***Oscillating spindle sander***—stationary power sander that uses a spindle that rotates and oscillates (moves up and down) at the same time. Most models have interchangeable abrasive drums or spindles in varying diameters. These sanders are useful for sanding curves and contours.

Sander Safety

Keep the following safety points in mind when using a sander:

- *Always* follow the safety and maintenance procedures outlined by the manufacturer.
- Proper eye protection should *always* be worn when operating a sander. Use safety glasses with side shields to fully protect your eyes.
- Portable and stationary sanders operate at high speeds and may easily snag loose clothing and long hair or cut power cords.
- *Always* wear a dust mask or respirator when using a sander. Use an attached dust collector when possible. Hearing protection is also recommended.
- Secure work with vise or clamps as needed.
- *Always* make sure the power switch is in the off position before plugging in the tool.
- Firmly hold the sander by its gripping handle(s), and turn on the sander before placing it in contact with material to be sanded.
- Use the appropriate sized grit for the job, and do *not* push down on the sander. Too much pressure may damage the work or cause the sander or workpiece to kick back.
- *Always* unplug the tool before changing worn paper, belts, or disks, or performing maintenance.
- Keep fingers clear of the moving belt, pad, or disk.
- Do *not* use sanders in wet conditions.

For most sanding work, you should begin with a coarse grit and work your way progressively through finer grits until you achieve the desired level of smoothness.

Grinders

Grinders and their accessories are used in metal working to smooth and shape metal, polish metal surfaces, and to remove small amounts of metal from surfaces and edges. Grinders are available in portable and stationary models. They operate at high speeds so special care should be taken in their operation, **Figure 7-53**.

Grinder Safety

Keep the following safety points in mind when using a grinder:

- *Always* follow the safety and maintenance procedures outlined by the manufacturer.
- *Always* wear safety glasses with side shields and a dust mask or respirator when using a grinder. Use an attached dust collector when possible. Hearing protection is also recommended.
- Wear gloves and a shop apron designed to stop small abrasive or workpiece fragments from penetrating your skin. Metals get very hot in the grinding process. Handle hot metal with pliers or gloves.
- Secure work with vise or clamps as needed.
- *Always* make sure the power switch is in the off position before plugging in the tool.
- Firmly hold the portable grinder by its gripping handles and turn it on before placing it in contact with the workpiece.
- Use the appropriate size and type of abrasive for the job. *Always* unplug the tool before changing worn abrasives or accessories.
- Use the proper grinding wheel for the job. Using the wrong wheel may cause the wheel to shatter.
- Use the properly sized accessories for the tool. Incorrectly sized accessories cannot be properly guarded or controlled.
- Portable and stationary grinders operate at high speeds and may easily snag loose clothing and long hair or cut power cords.
- Do *not* use a grinder that is too heavy for you to control.
- Do *not* use grinders in wet conditions.

vdimage/Shutterstock.com

Eimantas Buzas/Shutterstock.com

Figure 7-53. A portable angle grinder and small bench grinder.

Stationary Power Tools

You may find yourself working in an agricultural shop where stationary power tools are present. It is important to know the purpose and features of these tools along with the safety procedures for operating them.

Table Saws

Table saws are extremely useful tools in any workshop. They may also be extremely dangerous if safety rules and instructions are not followed. Table saws are designed for numerous types of cutting, and with proper use and maintenance, will provide years of service. The table saw gets its name from its design which includes a work "table" that houses the circular blade and provides a platform for resting the workpiece.

The table saw blade rises through a slot in the work table upon which the work material rests. The saw blade is driven by an electric motor and rotates toward the operator who "feeds" the work into the blade. Some agricultural shops have small tabletop table saws while others have large floor models. **Figure 7-54** shows the parts of the table saw.

Safety Note

Table saws are equipped with a safety shield that "rides" along the top of the material and protects the operator from the blade protruding from the top of the material. *Never* remove the safety shield.

It is important to keep hands and fingers out of the path of the saw blade. *Always* use a ***push stick*** to feed material into the saw. In the event the material slips, the saw will cut the push stick and not your fingers.

Crosscutting and Ripping

Table saws are useful for both crosscutting and ripping lumber:

- The miter gauge is used when making crosscuts. The gauge can be adjusted to feed the material into the saw at an angle.
- The rip fence is used when ripping lumber along the grain of the wood. The material is held against the rip fence and fed into the saw. The rip fence should be kept parallel to the blade so the stock does not bind and there will be less chance of kickback.

Blade Height and Angle Adjustment

Table saws have two main adjustments: blade height and blade angle. The adjustment knobs or wheels are typically located below the table surface. Raising the blade up or down in the slot allows the depth of the cut to be adjusted. Changing the blade angle allows you to make angled cuts. The adjustment wheels use a locking bolt to prevent them from turning while the machine is in use.

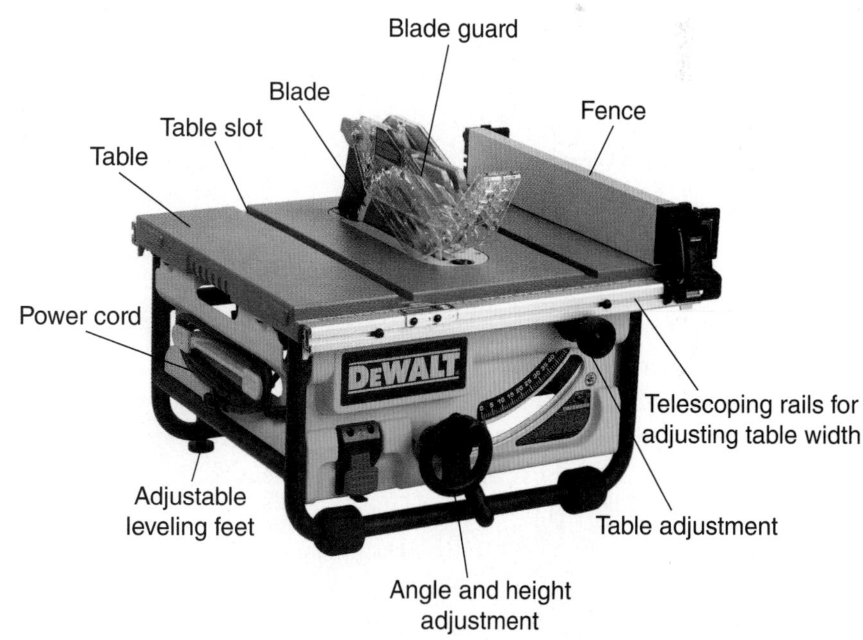

Provided courtesy of DeWALT Industrial Tool Co.

Figure 7-54. Portable table saws are available in a variety of sizes and capabilities.

Table Saw Safety

Table saws can be extremely dangerous when used improperly. In addition to your shop's safety rules, keep the following safety points in mind when using a table saw:

- *Always* follow the safety and maintenance procedures outlined by the manufacturer.
- Proper eye protection should *always* be worn when operating a table saw. Use safety glasses with side shields to fully protect your eyes. A full-faced safety shield may also be used.
- *Always* use hearing protection when working with power tools to prevent hearing loss.
- If you believe the saw needs maintenance, or if the safety guards do not appear to be working properly, *do not operate the saw!* Switch off the power, and notify your teacher or supervisor immediately.
- Do *not* wear jewelry, loose-fitting clothes, long sleeves, or gloves when operating a table saw. They all have the potential of being pulled into the spinning blade.
- *Always* check stock for loose knots, nails, and other metal pieces *before* cutting. This is especially important when working with recycled materials.
- Stand comfortably with your feet far enough apart for good balance, especially when feeding longer pieces of stock.
- Keeping the tabletop smooth and polished prevents rust formation and allows the work to move more freely across the table.
- *Always* keep the floor and tabletop clear of cut off pieces and piled up sawdust.
- Stay alert and pay attention to your work. If you are distracted by someone or something, switch off the saw and attend to the distraction.
- Do *not* remove safety guards and shields. Make sure they are in proper working order before using the machine.
- *Never* operate a table saw with the throat insert removed. The throat insert prevents wood from dropping down and getting caught on the blade.
- *Always* use the proper zero-clearance blade inserts. Without a blade insert, a piece of stock may fall into the cabinet and become a projectile.
- *Never* reach across the saw blade for any reason, even if the saw is switched off.
- *Always* stand slightly to one side of the blade. This helps to keep sawdust out of your face and prevents you from being struck by a piece of lumber if it kicks back out of the saw.
- Do *not* make free-hand cuts on a table saw.
- Use a push stick that is at least two feet long to keep hands and fingers away from the saw blade.
- Do *not* release work before it is past the moving blade. Releasing work before finishing the cut may cause kickback.
- *Never* use the fence and miter gauge together. Use the miter gauge for crosscutting lumber and the rip fence for ripping lumber.
- *Never* try to stop the spinning blade by holding something against it.
- Make adjustments to the machine components, shields, and guards *only* when the saw has been switched off and all moving parts are at a dead stop. If possible, disconnect the power so the machine cannot accidentally be turned on by someone else.
- When changing the blade height or angle, use the locking bolts to prevent the blade height and angle from changing during the sawing operation.
- Adjust the saw blade height so that it clears the top of the material by no more than 1/8″ to 1/4″.
- *Always* make sure the blade is spinning freely after making adjustments and *before* turning the saw back on.
- Sharp blades are safer to use than dull blades. Dull saw blades take more effort to push lumber into the saw, which may cause you to lose your balance.
- Do *not* turn on the saw with the blade engaged. Allow the saw motor to reach full speed before feeding work into the saw. This helps prevent binding and kickback.

(*Continued*)

- Use an assistant to help you cut large pieces of material. The assistant is only present to help you. You are in charge of the sawing operation. Make sure the helper understands the safety rules and is wearing the appropriate personal protective equipment.
- Once you have completed a sawing operation, switch off the power and wait next to the saw until the blade comes to a complete stop. You are *not* finished with the saw until all moving parts on the saw have come to a complete stop.
- After you have finished using the saw for the day, clean up sawdust and wood chips. Leave the saw in a clean condition, with all shields and guards in place. Lower the blade below table height after you finish cleaning.

Without these locking bolts, the blade angle or height may change while sawing, creating a potentially dangerous situation. At the very least, the cut will be inaccurate. Before turning these adjustment wheels, loosen the locking bolt. Once you have the correct blade height and angle, tighten the bolts.

The blade height should be set so that the saw blade protrudes 1/8″ to 1/4″ above the wood. While a higher blade height may improve the quality of the cut, it is unsafe to have the blade more than 1/4″ above the wood. High blade heights increase the potential for kickback because more of the saw blade is in contact with the wood. Safety is *always* the main concern, so adjust the blade height accordingly.

Radial Arm Saw

The ***radial arm saw*** is an excellent tool for fast and convenient crosscutting of lumber. The difference between the table saw and the radial arm saw is that the blade is on top of the lumber and all cutting occurs from the top down. As the saw moves along a track in the same direction as the saw's rotation, the saw exerts downward force on the lumber. This helps to hold the lumber against the guide fence at the rear of the saw. The radial arm saw may be described as a circular saw mounted on a sliding horizontal arm.

Portable miter saws are a version of the radial arm saw, **Figure 7-55**. The saw blade is positioned above a small platform on which the lumber rests, and is lowered into contact with the lumber. This saw is versatile and commonly found on farms and construction sites. The same safety

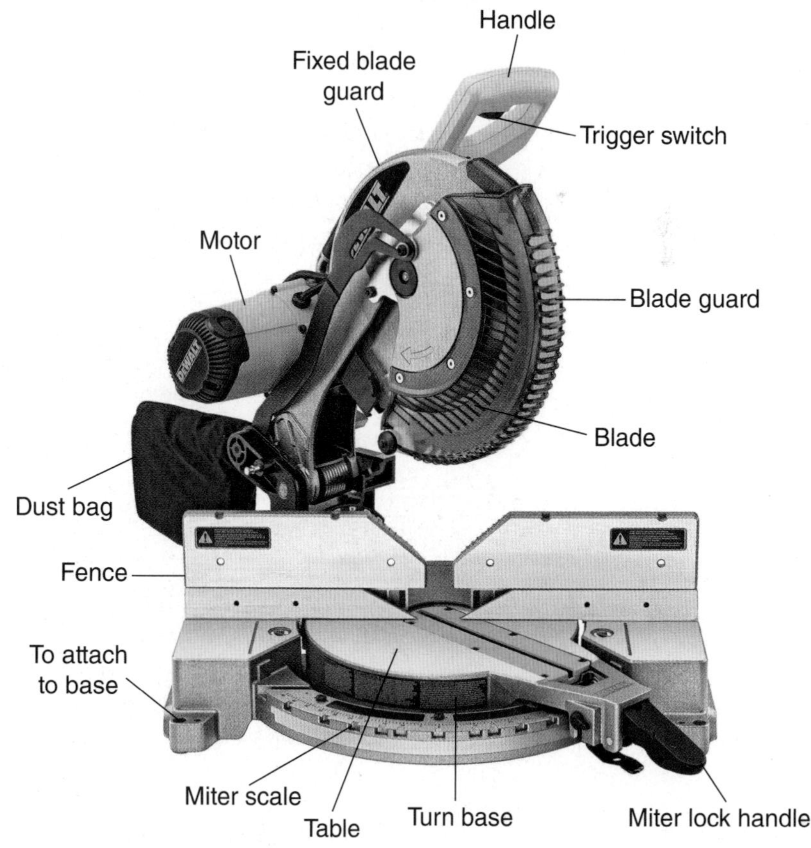

Provided courtesy of DeWALT Industrial Tool Co.

Figure 7-55. Due to its versatility, a portable miter saw will be found on almost any job site.

Radial Arm Saw Safety

When used properly, the radial arm saw is a safe and efficient tool. Keep the following safety points in mind when operating a radial arm saw:

- *Always* follow the safety and maintenance procedures outlined by the manufacturer.
- Proper eye protection should *always* be worn when operating a radial arm saw. Use safety glasses with side shields to fully protect your eyes.
- The lumber should be held firmly against the guide fence.
- The saw blade should be sharp. Sharp blades are safer to use than dull blades.
- All safety guards and shields must be in proper working order and in place on the saw.
- All clamps should be tight to prevent the lumber from vibrating loose and binding the saw blade.
- The saw should *always* be returned to the rear of the table after the cut is complete. If the saw does not have a retraction device that automatically pulls the saw back, then you must push it back to the rear of the table.

rules that apply to the radial arm saw also apply to the portable circular miter saw.

Pneumatic Tools

Pneumatic tools use compressed gas to operate. Compressed air is the most common gas used to power these tools, although other inert gases such as carbon dioxide may be used in special situations. Many of the same tools that are powered by electricity are available as pneumatic models. Paint sprayers, jackhammers, and air impact wrenches are examples of commonly used pneumatic tools, **Figure 7-56**.

Pneumatic Tool Safety

Keep the following safety points in mind when using pneumatic tools:

- *Always* follow the safety and maintenance procedures outlined by the manufacturer.
- Proper eye protection should *always* be worn when operating pneumatic tools. Use safety glasses with side shields to fully protect your eyes.
- Pneumatic tools exert a great deal of power and can easily "overdo" the job. Be careful to *not* overtighten fasteners because you may damage the threads on nuts and bolts.
- Do *not* use compressed air to clean off your clothes, arms, or any part of your body.
- Follow the same safety practices you would for battery-operated or electric power tools when using pneumatic tools.

Petroleum-Powered Tools

Petroleum-powered tools are commonly used in the turf and landscape industry where access to electricity is not practical. Lawn mowers, chainsaws, and string trimmers are examples of petroleum-powered tools. These tools are divided into two basic categories: two-cycle engines and four-cycle engines.

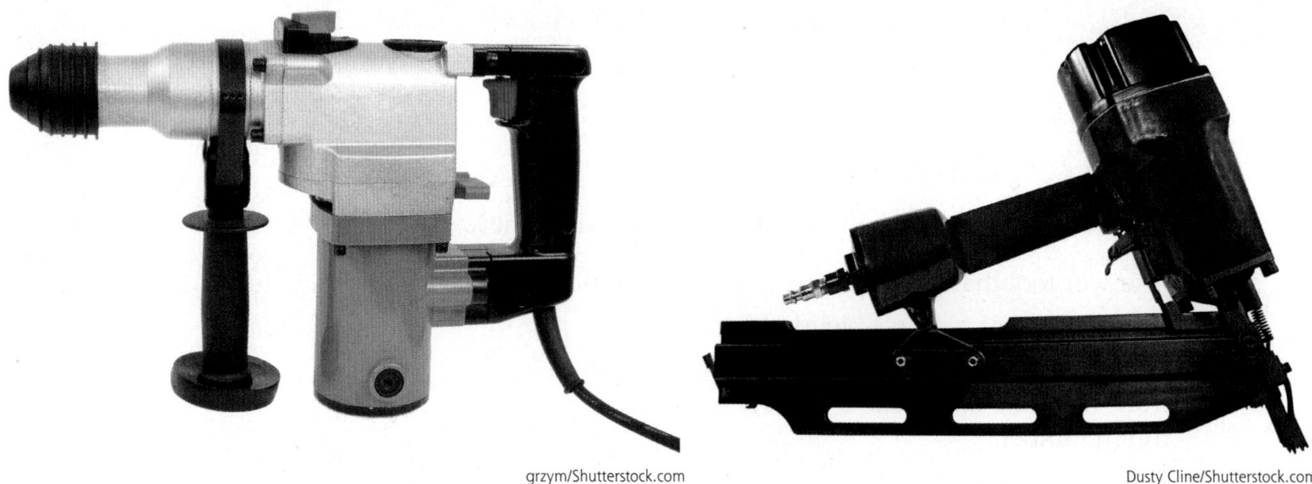

grzym/Shutterstock.com Dusty Cline/Shutterstock.com

Figure 7-56. Pneumatic tools require the use of a power source such as an air compressor.

Two-Cycle Engines

Two-cycle engines require blended fuel that is part oil and part gasoline. The oil must be mixed with the gasoline to be distributed throughout the engine for lubrication. Two-cycle engines are smaller and used where a lighter weight is desired. Tools such as string trimmers and chainsaws use two-cycle engines.

- *Always* follow the safety and maintenance procedures outlined by the manufacturer.
- Proper eye protection should *always* be worn when operating power tools. Use safety glasses with side shields to fully protect your eyes from flying debris such as rocks, pebbles, twigs, and wood chips.

Four-Cycle Engines

Four-cycle engines are larger, heavier, and more powerful. Self-propelled lawn mowers and push mowers use four-cycle engines.

Fuel Can Safety

Keep the following safety points in mind when using fuel cans:

- Fuel cans must be clearly marked for use with two-cycle or four-cycle engines. Four-cycle engines do not use mixed fuel. They have an oil sump where the lubricating oil is separate from the fuel. This oil level must be maintained for continued operation of the engine.
- Fuel cans must be stored in a cool, dry place.
- Keep fuel cans clean, and regularly inspect them for cracks and other damage.
- Promptly clean up spills, and properly dispose of cleanup materials.
- Use only oil that is labeled for use in two-cycle or two-stroke engines.
- Use the manufacturer's specified fuel-to-oil ratio.
- Mix fuel and oil in fuel cans. Do *not* mix gas and oil in the gas tank of the equipment.
- Do *not* mix more fuel than you will use in less than 30 days.

Lesson 7.3 Review and Assessment

Words to Know

Match the key terms from the lesson to the correct definition.

1. A power tool that uses compressed gas to operate.
2. Power tool that requires blended fuel.
3. Handheld woodworking tool ranging in length from 12″ to 18″ and used to smooth and square rough lumber.
4. Handheld power tool that uses a circular blade to cut through most dimensional lumber.
5. Handheld power tool that uses a thin blade that moves back and forth to make a cut.
6. Versatile handheld power tool that can be used to drill holes in wood, metal, and other materials; drive screws; turn nuts and bolts; and sand or grind material.
7. Strong, handheld power tool that uses a continuous loop of abrasive paper.
8. A stationary power tool that uses simultaneous motions to rotate an abrasive drum or spindle.
9. Handheld or stationary power tool used to smooth and shape metal.
10. Power tool that uses a rotating bit to shape the surfaces and edges of wood stock.
11. A lightweight, handheld power tool that uses two simultaneous motions to smooth surfaces or remove stock.
12. A tool commonly used in the turf and landscape industry where access to electricity is not practical.
13. A lightweight, handheld power tool that uses a square abrasive pad to smooth surfaces.
14. A wood-cutting power tool in which all cutting occurs from the top down.
15. A stationary power tool that uses a recessed blade and a built-in surface designed to support work material.

A. belt sander
B. circular saw
C. drill
D. grinder
E. jack plane
F. jig saw
G. orbital sander
H. oscillating spindle sander
I. petroleum-powered tool
J. pneumatic tool
K. radial arm saw
L. random orbital sander
M. router
N. table saw
O. two-cycle engine

Know and Understand

Answer the following questions using the information provided in this lesson.

1. List three general hand tool safety rules you should *always* follow.
2. List three types of measuring tools.

3. Briefly explain the purpose of each of the following types of hammers: nail hammer, ball-peen hammer, sledge hammer, tack hammer.
4. *True or False?* Using prying tools to remove fasteners will save unnecessary wear and tear on your hammers.
5. Explain why you should not use a screwdriver if the tip is rounded or chipped.
6. If you must use an SAE-sized wrench with a metric-sized fastener, what "tool" can you use to determine which size SAE wrench may be used?
7. Explain why it is important to use the correct wrench type and size for a job.
8. Explain why pliers should not be exposed to excessive heat or open flames.
9. Why is it important to keep clothing, long hair, and power cords away from the rotating drill bit on a drill press?
10. *True or False?* The teeth on a band saw blade should be pointing downward toward the table.
11. Explain why you should inspect wood stock for nails and other fasteners before cutting it on a band saw.
12. List three common types of hand saws and their intended purpose.
13. *True or False?* Starting cuts slowly will prevent a hand saw blade from jumping.
14. Explain when it is okay to remove the safety guard from a power tool.
15. Explain why you should always use personal protective equipment when working with hand and power tools.

Analyze and Apply

1. Research the history of nuts and bolts. How is their development and use tied in with the development of thread standards? Why were thread standards developed? Topics to cover in your research could include strength, materials, shapes, and sizes.
2. Choose one area of hand tools and research the history of their origin and changes over the years. Include examples and explanations of hand tools not covered in the text. For example, wrenches with interesting names and uses not covered in the text include: crowfoot, drum or bung, die wrench, lug wrench, flare nut, dogbone wrench, spoke wrench, chain whip, wing nut wrench, etc.

Thinking Critically

1. Identify other hand tools aside from the ones mentioned in this lesson. What are the uses of these tools?
2. Identify the types of materials used to make tools. Include materials used for handles and the main parts of the tools. Are any of the tools one solid piece? Are special processes used to treat the metals and woods used?
3. What types of wood make the best handles for hand tools? Why?

Lesson 7.4 Agricultural Design and Fabrication

Words to Know

American Society for Testing Materials (ASTM)
assembly drawing
border line
centerline
cutting-plane line
detail drawing
dimension line
elevation plan
extension line
floor plan
formal drawing
hidden line
leader line
level
National Fire Protection Association (NFPA)
object line
orthographic drawing
orthographic projection
phantom line
pictorial drawing
scale
sectional drawing
section line
site plan
sketch
sliding T-bevel

Lesson Outcomes

By the end of this lesson, you should be able to:

- Identify and demonstrate proper use of measurement and layout tools.
- Identify symbols and drawing techniques used to develop plans and sketches.
- Identify the major parts of a construction drawing.
- Identify the source and importance of industry construction and materials standards.
- Identify criteria in selecting materials in agricultural construction and fabrication.
- Prepare a bill of materials to accompany plans and sketches.

Before You Read

Arrange a study session to read the lesson aloud with a classmate. After you read each section, stop and tell each other what you think the main points are in the section. Take notes of your study session to share with the class. Continue with each section until you finish the lesson.

The design phase is the most important phase of any project, no matter how small or how large the project. Many projects fail or have numerous problems because of poor design and/or poor planning. Take the time to learn how to design and plan your projects. In the end, you will save money and time and prevent a lot of frustration.

The Design Process

Project design and fabrication is a multistep process. The first step in the design process is determining what need is to be fulfilled or what problem is to be solved. For example, the problem may be a small poultry operation that has outgrown a repurposed barn and needs a more efficient poultry house, **Figure 7-57**. Or, a particular horse is able to open the stall and barn doors and keeps letting itself out and escaping to a neighbor's pasture.

Research the Issue

The second step is to research the issue. Can the old barn be remodeled? Would it be more efficient to build a new structure? Should the old site be cleared or should it be built elsewhere? Will the new structure allow for expansion in the future? Would an alternative power source such as solar

konzeptm/Shutterstock.com

Figure 7-57. An open agricultural structure could be enclosed and finished for any number of uses. ***What types of circumstances would prohibit changing an existing structure to suit another use? For example, what would prohibit a producer from remodeling a building to house poultry?***

panels be worthwhile? In terms of the wayward horse, is the issue with the latches or the doors?

Creating a Plan

The third step is to create a plan for financing and building a new structure. Plans must be drawn up indicating the location of the building as well as its design. Basic areas of concern include electrical service, water supply, waste system, and the heating and ventilation system. If you are borrowing money to construct the building, the lender will need to see the design plans as well as the business plan. As for the escaping horse example, it may be necessary to install new latches or alter the doors in some way, or possibly both.

Refinement of Plan

The fourth step may include refinement of the plans as well as securing a contractor to prepare the site and build the structure. Situations change and problems may develop during construction. Most design plans contain some flexibility to allow for unforeseen circumstances.

Sketches and Drawings

Before you build any project, it is essential that you first develop a construction plan. A construction plan consists of a sketch of the project, a formal drawing, and a bill of materials. It may also include code restrictions, as well as other important information.

Sketches

A ***sketch*** is a rough drawing of the project you wish to construct. It captures the basic likeness of the project and is used to create a detailed and accurate drawing. Use graph paper to create the most accurate sketch possible. This will make it easier for you or someone else to complete the formal drawing.

Drawings

The ***formal drawing*** is more detailed than the sketch and drawn to scale (a specific ratio relative to the actual size of the place or object). It uses the basic ideas from the sketch to create a workable plan for constructing the project. To ensure precise dimensions and accurate angles, designers and architects use a computer drawing program and drafting instruments.

Hand Drawings

Although most architectural or mechanical drawings done today are created with computer programs, many people still use a drafting board and paper. If you are creating a hand drawing, use quality drafting tools to ensure accurate angles and lines. Some of the drafting tools you will need include steel rulers, a protractor, compass, triangle, T-square, drafting table/board, an architect or engineer's scale, drafting pencils and lead, and various stencils for lettering and symbols, **Figure 7-58**. You will also need drafting tape, paper, and erasers.

Preprinted drafting paper or blank paper may be used to create your drawing. Preprinted paper includes a blank title block as well as a frame and other guidelines to help you orient your drawing. Select paper large enough to clearly draw and label the components of the project.

Tape the paper to the drafting board, making sure that the edges of the paper are parallel to the edges of the board. If you are using blank paper, use the straightedge of a T-square to draw a border around the paper's edge, leaving a margin of 1/2″ to 1″. Make a title block along the bottom border. In this title

mihalec/Shutterstock.com

Figure 7-58. Quality drafting tools must be properly stored to ensure continued accuracy.

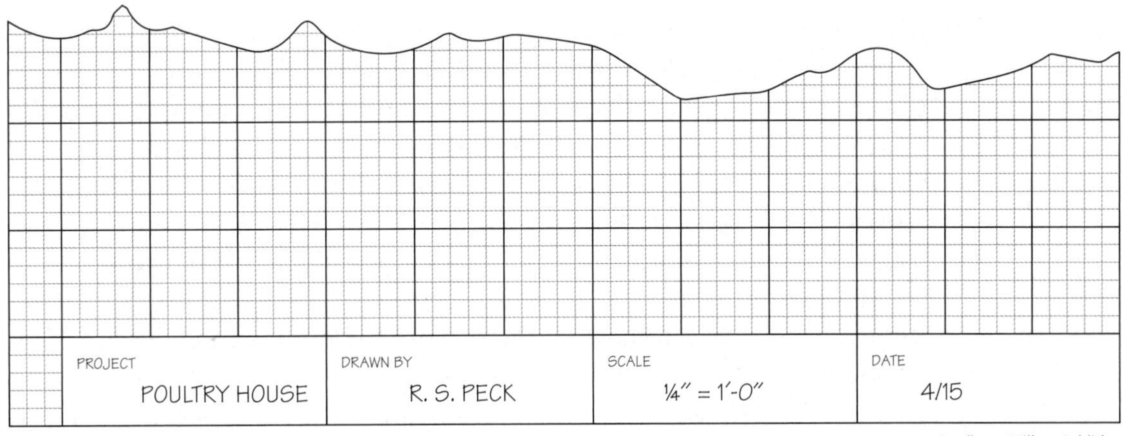

Goodheart-Willcox Publisher

Figure 7-59. The title block should contain the project title, your name, the scale used, and the date.

block, include your name, the name of your drawing, the scale you used, and the date you made the drawing. Refer to the example in **Figure 7-59**.

Types of Drawings

It is usually necessary to draw the project from a number of different views to show all the details. Therefore, several types of drawings are used in structural and mechanical drawings, including:

- ***Pictorial drawings***—show a likeness of the object or structure as viewed by the human eye.
- ***Assembly drawings***—show how the components of a project fit together. A common example would be an assemble-your-own type furniture piece.
- ***Detail drawings***—include view(s) of the product with dimensions and other important information (name of the object, quantity, drawing number, material used, scale, revisions).
- ***Orthographic drawings***—show all surfaces (top, front, sides, bottom) of an object projected onto flat planes, and set at 90° angles to one another. Orthographic drawings include detailed dimensions essential to constructing the project.
- ***Sectional drawings***—show how an object would look if a cut were made through the object.
- ***Orthographic projection***—when an object is drawn as a two-dimensional drawing.

Building plans use multiple drawings to provide detailed dimensions. If all dimensions for large-scale projects were included on one drawing, it would be too cluttered to be read. For larger projects, there are separate drawings for framing, plumbing, wiring, heating and ventilation, finishing, and even painting and wallpaper applications.

Scale

Before you begin to draw the project, determine the scale you wish to use. ***Scale*** describes the relationship between the size of the object in the drawing and the size of the actual constructed object. For instance, a scale of 1″ = 1′ means that 1″ in length on the drawing is equal to 1′ in length of the object being constructed.

Lines

Architects and engineers use specific line types to indicate dimensions or outlines, define edges, indicate internal or hidden edges, and to represent various types of materials. Lines used on formal drawings include:

- ***Border lines***—heavy, bold lines used to define the edges of the drawing surface.
- ***Object lines***—solid, black lines that show the outline of the object being drawn.
- ***Hidden lines***—thin, dashed, black lines used to outline edges and intersections that are hidden from view.
- ***Cutting-plane lines***—solid, black lines (same weight as object lines) used to indicate where the theoretical cut is taken for a sectional view. They may be a series of dashes or long dashes with intermittent short dashes.
- ***Centerlines***—thin lines with a crosshair at the center of the round object that are used to represent the diameter of a round object.
- ***Dimension lines***—thin lines (usually terminated with arrowheads at each end) used to indicate measurements.
- ***Extension lines***—thin lines that extend beyond the outline of a view to make it easier to read the dimensions.
- ***Section lines***—parallel, inclined lines used to indicate the cut surface of an object (in sectional view).
- ***Phantom lines***—thin lines used to indicate alternate positions of moving parts, repeated details (like threads), and motion.
- ***Leader lines***—lines with an arrow at one end. These lines are used to call out and explain or describe components of the drawing. See **Figure 7-60**.

There are also standard symbols used to represent everything from electrical outlets to plumbing fixtures as well as furnaces and doors, **Figure 7-61**.

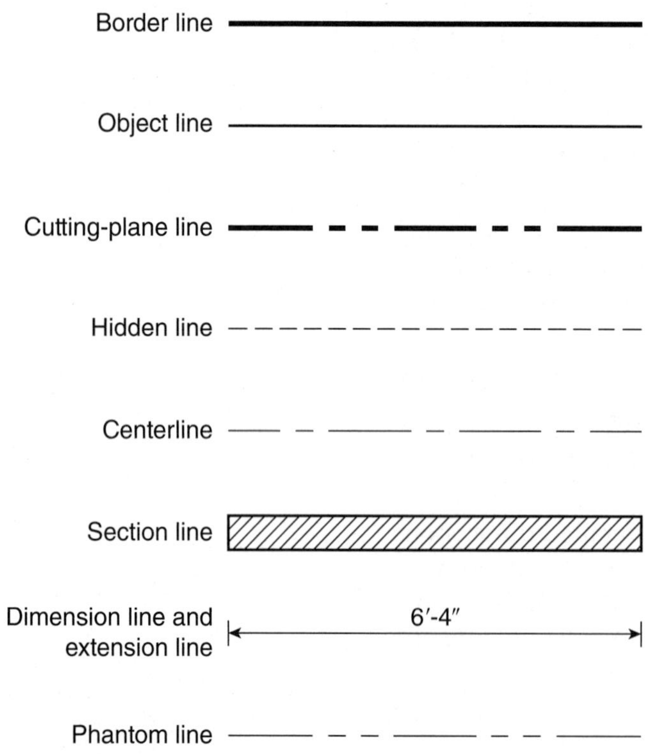

Goodheart-Willcox Publisher

Figure 7-60. The assortment of lines used on formal drawings is often referred to as the alphabet of lines. The example illustrated here is only a sample of those used.

Lettering

All letters, numbers, and symbols must be printed neatly to avoid errors and confusion. Use the standard letters, numbers, and symbols used in the industry.

Accuracy and Perspective

Drawings and sketches are drawn from a perspective that allows the length, width, and height of the object to be viewed. Accuracy is important because the drawn project plan will guide the purchase of materials and provide instructions for building the project. To improve your accuracy in drawing plans, practice sketching

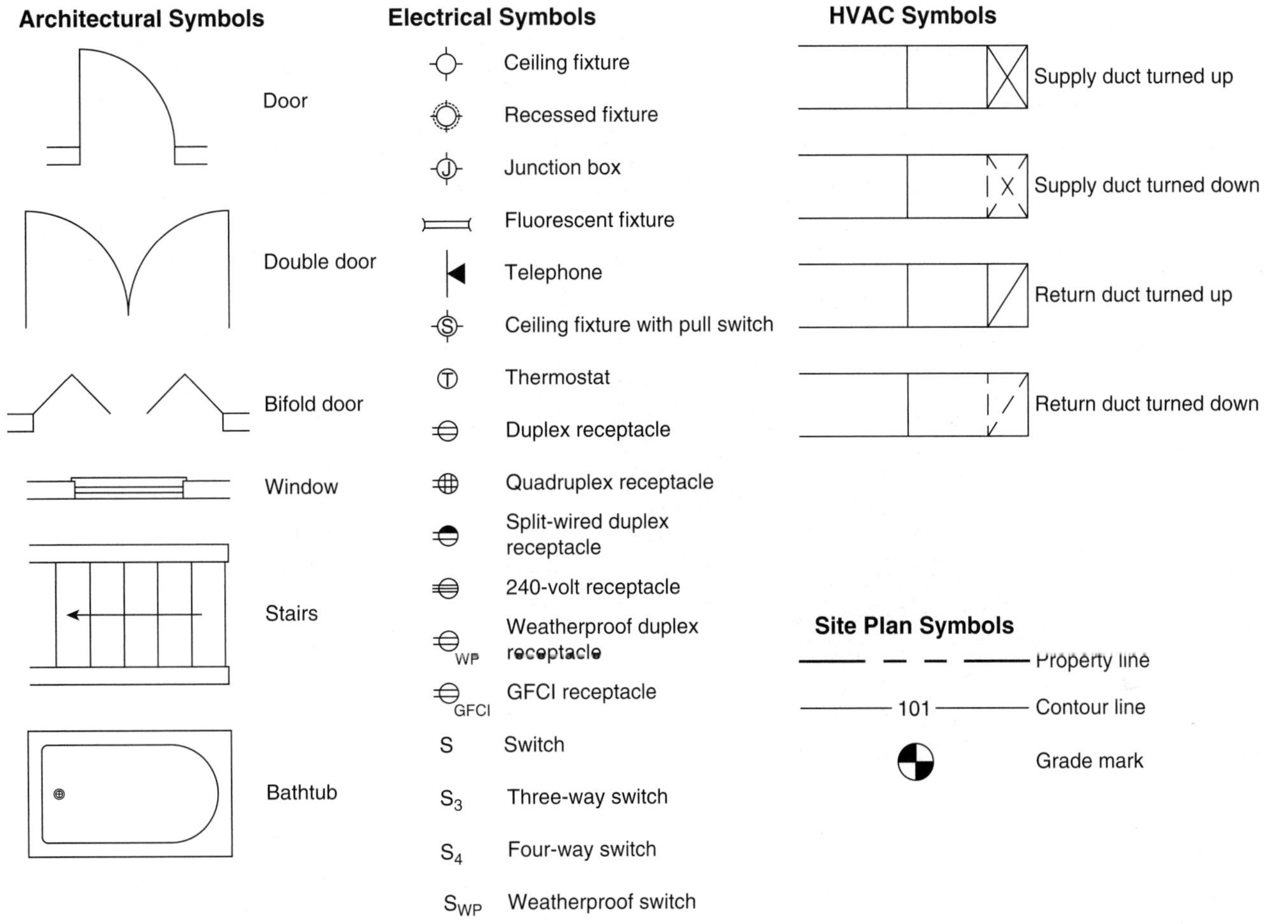

Figure 7-61. Architectural drawing programs usually include assortments of standard symbols such as those illustrated here.

lines until you can draw a reasonably straight line freehand. Practice drawing parallel lines as well as triangles, circles, and arcs. Use a straight-edged tool or ruler to make precise lines.

Proportion

Another key to a good drawing is using accurate proportion. That is, all of the parts of the object are drawn to the same scale. Practice freehand drawing of a series of the same object until you are able to draw them all to the same size and proportion.

Measuring

Agricultural projects require measurements to be taken before, during, and after the project is completed. If you are building a tack box, you will have to measure the proposed contents as well as the space in which the box will be kept. All structural projects also require measurements to be taken before, during, and after the project is completed. The property dimensions must be clear to ensure the structure meets all code restrictions and will not go over property lines. Measurements are also needed to determine how much of each material will be required to complete a project and how much it will cost.

Measuring Systems

If you use English units, you will need measurement tools with markings based on feet and inches. When using the metric system, you will need tools using the meter as the standard unit of measurement. This information should also be indicated on your drawings.

ShiningBlack/Shutterstock.com

Figure 7-62. Micrometers are precision instruments that come in various sizes and designs.

Tools

Measuring tools are the backbone of the design process. These tools range from a simple ruler to a precision micrometer, **Figure 7-62**. Many measuring tools are designed to perform more than one task. Basic measuring tools include:

- Rules
- Tapes
- Squares

Rules

Measuring rules are rigid and can be used for linear measurement and for drawing lines on drawings and building materials. Rules are made of aluminum, steel, or wood. The bench rule is useful in shop settings because it does not need to be unrolled or unfolded. Bench rules are short, usually between 2′ and 3′ in length. Their rigid construction and short length prevents them from being used in fieldwork. Rules may also be used for determining angles. Bench rules may also be referred to as yardsticks or meter sticks.

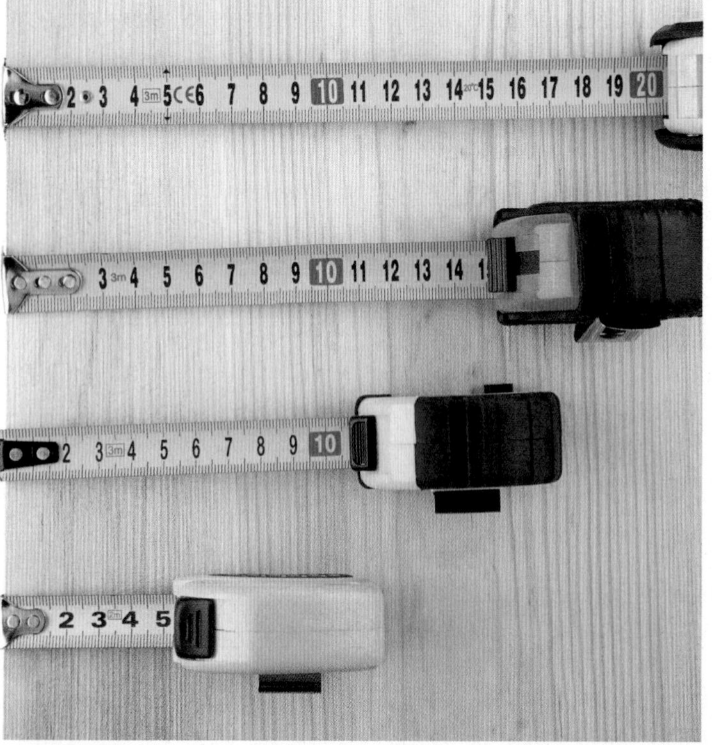

Seregam/Shutterstock.com

Figure 7-63. Measuring tapes are available in varying widths and lengths as well as in metric or English units.

Tapes

Measuring tapes are useful for both linear measurements and curved measurements. They are available in a wide variety of lengths, widths, and materials, **Figure 7-63**. The flexibility of some measuring tapes allows you to measure the circumference of a round or cylindrical object. These flexible measuring tapes use fabric with special coatings designed to prevent stretching, which would result in inaccurate measurements.

Squares

There are a variety of tools that fall under this category, including the framing square, try square, T-square, and combination square. Squares are used to measure angles, and square materials such as lumber, and to draw accurate lines on drawings and on construction materials, **Figure 7-64**.

D.B.Croom

Figure 7-64. Framing square, combination square, try square, and sliding T-bevel.

- The framing square is flat and made of steel or aluminum. The two major parts of the square are the blade and tongue. Framing squares are used in the laying out of rafters and stair stringers where a uniform angle measurement is needed. Framing squares are also used to square lumber.
- Try squares are smaller than framing squares. The try square is used to measure the accuracy of cuts in lumber. The try square is also referred to as a *try and miter square*. The term miter means *angle*. Some try and miter squares can test the accuracy of 45° angles.
- T-squares are used primarily for drawing horizontal lines. T-squares come in various lengths, including 18″, 24″, 36″, and 42″. The long shaft is called the *blade*, and the short shaft is called the *stock* or *head*.
- A combination square combines the tasks of measuring and leveling. The combination square is a 12″ steel or aluminum rule mounted in a housing

Hands-On Agriculture

Squares are often used as a straightedge for marking lumber for accurate saw cuts. The ***sliding T-bevel*** is a special type of square. It is used to lay out angles. The two parts of the sliding T-bevel are the handle and the blade. A set screw can be tightened to hold the blade at the desired angle. T-bevel squares are useful for transferring angles from one object to another.

Let's say you wish to transfer the angle of one board to another board. Set the T-bevel square to the angle of the first board and tighten the set screw. Then place the T-bevel square at the appropriate place on the other board and mark the angle. The second board is now ready for cutting.

D.B. Croom

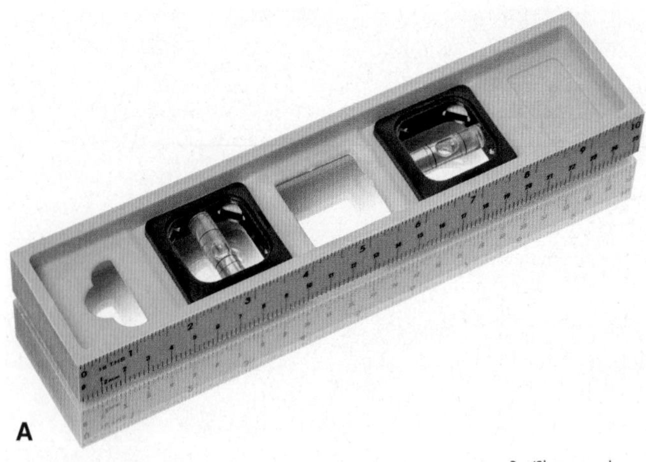

ra3m/Shutterstock.com serato/Shutterstock.com

Figure 7-65. A—A smaller level is useful for tight spaces. B—A nylon line is kept in place as a guide for the mason as he or she lays each row of bricks.

with a bubble level embedded in one edge. This level can be used to determine whether a horizontal surface is level or a vertical object is "plumb," or exactly vertical. The blade can slide through the frame, creating the ability to measure depth. A set screw can be tightened to hold the rule in place.

Levels

A ***level*** is a tool used to measure the degree to which an object is horizontal and the degree to which it is vertical. Levels come in a variety of lengths. The longer the level, the larger the area that can accurately be measured. Levels have bubble gauges that indicate the degree to which an object is level. Line levels are used when measuring over long horizontal distances. For example, when laying masonry block for a building foundation, the line level is attached to a nylon line that is stretched taught over the distance to be leveled. See **Figure 7-65**.

Chalk Line Reels

A chalk line reel is useful for marking straight lines over a long distance, **Figure 7-66**. A braided line is kept on a reel in a special housing filled with chalk powder. This powder coats the line. When the line is pulled from the reel and stretched tight across an object, the carpenter "snaps" the line by picking the line up at a point midway between the ends of the material and letting it snap back into place. This snapping action causes the chalk powder to form a straight line on the material.

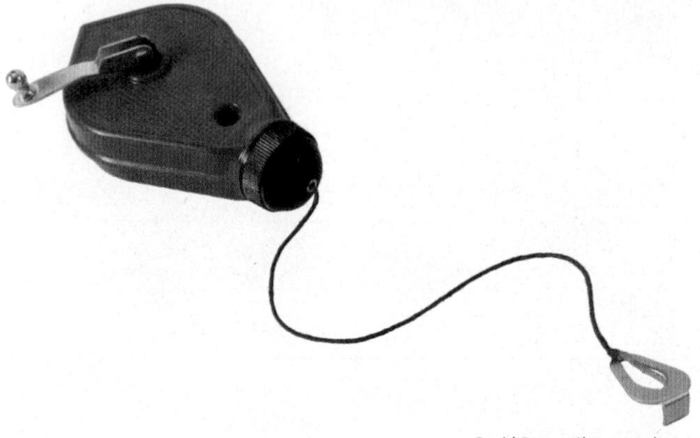

David Desoer/Shutterstock.com

Figure 7-66. A chalk line reel may be used to lay lines on most types of materials.

Construction Drawings

Construction drawings are those used by a construction crew to build the project. The larger and more complex the construction project, the more complex the drawing. There are several types of construction drawings:

- ***Site plans***—drawings of the land where the project will be constructed. These plans show property boundaries and the location of buildings on the site. Landscaping may also be portrayed on these plans.
- ***Floor plans***—these are common construction drawings. They represent the layout of rooms, doors, and other features within a building. Floor plans indicate where walls will be placed and provide a top view of rooms with the building.
- ***Elevation plans***—these drawings portray the building from one side, showing the vertical features of the structure.
- ***Assembly drawings***—for small projects, assembly drawings show how the structure or piece of equipment is to be constructed. These drawings may be in a step-by-step format, which shows each stage of the assembly.

Bill of Materials

Once the drawing is complete, determine the amount and cost of the materials you will need to build the project, **Figure 7-67**. This part of the project has three simple steps to follow:

Step 1—Select materials designed for the project you are building. If the project is an indoor project, select and use lumber created for indoor use. If the project is to be used outdoors, such as a picnic table or park bench, use lumber specially treated for outdoor use. If you are constructing wildlife feeders or nesting boxes for birds, be sure to use materials free of wood preservatives that could harm wildlife.

Step 2—Make a list of all of the parts of the object. Be thorough and get everything into one list. Make sure that you record the size and dimension of these parts. List the quantities of each part.

Step 3—Check the local hardware and supply stores for the items you need to build the project. Shop around to find the most economical source of materials.

Building a project can be a lot of fun, provided that you spent the time necessary to plan the project carefully. A well prepared plan and drawing can make the difference between a successful project and a project that falls short of your expectations.

BILL OF MATERIALS		
QUANTITY	ITEM / DESCRIPTION	COST
220 lin. ft.	Form boards, 18″ wide	
200 lin. ft.	#3 rebar	
40	1/2″ anchor bolts, 12″	
40	1/2″ washers (for anchor bolts)	
40	1/2″ nuts (for anchor bolts)	
180 lin. ft.	2 × 6 sill plate, pressure-treated	
180 lin. ft.	Sill sealer, rolled	

Goodheart-Willcox Publisher

Figure 7-67. A bill of materials is needed to keep track of costs for any project.

Industry Construction Standards and Materials Standards

Structures are only as good as the materials used to construct them and the quality of the construction. The ***American Society for Testing Materials,*** or ***ASTM,*** sets the industry standard on the quality of construction materials and their manufacture. Methods of construction are governed by local and state government standards as well as national organizations like ASTM and the ***National Fire Protection Association (NFPA)*** that develops electrical wiring codes followed by state and local governments. Building material and construction standards is developed through a series of rigorous tests. The goal of industry standards is to protect human life and property from substandard materials and inferior construction practices.

It is important to only use materials and construction processes that meet industry standards. Most construction materials will have a label that indicates whether or not they meet industry standards. State and local building inspectors periodically inspect new construction to determine whether it meets industry standards for safety and quality, **Figure 7-68**.

SpeedKingz/Shutterstock.com

Figure 7-68. Building inspectors determine whether the construction processes are being performed according to industry standards for safety and quality. A building inspector can halt construction if the work or materials are below industry standards.

Lesson 7.4 Review and Assessment

Words to Know

Match the key terms from the lesson to the correct definition.

1. A thin line with a crosshair at the center of the round object that is used to represent the diameter of a round object.
2. A parallel, inclined line used to indicate the cut surface of an object (in sectional view).
3. A solid, black line used to indicate where the theoretical cut is taken for a sectional view.
4. An organization that sets the industry standard on the quality of construction materials and their manufacture.
5. A thin line used to indicate alternate positions of moving parts, repeated details (like threads), and motion.
6. A drawing used to portray the building from one side, showing the vertical features of the structure.
7. A specific ratio relative to the actual size of a place or object.
8. A thin line that extends beyond the outline of a view to make it easier to read the dimensions.
9. A common construction drawing used to represent the layout of rooms, doors, and other features within a building.
10. A drawing designed to show a likeness of the object or structure as viewed by the human eye.
11. A heavy, bold line used to define the edges of the drawing surface.
12. A drawing that shows all surfaces (top, front, sides, bottom) of an object projected onto flat planes, and set at 90° angles to one another.
13. When an object is drawn as a two-dimensional drawing.
14. A rough drawing that captures the basic likeness of a project.
15. A more detailed drawing of a project that is drawn to scale.
16. A thin, dashed, black line used to outline edges and intersections that are hidden from view.
17. A drawing that includes view(s) of the product with dimensions and other important information.
18. A line with an arrow at one end that is used to call out and explain or describe components of the drawing.
19. A drawing created to show how the components of a project fit together.

A. American Society for Testing Materials (ASTM)
B. assembly drawing
C. border line
D. centerline
E. cutting-plane line
F. detail drawing
G. dimension line
H. elevation plan
I. extension line
J. floor plan
K. formal drawing
L. hidden line
M. leader line
N. object line
O. orthographic drawing
P. orthographic projection
Q. phantom line
R. pictorial drawing
S. scale
T. sectional drawing
U. section line
V. site plan
W. sketch

20. A solid, black line that shows the outline of the object being drawn.
21. A drawing created to show how an object would look if a cut were made through the object.
22. A drawing of the land where a project will be constructed.
23. A thin line (usually terminated with arrowheads at each end) used to indicate measurements.

Know and Understand

Answer the following questions using the information provided in this lesson.

1. What are the steps to the design process for an agricultural structure?
2. What is the purpose of a sketch?
3. What is the difference between a sketch and a formal drawing?
4. What are the different types of drawings?
5. What does a scale of 1″ = 10′ mean?
6. What is the purpose of phantom lines in a formal drawing?
7. Formal drawings need to be in proportion. What is proportion in a formal drawing?
8. What is ASTM and what does it do?
9. Why are there standards for construction materials and construction methods?
10. What does a building inspector do?

Analyze and Apply

1. Think of a structure you would like to build, and draw an informal sketch of the floor plan and elevation plan for the project. Share this with your instructor.
2. Examine some building materials made available to you by your agricultural education teacher. Can you find labels or marks that demonstrate that the material meets industry standards?

Thinking Critically

1. Develop a set of plans to construct a small project, such as a birdhouse or wooden toolbox. Informally sketch a series of assembly instructions that show how to assemble the project. Share your plans with another student and see if he or she can understand your assembly directions.

Lesson 7.5
Power Systems

Lesson Outcomes

By the end of this lesson, you should be able to:

- Identify power, structural, and technical systems used in agriculture.
- Give examples of power, structural, and technical systems in agronomy, horticulture, and nursery and landscaping.
- Examine power, structural, and technical systems in livestock and poultry production.
- Describe the impact of power systems in forestry and natural resources.

Words to Know

hydraulic system
internal combustion engine
pneumatic system
power
power system
work

Before You Read

Before reading the lesson, read through the *Know and Understand* questions at the end of the lesson. Use them to help you focus on the most important concepts as you read the lesson.

All production agriculture activities require the use of some form of power. A power system may be as simple as a lever and fulcrum or as complicated as the wheat threshing system in a self-propelled combine. Because it is not possible to list and explain every type of power system used in agricultural operations in the space of one lesson, this lesson will serve as a brief overview.

What Is Power?

Work is the application of force over a distance. For example, if you lift a bucket of water off the ground, you are performing work. The larger the force, the greater the amount of work. For example, it makes sense that it would require more force to lift a heavier bucket or to lift a bucket higher off the ground. The amount of work completed has no relation to the time required to complete the movement. That is, whether the bucket of water is lifted slowly or quickly, the amount of work required is the same.

Power is the rate of energy use. The greater the power, the more quickly work can be performed. For example, lifting a bucket of water requires a certain amount of power. If you want to lift the bucket twice as quickly, twice the power is needed. (The work remains the same, but it is being accomplished in half the time.)

Some degree of power is required to perform even the smallest of tasks. How that power is created and supplied varies by the energy source and the amount of power required.

Conny Sjostrom/Shutterstock.com

Figure 7-69. In the early days of agriculture, horses or oxen provided the energy required for work. Today, energy may be provided through power plants using such sources as biomass, solar energy, hydropower, or even wind energy.

Power Systems

A ***power system*** is a system that harnesses, converts, transmits, and controls energy to perform work. A power system receives energy from an energy system, **Figure 7-69**. Power systems used in agriculture include mechanical, electrical, hydraulic, and even pneumatic systems.

Mechanical Systems

Mechanical systems manage power to accomplish a task that involves force and movement. When a tractor is being used to till a field, its mechanical system is providing the force needed to pull the plow through the soil. Many mechanical systems are powered through internal combustion engines. In an ***internal combustion engine,*** power is generated as fuel is burned inside the engine. Hot expanding gases produced by the burning fuel drive the engine components to do work, **Figure 7-70**.

Electrical Systems

Electrical systems use a network of electrical components to supply, transmit, convert, and use electric power. An electrical system may be part of a machine or a component of a structural system. The ignition system used on most engines and the electric service in your home are both electrical systems.

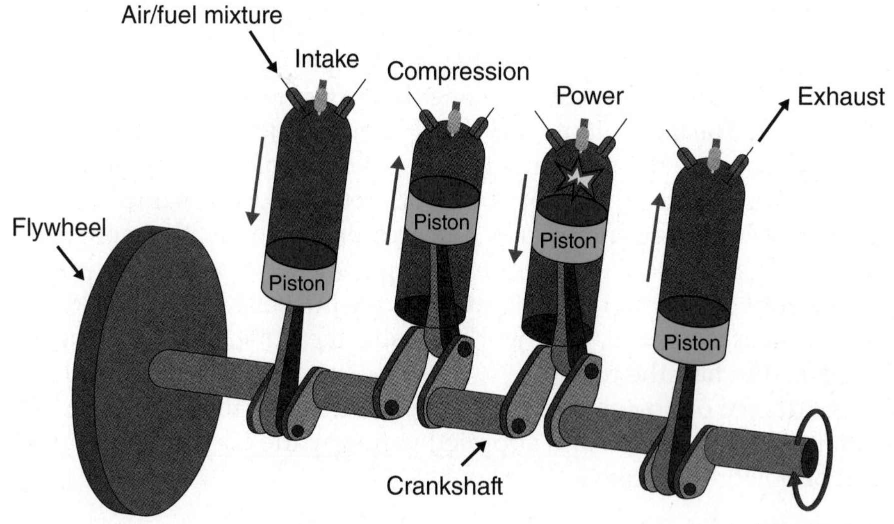

photoiconix/Shutterstock.com

Figure 7-70. An internal combustion engine generates power by burning fuel mixed with air inside the engine.

STEM Connection How an Internal Combustion Engine Works

One of the most important advances in power systems was the invention of the internal combustion engine. The internal combustion engine changes chemical energy from fuel into mechanical energy. The engine consists of a fixed cylinder and a piston. The piston fits snugly in the cylinder, but it still has the freedom to move up and down.

Most of the engines operate in a four-stroke cycle, meaning four piston strokes (up or down movements) are needed to complete a cycle. The four strokes are named according to their function and are known as the intake stroke, compression stroke, power (combustion) stroke, and exhaust stroke.

During the intake stroke, the piston moves downward, drawing air and fuel into the cylinder through the open intake valve. When the piston nears the bottom of the cylinder during the intake stroke, both valves close and the compression stroke begins. During the compression stroke, the piston moves upward in the cylinder to compress the air/fuel mixture. When the piston reaches the top of the cylinder during the compression stroke, the power stroke begins. During the power stroke, the compressed air/fuel mixture is ignited. The burning mixture produces heat and rapidly expanding gases. The expanding gases push down on the piston, causing it to again move downward in the cylinder. As the piston reaches the bottom of the cylinder at the end of the power stroke, the exhaust stroke begins. During the exhaust stroke, the exhaust valve opens and the upward-moving piston pushes the spent gases out of the cylinder, preparing it for another cycle.

The up-and-down motion of the piston is transferred to the crankshaft through a connecting rod. The crankshaft converts the up-and-down motion of the piston and connecting rod to a rotary motion to complete work like turning wheels or a power take off (PTO). Some small pieces of equipment, like lawn mowers, are equipped with single-cylinder engines, while large equipment requires engines with multiple cylinders. Large engines commonly have six or eight cylinders working together to provide energy output.

There are two basic types of internal combustion engines: spark-ignition engines and compression-ignition engines. The difference between spark-ignition engines and compression-ignition engines lies in the way each ignites the fuel. In a spark-ignition engine, a spark plug produces an electric arc that ignites the fuel, causing combustion. A compression-ignition engine relies on the heat generated during the compression stroke to ignite the fuel. Compression-ignition engines spray the fuel into the hot compressed air at a suitable, measured rate. Spark-ignition engines generally use gasoline as fuel, while compression-ignition engines use diesel fuel.

Understanding internal combustion engines can give you an important knowledge of how many systems in agriculture and other industries are powered.

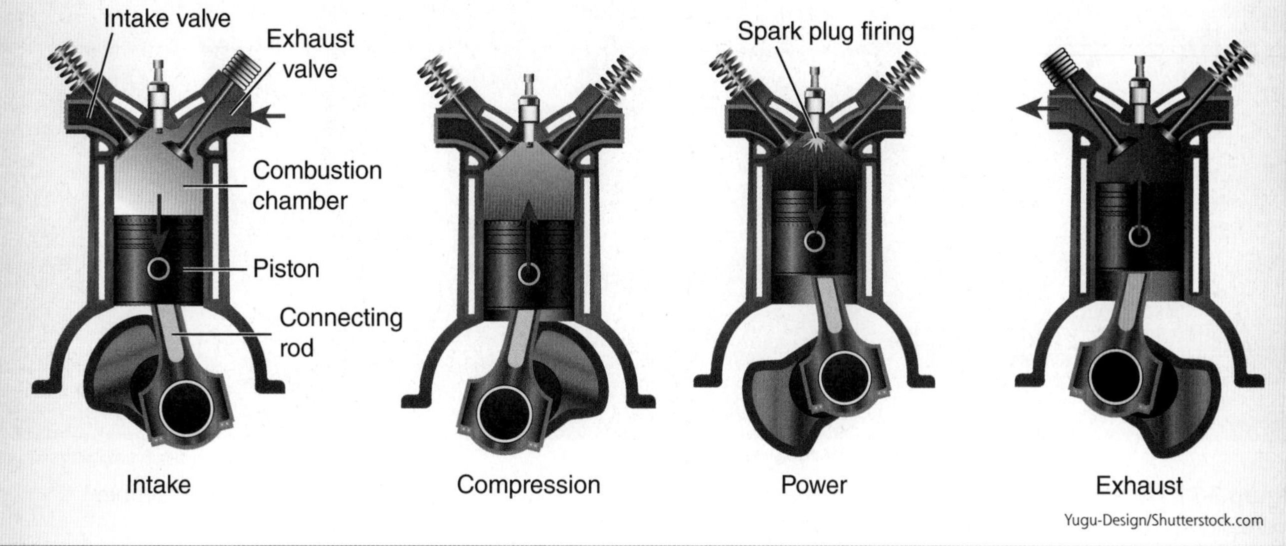

Yugu-Design/Shutterstock.com

Did You Know?

In summer months, agriculturalists rely on power to help regulate conditions for their animals. A power outage of less than an hour during a heat wave in North Carolina caused the deaths of tens of thousands of chickens and turkeys.

Electricity is used in agriculture for many purposes, and its constant supply is often vital to the survival of young and vulnerable plants, poultry, and livestock. Can you think of what would happen if there were a power outage in a greenhouse in the winter? The economic impact of losing that power could have devastating impacts on a greenhouse operation. Many agricultural industries cannot rely solely on the natural environment to give them the proper lighting, temperature, and moisture to produce their products. Electricity is used to control the environment in greenhouses, hatcheries, and livestock facilities, which allows many agricultural industries to remain productive. Electricity powers the machinery used to clean and sort produce and also powers cooling and storage facilities to prevent spoilage of food products.

Hydraulic Systems

Hydraulic systems use fluids to do work. In a hydraulic system, energy is transmitted and controlled using a fluid such as oil.

Hydraulic systems capitalize on some general properties of liquids:

- A liquid has no shape of its own—it conforms to the shape of its container.
- A liquid is basically uncompressible—when placed under pressure in a container, it applies the same amount of pressure to all sides of the container.

This transfer of pressure by a liquid is the basis of hydraulic systems. When a force is applied to the liquid at one location in the hydraulic system, it creates pressure throughout the system. For example, when you press a brake pedal, a plunger pushes against the hydraulic fluid, creating pressure. This pressure is transmitted by the fluid throughout the system. Most of the system, such as the hoses and valves, are not affected by this pressure. The pressure of the fluid moves a piston in the brake caliper, however, and this piston applies the brakes and causes the vehicle to slow.

Safety Note

The hydraulic fluid in a hydraulic power system is under very high pressure. A tiny leak in the system can allow fluid to escape at high velocity and with enough force to puncture your skin.

Some tractors use hydraulic systems for assist with braking and steering. Many tractor implements use hydraulic power to operate. Hydraulic systems are also used to power pumps and motors. See **Figure 7-71**.

Taina Sohlman/Shutterstock.com

Andrey_Popov/Shutterstock.com

Figure 7-71. Hydraulic systems enable large machinery to perform heavy tasks quickly and efficiently. A simple hydraulic system you may be familiar with would be a jack.

Pneumatic Systems

Pneumatic systems are similar to hydraulic systems. Both systems use a substance (a liquid in hydraulic systems and a gas in pneumatic systems) to transmit and control energy. Pneumatic systems use pressurized air or another pressurized gas to create the power needed to complete a task. A common example of a pneumatic power system is an air compressor and its components used to run pneumatic tools.

Did You Know?

Pneumatic power systems have been used since the eighteenth century. Many of the basic valves used in pneumatics today were developed in conjunction with the invention of the steam locomotive.

Combined Systems

Most agricultural machinery uses a combination of power systems. For example, a tractor may have an electronic ignition and use a hydraulic system to operate implements. If a PTO (power take off) is attached, the tractor's mechanical system is used to power the attachment, **Figure 7-72**. For someone who owns or manages an agricultural operation, it is necessary to understand the basic operation of many types of power systems. Most agriculturalists learn how to maintain and make basic repairs to the machinery used in their operation.

Structural Systems

Structural systems used in agriculture include construction systems such as framing and roofing, as well as building support systems such as plumbing; septic; wiring; and heating, cooling, and ventilation. These systems come together to make a building suited for specialized functions

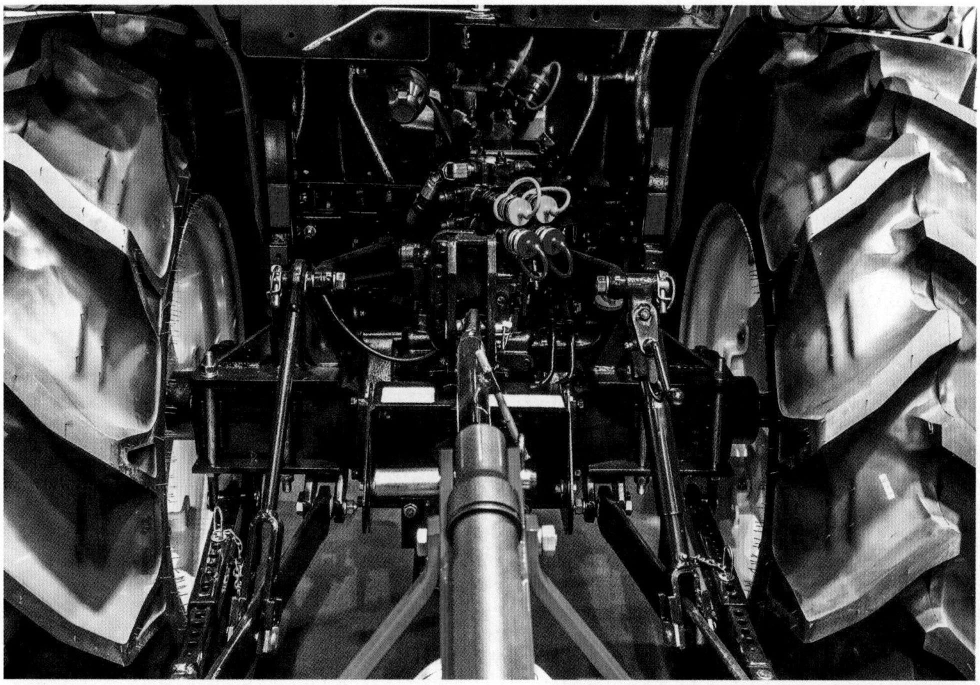

Kagai 19927/Shutterstock.com

Figure 7-72. The PTO (power take off) enables the agriculturalist to use one tractor for multiple purposes by changing the attached implement.

and purposes. More information about how structural systems come together can be found in *Lesson 6.1, Agricultural Structures*.

Technical Systems

Technical systems used in agriculture include computer applications, global positioning systems (GPS), global information systems (GIS), and monitor and control systems. Many production operations use security cameras to monitor facilities and ensure biosecurity protocol is followed. Computerized systems have become an integral part of agriculture.

Agricultural Machinery

Although there is specialized equipment for many agricultural applications, many agricultural machines are designed for versatility. For example, most tractors can be used to perform many tasks simply by attaching different implements. This versatility is essential because agricultural machines are quite often the most expensive portion of an agricultural endeavor, and purchasing multiple machines would be cost-prohibitive for most operations.

System Applications

There are many ways in which the different power systems benefit agriculture. Each industry has many different ways in which work needs to be completed. In fact, every single area of agriculture has specific power needs. Can you think of some ways that energy is harnessed in a specific area of agriculture?

The following sections list the many areas in which power, structural, and technical systems are used in agriculture. The lists are by no means all-inclusive as the needs of each individual operation vary.

Agronomy

In agronomy (soil management and crop production), power systems are used in every step of production:

- Soil preparation—tractors, plows (moldboard, chisel, and disk), and field cultivators.
- Planting and seeding—tractors, drills, and planters, **Figure 7-73**.
- Pesticide, herbicide, and fertilizer application—sprayers, granular applicators or spreaders, manure spreaders, tractors (with attachments), and even aircraft used to apply pesticides or scout crops.
- Irrigation and drainage—pumps, tractors (to move equipment), and drainage plows.

Fotokositc/Shutterstock.com

Figure 7-73. Seeding equipment is attached to a tractor and may be connected and powered with a PTO or hydraulics.

- Harvesting—tractors, forage harvesters, combines, cotton harvesters, mowers, rakes, threshers, and balers.
- Cleaning, drying, and storing—dryers, grinders, separators, augers, cotton gins, and ventilation.
- Processing—grinders, extractors, clarifiers, and packaging machinery.
- Transport—skid steers, tractors, trailer trucks, tankers, and trains and the equipment used to move the product from one implement or location to another.

Structural systems used in agronomy include storage facilities for product and equipment. Crops may be stored in barns, silos, and grain bins. Equipment may be stored in multipurpose barns or in buildings constructed specifically for equipment storage and maintenance.

Technical systems used in agronomy are often found in every step of crop production. Newer equipment includes computer systems to monitor each aspect of the equipment itself, as well as to tie into GPS and GIS used to monitor crops for herbicide, pesticide, and fertilizer needs. Older equipment may usually be retrofitted to use these systems. Crop telemetry and drones are also used in crop production. Technical systems are used to monitor air quality and moisture content in crop storage facilities.

Horticulture

In horticulture (the cultivation of fruit, vegetables, and ornamental plants), power systems and machinery are used for:

- Soil preparation—rototillers, posthole diggers, tractors, and rotovators.
- Planting—augers, diggers, and automated plug planters.
- Pesticide, herbicide, and fertilizer application—sprayers, granular applicators or spreaders, and tractors (with attachments).
- Irrigation—pumps and sprinkler systems.
- Transplanting—tree spades and other digging equipment, **Figure 7-74**.
- Pruning, trimming, and maintenance—chainsaws, hedge trimmers, and chipper/shredders.
- Harvesting and cleaning—grading machines and wash-dry-brush machines.
- Storing and processing—cooling sheds.

Structural systems used in horticulture vary depending on the operation and the final use of the product. For example, if produce is harvested for fresh-market sales, it may be necessary to remove field heat after harvesting by cooling it in a cooling shed and then transporting it directly to market. If the same product is harvested for use in other products (i.e. filling for baked goods), it will not only be cooled after harvest, but also washed, packaged, and refrigerated prior to transport. The structures will

Sue Smith/Shutterstock.com

Figure 7-74. Tree spades vary in size and ability.

require a variety of power and structural systems and the space needed to house the machinery.

Horticulture operations may use greenhouses to raise seedlings for transplant and off-season production of fruits and vegetables. Greenhouses require plumbing systems for irrigation; electrical service for lighting, cooling, and ventilation; and natural gas or propane systems for heating.

Technical systems used in horticulture include controls and monitoring systems for various applications including soil monitoring and irrigation timers.

Nursery and Landscaping

In nursery production and landscaping, power systems and machinery are used for:

- Soil preparation—rototillers, posthole diggers, and tractors.
- Planting—augers, diggers, and planters.
- Pesticide, herbicide, and fertilizer application—sprayers and granular applicators or spreaders.
- Irrigation—pumps, sprinkler systems, timers, and automated controls.
- Transplanting—tree spades and other digging equipment.
- Pruning, trimming, and maintenance—aerators, edgers, chainsaws, hedge trimmers, chipper/shredders, mowers, string trimmers, brush cutters, and leaf blowers/vacuums, **Figure 7-75**.
- Greenhouse construction, maintenance, and environmental control.

ThamKC/Shutterstock.com

Figure 7-75. A chipper/shredder is used to mulch woody plant materials from trimming, pruning, and tree removal. As long as the plant material is not diseased, it may be treated and used as mulch.

Structural systems used in nursery and landscaping operations include greenhouses and equipment storage and maintenance facilities. Equipment storage and maintenance facilities require electrical service, as well as heating and plumbing systems. Greenhouses require plumbing systems for irrigation, electrical service for cooling, and ventilation as well as natural gas or propane service for heating.

Technical systems used in nursery and landscaping include automated controls for irrigation. If an aquaponics system is in place, additional monitoring and control systems will also be used.

Livestock Production

In livestock production operations, power systems are used for:

- Environmental control—heating, cooling, and ventilation.
- Feeding—grinding equipment, as well as automated feed and water dispensers.
- Pest control—spray-dip machines and face and back rubbers.
- Harvesting, processing, and storage of meat products.

Structural systems used in livestock production include housing facilities and containment structures. The housing facilities will vary by the type of livestock being produced and whether the livestock is being raised for meat, dairy, or wool. Livestock facilities will also vary by the stages of animals being housed. For example, facilities for breeding will have different needs than those intended for finishing operations. Climate conditions will also influence the structural systems used in an operation. Livestock operations also need feed storage and handling areas, as well as systems for waste removal, storage, and processing.

Containment structures and their components will also vary by the type of livestock being raised. For example, gates and fences required for cattle containment are not the same as those required for poultry containment.

Technical systems used in livestock production include temperature controls in facilities and GPS for monitoring livestock location and condition, **Figure 7-76**. Closed-circuit camera systems may also be used to monitor facilities. In larger facilities, feed distribution schedules may also be computer-controlled.

USDA ARS

Figure 7-76. The GPS collar may be used to track the location of the cow. Some systems also allow the farmer or rancher to monitor the animal's physical condition.

Poultry Production

In poultry production operations, power systems and machinery are used for:

- Housing—heating, ventilation, and egg collection.
- Incubation and hatching—heating, ventilation, and turning eggs.
- Feeding—grinding and automated feeding and watering, **Figure 7-77**.
- Harvesting, processing, and storage of meat and eggs.

Structural systems used in modern poultry houses vary in design by use and type

terekhov igor/Shutterstock.com

Figure 7-77. Large-production poultry houses have automated feeding and watering equipment. This type of equipment enables the producer to efficiently feed a large number of birds and provide consistently fresh water.

of poultry being raised. Broiler (meat) production facilities have different requirements than egg production facilities. However, both types of facilities require electrical service, as well as heating, cooling, and ventilation systems. There must also be feed storage and handling areas, as well as waste removal, storage, and processing facilities.

Most poultry operations use elaborate technical systems to monitor and control each step of production. The strict biosecurity standards used in poultry production often require the use of alarm systems and closed-circuit cameras to ensure protocols are followed.

Dairy

In dairy operations, power systems and machinery are used for:

- Environmental control—heating, cooling, and ventilation. It is especially important to maintain a low-stress environment for dairy cows as it improves and maintains milk production levels.
- Milking and pasteurization—elaborate plumbing systems with pumps and pasteurization machinery.
- Product storage and transport—pumping systems and semitrucks to ship the product to processing facilities.
- Waste removal and processing—flushing, pumping, and processing of waste products.

Dairy operations require structural systems for multiple areas of production, including animal housing, forage and grain storage facilities, milking parlors, milk processing and storage areas, and space for large

Mark Yuill/Shutterstock.com

Mark Yuill/Shutterstock.com

Figure 7-78. Modern dairy production uses a variety of systems to monitor, control, and maintain milk quality.

equipment storage and maintenance. Larger operations may require waste storage and processing facilities.

Technical systems are used in every area of dairy production. Computerized systems are used to create and maintain milking schedules, monitor individual production levels, monitor milk temperature and quality, and monitor the health and well-being of individual animals, **Figure 7-78**. A computerized system may also be used to track each "batch" of milk from harvest to consumer sale.

Forestry

In forestry, power systems and machinery are used for:

- Soil and land preparation for field cultivation—tractors, plows, and field cultivators.
- Planting—tractors, drills, and augers, as well as seed and mechanical tree planters.
- Maintenance—tractors, chainsaws, log splitters, chipper/shredders, and stump grinders.
- Harvesting—log loaders, log stackers, stump cutters, log skidders, grinders, feller bunchers, chipper dumps, tree spades, stump grinders, forwarders, harvesters, slasher/loaders, and yarders, **Figure 7-79**.
- Processing—debarkers, delimbers, and sawmill machinery.

Structural systems used in forestry include those used in nursery production, in processing and storage of wood and its by-products, and large equipment storage and maintenance facilities. Many tree species cultivated for forestry are grown in greenhouse operations until they are considered viable for transplanting. As with the majority of agricultural structures, forestry structures require plumbing and electrical service in order to operate properly.

Technical systems used in forestry include GPS, GIS, and telemetry for monitoring forest and seedling conditions. Telematics and mobile applications may also be used to help maintain equipment.

TFoxFoto/Shutterstock.com

Figure 7-79. The power systems and machinery used in forestry operations enable producers to more efficiently harvest and transport trees. Much of this heavy work was once done with horses and oxen.

Pan Xunbin/Shutterstock.com

Figure 7-80. The structures used to contain stock must be designed for strength and durability.

Aquaculture

In aquaculture operations, power systems are used for:

- Housing, breeding, nurturing, and feeding applications.
- Maintaining water quality and supply.
- Temperature monitoring, control, and supply.
- Oxygen monitoring, control, and supply.
- Waste management and disposal.
- Transporting and harvesting.

Structural systems used in aquaculture vary by operation. Indoor operations require structures designed to withstand a moist environment and to provide adequate space and power for heating and cooling (water and air), lighting, and other specialty equipment. Structures used for indoor saltwater operations must also be designed to withstand the corrosive nature of salt. Materials such as plastic, fiberglass, or concrete are used for containment structures and must be designed to withstand the weight and pressure of the water, **Figure 7-80**. Outdoor structures include concrete runs, plastic or fiberglass tanks, and net or cage structures used in open bodies of water.

Technical systems are vital to successful aquaculture production. Computer-controlled systems are used to monitor water quality, temperature, and oxygen content. Fluctuations in any of these areas can easily kill most aquatic organisms.

Backup Power Sources

Many operations have backup power systems in case of power outages or system failures. The size of the operation and the vulnerability of its product will dictate the type and size of the backup system that should be in place. For example, a small computer system used for accounting and word processing operations may have a battery backup system that would allow the user to save and back up files during a power outage. A dairy operation would require a much larger backup power system, as a power outage could endanger the lives of newborn calves requiring heaters, interrupt milking, and jeopardize any refrigerated milk or other dairy products. Large backup power systems often use fuel-powered generators, **Figure 7-81**.

Visionsi/Shutterstock.com

Samuel Acosta/Shutterstock.com

Figure 7-81. Small gasoline-powered generators may be used for small operations. Larger, built-in systems are essential to operations where a power outage would endanger the well-being of plants or animals or jeopardize a product such as milk or meat.

Lesson 7.5 Review and Assessment

Words to Know

Match the key terms from the lesson to the correct definition.

1. A system that uses gas to transmit and control energy.
2. The rate at which work is performed or energy is applied.
3. A system that uses fluids to do work.
4. The application of force over distance.
5. A device that generates power by burning fuel with air inside the engine.
6. A system that harnesses, converts, transmits, and controls energy to perform work.

A. hydraulic system
B. internal combustion engine
C. pneumatic system
D. power
E. power system
F. work

Know and Understand

Answer the following questions using the information provided in this lesson.

1. How does work differ from power?
2. Why does the need for power increase when speed is increased?
3. Give an example of a mechanical system used in agriculture.
4. Identify three uses for electrical systems on a farm or ranch.
5. How does a hydraulic system perform work?
6. Why should you be cautious working around hydraulic systems?
7. What are three uses of technical systems on a farm or in an agricultural business?
8. Give an example of how structural systems and electrical systems work together in a combined system.
9. Identify three mechanical systems used in livestock production.
10. Waste storage and removal on a dairy farm may involve which types of systems?

Analyze and Apply

1. What are the major differences between pneumatic systems and hydraulic systems?
2. What are the similarities and differences between the various systems on a cattle ranch and a poultry farm?

Thinking Critically

1. Prepare a lecture to teach your classmates about the difference between work and power. Give real-life examples for each point you make.
2. Pest control in livestock and crops can be controlled by several methods. Can you identify a pest control method that uses a mechanical system and one that uses a technical system?

Chapter 7

Review and Assessment

Lesson 7.1

Energy Systems

Key Points

- Energy can come from renewable and nonrenewable sources.
- Solar energy, wind power, hydroelectric, and geothermal energy sources are expensive to establish, but economical in the long term.
- Biomass energy is derived from renewable resources.
- Oil, natural gas, and coal are three major fossil fuels.

Words to Know

Use the following list and the textbook glossary to review and study the *Words to Know* from *Lesson 7.1*.

air turbine
anaerobic digestion
biomass
biomass energy
coal
crude oil
derrick
diode
drag
drilling fluid
drilling mud
fossil fuel
geothermal energy
geothermal heat pump
ground-source heat pump
hydroelectric power plant
hydropower
in-situ leach (ISL) mining
in-situ recover (ISR) mining
kinetic energy
lift
nacelle
natural gas
nonrenewable energy
open-pit mining
photovoltaic cell (PV cell)
renewable energy
solar energy
solar power
strip mining
syngas
uranium
waste rock
wind energy
wind farm
wind power
wind turbine

Check Your Understanding

Answer the following questions using the information provided in *Lesson 7.1*.

1. Give two examples of renewable energy.
2. What is kinetic energy?

3. Explain how a wind turbine works.
4. In wind turbine construction, how do lift and drag work together to make the turbine blades work?
5. What are the advantages of hydropower systems?
6. What are some of the negative effects of geothermal plants on the environment?
7. How does anaerobic digestion of plant matter create energy?
8. What are some of the positive impacts of biomass power production?
9. What is the difference between open-pit mining and underground mining?
10. Why is natural gas a popular energy source?

Lesson 7.2

Biofuels

Key Points

- Biofuels are an alternative to traditional petroleum-based fuels.
- The most common biofuels are ethanol, biodiesel, and biogas.
- Biofuels can reduce carbon emissions as long as the energy it takes to produce the fuels does not surpass the energy created by the fuel.
- Fossil fuels will not last forever. Biofuels provide a solution to diminishing crude oil stocks.

Words to Know

Use the following list and the textbook glossary to review and study the *Words to Know* from *Lesson 7.2*.

alkyl esters
biodegradable
biodiesel
biofuel
biogas
cellulose
corn stover
crop biomass
distillation
dry mill process
ethanol
ethanol soluble
ethyl alcohol
feedstock
fermentation
glycerin
gross domestic product (GDP)
methanol
steepwater
stillage
transesterification
viscosity
wet mill process

Check Your Understanding

Answer the following questions using the information provided in *Lesson 7.2*.

1. What are three types of biofuels?
2. What is the most common source of ethanol?
3. How are native grasses converted to a fuel source?
4. Where is most of the world production of canola produced?
5. From what plant is canola derived?
6. Why are soybeans *not* used more often in biofuel production?
7. What are some natural sources of methanol?
8. *True or False?* The main goals behind the development of biofuels are to reduce fossil fuel dependency and reduce greenhouse gas pollutants.
9. How is the EPA involved in biofuel production and use?
10. How important is ethanol production to the U.S. economy?

Lesson 7.3

Agricultural Tools and Equipment

Key Points

- The proper identification of tools and their applications is essential for proper and safe tool use.
- The appropriate procedures for the use and operation of specific tools and equipment should be followed to ensure the most efficient and safe use of tools and equipment.
- Safety precautions must be followed when using tools to ensure the safety of the user and bystanders.

Words to Know

Use the following list and the textbook glossary to review and study the *Words to Know* from *Lesson 7.3*.

adjustable wrench
Allen wrench
aviation snips
back saw
ball-peen hammer
band saw
belt-and-disk sander
belt sander
blacksmith hammer
block plane
box end wrench
bull tongs
bush hammer
chisel
chuck
circular saw
claw hammer
club hammer
cold chisel
combination square
combination wrench
compass saw
coping saw
Crescent wrench

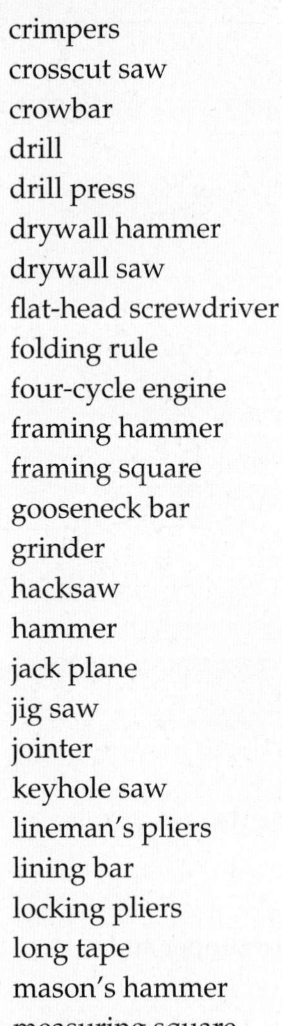

crimpers
crosscut saw
crowbar
drill
drill press
drywall hammer
drywall saw
flat-head screwdriver
folding rule
four-cycle engine
framing hammer
framing square
gooseneck bar
grinder
hacksaw
hammer
jack plane
jig saw
jointer
keyhole saw
lineman's pliers
lining bar
locking pliers
long tape
mason's hammer
measuring square
miter box
miter square
nail hammer
nail puller
nail set
needle-nose pliers
open end wrench
orbital sander
oscillating spindle sander
petroleum-powered tool
Phillips-head screwdriver
pincers
pinch bar
pin punch
pipe wrench
planes
pliers
pneumatic tool
pry bar
punches
push stick
radial arm saw
random orbital sander
reciprocating saw
ripping hammer
rip saw
router
rubber mallet
safety glasses
sander
screwdriver
shingler hammer
side-cutting pliers
sledge hammer
slip-joint pliers
slotted-head screwdriver
Society of Automotive Engineers (SAE)
socket wrench
soft-face hammer
speed square
strap wrench
table saw
tack hammer
tape measure
torque
try square
T-square
two-cycle engine
upholsterer's hammer
wire strippers
wood chisel
wooden mallet
wrecking bar
wrench
zigzag rule

Check Your Understanding

Answer the following questions using the information provided in *Lesson 7.3.*

1. What are some of the uses of a band saw?
2. Your teacher reminds you to never stand on the right side of a band saw while it is in operation. Why should you not stand on the right side of a band saw while it is in operation?
3. What are the differences between a compass saw and a coping saw?
4. What is the difference between a jack plane and a block plane?
5. How are routers similar to portable power drills?
6. What are the differences between a sander and a grinder?
7. Why is it important to use a push stick to feed material into a table saw?

8. Why is it unsafe to have more than 1/4″ of the saw blade protruding above the wood being sawn on a table saw?
9. Why should you never reach across a table saw blade, even when the saw is switched off and the blade is not moving?
10. Why should you not use compressed air to remove dirt, wood shavings, or metal shavings from your body and clothing?

Lesson 7.4

Agricultural Design and Fabrication

Key Points

- Every well constructed agricultural project begins with a well designed construction drawing and plan.
- Drawings of a project should be detailed enough that a person with the right skills could build the project using only the plans for direction.
- Aside from rough sketches, drawings should be as accurate as possible so that the final project is constructed with a high degree of accuracy.
- A bill of materials is a companion to the project drawings. Both are essential for accurate construction.

Words to Know

Use the following list and the textbook glossary to review and study the *Words to Know* from *Lesson 7.4*.

American Society for Testing Materials (ASTM)
assembly drawing
border line
centerline
cutting-plane line
detail drawing
dimension line
elevation plan
extension line
floor plan
formal drawing
hidden line
leader line
level
National Fire Protection Association (NFPA)
object line
orthographic drawing
orthographic projection
phantom line
pictorial drawing
scale
sectional drawing
section line
site plan
sketch
sliding T-bevel

Check Your Understanding

Answer the following questions using the information provided in *Lesson 7.4*.

1. What is the difference between a sketch and a formal drawing?
2. How are object lines and hidden lines similar in a formal drawing?
3. What is the difference between a dimension line and an object line?
4. What is the difference between a hidden line and a phantom line?
5. What does it mean to say that an object has been drawn to proportion?
6. What is the difference between a folding rule and a bench rule?
7. What tasks can you perform with a framing square?
8. What does the term *plumb* mean when measuring and leveling an object?
9. What are the steps to developing a bill of materials?
10. Why is it important to use only building materials that meet industry standards?

Lesson 7.5

Power Systems

Key Points

- Machines make the job of producing agricultural products easier and more efficient.
- It is essential that you understand the basics of machine systems in order to use that equipment safely, efficiently, and effectively.
- A variety of power, electrical, mechanical, structural, and technical systems are used in agricultural operations.

Words to Know

Use the following list and the textbook glossary to review and study the *Words to Know* from *Lesson 7.5*.

hydraulic system	pneumatic system	power system
internal combustion engine	power	work

Check Your Understanding

Answer the following questions using the information provided in *Lesson 7.5*.

1. How does an internal combustion engine convert power into work?
2. Give two on-farm examples of a mechanical system.

3. Give examples of power systems used in agronomic operations.
4. How are GPS and GIS used in crop production?
5. What types of support systems are used in greenhouses?
6. Why would housing facility needs differ for the same livestock species at different life stages?
7. Why would alarm systems and closed-circuit cameras be used with a poultry facility using strict biosecurity standards?
8. What structural systems would be essential to dairy production?
9. How are power systems used in forestry operations?
10. Why are backup power systems important in farming operations?

Chapter 7 Skill Development

STEM and Academic Activities

1. **Technology.** Conduct research to determine how lasers are being used to improve the accuracy of measurements in agricultural construction.
2. **Math.** Using a measuring device, practice measuring materials of various lengths and widths to develop your skills.
3. **Social Science.** Prepare a report on the historical development of woodworking tools used by a carpenter. Aside from the Internet, use woodworking, cabinetmaking, and carpentry books for your research. Write a report to present to your class using visual aids such as pictures and drawings of early tools. If possible, secure examples from a collector.

Communicating about Agriculture

1. **Reading and Writing.** Make a list of the jobs involved in locating and retrieving oil from underground to the delivery of gasoline and diesel fuel at the filling station pumps. Do not forget to include exploration, drilling, refining, transporting, selling, and all the supporting jobs.
2. **Writing.** Design a poster that encourages local farmers to recycle used oil and other used petroleum products.
3. **Reading and Writing.** Some proponents of biomass energy consider it to be a carbon-neutral resource. Research the topic and write a two-page report supporting or opposing this point of view.

Extending Your Knowledge

1. To help keep you and your coworkers safe, create a safety checklist for operators to use before and after they use the table saw. The checklist should be located in clear view of the table saw.
2. Using safe procedures approved by your teacher, crush raw or roasted peanuts to see if you can extract oil from them. Was this an easy process? How useful is peanut oil as a biofuel?

Chapter 8

Agricultural Mathematics

©iStock/XiXinXing

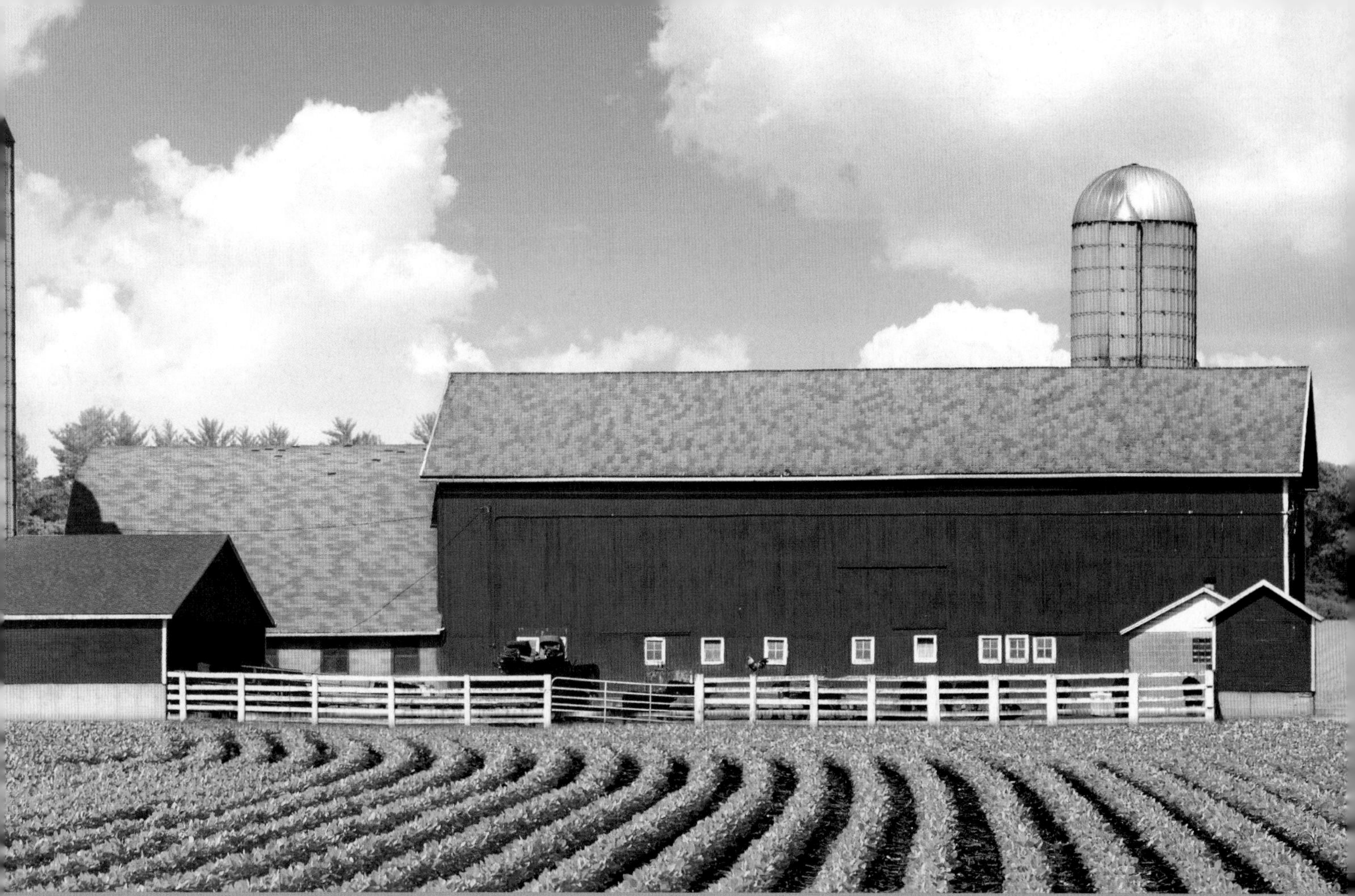

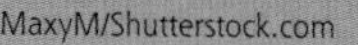
MaxyM/Shutterstock.com

©iStock/jordan rusev

©iStock/perkmeup

While studying, look for the activity icon to:

- **Practice** vocabulary terms with e-flash cards and matching activities.
- **Expand** learning with video clips, animations, and interactive activities.
- **Reinforce** what you learn by completing the end-of-lesson activities.
- **Test your knowledge** by completing the end-of-chapter questions.

Lesson 8.1

Agricultural Business Principles

Words to Know

accounts payable
amortization
asset
budget
business ethics
business plan
capital
capital loan
capital purchase
capital sale
cash flow statement
cash purchase
cash sale
chattel mortgage
collateral
comparative advantage
current asset
depreciation
expense
income
interest
inventory
liability
loan
market
net worth
niche market
noncash expense
noncurrent asset
profit
return on investment (ROI)
salvage value

Lesson Outcomes

By the end of this lesson, you should be able to:

- Develop project proposals and business plans.
- Determine the appropriate financial management principles for maintaining an agricultural business.
- Examine the principles behind accounting for financial records in agriculture.
- Analyze business productivity based on net worth, cash flow, and return on investment calculations.

Before You Read

Think about the title of this lesson, then write a list of five to ten topics that you think should be covered related to the title. After reading the lesson, revisit the list and examine the differences between what you thought would be covered and the actual lesson content.

Look at the items shown in **Figure 8-1**: green pastures filled with grazing cattle...acres and acres of corn...thousands of tomato plants in a greenhouse...a warehouse full of fencing materials...What do all of these things have to do with agribusiness? *Agribusiness* is the management of the profitability of agricultural goods. Without agribusiness, the agriculturalists producing commodities like cattle, corn, tomatoes, and fence panels, have no chance of making a profit.

With more than $7 trillion in sales of agricultural goods each year, agribusiness is big business in the United States. Each and every agribusiness, large or small, uses the same basic business principles used in nonagricultural businesses.

Vast amounts of literature have been written on business and how to make a business succeed. There are thousands of books on topics such as management, hiring practices, training, marketing, and financing. Many offer sound advice for owners and perspective entrepreneurs, but regardless of the size or type of business, all businesses begin with certain basic business principles. This lesson about agricultural business principles briefly explains how to plan for business ventures, and how to account for agricultural products and the income they produce.

Klaus Rademaker/Shutterstock.com AlinaMD/Shutterstock.com Gemenacom/Shutterstock.com Sergiy1975/Shutterstock.com

Figure 8-1. As discussed earlier in the text, there are many sectors in agriculture. Each operation must be run as a business to sustain itself and hopefully earn a profit. *How do these things relate to agricultural business?*

Business Planning

A ***business plan*** is the statement of business goals, along with the strategies and actions for achieving those goals, **Figure 8-2**. Businesses that are planning for future success often set up a business plan to guide and direct their decision-making process. In addition, many banks and lending institutions require the owners of a company to present a business plan when requesting a loan or business account. There are several components to keep in mind when developing your business plan. These components include:

- Business description including the market for the products
- Time investment analysis
- Financial plan

BlueSkyImage/Shutterstock.com

Figure 8-2. A good business plan includes how the business fits into an existing market, or how it will create a new section of the market before spending the money for startup. *Did your SAE require a business plan?*

Business Description

The first component of a good business plan is the description of the business. This should include a complete description of the time frame the business will be in development, when the business enterprise will start developing products or offering services, and the information on the longevity of the project. Understanding the timeline allows a business owner to make better decisions about how to manage time and resources.

gkuna/Shutterstock.com

Figure 8-3. There is a seasonality to when agricultural products are ready to go to market. Having a business plan allows agricultural businesses to make arrangements for their use of resources. *How long in advance do you think the producer had to plan this cotton crop?*

Seasonality

Many agricultural businesses have seasonality and the owners need to consider the time involved from the start of the enterprise to the sale of their product, and how this will affect their financial planning, **Figure 8-3**. For example, if you had a business selling pumpkins, you would need to begin your enterprise in the spring, and would spend the majority of your money before the summer months. You would not see any money from the project until October. Another example of seasonality would be if you had a guided hunting or hunting lease business. Almost all of your income and expenses would occur during the hunting season. Having a solid business plan could help you show potential investors or loan officers that the initial spending would have a return in the future.

Market

The business description should also include information on the market for your products. A ***market*** is the group of consumers that are likely to purchase a specific product. Considering who your market will be is an integral part of the planning. Think about items related to your consumer including:

- Is there a certain age that this product will appeal to?
- What level of income will these consumers likely have?
- How do consumers typically buy this type of product?

Minerva Studio/Shutterstock.com

Figure 8-4. Understanding what your product can offer over your competitors is an important part of developing a business plan. Knowing how your product stacks up to others that consumers can choose from will allow you to better market your product.

Part of understanding your market may include conducting an analysis of the comparative advantage of the specific business, **Figure 8-4**. A ***comparative advantage*** is a statement looking at potential competitors for a certain business, and how the business fits into the same market.

If you wanted to create a new business selling animal feed, you would likely want to do a comparative advantage that included how

FFA Connection Marketing Plan CDE

The Marketing Plan CDE gives a team of FFA members the opportunity to create a marketing plan for a real agricultural business in their area. In this CDE, students have the chance to analyze products available from a company and think of creative ways to increase the profitability of the company through marketing goods and services. Do you think you might like a career in advertising, business management, or owning your own company? If so, the Marketing Plan CDE might be a great fit for you.

Rawpixel/Shutterstock.com

images72/Shutterstock.com

Figure 8-5. This beeswax lip balm fills a niche by providing producers an outlet for leftover beeswax and consumers an all-natural and locally produced lip balm. *Can you think of other products made from the leftovers of other products?*

your business would compete with companies like Purina®, Nutrena®, and the feed available through your local producers' cooperative. Knowing how your company's offerings are better than your competitors' product is a key component in knowing how to market your product and make business decisions.

Niche Market

You may be able to find a place in the market where you have an advantage that makes your company much different than the competitors. In this case, the business is said to be in a ***niche market*** because it is able to offer something that the competitors do not. It is also the sole source for consumers to buy that type of product or service. Can you think of some examples of niche markets?

Many agricultural producers are able to find niche markets for the by-products of their main business. Some examples include a company that creates biodegradable pots from the manure residue on their dairy, and a honey producer who markets a line of lip balm and lotions from its leftover beeswax, **Figure 8-5**.

Hands-On Agriculture

Developing Niche Markets

How are agricultural niche markets created? The answer is simply by creative and forward-thinking agriculturalists. To give you some idea of how simple it is to create a niche market, please complete the following activity in groups of 2–3.

Step 1: Choose an existing agricultural product market (i.e. milk, honey, potatoes, etc.).

Step 2: Determine how you could take the product and make a version that is different from the mainstream market product.

Step 3: Develop a promotional poster for your niche market to share with the class. Make sure you come up with an original name, a thorough description of how this product is different, and to whom you would likely market this niche product.

Step 4: Share your niche market with the class.

Monkey Business Images/Shutterstock.com

Get together with friends to come up with a product and marketing plan.

Time Investment

Once the business plan has a well-developed description, the next step is to determine the amount of time that will need to be invested into the project in order for it to be successful. This is an important and often overlooked portion of a new business. The time that is spent working at a business has a value, even if you are not paying wages to the employees. Consider the following scenario:

Jason has a business building small greenhouses for people to keep in their backyard. He spends about $500 on the materials to put up a greenhouse, and sells them for $1000; it sounds like a really great return, right? What if he kept track of his hours and it takes him 120 hours of work to completely install the greenhouse? If you divide his $500 profit by the 120 hours of labor, he is actually earning $4.17 per hour. That does not sound so impressive, does it?

Having a good business plan can help business owners make decisions about which products to offer, the pricing, and other choices that will affect their bottom line.

Financial Plan

The next part of the business plan is a description of the financial plan for the business. This should include a detailed ***budget***, which is a list of the money a business plans to deal with over the course of the upcoming year. Having a budget allows you to make decisions that will ensure the business is going to be profitable, **Figure 8-6**. For example, if a planned expense is actually going to cost much more than the amount it is budgeted for, then costs should be cut somewhere else in order to ensure that the amount of profit will stay the same.

docstockmedia/Shutterstock.com

Figure 8-6. Can you imagine what would happen if a business decided to run without a budget? *What kinds of poor decisions do you think the company might make?*

SAE Connection Business Plan

When setting up your SAE, it is important to develop your SAE plan. This is the outline of the business plan for your SAE and, just like a business plan, it can help share the story of your SAE and allow you to make more informed decisions. Make sure you set up your SAE plan and look it over throughout your enterprise.

Make sure you enter your SAE plan into your record book to record the details of your business. Most online record books, like AET, have sections for you to input a business plan for each SAE enterprise.

Marlon Lopez MMG1 Design/Shutterstock.com

Capital

Business plans should also include how capital will be acquired and used. ***Capital*** is the term used to refer to the financial and material resources a business has available for use. At the beginning of the business, the plan needs to outline how machinery and equipment will be obtained, along with the plan for repayment if a loan is secured to pay for the equipment.

Accounting for Value

Once a business has a well-developed business plan, the enterprise can start working toward the business goals. How does a business owner know what the enterprise is worth? Before they can figure out a value, they should understand what the business has of value and any outstanding debts the business has incurred.

Paul Vasarhelyl/Shutterstock.com

Figure 8-7. Keeping an accurate count of inventory can not only help with ordering supplies, but it also helps businesses know their current worth. *How do businesses take inventory? How often?*

Assets

Assets are the items of value that are owned by the business. The complete list of assets a business enterprise has is called the ***inventory***, **Figure 8-7**. To fully calculate the total amount of assets that a company has, making an inventory list is a great starting point.

For accounting purposes, there are two basic types of assets. These types are categorized by how long the item is likely to be used by the business. ***Current assets*** are items that have a short-term usefulness for the business. ***Noncurrent assets*** have a longer useful life for the business.

Current Assets

Current assets include things that are consumable, or disappear over time. These items are generally in the business inventory less than one year. Animal feed is an example of a current asset because, in most cases, it is used up by the business in less than a year as it is fed to animals. Other examples of current assets include: fuel, seeds, fertilizer, and materials for packaging. Money that is owed to a business, called accounts receivable, is also considered a current asset because, in most cases, it is expected to come into the business within one year's time. Market animals in a livestock enterprise are also considered current assets. Can you think of a reason why market animals are considered current assets? Even though some market animals, like steers, may be over a year old, the length of their useful life as a show animal is less than a year, **Figure 8-8**. This places them in the inventory category of current assets.

TFoxPhoto/Shutterstock.com

Figure 8-8. Although market animals, like this steer, may be older than one year, their useful life is less than a year, and they are therefore classified as current inventory.

Noncurrent Assets

The other category of assets is noncurrent items. Noncurrent items are typically useful to the business for more than one year. Noncurrent items also generally have enough value to be considered capital for the business. Some examples of noncurrent items would be: buildings, heavy machinery, and equipment. Animals used for breeding are also considered noncurrent inventory, as they have an intended useful life to the business of longer than one year.

Noncurrent inventory loses value, or depreciates, over time. ***Depreciation*** is the reduction in the value of an asset with the passage of time. The concept of depreciation is easy to understand if you have ever purchased a vehicle. If you buy a new car today for $20,000, will it always be worth $20,000, **Figure 8-9**? The obvious answer is no. As you drive the car, as new models become available, and as the car experiences everyday wear and tear, the value is decreasing. Almost all noncurrent items depreciate over time,

Lisa F. Young/Shutterstock.com

Figure 8-9. The concept of depreciation is that noncurrent assets will lose value over time. *Do you think this car will be worth the same amount of money tomorrow as it is today?*

some more quickly than others, until they reach the end of their usefulness. The ***salvage value*** of a depreciable asset is the amount of money the asset is worth when all of the usefulness to the business is gone. In our car example, the salvage value would be the price of the scrap metal and parts from the car. For a breeding animal, the salvage value is the market price if that animal were sold to a processor.

Acquisition of Assets

All businesses need to consider how they will acquire assets. Assets can be acquired through both cash and noncash methods. Cash acquisition is fairly straightforward: the business exchanges its expendable cash for the asset. Noncash acquisition is a little bit more complicated. In this method of acquisition, the business owner does exchange cash for the items. Noncash purchases are important to building the inventory, especially for students working with SAE enterprises. To obtain assets in your SAE, you can trade labor or goods for the asset, receive them as a gift, or simply transfer a current item, like a young market animal, to a noncurrent asset, like breeding stock. The most important thing to consider is that there must be a record of where each asset came from and how it was acquired.

Liabilities

Liabilities are the debts or financial obligations of a business. These are debts owed to other people or vendors. Liabilities count against the overall value of a company. Having liabilities is not necessarily a bad thing; many businesses need to have additional funds for startup expenses or to purchase large capital items, **Figure 8-10**. There are two basic types of liabilities, accounts payable and loans. ***Accounts payable*** are charges that are typically due in full when received, and have a short time frame for payment. ***Loans*** are sums of money that are paid back to the lender over a certain length of time.

Taina Sohlman/Shutterstock.com

Figure 8-10. Large noncurrent items may require a business to pay over time, which means that they are counted as both an asset and a liability until they are paid in full.

Loans

Most loans allow you to pay back the money in equal installments over a period of time. ***Amortization*** is the process of splitting the total loan repayment into equal installments. There are many different types of loans, and understanding loans can allow you to understand why some types are better than others. A ***capital loan*** is a loan for the costs necessary to run a business. A ***chattel mortgage*** is a loan in which movable personal property is used to secure the loan. The property used to secure the loan is called ***collateral***. If a chattel mortgage is not repaid, the lender can take the collateral as payment.

Most vehicle loans are chattel mortgages where the vehicle is used as the collateral. Real estate mortgages are similar to chattel mortgages, but the property is used as collateral instead of movable property.

Interest

Why would financial institutions be willing to give loans to businesses? Loans are generally repaid with interest. ***Interest*** is the fee charged by the lender in exchange for loaning the money. There are different methods for calculating interest, depending on what the interest rate is and how often it is applied to the loan. Knowing how much interest you will pay on a loan can help you make a good decision on whether or not a loan is better than saving and paying for the item with cash. For example, if you get a $100 loan from your cousin and he is charging you 20% interest on the loan for a month, at the end of the month, you will owe him $120. Interest rates vary depending on the type of loan, amount of loan, and the estimated ability of the borrower to repay the loan amount. Paying close attention to interest rates can help you make a good decision about which loan is the best option for borrowing money.

Net Worth

One of the most important mathematical calculations that a business owner can complete is a net worth analysis. The ***net worth*** of a business is the actual value of the business. To calculate the net worth, the total liabilities are compared to the total assets. The remaining value is the actual value of the business, or net worth.

STEM Connection Calculating Net Worth

Franco has a lawn mowing business. On a separate sheet of paper, see if you can correctly identify his assets and liabilities and calculate his net worth. Franco's business includes:

- 2 lawn mowers worth $500 each
- 3 gas-powered trimmers/edgers worth $50 each
- 1 trailer worth $600
- A loan from his parents for $1,000 for the lawn mowers
- Bank account with $250
- 4 clients who owe him $50 each for last month's work
- A bill from the hardware store for $100
- $50 worth of gasoline
- Miscellaneous hand tools worth $100

Aigars Reinholds/Shutterstock.com

What is his net worth? Can you classify his assets as current or noncurrent?

evka119/Shutterstock.com

Figure 8-11. Income comes into a business from two sources, cash sales and capital sales. *Which type of income do you think is more helpful for the overall profitability of a business?*

Accounting for Profitability

Another area that agricultural business owners need to be concerned with is how they account for the profitability of their company. ***Profit*** is the amount of money that a business earns over its expenses. This type of accounting requires an accurate record of the exchange of money, along with information on available cash, and how much money the company is earning for each dollar it spends.

Exchange of Money

The basic components of accounting for profitability revolve around measuring the income and expenses of the business. ***Income*** is the money a business earns. An ***expense*** is any money that the business pays out.

Income

Income can come through a cash sale or through a capital sale, **Figure 8-11**. In a ***cash sale***, the company is receiving money for either services provided, like mowing a lawn, or for the sale of a current inventory item, like a market animal. ***Capital sales***, on the other hand, are based on the sale of a noncurrent inventory item, like a piece of equipment or a breeding animal.

Stockbyte/Thinkstock

Figure 8-12. When you decide to start your SAE, it is probable that your parents might help you with the initial startup funds. Make sure to keep track of their initial investment in your records. *Has someone agreed to sponsor your SAE financially? Do you have a plan for paying back the investment?*

Expenses

Expenses can also come in several different types. The first is a ***cash purchase***, where the business owner exchanges money for a current inventory item. An example would be a hay grower using his debit card to buy hydraulic fluid, which is a current inventory item. The next type of expense would be a capital purchase. ***Capital purchases*** are just the opposite of a capital sale, and involve spending cash to purchase a noncurrent inventory item.

There are also noncash expenses. As an agricultural education student, it is likely that your parents may help fund the initial financial requirements of your SAE project or another business venture you want to start, **Figure 8-12**. ***Noncash expenses*** occur when you receive a gift, or more likely, when you exchange labor

Career Connection Certified Public Accountant (CPA)

Managing the money for agricultural businesses

Job description: Certified Public Accountants, or CPAs, work to manage the operating finances of agricultural businesses. Larger agribusinesses may employ several CPAs just for their operations, while other CPAs will work for themselves and have a variety of agricultural businesses as their clients. They maintain financial records, prepare tax statements, and inform business owners about the financial health of their company.

Robert Kneschke/Shutterstock.com

Certified Public Accountants are state licensed to help businesses and individuals maintain their financial records. ***Why would it be a wise decision to hire a professional accountant?***

Education required: To become a CPA, you must meet the requirements for your state and pass a state board exam to become certified. Generally, CPAs are required to have a bachelor's degree in business or accounting and additional education to meet 150 semester-hours or credit. Additionally, most states require the completion of an ethics course and ethics exam.

Job fit: This job may be a fit for you if you have attention to detail and enjoy working with numbers, you like to help find areas for businesses to improve their profitability, and you are able to see patterns and identify places where patterns are not followed.

or products for the inventory that your business needs. Some examples of noncash expenses would be:

- Your grandpa giving you one of his calves to use as a show steer.
- Your mom agreeing to pay for all of the gas for your lawn mowing business if you mow the lawn at home.
- Your neighbor allowing you to use their field to grow corn in as long as you give them 50% of the corn at harvest time.

It is important from a recordkeeping perspective to keep track of the noncash purchases as well, as they can be critical to calculating the overall profitability of your business.

Cash Flow Statement

Another way to determine the profitability of a business is to look at the cash flow statement. A ***cash flow statement*** is a financial report that shows

FFA Connection Farm Business Management CDE

The Farm Business Management CDE allows you the opportunity to develop your skills in financial recordkeeping. This CDE involves individual FFA members taking a written exam covering financial management and accounting principles.

Cash Flow Statement for Katy's Guided Trail Rides		
Cash flow from operating activities		
Cash receipts from customers ($10,500 + $5,000)	$15,500	
Cash paid to suppliers and employees ($4,000 + $200)	(4,200)	
Net cash flow from operating activities		$11,300
Cash flow from investing activities		
Additions to equipment	(12,000)	
Net cash flow from investing activities		(12,000)
Cash flow from financing activities		
Proceeds from capital contributed	$15,000	
Drawings	(500)	
Proceeds from loan	5,000	
Payment of loan	(4,000)	
Net cash flow from financing activities		15,500
Net Cash Flow @mo/dy/yr		$14,800

Goodheart-Willcox Publisher

Figure 8-13. Keeping accurate records is essential when running a business. *Are there other categories Katy should include in her record book?*

how changes in income and expenses affect the amount of available cash a business has to work with at a given point in time. The statement breaks the income and expense into the operating, investing, and financing activities, and mathematically compares them to allow the owner to see where the money has come from and gone over the time period. Look at the example in **Figure 8-13**. Can you see how much money Katy spent to obtain supplies and pay her employees last year? Can you tell how much new equipment she invested in? What does the statement tell you about the overall status of Katy's Guided Trail Rides as a company?

Return on Investment

Another way to account for the profitability of a business is to look at the return on investment, or ROI. The ***return on investment (ROI)*** of a company is a mathematical calculation that allows the owners to see how much money they are making compared to the money that they are spending. To calculate the ROI, the business owner uses the following formula:

$$\frac{(\text{amount of income} - \text{total amount spent})}{\text{amount of income}}$$

For a business, it is definitely advisable to have a positive ROI. That means that you made *more* money than you spent. How high the ROI needs to be to make the company worth your time depends on your desired goals for the company. This goes back to the importance of setting business goals that we talked about in the beginning of this lesson.

Let us try one. Imagine that your friend Chaleesa has a pick-your-own pumpkin patch and corn maze business, **Figure 8-14**. Last year, Chaleesa made a total of $25,000 with her company. When she looked at her records, she determined that she spent a total of $12,500 on seed, ground lease, labor, marketing, and all her other expenses. What was her ROI? If you calculated her ROI at 0.50, great job. Essentially, she was able to double her money by putting it into her business.

Stu99/iStock/Thinkstock

Figure 8-14. Owning a business, even as simple as a corn maze, requires you to keep quality records so that you can make good business decisions. *How much do you think you could make with a corn maze in your area?*

Business Ethics

While most of the decisions that should be made for a business come from the information provided by mathematical calculations, there is another factor that business owners should consider. ***Business ethics*** is the practice of holding businesses accountable for making decisions that are in the best interest of their consumers, even if they are not what the numbers say is most profitable. Can you think of a company that you do not trust because of a negative ethical choice it made? Consider this example:

Company A purchases an inferior part for its product because it costs less than a better quality product. Management is aware the part is more likely to fail and create an unsafe situation for the consumer. Good business ethics would dictate that the company should decide to use a higher-priced, safer part, **Figure 8-15**.

Business ethics also include things like not investing money based on information that the general public does not know, discriminating against employees or consumers, and being held responsible for falsely advertising a product to the consumers. Having the same high standards of integrity for your business that you have for yourself will lead to increases in consumer confidence and loyalty, and will generally positively impact the finances of your company as well.

Minerva Studio/Shutterstock.com

Figure 8-15. Business ethics means making the right choice for consumers even if it means that the company will not make as much money as it would if it put customers in danger.

Lesson 8.1 Review and Assessment

Words to Know

Match the key terms from the lesson to the correct definition.

1. An item of value owned by a business.
2. A statement of business goals, along with the strategies and actions for achieving those goals.
3. The financial resources a business has available for use.
4. A loan in which movable personal property is used to secure the loan.
5. A list of the money a business plans to deal with over the course of the upcoming year.
6. Any money a business pays out.
7. A financial report that shows how changes in income and expenses affect the amount of available cash a business has to work with at a given point in time.
8. A statement looking at potential competitors for a certain business and how the business fits into the same market.
9. A business loan for the costs necessary to run a business.
10. Charges typically due in full when received; they have a short time frame for payment.
11. The actual value of a business determined by total liabilities compared to the total assets.
12. A purchase in which a business owner exchanges money for a current inventory item.
13. The amount of money that a business earns over its expenses.
14. A sale in which a business is receiving money for either services provided or for the sale of a current inventory item.
15. An item that has a longer useful life for a business.
16. The property used to secure a loan.
17. The process of splitting a loan repayment into equal installments.
18. Item that has a short-term usefulness for a business.
19. The group of consumers that are likely to purchase a specific product.
20. The money a business earns.
21. A mathematical calculation that allows the owners to see how much money they are making compared to the money that they are spending.
22. The fee charged by the lender in exchange for loaning the money.

A. accounts payable
B. amortization
C. asset
D. budget
E. business plan
F. capital
G. capital loan
H. cash flow statement
I. cash purchase
J. cash sale
K. chattel mortgage
L. collateral
M. comparative advantage
N. current asset
O. depreciation
P. expense
Q. income
R. interest
S. inventory
T. liability
U. market
V. net worth
W. noncash expense
X. noncurrent asset
Y. profit
Z. return on investment (ROI)

23. When a business receives a gift or exchanges labor or products for the inventory the business needs.
24. A complete list of a company's assets.
25. The reduction in the value of an asset with the passage of time.
26. A debt or financial obligation of a business.

Know and Understand

Answer the following questions using the information provided in this lesson.

1. What are the parts of a business plan?
2. Explain the purpose of a business plan.
3. In terms of an agricultural business, what does the term *seasonality* indicate?
4. Give two examples of comparative advantage.
5. A business is said to be in a(n) _____ when it has something to offer that the competitors do not.
6. Market animals in a livestock enterprise are considered _____ assets.
7. The _____ value of a depreciable asset is the amount of money the asset is worth when all of the usefulness to the business is gone.
8. A(n) _____ counts against the overall value of a company.
9. The mathematical calculation that business owners can complete to help them know the total value of their business is a(n) _____ analysis.
10. *True or False?* Capital sales are based on the sale of a current inventory item.
11. *True or False?* A positive ROI means a company made more money than it spent.

Analyze and Apply

1. Please sort the following Market Lamb SAE assets into two groups; current (C), or noncurrent (NC):
   A. Market lamb
   B. 50 lb of feed
   C. Halter
   D. Show box
   E. Soap
2. Kelby has a floral design business. He has a floral cooler worth $2000, tools and equipment worth $300, supplies worth $75, $200 worth of flowers on hand, and $120 in his business bank account. He owes his floral supplier $100 from last week's invoice, and he owes his parents $500 from the loan they gave him to start the business. His customers owe him a total of $80. What is his net worth?

Thinking Critically

1. Create a business plan for a small business that you would like to set up. Make sure you include the business description, market, comparative advantage of your product, seasonality, time investment, and an estimate of the cost to start this business venture. Ask your agriculture science teacher if you could potentially do this as a start for your own SAE.
2. Take an inventory of the things that you personally own and the liabilities you currently have to calculate your own net worth. Set two goals for increasing your net worth over the course of the upcoming year.

Lesson 8.2

Practical Mathematics in Agriculture

Lesson Outcomes

By the end of this lesson, you should be able to:

- Differentiate between the major fields of mathematics.
- Describe how mathematics, including calculation, measurement, and statistics is integral to agriculture.
- Recognize the application of arithmetic, algebraic, and geometric principles in agricultural settings.
- Use problem-solving skills to analyze and solve agricultural problems requiring basic mathematical calculations.

Words to Know

acre
algebra
applied mathematics
area
arithmetic
bushel
computational mathematics
conversion
conversion factor
formula
fraction
geometry
mathematics
mean
median
mode
percentage
ratio
standard deviation
volume

Before You Read

Read through the captions for the figures in this lesson and write down five questions you still have about the lesson topic after reading the captions. As you read the rest of the lesson, see if you can answer the questions you wrote. After reading, talk over your questions and answers with a partner.

Have you ever said that you *hated* math? Since at least the 1950s, studies have shown that mathematics is the subject that high school students like the *least*. Why? Is it because of the countless hours of math homework you have endured in your life? Is it because students have no idea why solving for *x* is important to their future? Is it because math just doesn't make sense to some high school students? Would it change your opinion of math to know that you probably do math every day outside of math class, without even realizing it, **Figure 8-16**?

This lesson is designed to help us explore the many ways that math and agriculture work together. We will look at the different types of mathematics, and then see how these concepts are used every day in agricultural applications. ***Mathematics*** is the study of numbers, equations, and geometric shapes and their relationships. Almost everything in agriculture relies on some degree of math.

There are four basic areas of mathematics: arithmetic, algebra, geometry, and applied mathematics. Let us take a deeper look at how these fields of math interact to help agriculturalists manage the production, processing, and marketing of agricultural goods.

Arithmetic

Arithmetic is the mathematics of counting quantities and manipulating numbers. It is the simplest category of mathematics. There are four basic arithmetic functions: addition, subtraction,

maurusone/istock/Thinkstock

Figure 8-16. *Have you ever sat in class and calculated how much time you had left until the bell?* If so, you've just discovered one of the hundreds of ways you use math every day without even thinking about it.

multiplication, and division. Each of these principles has rules and boundaries that you have likely learned in your math classes. This is probably the area of mathematics where you feel most comfortable. After all, you have been doing arithmetic in school for at least the last 10 years.

Agriculture uses arithmetic for a variety of purposes, as there are many different quantities in agriculture that need to be calculated. Agriculture involves having to manage products that have quantities that are fluid or dry and to determine what portions of a quantity are involved in various production systems, **Figure 8-17**.

Fluid Quantities

Many agricultural products are measured in fluid quantities. In the United States, fluid measurements are more commonly calculated using the measurements shown in **Figure 8-18**.

Blue Jean Images/Photodisc/Thinkstock

Figure 8-17. Arithmetic is the field of mathematics responsible for counting quantities. *How many pounds of rice and beans are there in this photo? How could we find out?*

STEM Connection Importance of Math in Agriculture

Agriculture simply could not function without mathematical calculations. Everything in agriculture, from the date to plant a seed or breed an animal, to the amount of profit expected to come from agricultural commodities is tied to math.

Developing strong math skills is essential to your success in any area of agriculture. Can you think of some ways that the following agricultural careers use math?

- Veterinarian
- Agricultural Engineer
- Production Agriculturalist

Think about any career that you might want to pursue in agriculture. Can you think of some ways that mathematics will be involved in that career? Every agricultural career uses mathematics in one way or another, even if you are only calculating a paycheck.

Sean Locke Photography/Shutterstock.com

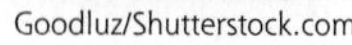

Goodluz/Shutterstock.com

Patrick Foto/Shutterstock.com

Examples of Calculating Fluid Quantities

We can start looking at the importance of arithmetic with our first practice problem:

1. If there are 16 fluid ounces in a pint, how many fluid ounces are there in 3 pints? Can you set up this equation? It should look like this:

$$16 \text{ (ounces in a pint)} \times 3 \text{ (pints)} = 48 \text{ fluid ounces}$$

Unit of Measurement	# fl. oz.	# cups
Fluid Ounce (fl. oz.)	1	1/8
Cup (c)	8	1
Pint (p)	16	2
Quart (qt)	32	4
Gallon (gal)	128	16
Conversions 1 fluid ounce = 29.57 ml 1 gallon = 3.79 l		

Goodheart-Willcox Publisher

Figure 8-18. Fluid measurements. There are both U.S. and metric measurements for fluids. *What do you need to do to change one to the other?*

Unit of Measurement	# ml	# l
Milliliter (ml)	1	0.001
Centiliter (cl)	10	0.01
Liter (l)	1,000	1
Kiloliter (kl)	1,000,000	1,000
Conversions (rounded to nearest .001) 1 ml = .034 fluid ounces 1 l = .264 gallon		

Goodheart-Willcox Publisher

Figure 8-19. Fluid measurements in metric. *Where in the world are metric measurements used?*

Let us try another one:

2. If we have 64 ounces of fluid, how many cups do we have? To set up this equation, we need to divide the number of ounces we have by the number of fluid ounces in a cup.

$$\frac{64 \text{ (ounces of fluid)}}{8 \text{ (ounces in 1 cup)}} = 8 \text{ cups of fluid}$$

Outside of the United States, fluids are measured using the metric system. See the table in **Figure 8-19**. This table begins with the milliliter (ml), which is a commonly used measure; 29.57 ml is equal to 1 fluid ounce. The other commonly used metric volume measurement is a liter, which is 1000 milliliters.

Conversions

Changing one unit of measurement to another is called ***conversion***. Conversions are important for agricultural companies, and are completed by multiplying or dividing a quantity by a ***conversion factor***. A conversion factor allows you to know *how many* of one unit of measurement are in another unit of measurement. For example, if you know that there are 8 ounces in a cup, you can easily use arithmetic to figure out how many ounces there are in 2 cups. In this case, the number of ounces per cup (8) is the conversion factor. To convert cups to ounces, you multiply by the conversion factor; to convert ounces to cups, you divide by the conversion factor.

Conversions are even more important to those agricultural producers who live in the United States because the United States has different units of measurement than many other countries.

Examples of Using Conversion Factors

Consider the following example:

1. Harold is a maple syrup producer who sells his highest grade product wholesale for $42 per gallon. He just received an order for 950 liters of maple syrup for a Canadian food processor. What information would you need to calculate this invoice? Did you say that you would need to know the conversion factor for liters and gallons? There are 3.79 liters in 1 gallon. Can you figure the order now? How many gallons would he have to send? How much would this order cost?

$$\frac{50 \text{ liters in the order}}{3.79 \text{ liters per gallon}} = 13.19 \text{ gallons in the order}$$

13.19 gallons in the order at $42 per gallon = $553.98 order total

Using arithmetic to calculate fluid quantities is vital to the agricultural commodities produced in a fluid form. In addition, fluid conversions are important in agriculture for calculating chemical ratios in fertilizers and pesticides, properly dosing medication for animals, and calculating the amounts of mechanical fluids like fuel and hydraulic fluid, **Figure 8-20**. There are more examples of fluid conversion practice problems at the end of the lesson.

rthoma/Shutterstock

Figure 8-20. One of the main supplies that agriculturalists need to be able to calculate properly is fuel. These fuel tanks allow equipment to run continuously from the farm. *How many gallons do you think each of these tanks holds?*

Dry Quantities

Just like calculating quantities of fluid volume is important, using arithmetic to calculate and convert dry weights is an important skill to master for agricultural producers and processors. The U.S. dry weights are shown in **Figure 8-21**.

One of the most common dry weight measurements for agricultural commodities is *cwt*, which equals 100 pounds. Another important dry weight measurement in agriculture is the bushel. A ***bushel*** is a unit of dry measurement equivalent in volume to 8 gallons. The weight for a bushel is different depending on the commodity, **Figure 8-22**. Corn, soybeans, wheat, oats, and barley are all commodities that are commonly sold by the bushel.

Outside of the United States, the metric system is used to calculate weights. The metric dry weights are shown in **Figure 8-23**. To understand the dry weight conversions, you need to know the common conversion factors: there are 453.59 grams in each pound, and about 2205 pounds in 1 metric ton.

One of the reasons being able to convert dry weights is important is for making business decisions for international trade.

Examples of Calculating Dry Quantities

Look at this example:

1. Laura is a soybean producer. She has been offered $10.25 per bushel to sell locally, or can sell for $398.00 per metric ton to be sold overseas. Which deal should she take? She needs to use arithmetic to figure out the price for equal units. To figure out the price per pound based on the bushel price, we just need to divide the price per bushel by the number of pounds of soybeans in a bushel. That equation looks like this:

$$\frac{\$10.25 \text{ per bushel}}{60 \text{ pounds of soybeans per bushel}} = \$0.1708$$

Did You Know?

The term *bushel* was originally used to measure the capacity of grain that would fit into a basket approximately 18.5″ in diameter and 8″ tall. Today, the U.S. bushel is based on a measure of exactly 2150.42 cubic inches.

Unit of Measurement	# pounds
Ounce (oz)	1/16
Pound (lb)	1
Hundred Weight (cwt)	100
Ton	2,000
Conversions (rounded to nearest .001) 1 oz = 28.350 g 1 lb = 453.592 g 1 lb = 0.454 kg	

Goodheart-Willcox Publisher

Figure 8-21. U.S. dry weights.

Commodity	# pounds (weight per bushel)
Barley	48
Corn (shelled)	56
Oats	32
Soybeans	60
Wheat	60

Goodheart-Willcox Publisher

Figure 8-22. Commodities have different weights for a bushel because it is based on volume rather than weight. ***How do you think moisture content could skew the bushel measurement?***

Unit of Measurement	# grams
Milligram (mg)	0.001
Gram (g)	1
Kilogram (kg)	1,000
Metric Ton	1,000,000
Conversions (rounded to nearest .001) 1 g = 0.035 oz 1 kg = 2.205 lb 1 metric ton = 2204.622 lb	

Goodheart-Willcox Publisher

Figure 8-23. Metric dry weights. Metric units of weight are used in many places in the world. ***Which of these weights is closest to a U.S. ton?***

To figure out the price per pound for the metric ton, we should know that there are about 2205 lb per metric ton; therefore, to calculate the price per pound based on the metric ton price, we can set up the equation like this:

$$\frac{\$398.00 \text{ per metric ton}}{2205 \text{ pounds per metric ton}} = \$0.1805$$

When we compare the prices, we can see that there is about a one cent per pound difference in the offers, which can really add up if Laura has hundreds of thousands of pounds of soybeans to sell.

Ratios

Another role that basic arithmetic plays in agriculture is in helping calculate ratios and percentages. A ***ratio*** is the comparative value of two or more quantities. For example, if you know that you need to mix cleaning solution for food processing equipment at a ratio of 5 to 1, you would need to put in five parts of water to one part of cleaning solution. Using basic math functions, you could calculate the amount of cleaning solution to add to any volume of water to reach the required concentration. Ratios are commonly used when mixing fuel additives, fertilizers, pesticides, and other spray application materials.

Examples of Calculating Ratios

Consider this example: You need to mix fuel and oil to the proper concentration for a weedeater. The owner's manual suggests a ratio of 50 parts fuel to 1 part oil. This means that 1/50 of the mixture should be oil. You have a 1 gallon (128 oz) container to mix the oil into. How much oil should you add?

$$\frac{128 \text{ oz in 1 gallon}}{50 \text{ parts fuel to oil}} = 2.56 \text{ oz of oil per gallon}$$

Fractions

One special thing to consider when using arithmetic in agriculture is that you may need to use arithmetic with fractions. ***Fractions*** are numerical expressions that represent one number divided by another to explain the part of a whole that is being measured. Each fraction is basically a ratio, for example, 1/2 is one out of two parts. The top number is called the numerator, and the bottom number is called the denominator. There are some basic rules about fractions that will help you when working with fractions that are shown in **Figure 8-24**.

Examples of Calculating Fractions

Some of the best examples of fractions in agriculture come with measuring lengths. For example, if you were building small wooden boxes for a project in class and needed to have 20 boards that were 3/4″ long, how many total inches of board would you need?

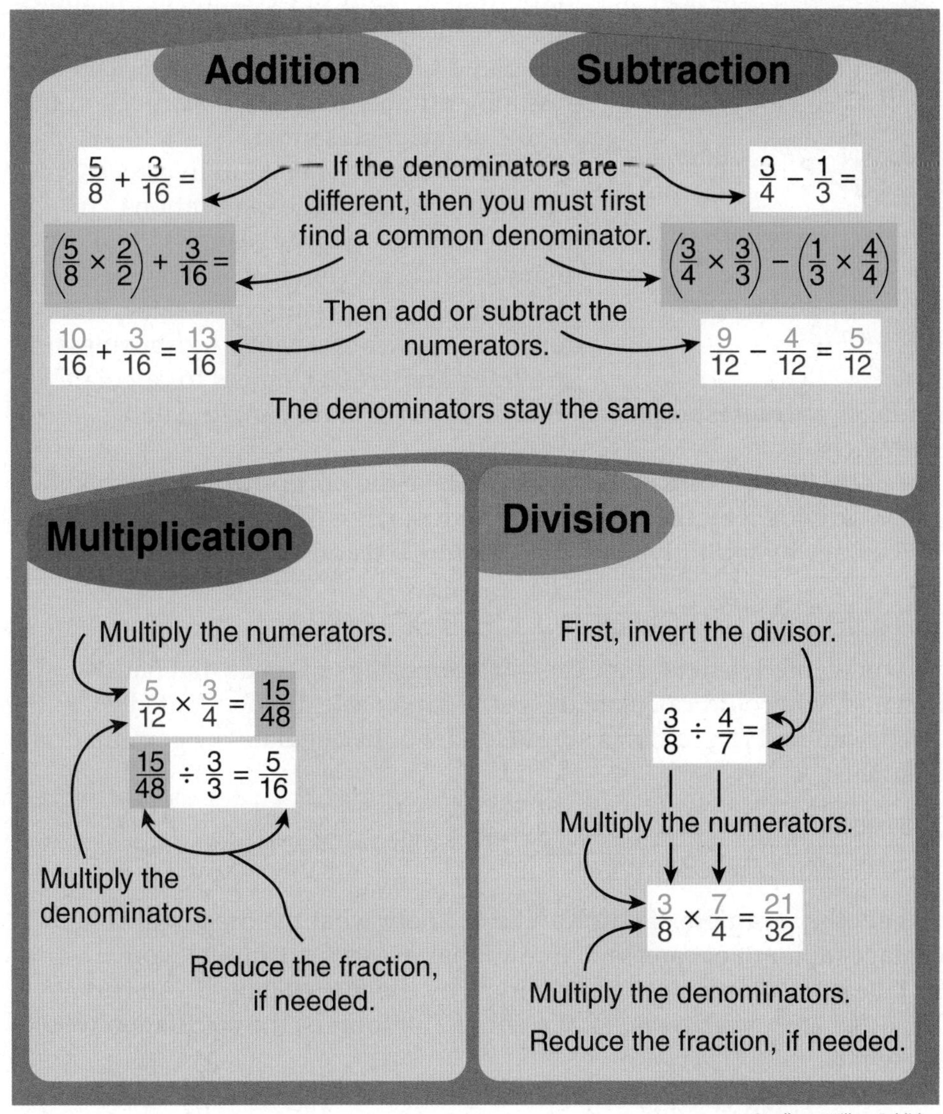

Figure 8-24. Rules about fractions. Knowing how fractions work can be incredibly helpful to understanding how to arithmetically calculate them. *What is the bottom number in a fraction called?*

Steve Heap/Shutterstock.com

Figure 8-25. Calculating a percentage shows you what portion one piece makes up of a larger whole. *What percentage would you guess the green section of this pie represents?*

To calculate this problem, you need to think back to your math class when you learned that multiplying fractions only involves multiplying the numerator, so:

$$20 \times \frac{3}{4} = \frac{60}{4}$$

You can then divide 60 by 4 to tell you that you would need 15″ of board to give you 20 pieces that were 3/4″ each.

Percentages

Understanding percentages is also important to many agricultural applications. A ***percentage*** is the number of parts compared to 100 parts, **Figure 8-25**. The resulting number tells you how much of the whole unit is represented.

Percentages are used in livestock operations to calculate feed rations, production success, and carcass yield. In plant science, percentages can be used to determine yield, germination rates, and fertilizer application rates. Agribusiness provides numerous chances to calculate percentages as interest on loans, return on investment, and available operating costs.

To calculate a percentage, you simply take the portion you want to know about, and divide it by the whole. In these cases, 1 = 100% and you can get the percentage by multiplying the decimal by 100 and adding a percent sign.

Examples of Calculating Percentages

If you wanted to determine the total percentage of seeds that had germinated, you could calculate the germination rate. If you planted 100 seeds and had 78 seedlings sprout, you would calculate the germination rate like this:

$$\frac{\text{78 seedlings sprouted}}{\text{100 seeds planted}} = .78 \text{ and } .78 \times 100\% = 78\% \text{ germination rate}$$

As another example, if you wanted to know what the dressing percentage (amount of carcass to live animal) was for an animal, you could take its carcass weight and divide it by the live weight. If you had a 1,400-pound steer with a carcass weight of 714 pounds, the calculation would look like this:

$$\frac{\text{714 lb of carcass}}{\text{1,400 lb live weight}} = .51 \text{ and } .51 \times 100\% = 51\% \text{ yield}$$

To explain the calculation, this means that 51% of the live animal was actually able to be used as retail meat products.

Algebra

Another category of mathematics is algebra. ***Algebra*** is the field of mathematics in which symbols are used to represent unknown values. Algebra involves using specific mathematically derived rules to find the unknown values. Algebra is important because there are many times when values in agriculture are unknown. Using algebraic principles allows agriculturalists to find the unknown values without having to guess. ***Formulas*** are predetermined mathematical patterns used to solve for the unknown values.

R.MACKAY PHOTOGRAPHY, LLC/Shutterstock.com

Figure 8-26. Algebra uses letters to represent unknown variables in the equation. It allows you to solve for the unknown values.

Examples of Using Algebra

Agriculture relies heavily on algebra, **Figure 8-26**. In crop science, algebra can be used to find the proper rate of flow for irrigation systems, determine the amount of damage weather can cause, and help calculate the rate of drying for crops that need to dry in the field. In animal science, algebra can help determine mixing feed rations, medications, and for estimating return on investment for feedlots. In agricultural engineering, everything from calculating horsepower to determining the speed of a PTO to calculating the proper wattage for an electrical system requires an algebraic formula. These formulas are highly specialized and once you know the formula, you can find the correct value by plugging in the known variables.

For example, if you were trying to determine how much horsepower a tractor would need for a given application, you would need to calculate the drawbar horsepower. The formula for calculating drawbar horsepower is

$$\frac{\text{total draft (in lb)} \times \text{speed (in miles per hour)}}{375}$$

If you knew that you had 7,500 lb of draft and wanted to go 5 mph, what would be the required drawbar horsepower?

Geometry

Geometry is the field of mathematics related to understanding spatial relationships and geometric shapes. Geometry is important to agriculture because there are applications for both two-dimensional and three-dimensional calculations.

Unit of Measurement	# feet
Inch	1/12′
Foot	1′
Yard	3′
Mile	5,280′
Unit of Measurement	**# meters**
Millimeter	0.001 m
Centimeter	0.01 m
Meter	1 m
Kilometer	1,000 m

Goodheart-Willcox Publisher

Figure 8-27. U.S. linear measurement and metric linear measurement. Linear measurements are used often in construction and fabrication situations. Metric linear measurements are used in many countries. ***Are you going faster at 50 miles per hour or 50 kilometers per hour?***

Calculating Length and Area

One of the principle reasons for using geometry in agriculture is to calculate ***area***, the size of a two-dimensional space. Both linear measurement in the United States and metric units of measurement are shown in **Figure 8-27**. There are 3.28′ in 1 meter, and 1.61 miles in 1 kilometer. Using linear measurements will tell you the distance between two points on a line.

To determine area, you need to first determine the shape of what you are measuring. Then, you can use the formula to calculate the area. The table in **Figure 8-28** will help you identify which formula is correct for the shape of your given space. Area is expressed in units squared. There are numerous agricultural applications for calculating area. Determining the size of a field, the amount of surface area for irrigation, and many agricultural mechanics applications require a knowledge of calculating area. One special unit of measurement for area in agriculture is an ***acre***, which is 43,560 square feet. Can you figure out how many square yards an acre would be? There are some additional practice problems about area for you to consider at the end of this lesson.

STEM Connection Pythagorean Theorem

One of the most important geometric formulas used in agriculture is the Pythagorean Theorem. This formula allows you to calculate the third side of a right triangle as long as you know the two other sides. The formula is:

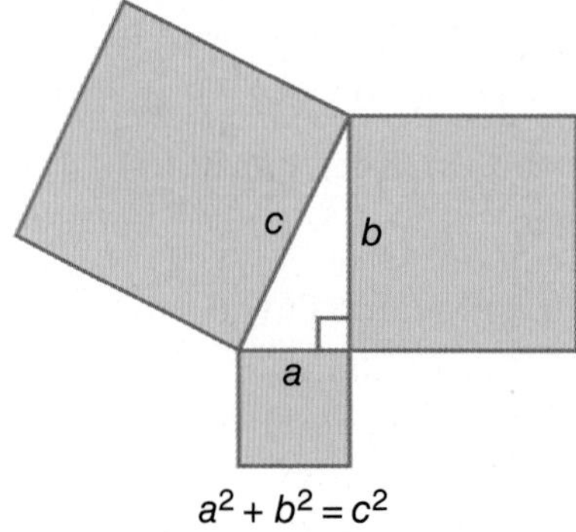

$$a^2 + b^2 = c^2$$

Understanding the Pythagorean Theorem is important for measuring triangles, squares, and any other shape that includes right triangles.

Two-Dimensional Plane Shapes	Formula for Area
Square 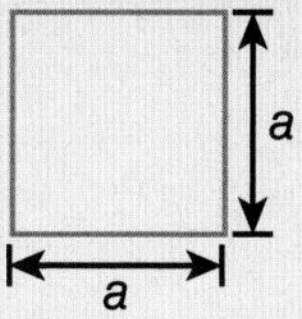	Area = a^2 or $a \times a$ Example: a = 5 cm Area = 5^2 = 25 cm^2
Rectangle 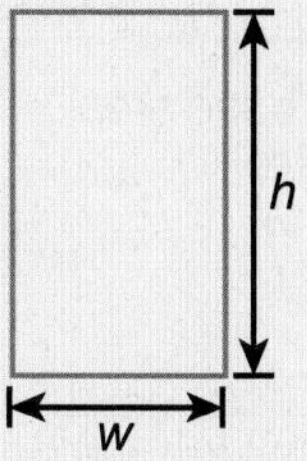	Area = $w \times h$ = width × height Example: w = 10 cm h = 20 cm Area = 10 × 20 = 200 cm^2
Triangle 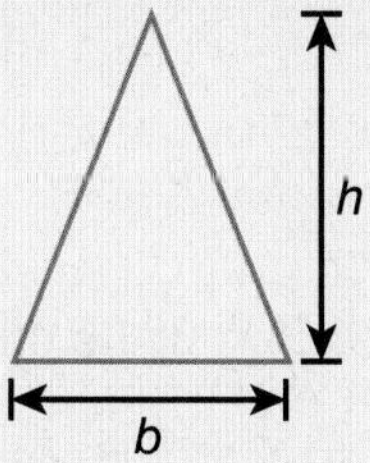	Area = $b \times h \times 0.5$ = base × height × 0.5 Example: b = 20 cm h = 15 cm Area = 20 × 15 × 0.5 = 150 cm^2
Circle 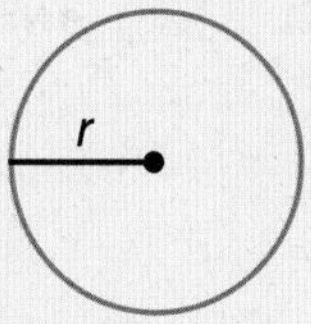	Area = $\pi \times r^2$ = pi × radius^2 Example: Use π = 3.14 r = 5 cm Area = 3.14 × 5^2 = 3.14 × 5 × 5 = 78.5 cm^2

Goodheart-Willcox Publisher

Figure 8-28. Finding the area of different shaped spaces or project pieces.

Calculating Volume

Now that we know how to calculate the amount of area a shape contains, we can consider the volume. ***Volume*** is the amount of space contained in a three-dimensional boundary. Once again, the volume depends on the shape of the container. Using the table in **Figure 8-29**, we can calculate the amount of cubic space a shape contains. Knowing the volume is important for agriculturalists who need to figure out how much room there is in a transportation vehicle, how much space is available for storage, or the total capacity for cylinders and silos. Using the table, let us try a volume problem together: Tyler is a grain producer. He has a cylinder-shaped silo that is 14 feet in diameter and 20 feet tall. What is the volume of his silo in cubic feet? If one bushel is 1.24 cubic feet, how many bushels of wheat could Tyler put in this silo?

Three-Dimensional Solid Shapes	Formula for Volume
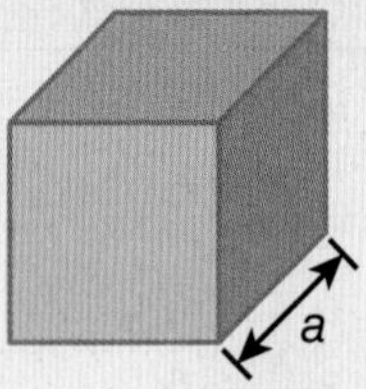 Cube	Volume = a^3 or $a \times a \times a$ Example: a = 5 cm Volume = 5^3 or $5 \times 5 \times 5$ Volume = 125 cm^3
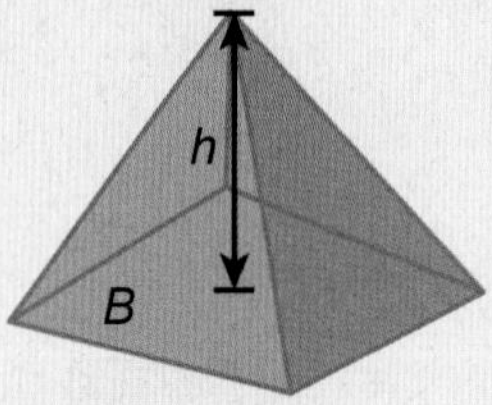 Pyramid	Volume = $B \times h \times 1/3$ = base area × height × 1/3 Example: B = 16 cm^2 h = 9 cm Volume = $16 \times 9 \times 1/3$ = 48 cm^3
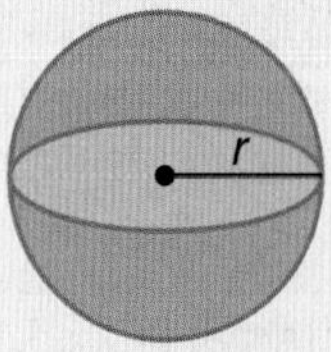 Sphere	Volume = $4/3 \times \pi \times r^3$ = 4/3 × pi × radius^3 Example: r = 4.5 cm Use π = 3.14 Volume = $4/3 \times 3.14 \times 4.5^3$ = 381.5 cm^3
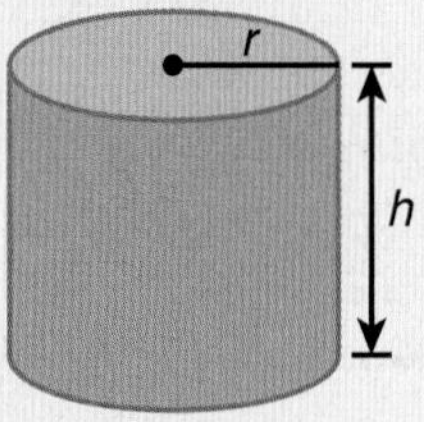 Cylinder	Volume = $\pi \times r^2 \times h$ = pi × radius^2 × height Example: r = 5 cm h = 10 cm Use π = 3.14 Volume = $3.14 \times 25 \times 10$ = 785 cm^3

Goodheart-Willcox Publisher

Figure 8-29. Use the formulas above to calculate volume.

Applied Mathematics

In addition to arithmetic, algebra, and geometry, there are also applied mathematical areas that are important to agricultural producers. ***Applied mathematics*** is the field of mathematics involved in the study of physical, biological, or sociological areas. Applied mathematics is math used by other areas. Agriculture relies on applied mathematics primarily in the areas of statistics and computers.

Statistics

Statistics is a concept that was covered in *Lesson 4.1* when we talked about agriculture as a social science. Statistics allows agriculturalists to make decisions based on incomplete information, by mathematically calculating potential outcomes. One of the core principles of statistics is understanding where the center of the existing information is so that you can use one number to represent a group of numbers. This can be done through several different methods. The ***mean*** is the arithmetic average of the numbers in a

Career Connection Agricultural Fabricator

Job description: Agricultural fabricators work to build the structures, equipment, and machinery required for agriculturalists. Many own their own business and custom build pieces based on an individual producer's needs.

Education required: Typically agricultural fabricators attend a certification program for welding or metalworking. Some may go to college to obtain a bachelor's degree in agricultural systems or agricultural engineering.

Job fit: This job may be a fit for you if you enjoy working with your hands, building things with metal or wood, and figuring out how to solve problems.

Vadim Ratnikov/Shutterstock.com

Agricultural fabricators use materials to build the equipment and structures that agricultural companies and producers need.

data set. To get the mean, you simply add all of the numbers and divide by the number of items you added. We can also look at the number in the center of the list when numbers are ordered from least to greatest, called the ***median***, or examine the number that occurs most often in a data set, which is called the ***mode***.

Looking at how far away a new number is from the center value of your total information can tell you important things about whether the number follows the same pattern. The average distance from the center value is called the ***standard deviation***. Understanding the concept of standard deviation is important for using statistics to make financial decisions in agriculture.

Statistics is a major factor in considering livestock breeding decisions. For example, in cattle, statistics can be used to figure a breed's average performance for birth weight, weaning weight, yearling weight, and allows producers to compare the performance of their individual animals. For example, if you calculated that the average weaning weight for a heifer calf from your operation was 516 lb and you had a cow with a calf that was 462 lb at weaning, it might help you make a decision about keeping her in your breeding program. Knowing the statistics for a potential bull allows breeders to decide which bull to breed to their cows.

Statistics is also important in making marketing and business planning decisions. Understanding what consumers are willing to spend, what factors play a role in their buying decisions, and what amount of output expense is likely to result in a better return on your investment are all factors that can be calculated with statistics.

Computational Mathematics

Computational mathematics is the area of math involved in developing computer programs. Most of this area is related to developing computer-coded algorithms for agricultural software programs and applications, **Figure 8-30**. There is a large demand for computer programmers who are able to understand the agricultural industry and computational mathematics.

Jevtic/iStock/Thinkstock

Figure 8-30. There is a high demand for the computer programmers who will develop the programs that will help the next generation of agriculturalists. *What agricultural applications do you think would be helpful for producers?*

Lesson 8.2 Review and Assessment

Words to Know

Match the key terms from the lesson to the correct definition.

1. The size of a two-dimensional space.
2. The arithmetic average of the numbers in a set of data.
3. A special unit of measurement for area in agriculture; equivalent to 43,560 square feet.
4. Numerical expression that represents one number divided by another to explain the part of a whole that is being measured.
5. The study of numbers, equations, and geometric shapes and their relationships.
6. The mathematics of counting quantities and manipulating numbers.
7. A unit of dry measurement equivalent in volume to 8 gallons.
8. The average distance from the center value.
9. The field of mathematics in which symbols are used to represent unknown values; involves using specific mathematically derived rules to find the unknown values.
10. A number that allows an understanding of *how many* of one unit of measurement are in another unit of measurement.
11. Predetermined mathematical patterns used to solve for unknown values.
12. The comparative value of two or more quantities.
13. The field of mathematics involved in the study of physics, biology, or sociology.
14. The number in the center of a list when numbers are ordered from least to greatest.
15. The area of math involved in developing computer programs.
16. The number that occurs most often in a data set.
17. The act of changing one unit of measurement into another.
18. The amount of cubic units that a three-dimensional space contains.
19. The field of mathematics related to understanding spatial relationships and geometric shapes.

A. acre
B. algebra
C. applied mathematics
D. area
E. arithmetic
F. bushel
G. computational mathematics
H. conversion
I. conversion factor
J. formula
K. fraction
L. geometry
M. mathematics
N. mean
O. median
P. mode
Q. percentage
R. ratio
S. volume

Know and Understand

Answer the following questions using the information provided in this lesson.

1. Why is it especially important for producers in the United States to know conversion factors?
2. Which area of mathematics is a land surveyor most likely to use in calculating field sizes?
3. Why do differences occur in the weight of bushels for different commodities?

4. If you needed to explain how many ounces of pesticide should be mixed per gallon, you would likely use which mathematical procedure?
5. What is the difference between area and volume?
6. Please give two examples of how a veterinarian might use mathematics.
7. Give an example of a time you would need to use a conversion factor.
8. What are the four basic arithmetic functions?
9. Give two examples of how algebra is used in animal science.
10. Which area of mathematics is the most commonly used in all agricultural applications?

Analyze and Apply

Practice problems by category:

Fluid Quantities (hint: use the information from Figures 8-18 and 8-19)

1. You have a dairy herd of 200 cows who produce an average of 7 gallons of milk each per day. Someone has asked you to determine the number of each container of milk your whole dairy produces each day:

   A. Gallons:

   B. Quarts:

   C. Pints:

   D. Cups:

   For the same dairy, how many liters of milk do you produce each day (remember that there are 3.79 liters per gallon)?
2. You want to buy a new liquid feed supplement for your show lambs. You know that they need 4 ounces of the fluid in standard measurement units, but you only have a milliliter measurement. How many milliliters of the supplement should you give each lamb? Remember that 1 oz is equal to 29.57 ml.

Dry Quantities (hint: use the information from Figures 8-21, 8-22, and 8-23)

3. You have a very specific fertilizer program for your prize-winning giant pumpkins. It involves giving them 2 pounds of your special fertilizer blend every week. The fertilizer provider can only mix up the fertilizer in kilograms. If you have 10 pumpkins that you need to fertilize for 6 weeks, how many kilograms of the fertilizer will you need?

   60 doses of fertilizer × 2 pounds per dose = 120 pounds of fertilizer needed

   120 lb × 0.454 kg per pound = _____ kilograms
4. There are two different producers who are looking to purchase your barley. One of them is willing to pay you $0.11 per pound, and the other is willing to pay you $6.00 per bushel. Which offer should you take?

Ratios and Percentages

5. If you had a pesticide ratio of 1 to 100 chemical to water, and you needed to mix up 500 gallons of spray, how much chemical would you add?
6. You know that in order for your flock of commercial sheep to be productive, you need to have a 140% lambing rate. If you have 346 ewes and had 474 lambs born, did you meet your required lambing rate?

7. If you have a market hog that was 297 lb at the show, and his carcass weighs 208 lb, what was his dressing percentage?

Geometry

8. You have access to a neighbor's barn for your livestock. He has allowed you to use the barn for free if you pay for the pipe to bring water to the barn from the well. You contact two different companies: one wants to charge you $1.50 per meter to put in the pipe, and the other one wants to charge you $0.50 per foot to put in the water line. Which company should you choose?
9. You are buying a new silo to hold grain. It is 32′ tall and 12′ in diameter. What is its storage capacity?

Statistics

10. You are trying to determine information about the market hogs you sold to participants at a local show. The following weights (in pounds) from the 10 market hogs you sold to exhibitors were:

    255
    265
    265
    272
    272
    272
    275
    282
    282
    295

    Please find the following statistics about this data:

    A. Average/mean weight:
    B. Mode:
    C. What would happen to the average if an 11th hog was weighed at 306 lb?

Thinking Critically

1. In order to use mathematics, it is important that you are able to calculate information from multiple sources. The following scenario is designed to test your ability to select different mathematical procedures to solve a real-world agricultural problem. Scenario: you are a local corn producer with 1000 acres of corn. Your yield this year is 150 bushels per acre. You plan on selling half of your corn to a local grain mill that will pay you $3.75 per bushel. How much money will you bring in? How much corn will you have left?
2. Create a brochure encouraging students interested in agricultural careers to take math classes. Highlight all of the places where agriculture and mathematics interact. Make sure your brochure explains how important it is for agriculturalists to have a solid mathematical foundation.

Chapter 8

Review and Assessment

Lesson 8.1

Agricultural Business Principles

Key Points

- Agricultural business functions on the same principles and relies on the same mathematical calculations as other business.
- Creating a business plan allows you to develop goals for acquiring the resources and the expected market for the products of a business.
- Understanding how inventory is obtained and recorded is important to calculating the overall net worth of a business.
- By tracking income and expenses, decisions can be made about the profitability and performance of an agricultural business.
- Relying on mathematical calculations and ethics ensure that business decisions are made with the best interest of both the business owners and the consumers.

Words to Know

Use the following list and the textbook glossary to review and study the *Words to Know* from *Lesson 8.1*.

accounts payable
amortization
asset
budget
business ethics
business plan
capital
capital loan
capital purchase
capital sale
cash flow statement
cash purchase
cash sale
chattel mortgage
collateral
comparative advantage
current asset
depreciation
expense
income
interest
inventory
liability
loan
market
net worth
niche market
noncash expense
noncurrent asset
profit
return on investment (ROI)
salvage value

Check Your Understanding

Answer the following questions using the information provided in *Lesson 8.1*.

1. Why is a business plan important?
2. Why is seasonality more important in agribusiness than in other businesses?
3. Why is it important to know the estimated time investment for a business?
4. List five items that could be considered capital by a business.
5. How does depreciation work?
6. What is the difference between a chattel mortgage and a real estate mortgage?
7. What is the difference between a current and noncurrent inventory item?
8. Please list three ways that noncash expenses can occur.
9. What does a positive ROI indicate?
10. What are the benefits of having good business ethics?

Lesson 8.2

Practical Mathematics in Agriculture

Key Points

- Even though many high school students do not like mathematics, there are a wide variety of applications for math principles in agriculture. By understanding the basic concepts of arithmetic, algebra, geometry, and applied mathematics, you will be better prepared for any career in the twenty-first century, especially an agricultural career.
- Understanding the basic mathematical procedures will allow the opportunity to apply them in an agricultural setting.

Words to Know

Use the following list and the textbook glossary to review and study the *Words to Know* from *Lesson 8.2*.

acre	applied mathematics	arithmetic
algebra	area	bushel

computational mathematics
conversion
conversion factor
formula
fraction
geometry
mathematics
mean
median
mode
percentage
ratio
standard deviation
volume

Check Your Understanding

Answer the following questions using the information provided in *Lesson 8.2*.

1. What are the four basic areas of mathematics?
2. Please list three units of measurement for fluid quantities and three units of measurement for dry quantities.
3. What is one of the most important reasons to be able to convert dry weights in agriculture?
4. What are the top and bottom numbers of a fraction called?
5. Why is it important to know mathematical formulas?
6. Please list five units of linear measurement.
7. What does using algebraic principles do for agriculturalists?
8. List two reasons geometry is important to agriculture.
9. How is a percentage calculated?
10. What does the area of statistics allow agriculturalists to do?

Chapter 8 Skill Development

STEM and Academic Activities

1. **Science.** There are many different applications for mathematics in science. Choose three science careers and conduct research to find the courses that someone would need to complete in order to get a degree in each area. For each of the three careers, list the math courses required along with how each class will help someone with that career be successful.
2. **Technology.** Research recordkeeping or inventory control software for small businesses. Determine which software might be best suited for your SAE. If possible, download a free trial of the software and use it to record the materials you have used in your SAE. Evaluate how well the software meets your needs. Design a recordkeeping or inventory control system, either on paper or using an electronic spreadsheet that you think would work better than the commercial one for your SAE. Use your system to enter the same information you entered using the commercial software. Which system works better for your SAE?

3. **Engineering.** Engineering is a combination of math and science principles. Contact a local engineer and ask how he or she uses mathematics every day on the job. Write a 1–2 paragraph summary highlighting the importance of math in engineering fields.
4. **Math.** Assuming a sales tax of 7.5%, calculate the sales tax and the total amount to be collected for each of the following sales.
   A. $185.62
   B. $223.95
   C. $179.04
   D. $56.47
   E. $1,108.52
5. **Social Science.** Research the sales tax in your area and find out how much of the tax is state sales tax and how much is levied by local taxing authorities, such as the city or county. Research the history of each component of the sales tax. When was it instituted? Has the amount changed over the years? What reasons were given for any increases?
6. **Language Arts.** Agricultural business plans often have to develop a marketing plan prior to getting funding. Write up a business plan for an innovative business that you would like to start.

Communicating about Agriculture

1. **Reading and Writing.** Written communication plans are essential for business success. Using the National FFA Marketing Plan CDE as a guide, choose an agricultural business in your area and write a business plan outlining how you would help them market their business to their target market.
2. **Reading and Writing.** Select an agricultural product, and then determine how you could improve upon it to create a niche market. Write a product description outlining the comparative advantage that your new product would have.
3. **Writing and Speaking.** Using the product you created in question 2, create a 5–10 minute presentation to pitch your idea to potential investors, similar to television shows where business hopefuls share their ideas in order to secure money from wealthy investors. Be prepared to share your presentation with the class.

Extending Your Knowledge

1. Prepare a team for the Marketing Plan CDE. Using the rules and regulations from National FFA, choose a local agricultural business and complete the development of a written business plan, outlining ways that you could examine its market and product and looking for places to expand the company.
2. Take your most recent math homework and write agricultural story problems for 10 of the questions on the assignment. For a very simplified example, if your math homework included the question 2 + 2, you could write the story problem: if one producer has 2 jars of honey and his neighbor has 2 jars of honey, how many jars of honey do they have altogether?

Chapter 9

Importance of Food

Lesson 9.1
Local Food Systems

Lesson 9.2
Global Food Systems

Lesson 9.3
Maintaining a Safe Food Supply

Lesson 9.4
Animal Feeds and Feeding

©iStock/123ArtistImages

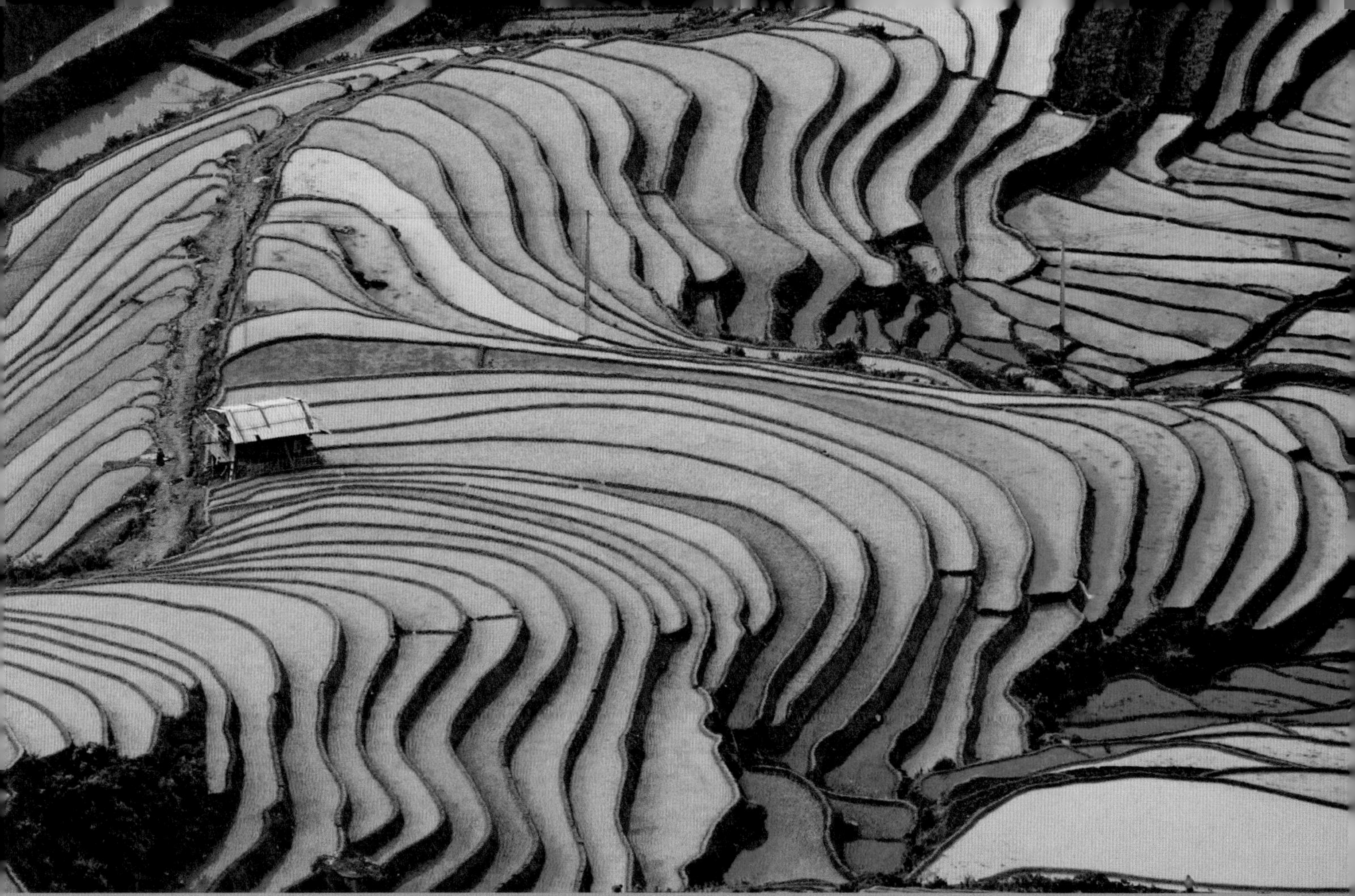

Jimmy Tran/Shutterstock.com

©iStock/Teun van den Dries

Wasu Watcharadachaphong/Shutterstock.com

While studying, look for the activity icon  **to:**

- **Practice** vocabulary terms with e-flash cards and matching activities.
- **Expand** learning with video clips, animations, and interactive activities.
- **Reinforce** what you learn by completing the end-of-lesson activities.
- **Test your knowledge** by completing the end-of-chapter questions.

Lesson 9.1

Local Food Systems

Words to Know

community garden
consumption
cooperative
disposal
distribution
exotic foods
farmers market
food miles
food system
harvesting
local food system
marketing
processing
production

Lesson Outcomes

By the end of this lesson, you should be able to:

- Define a local food system.
- Explain the advantages and disadvantages of local food systems.
- Examine methods for distributing local foods to consumers.

Before You Read

Before you read the lesson, read all of the table and photo captions. Share what you know about the material covered in this lesson just from reading the captions with a partner.

What did you eat last night for dinner? Have you ever considered how your food reaches your plate? Think about the food you have eaten recently and try to figure out where it came from. Did it come from a restaurant or grocery store? Think about it a little bit deeper. Let us say, for example, you had pepperoni pizza last night for dinner. What are the different components that make up this food? Where did each of those components come from? Take a look at **Figure 9-1**, which lists some of the most common ingredients in pepperoni pizza, along with the raw agricultural product that goes into that ingredient. How many different agricultural producers were involved in the production of your pepperoni pizza?

Ingredient	Typical Ingredient Contents
Dough (flour)	Bleached wheat flour, sugar, soy oil, salt, yeast, corn syrup solids
Pizza sauce and seasonings	Tomato puree (water, tomato paste), sugar, salt, garlic, onion, black pepper, oregano, basil, soybean or corn oil, parsley (the seasonings may be fresh or processed)
Sliced pepperoni	Pork and beef, salt, water, seasonings (the seasonings may be fresh or processed)

Goodheart-Willcox Publisher

Figure 9-1. The number of common ingredients in a pepperoni pizza may surprise you. Make a list of the ingredients and carefully evaluate each ingredient to determine how many producers were involved in its production. Do not forget about transportation and fuel.

In the United States, we are blessed with the safest and most abundant food supply in the world. A ***food system*** is the collaboration between producers, processors, distributors, consumers, and waste managers related to agriculture in a particular area. The size of the food system depends on the size of the community. In individual neighborhoods or small rural communities, the food system may be relatively small. In larger suburban and urban neighborhoods, the food system may be much larger. Food systems that are part of cities, counties, and states would be fairly large. A ***local food system*** is the systematic production of food that is consumed in close proximity to its origin, **Figure 9-2**. There is some disagreement about just how close to the source the food must be consumed to qualify. According to federal legislation, any food consumed within 400 miles of where it was produced can be considered local food.

James Horning/Shutterstock.com

Figure 9-2. Local food systems allow consumers to use food products that have not been transported more than 400 miles from the farm. *Have you ever bought food from a roadside stand?*

In this lesson, we will cover the cycle of local food system production and the factors surrounding the use of local food systems. We will also explore several specific types of local food systems and learn how these systems play into agriculture as a whole.

Food System Production Chain

In order for food to get to your table, it must make its way from a raw agricultural product through production channels before it reaches you, the final consumer. These channels form a production chain composed of seven basic steps.

Step 1: ***Production***—the act of growing a crop or raising animals for food. In this stage, the crop is cultivated or the animal is raised until the optimal harvesting stage.

Step 2: ***Harvesting***—the act of gathering raw agricultural products for consumer use. This step may be done by hand, but most often is done with the help of equipment.

Step 3: ***Processing***—the conversion of a raw product into a product ready for human consumption. Some foods require less processing than others. An apple, for example, may only be washed and packaged in this stage. However, a premade apple pie requires the apples to go through more processing stages before they are ready for human consumption.

Step 4: ***Distribution***—the transportation of products from their processing point to the place where they are available for sale. Once the products have been transported, they are marketed for sale to the public.

Step 5: ***Marketing***—includes everything that is done to convince consumers to purchase a product. Marketing takes place at the grocery store, in restaurants, on TV, and through various other media outlets.

Step 6: ***Consumption***—involves the actual use of the food product by the consumer. When you eat your piece of pepperoni pizza, you are taking your place in the production chain.

Step 7: ***Disposal***—the management of the unused portion of the product purchased for consumption. In our apple example, the waste management system that hauls away your trash (with the apple core) would be an example of this final step in the food production chain.

The production chain is basically the same for both local and global food systems. By knowing how the production cycle works, we can begin to see where local and global food systems differ. Which step in the production chain do you think is the most different between local and global food systems? Why do you think the step you selected is the most different?

The Use of Local Food Systems

In the early days of agriculture in our country, almost all food systems were local food systems. People would eat what was in season, and would rely on the items that they were able to produce for themselves, or that could be produced within traveling distance from their homes. There were several reasons people were required to eat most of their food through production in local food systems: refrigeration was not readily available, food preservation technologies were nonexistent, and there was no method of transportation for widespread distribution of food products, **Figure 9-3**.

Today, there is an increasing demand from some consumers to move toward a local food system again. This is due in part to a changing demand in the nature of U.S. food consumption. It is also influenced by the desire to decrease fossil fuel use and to establish a greater connection to the food we consume.

I. Pilon/Shutterstock.com

Figure 9-3. The various means of transportation and storage allow your local grocery store to provide foods from around the world. ***Where does your food come from?***

	2011	2012	2013
Sales ($ billion)			
At home	650.1	674.8	694.5
Away from home	589.6	623.9	653.0
Percent Change from Previous Year			
Sales			
At home	5.2	3.8	2.9
Away from home	5.8	5.8	4.7
Food sales exclude alcoholic beverages as well as home production, donations, and supplied and donated foods.			

ERS, USDA

Figure 9-4. Spending on food. In industrialized nations, food represents about 10% to 20% of consumer spending. In developing countries, food represents as much as 60% to 80% of consumer spending.

Nature of Consumption in the United States

Americans have changed the way they eat. Over the last few decades, eating habits of Americans have moved from eating mainly at home to eating mainly away from home. In 2013, the USDA economic research service estimated that Americans spent $1.35 trillion on food. To see how that food spending was broken down, look at **Figure 9-4**. About 52% ($694.5 billion) of that total was spent on home food consumption. The other 48%, or $653 billion, was spent away from home. Full-service and fast-food restaurants account for 75% of all away-from-home food sales, meaning that we are eating out, a lot. How many times this week have you eaten a meal that was purchased from a restaurant?

Did You Know?

The average U.S. adult purchases a restaurant or snack meal 5.8 times per week.

The changing nature of food consumption has led some consumer interest groups to conclude that eating out is having a negative effect on Americans' health, social well-being, and even family structure. This change has also led many consumers to desire food that is locally produced.

Decrease in the Use of Fossil Fuels

When we examined the food production chain, you were asked to think about the step in the chain that was most different between local and global food systems. Did you answer step four, distribution? If you did, you honed in on one of the biggest reasons that some consumers desire locally produced food. ***Food miles*** is the term used to describe the total distance that the ingredients of a food product travel before they reach the consumer, **Figure 9-5**. Foods

lucato/iStock/Thinkstock

Figure 9-5. How many miles do you think these onions will travel before they reach the consumer? Do you think that will increase or decrease if these onions are used to make prepared and frozen onion rings?

STEM Connection Calculating Food Miles

The food miles for a food product is the number of miles the ingredients for that food travels before reaching the consumer. Using online resources, find the closest place to your location for production of the ingredients for the following foods. Determine how many food miles the food product has for consumption in your area.

- Apples
- Bananas
- Pepperoni pizza (make sure you break down this food into all of its raw ingredients)

(left to right) Samokhin/iStock; Bedrin-Alexander/iStock; Boarding1Now/iStock

Did You Know?

Although corn is grown in all 50 states, only one U.S. state has the right climate and soil structure to grow coffee—Hawaii.

with lower food miles are produced using less transportation costs, and are therefore able to reach consumers with less fossil fuel use.

Exotic Foods

There are two types of foods that have particularly high food miles: exotic food and foods that are eaten out-of-season. ***Exotic foods*** are foods that cannot be produced in the area where they are consumed. Can you think of some examples of exotic foods? Tropical fruits, coffee, and cocoa are some of the most common exotic foods that are brought into the United States, **Figure 9-6**. These foods can travel thousands of miles before they reach your dinner plate and are part of the larger global food system that we will examine more in *Lesson 9.2*.

Out-of-Season Foods

Food that is consumed out-of-season is the other category of food with particularly high food mileage. Have you ever noticed your local grocery store has fresh vegetables in the middle of winter? In order to meet consumer demands, produce is shipped from regions where it is in season to areas where it is out-of-season. This provides consumers access to produce that is not easily stored like watermelon, sweet corn, and bell peppers at times of the year when they cannot be grown locally. Proponents of local food systems note that consuming foods when they are in season allows consumers to purchase the food without the added time required for transportation, and often allows the consumer to eat the food when it is at peak ripeness.

LauraKick/Shutterstock.com

Denis Tabler/Shutterstock.com

joannaunak/iStock/Thinkstock

elxeneize/Shutterstock.com

Figure 9-6. Foods that are grown in tropical climates are commonly imported and pose a drawback to local food systems.

Closer Connection to Agricultural Producers

Another reason consumers prefer food from a local food system is that a local food system is more likely to allow for a personal connection to their food. Purchasing food products that have been grown in another state or country, then processed by another company, and transported to a grocery store allows little opportunity for consumers to have a personal connection with the farmers and ranchers who produced the food, **Figure 9-7**.

Edler von Rabenstein/Shutterstock.com

Figure 9-7. Many consumers want to be able to put a face with the agricultural producers who have grown what they will be eating. ***Do you have a personal connection to the producers who grow your food?***

Have you ever been involved in growing or raising some of your own food? If so, you can understand the pride that comes with supplying your family with some of the nutrients they need. You also understand the feeling of knowing exactly what has happened to your food at every step before you ate it. If you have not had the opportunity to grow or raise your own food, have you ever thought about who is raising and processing your food? A local food system gives consumers greater opportunity to know who was involved in the production of their food along with the knowledge of where and how the food was produced. Consumer desire to be closely connected to their food leads to many local food systems that involve the consumers in all aspects of the production chain.

Drawbacks of Local Food Systems

As you can see, there are many positive aspects to local food systems. Can you think of drawbacks to a local food system? Drawbacks to local food systems are driven by geographic region, economic viability, and consumer demands.

Geographic Limitations

In colonial times, most of the population settled in areas where conditions were favorable for agricultural production. Today, many people live in areas where there is not enough agricultural land to support food production for the total population. The geographic drawback of local food systems is that it simply is not possible for some Americans to have all their desired foodstuffs produced by local producers, **Figure 9-8**.

Production and Shipping Costs

There are also economic drawbacks to local food systems. Although the cost of transportation for local foods is lower, small-scale producers are often less able to produce large quantities of food at a low cost. Large-scale producers growing and raising food in an area with optimal resources are usually able to produce more food at a lower cost. Even with the shipping costs factored in, it is sometimes less expensive than locally produced food.

JaysonPhotography/Shutterstock.com

Figure 9-8. Some Americans live in places that are not within 400 miles of agricultural production areas large enough to feed the entire population. *Do you see much agricultural production in this scene?*

Consumer Demand

Another drawback to local food systems is that some consumer demands cannot be met through local food systems alone. American consumers desire exotic and out-of-season foods, and are willing to support the global food systems that bring these foods to their table. Simply explaining the benefits of local food systems will not likely convince most Americans to give up products such as coffee, bananas, and cocoa.

Examples of Local Food Systems

There are many different ways that local foods reach consumers. For those who maintain their own garden, obtaining local food is literally as simple as walking outside. For livestock producers, obtaining locally grown meat is simply a matter of taking an animal to the local processing plant. The majority of Americans do not have their own garden to obtain fresh produce or livestock operations for meat, milk, and eggs. These consumers must rely on other nearby producers if they want to obtain local food products. Some common local food distribution systems that allow local products to reach consumers include farmers markets, community gardens, and agricultural consumer cooperatives.

Farmers Markets

A ***farmers market*** is a place where groups of agricultural producers gather to market their products directly to consumers. The concept of farmers markets is not new; in fact, street vendors are the most common type of retail sales outlets in the world, **Figure 9-9**. There are some farmers markets that

danielvfung/iStock/Thinkstock

Figure 9-9. Street markets, like this one in Hong Kong, are common throughout the world as a way to distribute raw agricultural products. They are becoming popular in the United States in the form of farmers markets.

operate from a set location. However, most farmers markets are seasonal and located in common community gathering places like downtown streets or squares and community parks. Check your local community guide to find out if there is a farmers market scheduled near you.

Community Gardens

People who do not have the chance to grow their own garden because of space, time, or knowledge limitations may choose to participate in a community gardening program. ***Community gardens*** are plots of land gardened collectively by community members. There are community gardens at schools, on rooftops in metropolitan areas, and on empty plots of land all around the country, **Figure 9-10**. There are even entire communities designed around the concept of maintaining a community garden large enough to grow enough produce for the neighborhood residents. Community gardens allow those who might not have experience in agricultural production to gain a hands-on appreciation for what goes on in the production of local food products.

Agricultural Consumer Cooperatives

Community gardens and farmers markets are great ways for consumers to gain access to locally produced fruits and vegetables, along with processed products like jams, jellies, and bread. What about meat, dairy, and egg

juripozzi/iStock/Thinkstock

Figure 9-10. Community gardens allow people to produce food even if they are not able to own their own land. *Can you think of some other advantages to a community garden?*

Career Connection Community Garden Program Manager

Job description: Community garden managers often work for nonprofit organizations to develop and build garden plots, manage planting and volunteer schedules, and coordinate the harvest of garden crops. They also may be involved in community outreach programs to discuss the importance of supporting local food systems.

Education required: Education in organizational leadership or community development is helpful if you want to pursue this career path.

Job fit: This job may be a fit for you if you enjoy working outdoors, like gardening, and have a desire to share your agricultural knowledge with others.

huePhotography/iStock

products? One of the most common solutions to allowing consumer access to these types of goods is through participation in an agricultural consumer cooperative. A ***cooperative*** is a farm or business that is owned and run jointly by its members. Agricultural consumer cooperatives allow consumers to buy into the agricultural business, and then receive products in exchange for their contributions, **Figure 9-11**.

Consumer cooperatives allow members to have a personal connection to local food sources. This type of local food system is common with meat products, dairy products, eggs, and is also commonly used with fruits and vegetables. One of the advantages for producers is that the consumers will often pay a monthly contribution whether they use their share of the local foods or not. For consumers, being part of a consumer cooperative allows them the assurance that they will have consistent access to quality food items.

lola1960/iStock/Thinkstock

Figure 9-11. A common type of agricultural consumer cooperative allows members to pick up a basket of in-season, locally grown fruits and vegetables every week, and exposes members to sometimes unfamiliar produce. *Can you name all of the produce in this sample basket?*

Lesson 9.1 Review and Assessment

Words to Know

Match the key terms from the lesson to the correct definition.

1. The actual use of the food product by the consumer.
2. Everything that is done to convince consumers to purchase a product.
3. The management of the unused portion of the product purchased for consumption.
4. The systematic production of food that is consumed in close proximity to its origin.
5. The transportation of products from their processing point to the place where they are available for sale.
6. The total distance that the ingredients of a food product travel before they reach the consumer.
7. A farm or business that is owned and run jointly by its members.
8. The act of growing a crop or raising an animal for food.
9. Foods that are not able to be produced in the area where they are consumed.
10. The collaboration between producers, processors, distributors, consumers, and waste managers related to agriculture in a particular area.
11. The conversion of a raw product into a product ready for human consumption.
12. A place where groups of agricultural producers gather together to market their products directly to consumers.
13. The act of gathering raw agricultural products for consumer use.
14. Plots of land that are gardened collectively by community members.

A. community garden
B. consumption
C. cooperative
D. disposal
E. distribution
F. exotic foods
G. farmers market
H. food miles
I. food system
J. harvesting
K. local food system
L. marketing
M. processing
N. production

Know and Understand

Answer the following questions using the information provided in this lesson.

1. Please list two reasons that consumers may desire food from local food systems.
2. Which food product is likely to have more food miles, an apple or a banana? Explain why.
3. List the steps in the food production cycle.
4. Give four examples of exotic foods.
5. Who owns a consumer cooperative?
6. Give three examples of local food production systems.

7. What are three drawbacks to local food production systems?
8. Which step in the food production cycle has the most differences between a local food system and a global food system? Explain how this step is different in each food system.
9. How many miles can food be produced from the consumer and still be considered local, according to legislation?
10. In which stage of the production cycle is every human who eats food involved?

Analyze and Apply

1. Think about your two favorite foods. List them, and then explain which one you think has the lowest food miles. Make sure you account for all of the ingredients in the food.
2. Conduct research and compile a list of local food distribution outlets in your area. Include in your list the contact information for the program director.

Thinking Critically

1. With a partner, develop your own business plan for a local food distribution system. Make sure you include how you will market your business to local consumers.
2. There are proponents of moving to a completely localized food system, much like the early American settlers had. What are your thoughts on this topic? Do you think the move toward local food systems would be good for agriculture? Write 1–2 paragraphs outlining your position.

Lesson 9.2

Global Food Systems

Words to Know

arable land
demand
export
foodborne illness
food desert
food insecurity
global food system
hunger
import
malnourishment
spoilage
staple crop
supply
tariff
trade surplus
urbanization
world hunger

Lesson Outcomes

By the end of this lesson, you should be able to:

- Define a global food system.
- Explain the factors surrounding the challenge of feeding the global population.
- Discuss the impacts of food production and distribution practices on world hunger.

Before You Read

Pair with a partner and discuss what you know about global food systems. Write down four questions that you have about global food systems before reading the lesson. After reading, discuss your questions with your partner and see if you can answer them. For any questions that remain, have a class discussion, and allow your classmates and agricultural science teacher to give their thoughts on the answers.

Look at a clock so that you can time yourself for 30 seconds. Now, take a full 30 seconds and think about the role that agriculture plays in feeding the world. You may find it interesting that in the 30 seconds you were contemplating agriculture's role, experts estimate that the global population increased by about 75 people. That is right, current population growth projections lead to an increase of 150 people per minute, 9,000 people per hour, 216,000 per day, 6,480,000 per month, and 77,760,000 people per year. With already well over 7 billion people to feed on the planet, feeding the world is a real challenge for the future of agriculture, **Figure 9-12**.

In this lesson, we will examine how global food production works. A ***global food system*** is the systematic production of food for distribution to consumers around the globe. We will examine the factors affecting global food systems and look at where major commodities are produced. Then, we will take a look at how global food systems work together to gain a better understanding of global hunger, and finally, examine food distribution so that you can gain a better understanding of the importance of an efficient global food system.

Factors Affecting Food Systems

Have you lived in another country? If you have, you may have noticed that the food systems for that country were different than the food system in the United States. Depending on the country, that system may have been only slightly different from what you were used to in this country, or may have been a drastic change from what most Americans are used to. If you have never lived out of the United States, you might be surprised to find that there are many places in the world where the concept of a single grocery store that sells all different food products simply does not exist, even in large metropolitan areas, **Figure 9-13**.

The truth is that food systems are different in different areas of our world, and all of these food systems working together make up the larger global food system. According to the Global Food Security Index, the United States is the most food secure country in the world. Before we can talk about the importance of the global food systems, we should first examine some of the factors that differ in food systems around the globe. The main factors affecting the global food system are the affordability, availability, and safety of the food that is produced.

Global Population Milestones

Population	Year
1 billion	1804
2 billion	1927
3 billion	1960
4 billion	1974
5 billion	1987
6 billion	1999
7 billion	2012
8 billion (projected)	2026
9 billion (projected)	2042
10 billion (projected)	2082

EPI fromUNPop

Figure 9-12. The global population is estimated to reach more than 10 billion people before the beginning of the next century.

Affordability of Food Products

The amount of money that people spend on food is drastically different in many food production systems. Americans spend the smallest percent of their income on food of any other country, with only 6.6% of the annual income spent on food costs, on average. The countries with the highest and lowest percent of

Iakov Filimonov/Shutterstock.com

Figure 9-13. It is common throughout the world to find stores that specialize in only one or two types of food. For example, this European meat market sells mainly cured, smoked, and cooked meats.

Country/Territory	Percentage of Income Spent on Food	Country/Territory	Percentage of Income Spent on Food
United States	6.7	Kuwait	18.5
Singapore	6.7	South Africa	19.2
Switzerland	8.9	Turkey	22.0
United Kingdom	9.3	Iran	25.0
Canada	9.5	Mexico	25.1
Australia	10.0	China	26.1
Ireland	10.4	India	29.6
Germany	12.0	Russia	30.5
Sweden	12.2	Vietnam	35.5
Japan	13.6	Peru	36.5
Taiwan	13.6	Guatemala	40.1
France	13.8	Philippines	42.4
Spain	13.8	Algeria	42.6
Italy	14.1	Kazakhstan	43.5
Brazil	15.7	Kenya	46.9
Greece	16.6	Pakistan	48.1
Poland	17.7	Nigeria	56.7

ERS, USDA

Figure 9-14. Percent of income spent on food.

annual household income spent on food are shown in **Figure 9-14**. How do you think the food system is different in these two groups?

Understanding the amount of income spent on food is not only important for food systems, but for the economy as well. In countries where there is a lower percentage of money spent on food, there is more money left to spend on housing, transportation, and nonessential items like recreational pursuits. As a general rule, countries that have lower overall income levels and provide less federal assistance to production agriculture have higher percentages of income spent on food products, **Figure 9-15**.

Availability of Food Products

Certain regions of the globe are not able to gain access to the types of food that they desire simply because these foods are not grown widely in their region. The five countries having the highest and lowest rankings for food availability are shown in **Figure 9-16**. The availability of food products is controlled by several factors, including how many agricultural commodities are produced, food loss from spoilage, and political corruption affecting the food system.

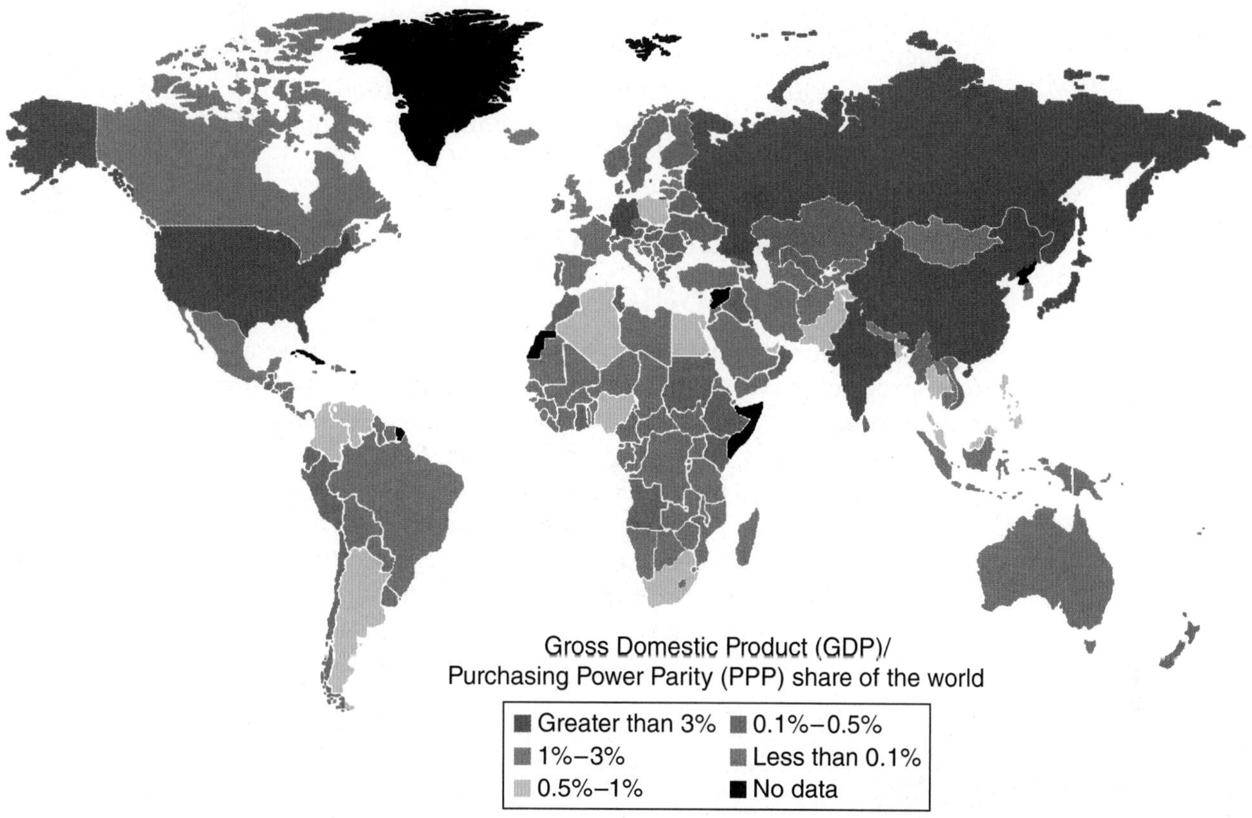

Share of World GDP PPP 2014, IMF by Digital Omniscience

Figure 9-15. It is clear that certain areas of the world have more poverty than others. A higher percentage of the income is spent on food in these areas.

Countries with the Highest Food Availability		Countries with the Lowest Food Availability	
Rank	**Country**	**Rank**	**Country**
1	United States	109	Chad
2	Austria	108	Burundi
3	Netherlands	107	Congo
4	Germany	106	Niger
5	Switzerland	105	Sudan

Data adapted from the Global Food Security index, 2015 evaluation of 109 reporting countries

Figure 9-16. Food availability varies by country.

Imagine for a moment you went to the grocery store and more than half of the shelves were empty. How would that affect the way you ate? Would you modify your diet in order to make use of the food that was available? For people in many countries, a lack of access to food has dramatically altered their eating habits, leading to widespread malnourishment and starvation.

You might be surprised to find that while the United States has the most available food supply on Earth, the future of food availability in our country is in danger. Our expansion of agricultural production is dwindling. This is due largely in part to an increase in ***urbanization***, which takes agricultural land out of production.

Food Product Regulations, Quality, and Safety

Many countries have their own regulatory agencies that regulate the food products that consumers purchase. In the United States, there are more regulations related to the production, processing, and distribution of food products than there are in any other nation. Most of the regulation in our country is managed by the USDA, the Food and Drug Administration (FDA), and the Food Safety and Inspection Service (FSIS). These agencies work together to ensure that food products are plentiful, high-quality, and safe for consumers. In fact, the careful attention to detail and diligent efforts to secure food safety by these agencies lead to the United States having the highest food safety ranking for all food systems in the world.

When food quality and food safety are evaluated, the United States ranks fifth in the world, behind Israel, France, Portugal, and Greece. The U.S. food quality ranking is lowered due largely in part to the lack of vegetal iron (iron which enters plants from iron-rich soil) in the average U.S. diet.

A lack of regulation for food products can lead to an unsafe food supply. ***Foodborne illnesses*** are ailments caused by improper handling, storage, or processing of food products. Have you ever experienced food poisoning? It is definitely not fun. Luckily for those who get ill from food in the United States, most of the cases of food poisoning are due to improper handling of food products during preparation and are easily treated. In developing countries, millions of people die annually from a lack of proper food inspection procedures, which can lead to widespread outbreaks of foodborne illness, **Figure 9-17**.

Image Point Fr/Shutterstock.com

Figure 9-17. Foodborne illness affects many people on a daily basis. *What precautions do you take at home to prevent food poisoning?*

Major Commodity Production

As we discussed in *Lesson 9.1*, for many years, people were limited to a local food system to supply their food needs. This was due to a lack of transportation that limited their ability to get food from other places on the planet. As early as 3000 BCE, humans expressed a desire to expand their food systems to include food

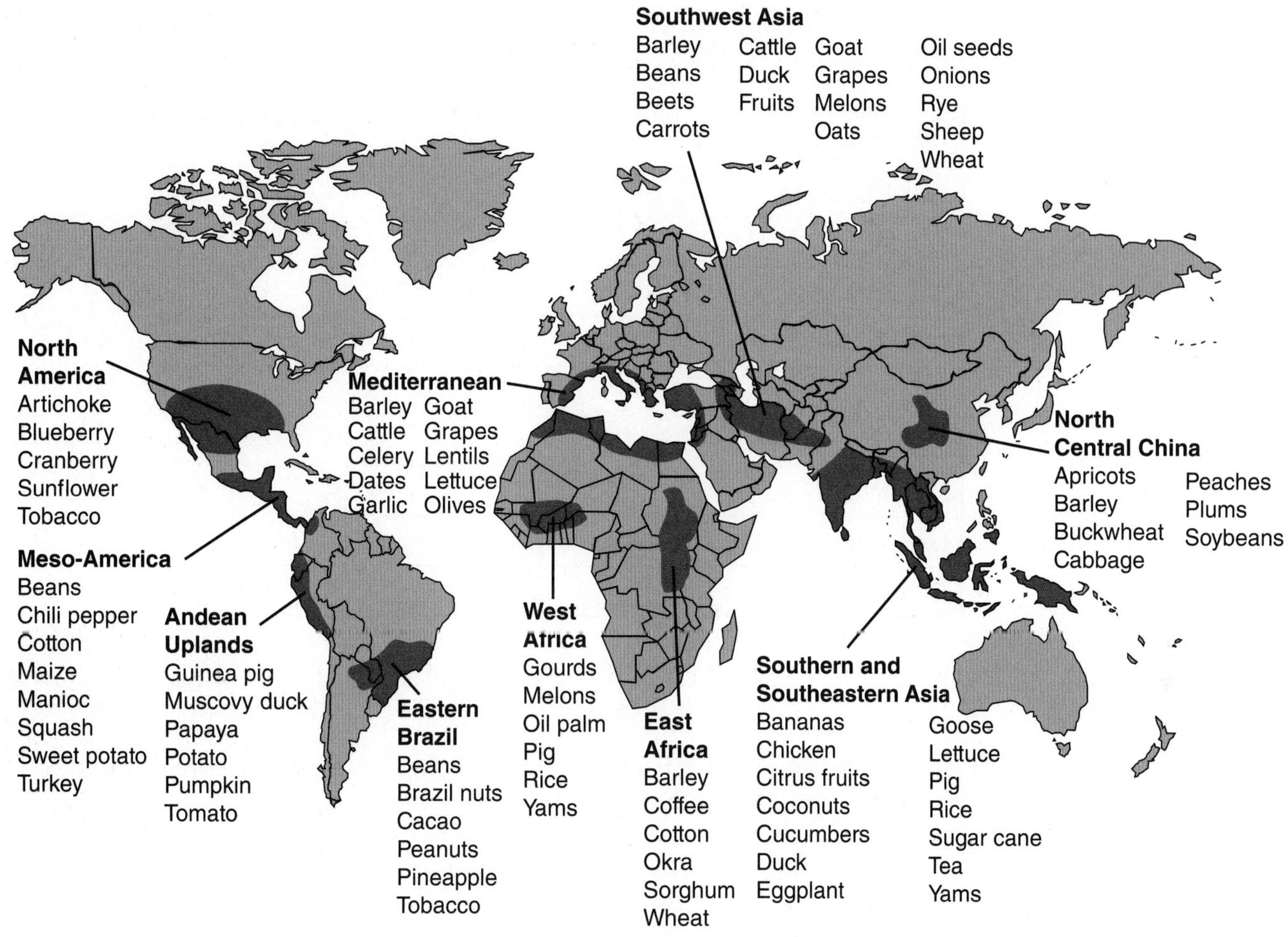

Figure 9-18. Areas where major agricultural commodities were first cultivated.

products that were not locally grown. In fact, the spice trade between Asia and Europe was one of the driving factors behind the "Age of Discovery," a historical time beginning in the fifteenth century that ultimately led global explorers like Christopher Columbus to expand their previous knowledge of the existing world.

Since that time, consumers have been able to sample food products from around the globe. The area where major agricultural commodities were first cultivated is shown in **Figure 9-18**. You can find commodities being cultivated today in areas that are thousands of miles away from their historical origin.

Did You Know?

While Christopher Columbus originally set sail to find an easier route to obtain spices and silk from Asia, he ended up expanding Europeans' diet by bringing back foods like corn, tomato, vanilla, tobacco, squash, cacao (chocolate), peanuts, and blueberries.

World Hunger

Did you eat something last night? Do you know there is food waiting for you at home when you get there this evening? ***Hunger*** is the scarcity of food in a particular region. ***World hunger*** is the combined scarcity of food for all countries in the world. Hunger leads to ***malnourishment***, which is the term used for a person not receiving the proper nutrients to sustain life. If you answered yes to the questions we just read, then you are not part of the more than 805 million people

	1990–1992	2012–2014
World	18.7%	11.3%
Developed regions	<5%	<5%
Developing regions	23.4%	14.5%
Africa	27.7%	20.5%
Sub-Saharan Africa	33.3%	23.8%
Asia	23.7%	12.7%
Eastern Asia	23.2%	10.8%
Southeastern Asia	30.7%	10.3%
Southern Asia	24.0%	15.8%
Latin America and Caribbean	15.3%	6.1%
Oceana	15.7%	14.0%

FAO The State of Food Insecurity in the World 2014

Figure 9-19. Percentage of malnourishment across the world.

worldwide who suffer from hunger and chronic malnourishment. Based on the numbers from the United Nations, one in nine people worldwide is chronically malnourished, a ratio that increases to one in eight when we only consider developing countries. **Figure 9-19** shows the percentage of malnourished people in each country. Which area seems to be the most critical? There are several issues that we need to examine in order to gain a more complete understanding of the scope of world hunger.

Global Population Growth

The need for more food production puts added pressure on the global battle to combat hunger. The fact of the matter is, based on the current rate of population growth, the agricultural industry will need to produce more food in the next 40 years than has been consumed in the entire history of humanity to this point. That is a huge task, especially considering that there is a decrease in the amount of agricultural land available each year, and that the percentage of the population involved as production agriculturalists is dwindling.

Innovations in agriculture will be the key to being able to keep up with the growing number of mouths that need to be fed. Scientists are working with agriculturalists to help produce varieties of crops that will grow faster, produce more, and allow for cultivating areas that we have previously not been able to use for agricultural production. As more and more people join the planet, it will become important to think about how to grow enough food to sustain them. Feeding the world in the future will require that agricultural production increases at exponential levels, and that we are able to produce more food for more people on less ***arable land*** (land that is suitable for growing crops).

Global Food Distribution

Did you know that even with the severity of world hunger, the agricultural industry currently produces enough food to feed the world population? At current rates of production, there is enough food produced annually to supply every person on the planet with around 3000 calories worth of food every day. Why, then, do you think that world hunger is still an issue? Did your answer involve something to do with the distribution of food? If so, you are right on track.

The challenge of solving world hunger is not only about producing more food to meet the demands of population growth, it is about getting food to the places that need food the most. The most critically malnourished areas in the world, including Sub-Saharan Africa and Southeastern Asia, are in areas where resources to produce sustainable amounts of food are limited. Solving the problem of world hunger means that we need to find a solution to get sustainable food sources into areas where resources are scarce.

Other factors that impede food distribution are conflict, political pressures, and poorly structured economic systems that prevent food from being equitably distributed to everyone in a particular region.

Conflict is both a cause and effect of hunger. People who do not have enough food are often more likely to resort to extreme measures in order to protect their ability to get food. During times of conflict, countries are less likely to focus on agricultural production, leading to a period of food shortage both during the conflict and once the hostility is over. These countries become more dependent on food products brought into the region.

Did You Know?

Some of the largest groups of malnourished people are living in refugee camps where they have sought safety from conflict in their region.

Imports and Exports

Food that is transported outside of a country has its own set of rules and regulations. When food is brought into a country, it is considered an ***import***, the term for a commodity brought into a country for consumption. Imported products are subject to the governmental regulations of the importing country, and are subject to tariffs. A ***tariff*** is a tax charged by an importing country on imported goods. Tariffs are an important factor in global food systems because they can drastically change the cost of imported goods.

A commodity that is shipped to another country for consumption is considered an ***export***. Like tariffs, countries can impose an export tax on the products leaving their country. These taxes can be used to discourage producers from exporting certain products. The global issue surrounding tariffs and export taxes can play a large role in global hunger, as both increase the cost of the product reaching the intended consumer.

When a country exports more than they import, the country has a ***trade surplus***. Countries that do not have a food trade surplus do not produce enough food to feed their population and are subject to global markets to supply their food products. This leaves those countries dependent on other countries to set their food prices. You can get an idea of how global food trade surpluses work by looking at the global grain trade shown in **Figure 9-20**.

Top 10 Net Exporters of Corn, Wheat, Rice, and Total Grain

Corn			Wheat		
Rank	Country	Quantity	Rank	Country	Quantity
		Million Tons			Million Tons
1	United States	21.6	1	United States	25.0
2	Argentina	19.5	2	Canada	18.1
3	Brazil	16.7	3	Australia	16.4
4	Ukraine	12.5	4	European Union	12.0
5	India	3.0	5	Russia	9.0
6	South Africa	2.5	6	Kazakhstan	7.0
7	Russia	2.1	7	India	6.5
8	Paraguay	2.1	8	Ukraine	6.1
9	Canada	1.0	9	Argentina	5.0
10	Zambia	0.5	10	Uruguay	1.0
Rice			**Total Grain**		
Rank	Country	Quantity	Rank	Country	Quantity
		Million Tons			Million Tons
1	India	10.4	1	United States	49.0
2	Vietnam	7.6	2	Argentina	32.0
3	Thailand	6.3	3	Australia	21.9
4	Pakistan	3.4	4	Canada	21.9
5	United States	2.6	5	Ukraine	20.7
6	Uruguay	1.1	6	India	17.5
7	Cambodia	0.8	7	Russia	12.9
8	Burma	0.7	8	Brazil	10.4
9	Argentina	0.7	9	Kazakhstan	7.2
10	Australia	0.3	10	European Union	7.0

Total grain includes barley, corn, millet, mixed grain, oats, rice, rye, sorghum, and wheat. Compiled by Earth Policy Institute from U.S. Department of Agriculture, *Production, Supply, & Distribution* (2012 Exports/2013 Imports)

USDA

Figure 9-20. Global grain trade.

Top 10 Net Importers of Corn, Wheat, Rice, and Total Grain

Corn			Wheat		
Rank	Country	Quantity	Rank	Country	Quantity
		Million Tons			Million Tons
1	Japan	15.0	1	Egypt	9.3
2	Mexico	8.9	2	Indonesia	6.3
3	South Korea	8.0	3	Brazil	6.0
4	European Union	7.0	4	Japan	5.6
5	Egypt	5.5	5	Algeria	5.2
6	Taiwan	4.3	6	Morocco	4.4
7	Colombia	3.5	7	South Korea	4.3
8	Iran	3.5	8	Iraq	3.7
9	Malaysia	3.1	9	Iran	3.6
10	Algeria	2.8	10	Mexico	3.4
Rice			**Total Grain**		
Rank	Country	Quantity	Rank	Country	Quantity
		Million Tons			Million Tons
1	Nigeria	3.2	1	Japan	24.0
2	Iran	1.8	2	Mexico	14.7
3	Indonesia	1.7	3	Egypt	14.2
4	Philippines	1.5	4	South Korea	12.9
5	Cote d'Ivoire	1.4	5	Saudi Arabia	12.6
6	China	1.4	6	Iran	10.0
7	Iraq	1.3	7	Indonesia	9.3
8	Senegal	1.2	8	Algeria	8.2
9	Saudi Arabia	1.1	9	China	8.1
10	European Union	1.1	10	Morocco	7.1

Total grain includes barley, corn, millet, mixed grain, oats, rice, rye, sorghum, and wheat. Compiled by Earth Policy Institute from U.S. Department of Agriculture, *Production, Supply, & Distribution* (2012 Exports/2013 Imports)

USDA

Figure 9-20. (*Continued*)

STEM Connection Profile of the Man Who Saved a Billion Lives

In 1970, a humble plant breeder who grew up on a farm in Iowa won the Nobel Peace Prize. Norman Borlaug was instrumental in beginning the technological changes to the way that food crops are produced. His belief was that through proper plant breeding and management to increase yields and potential in crops, the issue of world hunger could be addressed.

Borlaug is known as the Father of the Green Revolution and relied on scientific experimentation and careful selection of plant genetics to create a strain of wheat that would flourish in the then malnourished regions of India and Pakistan. Borlaug believed in a systematic approach to food production that allowed local farmers to become self-sufficient, using high-quality crops and the most advanced management practices. Because of his efforts, it was said that he started a movement in the 1960s that likely saved more than a billion people from starvation.

What do you think we can learn about the challenges of today's hunger issues from Dr. Borlaug?

Supply and Demand

In *Lesson 9.1*, we learned about the food production chain. This chain is essentially the same for both local and global food systems. Let us review the seven steps in the food production chain:

Step 1: Production

Step 2: Harvesting

Step 3: Processing

Step 4: Distribution

Step 5: Marketing

Step 6: Consumption

Step 7: Disposal

These seven steps illustrate the basic overview of the food system, but to understand the global food systems and how they differ around the world, there is more information we need to know. We know what happens in the global food chain, but we also need to understand what drives the production of food and other agricultural products. The simple answer to what makes the food production chain work? YOU.

Consumer demands drive the production of food and agricultural products. After all, if there is no one who wants to purchase the product, there is no reason to produce it. ***Demand*** is the measure of consumers' desire for a particular product. The factor that balances demand in an agricultural production system is supply. ***Supply*** is the amount of a specific product that is available at a particular price point. The difference between supply and demand is an important factor that plays a role in the production of food products.

Something important to consider when looking at supply and demand are the staple crops for people in a particular area. ***Staple crops*** are the plants that make up the majority of the calories consumed by people in a region. For example, in many of the Asian countries, rice is a main staple crop; for those in many African nations, the staple crop is the sweet potato. Understanding the staple crops of a region can play a key role in helping world hunger.

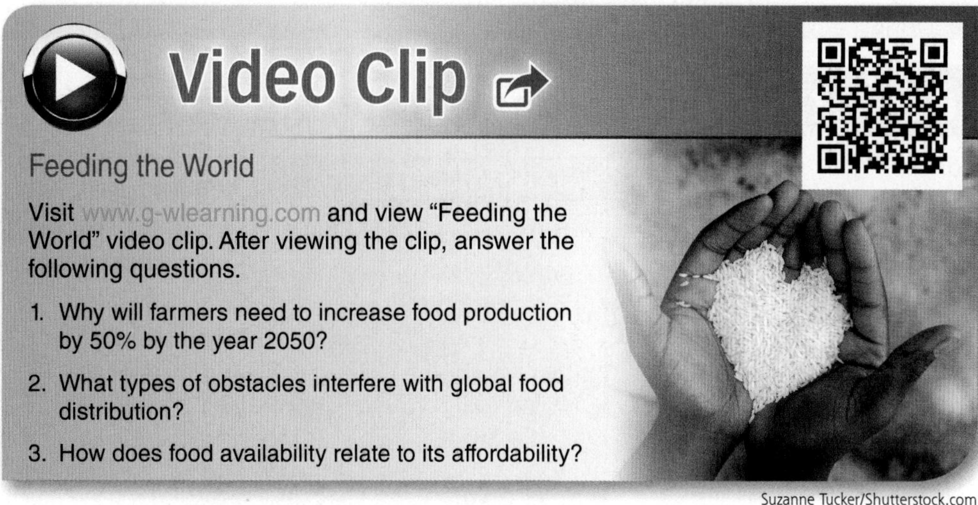

Suzanne Tucker/Shutterstock.com

How does global supply and demand play a role in world hunger? Areas of the globe that rely on importing their foods are commonly found in regions with the lowest annual income. In order to solve global hunger on a permanent level, supply and demand would need to be shifted to be primarily driven by dietary demands, not economic factors. Although strides have been taken by relief organizations and governments around the world to help provide food to malnourished populations, there is still a long way to go.

Transporting Food

Transporting food products has different challenges than transporting other goods. To start with, there are issues related to spoilage. ***Spoilage*** is the amount of a food product that cannot be consumed because of a decrease in quality or safety during storage. Most agricultural products that are transported are perishable, and a certain amount of spoilage is expected during transportation, **Figure 9-21**. Minimizing spoilage is important, especially for shipments of food to areas with high percentages of undernourished people. Proper climate controls, expedited shipping, and proper packaging can all help to reduce spoilage of food products during transportation.

There is also competition for commodities that need to be shipped, especially in areas where railways are used to transport goods. Often, lower-priced agricultural goods are not shipped in favor of more profitable commodities, like fossil fuels. Expediting the shipment of agricultural goods can help products reach consumers at peak quality.

©iStock/IMNATURE

Figure 9-21. A certain amount of spoilage is expected during transportation. For example, these oranges grow in a hot climate and are warm when they are picked. Unless they are being shipped locally for fresh consumption or processing, they would be cooled down quickly after picking to reduce spoilage.

Career Connection International Shipping Specialist

Shipping the food to the world

Dmytro Vietrov/Shutterstock.com

Job description: International shipping specialists work to advise companies about import and export requirements, coordinate worldwide shipping, prepare shipping documents, and arrange for proper product packaging and labeling.

Education required: Education in international business or logistics is helpful for someone in international shipping. Speaking a second language is also a benefit.

Job fit: This job may be a fit for you if you enjoy working with a global workforce, are able to manage and organize paperwork, and desire a fast-paced work environment.

©iStock/Snyderdf

Figure 9-22. Food insecurity affects people around the world. *Does your school or local youth group hold food drives to help local families dealing with food insecurity? Are you a member?*

The Local Issue of World Hunger

Food distribution is not just a problem in developing countries. There are those in developed countries, including the United States, who lack easy access to food. For Americans living below the poverty line, hunger is a real threat. Some experts estimate that as many as one in five American families are living with food insecurity. ***Food insecurity*** is a situation in which consistent food is unavailable year-round due to a lack of money or resources, **Figure 9-22**.

A factor leading to the issue of hunger in developed countries is that many people in these countries are living in food deserts. A ***food desert*** is an area with low income levels, where more than 500 people live more than one mile from the nearest grocery store. People living in food deserts are often living below the poverty line, lack their own transportation, and therefore have limited access to purchasing groceries. For these individuals, gaining access to consistent food is a daily challenge.

Solving the Issue of World Hunger

You may be thinking that world hunger is such a huge problem that there is nothing you can do to help, but in order to help feed everyone on the planet, it will require the efforts of many. So, how can you help make sure that the global food system runs efficiently? Here are a few ideas:

- Become an informed consumer. Understanding the role you play in the global food system is an important first step. Do not take media trends or claims about food products as fact without first conducting your own research into food issues.
- Support relief efforts that make a difference in the lives of the undernourished. Give your time, money, and resources to reputable relief organizations that are making a real impact on the lives of others. Do your homework, and make sure that you know exactly where your donations will go.
- Consider becoming involved in global food policy. The world needs dedicated policymakers who understand agriculture to help write the policy that will help feed the hungry, both in our country and around the globe.
- Volunteer to provide aid. Many organizations are looking for young people who want to help on a personal level. Research volunteer opportunities and consider spending some time working for a hunger relief organization, **Figure 9-23**.

Monkey Business Images/Shutterstock.com

Figure 9-23. Most hunger relief organizations rely heavily on volunteers to help collect and distribute food to those in need. *How can you help?*

Lesson 9.2 Review and Assessment

Words to Know

Match the key terms from the lesson to the correct definition.

1. The combined scarcity of food for all countries in the world.
2. An area with low income levels, where more than 500 people live more than one mile from the nearest grocery store.
3. The state in which a country exports more than they import.
4. A commodity brought into a country for consumption.
5. A situation in which consistent food is unavailable year-round due to a lack of money or resources.
6. The measure of consumers' desire for a particular product.
7. The amount of a food product that cannot be consumed because of a decrease in quality or safety during storage.
8. The systematic production of food for distribution to consumers around the globe.
9. A tax charged by an importing country on imported goods.
10. A commodity that is shipped to another country for consumption.
11. The scarcity of food in a particular region.
12. An ailment caused by improper handling, storage, or processing of food products.
13. When a person does not receive the proper nutrients to sustain his or her life.
14. The amount of a specific product that is available at a particular price point.

A. demand
B. export
C. foodborne illness
D. food desert
E. food insecurity
F. global food system
G. hunger
H. import
I. malnourishment
J. spoilage
K. supply
L. tariff
M. trade surplus
N. world hunger

Know and Understand

Answer the following questions using the information provided in this lesson.

1. What is the main difference between a local food system and a global food system?
2. What are the three main factors that affect a global food system?
3. Explain the difference between hunger and malnourishment.

4. If agricultural producers produce enough food for everyone on the planet, why does hunger still exist?
5. What is the difference between an import and an export?
6. Explain what a food desert is.
7. Does hunger exist only in developing nations? Explain your answer.
8. Why is shipping food products more difficult than shipping other goods?
9. What makes the global food system work?
10. Please list four things you can do to help the global food system run more smoothly.

Analyze and Apply

1. What agricultural innovation do you think will be most important in helping solve the challenge of global hunger? List your innovation along with a 1–2 sentence explanation of why you chose this particular innovation.
2. If someone came to you and said, "There is not an issue with feeding the world; we produce enough calories for everyone on the planet," what would you say to them to help them understand world hunger?
3. Explain how supply and demand of food products play a role in the global food system.

Thinking Critically

1. Imagine for a moment that you were put in charge of a charitable giving organization and were given the task to donate $100,000 to an organization that is involved with ending world hunger. Conduct research into hunger relief organizations and choose the one that you would give the money to. Write a 1–2 paragraph explanation as to why you chose this organization.
2. Create a flyer for an event that you would like to create to raise awareness of world hunger. Make sure you include the name of the event, the purpose, and any marketing slogan or facts that you feel would help others support your cause.

Lesson 9.3

Maintaining a Safe Food Supply

Words to Know

Animal and Plant Health Inspection Service (APHIS)
Centers for Disease Control and Prevention (CDC)
cross-contamination
Environmental Protection Agency (EPA)
Food and Drug Administration (FDA)
Food and Nutrition Services (FNS)
foodborne illness
food safety
Food Safety and Inspection Service (FSIS)
food security
Hazard Analysis and Critical Control Points (HACCP)
National Agricultural Library (NAL)
National Institute of Food and Agriculture (NIFA)
pathogen
shelf-stable

Lesson Outcomes

By the end of this lesson, you should be able to:

- Identify government agencies responsible for food safety.
- Describe the food safety and processing continuum.
- Understand common food safety and security issues.
- Understand Hazard Analysis and Critical Control Points (HACCP).
- Explore emerging technology in food safety.
- Explore career opportunities related to food safety.

Before You Read

Have you ever watched the news and wondered why it is such a big deal when there is a food recall? Research food safety in the United States to understand why a safe and wholesome food supply is critical to our society.

Maintaining and ensuring a safe food supply is something consumers in the United States expect. Anytime we go out to eat or purchase food in the grocery store, we expect the food we consume or purchase meets the highest standards and will be safe for consumption. The Centers for Disease Control and Prevention (CDC) estimates that one in every six people in the United States, which totals about 48 million people per year, gets sick from foodborne illnesses. A ***foodborne illness*** is an ailment caused by improper handling, storage, or processing of food products. Approximately 128,000 people are hospitalized as a result of these illnesses and 3,000 die as a result of the illness. There are more than 250 types of pathogens that can cause foodborne illness. A ***pathogen*** is an infectious or biological agent that causes disease or illness to its host.

Food safety and food security are two related and overlapping concepts. ***Food safety*** deals with handling, preparation, and storage of food to prevent foodborne illnesses. ***Food security*** deals with the ability to access and secure adequate food. In developing countries, more emphasis is placed on food security, as access to adequate food supplies is often limited, **Figure 9-24**.

Government Agencies Responsible for Food Safety

Many government agencies have a role in protecting and maintaining a safe food supply in our country. The ***Food Safety and Inspection Service (FSIS)*** is administered through the U.S. Department of Agriculture (USDA), and is responsible for food safety related to meat, poultry, and egg products. FSIS ensures products are safe for consumption, wholesome, and labeled and packaged correctly for consumers. The ***Food and Drug Administration (FDA)*** is responsible for ensuring that products other than meat, poultry, and eggs are safe, pure, and labeled correctly for consumers. The FDA is also responsible for the safety of drugs, medical devices, animal feeds and drugs, cosmetics, and radiation emitting devices. The ***Centers for Disease Control and Prevention (CDC)*** lead the way in identifying and investigating foodborne illnesses. The CDC plays a large role in local and state health departments and monitors surveillance of foodborne disease and illness outbreaks. The ***Environmental Protection Agency (EPA)*** regulates the environment in which our food is produced. The EPA works to protect public health by monitoring the use of pesticides on crops and promotes safe means of pest management. The U.S. Department of Homeland Security works with FSIS to ensure emergency response if our food supply is attacked, if there is a major disease outbreak, or if a natural disaster affects the food supply.

The USDA plays a large role in food safety and security in the United States. As previously mentioned, FSIS operates under the umbrella of USDA, but there are other agencies under USDA that assist with food safety and security in our country. The ***Animal and Plant Health Inspection Service (APHIS)*** plays a large role in the U.S. food safety network. APHIS is responsible for protecting consumers against plant and animal pests and diseases in our food supply. The ***Food and Nutrition Services (FNS)*** is the agency responsible for nutrition assistance programs administered through the USDA. These programs provide healthful and nutritious food for children and low-income adults who may not have access to an adequate food supply, **Figure 9-25**. This arm of USDA is most involved with food security in our country. The ***National Institute of Food and Agriculture (NIFA)*** promotes knowledge of agriculture and the environment through teaching, research, and extension

©iStock/dezombo

Figure 9-24. In many developing nations, food security is greatly affected by aid from hunger relief organizations. Raw commodity donations include corn, wheat, and rice.

Bob Nichols/USDA

Figure 9-25. School lunch programs are designed to ensure all students receive adequate nourishment.

Did You Know?

There are more than 100 land-grant universities in the United States. Which land-grant university are you closest in proximity to?

at Land-Grant Universities (Colleges of Agriculture) in the United States. The ***National Agricultural Library (NAL)*** gathers and shares U.S. and international agricultural information. It provides access to information and data related to agriculture to a global society. Each state has a department of agriculture that is actively involved in food safety and security for citizens of its respective state. These state agencies work closely with FSIS and other federal agencies to provide consumers with the safest food supply possible.

Food Safety and Processing Continuum

Our food can become contaminated with harmful microbes at any stage of the production process. As we have seen in earlier chapters, our food goes through several steps before it reaches our plate. Food is produced on farms or within some type of agricultural business. From the growing or production phase, it is harvested either by machine or human hands and is packed for transportation. The food products are then transported to a retail market for sale. Consumers buy the products and take them home for preparation and consumption. What is left over from consumption is then disposed of as waste. There are many opportunities to introduce harmful pathogens into our food supply. Whether it is from the raw plants or animals, or in packing, distribution, or marketing, the steps in the production cycle are all open for sources of contamination, **Figure 9-26**. Some foods are considered ***shelf-stable***, meaning that prior to opening, they have been packaged in a way that removes all the pathogens from them. That is why food, like ketchup, can stay on your shelf for months and be safe, but should be refrigerated after opening to prevent any bacteria that have entered the bottle from growing.

In order to prevent illness, the USDA has developed guidelines for safe handling procedures for food. The "Be Food Safe" campaign has a four-step process that will help prevent the risk of foodborne illness. The four steps are:

1. Clean
2. Separate
3. Cook
4. Chill

Clean and Separate

Always wash your hands with warm soapy water for 20 seconds prior to preparing food, and clean surfaces that are used to prepare food regularly. One of the biggest issues that plays a role in allowing pathogens to enter the food supply is cross-contamination. ***Cross-contamination*** is when harmful bacteria are transferred to other foods through utensils and cutting surfaces that have not been properly cleaned. Cross-contamination can also happen when grocery shopping if you do not separate your raw meat products from other foods. Anytime two food products come in contact with each other, there is the risk of cross-contaminating one with any pathogens existing on the other. For many people with food allergies, cross-contamination can be deadly. If allergens are cross-contaminated with safe food items, it can cause serious allergic reactions for food allergy sufferers. Separating foods and properly cleaning utensils may cut down on foodborne illness and possibly save lives!

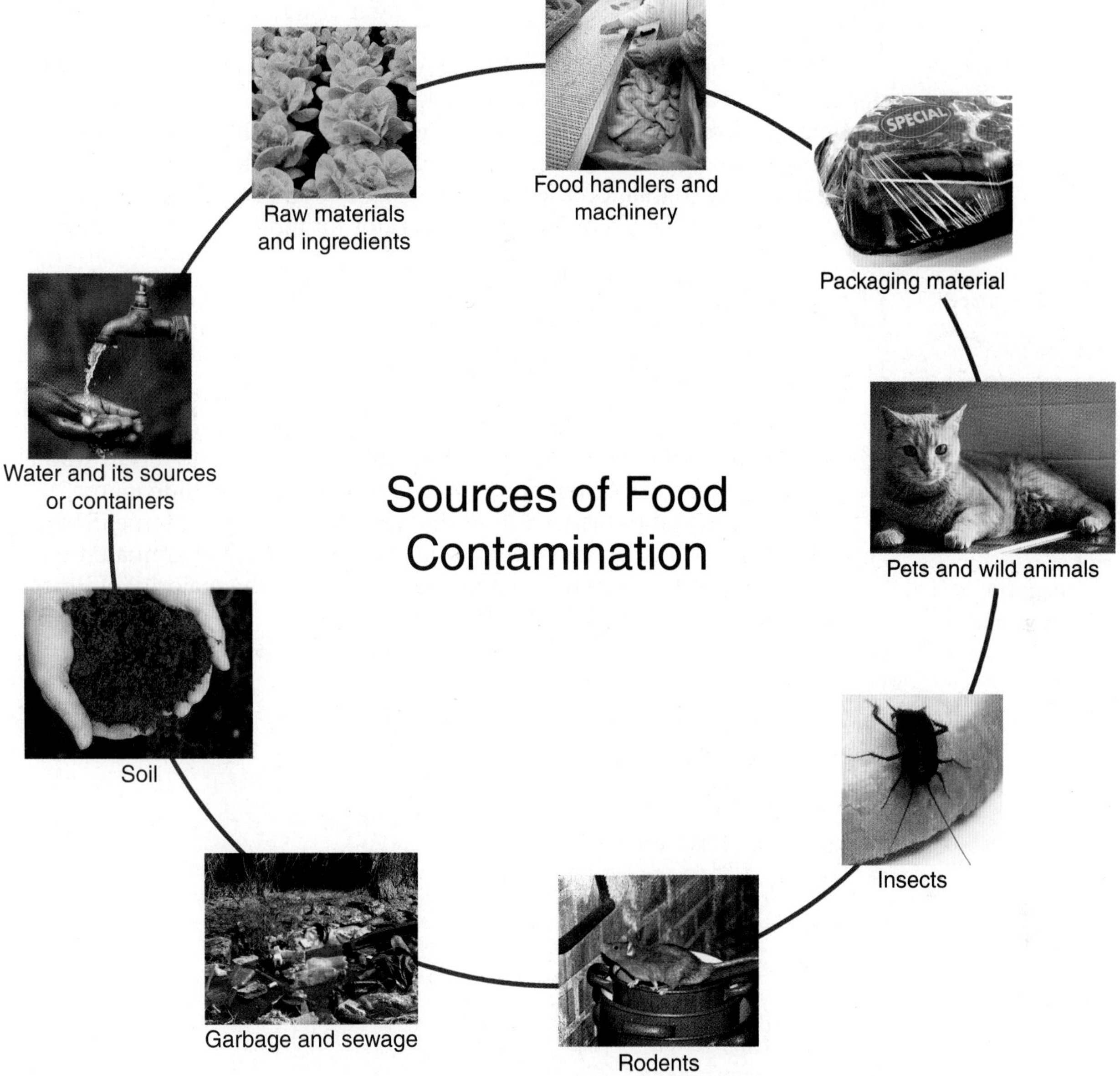

(clockwise from top center) ©iStock/Picsfive; ©iStock/onepony; ©iStock/sjallenphotography; ©iStock/IvelinRadkov; ©iStock/PeterHermesFurian; ©iStock/philly077; ©iStock/BartCo; ©iStock/Riccardo Lennart Niels Mayer; ©iStock/hansslegers

Figure 9-26. Food may be contaminated at any stage of its production.

Cook to Temperature

Cooking temperature is another important food safety concept. Cooking everything using cooking temperature guidelines will kill harmful bacteria and pathogens. Beef, pork, lamb, poultry, and all other meats from agricultural animals have established guidelines and temperatures to follow in order to ensure safe consumption. Raw meat and poultry should be cooked to a minimum internal temperature of 165°F and should be served above 140°F. Most foodborne illnesses can be avoided if meat is prepared to the appropriate temperature for the appropriate length of time.

Chill

The final step in the food safety recommendations, chill refers to proper storage of food after it has been prepared. By leaving food out at room temperature, bacteria can grow more rapidly and cause illness. After meals are complete, properly store and refrigerate all food items. The "Danger Zone" rule applies here in that food should always be above 140°F or below 40°F. Food should never be left out of the refrigerator for more than two hours.

Did You Know?

While it is a common belief that a large majority of foodborne illnesses come from meat and dairy products, research from the CDC has concluded that almost half of all foodborne illnesses can be traced to produce, especially leafy greens.

Common Food Safety and Security Issues

Foodborne illness, commonly referred to as *food poisoning*, is a common problem in our society, but it is very preventable. According to recent studies, there are more than 250 different foodborne diseases that include many bacteria, viruses, and parasites. Other diseases that cause foodborne illness are caused by harmful toxins or chemicals that may have contaminated the food. The CDC states that eight pathogens account for the majority of illness, hospitalization, or death in the United States.

The top five pathogens that cause foodborne illness are:

1. *Norovirus*
2. *Salmonella*
3. *Clostridium perfringens*
4. *Campylobacter spp.*
5. *Staphylococcus aureus*

The top five pathogens that cause hospitalization are:

1. *Salmonella*
2. *Norovirus*
3. *Campylobacter spp.*
4. *Toxoplasma gondii*
5. *E.coli (STEC) 0157*

The top five pathogens that cause foodborne illness and result in death are:

1. *Salmonella*
2. *Toxoplasma gondii*
3. *Listeria monocytogenes*
4. *Norovirus*
5. *Campylobacter spp.*

The majority of these pathogens are found in raw meat and poultry products. The symptoms of these conditions are similar, ranging from abdominal cramps to fever and life-threatening infections. All are caused by ingestion of the pathogen that would have been destroyed had the meat or poultry been cooked properly. Improvements in food safety over the last century have decreased the incidents of typhoid fever, tuberculosis, and cholera, which were all foodborne diseases of the late 1800s and early 1900s. New strains of pathogens appear regularly, but U.S. consumers can rely on the agencies of our government to ensure we have the safest food supply in the world.

FFA Connection Food Science and Technology CDE

The Food Science and Technology CDE is designed to let FFA members showcase their understanding of food development regulations and practices. In addition to formulating and designing their own food product, competitors in this event must identify food safety violations and develop a plan for ensuring safe food production practices. For more information, ask your FFA advisor how to participate.

Peredniankina/Shutterstock.com

Students who raise bees may make a variety of food and cosmetic products from the honey produced in their hives that could be used for a Food Science and Technology CDE.

Food security in the United States is related to people's ability to access an adequate supply of food in order to lead a healthy lifestyle. It is estimated that more than 14% of American homes were not able to secure adequate food last year. The percentage of Americans that were classified as having very low food security was 5.6% of the population. Although these numbers are low compared to many countries, it is troubling to think that a percentage of our population has trouble securing an adequate diet to maintain a healthy lifestyle. The USDA administers many programs [National School Lunch Program; Supplemental Nutrition Assistance Program (SNAP); Food Stamp Program; Special Supplemental Nutrition Program for Women, Infants, and Children (WIC)] to assist this percentage of the population to ensure that the nutritional needs of all Americans are met.

Hazard Analysis and Critical Control Points (HACCP)

In an effort to improve food safety in the United States, the FDA developed ***Hazard Analysis and Critical Control Points (HACCP)*** guidelines to serve as a management system for food safety involving control of biological, chemical, and physical hazards that may occur during agricultural product production, handling, manufacturing, distribution, and consumption. HACCP was initially used in the United States by Pillsbury and NASA to ensure food safety on space missions. Today, HACCP encompasses much more than food safety. It is being used in the pharmaceutical and cosmetic industries. HACCP is used in the red meat, poultry, seafood, and dairy industries to effectively monitor and audit safe food production practices. **Figure 9-27** shows a HACCP plan for restaurants.

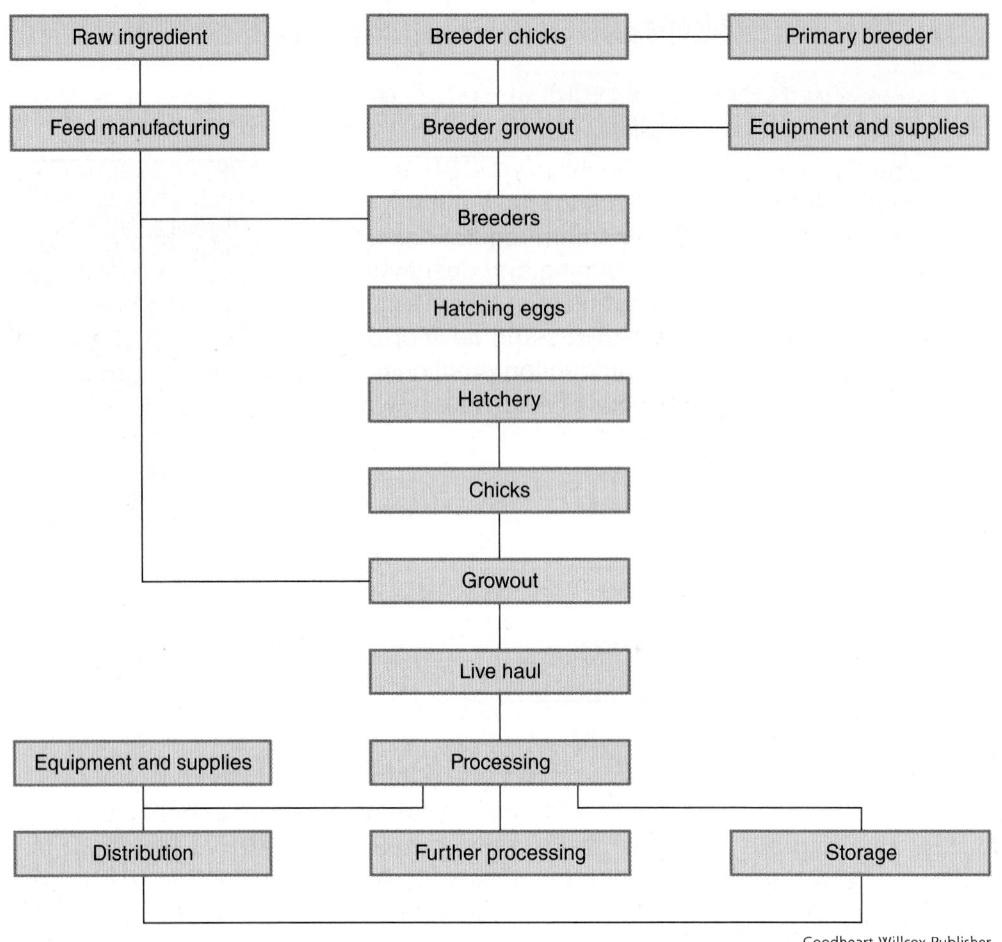

Goodheart-Willcox Publisher

Figure 9-27. HACCP guidelines require the processor to create a flow diagram covering each stage of the process.

Emerging Technology in Food Safety

At the turn of the nineteenth century, many advances were made in food safety from canning practices to pasteurization of milk. Today, food irradiation is becoming more popular than ever. It is similar to pasteurization in that it destroys harmful pathogens and bacteria that can lead to food spoilage or foodborne illness, **Figure 9-28**. Pasteurization relies on heat to destroy the pathogens while irradiation utilizes the energy of ionizing radiation. Irradiation uses gamma rays to penetrate deeply into the fruit, vegetable, or meat product being treated. Many consumers have concerns over whether or not irradiation affects the quality of their food. In fact, radiation doses vary across the types of food being irradiated. The FDA sets and monitors acceptable limits of radiation. Fruits only require one dose, while poultry requires three doses to be free of pathogens. While this new technology is promising, it is still in its early stages. Irradiation may prove useful in the long run of food safety in our country, but it will probably never replace the simple steps of washing, separating, cooking, and chilling we described earlier.

U.S. FDA

Figure 9-28. The FDA requires that irradiated foods bear the international symbol for irradiation.

Career Opportunities in Food Safety

There are many career opportunities associated with the food safety and processing industry. The USDA is the largest employer for this type of career. Food inspectors and consumer safety inspectors account for approximately 7,500 jobs nationally. Typically, these jobs require a bachelor's degree or one year of related experience in the food industry. These can range from inspecting meat, fruits, and vegetables to inspecting imported agricultural commodities.

There are many career opportunities which exist as part of a government agency; however, the food safety industry stretches far beyond the agencies of USDA. The meat, poultry, dairy, and fruit and vegetable industries also employ many people to inspect, grade, and ensure high-quality, safe food for U.S. consumers. From food production to processing facilities, to distribution and to the consumer, thousands of career opportunities are waiting for you in food safety.

Career Connection FSIS Veterinarian

Keeping animals and humans healthy

Job description: FSIS is the largest employer of veterinarians in the country. Approximately 1,100 veterinarians work to protect the public from foodborne illnesses. These veterinarians work in meat and poultry plants to inspect federal procedures. They oversee animal transport and handling, humane slaughter and harvesting, processing operations, and distribution and transportation of products to retail stores.

Education required: Completion of DVM certification. Includes bachelor's degree in a related field and a minimum of four years of education beyond a BS degree for professional licensing.

Job fit: This job may be a fit for you if you want to be a veterinarian but also want the stability of fairly regular working hours and governmental employee benefits.

USDA Food Safety and Inspection Service (FSIS)

Veterinarians work with FSIS inspectors to ensure humane animal treatment and safe processing of meat and poultry products.

Lesson 9.3 Review and Assessment

Words to Know

Match the key terms from the lesson to the correct definition.

1. A government agency responsible for ensuring that products other than meat, poultry, and eggs are safe, pure, and labeled correctly for consumers.
2. When harmful bacteria are transferred to other foods through utensils, cutting surfaces, or workers' hands or gloves that have not been properly cleaned.
3. A government agency responsible for nutrition assistance programs administered through the U.S. Department of Agriculture (USDA).
4. A government agency responsible for protecting consumers against plant and animal pests and diseases in our food supply.
5. An infectious or biological agent that causes disease or illness to its host.
6. The ability to access and secure adequate food.
7. A government agency tasked with regulating the environment in which our food is produced.
8. The handling, preparation, and storage of food to prevent foodborne illnesses.
9. An organization that promotes knowledge of agriculture and the environment through teaching, research, and extension at Land-Grant Universities (Colleges of Agriculture) in the United States.
10. An agency administered through the USDA that is responsible for food safety related to meat, poultry, and egg products.
11. Food safety guidelines developed by the FDA to serve as a management system for food safety involving control of biological, chemical, and physical hazards that may occur during agricultural product production, handling, manufacturing, distribution, and consumption.
12. A government agency responsible for identifying and investigating foodborne illnesses.
13. An ailment caused by improper handling, storage, or processing of food products.

A. Animal and Plant Health Inspection Service (APHIS)
B. Centers for Disease Control and Prevention (CDC)
C. cross-contamination
D. Environmental Protection Agency (EPA)
E. Food and Drug Administration (FDA)
F. Food and Nutrition Services (FNS)
G. foodborne illness
H. food safety
I. Food Safety and Inspection Service (FSIS)
J. food security
K. Hazard Analysis and Critical Control Points (HACCP)
L. National Agricultural Library (NAL)
M. National Institute of Food and Agriculture (NIFA)
N. pathogen
O. shelf-stable

14. An organization that gathers and shares U.S. and international agricultural information.
15. Food packaging that removes all pathogens and enables extended storage before opening.

Know and Understand

Answer the following questions using the information provided in this lesson.

1. How many different pathogens can cause foodborne illnesses?
2. Which governmental agency regulates the food safety of meat, poultry, and egg products?
3. Which governmental agency regulates food products that are not meat, poultry, or eggs?
4. Which governmental agency would be responsible if an earthquake or flood affected the food supply?
5. What are the food safety steps outlined in the USDA "Be Food Safe" campaign for consumers?
6. Which pathogen is responsible for the most foodborne illnesses in the United States?
7. What is the term used for a plan to manage food safety practices?
8. What role do veterinarians play in food safety in the United States?
9. What is cross-contamination?
10. What is the most common term for foodborne illness?

Analyze and Apply

1. What can restaurant owners and kitchen managers do to prevent foodborne illnesses resulting from their kitchen? List 10 recommendations you would give to these individuals to keep the food they produce safe for consumers.
2. Which governmental agency do you think is the most important in maintaining a safe food supply? Please explain your answer in 1–2 sentences.

Thinking Critically

1. Imagine for a moment that you are the owner of a food processing plant and *Listeria monocytogenes* has just been discovered in food products produced at your plant. What actions would you take to correct the situation? Write a press release that you could distribute to local media outlets to share this plan with consumers.
2. As a consumer, there are steps that can be taken to prevent foodborne illnesses at home. Create a poster highlighting at least five important things consumers can do at home to keep their food safe.

Lesson 9.4 Animal Feeds and Feeding

Words to Know

abomasum
acidosis
appetite
avian digestive system
balanced ration
bloat
caeca
carbohydrate
cecum
chemical digestion
cloaca
colic
concentrate
crop
cud
digestion
enzymes
fat
fiber
gizzard
honeycomb
legume
lipid
mechanical digestion
metabolism
mineral
modified monogastric digestive system
monogastric digestive system
nutrient
nutrition
omasum
palatability

(Continued)

Lesson Outcomes

By the end of this lesson, you should be able to:

- Identify the six major nutrients and explain their role in animal feeding.
- Determine the components of an animal ration.
- Explain the importance of feeding high-quality feeds.
- Compare and contrast the functions of different animal digestive systems.
- Examine some common digestive disorders.

Before You Read

Obtain sticky notes that you can use when you are reading the lesson. As you read, place a sticky note with questions you have that are not answered by the text. After you complete the lesson, get with a partner and discuss the questions you wrote on the sticky notes. See if you can answer the questions for each other. If not, bring your unanswered questions to the class for discussion.

Have you heard the expression, "You are what you eat?" Do you think the same thing is true for animals? The food that we feed to our animals plays a huge role in their growth and development. Because there are big differences in agricultural animals, we need to make sure we develop a complete understanding of how food is broken down, what nutrients are contained in foods, and what each type of nutrient does for the development of animals.

Nutrition

Nutrition is the process by which an organism obtains food which is used to provide energy and sustain life. Understanding how to properly feed animals requires an understanding of nutrients and how they work to fuel an animal's body. It also means understanding foodstuffs, animal preferences, and tailoring a feeding program to each individual animal, **Figure 9-29**. In addition, having a good knowledge of animal nutrition involves an understanding of digestive systems and the components that make up feed.

As we move through this lesson, we will first examine the major nutrients required for animal life, and the micronutrients important for giving animals complete nutrition. Next, we will examine the components of feeds and feedstuffs and take a look at how they work together to help animals develop. Finally, we will take a look at the process of digestion and how digestive systems differ in different species.

Words to Know

(*Continued*)

protein
proventriculus
ration
reticulum
roughage
rumen
ruminant
ruminant digestive system
scours
total digestible nutrients (TDN)
vitamin

Nutrients

A ***nutrient*** is a substance that provides nourishment essential for growth and the maintenance of life. There are six major nutrients that all animals need in order to survive.

- Water
- Carbohydrates
- Fats
- Protein
- Vitamins
- Minerals

The major nutrients work together to supply the animal with the chemicals that are needed to sustain its life. Although every animal needs all of the nutrients, each nutrient serves a very different purpose in the animal's well-being. You can see how much of an animal is composed of five of the nutrients by looking at **Figure 9-30**.

Water

Water is the single most important nutrient for any animal. It is the most limiting nutrient, meaning that if the animal is without water, it will perish the fastest. It makes sense, because most of an animal's body is composed of water. Making sure that animals have good access to quality water is the most important thing you can do when feeding animals. Without water, it does not matter how good you are serving their other nutrient needs, **Figure 9-31**.

©iStock/Jevtic

©iStock/Jevtic

Figure 9-29. A grasp of animal feed and feeding can help you make the connections that will lead to an understanding of the digestive process of animals. ***Have you ever thought about how feed (left) translates to useable products from animals (right)?***

	Water	Protein	Fats	Minerals	Carbohydrates
Dairy cow	57	17.2	20.6	5.0	0.2
Cow milk	87	3.5	4.0	0.8	4.7
Cow blood	82.2	16.4	0.6	0.7	0.1
Cow muscles	72	21.4	4.5	1.5	0.6
Swine	58	15.2	23.8	2.8	0.2
Sheep (ewe)	60	16.6	19.8	3.4	0.2
Chicken	66	21.0	9.5	3.5	0.2
Chicken egg	66	13.0	10.0	11.0	—

Goodheart-Willcox Publisher

Figure 9-30. Animals really are made up of the main nutrients. This figure shows the composition of each animal in five of the six nutrient areas.

Dmitry Kalinovsky/Shutterstock.com

Figure 9-31. Access to clean water is the most important nutrient for animal producers to be mindful of because most of an animal's body is water. *Can you think of how the design of this water system plays a role in the production of these pigs?*

Water aids in transportation of nutrients, chemicals, and waste to every cell in an animal's body. Water also aids in digestion and allows animals to regulate their body temperature.

There are some considerations to think about when considering water for animals:

- Provide clean, free-flowing, or regularly changed water for your animals. Stagnant water can harbor bacteria and other pathogens that are harmful to animals.

- Try to avoid giving an animal which has been recently exercised water that is very cold. Making sure your animal has had time to cool down before giving it water can prevent complications.
- Not all water tastes the same to animals. If you will be traveling with your animal and are concerned about it not wanting to drink the water in the new location (a stockshow, for example), you may want to consider flavoring the water at home and at your destination so it tastes the same to the animal. If possible, you may want to bring water from the same source the animal is used to when you go to your new location.

STEM Connection The Chemistry of Carbohydrates

Animal nutrition is deeply rooted in chemistry. A perfect example of this is the complexity of carbohydrates. Carbohydrates are all molecules made up of the same structure, a simple glucose molecule, **Figure A**. Simple sugars, such as glucose, fructose, and lactose, are made up of only one or two of these sugar molecules. The body finds them easy to break down and uses them for readily available energy. Have you ever had a sugar rush? If so, it is because your body broke down the sugar molecules easily.

The more rings of sugar, the more complicated it is for your body to break down the carbohydrate because each bond that is broken requires time, energy, and enzymes to break apart. Starches and cellulose, **Figure B**, have

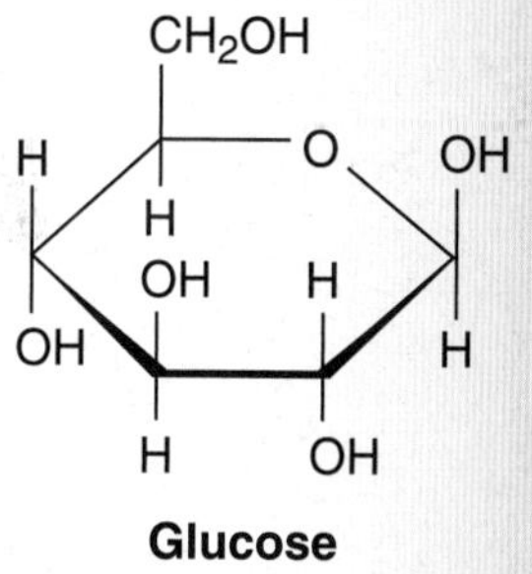

Glucose

Figure A.

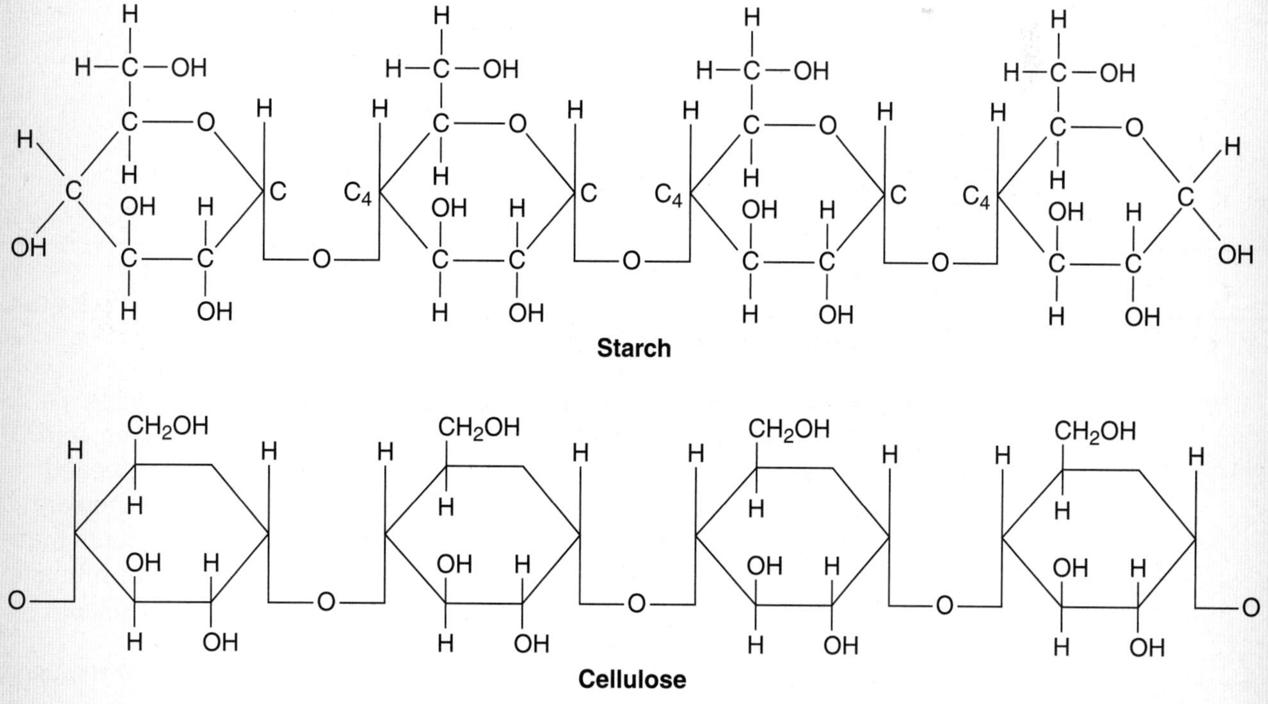

Starch

Cellulose

Figure B.

(Continued)

more molecules to take apart. This means they have more potential energy but are also more complicated to digest.

Lignin, **Figure C**, one of the most complicated molecules, is not digestible unless microbial fermentation comes into play. This is due to the multiple bonds that bind this molecule together. Lignin is so complicated that if you eat it, your body will not be able to break it down.

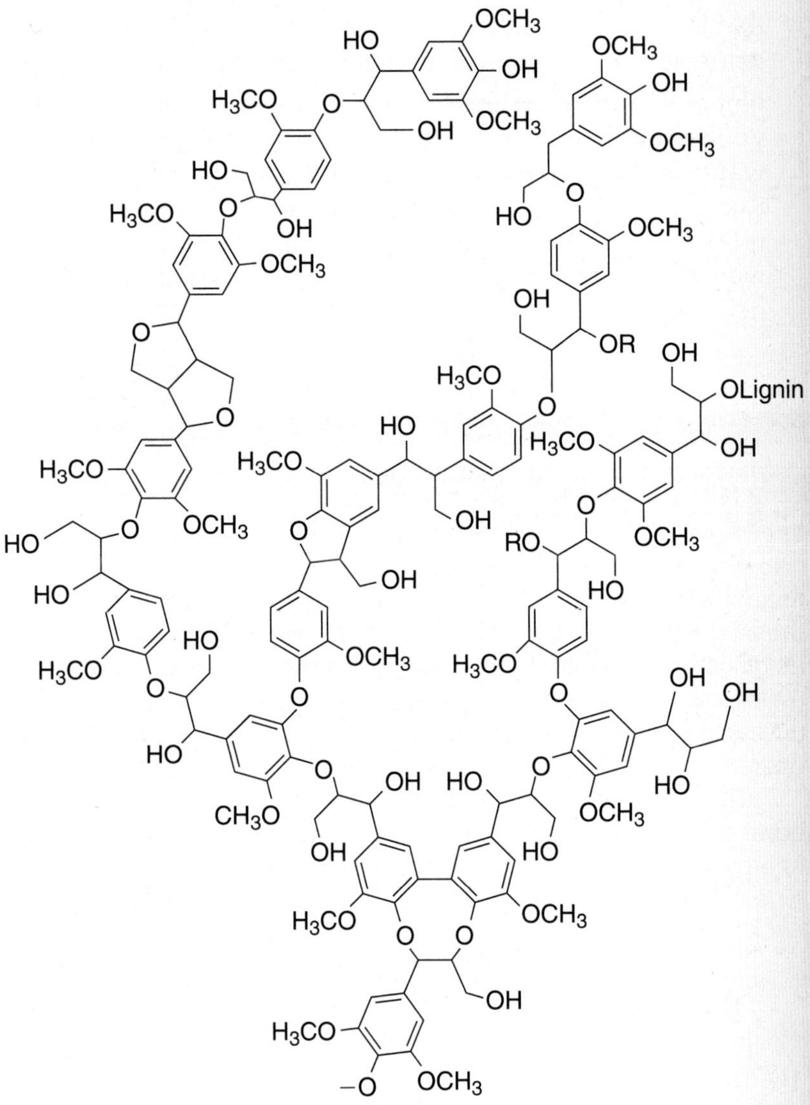

Lignin

Figure C.

Carbohydrates

Carbohydrates are nutrients that give the animal readily available energy. A carbohydrate is easily broken down in the digestive system, and the breakdown of carbohydrates provides the animal with energy needed for movement, life functions, and maintaining body heat. Because they are easily broken down, carbohydrates are the most quickly used nutrients that enter the digestive system. Unused carbohydrates are stored by the animal's body in the form of fats. Because of this, very little of the animal's actual composition is made of carbohydrates.

Sugars and starches are types of carbohydrates. Approximately 70%–80% of all grains that animals eat are carbohydrates. The most common source of carbohydrates in animal feed worldwide is corn, **Figure 9-32**.

©iStock/wlfella

Figure 9-32. Grain corn is the most common source of carbohydrates in livestock animals. Corn is high in energy, and what animals do not need for everyday energy, they convert into fat, which is stored around and within muscle fibers. ***How does feeding corn affect meat quality?***

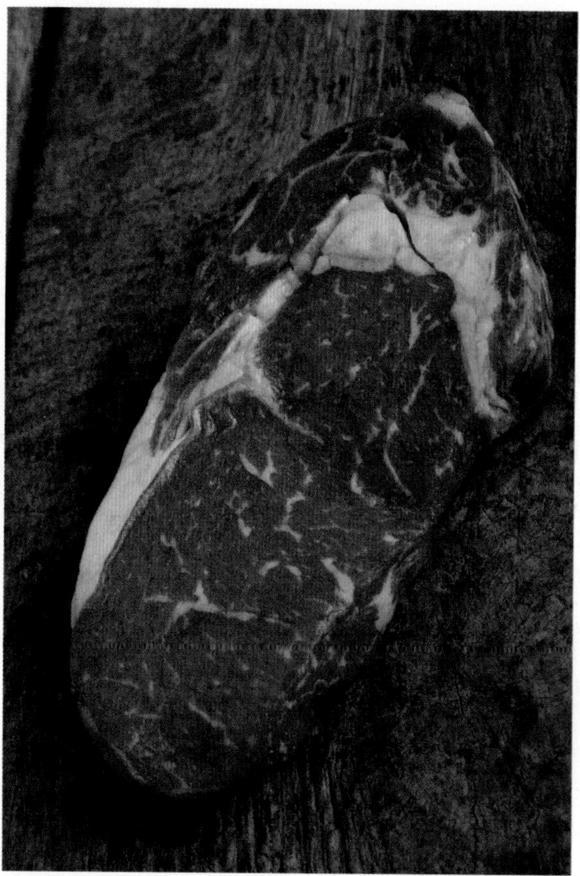

©iStock/neiljlangan

Figure 9-33. Marbling is the term for the small fat deposits made within muscle fibers in meat-producing animals. ***Can you think of reasons that you would want to increase marbling in the meat?***

Fats

Fats, which are also called ***lipids***, are nutrients that provide long-term storage of energy for animals. Fats contain 2.25 times more energy per gram than carbohydrates and can be stored in the body as energy reserves. Fats are also essential in the production of hormones required for life processes and in the cell membranes of new animal cells. For animals that are used for meat production, having an adequate amount of fat before harvesting is important in producing high-quality meat, **Figure 9-33**.

Animal feeds that have high amounts of fats include the meal of any seed (cottonseed, sunflower, safflower, linseed), animal by-products, and high-fat distillers' grains. Because they have evolved on a low-fat diet, it is typically recommended not to exceed 5% fat in the diet of animals with a ruminant digestive system.

Protein

Protein is the nutrient responsible for building and repairing cells in an animal's body. Animals use protein in building muscle and all other body

©iStock/cta88

Figure 9-34. Alfalfa is a legume plant that can take nitrogen from the soil and convert it into plant protein. *Why would feeding alfalfa hay to meat animals be more productive than feeding them grass hay?*

structures, making it an important factor in feed for meat animals needing to produce large amounts of muscle. Milk is also high in protein and feeding high levels of protein to lactating animals is essential.

Protein is made up of long chains of smaller molecules, called amino acids. Ruminant animals can produce all of the amino acids that they need, but other animals need to consume some of their amino acids. There are two different ways to measure the protein in feeds: crude protein measures the amount of protein based on the nitrogen content of the feed, and digestible protein adjusts the total to include only the amount the animal can digest.

Animal feeds that are high in protein come from animal and plant sources. Animal sources of protein feed include fish meal, bone meal, poultry litter, and dried food processing products. Plants that are used as protein sources come from the legume family. ***Legumes*** are plants that have a symbiotic relationship with bacteria, allowing them to take nitrogen from the soil and uptake it as a protein source, **Figure 9-34**. These protein-rich plant feeds include alfalfa and soybeans.

Vitamins

Vitamins are a group of organic compounds that are essential for life. Organic compounds all contain the element carbon. Vitamins are involved in the regulation of metabolism. ***Metabolism*** is the sum of all the chemical processes conducted by the body to sustain life. Vitamins impact reproduction, skin and coat quality, and immune function.

Vitamins can be classified as either fat-soluble, meaning they break down in fat, or water-soluble, meaning they break down in water. Vitamins A, D, E, and K are fat-soluble vitamins that can be stored in the liver or other fatty tissue within an animal. Vitamin C and all of the B complex vitamins are water soluble and should be included as a part of an animal's feed daily. Ruminant animals can typically rely on their digestive microbes to produce all of the B vitamins that they will need for survival.

Do you take your vitamins daily? Some animals do not need to. Grazing animals typically do not need much vitamin supplementation. Animals that are not grazing may benefit from the addition of a vitamin supplement in their feed.

Minerals

Minerals are inorganic compounds that are required by animals to sustain their life functions. These compounds differ from vitamins because they do not contain the element carbon. Minerals are used to form structures like bones and teeth, transmit nerve impulses, produce enzymes, and carry oxygen throughout the body.

Minerals can be divided into two groups based on the amount that an animal needs in its diet. Macrominerals are needed in large amounts and include:

- Calcium
- Phosphorous
- Potassium
- Sodium
- Sulfur
- Chloride
- Magnesium

The microminerals are needed in smaller amounts and include copper, chromium, iron, cobalt, iodine, manganese, selenium, molybdenum, zinc, and nickel.

Mineral deficiencies are very regional in nature. The best way to ensure that animals are getting the correct amount of minerals is to provide access to mineral blocks or to use loose minerals added to feed, **Figure 9-35**.

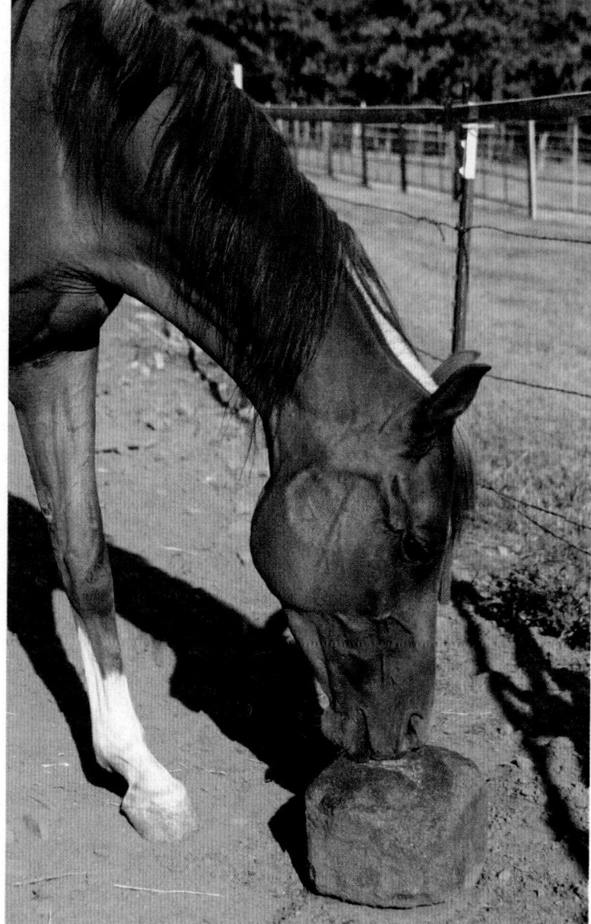

©iStock/sfmorris

Figure 9-35. Why is this horse so interested in that block? It contains minerals and vitamins necessary for proper health. Animals will have cravings for the minerals and vitamins and can easily get them into their diet with mineral blocks or tubs placed in pastures or paddocks.

Nutrient Requirements

The amount of each nutrient that an animal requires is based on several factors, including:

- **Age**—younger animals typically need more nutrients per pound of body weight than mature animals and have greater protein needs.
- **Species**—the different digestive systems of animal species and their ability to break down feeds makes a huge difference in the nutritional requirements of animals. As a general rule, ruminants can consumer higher amounts of complex carbohydrates.
- **Intended use**—the use of an animal plays a big role in its nutritional requirements. Animals that are expending a lot of energy, are pregnant, or are lactating have much different nutrient requirements than animals who are being fed to maintain their weight.
- **Individual genetics**—just like people, each animal has its own unique ability to digest and absorb nutrients. Animals of the same species, age, and use may need to be fed differently, depending on how they process the food. Paying careful attention to the animals you are feeding will allow you to tailor your feeding program to meet their individual needs.

Types of Feed

Animal feeds are made up of different amounts of each nutrient. Typically, there are two broad classifications of animal feed:

- Concentrates
- Roughages

These different categories of feed are based on the amount of fiber, energy, and total digestible nutrients (TDN). ***Fiber*** is the term used for very complex carbohydrates, which require a lot of digestive effort to break down into absorbable nutrients. Cellulose and lignin are examples of fiber sources in animal feeds. ***Total digestible nutrients (TDNs)*** are the sum of the digestible fiber, protein, lipids, and carbohydrates in a feed.

Concentrates

Concentrates are feeds that are high in energy and TDN and low in fiber content. Concentrates generally contain less than 18% fiber once all water is removed from the feed. Can you think of some animals that might require high levels of concentrates in their diet? Any animal that is using a lot of energy is likely to need a diet high in concentrates. Some feedstuffs that are considered concentrates include: grains (except rice), oil millings, and other energy or protein concentrates including processed livestock products and industrial products.

Livestock that are typically fed high-concentrate feeds include:

- Dairy cattle
- Beef cattle
- Swine
- Layers
- Broilers
- Performance horses

Roughages

A ***roughage*** is a feed that is low in energy and TDN and high in fiber content. As the opposite of a concentrate, roughages typically have more than 18% fiber once completely dried. Roughages are tough to digest. In fact, animals which do not have a digestive system with microbial digestion, like pigs, cannot digest some of the more complicated sources of fiber, like lignin. Feeds that are high in roughages include: legume hays, grass hays, rice grain, beet pulp, straw, silage, and fresh grass, **Figure 9-36**. Animals that are fed diets high in roughages will typically not grow or gain muscle mass as quickly as those animals that are on a high-concentrate diet.

Livestock that are typically fed high roughage feeds include:

- Cows in cow-calf operations.
- Horses being fed a maintenance diet.
- Sheep used for wool production.

©iStock/StephM2506

Figure 9-36. Grazing animals are typically consuming large amounts of roughages. Roughages are low in TDN and high in fiber. *What percentage of these sheep's diet do you think is made up of roughages?*

Career Connection Animal Nutritionist

Job description: Animal nutritionists work for livestock producers, commercial feedlots, feed companies, and research institutions to help determine the role of nutrients in animal production. They work to balance rations, calculate feed costs, and develop new feeding programs tailored to specific animals' needs.

Education required: Most animal nutritionists have at least a master's degree in animal science and nutrition, ruminant nutrition, or a related field.

Job fit: This job may be a fit for you if you love chemistry, working both with animals and in a laboratory setting, and enjoy conducting research.

nandyphotos/iStock

As you can see, animals that need to be actively growing or performing typically do not perform as well when fed solely roughages.

Animal Feeding Considerations

The amount of feed that an animal consumes in a day is called a ***ration***. A ration that contains all of the necessary nutrients for an animal's intended growth and development is called a ***balanced ration***. Balancing rations is a highly specialized skill and requires a lot of knowledge about how livestock nutrition and digestion works. In addition to having a balanced ration, properly feeding animals requires several factors be taken into consideration.

Animal Preferences

Do you have a favorite food? Are there certain foods you just will not eat because you do not like the way they taste or their texture in your mouth? Animals have food preferences, too. It is the same concept with you preferring a piece of cake over a piece of broccoli. The term used to describe how desirable an animal finds a foodstuff is called ***palatability***. Feeding more palatable foods will increase the amount of feed that an animal will eat, but does not necessarily mean that the feed is more nutritious, **Figure 9-37**.

©iStock/RGtimeline

Figure 9-37. Have you ever refused to eat, not because you were not hungry, but because you did not like the food? Animals have the same issue. A big consideration in livestock feeding is making sure the food is palatable so that the animals will eat it.

Animal	Body Weight Consumed per Day in Feed
Beef cattle	2%–4%
Horses	1.5%–2.5%
Swine	2.5%–4%
Chickens	4%–6%
Sheep	2%–5%
Goats	2%–5%

Goodheart-Willcox Publisher

Figure 9-38. This figure shows a rough guideline of approximately how much feed per day livestock animals consume, as a function of their body weight. For example, if you had a 1,000-lb horse, you would expect to feed it 15–25 lb of feed per day. It is important to remember that this will vary, depending on the life stage and desired output for your animal.

Appetite is the term for how much an animal desires food of any type. A lack of appetite is something to pay attention to, as it often indicates poor health.

Another animal consideration is the fill that the feed will provide to the animal. Animals need to eat a certain percentage of their body weight each day to feel full, **Figure 9-38**. A ration that has too little volume will leave the animal feeling hungry, and they will not eat more than they can comfortably consume in a day.

Feed Quality

Animal feeds vary greatly in their nutritional content based on several factors. A field of alfalfa hay will often have different nutritional value if the farmer chooses to cut it a day or two earlier or later than the peak of its harvest time. Storing feed improperly can also decrease the nutritional value, **Figure 9-39**.

Different methods of processing feed can also play a role in the nutrients that an animal obtains from the feed. Think about this: whole corn has a lower nutritional content than cracked, rolled, or steam-flaked corn. Why do you think this is so? The corn could come from the exact same cob and still vary in the nutrients that an animal can get from it. That is because the processed corn

©iStock/CarolBeckman

Figure 9-39. Proper storage of feed can help retain its nutrient value. This feed storage area allows the feed to be kept out of the elements and stored in bulk quantities. *What do you think these animal feed are?*

has already been partially mechanically broken down, making the digestion process easier for the digestive system of the animals.

Feed Costs

Feeds that are similar in nutritional content can be very different in their cost to the producer. For this reason, agriculturalists often look to the feeds that are most readily available in their area and have a lower cost. For example, cottonseed meal is used as a great source of both protein and fats in the diets of many livestock animals in the areas where cotton is produced, and is not found in many feeds that are produced outside of cotton-producing areas.

Because of the desire to feed livestock the most nutritious and cost-effective feed, livestock are sometimes fed things that may seem odd to the average consumer. Some examples of this are poultry litter, leftovers from large commercial bakeries, and food industry waste products that are leftover from processing human foodstuffs. As an added bonus, this practice allows livestock to play an important role in recycling.

Did You Know?

Have you ever watched someone who "chewed like a cow?" Cows (and other ruminants) chew their food by moving their lower jaw in a circular motion. This is because ruminants do not have incisors (front teeth) on the top of their mouth. Instead, they have a hard dental pad that they push food against to mechanically break it down prior to swallowing.

How Digestion Works

Digestion is the process of breaking down food through mechanical and chemical means and absorbing nutrients into the bloodstream. In its most basic explanation, during digestion, foodstuffs are broken down into molecules small enough to pass into the bloodstream of the animal. Digestion occurs through two processes:

- ***Mechanical digestion***—the physical separation of the foodstuff into smaller and smaller pieces. Chewing and churning actions in the digestive system help with this process, **Figure 9-40**.
- ***Chemical digestion***—the addition of digestive chemicals that break chemical bonds in foods at a molecular level. The chemicals that help break down food are called ***enzymes***.

The Basics of Digestion

Food goes on a very similar journey in all of the mammal digestive systems. The process is outlined below:

Step 1: Food is taken in through the mouth, where the teeth provide mechanical breakdown of the feeds.

Step 2: The food then travels down the esophagus to the stomach, where, depending on the type of system, the food is broken down by digestive enzymes.

Step 3: Next, the food passes into the small intestine, which contains

©iStock/Jevtic

Figure 9-40. Chewing is the process of mechanically breaking down feed into smaller parts. Chewing food increases the surface area of the ingested feed, allowing more surface area for microbes and digestive enzymes to work.

many folds and a lot of surface area. This is where most of the broken down nutrient molecules enter the bloodstream.

Step 4: The next section, called the large intestine, allows for continued absorption of nutrients and is where much of the water from the digestive system is reabsorbed into the body.

Step 5: Finally, excess waste product leaves the animal body through the rectum and anus.

There are modifications to basic digestion based on the type of digestive system that the animal has. Let us explore these digestive systems a little more in depth.

Types of Digestive Systems

Different animals can digest different kinds of feed. For example, cattle can eat a protein-rich hay their entire lives and still be able to sustain their life functions, but if we fed a market hog the same hay as its only feed, it would not be able to survive. The differences in type of digestive system are driving factors in the proper feeding of animals.

Ruminant Digestive Systems

Traditional livestock animals including cattle, sheep, and goats, along with nontraditional livestock animals like llamas, alpacas, bison, deer, and elk, are all ruminant animals. A ***ruminant*** animal has a ***ruminant digestive system*** with four compartments to its stomach, **Figure 9-41**:

- ***Rumen***—this is the largest section of the ruminant stomach, which contains microbes that process feed and help break down fiber.
- ***Reticulum***—also called the ***honeycomb,*** this chamber serves as an area to sort feed coming from the mouth and remove any foreign objects from the feed prior to passage into the rumen.

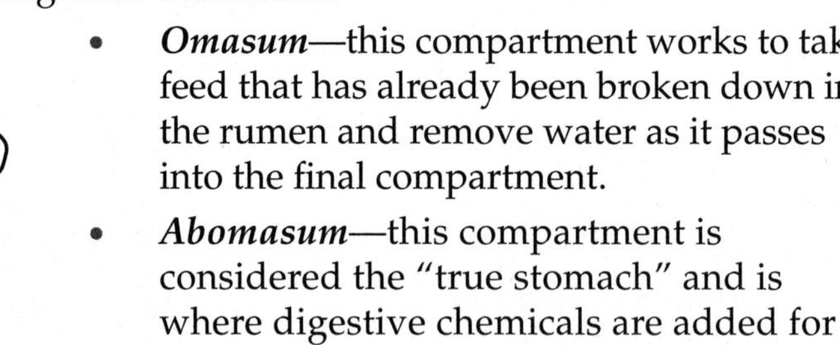

- ***Omasum***—this compartment works to take feed that has already been broken down in the rumen and remove water as it passes into the final compartment.
- ***Abomasum***—this compartment is considered the "true stomach" and is where digestive chemicals are added for the chemical breakdown of food.

Ruminant animals have microbes that live in their rumen and help to break down roughages. Because of this symbiotic relationship, ruminant animals are able to process roughages like cellulose and lignin that other species of animals get no nutritional value from. They can also use their chemical digestion processes to produce all of the fat-soluble vitamins and essential amino acids necessary to sustain their life functions.

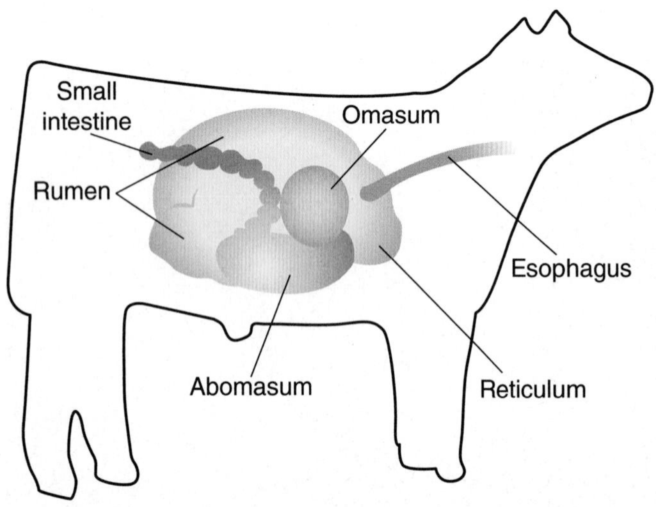

Goodheart-Willcox Publisher

Figure 9-41. The ruminant digestive system contains four compartments. The inside of each compartment has a different texture. *How do you think the different textures help each compartment serve its intended role in digestion?*

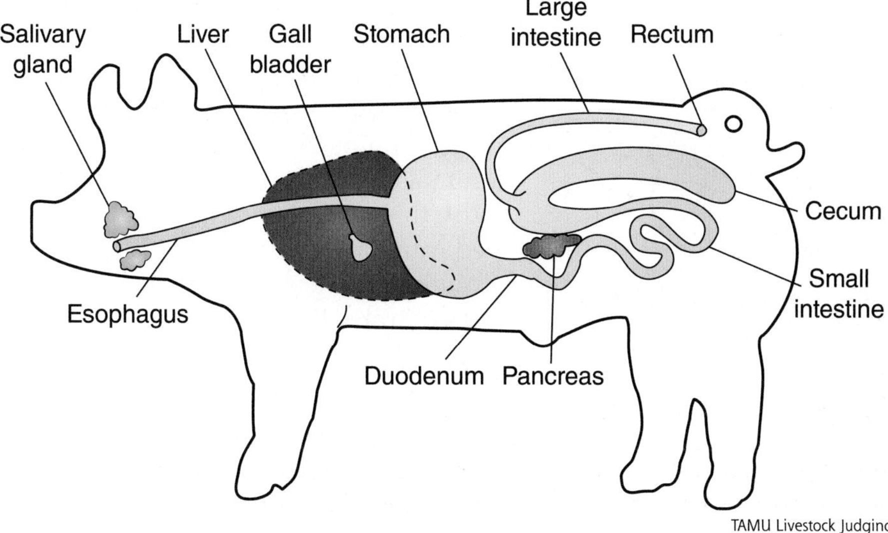

TAMU Livestock Judging

Figure 9-42. The monogastric digestive system has no area for microbial breakdown of feed. *What does this mean in terms of the feeds that can be fed to these animals?*

The partial breakdown of food in the rumen is also the reason that ruminants chew their cud. ***Cud*** is partially digested food that is regurgitated into the mouth and rechewed so that it can pass to the rest of the digestive system.

Monogastric Digestive System

A ***monogastric digestive system*** is a digestive system that has a single compartment to the stomach and is unable to process cellulose. Pigs, dogs, cats, and humans all have a monogastric digestive system, **Figure 9-42**. Animals with a monogastric digestive system do not perform well when fed a high-roughage diet because of the lack of microbial processes to help break down complex carbohydrates.

Modified Monogastric Digestive System

A ***modified monogastric digestive system*** is a system which has a single compartment stomach and an enlarged cecum for microbial digestion. The ***cecum*** is a pouch-like area found at the beginning of the large intestine, where digestive microbes are housed to break down roughages, **Figure 9-43**. Horses

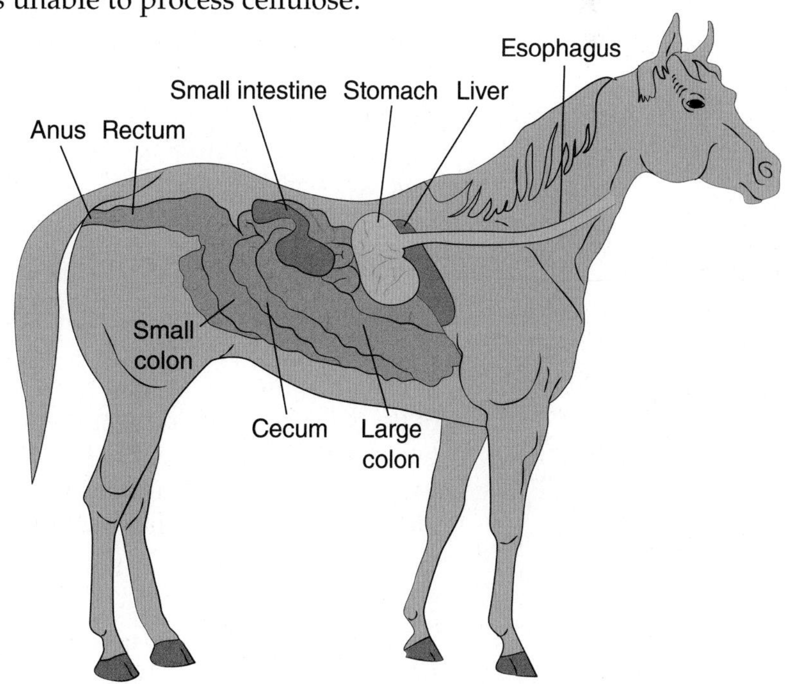

Goodheart-Willcox Publisher

Figure 9-43. Horses and rabbits have modified monogastric digestive systems. *What is the function of the enlarged cecum in this system?*

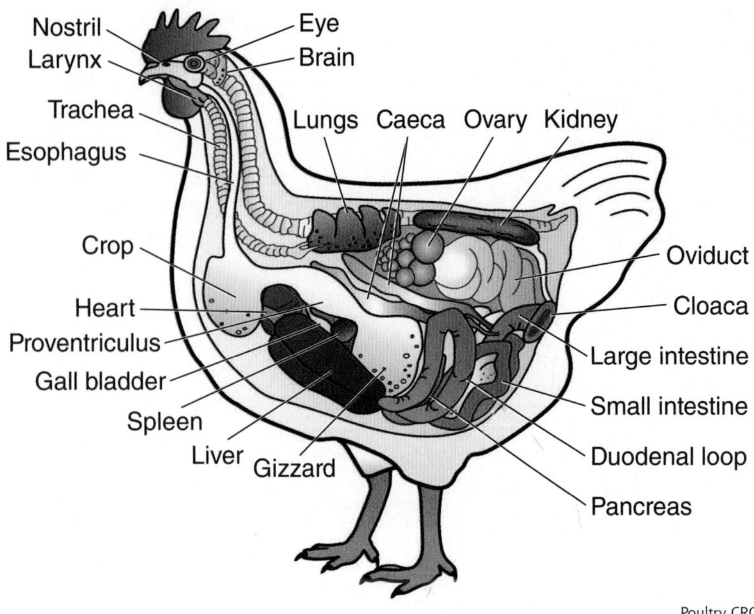

Figure 9-44. The avian digestive system works in a very different manner than the mammal systems we have already looked at. *Does this system look at all like the others? What differences can you see?*

and rabbits have this type of digestive system.

Animals that have a modified monogastric system can break down fiber, but not as efficiently as ruminant animals. For this reason, animals with this type of system need to be fed roughages with higher TDN values and are more likely to require concentrates than ruminants.

Avian Digestive System

The ***avian digestive system*** is the digestive system used by poultry and all birds, **Figure 9-44**. This system is very different from any of the mammal digestive systems. In the avian system, the food is passed from the beak into the esophagus, where it then travels into a storage area at the base of the esophagus called the ***crop***. From there, food travels into a muscular pouch called a gizzard. The ***gizzard*** is an area where muscular contractions, combined with pebbles or rocks consumed by the bird, work to mechanically break down the food. From here, the food travels into a small area called the ***proventriculus***, where digestive enzymes for chemical breakdown are added. Nutrients are then absorbed into the bloodstream in the small intestine. Near the large intestine, the waste products from the kidneys enter the digestive system through the ***caeca***, and finally, waste leaves the avian body through an opening called the ***cloaca***. The cloaca is also the opening through which eggs are laid.

These vast differences in the digestive system of avians are important to consider when developing avian rations.

Did You Know?

Chickens do not urinate. Rather than having liquid urine, birds process their liquid waste and convert it into a high-urea, low-moisture waste product that is incorporated into their feces.

Digestive Disorders

Not feeding animals properly can result in a wide array of digestive disorders. Digestive disorders are one of the leading causes of death in all livestock animals. Can you think of some animal digestive disorders? One of the most common digestive problems is nutrient deficiencies. This occurs when one of the nutrient needs of an animal is not being met and has a wide variety of negative health implications. The following are just a few of the other most common disorders in livestock animals.

Acidosis

Acidosis is caused when the pH level of a ruminant animal remains acidic for an extended period of time. This disorder is usually due to a rapid intake of highly digestible concentrates, and symptoms

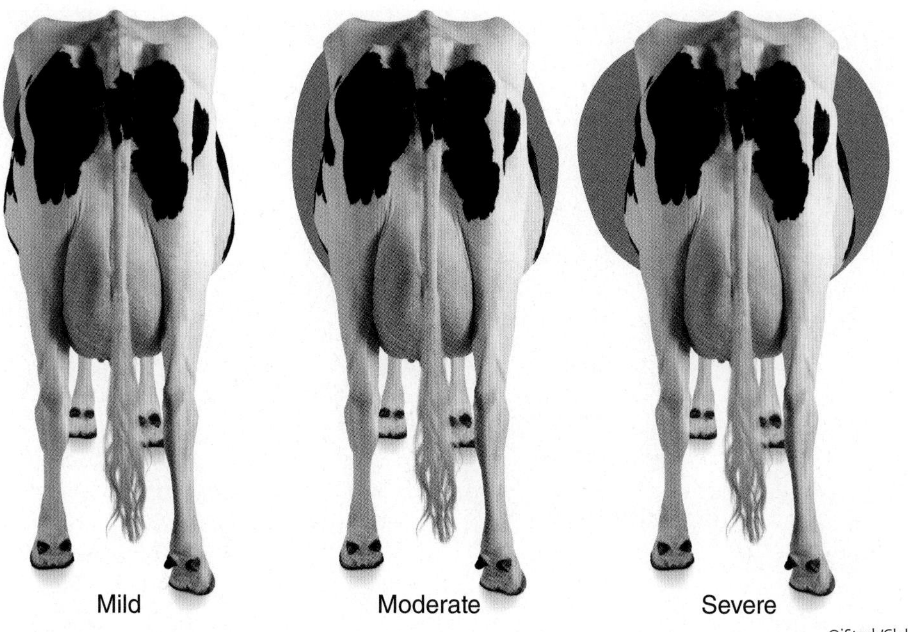

Figure 9-45. Bloat will initially be seen on the left side of the animal as gases build up in the rumen. As the bloat becomes more severe, the animal will have a very swollen left side and the bloat will eventually be visible from the right side as well, as shown in this dairy cow example.

include an animal that will not eat or has acute diarrhea. Prevention of acidosis includes making sure that animals are fed a ration that does not contain concentrates with TDN levels that are too high for the animal to process.

Bloat

Bloat is a condition in livestock animals where excessive gases build up in the rumen, **Figure 9-45**. This can be caused by an obstruction of the path for gases to escape or, more commonly, when a buildup of froth occurs on the top of the ruminant liquid, preventing gas from being belched out. Bloat is common in animals that have eaten rapidly growing forages, like cattle on pastures with lots of clover in the spring. Animals with bloat are often diagnosed because of a swollen left side, the side of the animal where the rumen sits. To prevent bloat, cattle can be given anti-foaming supplements.

Colic

Colic is a digestive disorder that is the number one cause of death in horses. The word *colic* basically means any digestive pain in horses, **Figure 9-46**. Colic can be caused by poor quality feed, twisted intestines, ingestion of sand, or drinking excessive amounts of water when the animal is hot. Horses with colic will typically be restless, bite or kick at their belly, sweat, and have a lack of gut noises. Treatment for colic depends greatly on the severity and cause of the condition.

©iStock/trueblu

Figure 9-46. Is this horse colicking? Careful attention to your horse will let you know if the horse is just rolling or if something more serious is going on. If the rolling is combined with sweating, signs of discomfort, and biting or kicking at the stomach, the horse could be colicking.

Scours

Scours is a digestive disorder characterized by extreme diarrhea in young animals. This disorder causes more loss of young animals than any other condition. There are many different causes for scours including bacterial and viral digestive infections. Generally, young animals that have incredibly bad scours become very dehydrated. Keeping the animal hydrated is important to treating scours.

Lesson 9.4 Review and Assessment

Words to Know

Match the key terms from the lesson to the correct definition.

1. An animal digestive system with four compartments to the stomach.
2. A condition caused when the pH level of a ruminant animal remains acidic for an extended period of time; usually due to a rapid intake of highly digestible concentrates.
3. A storage area at the base of the esophagus in an avian digestive system.
4. An animal whose digestive system has four compartments to its stomach.
5. The addition of digestive chemicals that break chemical bonds in foods at a molecular level.
6. The nutrient responsible for building and repairing cells in an animal's body.
7. The opening through which waste leaves the avian body; also the opening through which eggs are laid.
8. A digestive system that has a single compartment to the stomach and is unable to process cellulose.
9. The process of breaking down food through mechanical and chemical means and absorbing nutrients into the bloodstream.
10. Very complex carbohydrates, which require a lot of digestive effort to break down into absorbable nutrients.
11. The amount of feed that an animal consumes in a day.
12. A condition in livestock animals where excessive gases build up in the rumen.
13. The sum of all the chemical processes conducted by the body to sustain life.
14. A pouch-like area found at the beginning of the large intestine (in a modified monogastric digestive system) where digestive microbes are housed to break down roughages.
15. The chemicals that help break down food.
16. An inorganic compound that is required by animals to sustain their life functions.
17. The physical separation of the foodstuff into smaller and smaller pieces.
18. A digestive disorder characterized by extreme diarrhea in young animals.
19. A digestive system that has a single compartment stomach and an enlarged cecum for microbial digestion.
20. A digestive disorder that is the number one cause of death in horses.

A. acidosis
B. balanced ration
C. bloat
D. cecum
E. chemical digestion
F. cloaca
G. colic
H. crop
I. cud
J. digestion
K. enzymes
L. fiber
M. mechanical digestion
N. metabolism
O. mineral
P. modified monogastric digestive system
Q. monogastric digestive system
R. nutrition
S. protein
T. proventriculus
U. ration
V. roughage
W. ruminant
X. ruminant digestive system
Y. scours
Z. total digestible nutrients (TDN)

21. The process by which an organism obtains food which is used to provide energy and sustain life.
22. A small area in an avian digestive system where digestive enzymes for chemical breakdown are added to the food.
23. The sum of the digestible fiber, protein, lipids, and carbohydrates in a feed.
24. Partially digested food that is regurgitated into the mouth and rechewed so that it can pass to the rest of the digestive system.
25. A feed that is low in energy and total digestible nutrients (TDN) and high in fiber content.
26. A ration that contains all of the necessary nutrients for an animal's intended growth and development.

Know and Understand

Answer the following questions using the information provided in this lesson.

1. What are the six major categories of nutrients?
2. Which nutrient is the most important for animal survival?
3. Please list four factors that can affect an animal's nutrient requirements.
4. What are the two main types of feed? How are they different?
5. Please list the five steps in digestion.
6. What is a ruminant animal?
7. Which section of the ruminant stomach is most important in the digestion of fiber? Why?
8. List three animals with a monogastric digestive system.
9. Why can modified monogastrics digest fiber?
10. Please list four common digestive disorders in livestock animals.

Analyze and Apply

1. Explain to your neighbor, who is a first-time hog buyer, why he cannot just feed his pigs hay.
2. Which animal do you think would need more protein, a market steer or a racehorse? Please explain your answer.

Thinking Critically

1. Search online for feed labels from two common livestock feeds. By analyzing the components of the label, determine what type of animal would likely benefit from eating this product. Be sure to base your knowledge on what you know about the role nutrients play in an animal's body.
2. Please write 1–2 paragraphs explaining why modified monogastrics have to eat feed that is higher in TDN than ruminant animals.

Chapter 9

Review and Assessment

Lesson 9.1

Local Food Systems

Key Points

- The food production system includes all of the steps from production through disposal of waste from food products.
- A local food system is one that produces food within 400 miles of where it is consumed.
- Many consumers desire locally produced foods because they want to decrease the amount of fossil fuels used, have access to fresher products, or want a closer connection to their food.
- There are many distribution systems for getting local foods to consumers including farmers markets, community gardens, and agricultural consumer cooperatives.

Words to Know

Use the following list and the textbook glossary to review and study the *Words to Know* from *Lesson 9.1*.

community garden
consumption
cooperative
disposal
distribution
exotic foods
farmers market
food miles
food system
harvesting
local food system
marketing
processing
production

Check Your Understanding

Answer the following questions using the information provided in *Lesson 9.1*.

1. Please list the steps in the food production chain in order.
2. How does a local food system differ from a global food system?
3. Give one advantage and one drawback to using local food systems.
4. Why were local food systems more common historically than they are now?
5. Explain the concept of food miles and why they are important to local food systems.
6. What is the purpose of an agricultural consumer cooperative?
7. How does a farmers market work?

8. What is a community garden?
9. What is the purpose of processing in a food production chain?
10. What percentage of American food spending is spent on foods consumed outside the home?

Lesson 9.2

Global Food Systems

Key Points

- The global food system is the systematic production of food for distribution to consumers around the globe.
- The key factors that play a role in the global food system are the affordability, availability, and safety of food products.
- More than one out of nine people on the planet is chronically malnourished.
- Feeding the world will become increasingly important, as agriculture is expected to produce more food in the next 40 years than it has in the history of humanity.
- Distribution of food products is important in understanding world hunger.
- World hunger is not only an issue in developing countries; hunger exists in our own country as well.

Words to Know

Use the following list and the textbook glossary to review and study the *Words to Know* from *Lesson 9.2*.

arable land
demand
export
foodborne illness
food desert
food insecurity
global food system
hunger
import
malnourishment
spoilage
staple crop
supply
tariff
trade surplus
urbanization
world hunger

Check Your Understanding

Answer the following questions using the information provided in *Lesson 9.2*.

1. What are the factors that determine the nature of the global food system?
2. Please list three factors that contribute to food availability.

3. Please list two factors that contribute to an increase in average spending on food in a country.
4. What is the impact of developing nations lacking proper food inspection regulations?
5. How does conflict affect world hunger?
6. Explain what a trade surplus is.
7. Why are countries that do not have a trade surplus more vulnerable to hunger?
8. How many American families are dealing with food insecurity?
9. What factor is playing a role in the future of food availability in the United States?
10. Which country spends the smallest percentage of its income on food?

Lesson 9.3

Maintaining a Safe Food Supply

Key Points

- Food safety and food security are huge issues in our country and around the globe.
- The U.S. government is actively involved in making sure that we have a safe and wholesome food supply.
- Although numerous pathogens threaten our food supply, there are measures, policies, and procedures in place to deal with illnesses, outbreaks, diseases, and disasters related to food safety and security.
- Technology improves daily and with modern advances, food safety and security may continue to decrease in our society.

Words to Know

Use the following list and the textbook glossary to review and study the *Words to Know* from *Lesson 9.3*.

Animal and Plant Health Inspection Service (APHIS)
Centers for Disease Control and Prevention (CDC)
cross-contamination
Environmental Protection Agency (EPA)
Food and Drug Administration (FDA)
Food and Nutrition Services (FNS)
foodborne illness
food safety
Food Safety and Inspection Service (FSIS)
food security
Hazard Analysis and Critical Control Points (HACCP)
National Agricultural Library (NAL)
National Institute of Food and Agriculture (NIFA)
pathogen
shelf-stable

Check Your Understanding

Answer the following questions using the information provided in *Lesson 9.3*.

1. What are the three factors that cause foodborne illnesses?
2. Please list the governmental agencies that play a role in food safety in the United States.
3. What are the four steps that the USDA says should be followed to help consumers keep food products safe for consumption?
4. What temperature should raw meat and poultry be cooked to in order to be safe for consumption?
5. What temperature is considered the "danger zone" for bacterial growth?
6. List five pathogens that are common causes of foodborne illnesses.
7. What is a HACCP plan?
8. How does the process of irradiation relate to food safety?
9. Explain why cross-contamination is a big issue in food safety.
10. What do you feel is the single most important thing consumers can do to keep their food products at home safe to eat?

Lesson 9.4

Animal Feeds and Feeding

Key Points

- There are six essential nutrients: water, carbohydrates, fats, protein, vitamins, and minerals.
- Each animal will have different nutrient requirements based on their age, species, intended use, and individual genetics.
- Feeds can be classified as concentrates or roughages.
- Ruminant animals have a large compartment in their stomach for microbial breakdown of feed. This allows them to digest fiber.
- Modified monogastric animals can digest roughages because they have a functional cecum for microbial breakdown.
- The avian digestive system is vastly different from the digestive systems found in mammals.
- Proper feeding is important in the health and maintenance of animals and can prevent digestive disorders.

Words to Know

Use the following list and the textbook glossary to review and study the *Words to Know* from *Lesson 9.4.*

abomasum
acidosis
appetite
avian digestive system
balanced ration
bloat
caeca
carbohydrate
cecum
chemical digestion
cloaca
colic
concentrate
crop
cud
digestion
enzymes
fat
fiber
gizzard
honeycomb
legume
lipid
mechanical digestion
metabolism
mineral
modified monogastric digestive system
monogastric digestive system
nutrient
nutrition
omasum
palatability
protein
proventriculus
ration
reticulum
roughage
rumen
ruminant
ruminant digestive system
scours
total digestible nutrients (TDN)
vitamin

Check Your Understanding

Answer the following questions using the information provided in *Lesson 9.4.*

1. Please explain the function of each of the six major nutrient categories.
2. What is the difference between a vitamin and a mineral?
3. List and define the two main categories of animal feeds.
4. Why is it important for animal feed to be palatable?
5. What are the two types of digestive processes?
6. List the four types of digestive systems covered in the lesson and give two examples of animals with each type of system.
7. Explain which digestive systems can break down cellulose and which structure each of them uses to accomplish the cellulose processing.
8. Please list and give the function of the four compartments of the ruminant stomach.
9. Please list three considerations that should be taken into account when selecting animal feeds.
10. Please define digestion.

Chapter 9 Skill Development

STEM and Academic Activities

1. **Science.** Digestive chemicals. Regardless of which digestive system animals have, they all rely on digestive chemicals to break down foods. Foods with less complex molecules are easier to break down. For this lab, take the following foods: sugar cube (sugar), dried pasta (starch), and hay (cellulose). Place each of the foods into a small clear plastic cup and pour cola over them. The cola simulates digestive chemicals in animal stomachs. Make a hypothesis about which will break down more quickly. Observe the foods for several days and see what happens to them in the presence of the acid. Explain the results in 3–4 sentences.
2. **Technology.** Certain Genetically Modified Foods (GMOs) have been developed with a focus on increasing their shelf life. Conduct a search to find information on one of these crops and explain in 2–3 paragraphs why this technology could be used to help decrease world hunger.
3. **Engineering.** Food packaging is an important thing to consider in food safety. Design a container that could be used to take leftovers home from a restaurant while at the same time informing consumers about the safe handling of the leftovers. Draw and label the features and writing on your packaging that will help aid in food safety.
4. **Math.** Determining the value of a foodborne illness. What is the potential for cost due to a foodborne illness? Consider the following scenario of a very mild foodborne illness: A local hamburger restaurant averages sales of $1,000 per day. A malfunctioning grill leads to a batch of 100 hamburgers that are not cooked properly and have foodborne pathogens in them. Sixty people become ill and spend an average of $250 on medical-related expenses and loss of wages because of the illness. The news shares the story and the sales of the restaurant decrease by 60% for the next 100 days until the owners can earn back consumer trust. How much has this food safety issue cost?
5. **Social Science.** How safe do consumers feel their food is? Conduct a survey of 50 people asking their position on food safety. Develop 1–10 food safety questions that you could ask, for example, "On a scale from one to 10, how safe do you feel the food you purchase from the grocery store is?" Remember that you can choose to ask questions about the food products consumed at home or out of the home; you could ask about food products from different segments of the agricultural industry (i.e. produce, meat, or dairy). Be creative and ask questions that you are interested in finding the answers to. When you have completed your survey, share your results in a graphic form (bar charts, pie charts, etc.) with the rest of the class. How do your results compare to others?
6. **Language Arts.** Conduct an Internet search for *The Jungle* by Upton Sinclair, which outlines the unregulated conditions in the meat processing plants in the country at the turn of the century. Determine the impact that this book had on federal meat processing plant regulations. Write a one-page summary of how this work of literature changed federal policy.

Communicating about Agriculture

1. **Speaking.** In a group of 3–4, create a short video about the importance of addressing hunger on a local level.
2. **Speaking and Writing.** Visit a local elementary school to teach them about safe food handling procedures. Make sure you develop handouts and presentations that are geared toward your intended audience.
3. **Reading and Speaking.** In a group, select one of the digestive systems and create a poster explaining the parts, function, and animals that have your selected type of digestive system. Share it with the class as a way to review digestive system information.

Chapter 10

Large-Animal Production

Santa Gertrudis Breeders International, Kingsville, Texas

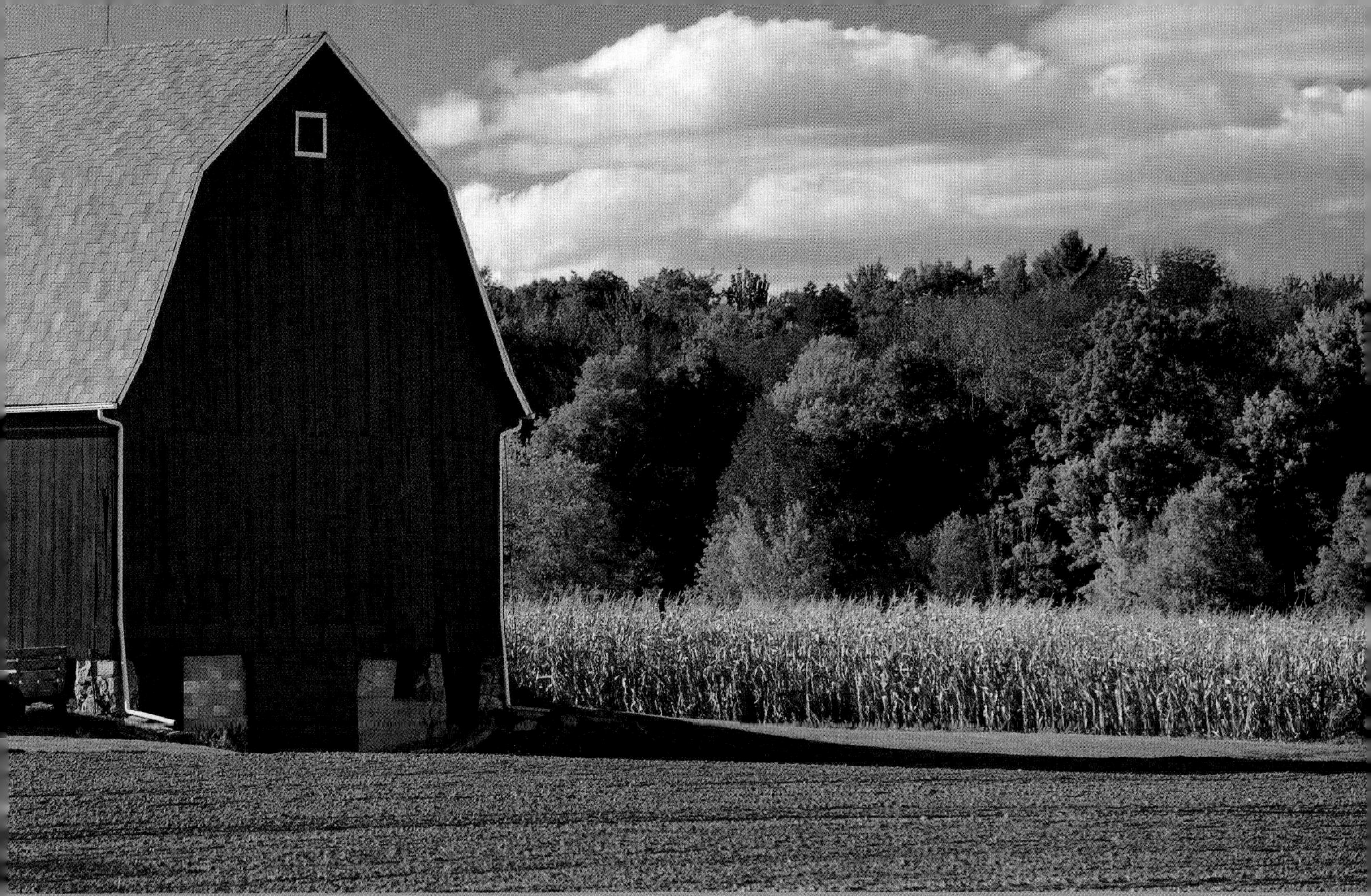

John McCormick/Shutterstock.com

sw_photo/Shutterstock.com

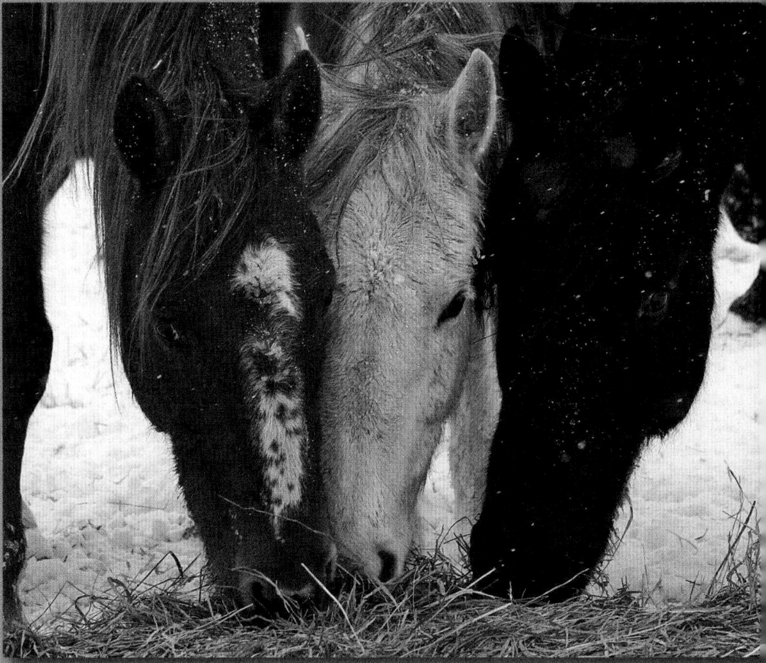

Bureau of Land Management, Jana Wilson

While studying, look for the activity icon to:

- **Practice** vocabulary terms with e-flash cards and matching activities.
- **Expand** learning with video clips, animations, and interactive activities.
- **Reinforce** what you learn by completing the end-of-lesson activities.
- **Test your knowledge** by completing the end-of-chapter questions.

Lesson 10.1

Beef Industry

Words to Know

artificial insemination (A.I.)
backgrounding operation
beef
beef cattle
Bos Indicus
Bos Taurus
bovine
breed
breed association
brisket
bull
calf
calving
chuck
commercial cattle
cow
cow-calf operation
cud
cutability
feedlot
flank
gestation
heifer
herd
loin
offal
plate
polled
primal cut
purebred
purebred breeder
rib

(Continued)

Lesson Outcomes

By the end of this lesson, you should be able to:

- Understand the size and the scope of the beef industry.
- Understand common terminology of the beef industry.
- Identify the major cuts of beef.
- Describe the components of the beef industry.
- Understand commonly used production systems in the beef industry.
- Describe different breeds of beef cattle in the United States.

Before You Read

Do an Internet search for breeds of beef cattle in the United States. Pick three breeds you would like to research and learn more about. Research at least one breed of which you have not heard of before.

The term ***beef cattle*** refers to cattle used to produce beef. ***Beef*** is the meat humans consume from harvested cattle. Examples of beef would be hamburger, steak, and roast beef. Beef cattle are not indigenous to the United States. Many historians report that when Columbus came to America, there were no domesticated animals. Cattle and other domesticated animals came to America on Columbus' second voyage in 1493.

Beef Industry in the United States

Today, the beef industry in the United States is a large, multibillion dollar industry that contributes greatly to our economy. According to the USDA, the beef cattle industry generates $44 billion in economic impact, and there are approximately 87.7 million head of cattle in our country. The average size cow herd in the United States is 40 head. The top five states (in order) in beef cattle production are Texas, Nebraska, Kansas, California, and Oklahoma, **Figure 10-1**.

U.S. Beef Exports

In 2014, the U.S. beef industry exported 5.6 billion pounds of beef with the top export markets being Canada, Japan, Mexico, South Korea, and Hong Kong. In 2013, the average person spent $288.17 on beef in the United States.

Careers

There are many careers within the beef industry. It is easy to become a rancher or cattle producer if your family raises cattle, **Figure 10-2**. What if you grow up in the city or have little experience with cattle? There are many opportunities to become engaged in the beef industry through breed associations, cattle shows, working on a farm or ranch that raises cattle, working at a feedlot, or being a large-animal veterinarian—the opportunities are endless!

Working on a farm, ranch, or at a feedlot requires some education. Working for a breed association usually requires a college degree, and being a veterinarian requires approximately eight years of college.

Words to Know

(Continued)
round
ruminant
seedstock
seedstock cattle producer
steer
stocker operation
subprimal cut

Common Beef Cattle Terms

As with most animals, different terms are used to describe cattle during different stages of their lives:

- ***Calf***—term to describe young
- ***Bull***—young male
- ***Steer***—castrated male
- ***Heifer***—young female
- ***Bull***—mature male
- ***Cow***—mature female
- ***Calving***—act of giving birth in cattle

Did You Know?

In today's society, the average person consumes approximately 70 pounds of beef per year.

Figure 10-1. The top states in beef cattle production. ***Are there certain climates or topography that make an area better suited for raising beef cattle?***

Diamond K Ranch, Hempstead, TX

Figure 10-2. Careers in the beef cattle industry often begin with raising and showing cattle as part of an SAE or 4-H program.

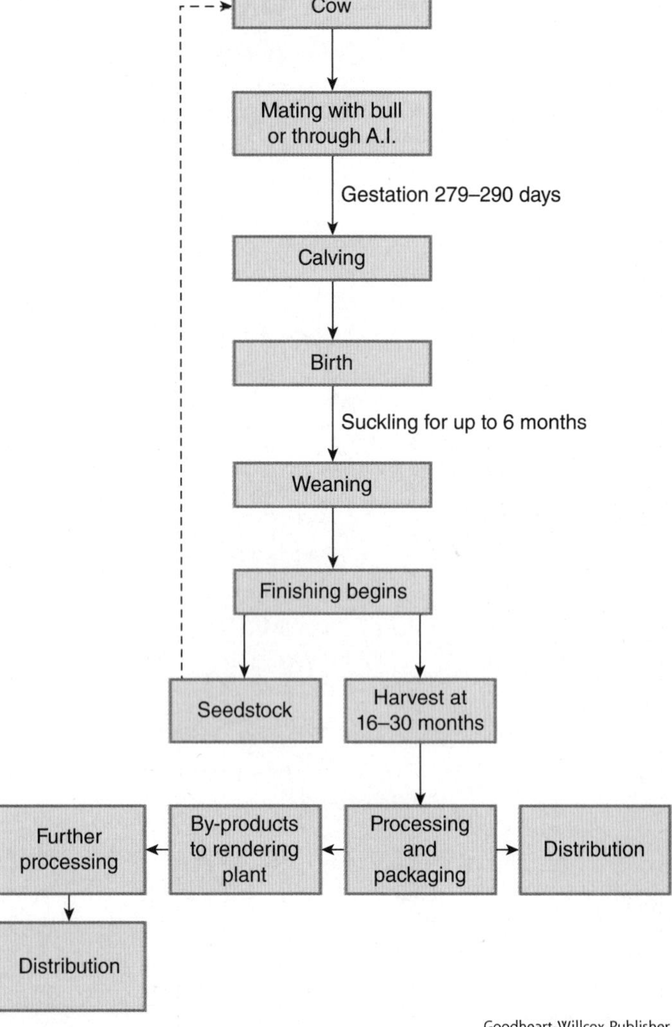

Goodheart-Willcox Publisher

Figure 10-3. The production cycle of beef cattle. After studying the production cycle, visit www.g-wlearning.com to test your knowledge.

Production Cycle of Beef Cattle

Cows are bred through natural means (exposure to a bull) or through artificial insemination. ***Artificial insemination (A.I.)*** is a process where sperm is collected from a bull, processed and frozen for storage, and then thawed and placed in the reproductive tract of the cow. Once the cow is pregnant, she will carry the calf in ***gestation*** (pregnancy) for approximately 279–290 days. The gestation period varies, depending on the breed of cattle. The term *calving* is used to describe a cow who is in the process of giving birth.

When the calf is born, it will nurse its mother until six months of age when it is weaned or separated into a different pen or pasture from the cow. At weaning, calves of both sexes are sent to backgrounding operations to graze and gain weight prior to being sent to a feedlot. Some calves are kept as potential replacement breeding animals. These bulls and heifers are usually grazed and fed until an acceptable breeding age. Bulls usually reach breeding age at two years. Heifers are usually bred when they are 15–18 months of age so they can calve at approximately two years of age, **Figure 10-3**.

The Beef Cattle Industry

There are many different segments of the beef cattle industry. The following are examples of different types of beef cattle operations in the United States. Some of these examples may be common in your geographic region.

Cow-Calf Operations

Most beef cattle are born on ***cow-calf operations***. These operations are commonly seen as you travel down highways or country roads. Cows are bred each year to produce calves for the market, **Figure 10-4**. These operations are usually family-owned with families providing the daily care and maintenance of the cow herd. These producers profit mainly from the sale of calves at weaning age, which is usually around 6–8 months of age.

Red Angus Association

Figure 10-4. Calves nurse for six months before weaning. *Do nursing cows have additional nutritional needs?*

Stocker Operations

When calves are weaned from their mothers (6–8 months), they are often transferred to ***stocker*** or ***backgrounding operations***. These operations use grazing pastures for cattle to provide moderate weight gain with minimal grain being added to their diet. This stage of production usually lasts from when calves are weaned until they are 12–18 months. At this time, they are taken to a feedlot.

Feedlot Operations

In ***feedlots***, large numbers of cattle are grouped based on size, genetics, and consistency in order to maximize profit for the feedlot. Cattle are not allowed to graze at feedlots. They are fed high-grain diets with roughage (hay) to increase fat deposition which, in turn, increases the quality of beef. Cattle usually spend four to six months in the feedlot. Most feedlots are located in the high plains of the United States. The Texas panhandle, Nebraska, Kansas, Iowa, and Colorado are the leading states in feedlot production, **Figure 10-5**.

Santa Gertrudis Breeders International, Kingsville, Texas

Figure 10-5. Cattle on feedlots are grouped to maximize profit for the feedlot. *How does grouping the animals by size, genetics, and consistency help maximize profit for the feedlot?*

Seedstock Operations

Seedstock cattle producers raise cattle that are typically registered with a ***breed association***. They may be referred to as ***purebred breeders***. (A ***purebred*** animal is an animal species or breed achieved through the process of selective breeding.)

They produce ***seedstock***: bulls, heifers, and cows that are used in registered breeding herds as well as commercial cattle operations. These cattle are usually registered in a breed association with a documented pedigree and performance data (birth date, birth weight, weaning weight, yearling weight, maternal/milk or carcass data) that is useful to purebred and commercial cattlemen.

Associations

Each breed of cattle has an association for cattle producers who raise that specific breed. Most breed associations have websites and promotional materials about their breed available so you can learn more about them. Most states have a cattle producers' or cattleman's association. The National Cattleman's Beef Association (NCBA) is a national organization that promotes beef cattle and the beef industry in the United States.

Commercial Operations

Commercial cattle are typically crossbred (composed of two or more breeds) and are not registered with a breed association. The goal of most commercial cattle producers is to produce cattle to sell to stocker operations and ultimately to feedlots to produce beef.

Anatomy of Beef Cattle

In order to understand many of the principles behind the growth and physiology of beef cattle, it is important to know the basic parts of the animal. **Figure 10-6** illustrates the parts of a beef steer. Knowledge of these parts becomes even more critical when participating in FFA Career Development Events, such as Livestock Evaluation or Meats Evaluation.

Beef cattle are ***ruminants***, which means they can digest high amounts of roughage or forage, usually grass or hay. Their stomach is divided into four

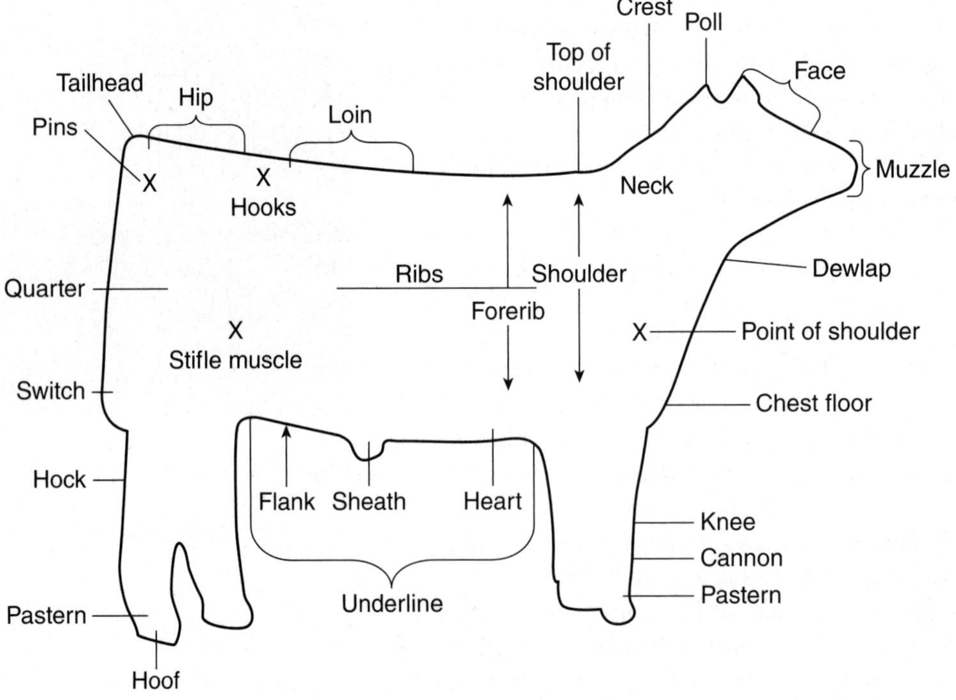

Figure 10-6. Aside from the distinctive physical variations of each breed, all beef cattle share the same basic anatomy. After studying the anatomy image, visit www.g-wlearning.com to test your knowledge.

American Hereford Association

Figure 10-7. A good understanding of natural herd behavior may help workers stay safe and prevent harm to the cattle. *Aside from reading reputable publications, how can you develop your own understanding of herd behavior? Who would be a good resource?*

compartments: rumen, omasum, abomasum, and the reticulum. Ruminants like cattle are commonly seen chewing their ***cud***, which is a regurgitation of the roughage and feed they have eaten to aid in further digestion.

Herd Behavior

Cattle stay together in groups called ***herds***, **Figure 10-7**. It is important to understand how herd animals behave as a group when working with livestock. Each animal will imitate the behavior of other animals in the herd, particularly the nearest ones. This behavior causes the herd to function as a single entity. For example, the herd will move in one direction away from perceived danger, such as a predator. This movement is not planned and occurs spontaneously as lead animals begin the movement and each subsequent animal makes the same move. This is important to understand because any real or perceived danger may cause a herd to panic, creating a dangerous situation for workers and cattle alike.

It is also important to understand the natural hierarchy that forms in a herd. Lead animals establish their position and usually maintain it for extended periods of time. The herd composition continually changes as animals are added or culled (removed from the herd). This means the social structure changes each time the herd changes. Careful introduction of new cattle will help reduce the aggressive behavior and prevent potentially dangerous situations for workers and cattle.

Maintaining Herd Health

Since cattle are herd animals, sickness and disease may easily spread throughout a herd if the animals are not properly vaccinated or if they live in unsanitary conditions. Keep the following points in mind when raising cattle:

- Establish a complete herd health program and consistently implement regular vaccination and parasite control programs.

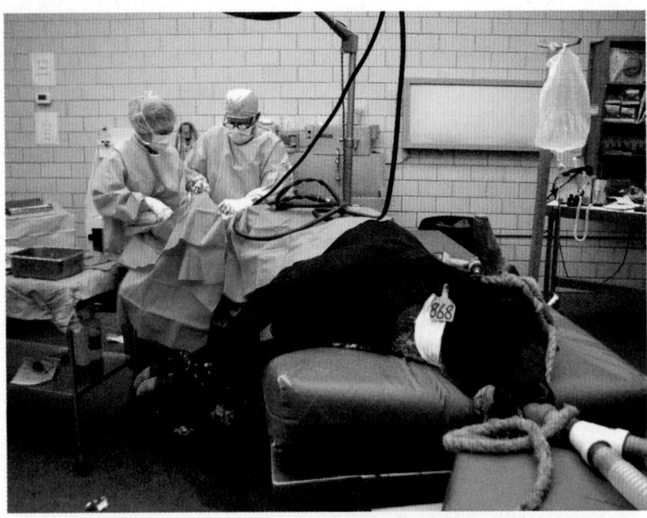

Elgin Veterinary Hospital, Bovine Division, Elgin, Texas

Figure 10-8. Although large-animal vets schedule regular visits to examine and treat animals, animals must sometimes be transported to veterinary facilities for surgery or other specialized treatment. Here, Dr. Gary Warner is performing a complex orthopedic procedure on a bovine athlete (bucking bull).

- Have a good working relationship with a large-animal veterinarian to ensure proper herd health management, **Figure 10-8**.
- Purchase replacement cattle only from reputable breeders who maintain high herd health standards.
- Observe new animals introduced into the herd for possible diseases and parasites.
- Maintain clear, complete, and accurate records for calving, breeding, weaning, and vaccination.
- Monitor cows during calving, especially first-calf heifers, to minimize potential harm to or loss of calves and cows.

Handling Cattle

There is a great deal more information regarding breeding and handling of beef cattle than can be included in this chapter. Many of these concepts will be learned in other agricultural classes or in a real-world career setting working with beef cattle. The following list of tips is not all-inclusive, but each point is important enough to remember when working with or around beef cattle.

- Cattle should be selected according to their intended use, the geographical area and environment in which they will be raised, and the breeder's personal preference of the traits he or she wants to produce with the cattle.
- Maintain human contact with the beef herd to make it easier to handle cattle when necessary.
- Break young calves to lead at an early age if they are going to be exhibited in shows and fairs.
- Maintain the correct ratio of cows to bulls for breeding purposes.
- Provide sufficient access to clean water at all times and provide shelter from harsh weather conditions when possible.
- Use safe actions when working around cattle. Avoid situations that startle or frighten them and always employ enough help when needed.

Cuts of Beef

When beef cattle are harvested for meat, a beef carcass is cut into large primary pieces called primal cuts. The ***primal cuts*** of beef are chuck, rib, loin, round, brisket, plate, and flank, **Figure 10-9**.

- ***Chuck***—primal cut from the shoulder and neck of beef cattle. Flavorful, but may be tough and fatty and contain excess bone and gristle. Cuts of meat from the chuck include: 7-bone roast, arm pot roast (bone-in/boneless), blade roast, eye roast (boneless), eye steak (boneless), mock tender roast, mock tender steak, petite tender, shoulder pot roast (boneless), and top blade steak (flat iron).

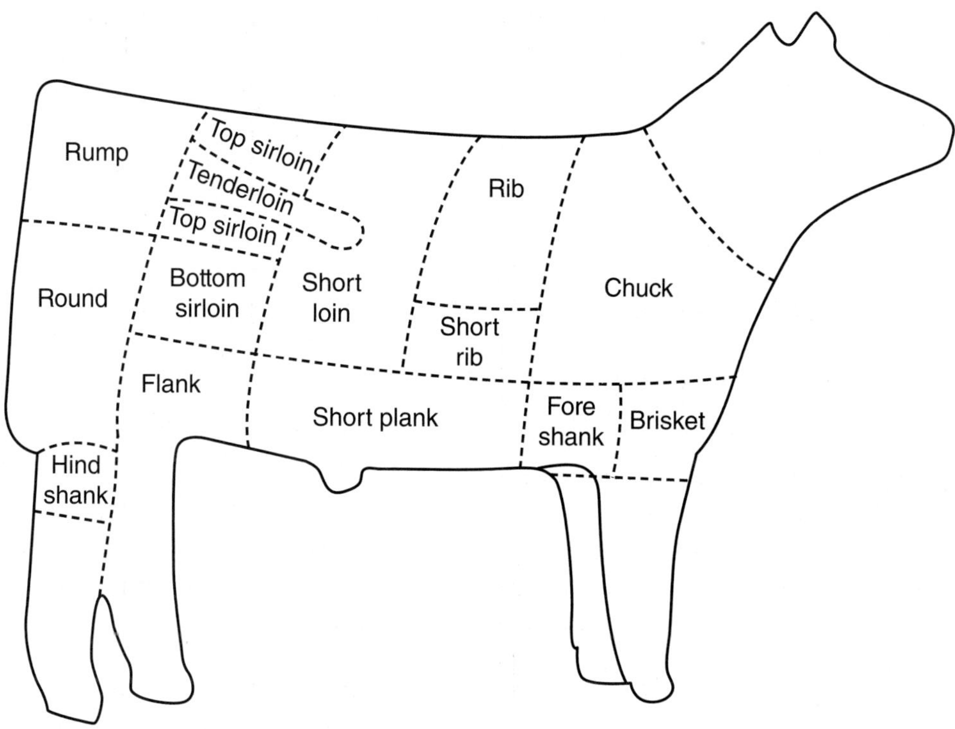

Figure 10.9 The primal cuts of beef include the chuck, rib, loin, round, brisket, plate, and flank. After studying the image, visit www.g-wlearning.com to test your knowledge.

- ***Rib***—tender primal cut made from the center section of rib (specifically the sixth through twelfth ribs). Cuts of meat from the rib include: rib or ribeye roast (large end), rib or ribeye roast (small end), rib steak (small end), rib eye steak (boneless or lip-on), back ribs, and short ribs.
- ***Loin***—primal cut from the section between the ribs and the round, and above the flank. Cuts of meat from this area include: the porterhouse steak, T-bone steak, tenderloin roast, tenderloin steak, top loin steak, top sirloin cap steak, top sirloin steak (cap off), top sirloin steak, and tri-tip roast.
- ***Round***—primal cut of meat from the hindquarters of beef cattle. It is lean and may be tough. Cuts from the round include: bottom round roast, rump roast, eye round roast/steaks, top round steak, round steak, round tip roast, round tip steak, and top round roast.
- ***Brisket***—fairly tough, boneless primal cut that lies over the sternum, ribs, and connecting cartilage of beef cattle. Cuts of meat from this section include: corned beef, flat half, and whole brisket.
- ***Plate***—tough, fatty primal cut of meat from the front belly of beef cattle, just below the rib cut. Cuts of meat from this section include: short ribs and skirt steak.
- ***Flank***—a primal cut from the abdominal muscles of beef cattle. Cuts of meat from this section include the long, flat flank steak.
- ***Offal***—the edible offal of cattle, including the liver. Offal is the entrails and internal organs of animals processed for food.

The steaks, roasts, and hamburger meat that are commonly used as a source of protein in our diets are cut mainly from the round, loin, rib, and chuck. The cuts made from the primal cuts are called ***subprimal cuts***. The way the subprimals are cut determines the quality of the final cut and its cost. Less expensive cuts of beef are cut from the brisket, plate, and flank.

Breeds of Beef Cattle

Did You Know?

There are more than 50 breeds of cattle in the United States and over 800 breeds recognized worldwide.

The species name for cattle is ***bovine***. Within the bovine species, there are two subspecies, ***Bos Taurus***, which describes the breeds that originated in Europe (England, Scotland, France, Germany, and Italy), and ***Bos Indicus***, which describes cattle from more tropical countries (Asia, Africa, and India), **Figure 10-10**. A ***breed*** is a specific group of cattle that has similar appearance, characteristics, and behaviors that distinguish it from other cattle in the same species.

Bos Indicus breeds of beef cattle have Brahman influence in their pedigree. They are typically well suited to warmer climates and have a higher resistance to diseases and insects. They are very popular in the southeastern part of the United States because of their heat tolerance and ability to thrive on limited forage.

There are many other breeds of beef cattle other than those covered in the following sections. Several of the breeds discussed in the following sections also have miniature breeds that have evolved such as Miniature Herefords, Lowline Angus, and Minature Zebu to name a few. To find out more about breeds of beef cattle, contact the specific breed association.

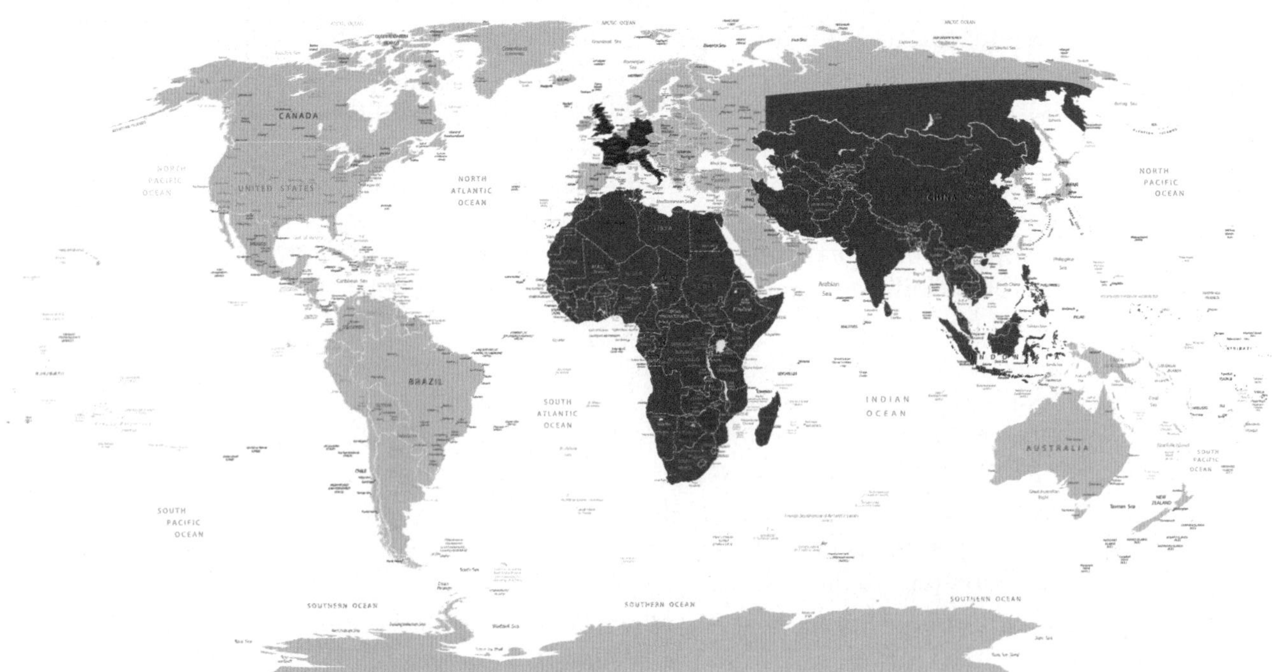

ekler/Shutterstock.com

Figure 10-10. The subspecies *Bos Taurus* and *Bos Indicus* each originated in specific areas of the world. ***Are there certain characteristics or traits that developed in each subspecies that enabled them to thrive in their respective environments?***

Bos Taurus Cattle Breeds

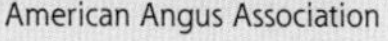

American Angus Association

American Angus Association

Angus. Angus cattle originated in Scotland in the shires of Aberdeen and Angus. They are black in color and naturally ***polled*** (hornless). Angus are known for their maternal ability as well as carcass quality. Angus cattle are found in all 50 states in the United States, and the American Angus Association's branded beef program, Certified Angus Beef (CAB), is world renowned.

Thomas Ranch, Harrold, South Dakota

Thomas Ranch, Harrold, South Dakota

Charolais. Charolais is one of the oldest French breeds of cattle. Charolais are solid white in color and are known for high growth rates and heavy muscling. Most Charolais cattle are naturally horned, but some are polled. They are commonly used in crossbreeding programs to produce smoky-gray colored calves that perform well in stocker and feedlot operations.

American Chianina Association

American Chianina Association

Chianina. Chianina cattle originated in Italy. They are one of the oldest breeds of cattle in the world. Chianinas are white in color with black skin pigment. In the United States, Chianinas are predominantly used in crossbreeding programs and will be black in color. They are known for exceptional growth. The terms Chi (pronounced key), Chi-Angus, and Chi-Maine are all associated with modern day Chianina-influenced cattle.

Bos Taurus Cattle Breeds (*Continued*)

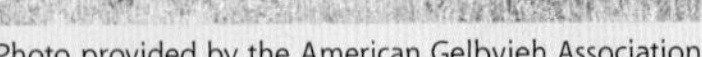

Photo provided by the American Gelbvieh Association

Photo provided by the American Gelbvieh Association

Gelbvieh. Gelbvieh cattle came from Germany. They are a reddish-yellow color and are known for having excellent maternal characteristics. Many Gelbviehs in the United States today are black in color as a result of crossbreeding programs. The term *balancer* is used to describe Gelbvieh cattle that have 25%–75% Gelbvieh in their pedigree, with the remaining portion of the breed makeup being Angus or Red Angus.

Provided by American Hereford Association

Provided by American Hereford Association

Hereford/Polled Hereford. Hereford cattle originated in Herefordshire, England. Herefords are easily recognizable with their white face, red body, and white belly and legs. There are more Hereford cattle registered than any other breed of cattle. They are known for their longevity and being docile. Polled Herefords were developed in Iowa by Warren Gammon. He sought out naturally polled Hereford cattle and started developing them as a breed. In 1995, the American Hereford Association and the American Polled Hereford Association merged to create the American Hereford Association, which today registers polled and horned Herefords.

North American Limousin Foundation

North American Limousin Foundation

Limousin. Limousin cattle originated in France. When they were imported into the United States in the late 1960s, they were yellow to red in color. These cattle were known for their cutability and heavy muscling. (***Cutability*** is the proportion of lean, salable meat yielded by a carcass.) Although most Limousin cattle are black in color, some red genetics still exist. Today, Lim-Flex cattle are gaining popularity. Lim-Flex must be 25%–75% Limousin and 25%–27% Angus or Red Angus and may have 12.5%, or 1/8, of their breed composition being another breed.

BOSS Beauty 84M

DMCC Limited Edition

Maine Anjou. Maine Anjou cattle originated in France and were the result of mating English Shorthorn bulls with Mancelle cows. They were originally bred to be draft animals. The first Maine Anjou semen was imported into the United States in 1970. These cattle were dark red and white in color. Today, most Maine Anjou cattle are black as a result of crossbreeding. Maine Anjou cattle are fast growing and produce high-quality carcasses. Over the last couple of decades, they have gained popularity in the club calf sector of the beef industry.

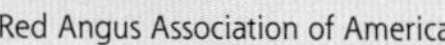

Red Angus Association of America

Red Angus Association of America

Red Angus. The Red Angus breed originated as a result of a recessive gene in Angus cattle. They have the same origin as Angus cattle, which are black in color. In the mid 1950s, a group of breeders chose to start raising and registering Red Angus cattle as its own breed. Red Angus have many of the same traits as Angus cattle; they are naturally polled, but due to their red color, are much more heat tolerant than black Angus cattle.

V8 Ranch, Wharton, TX

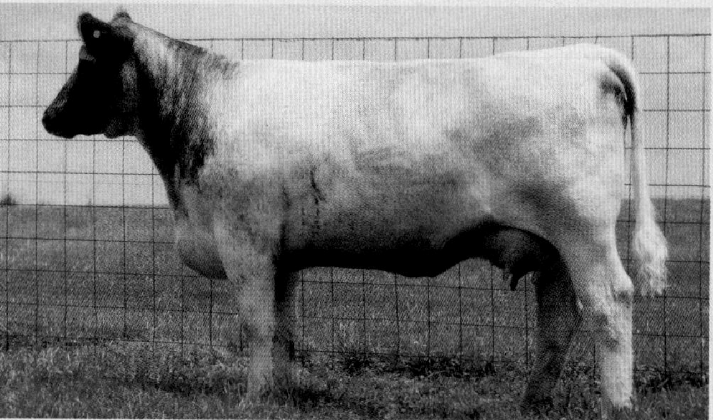

V8 Ranch, Wharton, TX

Shorthorn. The Shorthorn breed originated in northern England. They were a dual-purpose breed, used for milk and meat, and were originally known as Durhams. They are one of the oldest breeds of cattle. They are red, white, or roan in color and are horned. They are known for their maternal characteristics and their usefulness in crossbreeding programs. ShorthornPlus is a program designed to recognize and market Shorthorn-influenced genetics with animals that have 1/4 to 7/8 Shorthorn in their pedigree.

Bos Taurus Cattle Breeds (*Continued*)

Photo courtesy of the American Simmental Association, www.simmental.org

Photo courtesy of the American Simmental Association, www.simmental.org

Simmental. Simmental cattle originated in Switzerland and date back to the Middle Ages. Originally, they were a dual-purpose breed used to produce milk and meat. Since arriving in the United States in the late 1960s, they are noted for their fast growth, heavy muscle, and cutability. Simmentals in this era were typically red to yellow with white faces and were sometimes spotted. Today, most Simmentals are black or black with a white face. The terms *SimAngus* and *SimSolutions* refer to Simmental-influenced cattle that are crossed with other breeds, typically Angus.

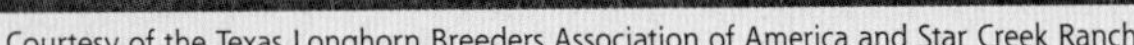

Courtesy of the Texas Longhorn Breeders Association of America and Star Creek Ranch

Courtesy of the Texas Longhorn Breeders Association of America and Star Creek Ranch

Texas Longhorn. Texas Longhorns are descendants of the Spanish cattle brought over on Columbus' second voyage in 1493. Many of these cattle migrated north from Mexico and became acclimated to the environment of the southwestern United States. Texas Longhorns have horns that curve upward and may spread up to four to five feet. They are highly fertile and adapt to harsh conditions. They can survive on little forage, have excellent calving ease, and are known for their longevity.

Bos Indicus Cattle Breeds

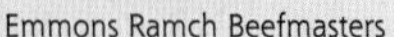

Emmons Ramch Beefmasters

Emmons Ramch Beefmasters

Beefmaster. Beefmaster cattle originated in south Texas in the 1930s. Tom Lasater developed this breed by crossing Hereford, Shorthorn, and Brahman cattle. There are no breed standards for color. Many Beefmasters are red, gold, or yellow in color, but some are black or even spotted. Beefmasters are hardy cattle that thrive in harsh conditions. They have excellent maternal ability and fertility.

Courtesy of the United Braford Breeders, Inc.

Courtesy of the United Braford Breeders, Inc.

Braford. Braford cattle were developed by crossing Brahman cows with Hereford bulls. This started at a ranch in Florida in the late 1940s. By crossbreeding these two breeds, cattlemen were able to use positive traits of each breed to form a breed that was well suited for its environment. Brafords have a Hereford-type color pattern: they are dark red with a white face and underline. They are typically 5/8 Hereford and 3/8 Brahman. Brafords are known for their maternal ability, high growth rate, and adaptability to a given environment.

Diamond K Ranch, Hempstead, TX

Diamond K Ranch, Hempstead TX

Brangus. Brangus cattle were derived by crossing Angus and Brahman cattle to arrive at a 3/8 Brahman and 5/8 Angus cross. The early crossing of these two breeds can be traced back to 1912 at a USDA Experiment Station in Louisiana. Brangus cattle are solid black and polled. They are known for maternal ability, rapid growth, and adaptability.

Red Brangus. Red Brangus cattle originated in Texas in the late 1940s and are a result of crossing Brahman and Angus cattle. The unique thing about Red Brangus is that the breed percentages of Brahman and Angus can vary. Cattle can be registered with a 5/8 Brahman and 3/8 Angus cross, 1/2 Brahman and 1/2 Angus, or 3/8 Brahman and 5/8 Angus. This allows for geographic adaptability of a certain composite of breeds. Red Brangus cattle are hardy, red, polled, and well suited to a variety of conditions.

Bos Indicus Cattle Breeds (*Continued*)

V8 Ranch, Wharton, TX; American Braham Breeders Association

V8 Ranch, Wharton, TX; American Braham Breeders Association

Brahman. Sometimes referred to as Zebu cattle, Brahman cattle were imported into the United States from India in the late 1800s and early 1900s. Brahman cattle are easily distinguished from other breeds: they have a large hump over their shoulders, black pigment, and long droopy ears. They also have excess skin hanging from their neck and throat region. Brahmans may be light gray, dark gray, red, or nearly black in color. Light gray is the most common color. They are typically horned and are used extensively in crossbreeding programs. Brahman cattle are known for their maternal ability and growth rate.

La Muneca Ranch, Linn, TX

La Muneca Ranch, Linn, TX

Simbrah. Simbrah cattle are a result of crossing Brahman and Simmental breeds. This breed evolved in the 1960s in the gulf coast region of the United States. Simbrahs are typically red with white markings, but there is no set color pattern for this breed. They combine positive traits of the Brahman and Simmental breeds and make excellent mothers, while having high performance and growth with added muscling. The Simbrah registry is housed at the American Simmental Association.

Santa Gertrudis Breeders International, Kingsville, Texas

Santa Gertrudis. Santa Gertrudis cattle were developed on the King Ranch in south Texas. The breed is composed of 3/8 Brahman and 5/8 Shorthorn. Santa Gertrudis cattle are dark cherry red in color and may be horned or polled. This breed was developed to survive in the harsh conditions of south Texas where native grasses are often sparse. Santa Gertudis are hardy, disease-resistant cattle that easily adapt to many different production scenarios.

Lesson 10.1 Review and Assessment

Words to Know

Match the key terms from the lesson to the correct definition.

1. A term to describe young and mature male beef/dairy cattle.
2. A primal cut of meat from the rear region of beef cattle.
3. Primal cut from the shoulder and neck of beef cattle; flavorful, but may be tough and fatty and contain excess bone and gristle.
4. A specific group of cattle that has similar appearance, characteristics, and behaviors that distinguish it from other cattle in the same species.
5. A beef cattle operation in which young calves are allowed to graze pastures for moderate weight gain, before moving to feedlots.
6. The term used to describe a cow who is in the process of giving birth.
7. A term to describe young beef/dairy cattle.
8. A fairly tough, boneless primal cut that lies over the sternum, ribs, and connecting cartilage of beef cattle.
9. A cattle operation in which large numbers of cattle are grouped based on size, genetics, and consistency in order to maximize profit; cattle are fed high-grain diets and not allowed to graze.
10. Usually crossbred beef cattle produced for sale to stocker operations and feedlots.
11. An operation used to breed and birth beef cattle for sale.
12. Cattle that lack horns by natural or artificial means.
13. A term to describe young, female beef/dairy cattle that have not been bred.
14. The cattle breeds that originated in Europe.
15. A term to describe mature female beef/dairy cattle.
16. A group of cattle of a single kind that is kept together for a specific purpose.
17. The cattle breeds that originated in tropical countries such as Asia, Africa, and India.
18. The proportion of lean, salable meat yielded by a carcass.
19. A tender primal cut made from the center section of rib.

A. backgrounding operation
B. *Bos Indicus*
C. *Bos Taurus*
D. breed
E. brisket
F. bull
G. calf
H. calving
I. chuck
J. commercial cattle
K. cow
L. cow-calf operation
M. cutability
N. feedlot
O. flank
P. gestation
Q. heifer
R. herd
S. loin
T. plate
U. polled
V. primal cut
W. rib
X. round
Y. seedstock
Z. steer

20. A primal cut from the section between the ribs and the round, and above the flank of beef cattle.
21. The process of carrying young in the womb between conception and birth; pregnancy.
22. Castrated beef/dairy cattle.
23. Term used to describe the large primary pieces into which a beef carcass is divided.
24. Bulls, heifers, and cows that are used in registered breeding herds as well as commercial cattle operations.
25. A tough, fatty primal cut of meat from the front belly of beef cattle, just below the rib cut.
26. A primal cut from the abdominal muscles of beef cattle.

Know and Understand

Answer the following questions using the information provided in this lesson.

1. Which beef cattle breeds are indigenous to the United States?
2. What is the approximate gestation period for beef cows?
3. At what age are heifers usually bred?
4. At what age are calves weaned and sold in cow-calf operations?
5. Identify the type of diet beef cattle enjoy while living in feedlot operations. Explain why they are fed this type of diet.
6. List the types of performance data kept on record for purebred cattle.
7. Explain why cattle chew cud.
8. Explain why it is a good idea to maintain some level of human contact with a herd of beef cattle.
9. List the primal cuts of beef.
10. Explain the difference between *Bos Taurus* and *Bos Indicus*.

Analyze and Apply

1. How has the beef industry changed since it first began in the United States? Research the development of the beef industry in the United States. Find at least five major events that affected the course of the industry. How were these events significant?
2. The beef industry is strong in many other countries around the world. Aside from the United States, which countries produce and export a significant amount of beef and beef products? Who purchases or imports the most beef from these countries?
3. How large is the organic beef industry in the United States? How are these animals raised? What are they fed? How is it determined whether or not the cattle have been raised organically?

Thinking Critically

1. What are some of the ethical issues surrounding the beef industry? Research these issues and choose a specific area that interests you. Use your research to write a brief argument for or against a specific issue.
2. The waste from cattle operations produces a considerable amount of methane. Is this methane production detrimental to the environment? How can methane produced from animal waste be used productively?

Lesson 10.2

Dairy Industry

Words to Know

artificial insemination (A.I.)
butter
buttermilk
cheese
condensed milk
cream
evaporated milk
frozen yogurt
homogenization
ice cream
mastitis
milk
milking parlor
parturition
somatic cell
sour cream
teats
udder
yogurt

Lesson Outcomes

By the end of this lesson, you should be able to:

- Understand the size and scope of the U.S. dairy industry.
- Understand common terminology of the dairy industry.
- Identify common dairy products.
- Describe the components of the U.S. dairy industry.
- Understand commonly used production systems in the dairy industry.
- Describe different breeds of dairy cattle in the United States.

Before You Read

Take a moment to compare and contrast the differences and similarities between dairy cattle and beef cattle. Share the results with your classmates and teacher.

Dairy cattle are cattle raised for their ability to produce large amounts of ***milk***. According to the USDA, there are approximately 9.2 million head of dairy cattle in the United States. Dairy cattle are found in all 50 states of the United States. More than 19% (1.78 million) of the total number of dairy cows are found in the state of California. The top five states (in order) in dairy production are California, Wisconsin, New York, Idaho, and Pennsylvania, **Figure 10-11**.

Dairy Industry in the United States

The U.S. dairy industry contributes approximately $140 billion to our economy. There are more than 50,000 dairies in the United States and more than 95% are family owned and operated. The average herd size for a dairy is 115 cows. Three-fourths of U.S. dairies have less than 100 cows, **Figure 10-12**. Eighty-five percent of milk produced in our country comes from dairies that have more than 100 cows. Dairy farms in the United States produce approximately 23 billion gallons of milk per year. The average dairy cow will produce almost 7 gallons of milk a day. That totals more than 2,500 gallons per year.

The main commodity produced from dairy cattle is milk and milk products. However, when dairy cattle are harvested, they also produce beef. Dairy calves account for approximately 14% of beef production in the United States. One in every five pounds of beef consumed in our country comes from dairy cattle.

Mirec/Shutterstock.com

Figure 10-11. Surprisingly, the top five states in dairy production are not all in the same area of the country. *What makes a particular area more conducive to dairy cattle production than another?*

Wasan Srisawat/Shutterstock.com

Figure 10-12. Dairy operations are high-maintenance, labor-intensive operations. *What are some of the jobs or careers involved in dairy cattle production?*

Africa Studio/Shutterstock.com

Figure 10-13. Cheese is only one of the many dairy products most people consume on a regular basis. *Have you ever made cheese? Was it difficult? Do you think any one kind is more complicated to produce?*

Dairy Products

Some of the most commonly consumed dairy products are types of fluid milk: fat-free or skim milk, low fat (1%) milk, reduced fat (2%) milk, and whole milk. Other commonly consumed dairy products are made from milk and include:

- ***Cream***—the thick white or pale yellow fatty liquid that rises to the top of milk that is left to stand (unless it is homogenized). ***Homogenization*** is a process in which the milk's fat droplets are emulsified and the cream does not separate.
- ***Yogurt***—a semisolid food made of milk and milk solids fermented by two added cultures of bacteria (*Lactobacillus bulgaricus* and *Streptococcus thermophilus*).
- ***Butter***—a soft yellow or white food churned from milk or cream and used to spread on food, in cooking, and in baking.
- ***Sour cream***—cream that has been fermented with certain kinds of bacteria. The bacterial culture sours and thickens the cream.
- ***Buttermilk***—the slightly sour liquid left over after butter has been churned. It is used in baking and consumed as a drink.
- ***Ice cream***—a sweet, frozen food made from no less than 10% butterfat and made in a myriad of flavors.
- ***Frozen yogurt***—a frozen dessert that is lower in fat than ice cream and contains yogurt cultures that may or may not be active.
- ***Cheese***—a solid food made from milk curd that is produced in a range of flavors, textures, and forms. The milk curd is pressed together and may be seasoned, or allowed to age or ripen with bacterial cultures, **Figure 10-13**.

STEM Connection Is It a Dairy Product?

Many of us assume all the dairy products we purchase contain milk or cream. But do they? Make a list of all the dairy products you can think of, and then visit your local supermarket. Look at the ingredients labels. Do they all contain milk or cream? Were you surprised by your findings? What ingredients surprised you the most?

What other foods contain dairy products? Look at the labels of some of your favorite snack foods. Many flavored crackers, potato chips, and other snack foods contain dairy products as well.

nevodka/Shutterstock.com

- ***Evaporated milk***—a milk product made by removing about 60% of the water from ordinary milk. It may be used as a substitute for milk or cream.
- ***Condensed milk***—a heavily sweetened milk product made by removing about 60% of the water from ordinary milk. It may be used in baking and desserts.

Milk is used in many milk-based desserts including pudding (instant and cooked), custard (frozen and baked), and ice milk. It is used with breakfast cereals (cold and hot), in soups, and as the base for many types of sauces. It is easy to see the dairy industry's influence on our food supply with the wide variety of products that come from milk.

Did You Know?

The National FFA Organization has a career development event called Milk Quality and Products. This competition is designed to teach students about milk quality and safety along with manufacturing and marketing of milk and dairy products. Ask your FFA advisor for more information.

Common Dairy Cattle Terms

The terminology used to describe dairy cattle is identical to the terminology used to describe beef cattle during different stages of their life:

- Calf—term to describe young dairy cattle
- Bull—young male
- Steer—castrated male
- Heifer—young female
- Bull—mature male
- Cow—mature female
- Calving—act of giving birth

Production Cycle of Dairy Cattle

The production cycle of dairy cattle is similar to that of beef cattle; however, some aspects of the production cycle are vastly different, **Figure 10-14**. ***Parturition***, the act of giving birth, is what causes milk production in cows. When dairy cattle give birth, their calves are only allowed to stay with the cow and nurse for a short period of time, usually a few hours. The newborn calves are fed milk or milk replacer from a bottle until they reach 6–8 weeks of age.

Heifers are usually retained and placed into a replacement program where they are grown out to 15 months of age and bred. They will typically have their first calf at two years of age. Bull calves are usually grown out in similar fashion; most will be castrated and grown out for beef production. Only elite bull calves with superior genetics are selected to keep as bulls. Most dairy producers use artificial insemination (A.I.) to breed their cows and do not keep a bull at their farm. ***Artificial insemination (A.I.)*** is a process where sperm is collected from a bull, processed and frozen for storage, and then thawed and placed in the reproductive tract of the cow. A major advantage of A.I. in dairy cattle is that producers can breed their cows to the top bulls in the industry without having to ship either of the animals.

Culling the Dairy Herd

A dairy cow will usually stay in the production herd about five years. There are several factors that can lead to cows being culled from the dairy herd:

- ***Mastitis*** is inflammation of the mammary glands, which is usually caused by bacteria.

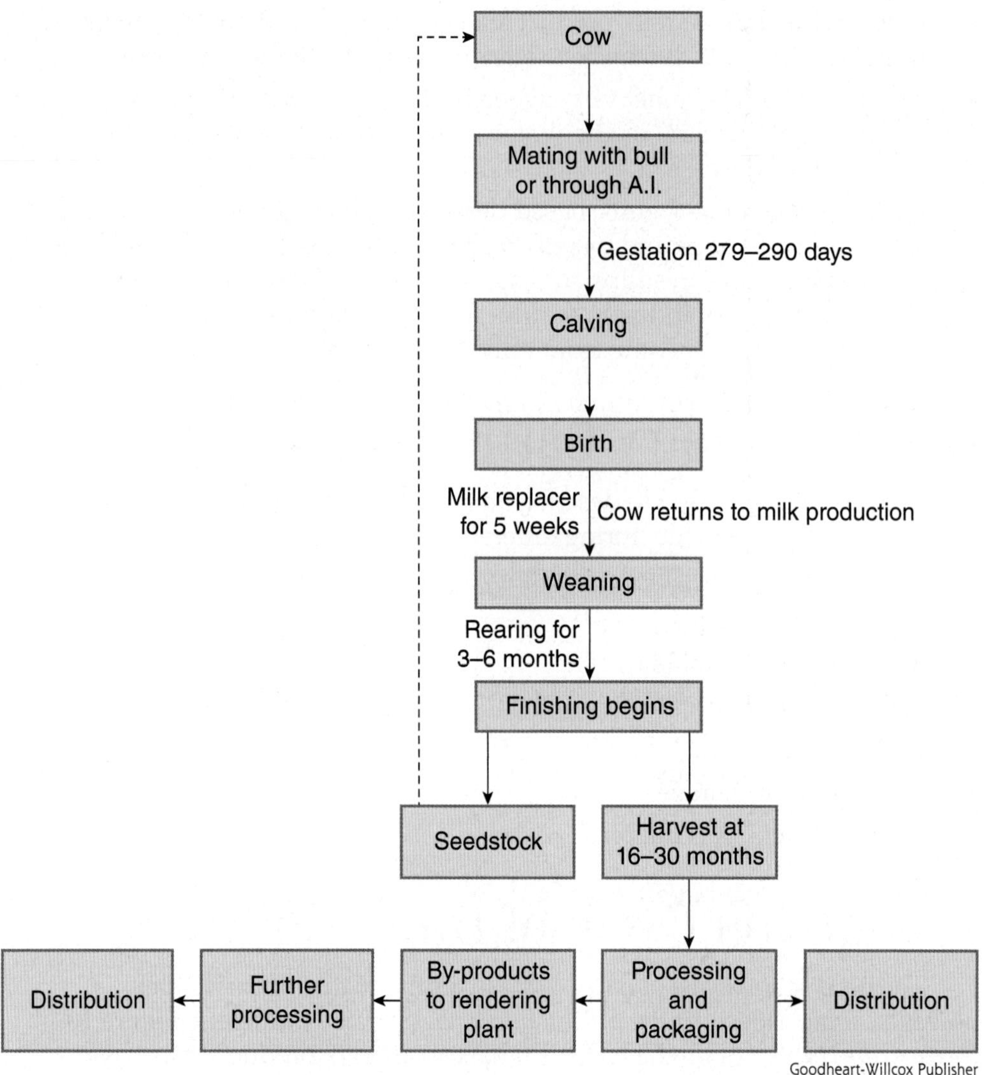

Figure 10-14. The production cycle of dairy cattle production is similar to that of beef cattle production. ***What are the main differences between the two cycles?*** After studying the production cycle, visit www.g-wlearning.com to test your knowledge.

- Inflammation of the udder is swelling and discoloration of the udder due to infection or injury. The ***udder*** is the baglike mammary gland of cattle (and other livestock) that has two or more teats hanging near the hind legs. The ***teats*** are the nipples of the mammary gland from which the milk is extracted. See **Figure 10-15**.
- Lameness is when the cow is having trouble moving on its feet and legs. Lameness becomes a serious issue for dairy cattle because they walk mainly on concrete to enter the milking parlor.

Milking Parlor

The ***milking parlor*** is the section of a dairy where the cow is moved to in order for milking to occur. There are many different types and designs of milking parlors, **Figure 10-16**. When the cow moves into the milking parlor, there is a multistep process that occurs prior to and after milking:

- The cow's udder and teats must be cleaned and dried prior to milking.
- The milking equipment must be working properly and must be attached to the udder correctly to ensure maximum production.
- After the cow is milked, her teats must be disinfected with a teat dip. This prevents the spread of mastitis and other afflictions from cow to cow.
- Once the milk leaves the cow, it must be cooled to a temperature below 45°F within a two-hour period.
- The milk is stored in tanks before the truck comes to transport it to a processing facility.
- The milk is tested before processing to make sure it meets health and safety standards.

smereka/Shutterstock.com

Figure 10-15. It is vital for dairy operations to keep the udders and teats of their herd in top condition in order to maintain maximum milk production. *What steps are taken to keep the udders free of infection or from physical damage? Are these steps performed on a daily basis?*

Dairy farmers are required to meet specific standards and are paid a premium for milk that meets the industry standards. The requirements involve bacteria counts, ***somatic cell*** (white blood cell) counts, no added water, and no antibiotics present in the milk.

Dairy cows are typically milked twice a day. Some producers have seen increased production when milking three times a day, but this is time-consuming and cost-prohibitive for most dairy operations.

Pavel L Photo and Video/Shutterstock.com

Picsfive/Shutterstock.com

Figure 10-16. Today's milking parlors are complex operations requiring daily cleaning and maintenance. *How would extensive knowledge of plumbing benefit someone working in a dairy operation?*

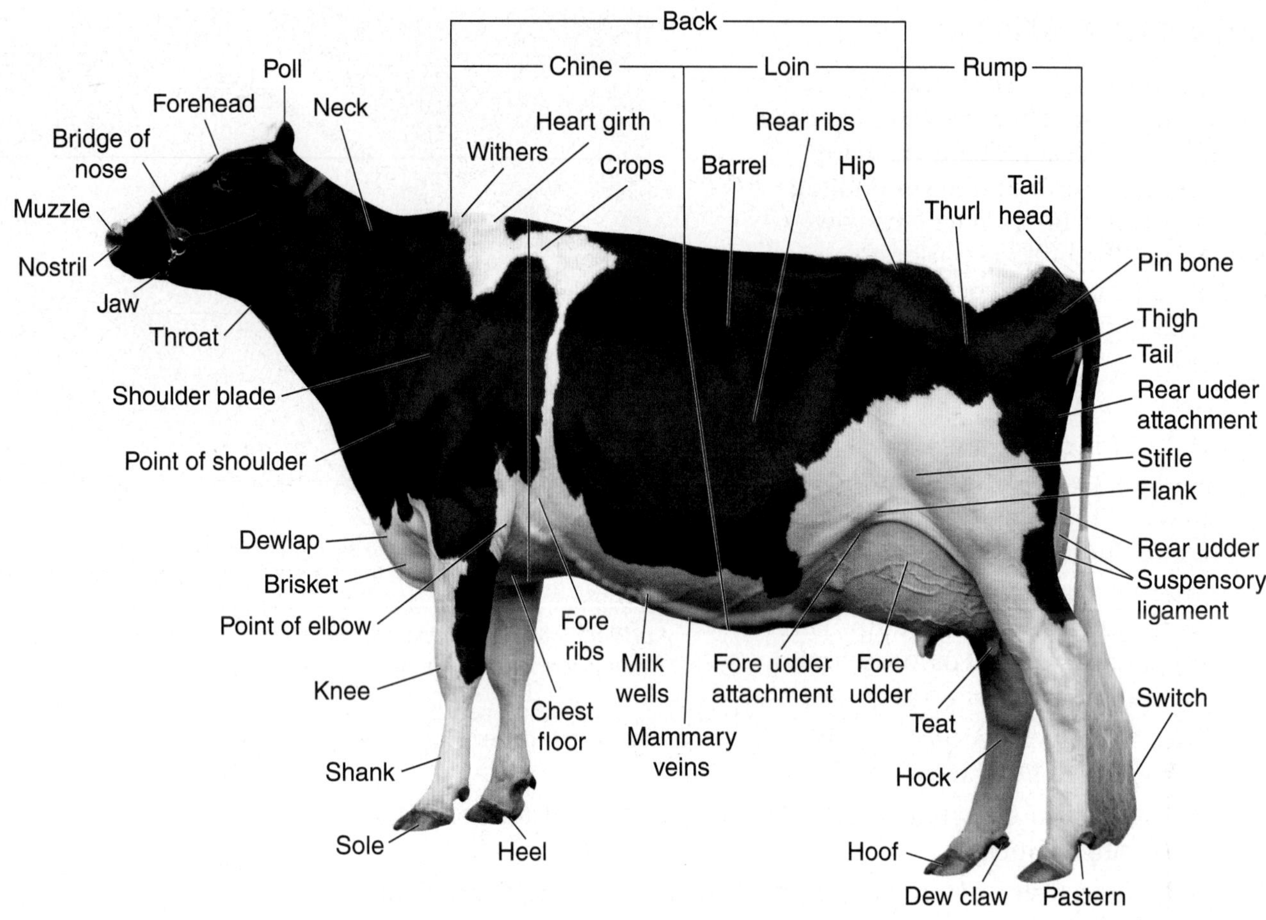

Figure 10-17. Dairy cow anatomy. After studying the image, visit www.g-wlearning.com to test your knowledge.

Anatomy of Dairy Cattle

Learning the basic parts of dairy cattle is a fundamental step in learning to evaluate and judge dairy cattle, **Figure 10-17**. To further develop your knowledge and understanding of the confirmation and makeup of dairy cattle, talk to your teacher about the Dairy Cattle Evaluation and Management Career Development Event and the Dairy Handlers event in FFA.

Maintaining Herd Health

Since cattle are herd animals, sickness and disease may easily spread throughout a herd if the animals are not properly vaccinated or if they live in unsanitary conditions. Keep the following points in mind when raising cattle:

- Establish a complete herd health program and consistently implement regular vaccination and parasite control programs.
- Have a good working relationship with a large-animal veterinarian to ensure proper herd health management.
- Purchase replacement females only from reputable breeders who maintain high herd health standards.

- Observe new animals introduced into the herd for possible diseases and parasites.
- Maintain clear, complete, and accurate records for calving, breeding, and vaccination.
- Monitor cows during calving, especially first-calf heifers, to minimize potential harm to or loss of calves and cows.

Handling Dairy Cattle

There is a great deal more information regarding breeding and handling of dairy cattle than can be included in this chapter. Many of these concepts will be learned in other agricultural classes or in a real-world career setting working on a dairy or in the dairy industry. The following list of tips is not all-inclusive, but each point is important enough to remember when working with or around dairy cattle.

- Dairy cattle should be selected according to their intended use, the geographical area and environment in which they will be raised, and the breeder's personal preference of the traits he or she wants to produce with the dairy cattle.
- Remain calm and handle dairy cattle with ease when moving them in and out of the milking parlor to avoid slips, falls, and feet and leg injury.
- Extra caution must be exercised if the dairy producers uses bulls rather than artificial insemination. Dairy bulls are very mean and aggressive.
- Provide sufficient access to clean water at all times and provide shelter from harsh weather conditions when possible, **Figure 10-18**.
- Use safe actions when working around dairy cattle. Avoid situations that startle or frighten them, especially when approaching them in the milking process.

Did You Know?

Some of the top-producing Holstein cows have been known to produce more than 72,000 lb of milk a year! That is under ideal conditions as far as nutrition and milking three times a day versus two.

Catherine311/Shutterstock.com

Figure 10-18. It is essential for any living organism to have access to a clean and safe water supply. *How would insufficient access to a safe water supply affect a dairy cow's milk production? How would it affect the dairy operation's profits?*

Breeds of Dairy Cattle

The species name for dairy cattle is bovine. Dairy cattle are classified under the subspecies *Bos Taurus*. A breed is a specific group of cattle that has similar appearance, characteristics, and behaviors that distinguish it from other cattle in the same species. Let us take a look at the most popular dairy breeds in the United States and their characteristics.

Breeds of Dairy Cattle

Courtesy of Hoard's Dairyman

Ayrshire. Ayrshire cattle originated in the county of Ayr in Scotland prior to the 1800s. They were first imported into the United States in 1822. Ayrshires are red and white in color and are known for their udder quality, efficiency, and longevity. Ayrshire are moderate in size with mature cows weighing around 1,200 lb. They produce high-quality milk, which is popular in manufactured products.

Courtesy of Hoard's Dairyman

Brown Swiss. Probably the oldest of all dairy breeds, the Brown Swiss originated in Switzerland. They were developed in the Swiss Alps and are known for their ability to adapt to different environments. Brown Swiss were imported into the United States in 1869. The Brown Swiss illustrated here is predominantly brown in color with a black nose encircled by a white ring. Brown Swiss are large-framed cattle. Mature cows will weigh around 1,400–1,500 lb. They are known for their docile temperament. The high protein content of their milk makes it highly desirable in the cheese industry.

Courtesy of Hoard's Dairyman

Guernsey. Guernsey cattle can be traced back to the Isle of Guernsey off the coast of France. The Guernsey breed was developed by French monks from crossbreeding and can be traced back over 1,000 years. They were first imported in the United States in 1831. Guernseys are fawn colored with white markings. Mature cows are smaller to moderate in size, weighing about 1,100–1,200 lb. They produce milk that is high in fat content and golden in color. Guernsey cattle mature early, adapt well, and are very docile.

Courtesy of Hoard's Dairyman

Holstein. Holstein is the most popular breed of dairy cattle in the United States and is the world's largest dairy breed. They make up more than 90% of the dairy cattle in this country. Holstein cattle originated in the Netherlands over 2,000 years ago. One of the provinces they were developed in is called *Friesland*. In other countries of the world, Holsteins are called Friesians. They are large-framed cattle with mature cows weighing more than 1,400 lb. Holsteins are known for their excellent milk production. The average Holstein cow produces more than 23,000 lb of milk a year.

Courtesy of Hoard's Dairyman

Jersey. Jersey cattle are the second most popular dairy breed in the United States. Jersey cattle originated on the Isle of Jersey in the English Channel off the coast of France. They are one of the oldest dairy breeds and were imported into the United States in the early 1800s. Jerseys are the smallest breed of dairy cattle in terms of size. Mature cows are usually 1,000 lb or lighter. They are cream to fawn color and sometimes almost black. Their points are black and they have a distinctive dished face. Jerseys are known for the exceptional fat content in their milk. They rank first among all dairy breeds in fat content, which makes their milk popular for products such as ice cream.

Courtesy of Hoard's Dairyman

Milking Shorthorn. Shorthorns date back to the early 1500s and are probably the oldest breed of cattle. The Milking Shorthorn is a portion of the Shorthorn breed, which was a dual-purpose breed. Milking Shorthorns were developed in Northeastern England in the valley of the Tees River. They were imported into the United States in the late 1700s. Milking Shorthorns come in an array of colors. Milking Shorthorns can range from red, to red and white or roan, much like Shorthorn beef cattle. Cattle are medium in size with mature cows weighing approximately 1,400 lb. Milking Shorthorns are known for their versatility and being docile. They are also appreciated for their easy calving and longevity.

Courtesy of Hoard's Dairyman

Red and White (Holstein). Red and White dairy cattle are simply red Holsteins. The Red and White Dairy Cattle Association was started in 1964 in the United States. Their breed association is strong and growing. They have basically the same characteristics as Holsteins except for their variation in color. They are a larger breed in stature, like Holsteins, with mature cows weighing around 1,500 lb. They have excellent milk production.

Lesson 10.2 Review and Assessment

Words to Know

Match the key terms from the lesson to the correct definition.

1. A fluid rich in protein and fat, secreted by female mammals for the nourishment of their young.
2. The slightly sour liquid left over after butter has been churned. It is used in baking and consumed as a drink.
3. The section of a dairy where the cow is moved to in order for milking to occur.
4. A heavily sweetened milk product made by removing about 60% of the water from ordinary milk.
5. A solid food made from milk curd that is produced in a range of flavors, textures, and forms.
6. A milk product that has been fermented with certain kinds of bacteria; the bacterial culture sours and thickens the cream.
7. The act of giving birth.
8. The thick white or pale yellow fatty liquid that rises to the top of milk that is left to stand (unless it is homogenized).
9. A semisolid food made of milk and milk solids fermented by two added cultures of bacteria.
10. A milk product made by removing about 60% of the water from ordinary milk.
11. A soft yellow or white food churned from milk or cream and used to spread on food, in cooking, and in baking.
12. The baglike mammary gland of cattle that has two or more teats hanging near the hind legs.
13. A frozen dessert that is lower in fat than ice cream and contains yogurt cultures that may or may not be active.
14. A process where sperm is collected from a bull, processed, and frozen for storage, and then thawed and placed in the reproductive tract of the cow.
15. The nipples of the mammary gland from which the milk is extracted.
16. A white blood cell.
17. A process in which the milk's fat droplets are emulsified and the cream does not separate.

A. artificial insemination (A.I.)
B. butter
C. buttermilk
D. cheese
E. condensed milk
F. cream
G. evaporated milk
H. frozen yogurt
I. homogenization
J. ice cream
K. mastitis
L. milk
M. milking parlor
N. parturition
O. somatic cell
P. sour cream
Q. teats
R. udder
S. yogurt

18. A sweet, frozen food made from no less than 10% butterfat and made in a myriad of flavors.
19. The inflammation of the mammary glands which is usually caused by bacteria.

Know and Understand

Answer the following questions using the information provided in this lesson.

1. *True or False?* Less than 65% of dairies in the United States are family owned and operated.
2. What is the main commodity produced from dairy cattle?
3. List five common products made from milk.
4. When dairy cattle give birth, their calves _____.
   A. nurse until they reach 6–8 weeks of age
   B. stay and nurse for only a few hours
   C. are placed in a nursery with nursing cows
   D. None of the above.
5. Explain why dairy farmers commonly use artificial insemination to breed their cows.
6. Identify three reasons cows are culled from the dairy herd.
7. Explain why a cow's teats are disinfected after milking.
8. Once the milk leaves the cow, it must be cooled to a temperature below _____ °F within a two-hour period.
9. List three ways in which you can help maintain herd health.
10. List three points of proper handling of dairy cows.

Analyze and Apply

1. Convert pounds of milk to gallons of milk.
2. Convert gallons to glasses of milk.

Thinking Critically

1. Milk is used to make many food products. Develop a list of common foods that have milk as a major ingredient. What percentage of your daily diet includes milk and dairy products? Approximately how many pounds of milk or dairy products do you consume in a day? In a month? In a year?
2. There are many different kinds of milk: whole milk, 2% milk, 1% milk, skim milk. With a partner, make a chart with the different kinds of milk, what they are used for, how much fat is in each type, and how much price difference there is between each type.

Lesson 10.3

Equine Industry

Words to Know

allelomimetic behavior
bars
broodmare
cecum
coldblood
colt
conformation
draft horse
English riding
equine
equine assisted therapy
equines
Equus asinus
Equus caballus
farrier
feral
filly
foal
foaling
frog
gelding
hand
hinny
hotblood
imprinting
jack
jennet
jenny
light horse
long-eared breed
mare
modified monogastric
mule
pony

(Continued)

Lesson Outcomes

By the end of this lesson, you should be able to:

- Understand the size and scope of the equine industry.
- Define common terms used in the equine industry.
- Describe the components of the equine industry in the United States.
- Differentiate between different types of equine animals based on physical characteristics.
- Identify and describe common breeds of equine animals.

Before You Read

Write down 10 facts you think you already know about the lesson topic. Discuss these facts with a partner. After reading the lesson, revisit your list and add any new information you found about the facts to the list. Have a conversation with your partner about how your knowledge of the topic changed after reading this lesson.

Which livestock animal has historically been used as a food source, a power source, a weapon, a tool, a mode of transportation, a sporting event, a hobby, and even a companion? If you guessed the horse, you are absolutely correct. Winston Churchill once said, "The dog may be man's best friend, but the horse wrote history." Human history is closely intertwined with horses. Although the advent of the automobile, electric streetcars, and mechanized agricultural machinery once threatened the future need for the horse, the horse industry is still incredibly important to both horse owners and the economy in the United States.

Equine Industry in the United States

Equines are animals in the family *Equidae*, which includes single-toed hooved animals like horses, asses, and zebras, **Figure 10-19**. Horses make up the largest population of equines in the United States, followed by asses, and less common equines such as zebras. According to the American Horse Council (AHC), there are well over nine million horses in the United States. The top five states by number of horses are:

- Texas
- Oklahoma
- Kentucky

- Tennessee
- Missouri

The majority of horses in the United States are kept by private owners for personal use such as pleasure riding, trail riding, and showing/competition. Horses are also kept on breeding farms and in riding and training stables related to their specific use. According to the AHC, the equine industry directly employs almost half a million Americans and provides $39 billion in direct sales. The equine industry is also a factor in more than $100 billion of revenue in the United States every year, **Figure 10-20**.

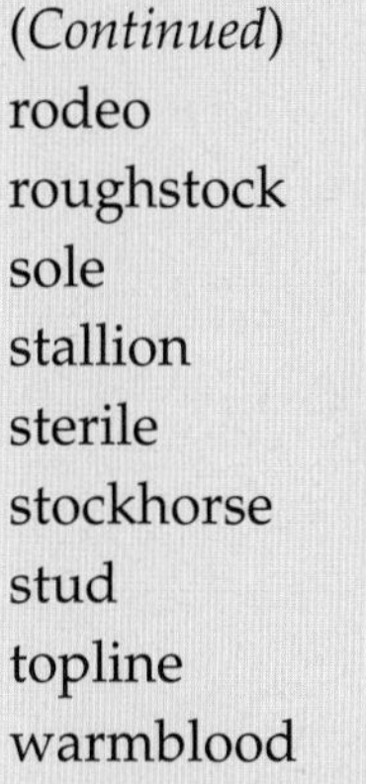

Words to Know (*Continued*)
rodeo
roughstock
sole
stallion
sterile
stockhorse
stud
topline
warmblood
western riding
white line

Common Equine Terms

When talking about equines, there are different terms for the animals based on gender, age, and subcategory:

- ***Foal***—young equine of either gender
- ***Filly***—female horse two years or younger
- ***Colt***—male horse two years or younger
- ***Gelding***—castrated male equine
- ***Mare***—mature female horse

Did You Know?

New Jersey is the state with the most horses per capita.

jacotakepics/Shutterstock.com

Stefano Pellicciari/Shutterstock.com;

Fireglo/Shutterstock.com

Figure 10-19. The equine family includes all horses, asses, donkeys, and even zebras. *What characteristics do these animals have in common?*

jessicakirsh/Shutterstock.com

Figure 10-20. The equine industry in the United States is a lucrative industry. Major events like the Kentucky Derby draw thousands of spectators and contribute to the more than $100 billion the equine industry helps generate each year.

- ***Broodmare***—mature female horse kept for breeding
- ***Stallion***—mature male horse
- ***Jenny*** or ***jennet***—female ass or donkey
- ***Jack***—male ass or donkey

Did You Know?

The term *stud* is often misused to describe a mature male horse. The word ***stud*** is actually used to identify the place a stallion is kept for breeding. For example, if you had a stallion and were using him for breeding purposes, your farm or ranch would be the place the stallion was at stud.

Production Cycle of Horses

While production operations exist for all equines, horses are by far the most common equines in production in the United States. Unlike cattle, horses are not bred for meat or milk production in the United States. Horses are bred primarily to continue bloodlines and to produce quality horses for specific uses, such as racing and showing, and to produce good stockhorses for ranch work or roughstock for rodeos.

The production of horses begins through natural breeding or through artificial insemination of the mare. Depending on the breed, a mare will carry the foal for 335 to 370 days. Larger horse breeds have a longer gestation than smaller or pony breeds. Most producers calculate foaling dates at an average of 342 days. ***Foaling*** is the term for the process of giving birth in all equine animals, **Figure 10-21**. A foal will nurse for 3–6 months before being weaned. The continued growth, training, and development of an individual horse will depend heavily on the intended use for the horse.

R Bar T Quarter Horses

Figure 10-21. A—This mare has just completed foaling. B—Nursing mares have very demanding nutritional needs. Make sure that you are feeding mares high-protein feed so that they do not lose muscle mass when nursing. ***What are some of the considerations that the producer should be mindful of to take care of this mare and her new foal?***

The Horse Industry

The most common sectors of the horse industry in the United States are:

- Horse showing
- Ranch work
- Rodeos
- Racing
- Recreational riding
- Other horse employment

Did You Know?

An easy way to remember the gestation of a horse is that it is about 11 months and 11 days for most horse breeds.

Bureau of Land Management

Horse Showing

Over the course of history, horsemen have tried to hone and compare their riding and fighting skills through show competitions. Think back to the times of jousting knights in armor where men competed for honor and position, as well as monetary compensation. Today's equestrians compete to show their horse training and handling skills, as well as the quality of their animals. Some equestrians train to reach one of the ultimate competitions, the Olympics.

Horse shows may be breed association championship shows or open to horses of any breed. There are also horse shows for different age and skill levels for both horses and riders. Many different associations sponsor horse shows. Almost anywhere in the country, you are certain to find a show that will match both your own and your horse's skill level and desired discipline.

Horse shows encompass both structural and performance disciplines. Horses are shown at halter based on their conformation to breed standards and based on specific criteria while performing. Some equitation events involve the judging of the rider's ability; other events are based more on the horse's ability. As you can see, there are horse-based events for every level of interest and skill.

The two widely recognized horsemanship disciplines in which horses are shown are English riding and western riding.

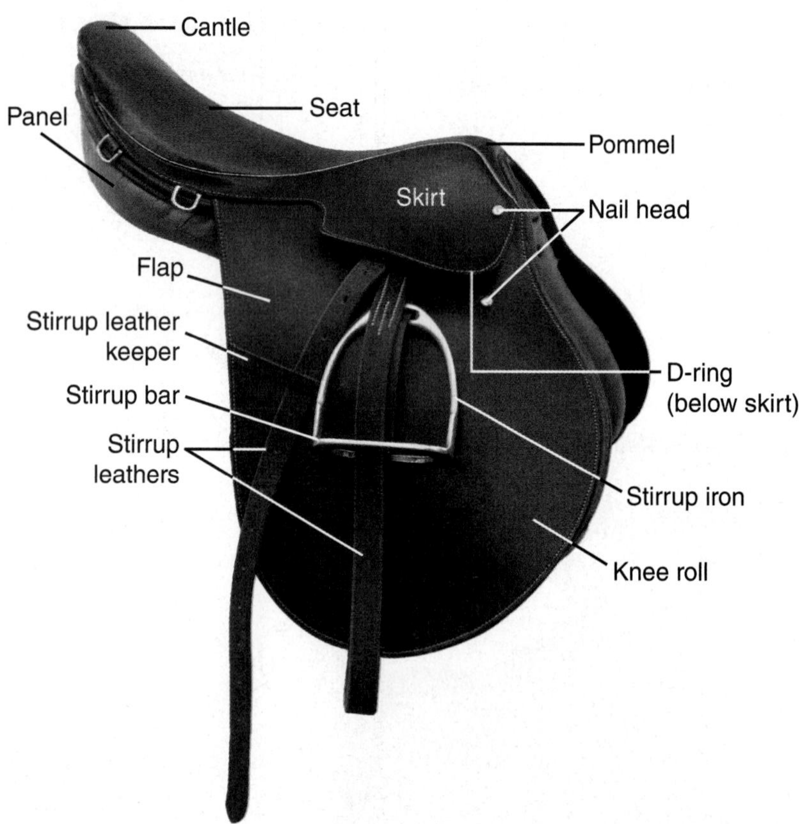

Figure 10-22. English riding and showing occurs with the use of an English saddle, like this one. There are many different modifications that can be made based on the specific use of the saddle. ***Can you think about how a dressage saddle might differ from a saddle used for jumping?*** After studying the image, visit www.g-wlearning.com to test your knowledge.

English Riding

In ***English riding***, horses are ridden in an English saddle, **Figure 10-22**. One of the biggest differences between English and western riding is the use of the posting trot. This action requires the rider to rise up and down in rhythm with the two-beat gait of the trot. There are also differences in how the reins are held; English riding is completed exclusively with two hands on the reins. Some events in English riding are based on the skills horses and riders used for fox hunting in the English countryside. Horse show events using English riding include:

- Dressage
- Hunter under saddle
- Hunt seat equitation
- Working hunter

Stadium and cross-country jumping are also events considered to be within the English riding discipline, **Figure 10-23**. In addition, there are also classes in which the horses are judged on their ability to pull a carriage.

Margo Harrison/iStock/Thinkstock

Mikhail Kondrashov/iStock/Thinkstock

Figure 10-23. Both dressage and stadium jumping are considered English horse showing events. These events both demand careful communication between horse and rider to complete the required tasks.

Western Riding

Western riding is the discipline of horse showing in which horses are shown in a western saddle, **Figure 10-24**. Western riding disciplines are designed so that the horse can be shown with the reins in one hand. Traditionally, this allowed the rider's free hand to be used for handling ropes and other tools while working. Most of the western riding events are based on skills required by stockhorses used to move cattle. Western riding horse showing includes events like:

- Trail riding
- Western pleasure
- Western riding

FFA Connection Horse Evaluation CDE

Do you think you would be a great horse show judge? Would you like to learn more about the specifics of conformation, equitation, and judged horse show events? The Horse Evaluation CDE through the National FFA Organization allows you the chance to judge both conformation and riding classes. This CDE is a great way to examine breed ideals and test your knowledge of the proper way to evaluate horses at halter and under saddle. You have the opportunity to place classes and then defend your placings through oral reasons.

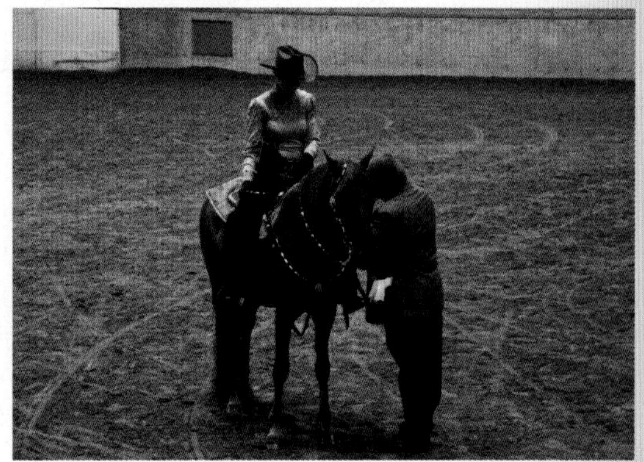
EP photo/Shutterstock.com

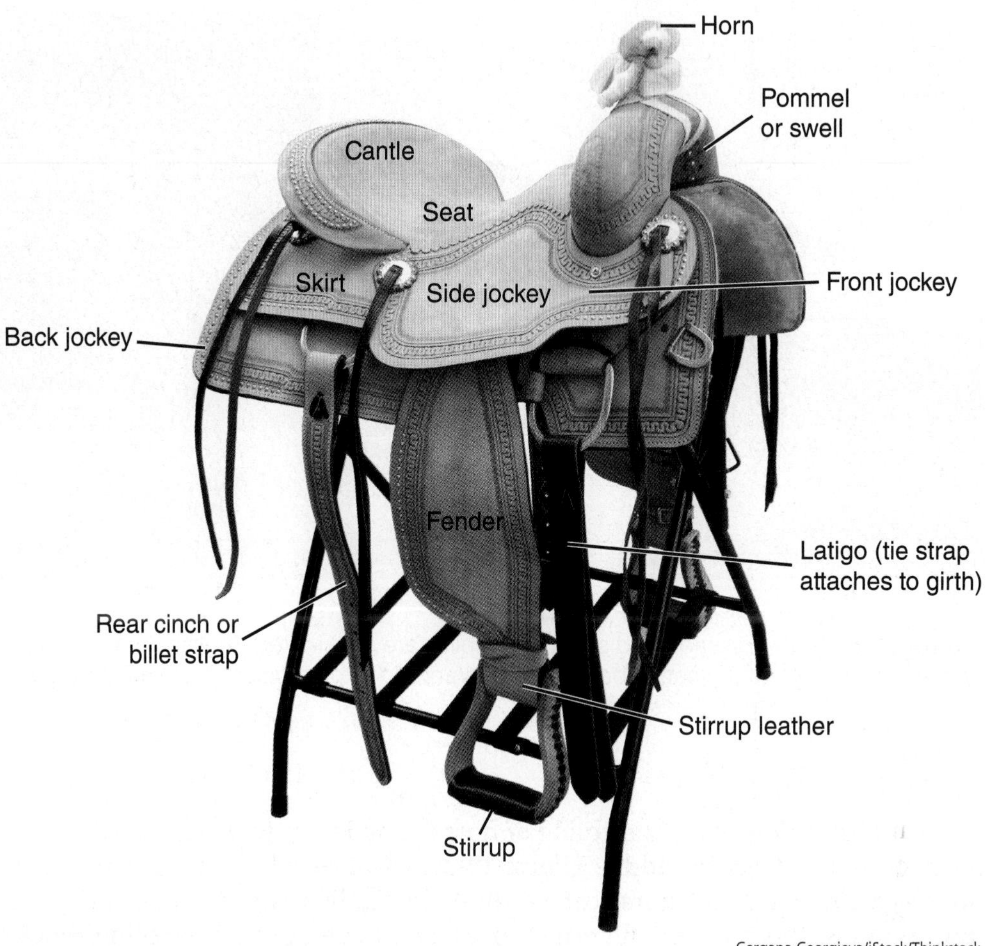

Gergana Georgieva/iStock/Thinkstock

Figure 10-24. Western riding happens in a western saddle, like this one. *How do you think the differences between this saddle and the English saddle help with western riding events?* After studying the image, visit www.g-wlearning.com to test your knowledge.

- Reining
- Working cow horse
- Cutting (**Figure 10-25**)

There are also timed western events which are placed by completing a task in the fastest time such as pole bending, stake race, team penning, and barrel racing.

Ranch Work

Horses have been a tool used by ranchers since the era of the American West. Even with modern technology and machinery, horses are still used in many beef cattle operations as an efficient way to gather, sort, and work with cattle. Horses are more efficient, less likely to spook cattle, and are able to go into areas where many motorized vehicles are unable to gain access. It requires a specific level of athleticism, speed, and instinctive ability to work with cattle to be a good ranch workhorse. Horses that have been selected for these ranch working traits are called ***stockhorses***.

Candia Baxter/Shutterstock.com

Venessa Nel/Shutterstock.com

Figure 10-25. Western horse shows include events like western pleasure (left) and cutting (right). Some events require the horse and rider to demonstrate their skills in handling cattle.

Thousands of horses are used every day to help with ranch work around the country, **Figure 10-26**.

Rodeos

Another sector of the horse industry is rodeos. A ***rodeo*** is a sporting event comprised of events based on ranch skills that cowboys and horses used to work cattle in the American West, **Figure 10-27**. The first rodeos began when groups of neighboring ranch hands met to see who had the

Did You Know?

Three different towns all claim to be the site of the first rodeo:

- Santa Fe, New Mexico, 1847
- Deer Trail, Colorado, July 4, 1869
- Pecos, Texas, 1883

Actually, much of the current rodeo structure is based on the rodeo held in Prescott, Arizona, on July 4, 1888. The Prescott Rodeo is still an annual event.

Lenice Harms/Shutterstock.com

Figure 10-26. Even in today's world of modern equipment and technology, horses are still used as the primary means for moving and sorting cattle on many ranches. *Do you think it has more to do with tradition or efficiency? What benefits could you see to using horses?*

David Thoresen/Shutterstock.com

Kobby Dagan/Shutterstock.com

Figure 10-27. Skills used to manage and give medical attention to cattle on ranches directly translate to rodeo events. *Can you see the similarity between the actions of the ranch hands (left) and the team ropers (right)?*

best ranch skills. Today, horses are used in rodeo events as both roughstock and saddle horses. ***Roughstock*** animals are the bucking animals cowboys are scored on riding during rodeo events. The two roughstock events that use horses are bareback riding and saddle bronc riding.

Horses in rodeos are also used under saddle for rodeo events including: steer wrestling, tie-down roping, steer roping, team roping, breakaway roping, goat typing, pole bending, and barrel racing. Rodeo is a very competitive industry, with levels of competition ranging from youth rodeos to high school and collegiate rodeos, and adult rodeos for both part-time and full-time cowboys and cowgirls.

Neale Cousland/Shutterstock.com

Figure 10-28. The majority of the horse racing in the United States is done with the Thoroughbred breed. It is called the "Sport of Kings" although sultans, czars, and even Queen Elizabeth have been known to get in on the action.

Racing

Horse racing, or the "Sport of Kings," in the United States is a multibillion dollar industry, **Figure 10-28**. The Thoroughbred breed is often seen as the ultimate racing horse. Consequently, the most notable races in the country, the Kentucky Derby, the Preakness Stakes, and the Belmont Stakes, are all Thoroughbred races. Other breeds are also involved in racing, but not to the scale of the Thoroughbred.

Race horses undergo a closely monitored exercise and nutrition program to ensure that they are in top physical condition before they race. While most horse races occur on a flat track with a single rider called a jockey, there are also jumping races called steeplechases, and harness racing where horses pull a small buggy called a sulky, **Figure 10-29**.

Did You Know?

The most prestigious title in the Thoroughbred racing industry is the Triple Crown. To win the Triple Crown, a horse must win the Kentucky Derby, the Preakness Stakes, and the Belmont Stakes.

Andreas Glossner/iStock/Thinkstock

Figure 10-29. Chariot racing began in the Greek coliseums and continues today as an exciting (and dangerous) type of horse racing called harness racing. The small buggies are called sulkies.

Recreational Riding

Many of the horses in the United States are kept for recreational riding. The owner may merely desire to use the horse for personal trail riding or other recreational activities like hunting. Many people join riding clubs so they can ride with other horse enthusiasts and share their love of riding and sometimes the expense of transporting their horses to riding venues across the country. Recreational riding is available in many national parks and forests, with both managed and unmanaged trail opportunities.

Equine Assisted Therapy

Many horse owners use their horses in ***equine assisted therapy*** programs. These programs use equine activities to improve the physical, occupational, or psychological health of the riders. Studies have found that equine assisted therapy can be helpful for individuals with conditions like ADHD, autism, cerebral palsy, depression, anxiety, developmental delays, and post-traumatic stress disorder, **Figure 10-30**.

Tom Ervin/Getty Images News/Thinkstock

Figure 10-30. Studies have found that there are many benefits to using horses for therapy. This little boy is autistic and suffers from cerebral palsy. ***How do you think being involved with horses can enrich his life?***

Guided Riding Experiences

Many people desire the opportunity to ride horses, even those who do not have the desire to own their own horses. Of these individuals, there are many novice horsemen who choose to have a guided horseback experience. Dude ranches provide this opportunity for those who want the chance to experience horseback ranch work. Guided trail rides are available for many people who want to experience riding in a particular location, and are very common in vacation destination locations. Horse-oriented day camps are also popular for giving youth an equine experience. Novice riders who want to have a prolonged engagement with horses can sign up for riding lessons that provide a lesson horse for the student to use.

Other Horse Employment

Another section of the equine industry in the United States is the use of horses as tools for completing a specific job. Even though the notion of using horses for our main source of power and transportation is a thing of the past, horses are still used in police and military work, and for pulling wagons and carriages.

Horses can be effective police officers. Especially in large cities, police officers ride horses to allow them a way to move through crowds. The height of a rider on a horse gives the officer a better vantage point for watching a situation, and the use of a horse provides the maneuverability of being on foot with the speed of having a motorized vehicle. Military cavalry units still exist, although most of their duties now are related to conducting ceremonies, rather than being used in battle.

Some horses also go to work every day as carriage horses in major metropolitan areas or at resorts and agricultural farm tours across the country.

Anatomy of an Equine

The anatomy and physiology of horses and other equines is essentially the same. Equines come in all different sizes and are measured in four-inch increments called ***hands***. The proper way to express a height in hands is the whole number of hands, followed by a decimal and any remaining inches. For example, if a horse was 50″ tall, it would be 12.2 hands tall. An equine is measured to the top of its withers.

Conformation

Conformation classes at horse shows are events in which horses are evaluated without a rider based on their structure in comparison to breed ideal anatomy. A horse's ***conformation*** is the correctness of its bone structure, musculature, and body proportions in relation to each other. Before you can accurately evaluate horses, you must first be acquainted with the external parts of a horse, **Figure 10-31**.

FFA Connection Horse Evaluation CDE

Are you interested in analyzing the proper conformation of horses? Do you want the opportunity to learn how to evaluate proper horsemanship? The Horse Evaluation CDE allows teams of FFA members to evaluate horses for conformation at halter and under saddle. Team members also complete a team activity related to the proper management and maintenance of animals in the equine industry.

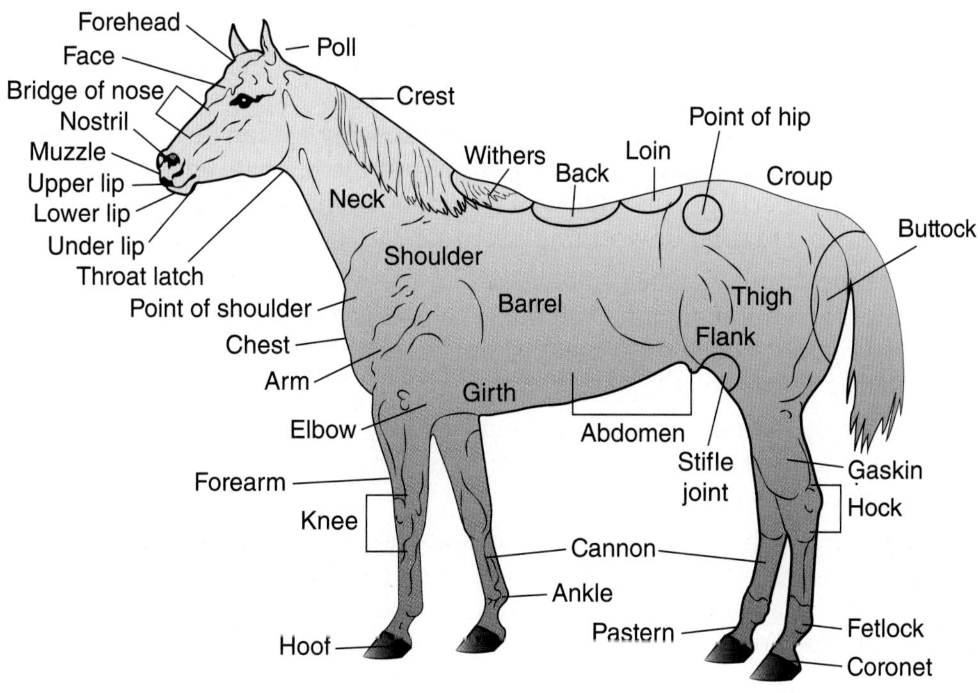

Figure 10-31. Understanding the parts of the external anatomy is an important part of being able to talk about correct conformation of horses. ***Can you name all the parts of this horse?*** After studying the image, visit www.g-wlearning.com to test your knowledge.

Equine Digestion

The digestion of equines is different from the digestive systems of other livestock animals. Equines are considered ***modified monogastrics***, which means that they have a stomach with a single compartment, but also have a special compartment in their intestine called the cecum, which helps them digest roughages, **Figure 10-32**. The ***cecum*** is a pouch-like area found at the beginning of the large intestine, where digestive microbes are housed to break down roughages.

Maintaining Herd Health

There are some special considerations that equine producers need to take to ensure that their equine animals remain in good health. Keep the following things in mind when raising and training horses:

- Maintaining current and complete immunizations is important for all horses, especially those who will be in contact with other horses at shows and rodeos.
- Some equine diseases are transmitted by flies and mosquitoes. Establish

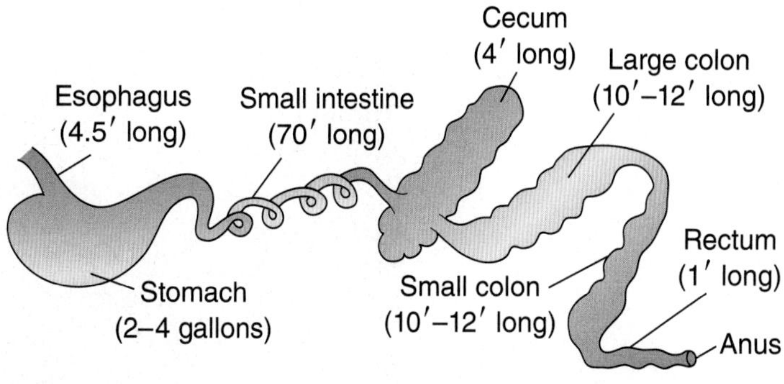

Figure 10-32. Horses only have one compartment to their stomach. In order to allow them to digest roughages, they have evolved an enlarged cecum. The cecum works similar to the rumen in ruminant animals and is in the same basic place on their digestive system as your appendix is on yours. After studying the image, visit www.g-wlearning.com to test your knowledge.

a pest management plan that will reduce these insects in your stable area, and use pest control on your horses during peak fly and mosquito season.

- Internal parasites, like roundworms and strongyles, can be a big problem for grazing animals. Grazing animals ingest the parasite eggs while grazing, especially when they eat roughage to the ground. Keep your horses on a regular deworming schedule.
- The most common cause of death in equines is *colic*, which is a general term for abdominal pain. Any horse showing signs of colic (i.e., sweating, refusing to eat, kicking at belly) should be carefully monitored and veterinary care sought if symptoms continue, **Figure 10-33**.
- Ensure that horses are fed high-quality feeds that are free from both mold and dust, which can cause serious digestive and respiratory complications for horses.
- Traveling with horses has certain health restrictions. Before crossing the state line with your horses, you will likely need a health inspection from a veterinarian and a blood test to ensure they are free from Equine Infectious Anemia.

Did You Know?

There are very specific ages at which horses' teeth come in, are replaced by permanent teeth, and wear through old age. Once you understand the ways that horse teeth wear, you can determine a horse's age by examining its teeth. Hence the phrase, "Never look a gift horse in the mouth."

Hoof Care

There is an important saying in equine science: *no hoof, no horse.* Managing equine health rests heavily on caring for hooves and understanding the importance of the structures in the horse's feet.

Dan Kitwood/Getty Images News/Thinkstock

Figure 10-33. Colic is the number one cause of death in horses and often requires surgical intervention. Even with surgery, recovery is not certain. If you are concerned that your horse has abdominal pain, seek veterinary attention immediately.

Career Connection Equine Chiropractor

Straightening out the horse industry

Job description: Equine chiropractors work to diagnose and treat muscular and skeletal ailments in horses. Generally, they work with performance horses who need to be in top physical condition. Much like a human chiropractor, equine chiropractors perform massage, rehabilitation therapy, and spinal manipulation—they just do it with much larger patients.

Education required: Several training programs offer one-year or two-year certification programs in equine chiropractic services. Some specialized chiropractors are also licensed veterinarians.

Job fit: This job may be for you if you enjoy working with horses and horse owners, and you do not mind putting all your weight into the job, literally.

Leslie Trimble

Leg Conformation

The lower leg of the horse has many small bones that support the entire weight of the animal. The internal structures of the lower leg are shown in **Figure 10-34**. Proper conformation will ensure that these bones are not put under additional stress. Selective breeding of horses with sound legs can be a huge benefit to offspring. Conformational defects in any of the structures in the lower leg can lead to the horse being unsound and affect the usefulness of the animal.

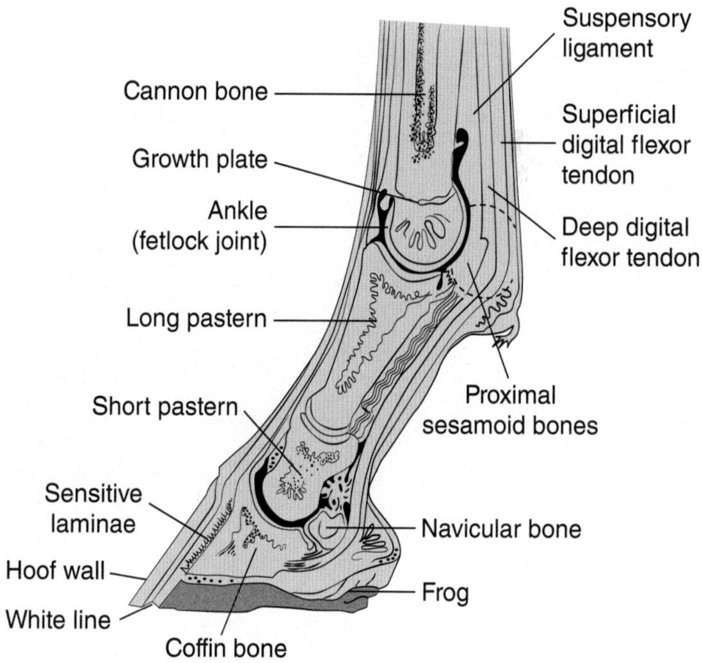

Goodheart-Willcox Publisher

Figure 10-34. There are many small bones and connective tissue that make up the internal structure of the leg in horses. Many of these bones are associated with specific problems. For example, navicular disease is caused by inflammation and degeneration of the navicular bone, and laminitis affects the coffin bone. After studying the leg structure image, visit www.g-wlearning.com to test your knowledge.

Hoof Composition

The hoof may appear to be solid, but it is actually made up of several different components, **Figure 10-35**. The outer layer is hard and composed of semi-hard material similar to your fingernails. It is constantly growing and, like your fingernails, may be very thin or very thick. The outer layer attaches to the outer wall of the hoof. The point at which they attach is called the ***white line***. ***Farriers*** (craftspeople who trim and shoe

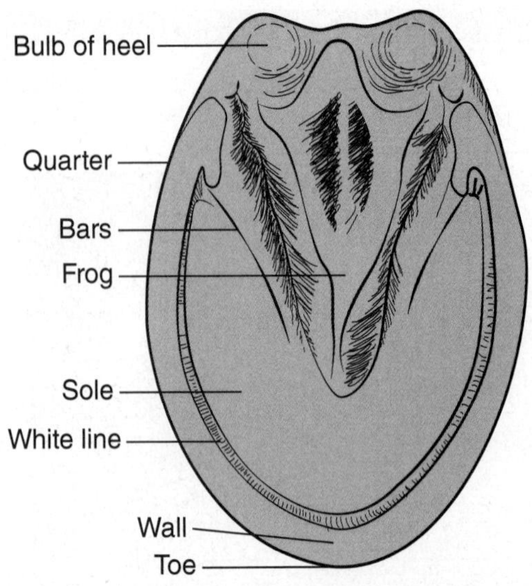

Goodheart-Willcox Publisher

Figure 10-35. The bottom of a horse's hoof is designed to protect the internal structures and aid the body in pumping blood back to the horse's heart. ***What are some considerations you should be mindful of when caring for hooves?*** After studying the hoof image, visit www.g-wlearning.com to test your knowledge.

horses) use the white line as a guide for trimming and for setting nails.

The bottom of the hoof is called the ***sole***. The sole is thick, hard, and somewhat flexible and protects the more sensitive part of the sole directly beneath the bones of the foot. The horse's sole protects the horse much like the sole of your shoe protects your foot. The ***frog*** is a triangular area in the middle of the sole. The frog helps circulate blood in the foot and from the bottom of the leg to the heart. The bars are located on either side of the frog. The ***bars*** are depressions in the sole of the hoof which provide stability for the walls and allow the frog to expand as the hoof impacts the ground.

A horse's hooves are designed to help absorb impact; however, there are risks from placing too much pressure on a section of the hoof. Placing too much pressure on a section of the hoof can cause bruising to the internal portion of the hoof. Excess pressure may be due to running on extremely hard surfaces or hard objects, such as rocks, impacting the sole. To avoid undue pressure, never run your horse on pavement, and take extra care to remove rocks and other hard objects from arenas and areas where horses are ridden.

Hoof Maintenance

The makeup of the hoof requires horse owners to take special care to ensure that hooves are appropriately maintained. Some of the considerations for hoof care include:

- Ensuring that your horses' hooves are kept trimmed and well maintained. Hooves should be trimmed or reshod every 6–8 weeks, **Figure 10-36**.
- Clean hooves with a hoof pick every time you groom your horse. Keeping the bars clean will allow the frog to function properly.
- Proper hoof moisture should be maintained to prevent cracking or fungal growth. Keep stalls clean to prevent ammonia exposure to hooves.

Olga_i/Shutterstock.com

Figure 10-36. Hoof health is one of the most important parts of managing equine health. The triangular area at the back of the hoof is called a frog and functions like a pump to return blood from the hoof to the heart.

Horse Behavior

The behavior of horses is a topic that has been studied in depth. Horses are driven by several instincts that guide much of their activities. These behaviors include:

- Fight or flight—horses will naturally try to remove themselves from situations that cause them fear, but will stand and fight if they feel cornered.

- Allelomimetic behavior—horses will copy what other horses do. ***Allelomimetic behavior*** is a range of activities in which the performance of a behavior increases the probability of that behavior being performed by other nearby animals. This instinctive herd mentality was originally developed to help horses stay in large groups to prevent predator animals from singling out individuals.
- Herd dominance—it is important to understand the natural hierarchy that forms in a herd. In a wild herd, lead animals establish their position and usually maintain it for extended periods of time. In domestic herds, the herd composition continually changes as animals are culled or added. This means the social structure changes each time the herd changes, which may lead to more instances of aggressive behavior as horses vie for the lead position. Careful introduction of new horses will help reduce the aggressive behavior and prevent potentially dangerous situations for workers and horses alike.

Handling Horses

Horses are instinctively prey animals who have a natural fear of humans. Most horses receive some training shortly after birth, called ***imprinting***, to ensure that they are accustomed to human interaction and will be safer to handle when they are older. Imprinting generally includes the foal learning to be haltered and led, along with tasks like allowing their feet to be picked up and allowing humans to touch their mouth, ears, and girth area, **Figure 10-37**.

Charlene Bayerle/Shutterstock.com

Figure 10-37. Working with foals as soon as possible after birth will help them overcome their instinctive fear of humans. Teaching a 100-lb foal to pick up its feet is a lot easier than teaching a 600-lb yearling or 1,000-lb 2-year-old.

Right eye
Both eyes
Blind
Left eye

Goodheart-Willcox Publisher

Figure 10-38. Horses have broad peripheral vision. They can see with both eyes out in front of them and to both their left and right sides with each respective eye. They also have a large blind spot behind them.

Horses are large and often unpredictable animals; in order to ensure both your own and the horse's safety, keep the following tips in mind when working with or around horses:

- Always approach a horse from a slow and consistent speed; quick movements can startle or spook the horse.
- Work with horses from an early age to help them become more comfortable around humans.
- Remember that horses have a well-developed "fight or flight" instinct. Never put a horse in a situation where it feels cornered.
- Horses have good peripheral vision, but have distinct blind spots, **Figure 10-38**. Avoid standing in these blind spots to prevent spooking the animal.
- It is a well-known fact that horses kick. Do *not* walk directly behind a horse within reach of its hind legs, and always ensure the horse is aware of your presence.
- When tying a horse, use a quick-release knot. This will allow you to prevent yourself and/or the horse from becoming injured should it be spooked.

Wild Horses and Burros

Did You Know?

More than 49,000 wild horses and burros currently roam on Bureau of Land Management (BLM) rangelands.

In the United States, there are still horses that some call *wild*, **Figure 10-39**. These "wild" horses are descendants of animals that escaped from the Spanish explorers in the fifteenth century. These horses are not actually classified as wild by scientific terms because they come from animals that have previously been domesticated. The correct term for animals that have historically been domesticated and now live in the wild with no human assistance is ***feral***. Careful observation of these feral animals has helped horse behaviorists more accurately understand the instinctive behaviors of horses.

NIKKOS DASKALAKIS/Shutterstock.com

Figure 10-39. These feral horses, in the high mountain deserts of Nevada, live in groups dominated by both an alpha mare and a stallion. Horses like this live without human assistance or interaction and can allow for examination of natural horse behavior.

Feral Horses and Burros in Western States

Prior to 1971, the feral horse herds were allowed to roam largely across western states without human interaction. The 1971 Wild Free-Roaming Horses and Burros Act set forth regulations and policies for the management of these herds. The Bureau of Land Management (BLM) currently has the responsibility for managing the feral horse population in the United States. Managing feral horses and burros is often a source of contention in the states where these animals roam. Herd numbers often exceed the number of animals that ranges can adequately support, and ranchers are concerned with cattle having to compete for grazing.

In order to reduce the number of animals on the rangeland, the BLM periodically conducts roundups of the animals and sorts out some horses and burros for public adoption, **Figure 10-40**. Many of these horses are also kept on feedlots or private ranches. Populations are also managed by treating mares with fertility control injections.

BLM Nevada Diamond Gather

Figure 10-40. The BLM holds adoptions for the excess feral horses and burros rounded up from rangeland. In order to adopt a horse, you must meet minimum housing criteria and pay a $125 adoption fee. *Do you know someone who has adopted a "wild" horse?*

East Coast Feral Horses

When most people think of wild horses in the United States, they picture those running across western states, not running free on an island on the East Coast. The Assateague horses are said to be descendants of horses owned by seventeenth century owners seeking to avoid fencing laws and livestock taxation. Others paint a more romantic story of the horses as descendants of shipwreck survivors. Feral horse herds can also be found on the Outer Banks of North Carolina. Populations in these areas are controlled through roundup and adoption operations, much like their Western counterparts.

Breeds of Equines

The species name given to all horses, donkeys, and asses is ***equine***. Within the equine species, there are two subcategories, *Equus caballus* and *Equus asinus*. ***Equus caballus*** is the scientific name for wild and domesticated horses, and ***Equus asinus*** is the scientific name for donkeys and asses.

Did You Know?

The current claim for the tallest horse in the world belongs to a Belgian gelding named Big Jake. Big Jake weighed 240 lb at birth and stands almost 20.3 hands tall. He currently weighs more than 2,600 lb and consumes about 80 lb of hay and 40 qt of oats every day.

Types of Equines

The *Equus asinus* equines are often called the ***long-eared breeds***. These asses and donkeys are characterized by their long ears, larger heads in proportion to their body, and flatter backs and croups. Horses, the subcategory *Equus caballus*, are broken down into three different categories based on size. ***Pony*** is the classification typically given to horses under 14.2 hands tall and weighing up to 900 lb. ***Light horses*** are around 14 to 16 hands tall and weigh generally between 900 and 1,400 lb at maturity. ***Draft horses*** are typically 16 to 18 hands tall and may weigh upward of 2,000 lb. They are characterized by their heavy bone and muscular build. It is important to note that height alone is not the only indicator of which subcategory a horse belongs in. For example, in cases where a light breed horse matures at under 14.2 hands tall, it is not necessarily considered a pony.

STEM Connection Mules: Hybrid Animals

You may wonder where mules fit into all of this equine stuff. Mules are actually a hybrid animal, combined by crossing a horse with an animal from the ass family. When a mare is bred to a jack, the offspring is called a ***mule***. You can also breed to get a ***hinny***, the product of a female donkey and a stallion.

There are some interesting scientific principles behind these hybrid animals. You see, even though they have a distinct male or female gender, all mules (and hinnies) are ***sterile***, meaning they cannot produce offspring. This is because of some very specific things that happen in the creation of their DNA.

During the process of meiosis, sperm and egg cells are created. These cells, also called sex cells, have half the regular number of chromosomes for that species. For example, horses have 64 chromosomes, meaning that a horse egg or sperm cell only has 32 chromosomes. In donkeys, there are 62 total chromosomes, leading to sperm and egg cells that have 31 chromosomes. When donkeys and horses mate, the result is a zygote (newly fertilized egg) that has 63 chromosomes. Because the DNA structure of horses and donkeys is so closely related, this zygote grows into a complete organism with 63 chromosomes. Mules function well until it comes time for them to create sex cells of their own. In meiosis, there is no way to split an uneven number of chromosomes, leaving mules and hinnies unable to create viable sperm or egg cells.

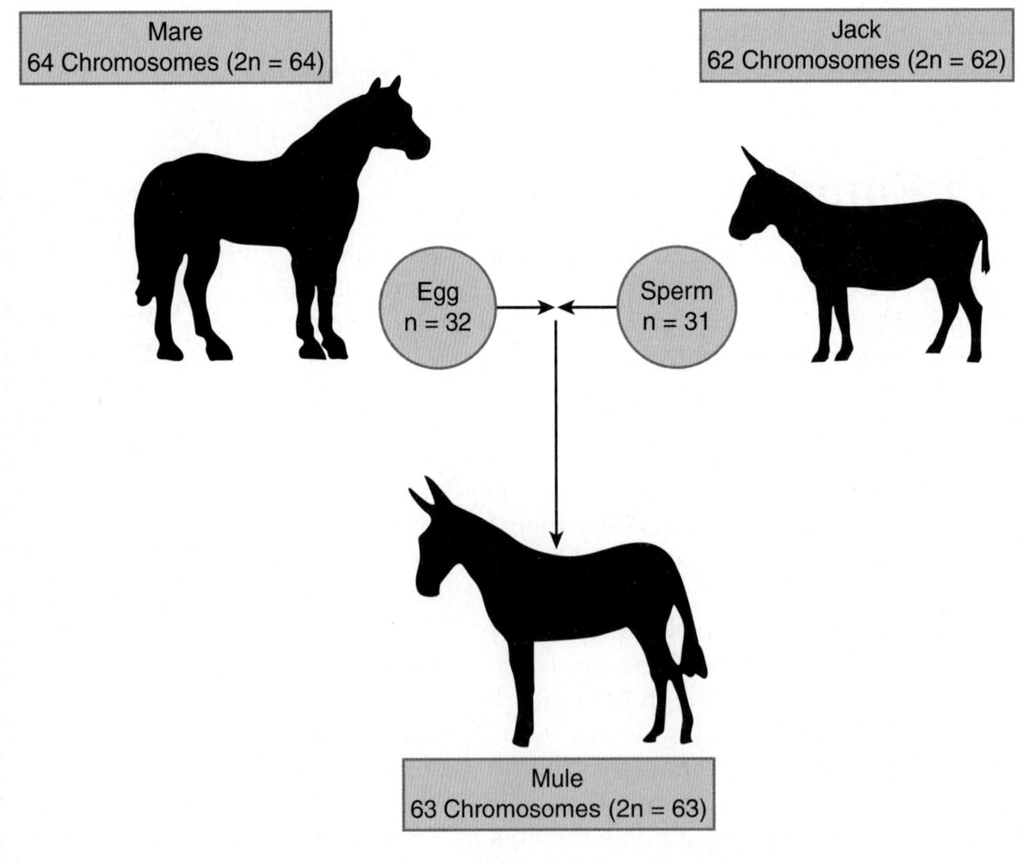

Long-Eared Breeds *(Equus asinus)*

kubais/Shutterstock.com

American Mammoth Jack. The American Mammoth Jack originated from a combination of breeds of donkey imported into the United States. These animals may be solid or have large white patches throughout their coat. Although they are called "mammoth," they are not large enough to be considered a draft breed. Typical heights are 13.5 hands for jennets and 14 hands for jacks.

Dennis W. Donohue/Shutterstock.com

Miniature Donkey. Miniature donkeys originated in the Italian islands of Sardinia and Sicily. These donkeys are small in stature, gray or grayish-tan in color, and typically under 9 hands at the withers. They are most frequently kept as companion animals and used for weed control.

Critter Haven Pandemonium, photo provided courtesy of Diane Hildebrand/Critter Haven Farms

Spotted Ass. The spotted ass is a color-based breed association incorporating any ass or donkey having a coat with spots of color and white. These asses may come in a variety of sizes and varied colors.

DragoNika/Shutterstock.com

Standard Donkey/Burro. The Standard Donkey is characterized by its gray color and small stature, standing approximately 10 to 12 hands tall with a slender build. *Burro* is the name for a Standard Donkey that originated in North America. Burros exist in both domesticated and feral groups. Standard Donkeys and Burros are still used for transportation and as cart animals in many parts of Mexico and Latin America.

Pony Breeds

Ponies are the smallest recognized breeds of horses and include miniature horses.

Vera Zinkova/Shutterstock.com

American Miniature Horse. The miniature horse is the smallest horse breed. In order to be considered a miniature horse, the animal must stand less than 8.2 hands at the withers. The breed standard requires these horses to have the same proportional conformation as their larger light breed counterparts. These horses originated from the Argentine Falabella horses, who were crossbred beginning in the fifteenth century to create a distinctly small breed of horse. While mainly used as companion animals, these horses are also used in shows and are showing promise as therapy animals, perhaps even replacing seeing-eye dogs for some individuals.

Vera Zinkova/Shutterstock.com

Shetland. The Shetland Pony originated from the cart ponies used in the Shetland Isles to haul peat and coal. Their small size allowed the ponies to easily enter mines and pull coal carts. Shetlands range from 7 hands to a maximum height of 10.2 hands. They are heavy boned, muscular in build, and have thick necks and strong ***toplines*** (the muscles going over the haunches, back, and neck of a horse). Today, they are used primarily for youth riders, but are also shown in harness driving classes and used for pleasure driving.

Photo courtesy of Pony of the Americas Club, Inc./Impulse Photography

Pony of the Americas (POA). The Pony of the Americas (POA) is a breed that originated in the mid-1950s in Iowa. These ponies were developed as a combination of Shetland Ponies and Arabian and Appaloosa breeds. Today, the breed also incorporates Quarter Horse and Welsh Pony genetics. Characteristics of the breed include heights between 11.2 and 14 hands, muscular builds with slightly dished heads, and white *sclera* (the white outer layer of the eyeball) around their eyes. They are most commonly found with a spotted coat pattern. POAs are used in many disciplines and are commonly trained for use by youth in competitions.

Light Horse Breeds

Light horses are the horses that most people think about when they picture a horse. These horses are most commonly used as saddle horses. There are literally hundreds of breeds of light horses. Some of the most common breeds include the American Mustang, American Paint Horse, American Quarter Horse, Appaloosa, Arabian, Dutch Warmblood, Missouri Fox Trotter, Morgan, Tennessee Walking Horse, and the Thoroughbred.

Dennins W. Donohue/Shutterstock.com

American Mustang. American Mustangs come from the feral horses in the American West. To be considered part of this breed, the horse must have come directly from the range or have parents who have come directly from the range. They are typically 14.2 to 15.2 hands tall and have heavy bone and large heads. Conformation varies greatly, depending on the genetics in the area where individual horses are found.

Courtesy of APHA/Paint Horse Journal

American Paint Horse. The American Paint Horse breed originated in the United States as a registry for horses who have a coat pattern with large sections of white coloring. They have muscular hindquarters and strong toplines, combined with athletic ability and a willing disposition. They typically range from 14.2 to 16 hands tall. There are bloodline restrictions and horses must have a distinctive stockhorse body type. Horses who are the offspring of American Paint Horse parents, but who do not meet the coloring requirements, can also be registered in this breed. The American Paint is used in a wide spectrum of activities including showing, ranch work, racing, and pleasure riding.

Telasecret, AQHA World Champion Producer, Courtesy Clark Rassi Quarter Horses

American Quarter Horse. The American Quarter Horse Association is the largest breed registry in the United States, and the Quarter Horse is the most popular breed in the nation. The breed got its name from its ability to quickly sprint over a quarter mile. These horses are 14.2 to 16 hands tall and are characteristically eye appealing. They typically have a blending of an attractive head with a refined throatlatch and trim neck. Conformation standards include a long shoulder; deep girth; strong back, loin, and hip; and uniform muscularity in their hip, gaskin, forearm, and chest. Quarter Horses are considered one of the most versatile breeds. They excel at ranch work, both English and western showing, racing, and pleasure riding.

Darrell Dodds courtesy of the Appaloosa Horse Club

Appaloosa. The Appaloosa breed originated from the horses of the Nez Perce Native Americans in the Pacific Northwest. This breed is characteristically 14.1 to 15.3 hands tall. Coloring often includes a spotted coat pattern on either their entire body or over their hindquarters. They are often seen with mottled or spotted colored skin, white sclera around their eyes, and striped hooves. Horses without spotted coloring can still be registered in this breed if their parentage is verified. Appaloosas are used for general riding, showing, and ranch work.

Light Horse Breeds (*Continued*)

Original breed standard painting by Gladys Brown Edwards; printed with permission by Arabian Horse Association

Arabian. Arabian horses are descendants of the horses ridden in ancient Arabia. These horses have been carefully bred over thousands of years to create a horse that is refined in nature and exhibits the distinct characteristics of the breed. Arabians have a characteristic dished profile, large wide-set eyes, a broad forehead, small curved ears, and large nostrils. They also have a long arched neck, short back, refined bone structure, and high tailset. Arabians are used in a variety of events including showing, endurance racing, and pleasure riding.

Sunrise Tradition, Courtesy of P Bar T Fox Trotters, Independence, MN

Missouri Fox Trotter. Missouri Fox Trotters were developed in the Ozark Mountains to breed horses that could haul logs and work cattle. They are a combination of Arabian, Morgan, Standardbred, and Saddlebred horses. Missouri Fox Trotters are between 14 and 16 hands tall and weigh 900–1,200 lb. They should have symmetrical features with large bright eyes and a tapered muzzle. Their body should have a deep girth and well-sprung ribs. Fox trotters have a distinct fox trot gait. In this gait, the horse walks on its front legs while trotting on its hind legs. These horses are used primarily for pleasure riding, but are also used in showing and racing.

Breed standard painting courtesy of the American Morgan Horse Association/Jeanne Mellin

Morgan. The Morgan horse originated in the late eighteenth century when a man named Justin Morgan began promoting his stallion. This horse was the foundation of the breed and was known for a combination of speed and strength. The ideal Morgan horse stands between 14.2 and 16 hands, has a straight face, well-rounded jowls, and a deeper throatlatch. Morgans should also have an angular shoulder, a compact body, and correctly angled hocks. Some Morgans are gaited, meaning they have a specific way of moving different from non-gaited breeds.

Courtesy of The Jockey Club/Susan Martin

Thoroughbred. The Thoroughbred breed originated in England from two types of Arabian horses and Turkish horses. The horses were bred in an effort to create a horse that could carry weight with sustained speed over extended distances. These horses range from 15.2 to 17 hands tall, and are most often a solid dark color or gray. Thoroughbreds have a well-chiseled head, long neck, high withers, deep chest and hindquarters, lean body, and long legs. In the United States, Thoroughbreds are widely used in racing, as well as in English horse showing and pleasure riding.

Warmbloods

The term ***warmblood*** is used to define horses that are a combination of larger draft horses, which are considered ***coldbloods***, and more spirited light breeds like the Arabian and Thoroughbred, which are considered ***hotbloods***. The result is a horse that is large in stature, but retains the athleticism of a light horse. Dutch Warmbloods are at least 15.3 hands tall at the withers, and are often found in heights up to 17.2 hands. These horses are most commonly used for show jumping, dressage, cross-country jumping, or the combination of all those events, called Three Day Eventing. Other warmblood breeds of note include the Hanoverian, Trakehner, and Holsteiner.

Trakehner dien/Shutterstock.com

Hanoverian Abramova Kseniya/Shutterstock.com

Eric Isselee/Shutterstock.com

Dutch Warmblood

Draft Breeds

Draft horses are the historical powerhouses of the horse world. These animals have primarily been used to pull carts and farm equipment. Their large size defies their general calm disposition.

Hyjak Legacy Lady - Hyjak Legacy's Supreme Lady, M124671, Registered Belgian mare shown by Oak Haven Belfians, DOB: 3/10/2001

Belgian. Belgian horses originated in Belgium in the 1800s. These horses were bred specifically to be thicker bodied and heavier boned. They stand between 16 and 17.3 hands and have thick muscular chests, heavy bone, and strong loins and hindquarters. They are most commonly seen with a reddish brown (sorrel) body and a gold or tan (flaxen) mane and tail. These horses are often used for pulling carts and wagons, and are also used in horse weight-pulling competitions.

Muskoka Stock Photos/Shutterstock.com

Clydesdale. Clydesdales originated in Scotland where they were used as farm horses. These horses are generally brown with a black mane and tail (bay), and have white markings on their legs and face. They are also well-known for having excess hair, or feathering, on their lower legs. Clydesdales typically mature between 17 and 18 hands, and weigh between 1,600 and 2,400 lb. They are used for pulling carts and wagons and in horse-pulling competitions, although some people use them for pleasure riding as well.

On Behalf of Horse Photos

Percheron. Percheron horses originated in France, where they were used as war horses, for farm work, and for carriage pulling. They are usually black or gray in color, although they may be found in other colors. They typically stand between 16.2 and 17.3 hands tall and weigh an average of 1,900 lb. These horses are known for exceptional muscling in their lower thighs, allowing them to pull heavy loads over long distances. These horses are still used in farming applications in some areas, as well as being used as cart and wagon horses.

2014 National Grand Champion Shire Mare, courtesy of Jenson Shires, Blair, NE

Shire. Shire horses are a minimum of 16.2 hands and average 17.1 hands. The ideal Shire should have a broad head, sloping shoulders, deep girth, and a strong short back. They should have heavy bone and large feet. Shires are most desirable in brown, bay, black, or gray. These horses are especially known for their ability to move large amounts of weight over short distances and are commonly used in horse-pulling competitions.

Lesson 10.3 Review and Assessment

Words to Know

Match the key terms from the lesson to the correct definition.

1. The scientific name for wild and domesticated horses.
2. A female horse two years or younger.
3. A young equine of either gender.
4. The scientific name for donkeys and asses.
5. The correctness of an animal's bone structure, musculature, and body proportions in relation to each other.
6. The process of when an equine is giving birth.
7. A castrated male equine.
8. Training given to newborn equines to ensure they are accustomed to human interaction and will be safer to handle when they are older.
9. A mature female horse.
10. The species name given to all horses, donkeys, and asses.
11. The bucking animals that cowboys attempt to ride in rodeos.
12. The equine offspring of a female donkey and a stallion.
13. A male horse two years or younger.
14. A female ass or donkey.
15. The classification given to horses measuring 16 to 18 hands tall and weighing upward of 2,000 lb.
16. A mature male horse.
17. The classification typically given to horses around 14 to 16 hands tall and weighing between 900 and 1,400 lb at maturity.
18. Animals that have historically been domesticated and now live in the wild with no human assistance.
19. A male ass or donkey.
20. Horse that has been selected for ranch working traits such as a specific level of athleticism, speed, and instinctive ability to work with cattle.
21. The equine offspring of a mare bred to a jack.

A. colt
B. conformation
C. draft horse
D. equine
E. *Equus asinus*
F. *Equus caballus*
G. feral
H. filly
I. foal
J. foaling
K. gelding
L. hinny
M. imprinting
N. jack
O. jenny
P. light horse
Q. mare
R. modified monogastric
S. mule
T. pony
U. rodeo
V. roughstock
W. stallion
X. stockhorse

22. A type of digestive system with a stomach that has a single compartment and a special compartment in the intestine called the cecum.
23. The classification typically given to horses under 14.2 hands tall and weighing up to 900 lb.
24. A sporting event comprised of events based on ranch skills that cowboys and horses needed to have in order to work cattle.

Know and Understand

Answer the following questions using the information provided in this lesson.

1. *True or False?* A foal will typically nurse for 3–6 months before being weaned.
2. List the five most common sectors of the horse industry in the United States.
3. Identify four show events that are commonly found in English riding.
4. List five western horse showing events.
5. Explain why it is sometimes preferable to use horses instead of motorized vehicles in cattle operations.
6. The bucking animals used in rodeo events are referred to as _____ animals.
7. *True or False?* Equine activities are often used in therapy to improve the physical, occupational, or psychological health of the riders.
8. *True or False?* An equine is measured in feet to the top of its withers.
9. Explain how the digestive system of equines differs from the digestive system of other livestock animals.
10. List and explain three actions you can take to help maintain equine health.
11. Explain the fight or flight behavior common to horses.
12. Explain why breeders use imprinting on newborn foals.

Analyze and Apply

1. **Math.** Horses are measured in hands (4″ each). Calculate the following horse heights in hands. (Remember that you express the hands in whole numbers, then the number of remaining inches, i.e. 12.3 hands.)

   A. 48″

   B. 56″

   C. 62″

   D. 5′

   E. 6′ 1″

   F. 6′ 7″

2. Horses are driven by a combination of their instincts and training, and managing horses is a very intense industry. Imagine you are the owner of a horse boarding facility. Using the information from this lesson and your previous knowledge, write a list of ten "Barn Rules" that you would post in your facility and require all horse owners to follow.

Thinking Critically

1. There is a lot of controversy about feral horse populations in the United States. Conduct research on the Internet to find out why some organizations oppose the current management plan being used by the Bureau of Land Management (BLM). Compile a list of five arguments that opposition to the BLM has for the current Wild Horse Management Plan.
2. Horse slaughter is another controversial issue in the United States. Conduct research on horse slaughter in the United States and write 2–4 paragraphs explaining your personal opinion on whether horse slaughter should be legal in the United States. Use factual statements to back your opinion and cite your references.
3. Think about the wide difference in breeds of horses. Why do you think there is the need for so many different breeds?
4. What steps should be taken when introducing a new horse onto your ranch or farm?

Chapter 10

Review and Assessment

Lesson 10.1

Beef Industry

Key Points

- The beef industry is vital to our economy and food supply.
- Beef production systems range from simple, family-owned operations to operations boasting thousands of head of cattle.
- The beef production systems work together to supply a constant and safe supply of beef in the United States and worldwide.
- No two breeds of beef cattle are the same, and each brings its own set of strengths and weaknesses to the industry.
- Through research and development, beef breeds have evolved to fit all climates, terrains, and geographic areas of the world.
- Raising beef cattle is the single largest segment of American agriculture.

Words to Know

Use the following list and the textbook glossary to review and study the *Words to Know* from *Lesson 10.1*.

artificial insemination (A.I.)
backgrounding operation
beef
beef cattle
Bos Indicus
Bos Taurus
bovine
breed
breed association
brisket
bull
calf
calving
chuck
commercial cattle
cow
cow-calf operation
cud
cutability
feedlot
flank
gestation
heifer
herd
loin
offal
plate
polled
primal cut
purebred
purebred breeder
rib
round
ruminant
seedstock
seedstock cattle producer
steer
stocker operation
subprimal cut

Check Your Understanding

Answer the following questions using the information provided in *Lesson 10.1*.

1. At weaning, calves of both sexes are sent to _____ operations to graze and gain weight.
2. Most beef cattle are born on _____ operations.
3. Calves weaned from their mothers are often transferred to _____ operations.
4. Explain why cattle are fed high-grain diets with roughage in feedlot operations.
5. Explain why it is important to understand how herd animals behave in a group.
6. List four ways in which you can help maintain herd health.
7. List three points to keep in mind when handling cattle.

Lesson 10.2

Dairy Industry

Key Points

- The dairy industry is vital to our economy and food supply.
- Dairy production operations range from small, family-owned operations to operations with a couple of hundred dairy cattle.
- The main commodity of the dairy industry is fluid milk.
- When dairy cattle are harvested, they also produce beef.
- There are innumerable products made from milk, ranging from cream, yogurt, and butter to cheeses and frozen desserts.
- The terminology used to describe dairy cattle is identical to that used to describe beef cattle.
- No two breeds of dairy cattle are the same, and each brings its own set of strengths and weaknesses to the industry.
- Through research and development, dairy breeds have evolved to fit most climates, terrains, and geographic areas of the world.

Words to Know

Use the following list and the textbook glossary to review and study the *Words to Know* from *Lesson 10.2*.

artificial insemination (A.I.)	buttermilk	condensed milk
butter	cheese	cream

evaporated milk
frozen yogurt
homogenization
ice cream
mastitis
milk
milking parlor
parturition
somatic cell
sour cream
teats
udder
yogurt

Check Your Understanding

Answer the following questions using the information provided in *Lesson 10.2*.

1. The process in which the milk's fat droplets are emulsified and the cream does not separate is referred to as _____.
2. The act of giving _____ is what causes milk production in cows.
3. Heifers are usually retained and placed into a replacement program where they are grown out to _____ months of age and bred.
4. Dairy cows are culled from the dairy herd due to _____.
   A. lameness
   B. inflammation of the udder
   C. mastitis
   D. All of the above.
5. List four specific criteria milk must fulfill to meet industry standards.
6. Dairy cows are typically milked _____ a day.
7. List three factors that must be considered when selecting dairy cattle.
8. Explain why extra caution must be exercised if bulls are used for breeding rather than artificial insemination.

Lesson 10.3

Equine Industry

Key Points

- The equine industry in the United States includes more than nine million animals and contributes more than $100 billion in revenue annually.
- Horses are selectively bred to have characteristics that meet their desired discipline.
- Horses are used for showing, ranch work, rodeo, racing, recreational riding, and other highly specialized uses.
- Behavior plays an important role in handling and managing horses. Using caution around horses can prevent injuries to you or the horses you are working with.

- Feral horses still exist in the United States. Hundreds of breeds of domestic horses have been developed in order to suit the needs of humans.

Words to Know

Use the following list and the textbook glossary to review and study the *Words to Know* from *Lesson 10.3*.

allelomimetic behavior	feral	mare
bars	filly	modified monogastric
broodmare	foal	mule
cecum	foaling	pony
coldblood	frog	rodeo
colt	gelding	roughstock
conformation	hand	sole
draft horse	hinny	stallion
English riding	hotblood	sterile
equine	imprinting	stockhorse
equine assisted therapy	jack	stud
equines	jennet	topline
Equus asinus	jenny	warmblood
Equus caballus	light horse	western riding
farrier	long-eared breed	white line

Check Your Understanding

Answer the following questions using the information provided in *Lesson 10.3*.

1. *True or False?* The majority of horses in the United States are kept by racing stables.
2. Horses are bred primarily to _____.
   A. produce quality horses for racing and showing
   B. continue bloodlines
   C. produce individuals with superior traits for their given discipline
   D. All of the above.
3. How long is the gestation period for a mare?
4. *True or False?* English riding disciplines are designed so that the horse can be shown with the reins in one hand.
5. Identify three common types of horse racing.
6. Explain why horses can be effective transportation for police officers.
7. Explain the main difference between an equine's digestive system and that of most other livestock.

8. How often should a horse's hooves be trimmed or reshod?
9. Explain why it is important to establish a pest management plan for use with and around equines.
10. What is the most common cause of death in equines and what are its symptoms?

Chapter 10 Skill Development

STEM and Academic Activities

1. **Science.** Cheese is made from a chemical process that alters the milk protein using enzymes. What makes different varieties of cheese different? Choose three types of cheese to research and write a paragraph on each, describing the specific process for making that type of cheese. Then, compare and contrast your three cheese types so that you can explain how they are different. Be prepared to share your findings with the class.
2. **Technology.** Milking machines are incredibly fascinating machinery. To find out how the milking machine mirrors the actual suckling of a calf, conduct some research or ask a local dairyman. Draw a diagram that shows the parts of the milking machine that are required to ensure that dairy cows are milked efficiently.
3. **Engineering.** Cattle handling facilities are incredibly diverse. Imagine that you are the owner of a cow-calf operation and need to design a cattle handling area. In groups of two or three, research the common components of a cattle handling area, and design your layout. Make sure you include a squeeze chute or calf table, holding pen, and any other pens or structures that you think would be important.
4. **Math.** Working with a partner, measure the stalls in an equine stable. Calculate the number of square feet for each stall. Obtain prices for various types of bedding used for equines. Using the total square footage of the stalls, calculate how much bedding will be needed to provide adequate bedding for each stall. Calculate the cost of each type of bedding. Determine the cost for one month's bedding for all the stalls. If possible, include additional costs such as shipping and delivery.
5. **Social Science.** Why are businesses, homes, and livestock facilities usually in separate areas of a community? Are there instances where livestock may be kept on property that is not zoned as agricultural land? Interview the person in your community who is in charge of planning or zoning. Find out why zoning is considered important for the community and how a property owner can seek a zoning change. Obtain a copy of the community's zoning map and use it for a visual aid as you report your findings to the class.
6. **Language Arts.** Many stories, books, and poems have been written about horses. Choose one of the following titles, or one of your own favorites, to read and write a detailed book report. Use a standard format from your agriculture or English teacher. Create a diorama of your favorite or most memorable scene. Books to consider: *Black Beauty* by Anna Sewell, *The Black Stallion* by Walter Farley, *The Horse Whisperer* by Nicholas Evans, *A Horse Called Wonder* by Joanna Campbell.

Communicating about Agriculture

1. **Reading and Speaking.** Research the products and services available for raising either cattle or equines. Collect promotional materials for a variety of products and services from product manufacturers. Analyze the data in these materials based on the knowledge gained from this chapter. With your group, review the list for words that can be used in the subject area of raising livestock. Practice pronouncing the word, and discuss its meaning. As a fun challenge, work together to compose a creative narrative using as many words as you can from your new list.
2. **Speaking and Listening.** Divide into groups of four or five students. Each group should choose one of the following types of beef cattle operations: backgrounding operation, cow-calf operation, and stocker operation. Using your textbook as a starting point, research your topic and prepare a report on the operation. Include topics such as costs, land needs, structural needs, and employees, as well as the types of challenges the operation faces. As a group, deliver your presentation to the rest of the class. Take notes while other students give their reports. Ask questions about any details that you would like clarified.
3. **Reading and Writing.** The ability to read and interpret information is an important workplace skill. Presume you work for a local, well-known dairy operation. Your employer is considering upgrading the milking parlor with new milking equipment. Your supervisor wants you to evaluate and interpret some research on several new milking systems. Locate at least three reliable resources for the most current information on new milking systems. If possible, contact representatives from the manufacturers, and (after explaining your project) ask them about their products and additional costs, such as delivery and installation and employee training on the new equipment. Write a report summarizing your findings in an organized manner.

Extending Your Knowledge

1. Buying a horse is a huge financial responsibility. Even if you already have the horse and the equipment needed, there are additional costs such as entry fees for shows or rodeos, and medical fees. Figure the operating costs to keep a horse for one year. Research the cost for horse-related expenses in your area, and find the total. You will need to provide:
   A. Housing—contact a local boarding facility to determine the cost per month for stall, paddock, or pasture rent.
   B. Horse shoeing—calculate the number of trimmings or shoeings your horse will require per year (once every 6–8 weeks).
   C. Feed—light breed horses will eat on average one ton of hay every three months. Find this cost plus the cost of any feed additives.
   D. Vaccinations—most horses will receive annual and semiannual vaccinations. Contact an animal health provider to calculate this cost. Remember that this cost does *not* include other common costs like additional veterinary bills or paperwork for out-of-state travel.

Chapter 11

Small-Animal Production

Pattakorn Uttarasak/Shutterstock.com

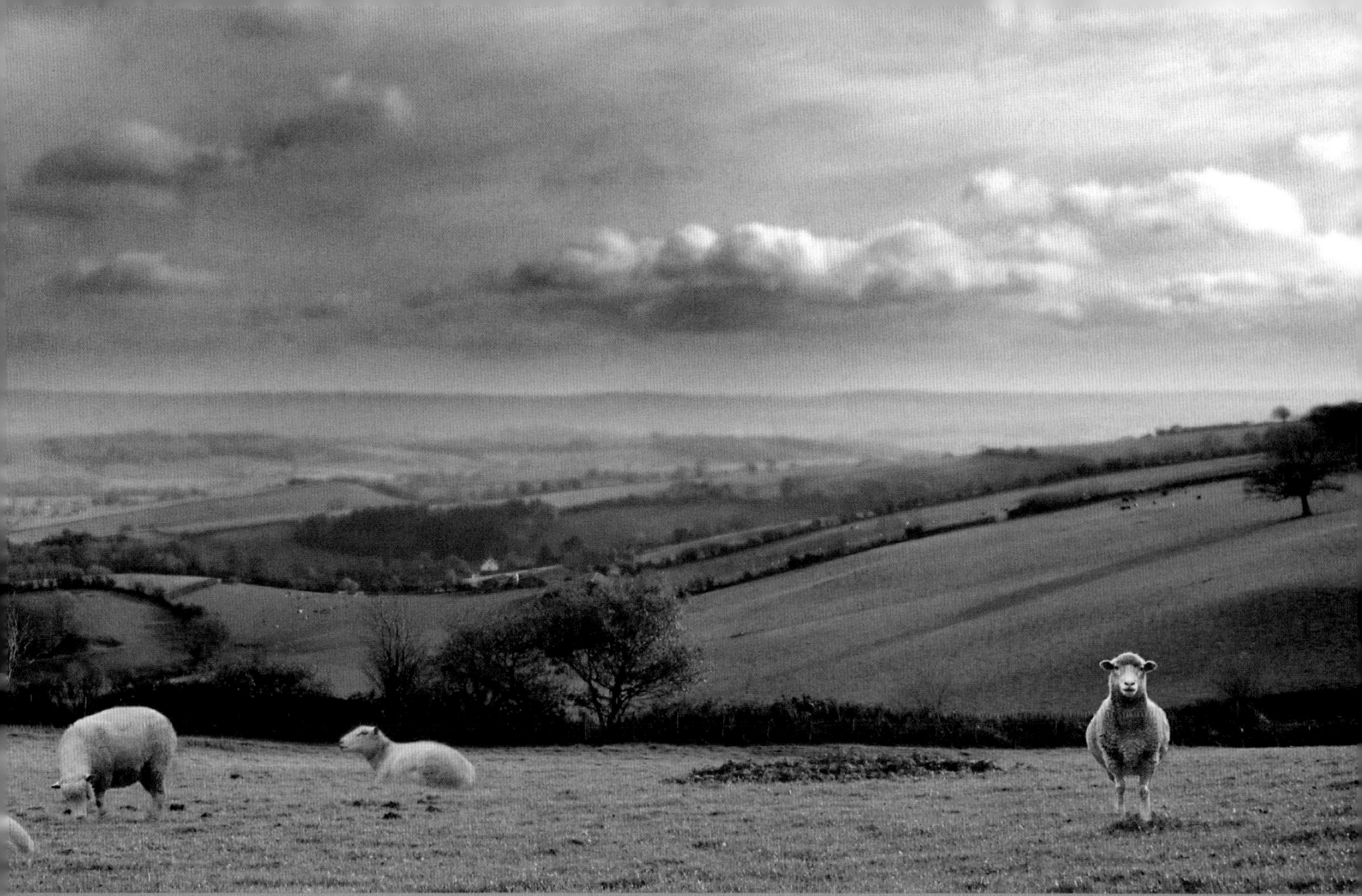

Pixel Memoirs/Shutterstock.com

Pichugin Dmitry/Shutterstock.com

aaltair/Shutterstock.com

While studying, look for the activity icon to:

- **Practice** vocabulary terms with e-flash cards and matching activities.
- **Expand** learning with video clips, animations, and interactive activities.
- **Reinforce** what you learn by completing the end-of-lesson activities.
- **Test your knowledge** by completing the end-of-chapter questions.

Lesson 11.1

Poultry Industry

Words to Know

avian
black-out house
boneless/skinless breast
breast quarter
breeder farm
broiler
caeca
candling
capon
chick
chicken
cloaca
cockerel
crop
drake
drumstick
duck
duckling
feet
flock
free-range system
gaggle
game bird
gander
gang
giblets
gizzard
goose
gosling
grit
hatchery
hen
hybrid
intensive housing system
layer
leg quarter

(*Continued*)

Lesson Outcomes

By the end of this lesson, you should be able to:

- Understand the size and scope of the U.S. poultry industry.
- Understand common terminology of the poultry industry.
- Describe the components of the U.S. poultry industry.
- Understand commonly used poultry production systems.
- Describe different types of poultry produced in the United States.

Before You Read

The term *poultry* includes several species of birds. Research the species of birds that fall under the category of poultry and compare and contrast the differences.

People have been raising poultry for meat, eggs, and feathers for thousands of years. It is estimated that chickens were domesticated as early as 2000 BCE. Turkeys have been domesticated for more than 2,000 years and have even been held in high esteem in some cultures. Depending on the culture, ducks and geese have even held their place as royal birds. Today, poultry serves as a primary source of meat and eggs for human consumption. The top five countries in poultry production are (in order): China, United States, Brazil, Russia, and India, **Figure 11-1**.

The Poultry Industry

Although the poultry industry includes turkeys, ducks, geese, and some game birds, the majority of poultry production in the United States is chicken production for both meat and eggs. Many of the production practices used for chickens apply to the commercial production of other poultry, with some variations, of course.

Common Chicken Terms

The following terms are used to describe chickens during different stages of their life:

- *Chick*—baby chicken (remains a chick until its body is covered with feathers)
- *Cockerel*—male chicken under one year old
- *Hen*—adult female chicken over a year old

- ***Pullet***—a female chicken under one year old
- ***Rooster***—adult male chicken over a year old (may also be called a cock)
- ***Capon***—castrated male chicken
- ***Broiler***—chicken raised strictly for meat
- ***Layer***—chicken raised for laying eggs
- ***Flock***—a number of chickens feeding and living together
- ***Chicken***—a domestic fowl kept for its meat or eggs; the meat from the bird

Broiler Production

The United States is the largest producer of broiler chickens in the world. In the past 50 years, the American broiler industry has evolved into a network of efficient, integrated businesses that supply customers with nutritious, quality meat, **Figure 11-2**. People in the United States consume more chicken (nearly 84 pounds per person, per year) than people in any other country. Chicken is the number one protein consumed in the United States, and American consumers spend approximately $70 billion per year on chicken.

The top states in broiler production are (in order): Georgia, Arkansas, Alabama, North Carolina, and Mississippi. Nearly 50 billion pounds of broiler meat are produced in the United States per year.

Vertically Integrated Production

Commercial broiler production began on a small scale in the 1920s and 1930s. The first producers only bred and raised the chickens, and had little to do with the feed production and processing of the birds. Eventually,

Words to Know

(*Continued*)
litter
molted
offal
paws
poult
proventriculus
pullet
rafter
rooster
semi-intensive housing system
sexually dimorphic
split breast
tenderloin
thigh
tom
vertical integration
whole bird
whole fryer
wing
WOG

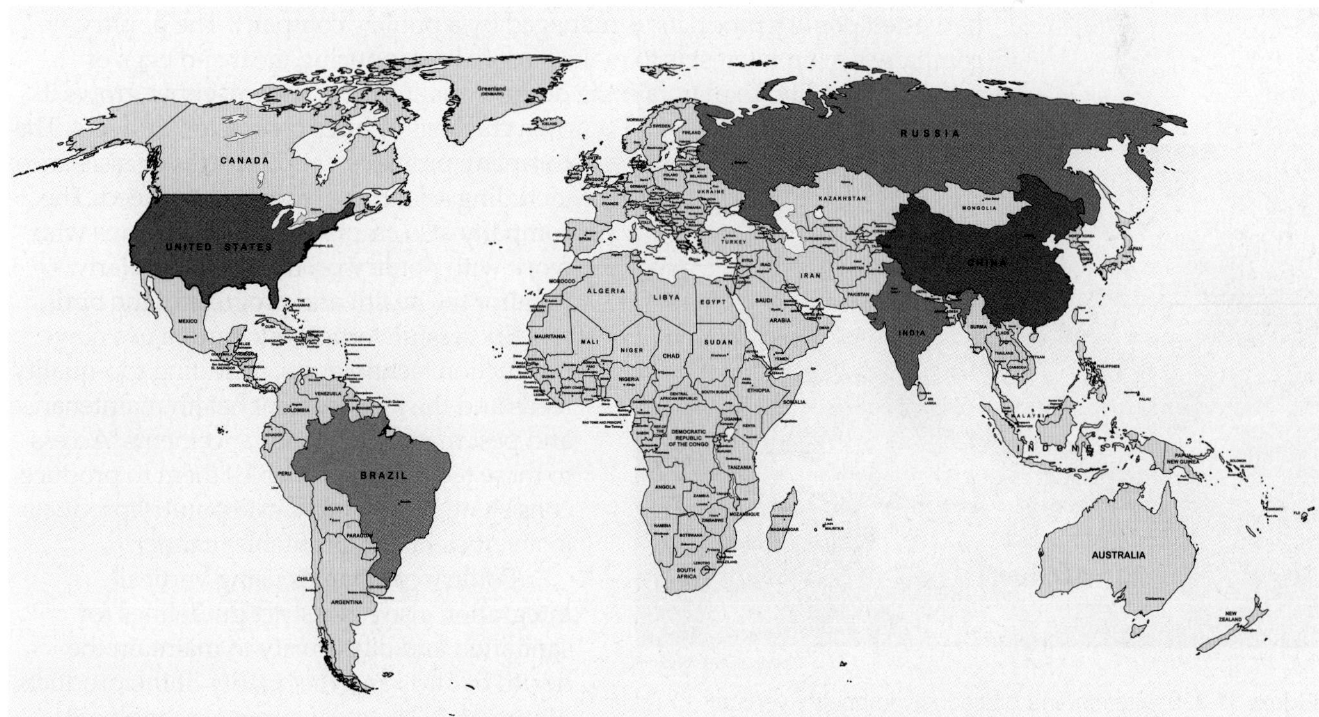

Bardocz Peter/Shutterstock.com

Figure 11-1. Many successful poultry operations are located throughout the world. The top five producing countries are (in order) China, United States, Brazil, Russia, and India.

Kharkhan Oleg/Shutterstock.com

Figure 11-2. Most broiler production operations have multiple poultry houses containing thousands of birds. *Is there a space requirement per broiler in a poultry house? If so, how much space is required?*

entrepreneurs combined the feed mill, hatchery, and processing operations, resulting in the beginning of the integrated poultry industry we have today.

The majority of the poultry industry in the United States uses ***vertical integration***. This means that every step in the production cycle, from birth to harvested poultry products, is managed by a poultry company. The poultry company's main interest is to make a profit by producing meat and eggs of consistent quality. To maintain the desired quality, the company either grows its own chickens or hires contracted growers. The company provides various forms of assistance, including scientifically formulated feed. The company also employs field technicians who work with poultry producers to regularly monitor the health and progress of the birds.

Picsfive/Shutterstock.com

Figure 11-3. Sanitation and biosecurity in poultry vertical integration are extremely important to maintain the health of the birds, and to ensure safe food products. *What measures are taken to prevent contamination or deterioration of the product in the harvesting and processing operations?*

Successful vertical integrators use new production technologies, including top-quality feeds and the most current health maintenance and pest management advancements. Access to these technologies enables them to produce consistent, high-quality, safe poultry products in an efficient and profitable manner.

Poultry companies using vertical integration also have strict guidelines for sanitation and biosecurity to maintain the health of birds and the quality of the products, **Figure 11-3**. The most current technologies are also used to monitor the heating, cooling, and ventilation of poultry houses to provide optimum conditions for the birds.

Egg Production

Egg production (for consumption) is another important aspect of the U.S. poultry industry. As with broiler production, commercial egg production in the 1920s and 1930s consisted of small farm operations slowly building up flocks of birds as consumer demand increased. Producers faced many problems, including weather extremes, predators, pecking order issues, diseases, and common parasites. Hens were laying around 150 eggs per year and had high mortality rates.

Did You Know?

Beak trimming is commonly done in large egg production operations to prevent the birds from pecking each other to death or cannibalism. Beak trimming has been questioned as an animal rights concern, but research has shown that it reduces stress and lowers mortality rates in commercial poultry production.

Early Advancements

As research into poultry production advanced, producers began moving the layers into indoor housing to help protect them from extreme weather conditions and predators. Egg production did improve somewhat with the new facilities, and the hen mortality rate was reduced. However, greater improvements in egg production and bird health did not occur until the late 1940s, when growers began using raised, wire-floor housing. The wire housing eliminated both hen and egg contact with waste, and allowed more uniform feeding throughout the flock. The wire caging system also lent itself to increased automation of egg collection.

Egg Production Today

Today, the poultry industry produces around eight billion eggs annually. These eggs are laid by the approximately 275 million laying hens in production at any given time. Each hen produces an average of 200–300 eggs in its commercial lifetime. If ***molted*** (forced shedding of feathers), the hens may produce more than 500 eggs. Careful genetic selection has been used to produce these prolific layers.

Egg production in our country primarily takes place with large companies holding a vast share of total egg production. There are approximately 172 egg-producing companies in the United States that have more than 75,000 hens. These companies represent 99% of the laying hens in the United States, **Figure 11-4**.

The top states in egg production are (in order): Iowa, Ohio, Indiana, Pennsylvania, and California. These five states represent more than 50% of

Did You Know?

Other than shell color, there is no nutritional difference between colored chicken eggs. The shell color varies by the breed of chicken.

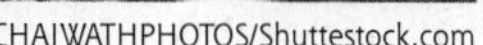

CHAIWATHPHOTOS/Shuttestock.com

spflaum/Shutterstock.com

Figure 11-4. Perishable products such as eggs require prompt harvesting, processing, and shipping to ensure freshness. *How much time is allotted between harvesting and consumption?*

the commercial egg producing hens in the country. The poultry industry also uses vertical integration systems for egg production.

Production Cycle of Poultry

Although the production cycle described here is primarily that of commercial chickens, it is, or would be, similar for most other poultry. It is simply not possible to include variations specific to each type and breed of chicken, turkey, duck, goose, or game fowl in the space of one lesson. See **Figure 11-5**.

Breeder Farms

For both broilers and layers, the production process starts at a ***breeder farm*** where pullets about 20 weeks old are placed with roosters to produce fertilized eggs. The poultry house used for breeding varies slightly in design, according to

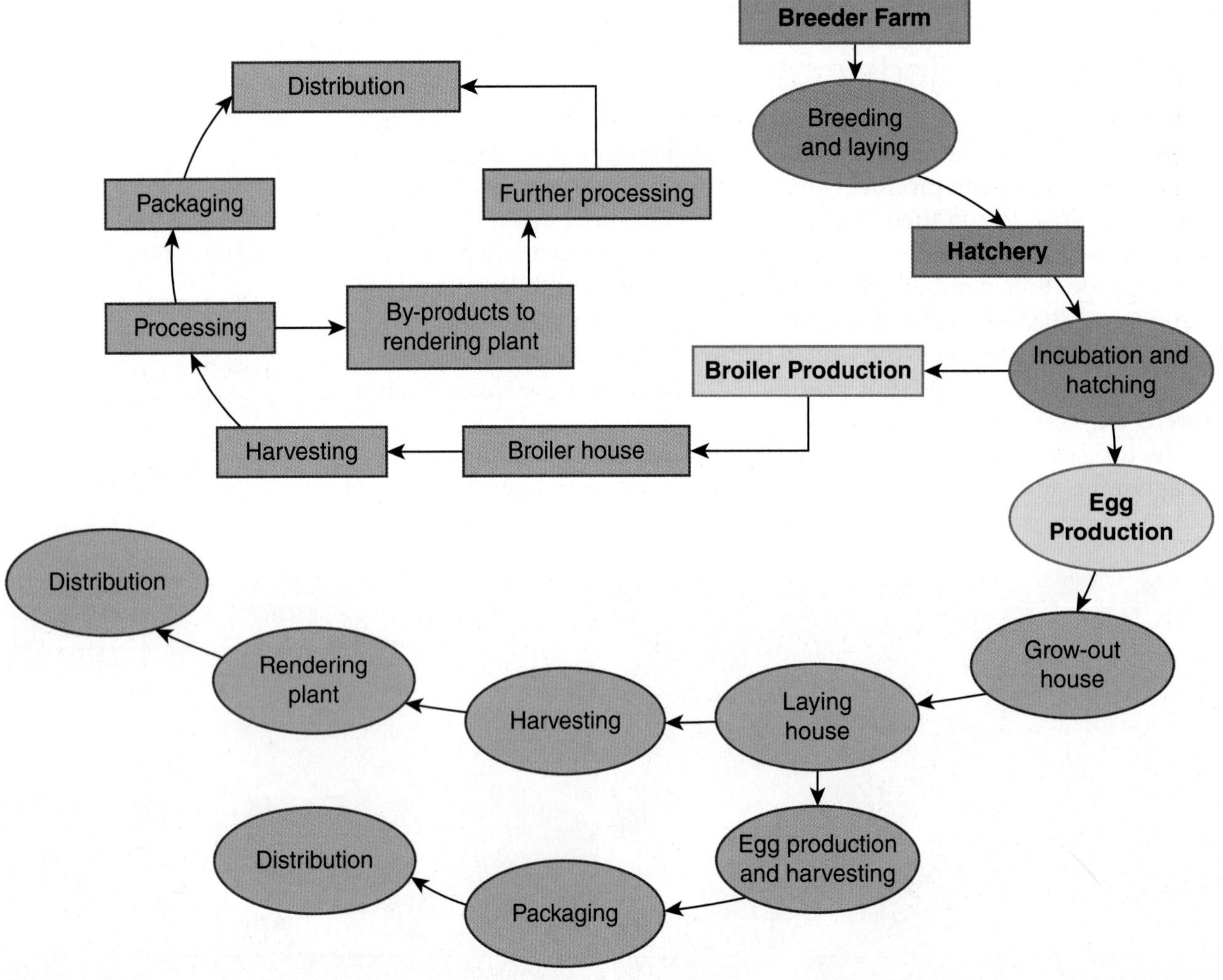

Figure 11-5. The production cycle is basically the same for all types of poultry. *What differences are there between smaller and larger operations? What differences in the production cycle might there be between the types of poultry?* After studying the production cycle, visit www.g-wlearning.com to test your knowledge.

the type of poultry being raised. For example, poultry houses for breeder broilers and ducks may have an area in the middle of the house covered with litter to act as a mating area, whereas the entire floor in turkey houses may be covered in litter for the same purpose. All large production houses will have nesting boxes of some type to facilitate egg laying and collection.

Avatar_023/Shutterstock.com

Figure 11-6. Today's modern hatchery facilities use computer systems to help monitor and control temperature and airflow, and to control the timing for turning eggs. *What types of jobs or careers are available in poultry production operations?*

Hatcheries

The fertilized eggs are gathered and taken to a hatchery, **Figure 11-6**. A ***hatchery*** is a carefully designed and specially constructed facility. Careful control of each stage of the operation is essential for optimum production. Most hatcheries have distinct areas for each of the following operations:

- Fertilized egg arrival and processing—this may include egg storage areas and containers, a grading and selection area, a fumigation area, an area for discarding eggs, and facilities for washing and storing storage trays and other equipment.
- Egg incubation—this area includes setter rooms, ***candling*** (using light to check embryo development through the shell) and transfer rooms, hatcher rooms, and an area/method for discarding eggs. All operations use automated incubators to aid in turning and rotating eggs to mimic natural brooding habits. Temperature and humidity are closely monitored and strictly controlled in the incubation areas to ensure optimum hatching numbers and chick health. See **Figure 11-7**.
- Newly hatched chicks—handling room for sexing and vaccination, holding area for chicks, indoor (ideally) loading area, disposal area/method, and an area for cleaning dirty trays and other reusable equipment.

branislavpudar/Shutterstock.com

Avatar_023/Shutterstock.com

Figure 11-7. Large poultry production facilities have walk-in incubation rooms in which movable racks house the eggs and the baby chicks. *How are eggs turned on these racks?*

Kharkhan Oleg/Shutterstock.com

Figure 11-8. Once the chicks are hatched, they are sorted and shipped to poultry production facilities.

- Technical operations—separate rooms for electrical controls, water controls, and ventilation controls. These rooms are best located on an outside wall to prevent unnecessary entry of outside personnel. Larger operations may also have employee showering and changing areas and separate employee areas for egg handling personnel and chick handling personnel.
- Waste processing and removal—these areas may be located separate from each operational area or incorporated within each area. Stringent protocol is usually used for waste processing and removal to maintain the most sanitary conditions possible.

Once the fertilized eggs arrive and are processed at the hatchery, they will be incubated for approximately 21 days. Turkeys and ducks typically require 28 days. After hatching, the young chicks are vaccinated, sorted by sex, and transported to grow-out facilities, broiler houses, or laying houses, **Figure 11-8**. The young chicks are fed very efficient feed and grow to four to seven pounds within 45–50 days. Around 63 days of age, broilers are sent to the processing plant to be harvested.

FFA Connection Candling Eggs

Candling is a process used to examine the interior quality of eggs without breaking the shell. In the FFA Poultry Evaluation Career Development Event, contestants candle eggs to determine the grade of the eggs. The eggs are held up to a bright light that illuminates the egg and allows the contestant to see air cell depth and imperfections in the eggs. The candling process also helps the contestant determine the air cell size of the egg. AA is the best grade an egg can receive on interior quality. This means the air cell size is less than 1/8″ deep, or about the size of a dime. A grade A egg has an air cell size of 3/16″, or approximately the size of a nickel. Grade B eggs have no size limit on air cell, but typically anything larger than 3/8″, or the size of a quarter, is a grade B egg. Eggs that have embryo development or blood spots are graded as inedible. Candling is also used to check the viability of fertilized eggs.

Download and print a practice air cell gauge from the USDA. Although participants are not allowed to use the gauge in competition, practicing with one will help you develop a keener sense of egg grades.

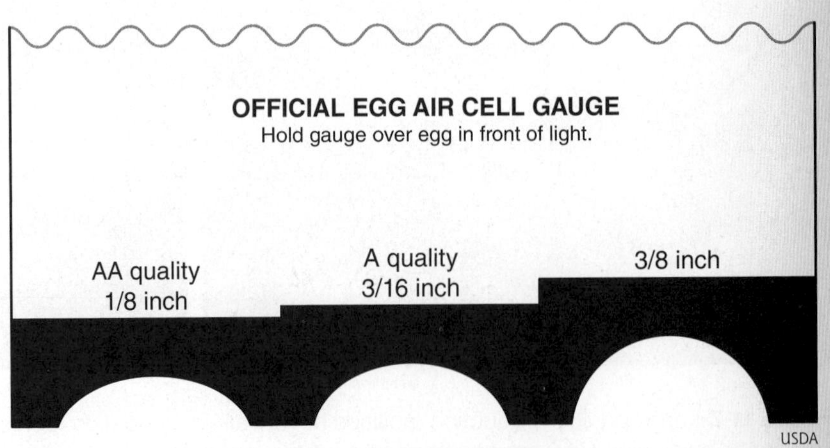

USDA

Lighting

As more research was performed on commercial poultry production, producers learned how lighting plays an important role in bird growth, development, and maturity. In natural environments, the seasonally changing amount of daylight affects a bird's development and the amount of eggs it will lay. Commercial producers manipulate the number of hours poultry are exposed to light in order to speed growth and development, and maintain higher levels of egg production. Poultry houses in which the lighting is controlled completely by the producer may be referred to as ***black-out houses***.

Did You Know?

More than 90% of the chickens raised for human consumption in the United States are produced by independent farmers who contract with a poultry company to raise chickens.

Anatomy of Poultry

There are many external parts on poultry. These parts are very similar across all types of poultry, **Figure 11-9**. The FFA Poultry Evaluation Career Development Event is a good way to learn the external and internal parts of all types of poultry.

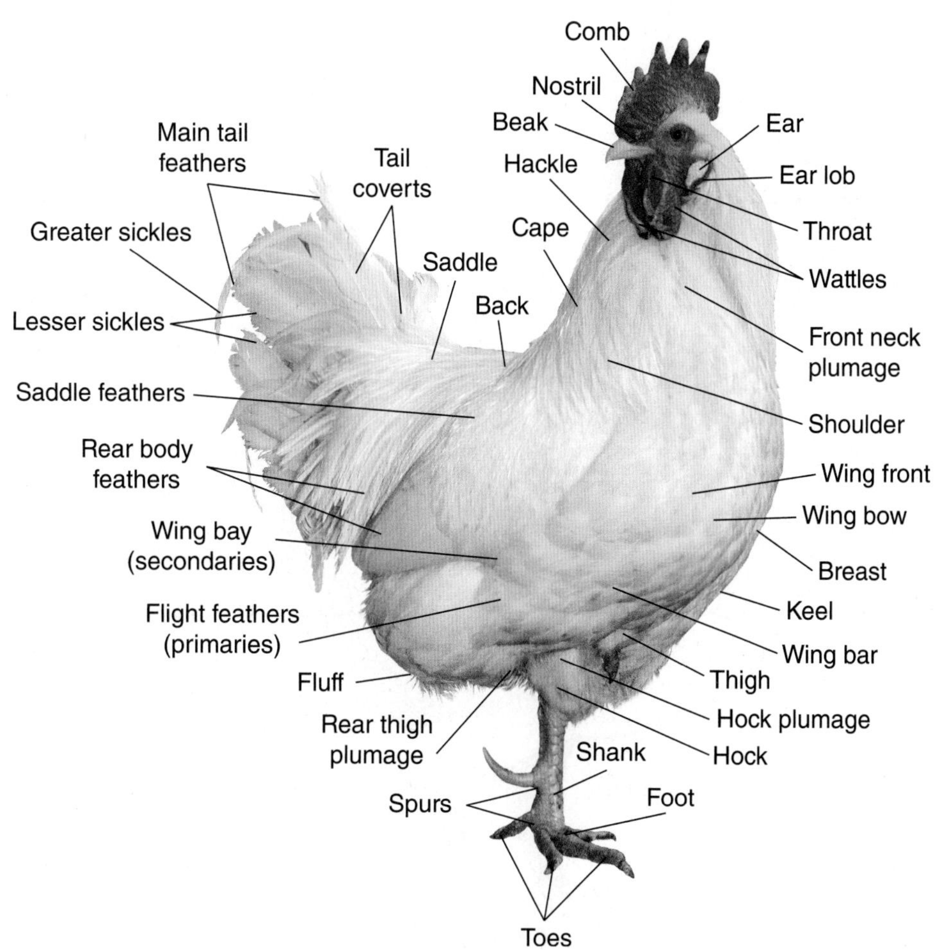

Aksenova Natalya/Shutterstock.com

Figure 11-9. Chickens, ducks, geese, and turkeys vary greatly in appearance. However, they do share the same basic anatomy. After studying the image, visit www.g-wlearning.com to test your knowledge of poultry anatomy.

Poultry Digestive System

The poultry digestive system is the same ***avian*** (of, or relating to, birds) digestive system used by all birds, **Figure 11-10**. This system is very different from mammal digestive systems. In the avian system, the food is passed from the beak into the esophagus, where it then travels into a storage area at the base of the esophagus called the ***crop***. From there, food travels into a muscular pouch called a gizzard. The ***gizzard*** is an area where muscular contractions, combined with pebbles or rocks, work to mechanically break down the food. When they have access to the ground outside, chickens eat pebbles and other hard objects (***grit***) that are used in the gizzard to grind up food. Grit is not necessary for commercially raised poultry, as the feed is already ground. Sometimes it is fed as a mash, or it is formed into pellets.

From the gizzard, the food travels into a small area called the ***proventriculus***, where digestive enzymes for chemical breakdown are added. Nutrients are then absorbed into the bloodstream in the small intestine. Near the large intestine, the waste products from the kidneys enter the digestive system through the ***caeca***, and finally, waste leaves the avian body through an opening called the ***cloaca***. The cloaca is the same exit for eggs.

Did You Know?

Chickens swallow their food whole because they have no teeth.

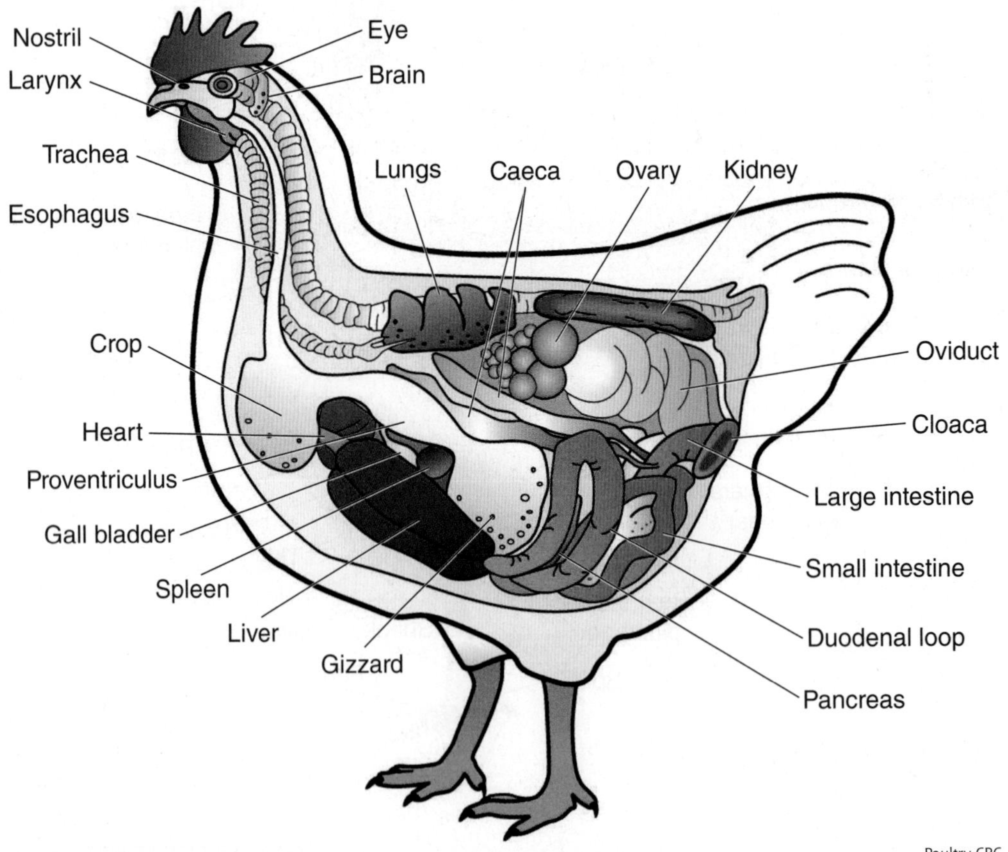

Figure 11-10. The avian digestive system is relatively simple. After studying the image, visit www.g-wlearning.com to test your knowledge.

Major Cuts of Poultry

Poultry may be sold as whole, dressed birds, as halved birds, or packaged by the individual parts. The primal cuts of poultry are described and illustrated on the following pages. Although the parts illustrated are from a broiler chicken, they are the same for all poultry, varying only in size, meat color, and possibly skin color and texture.

de2marco/Shutterstock.com

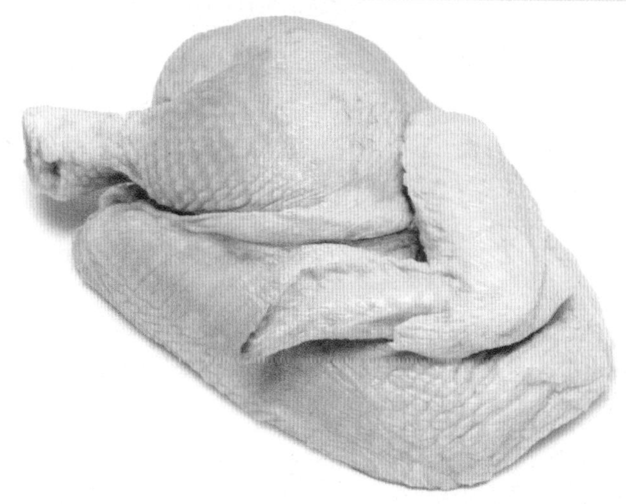

NRT/Shutterstock.com

Whole bird or **whole fryer.** Although still sold packaged this way, many consumers prefer to purchase individual pieces. A ***whole bird*** may be sold fresh or frozen as a ***whole fryer***, broiler, or roaster. The price per pound for whole birds is usually less than the price per pound of individual cuts. It may also be sold cut up as an eight-piece cut or whole cut chicken. A whole bird usually includes the giblets packaged in a small bag left inside the bird. A whole fryer without the giblets is referred to as ***WOG***. The chicken may also be sold as a half bird.

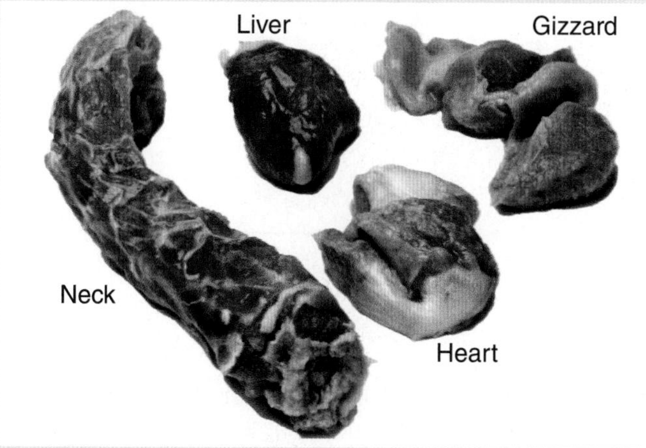

Vizual Studio/Shutterstock.com

Giblets. The ***giblets*** are the edible offal of the chicken, including the *gizzards*, liver, and heart. It may also include the neck. ***Offal*** is the entrails and internal organs of animals processed for food.

©istock/travellinglight

Tenderloin. White meat that may be slightly more tender than the breast. The ***tenderloin*** is a small muscle that sits adjacent to the breast.

Major Cuts of Poultry (*Continued*)

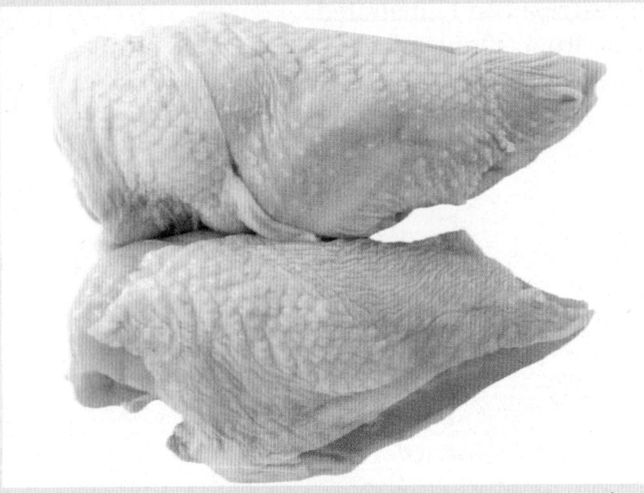

taro911 Photographer/Shutterstock.com

Split breast. The ***split breast*** is a breast quarter (bone in) with the wing removed. It may or may not come with the back portion. A ***breast quarter*** is a bone-in cut (with skin attached) that includes a portion of the back, the breast, and the wing.

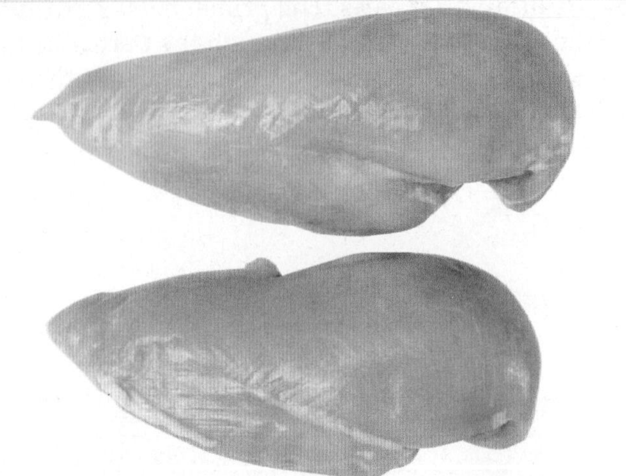

Nils Z/Shutterstock.com

Boneless/skinless breast. The ***boneless/skinless breast*** is one of the most popular cuts of chicken composed of white meat. The lean meat is from a split breast which has been deboned and skinned.

Madlen/Shutterstock.com

Wing. The ***wing*** is the least expensive but most popular cut, with little meat (white). It may be sold whole or divided into three sections: the drumette, midsection, and tip. The drumette is the portion between the shoulder and the elbow. The midsection is the portion between the elbow and the tip.

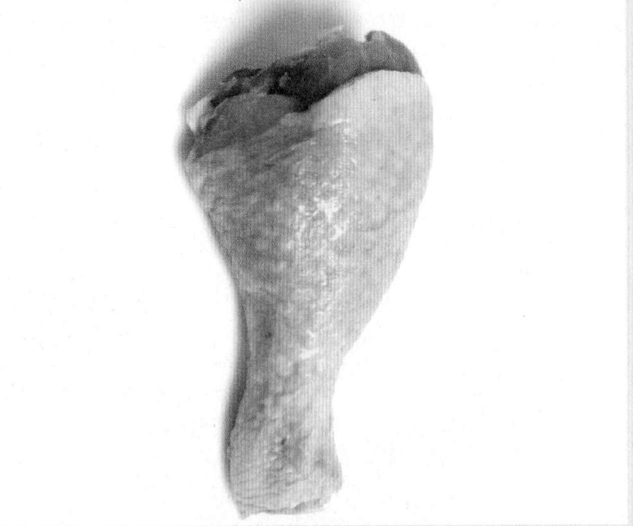

HomeStudio/Shutterstock.com

Drumstick. The ***drumstick*** is the lower portion of the leg quarter (between the knee joint and the hock). The drumstick is a less expensive cut sold with the bone in.

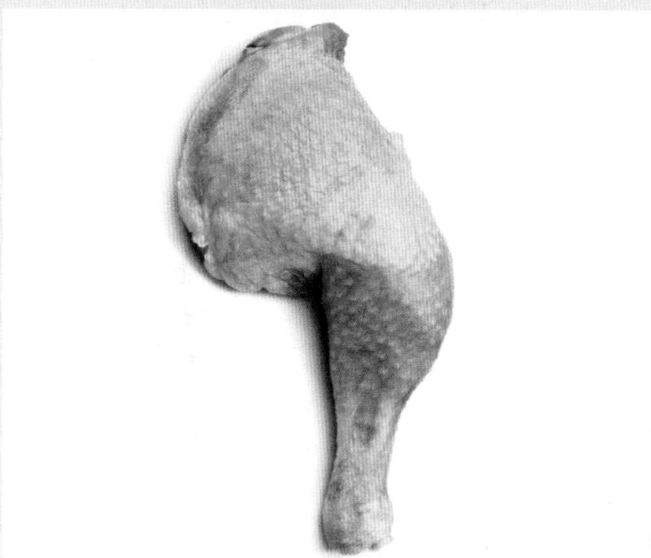

Petr Malyshev/Shutterstock.com

Leg quarter. The ***leg quarter*** is both the drumstick and the thigh when still connected.

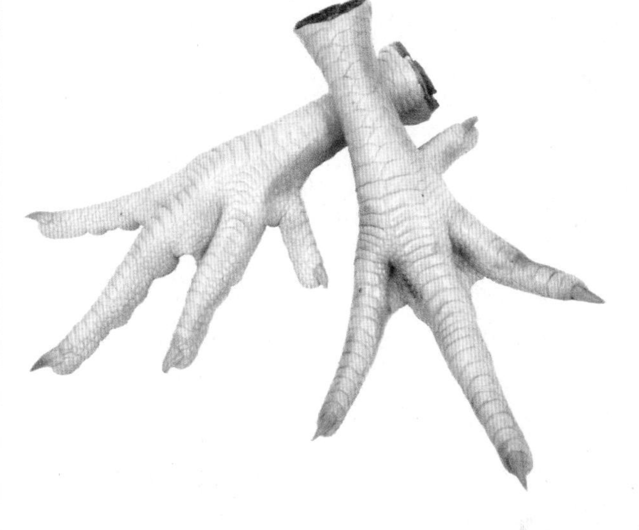

Mr. SUTTIPON YAKHAM/Shutterstock.com

Feet. The ***feet***, or ***paws***, are often used to make stock. They are also harvested and sold for a premium to export markets such as China.

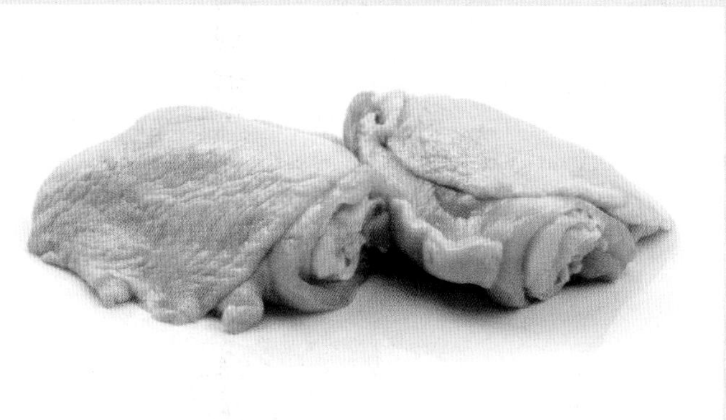

Ratchapol Yindeesuk/Shutterstock.com

MSPhotographic/Shutterstock.com

Thigh. The ***thigh*** is the portion of the chicken leg cut above the joint of the knee. The thigh is dark meat that tends to be firmer than the breast meat and may contain more fat. It may require longer cooking time than a breast. Thighs may be purchased boneless/skinless or bone-in with the skin attached.

Poultry is also sold as ground meat and often used as a substitute for hamburger meat. One section of the FFA Poultry Evaluation Career Development Event focuses on the identification of processed parts of the poultry carcass. Being able to identify the cuts of poultry will make you a better informed consumer.

Modfos/Shutterstock.com

Figure 11-11. Poultry production facilities use biosecurity methods to prevent the introduction of diseases into their poultry houses. *If disease spreads through a facility, what steps must be taken to dispose of the birds if they cannot be treated successfully?*

Maintaining Flock Health

The use of vertical integration has alleviated most health concerns and issues in modern poultry production. Field technicians, along with veterinarians hired by poultry companies, maintain strict health and biosecurity requirements to ensure healthy flocks. A few things to remember about flock health:

- Follow all biosecurity regulations of the poultry company, **Figure 11-11**.
- Eliminate exposure of flocks to infectious agents.
- Maintain access to feed and clean water at all times.
- Closely monitor the ventilation system of poultry houses.
- Practice proper removal and storage of litter from poultry houses.

Housing Systems

The three common methods used to house commercial poultry are free-range, semi-intensive, and intensive.

Free-Range Systems

The free-range system is the oldest system of housing poultry. In ***free-range systems***, the birds are allowed to roam and forage for insects, seeds, and herb material during the day, **Figure 11-12**. As on most farms,

Did You Know?

No hormones are used in poultry production in the United States. The Food and Drug Administration (FDA) prohibits the use of any hormone in poultry production.

Ivonne Wierink/Shutterstock.com

Figure 11-12. Free-range chickens spend the majority of their time outdoors. *What types of natural predators endanger free-range chickens?*

both large and small, the birds are brought indoors to roost at night and to protect them from predators. The downside to free-range production is the lack of environmental control. The birds are exposed to pathogens, parasites, and wildlife that can easily transmit otherwise controllable illnesses and diseases. They are also more susceptible to predators. The amount of light exposure the hens need for optimum laying is also out of the producer's control. Today, less than one percent of all of the chickens raised in the United States are raised free-range.

Semi-Intensive Housing Systems

In ***semi-intensive housing systems***, the birds are allowed some access to outdoor areas for fresh air and foraging. Free space in this system is limited, but outdoor runs may be attached to the poultry house to provide 20–30 square yards of space for birds to move around outside. The feeding and watering systems are located inside. The birds are also kept inside overnight to protect them from extreme weather and predators. This type of system is more typical for small-scale poultry production, or for producers who want to raise chickens in a more controlled environment than a free-range system. It is also commonly used in duck production. See **Figure 11-13**.

chinahbzyg/Shutterstock.com

Figure 11-13. Semi-intensive housing facilities allow the birds to spend part of the day outdoors.

Intensive Housing Systems

Intensive housing systems are total confinement operations where the birds stay inside a house or a cage their entire life, **Figure 11-14**. This is the most common housing method used in commercial poultry production in the United States. Intensive housing systems allow close management of birds being raised for vertically integrated operations. Producers can effectively use biosecurity systems to prevent the flock's exposure to disease and illness, carefully monitor food and water distribution, and control light exposure for optimum laying. Intensive management systems have been criticized as being cruel to the animals. Without these systems, however, poultry production would struggle to meet consumer demands

Gualberto Becerra/Shutterstock.com

Figure 11-14. Most large-scale poultry production uses intensive housing. It is the most practical method of maintaining such a large number of birds.

STEM Connection Scientific Housing

At first glance, poultry houses look like a big barn with a tin roof and a lot of chickens inside. They are actually highly technical and scientific housing units designed to maximize production and efficiency of poultry. Poultry houses are climate controlled to provide the proper temperature for the stage of growth and maturity of the chickens. The feeding and watering systems are automated to provide optimum nutritional requirements. The optimum depth of shavings is applied to the floor of the house to provide proper bedding and sanitation for the birds. Medication and treatment of birds is administered through the watering system to ensure coverage for all birds and consistency of treatment. The modern poultry house is much more specialized and scientific than your grandfather's chicken coop!

for good quality poultry products at reasonable prices. Research has shown that intensive management systems for poultry production have decreased poultry mortality in commercial production by almost 500% over the last several decades.

Did You Know?

Paleontologists have dubbed the *Anzu wyliei* (an oviraptorosaur) the "chicken from hell." These dinosaurs were among the largest "feathered" dinosaurs ever found in North America. They were 11′ long and weighed about 500 pounds.

Litter

Intensive housing systems produce a large amount of waste from the chickens called litter. Poultry ***litter*** contains manure, molted feathers, and bedding materials (usually straw, rice hulls, or wood shavings) from the poultry house. The litter's nutrient content varies depending on how the litter is removed from the poultry house and how it is stored. Most poultry litter is high in nitrogen and contains phosphate, potash, and other nutrients in varying amounts. Poultry litter can be a low-cost fertilizer for crops, used to feed cattle, and also used as a biomass energy source.

Did You Know?

Many historians attribute the wide distribution of chickens to the now banned sport of cockfighting.

Chickens

Although the term *poultry* includes all types of domesticated fowl, most of the poultry raised commercially throughout the world are chickens. It is estimated that there are more than 24 billion chickens in the world. That is more than three times as many people living on Earth!

Chicken Breeds

Today's domesticated chickens, *Gallus gallus domesticus*, are descended from the red jungle fowl, *Gallus gallus*, originally from the areas around Southeast Asia. There are many shapes, sizes, and colors of chickens around the world, but the ones most commonly used in commercial broiler production are ***hybrids*** (cross between two breeds) descended from original crosses between the Cornish and the Plymouth Rock. Producers primarily sought birds with high meat production, but they also sought birds with fast growth rates and high feed conversion ratios. For example, today's broiler chickens grow rapidly (reaching slaughter weight in 35–49 days) and convert feed at 1.8–2 lb of feed per pound of gain. In the 1920s, the feed conversion ratio was closer to 5 lb. The most common breed of chicken used to produce eating eggs is the White Leghorn.

Chicken Breeds

Jeannette Beranger/The Livestock Conservancy

Cornish. The Cornish breed was developed in England in the early 1800s to produce a superior fighting chicken. The Cornish found its way into commercial production because of its large breast and high level of meat production. Adult birds weigh between 8–10.5-lb. Depending on the variety, plumage may be white, white with red, or dark with some color variations. Unlike most chicken breeds, the male and female birds are not ***sexually dimorphic*** (different in color and size) but look very much alike.

Courtesy of Les Farms, Canada

Plymouth Rock. The Plymouth Rock breed is a dual-purpose bird found on many small farms in the United States and Europe. They are heavy, long-lived birds well-known for both their egg laying and meat production. The birds weigh between 7–9.5-lb, and their plumage is distinct between varieties. The most common varieties are the Barred Rock and the White Rock. The White Rock is used for the female side of today's commercial broiler cross.

Jeannette Beranger/The Livestock Conservancy

White Leghorn. The White Leghorn is said to have been introduced to the United States around 1852 in Mystic, Connecticut. They are hardy birds, known for their active foraging and small appetites as well as their prolific egg laying and fertility. Leghorns mature quickly but are not large birds, weighing between 4–6 lb. The White Leghorn is used for large-scale egg production and is also preferred by many researchers for avian biological research.

Chicken Breeds (*Continued*)

Jeannette Beranger/The Livestock Conservancy

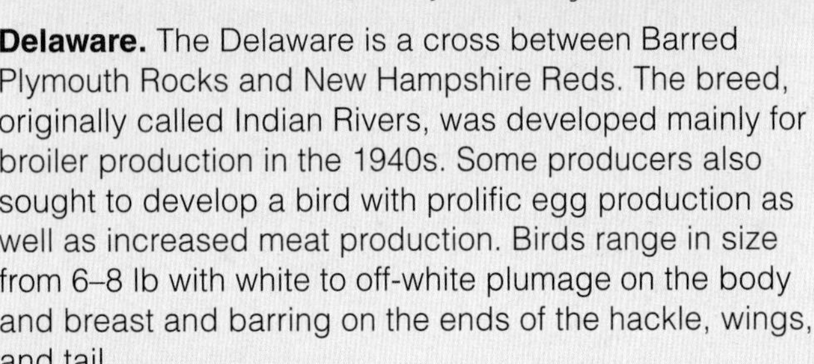

Delaware. The Delaware is a cross between Barred Plymouth Rocks and New Hampshire Reds. The breed, originally called Indian Rivers, was developed mainly for broiler production in the 1940s. Some producers also sought to develop a bird with prolific egg production as well as increased meat production. Birds range in size from 6–8 lb with white to off-white plumage on the body and breast and barring on the ends of the hackle, wings, and tail.

Tony Magdaraog/Shutterstock.com

Jeannette Beranger/The Livestock Conservancy

Rhode Island Red. The Rhode Island Red was developed in Massachusetts and Rhode Island in the late 1800s from crosses between the Malay, Shanghai, Java, and Brown Leghorn. They are hardy birds, known to be resistant to illness, good foragers, and weigh between 7–8 lb. Reds are excellent egg layers and may lay between 200–300 extra-large brown eggs annually. Their plumage varies and ranges from rust colored to darker maroon bordering on black.

American Game. The American Game breed was created originally for cockfighting through crossbreeding of various European and Oriental game breeds. The American Game is a beautiful, hardy bird and is raised primarily for showing. American Game birds are pleasant birds, but they have retained aggressive behaviors, and stags must be separated as they mature to prevent fighting and cannibalism. The breed is not meaty and are marginal egg layers.

Turkeys

Turkey production in the United States has increased more than 100% since the 1970s. It is estimated that the United States produces nearly 250 million turkeys per year. Turkey is the fourth most popular protein source of Americans behind chicken, beef, and pork. The average consumer in our country consumes about 16 lb of turkey per year. The top states in turkey production are (in order): Minnesota, North Carolina, Arkansas, Indiana, and Missouri.

Management Systems

Turkeys are mainly raised in semi-intensive or intensive management systems, **Figure 11-15**. Being much larger in mature size than chickens, turkeys require more space for laying and growing. Turkeys are also very susceptible to disease outbreaks and do well in intensive management systems where stringent biosecurity systems are maintained.

Common Turkey Terms

The following terms are used to describe turkeys during different stages of their life.

- ***Poult***—newly hatched turkey not fully covered with feathers
- ***Hen***—female turkey
- ***Tom***—male turkey
- ***Rafter***—a group of turkeys. It may also be called a ***gang***

Did You Know?

The average turkey egg is 50% larger than a chicken egg but contains nearly twice as many calories and grams of fat and four times as much cholesterol.

Egg Production

Although turkeys are raised commercially for meat production, it would be impractical to use turkeys for commercial egg production. Turkeys may produce four to five eggs per week whereas layers (chicken) average one egg per day. Chickens produce eggs at an earlier age (between four and five months) than turkeys (around 7 months), and require less space than turkeys. In commercial egg production, a turkey would require around three square feet, enough space to accommodate eight chickens. Turkeys also require more food than chickens.

Did You Know?

As larger turkeys were produced, they lost their ability to mate naturally. Today's commercial turkeys are all conceived artificially.

terekhov igor/Shutterstock.com

Leonid Shcheglov/Shutterstock.com

Figure 11-15. Turkey production operations are similar to those used for chickens. *What types of modifications might a poultry operation need if the producer were to switch from chicken to turkey production?*

Turkey Breeds

All domesticated turkeys are descended from wild turkeys indigenous to North and South America. Through Spanish explorers, turkeys found their way to Europe and then back to the New World through early American settlers. Some turkeys were bred for meat production and others to show off the bird's beauty. Today, most of the turkeys raised for commercial production are Broad Breasted White turkeys. The Broad Breasted White turkey is used in commercial production because of its size and the amount of meat it produces.

Since turkey breeds raised for commercial production are limited, many turkey breeds are disappearing and considered endangered. Some of these turkeys are raised in small operations to maintain the breeds and produce birds that fulfill breed standards. The following are a few of the breeds listed in the American Poultry Association's (APA) Standard of Perfection: Bronze, Narragansett, Bourbon Red, Royal Palm, and White Holland.

Jeannette Beranger/The Livestock Conservancy

Bronze. Bronze turkeys were produced by crossing Black European turkeys with American wild turkeys. The Bronze turkeys were larger and more robust than the European birds and tamer than the wild turkeys. The plumage of these birds is brown with highlighted shades of copper and blue-green. Larger varieties of Bronze turkeys (Broad Breasted Bronze and Broad Breasted White) dominated the commercial turkey industry for two decades after their development. Toms weigh approximately 25 lb, and hens weigh about 16 lb. The Standard Bronze is listed as critical by The Livestock Conservancy.

Jeannette Beranger/The Livestock Conservancy

Narragansett. Narragansett turkeys are unique to North America and are named for Narragansett Bay in Rhode Island. Narragansett turkeys are descended from a cross between wild turkeys of the eastern portion of the United States and the domestic turkey. The bird's plumage is black, gray, tan, and white. It resembles the Bronze turkey but has dull black or gray feathers in place of the Bronze turkey's copper feathers. The Narragansett turkey has a black beard and mostly featherless head and neck that range in color from red to bluish white. The Narragansett turkey was not produced commercially until it regained popularity in the twenty-first century. Toms weigh between 22–28 lb, and hens weigh between 12–16 lb.

Jeannette Beranger/The Livestock Conservancy

Bourbon Red. The Bourbon Red turkey is an attractive domestic breed named for Bourbon County in Kentucky. The breed was developed from crosses between Buff, Bronze, and White Holland turkeys by J.F. Barbee in the late 1800s. The plumage ranges from brownish to dark red with white flight and tail feathers featuring soft red feathers near the end. The Bourbon Red was known for its heavy breast and richly flavored meat and was commercially produced through the 1930s and 1940s. As with many other turkey varieties, it was unable to compete with the broad-breasted varieties on a commercial scale. Toms weigh approximately 23 lb, and hens weigh about 14 lb. The breed is currently listed as on "watch" status by The Livestock Conservancy.

Jeannette Beranger/The Livestock Conservancy

Royal Palm. The Royal Palm was developed primarily for its unique appearance with mostly white plumage and broad bands of metallic black feathers. The Royal Palm lacks the size for commercial production. Toms weigh between 16–22 lb, and hens weigh between 10–12 lb. The breed was developed in Lake Worth, Florida, in the 1920s as a cross between Black, Bronze, Narragansett, and native turkeys. The breed is currently classified as on "watch" status with The Livestock Conservancy.

Jeannette Beranger/The Livestock Conservancy

White Holland. The White Holland turkey was produced from crosses of white European turkeys and native turkeys. White Holland turkeys were crossed with Broad Breasted Bronze turkeys to create the Broad Breasted White turkey, the most common turkey breed in the world today. Toms weigh between 30–36 lb, and hens weigh between 18–20 lb. The White Holland is known for its hardiness, smaller breast, and longer legs. The breed is currently listed as threatened by The Livestock Conservancy.

Ducks

Most people think of ducks as wild birds, but domesticated duck production is a growing segment of the poultry industry. Ducks are raised for meat, eggs, feathers, as pets, and for showing, **Figure 11-16**. According to the Census of Agriculture, there are about 30 million ducks raised annually in the United States for processing. The top states in duck meat production are (in order): Indiana, California, and Pennsylvania.

Although duck production is much smaller than commercial chicken production, it is definitely finding its niche. People purchase duck eggs because they are larger than chicken eggs with larger yolks and provide a richer, more moist texture to foods such as pastas and baked goods. Duck eggs are also an important ingredient in a number of traditional Asian dishes. Partly due to size, duck eggs have more nutrients and higher concentrations of certain vitamins and minerals, as well as more protein, fat, and cholesterol than chicken eggs.

Management Systems

Commercial duck management systems are similar to those used for chickens and turkeys. However, waterfowl do have some special considerations.

- They require flooring that will not abrade the smooth skin of their webbed feet or cause them to trip and stumble.
- Ducks also drink and excrete more water than land fowl, and the ventilation and heating system must be adjusted to remove the extra

mmphotographie.de/Shutterstock.com

Figure 11-16. Duck down is popular as a natural insulator in cold-weather outerwear. *How common is it for down feathers to be mixed with regular feathers when used as insulation in clothing or bedding?*

moisture and maintain proper temperatures. It is also necessary to carefully monitor litter as their droppings contain over 90% moisture.

- Waterers and feeders must be designed to accommodate the size of a duck's bill. The feeder must provide sufficient room for the larger bill and their shoveling eating motion.

Free-range systems are typical in smaller operations, often allowing ducks access to ponds for swimming and breeding. Ducks are often seen within flocks of other fowl in smaller operations. Semi-intensive housing systems may also allow ducks daily access to open water and foraging. Intensive housing systems are typically used in larger scale duck production for better control and prevention of illnesses and disease.

Common Duck Terms

The following terms are used to describe ducks during different stages of their life.

- *Duckling*—baby duck of either sex
- *Duck*—female duck
- *Drake*—male duck
- *Flock*—a group of ducks

Egg Production

Although duck egg production is not done on as large a scale as chickens in the United States, ducks can be very proficient layers. Depending on the breed, ducks may lay from 100 to 300 eggs per year. Duck eggs are considered relatively easy to hatch when compared to other fowl since they are more forgiving of variations in temperature and humidity.

Did You Know?

The average life span of domestic ducks is 10 years, while ducks in the wild average between 2 to 5 years.

Duck Breeds

The majority of domesticated ducks are descended from the wild Mallard. The one well-known exception being the Muscovy, which is originally from Central and South America and almost the size of a goose. The breed most widely used in U.S. commercial production is the Pekin because of its generous size, rapid development, and egg-laying proficiency. A few popular breeds include the Aylesbury, Campbell, and Runner. As with turkey breeds, due to the limited number of duck breeds used in production systems, many domesticated duck breeds are also nearing extinction. Breeds may be chosen for temperament, size, egg production, mothering ability, egg size, meat production, and even foraging abilities.

Duck Breeds

Ian Lee/Shutterstock.com

Pekin. The Pekin duck is the primary breed used for commercial meat production in the United States. (Other breeds, such as the Khaki Campbell or breed crosses, are generally used for commercial egg production.) The breed was first introduced to the United States in 1873 when a small number were imported into Long Island from China. The Pekin is popular in commercial production because it is very large, develops rapidly, is very hardy, and is an excellent layer. Mature birds weigh between 8–11 lb in captivity and have substantial meat on the breast, leg, and thigh. An adult Pekin may lay an average of 200 eggs per year if they are kept from brooding. Their plumage is primarily white but may have a yellowish tinge, and they have distinctive orange bills and legs.

Jeannette Beranger/The Livestock Conservancy

Aylesbury. The Aylesbury is similar to the Pekin duck and is currently the primary duck used in commercial production in Great Britain. Like the Pekin, Ayelsbury ducks are large birds that mature quickly and are proficient egg layers. Mature birds weigh between 8–9 lb, and an adult Aylesbury may lay an average of 125 eggs per year. Aylesbury ducks have white skin; white plumage; yellow feet; and long, straight, light-pink bills.

Jeannette Beranger/The Livestock Conservancy

Campbell. The Campbell, an exceptional egg layer, was developed in the late 1800s in England through crossing of Fawn and White Indian Runners and the Rouen. Although there are four color varieties in the United States (Khaki, White, Dark, and Pied), only the Khaki Campbell is recognized by the American Poultry Association (APA). As with many fowl breeds, the drake and duck have slightly different coloring. The drake is more colorful with a green bill, orange legs and feet, and a combination of brown-bronze and warm-khaki plumage. The duck has a green bill, brown legs and feet, and a combination of brown and khaki plumage. Campbells weigh between 4.5–5.5 lb, and begin laying eggs around 5–7 months of age. Campbells may lay between 200–300 eggs per year.

Geese

Geese production in the United States is declining. More geese are being raised but fewer are being sold. The most recent estimates state that there were approximately 339,000 geese raised in the country. The top states in geese production are (in order): Texas, Minnesota, and South Dakota. We consume even less goose than we do duck in the United States. The majority of geese products produced in the United States are exported to Japan. The majority of goose meat consumed annually is produced in China, followed by the Ukraine, Egypt, and Hungary.

Geese are hardy, generally disease free, and long-living birds. In general, they do well on foraging without supplementary feed. It is believed this is due (in part) to their ability to extract protein from the herbage they forage. They have long been standard on many farms, kept for eggs, meat, feathers, as companion animals, and even as guard "dogs." They have been used in the United States for controlling weeds in crops such as asparagus, tobacco, beans, and potatoes, and also to control water hyacinth from taking over waterways. They may be kept as free-range birds requiring minimal protection from the elements, **Figure 11-17**.

Did You Know?

Goslings grow rapidly but are not fully mature until two years old.

Common Geese Terms

The following terms are used to describe geese during different stages of their life.

- *Gosling*—baby goose of either sex
- *Goose*—female goose
- *Gander*—male goose
- *Gaggle*—a group of geese

Did You Know?

Unlike European geese, Asian geese have a large knob at the base of the bill.

Reproduction

Most geese are able to reproduce naturally. The gander steps on the female from the side, and stands on her to mate. Overweight birds may have difficulty breeding naturally.

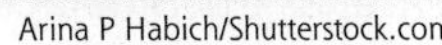

Arina P Habich/Shutterstock.com

Photographee.eu/Shutterstock.com

Figure 11-17. Free-range housing is used in most geese production operations because geese are hardy poultry and generally disease free.

Goose Breeds

Roman Samokhin/Shutterstock.com

Embden. The Embden breed was introduced to the United States around 1820 and is currently one of the most commonly used for commercial production. They grow fast, are large, and are very hardy birds. Mature birds weigh between 18–20 lb and produce more attractive carcasses than other darker-feathered breeds. They are good layers and produce large eggs. The Embden has white plumage and a yellow-orange bill with yellow-orange legs and feet.

Eric Isselee/Shutterstock.com

Toulouse. The name Toulouse is used for several types of geese descended from the European Greylag breed. The Production Toulouse is more common and used for both meat and egg production. The Production Toulouse are large birds most suitable for small farm flocks. Mature birds weigh between 18–20 lb and produce 25–40 eggs annually. The most common colored plumage is grey, but there is also a buff-colored variety. The bill, legs, and feet are orange.

Microcosmos/Shutterstock.com

African. The African goose is a large, heavy-bodied bird weighing between 18–22 lb. The mature African goose has a distinctive large knob on its forehead and a smooth, crescent-shaped dewlap hanging from its lower jaw and upper neck. African varieties found in North America include the Gray or Brown African and the White African. The African is considered a premier roasting goose with high-quality, lean meat. African geese are not considered prolific layers but may lay better than the Embden.

zikovec/Shutterstock.com

Chinese. Chinese geese have long, slender bills with large, rounded erect knobs on their foreheads. White Chinese geese have blue eyes, white plumage, and bright orange bills, knobs, and feet. Brown Chinese geese have brown and fawn-colored plumage. Their bill and knob are glossy black, and their legs and feet are orange. The Chinese is the most prolific layer (40–100 eggs annually) of all the breeds and also has less "greasy" meat as they have less fat on them. They are excellent foragers and are often used to control weeds in various crops and in open acreage. Mature geese weigh between 11–14 lb.

Game Birds

Game birds include a number of different fowl species and subspecies that are legally hunted. They may be native or nonnative to the United States. Many game birds raised commercially are released and hunted, but they may also be raised for meat or egg production. Common game birds raised for hunting in the United States include quail, pheasants, pigeons, ducks, wild turkeys, and the Chukar partridge.

The game bird industry generates most of its revenue from hunting and recreation. It is estimated that the game bird industry generates more than $1.6 billion in the United States.

Career Connection Metzer Farms

Duck, Goose, and Game Bird Hatchery

When John Metzer graduated from the University of California at Davis in 1978 with a degree in Animal Science, he returned home to find his father's duck hobby had grown substantially. His father began keeping ducks to help rid his sheep herd of liver fluke, a parasite of sheep. The ducks loved to eat the snails that ingested the fluke eggs, thereby ending the cycle. The snail-eating ducks led to duckling and egg sales, which led to John's dreams of Metzer Farms.

Today, the main products sold are day-old ducklings and goslings. Customers include hobbyists, feed stores, and commercial meat and egg producers located in the United States and around the world. Metzer Farms raises more than 25 breeds of ducks, geese, guineas, pheasants, and wild turkeys. Other products sold include duck hatching eggs, blown eggs for decorators, *balut* (partially incubated duck eggs), salted duck eggs, fresh duck eggs, and grown Mallards for dog trials and hunting preserves.

The Metzer family controls every aspect of the business—from breeding to shipping—to ensure the highest quality birds and best customer service. Metzer Farms is located in California's Salinas Valley, just 20 miles from the Pacific Ocean. The valley's mild climate is ideal for raising waterfowl.

Metzer Farms

John and Marc Metzer.

Lesson 11.1 Review and Assessment

Words to Know

Match the key terms from the lesson to the correct definition.

1. A poultry production facility in which pullets are placed with roosters to produce fertilized eggs.
2. A baby chicken.
3. A group of turkeys.
4. A chicken raised strictly for meat.
5. A baby goose of either sex.
6. A male duck.
7. The opening through which waste leaves the avian body.
8. Poultry meat from a split breast which has been deboned and skinned.
9. A female chicken under one year old.
10. A bone-in poultry cut (with skin attached) that includes a portion of the back, the breast, and the wing.
11. A female goose.
12. A male turkey.
13. A poultry production system in which the birds are allowed to roam and forage for insects, seeds, and herb material during the day.
14. An adult male chicken over a year old (may also be called a cock).
15. A group of geese.
16. A castrated male chicken.
17. A whole fryer (chicken) packaged without the giblets.
18. A baby duck of either sex.
19. A newly hatched turkey not fully covered with feathers.
20. A male goose.
21. A male chicken under one year old.
22. The edible offal of the chicken, including the gizzard, liver, and heart.
23. A poultry breast quarter with the wing removed.
24. A total confinement poultry production system in which the birds stay inside a house or a cage their entire life.
25. A poultry cut of meat composed of both the drumstick and thigh when still connected.
26. The bedding materials (including manure and molted feathers) from a poultry house.

A. boneless/skinless breast
B. breast quarter
C. breeder farm
D. broiler
E. capon
F. chick
G. cloaca
H. cockerel
I. drake
J. duckling
K. free-range system
L. gaggle
M. gander
N. gang
O. giblets
P. goose
Q. gosling
R. intensive housing system
S. leg quarter
T. litter
U. poult
V. pullet
W. rooster
X. split breast
Y. tom
Z. WOG

Know and Understand

Answer the following questions using the information provided in this lesson.

1. Which birds are included in the poultry industry?
2. *True or False?* Chicken is the number two protein consumed in the United States.
3. Briefly explain vertically integrated poultry production.
4. Explain the biosecurity concept as it applies to poultry production.
5. On average, how many eggs does a hen lay in its commercial lifetime?
6. Why do hatcheries have distinct areas for each operation of chick production?
7. Explain why commercial producers manipulate the number of hours poultry are exposed to light.
8. *True or False?* The avian digestive system is similar to that of animals with a modified monogastric digestive system.
9. List three actions you can take to ensure healthy poultry flocks.
10. The most common housing method used in commercial poultry production is _____.

Analyze and Apply

1. Exterior egg grading is a portion of the FFA Poultry Evaluation Career Development Event. The eggs shells are evaluated for stains, dirt or foreign material, shape, shell texture, ridges, shell thickness, and body checks. Using the USDA standards for exterior quality of eggs, work with a partner to grade the exterior quality of a dozen eggs.
2. Research various methods of creating your own candler. Acquire the needed materials and practice candling eggs.

Thinking Critically

1. The top four companies in poultry production are Tyson, Pilgrim's Pride, Perdue, and Koch. Pick a poultry company, research its website, and report on how it uses vertical integration in its company. Are all vertically integrated poultry companies the same?
2. Take the list of parts for a poultry carcass and label a ready-to-cook fryer. How many different poultry products come from each major part (breast, thigh, legs)? Use the ready-to-cook birds to analyze the standards for ready-to-cook poultry based on weight categories.

Lesson 11.2

Swine Industry

Words to Know

American Berkshire Association (ABA)
barrow
boar
Boston butt
Boston shoulder
castrate
certified pedigreed swine (CPS)
confinement operation
export
farrow
farrowing
farrowing crate
farrow-to-finish operation
farrow-to-wean operation
feeder pig operation
finishing operation
gilt
ham
litter
loin
market hog
maternal breed
monogastric
National Swine Registry (NSR)
nursery
picnic shoulder
pig
piglet
porcine

(*Continued*)

Lesson Outcomes

By the end of this lesson, you should be able to:

- Understand the size and scope of the swine industry.
- Understand common terminology used in the swine and pork industry.
- Describe the components of the swine industry.
- Understand production systems in the swine industry.
- Describe different breeds of swine in the United States.

Before You Read

Scan this lesson and look for information presented as fact. As you read this lesson, try to determine which topics are fact and which are the author's opinion. After reading the lesson, research the topics and verify which are facts and which are opinions.

Swine is the term used to describe pigs and hogs at various stages of growth and development. The introduction of swine in the United States can be traced back to Vikings from Scandinavia around the year 1000 CE. Christopher Columbus was reported to have brought eight head of swine with him on his 1493 voyage. Hernando De Soto explored Florida and other regions of what is now the southeastern United States in 1539. De Soto brought 13 sows with him on his exploration and is given much of the credit for starting the swine industry in the United States.

Swine Industry in the United States

According to the 2012 Census of Agriculture, the U.S. swine industry generated $22.5 billion in sales, which accounts for six percent of total agricultural sales in the United States. Pork exports have increased dramatically over the last few years. Japan is the largest importer of U.S. pork, accounting for nearly one-third of all pork product exports. (An ***export*** is a commodity or service sold abroad. Any of the meat products produced in the United States and sent to other countries for consumption is considered an export.)

Today the swine industry is largely focused in the Corn Belt, which is located in the upper Midwest, **Figure 11-18**. The top five states in the United States in swine production are: Iowa, North Carolina, Minnesota, Illinois, and Indiana.

Pork is second behind beef in the United States at $170.65 per person in per capita meat expenditures. Pork is the most widely consumed meat in the world. It accounts for 37% of all meat and poultry consumed worldwide.

Common Swine Terms

The following terms are used to describe swine during different stages of their life:

- ***Pig*** or ***piglet***—term to describe young
- ***Boar***—young, noncastrated male
- ***Barrow***—castrated male
- ***Gilt***—young female
- ***Boar***—mature male
- ***Sow***—mature female
- ***Pork***—type of meat
- ***Farrowing***—act of giving birth

Words to Know

(Continued)
pork
pork butt
purebred swine
sausage
seedstock
sow
spareribs
swine
terminal breed
vertical integration
weaned

Mirec/Shutterstock.com

Figure 11-18. In the United States, most swine production facilities are located in the upper Midwest. *What might the reasons be for this concentration of swine operations?*

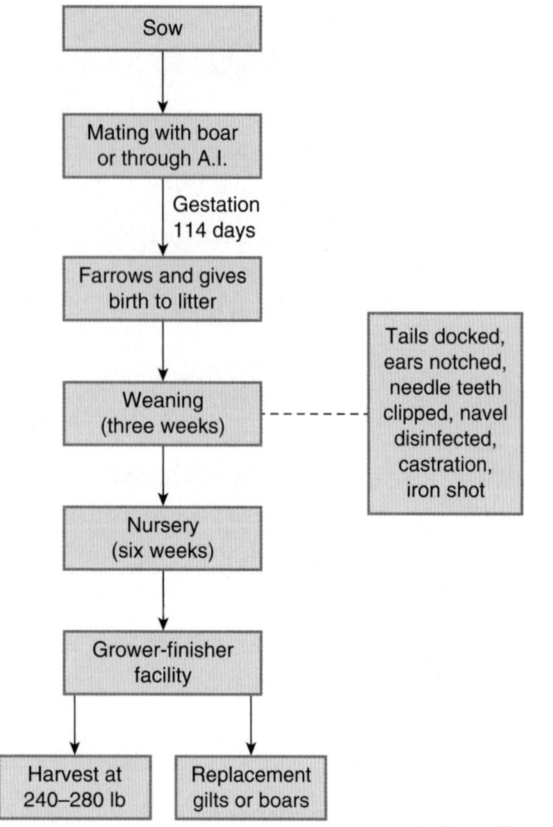

Goodheart-Willcox Publisher

Figure 11-19. The production cycle of swine. After studying the production cycle, visit www.g-wlearning.com to test your knowledge.

Production Cycle of Swine

The production cycle of swine starts when a boar breeds a sow or gilt. Breeding may occur naturally or through artificial insemination. The sow or gilt is in gestation (pregnancy) for 114 days. The female ***farrows*** and gives birth to a ***litter*** of piglets (usually 8–12 pigs). The piglets stay with their mother until they are three weeks old and are ***weaned*** (no longer nursing). During this three-week period, the pigs are processed for continued growth and health. This involves having their tails docked, ears notched, needle teeth clipped, navel disinfected, and receiving an iron shot. Boar pigs that will not be saved as replacements are usually ***castrated*** at this time. The weaned pigs are placed in a ***nursery*** for approximately six weeks. At nine weeks of age, they move to a grower-finisher facility. Once they reach 240–280 pounds, they are sent to harvest as ***market hogs*** unless they are being kept as replacement gilts or boars for purebred or commercial operations. Refer to **Figure 11-19**.

STEM Connection Xenotransplantation

The term *xenotransplantation* refers to transplantation of organs and other parts between two different species. Swine are commonly used for this procedure in human medicine. Pig and cattle heart valves have been used to replace human heart valves, and pig skin has been used as a replacement for human skin. Research is being conducted daily to explore the possibility of using organs from animals to replace human organs for people who suffer from an organ failure. This could have a profound effect on the medical and agricultural industries.

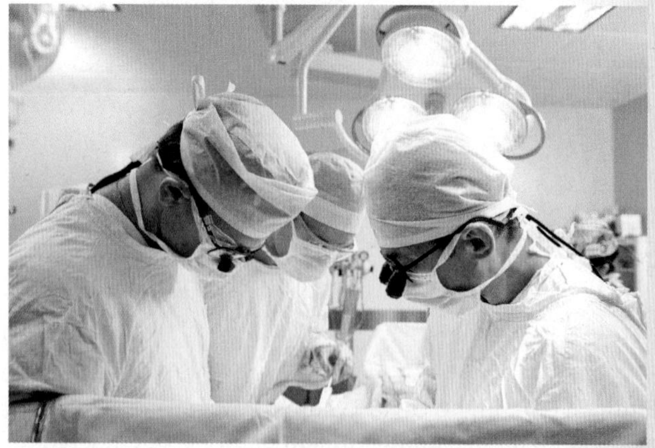

Dmitry Kalinovsky/Shutterstock.com

Mark William Richardson/Shutterstock.com

Figure 11-20. Although outdoor facilities are used in some swine production operations, it is not common in the United States. ***What would be the greatest challenges a producer would face when raising a large number of swine in this type of facility?***

Swine Production Systems

Swine production systems can be divided into two categories, outdoor and indoor.

Outdoor Facilities

Outdoor facilities consist of pastures with huts or sheds for sows to farrow under, **Figure 11-20**. There is no way to control environmental factors in outdoor facilities, and it is difficult to efficiently handle waste from the hogs. This is especially true for larger operations. Today, especially in the United States, most swine are raised in indoor facilities.

Did You Know?

3 months 3 weeks 3 days—this is an easy way to remember the gestation length (length of pregnancy) of hogs. 90 days (3 months) + 21 days (3 weeks) + 3 days = 114 days for a sow to have a litter of pigs.

Indoor Facilities

Indoor swine facilities are referred to as ***confinement operations***. In a confinement operation, the pigs are bred, born, and raised indoors. A confinement operation uses slatted floors that allow the manure and other waste to drop through the floor and easily be transferred to a separate processing facility, **Figure 11-21**.

Although indoor confinement operations are seen by some as cruel and inhumane, the confinement system for swine production actually engages the producer in better animal welfare practices and has become the standard for swine husbandry. Indoor production also takes the environment out of the equation because swine producers are able to climate control their swine facilities to the desired temperature needed for any stage of production. It also limits exposure to pathogens and other factors that may cause illness.

A_Lesik/Shutterstock.com

Figure 11-21. Modern confinement operations are efficient and the most secure manner in which to raise large numbers of swine for meat production. ***How much space is allotted per animal in a confinement operation?***

Farrowing Crates

Indoor systems use ***farrowing crates*** for the sows to birth and nurse the piglets, **Figure 11-22**. Using farrowing crates decreases the chances that the sow will lay down on the piglets and kill them. This is a common problem in outdoor production. The farrowing crate is designed with room for the sow to stand or lie down, and move from side to side. Each farrowing crate has its own feed and water supply, and allows the producer to control the temperature for the sow and pigs separately. (The sow prefers a lower temperature than what is needed to keep the piglets warm.)

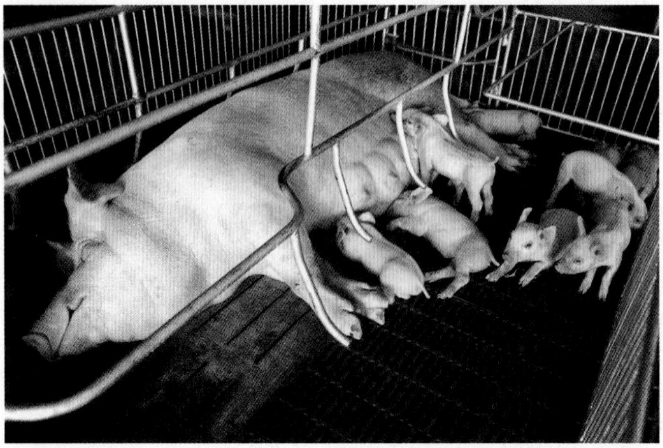

Tawin Mukdharakosa/Shutterstock.com

Figure 11-22. Farrowing crates allow the producer to reduce piglet mortality (death rate), and to control the temperature of adjoining areas. ***What are the ideal environment temperatures for nursing sows and newborn piglets?***

Vertical Integration

Indoor production of swine has moved toward ***vertical integration***, similar to what we see in the poultry industry. Vertical integration is where a major company, such as Tyson, Cargill, or Smithfield, controls the production process from birth to a finished pork product. The company controls the production of the animal as well as the marketing and sales of the end product.

The Swine Industry

The swine industry can be divided into two distinct categories, commercial and purebred

Career Connection Livestock Veterinarian

Job description: A livestock veterinarian provides preventive care and treats the injuries and diseases of animals that are bred for food and other products. Some veterinarians specialize in certain types of livestock, such as cattle, horses, swine, sheep, or goats.

Education required: A bachelor of science degree in chemistry, physics, biochemistry, biology, animal biology, or zoology, and a doctor of veterinary medicine (DVM or VMD) degree from an accredited college of veterinary medicine.

Job fit: This job may be a fit for you if you enjoy working with large animals, living and working in a rural area, and do not mind long hours and being on call.

Dmitry Kalinovsky/Shutterstock.com

swine operations. Commercial swine operations can be divided into farrow-to-finish, feeder pig, and finishing operations.

Farrow-to-Finish Operations

In ***farrow-to-finish operations***, the swine herd is maintained with sows farrowing year-round. The pigs are raised to market weight (240–280 pounds). Market hogs are sold to a processor or packer to harvest the hogs for meat production.

Feeder Pig Operations

In ***feeder pig operations***, piglets are weaned between five and eight weeks of age. At this time, pigs will weigh approximately 15–25 pounds. These pigs are sold to finisher operations to be fed and marketed at 240–280 pounds. Feeder pig operations may also be referred to as ***farrow-to-wean operations***.

Did You Know?

Bacon is one of the oldest processed meats in history. The Chinese began salting pork bellies as early as 1500 BCE.

Finishing Operations

In ***finishing operations***, feeder pigs are purchased weighing approximately 25 lb and are fed to desirable market weight. Market hogs reach 240–280 pounds at approximately five to six months of age. Finished hogs are then sold to a packer for pork production.

Purebred Operations

Purebred swine producers raise seedstock for other breeders to purchase and use as breeding animals. ***Seedstock*** are the boars, gilts, and sows used in breeding operations. Purebred swine producers make up a very small portion of the total swine industry. ***Purebred swine*** (or any type of animal) are those with documented pedigrees that are usually registered with breed organizations and national registries. Less than one percent of hogs produced in the United States are by purebred producers. In 1994, the ***National Swine Registry (NSR)*** was formed to consolidate many of the former breed associations that existed for purebred swine. ***Certified pedigreed swine (CPS)*** and the ***American Berkshire Association (ABA)*** are also swine breed organizations committed to the promotion and success of the purebred swine industry.

Anatomy of Swine

In order to understand many of the principles behind the growth and physiology of hogs and pigs, it is important to know the basic parts of the animal. **Figure 11-23** illustrates the parts of a market hog. To learn more about the parts of a hog and how swine are selected, refer to the FFA Livestock Career Development Event handbook. To learn more about the cuts of pork or the meats industry, focus on the Meats Evaluation Career Development Event sponsored by FFA.

Maintaining Swine Health

Swine are ***monogastric***, which means they have a single-chambered stomach just like humans, **Figure 11-24**. Typically they eat a high-energy diet containing a large amount of corn. Basic nutrient needs of swine are energy, protein, minerals, vitamins, and of course, water. Water can never be overlooked as an important nutrient for any livestock, but it is critical to the overall health and well-being of swine.

Since swine are kept indoors in close quarters, it is imperative to maintain their health. Sickness and disease can easily spread throughout the herd if animals are not properly vaccinated or if they live in unsanitary conditions. Keep the following points in mind when raising swine:

- Establish a complete herd health program from processing litters of pigs to sale of market or breeding swine.

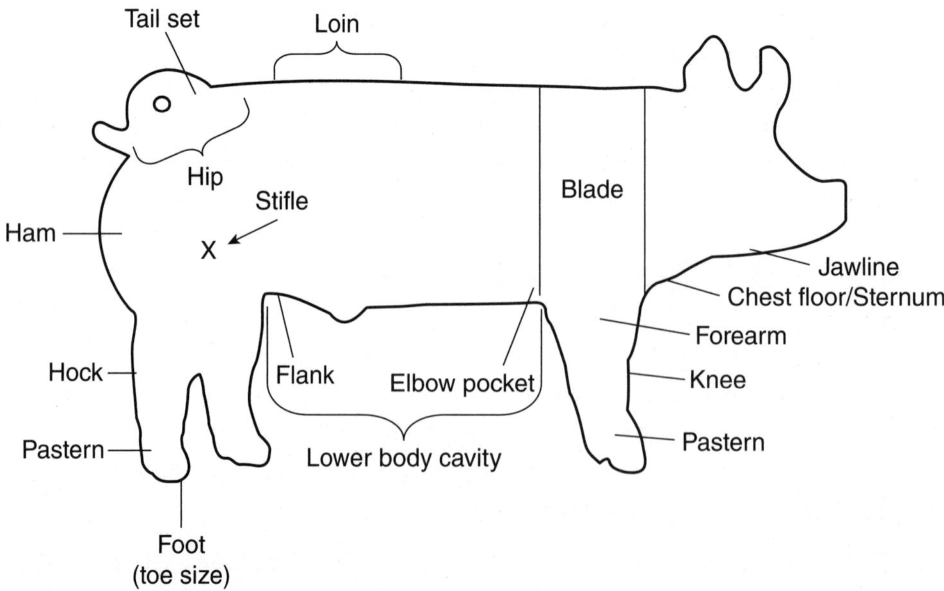

TAMU Livestock Judging

Figure 11-23. Study the illustration to learn the anatomy of swine. After studying the image, visit www.g-wlearning.com to test your knowledge.

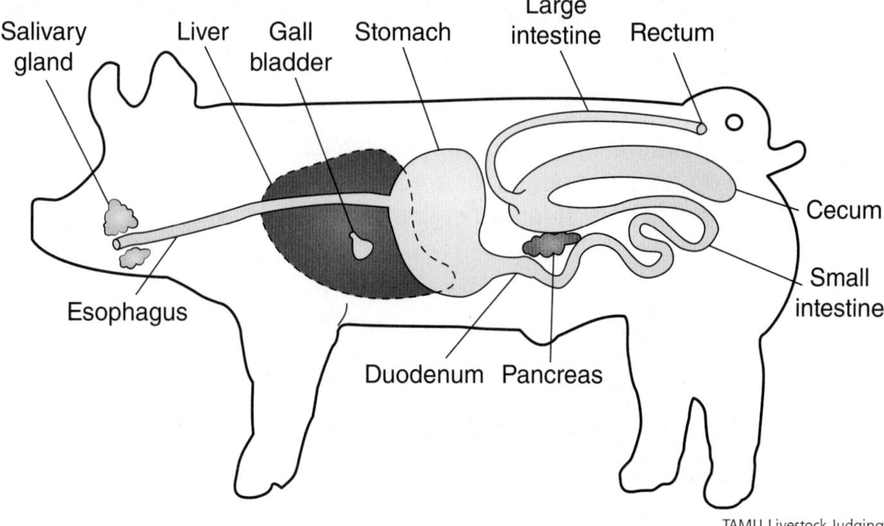

Figure 11-24. Swine are monogastric, which means they have a single-chambered stomach just like humans. Many of a swine's organs are very similar to human organs, with only slight variations.

- Work with your veterinarian to establish a vaccination program to prevent diseases.
- Purchase swine only from reputable breeders who maintain high herd heath standards.
- Observe new animals for possible diseases and parasites and quarantine if necessary.
- Maintain clear, complete, and accurate records for farrowing, breeding, weaning, and vaccination.
- Ensure farrowing occurs at the appropriate times and that sows farrow in clean facilities.
- Provide unlimited access to fresh, clean water at all times.
- Clean and disinfect all facilities and equipment after a group of swine is sold or moved to another facility.
- Provide adequate protection from extreme temperatures, especially heat. Swine do not sweat, and care must be taken to prevent overheating.
- To prevent the spread of disease, biosecurity measures are also used in swine production. Each operation has its own protocol, which may be as complicated as shower-in-shower-out restrictions, or as simple as putting on clean rubber boots and stepping in a disinfectant tub before entering the facilities. All protocol should be respected and followed.

Handling Swine

Swine are unpredictable and not every action can be foreseen, but there are precautions that can be taken to reduce hazards and injuries to both producers and animals. Keep the following points in mind when handling and transporting swine.

- As with most livestock, swine become extremely stressed when being transported or moved. The key to smooth transport is to do it as quickly, quietly, and efficiently as possible.

STEM Connection

You can tell a lot by looking at a hog's ears. All breeds of swine that have "shire" in their name have erect ears (their ears stand up). Hog breeds that do not have genes for erect ears have droopy ears. What part does genetics play in determining whether a hog's ears stand up or flop over?

Duroc

Hampshire

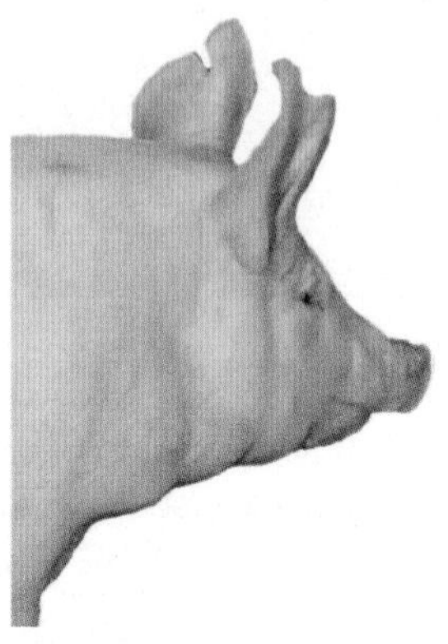

Yorkshire

Courtesy of the National Swine Registry

Compare the ears of the three breeds. *Can you see the differences? What are the similarities?*

- Pigs have a range of vision that is almost 360°. To prevent distractions when moving swine, use solid sidewalls (boards of hurdles) to help keep them focused.
- Pigs tend to move from dimmer areas to well-lit areas. Use spotlights to help guide the animals up ramps or into other pens.
- Use moving aids such as panels or hurdles when appropriate.
- Stay calm and handle swine gently to prevent stress. Some breeds of swine are more susceptible to stress than others. Use caution when handling swine prone to stress; it can lead to sudden death.
- Avoid loud noises, maintain clear walkways, and do not overcrowd chutes and pens.
- Animals may be especially aggressive during breeding. Use sorting panels or hurdles, and do *not* get between the sow and the boar.
- Sows may be increasingly aggressive during different points of the reproductive cycle. Always use caution when handling aggressive sows.

Cuts of Pork

The primal cuts of pork are ham, loin, Boston shoulder, and picnic shoulder, **Figure 11-25**. Other common cuts of pork are ***spareribs***, jowl, ham hock, and side (bacon). ***Sausage*** is made with ground pork, which may come from any of the major cuts.

- ***Ham***—fresh cuts from the ham/leg include ham, center slice, rump portion, and shank portion. Cuts from smoked ham include boneless ham, center slice, rump portion, and the shank portion. The tip roast (boneless) and top roast (boneless) are also cut from the ham/leg primal cut.

- *Loin*—tender primal cut from the upper back and side. Cuts of pork from the loin include blade chops, boneless blade chops, blade roast, butterflied boneless chops, center loin roast, center rib roast, country style ribs, loin chops, rib chops, sirloin chops, sirloin cutlets, and the sirloin roast. Cuts from the smoked pork loin include chops or rib chops, tenderloin, top loin chops, and the top loin roast. Ribs are cut from the loin back.
- *Boston shoulder*—tougher primal cut which contains muscles and sinew. Cuts of meat from the shoulder include the arm roast, arm steak, blade Boston roast, blade steak, smoked picnic (whole), and the arm picnic (whole). Although it comes from the upper shoulder of the hog, it may also be referred to as ***pork butt*** or ***Boston butt***. It is often used to make ground pork or sausages.
- *Picnic shoulder*—tough primal cut which is frequently cured or smoked. Often used to make ground pork or sausage meat.

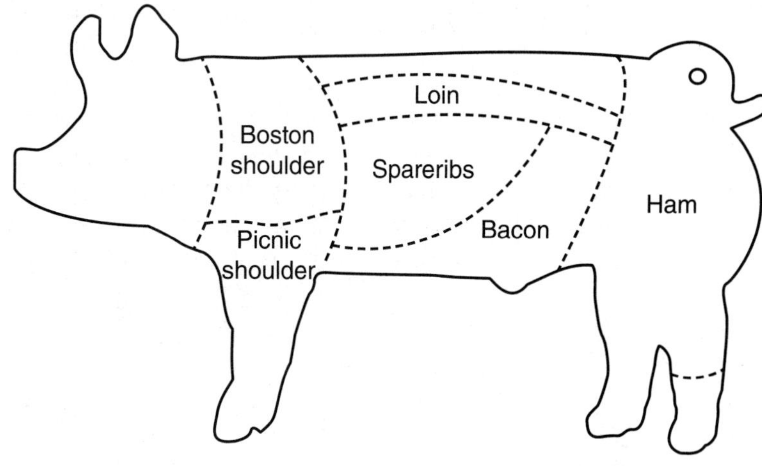

Figure 11-25. Study the illustration to learn the primal cuts of swine. After studying the image, visit www.g-wlearning.com to test your knowledge.

Hands-On Agriculture

Ear Notching

Swine are identified using ear notches. Look at the placement for each number on the right and left ears on the illustration. The right ear is used for the litter number and the left ear is used for the pig number within the litter. Using a cardboard cutout of a pig's head, practice ear notching your own pig with a litter and pig number.

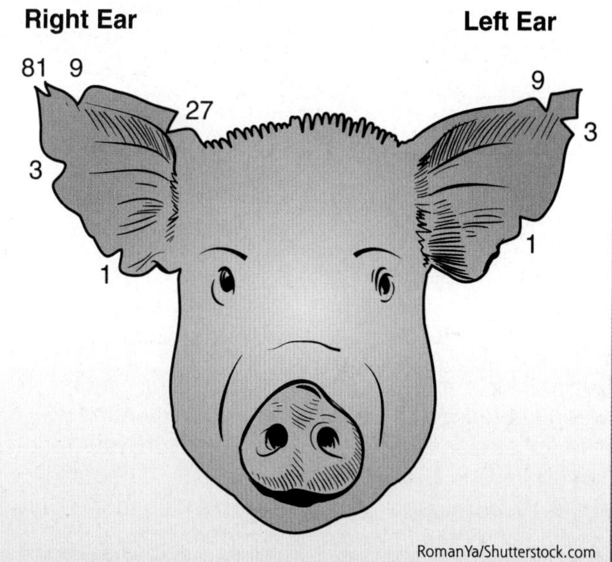

Swine Breeds

The species name for swine is ***porcine***. The scientific name for swine is *Sus scofa*. Swine may have been domesticated as early as 9,000 years ago in China. There are more than 70 breeds of swine worldwide. The following is a list of the most common swine breeds found in the United States. As you read the descriptions of the breeds of swine, some will be described as terminal breeds or as maternal breeds. ***Terminal breeds*** are breeds known for their muscling, carcass quality, and cutability. Generally speaking, the colored breeds of hogs are terminal breeds. ***Maternal breeds*** are known for their mothering ability, litter size, and longevity. Breeds that are solid white in color are typically recognized as maternal breeds.

Courtesy of the National Swine Registry

Berkshire. Berkshire hogs originated in England. Berkshires were imported into the United States in 1823 and were used to improve all breeds of swine. A Berkshire is black with six white points: all four feet, a white face or forehead, and on the tail. Their ears are erect and their face is slightly dished. Berkshire hogs are widely recognized for their carcass quality. They are also known for their fast growth and for being prolific.

Courtesy of the National Swine Registry

Chester White. Chester White hogs originated in Chester County, Pennsylvania, in the late 1800s. The Chester White hog illustrated here is solid white with floppy, medium-size ears. Chester Whites must not have any other color on their skin larger than a silver dollar or any colored hair. Chester Whites are a maternal breed known for their mothering ability, soundness, and durability.

Courtesy of the National Swine Registry

Duroc. Duroc hogs originated in the United States. The breed started in New York and New Jersey with two different strains of red hogs: Red Durocs in New York and Jersey Reds in New Jersey. The first Duroc hog show was at the 1893 Chicago World's Fair. This helped the Duroc breed gain popularity nationwide. Durocs are usually dark red but can range from almost yellow to a mahogany color. As you can see here, their ears droop forward. Durocs are known for the fast growth, muscling, and terminal sire ability.

Courtesy of the National Swine Registry

Hampshire. Hampshire hogs originated in southern Scotland and northern England. They were originally known as the *Old English Breed*. Hampshires were imported into the United States between 1825 and 1835. During this time period, the Hampshire breed was referred to in many ways, including: Thin rinds (due to their thin skin), McKay hogs, McGee hogs, Saddlebacks, and Ring Middles. Hampshire hogs are black with a white belt encircling their body, including both front legs and feet. Hampshires are known for their muscling, carcass quality, and durability.

Courtesy of the National Swine Registry

Landrace. Landrace hogs originated in Denmark and were imported into the United States in the 1930s. The USDA purchased 24 Danish Landrace from Denmark to use in swine research studies in the United States. The Landrace is solid white with large droopy ears. They are noted for their farrowing ability and large litter size. Landrace are very prolific and cross well with many other breeds. Landrace is a very popular maternal breed of swine.

Courtesy of the National Swine Registry

Poland China. Poland China hogs originated in Butler and Warren counties in Ohio. They are descendants of Big China hogs, which were popular in the northeastern United States in the early 1800s. The Poland China is black with six white points: the face, all four feet, and the switch of the tail. Polands have droopy ears and must not have evidence of a belt forming from their white points. Polands are long bodied, lean, big framed, and muscular. Poland Chinas lead the U.S. pork industry in pounds of hog per sow per year.

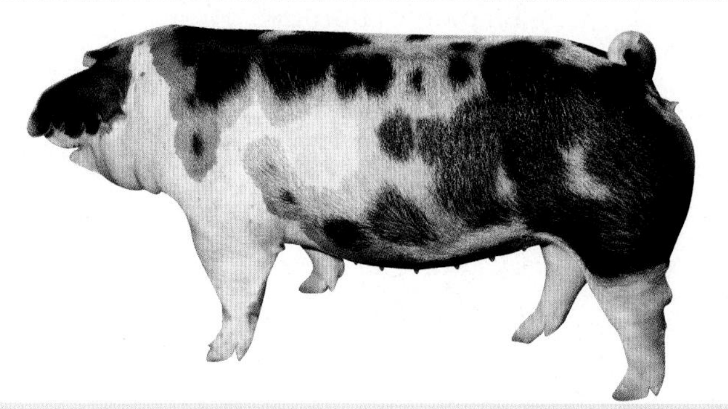

Courtesy of the National Swine Registry

Spotted. Spotted hogs are descendants of the original Poland China, which was called the Warren County Hog in the late 1880s. Swine producers from Indiana and Ohio continued to develop the breed and introduced Glouster Old Spots from England. As the name of the breed indicates, they are spotted. Their ears are droopy, and their head cannot be solid black from the ears forward. Spots are known for their feed efficiency, rate of gain, and carcass quality.

Courtesy of the National Swine Registry

Yorkshire. Yorkshire hogs originated in York County in England. They are also known as English Large Whites. Yorkshires were first imported into the United States in 1830 in Ohio. Yorkshires are the most recorded breed of swine in the United States and are found in nearly every state. Yorkshires are solid white with erect ears. They are a maternal breed known for their mothering ability, litter size, and length. Modern-day Yorkshires are known as the Mother Breed.

Lesson 11.2 Review and Assessment

Words to Know

Match the key terms from the lesson to the correct definition.

1. Swine production system in which the swine herd is maintained with sows farrowing year-round; pigs are sold at a market weight between 230–280 pounds.
2. A young or mature male swine.
3. A primal cut of pork meat from the upper hind leg of swine.
4. A swine production system in which feeder pigs are purchased weighing approximately 25 pounds and are fed to desirable market weight.
5. A castrated male swine.
6. Swine production system in which weaned piglets (5–8 weeks) are sold to finisher operations to be fed and marketed.
7. A young female swine.
8. A mature female swine.
9. A tougher primal cut of pork meat from the shoulder of swine that contains muscles and sinew.
10. A common cut of pork cut from below the loin and above the stomach of swine.
11. A tough primal pork cut which is frequently cured or smoked. Often used to make ground pork or sausage meat.
12. A tender primal cut of pork meat located on the upper back and side of swine.
13. A meat product commonly made from ground pork.
14. A swine breed known for its mothering ability, litter size, and longevity.
15. Confined area in which weaned piglets are placed for approximately six weeks before being moved to a grower-finisher facility.
16. Term used to describe young swine.
17. When a major company controls the production process from birth to a finished pork product.
18. Swine breeds known for muscling, carcass quality, and cutability.

A. barrow
B. boar
C. Boston shoulder
D. farrow-to-finish operation
E. farrow-to-wean operation
F. finishing operation
G. gilt
H. ham
I. loin
J. maternal breed
K. nursery
L. picnic shoulder
M. piglet
N. sausage
O. sow
P. spareribs
Q. terminal breed
R. vertical integration

Know and Understand

Answer the following questions using the information provided in this lesson.

1. *True or False?* Christopher Columbus is given much of the credit for starting the swine industry in the United States.
2. At what weight are pigs sent to harvest as market hogs?
3. Explain why indoor facilities provide the producer more control over swine production.
4. Why are farrowing crates used for sows to birth and nurse piglets?
5. *True or False?* Vertical integration requires the use of both indoor and outdoor production facilities.
6. *True or False?* Swine are monogastric, which means they have a single-chambered stomach.
7. List four important factors for maintaining swine health.
8. Why is it important to minimize stress when handling and transporting swine?
9. List at least eight common cuts of pork.
10. Explain the difference between terminal and maternal breeds.

Analyze and Apply

1. The hog-corn price ratio is a good way to measure profitability in the swine industry. The hog-corn price ratio is calculated by taking the price per hundred weight at which market hogs are selling, and dividing it by the price of corn per bushel. If market hogs are selling for $60 per hundred weight, and corn is bringing $3.00 per bushel, what is the hog-corn ratio?

Thinking Critically

1. Swine herd management and biosecurity are vital to the success of swine producers. With a partner, create a list of swine herd health precautions.
2. Pork is a major source of protein for many people. Using the diagram of a market hog, list the various types of meat that come from the major cuts of swine.

Lesson 11.3

Sheep Industry

Words to Know

crossbred-wool breed
ewe
ewe lamb
farm flock
fine-wool breed
fleece
flock
greasy wool
hair sheep
lamb
lambing
leg
loin
long-wool breed
medium-wool breed
mutton
ovine
purebred producers
rack
ram
ram lamb
range production
seasonal-use production
shoulder
stocking rate
wether

Lesson Outcomes

By the end of this lesson, you should be able to:

- Understand the size and scope of the U.S. sheep industry.
- Understand common terminology of the sheep industry.
- Identify the major cuts of lamb and mutton.
- Describe the components of the U.S. sheep industry.
- Understand commonly used production systems in the sheep industry.
- Describe different breeds of sheep in the United States.

Before You Read

Conduct research on the differences between wool and meat breeds of sheep. Why are certain breeds better suited for specific regions of the United States?

Sheep are probably the most widespread species of domesticated livestock in the world. The first domesticated sheep arrived in the Americas in 1493 with Columbus' second voyage. Cortes brought sheep on his 1519 voyage to Mexico. It is believed that these sheep are the ancestors of the flocks that spread throughout Mexico and the southwestern part of the United States.

Sheep Industry in the United States

Sheep are produced in every state in the United States. According to the USDA, in 2014, there were approximately 5.21 million head of sheep in the United States, **Figure 11-26**.

Most of the sheep raised in the United States are dual purpose. This means they are raised for two purposes: meat and wool. However, in recent years, there has been a dramatic increase in the production of hair sheep, seeing dual-purpose breeds decline in many areas of the United States. The sheep industry accounts for less than one percent of total livestock industry income in the United States. Per capita lamb and mutton consumption is very low in the United States. In the 1960s, the average person consumed about 5 pounds of lamb or mutton per year. Today, that figure is around one pound per year per person. This is drastically different from other countries like Australia and New Zealand where per capita lamb and mutton consumption is around 25 pounds per person.

Mirec/Shutterstock.com

Figure 11-26. The top five states (in order) in number of sheep and lambs are Texas, California, Colorado, Wyoming, and Utah.

Sheep around the Globe

Worldwide, the average person consumes more than 4 lb of lamb and mutton per year. Lamb and mutton account for less than 5% of total meat consumption in the world. The United States continues to import nearly half of its lamb and mutton supply from Australia and New Zealand. There continues to be demand for lamb and mutton from certain ethnic communities. Middle Eastern, Caribbean, and African consumers in larger metropolitan areas are a major market for lamb and mutton products.

Wool Production

The wool industry in the United States is small compared to other countries. The top five wool producing countries in the world are Australia, China, New Zealand, Russia, and Argentina. The top five states in the United States in wool production are the same as the top five sheep and lamb producing states with Wyoming producing slightly more than Colorado.

Fleece

Typically wool is harvested in spring when sheep are in full fleece. The ***fleece*** is the woolly coat of the sheep, **Figure 11-27**. The United States produces more than 30 million pounds of ***greasy wool*** (wool in its natural

RTimages/Shutterstock.com

Figure 11-27. One sheep produces anywhere from 2 lb–30 lb of wool annually. Harvesting fleece is labor-intensive, as sheep must be shorn by hand. *Have you ever sheared fleece or spun wool? Does the color of the fleece affect its value?*

state) per year. There is great variation in the weight of fleece produced from region to region in the United States. Fleece produced in the western part of the United States might weigh double what fleece from the eastern United States weighs. Wool is a versatile fiber that is comfortable, flame resistant, biodegradable, and washable. Wool has many uses in the textile, fabric, and carpet industries.

Common Sheep Terms

The following terms are used to describe sheep during different stages of their life.

- ***Lamb***—term to describe young
- ***Ram lamb***—young male
- ***Wether***—castrated male
- ***Ewe lamb***—young female
- ***Ram***—mature male
- ***Ewe***—mature female
- ***Lamb***—meat from sheep less than one year old
- ***Mutton***—meat from sheep older than one year
- ***Lambing***—act of giving birth

Production Cycle of Sheep

Sheep are very gregarious (very sociable) and stay in a group called a ***flock***. They are seasonal breeders, which means they are typically influenced by the length of the day. Optimal breeding season for most sheep is in late fall when the days start to get shorter. The female sheep (ewes) are mated to rams. Mating may take place naturally or sheep may be impregnated through artificial insemination. A ewe is in gestation (pregnancy) for approximately 150 days. The baby sheep are called lambs. It is very common for ewes to have twins or even triplets. The lambs stay with their mother until they are about five to six months old. At this point, they weigh between 80–120 lb. At this time, wethers (castrated males) can be sold as market lambs or sent to a feedlot for finishing. Rams and ewes will either be sold as market lambs or placed back into a breeding herd as replacements. The life expectancy of sheep is 6–7 years, but they can live to be as old as 20 years. Refer to **Figure 11-28**.

Did You Know?

The first cloned mammal in the world was a sheep named Dolly. In 1996, British scientists made headlines with this scientific breakthrough.

The Sheep Industry

In the United States, sheep are primarily raised in two types of production systems:

- Range production.
- Farm flocks.

Range production of sheep occurs mainly in the western states, **Figure 11-29**. These are typically large, commercial (non-registered) flocks of sheep that have more than 1,000 head of ewes. ***Farm flocks*** are smaller operations that typically have less than 100 ewes and are a secondary enterprise for most of those producers.

Within range and farm flock production, there are ***purebred producers*** who raise replacement breeding sheep (rams and ewes) to market to other purebred and commercial sheep producers. Commercial sheep producers raise lambs to go to market. They may choose to lamb at different stages of the year, depending on weather conditions, forage, and other production systems. Their main goal is to raise pounds of lamb to send to the market. In some regions of the United States, ***seasonal-use production*** systems are used. Seasonal-use systems can make use of farm-produced roughage such as winter pasture or crop residue to serve as a source of nutrition for bred or open ewes. This is typically done in a range setting and can vary, depending on forage, weather, and additional feed resources.

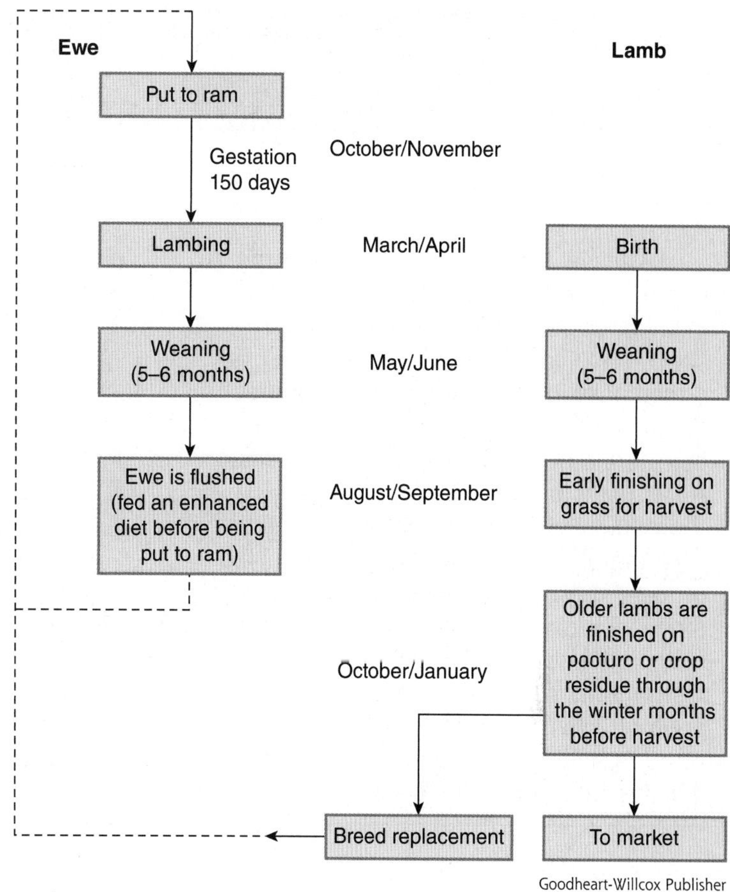

Goodheart-Willcox Publisher

Figure 11-28. The sheep production cycle. After studying the production cycle, visit www.g-wlearning.com to test your knowledge.

irynal/Shutterstock.com

Figure 11-29. Range production is used throughout the world. In the United States, range production is limited to the western region of the country. *Why is the topography in the western portion of the United States more conducive to range production of sheep?*

Did You Know?

Sheep are very susceptible to predators. Coyotes, foxes, wild dogs, wolves, mountain lions, and eagles are all capable of killing and eating young lambs. Some ranchers use donkeys, llamas, and certain breeds of dogs to live in the sheep flock to guard against predators.

STEM Connection Laparoscopic Artificial Insemination

Budimir Jevtic/Shutterstock.com

Artificial insemination (A.I.) is common in the livestock world. The process of A.I. in sheep is very different than in other species of livestock. *Laparoscopic A.I.* is a minimally invasive surgical procedure that involves placing semen directly into the uterine horn of the ewe. Laparoscopic A.I. is used in sheep because their reproductive tract is structured differently than that of other ruminant livestock. This process requires sedation of the ewe, a small incision in their abdomen, a special breeding cradle, and specialized surgical instruments with a laparoscope or light source. This is usually done by a veterinarian or trained professional. Do you think the requirement to perform A.I. laproscopically has an impact on the use of A.I. in sheep? If you thought about how this would limit the number of artificially inseminated lambs born, you would be exactly right!

Anatomy of Sheep

In order to understand many of the principles behind the growth and physiology of sheep, it is important to know the basic parts of the animal. **Figure 11-30** illustrates the parts of a market lamb. Knowledge of these parts becomes even more critical when participating in FFA Career Development Events such as Livestock Evaluation or Meats Evaluation.

Sheep are small ruminants, which means they can digest high amounts of roughage or forage, usually grass or hay. Their stomach is divided into four compartments: rumen, omasum, abomasum, and reticulum. Ruminants like sheep commonly chew their cud, which is a regurgitation of the roughage and feed they have eaten to aid in further digestion.

Maintaining Flock Health

Maintaining flock health is essential to keep sheep production affordable and profitable. The following brief list of tips for maintaining flock health is not all-inclusive, but it does contain basic steps to ensure a healthy flock.

- Establish and implement a flock health management plan. Work with your veterinarian to establish a parasite control program and flock health management plan.
- Keep accurate records that include: breeding and lambing, feeding, disease prevention, purchase of new animals, and sale of livestock.

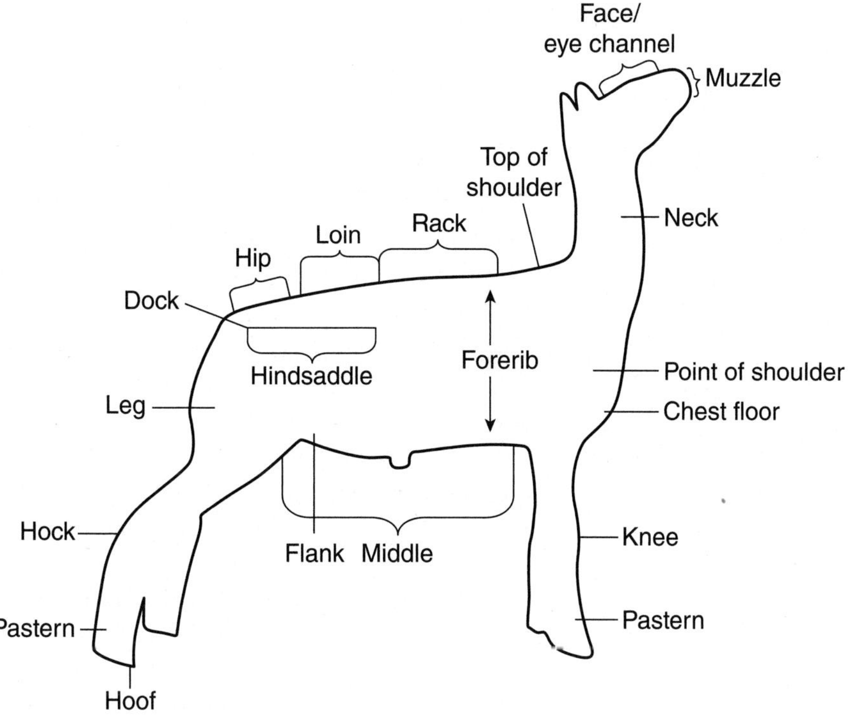

TAMU Livestock Judging

Figure 11-30. Use the above illustration to study the basic anatomy of sheep. After studying the image, visit www.g-wlearning.com to test your knowledge.

- Restrict uncontrolled traffic and contact of flock with people and sheep of unknown origin. Quarantine new sheep if necessary to maintain flock health.
- Maintain proper environments. This includes a comfortable climate, dry and clean bedding, natural lighting, and adequate space and stocking rate. ***Stocking rate*** is the number of animals on a given amount of land over a certain period of time.
- Provide proper feeding, practice proper handling, and use proper care.

Did You Know?

Because sheep are aggressive foragers, they are very susceptible to internal parasites. Sheep producers must have a procedure for regularly deworming lambs and sheep to maintain the health of the flock.

Handling Sheep

There are precautions that can be taken to reduce hazards and injuries to both producers and sheep. Keep the following points in mind when handling and transporting sheep.

- Sheep are gregarious, which means they like to stay together in a flock or herd. It is difficult to single out one or two sheep without the rest of the flock wanting to follow along.
- Carefully plan and design your handling facilities. This includes the size of pens and runs as well as closely boarding the sides so the sheep cannot see through and get distracted. Make the runs narrow enough to prevent the sheep from turning around and blocking the flow.

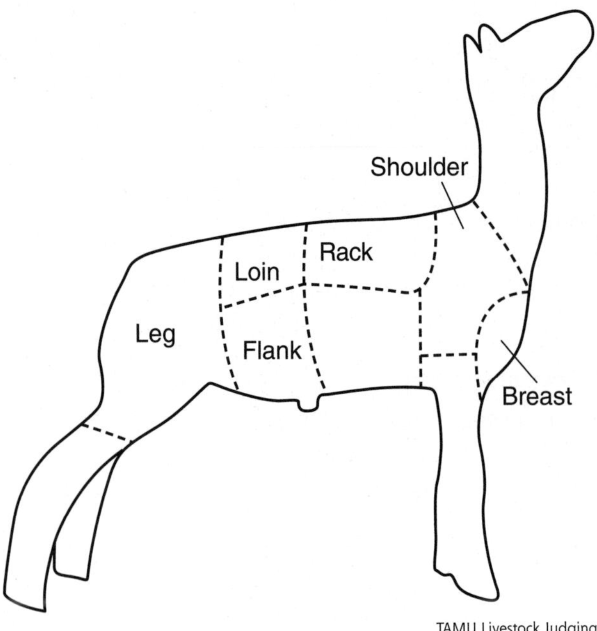

Figure 11-31. Use the illustration to learn the primal cuts of lamb. After studying the image, visit www.g-wlearning.com to test your knowledge.

- As with swine, sheep tend to move from dimmer areas to well-lit areas. Use spotlights to help guide the animals up ramps or into other pens.
- Sheep do not like visual dead ends and will not move freely into one. Make sure there is always a clear way ahead.
- Learn how to properly handle sheep for procedures like feet trimming, vaccinating, and shearing.
- Use the sheep's flight zone to help manage their movements. The term *flight zone* is used to describe how close you can get to sheep before they start moving. Sheep are fast and can leap as high as 3′ to 4′ in the air. Caution must be taken when working them in tight spaces with short fences.
- Dogs can be extremely useful when working with sheep. However, they must be well-trained. Sheep view dogs as predators, so they must become familiar with them in the working and handling process.

Major Cuts of Lamb and Mutton

Unlike beef, which is cut into sides, lamb is divided into sections called the foresaddle and hindsaddle. These sections are then cut into their main primal cuts. The primal cuts of lamb and mutton are shoulder, rack, loin, and leg, **Figure 11-31**. Less expensive cuts include the breast, neck, and shank. Meat from these cuts is commonly made into ground lamb.

- ***Shoulder***—tender primal cut that is usually deboned and cooked as a roast. The shoulder may also be cut into chops. Cuts of meat from the shoulder include arm chops, blade chops, and the square cut.
- ***Rack***—the rack is actually the rib primal cut. Cuts of meat from the rack include rib chops, rib chops Frenched, rib roast, and rib roast Frenched.
- ***Loin***—tender primal cut. Cuts of meat from the loin include loin chops and the loin roast.
- ***Leg***—tender primal cut that may be cooked as a roast or cut into chops. Cuts of meat from the leg include American style roast, center slice, Frenched style, leg roast (boneless), sirloin chops, and the sirloin half.

Sheep Breeds

The species name for sheep is ***ovine,*** and the scientific name is *Ovis aries*. There are more than 150 different breeds of sheep with several other closely related species worldwide. There are many ways to categorize sheep. One of the most common ways is by the type of wool they have. The categories are fine-wool, long-wool, medium-wool, crossbred-wool, and hair sheep.

Fine-Wool Breeds

Fine-wool sheep produce wool with tightly crimped, small-diameter fibers. The wool is desirable as it produces very soft, delicate garments. Fine-wool sheep account for nearly half of the world's sheep population.

Courtesy of Forbes & Rabel Rambouillet, Kaycee, Wyoming

Rambouillet. All of the ***fine-wool breeds*** were developed from the Spanish Merino, which was first imported into the United States in the early 1800s. These breeds are primarily used for wool production and have less value as a meat animal. They have the ability to survive in harsh range conditions with little or poor-quality forage. Merinos and Rambouillets are the most popular fine-wool sheep breed.

Long-Wool Breeds

Long-wool breeds common in the United States are Costwold, Leicester, Lincoln, and Romney. All of these breeds originated in England and produce long, coarse wool fiber. These breeds are larger in size than medium- or fine-wool breeds. They are large framed and typically used as wool-producing breeds.

Medium-Wool Breeds

Sheep that are raised for lamb and mutton are usually ***medium-wool breeds***. Their wool or fleece is a secondary product and is not as valuable as fine-wool fleece. Popular medium-wool breeds are Dorset, Hampshire, Shropshire, Southdown, and Suffolk. Other medium-wool breeds include the Cheviot, Finnsheep, Montadale, Oxford, and Tunis.

Courtesy Continental Dorset Club

Courtesy Continental Dorset Club

Dorset. The Dorset originated in England and were imported into the United States in the late 1800s. Dorsets are medium-sized and solid white. They may be horned or polled. They produce a heavy-muscled carcass, and the ewes can breed out of season to produce fall-born lambs.

Medium-Wool Breeds (*Continued*)

Courtesy of the American Hampshire Sheep Association

Hampshire. The Hampshire breed originated in southern England. They were first imported into the United States in the 1840s. Hampshires are large with a solid white body, a black head, and black legs. They have a wool cap on top of their head with wool extension going down their legs. They cross well with other breeds to produce fast-growing market lambs.

Courtesy of the American Southdown Association

Southdown. Southdowns were developed in Sussex, England, in the late 1700s and early 1800s. They were imported into the United States in the 1820s to Pennsylvania. Southdowns are moderate to small-sized sheep that are early maturing. Their body is white, and their face and lower legs are a grayish brown color. They are adaptable to varied climates but work best in farm flock production scenarios.

Courtesy of the United Suffolk Sheep Association

Suffolk. *(Left)* Suffolk sheep were developed in southern England by crossing Southdown rams with Norfolk Horned ewes. They were imported into the United States in 1888. Suffolks are white with a black head and legs and have no wool extension on their points. They are a large breed known for their growth and muscling. They are widely used in crossbreeding programs and produce excellent market lambs.

Shropshire. *(Not illustrated)* Shropshires were imported into the United States in the late 1850s. They are moderate in frame size and weight and one of the smaller breeds of medium-wool sheep. Their face and legs are dark with heavy wool extension on their face and legs. Shropshires have good maternal characteristics and are recognized as one of the better dual-purpose sheep breeds.

Crossbred-Wool Breeds

In the United States, long-wool breeds are frequently crossed with fine-wool breeds to produce ***crossbred-wool breeds*** such as Columbia and Corriedale. Other crossbred-wool breeds include the Panama, Romeldale, Targhee, Tailess, and Southdale. These breeds are less common in the United States.

Courtesy of Jarvis Sheep Company

Columbia. Columbia sheep are a result of breeding Lincoln rams to Rambouillet ewes. They originated in the United States and were bred to be adapted to the mountainous regions of the western United States. They are the largest of the crossbred-wool breeds and are used extensively in crossbreeding programs.

American Corriedale Association

Corriedale. The Corriedale originated in New Zealand as a result of crossing Lincoln and Leicester rams with Merino ewes. They were imported into the United States in the early 1900s. They were very suitable for the climate and terrain of the western range. They are medium to large in size and produce quality wool along with acceptable carcasses, making them a good dual-purpose breed.

Hair Sheep Breeds

Hair sheep have gained popularity in the United States over the last few years. Hair sheep are lower maintenance because they are more parasite-resistant, more heat-tolerant, and provide multiple lambing opportunities per year. The primary difference between hair sheep and wool sheep is the ratio of hair to wool fibers. Hair sheep have more hair fibers and less wool fibers than wool sheep. Another valuable by-product that comes from hair sheep is their hide. Dorper skin is the most sought after sheepskin in the world and is used to make high-quality shoes, handbags and other leather garments. The two most popular breeds of hair sheep are Katahdin and Dorper.

Courtesy of the American Dorper Sheep Breeders Society, Inc.

Dorper. *(Left)* Dorper is a South African mutton breed developed from Dorset Horn rams and Black Headed Persian ewes. Dorpers are hornless and very fertile. Typically, they will be white in color with a solid black head. Dorpers are not seasonal breeders and can breed and lamb year-round. Dorpers are hardy, thick, heavily muscled animals that are well-suited for mutton production.

Katahdin. *(Not illustrated)* Katahdins were developed in the United States by Michael Piel in Maine. He imported three African Hair Sheep in 1957 to start his breeding program. Katahdins are moderate-sized white sheep that are docile and easy to handle. They are hardy and easily adapt to most climates.

Lesson 11.3 Review and Assessment

Words to Know

Match the following key terms with the proper definition.

1. A tender primal cut of lamb or mutton that is usually deboned and cooked as a roast.
2. The meat from sheep less than one year old.
3. A young female sheep.
4. The meat from sheep older than one year.
5. A mature male sheep.
6. Sheep primarily used for finer wool production and have less value as a meat animal.
7. Typically large, commercial flocks of sheep that have more than 1,000 head of ewe.
8. Breeds of sheep that have a higher ratio hair to wool fibers than wool sheep.
9. A mature female sheep.
10. A young male sheep.
11. Large sheep that are primarily raised for their long, coarse, wool fiber.
12. The rib primal cut of lamb/mutton.
13. A castrated male sheep.
14. Sheep primarily used for lamb and mutton and produce a less valuable fleece than fine-wool breeds.

A. ewe
B. ewe lamb
C. fine-wool breed
D. hair sheep
E. lamb
F. long-wool breed
G. medium-wool breed
H. mutton
I. rack
J. ram
K. ram lamb
L. range production
M. shoulder
N. wether

Know and Understand

Answer the following questions using the information provided in this lesson.

1. What does it mean when a sheep is raised as dual purpose?
2. *True or False?* The sheep industry accounts for more than 10% of total livestock industry in the United States.
3. *True or False?* Lamb and mutton account for less than 5% of total meat consumption in the world.
4. The United States produces more than _____ million pounds of greasy wool per year.
5. When is the optimal breeding season for most sheep?
6. How long is a ewe in gestation?
7. Explain the difference between range production of sheep and farm flocks.
8. *True or False?* Sheep are ruminants, with four stomach compartments.
9. List three things you should do when handling or working with sheep.
10. What are the four primal cuts of lamb and mutton?

Analyze and Apply

1. Meat breeds, wool breeds, haired sheep? There are many uses for sheep, regardless of type. Work with a partner to develop a chart that displays breeds, type of breed, main product (meat, wool) derived from that breed, and where they are located in the United States.
2. Sheep and lamb numbers have steadily declined over the last decade. Research the factors contributing to the decline, and create a marketing plan for increasing sheep and lamb numbers in your community. How could they be used? Who would your consumers be? How would this impact the local economy?
3. There is great variation in the weight of fleece produced from region to region in the United States. Other than the breed, what would cause this variation?

Thinking Critically

1. Sheep have been widely used in research for decades. More recently, sheep have been used extensively in cloning. Research the various ways sheep have been used in the scientific community and how they contribute to advancements in this area.
2. Most sheep producers are not happy with a 100% lamb crop. How is it possible to have over a 100% lamb crop? What is an acceptable lambing rate?

Lesson 11.4 Goat Industry

Words to Know

billy
browse
buck
cabrito
cape
caprine
chevon
doe
flock
herd
kid
kidding
mohair
myotonic goats
nanny
tribe
trip
wether

Lesson Outcomes

By the end of this lesson, you should be able to:

- Understand the size and scope of the U.S. goat industry.
- Understand common terminology of the goat industry.
- Describe the components of the U.S. goat industry.
- Understand commonly used production systems in the goat industry.
- Describe different breeds of goats in the United States.

Before You Read

There are many different uses and classifications of goats in our country. Investigate the many ways that goats are a part of our meat, dairy, and fiber industries in the United States. Compare and contrast their differences in a table.

Goat Industry in the United States

The goat industry is one of the fastest growing animal industries in the United States. This is mainly due to the changing makeup of our society. As our population becomes more ethnically diverse, the demand for goats increases. Most experts believe that the estimated number of total goats in the United States is low. The census of agriculture in 2012 reported that there are approximately 2.6 million goats in the United States. This number includes kids, milk goats, meat goats, and haired goats.

The USDA readily admits that it is difficult to produce an accurate count of goats in the United States for many reasons. The primary reason is that most goats are produced on very small agricultural operations that may be excluded from market reporting. The size and scope of 4-H and FFA goat projects are equally as hard to capture. The sale of these projects would only be reported if the 4-H or FFA member lived on a farm that produced at least $1,000 in agricultural sales for the year.

Aggie 11/Shutterstock.com

Figure 11-32. Goats are the prominent meat source in countries that cannot support other bovines such as cattle. Goats are agile creatures content to browse for food such as twigs, leaves, and shoots of plants in lieu of grass.

Determining the top states in total goat production is difficult due to the varying nature of type and purpose of the specific goat breeds, **Figure 11-32**. Texas is the leading state in goat production; from there, it is difficult to determine the other leading states due to agricultural reporting challenges.

Goats around the Globe

Goats are one of the oldest domesticated animals in the world. For centuries, they have been used for their meat, milk, hair, and hide. More than 80% of the world goat population lives in Asia and Africa. The top five countries in the world in number of goats are (in order): China, India, Pakistan, Bangladesh, and Nigeria. Worldwide, more people eat meat and drink milk from goats than any other animal.

It is important to recognize that the U.S. goat industry has been impacted by the increase of Hispanic, Muslim, Caribbean, and Chinese consumers. The demand for goats is especially high around religious holidays for these ethnic groups, and each has its own preference for age, sex, weight, and maturity of the goat.

Did You Know?

Goats are members of the family bovidae and are closely related to cows and antelopes. However, only cattle are referred to as bovines.

Common Goat Terms

The following terms are used to describe goats during different stages of their life.

- ***Billy***—a male goat of any age
- ***Buck***—a male goat of any age (industry-preferred term)
- ***Cabrito***—Spanish word for young, milk-fed goat
- ***Doe***—a female goat of any age (industry-preferred term)
- ***Kid***—a young goat of either sex
- ***Nanny***—a female goat of any age
- ***Wether***—a castrated male goat

Production Cycle of Goats

The production cycle of goats is very similar to that of sheep, with a few minor exceptions, **Figure 11-33**.

Goats stay in a group called a ***flock*** or a ***herd***. In other cultures, a grouping of goats may be called a ***tribe*** or a ***trip*** of goats. Goats are seasonal breeders, which means they are typically influenced by the length of the day. Optimal breeding season for goats is in late fall when the days start to get shorter. Female goats (does or nannies) are naturally mated to male goats called bucks or billies. The doe is in gestation (pregnancy) for approximately150 days before ***kidding*** (giving birth). The baby goats are called kids. It is very common for does to have twins or sometimes triplets.

The kids stay with the doe until they are 10–12 weeks old. At this point, they weigh between 30−60 lb. Also at this time, wethers can be sold as market goats or sent to a feedlot for finishing. Bucks and does will either be sold as market goats or placed back into a breeding herd as replacements. Replacements are used in the breeding operation for animals that have been culled (removed) from the herd due to age, illness, and physical condition. The life expectancy of goats is 8–12 years, but they may live to be as old as 15.

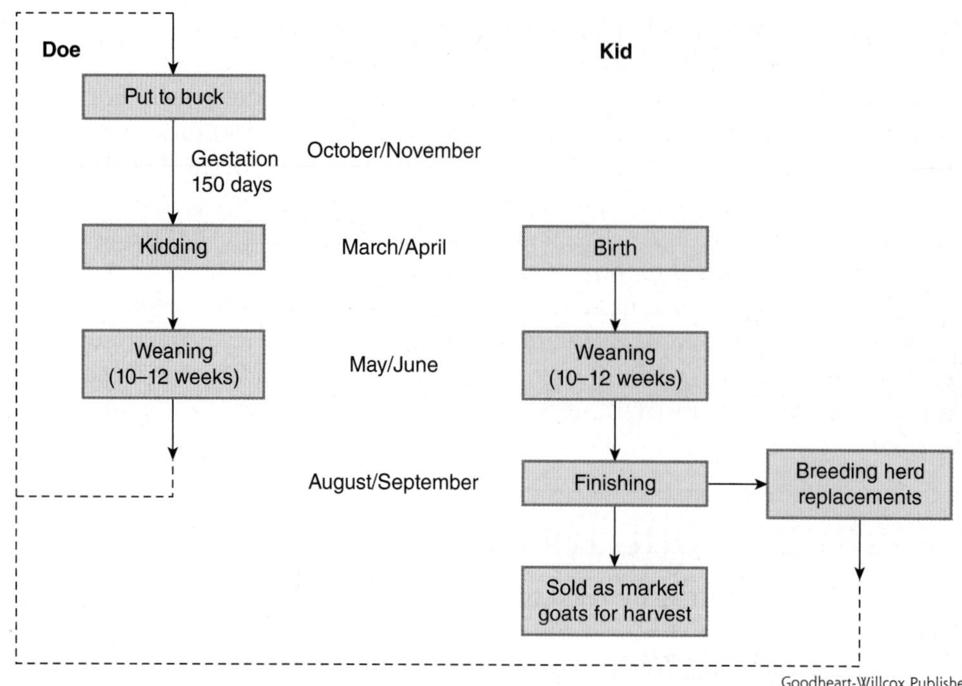

Figure 11-33. The production cycle of goats is similar to that of sheep. After studying the production cycle, visit www.g-wlearning.com to test your knowledge.

The Goat Industry

Goats are multipurpose animals that can be used for milk, meat, or fiber. Most goats are raised in small flock or herd settings. The average farm herd size for goats is around 30 head. The goat industry in the United States is divided into four main sectors:

- Meat goats
- Dairy goats
- Fiber or hair goats
- 4-H and FFA project goats

Meat goats are currently the largest sector of the U.S. goat industry, followed by dairy goats. Over the last 20 years, fiber or hair goat production has declined in the United States while the number of 4-H and FFA goat projects has increased dramatically. In most states, 4-H and FFA show goat numbers have surpassed and, in many cases, doubled or tripled that of sheep.

Did You Know?

Goats have often been used as a natural means of weed control.

Breed Associations

Most goat breeds have their own breed association similar to other species of livestock. Meat goats, dairy goats, and Angoras may be registered within their respective breed associations, and bucks and does that are kept as replacements in the purebred operations are often registered.

Commercial Production

Commercial goat production is raising goats strictly for market purposes. This is achieved by breeding bucks to does and selling the offspring at weaning as market animals. These commercial animals are usually crossbred, meaning they are a mixture of more than one breed of goat.

Anatomy of Goats

Because goats are multipurpose animals, it is important to recognize the differences between meat and dairy goats. Meat goats have been incorporated in the National FFA Livestock Career Development Event and are a component of most state and collegiate livestock judging competitions. **Figure 11-34** shows the parts of a meat goat. To learn more about selection and judging of meat goats, ask your teacher about the FFA Livestock CDE.

Meat Goats and Dairy Goats

The differences between meat and dairy goats are similar to that of beef and dairy cattle. Although dairy goats can be harvested for meat production, their main purpose is to produce milk and milk products. **Figure 11-35** shows the parts of a dairy goat with additional views of the parts of the udder.

Goats are small ruminants, which means they can digest high amounts of roughage or forage, usually grass or hay. Their stomachs are divided into four compartments: the rumen, omasum, abomasum, and the reticulum.

Ruminants like goats commonly chew their cud, which is a regurgitation of the roughage and feed they have eaten, to aid in further digestion. Goats prefer to ***browse*** rather than graze like sheep and cattle. They prefer to eat leaves, twigs, and shoots of plants in rangeland over grazing.

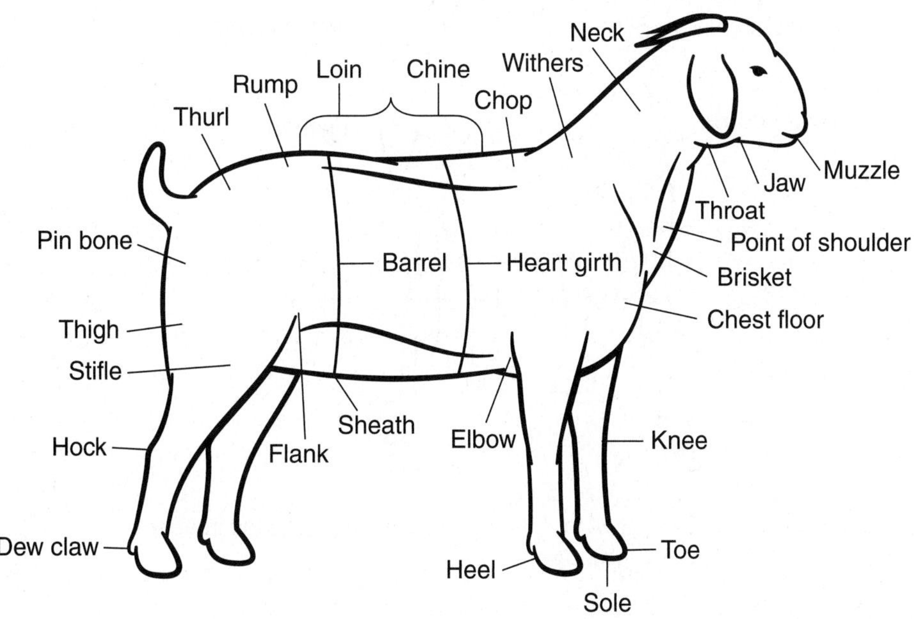

Vlad Klok/Shutterstock.com

Figure 11-34. Study the anatomy of the meat goat illustrated above. After studying the image, visit www.g-wlearning.com to test your knowledge.

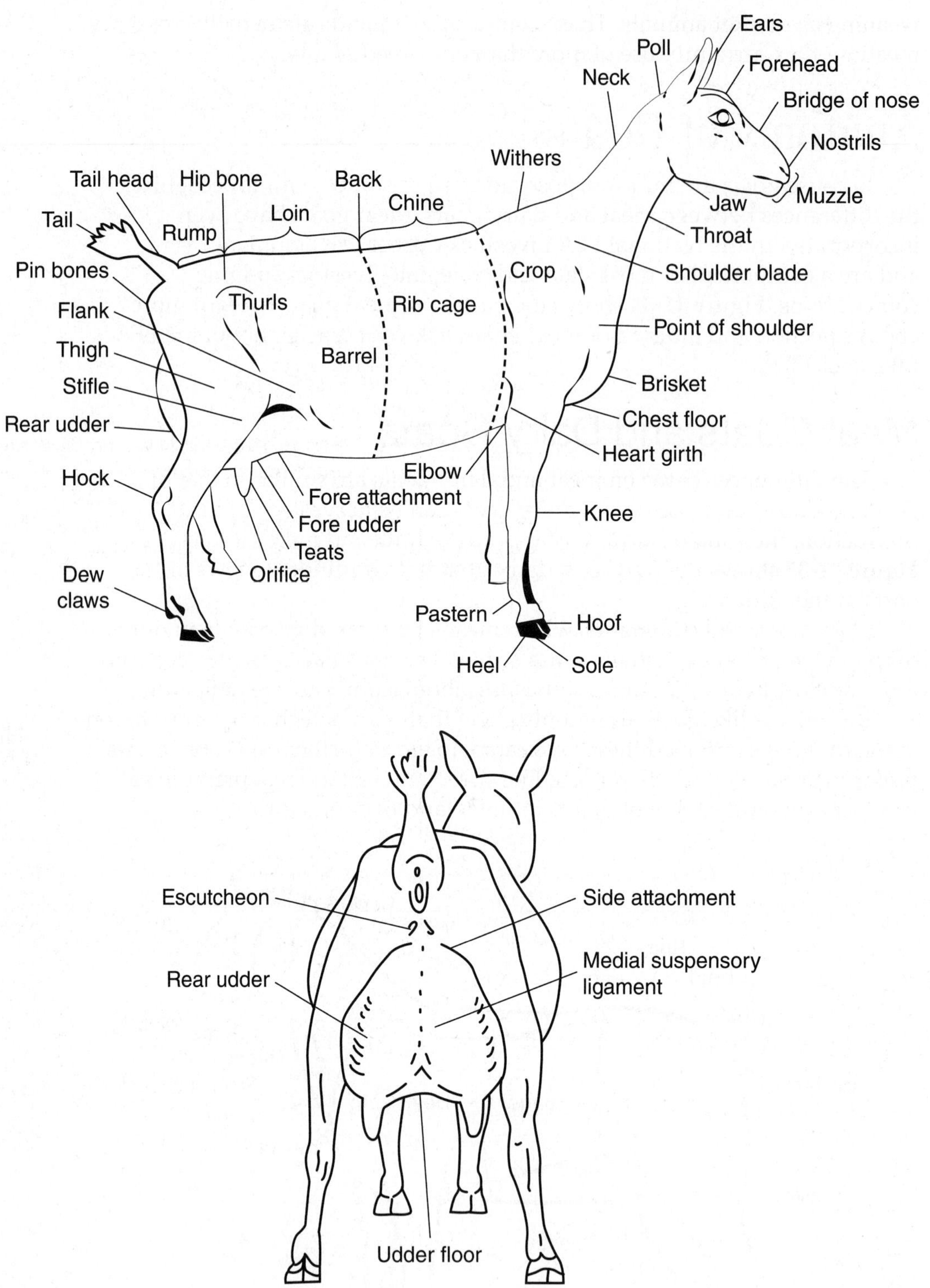

Figure 11-35. Study the anatomy of the dairy goat. After studying the image, visit www.g-wlearning.com to test your knowledge.

Maintaining Herd Health

Maintaining herd health is essential to keep goat production affordable and profitable. The following brief list of tips for maintaining herd health is not all-inclusive, but it does contain basic steps to ensure a healthy herd of goats.

- Establish and implement a herd health management plan. Work with your veterinarian to establish a parasite control program and a herd health management plan, **Figure 11-36**.
- Keep accurate records that include: breeding and kidding, feeding, disease prevention, purchase of new animals, and sale of goats.
- Restrict uncontrolled traffic and contact of herd with people and goats of unknown origin. Quarantine new goats if necessary to maintain herd health.
- Maintain proper environments. This includes adequate space and stocking rate, comfortable climate, dry and clean bedding, and natural lighting.
- Provide proper nutrition and roughage, practice proper handling, and use proper care.

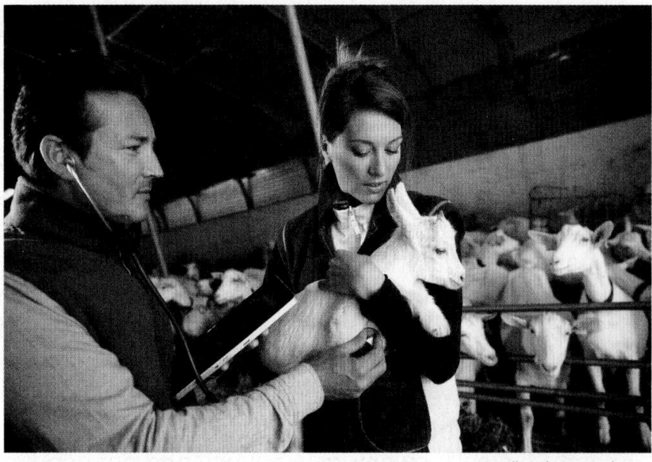

Goodluz/Shutterstock.com

Figure 11-36. Many veterinarians are not as familiar with goat health issues as they are with other livestock issues because it is a newer area of interest in the United States. It is important to find a veterinarian who is knowledgeable about goats and their health issues if you plan on raising goats.

Handling Goats

There are precautions that can be taken to reduce hazards and injuries to both producers and goats. Keep the following points in mind when handling and transporting goats.

- Goats are more difficult to handle than sheep due to their stubborn nature. When frightened, they lie down or sulk and make it difficult to work through a chute.
- Carefully plan and design your handling facilities. This includes the size of pens and runs as well as closely boarding the sides so the goats cannot see through and get distracted. Make the runs narrow enough to prevent the goats from turning around and blocking the flow.
- Goats tend to move together as a family. The older goats usually lead the way.
- Most breeds of goats have horns. Extra caution must be taken when handling any type of livestock with horns.
- Learn how to properly handle goats for procedures like hoof trimming and vaccinating.
- Use the goat's flight zone to help manage its movements. The term *flight zone* is used to describe how close you can get to goats before they start moving.
- Goats like other ruminants prefer to move toward light, and not dark, places. Keep chutes and working facilities in well-lit areas.

Did You Know?

Goats and sheep are both susceptible to *enterotoxemia*, also known as *overeating disease*. This is a condition that young lambs and goats can be affected by and results in sudden death. This can be prevented with a proper vaccination schedule and good management.

Did You Know?

Cashmere is the soft down or winter undercoat of a goat. Cashmere does not come from a specific breed of goat—it is harvested when goats shed their hair naturally. One goat will only produce about 1 lb of cashmere. The cashmere industry in the United States is one of the least economically important segments of the goat industry.

STEM Connection Myotonia Congenita

Myotonic goats, also known as Tennessee fainting goats or Wooden Leg goats, are a breed of meat goat with very unique characteristics. Myotonic, or fainting, goats do not actually faint or pass out like many people believe. They are born with a hereditary genetic disorder known as *myotonia congenita*. When the goat feels stressed or threatened, their muscles tense up and they collapse. There is no pain involved for the goat, and they can recover from this quickly. Fainting goats are listed as a threatened species and are typically owned as a novelty and not used for meat production.

col/Shutterstock.com

Cuts of Goat

Meat from an adult goat is referred to as ***chevon***. This meat may be darker in color and less tender than meat from younger goats. *Kid* or *cabrito* are terms used to describe young milk-fed goats under a year old weighing 25–50 lb. This meat is typically more tender and leaner than that from older goats. Chevon or kid is usually sold as an entire carcass or in quarters or sixths. The primal cuts are similar to that of sheep, including the breast, leg, loin, neck, rack, shank, shoulder, and sirloin, **Figure 11-37**.

Retailers have not seen the demand for retail cuts of this meat, but goat is USDA inspected like beef, pork, and mutton.

Leg

Loin

Flank

Rib

Shoulder

Vlad Klok/Shutterstock.com

Figure 11-37. The meat cuts of goats are similar to that of sheep. Most consumers order a whole carcass when they purchase goat meat. For this reason, retailers have yet to experience a demand for retail goat cuts. After studying the image, visit www.g-wlearning.com to test your knowledge.

Goat Breeds

The species name for goats is ***caprine,*** and the scientific name is *Capra aegagrus hircus.* There are more than 200 different breeds of goats with an estimated population of 450 million worldwide. The breeds of goats can be broken down into three categories: meat goats, dairy goats, and Angora goats. Let us examine the breeds within each of these categories.

Meat Goat Breeds

Meat goat breeds are known for their ability to produce heavy-muscled carcasses with acceptable yield and quality. The following breeds are some of the most recognized breeds of meat goats in the United States.

Rachelsie Farm, Inc.

Boer. Boer goats originated in South Africa in the early 1900s. Boer goats were not introduced into the United States until 1993 but have increased in popularity to become the most popular meat goat breed in the United States. Boer goats are white with a red head and a white blaze down the middle of their face. The red or brown coloring extending down from the head to the shoulder region is called a ***cape.*** Other color patterns include paint, red, spotted, and solid white. Boer goats have changed the meat goat industry in the United States because of their ability to be feed-efficient and cross well with other breeds.

MJ Ironwater Acres

Kiko. Kiko goats originated from the feral native goats of New Zealand and were developed by the Goatex Group LLC. Kikos are very hardy and vigorous. They have excellent maternal characteristics and can raise twins or triplets without human intervention. Kikos are born very small but grow rapidly to mature into a large-framed goat. There are no breed standards for color or size.

Morgan Frederick, Three Mill Ranch

Spanish. Spanish goats are descendants of feral goats brought over by the early explorers of the United States. They are commonly referred to as brush goats because of their hardy nature and ability to survive on limited or poor-quality forages such as brush or weeds. Spanish goats vary greatly in color and are horned. They are very prolific and can survive with little care. They are small compared to other breeds of meat goats but are agile and hardy. Spanish goats cross well with Boer goats to produce good commercial meat goats.

Dairy Goat Breeds

Dairy goats are vital to the total dairy industry in the United States. Goat milk can be used to make the traditional dairy products we think about that come from cow's milk, such as: cheese, butter, yogurt, and sour cream. Goat milk is slightly higher in fat content than cow's milk. The lactation period for a dairy goat is about 280 days. Peak milk production in does occurs approximately four to six weeks after kidding.

Redwood Hill Farm

Alpine. Alpine goats originated in France and were imported into the United States in the 1920s. Alpines have erect ears and varying color patterns from black to white. Alpines are known for their excellent milking ability and large frame size. An average Alpine doe can produce more than 2,300 pounds of milk per year.

Redwood Hill Farm

LaMancha. LaMancha goats originated in Oregon and were the result of crossing short-eared goats with Nubians. Their ears are very small—so small that it may appear that they do not have ears at all. LaManchas are known for being docile and steady producers of milk with high fat content. LaManchas tend to be smaller framed than most dairy goat breeds, and colors vary.

Redwood Hill Farm

Nubian. Nubian is the most popular breed of dairy goat in the United States. Nubians originated in Africa; however, the modern-day Nubian goat is a result of crossing Nubian bucks with different British dairy goat breeds. They can be any color. Nubians produce the highest fat content milk of any dairy breed. Nubians are medium to large in size and short haired.

Redwood Hill Farm

Saanen. Saanen goats originated in Switzerland and were one of the first dairy goat breeds imported into the United States in the early 1900s. Saanens are solid white or light cream colored with upright ears. They are known as the "Holstein" of the dairy goat world. Saanen is the most widely distributed breed of dairy goat in the world. They produce high volumes of milk, with does producing more than 2,500 pounds of milk per year.

Dairy Goat Breeds (*Continued*)

Rachelsie Farm, Inc.

Oberhasli. (*Above*) The Oberhasli breed originated in Switzerland and was imported into the United States. Until 1979, they were known as Swiss Alpines. Their color is Chamoise, which is a bay color with a black dorsal stripe and a black underline. Their heads are nearly all black with black below their knees. Does may be nearly all black, but this color pattern is not acceptable for bucks. Oberhaslis are moderate to small in size and are moderate to high in milk production with moderate milk fat content.

Toggenburg. (*Not illustrated*) Toggenburgs are another Swiss breed of dairy goats. They were imported into the United States around 1893. Toggenburgs are light fawn to dark chocolate in color with white ears and white on their lower legs. They also have white on the side of their tail and two white stripes down their face from the eye to the muzzle. Toggenburgs are the oldest registered breed of any type of livestock in the world. They are not as docile as other breeds of dairy goats and are known as moderate milk producers that are small to moderate in size.

Angora Goat Breeds

Smith Family Angoras & Erbstruck Ranch

Smith Family Angoras & Erbstruck Ranch

Angora. Angora goats are a very old breed that originated in Turkey. They were imported into the United States in the mid-1800s. Angora goat production in the United States occurs mainly in Texas, where 90% of all Angora goats in the country are located. The fleece from Angora goats is called ***mohair***. Mohair can be classified into three types: kid, young adult, and adult. The younger the goat, the finer or higher quality mohair is produced. Angoras are shorn twice per year, producing about 5 lb of fleece each time they are shorn. Angora goats like dry, mild climates and can be kept in small farm herds or large rangeland flocks like sheep. Their popularity has dwindled in the United States since the early 1990s when government subsidies for mohair were eliminated.

Lesson 11.4 Review and Assessment

Words to Know

Match the key terms from the lesson to the correct definition.

1. The fleece from Angora goats.
2. A group of goats.
3. The red or brown coloring extending down from the head to the shoulder region of a Boer goat.
4. Spanish word for young, milk-fed goat.
5. A female goat of any age.
6. A male goat of any age.
7. The preferred eating method of goats in which they eat leaves, twigs, and shoots instead of grazing on grasses.
8. The act of a goat giving birth.
9. A castrated male goat.
10. The meat from a mature goat.
11. A young goat of either sex.
12. The species name for goats.
13. A breed of meat goat that appears to faint when startled.

A. browse
B. buck
C. cabrito
D. cape
E. caprine
F. chevon
G. doe
H. flock
I. kid
J. kidding
K. mohair
L. myotonic goat
M. wether

Know and Understand

Answer the following questions using the information provided in this lesson.

1. Explain why the demand for goat meat has increased in the United States.
2. *True or False?* A grouping of goats may be called a tribe of goats.
3. *True or False?* The kids stay with the doe until they are weaned at 4–6 weeks old.
4. List the four main sectors of the goat industry in the United States.
5. Which of the four main sectors of the goat industry in the United States is the largest?
6. *True or False?* Goats are modified monogastric.
7. *True or False?* The fleece from myotonic goats is called mohair.

Analyze and Apply

1. Goats are essential to the dairy industry. Compare and contrast dairy products made from cow's milk and goat's milk. Are they different? Similar? The same? How do consumers view goat milk vs. cow milk?
2. Meat goats are one of the fastest growing livestock enterprises in the United States. With a partner, develop a list of pros and cons for goat production in your area. Are they more popular than other small ruminants? How could they be incorporated into other livestock production systems?

Thinking Critically

1. Goat meat is one of the most widely consumed meats in the world. Why is goat meat more popular in other countries than in the United States? What factors have limited the popularity of goat in U.S markets? Grocery stores?
2. Many people view goats and sheep as very similar livestock animals. The two species are vastly different. Divide into pairs or groups. Research the two industries and debate the pros and cons of sheep production and goat production. Determine which is more suitable for your given area.

Chapter 11

Review and Assessment

Lesson 11.1

Poultry Industry

Key Points

- The poultry industry is vital to the economy and food supply in the United States.
- There are various species and subspecies of birds that are classified as poultry.
- Both the meat and eggs produced in the commercial poultry industry are essential protein sources for the American consumer.
- The United States is second in the world in poultry production. As demand for poultry products increases in our country, the poultry industry will continue to be a very influential part of our food system nationally and globally.

Words to Know

Use the following list and textbook glossary to review and study the *Words to Know* from *Lesson 11.1*.

avian
black-out house
boneless/skinless breast
breast quarter
breeder farm
broiler
caeca
candling
capon
chick
chicken
cloaca
cockerel
crop
drake
drumstick
duck
duckling
feet
flock
free-range system
gaggle
game bird
gander
gang
giblets
gizzard
goose
gosling
grit
hatchery
hen
hybrid
intensive housing system
layer
leg quarter
litter
molted
offal
paws
poult
proventriculus
pullet
rafter
rooster

semi-intensive housing system
sexually dimorphic
split breast
tenderloin
thigh
tom
vertical integration
whole bird
whole fryer
wing
WOG

Check Your Understanding

Answer the following questions using the information provided in *Lesson 11.1*.

1. *True or False?* People in the United States consume more chicken than people in any other country.
2. *True or False?* The poultry company's main interest is to produce a profit by producing meat and eggs of consistent quality.
3. What is the purpose(s) of beak trimming in commercial egg production operations?
4. What were some of the problems early poultry producers faced in the 1920s and 1930s?
5. How did growers in the 1940s improve egg production and bird health?
6. What is the average number of eggs a hen produces in its commercial lifetime?
7. *True or False?* Approximately 172 egg-producing companies in the United States represent 99% of the laying hens in the United States.
8. *True or False?* Turkey is the second most popular source of protein for Americans.
9. Why is it impractical to use turkeys for commercial egg production?
10. Identify three considerations specific to duck or other waterfowl production.

Lesson 11.2

Swine Industry

Key Points

- The U.S. swine industry is a complex system made up of many different segments.
- The ultimate goal of the swine industry is to produce pork for consumers in the United States and for a global society.

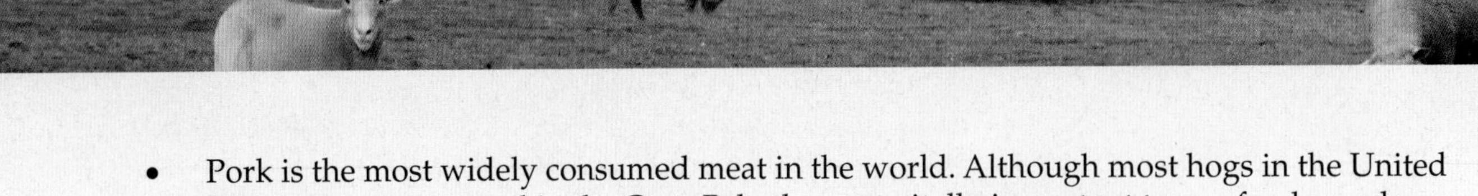

- Pork is the most widely consumed meat in the world. Although most hogs in the United States are concentrated in the Corn Belt, they are vitally important to our food supply and economy.
- The swine industry depends on many other agricultural industries for its success.
- Other agricultural commodities such as corn and soybeans are used as key ingredients in swine feed. As the price of these commodities rises, so does the price to produce pork.
- Animal agriculture is only one part of a huge industry that feeds our nation and our world.

Words to Know

Use the following list and the textbook glossary to review and study the *Words to Know* from *Lesson 11.2*.

American Berkshire Association (ABA)
barrow
boar
Boston butt
Boston shoulder
castrate
certified pedigreed swine (CPS)
confinement operation
export
farrow
farrowing
farrowing crate
farrow-to-finish operation
farrow-to-wean operation
feeder pig operation
finishing operation
gilt
ham
litter
loin
market hog
maternal breed
monogastric
National Swine Registry (NSR)
nursery
picnic shoulder
pig
piglet
porcine
pork
pork butt
purebred swine
sausage
seedstock
sow
spareribs
swine
terminal breed
vertical integration
weaned

Check Your Understanding

Answer the following questions using the information provided in *Lesson 11.2*.

1. *True or False?* The hog and pig industry accounts for 6% of total agricultural sales in the United States.
2. At what age are piglets weaned? List five processing actions that are taken while the piglets are still nursing.
3. *True or False?* In farrow-to-finish operations, the swine herd is maintained with sows farrowing once per year.

4. *True or False?* In feeder pig operations, piglets are weaned between ten and twelve weeks of age.
5. What percentage of hogs produced in the United States are by purebred producers? For what purpose are these hogs raised?
6. _____ is critical to the overall health and well-being of swine.
7. Why is it so important to provide hogs adequate protection from extreme heat?
8. Why is it important to use solid sidewalls and spotlights when moving swine?
9. What purpose do sorting panels or hurdles serve during breeding?
10. *True or False?* There are more than 70 breeds of swine worldwide.

Lesson 11.3

Sheep Industry

Key Points

- Although the sheep has declined in the United States over the last few years, it is still a very important segment of animal agriculture.
- The sheep industry provides us with lamb and mutton to eat as well as wool and leather to use as clothing.
- Sheep have been around since biblical times and have provided a source of food and clothing for many civilizations.
- Today, scientists continue to use sheep in groundbreaking medical research and in the field of cloning.
- The sheep industry has and will continue to play a major role in all walks of daily life.

Words to Know

Use the following list and the textbook glossary to review and study the *Words to Know* from *Lesson 11.3*.

crossbred-wool breed	fleece	lambing
ewe	flock	leg
ewe lamb	greasy wool	loin
farm flock	hair sheep	long-wool breed
fine-wool breed	lamb	medium-wool breed

mutton
ovine
purebred producers
rack
ram
ram lamb
range production
seasonal-use production
shoulder
stocking rate
wether

Check Your Understanding

Answer the following questions using the information provided in *Lesson 11.3*.

1. *True or False?* Sheep are produced in every state in the United States.
2. At what age and weight are lambs removed from the ewes?
3. What is the life expectancy of sheep?
4. *True or False?* Commercial sheep producers may choose to lamb at different stages of the year, depending on weather conditions, forage, and other production systems.
5. Why do ruminants like sheep chew their cud?
6. List three basic steps a producer can take to help ensure a healthy flock.
7. What is a sheep's flight zone, and how can you use it when handling or moving sheep?
8. *True or False?* Fine-wool breeds are primarily used for wool production and account for nearly half of the world's sheep population.
9. *True or False?* Medium-wool breeds are primarily used for wool production and have less value as a meat animal.
10. *True or False?* Long-wool breeds are typically used as wool-producing breeds.

Lesson 11.4

Goat Industry

Key Points

- Goats are one of the oldest domesticated animals in the world.
- In many parts of the world, goats are the sole source of meat, milk, and dairy products, especially in climates and terrains where cattle cannot survive.
- Goats are multipurpose animals that have stood the test of time and are vital to agriculture in our country and around the world.

Words to Know

Use the following list and the textbook glossary to review and study the *Words to Know* from *Lesson 11.4*.

billy
browse
buck
cabrito
cape
caprine
chevon
doe
flock
herd
kid
kidding
mohair
myotonic goats
nanny
tribe
trip
wether

Check Your Understanding

Answer the following questions using the information provided in *Lesson 11.4*.

1. Approximately how many goats are in the United States?
2. Identify the top five countries in number of goats.
3. What demographic increase has impacted the U.S. goat industry the most in recent years?
4. When is the optimal breeding season for goats? How long is the gestation period?
5. At what age and weight are the kids removed from the doe's care?
6. What are the four main sectors of the goat industry in the United States?
7. *True or False?* Goats prefer to graze like sheep and cattle rather than browse.
8. List three actions you may take to help ensure a healthy herd of goats.
9. *True or False?* Goats are more difficult to handle than sheep due to their stubborn nature.
10. When does peak milk production occur in does?

Chapter 11 Skill Development

STEM and Academic Activities

1. **Science.** Through careful observation, we can learn a lot about animal behavior. Understanding animal behavior helps farmers and ranchers reduce animal stress and catch problems before they become unmanageable. Carefully observe livestock animals or a pet for a week to 10 days. Use an ethogram (a description of an animal's behavior) to record your observations. Areas of observation to include in your ethogram should cover feeding, aggressiveness, vocalization, movement, grooming, scratching, sleeping, and

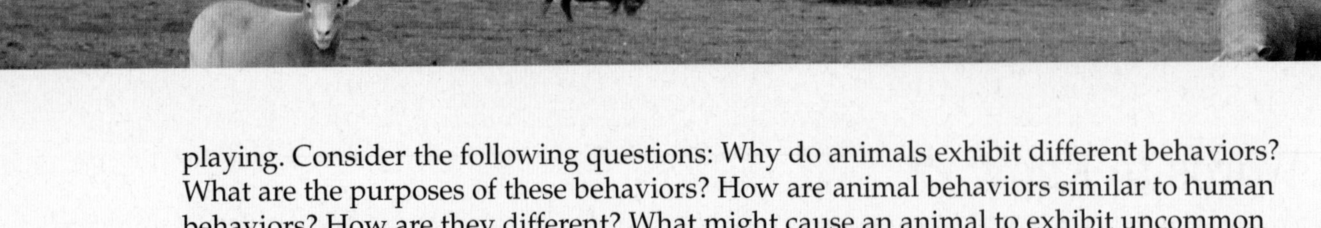

playing. Consider the following questions: Why do animals exhibit different behaviors? What are the purposes of these behaviors? How are animal behaviors similar to human behaviors? How are they different? What might cause an animal to exhibit uncommon behavior? How can your observations help raise healthier, happier animals? Be prepared to share the behavior of your animal with the class.

2. **Technology.** Choose one of the animal industries from this chapter: swine, sheep, goats, or poultry. For the industry you have selected, please conduct research to identify a piece of equipment that is commonly used in modern production systems. Learn how this piece of equipment helps producers increase the quality or efficiency of their operation. Collect information on what the equipment is, how it works, and why it is important to the industry.
3. **Engineering.** Design, build, and use a candler using common household items. Items you might use include: a can light, cardboard, fluorescent bulb, and scissors; cardboard tube, black electrical tape, and a bright light; a metal can with a lid, a light fixture, a bulb, sheet metal screws, and some rubber gasket material or felt. Research the timing and temperature issues involved before candling eggs.
4. **Mathematics.** Poultry are kept in climate- and light-controlled houses. By increasing the amount of light, laying hens will increase their production. During the winter, many places in the United States receive around 10 hours of natural sunlight per day; in the summer, they can receive 12–14 hours of natural light. Poultry houses can increase production by increasing the amount of light to 16–18 hours. But how much impact does the increased light have on the bottom line? At 12 hours of light, hens will produce an egg about every 28 hours; at 16 hours of light, hens will produce an egg about every 24 hours. If you had 100,000 hens, and eggs were sold wholesale at $2.00 per dozen, how much more income would you have each week (168 hours) by changing from 12 to 16 hours of light in the poultry house?
5. **Social Science.** There is a lot of controversy and misinformation about the terms used to label poultry. Research the following terms (free range, farm-raised, natural, organic, no hormones added, antibiotic free, all-vegetable diet, made in the U.S.A.) and write a two- to three-page report explaining the pros and cons of labels used in poultry production. Which labels are regulated and which labels are not? What labels would increase your likeliness to purchase a specific poultry product?

Communicating about Agriculture

1. **Reading and Writing.** There are many duck breeds in the United States and around the world. Choose one of the following breeds and write a two-page written report. Use a standard format and cite your references. Include the following information in your report: history, country of origin, temperament, egg-laying proficiency, weight, plumage, and current status (endangered, watch list, etc.). Include images and any unusual or interesting facts.

   Ancona, Australian Spotted, Buff or Orpington, Cayuga, Dutch Hookbill, Magpie, Saxony, Silver Appleyard, Swedish, or Welsh Harlequin.

2. **Speaking.** Debate the topic of hormones and antibiotics as they apply to animal production. Determine which type of livestock or small-animal production you will discuss. Divide into two groups. Each group should gather information in support of either the pro argument (hormones and antibiotics are necessary for animal production) or the con argument (the hormones and antibiotics used in animal production are detrimental to the animals and those who consume them). Use material from the chapter, as well as other resources, to support your side of the debate and to clarify word meanings as necessary. Do additional research to find expert opinions, costs associated with hormones and antibiotics, and other relevant information.

Chapter 12

Other Animal Production

©iStock/EyeEmCLOSED

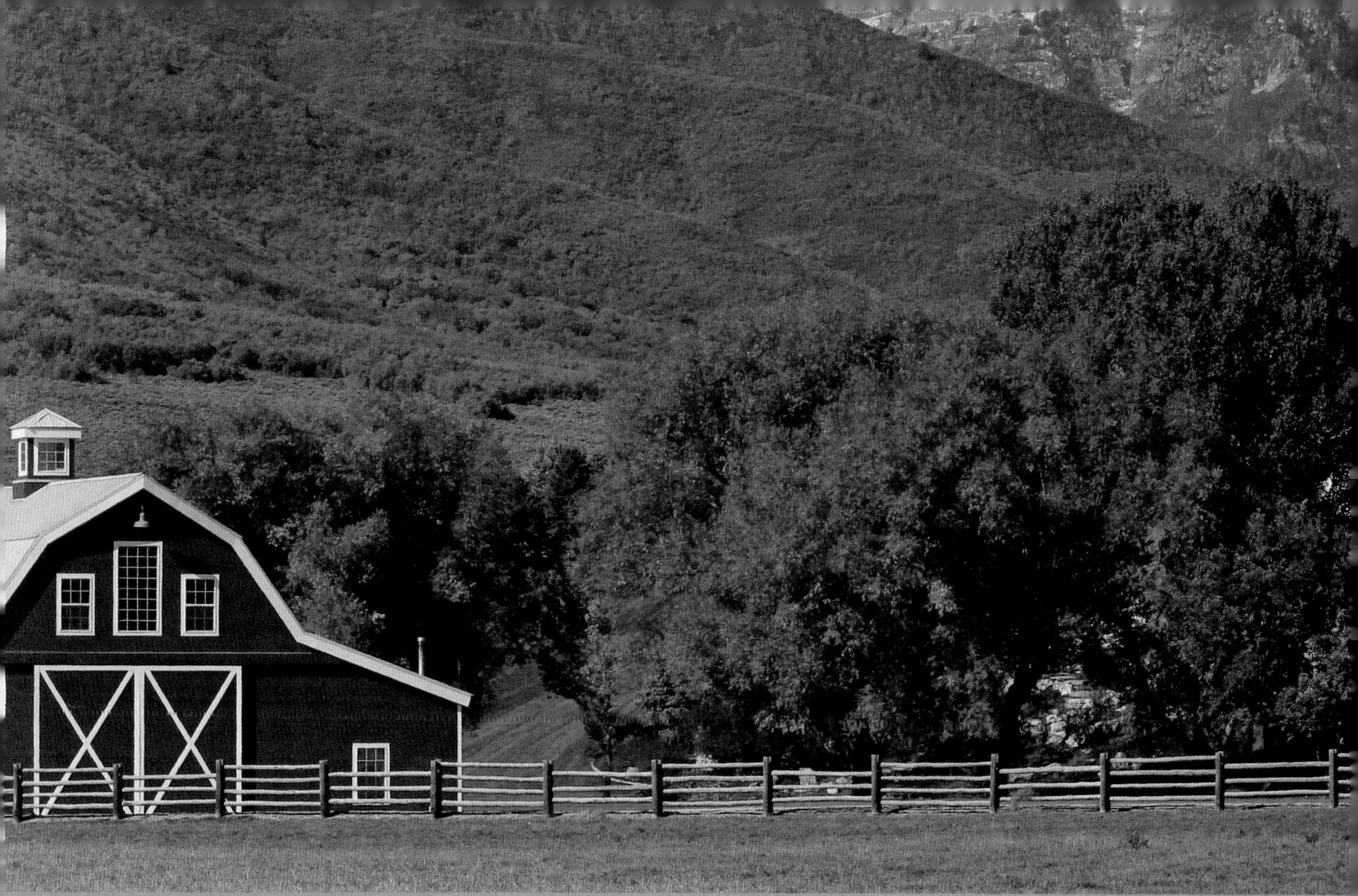

Johnny Adolphson/Shutterstock.com

BrendaLawlor/iStock/Thinkstock

©iStock/ultrapicture

While studying, look for the activity icon to:

- **Practice** vocabulary terms with e-flash cards and matching activities.
- **Expand** learning with video clips, animations, and interactive activities.
- **Reinforce** what you learn by completing the end-of-lesson activities.
- **Test your knowledge** by completing the end-of-chapter questions.

Lesson 12.1

Companion Animals

Words to Know

buck
companion animal
doe
domesticated animal
euthanasia
feral animal
fryer
hutch
kindling
kit
lapin
litter
neuter
spay
whelping
wild animal
zoonoses

Lesson Outcomes

By the end of this lesson, you should be able to:

- Understand the size and scope of the companion animal industry.
- Understand common terminology of the companion animal industry.
- Describe the components of the companion animal industry.
- Describe different types of companion animals in the United States.
- Develop basic skills for maintaining healthy pets.
- Explore career opportunities in the companion animal industry.

Before You Read

Companion animals are an important part of the animal industry. Make a list of ways you think companion animals are used in society. After you complete your list, research the ways companion animals are used in our society. Did you find any new or unusual ways they are used?

Companion animals are the animals that live, play, and sometimes work with people on a daily basis, **Figure 12-1**. Most people refer to them simply as *pets*.

Andresr/Shutterstock.com

Figure 12-1. Many people consider their pets members of their family.

Humans initially took on companion animals in relationships that benefited the animal and human in terms of survival, but today's pets mean so much more. The importance of companion animals in the United States is evident by the billions of dollars spent annually on pet food and care. It is also evident in the sheer volume of pet videos and images found on the Internet, as well as all the amusing animal images people text and email to each other on a daily basis.

Domestication of Companion Animals

Domesticated animals are the most common types of animals kept as pets or companion animals. A ***domesticated animal*** is any animal that has been tamed from its wild form to a domesticated form for the benefit of humans. In order for an animal to be considered domesticated, it must be different from its wild variety in five ways:

- The animal must have a structural difference.
- The animal must behave differently. For animals, their fear or aggressiveness to humans is diminished or eliminated.
- The animal must rely on humans for sustenance, **Figure 12-2**. Most pets would not survive long if they were left to fend on their own in the wild. They depend on their owners for shelter and food, as well as affection and companionship.
- The animal's reproduction must be subject to human control.
- There must be a clear purpose for humans to use and cultivate the animal.

Domestication of animals occurred at roughly the same time as the earliest crop domestication, with one exception, the dog. Most anthropologists agree that the dog was domesticated as early as 30,000 years ago. Evidence suggests nomadic tribes kept dogs with them for protection, hunting, and companionship.

Wild or Exotic Animals as Pets

Some people keep wild animals as pets. A ***wild animal*** is an animal that lives in a natural setting, must get food and water on its own, and is not cared for by humans. Keeping wild animals is discouraged, and sometimes illegal, unless the owner is qualified to feed and safely handle the animal. Many local and state jurisdictions require special licensing to house a wild or exotic animal. For example, there are licensed sanctuaries that take in wild animals that were raised by people as novelty pets, or are injured in some way. This includes animals such as large cats (tigers, lions, cougars), bears,

Susan Schmitz/Shutterstock.com

Figure 12-2. Most domesticated animals would not survive if left in the wild to fend on their own. For example, breeds like the English Bulldog often have breathing issues due to the shortened profile that has been developed as a desirable show trait. *How would these breathing issues affect the dog's ability to survive in the wild?*

Did You Know?

Some of the most common forms of zoonotic diseases are toxoplasmosis, cryptosporidiosis (crypto), and Lyme disease.

wolves, raccoons and opossums, birds of prey, and even monkeys. These animals would not survive if released into the wild, but they are not really "tame" or domesticated. Many of these animals retain aggressive or wild behaviors, and they all have special nutritional needs that few people are trained to handle. Fortunately, some of these rescued animals can be used in programs to educate the public on the benefits of leaving wild animals in the wild.

Zoonotic Diseases

Some exotic species may also carry contagious zoonotic diseases, or zoonoses. ***Zoonoses*** are diseases and infections that can be transmitted from animals to humans. Common diseases that can be transmitted are rabies, toxoplasmosis, ringworm, roundworms, and Rocky Mountain Spotted Fever, **Figure 12-3**. All of these are preventable with proper healthcare and vaccination of pets and proper hygiene and handling of pets and their waste. Zoonotic diseases may be transmitted by breathing in droplets of contaminated air, through insect bites, by ingestion, direct contact (through mucus membranes, open cuts, or wounds), and through objects carrying pathogens (brushes, shavings, obstetrical chains).

To prevent transmission of these diseases, keep the following points in mind when handling companion animals:

- Be aware of places where you might come in contact with an infected animal or insect. This includes: nature parks (woods, beaches, fields, deserts), animal displays, petting zoos, pet stores, pony rides, farms, and county or state fairs.
- Always wash your hands with soap and clean water for at least 20 seconds after being around animals or their areas of confinement.
- Use masks and gloves during animal birthing or when cleaning their areas of confinement.
- Use insect repellent on any exposed skin. Follow the instructions and reapply as necessary.

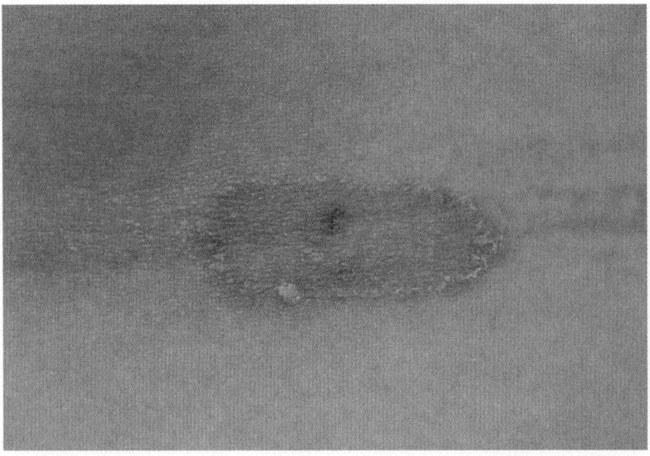

schankz/Shutterstock.com

Dariusz Majgier/Shutterstock.com

Figure 12-3. Diseases and infections like ringworm (left) and Rocky Mountain Spotted Fever (spread by ticks such as the Rocky Mountain Wood tick, right) are considered zoonotic diseases. They are easily transmitted from animals to humans. *What other zoonotic diseases can you identify? Are they all treatable?*

- Use products containing permethrin on your clothing and gear. It repels and kills ticks, mosquitoes, and other arthropods.
- Eliminate any standing water to prevent mosquito breeding. This includes regularly changing the water in water bowls and birdbaths.
- Carefully check your body and your companion animals for ticks after you have been in areas of exposure.

Did You Know?

One of the most common zoonotic diseases is ringworm. This fungus is easily transmitted from infected animals to humans. Pay careful attention to your animals to keep ringworm under control, and make sure you are disinfecting grooming tools and washing your hands properly to prevent infection.

Feral Animals

Feral animals are animals living in the wild but descended from domesticated counterparts. Examples include the "wild" horses in the western part of the United States and various herds of horses on the East Coast, **Figure 12-4**.

Many farmers encounter problems with feral dogs that terrorize and sometimes kill poultry and small livestock. People often let unwanted dogs go "free" near open farm fields instead of finding new homes for them or taking them to a shelter. The "freed" animals that survive may form packs that live by scavenging from humans and killing vulnerable livestock and poultry. Farmers must often kill these animals to protect their own animals and to help prevent the spread of animal diseases such as rabies. In developing nations without centralized animal control systems, these feral dogs can create a serious public safety concern.

Ranchers and farmers in the United States also encounter multiple problems with feral pigs. In addition to eating crops when available, feral pigs also damage or tear down fencing, damage and kill a wide variety of tree seedlings, compete with native wildlife for hard mast, and cause extensive damage to delicate wetland areas. Feral pigs have been reported in at least 45 states, with most of their range expansion due to illegal translocation of pigs for hunting.

Stephen Bonk/Shutterstock.com

Figure 12-4. Feral horses found on the East Coast and western parts of the United States are descended primarily from horses brought over by Spanish explorers.

STEM Connection Science

Russian scientist Dmitri Konstantinovich Belyaev began a research project in 1959 to determine if the silver fox would become tame, or domesticated, over time as a result of selective breeding. Animal behavior stems from biology, and changes in the tameness or aggression of an animal is a physiological change. Belyaev selected foxes that were the friendliest toward humans for breeding. Forty years after Belyaev began, the project has yielded a unique group of domesticated foxes. Through genetic selection alone, the research group has created a population of tame foxes that are much different in temperament, behavior, and physical traits from their wild counterparts. The study has continued for more than 40 years and is still in progress.

Ivan Protsiuk/Shutterstock.com

Many of the domesticated foxes in the study had domesticated animal traits like floppy ears and changes in their coat coloring.

Companion and Small-Animal Industry

Companion animals play an important role in many American families. More than 71 million American households have a pet. That is about 62% of all households in the United States. What companion animals have played a role in your life?

Economic Impact

The economic impact of companion animals is substantial. Americans annually spend billions of dollars on pet care. In the United States, the amount spent increases each year and is currently around $59 billion, almost half which is spent solely on pet food, **Figure 12-5**. This is mainly attributed to the phenomenon that people "humanize" their pets, which means they treat them like people. Pet owners are increasingly interested in feeding their pets quality foods to help them lead healthier and longer lives.

U.S. Market Sales

	2014 Sales	2013 Sales
Food	$22.62 billion	$21.57 billion
Supplies/over-the-counter medicines	$13.72 billion	$13.14 billion
Veterinary care (includes insurance, routine care, prescription medications)	$15.25 billion	$14.37 billion
Pet services (grooming, boarding, pet sitting, training)	$4.73 billion	$4.41 billion

American Pet Products Association (APPA)

Figure 12-5. The amount of money pet owners spend on feeding and caring for their pets has steadily increased each year.

Veterinary Expenses

Years ago, most people did not spend more than absolutely necessary on the veterinary care of their companion animals. Other than what was required by law, most pet owners simply could not afford preventive medical care for their pets. Today, veterinary medical treatments account for more than $15 billion per year spent on companion animals, with another $13 billion being spent on pet supplies and over-the-counter medications and treatments, **Figure 12-6**. Pet owners give their pets life-sustaining medical treatment and pay thousands of dollars on surgeries to retain their companion's quality of life. People also pay to euthanize their pets and may pay to have their pets cremated or buried in pet cemeteries.

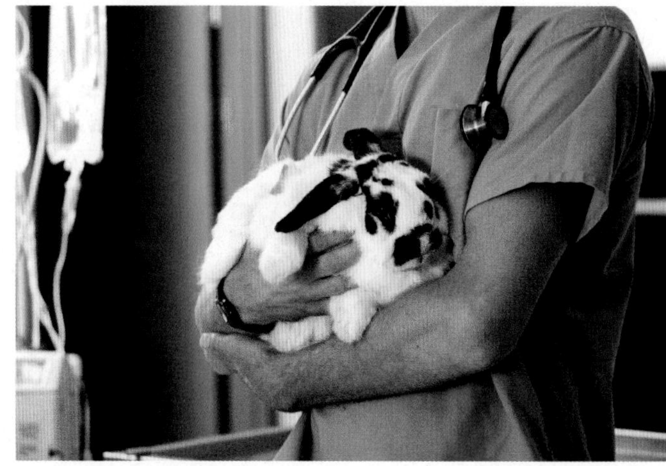

Tyler Olson/Shutterstock.com

Figure 12-6. People will spend thousands of dollars on veterinary services to retain their companion's quality of life.

Live Animal Purchases

More than $2 billion is spent annually on live animal purchases. Although there is an increased awareness of the need to adopt abandoned pets and to discourage disreputable breeders from running "puppy mills," the sale of dogs and cats increases each year. This is due in part to the fact that pets are increasingly seen as humanized members of the family, and people have more discretionary income to spend on their companions. Another reason for the continued spending on purchasing live animals is the desire to have a purebred animal who can be registered with a breed association.

Pet Adoptions

Although the adoption of pets itself may not figure into the sales of live animals, many of these animals join families and lead long lives that require spending on goods and materials to support their care. The increasing numbers of nonprofit, breed-specific organizations who take in and place unwanted purebred animals in new homes has also increased the number of adopted pets. For example, for years, Greyhounds no longer useful for racing were euthanized at a relatively young age to rid the owners of their expense. Today, many of these animals are rescued from racetracks, trained, and adopted into new homes.

To help reduce the number of unwanted pets, many pet store chains do not sell live animals and sponsor adoption days in conjunction with local shelters. Many also sponsor clinics where pet owners get discounted prices for vaccinating and spaying or neutering their pets. Most breeders selectively sell their animals. For example, if the animal is being sold as a pet, and not to be used for breeding, some breeders require a portion of the purchase price be used to spay or neuter the animal. This is especially true for show animals, so the breeder can control bloodlines.

Overall

The companion animal industry contributes greatly to our national economy and provides thousands of jobs ranging from pet grooming and pet daycare to pet food sales and manufacturing. The companion animal industry spending has continued to increase since the American Pet Products Association (APPA) started keeping statistical records of spending and ownership.

Owner Obligations and Benefits

Most pet owners feel obligated to provide their pets the same comfortable and happy lives they and their family members lead. These owners gladly accept the costs and responsibilities of pet ownership and are the driving force behind the economic success of the companion and small-animal industry. Whether you are new to pet ownership, adding companion animals to your family, or contemplating breeding animals for profit, there are a number of basic factors to keep in mind. These include:

- Time commitment
- Legal restrictions
- Financial commitment (purchasing, feeding, housing, medical care)

Time Commitment

Many people fall in love with a prospective pet before they give the slightest thought to the time involved in the pet's care. Many pet owners give up new pets when the animal fails to meet their expectations and develops bad or unacceptable habits. What they fail to understand is the amount and quality of time spent training a pet in the first year of its life will often determine much of the animal's behavior for the rest of its life, **Figure 12-7**.

Elena11/Shutterstock.com

Figure 12-7. Consistent and quality training will help your companion develop good habits and provide ample stimulation.

The amount of time needed may vary by the type and age of the animal, as well as what purpose the animal will serve. For example, young puppies and kittens require almost as much time and attention as small children. They need to be fed and watered daily, and they must be taught to use the litter box or to go outside to relieve themselves. Most cats will quickly adapt to using a litter box, but most puppies need more time and attention to learn to control their bladder and bowel movements. This often requires taking the puppy outside on an *hourly* basis.

The time needed for house training is only one factor to consider when contemplating your time commitment. Depending on the type of animal, you will need to teach your pet everything from staying off the furniture or kitchen counters, to not eating the tray of brownies you left to cool on the kitchen table. Although adopting older, house-trained animals may eliminate some training time, you may have to retrain a pet to eliminate bad habits. Pets that do not live within the owner's home may not require house training, but they will need to be trained to meet the owner's expectations and needs.

Legal Restrictions

Legal restrictions and obligations include everything from required vaccinations (such as rabies) to the numbers and types of animals you may own. Depending on your location, there are restrictions to the number of animals (including dogs) you can have without a special license or modifications to their housing facilities, **Figure 12-8**. Some animals, especially those considered livestock, are prohibited in most urban or suburban locations.

Jayme Burrows/Shutterstock.com

Figure 12-8. It is important to understand and comply with local codes regulating the types and numbers of animals you may house on your property. Carefully designed and constructed facilities will ensure the welfare of its inhabitants.

It is important to research and understand the local codes and restrictions regarding animal ownership *before* committing to animal ownership. If not, you may find yourself in a situation where you are required to give up your animals or move to a different location.

Financial Commitment

People do not often consider the financial obligations involved with animal ownership beyond the initial cost of the animal itself. If you are purchasing a purebred animal for companionship or breeding, the initial expense may be very high. If you are acquiring an animal for companionship, and are not interested in conducting a breeding operation, you may get the animal for free or for minimal cost at a shelter. A companion animal living in your home will not necessarily need its own bed, let alone a kennel. However, if you are raising companion animals for show or sale, you will need facilities to accommodate the animals from birth to sale.

In addition to the initial cost of the animal(s), and the costs involved with feeding and housing the animal(s), other costs to consider include: bedding, cleaning supplies, training supplies, toys for stimulation and teething relief, training/trainer fees, and boarding costs. You may also incur costs for showing that include entry fees and hotel and travel expenses.

Medical Costs

All pets require some veterinary care. Some vaccinations are required by law, whereas others may be required in order to travel with or ship your animals across state or international borders. If you are breeding companion animals for sale, you may be able to negotiate fees up front with your veterinarian. For example, if you are breeding dogs, you may want the puppies' dew claws removed, their ears or tails docked, and you may need to schedule regular caesarians because your particular breed is small and known for delivery problems. Many vets will accommodate reasonable requests as long as the breeder is committed to their services.

As you can see, having a pet requires a financial commitment to the animal's health and well-being. Having multiple animals requires additional financial commitments. The cost estimates in **Figure 12-9** will give you a rough idea of how costly a commitment owning a companion animal may be.

Health, Welfare, and Handling

As you can see, your companion animals depend on you for their health and welfare. As the owner, you must ensure your animals are safely housed and fed appropriately. You must also learn proper handling and restraint techniques. This will reduce stress for the animal and may prevent injury to the humans caring for the animal, and to the animal itself. Keep the following points in mind when working with companion animals:

- Establish a good relationship with your veterinarian. He or she will advise you about proper vaccinations, health management plans, and daily care of your pet.
- Certain veterinarians specialize in exotic pets (reptiles, birds, other small animals); if you have an exotic pet, find a suitable vet.

Expense	First Year	Subsequent Years
Adoption	\$0–\$500	N/A
Purchase	\$0–\$1,000+	N/A
Food	\$120–\$500	\$120–\$500
Dental/chew toys	\$20–\$200	\$20–\$200
Routine veterinary exam	\$45–\$200	\$45–\$200
Vaccinations	\$60–\$150	\$60–\$150
Vaccine boosters	N/A	\$18–\$25
Emergency veterinary care	\$0–\$2,000+	\$0–\$2,000+
Heartworm test	\$40–\$50	\$40–\$50
Heartworm prevention	\$24–\$120	\$36–\$132
Fecal exam	\$25–\$45	\$25–\$45
Flea/tick prevention	\$200–\$500	\$200–\$500
Spaying/neutering	\$35–\$200	N/A
Teeth cleaning	\$0–\$500	\$0–\$500
Boarding (per day)	\$15–\$50	\$15–\$50
Grooming	\$35–\$500	\$35–\$500

Goodheart-Willcox Publisher

Figure 12-9. The information included in this chart pertains to individual dog ownership. It is a rough estimate of what it would cost a pet owner to purchase and care for a new pet. Costs will vary according to the pet's individual needs, especially if the animal is prone to health problems. *What does it cost your family to care for your family pets?*

- Always use proper restraint equipment when visiting the veterinarian. This includes leashes and harnesses, carriers, and cages. This goes for all pets, from the smallest to the largest.
- Be aware of other animals and other smells in a clinic environment. This may excite or confuse some pets. Some veterinarian clinics have divided waiting areas for different types of pets. Some will even make special arrangements to bring pets in through a separate door to avoid contact with other patients.
- If an animal has been raised outside, it will behave differently than a house pet, especially when restrained in any way.
- Normally friendly animals may act aggressive to other animals in a strange setting.
- Always place animals on an examining table, never sit on the floor while trying to restrain an animal, **Figure 12-10**.

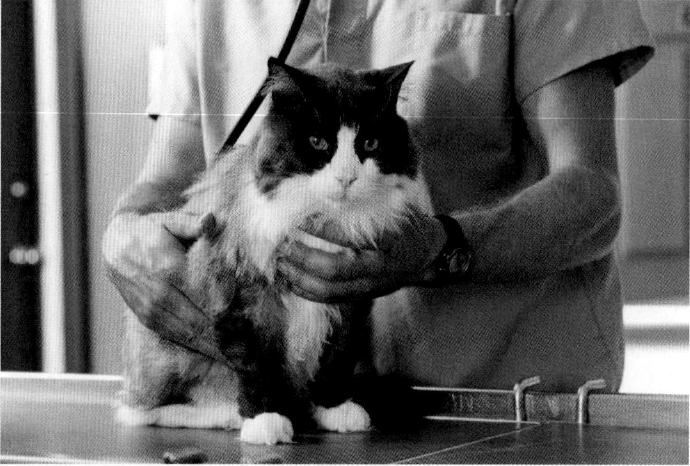

Tyler Olson/Shutterstock.com

Figure 12-10. It is important to properly handle animals when they are being examined or treated to prevent injury to the handler and the animal.

- Always use a low soothing tone when working with companion animals. Do not yell or speak in a high-pitched voice when working with pets.
- Restrain animals when necessary. This may include using your hand, a towel, control pole, nets, muzzle, or possibly drugs to calm the animal enough for treatment.
- Never put your face near the face of a dog or cat—they may lash out or bite. Reinforce this point with small children when they are around your companion animals.
- Cats and other small animals may have sharp claws as well as teeth. Wear protective gloves as necessary.
- Prevent inappropriate feeding and consumption of inappropriate foods. For example, both chocolate and grapes may be toxic to dogs. Ask your vet for a list of inappropriate foods or search online for the information.
- Prevent exposure to toxic substances, such as antifreeze and certain houseplants. Ask your vet for a list of toxic substances or search online for the information.
- Prevent obesity of your companion animals. Do not overfeed pets, do not feed pets human food, and always ensure they have adequate exercise.
- Practice safe handling and sanitation to prevent the transmission of zoonotic diseases.
- Learn how to properly treat bites and scratches, and when to seek medical care to prevent health problems.

Breeding

Should you decide to raise companion animals for sale, you will have to understand the animals' breeding habits, gestation, and parturition. You must also understand the concept and responsibility of euthanasia in order to make the sometimes difficult decisions required to be a reputable breeder.

To better understand the breeding habits of a particular type of companion animal, research the topic and speak with other breeders for advice. Some breeds do well mating naturally, whereas others may need to be artificially inseminated. Having this knowledge before you get started will help you decide which animals you want to breed, and save you a lot of frustration.

Gestation and Parturition

Gestation time varies greatly by animal types, breeds, size, and age. Knowing the gestation period of your companion animals will help you time breeding and birthing (parturition). It will also make it easier to keep track of the pregnancy progress and schedule help for parturition if needed.

Neutering and Spaying

Neutering and spaying are used to prevent animals from reproducing. ***Neutering*** is used to describe the castration of a male animal, whereas ***spaying*** is used to describe the removal of the ovaries and/or uterus (ovariohysterectomy) of a female animal. Studies have shown that, in addition to preventing reproduction, neutering or spaying has health benefits for animals. It may reduce the risk of certain cancers and tumor development, as well as the onset of some diseases. However, it may also increase an animal's risk of obesity. Behavioral benefits include less desire to roam, reduction or elimination of spraying and marking, and a decrease in aggressiveness.

Euthanasia

Euthanasia is the act of humanely putting an animal to death through chemical means, or allowing it to die by withholding extreme medical measures. Many pet owners choose to have a companion animal "put down" or "put to sleep" if the animal has an incurable and especially painful illness, or if the animal can no longer eat, drink, and relieve itself without help. Unfortunately, many healthy and young companion animals are euthanized because of overpopulation and lack of resources.

Companion Animals and Society

Companion animals of all shapes and sizes can physically, psychologically, and socially enrich the lives of their caretakers. Have you experienced some of the benefits of companion animals?

Psychological Benefits

Reducing depression, alleviating loneliness, aiding in the bereavement process, and overcoming illness have all been attributed as psychological benefits of pets. Caring for pets teaches children responsibility in caring for another living being, and provides elderly people with companionship, **Figure 12-11**. Service animals may also help alleviate anxiety and other debilitating issues, in addition to keeping their owners from potentially harmful or hazardous situations.

Physical Benefits

Research has shown that people who own companion animals have a healthier lifestyle. Walking and playing with companion animals increases

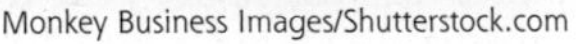

Monkey Business Images/Shutterstock.com

Monkey Business Images/Shutterstock.com

Figure 12-11. Companion animals offer psychological, physiological, and social benefits to their owners on a daily basis.

the amount of physical exercise a pet owner performs. Studies have also shown that regular interaction with pets can reduce stress, lower blood pressure, decrease heart rate, and improve cardiovascular health. Doctors often recommend pet ownership or interaction to patients with conditions that would benefit from the moderate exercise that results from feeding and exercising a pet.

Social Benefits

The main social benefit of pet ownership is the facilitation of interaction with other people. People will usually stop and speak to one another if their pets show a mutual interest. Public dog parks allow pets to freely interact and allow their owners the opportunity of social interaction. Service animals also allow freedom and socialization to those who may not otherwise be able to get around by themselves.

Choosing Companion Animals

Whether you are choosing a companion animal to keep as a pet or to breed and sell, there are many considerations you must make. Some factors to keep in mind when choosing companion animals include:

- Allergies and asthma—these conditions may be aggravated by dander, fur, and feathers. You must consider your own physical well-being, as well as that of others living in the same household.
- Stress—undesirable behaviors such as constant barking may induce stress in the owner, in other members of the household, and of nearby neighbors.
- Physical well-being—pets that are too large or strong for you to handle when they are full grown may cause you or others physical harm. This is especially a concern for older owners who may be tripped or knocked down.

Did You Know?

There are no 100% hypoallergenic domestic cats, but there are some breeds that produce less of the protein (Fel d 1) that is thought to be the culprit of most cat allergies.

Other factors to keep in mind when choosing companion animals include: breed characteristics (especially temperament); physical, housing, and exercise needs; breeding and birthing issues; known health issues; and licensing and ownership restrictions. Although there are exceptions to the rule, most breeds are true to their heritage, and you should carefully research any animal you are considering for purchase. This holds true for purebred animals as well as crossbred animals. Careful research early on will save you frustration and expenses in the long run. The most common types of companion animals include dogs, cats, rabbits, cavies, reptiles, birds, fish, and horses.

Dogs

Dogs are the most common companion animal in the United States. It is estimated that there are approximately 83 million dogs in U.S. households. Today's domesticated dog is descended from animals existing more than 15,000 years ago. Today's dogs vary greatly in size, temperament, appearance, and composition. They range from 2 lb to over 200 lb, and can be small enough to fit in your hand, or large enough to look you in the eye when standing on their hind legs, **Figure 12-12**. Some have thick, long furry coats that shed constantly while others are referred to as hairless, with little, if any hair. The scientific term *Canis lupus familiaris* or *Canis familiaris*, refers to today's domesticated dogs.

Eric Isselee/Shutterstock.com

Figure 12-12. Dogs come in many shapes and sizes. It is important to understand which breed will best suit your needs before acquiring any companion animal.

Gestation and Parturition

The average gestation period for most dogs ranges from 58 to 65 days. Most dogs require little help when ***whelping*** (giving birth), whereas some dogs require help delivering, cleaning, and stimulating their puppies. It is not uncommon for many smaller dogs to require caesarian sections to safely deliver their puppies. Although dogs usually welcome their litters and take good care of the pups, some will refuse to care for their offspring and may even kill the puppies. This is more common in first-time mothers, so extra care and attention must be given to those birthing first litters.

Life Expectancy

The life span of most dog breeds is between 10 and 13 years. However, some breeds may average between 15 and 20 years. It is reported that smaller dogs tend to live longer than larger dogs, but longevity is greatly affected by lifestyle and good healthcare.

Breed Classification

Dog breeds are divided into seven major groups: sporting, hounds, terriers, working, herding, toy, and nonsporting. A dog breed fits into these divisions according to its use and size.

Sporting Breeds

Sporting dogs are very active and alert. They are excellent for people who enjoy the outdoors, hunting, and other field activities. These breeds require space to run and plenty of exercise. Their coats vary from smooth easy-care fur to longer, thicker fur that requires grooming. Sporting breeds include retrievers, setters, pointers, and spaniels. Common breeds found in the United States are Labrador retrievers, Irish setters, and Brittany spaniels, **Figure 12-13**.

Kirk Geisler/Shutterstock.com

Figure 12-13. Sporting dogs enjoy working to please their handler. They require a great deal of structured discipline and plenty of exercise.

cynoclub/Shutterstock.com

Irina oxilixo Danilova

Figure 12-14. Most people do not picture the elegant Afghan or adorable Dachshund as hound dogs. The Afghan was originally used for hunting large prey in the deserts and mountains of Afghanistan, and the Dachshund was bred to scent, chase, and flush out burrow-dwelling animals.

Hounds

Most hound breeds have been developed to make excellent hunting dogs. They are also used as search animals because of their acute scenting abilities to follow a trail. Many people picture a Basset or Bloodhound when thinking of hound dogs, but hounds vary greatly in size, temperament, and appearance. Most hound dogs have smooth, easy-care coats. Common breeds of hounds are Beagles, Basset Hounds, Bloodhounds, Black and Tan Coonhounds, Afghan Hounds, Dachshunds, and Whippets, **Figure 12-14**.

Terrier Breeds

Terriers range from fairly small to fairly large dogs. Terriers are known for their energetic and feisty personalities. Terriers do not like other animals, and often do better as a single family pet. Their ancestors were initially bred to hunt and kill vermin. Most terriers have wiry hair that requires extra grooming. Common breeds of terriers are Russell Terriers, Rat Terriers, Fox Terriers, Welch Terriers, and Bull Terriers, **Figure 12-15**.

Working Breeds

Dogs in the working group are typically larger animals that require a structured home life and firm, consistent training. Working dogs have been bred to perform a variety of tasks, including guarding, pulling sleds, and rescuing people. As pets, they require a lot of exercise, mental stimulation, and human interaction. Some have easy-care coats, but many have thick fur that requires consistent grooming. Working dogs have proven to be valuable in our society because of their strength and intelligence. Common breeds of working dogs are Boxer, Doberman Pinscher, Rottweiler, and St. Bernard.

Herding Breeds

Herding dogs are raised for herding and protecting livestock against predators. Herding dogs have the unique ability to move other animals like sheep and cattle in a group. At times, they will use their herding instincts on people, especially small children. Herding dogs are very intelligent and make excellent pets because they are responsive to training. They also require consistent exercise, mental stimulation, and human interaction. Their coats vary from short, easy-care fur to long, thick fur that requires regular grooming. Common breeds of herding dogs are Australian Cattle Dog (Blue or Red Heeler), Australian Shepherd, Border Collie, German Shepherd, and the Pembroke Welsh Corgi.

Eric Isselee/Shutterstock.com

Figure 12-15. Terriers range in size from toy Yorkshire Terriers, to spunky Russell Terriers, to large Staffordshire Terriers. Although they are very different in size and appearance, most terriers require ample attention and exercise.

Toy Breeds

Toy dog breeds are very small in size and work well as pets in urban settings. They are popular in large cities where space is limited and apartment living is common. Toy breeds have their own unique personalities and can be very tough given their small stature. Common toy breeds are Chihuahua, Pekingese, Pomeranian, Poodle, Pug, and Yorkshire Terrier, **Figure 12-16**.

E.J. Schreiner

Figure 12-16. Toy breeds like the Chihuahua are popular in both city and rural homes. They are loyal animals and often bond with one member of the family.

Did You Know?

The American Kennel Club (AKC) has excellent resources available for researching group and breed recognition.

Non-sporting Breeds

The non-sporting group of dogs is a catchall for breeds that do not fit under any of the other major categories. The breeds in this group have varied personalities, shapes, sizes, and temperaments. This group includes breeds like the Bulldog, Chow Chow, and Dalmatian.

Cats

It is estimated that there are more than 95 million cats kept as pets in the United States. Although dogs are found in more households, cats outnumber dogs in total number of pets because most cat owners have more than one cat. It is estimated that cats have been semidomesticated and living with humans for around 9,000 years. Scientists claim cats are only semidomesticated because they share almost all their genes with their wild counterparts. Domestic cats still have many of the hunting, sensory, and digestive traits as their wild kin. The human influence on cat evolution is most likely limited to fur color and pattern and, to some extent, their tameness. Domestic cats are all members of the species *Felis catus*.

It is surmised that cats were welcomed into human communities because of their ability to kill mice and other rodents. This was especially welcome in ancient times. Cats are still kept around homes and farms, as well as around grain storage facilities to control rodent populations. Many farms and stables have what they refer to as "barn cats" for this purpose. For the early Egyptians, cats went beyond being pets. The early Egyptians worshipped a cat goddess and even mummified their pet cats for their journey to the next world.

Did You Know?

Much like their wild kin, cats sleep from 12 to 16 hours per day. In the wild, they store up energy that enables them to quickly sprint after their quarry.

Gestation and Parturition

The gestation period for most cats is 66 days. Cats may reach puberty between four and 12 months and can have up to three litters a year. The litter size ranges from one to six kittens, depending on the breed and health of the cat. Since female cats cycle (*come into heat* or have *estrus*) early and so frequently, it is important to have them spayed if you do not have plans to breed them. Most cats deliver their kittens without complications, and prefer little interaction during the birthing process. As with most animals, first-time mothers should be closely observed before, during, and after birthing her kittens.

Life Expectancy

Cats who live indoors have a life expectancy of 12 to 20 years, whereas outdoor cats have a life expectancy of one to five years. There are many environmental issues which play a role in the decreased lifespan for outdoor cats.

Breed Classification

Most cats are very clean and quiet. They are highly intelligent and can be trained easily to use litter boxes in a home setting. There are more than 60 breeds of cats recognized in the United States. Although there is less breed variation in cats than in dogs, the breeds do vary in temperament, making some more or less suitable to certain home situations. It is best to research breeds before deciding on which type you would like to own or breed. Also like dogs, their fur ranges from long to nonexistent. Most cat breeds are named after a region of origin. Cats are commonly divided into two groups, longhair breeds and shorthair breeds.

Longhair Breeds

Longhaired cats are very popular in the United States as pets and for show. They do require more grooming and maintenance than shorthaired breeds. If left untended, a longhaired cat's fur is prone to matting and knotting, and the animal will likely ingest more hair during grooming. Most veterinarians and breeders advise against shaving cats unless it is medically necessary. It is usually very stressful for the animal and may cause the cat to lash out or even become depressed. An animal's fur insulates it from the cold and the heat, and helps it regulate its body temperature. It also serves as protection from the sun.

Longhaired cats are soft and people love to pet them; however, if you do not have the time to dedicate to grooming a longhaired cat, your better option would be to get a shorthair breed. Some popular longhair breeds are the Persian, Himalayan, Ragdoll, British Longhair, Birman, Norwegian Forest Cat, Turkish Angora, Turkish Van, Balinese, and Siberian, **Figure 12-17**.

Shorthair Breeds

Shorthair breeds may be better suited for some pet owners as they require less grooming and maintenance. They may also be a better choice for individuals who suffer from pet dander allergies. Popular shorthair breeds include the British Shorthair, Siamese, Russian Blue, Domestic Shorthair, and Sphynx. See **Figure 12-18**.

Eric Isselee/Shutterstock.com

Figure 12-17. Longhair cat breeds like the Persian and Turkish Angora have beautiful fur coats that require a lot of maintenance.

Axel Bueckert/Shutterstock.com Eric Isselee/Shutterstock.com

Figure 12-18. Shorthair cat breeds include Siamese with their distinctive markings and the hairless Sphynx.

Other Breeds

There are other classifications of cats such as curly-coat breeds, hypoallergenic breeds, and even hairless cat breeds. There are many shapes, colors, sizes, and variations of cat breeds. Finding the right breed of cat for your situation may take some research, but there are many credible sources to help you find the breed of cat for your family.

Rabbits

People raise rabbits for show, as pets, and for fur, fiber, wool, and meat production. Rabbits are clean, inexpensive to purchase and keep, and require little space. They do require a consistent supply of clean water and food and should never be fed moldy hay or alfalfa. Their ***hutches*** (cages) may be small, but they should be thoroughly cleaned on a weekly basis (minimally), **Figure 12-19**. The waste and hutch litter may be used for composting or fertilizer. Many rabbit breeders place worm trays below the hutches to raise fishing worms in the composting waste. Most rabbits can handle colder weather, but do not do well in high heat or damp environments.

Stephen Rees/Shutterstock.com

Figure 12-19. Rabbits are commonly kept in hutches like this to protect them from predators on the ground and give owners easier access.

Common Terms

The following terms are used to describe rabbits:

- ***Buck***—male rabbit
- ***Doe***—female rabbit
- ***Lapin***—castrated male rabbit
- ***Kit***—baby rabbit
- ***Litter***—group of newborn rabbits
- ***Fryer***—young rabbits used for meat (8–10 weeks old)

Maintaining Rabbit Health

The key to keeping rabbits healthy is maintaining a high level of cleanliness. Most modern hutches are designed to help make this easier. You can also ensure healthier animals by providing good ventilation, closely observing individual behavior, and quarantining new animals or those you think might be ill.

Gestation and Parturition

The rabbit gestation period averages around 31 days. The short cycle allows rabbits to breed almost continuously, unless they are prevented from doing so. If you are not breeding your rabbits, they may be spayed or neutered by a regular veterinarian. Shortly before giving birth (***kindling***), the doe will prepare her nest with fur plucked from her body. Rabbits give birth to their offspring with little if any problems, and are best left undisturbed until the young are a day old. The kits may be weaned around four weeks, and sold as fryers around eight weeks of age.

Breed Classification

Rabbit breeds are characterized by: size, shape, ears, fur texture, sheen, and color. As when choosing any livestock, the end purpose will help determine which breed will best suit your needs. If your rabbit(s) is going to be a pet, you can choose whatever appeals to you. If you decide to raise rabbits for show and/or breeding, you will want a registered or pedigreed animal. If you are interested in showing or raising purebred rabbits for sale, do your research and find a reputable breeder. Organizations such as the American Rabbit Breeders Association (ARBA) provide excellent resources and contacts. Associations such as the ARBA also usually provide a membership list when you join. Such a list may be used to find local breeders selling stock and offering breeding advice. The ARBA currently recognizes 48 unique rabbit breeds.

Commercial Production

As with commercial broiler production, you will want to use breeds that produce larger quantities of meat. Meat rabbit breeds known for rapid weight gain and large litters include New Zealand Whites, Californians, and Production Whites (commercial crosses). If you are raising rabbits primarily for fur or wool production, you might choose one of the Angora

Eric Isselee/Shutterstock.com Nikolai Tsvetkov/Shutterstock.com

Figure 12-20. Angora rabbits are beautiful but require additional efforts by the owner to ensure they do not develop matted fur or excessive hairballs caused by grooming their long fur on their own.

breeds, the Jersey Wooly, or the Chinchilla. Like longhair cats, these rabbits require regular grooming to maintain their coats, and may even require special nutrition supplements to prevent fur-growing problems. See **Figure 12-20**.

Other Small Animals

Other small animals considered companion animals include hamsters, gerbils, cavies (guinea pigs), potbellied pigs, ferrets, chinchillas, mice, rats, and tarantulas, **Figure 12-21**. Most of these small companion animals are popular because they can be kept in cages or pens that require little space. However, they still require time and attention like any other pet. It is not possible to cover each type of animal, let alone its needs, in the space of this lesson. If you are interested in raising any type of companion animal, it is always best to begin with adequate research. It is also recommended you seek the advice of experienced breeders.

Photok.dk/Shutterstock.com Irina oxilixo Danilova/Shutterstock.com Mirek Kijewski/Shutterstock.com

Figure 12-21. There are many other small animals used as companions for humans, each with their own needs and benefits. Small animals that make good companion animals include guinea pigs, chinchillas, and tarantulas. ***Which of these companion animals do you think would be best for you?***

Reptiles

Reptiles have become very popular as household pets. Approximately 11.5 million reptiles like snakes, lizards, turtles, frogs, and other amphibians are reported as companion animals in the United States. Unfortunately, many of these animals are abandoned once the novelty wears off, or the owner realizes the level of care their new pet requires. As with any companion animal you are considering raising or owning, you need to research the reptile's diet and habitat, as well as available veterinary care, *before* you purchase it. Make sure the reptile is suitable for you to keep. Know the cost and time involved in caring for the reptile, and keep the following points in mind as you consider different types of reptiles:

- Habitat—consider the amount of space you have to dedicate to the enclosure. You must also know and understand the reptiles native habitat, and be able to replicate it accurately. This includes everything from the substrate(s), plants, humidity level, and temperature range. And, as the reptile grows, you will have to increase the size of its enclosure.
- Electricity—most reptiles require a consistent source of heat. Some require a heating pad below the substrate, as well as overhead lighting. You will need to purchase the lighting equipment and bulbs on a regular basis to maintain proper temperatures.
- Feeding—each reptile has a unique diet. You must understand what the animal eats and how often it eats. Some reptiles require live or freshly killed food. If you cannot locate a supplier that will humanely kill the prey, you will have to do it yourself. You may also need to keep live worms or crickets for regular feeding, or frozen mice, rabbits, or rats. For herbivore reptiles, you must supply fresh foods specific to their natural diets. These foods must be regularly prepared in such a way as to enable the reptile to maximize its intake and digestion.
- Veterinary care—not all vets are familiar with reptile medicine. It is not unheard of for responsible reptile keepers to drive four or five hours to see a vet who specializes in or is familiar with reptile medicine.
- Handling—many reptiles are easily harmed or extremely stressed when handled improperly. People unfamiliar with handling such animals often grip too tightly or quickly release the animal when it squirms too much or defecates on their arm or hand. Reptiles must be handled regularly for taming. Careful sanitation practices must also be used to prevent transmission of zoonotic diseases.
- Cleaning and sanitation—keeping the enclosure clean and sanitary is extremely important to maintaining the health of reptiles. You will need to regularly change the substrate and have a regular supply of cleaning supplies such as sponges, disinfectants, and rubber gloves. You will also need a holding enclosure to keep your reptile while you clean its regular enclosure.
- Life expectancy—some reptiles have short life spans, whereas others may live for 10 to 150 years. Remember, most well-kept captive animals tend to live longer than their wild counterparts.

Some of the most common reptiles kept as pets or companions are lizards, snakes, turtles, and tortoises. See **Figure 12-22**. It is important to

mashe/Shutterstock.com

Olesia Bilkei/Shutterstock.com

Figure 12-22. Most experts do not recommend having turtles and tortoises as pets because their enclosure setup and care are much more complicated than most people understand. Larger lizards require large enclosures. Many also have sharp claws and spiny, powerful tails that will easily scratch your skin, no matter how tame the animal.

purchase or obtain captive-bred reptiles. Most wild-caught reptiles have internal and external parasites, respiratory infections, and even shell infections. If you decide to raise reptiles, visit reputable breeders before investing your time and money. There are also many reptile rescue operations in existence looking for good homes for abandoned pets.

Did You Know?

Birds are highly sensitive to air quality, and exposure to some toxic inhalants can cause premature or immediate death.

Birds

It is estimated that more than 20 million birds are kept as pets in the United States. People often keep birds as companion animals because they have limited space or resources for other, higher maintenance animals. Some smaller bird species can be housed in cages the majority of the time, and require very little human interaction, especially if they have a companion, **Figure 12-23**. Other bird species are extremely intelligent and very demanding of time and attention. If you are contemplating owning or raising birds, it is very important to understand the physical and emotional needs of the species you are considering. Before you decide to invest in bird ownership, research the species to understand its specific needs, and keep the following general information in mind:

- Time commitment—many birds require time for social interaction. Leaving a bird caged for extended periods of time will often lead to behavioral problems that are difficult to break. Many of these problems are self-destructive and include feather plucking or self-mutilation.
- Habitat—unlike reptiles, you do not need to replicate a bird's natural habitat in order for the bird to thrive. You should, however, provide the largest cage possible, and provide plenty of toys for stimulation. Most bird owners keep the cage open so the bird can enter and exit at will, and enclose the bird only at night. These birds may or may not have clipped wings. Many of the larger birds quickly destroy toys and require a constant supply of new items. Birds with clipped wings need ample space and climbing structures in order to get adequate exercise.
- Feeding—some birds thrive on commercial bird foods, whereas others require varied diets including fresh fruit and vegetables on a daily basis. Many birds enjoy throwing food they do not like or care to eat.

Pichugin Dmitry/Shutterstock.com

Figure 12-23. Some breeds of bird do much better when kept with other birds. Research the type of bird you are getting to determine if buying more than one will increase the happiness of your animal(s).

- Veterinarian care—not all vets are familiar with avian care. It is highly advisable to locate an avian vet before you purchase your bird(s), and establish a relationship with the vet. Annual well-bird examinations are highly recommended to maintain your bird's health.
- Handling—most captive-bred species do not mind being handled. However, you must usually establish trust over time, especially if you have acquired an older bird. Speak with your vet and fellow breeders to learn how best to hold your birds without stressing or harming them. Birds, especially the larger species, have sharp claws and strong beaks, and may need to be handled with gloves to prevent personal injury.
- Cleaning and sanitation—many species are messy and like to throw food and other items for their own amusement. Expect to clean cages on a daily basis and replenish water regularly. Careful hand washing is also recommended after handling birds to prevent the transmission of zoonotic diseases such as avian tuberculosis.
- Life expectancy—with proper care, some birds may live only a few years, whereas others may live to be more than 50 years old.
- Allergies and asthma—birds continually shed feather dust (particles of feathers) that may aggravate allergies or asthma in some people.

As with any companion animals, make sure you are dealing with a reputable breeder and purchasing healthy, legal birds. You should also consider obtaining your bird through a reputable bird rescue group.

There are many types of birds suitable for indoor cage rearing, including finches, parakeets, cockatiels, and quakers and conures. Larger parrot species include Amazons, Macaws, African Greys, and Cockatoos.

Kuttelvaserova Stuchelova/Shutterstock.com

Figure 12-24. Zebra Finches are a favorite household bird. They require minimal expertise to raise successfully.

Finches

Finches are one of the most common cage birds, **Figure 12-24**. They are easy to raise and an affordable choice for people who want to raise birds. They do not like to be handled, but are an easy, low maintenance indoor bird.

Parakeets

Parakeets, also known as budgies, make excellent caged pets, **Figure 12-25**. They are very colorful, friendly, and can be inexpensive to purchase. They

panbazil/Shutterstock.com

Figure 12-25. Parakeets are a common breed of bird. Their colorful feathers and curious personalities make them enjoyable to watch.

become attached to their owner, but can be difficult to tame at first. Male parakeets can learn to mimic familiar sounds and even a word or two.

Cockatiels

Cockatiels are lively and have a unique personality. They are large enough in size that look like a real parrot. Cockatiels can be the ideal first bird for pet owners because they are affordable and they have the ability to talk, **Figure 12-26**. Males can mimic familiar sounds, whistle songs, and even say words and phrases. They are very intelligent birds that require more out-of-cage time than parakeets.

Antonio Guillem/Shutterstock.com

Figure 12-26. Cockatiels enjoy socializing with humans, making them great companions.

Quakers and Conures

Quakers and conures are medium size parrots known for their ability to mimic phrases, **Figure 12-27**. They are excellent talkers and beautiful birds, but are needy and will crave human interaction. They are more expensive than parakeets and cockatiels and will require about three hours per day of out-of-cage time. Quakers and conures can live up to 35 years.

Lorcel/Shutterstock.com

Figure 12-27. Conures are smaller than large breeds of parrots and easily mimic the sounds of the human voice. *Have you ever had a conure mimic you?*

Large Parrot Species

There are several types of big parrots that can be household pets. Amazons, Macaws, African Greys, and Cockatoos are all large parrots that can be household pets, but require much more attention and commitment than the smaller varieties of birds, **Figure 12-28**. These birds become very attached to one person and can become temperamental and even destructive in your home. These parrots are beautiful birds; they are the most expensive of all companion birds and can outlive humans. The commitment and patience of the owners must be high to properly keep a large parrot.

Aquatic Pets

There are 145 million freshwater fish and 13.6 million saltwater fish kept as companion animals in the United

Eric Isselee/Shutterstock.com

Nature Art/Shutterstock.com

Figure 12-28. These large parrots often live longer than their owners and require special care to ensure they receive the correct diet, housing, and socialization to live a long and productive life.

States. There is debate about fish being classified as companion animals because the interaction between humans and fish is limited at best. Fish, however, do require consistent care to maintain their environments and their health.

The initial investment for raising fish may be quite expensive. It is better to begin with a small tank and several inexpensive fish. As you learn to care for your aquatic companions, slowly expand your investment. You will need to decide if you want to raise saltwater or freshwater fish, or warm or cold water fish. Each type of environment requires specific equipment and maintenance, and each species has specific environmental and nutritional needs. You will also need to learn about diseases and treatment and the reproductive processes of each species should you plan on breeding.

Some popular pet fish include goldfish, tetras, angelfish, bettas, clown fish, koi, and guppies to name a few, **Figure 12-29**. There are hundreds of types of fresh and saltwater fish that are appropriate for aquarium raising.

FamVeld/Shutterstock.com

Figure 12-29. Fish kept as companion animals are a multimillion dollar industry. People spend thousands of dollars on elaborate tanks and exotic fish to have in their homes and offices. *Do you have a favorite type of exotic fish?*

Horses

There is great debate in our country over whether horses are considered companion animals or livestock. With approximately 8.3 million horses in our country being reported as companion animals, the case can be made that horses are truly companion animals. They bring enjoyment, recreation, and companionship to many people in the United States and around the world. Horses are being more widely used as therapeutic animals for the disabled and even miniature horses have been approved in the Americans with Disabilities Act (ADA) as service animals. Refer to *Lesson 10.3* for more information about horses.

Careers in the Companion Animal Industry

As the companion animal industry continues to grow and expand, new career opportunities continue to arise. The pet care industry employs people who run pet boarding services, veterinarian clinics, animal shelters, pet stores, pet training facilities, and pet grooming services. People involved in these lines of work do things like feed and care for animals, bathe, brush, and groom pets, and perform basic sanitation tasks associated with running a pet care facility. People are also needed for training and working with therapy dogs (facility therapy dogs, animal-assisted therapy dogs, therapeutic visitation dogs).

The pet food industry is a multibillion dollar industry with careers in development, manufacturing, distribution, sales, and marketing. Entrepreneurial opportunities include everything from dog walking services to specialty pet boutiques.

In addition, research facilities require a consistent supply of laboratory animals for use in biomedical sciences.

FFA Connection Veterinary Science CDE

National FFA has a Career Development Event called Veterinary Science. This event prepares students for careers in the field of veterinary science, whether as a veterinarian or a veterinarian technician. Students learn clinical procedures, anatomy and physiology of animals, veterinarian medical terminology, identification of animals and symptoms, and safety precautions used in dealing with animals. For more information, contact your FFA advisor.

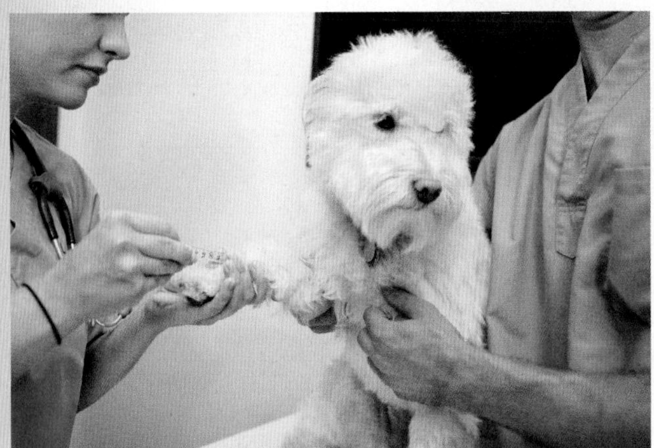
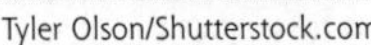

Tyler Olson/Shutterstock.com

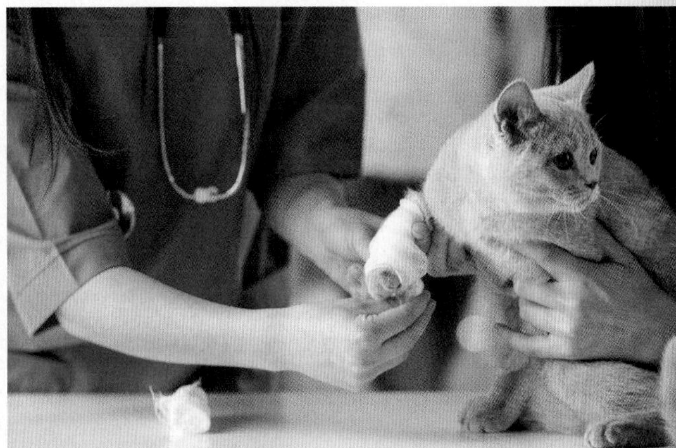

VGstockstudio/Shutterstock.com

Lesson 12.1 Review and Assessment

Words to Know

Match the key terms from the lesson to the correct definition.

1. A female rabbit.
2. A castrated male rabbit.
3. Any animal that has been tamed from its wild form to a domesticated form for the benefit of humans.
4. The animals that live, play, and work with people on a daily basis.
5. An animal living in the wild but descended from its domesticated counterparts.
6. The act of a dog giving birth.
7. A young rabbit raised for meat.
8. The act of humanely putting an animal to death through chemical means, or allowing it to die by withholding extreme medical measures.
9. The removal of the ovaries and/or uterus (ovariohysterectomy) of a female animal to prevent breeding and reproduction.
10. A male rabbit.
11. The castration of a male animal to prevent breeding and reproduction.
12. A type of cage typically designed for keeping rabbits.
13. An animal that lives in a natural setting, must get food and water on its own, and is not cared for by humans.
14. The act of a rabbit giving birth.
15. A baby rabbit.
16. Diseases and infections that can be transmitted from animals to humans.
17. A group of newborn rabbits.

A. buck
B. companion animal
C. doe
D. domesticated animal
E. euthanasia
F. feral animal
G. fryer
H. hutch
I. kindling
J. kit
K. lapin
L. litter
M. neuter
N. spay
O. whelping
P. wild animal
Q. zoonoses

Know and Understand

Answer the following questions using the information provided in this lesson.

1. Identify three ways an animal must differ from its wild counterpart in order to be considered domesticated.
2. *True or False?* The dog was domesticated roughly at the same time as the earliest crop domestication.
3. From your own point of view, why is it best *not* to capture and attempt to raise wild animals?
4. What are three actions you can take to prevent the transmission of zoonotic diseases?
5. What problems are caused by feral dogs and feral pigs?

6. How do pet adoptions contribute to the economy?
7. How do legal restrictions and obligations affect animal ownership?
8. Why is it important to find a vet who specializes in exotic pets if you plan on raising reptiles or birds?
9. List three known benefits of companion animals to humans.
10. What are five factors that should be considered when choosing companion animals?
11. What are the seven major groups of dog breeds? What basic criteria is used to determine which group a breed belongs to?
12. Why do some scientists claim cats are only semidomesticated?
13. What steps can you take to ensure healthy rabbits?
14. What are three things you should consider *before* you purchase any type of reptile?
15. Why is it important to understand the physical and emotional needs of the bird species you are considering owning?

Analyze and Apply

1. Shadow a veterinarian or a veterinarian technician to see the work he or she performs on a daily basis. Is this a position you would like? Report your findings to your classmates and teacher.
2. Research zoonoses and identify at least five diseases transmitted from animals to humans. Can animals contract diseases from humans?

Thinking Critically

1. Many people will list dozens of reasons they have companion animals. What are three good reasons to *not* own a particular type of pet, or any pet at all?
2. Consider the following practices commonly used with companion animals: spaying and neutering, declawing, docking ears and tails, and preserving thoroughbred lines. Are these practices morally acceptable? Explain your answer.

Lesson 12.2
Aquaculture Industry

Lesson Outcomes

By the end of this lesson, you should be able to:

- Understand the size and the scope of the aquaculture industry.
- Understand common terminology of the aquaculture industry.
- Describe the components of the aquaculture industry.
- Describe different types of fish and aquatic animals in the United States.
- Explore career opportunities in the aquaculture industry.

Words to Know

aeration
anadromous
aquaculture
concentrated aquatic animal production (CAAP)
crustaceans
dissolved oxygen (DO)
fingerling
fishery
fish farming
freshwater aquaculture
fry
hatchery
intensive recirculating aquaculture system (IRAS)
mariculture
marine aquaculture
marine cage
net pen
pond
raceway
salinity
silo
spawn
stocking rate

Before You Read

Aquaculture is a much larger industry worldwide than in the United States. Do some research to determine why aquaculture is so important to our global society.

You may be wondering what aquaculture has to do with agriculture. After all, fish are raised in water and not on farmland. And, do people really "farm fish"? As a matter of fact, today's aquaculture is considered a form of farming that not only contributes to food production, but also plays a major role in the preservation of many aquatic plant and animal species. Aquaculture also contributes greatly to the commercial fishing industry.

This lesson briefly discusses the different types of aquaculture, facilities, and aquatic life raised through aquaculture, as well as the benefits and problems associated with the farming of aquatic plants and animals.

Global and U.S. Aquaculture Production

The leading countries in aquaculture production are (in order): China, India, Vietnam, Indonesia, and Thailand, **Figure 12-30**. The continent of Asia accounts for 89% of the total world aquaculture production with China being responsible for 62% of that total production.

Even though the aquaculture industry is alive and well in the United States (U.S. aquaculture production contributes about $1.2 billion to our economy annually), it only meets 5% to 7% of the U.S. demand for seafood. That means more than 90% of the seafood we eat in the United States comes from other countries. The most common seafood items imported into the United States are: shrimp, Atlantic salmon, tilapia, and shellfish (scallops, mussels, clams, and oysters). Asian countries supply the majority of the imported shrimp, and Canada, Norway, and Chile supply most of the imported Atlantic salmon.

Country	Harvested Tons	Country	Harvested Tons
China	32,414,084	Egypt	539,748
India	2,837,751	Myanmar	474,510
Vietnam	1,437,300	United States	471,958
Indonesia	1,197,109	South Korea	436,232
Thailand	1,144,011	Taiwan	304,756
Bangladesh	882,091	France	258,435
Japan	746,221	Brazil	257,783
Chile	698,214	Spain	221,297
Norway	656,636	Italy	180,943
Philippines	557,251	Malaysia	175,834

Goodheart-Willcox Publisher

Figure 12-30. Aquaculture is practiced throughout the world. These are the top 20 producers in the world.

Did You Know?

The United States ranks third in the world in seafood consumption behind China and Japan. The average American consumer eats approximately 16 lb of fish and shellfish per year. Americans, as a whole, consume nearly 5 billion lb of seafood per year.

Types of Aquaculture

According to the National Oceanic and Atmospheric Administration (NOAA), ***aquaculture*** is defined as the breeding, rearing, and harvesting of plants and animals in all types of water environments including ponds, rivers, lakes, oceans, and land-based facilities. The acronym ***CAAP (concentrated aquatic animal production)*** may also be used when describing aquaculture production. The term ***fishery*** may also be used to describe different types of aquaculture operations.

Aquaculture is sometimes simply referred to as ***fish farming***. However, that reference is not always accurate as aquaculture producers farm many types of freshwater and marine species of fish, shellfish, and plants. Aquaculture produces food fish, sport fish, bait fish, ornamental fish, crustaceans, mollusks, shellfish, algae, sea vegetables, and fish eggs. Some also consider the production of alligators, crocodiles, turtles, and other amphibians as part of the aquaculture industry, **Figure 12-31**. Each one of these aquatic species has specific environmental requirements, and producers must closely mimic their natural environments to have successful production systems.

Eve Wheeler Photography/Shutterstock.com

Figure 12-31. Some consider the production of alligators as part of the aquaculture industry.

The two main divisions of aquaculture are marine aquaculture and freshwater aquaculture.

Marine Aquaculture

Marine aquaculture is the production of species of aquatic plants and animals that live in the ocean (saltwater). In the United States, marine aquaculture produces mainly oysters, clams, mussels, shrimp, and salmon. Marine aquaculture may take place in constructed on-land systems or in the ocean in containment systems, **Figure 12-32**. The farming of marine organisms may also be referred to as ***mariculture***.

The leading states for marine aquaculture are (in order): Maine, Washington, Virginia, Louisiana, and Hawaii. Some U.S. marine producers also raise species such as cod, yellowtail tuna, and seabass. Ongoing research and continued advances in technology and management practices are continually expanding U.S. aquaculture's potential role in producing more species on a commercial basis.

Vladislav Gajic/Shutterstock.com

135pixels/Shutterstock.com

Figure 12-32. Marine aquaculture commonly uses floating cages anchored in coastal waters to contain fish. Oysters may also be cultivated in contained environments in coastal waters. ***Can you see the benefits of using natural environments to cultivate marine life?***

Freshwater Aquaculture

Freshwater aquaculture is the production of aquatic life that is native to rivers, lakes, and streams, or spends the majority of its life in freshwater. In the United States, freshwater aquaculture for food production is dominated by catfish farming. Trout, bass, and tilapia are also raised in freshwater aquaculture facilities in the United States. Freshwater aquaculture may take place in constructed on-land systems or in freshwater habitats such as ponds or lakes, **Figure 12-33**. The top states for freshwater aquaculture production are Mississippi for catfish and Idaho for trout.

Did You Know?

Oysters, clams, and mussels account for about two-thirds of total U.S. marine aquaculture production, according to NOAA.

Aquaculture Systems

Aquaculture production systems may be as simple as a few tanks and a filter in someone's barn, or as large as those found along the coasts of many countries. Indoor and outdoor production systems share many common components, but each also have design and operation issues specific to their environment.

Indoor Production Systems

Indoor production of fish has gained popularity over the last few decades. Many aquaculture producers have chosen indoor systems because

Pan Xunbin/Shutterstock.com

Artens/Shutterstock.com

Figure 12-33. Freshwater aquaculture may use on-land containers or freshwater habitats such as ponds or lakes. *What are the advantages of each system? What are the disadvantages?*

of the amount of environmental control it allows. The producer can more easily prevent and control disease and pest problems, as well as ensure the optimum distribution of food and other nutrients. The producer can also more readily isolate or quarantine specimens that are new or may be exhibiting symptoms of disease. Indoor systems are used primarily for fish hatcheries, to raise ornamental fish, and to raise fish for consumption. Indoor production systems may be freshwater or marine systems.

Containment Structures

Indoor aquaculture production systems must have containment systems capable of holding and interchanging the water in which their "product" can live and reproduce. The type of containers used depends on the species being cultivated. For example, oysters and mussels require a different environment than tilapia or trout. Marine systems also require different types of containers than freshwater due to the corrosive nature of saltwater. The producer must understand the specific needs of the species he or she intends to raise before investing in containment structures.

The containment structures most common in indoor production systems include round tanks or raceways. Round tanks are commonly used in hatcheries due to the natural self-cleaning action. As the water swirls around the tank, solids are drawn to the center where the outlet is typically located. Round tanks are commonly made of fiberglass, steel, or concrete. These aboveground containers are also referred to as ponds. Raceways are straight-sided channels typically made of poured concrete or concrete blocks.

Hatcheries

Hatcheries are production facilities focused on breeding, nursing, and rearing fish, invertebrates, and aquatic plants to fry, fingerling, or juvenile stages, **Figure 12-34**. Many hatcheries are indoor production systems to enable better environmental control for the more vulnerable young.

Fry are young fish ready to begin eating on their own. ***Fingerlings*** are small, young fish under one year. Hatcheries supply aquaculture production with seed fish and other aquatic species to replenish their stock. Hatchery stock is also vital to rebuilding oyster reefs, increasing wild populations, and rebuilding threatened and endangered species.

siripong panasonthi/Shutterstock.com

Figure 12-34. The tilapia fry from this hatchery will supply seed fish to various fish farms. Aquaculture hatcheries provide the same type of services as livestock operations that raise seedstock animals.

Outdoor Production Systems

Outdoor aquaculture production systems include small-scale, freshwater fishing pond production and large coastal marine production systems. Outdoor production systems may cultivate fish, shellfish, mollusks, as well as aquaculture plants such as seaweed. In some situations, outdoor aquacultural production may be combined with crop production, as is the case with crawfish grown in flooded rice fields. Outdoor production systems may use various types of containment, depending on the species and whether the system is freshwater or saltwater. The types of containment used include:

- ***Ponds***—ponds are bodies of standing water that are usually smaller than a lake. A pond may be a natural or built structure. Large, aboveground aquaculture tanks may also be referred to as ponds.
- ***Raceways***—are straight-sided channels typically made of poured concrete or concrete blocks. One of the biggest advantages of raceway tanks is the efficiency of water exchange. Raceways may be constructed above- or belowground, **Figure 12-35A**.
- ***Silos***—deep round aquaculture tanks that are usually belowground.
- ***Net pens***—mesh enclosures suspended in a body of water from a rigid structure on the water's surface. Net pens allow the water to pass freely between the fish and surrounding water resource, **Figure 12-35B**.
- ***Marine cages***—aquaculture enclosures with rigid frames on all sides. These cages are suspended in a body of water. Marine cages also allow water to pass freely between the fish and surrounding water resource.

somrak jendee/Shutterstock.com Eugene Sergeev/Shutterstock.com

Figure 12-35. A—Raceway on a trout farm; B—Net pen/marine cage.

- Ropes, lines, and stakes—methods used to cultivate shellfish. The ropes or lines are suspended in deeper waters from rafts or buoys. The stakes are inserted into the seabed in intertidal areas. The shellfish cling to the ropes or stakes. See **Figure 12-36**.

Indoor and outdoor aquaculture systems share many issues. These issues include water quality and availability, waste accumulation, dissolved oxygen levels, water temperature, disease and pest control, and mechanical systems. The largest variant between indoor and outdoor aquaculture systems is the amount of control a producer may have over these issues.

Water Quality

Water quality is one of the most important aspects of any aquaculture operation as most cultivated species depend on water to breathe, feed, grow, excrete wastes, and reproduce. The quantity and quality of water available

VTT Studio/Shutterstock.com TTphoto/Shutterstock.com

Figure 12-36. Shellfish are cultivated in numerous ways. These ropes are suspended in coastal waters for mussels and other shellfish larvae to settle on and grow. Oysters may be raised in beds or cages located in tidal waters.

is often the determining factor for the location of an aquaculture facility. Common sources of water for indoor and outdoor operations are rivers, lakes, ponds, springs, wells, groundwater, and municipal water. Depending on the location and the species being cultivated, marine operations may use freshwater sources (with treatment) or water from the ocean.

Contained Systems

Maintaining water quality is one of the largest challenges for indoor freshwater and marine systems as well as outdoor systems using tanks or a series of ponds. Both freshwater and marine systems must regulate water temperature, water flow, and the filtering of waste, as well as maintain proper water levels throughout the system. There must also be a ready supply of clean or prepared water. Marine systems must also regulate and maintain the proper salinity of the water. ***Salinity*** is the dissolved salt content of the water.

Water treatment may be expensive. If the water requires extensive processing before it is suitable for aquaculture production, it may be difficult to establish a profitable operation.

Open-Water Systems

Open-water systems include those in coastal waters as well as in freshwater lakes, ponds, and rivers, **Figure 12-37**. The fish, shellfish, mollusks, or plants being raised are contained within the body of water and use the natural ecological cycle for much of their sustainment. Producers have little or no control of the water source in open-water systems. The aquatic life being cultivated is exposed to both natural and human contaminants, including disease, pests, predators, chemicals, and other pollutants.

sezer66/Shutterstock.com

Figure 12-37. Coastal aquaculture systems are common in tropical waters.

Freshwater Ponds

Freshwater ponds, such as those found on farmland, are usually fed by some source of underground or aboveground water source and/or through runoff. The water quality of a pond is not easily controlled and it may accumulate pesticides, herbicides, and fertilizers from surrounding cropland. A pond may also accumulate fecal contamination from livestock manure if livestock defecate around the pond or directly in the water. The water in ponds surrounded by croplands and accessible to livestock must be closely monitored and possibly treated to offset potential problems. Ponds used for aquaculture must also be able to maintain a suitable water level.

Water Temperature

Aquatic organisms take on the approximate temperature of their environment and very few will tolerate rapid or extreme changes in temperature. Indoor aquaculture systems and those with aboveground containment structures must have an effective means of maintaining the proper water temperature. Open-water systems are subject to many uncontrolled factors that may affect water temperature. These factors include fluctuating day and night temperatures, rainfall, runoff water temperature, and water depth.

Different species thrive naturally in specific water temperature and it is essential the producer understand the specific temperature needs of the species he or she is cultivating. Water temperature affects many aspects of aquaculture production, including oxygen, carbon dioxide, and nitrogen levels in the water. It will also affect the growth of bacteria and aquatic pests, as well as the behavior, feeding, growth, and reproduction of many aquatic species.

The means used to control water temperature will depend on the facility and the species being raised. Indoor production systems must also maintain the temperature of the facility itself.

Did You Know?

Low-dissolved oxygen levels are responsible for more fish kills, either directly or indirectly, than all other problems combined.

Dissolved Oxygen

The simplest definition of ***dissolved oxygen (DO)*** is that it is the amount of gaseous oxygen (O_2) dissolved in the water. Dissolved oxygen levels are an important assessment of any body of water intended to support living organisms. A dissolved oxygen level that is too high or too low can harm aquatic life and affect water quality. The aquaculture producer must ensure the DO level is properly maintained regardless of the type of production system he or she is using.

Aeration

Oxygen enters water by direct absorption from the surrounding atmosphere, through rapid movement of the water, and through plant photosynthesis. In most natural waters, the oxygen supply exceeds the amount used, and the aquatic life does not have difficulty obtaining enough oxygen. In aquaculture production systems, however, there may be more aquatic life consuming the oxygen faster than it can be replenished. Most aquaculture producers will need to use a means of mechanical ***aeration*** (adding oxygen to the water) to maintain the oxygen saturation of the water. This holds true for many open-water systems as the natural cycle may not generate enough DO for the contained aquatic life. If the dissolved oxygen level remains low for an extended period, the aquatic life will consume less feed, grow more slowly, be more susceptible to disease, and possibly suffocate and die.

Types of mechanical aeration include agitators, paddlewheels, airlift pumps, liquid oxygen injection, and air diffusers, **Figure 12-38.** Aquaculture producers must understand the specific oxygen needs of the species they are cultivating in order to maintain proper levels.

smuay/Shutterstock.com

Figure 12-38. Mechanical aeration is often needed to increase oxygen saturation of the water. Floating aeration turbines are commonly used on larger bodies of water.

Wastewater

All aquatic organisms excrete waste. In contained systems, this waste must be removed regularly to keep the water clean and minimize high levels of ammonia and other contaminants. For the most part, open-water systems use the natural ecology to control waste, especially in coastal systems. Although the natural ecosystem helps control waste, actions such as overstocking may upset the natural cycle and the producer may need to intervene. Close monitoring and testing of water quality will indicate the need for intervention and treatment.

Filtration and Disposal

Indoor aquaculture facilities cannot freely discharge wastewater into the water supply, nor can they freely discharge onto land. Aquaculture wastewater contains fecal matter, urine, food particles, and nutrients. It may also contain chemicals used to prevent disease or pest problems and those used to treat aquatic diseases. All these elements could be detrimental to the environment if wastewater were discharged into the water supply before treatment. For this reason, the EPA has established guidelines for aquaculture wastewater disposal.

Many facilities using intensive recirculating aquaculture systems (IRAS) use filtration systems similar to those used in municipal waste systems. An ***intensive recirculating aquaculture system (IRAS)*** is an aquaculture system that filters water from the tanks so it can be reused within the tanks. The design and functions of a production system are determined by the species being cultivated as well as the type of facility. For example, marine species and freshwater species excrete highly different

STEM Connection

The direction water, ammonia, and salt move into and out of marine fish and freshwater fish is very different. This movement affects the waste content of the water in aquaculture systems. Marine fish drink large amounts of water and excrete small amounts of concentrated urine. Freshwater fish do not drink water, but excrete large amounts of diluted urine.

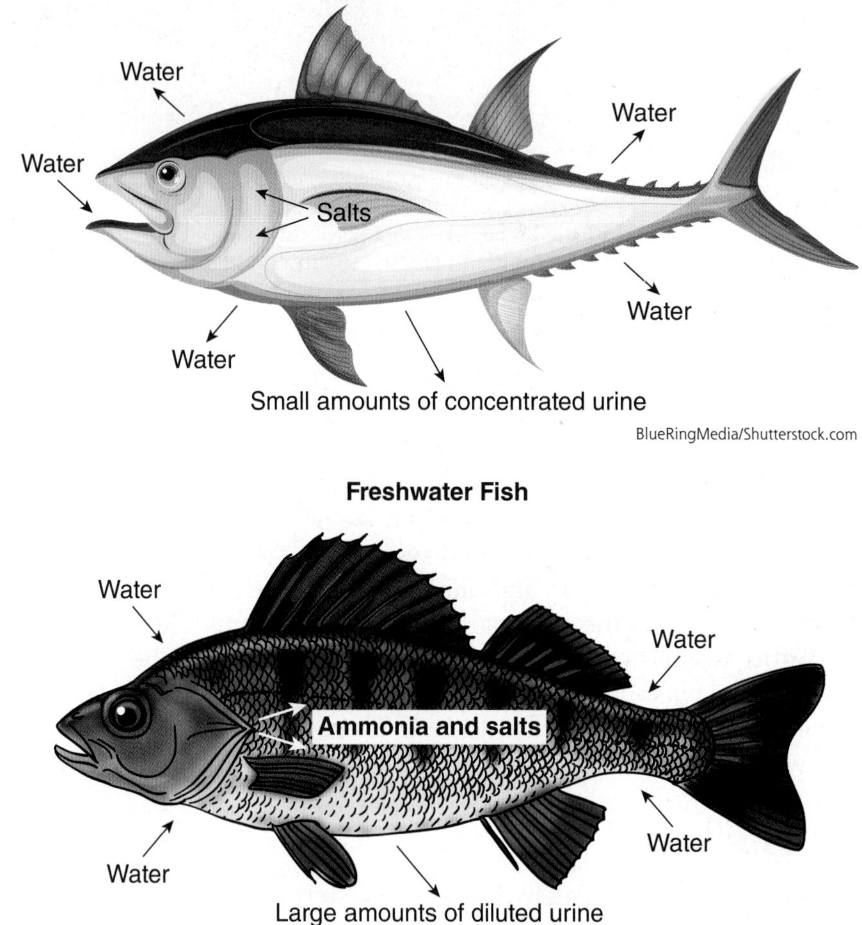

BlueRingMedia/Shutterstock.com

IRINA OKSENOYD/Shutterstock.com

urine concentrations, and the wastewater from each type of facility will require different types of filtration.

Disease and Pest Control

Aquatic organisms raised in indoor facilities are susceptible to bacterial, fungal, and parasitic infections. This is especially true during times of stress. Most problems can be prevented by properly maintaining water quality and regular disease monitoring of stock. Proper quarantine procedures of new stock is also essential for effective disease and pest control. Veterinarian-prescribed treatments may be effectively used in the event of a disease outbreak.

Inhabitants of outdoor and open-water aquaculture systems are susceptible to the same diseases and pests that affect wildlife. In addition, the transfer of viruses and parasites between farmed and wild species as well as among farmed species also presents a risk to wild populations and other farms. As with indoor production systems, most problems can be prevented by properly maintaining water quality and regular disease monitoring of stock. Proper quarantine procedures of new stock are also essential for effective disease and pest control. Some countries prohibit the import of live seed stock to help prevent the spread of disease.

Lynda Richardson, USDA NRCS

Figure 12-39. These giant vats of algal-infused saltwater serve as the habitat for oysters spawned in this Virginia oyster hatchery. The vats, tubing, and piping are all designed to withstand the corrosive saltwater.

Mechanical Systems

The types of mechanical systems an indoor aquaculture system uses depends on the specific needs of the species being cultivated. For example, marine production systems must have a means of regulating the salinity of the water and the mechanical systems must be designed to withstand the corrosive nature of saltwater, **Figure 12-39**. Mechanical systems used in indoor aquaculture systems include generators, water pumps, aerators, power washers, lighting, heating, cooling, and ventilation, and various types of monitoring and control systems.

Outdoor aquaculture systems use many of the same mechanical systems as indoor facilities. The type of equipment needed will depend on a variety of factors, including the size of the operation, the species being cultivated, and the water source. It will also depend on the configuration and types of containment being used.

Stocking Rate

Stocking rate refers to the number of gallons of water per fish in an indoor system or the number of fish per the amount of space in an outdoor system. Because many fish will grow based on the amount of available area, understanding stocking rates is important to both indoor and outdoor aquacultural producers. Stocking rates vary by species as well as by the amount of space. If low stocking rates are used, feed and fertilizer amounts are limited, but can increase greatly as the density of the stock increases. When the density or stocking rate increases, so does the intensiveness of the operation and the need for adding aeration to the system.

There are tables and equations available from various sources including the Fish and Wildlife Service (FWS), NOAA, and most extension programs to determine healthy stocking rates. These organizations, as well as many others, can also help determine healthy ratios of species and which species should be removed. Proper stocking rates will allow for the most efficient production system.

Common Types of Aquatic Animals

There are many different species of aquatic animals raised for aquaculture production. The following is a list of the most common species raised in the United States.

Fish

There are many types of fish grown in aquaculture environments in the United States, including those grown for ornamental, feed, and conservation purposes. The most common freshwater fish raised in the United States for human consumption are catfish, trout, and tilapia.

Did You Know?

Unlike other fish, catfish have no scales. Some species of catfish have stingers in their fins that can become a hazard when handling them.

Catfish

Catfish are bottom feeders—they usually stay close to the bottom of the body of water they are inhabiting. Catfish vary greatly in size. They can range from a few pounds to more than 100 pounds in extreme cases. Most catfish are raised commercially in the southeastern United States in Alabama and Mississippi, **Figure 12-40**.

Trout

Trout are more commonly found in cooler, clear water streams and lakes. They are well distributed throughout North America. Trout are very closely related to salmon and are classified as an oily fish. They feed off other fish and soft invertebrates like flies, mayflies, mollusks, or shrimp. As seen in **Figure 12-41**, they blend in with their surroundings and have the ability to camouflage themselves.

Kletr/Shutterstock.com

Scott Bauer, USDA

Figure 12-40. Catfish are the most commonly produced aquaculture fish in the United States. They have no scales and long sensing whiskers to help them search for food. The color differences between these Louisiana catfish ponds can be correlated to the number and type of algae present within the ponds.

Kletr/Shutterstock.com

Figure 12-41. Trout are the second most commonly produced fish in the United States. They are grown in cooler waters. ***How many trout species are raised in the United States?***

Tilapia

Tilapia is the second most widely cultured fish species in the world behind carp. Tilapia date back to biblical times and are often referred to as St. Peter's fish because they are found in the Sea of Galilee. There are nearly 100 species of fish classified as tilapia. Tilapia thrive in warmer waters but can survive in cooler waters above 50°F. **Figure 12-42** shows a

Ammit Jack/Shutterstock.com

Figure 12-42. Tilapia are a versatile aquaculture fish that can be produced in cooler or warmer waters.

bikeriderlondon/Shutterstock.com

Figure 12-43. Salmon travel from the ocean to freshwater streams for natural spawning.

tilapia with a small head, football shape, tall profile, and bluish gray color. They feed off soft invertebrates, organic matter, and plankton.

Salmon

Salmon are very adaptable and can survive in freshwater and saltwater. Salmon are ***anadromous,*** which means they are born in freshwater, migrate to sea, and return to freshwater to ***spawn*** (reproduce), **Figure 12-43**. Salmon are found in the Atlantic and Pacific oceans and are also found in the Great Lakes. They vary in color from grayish silver to brownish red color, depending on species. Young salmon feed off insects, soft invertebrates, and plankton while adult salmon feed off other fish, shrimp, squid, and eels. Salmon accounts for approximately 25% of the aquaculture industry in the United States.

Shellfish

Two-thirds of the marine production in the United States is molluscan shellfish like oysters, clams, scallops, and mussels. This is the largest sector of U.S. marine aquaculture. **Figure 12-44** shows common types of shellfish produced in the United States.

> **Did You Know?**
>
> According to NOAA, about 40% of the salmon caught in Alaska and 80%–90% in the Pacific Northwest start their lives in a hatchery—contributing more than $270 million to commercial fishery.

Crustaceans

Crustaceans are a very large group of arthropods with more than 67,000 species. Crustaceans have an exoskeleton that they molt. Crustaceans are a favorite of many consumers. Decapod crustaceans like crabs, lobster, shrimp, crawfish, and prawns are delicacies in many cultures including ours,

Figure 12-44. Shellfish production requires careful monitoring of the environment and often takes years before the shellfish reach harvestable size.

Figure 12-45. Shrimp farming is done mainly in earth-bottom ponds built close to the sea.

The Future of Aquaculture

Research shows that seafood is a healthful choice in our diets. Seafood is an excellent source of protein, and it is recommended Americans eat two serving per week. Seafood is low in fat and sodium and high in omega-3 fatty acids, which have been shown to improve health. Studies have also shown that eating seafood regularly can help prevent or reduce the onset of many major illnesses. These health benefits have greatly increased the demand for seafood in the United States and created great opportunity for new aquaculture operations as well as the expansion of existing production systems.

Today's aquaculture is also improving the environment in our country and abroad. As technology advances, some coastal systems are raising fish in enclosed areas in the ocean and turning them out to the

Chris Parypa Photography/Shutterstock.com bonchan/Shutterstock.com
Thawornnurak/Shutterstock.com Mario Savoia/Shutterstock.com

Figure 12-45. Shrimp account for approximately 10% of the U.S. aquaculture industry. Crustaceans are often seen as one of the biggest delicacies in the American diet. *Do you like crab, lobster, shrimp, or crawfish?*

wild for traditional harvesting. This contributes greatly to the commercial fishing industry. The U.S. aquaculture industry is also striving to be more sustainable while increasing production. Strict guidelines for aquaculture have been implemented regarding net safety and the tracking of feed and chemicals needed to maintain health. This environmental awareness is essential if the U.S. aquaculture industry is to continue to grow and thrive.

Another benefit of aquaculture is the efficiency of most fish and aquatic species. Typical livestock require many more pounds of feed to produce a pound of meat used for human consumption. Cattle require 8 lb of feed to produce 1 lb of beef, swine require about 6 lb of feed for 1 lb of pork, while fish require approximately 1.2 lb of feed to produce 1 lb of fish. Fish are the most feed-efficient animal raised in production agriculture in the United States.

Careers in the Aquaculture Industry

There are many careers associated with the aquaculture industry. Most center around rural coastal communities, **Figure 12-46**. Most aquaculture jobs are year-round, hourly wage earning positions. There are many careers

photomatz/Shutterstock.com

Figure 12-46. Commercial fishing is one of the deadliest jobs in the United States. This is particularly true for Alaskan shellfishing.

at local docks, boatyards, and fish processing facilities. This type of position usually does not require a college degree. For aquaculture industry positions requiring a college degree, a person may become a marine biologist, an aquaculture manager, or a wildlife biologist. The level of education and the wages will vary greatly, depending on level of education and years of experience in the field.

Many universities have undergraduate and graduate programs in aquaculture and fisheries. Explore the many career opportunities available in the aquaculture industry.

FFA Connection Aquaculture Proficiency Award

There is an FFA proficiency award category for Aquaculture. This award is given to students who are actively involved in the aquaculture industry for their SAE. If you are interested in participating in this award program, ask your FFA advisor for information on how you can develop an SAE in this award category.

Lesson 12.2 Review and Assessment

Words to Know

Match the key terms from the lesson to the correct definition.

1. A deep, round aquaculture tank that is usually belowground.
2. The number of fish per the amount of space in an outdoor system, or the number of gallons of water per fish in an indoor system.
3. A mesh enclosure suspended in a body of water from a rigid structure on the water's surface.
4. Young fish ready to begin eating on their own.
5. An aquatic animal born in freshwater that migrates to sea and returns to freshwater to reproduce.
6. Small, young fish under one year old.
7. A body of standing water that is usually smaller than a lake.
8. Straight-sided channels typically made of poured concrete or concrete blocks used to house aquatic life.
9. The addition of oxygen to the water of an aquaculture system to ensure proper oxygen levels for all species of aquatic life.
10. An aquaculture system that filters water from the tanks so it can be reused within the tanks.
11. The release or deposit of eggs by a fish, frog, mollusk, crustacean, or other such organism.
12. The amount of gaseous oxygen (O_2) dissolved in the water.
13. The breeding, rearing, and harvesting of plants and animals in all types of water environments, including ponds, rivers, lakes, oceans, and land-based facilities. May also be referred to as concentrated aquatic animal production (CAAP) or fish farming.
14. A large group of arthropods who have exoskeletons that they molt. The group includes crabs, lobster, shrimp, crawfish, and prawns.
15. An aquaculture production facility focused on breeding, nursing, and rearing fish, invertebrates, and aquatic plants to fry, fingerlings, or juvenile stages.
16. The dissolved salt content of the water.
17. An aquaculture enclosure with a rigid frame on all sides that is suspended in a body of water.
18. The production of aquatic life that is native to rivers, lakes, and streams, or spends the majority of its life in freshwater.
19. An aquaculture operation or the practice of aquaculture.
20. The production of species of aquatic plants and animals that live in the ocean.

A. aeration
B. anadromous
C. aquaculture
D. crustaceans
E. dissolved oxygen (DO)
F. fingerling
G. fishery
H. freshwater aquaculture
I. fry
J. hatchery
K. intensive recirculating aquaculture system (IRAS)
L. marine aquaculture
M. marine cage
N. net pen
O. pond
P. raceway
Q. salinity
R. silo
S. spawn
T. stocking rate

Know and Understand

Answer the following questions using the information provided in this lesson.

1. Which country produces the most seafood on a global level?
2. *True or False?* The U.S. aquaculture industry produces enough seafood to satisfy the desires of American consumers.
3. What is a main reason that a producer may choose an indoor aquaculture production system?
4. How does a hatchery contribute to the environment?
5. What is the biggest advantage of a raceway tank?
6. Why is water quality important in aquaculture production?
7. What aspects of aquaculture production are affected by water temperature?
8. Please explain why aquaculture producers use mechanical aeration.
9. Why do some countries prohibit the import of live seed stock?
10. Please explain how higher stocking rates affect an aquaculture production operation.
11. How does the anadromous life cycle of salmon work?
12. How does the feed efficiency of fish compare to other livestock animals?

Analyze and Apply

1. In many cool water aquaculture operations, whirling disease is a big concern. Conduct research on whirling disease and write 1–2 paragraphs explaining what this disease is and how aquaculturalists are trying to prevent it.
2. Many of the aquaculture production systems in the United States are owned by fish and wildlife services as a means of stocking fish for recreational fishing. Does stocking public lakes and ponds happen in your area? Find out if your local officials stock any of the bodies of water near you, and if so, find out which species and how many are stocked each year.
3. Often aquacultural production takes place in a symbiotic relationship with plants and producers. Find an example of aquaculture playing a role in the development and growth of crops.

Thinking Critically

1. Compare and contrast the aquaculture environments needed to grow the two most commonly produced fish in the United States, trout and catfish. Create a table which shows the optimal water temperature, oxygen levels, feed, and growth cycle for both types of fish.
2. The process of growing fish in floating cages and turning them into the wild for fishing is increasing in popularity. Give three reasons this would be beneficial for fish populations.
3. An aquaculture industry you may not have thought about is farm-raised pearls. Why might these pearls have less value from those found in the wild? How could you market farm-raised pearls to consumers to maximize your profit?

Lesson 12.3

Nontraditional Animal Industries

Words to Know

antler
apiculture
buck
Chronic Wasting Disease (CWD)
colony
Colony Collapse Disorder (CCD)
cria
dam
doe
drone
fawn
gelding
lanolin
nontraditional animal industry
stud
velvet
venison

Lesson Outcomes

By the end of this lesson, you should be able to:

- Define nontraditional animal industries.
- Explain how animals outside traditional livestock and companion animals play a role in agriculture.
- Describe the production cycle of different nontraditional animal industries.

Before You Read

Read the photo captions from this lesson. After you read the captions, write down a list of five questions you have about the content of this lesson. Once you have completed your reading, see if you can answer the five questions using information found in the text.

When someone says the phrase "animal production," what animals come to your mind? Most of us would think about standard livestock animals, including cattle, sheep, horses, goats, poultry, and swine. If someone asked you to list more animals that are produced in the United States, you might come up with fish grown on fish farms or companion animals like dogs, cats, and birds. While these are the most commonly produced animals in the United States, there are many other animals that are raised in a production cycle in our country, **Figure 12-47**.

A ***nontraditional animal industry*** is an industry that produces products from animals not generally considered livestock or companion animals. These industries are important to the U.S. economy and produce products that are marketable to consumers. The specialized nature of the production cycle for each of these animals requires producers to have highly specialized knowledge about the species they are producing.

Nontraditional animal production accounts for approximately $6 billion in annual revenue. The most common nontraditional animals produced in the United States are:

- Bees
- Llamas and alpacas
- Deer and elk
- Bison
- Alligators
- Mink

Jevtic/iStock/Thinkstock

Figure 12-47. A lot of animals are used for agricultural production that do not fall into the traditional livestock animal classifications. *Are these the animals you picture when you think about animal agriculture?*

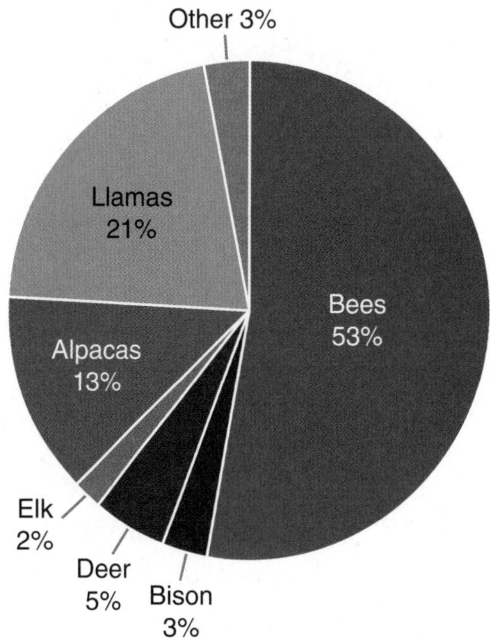

Goodheart-Willcox Publisher

Figure 12-48. This graphic shows the breakdown of the percentage of producers for the most common nontraditional animal industries in the United States.

According to the USDA, there are approximately 75,000 farms in the United States that produce nontraditional animals. The breakdown of what percent of nontraditional animals falls into each category is shown in **Figure 12-48**.

Bee Production

The production of bees is an important agricultural endeavor. In fact, there are more farms producing bees than any other nontraditional animal, **Figure 12-49**. The process of keeping bees for production is called ***apiculture***. Bees are produced primarily for crop pollination and honey production. Bee producers will often rent their hives to agricultural producers so the bees can pollinate their crops. Beekeeping also results in the production of honey, an important agricultural commodity.

predrag1/iStock/Thinkstock

Figure 12-49. Bees are the most commonly produced nontraditional agriculture animal. *Have you ever seen a grouping of beehives, such as the one shown in this sunflower field?*

Scope of Bee Production

In the United States, there are more than 2.5 million managed beehives. Each hive may contain upward of 50,000 bees. That translates to about 125 billion bees currently in production in our country. These hives are responsible for the pollination of around 30% of the food consumed in North America. Bees are also responsible for pollinating much of the nation's livestock feed. The value of pollination from bees in the United States is estimated at more than $16.5 billion annually. Some crops are more reliant on bees for pollination than others. For example, almond trees are 100% pollinated by bees. Bees are important for pollination of up to 30% of the human food products in North America, **Figure 12-50**.

In addition to their value from pollination, bees provide more than 150 million pounds of honey annually. Most of the honey produced in the United States is consumed here. However, the U.S. demand for honey also results in honey imports of around $375 million annually. Beeswax, a by-product of the honey industry, is estimated to be worth about $7 million in sales each year.

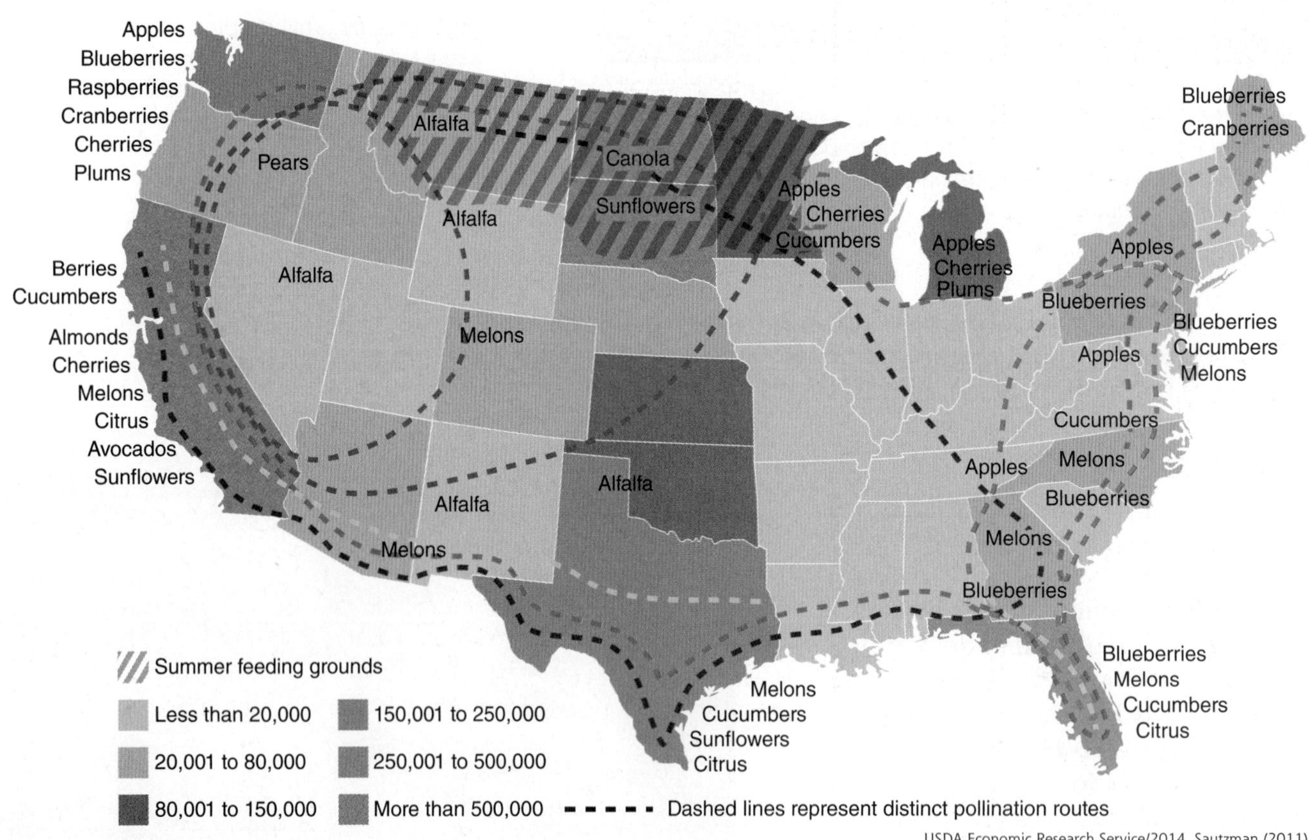

Figure 12-50. Commercial beehives are transported around the country to ensure the successful cultivation of many U.S. crops. The USDA estimates that 60% to 75% of the commercial hives are employed for California's almond bloom in early spring. After the almond bloom, some beekeepers remain in California, while others depart for northern states to pollinate orchard and berry crops, and others migrate to southern and eastern states to pollinate a variety of field crops. ***Which of these crops do you think would be most affected by a decrease in global bee populations?***

Reluk/Shutterstock.com

Figure 12-51. *Can you spot the queen bee in this group of bees?* The queen is the largest bee and the mother to all offspring produced during the time she is queen. Some apiculturalists will mark their queen for easy identification, such as with the yellow dot on the queen in this picture.

Production Cycle in Bees

Bees are kept in groups called ***colonies***. Each colony dwells within a single hive with a single queen, a large number of female worker bees, and several thousand male bees, called ***drones***. The queen is the only bee in the hive that is capable of laying eggs. Each queen will lay between 1,500–2,000 eggs a day during her lifetime (3–5 years). The queen has the ability to lay worker or drone eggs, but will only lay eggs that produce workers if she has left the nest to mate with drones from another hive, **Figure 12-51**.

Bee eggs hatch after 3 days and, depending on whether the larvae is a queen, worker, or drone, are in the larval stage for 8–13 days. The worker bees gather pollen, feed the larvae, and care for the queen and drones. After the larval stage, the cell is capped and the larvae develops into a pupa before becoming fully developed.

Did You Know?

It is true that bees die after stinging, as a large portion of their anatomy (the barb on their stinger) is torn away after they sting. The same is true for the male reproductive organ on drones, meaning they can mate only once.

Hive Placement

When renting their hives, apiculturalists will place their hives in close proximity to the crops requiring pollination. Careful analysis is taken to ensure that there are enough bees to pollinate the desired crops. The number of hives needed per acre depends in part on how attractive the crop is to the bees and the number of wild bees in the area. Accessibility to interior areas of the fields will also determine the number of hives needed and their strategic placement. Although worker bees may travel in a 3–5 mile radius around the hive, they prefer to work within 300′ of their hive.

Timing

It is also important to consider the timing of hive placement. It is best to move the hives into location after some flowering has begun. If you move them in before the crop blooms, the bees may search for other plants and ignore the intended crop when it blooms. The timing of pesticide applications must also be clearly worked out between the grower and the beekeeper.

Honey and Beeswax Harvesting

Worker bees collect pollen, bring it back to the hive, and store it in the cells of the honeycomb. The honeycomb is made of beeswax. Once a cell is full of honey, the cell is capped and the honey is ready for harvesting, **Figure 12-52**. The flavor and color of honey is determined by the various plant species visited by the bees. The pollen from some plants, like clover and fruit trees, produce much more desirable honey products.

Managing beehives involves periodically opening the hives to check for colony health and to collect honeycombs for honey extraction. The use of protective clothing and smokers are important in ensuring the safety of the apiculturalist and the bees during hive opening, **Figure 12-53**. At harvest, the hives are opened, and the frames containing the honeycomb are removed. The honey is separated from the beeswax, and the frames are replaced to allow the bees to begin the production cycle once again.

Pests, Disease, and Other Threats

Threats to the apiculture industry exist in the form of diseases, parasites, habitat loss, pesticides, and air pollution.

- Diseases—Colony Collapse Disorder (CCD) has wiped out millions of beehives in the past few years. ***Colony Collapse Disorder (CCD)*** is a disease that affects both wild and domesticated hives of honeybees.

greenantphoto/iStock/Thinkstock

Figure 12-52. Cells containing honey are capped once they are full to seal in the honey until harvested. *Can you see the capped cells in this hive?*

Researchers suspect pesticides, parasites, poor nutrition, and fungicides as possible causes for this disorder.

- Parasites—parasitic mites can cause widespread death of bees in a colony.
- Habitat loss—the destruction and crowding out of native plant species reduces food sources.
- Pesticides—pesticide misuse and drift from aerial spraying contaminates pollen grains, which in turn weakens or destroys colonies as the contaminated pollen is brought into hives.
- Air pollution—problematic for pollinators that rely on scent trails to find flowers.

JupiterImages/BananaStock/Thinkstock

Figure 12-53. Do you know the purpose of the smoker in apiculture? The smoke is used to slow down the reaction communication between bees in the hive, allowing them to be calmer during the opening of the hive. Protective clothing protects the beekeeper from being stung while working with bees.

Llama and Alpaca Production

Do you know someone who owns something made of alpaca or llama wool? Llamas and alpacas were historically used as pack animals in the Andes Mountains, and made their way to North America where they are typically used for wool production, **Figure 12-54**.

While these two animals look very similar, there are some differences to note. First, the wool of llamas and alpacas is different. Llamas have a hollow coarser wool on the top, and a thinner undercoat of guard hairs underneath. This is in contrast to alpacas, which have one, finer textured wool on their entire body. Because of this, alpaca wool is softer and more desirable for use in the clothing industry. Alpacas are about half the size of llamas, weighing in at around 100 lb

Tina_Gambell/iStock/Thinkstock

brackish_nz/iStock/Thinkstock

Figure 12-54. Llamas and alpacas are very similar in structure but are different sizes. *What other differences exist between llamas (left) and alpacas (right)?*

Levranii/iStock/Thinkstock

Figure 12-55. Newborn llamas and alpacas are called cria. The dam will fiercely protect her young from anything they think is a threat. This means that ensuring llamas are comfortable around humans prior to giving birth is helpful in managing llama production.

to 175 lb, and are more likely to be a single color, whereas llamas are often multicolored.

Scope of Llama and Alpaca Production

There are approximately 9,300 alpaca and 15,300 llama farms and ranches in the United States. These farms and ranches are responsible for producing more than $37 million in sales annually. Llamas and alpacas are raised as breeding stock for companion animals, and for the production of wool fiber. In some rare instances, llamas are kept with sheep as protectors from predators. The large and aggressive nature of these animals makes them a perfect "bodyguard" for sheep, which lack the size to fend off predators on their own.

Did You Know?

Llamas and alpacas have a modified ruminant stomach. This means they have a stomach with three compartments, unlike the four compartments of most ruminant animals. They also have an attached tongue, which means they can only stick out their tongue about a half-inch from their mouths.

Production Cycle in Llamas and Alpacas

The production cycles of llamas and alpacas is similar to that of sheep and goats. Typically, females, called ***dams***, will give birth to a single offspring, called a ***cria***, **Figure 12-55.** Male llamas and alpacas are called ***studs***. They are called ***geldings*** if they have been castrated. Llamas and alpacas produce well when raised on pasture or other forages. One of the benefits of llama and alpaca fleece compared to wool from sheep is that they do not produce lanolin, making their fleece less allergenic than wool. (***Lanolin*** is the natural grease found on sheep's wool.) These animals are highly sociable, and are relatively easy to train to halter and lead.

Fleece Production

Alpacas are generally shorn once a year. A mature alpaca will produce 5 lb to 8 lb of fleece annually. The value of the wool is determined in part by color, fiber length, tensile strength, and softness. White is the most common color of alpaca fleece, and it is also the most desirable because the white fiber can be easily dyed. As an alpaca gets older, its wool fibers become thicker in diameter and more coarse. For this reason, fleeces from younger animals typically have higher value. The coarser guard hair in their fleece can be used for rugs, wall hangings, and lead ropes. The undercoat of a llama may be used for the production of garments. Depending on the type of animal, llamas will typically produce 2 lb to 6 lb of fleece per year.

Restraining for Shearing

Alpacas and llamas are larger and stronger than sheep and require some method of restraint while being shorn. The most common method of restraint is to lay the animal down, stretched between two anchor points, **Figure 12-56.** The feet are tied to the anchor points, allowing the shearer access to the brisket of the animal. A tarp is placed below the animal to collect the fleece.

Chris Warham/iStock/Thinkstock

Figure 12-56. Llamas and alpacas are restrained for shearing to ensure the safety of both the animal and the shearer.

Control of the animal is essential to ensure the safety of the animal and the shearer. With practice and experience, the time an animal must be restrained is minimal, as an alpaca can be sheared in less than 10 minutes.

Guard Animals

Llamas have a similar production cycle to alpacas, although llamas are used for a broader variety of purposes. Some llamas are used in sheep and goat production as guard animals. This process has become increasingly common in the western United States since the 1980s. Typically, a gelded male is introduced to the flock and will bond with the other animals. If these animals are threatened by predators like coyotes or feral dogs, the large size and aggressive nature of the llama is often successful at chasing the predators away.

In some regions in the world, and even here in the United States, llamas are used for pack animals. They are sure-footed, small, and are able to travel on the same trails and paths as hikers. The structure of a llama makes it able to safely carry about 1/3 of its body weight.

Did You Know?

Llamas and alpacas (especially males) may develop "berserk llama syndrome," an attachment disease that can be very dangerous for owners.

Deer and Elk Production

Although most deer and elk are considered wildlife animals and live with no human assistance in many areas of the country, these animals are also raised in captivity as an agricultural commodity. There are several reasons for these animals to be produced by agricultural producers:

- Meat harvesting
- Recreational hunting
- Production of antlers and velvet

Deer kept in captivity in the United States are generally either mule deer or white-tail deer. Elk kept in captivity in the United States are almost exclusively Rocky Mountain Elk, a specific variety of wild elk typically found in the region west of the Rocky Mountains.

The United States has a large number of deer and elk farms. There are around 4,000 farms in the country raising more than 230,000 deer in captivity. The more than 1,100 elk farms raise around 40,000 animals. The production of these animals produces more than $3 billion in revenue to the U.S. economy each year.

Deer Production

Domestic deer are generally kept in forage pastures or fed forage feed. Mature female deer, called ***does***, have 2–3 offspring called ***fawns*** in the early spring. Proper nutrition of breeding females can increase the likelihood of multiple offspring. Reproduction in deer farms is managed to select individuals with superior structure and antlers. Seasonally, male deer (***bucks***) grow ***antlers***, bone structures that grow as an extension of the skull in members of the deer family. These antlers have a protective covering that has a rich blood supply. This vascular covering of antlers is called ***velvet***, **Figure 12-57**. Sales of velvet from deer antlers, as well as from cast (shed) antlers, make up a portion of the income from this type of agricultural operation.

Private Hunting

Do you like to hunt? One of the largest uses for commercial deer herds is private hunting. Managers of private deer farms typically have large tracts of land, and sell trophy hunting permissions to interested hunters. Hunting preserves are required to have a minimum of 80 connected acres of land for operation. Hunters will pay premium prices for the opportunity to hunt animals with superior antlers, and deer farms raise animals that are specifically bred to be trophy quality.

Lightwriter1949/iStock/Thinkstock

Figure 12-57. One of the most important products from deer and elk farming is velvet. *Does it look painful for this buck to sluff off his velvet?* The blood supply to the velvet is gone prior to it falling away, making the process itchy but not painful.

Meat Production

Deer meat is called ***venison***. There is a growing market for venison produced on deer farms in the U.S. market. The primary meat cuts of venison are similar to those of beef cattle.

Game Fencing

Deer and elk will not be contained with the same type of fencing suitable for cattle or sheep. A white-tail deer can easily clear an 8′ fence on level ground, so standard cattle fencing would not even pose a challenge. There are various types of effective fencing available. The type of fencing and installation design used will depend on the fencing budget and lay of the land. For example, a 7′ or 8′ fence would suffice in most situations;

however, you may need a 10′ or 11′ fence on sloping ground. An agricultural fencing contractor or supplier can provide advice and suggestions for each specific situation.

Elk Production

All domestic elk kept on elk farms in the United States must come from private herds. The sale of wild animals into captivity ended in 1960. Because of generations of selective breeding for animals that have superior structural traits and calm behavior, they are not considered wildlife, but rather tame wild herds.

Elk are able to convert large amounts of high fiber feed into muscle, making them very efficient meat-producing animals. This means that elk are able to convert very low-quality feeds into useful muscle. The terminology for elk are the same as cattle: bull, cow, and calf. Hunting preserves with domestic elk are common. The animals are raised in areas that allow them access to large tracts of land, **Figure 12-58**. Hunters may purchase hunting rights for a specific animal, time, or experience.

benjaminjk/iStock/Thinkstock

Figure 12-58. Elk are raised on large tracts of land and genetically selected to be trophy sized at maturity. ***Have you ever been on a deer or elk farm?***

Velvet Production

Velvet is a more income-generating commodity in elk than it is in deer. Elk velvet is prized as a nutritional supplement in many Asian markets. A mature bull elk can produce more than 16 lb of velvet annually. In farms where velvet is the main product, the antlers of the elk are removed after they have reached their full growth size for that year. The bony portion of the antlers are then sold for decorative purposes.

Meat Production

Elk meat is sought after as a low-fat and low-cholesterol alternative to beef, **Figure 12-59**. Elk produced for meat are fed similarly to cattle, ensuring that their meat will not have the "gamey" flavor of wild elk. The meat flavor of farmed elk can become similar to the taste of beef through careful control of their nutrition.

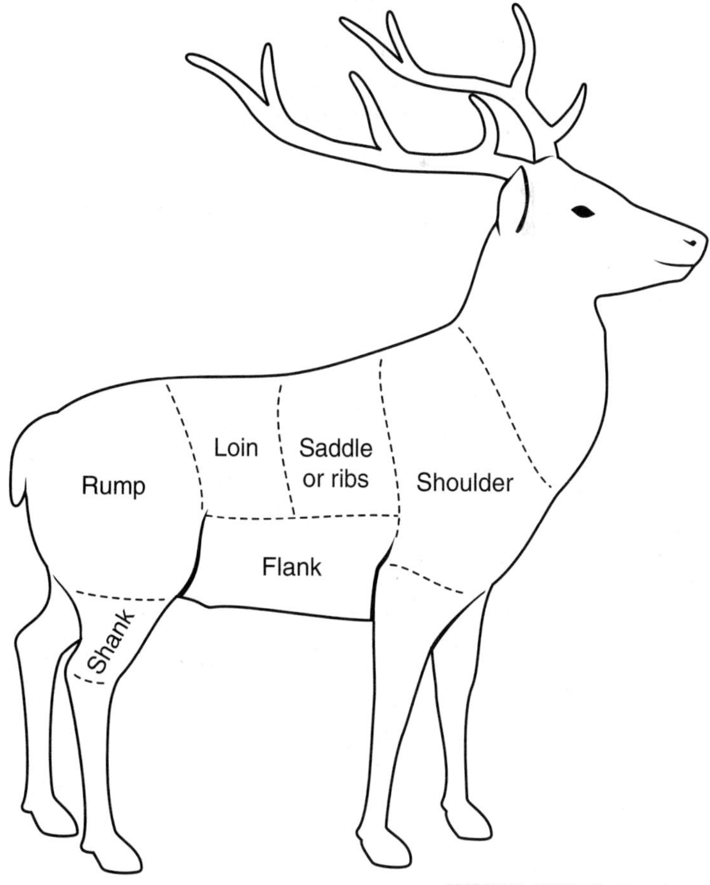

230873560 ONiONA/Shutterstock.com

Figure 12-59. The primary meat cuts of venison are similar to those of beef cattle.

Did You Know?

There is no quick test for detecting CWD. To detect CWD, federally approved labs will examine brain samples and lymph nodes only from freshly killed animals.

Deer and Elk Diseases

A big concern in the domestic production of deer and elk is the threat of ***Chronic Wasting Disease (CWD)***. CWD in elk and deer is similar to mad cow disease in cattle. This disease can be found in wild populations of deer and elk and can then be transmitted to domestic populations. Care should be taken by these producers to ensure that any wild herd near their farm or ranch are kept away from the domestic animals.

Bison Production

Have you eaten a bison burger? How about a bison steak? Bison are commonly produced on farms in the United States and are seen as a viable alternative to beef. Raising bison, which you may think of as buffalo, is very similar to raising cattle because both animals have very similar production cycles, nutritional needs, and structure, **Figure 12-60**.

Lunnderboy/iStock/Thinkstock

Figure 12-60. Raising bison produces low-fat meat with much of the same required management as cattle. *Have you ever considered how similar bison and cattle are in structure and function?*

Scope of Bison Production

About 7.5 million pounds of bison meat is sold annually in the United States, coming from around 15,000 animals that are harvested. Total domestic bison numbers in the country are estimated at about 200,000 animals on about 4,500 farms and ranches. The primary product from bison is meat. Bison meat is lower in calories per ounce than beef, chicken, and turkey, **Figure 12-61**. Combining this fact with the relatively large size of a bison carcass has driven the production of bison.

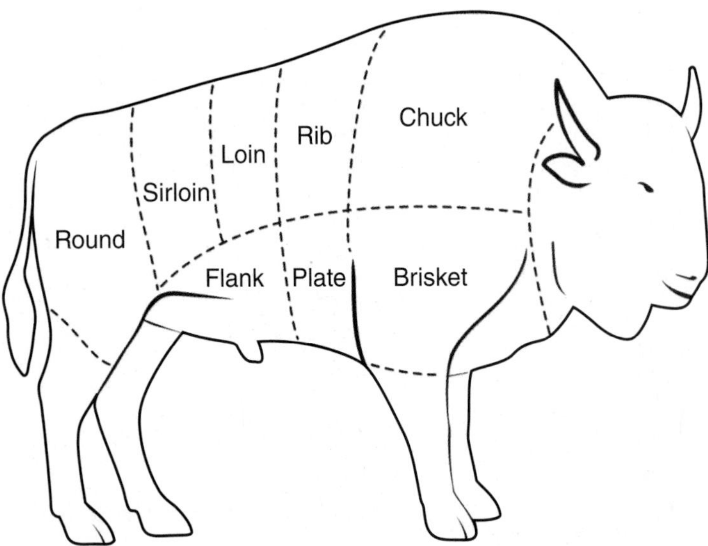

ONiONAstudio/iStock/Thinkstock

Figure 12-61. Do you remember the cuts of beef cattle from *Lesson 10.1*? *Can you see any similarities with the cuts of bison meat?*

Production Cycle of Bison

Bison are produced in a cycle similar to beef cattle, and the terminology for bison is the same: bull, cow, and calf. Bison have gestation of about nine and a half months, and calves weigh around 40 lb or 50 lb at birth. Calves are weaned at around six months old, weighing between 350–450 lb. Animals are then raised on a high-forage diet until they are between 2 1/2 and 4 years of age. Like beef, bison can be finished on a grass-only diet, or can be finished by feeding a high-nutrient diet, generally including corn, for the last 90 days of their production. Factors that affect meat quality in bison include: age of animal, feeding program, and animal stress. Animals that are under stress during the harvesting process will have additional adrenaline that can cause a more "gamey" flavor in their meat.

While the production of bison is similar to beef cattle, there are some noticeable differences. One of the biggest differences is that cattle have been subject to years of selective breeding to create animals who are more docile. In contrast, bison are not considered domesticated or even tame animals. Bison can be unpredictable and aggressive, and care should be taken to ensure that they are kept in facilities that can handle their large size and potentially aggressive character, **Figure 12-62**. Domestic bison are also susceptible to many of the same diseases that cattle suffer from and should be kept on a structured vaccination program to prevent illness.

Did You Know?

Although bison look large and cumbersome, they are able to run at speeds of up to 40 miles per hour and can jump over 6′ vertically.

Alligator Production

Do you think it sounds like fun to spend time working with 15-foot reptiles that could easily tear you apart, **Figure 12-63**? For those who raise alligators as production animals, this is a reality. Alligators in the United States have a long history with humans. They were once hunted almost to extinction, but through proper management practices, alligators have made a comeback in the wild and as a viable agricultural commodity. Alligator production may take place on alligator farms or ranches. You may wonder what the difference is between an alligator farm and ranch. Alligator farms are closed systems, which include the breeding of mature alligators to produce new animals. Alligator ranches, on the other hand, rely on harvesting eggs from the wild and hatching them in captivity.

mdulieu/iStock/Thinkstock

Figure 12-62. Bison have not been bred for docility and can become incredibly aggressive when provoked. Bison farming requires that fences, gates, and holding facilities are built to withstand the efforts of an angry bison.

Scope of Alligator Production

While a large portion of the alligator products in the United States come from the harvesting of wild alligators, some of the market share comes from alligator farms and ranches. Commercial alligator farms and ranches exist in both Louisiana and Florida, the two states with the highest alligator populations. In Louisiana, there are about 400,000 alligators being raised on alligator ranches.

BrendaLawlor/iStock/Thinkstock

Figure 12-63. Alligators are the most common predator animal that is farmed in the United States. *Do you think it would be fun to raise animals that could potentially eat you?*

amnarj2006/iStock/Thinkstock

Figure 12-64. The hides of alligators are made into the most expensive type of leather. *Can you think of some properties of alligator hide that would make their leather superior for designer goods?*

The most valuable product from an alligator is the hide, which can sell for more than $25 per square foot. In addition to the hide, alligator meat is a desirable product coming from this industry, **Figure 12-64**. Alligator farms also play an important role in the conservation of wild alligators. Young alligators are especially susceptible to being eaten by birds, fish, and other alligators. Many alligator farms remove eggs from the wild, hatch them, and grow them until they are big enough to have a better chance at survival, and then return them to the wild.

Production Cycle in Alligators

It is possible to breed and harvest eggs from captive alligators on an alligator farm. The majority of eggs from alligators raised in captivity are harvested from wild nests. This practice can be dangerous, as female alligators will aggressively defend their nests if they feel threatened. The temperature that the eggs are kept at during incubation determines the sex of the hatchlings. Eggs hatch 63 days after they are laid, **Figure 12-65**.

Alligators are kept in shallow water, with access to areas outside of the water. Most alligators are kept in heated houses until they are approximately

Eduard Lysenko/iStock/Thinkstock

Figure 12-65. Alligators on farms and ranches are generally raised to about 4′ long and then returned to the wild or processed for hides. *Can you believe this little guy can grow into an alligator that reaches nearly 20′ in length?*

4′ in length, which may take up to 15 months. At this time, regulations may require the alligator ranch to return some of the alligators to the wild if they were collected from wild eggs. The rest of the population is processed in order to harvest the hide and meat from the animals. Hides are required to have an identifying tag that accompanies the hide until the final wholesale destination, ensuring that alligator hides sold are taken through the proper regulatory channels.

Career Connection Hunting Preserve Manager

Job description: Hunting preserve managers are responsible for maintaining the herd health, hunter experience, and the environment of private hunting preserves. Duties range from managing the animals to ensuring that hunter clients have a favorable experience. Hunting preserve managers must ensure hunters do not disturb or kill animals that are not in season or protected or endangered.

Education required: While many hunting preserve managers learn on the job, others have education in wildlife management and the hospitality industry.

Job fit: This job may be for you if you enjoy working with both wildlife animals and people and are able to pay careful attention to regulations that govern hunting preserves in the United States.

©iStock.com/robertcicchetti

Are there instances when protected or endangered animals can be hunted?

Lesson 12.3 Review and Assessment

Words to Know

Match the key terms from the lesson to the correct definition.

1. The meat from deer.
2. A disease similar to mad cow disease found in wild populations of deer and elk that can be transmitted to domestic populations.
3. The process of keeping bees for commercial production.
4. A male llama or alpaca.
5. A disease that affects both wild and domesticated hives of honeybees.
6. A worker bee.
7. The vascular covering of deer or elk antlers.
8. A castrated llama or alpaca.
9. An industry that produces products from animals not generally considered livestock or companion animals.
10. Bone structures that grow as an extension of the skull in members of the deer family.
11. A female llama or alpaca.
12. The natural grease found on sheep's wool.
13. A mature female deer.
14. The offspring of a llama or alpaca.
15. A mature male deer.
16. The offspring of a mature female deer.
17. A large group of bees that live and work together in a beehive.

A. antlers
B. apiculture
C. buck
D. Chronic Wasting Disease (CWD)
E. colony
F. Colony Collapse Disorder (CCD)
G. cria
H. dam
I. doe
J. drone
K. fawn
L. gelding
M. lanolin
N. nontraditional animal industry
O. stud
P. velvet
Q. venison

Know and Understand

Answer the following questions using the information provided in this lesson.

1. What is the most common nontraditional animal produced in the United States?
2. Please list three threats to the health of the bees in a hive.
3. What are the two main differences between llamas and alpacas?
4. Explain why llamas make effective guard animals.
5. What are the three main reasons producers keep deer and elk?
6. How could deer and elk producers help reduce the threat of Chronic Wasting Disease entering their herd?
7. How does the calorie composition of bison compare to other meats?

8. What can producers do to decrease the chance of "gamey" flavor in bison meat?
9. What is the difference between an alligator farm and an alligator ranch?
10. What production factor determines the sex of a baby alligator?

Analyze and Apply

1. If you were going to produce one of the nontraditional animals in this lesson, which one would you pick? Please explain your answer in 1–2 sentences.
2. How does raising nontraditional animals differ from raising traditional livestock animals? Give at least three examples of how these industries are different.

Thinking Critically

1. Select a nontraditional animal that is used for production and not included in this lesson. Find the scope of this industry in the United States, and write a paragraph including the production cycle for this nontraditional animal.
2. Please share your opinion on whether you feel alligator ranchers should be able to take eggs from wild nests and raise them in a controlled environment. Write 1–2 paragraphs outlining your opinion, and give 3–4 statements that support your position.

Lesson 12.4

By-Products from Animal Industries

Words to Know

by-product
dressing percentage
gelatin
hide
leather
pelt
primary product
rendering
tanning
variety meat
xenotransplantation

Lesson Outcomes

By the end of this lesson, you should be able to:

- Differentiate between primary products and by-products.
- Examine the portion of livestock animals that are used for primary and by-products.
- List common uses for animal by-products.

Before You Read

Read the headings for each of the sections of this lesson. Write down one question you have about each of the main headings. After reading the lesson, see if you can answer the questions you wrote. If you find a question you cannot answer, pair up with a partner and use the Internet to research your topic.

What do you think common items such as lotion, shoes, mascara, asphalt, gummy bears, and fireworks have in common? Did you guess that they are all made with animal products? Most of the time, we only think of the main product an animal is raised to produce, which is called the ***primary product***, but what about the other parts of the animal?

In this lesson we will examine animal by-products. A ***by-product*** is something that is produced incidentally in the production of a primary product. The industry involved in taking the by-products and converting them into useful goods is called the ***rendering*** industry. Industry experts report that more than 98% of an animal harvested for human consumption is used to manufacture useful products, **Figure 12-66**. Let us take a look at how much of an animal is primary product compared to by-product, what parts of the animal are used in making products, and what everyday items come from animal by-products.

Primary Products vs. By-Products

In order to examine animal by-products, we should first take a look at how much of an animal becomes primary product and how much is by-product. Even though the intended use of the animals is the product listed, there are many other parts of the animal that can be useful after the animal is harvested. Even animals like dairy cattle and laying hens, who produce products that do not require them to be harvested, will eventually reach the end of their production cycle. When that happens, these animals are also often harvested for useful by-products as well.

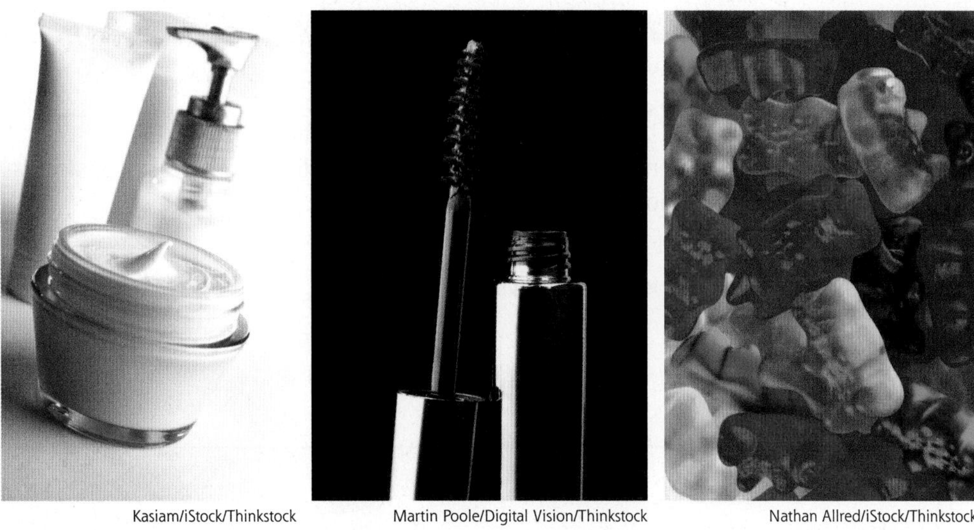

Kasiam/iStock/Thinkstock Martin Poole/Digital Vision/Thinkstock Nathan Allred/iStock/Thinkstock

Figure 12-66. *When you see items like this, does your mind automatically think about animal products?* Each of these items contains animal by-products.

The ***dressing percentage*** is the ratio of live animal weight to carcass. This is a good starting point in determining how much of a meat animal is used for meat, and how much is left over as a by-product. The average dressing percentages for common meat animals is shown in **Figure 12-67**, along with the percentage of those animals that becomes by-product.

If we are losing almost half of some of our meat animals, what happens to the rest of the animal? The image in **Figure 12-68** outlines the average amount of products that would come from a 1,000-pound steer. As you look at the items on the figure besides the retail meat, you will see that the other portion of the animal can be broken down into these broad categories:

- Variety meats
- Hide, hair, fur, or wool
- Bone, horn, and hooves
- Fats and fatty acids

Species	Dressing Percentage	By-Product Percentage
Beef cattle	60%	40%
Broilers (chickens)	71%	29%
Dairy cattle	55%	45%
Lambs and goats	50%	50%
Swine	70%	30%
Turkeys	79%	21%

Goodheart-Willcox Publisher

Figure 12-67. The amount of meat from an animal compared to the amount of by-products varies by species. Selective breeding has helped producers increase the dressing percentage of species of livestock through the years.

STEM Connection Dressing Percentage

Dressing percentage allows you to calculate how much of a live meat animal will become a harvested carcass. To calculate dressing percentage, use the following formula:

Carcass weight ÷ Live weight

See if you can calculate the dressing percentage for the following animals:

1. A 140-lb lamb with an 82-lb carcass
2. A 1,300-lb steer with a 680-lb carcass
3. A 290-lb hog with a 208-lb carcass

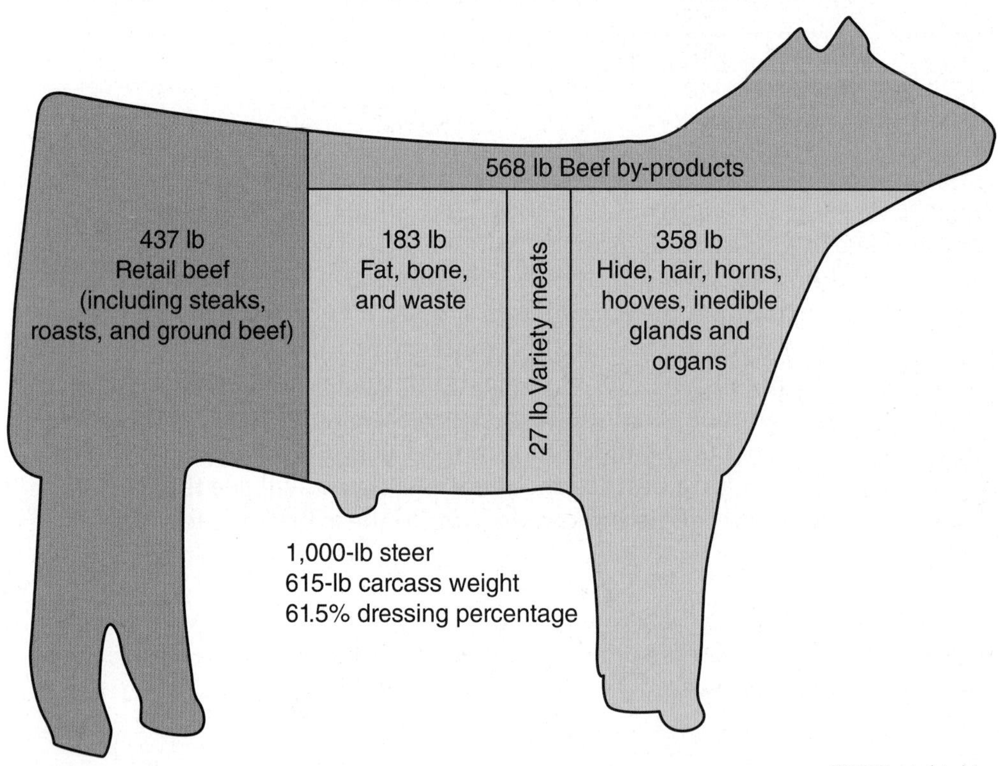

TAMU Livestock Judging

Figure 12-68. Many products are derived from a steer. *How many of them are primary products?*

Did You Know?

Unless specifically labeled as "with variety meats" or "with meat by-products," the USDA prohibits the use of variety meats in hot dogs. Most hot dogs in the United States are made only from regular retail cuts mixed with seasoning and spices.

Let us take a closer look at how these different portions of the animal become useful by-products.

Variety Meats

Steaks, pork chops, and chicken wings are what you might think of when we hear the word *meat*, but they are not the only meat on the animal. ***Variety meat*** is meat that comes from a nonskeletal muscle, **Figure 12-69**. Some common examples of variety meats include: liver, heart, sweetbreads (thymus gland), and tongue.

bonchan/iStock/Thinkstock

Figure 12-69. Variety meats, like the beef heart used in this Peruvian anticuchos dish, are common protein sources in many recipes. *Does this look tasty to you?*

Have you ever tried a variety meat? These types of meat make up about 2.7% of the total live weight of beef cattle. Even though they sometimes get a bad reputation, some of these meats are seen as delicacies in certain fine dining establishments.

Hide, Hair, Fur, and Wool

Sometimes, what is considered a primary product in one species of animal is actually considered a by-product of other animals. Hide, hair, fur, and wool is a great example. One of the most valuable by-products of animal agriculture is the hide of harvested animals. The ***hide*** is the term for the skin of an animal. Can you think of some animals that are raised primarily for their hide, hair, fur, or wool? Wool breed sheep, goats that are raised for mohair, and fur-bearing animals like mink are raised primarily for their hide, **Figure 12-70**. Other animals, like beef cattle, hogs, and meat breed sheep and goats are raised for another purpose. Even animals that are not raised primarily for their hair or wool have a hide that can be used as a by-product.

In order to preserve the animal's hide, it undergoes a process called tanning. ***Tanning*** is a chemical or mechanical process whereby the hide of an animal is preserved. If the hair is removed during the tanning process, the final product is called ***leather***. If

Timur Arbaev/iStock/Thinkstock

Figure 12-70. The hides of animals like the mink are considered the primary product of mink production. The hide of other animals is a by-product.

Did You Know?

Italian leather is considered some of the most valuable in the world, partly because many famous Italian leathermakers only tan one hide at a time and will produce leather only from the highest quality hides.

apichat naweewong/iStock/Thinkstock

Figure 12-71. When the hair is left on the hide, the resulting hide is called a pelt. *Can you list some of the animals whose pelts are more valuable than others?*

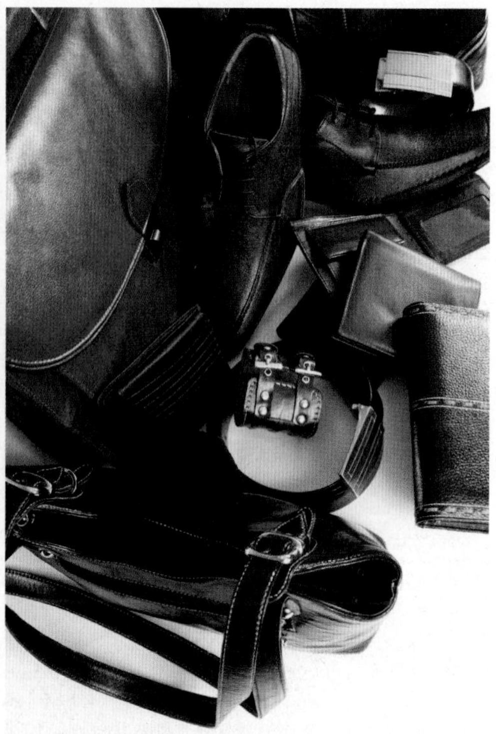

Youssouf Cader/iStock/Thinkstock

Figure 12-72. Leather goods include many clothing and accessory items. In fact, high-quality leather goods are one of the most value-added agricultural product. *What is the most money you have ever paid for a leather handbag or shoes?*

the hair, wool, or fur is left on the hide, the tanned product is called a ***pelt***, **Figure 12-71**.

What kinds of things can be made from the hide of an animal? Anything made out of leather is considered a by-product of the animal industry, **Figure 12-72**. In addition, the hair and hide of animals that have been removed from the hide can be used to make brushes and felt. The hide can also be used to create the bases for many pastes and ointments, along with the binders needed for plaster, paint, and asphalt. Another place you might find the hide or hair of animals is in the materials used to insulate homes and businesses. Did you ever think you might find material made from cattle hide inside your walls?

bhofack2/iStock/Thinkstock

Figure 12-73. Gelatin is an important component of marshmallows. *Have you ever thought about the role of animals in your favorite sweet treat?*

Bone, Horn, and Hooves

Think about the last time you ate marshmallows, gelatin dessert, or gummy bears. Did you think about how animal by-products played a role in bringing you those tasty treats, **Figure 12-73**? There are many uses for the bone, horn, and hooves of an animal. One of the most important uses is in the production of gelatin. ***Gelatin*** is an odorless, colorless, sterile protein powder that is used in many food products. Gelatin is produced by breaking down the collagen proteins found in the connective tissue of animals.

Another application for the bones, horns, and hooves is in adhesives like glue and paste. These products are made in a similar process to the one used to create gelatin. The main difference is that adhesives are not dried in the same process as gelatin powder.

Do you know someone who has a fine china collection, **Figure 12-74**? Have you ever heard of someone refer to the term *bone china*? Animal bones can be heated to a very high temperature and ground into a fine powder that is a great material for creating high-quality chinaware. This process began in the early 1800s and is still being used today.

pamela_d_mcadams/iStock/Thinkstock

Figure 12-74. Bones are heated until they have no moisture, then ground into powder that is used to make bone china. *Does this look like a pile of bones to you?*

Fats and Fatty Acids

The fats and fatty acids coming from animals have many different uses. Animal fats are commonly found in cosmetics, including makeup, shampoo, conditioner, and lotion, **Figure 12-75**. Lanolin, a by-product of the wool industry, is used for lotions, leather conditioner, and is also used as a commercial grade lubricant for heavy machinery.

ajafoto/iStock/Thinkstock

Figure 12-75. Even though animal fats are common in many cooking applications, they are also used in chemicals, industrial lubricants, and in the manufacturing of plastics. *Do you know someone who likes to cook with lard?*

You might also find these fatty acids in places that seem completely removed from that animal industry. In order to produce high-quality plastics, fatty acids can be added to increase desirable traits. That means that you can find animal fats in almost all plastic products. Who knew that you were using an animal by-product when you brought your groceries home in a plastic grocery bag? Animal by-products can also be found in vinyl records, photography processing liquids, crayons, and even perfumes and fabric softener.

Fats and fatty acids are also commonly used in the production of rubber tires because they have the ability to decrease friction and keep the tires cooler as they travel down the road.

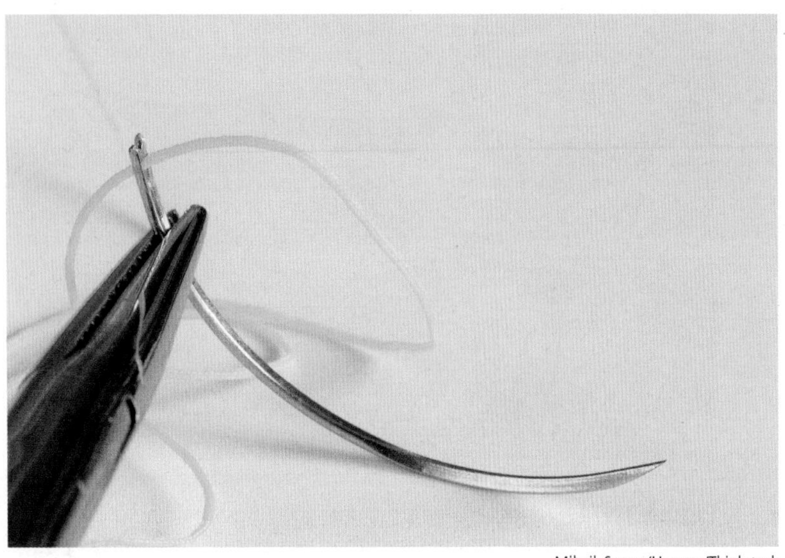

Mihail_Syarov/Hemera/Thinkstock

Figure 12-76. Disposable stitches are often made using catgut, a sterile material that comes from the intestines of ruminant animals. *Can you think of some reasons why this product would degrade in the body?*

Common Uses for Animal By-Products

There are many reasons that animal by-products are used in common items. One of the most important reasons to use animal by-products is in order to use all of the animal, rather than wasting everything but the primary products. We have already mentioned many of the common household items that use animal by-products. Can you think of any by-product uses that we might have missed? Consider for a minute all of the different ways that animal products could be used in human lives. Did you think about the uses that exist in the medical field, industry, agricultural products, and household items? Let us take a minute and see how by-products are used in these areas to make human life easier.

Did You Know?

Catgut is not actually made from cats and never has been. Catgut comes from the healthy intestinal tissues of ruminant animals, most commonly sheep, although catgut can also come from cattle and goats.

By-Products in Medicine

Humans and animals are similar in structure. In fact, you share 84% of your DNA sequence with your dog, 90% of the DNA of a mouse, and somewhere near 95% of the DNA sequence with swine. Because of our close biological connection to animals, there are medical applications for humans with regard to animal by-products. For example, a material made from animal intestines, called *catgut*, is used for disposable sutures, **Figure 12-76**.

Pharmaceuticals

Human pharmaceuticals are often produced from animal by-products. Pig insulin was the most common type of insulin given to human diabetics until biotechnology evolved to be able to recreate a human insulin product. Estradiol, a human female reproductive hormone, is produced in pregnant horses and then refined to a product that can be used to regulate the reproductive cycle of women. The adrenal glands of animals can also be used to harvest cortisone for use in humans.

Xenotransplantation

Another use for animal by-products in human medicine is through the process of xenotransplantation. ***Xenotransplantation*** is the process of grafting or transplanting tissues and organs from one species to another, **Figure 12-77**. It sounds a little bit like a science fiction movie, but doctors have been transplanting pig heart valves into human patients since the late 1960s. Doctors can also use cow heart valves as heart valve replacements.

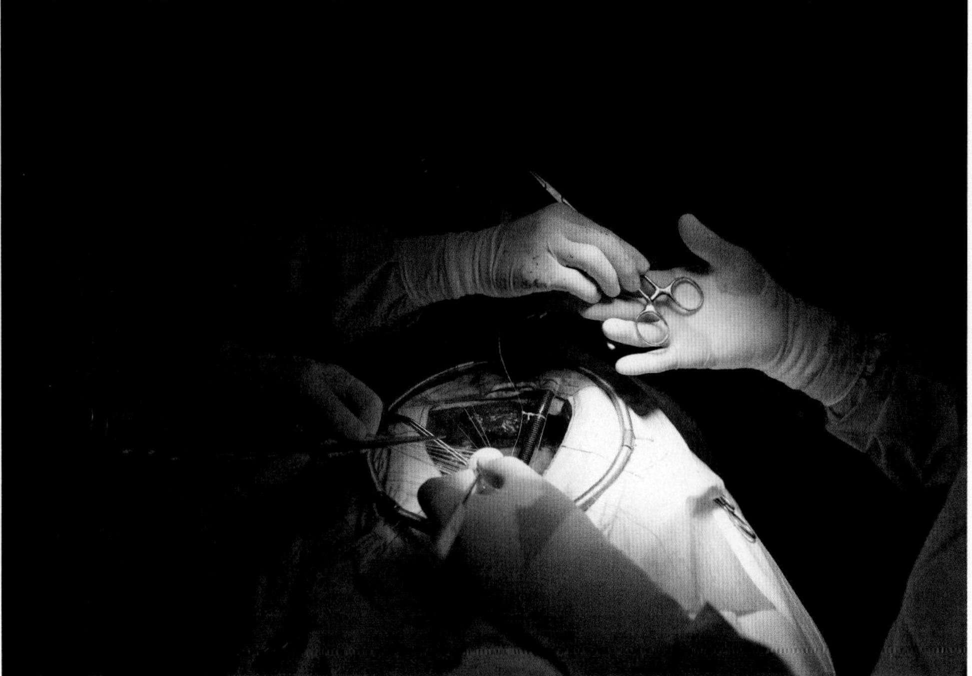

tanyaross71/iStock/Thinkstock

Figure 12-77. Heart surgeons perform more than 100,000 heart valve replacement surgeries each year. Artificial valves are an option, but many patients choose a valve replacement from a pig or cow heart. ***Which type of replacement valve would you choose?***

This choice may be more common than you think. There are literally thousands of Americans who are living their daily lives with the aid of animal heart valves.

Xenotransplantation is also used for skin grafts in many patients. All animal tissues used in humans go through special treatment procedures that decrease the risk of rejection. Scientists are working to increase this technology in the hope that one day, xenotransplantation of entire organs might be possible.

Industry

Take a moment and think about all of the moving parts in a large piece of industrial equipment. In order for that equipment to function properly, all of the moving parts need to be able to sustain constant movement without friction. To help with that, lubricants are used. Most lubricants, like the grease used on the roller bearing in **Figure 12-78**, are made using animal by-products. Lubrication is the largest industrial use for rendered animals.

In addition to lubrication, animal by-products are used in solvents, antifreeze, and in the production of chemicals like herbicides and pesticides. The water-resistant properties of fatty acids make animal by-products a good addition to industrial chemicals that need to be applied in wet environments. Even items like industrial-strength cleaning products often have the addition of animal by-products like fatty acids, glycerin, or gelatin.

simazoran/iStock/Thinkstock

Figure 12-78. Animal fats are used in the production of industrial lubricants, such as the grease being applied to this roller bearing. *What other types of industrial chemicals do you think contain animal by-products?*

Did You Know?

It is illegal for any ruminant animal feed to contain by-products from ruminant animals. This regulation is in place to prevent the transmission of encephalopathic diseases such as scrapie in sheep and mad cow disease (BSE).

Agricultural Products

Agriculture is the industry that produces animal by-products; it only makes sense that some of these by-products are used by the agricultural industry. Many animal feeds contain the by-products of other animals. The food fed to fish in aquaculture settings, for example, is commonly made from blood and bone meal. This meal is made by drying the blood and bones of other animals and grinding it into a fine powder. Other animal feeds include animal by-products as well.

By-products are also used in the production of herbicides and pesticides, and many farmers use animal by-products as fertilizer for their crops. When agricultural producers use the by-products of animal agriculture, the entire production cycle comes full circle.

Household Items

Now that we have taken some time to examine the many uses for agricultural by-products, think about all of the places where you use animal by-products every day. There are literally hundreds of ways that these products play a role in your daily life. When you were getting ready for school this morning, did you use soap, shampoo, conditioner, lotion, or any kind of cosmetic? If so, then chances are that animal by-products played a role in helping you get ready for the day. Some of the other household uses for animal by-products are shown in **Figure 12-79**.

Everyday Products Containing Animal By-Products	
Blankets	Mattresses
Brake fluid	Packaging ink
Brushes	Paint
Candles	Plastic bags
Chewing gum	Plywood (glue)
Cosmetics	Rubber tires
Couch cushions	Saddle soap
Crayons	Shampoo
Deodorant	Shaving cream
Hair conditioner	Synthetic fabrics
Lotion	Toothpaste

Goodheart-Willcox Publisher

Figure 12-79. There are many everyday products you would not necessarily think contain animal by-products. *Which of these products surprises you most?*

Career Connection Rendering Plant Operator

Managing the leftovers of animal agriculture

Job description: Rendering plant operators are responsible for taking the whole carcass or parts of animals that are left over from the meat industry and repurposing their parts. The job generally includes the production of by-products like bone and blood meal.

Education required: Most rendering plant operators receive on-the-job training from someone already in the industry. The highly specialized nature of this industry means that many times, the company is passed down from generation to generation.

Job fit: This job may be for you if you enjoy working outdoors, are interested in complete recycling of agricultural products, and do not mind getting a little dirty.

Lesson 12.4 Review and Assessment

Words to Know

Match the key terms from the lesson to the correct definition.

1. The skin of an animal.
2. Meat that comes from a nonskeletal muscle.
3. Tanned animal hide that has had the hair removed.
4. The process of grafting or transplanting tissues and organs from one species to another.
5. A chemical or mechanical process whereby the hide of an animal is preserved.
6. The ratio of live animal weight to carcass.
7. A tanned animal hide that has been processed with the hair on.
8. Something that is produced incidentally in the production of a primary product.
9. An odorless, colorless, sterile protein powder that is used in many food products.
10. The main purpose for which an animal is produced.
11. The taking of by-products and converting them into useful goods.

A. by-product
B. dressing percentage
C. gelatin
D. hide
E. leather
F. pelt
G. primary product
H. rendering
I. tanning
J. variety meat
K. xenotransplantation

Know and Understand

Answer the following questions using the information provided in this lesson.

1. *True or False?* More than 98% of an animal harvested for human consumption is used to manufacture useful products.
2. What is the primary product of the following animal industries: beef cattle, swine, and laying hens?
3. What percentage of the average beef animal is variety meats?
4. Give three examples of variety meats.
5. Explain why someone who wants to live a vegan lifestyle should not eat gummy-type candies.
6. How do animal by-products help in human medicine?
7. What are two common examples of xenotransplantation?
8. List three applications for animal by-products in industry.
9. Explain how animal by-products and plastic grocery bags are related.
10. Explain the difference between a pelt and leather.

Analyze and Apply

1. More dressing percentages. Using the formula from the lesson, calculate the dressing percentage of the following animals (round to the nearest tenth of a percent):
   A. Market Steer: live weight 1,240 lb, carcass weight 723 lb
   B. Market Hog: live weight 265 lb, carcass weight 196 lb
   C. Market Goat: live weight 160 lb, carcass weight 83 lb
   D. Market Lamb: live weight 130 lb, carcass weight 74 lb

Thinking Critically

1. Vegans are people who desire to live their lives without consuming or using any animal products. Explain to a vegan some of the things in addition to food that they would need to avoid to live a truly vegan lifestyle.
2. The rendering industry often gets the reputation for being smelly and messy. Now that you understand more about by-products, create an informational pamphlet that highlights the importance of animal by-products in society.
3. Which of the categories of animal by-products do you feel is most important in our society? Write a 1–2 paragraph response that outlines your reasoning.

Chapter 12

Review and Assessment

Lesson 12.1

Companion Animals

Key Points

- Companion animals and pets are important to the families they belong to for emotional and psychological reasons. Pets are a source of happiness, stress relief, and well-being for many Americans.
- The companion animal industry in the United States is huge and growing. It is a major contributor to our economy and provides countless career opportunities.
- Companion animals are more than dogs and cats in our society. There are many pets that bring enjoyment and pleasure to many American households.
- Companion animals require a large time commitment from their owners. From daily feeding and watering, to exercise and grooming, to providing veterinarian care, companion animals are a large responsibility.
- Safe handling of pets and companion animals is a necessity. Animals can revert back to their wild heritage, and caution must be taken when dealing with injured or sick companion animals.
- It is important for companion animal owners to have a relationship with a veterinarian skilled in dealing with their specific pet species.
- There are numerous careers in the companion animal industry. Some require minimal education while others require many years beyond college.

Words to Know

Use the following list and the textbook glossary to review and study the *Words to Know* from *Lesson 12.1*.

buck
companion animal
doe
domesticated animal
euthanasia
feral animal
fryer
hutch
kindling
kit
lapin
litter
neuter
spay
whelping
wild animal
zoonoses

Check Your Understanding

Answer the following questions using the information provided in *Lesson 12.1*.

1. *True or False?* Domesticated animals rely on humans for sustenance.
2. *True or False?* Wild animals will not retain aggressive or wild behaviors after they have lived in captivity.
3. How are zoonotic diseases transmitted from animals to humans?
4. What is the difference between a feral animal and a wild animal?
5. What type of economic impact does the companion animal industry have in the United States?
6. What three basic factors should be considered before you become a pet owner, add another pet to your family, or begin breeding animals for profit?
7. Identify three ways you can ensure the health and welfare of your companion animals.
8. Why is it important to know the gestation period of the species of animal you are breeding?
9. What are the benefits of neutering or spaying companion animals?
10. Why is it important to know the life expectancy of an animal you are considering owning?

Lesson 12.2

Aquaculture Industry

Key Points

- Aquaculture is very important in the United States and globally.
- The production of fish and other aquatic animals is economically significant, and the sustainability of aquatic ecosystems for fish and aquatic plant production makes this industry viable and complex.
- In order for the aquaculture industry in the United States to meet consumer demands for seafood, this industry must continue to expand and grow.

Words to Know

Use the following list and the textbook glossary to review and study the *Words to Know* from *Lesson 12.2*.

aeration
anadromous
aquaculture
concentrated aquatic animal production (CAAP)
crustaceans
dissolved oxygen (DO)
fingerling
fishery

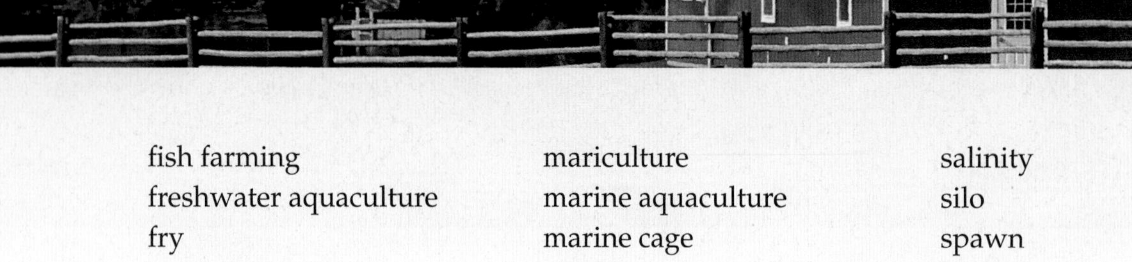

fish farming
freshwater aquaculture
fry
hatchery
intensive recirculating aquaculture system (IRAS)
mariculture
marine aquaculture
marine cage
net pen
pond
raceway
salinity
silo
spawn
stocking rate

Check Your Understanding

Answer the following questions using the information provided in *Lesson 12.2.*

1. What are the top three countries in aquaculture production?
2. *True or False?* Less than 10% of the seafood consumed in the United States comes from other countries.
3. What types of aquatic life are raised in aquaculture production systems?
4. What is the main difference between marine and freshwater aquaculture production?
5. What is the main reason for a producer to select an indoor production system over an outdoor production system?
6. What types of containment systems are used in indoor production systems?
7. What types of containment systems are used in outdoor production systems?
8. Why is water quality one of the most important aspects of any aquaculture operation?
9. What factors must be controlled in contained aquaculture systems?
10. How may freshwater ponds become contaminated?
11. Why do aquatic organisms have difficulty with rapid or extreme changes in water temperature?
12. List three aspects of aquaculture production affected by water temperature.
13. Why is aeration important in aquaculture production systems?
14. Why are aquaculture production systems prohibited from discharging untreated wastewater into the water supply?
15. Explain how fish are feed-efficient animals.

Lesson 12.3

Nontraditional Animal Industries

Key Points

- Nontraditional animals are raised and marketed as animal agricultural products.
- There is a wide array of nontraditional animals that are produced for human consumption.
- Bees are the most prolific nontraditional agricultural animal.
- Each nontraditional animal has its own considerations for production and marketing.

Words to Know

Use the following list and the textbook glossary to review and study the *Words to Know* from *Lesson 12.3*.

antler
apiculture
buck
Chronic Wasting Disease (CWD)
colony
Colony Collapse Disorder (CCD)
cria
dam
doe
drone
fawn
gelding
lanolin
nontraditional animal industry
stud
velvet
venison

Check Your Understanding

Answer the following questions using the information provided in *Lesson 12.3*.

1. What is a nontraditional animal industry?
2. Explain the roles the queen, drone, and worker bees play in a beehive.
3. List three things that a beekeeper using bees for pollination should consider when placing bees in the fields.
4. Why is the fleece from younger llamas and alpacas more desirable?
5. Why is the velvet from deer and elk valuable?
6. What disease is the most concerning to a deer and elk producer?
7. Bison are produced in a similar fashion to beef cattle. Please give two differences between raising beef cattle and raising bison.
8. What nutritional advantages come with the consumption of bison meat?
9. Alligator production is conducted for which primary product?
10. Explain how alligator ranching works as a conservation method.

Lesson 12.4

By-Products from Animal Industries

Key Points

- A by-product is produced incidentally in the production of a more desirable primary product.
- As much as 50% of some livestock animals is considered by-products.
- Leather is the most valuable by-product from most livestock animals.
- Gelatin is found in many food products and is made from the bone, horns, and hooves of animals.

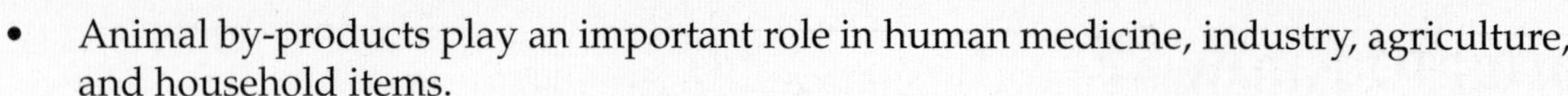

- Animal by-products play an important role in human medicine, industry, agriculture, and household items.

Words to Know

Use the following list and the textbook glossary to review and study the *Words to Know* from *Lesson 12.4*.

by-product
dressing percentage
gelatin
hide
leather
pelt
primary product
rendering
tanning
variety meat
xenotransplantation

Check Your Understanding

Answer the following questions using the information provided in *Lesson 12.4*.

1. Explain the difference between a primary product and a by-product.
2. What does an increase in dressing percentage mean for the amount of by-products from an animal?
3. How do variety meats differ from traditional cuts of meat?
4. Give three examples of by-products that come from hide and hair.
5. Give three examples of by-products that come from bone, horn, and hooves.
6. Give three examples of by-products that come from fats and fatty acids.
7. What role do animal by-products play in human pharmaceuticals?
8. Explain the concept of xenotransplantation.
9. Why are there limitations to feeding ruminant animals feed products derived from other ruminant animals?
10. Explain in 1–2 sentences why it would be difficult to eliminate all animal products from your life?

Chapter 12 Skill Development

STEM and Academic Activities

1. **Science.** Aquaculture relies on scientific measures to test and measure the amount of available oxygen, nutrients, and pH level in water sources. As a class, obtain water samples from several local water sources (ponds, lakes, streams, etc.). Test the water to determine the amount of oxygen, dissolved nitrogen, and the pH. As a class, create a chart and determine which of the water sources would be the most likely to support various types of aquatic life.

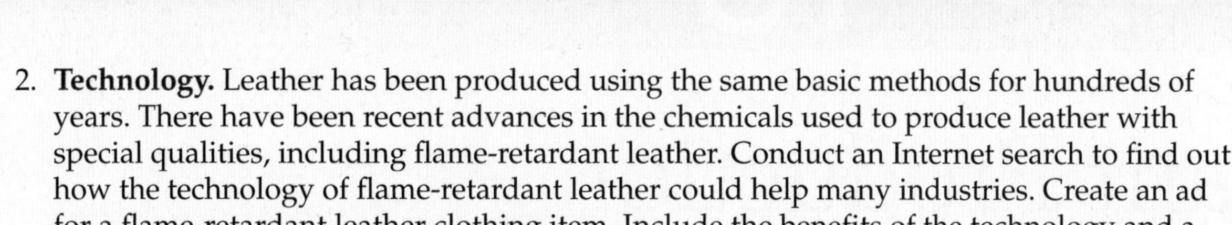

2. **Technology.** Leather has been produced using the same basic methods for hundreds of years. There have been recent advances in the chemicals used to produce leather with special qualities, including flame-retardant leather. Conduct an Internet search to find out how the technology of flame-retardant leather could help many industries. Create an ad for a flame-retardant leather clothing item. Include the benefits of the technology and a quick description of how this technology works.
3. **Engineering.** Although working beef cattle and bison are similar, the equipment required for working cattle is not strong enough to handle larger and more unpredictable bison. Select either a chute, holding pen, or feeding manger used for beef cattle. Explain and draw how you would adapt the standard beef cattle equipment to be suitable for bison. Make sure to label the changes you have made to the standard equipment.
4. **Math.** The real cost of pet ownership. The purchase of a pet is likely the smallest portion of owning an animal. Calculate the overall cost of ownership for a year in the following situation. Randy has five cats. He feeds them a total of 1 lb of food a day. He loves his cats and feels it is important to feed them high-quality feed that costs $21.00 for a 20-lb bag. He takes them to the vet annually for a checkup, which costs $75 each. He also spends on average $30 a month for toys and treats. During the holidays, he buys each of them a present for an average of $10 each. About how much does Randy spend on his cats annually?
5. **Social Science.** Cat person or dog person? There is an ongoing discussion between those who like cats and those who like dogs. Have you ever stopped to wonder if there are factors that contribute to someone being a cat person or a dog person? Conduct a survey of people in your community. Ask them if they consider themselves a cat person or a dog person. Collect data related to their age, occupation, gender, and other demographic characteristics. Compile your data to see how age, gender, and occupation relate to someone being a cat person or a dog person.
6. **Language Arts.** There are many literary works based on companion animals. Select and read a book where the main character is a companion animal. Write a report summarizing the book and explaining the sections in the book where the author attributed human characteristics to the animal.

Communicating about Agriculture

1. **Reading and Speaking.** Create a PowerPoint™ about the production of a nontraditional animal industry that was not covered in *Lesson 12.3*. Make sure to cover the production cycle, uses of the animals, equipment needed, and feeding considerations. Share your completed PowerPoint™ with the class.
2. **Reading and Writing.** A major animal welfare concern related to pet ownership is the issue of leaving dogs in hot vehicles during the summer months. Create a poster that serves as a public service announcement warning dog owners about the consequences of leaving a dog in a hot vehicle.
3. **Reading and Speaking.** The practice of sustainable fishing goes hand-in-hand with the principles of aquaculture. Research how fishermen are trying to fish responsibly. Create a 30-second radio broadcast explaining why sustainable fishing practices are important for our future. Be prepared to read your radio broadcast to the class.

Chapter 13

Plant Production

Lesson 13.1
Plant Anatomy and Physiology

Lesson 13.2
Cereal Grain Production

Lesson 13.3
Oil Crop Production

Lesson 13.4
Fiber Crop Production

Lesson 13.5
Fruit and Vegetable Production

Lesson 13.6
Ornamental Horticulture

©iStock/Shaiith

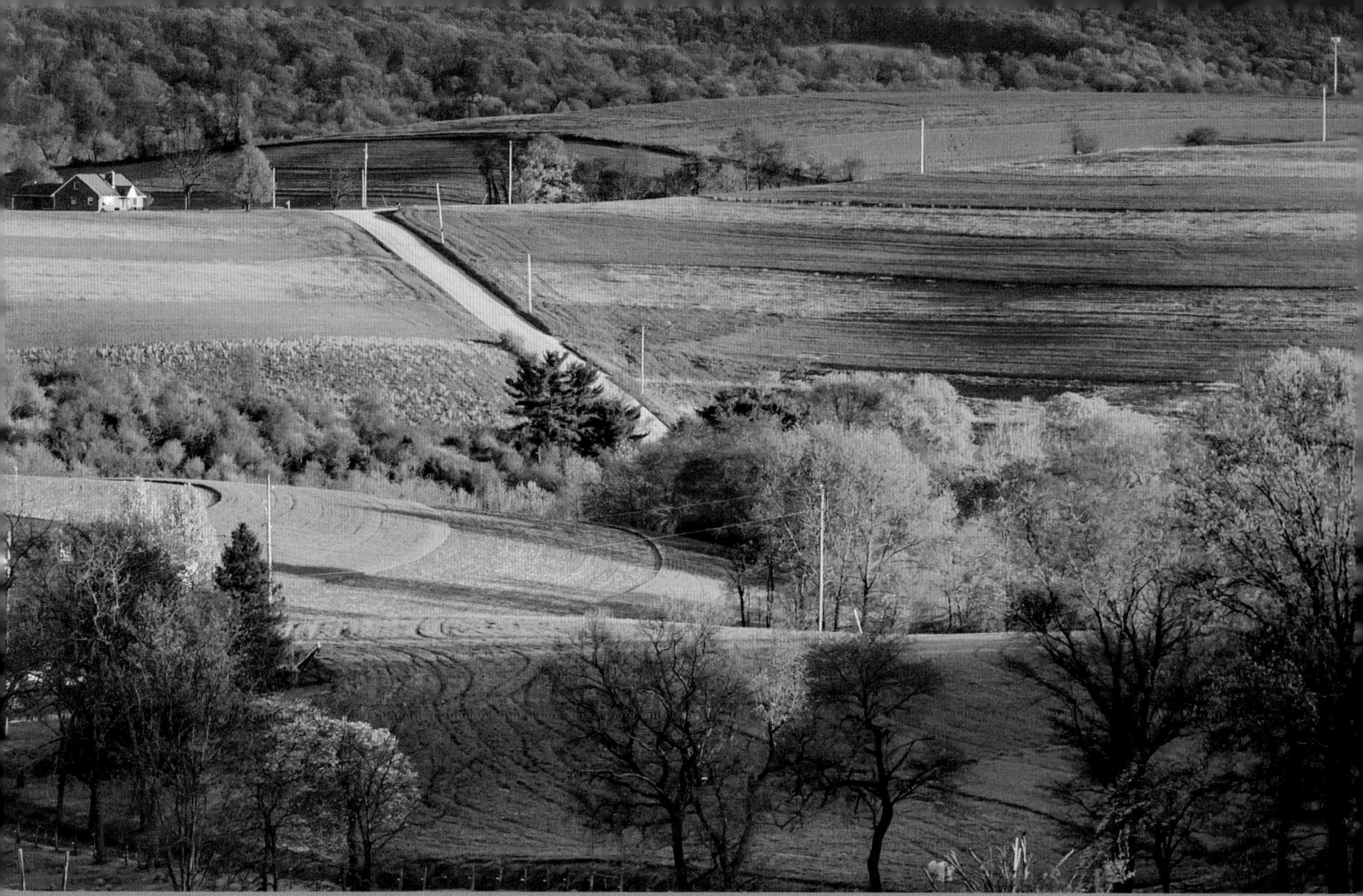

Jon Bilous/Shutterstock.com

©iStock/Romariolen

©iStock/sssimone

While studying, look for the activity icon to:

- **Practice** vocabulary terms with e-flash cards and matching activities.
- **Expand** learning with video clips, animations, and interactive activities.
- **Reinforce** what you learn by completing the end-of-lesson activities.
- **Test your knowledge** by completing the end-of-chapter questions.

Lesson 13.1

Plant Anatomy and Physiology

Words to Know

annual
anther
asexual reproduction
autotroph
biennial
blade
bulb
cambium
cellular respiration
chlorophyll
chloroplast
composite flower
corm
cotyledon
crop
cuticle
dark reaction
dicot
dicotyledon
disk flower
double fertilization
endosperm
epidermal tissue
fertilization
fibrous roots
filament
ground tissue
growth cycle
guard cell
herbaceous stem
imperfect flower
internode
lateral root
leaf
legume
lenticel
light reaction
meiosis
meristem
mesophyll
micropropagation
midrib
mitosis
monocot
monocotyledon
node
nodule
ornamental plant
ovary
ovule
palisade layer
peduncle
perennial
perfect flower
petal
petiole
phloem
photosynthesis
pistil
pistillate flower
pollen
pollen tube
pollination
primary root
ray flower
receptacle
rhizomes
root
root cap
root hair
runner
sepal
sexual reproduction
spongy mesophyll
stamen
staminate flower
stem
stigma
stolon
stomata
style
symbiotic relationship
taproot
tuber
vascular bundle
vascular tissue
vein
weed
woody stem
xylem
zone of elongation

Lesson Outcomes

By the end of this lesson, you should be able to:

- Explain the structure and function of basic plant anatomy.
- Classify plants based on developmental type, growth cycle, and use.
- Explain the basic process of photosynthesis and its importance.
- Compare and contrast asexual and sexual plant reproduction.

Before You Read

Find a partner and read the lesson title. Tell your partner what you have experienced or already know about this topic. Based on your prior knowledge, write three questions about what you would still like to learn about plant anatomy and physiology. After reading the lesson, share two things you have learned with your peer.

Scientists estimate that there are more than 300,000 different species of plants on Earth. These plants are diverse in how they grow and reproduce. Take a moment to think about some of the weird plants you know about. Did you think of the Venus Flytrap, **Figure 13-1A**, that digests insects? What about the Touch-Me-Not, **Figure 13-1B**, which will fold up its leaves if you touch it? Did you remember the Corpse Flower, **Figure 13-1C**, whose large blossom smells like rotting flesh? The truth is, plants have a unique ability to adapt to their surroundings. No matter how different plants may appear, they all have the same basic systems and structures.

This lesson examines how plants work and the different structures they use to grow and reproduce. Once we understand the plant parts, we will examine the way plants are classified, and identify plants based on their characteristics. Finally, we will examine some of the ways plant parts work together to support the life of the plant.

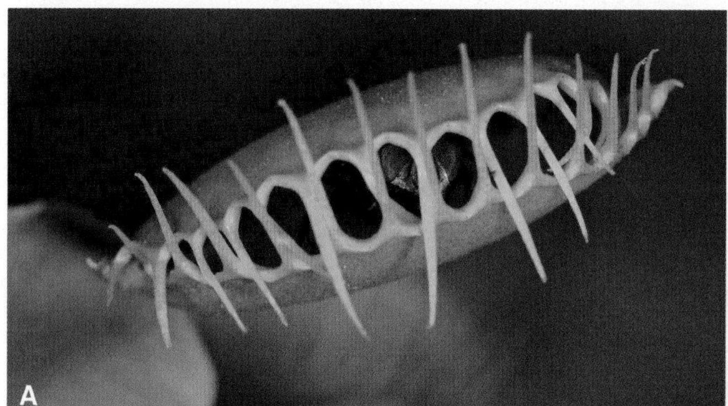

Marco Uliana/Shutterstock.com

kunanon/Shutterstock.com

Andrei Medvedev/Shutterstock.com

Figure 13-1. A—Venus Fly Trap (*Dionaea muscipula*) traps and digests animals as a source of food. B—Touch-Me-Not Plant (*Mimosa pudica*) moves away from touch. C—When the Corpse Flower (*Amorphophallus titanum*) blooms, it smells like rotting flesh. ***Have you ever seen (or smelled) this plant up close? How often does it bloom?***

Did You Know?

The spines on the epidermis of some desert plants serve a purpose besides protecting the plant from being eaten. In very hot climates, the spines provide shade, which can protect the plant from harmful sunrays.

Plant Structure

Just like we have organs that make our bodies work, plants have structures to carry out their life functions. Plants are composed of specialized tissues that combine to make up the basic plant parts. Plants have three main types of tissues: epidermal, ground, and vascular. These tissues function to give the plant its shape and provide structure, **Figure 13-2**.

Epidermal Tissue

Epidermal tissue is the outside covering of a plant. The epidermal tissue works like your skin. It protects internal parts of the plant. Sealing in water is one of the major functions of the epidermal tissue. Some plants have a waxy substance on the epidermal tissue to help the plant retain water. These plants often grow in dry, desert-like conditions.

Ground Tissue

Ground tissue is the tissue that makes up the largest part of most plants. The ground tissue includes all the plant parts that perform photosynthesis. This tissue houses the chloroplasts, which we will discuss later in the lesson. Ground tissue also holds the materials created through photosynthesis and provides the bulk of the structure that supports the plant.

Vascular Tissue

Vascular tissues transport materials around the plant, much like blood vessels transport materials in your body. These tissues are structured as a

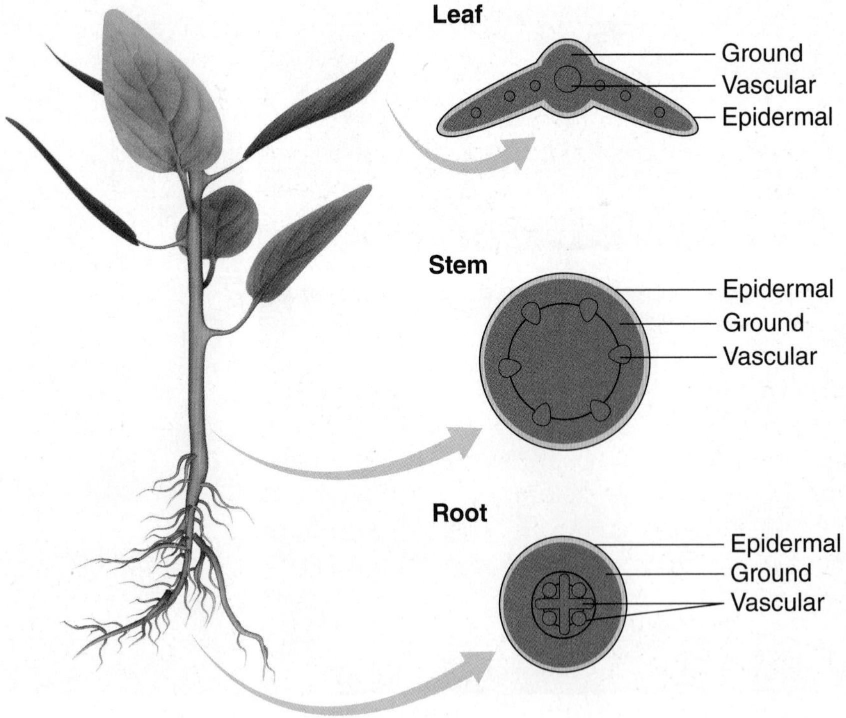

Sciencepics/Shutterstock.com

Figure 13-2. Plants have three types of tissue: epidermal, ground, and vascular. These tissues are found in all plant parts.

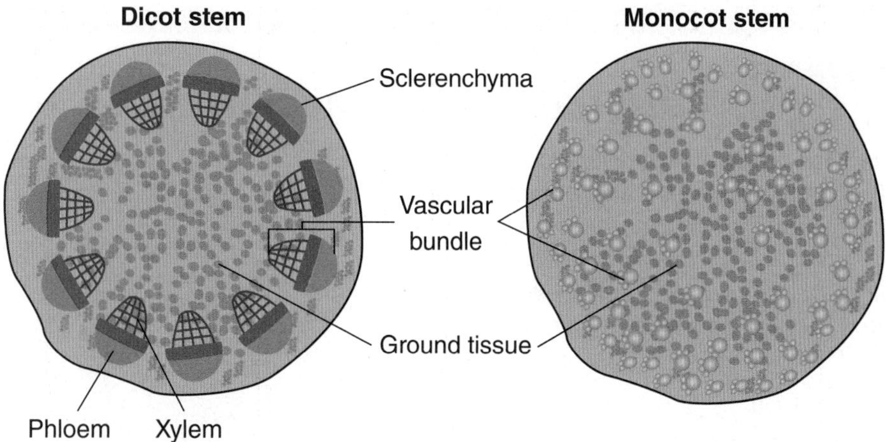

Figure 13-3. Vascular bundles are a combination of xylem and phloem together running throughout the plant.

series of long tubes for water and nutrients to flow through. There are two main types of vascular tissue: xylem and phloem.

Xylem and Phloem

Xylem is the tissue which is responsible for carrying water and nutrients up the plant. Material in the ***phloem*** runs the opposite direction, carrying nutrients produced in the plant toward the roots. Both the xylem and the phloem are made by a layer of cells called the ***cambium***. It is easy to see vascular tissues in the leaves of plants. Leaf veins are the two types of vascular tissue grouped together into structures called ***vascular bundles*, Figure 13-3**.

Did You Know?

The xylem of many vine plants contains large quantities of filtered water that may be safe to drink in survival situations. In many jungle or swamp areas, the water in vines is much safer than water found in ponds and streams.

Plant Parts

Working together, plant tissues make up the four main parts of a plant. These parts fulfill all of a plant's life functions. The four main parts of plants are: roots, stems, leaves, and flowers, **Figure 13-4**. These parts work together to ensure the plant can get nutrients, store energy, support itself, grow and develop, and reproduce.

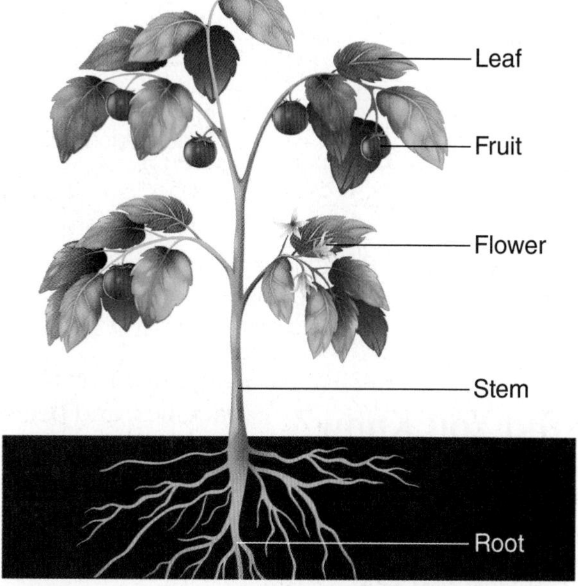

Figure 13-4. Main parts of a plant. *Are there any houseplants in your home? Can you identify the main parts of each plant?*

Roots

The ***roots*** of a plant absorb water and nutrients from the environment and store energy for the plant. They also make sure that the plant is firmly anchored into the ground. The structure of roots is shown in **Figure 13-5**.

The outside of the roots is made of epidermal tissue. Water and nutrients pass through the epidermal tissue and enter the plant. The main portion of the root is called the ***primary root***. The primary root includes the mature section of the root and the section of the root called the zone of

STEM Connection Calculating Surface Area

It is important for plants to have as much of their roots touching the soil as possible, for maximum ability to absorb water and nutrients. The area on the surface of the root is called the surface area and is calculated with a mathematical formula. To simplify the calculation, let us imagine the structure of a root like a cone with a cylinder on top, as shown on the right.

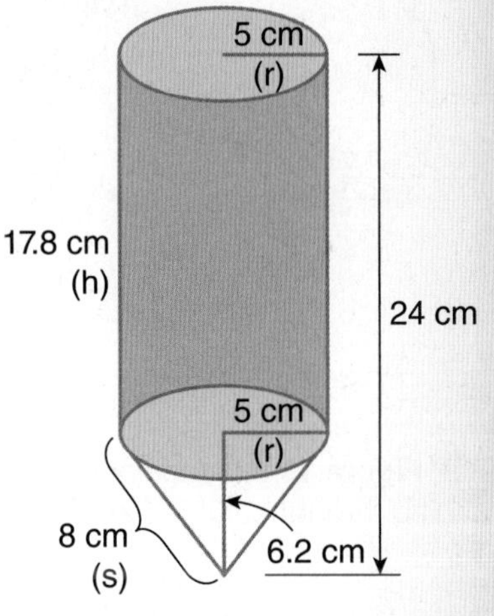

The formula for calculating the surface area of the cylinder portion of the root is:

$$2 \times \pi \times r\,(r + h)$$

To explain the formula, the surface area of the cylinder portion is 2 times the radius (halfway across the diameter) times π (3.14).

To calculate the cone portion of the root, the equation is:

$$\pi rs$$

or, π times the radius times the length of the slant.

By adding the two surface areas together, we can get the surface area of the root. Plants add root hairs to increase the surface area. Root hairs are small in diameter, but essentially long cylinders running out from the root, greatly increasing the overall surface area.

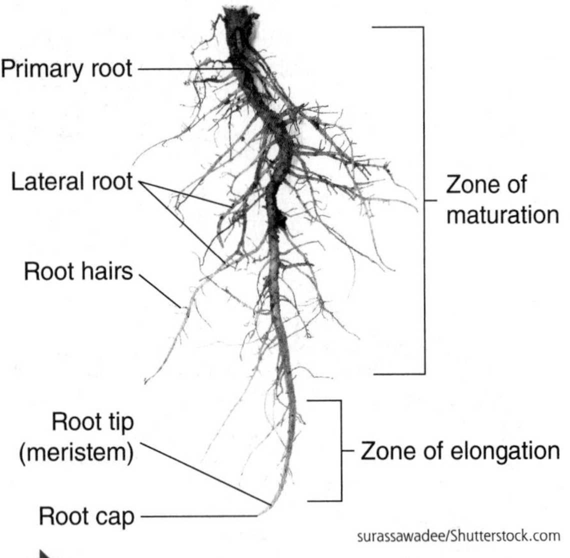

surassawadee/Shutterstock.com

Figure 13-5. Roots have different external parts to ensure they can grow and get the most contact with their surroundings.

elongation. The ***zone of elongation*** is the rapidly growing section of the root, where cells are rapidly dividing through *mitosis*. The very bottom of the root tip is called the ***meristem***. To protect the meristem from rocks and other harmful things in the soil, many plants have a hardened portion covering it called the ***root cap***.

Roots often have large branches growing out from the primary root called ***lateral roots***. The epidermal tissue often produces outshoots called ***root hairs***. Root hairs increase the area of the roots touching the soil, increasing the amount of water and nutrients that can be brought into the plant.

Root Types

Plants have different types of roots, **Figure 13-6**. ***Fibrous roots*** are roots that are heavily branched and spread out underground from the stem. Petunias, beans, and peas are all examples of plants that have fibrous roots. ***Taproots*** are a system of roots with one large primary root and few, if any, lateral roots growing from it. Carrots, radishes, and beets are excellent examples of taproots.

Some plants also have specialized roots that serve a specific purpose. Many plants have evolved to store much of their energy in their roots, making them a valuable food source. Taproot food plants include parsnips, rutabagas, turnips, carrots, beets, and radishes, **Figure 13-7**.

Did You Know?

The plant that grows the longest roots in the world is winter rye. Although the roots stay fairly condensed, the branched roots from a single plant can reach more than 380 miles in length.

Legumes

Legumes are plants that have developed an additional use for their roots. Plants that are legumes, like soybeans, alfalfa, and peas, have small bumps, or ***nodules***, on their roots which house a special form of bacteria. This bacterium has the ability to take nitrogen from the environment that plants cannot use, and convert it into a useable form. This ***symbiotic relationship*** means that legumes have developed a partnership with the bacteria that provides them with all the nitrogen they need. These plants are often used in rotation with other crops, because after a legume has been harvested, leftover usable nitrogen remains in the soil.

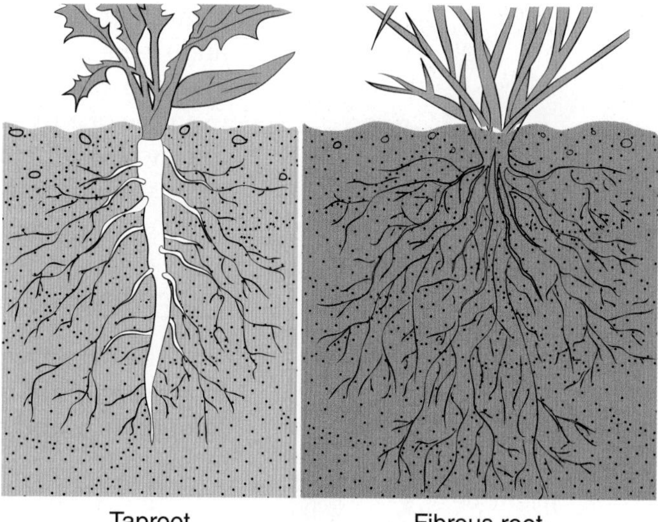

Goodheart-Willcox Publisher

Figure 13-6. Taproots have branches from a central root, whereas fibrous roots branch from the stem. *Is one type of root more effective than the other?*

Stems

A plant ***stem*** is the main trunk of a plant that gives the plant structure and shape. Plant stems also support the leaves and have some ability to make food for the plant.

Stem Types

There are two main types of stems, herbaceous and woody. ***Herbaceous stems*** are found in plants that do not produce wood, such as daisies, grasses, and most crop plants. They are more flexible and less rigorous than woody stems. Herbaceous stems continually grow upward, and have their xylem and phloem together in a bundle. Examples of herbaceous plants include corn, soybeans, tomatoes, and potatoes. ***Woody stems*** are found in plants that produce wood, such as trees and most shrubs. Plants with woody stems have stems that are constantly growing both upward and outward. The xylem and phloem in woody stems grow in large rings around the stem. This growth pattern is what creates a tree's annual rings.

Cary Bates/Shutterstock.com

Figure 13-7. Many roots are commonly eaten as food. *How many other roots do people eat? Are they related?*

Stem Structure

The external structure of a stem is shown in **Figure 13-8**. Stems have small openings that control air and water leaving the plant called ***lenticels***. Along the stem are places where a leaf, flower, or main branch of the stem will emerge; these places are called ***nodes***. The section of a stem without nodes is called the ***internode***. The internode distance can be important in determining plant height and shape.

Modified Stems

Stems have the ability to serve special functions as well. Many types of stems grow underground and provide us with important food sources. Did you know that the potato is not actually a root? It is a modified stem called a ***tuber***. Other modified stems include ***corms***, which are thick underground

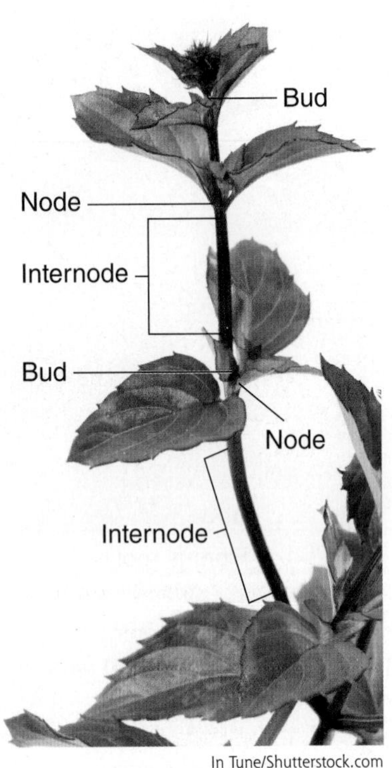

In Tune/Shutterstock.com

Figure 13-8. The external structure of a stem.

stems. Gladiolus and banana trees are examples of corms. ***Stolons*** and ***runners*** are stems that grow above the ground horizontally and connect plants or root themselves in a new place. A very good example of a plant that uses stolons is the strawberry plant. ***Rhizomes*** are another type of modified stem; they are similar to stolons, but grow underground and send plant shoots upward. Examples of common plants with rhizomes are bamboo, irises, and lilies.

Leaves

The primary purpose of a ***leaf*** is to collect sunlight for the plant to perform photosynthesis. Leaves also work to control the plant temperature and evaporation of water from the plant.

Leaf Structure

The internal structure of a leaf is arranged in layers, **Figure 13-9.**

- The top layer of the leaf is a waxy covering called the ***cuticle***. The cuticle seals the leaf and prevents the leaf from losing water through evaporation.

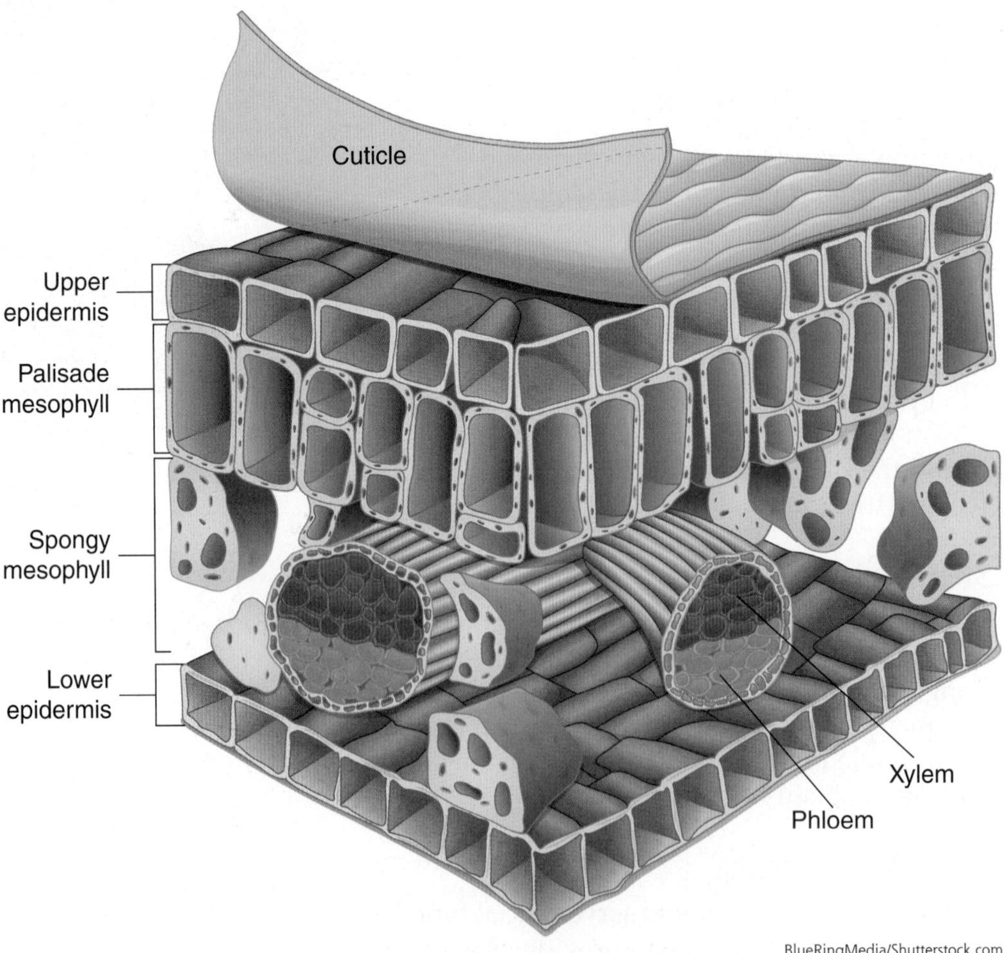

BlueRingMedia/Shutterstock.com

Figure 13-9. The interior of the leaf is sectioned into organized layers to maximize photosynthetic abilities.

- Underneath the upper epidermal tissue of a leaf is a layer of very tightly organized cells called the ***palisade layer***. The cells in the palisade layer have high concentrations of chloroplasts. ***Chloroplasts*** are the structures in the plant cells that collect sun energy and change it into nutrients for the plant.
- The layer beneath the palisade layer of a leaf is called the ***mesophyll***, or ***spongy mesophyll***. This layer has loosely arranged cells that allow gases to pass through for use in photosynthesis.
- The bottom layer (lower epidermis) of a leaf has kidney-shaped ***guard cells***, which create openings called ***stomata***. These stoma expand and contract to open and close the leaf, allowing or preventing gases and water from leaving the leaf. Tree trunks do not contain stomata. They do contain *lenticils*, which operate much like stomata.

Figure 13-10. The external structure of the leaf includes vascular bundles and attachment.

Leaf Shapes and Sizes

Leaves come in many shapes and sizes to serve the needs of each plant. The basic exterior structure of leaves includes their attachment to the stem and the type of vein shape they have, **Figure 13-10**. The leafstalk, which connects the leaf to the stem, is called the ***petiole***. The petiole leads into the ***blade***, which is the main part and shape of the leaf. Leaves contain vascular bundles called ***veins***, which run parallel to each other or branch off a central vein called the ***midrib***.

Like all plant parts, leaves can have modifications to make them more specialized. The most common type of modified leaves are bulbs. ***Bulbs*** are thickened leaves that grow underground attached to a shortened stem and roots. Food plants that contain edible bulbs include onions, shallots, and garlic.

Flowers

The primary purpose of the flower is to produce the male and female sex cells for sexual reproduction in plants.

Flower Structure

Flowers have four basic components: petals, sepals, pistils, and stamen, **Figure 13-11**.

- The ***petals*** serve the plant by protecting the reproductive structures and, for many plants, help attract insects or birds to pollinate the flower.

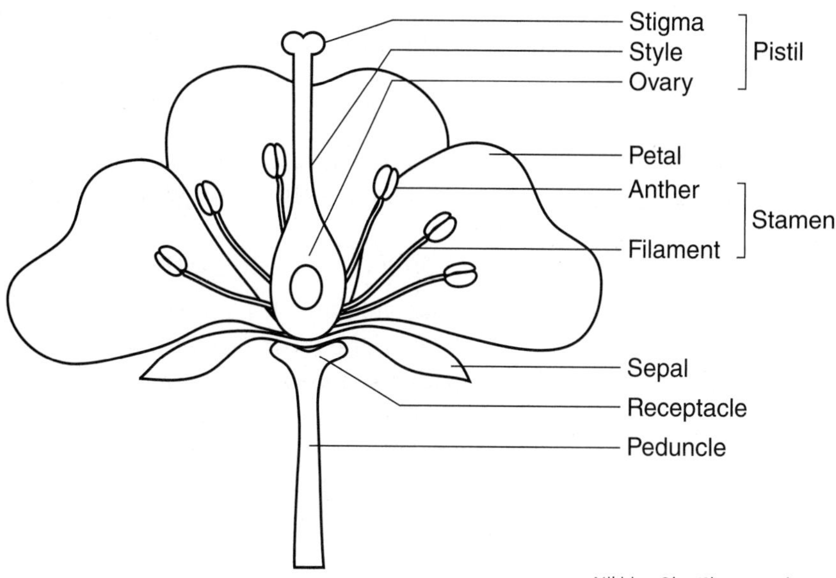

Figure 13-11. Flowers have four basic components: petals, sepals, pistils, and stamen. Flowers with all four parts are considered perfect flowers.

STEM Connection Growing Plants with Hydroponics

Some plants can be grown in aquatic media or water. Hydroponic production of plants is a rapidly growing area of agriculture. It is especially popular in cities and areas where fertile soil is lacking. In a hydroponic system, plants grow in a mineral-enriched water solution in a greenhouse or other structure. The plant needs support from its roots, so the roots are placed in floating pots. The pots contain glass wool, perlite, or gravel. These materials allow the roots to develop and absorb necessary water and minerals from the flowing water solution under the pots. Advantages of hydroponics include the ability to control plant nutrients, reduced water consumption, and fewer weeds and pests. Disadvantages include the initial cost of setting up a hydroponic system, high technical knowledge requirements, and careful monitoring of water quality to ensure optimum plant growth. Can you think of some structures of plants that would make them better suited to growth in a hydroponic system?

stoonn/Shutterstock.com

- ***Sepals*** are small leaves at the base of a flower which help protect the flower when it is in bud form.

  The reproductive parts of the flower are the pistil and the stamen.
- The ***pistil*** is the female part of the flower and is made up of a ***stigma*** (which collects pollen), and the ***style*** that allows pollen to travel to the ***ovary***, which is the site of fertilization in plants. Once fertilized, the ovary develops into the fruit, and the ovules become the seeds.
- The ***stamen*** is the male portion of a flower. A stamen is comprised of an ***anther***, where pollen is produced, and a ***filament***, the stalk that holds the anther.
- The wide bottom portion of a flower is called the ***receptacle***, and the ***peduncle*** is the stem which supports the flower.

Flower Classification

Flowers are classified based on the parts they have. If a flower contains all four of the basic flower parts (petals, sepals, pistils, and stamen), it is considered a ***perfect flower***. If it is missing one or more of the basic parts, it is called an ***imperfect flower***. Flowers can also be classified by the type of reproductive structures they have. ***Pistillate flowers*** have only female reproductive parts (stigma, style, ovary). ***Staminate flowers*** only have male reproductive parts (anther, filament). That means that staminate flowers cannot produce seeds or fruit.

Some plants have developed the ability to group many flowers close together into one structure. ***Composite flowers*** are made up of many small flowers in one large "flower" structure, **Figure 13-12**. Some scientists believe

Ian Grainger/Shutterstock.com

Kevin H Knuth/Shutterstock.com

weerayut ranmai/Shutterstock.com

Figure 13-12. Types of composite flowers. A—Coreopsis. B—Queen Anne's Lace. C—Gazania.

that these plants are the most recently evolved type of flower. Composite flowers can be made of ray flowers, disk flowers, or a combination of the two.

Ray flowers are small, narrow flowers that usually have no stamens, and one long petal. Examples of plants that have composite flowers with ray flowers only are dandelions, chicory, and endive. ***Disk flowers*** are wider, flat flowers with large stigmas. Some plants that have disk only composite flowers are thistles and burdock. Combination flowers have ray flowers on the edges and disk flowers in the center. Any flower that has an "eye" (i.e. sunflowers, daisies, chrysanthemums, or asters) is considered a combination composite flower, **Figure 13-13**.

Nattika/Shutterstock.com

Figure 13-13. A single flower, such as this gerbera, may have both disk and ray flowers.

Plant Classification

With such a large number of plant species on Earth, scientists have identified structures in plants that are similar and classified plants based on the structures they have, the way they grow, and their uses.

Structure

One of the largest classifications of plants is based solely on structure, and differences are obvious from the time that the plant comes through the surface of the soil. The ***cotyledon*** is the very first seed leaf produced by a plant. Plants that have one seed leaf when they first emerge are called ***monocotyledon*** or ***monocots***. Plants that have two seed leaves when first emerging are called ***dicotyledons*** or ***dicots***. Major differences in structure exist between monocots and dicots, **Figure 13-14**.

Growth Cycle

Another way plants are classified is by their growth cycle. A ***growth cycle*** is the amount of time it takes for a plant to germinate, grow to maturity, and

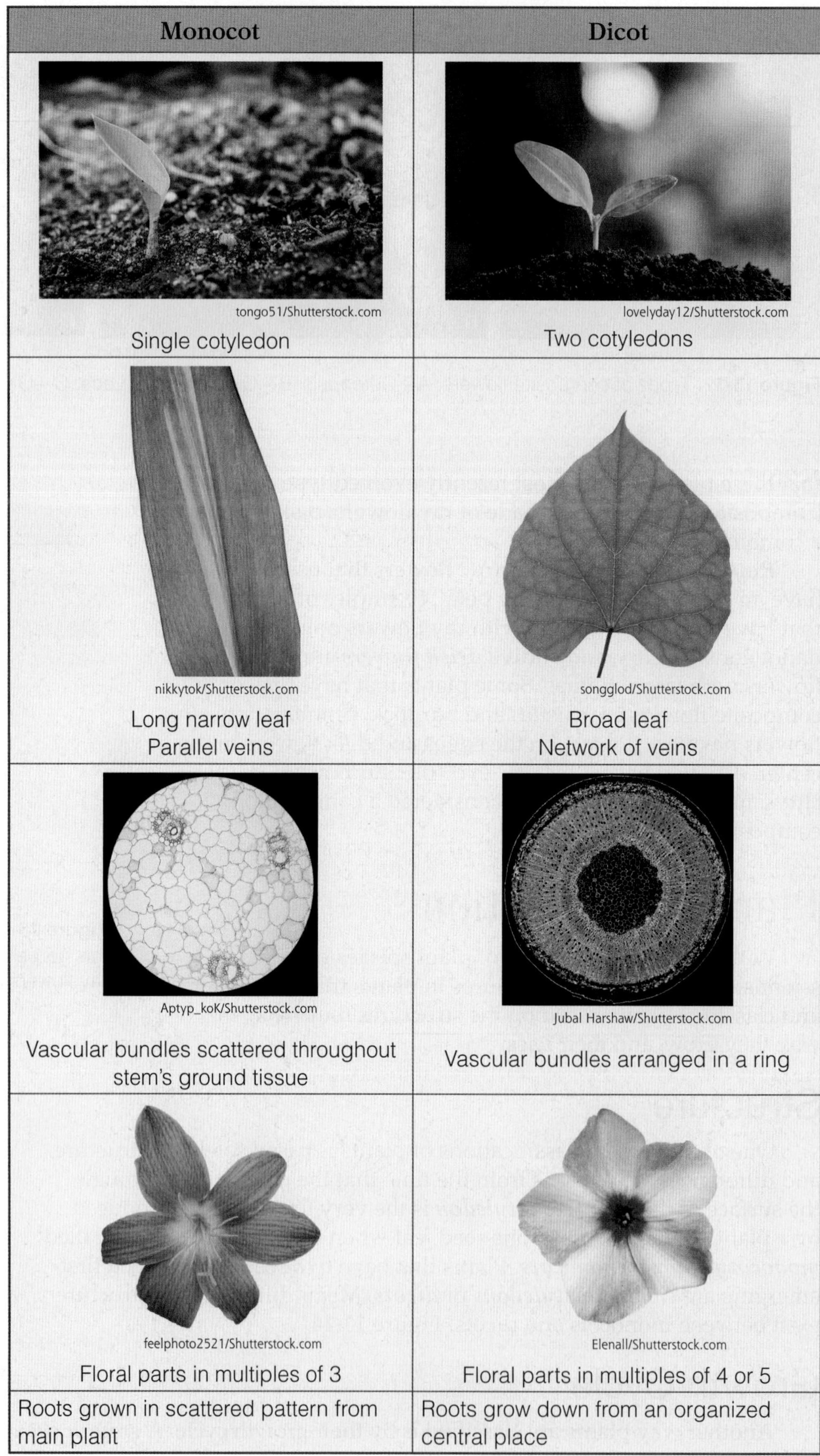

Monocot	Dicot
tongo51/Shutterstock.com Single cotyledon	lovelyday12/Shutterstock.com Two cotyledons
nikkytok/Shutterstock.com Long narrow leaf Parallel veins	songglod/Shutterstock.com Broad leaf Network of veins
Aptyp_koK/Shutterstock.com Vascular bundles scattered throughout stem's ground tissue	Jubal Harshaw/Shutterstock.com Vascular bundles arranged in a ring
feelphoto2521/Shutterstock.com Floral parts in multiples of 3	Elenall/Shutterstock.com Floral parts in multiples of 4 or 5
Roots grown in scattered pattern from main plant	Roots grow down from an organized central place

Figure 13-14. Differences in monocots and dicots.

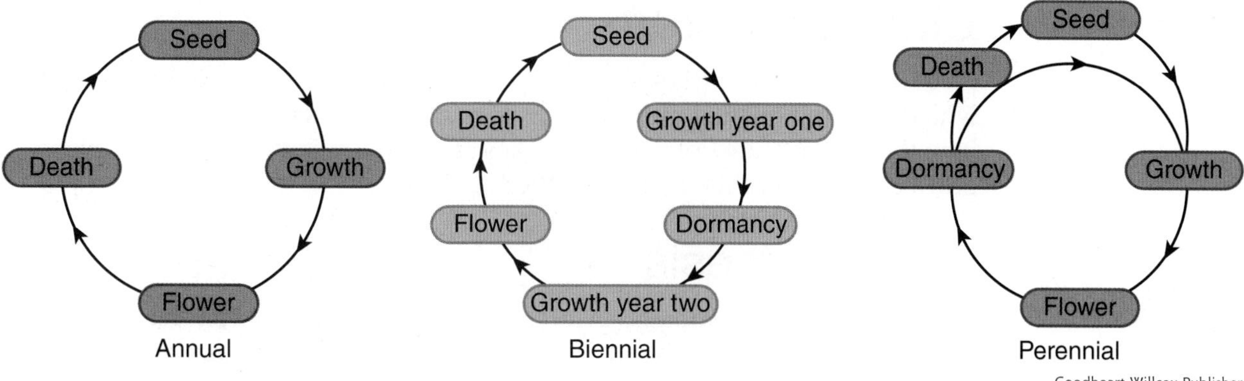

Figure 13-15. Annual, biennial, and perennial growth cycles. *Is it possible to extend the life of an annual by cultivating it indoors?*

reproduce. Plants are classified into one of three growth cycle categories, **Figure 13-15**:

- ***Annuals***—plants that complete their growth cycle in one growing season, and all roots, stems, and leaves die at the end of each growing season.
- ***Biennials***—plants that live for two years to complete their growing cycle. Typically, year one for a biennial includes growing a grouping of leaves near the ground, and the second season results in stem elongation and flowering.
- ***Perennials***—plants that grow for many growing seasons. In many cases, the top portion of the perennial, including the flowers and leaves, dies back at the end of each growing season and is replaced with new growth at the beginning of the next growing season.

Some plants can be perennials in certain climates, but grow as an annual in other regions where the environment is not as favorable. **Figure 13-16** shows common crops and their classification by growth cycle.

Plant Use

An additional way to classify plants is based on their use. This type of classification includes crops, ornamental plants, and weeds. ***Crops*** are any plant grown in the care of humans as a food source. ***Ornamental plants*** are those grown and cared for by humans to increase the beauty of their surroundings. A ***weed*** is defined as any plant that is not cultivated by humans and is growing in competition with desired or cultivated plants. Even plants that are considered to be domesticated varieties of plants can be weeds if they are in an area where they compete with other desired plants.

Growth Cycle	Crops
Annuals	Cereal grains, oil crops, fiber crops
Biennials	Carrots, onions, parsley, pineapples
Perennials	Tree fruits, nut trees, wheatgrass

Goodheart-Willcox Publisher

Figure 13-16. Common crops and their classification by growth cycle.

Plant Functions

The structures of plants work together to perform the various functions which sustain a plant's life and help them produce offspring. Understanding how these plant functions work is a key factor in understanding how plants grow, obtain nutrients, and reproduce.

Photosynthesis

Every bit of the food energy on Earth originally comes from plants harnessing energy from the sun through photosynthesis. ***Photosynthesis*** is the process of converting light energy to chemical energy and storing it in the form of sugar. Organisms that can perform photosynthesis are called ***autotrophs,*** meaning they make their own nutrients. All plants, along with some bacteria and some protists, are autotrophs. Photosynthesis begins with the harnessing of light energy by plant cell organelles called chloroplasts. Many chloroplasts are found in the organized and tightly packed top layer of leaves, called the palisade layer. Chloroplasts contain a green substance called ***chlorophyll,*** which absorbs sunlight to help complete the process of photosynthesis.

The basic chemical formula of photosynthesis is:

$$6CO_2 + 6H_2O \rightarrow C_6H_{12}O_6 + 6O_2$$

To read this formula, we need to rely on our knowledge of chemical symbols. Each element is represented by a letter, and the numbers show how many atoms of each chemical are found in a molecule (more than one atom bound together). There are only three elements involved in photosynthesis: carbon (C), hydrogen (H), and oxygen (O). These elements combine to make 4 molecules:

- CO_2 (carbon dioxide): molecule with one atom of carbon and two atoms of oxygen.

STEM Connection Chemistry Vocabulary

Knowing how to read chemical expressions is important to all types of scientists. Through their understanding of the similarities and differences of chemicals, scientists are able to examine complex chemical reactions.

One of the first steps in understanding chemical reactions is knowing how to read the chemical expression of a molecule. A chemical expression uses the symbols from the periodic table of elements. A chemical expression includes the chemical symbol for each element found in the molecule, along with the number of atoms of that molecule. (See the image on the upper right.) Test your knowledge by writing down the elements and number of atoms in each of the following chemical expressions:

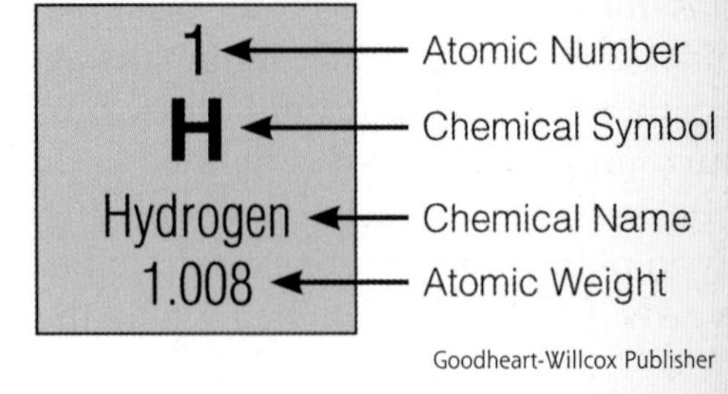

Goodheart-Willcox Publisher

1. Table Salt (sodium chloride): NaCl
2. Water (dihydrogen monoxide): H_2O
3. Simple Sugar (glucose): $C_6H_{12}O_6$
4. Baking Soda (sodium bicarbonate): $NaHCO_3$

ANSWERS: 1. (1 Sodium, 1 Chlorine) 2. (2 Hydrogen, 1 Oxygen) 3. (6 Carbon, 12 Hydrogen, 6 Oxygen) 4. (1 Sodium, 1 Hydrogen, 1 Carbon, 3 Oxygen)

STEM Connection Artificial Leaves

Scientists are developing an artificial leaf that can convert sunlight into useable energy using the same basic chemical processes as real leaves. This technology is more efficient than current solar electricity methods, and researchers are hoping to be able to create an entire power plant using artificial leaf technology.

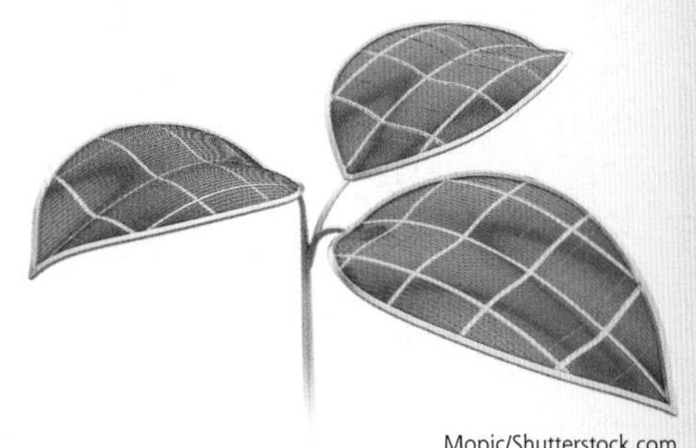

Mopic/Shutterstock.com

- H_2O (water): molecule with two hydrogen atoms and one oxygen atom.
- $C_6H_{12}O_6$ (glucose): molecule with six carbon atoms, 12 hydrogen atoms, and six oxygen atoms.
- O_2 (oxygen): two oxygen atoms bound together. The chemical structure of oxygen makes it more common to find oxygen in molecules of two atoms.

The photosynthesis formula begins with six molecules of carbon dioxide (CO_2) and six molecules of water (H_2O). These 12 molecules undergo a reaction, converting them into one molecule of glucose ($C_6H_{12}O_6$), along with six molecules of oxygen (O_2). Photosynthesis is conducted in two portions: light reactions, which require light, and dark reactions, which can be completed with or without light present. ***Light reactions*** involve converting light energy to chemical energy. Oxygen is also released during this stage. ***Dark reactions*** take the energy molecules produced and convert them into sugar molecules for energy storage.

The oxygen produced through photosynthesis is released into the environment. This is one of the reasons photosynthesis is so important to the planet. Ensuring that there are enough healthy plants to convert CO_2 to O_2 is essential to making sure that our planet is in balance.

There are several factors that contribute to the rate of photosynthesis in a plant. These factors include:

- Amount of light.
- Intensity of light.
- Availability of CO_2.
- Temperature.
- Amount of water.
- Wavelength of light.

Photosynthesis is essential to life on Earth. It is the process by which energy enters living organisms. In order to use the energy, a process opposite from photosynthesis occurs, called cellular respiration. ***Cellular respiration*** is the breaking down of glucose molecules in all living organisms to release energy, and is the inverse reaction to photosynthesis:

Photosynthesis

$$6CO_2 + 6H_2O + \underbrace{\text{energy}}_{\text{energy in}} \rightarrow C_6H_{12}O_6 + 6O_2$$

Cellular respiration

$$C_6H_{12}O_6 + 6O_2 \rightarrow 6CO_2 + 6H_2O + \underline{\text{energy}}$$

energy out

Plant Reproduction

In order for plants to grow and thrive on Earth, they need to have the ability to reproduce. Plants are different from vertebrate animals because they have the ability to reproduce in two ways.

- Asexual reproduction using mitosis.
- Sexual reproduction using meiosis.

Recall your knowledge of cellular reproduction as you look at **Figure 13-17**, which illustrates the differences between mitosis and meiosis.

Asexual Reproduction

Asexual reproduction involves mature plant tissues growing into an exact copy of the plant. This type of reproduction uses ***mitosis***, where one cell is made into two identical cells. Therefore, plants reproduced using asexual reproduction are genetically identical and have the exact same DNA. There are many types of asexual reproduction in plants including cuttings, grafting, division, and layering, **Figure 13-18**.

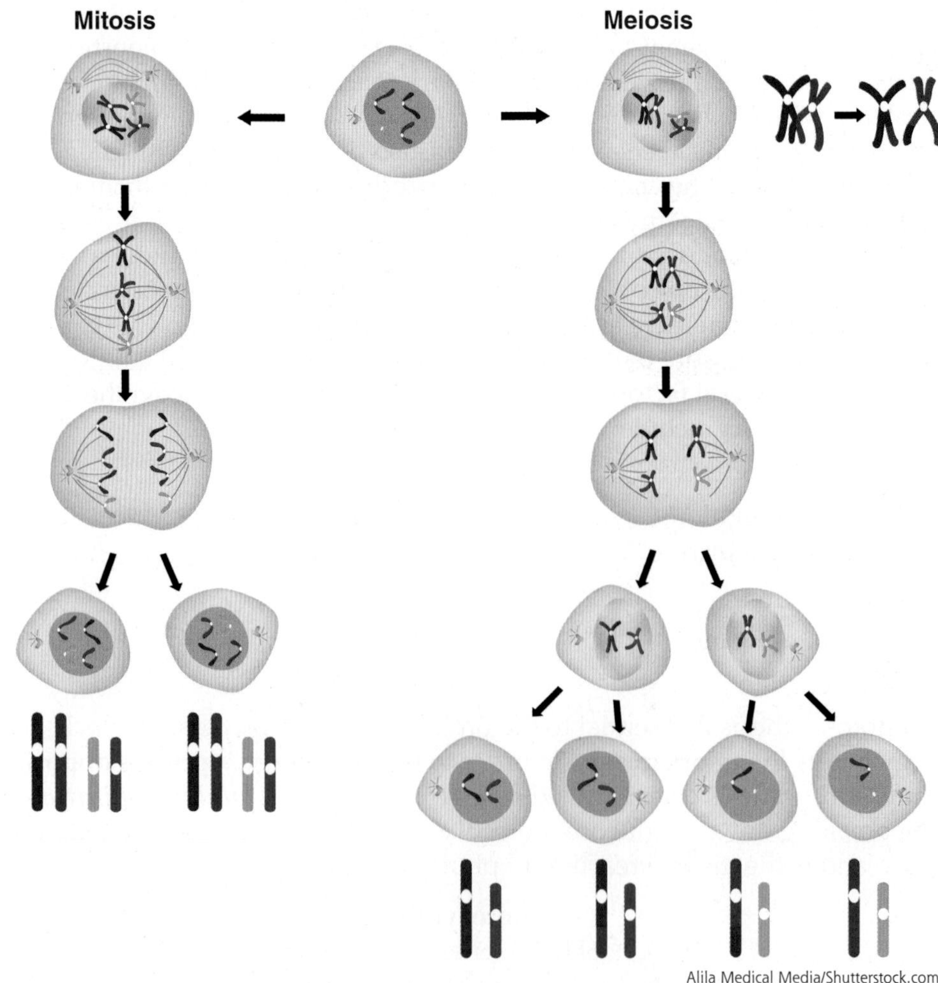

Figure 13-17. Comparison of mitosis and meiosis.

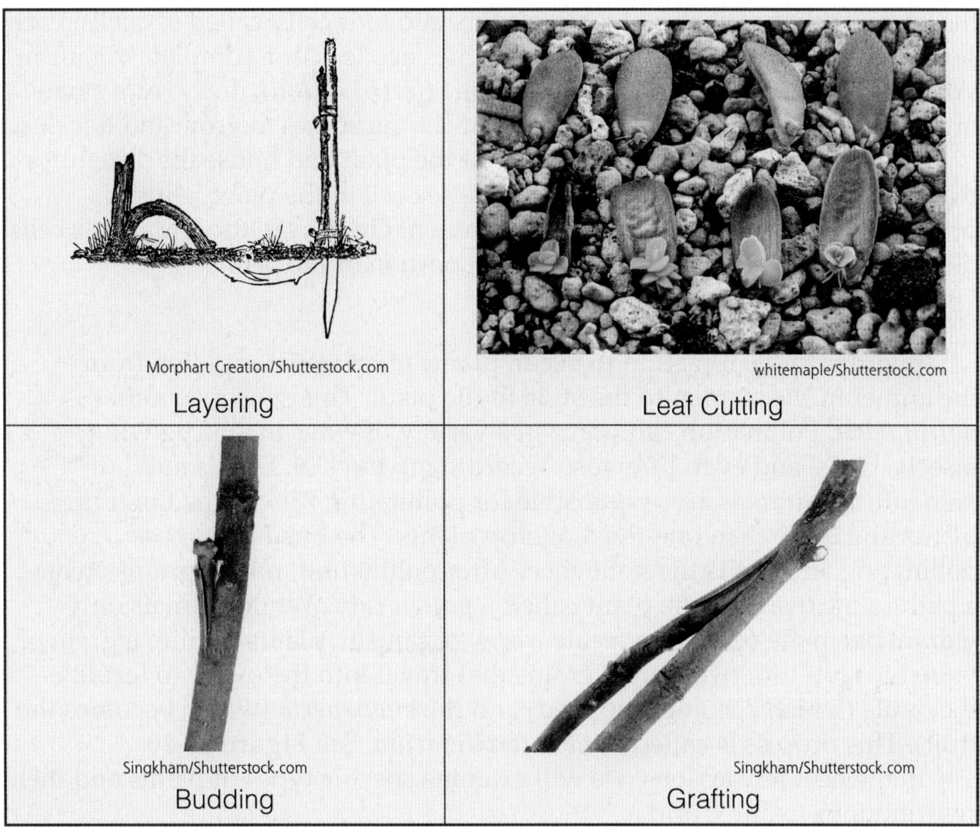

Figure 13-18. Types of asexual plant propagation.

Micropropagation

One of the most technology-enhanced types of asexual plant reproduction is called micropropagation. ***Micropropagation*** is the process of taking a single plant cell and, using a special nutrient-growing medium, growing an entire plant. This process is used in large-scale production of plants, as it provides a way to create thousands of identical copies of a single plant, **Figure 13-19**. This technology is especially helpful in the production of plant cells created through genetic engineering.

Sexual Reproduction

Sexual reproduction, in both plants and animals, involves the creation of sex cells through meiosis. ***Meiosis*** is a process in which one parent cell with

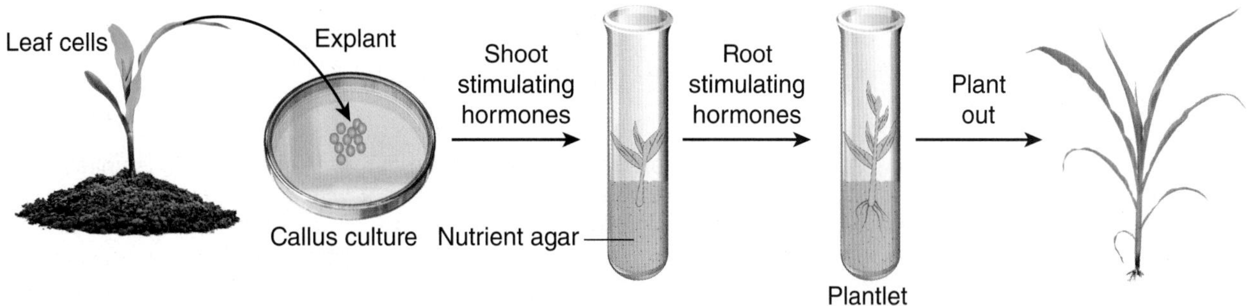

Figure 13-19. Process of micropropagation, a type of asexual reproduction in plants.

the normal number of chromosomes splits into four cells, called sex cells. Each sex cell has half the number of chromosomes needed for a complete organism. When a male and a female sex cell combine (*fertilization*), they create a new organism that has the required number of chromosomes to grow and develop.

Flowers are the reproductive parts of the plant and house the structures that make the sex cells. ***Pollen*** is the male sex cell in the plant, which is produced by the anther portion of the stamen. ***Ovules*** are the female sex cells of plants; they are produced in the ovary portion of the pistil.

Pollination

In order for fertilization to occur, plants must transfer pollen from the anther in the stamen to the style in the pistil. This process is called ***pollination***. Pollination can occur in a variety of ways including wind, insects, birds, and even humans. According to the U.S. Department of Agriculture, animals are responsible for pollinating 75% of the flowering plants and more than one-third of crop plants. The busiest of these pollinating animals is the honeybee. After pollination, pollen grains create a path to the ovary of the plant called a ***pollen tube***. While animals only require one male sex cell to create a new organism, plants require a grain of pollen to split into two pollen grains that travel into the ovary to fertilize the ovule (which becomes the seed) and the ***endosperm*** (which becomes the fruit). This process is called ***double fertilization***. See **Figure 13-20**.

In the next few lessons, we will examine specific types of plants and their contributions to our world.

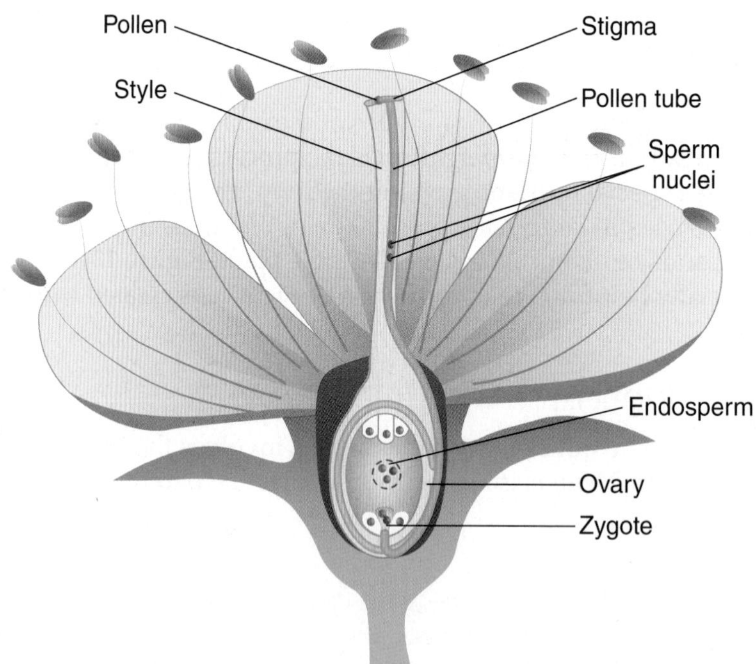

Figure 13-20. In order for seeds to grow, one pollen grain must split and fertilize the ovule and the endosperm.

Hands-On Agriculture

Plant Reproduction

©iStock/jurisam

There are advantages and disadvantages to each type of plant reproduction. Why would it be helpful to have multiple plants with the same genes? Why may it be good to have plants created that have genetics different from the parent plants?

Find a partner. Discuss your answers to the questions above with your partner. Come up with at least two answers for each of the questions. When time is up, share your answers with the class and discuss the main advantages and disadvantages of each type of plant reproduction.

With your partner, use your knowledge of the differences between asexual and sexual reproduction in plants to respond to the following scenario: You are a horticulturalist wanting to grow and market roses. What would be some of the reasons you would choose to propagate your plants using asexual reproduction methods? Explain reasons that you may choose to reproduce your plants sexually instead.

Lesson 13.1 Review and Assessment

Words to Know

Match the key terms from the lesson to the correct definition.

1. A part of photosynthesis in which light energy is converted to chemical energy; oxygen is also released during this stage.
2. A part of photosynthesis in which the energy molecules produced by light reactions are converted into sugar molecules for energy storage.
3. A plant stem that grows above the ground horizontally and connects plants or roots in a new place.
4. A plant that has one seed leaf when it first emerges.
5. A plant that has two seed leaves when it first emerges.
6. A section of a plant stem between nodes; this distance is often essential in determining plant height and shape.
7. A section of a plant stem where a leaf, flower, or main branch of the stem will emerge.
8. A thick modified underground plant stem with scaly leaves and buds from which new plants grow.
9. A tissue produced inside the seeds of a flowering plant around the time of fertilization; it surrounds the embryo and provides nutrition in the form of starch.
10. A type of flexible plant stem that continually grows upward and has its vascular system in a bundle.
11. An organism that can perform photosynthesis to provide its own nutrients.
12. Kidney-shaped cells in the lower epidermis of a leaf that create openings that allow or prevent gases and water from leaving the leaf.
13. Openings in the lower epidermis of a leaf that allow or prevent gases and water from leaving the leaf.
14. The breaking down of glucose molecules in all living organisms to release energy.
15. The female part of the flower made up of the stigma, style, and ovary.
16. The main trunk of a plant that gives the plant structure and shape.
17. The male portion of a flower which is comprised of an anther, where pollen is produced, and a filament, the stalk that holds the anther.
18. The male sex cell in a plant.

A. autotroph
B. cellular respiration
C. corm
D. cotyledon
E. dark reaction
F. dicot
G. endosperm
H. guard cell
I. herbaceous stem
J. internode
K. light reaction
L. meristem
M. monocot
N. node
O. petiole
P. phloem
Q. photosynthesis
R. pistil
S. pollen
T. root
U. runner
V. sepals
W. stamen
X. stem
Y. stomata
Z. xylem

19. The part of a plant that functions as a means of water absorption, food storage, and/or as a means of anchorage and support.
20. The portion of a plant's structure that connects the leaf to the stem.
21. The process plants use to convert light energy to chemical energy and store it in the form of sugar.
22. The small leaves at the base of a flower which help protect the flower when it is in bud form.
23. The very bottom of a plant's root tip.
24. The very first seed leaf produced by a plant.
25. Vascular plant tissue responsible for carrying nutrients produced in the plant down toward the roots.
26. Vascular plant tissue responsible for carrying water and nutrients up the plant.

Know and Understand

Answer the following questions using the information provided in this lesson.

1. What are the four basic plant parts?
2. What is the primary function of vascular tissue in plants?
3. What is the function of the xylem?
4. Name the two types of roots.
5. The nodules on the roots of legumes contain _____ that convert _____ from the soil into a usable form for plants.
6. What are the five types of modified stems?
7. The leaf is the primary site of _____, which produces food for the plant.
8. _____ is the green substance inside the cells that absorbs sunlight.
9. The part of the leaf that connects the blade to the stem is called the _____.
10. The pistil is the _____ part of the flower.
11. What are the two parts of the stamen?
12. Plants that have one seed leaf when they first emerge are called _____.
13. What is the difference between an annual and a perennial?
14. Which chemical elements are found in photosynthesis?
15. What factors would a greenhouse grower adjust to increase the rate of photosynthesis in his or her greenhouse plants?
16. Which type of propagation in plants involves pollen and ovules?
17. For sexual reproduction to occur in plants, what process must also take place?

Analyze and Apply

1. Obtain plants and dissect them into their parts.
2. Under a microscope, cut a cross section of different plant parts and see if you can identify the internal structures.
3. Go on a nature walk outside of the school. Identify every plant you see as either a monocot or a dicot.

Thinking Critically

1. Which of the four main parts of plants do you think are the most essential for survival? Please support your argument with at least three sources of evidence.
2. Explain why trees have annual rings.
3. What are some differences you would expect between flowers that are pollinated by wind and those that are pollinated by insects?
4. Are all perfect flowers complete? Are all complete flowers perfect? Explain your answers.
5. Many scientists believe that carbon dioxide in the atmosphere is increasing. How does this affect the planet's plant population?
6. What would be some reasons that an agricultural producer would want to reproduce plants using asexual reproduction?
7. What would be some of the reasons an agricultural producer might want to create seeds through sexual propagation?

Lesson 13.2 Cereal Grain Production

Lesson Outcomes

By the end of this lesson, you should be able to:

- Describe the morphological characteristics of cereal grains.
- Identify cereal grains by common names and scientific names.
- Explain the production practices of cereal grain crops.

Words to Know

barley yellow dwarf
boot stage
bran
creep feeding
crown rust
dent corn
embryo
endosperm
ergot
flint corn
flour corn
germ
Helminthosporium leaf blotch
hull-less barley
integrated pest management (IPM)
jointing
kernel
lodge
loose smut
moisture content
panicle
popcorn
pseudo-cereals
seedhead
six-row barley
smut
sweet corn
tillering
tillers
two-row barley
windrow

Before You Read

The next time you eat a sandwich at lunchtime, look closely at the bread. What is the bread made from? Which plant did it come from? Is it wheat bread or rye bread? How many different foods in your diet come from a cereal grain? Cereal grains are of major importance as a food source around the world.

For centuries, humans have eaten plants called cereals. You might think of cereals as what you eat for breakfast, but they are actually a classification of plants. Cereal grains belong to the grass family *Poaceae*. Cereal grains provide an excellent source of vitamins, minerals, protein, carbohydrates, and fats in an unprocessed form. Cereals are an excellent source of almost every essential nutrient that we need and that livestock need for growth. More cereal grains are produced than any other crop, and cereal grains compose the majority of the food source for most developing nations.

Some crops are similar to cereal grains, but are not actually grains at all. Quinoa and buckwheat are two examples of these and are known as ***pseudo-cereals***. They have the same properties as cereal grains, but do not produce a grain-type fruit. **Figure 13-21** shows the major cereal grains.

The fruit of a cereal grain is called a ***kernel***. It is composed of a hard outer coating that in most cereal grains is called the ***bran***. The softer inner portion of the kernel is called the ***endosperm***. The ***germ*** is located in or next to the endosperm and serves as the portion of the seed that actually forms the living plant. Another name for the germ of the kernel is ***embryo***. The germ "feeds" off the endosperm until the roots, stems, and leaves take hold and use soil nutrients for growth, **Figure 13-22**.

Let us take a look at the most common cereal grains grown in the United States.

Corn

With almost 80 million acres in production, corn (*Zea mays*) is the most widely produced feed grain in the United States. Corn forms a main ingredient in animal feed and human food and is widely used in industry.

kzww/Shutterstock.com; schankz/Shutterstock.com; zcw/Shutterstock.com; Garsya/Shutterstock.com; Madlen/Shutterstock.com; klaikungwon/Shutterstock.com; schankz/Shutterstock.com; Ratmaner/Shutterstock.com

Figure 13-21. Images of the major cereal grains are shown above. ***What similarities do you notice about these cereal grains?***

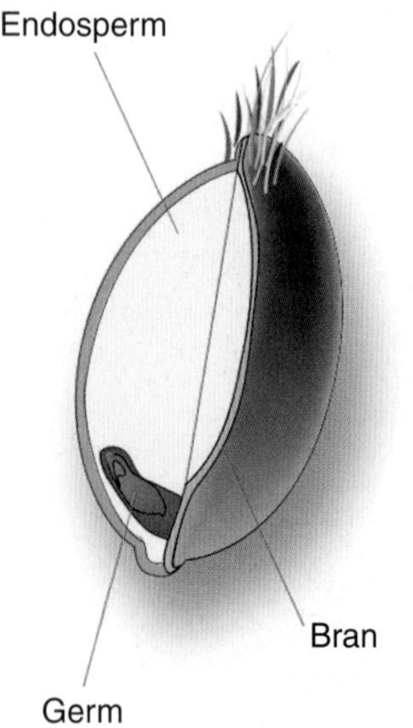

Jkwchui

Figure 13-22. All cereal grain kernels have three main parts: the bran, endosperm, and germ.

Distillers produce ethanol for fuel from corn on a large scale. Most often, they blend the ethanol from corn with fossil fuels to produce gasoline. Corn is native to Central America and is the only grain that originated in the Western Hemisphere.

There are several types of corn, including dent, flint, flour, sweet, and popcorn. The type of corn may be determined by examination of the corn seed.

Dent Corn

The corn you see growing as you drive along a highway is most likely ***dent corn***, actually number two, yellow dent corn. When you look at a dent corn kernel, you will see a small dent in the top, **Figure 13-23A**. Dent corn has a high starch content and is used as the base ingredient for cornmeal flour. Starch from dent corn is used for manufacturing plastics and fructose. Dent corn is also widely used in animal feed.

Flint Corn

Flint corn kernels have a hard outer layer that protects the soft endosperm. Flint corn has less soft starch than dent corn and a very low water content, making it more resistant to freezing than other varieties. Flint corn is also commonly used to make hominy. Although the hard outer layer makes it more difficult to process, processers use flint corn primarily in animal feed. Flint corn may also be called calico corn or Indian corn, **Figure 13-23B**.

Flour Corn

Flour corn has a high soft starch content and is primarily used for making corn flour. Corn flour is milled from the whole kernel and is ground finer than cornmeal.

Nancy Bauer/Shutterstock.com

Arina P Habich/Shutterstock.com

Figure 13-23. A—Dent corn has an obvious indentation in each kernel. B—Flint corn, or Indian corn, is colorful and used for decorative purposes. ***What time of year do you see flint corn displayed?***

Sweet Corn

Sweet corn varieties have a high sugar content. Unlike most field corn, sweet corn must be harvested when immature to be eaten as a vegetable. If harvested when mature, the kernels will be tough and starchy. Sweet corn may be eaten directly from the field.

Did You Know?

One acre of land can produce 14,000 pounds of sweet corn.

Popcorn

Popcorn kernels are small, roundish seeds with a hard outer coating that seals in the corn's starch, oil, and water. When popcorn kernels are heated, the water inside expands and becomes steam. This steam causes the starch to melt into a thick liquid mixture. When the inside temperature of the kernel exceeds 247°F, the pressure inside the kernel exceeds 135 pounds per square inch before bursting open. When the kernel bursts open, the steam is released and the starch both expands and cools into the puffy white shape we enjoy as popcorn, **Figure 13-24**.

Farmers grow popcorn in only a few areas of the United States. It looks like droopy corn when growing in the field. Popcorn kernels are rounder and smaller than dent corn kernels.

Wheat

Wheat is a major grain crop in the United States. More than 2.1 billion bushels of winter, durum, and spring wheat are grown in the United States, **Figure 13-25**. Nearly 150,000 farms produce wheat in this country. Most wheat is grown in the Great Plains of the United States. Winter wheat makes up the largest percentage of all wheat grown in the United States at 1.5 billion bushels annually.

Coprid/Shutterstock.com

Figure 13-24. Popcorn seeds are small, oval, and have a hard seed coat. ***Does popcorn come in other colors beside white?***

Zeljko Radojko/Shutterstock.com

Figure 13-25. Fields of golden wheat nearing harvest are often the subject of photographers around the world.

Globally, farmers grow more wheat on more land than any other crop. Wheat was grown in the Middle East as early as 5000 BCE. Wheat adapts well to many climates, including those that are too dry for corn or rice. Harvested wheat stores and travels well, and humans have relied on wheat as a main food source for thousands of years.

Wheat is the principle source of human food in the United States. One bushel of wheat contains about one million individual kernels. This one bushel can produce up to 40 lb of flour. There are six major types of wheat: hard red winter wheat, hard red spring wheat, hard white wheat, durum wheat, soft red winter wheat, and soft white wheat, **Figure 13-26.**

Rice

Rice is one of the most important grains in the human diet, **Figure 13-27.** It is widely grown around the world. The United States produces more than 200 million bushels of rice annually on 2.7 million acres. Rice is produced primarily in the states along the Mississippi Delta region. Arkansas is a major rice producer in the United States.

Wheat Type	Areas Grown in United States	Products Created
Hard red winter wheat, most commonly produced in the United States	Grown in the Great Plains (Colorado, Kansas, Nebraska, Oklahoma, Texas, Montana, South Dakota, North Dakota, Wyoming)	General all-purpose flour, most breads
Hard red spring wheat, highest protein content of any wheat	Grown in the upper Midwest	Artisan and fine breads, bagels, and rolls Laundry detergents
Hard white wheat, milder and sweeter flavor than other varieties; newest variety to be grown in the United States	Grown in Kansas and Colorado	Specialty-baked products such as hard rolls, tortillas, pan bread, flat breads, and oriental noodles
Durum wheat, hardest kernel of all wheat varieties	Grown in North Dakota	Pastas
Soft red winter wheat, lower protein content than other varieties	Grown in the eastern Midwest	Cake and pie crust flour, biscuits, and muffins
Soft white wheat	Grown in the Pacific Northwest	Pastries, crackers, pancake mix, and puffed breakfast foods

Goodheart-Willcox Publisher

Figure 13-26. Wheat is a versatile cereal grain. Farmers grow many types of wheat in the United States for a variety of uses.

Oats

The United States produces almost 66 million bushels of oats each year on 35,000 farms. Oats are the third most important grain crop for livestock production in the United States. Oats taste mildly sweet, which makes them useful for human breakfast cereals. Unlike other grains, oats are typically processed as a whole grain (without removing the bran, or seed coat, and germ), **Figure 13-28**. Whole-grain diets contain high levels of fibers, which are very healthful. Oats are prized for their ability to lower cholesterol in humans, which helps reduce the risk of heart disease.

Varts/Shutterstock.com

Figure 13-27. Rice grows in fields that are flooded for much of the growing season. Flooding helps control weeds and insect pests in the growing rice.

Livestock Feed

When feeding beef cattle, farmers include oats in the feed for the growing and finishing phases of production. Oats are a good supplement for forage and can be used in creep feeding. ***Creep feeding*** is a method of supplementing the diet of young livestock by offering feed to animals who are still nursing. Younger ruminants adequately break down oats in the gut. Oats contain higher levels of protein content than corn.

Oats are also a favorite food for horses. Oats are bulky enough to prevent digestive problems, yet contain a high protein content. Farmers feed whole or cracked oats to horses to reduce the levels of dust in their feed ration.

Timmary/Shutterstock.com

Figure 13-28. Oats have a slightly sweet taste, so they are processed into breakfast cereals. *Which of the cereals in your cupboard contain oats?*

Barley

Barley (*Hordeum vulgare*) is a plant in the grass family. Humans have cultivated barley for longer than most other grains. More than 200 million bushels are produced in the United States each year. Barley ranks fourth in world production among all cereal grains. Barley is used for human food and livestock consumption. It is used to prepare soups and stews and in the production of bread and distilled alcohol beverages. Barley is also an excellent livestock feed.

Types of Barley

There are three basic types of barley: two-row barley, six-row barley, and hull-less barley. You can tell the difference between these three types by examining the seedhead. The ***seedhead*** is the part of the plant made up of clusters of seeds. In cereal grains, the seedhead grows at the top of the plant. The seedhead may also be referred to as a ***panicle***.

Richard Griffin/Shutterstock.com

Figure 13-29. Farmers produce three kinds of barley: two-row, six-row, and hull-less. *What kind of barley is pictured in this image?*

chinahbzyg/Shutterstock.com

Figure 13-30. Millet is grown to produce bird feed and, to limited extent, human food.

Africa Studio/Shutterstock.com

Figure 13-31. Rye seeds form in the seedheads, or panicles, of the rye plant. *Which other cereal grains grow in rows of seeds in the seedhead?*

Two-row barley forms two rows of seeds opposite from each other on the seedhead. ***Six-row barley*** grows with six rows of seeds along the main shaft of the seedhead. ***Hull-less barley*** is the third type, but it is not actually without a hull as the name implies. The hull around the seed on this type of barley comes off easily in the threshing process. See **Figure 13-29**.

Millet

Many people recognize millet as these little round seeds in bird feed, **Figure 13-30**. Millet is a good source of magnesium and phosphorus in human food, but is most valuable as a forage and grain crop for livestock. Millet is an emerging crop in the United States, with more than 224,000 acres in production and 3.6 million bushels produced every year. Millet has about the same amount of protein as wheat and corn, but is not as popular as these two crops as a small grain.

Rye

Rye used for grain is produced on 265,000 acres in the United States, **Figure 13-31**. More than 6.6 million bushels are produced annually. Like other cereal grain crops, it has many uses. Rye can be used as a cover crop and forage crop, and it can be used to make certain types of bread and other human foods. Rye has a stronger flavor than most other cereal grains. Rye bread is easily identified by its dark color.

Grain Sorghum

Grain sorghum, or milo, is produced as a grain crop and a forage crop, **Figure 13-32**. As a grain crop, more than 264 million bushels are produced annually on 5.1 million acres in the United States. Grain sorghum serves as a good source of nutrition for humans and livestock. Grain sorghum has about the same nutritional value as corn, with slightly less fat and protein. Livestock find cracked grain sorghum kernels a highly palatable feed source. Farmers value grain sorghum as a forage crop. Livestock are often allowed to graze on grain sorghum stubble after the grain is harvested. Grain sorghum is sometimes harvested for silage, although other varieties of cereal grains produce higher forage yields.

Young sorghum plants resemble corn, but sorghum soon produces multiple tillers and branches after germination. Grain sorghum grows to a maximum height of 2′–4′. Grain sorghum kernels are located in a seedhead at the top of the main stem of the plant.

zhuda/Shutterstock.com

Figure 13-32. Field of grain sorghum nearing maturity makes for an appetizing meal for hungry birds.

Production

Farmers plant, grow, and produce most cereal grains in a similar manner. The principal exception is rice, which is grown much differently from the other cereal grains. Let us take a look at how cereal grains are produced.

Planting

As with every other crop, farmers must select the best variety of seed for their geographic region. Growers consider disease resistance, plant hardiness, and date to maturity when selecting the best species and varieties to plant. Purchasing seeds treated with fungicide is an effective method

FFA Connection Agronomy CDE

The National FFA Organization allows students who are interested in agronomy to participate in the Agronomy CDE. The objectives of this event include:

- Demonstration of basic knowledge of agronomic sciences.
- Identification of agronomic crops, weeds, seeds, insects, diseases, plant nutrient deficiencies, plant disorders, crop grading and pricing, and equipment.
- Evaluating scenarios and developing a crop management plan including crop selection, production, and marketing.
- Demonstration of understanding of sustainable agriculture and environmental stewardship through the use of integrated pest management (IPM) and best management practices.

Goodluz/Shutterstock.com

If you are interested in joining or starting an Agronomy CDE Team, talk to your agricultural science teacher/FFA advisor.

for protecting seeds from soilborne diseases. Farmers plant certified seed treated with a fungicide at the recommended seeding rates to improve germination rates and ensure the best yields.

Seedbed preparation controls weeds while conserving soil moisture. Weed control is important in the early stages of growth. The seed also needs sufficient moisture for germination. In no-till planting, farmers adjust the planter so that it is able to cut through the surface crop residue and plant the seed at the appropriate depth. Starter fertilizers kick start the growth process and improve early plant growth.

Corn

Typically, corn is planted in rows 30″–32″ apart to obtain seeding rates between 24,000 to 30,000 plants per acre. Growers plant corn between 10 and 14 days before the last killing frost of the spring. Soil temperatures should be above 50°F (10°C).

Wheat

Farmers plant wheat using a wheat drill. This machine plants wheat in rows that are approximately 7″ apart. The best stands of wheat start with quality seed with an 85% germination rate. Seedlings emerge within five to seven days of planting. Stems and roots rapidly develop as the plant reaches the joint stage. ***Jointing*** marks the point in the wheat plant's life when vegetative growth transforms into reproductive growth. Eventually, the seed head forms and is protected by the flag leaf at the top of the stem. This is known as the ***boot stage***. The heading stage occurs next, when the seed head emerges from the flag leaf.

Did You Know?

Brown rice is unpolished whole grain rice that is produced by removing only the outer husk. It becomes white rice when the bran layer is stripped off in the milling process.

Rice

Rice is planted when the soil temperature at planting depth reaches 50°F (10°C) in the spring. Seedlings emerge in five to 28 days, depending on the weather. The young rice plants will produce one new leaf per week under optimum growing conditions. Tillering will occur sometime around the fifth or seventh week of the growing season. ***Tillering*** is the production of tillers. ***Tillers*** are extra stems rising out of the root system at ground level.

Rice should be planted at a depth of 1/4″–1 1/2″ using a grain drill. There are two additional methods for planting rice. Seeds may be spread using a broadcast spreader and then lightly tilled into the soil. Farmers may also broadcast seeds into a flooded field. The natural settlement of silt covers the seeds and encourages germination. Rice competes favorably with weeds when planted in rows 4″–10″ apart.

Once the plants produce the fifth leaf, they enter the tiller stage. At this stage, the rice fields can be flooded. Weeds cannot compete well with rice plants in standing water, so the rice can outcompete weeds rather easily. The major insects that attack rice plants are held to a minimum because they do not thrive in water as rice does. Diseases in rice are controlled by selecting disease-resistant varieties. Scouting fields at regular intervals catches disease infestations before they become a major problem. Once a disease infestation is located, pesticides can be applied to control it.

Oats

The best dates for planting oats depend upon the variety, geographic region, and season. Spring planting usually begins in late February and ends in mid-April. Fall plantings are typically in September. Spring oats are usually harvested between July and August. Fall oats are typically harvested in June. Oats need a smooth, firm seedbed in order to obtain a good stand. Oats used for grazing will have higher planting rates per acre than oats produced for grain. Oats are planted with a grain drill or, in some cases, broadcast with a spreader and incorporated into the soil. Farmers plant oats at a depth of 3/4″–1 1/2″.

Did You Know?

Oats are a natural beauty product. An oatmeal bath may be used to soothe skin inflamed from conditions such as chickenpox, eczema, and sunburn.

Barley

Farmers grow barley as either a spring crop or a fall crop. Barley is a cool season crop and enjoys well-drained soils and low humidity. As a spring crop, barley is often planted when soil temperatures are between 55°F and 75°F (12.8°C and 24°C). Planting depth is 1″–1 1/2″. Farmers plant barley by using a grain drill, **Figure 13-33**. They strive for the optimum average for planting of 750,000 plants per acre.

Leonid Ikan/Shutterstock.com

Figure 13-33. A grain drill is commonly used to plant cereal grains in narrow rows. The main difference between a planter and a drill is that the drill plants seeds in narrower rows.

Millet

Millet is grown primarily on less fertile soils than similar higher value grains, such as wheat or barley. Millet is mostly grown for forage, so nitrogen and phosphorus must be supplied in recommended quantities to encourage foliage production.

Farmers plant millet in the same manner as other small grains, with a grain drill. Typically, millet is planted in rows spaced 6″–8″ apart to reduce weed competition. This crop needs warm soils to germinate, so planting occurs after the cold and wet winter months.

Rye

Planting and cultivation practices are similar to wheat, barley, and other cereal grains. Farmers plant rye in the fall as a cover crop during the winter months. As a ground cover, rye prevents soil erosion and limits weed growth. In the spring, farmers either harvest rye for grain or till it into the soil to provide organic matter for the next crop. It is winter hardy and will survive even when covered with snow. It grows vigorously in a variety of soils and can outcompete weeds. Rye is susceptible to ***ergot***, a type of fungus that causes severe illness in humans, **Figure 13-34**.

PHOTO FUN/Shutterstock.com

Figure 13-34. Ergot.

Natalia D./Shutterstock.com

Figure 13-35. Conservation tillage methods such as no-till or minimum tillage reduce the need for cultivation and labor while conserving soil.

Grain Sorghum

Grain sorghum likes a warm climate for growth. Grain sorghum is planted when the soil temperatures reach 60°F (15.6°C) to ensure good germination rates. The seedbed should be prepared in a similar method used to plant corn, although a roller-packer can increase the seed-to-soil contact and improve stands. Sorghum is typically planted at depths between 1″–1 1/2″ in rows 30″–34″ apart. Some growers experiment with narrower rows to take advantage of moisture and sunlight and to increase the crop's advantage against weeds. However, cultivation is difficult in these narrower rows, which may only be 10″ apart.

Tillage Operations and Cultivation

Farmers choose between conventional tillage, minimum tillage, and no-till when planting cereal grains, **Figure 13-35**. Tillage should preserve the natural structure of the soil and reduce wind and water erosion. Minimum tillage and no-till operations use cover crops to reduce soil compaction and to increase the crop's advantage over weeds.

Most small grains adapt well to almost any tillage system. If planting in a no-till operation, farmers try to distribute the straw and chaff as evenly as possible when harvesting the previous year's crop. This improves germination rates. Because small grains like oats, wheat, and barley are seeded using a grain drill, a level and smooth seedbed works best. Many farmers use no-till planting. Soils in a no-till system are typically wetter and cooler than in conventional tillage. This allows for good germination rates.

Fertilizer Needs and Applications

The results of soil testing should determine the fertilizer needs of all cereal grains. Nitrogen is the nutrient most frequently lacking in cereal grain production, especially for corn. Growers add a nitrogen-based fertilizer at the appropriate time during the growing stages of the corn crop. Of the three main fertilizer components (nitrogen, phosphorus, and potassium), nitrogen is the most mobile. If ammonium-based fertilizers are used to supply nitrogen for crop needs, then these fertilizers should be applied when the soil and air temperatures are near 50°F (10°C). At warmer temperatures, nitrogen should be applied just prior to rain or irrigation to help it percolate into the soil. Nitrogen is often applied after seedling emergence and once again later in the growing season. Nitrogen fertilizer can be broadcast on the surface, surface-banded, or subsurface banded.

Phosphorus helps the plant store and transfer energy to create the reproductive structures. Potassium contributes to strong stalk development.

Weak stalks will fall over, or ***lodge***, and result in harvest losses. Potassium reduces lodging. Phosphorus and potassium are applied in granular form at the rates recommended by soil tests. Phosphorus and potassium fertilizers are needed early in the growing season. Additional micronutrients may be needed as well. As always, growers should consult soil test results for recommended rates of application.

The pH of the soil is important for all crops. For example, oats yield well at a pH between 6 and 6.5. Lime is used to adjust pH levels and should be applied well before planting time.

Pest Control

Unfortunately, insects and other pests rely on crops for food. Insects and other pests damage and destroy millions of dollars of agricultural crops each year. Pest control is one of the primary challenges for farmers. Let us take a look at the major pests for each of our crops.

Corn

Corn grows quickly and vigorously when adequate nutrients are available and competition with other plants is minimal. Common broadleaf weeds that compete with corn for sunlight, water, and nutrients are pigweeds, velvetleaf, cocklebur, and smartweed. Grassy weeds such as large crabgrass, the foxtails, Johnson grass, and fall panicum also compete with corn.

Common insect pests in corn are rootworms, wireworms, cutworms, chinch bugs, thrips, European corn borer, and fall armyworm. Corn rootworms cause the most damage to growing corn crops. Larvae of the rootworms feed on the stalk, which reduces the plant's ability to transport nutrients and weakens the stalk. European corn borer and cutworms also feed aggressively on the growing corn plant. Growers control these insect pests with a variety of IPM strategies, including scouting, genetically modified seeds, biological controls, and pesticides.

A variety of diseases attack corn plants at every stage of plant growth. Fungi in the soil can cause seed rot. Stalk and root rots can cause lodging. Fungi can attack ears of corn and cause rot. Leaves are susceptible to various rusts and blights. Corn ***smut*** is a widely distributed fungus that attacks corn almost everywhere it is grown, **Figure 13-36**. There are other weeds, insects, and diseases in each specific region of the country that may also reduce crop yields through competition.

Did You Know?

Corn smut has long been considered a delicacy in Mexico. It is commonly served diced with a mixture of onions and corn kernels.

Nataliia Melnychuk/Shutterstock.com

Figure 13-36. Corn smut is a fungus that damages ears of corn.

Integrated Pest Management (IPM)

Integrated pest management (IPM) is the key to reducing pest damage in corn and many other crops. ***Integrated pest management (IPM)*** is an ecosystem-based method of pest control. It focuses on long-term prevention of pests and damage through a combination of techniques including modification of cultural practices, biological control, and the use of resistant varieties.

Wheat and Oats

Weed control in wheat and oats is similar to other small grain crops. Oats are susceptible to ***barley yellow dwarf***, which causes the leaves to turn a reddish color from the tips downward. Aphids spread this disease from plant to plant. This disease can be prevented somewhat by postponing planting until later in the planting period. ***Crown rust*** is another disease that can cause serious yield loss. Fungicides are a proven method of control for crown rust. ***Helminthosporium leaf blotch*** and ***loose smut*** are two fungi that attack kernels, causing severe losses. Using fungicide-treated seeds and resistant varieties help reduce the effect of these diseases.

Barley

Barley competes well against crop pests because it grows vigorously. However, weed infestations can lower yields and make harvesting operations difficult. Weeds like wild oat, wheatgrass, Russian thistle, and field pennycress can threaten crop yields. Farmers use integrated pest management (IPM) techniques to control weeds and pests. Common IPM techniques used with barley include using the correct barley seeding rates, crop rotation methods, and application of herbicides at recommended rates. Plant diseases such as damping off, bacterial kernel blight, barley stripe, ergot, rusts, and seedhead diseases can be controlled through crop rotation, irrigation management, and planting disease-resistant varieties. Insects such as cereal leaf beetles, aphid cutworms, and wireworms can be controlled by insecticides and by planting seed adapted to local growing conditions.

Millet

Integrated pest management is also important for controlling pests in millet. Using weed-free seed sources and controlling weeds along field borders will reduce weed competition. Narrow row spacing and careful variety selection also helps reduce weed competition and damage from other pests.

Grasshoppers and armyworms are two major insect pests for millet, **Figure 13-37**. These are controlled through the use of approved pesticides. Few diseases affect millet in the United States. Farmers apply fungicides to seeds at planting to reduce disease infestations.

Grain Sorghum

Birds are a major pest for grain sorghum. The birds love to eat the seeds readily available to them in the panicle. As a matter of fact, birds love grain sorghum so much that it is a major component of bird feeds.

Harvesting

Growers look forward to harvesting agricultural crops. At harvest time, the farmer realizes the fruits of his or her labors. They gather the yield and market crops to generate income for their families and communities. Harvesting crops is a high-tech, precise endeavor in today's agricultural world.

Eric Isselee/Shutterstock.com

science photo/Shutterstock.com

Figure 13-37. Grasshoppers and armyworms are two major pests for millet. Armyworm is also a serious threat to a corn crop.

Corn

Farmers harvest corn when kernels reach 15% to 18% ***moisture content***, which is the percentage of moisture by weight found in the crop. Corn that is harvested too wet or too dry is susceptible to damage during harvest. Wet crops may grow mold or even generate sufficient heat to catch fire. Crops that are too dry become brittle and are prone to cracking during harvest. Grain combines are equipped with a corn head that cuts the cornstalks, removes the ears, and then funnels the ears into the body of the harvester.

Wheat

Wheat is harvested when wheat kernels reach 12% to 13% moisture content. Grain combines have special headers designed for cutting wheat.

Rice

As the rice plants approach maturity, the fields are drained. Harvest occurs when moisture in the rice kernels drops to 20% or below. Rice is harvested using a grain combine in the same manner as other cereal grains.

Oats, Barley, and Rye

Oats, barley, and rye are harvested when grain moisture levels drop to 13%. If oats are intended for seeding and the grain must be dried after harvesting, dryer temperatures should be set no higher than 110°F (43°C) so seed germination is protected. Oats produced for feed or food sources can be dried at 200°F (93°C).

Farmers and food processors use very high-quality barley for malt production. Growers should make certain that harvesting, storage, and grain handling equipment is properly adjusted to minimize damage to kernels.

Millet

Seed millet is harvested in a different manner than most other small grains. The plants are cut and windrowed when the seedheads are nearing ripeness. (***Windrows*** are long lines of raked hay or sheaves of grain laid out to dry in the wind.) The seedheads continue to dry in the windrows before being picked up by a harvester with a special pickup attachment.

Grain Sorghum

Grain sorghum is harvested much like wheat. A grain combine threshes the seed. Sorghum should be harvested when the seed moisture content drops between 20% and 25%. Under optimum growing conditions, grain sorghum can reach yields of 100 bushels or more per acre.

Career Connection Crop Consultant

Job description: A crop consultant is a person who works to provide advice to growers about growing their crop. With their extensive knowledge of soils, plants, nutrients, and growing conditions, crop consultants can provide input on all aspects of growing a crop to produce a high yield. They may gather their own data or use existing data. If they gather their own data about fields and crops, then they would likely use soil tests, crop monitoring systems, and even GPS. Crop consultants must have good agronomic and science knowledge.

Crop consultants must also possess acute communication skills to convey that knowledge to growers and farmers. Because of this communication and intimate knowledge of the crop, the crop consultant becomes an integral part of a farmer's production team. Their goal is to help the farmer raise the best crop possible within economic and environmental parameters and to help the farmer adopt new management systems that work for his or her land.

Education required: Certification programs in addition to a college degree in agricultural education, agronomy, crop and soils science, or other plant-related major.

Job fit: This job may be a fit for you if you enjoy collecting and analyzing data, understand cropping systems, and have good communication skills.

igor.stevanovic/Shutterstock.com

Lesson 13.2 Review and Assessment

Words to Know

Match the key terms from the lesson to the correct definition.

1. A cereal grain with an easy-to-remove hull.
2. A method of supplementing the diet of young livestock by offering feed to animals who are still nursing.
3. A type of corn with a high, soft starch content that is primarily used for making corn flour.
4. A type of fungus to which rye is highly susceptible.
5. Cereal grain with six rows of seeds along the main shaft of the seedhead.
6. A variety of corn that has a hard outer layer on each kernel, a lower starch content than dent corn, and a very low water content.
7. A variety of field corn having yellow or white kernels that become indented as they ripen.
8. Cereal grain with two rows of seeds opposite from each other on the seedhead.
9. A variety of corn with a high sugar content. Most commonly used for human consumption.
10. Condition when wheat stalks fall over.
11. Crops with the same properties as cereal grains, but that do not produce a grain-type fruit.
12. Extra stems rising out of the root system of a rice plant at ground level.
13. Period of growth when the seedhead forms and is protected by the flag leaf at the top of the stem of a wheat plant.
14. The fruit of a cereal grain.
15. The hard outer coating of cereal grains.
16. The part of the plant made up of clusters of seeds.
17. The point in a wheat plant's life when vegetative growth transforms into reproductive growth.
18. The portion of a cereal grain's seed that forms the living plant.
19. Long lines of raked hay or sheaves of grain laid out to dry.
20. The soft inner portion of a grain's kernel.

A. boot stage
B. bran
C. creep feeding
D. dent corn
E. endosperm
F. ergot
G. flint corn
H. flour corn
I. germ
J. hull-less barley
K. jointing
L. kernel
M. lodge
N. pseudo-cereals
O. seedhead
P. six-row barley
Q. sweet corn
R. tillers
S. two-row barley
T. windrow

Know and Understand

Answer the following questions using the information provided in this lesson.

1. What is the fruit of a cereal called?
2. Where did corn originate?

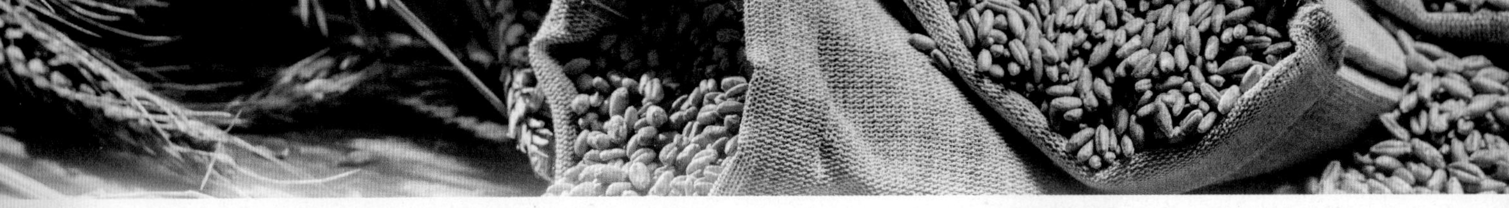

3. What are the six types of corn?
4. What is the white puffy part of popcorn?
5. Which kind of wheat is the most popular in the United States?
6. Which state produces the most rice?
7. Which food nutrient is higher in oats than in corn?
8. *True or False?* Barley has been grown by humans for longer than most other grains.
9. What is the primary purpose of millet production?
10. If bread is dark brown, which cereal grain is a likely component?
11. Where are the kernels of grain sorghum located?
12. What fungus attacks rye?
13. How does a farmer know how much fertilizer to apply for each crop?
14. Which plant nutrient helps the plant store and transfer energy to create the reproductive structures?
15. What is one corn insect pest?
16. At which stage is corn susceptible to disease?
17. How is barley yellow dwarf spread through wheat and oat crops?
18. What types of weed threaten barley crops?
19. *True or False?* In the United States, millet is affected by numerous diseases.
20. *True or False?* All crops are harvested and stored at the same moisture content.

Analyze and Apply

1. Draw and diagram the component parts of wheat and corn kernels. Identify and label each part.
2. Research the origins of the various cereal grain species. Where did they originate, and how did they get to the United States?
3. Why is row spacing important in helping the growing crop outcompete weeds in the field?

Thinking Critically

1. Why are cereal grains such important crops around the world? What makes cereal grains desirable as a food source for so many people?
2. What has been the impact of biotechnology on the production of cereal grains?
3. As you think about global climate change, what are the challenges of this climate change on the production of cereal grains? What challenges do farmers face and how do they combat those?

Lesson 13.3

Oil Crop Production

Lesson Outcomes

By the end of this lesson, you should be able to:

- Identify oil crops and their uses.
- Explain the planting and cultivation practices that produce optimum yields of these oil crops.

Words to Know

cash farm receipt
inoculation
legume
pegs
rust

Before You Read

Before you read the lesson, read all of the image captions. Share what you know about the material covered in this lesson just from reading the captions with a partner.

Do you burn candles in your home? Did you know that they likely originated as a plant? Have you noticed the ethanol sticker on the gasoline pump when you fill your car? Did you know ethanol is derived from a plant? Have you examined the ingredients of your lip balm? Where do you think many of those ingredients got their start? That is right, as plants!

Oil crops are essential for human and livestock nutrition and industrial products. Oils extracted from plants flavor our foods, power our automobiles, supplement livestock feed, and aid in the manufacturing of paints and plastics. As we continue to reduce our dependence on fossil fuels, oil crops become increasingly important for fuel and manufacturing in addition to food. Let us find out which plants produce the oils that we encounter in our daily lives.

Corn

Corn production as a grain crop has been discussed in *Lesson 13.2*; however, it is important to note that corn is the largest oil crop in the United States. One bushel of corn produces 1.5 lb of oil, **Figure 13-38**. Corn oil is used as a cooking oil, in margarines, as feedstock, and in the production of biodiesel, soap, paint, and insecticides. It is less expensive than other types of vegetable oils.

Evan Lorne /Shutterstock.com

Figure 13-38. Corn oil is used as a cooking oil.

Steve Collender/Shutterstock.com

Figure 13-39. Soybeans respond favorably to good management practices. Soybean seed pods dry and turn brown just before harvest in early fall.

sima/Shutterstock.com

Figure 13-40. Soybeans produce their own nitrogen fertilizer but need to be inoculated with a bacterium that helps the plant do this. Soybeans flower as the nights get longer starting in late June. Each flower has the potential to produce a soybean pod with several soybeans inside.

Soybeans

Soybeans arrived later on the American agricultural scene than many other crops, **Figure 13-39**. Soybeans were first planted in Georgia in 1765 and have since grown to be the second largest crop in the United States. More than 75 million acres are planted annually with gross receipts of more than $42 billion. Most soybeans are processed for their oil. You can find soybean oil in cooking oils, margarine, soy candles, soy ink, soy crayons, many lubricants, and even biodiesel.

Soybeans belong to a special class of plants called legumes. ***Legumes*** are plants that contain special bacterial nodules on their roots that fix nitrogen from the atmosphere into usable forms in the soil. Soybeans produce seed pods that arise out of its flowers. Soybean plants with more flowers produce more beans and a larger harvest, **Figure 13-40**.

Sunflowers

More than 1.4 million acres of sunflowers are produced every year in the United States, **Figure 13-41**. These acres yield 2 billion pounds of sunflower seeds per year. Sunflowers are grown in the United States for three major reasons:

- To provide a high-quality vegetable oil for human nutrition and livestock nutrition.
- As a major ingredient in bird feed.
- For habitat restoration (planted in wild food plots).

Sunflowers can yield as much as 1,400 pounds per acre. High yields are dependent on the grower's management ability. Sunflower varieties typically mature in 85 to 95 days.

Canola (Rapeseed)

Canola production has increased in the past 40 years, making canola oil the second largest oil crop in the world. Canola meal is the second largest feed meal in the world, surpassing all others except soybean meal. The United States canola crop is valued at more than $270 million.

Canola oil and meal does not come from a canola plant. The name "canola" was coined by the western Canadian oilseed crushers association. Canola oil is the oil extracted from the rapeseed plant. Today, most production is derived from edible rapeseed varieties. Rapeseed is part of the family *brassicaceae*, the mustard family. See **Figure 13-42**.

Canola Variety Selection

The most important element in the production of canola is the selection of an appropriate variety. Farmers select a variety that:

- Demonstrates high yields in their geographic region.
- Demonstrates strong stalk strength to avoid lodging.
- Exhibits disease tolerance with strong seedling vigor that will result in higher yields at harvest.

Peanuts

More than one million acres of peanuts are planted each year in the United States, resulting in more than $1 billion in gross farm receipts. (A ***cash farm receipt*** is the amount of money earned by the farmer or grower for his or her crop when he or she sells it to the marketplace.) Peanuts are somewhat of an oddity. Although the seeds are called nuts, peanuts are not true nuts. Peanuts are more similar to soybeans than the tree nuts we consume. Peanuts are produced underground, unlike any other major nut crop plant. The flowers produce ***pegs*** (stems) that go into the soil where the seedpod is formed, **Figure 13-43**. Like soybeans, peanut plants collect nitrogen from the air and convert

Popkov/Shutterstock.com

Figure 13-41. More than 1.4 million acres of sunflowers are produced every year in the United States. Sunflowers grown for seed harvest should not be spaced too close together.

Daniel Prudek/Shutterstock.com

Figure 13-42. Today, most canola production is derived from edible rapeseed varieties.

sunsetman/Shutterstock.com

Figure 13-43. Peanut plant.

it into a form suitable as fertilizer for the growing plant.

There are four major market types of peanuts: Virginia, runner, Valencia, and Spanish. The Virginia and runner types are different from Spanish types because they bloom and produce fruit on the upright main stem.

Planting Oil Crops

Before the planting season begins, farmers examine their planting equipment and make repairs or adjustments. Planters deposit seeds into the furrow at precise rates so careful calibration is needed to ensure planting occurs at the correct rates.

Most seeds require warm soil temperatures in order to germinate. Peanuts germinate when the soil temperature is at least 65°F (18°C). Soybeans like well-drained, warm soils between 55°F and 60°F (12.8°C and 15.6°C) to germinate. Sunflower seeds germinate when the soil temperatures reach 42° to 50° (5.6°C and 10°C). The warmer the soil, the more efficient the germination rate. Canola is a cool season crop and will germinate when soil temperatures are higher than 38°F (3.3°C).

Corn

Farmers plant corn in long rows that are generally 30″ apart. Corn is planted when soil temperatures warm above 50°F (10°C). Like all agronomic crops, corn should be planted as early in the growing season as possible to maximize yields. Most corn seed is treated with fungicides and insecticides to protect the corn seed until it germinates.

Soybeans

Farmers plant soybeans in narrow rows with high populations per acre. Rows are generally less than 30″ apart. Narrow row widths of 15″ to 20″ increase yields and reduce competition from weeds. Soybeans should be planted as early in the season as possible for optimum yields. Seeds should be treated with a fungicide that prevents disease from becoming a problem. Soybeans are planted 1 1/2″ deep.

Sunflowers

Sunflowers perform best when planted in rows 20″ to 30″ apart. Sunflowers that are grown for seed should not be spaced too close together. The plant adapts to a wider plant spacing by growing larger seedheads with larger seeds. For a food crop, larger seeds are more desirable. Soil sampling and testing will inform the grower of the fertility requirements needed for sunflowers. Farmers consult crop production guides from cooperative extension services and seed companies for the best recommendations on fertilizer application.

Canola

Narrow row spacing of 6″ to 9″ will help canola plants compete effectively against weeds. On average, canola will take 17 to 21 days to germinate, depending on soil temperature and moisture. After germination, it may take as many as 10 days for the seedlings to emerge from the soil. Fertilizer rates are determined based on soil test recommendations.

viphotos/Shutterstock.com

Figure 13-44. Peanuts perform well in 36″ to 38″ rows.

Peanuts

Growers must employ good management to produce the best peanuts. Peanuts perform best on a slightly raised bed. Planting occurs in late spring to avoid the risk of infestation by tomato spotted wilt. Farmers plant seeds at 1 1/2″ deep on average. Peanuts perform well on 36″ to 38″ rows, but twin row plantings at 7″ apart on 36″ row widths can reduce the risk of tomato spotted wilt, **Figure 13-44**.

Did You Know?

Canola is an herbaceous plant that can range from 3′ to 6′ in height.

Crop Stress

Managing crop stress is the biggest job of the producer. Any number of things can cause stress on the plant; the following represent the most important factors:

- Soil conditions—soil conditions in the field may be either too wet or too dry for optimal crop production. Some growers irrigate their land. However, irrigation is expensive, which reduces profits for the grower.
- Weather—weather is naturally volatile and unpredictable for any length of time beyond a few days. Farmers constantly monitor weather conditions to plan cultivation, spraying, and harvest.
- Pests—pests are constantly present in the field. Farmers scout their fields for insects, weeds, fungi, and other pests. They must understand stages of development of all pests to know when to properly treat and control the pests.
- Nutrient deficiencies—all soils contain different levels of plant nutrients and also release those plant nutrients differently to each plant or crop. Farmers must sample the soils in their fields periodically for nutrients and pH. Once they have this information, they can then apply fertilizers or lime to adjust the chemical properties of their soils.

Fertilizer Needs and Applications

Farmers fertilize fields based on soil test recommendations. The macronutrients of most interest are nitrogen, phosphorus, and potassium. Copper, zinc, calcium, and other micronutrients are also important, but seldom are limiting factors for high yields.

Corn

Fertilizing corn is primarily a function of adding sufficient nitrogen for the growing crop. In order to produce a good corn crop, most soils require 1–1.2 lb of nitrogen per bushel of yield goal for a corn crop. If a farmer wants to yield 150 bushels of corn per acre, then the soil needs to provide at least 150 lb of nitrogen per acre. If the soil lacks that availability of nitrogen, then the farmer must apply nitrogen in the form of urea or anhydrous ammonia. Corn also requires phosphorus and potassium, the other two plant macronutrients.

Soybeans

Soybeans produce their own nitrogen fertilizer, but need to be inoculated with a bacterium that helps the plant do this. ***Inoculation*** is the process of introducing beneficial bacteria to the soybean seeds prior to planting. *Bradyrhizobium japonicum* is placed on the seed. Soybeans can convert 75% of nitrogen from the air into a form of nitrogen-based fertilizer when *Bradyrhizobium japonicum* is present in the soil. Soil pH of 6.5 is important for soybean growth.

Sunflowers

Fertilizing sunflowers requires moderate amounts of nitrogen, phosphorus, and potassium. For example, a 2,000 pound-per-acre yield goal would require 100 lb of soil nitrogen, 20 lb of phosphorus, and 30 lb of potassium. Depending on the yield goals, farmers apply more or less of these nutrients. Sunflowers grow well in pH neutral or slightly alkaline soils.

Canola

Fertilizing canola requires moderate amounts of nitrogen, phosphorus, and potassium. For example, yields are highest when the available soil nitrogen is between 100–120 pounds per acre. Most soils require the addition of 35 lb of phosphorus and 50 lb of potassium for a healthy canola crop. Of course, farmers always conduct soil tests to determine the correct fertilizer applications for their soils. Depending on the yield goals, farmers apply more or less of these nutrients. Canola grows well in pH of 5.5 to 8.3.

Peanuts

A pH range of 5.8 to 6.2 is suitable for peanuts. Peanuts are a legume and are able to fix nitrogen for use as a fertilizer. Because of this, farmers inoculate peanut fields with nitrogen-fixing bacteria such as *Bradyrhizobium japonicum*.

Pest Control

A major part of crop production is pest control. Growers constantly monitor their crops for infestations of common pests. Pests generally take the form of insects, weeds, or fungi.

Corn

Corn growers scout their fields for insect pests, sometimes even prior to planting. Major insect pests in corn include corn borer, corn rootworm, and cutworm. Both the larval and adult stages of these insects feed on the corn plant. When insects feed on the corn plant, the plant's yield potential is reduced.

Common fungal pests in corn include various seedling rots, rusts, and gray leaf spot. Corn seeds are treated with fungicides to protect the germinating corn seed from attack of seedling fungal rots. Rusts and gray leaf spot are fungal diseases affecting the corn leaf.

Soybeans

Farmers scout the fields of growing soybeans frequently to identify pests and fertility problems. They apply pest control measures as needed. The soybeans transition from the growing vegetative stage into the reproductive stage as the day length gets shorter in the fall. The reproductive process begins with the emergence of flowers at any node on the main stem of the plants. These first flowers will appear five to six weeks after planting.

During the flowering stage, farmers scout carefully for pests. Soybean aphids can stunt plant growth. Spider mites can damage leaves and stunt growth also. Japanese beetles and bean leaf beetles are major pest problems in soybeans and cause major damage if not controlled. Among major pests, the soybean cyst nematode causes the most serious problem for soybean growers. Nematodes damage the ability of the roots to perform their duty. Growers control pests using the recommended pesticides for their area and follow all pesticide label directions for their safe use.

Sunflowers

Insects are a major problem for sunflowers. The sunflower moth larvae burrow into seeds and feed on them. The carrot beetle burrows into the soil and feeds on the plant root system. Insects are not the only problem. Downy mildew causes stunting in plants. ***Rust***, another fungus, infects and damages young seedlings.

Weeds can be a nuisance as well, especially in the sunflower seedling stage when the crop cannot compete effectively against weeds. Cultivation and the safe application of the appropriate pesticides are two effective methods for managing pests in sunflowers.

Canola

Weed control in canola is best accomplished by planting and maintaining a uniform stand. Plants that are planted at the appropriate row spacing will outcompete weeds for nutrients and sunlight. Canadian thistle and wild oat are two weeds that are troublesome in canola. These can be controlled by the safe application of appropriate herbicides. A variety of insects can cause damage to canola at almost any point in the growth cycle, **Figure 13-45**.

Seedling to flowering:

- Flea beetles
- Cutworms

InsectWorld/Shutterstock.com

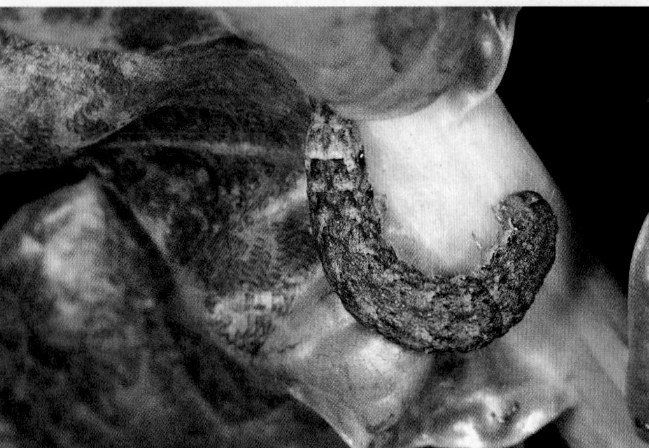
Sarah2/Shutterstock.com

Lostry7/Shutterstock.com

Figure 13-45. Flea beetles (top), cutworms (middle), and lygus bugs (bottom) are all major pests for canola, or rapeseed.

- Grasshoppers
- Lygus bugs

Flowering to pod development:

- Aphids
- Blister beetles
- Diamondback moths
- Grasshoppers
- Lygus bugs

A number of diseases may damage canola, but two major ones are blackleg and sclerotinia stem rot. Pesticides are the most effective methods of dealing with pest infestations.

Peanuts

Weed management requires herbicides and cultivation to reduce weed competition. Peanuts do not compete well with weeds. Fields should be maintained weed-free for 45 days after planting. Cultivation helps when herbicides are not effective. Farmers avoid throwing cultivated soil onto peanut foliage to prevent the spread of disease. All peanut seed is treated with a fungicide to control for disease.

A major disease of peanuts is the tomato spotted wilt virus, which is transmitted to peanuts by thrips. This disease reduces yield by damaging the seed pods. Tomato spotted wilt virus can be controlled by selecting resistant varieties, planting in late spring, planting at optimum and uniform seeding rates, and treating with insecticides to control the thrips that spread the disease, **Figure 13-46**.

Harvesting

Harvesting occurs after the crop matures and the moisture content drops. Harvest for most agronomic crops occurs in the fall of the year. Most crops require over 100 days from planting to maturity.

Corn

Corn is harvested when the moisture drops below 18% relative moisture. The corn should feel hard when held in your hand. Farmers harvest corn using a machine called a combine. A combine cuts the dried cornstalks, removes the ears from the stalks, and removes the corn kernels from the ears. Corn kernels are stored temporarily in the bulk tank on the combine. Once the bulk tank is full, then the farmer unloads the corn into a grain truck, semi, or bulk wagon.

Soybeans

Soybeans respond to night length. Plants begin to flower as soon as the nights begin to get longer, just after the first few days of summer. After the flowers have given way to the development of the seedpod, the plant enters the maturity stage. The seed pods fill, mature, and dry down. Soybeans are harvested between 12% and 15% moisture content. Harvest is accomplished with a reel header on a combine harvester. Soybeans are transferred from the combine into trucks for transportation to market or to on-farm storage.

Katarina Christenson/Shutterstock.com

Figure 13-46. Thrips spread tomato spotted wilt virus in peanuts.

Sunflowers

Farmers harvest when the seeds reach 18% to 20% moisture content to reduce shattering and bird damage. Farmers often dry the seeds once they arrive in storage. A grain combine with a sunflower head attachment can do the job of harvesting, **Figure 13-47**.

Did You Know?

Sunflowers can be used to extract toxins, such as lead, from contaminated soils. This natural process is called *phytoremediation*.

Canola

Canola matures when the seedpods turn brown. This typically happens in mid-May. The seedpod walls become very loose and brittle at maturity. This allows for the seedpod to shatter and dump seeds on the ground. Keep in mind that all plants have developed methods to spread their seeds naturally. When growing canola, the farmer must time his or her harvest with the canola plant's natural adaptive traits and harvest these seeds before they land on the ground. Monitoring and harvesting at the right moisture levels will help avert this problem, **Figure 13-48**.

smart.art/Shutterstock.com

Figure 13-47. A grain combine with a sunflower head attachment used to harvest sunflowers.

Peanuts

Peanuts pollinate and produce seed pods in the hot summer months. The fertilized flower produces an elongated ovary that

v.schlichting/Shutterstock.com

Figure 13-48. Canola is harvested like small grains. Canola needs to be harvested at the right moisture level.

continues to grow and travel toward the soil surface. This is called the *peg* and it is first visible about one week after the flower is fertilized. In about 12 days, the pegs enter the soil to a depth of 1″–2″. The seedpod develops from the tip of the peg.

When the seedpods are finished developing and ready to harvest, the farmer uses a digger-shaker machine to prepare the peanuts for drying in the field. This machine digs beneath the plants, severs the roots, then flips and shakes the peanut plants to remove excess soil, **Figure 13-49**. The plants are left in a windrow until the seedpods dry enough to harvest. After threshing, the peanuts are cured or dried in trailers before being transported to a processing facility.

AFNR/Shutterstock.com

Wellford Tiller/Shutterstock.com

Figure 13-49. A peanut digger-shaker is used to pull the plants from the ground and shake off excess soil. Once the seedpods are dry enough, a peanut harvester is used to collect the pods from the windrows.

Career Connection Grain Inspector

Job description: A grain inspector inspects grain when it arrives at the grain elevator from the farmer and then again when it is loaded onto barges and ships leaving and entering U.S. shipping ports. Grain must be inspected for insects, fungi, and other pests. Grain inspectors also monitor the grain for moisture content and foreign matter. Foreign matter may include stones, sticks, and other materials that are not grain, but contribute to the weight of the shipment. Other abnormalities may be present in grain, so grain inspectors also need to be able to see, feel, and/or smell abnormalities in the grain.

Education required: Certification programs in addition to a high school degree.

Job fit: This job may be a fit for you if you enjoy collecting and analyzing grain data and working both indoors and outside.

USDA/ARS

This USDA electrical engineer is testing the accuracy of a new *insector probe receptacle*. The probe is designed to monitor insect infestations in grain bins. The engineer will count the insects caught in the receptacle to determine the probe's accuracy.

Lesson 13.3 Review and Assessment

Words to Know

Match the key terms from the lesson to the correct definition.

1. The flower on a peanut plant produces this type of stem that goes into the soil where the seedpod is formed.
2. A plant disease caused by fungus.
3. Plants that contain special bacterial nodules on their roots that fix nitrogen from the atmosphere into usable forms in the soil.
4. The process of introducing beneficial bacteria to the soybean seeds prior to planting.
5. The amount of money earned by the farmer or grower for his or her crop when he or she sells it to the marketplace.

A. cash farm receipt
B. inoculation
C. legumes
D. peg
E. rust

Know and Understand

Answer the following questions using the information provided in this lesson.

1. What is the second largest cash crop in the United States?
2. Which household products contain soy oil?
3. What are three uses for sunflower crops?
4. What is the appropriate name for the plant that produces canola oil?
5. Canola belongs to the _____ family.
6. *True or False?* Peanuts are true nuts.
7. *True or False?* Peanuts are a legume like soybeans.
8. What are three major pests for soybeans?
9. Are insects a problem when growing sunflowers? Identify two common pests.
10. At what growth stage of canola are Lygus bugs a problem?
11. Tomato spotted wilt virus is transmitted to peanuts by _____.
12. What triggers soybean plants to begin to flower?

Analyze and Apply

1. Research the production of oil crops in the United States. Find the top five states in terms of production for each oil crop. Map these top-producing states on a map of the United States.
2. Why would a farmer rotate his cropping system by planting corn in the field one year, followed by soybeans the next year? What do soybeans provide to the soil that helps corn grow?

Thinking Critically

1. What are the advantages to gasolines and diesel fuels produced from plants? What are the disadvantages?
2. What are some of the by-products of the oil crops that we also use throughout our homes? Research by-products of oil crops and then determine which are found in your home. Which ones surprised you? Why?
3. Which states produce the most of each kind of oil crop? What information in this lesson would provide clues about where oil crops are planted? Use data and evidence to justify your reasoning.

Lesson 13.4

Fiber Crop Production

Words to Know

Bacillus thuringiensis (Bt)
boll
bract
defoliant
doffers
linen
linseed oil
lint
square stage
windrow

Lesson Outcomes

By the end of this lesson, you should be able to:

- Describe the basic production practices for growing and harvesting cotton.
- Explain the process of preparing cotton for textile processing.
- Describe the basic production practices for growing and harvesting flax.

Before You Read

Check the label on your clothing. Is cotton one of the products used in the manufacture of what you are wearing? Have you ever wondered how cotton bolls become the fabric you wear?

Farmers grow a number of fiber crops in the United States, including cotton, flax, and kenaf. These crops are used to make fabric, paper, rope, and other specialized products used in industry. Some fiber crops are used as food additives, and to make soap, cosmetics, paints, and detergents. Some are also used for livestock feed. This lesson focuses on the role of fiber crops as a source of fiber used in the manufacture of various products. Let us look at the production process of fiber crops, beginning with the king of all fiber crops produced in the United States: cotton.

©iStock/pelicankate

Figure 13-50. Cotton remains a major crop in the United States.

Cotton

Cotton is, and has been, a major crop in the United States since colonial times. More than 20 million bales are produced on 14 million acres. The average value of cotton produced in the United States exceeds $3 billion annually. Not bad for a shrubby little plant, **Figure 13-50**.

The cotton plant has many uses, including:

- The cotton seed is ground up and used as a livestock feed additive.
- Oil is extracted from the seeds and used in foods, soap, cosmetics, paint, and detergents.

- Cotton fiber is used to manufacture fabrics for clothing, automotive tire cord, cardboard, and paper.

Did You Know?

The white or yellow fibers produced by the cotton plant are called ***lint***. Even the tiny lint fibers present on cotton seed are removed and used in textile production.

Planting

Cotton is a warm season crop planted in late spring. The seedbed is formed into rows 36″ to 40″ apart. Cotton planters open a furrow, place the seed in that furrow at a depth between 3/4″ and 1 1/2″ deep. The planter then covers and packs the soil to provide good seed-to-soil contact. Seedlings emerge five to seven days after planting. Six weeks after emergence, the plant enters the square stage where flowers begin to form on the plant. The ***square stage*** occurs when three bracts form a triangle to protect the growing flower bud. A ***bract*** is a small cluster of flowers found in the axil of a stem. The flowers are pollinated, and the ovary begins to grow into a round green pod called the boll, **Figure 13-51**.

Fertilizer Needs and Applications

Cotton grows slowly in the first 40 to 60 days of growth. Once the plant reaches the square stage, growth begins to occur rapidly. The crop needs to receive adequate fertilizer applications at this critical growth stage. Farmers follow soil test recommendations for fertilizer application.

Pest Control

Pests are a major concern in cotton production. Insects, weeds, and disease can reduce yields significantly. In the past, the boll weevil was a major pest in cotton, destroying thousands of acres of cotton in a growing season, **Figure 13-52**. (The ***boll*** is the rounded seed capsule of the cotton plant that contains the ovary.) It has now been eliminated as a serious threat through scouting and the rigorous application of boll weevil control measures.

A

Jerry Horbert/Shutterstock.com

B

itman_47/Shutterstock.com

Figure 13-51. A—Cotton is usually planted in rows 36″ to 40″ apart. B—This cotton plant shows two cotton squares (top and right), a fully formed flower (middle), and a boll emerging from a square (bottom).

Leong9655/Shutterstock.com

a katz/Shutterstock.com

Figure 13-52. A—Years ago, the boll weevil almost destroyed the U.S. cotton industry. Successful measures to track and eradicate the insect have eliminated the threat. B—Boll weevil traps are placed at the edges of fields to attract the insects. By examining traps, growers can determine the extent of the infestation, the direction of migration, and what measures need to be taken.

Did You Know?

Cotton keeps the body cool in summer and warm in winter because it is a good conductor of heat.

The eradication of the boll weevil has reduced the amount of pesticides needed to protect cotton crops. Insects and disease can be controlled by selecting resistant varieties, using biological controls, and through the application of pesticides as needed. Biological controls are living organisms that control pests. In cotton, ***Bacillus thuringiensis (Bt)*** is a bacterium that kills certain insect species that prey upon cotton. *Bt* is also found in corn, tomatoes, potatoes, and canola. This bacteria lives within the cotton plant, saving the grower the time and money of controlling pests by other methods.

Harvesting

Approximately 10 weeks after flowering, the bolls split apart and the cotton fibers emerge from the boll, **Figure 13-53**. Once the bolls open, a defoliant is applied to the field. A ***defoliant*** is a chemical herbicide that kills the cotton plant. This causes the leaves to fall off the plant, **Figure 13-54**. The leaves need to be removed so that the harvesting equipment can efficiently pick only the cotton from the bolls. The leaves may also stain cotton fibers.

Heidi Brand/Shutterstock.com

Figure 13-53. This boll is nearing maturity.

The most prevalent system for harvesting cotton uses a picker to remove bolls from the plant. It is important to harvest cotton when the bolls are dry. Harvesting should begin after the morning dew has evaporated and before the dew falls at the end of the day, and the boll moisture level is at 12%. Cotton must be dry when harvested for a variety of reasons, including:

- Wet cotton may get mildewy and/or stained.

Steven Frame/Shutterstock.com

Figure 13-54. This cotton field has been defoliated to remove leaves that hamper the harvesting process.

- Wet cotton will get stuck and jam up the harvest machinery.
- Wet cotton bales may heat up and start on fire.
- The cottonseed coat may weaken with moisture and sprout within the boll or break into chips in the gin.

As the cotton picker moves through the field, spindles with sharp teeth pull cotton fibers from the boll. ***Doffers*** are rubber pads with ridges or fingers that remove the fibers from the spindle. The cotton lint is blown by air pressure from a fan into the basket at the rear of the picker. Once the basket is full, it is dumped into a module builder. The cotton is packed firmly into the module builder, making a large compact bale approximately 30′ long by 10′ wide by 12′ high. A specialized truck picks up the module and transports it to the cotton gin. See **Figure 13-55.**

Ginning Cotton

The cotton gin gets its name from a shortened version of the phrase, *cotton engine*. Over time, the word *engine* was shortened to gin. The modern cotton gin uses sophisticated processes to remove debris and leaves from the cotton fibers, and to separate the cotton seed from the fibers.

Ginning cotton was a slow and tedious process before the invention of the cotton gin. A modern cotton gin can do the job much faster today. When the cotton arrives at the gin, it is sucked out of the module and run through a dryer to ensure that it is being processed at the optimum moisture level. The lint enters a thresher that removes the larger bits of plant material that may be mixed in with the lint. The cottonseed is then removed from the lint, and circular saw blades remove the small lint fibers from the seed. The seed is

Safety Note

Cotton harvest can be hazardous. Fires may start on cotton pickers and consume the machinery in a short period of time. It is important that machinery be properly maintained and cleaned to prevent the buildup of cotton lint, which is highly flammable. Dry and windy conditions, lack of dew formation, and static electricity increase the potential for fire.

Did You Know?

A cotton boll was the background of the New Farmers of America emblem.

Did You Know?

The fiber from one 500-lb (227-kg) cotton bale can produce 215 pairs of jeans, 250 single bed sheets, 1,200 T-shirts, 2,100 pairs of boxer shorts, 4,300 pairs of socks, or 680,000 cotton balls.

VanHart/Shutterstock.com Lindasj22/Shutterstock.com Lindasj22/Shutterstock.com Lindasj22/Shutterstock.com Lindasj22/Shutterstock.com

Figure 13-55. A—The spindles on the cotton picker have sharp teeth that pull cotton fibers from the boll. B—Cotton picker moving through a field. C—Cotton being dumped from the picker basket into a module builder. D—The module builder tamps down the cotton into a compact module. E—The module builder is removed, and the cotton module is covered with a tarp to prevent rain and dew from penetrating. A specialized truck picks up the module and delivers it to a cotton gin for processing.

STEM Connection Calculating Actual Nitrogen in Fertilizer

Fertilizer can be purchased in bags with an N-P-K analysis that might be 10-10-10. What those numbers mean is that the bag of fertilizer is 10% nitrogen, 10% phosphorus, and 10% potassium, always in that order. The numbers can change, but they always represent the percentage of elemental fertilizer by weight. So, if a bag weighs 50 lb and is 10-10-10, then the bag contains 10% or 5 lb each of nitrogen, phosphorus, and potassium. Using this information, calculate the amounts of each elemental fertilizer in the following:

A) 100 lb of 8-12-20
B) 50 lb of 15-30-15
C) 2,000 lb of 47-0-0

egilshay/Shutterstock.com

stored in binds for further processing. The clean lint is formed into 500-pound bales and packages for shipping to textile mills, **Figure 13-56**.

muratart/Shutterstock.com

Figure 13-56. Processed cotton bales awaiting shipment to textile mills.

Flax

Flax is both a food and fiber crop that has been grown in the United States since colonial times, **Figure 13-57**. The early colonists grew flax primarily to produce fiber for clothing and linens, and for its oil to be used in paints and stains. Today, flax is grown for a variety of purposes, including:

- Flaxseed is used in breakfast cereals and bread, and in quality pet foods, livestock feed, and poultry feed. Flaxseed oil is also used in a variety of food products.
- Flaxseed oil is used to make linoleum, a nonallergenic and biodegradable flooring product. It is also used in oil-based paints and stains. Flaxseed oil is commonly referred to as ***linseed oil***.
- Flax fiber is used to make ***linen***, a material used in bed linens, napkins and tablecloths, and clothing, **Figure 13-58**. It is also used to make specialty papers like parchment paper.

Did You Know?

Flax is also known as *common flax* or *linseed*.

©iStock/Smithore

Figure 13-57. Flax field in bloom.

Figure 13-58. Linen made from flax fibers.

- Flax fibers are blended with cotton or other fibers to make medical products such as bandages.
- Flax straw not suitable for fiber production is often sold for biofuel production.
- Flaxseed meal, a by-product from production processes, is used as livestock feed.
- Flax is also planted and grown in areas as a means of erosion control, and can also be used as a fire suppressant species in green strip plantings.

U.S. Production

In the United States, flax is grown primarily in North Dakota, followed by South Dakota, Montana, and Minnesota. According to the USDA National Agricultural Statistics Service (NASS), the total U.S. flax production in 2012 was nearly 5.8 million bushels valued at nearly $78.3 million. The United States exported flaxseed valued at more than $16 million, and linseed oil at $46.2 million (2012). Most commercial operations in North America produce flax for its seeds.

Figure 13-59. As part of their Environmental Quality Incentive Program (EQIP) contract, brothers Robert and Mike Rausch have incorporated cover crops such as flax to their rotational cropping system. *What is an Environmental Quality Incentive Program (EQIP)?*

Planting

Flax should be seeded in early spring (late April–early May), as early planting dates allow the plants to flower and begin seed set in the cooler part of summer. The seedbed should be prepared and seeded in 24″ or 36″ rows, depending on the seeding rate used. Properly spaced rows will allow mechanical weed control. Seeds should be planted at a depth between 3/4″ and 1 1/2″ with a grain drill and press attachment. The press attachment helps compress soil against the seed and will promote rapid and more uniform germination.

Deep loams containing a large proportion of organic matter are the most suitable for growing flax. Flax will not thrive in heavy clays or gravelly or dry, sandy soils. Poorly drained land or land subject to excessive drought or erosion should not be used. Fields suitable for corn and soybeans are also suitable for flax, **Figure 13-59**. To the north and west of the Corn Belt, flax usually is grown in fields suitable for wheat or barley.

Flax plants will reach 3.9″–5.9″ (10 cm–15 cm) in height within eight weeks of sowing, and will grow several centimeters per day under optimal growth conditions, reaching 28″–31″ (70 cm–80 cm) within 15 days.

Fertilizer Needs and Applications

Farming flax requires few fertilizers or pesticides. Soil test levels, yield goals, and previous crops in the rotation will often determine the need for fertilizers. Farmers may also obtain the most current, best-management practices regarding fertilizer application through university extension programs and government organizations.

Pest and Disease Control

Although flax is not usually plagued with insect problems, insects that may affect crops include cutworms, potato aphids, Lygus bugs, and grasshoppers. Chemical control of insects is available for most pests; however, application times may be limited due to pollination needs and harvesting dates.

Disease problems are minimal with flax. However, fungus, such as rust and wilt, have been noted for some native species. Rust and wilt can be best managed by growing resistant varieties.

Harvesting

As stated earlier, flax is both a food and fiber crop. Some farmers may harvest only the seeds for food and oil production, whereas others also harvest the straw for fiber production. A combine harvester may be used to cut only the heads of the plant (for seeds) or the whole plant. The amount of weeds in flax straw greatly affects its market value. Therefore, the amount of weeds in the crop may determine whether or not the straw is harvested for fiber production. If the flax straw is not harvested for fiber production, it may be sold as biofuel.

Did You Know?

Flax fibers are taken from the stem of the plant and are two to three times as strong as those of cotton, but less elastic.

Flaxseed is generally harvested in late July to mid-August by windrowing before seed shatter. ***Windrows*** are long lines of raked hay or sheaves of grain laid out to dry in the wind, **Figure 13-60**. Flax is best harvested as the base begins to turn yellow. If harvested while still green, the seed will not be useful and the fiber will be underdeveloped. Although harvesting by hand may maximize the fiber length, flax harvesting in the United States is performed with mowing equipment, similar to hay harvesting.

Ruud Morijn Photographer

Figure 13-60. Flax may be harvested and laid out in windrows to dry in the same manner as hay.

Processing

Flaxseed is processed by cold pressing to obtain flaxseed oil for human consumption. The flaxseed meal remaining after cold pressing is often used in livestock feed. For industrial purposes, solvent extraction is often used to remove the flaxseed oil.

jocic/Shutterstock.com

Tukaram Karve/Shutterstock.com

Figure 13-61. Kenaf and jute are used to make twine and coarse fabrics such as burlap. These natural materials allow proper air circulation through the sacks of stored grains and seed.

Other Fiber Crops

Other crops grown for fiber include hemp, ramie, bamboo, kenaf, and jute, **Figure 13-61**. Although some are suitable for growing in the United States, most prefer more tropical climates. In addition, these fiber crops are not profitable enough to justify using land currently being used to grow other types of crops.

Career Connection Cotton Cooperative Manager

Job description: A cotton cooperative manager may or may not be a cotton farmer. The cotton gin is an expensive piece of machinery, costing hundreds of thousands of dollars. And the cotton gin is used only at harvest time. So, in some areas, several farmers will form a cooperative to manage the cotton gin and collaborate on volume sales of cotton. In general, one of the farmers is hired to manage the cotton gin. The manager maintains the gin, schedules farmers to use the gin, and stores the cotton bales.

©iStock/photosbyjim

Education required: High school degree required, college degree with courses in management and finance preferred.

Job fit: This job may be a fit for you if you enjoy working with people, managing schedules, and handling finances.

Lesson 13.4 Review and Assessment

Words to Know

Match the key terms from the lesson to the correct definition.

1. The white or yellow fibers produced by the cotton plant.
2. Growing stage in cotton plants in which three bracts form a triangle to protect the growing flower bud.
3. A bacterium that kills certain insect species that prey on cotton.
4. A chemical herbicide used to kill the cotton plant (or other plant) and make the leaves fall off to ensure a cleaner harvest.
5. Rubber pads with ridges or fingers that remove the fibers from the spindle of a cotton picker machine.
6. Oil from flaxseed used in both food and other products.
7. A natural material made from flax used in bed sheets, napkins, tablecloths, and clothing.
8. A long line of raked hay or sheaves of grain laid out to dry.
9. The rounded seed capsule of the cotton or flax plant.
10. A small cluster of flowers found in the axil of a stem.

A. Bacillus thuringiensis (Bt)
B. boll
C. bract
D. defoliant
E. doffers
F. linen
G. linseed oil
H. lint
I. square stage
J. windrow

Know and Understand

Answer the following questions using the information provided in this lesson.

1. List five common products made from the cotton plant.
2. Cotton fibers after harvest are called _____.
3. What is the part of the cotton plant where the ovary grows?
4. What is the major insect pest of the cotton plant?
5. *True or False? Bacillus thuringiensis* lives in the cotton plant.
6. Explain why a defoliant is used on cotton crops before harvesting.
7. What is a harvested bundle of cotton called?
8. What are three common products made from the flax plant?
9. What main factor determines if flax straw is harvested for fiber production?
10. *True or False?* Cotton fibers are stronger than flax fibers.

Analyze and Apply

1. Research the history of cotton production in the United States. How did cotton come to be a major crop?
2. Research the world production of cotton. How important is this crop on a global scale?

Thinking Critically

1. How has cotton played a role in our nation's history?
2. Where is cotton grown in the United States? Why do you think this might be the case?
3. Compare and contrast linen and cotton products. What are some of the benefits of linen over cotton garments?

Lesson 13.5

Fruit and Vegetable Production

Lesson Outcomes

By the end of this lesson, you should be able to:

- Identify fruits that are grown in the United States.
- Map where these fruits are grown in the United States.
- Define and identify brambles.
- Identify vegetables that are grown in the United States.
- Map where these vegetables are grown in the United States.
- Explain good management practices for fruit production.
- Explain good management practices for vegetable production.

Before You Read

Do a survey of your community to determine where vegetables are grown on a commercial basis. Research one of the vegetable crops to learn more about how it is produced.

Words to Know

apiarist
bramble
bulb vegetable
canes
cash farm receipt
crop rotation
damping off
cycs
fresh-market consumption
fruit
fruit vegetable
grafting
leafy vegetable
plant hardiness
pollination
pomology
rootstock
root vegetable
scion
seed treatment
stalk vegetable
traceability
tuber vegetable
value-added
vector

The average American eats around a ton of food each year; yes, 1,996 lb of food. Of that ton of food, the average American eats 415 lb of vegetables, 270 lb of fruits, and 24 lb of nuts. And, many of us should be eating even more vegetables for a healthful diet!

The USDA's Census of Agriculture shows that American farmers produce more than 100 different fruits and vegetables. Some of these are annual crops and others are perennial crops. Some fruits and vegetables are grown for ***fresh-market consumption***, meaning that they are consumed almost immediately without much processing other than washing, **Figure 13-62**. Other crops are grown and processed into canned, frozen, dehydrated, juiced, sauced, mixed, blended, and other value-added products. ***Value-added*** means that the grower or processor does something to the crop to make it more valuable to a consumer. A prime example of value-added produce is fruits and vegetables that have been cut, peeled, sliced or diced, and packaged in convenient quantities.

Did You Know?

Potatoes provide the highest cash farm receipts for vegetables. They are followed by lettuce, tomatoes, mushrooms, and onions.

Commercial Production

Commercial vegetable, fruit, and nut production is important to the U.S. economy, adding more than $46 billion in U.S. cash farm receipts annually. A ***cash farm receipt*** is the amount of money earned by the farmer or grower for his or her crop when he or she sells it to the marketplace. This represents

Goran Bogicevic/Shutterstock.com

Figure 13-62. Fresh-market fruits and vegetables have grown in popularity with the increase in prevalence of farmers markets in many parts of the country.

about 30% of all cash farm receipts from crops. Most of this production occurs on small farms.

While world consumption of fruits and vegetables has trended upward over the past 20 years, American production has slowed. Several factors contributed to this slowdown. Weather and disease problems have devastated some crops. Farmers in foreign countries have improved their production and processing practices, so American farmers face increased global competition in fruit and vegetable production. As a matter of fact, since about 1999, we have imported a greater volume of fruits and vegetables than we exported. As cities continue to grow, housing additions and shopping malls crowd out fruit and vegetable fields. So, fruit and vegetable farmers face competition over land with urban developers.

Vegetables

Vegetables can be grown almost anywhere. They can be grown in large fields in rural communities. They can be grown in small gardens in the cities, even in windowsills and on balconies. Vegetables can be produced by both small farm operators and large farm operators. Small farm operators may operate fruit and vegetable stands or sell products at food markets. Large farm operators typically sell their products directly to a processor in large quantities. Many vegetables can be eaten raw directly from the garden after careful washing.

Did You Know?

Grapes provide the highest cash farm receipts for fruit and tree nut production. They are followed by oranges, apples, almonds, and strawberries. Grapes and almonds have shown the most growth in cash farm receipts since 1990. Can you think of reasons why?

Most of the fresh commercially produced vegetables in the United States are grown in California. About half of the production of the major vegetable crops are grown there. We produce more than 1.7 million acres of fresh commercial vegetables in the United States at a value of more than $16 billion. Vegetable production is big business!

Vegetables need to be grown in the environment for which they are best suited. Vegetables are sensitive to climate and temperature variations. They often require a lot of water to grow and produce. Plant hardiness zones were created to guide farmers in growing the vegetable crops best suited for their particular region in the United States. ***Plant hardiness*** refers to the plant's ability to grow in cold or warm environments. The chart in **Figure 13-63** shows the plant hardiness zones in the United States.

Soil Management

Soil conditions are very important to the grower. Soil with good structure, moisture, and nutrients is essential for optimum plant growth. Soil management includes managing the amount of organic matter in the

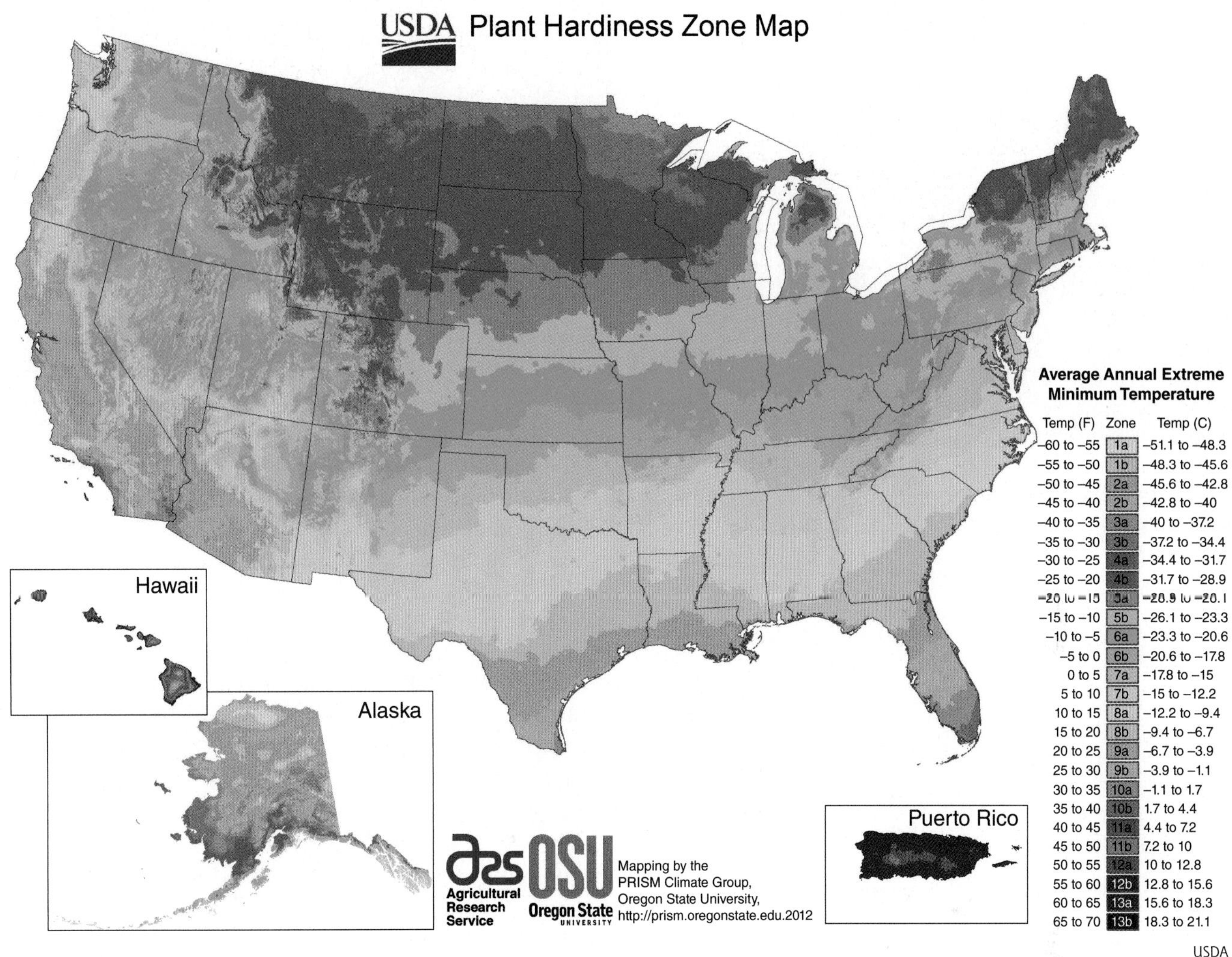

USDA

Figure 13-63. The USDA updates the hardiness zones as climate changes across the country and around the world. Visit the USDA website to use the interactive version and view hardiness zones from around the world.

soil for natural nutrient supply, and for protecting soil moisture. Careful tillage protects the soil structure, whereas excessive tillage breaks down soil particles and changes the soil's ability to hold moisture. Farmers lime and fertilize the soil according to the recommendations of a soil test. Over-fertilization is a waste of money, could lead to off-site pollution of water resources, and may inhibit plant growth.

Fertilizer Application

In small operations, fertilizers can be spread by hand. In larger operations, fertilizers are applied at planting and via surface applications once the crop has emerged. For those crops grown in greenhouses using hydroponics, fertilizers are injected directly into the water flow. Fertilizers should be used sparingly. Overdependence on manufactured fertilizers to boost yields has caused pollution from runoff. Increased fertilizer use means more time in the fields, increasing equipment and labor costs.

Did You Know?

If you look at a map of land used for vegetable production in the United States, you will see that most vegetable land is near the edges of the United States. Both coasts and around the Great Lakes have a large amount of vegetable land. Why do you think this might be the case? What explanations can you provide?

B Brown/Shutterstock.com

Narinto/Shutterstock.com

Figure 13-64. Alfalfa is a good cover crop for vegetable fields. It can be harvested for livestock feed or tilled in to improve soil conditions. Plastic mulch is often used in vegetable crop production.

Cover Crops and Mulches

Cover crops provide protection against wind and water erosion. Mulches can be made from natural materials, such as crop residue, or from manufactured products such as plastic, **Figure 13-64**. Ladino clover and alfalfa are good cover crops that can be tilled into the soil to increase the nutrients in the soil for following vegetable crops.

When using plastic mulch, growers till the seedbed prior to laying down the mulch in order to prepare a fine seedbed. Plastic mulch works best at warming the soil when the plastic is in direct contact with the soil surface. One disadvantage of plastic mulches is their disposal at the end of the season. Many landfills do not accept plastic mulches. There are some biodegradable mulches that break down over time in the field.

SeDmi/Shutterstock.com

Figure 13-65. This potato plant sprouted from a piece of "seed potato." It has already begun to produce potatoes.

Seeding and Transplanting

Some vegetables are large enough and hardy enough to be planted directly in the field. Sweet corn, pumpkins, and lima beans are good examples of these.

Other crops are seeded through vegetative means. Potatoes are a good example of a crop that is planted in this way. White potatoes are actually stem tissue that grows underground. The nodes on these potatoes are sometimes called the ***eyes*** of the potato. They are where new growth forms. Potatoes are sliced up and planted so that the nodes can sprout and form new potatoes, **Figure 13-65**.

Transplanting

Some vegetable seeds are too small to plant directly in the field. Two examples are tomatoes and peppers. Vegetables are planted in small containers in a greenhouse. These plants are watered and maintained until they reach the age and size that allows for successful transplanting. Greenhouses provide the uniform temperatures needed for successful seed germination.

Pest Control

Tree fruit, tree nut, bramble, and vegetable growers use integrated pest management (IPM) to control harmful insects, weeds, diseases, and fungi in their crops. The first step in integrated pest management is scouting the crop to identify and monitor pests in the crop. Farmers will always have a few weeds, harmful insects, and diseases in their crops. But, with each crop, there is a threshold above which it is necessary to use some means of controlling the pest. Pests can include weeds, insects, fungi, bacteria, viruses, and diseases.

Disease Control

Seed treatments effectively control disease at planting. Growers use approved seed treatments and follow all recommended safety precautions. ***Seed treatments*** are pesticides that are applied to the seed to provide pest control to the growing seedling. Because of seed treatments, seed should be handled with caution. Transplants are susceptible to diseases such as ***damping off***, which kills young plants before they can mature in the field. Damping off can be controlled by planting seeds in fresh soil mixes and by avoiding overwatering the young seedlings in the greenhouse, **Figure 13-66**. Diseases in transplants can also be controlled by keeping greenhouses clean, using certified seed treated with a fungicide, and by the approved use of fungicides.

Insect Control

For fruits and vegetables, insects pose a serious threat. Insects that feed on fresh-market fruits and vegetables make the fruit or vegetable unmarketable because of the insect feeding. Insects are vectors for diseases, viruses, and fungi. A ***vector*** is any organism that transmits disease, viruses, or fungi from one host organism to another.

Growers control insects primarily through the use of biological or chemical controls. Pesticides are applied to kill unwanted insects in the field. Growers must follow label directions explicitly to avoid chemical contamination of the produce. Growers must also keep accurate records of the exact timing and amounts of pesticide applications. If the produce is grown for markets other than retail, then the next entity that purchases the crop must also certify the pesticide records of the grower. This is called ***traceability***.

Humannet/Shutterstock.com

Figure 13-66. Damping off kills seedlings before they can grow and mature.

Cultural Practices

Selecting a good variety can reduce insect damage. Research on vegetables has begun to yield varieties that may be resistant to insect damage. Some varieties of corn, for instance, have husks that completely protect the growing ear of corn from corn earworm damage. Some vegetable varieties allow for earlier planting and can therefore avoid damage by late-season insects.

Crop rotation is another useful method for reducing insect damage. Rotating crops prevents the buildup of certain insect species that can damage crops. Plant residue left over from the previous crop should also be destroyed to kill insects that might overwinter in this vegetative material.

By practicing good tillage at the proper times during the growing season, growers can reduce the amount of damage from insects. Insects live in weeds alongside crops. If the weeds are removed, insect populations can be reduced. Thus, good sanitation and care of even the areas surrounding the fields reduces insect populations.

Conserving natural enemies of insects is one form of control that vegetable growers should utilize. Some insects feed upon more damaging insects in vegetable crops. Growers reduce the amount of insecticides applied to vegetables in order to increase the number of beneficial insects, **Figure 13-67**. Beneficial insects include earthworms, many types of predatory mites and wasps, and red ladybugs.

Weed Control

The key to effective disease and insect control is a good weed control program. Weeds compete with vegetables for water, nutrients, and sunlight. They also harbor insects and disease. To control weeds, growers do not plant vegetables on land with historically high levels of weed infestation. Growers plant their most competitive crops in weed-infested fields and plant the least competitive crops in cleaner fields. Rotating crops allows growers to effectively use herbicides to control difficult weed problems. Applying mulches also holds down weed germination and emergence. By combining several weed control methods in tandem with the appropriate herbicides, growers can reduce weed competition in vegetable crops. The goal is not to remove all weeds entirely, but rather to reduce the effect that weeds have on the growing vegetable crops.

The Gallery/Shutterstock.com

Figure 13-67. Insects can quickly damage a vegetable crop unless preventive measures are taken.

Pollination

Pollination is a fertilization process that occurs when pollen from the male portions of the flower is transferred to the female portions of the flower. Vegetables are produced as a result of this fertilization process. For many vegetable species, this process is essential for production. Cucumber, cantaloupe, pumpkin, squash, and watermelon are examples of crops that depend heavily on pollination for growth and development. Pollination may occur by wind or by insects.

At least 90 vegetable crops depend on bees to pollinate their flowers. Honeybees are often thought of as the most frequent pollinator, but there are other species of bees, including bumblebees and orchard bees, that pollinate vegetable crops. In tomato plants, the vibration made by bumblebees as they move from flower to flower often triggers the release of pollen onto the female flowers.

Fruit orchard owners and vegetable growers either rent honeybee hives from local beekeepers or manage their own hives. Growers can also improve the pollination services of native and nonnative bees by improving bee habitats, **Figure 13-68**.

Types of Vegetable Crops

Vegetable crops are typed by the edible portion for which they are grown.

- ***Root vegetables*** are those grown for their edible roots. Examples include beets, carrots, and radishes. Other less common root vegetables include parsnips, rutabagas, yucca, kohlrabi, garlic, horseradish, and ginger. Root vegetables contain high amounts of antioxidants, iron, and vitamins A, B, and C. They also generally contain high levels of carbohydrates and fiber.
- ***Fruit vegetables*** are those that produce a fruit, such as bell peppers, squash, and cucumber. Melons also fall into this category. They include watermelons, cantaloupe, and muskmelons. These are classified as fruit

Ratikova/Shutterstock.com

Figure 13-68. Honey bees pollinate vegetable crops. Vegetable growers will either maintain their own bee hives or rent them from local beekeepers. *At which plant growth stage should hives be placed near a crop?*

vegetables because they are consumed as vegetables, but also contain seeds in the part that is consumed.

- ***Tuber vegetables*** are those grown for their underground stems. Examples of these are potatoes and sweet potatoes. Tuber vegetables are generally high in starch content and minerals.
- ***Leafy vegetables***, such as lettuce, cabbage, and kale, are grown for the purpose of having their leaves consumed. You can think of salads as examples of leafy vegetables. Leafy vegetables contain high concentrations of vitamins A, C, and K; fiber; and vegetable proteins. Because of these nutrients, many doctors believe leafy vegetables are some of the most healthful foods you can eat.
- Asparagus, rhubarb, and celery have edible stalks. These are called ***stalk vegetables***. Stalk vegetables provide high amounts of fiber when consumed in the diet.
- ***Bulb vegetables*** produce belowground plant material that can be consumed. Examples of bulbs are leeks and onions.

Maksud/Shutterstock.com

Figure 13-69. Good harvesting methods avoid damaging the vegetables.

Harvesting

Harvesting the vegetable crop requires careful handling, packaging, and storage in order to get the vegetables to market at their optimum quality. Poor harvesting methods will almost always reduce the value of the harvested crop, **Figure 13-69**. Fresh fruits and vegetables are perishable products. It is important that harvest be done quickly while protecting the quality of the harvested product.

Vegetables are picked in the fields and then cleaned in packing sheds. Some sheds have cold storage available to cool down the produce and prevent early spoilage.

Growers handle harvested vegetables carefully to prevent bruising and damage to the vegetable's skin. Bruise damage reduces the shelf life of vegetables. Field containers used to harvest vegetables should be free of sharp and rough edges. Growers avoid dumping vegetables onto hard surfaces.

Vegetables are picked during the warm parts of the day. Heat can build up in vegetable containers and cause damage. This heat needs to be removed as soon as possible to increase the shelf life of the produce. Flowing cool water is often used to cool vegetables. This water is tested for bacteria and other contaminants regularly.

Richard Thornton/Shutterstock.com

Figure 13-70. Fruit and nut production adds more than $18 billion to the U.S. economy annually.

Fruits

Pomology is the branch of horticulture that studies the production of fruit and tree nut crops. A ***fruit*** is the ripened ovary surrounding the seed of the plant. Common fruits include apples, oranges, avocados, apricots, peaches, pears, plums, cherries, and nectarines, **Figure 13-70**. Common tree nuts grown in the United States include almonds, pecans, and walnuts.

About half of the annual U.S. fruit crop goes into the fresh market. The other half is processed into foods such as canned fruit, fruit juice, and dried fruit. The top five fruits consumed in the United States are oranges, grapes, apples, bananas, and pineapples.

Tree Fruit Farms

When considered collectively, more than half of the fruit and vegetable farms in the United States are tree fruit farms. This may be due in part to the fact that tree fruits are perennial crops and, once established, are cultivated for years. A vast amount of fruit is grown and harvested on the west coast. In fact, California produces more grapes, strawberries, and peaches than any other state. California also produces more than half of all fresh fruit consumed in the United States. Florida farmers produce the most citrus fruit, and more apples are produced in Washington than in any other state.

Did You Know?

Eight average strawberries contain more vitamin C than an average orange, and one large yellow bell pepper contains almost five times as much vitamin C as an orange.

Orchards

Fruit crops are generally grown in orchards. These are planted and maintain their productivity for decades. Apple trees generally produce fruit

Anastasiia Malinich/Shutterstock.com

Figure 13-71. Grafting involves splicing a scion from one variety of tree to the rootstock of another variety of tree, as was done with this sweet cherry tree.

for 15 to 20 years. Orange trees will produce fruit for even longer, sometimes up to 50 years or more.

Grafting

Most fruit trees are grafted. ***Grafting*** is the technique of splicing the shoot of one plant onto the rootstock of another plant, **Figure 13-71**. ***Rootstocks*** are the roots and growing stems that are cut off just a few inches aboveground. The shoot is also called a ***scion***. Why do orchard owners graft trees? First, some tree fruits, such as apples, produce wild types when seeds are planted, so the offspring do not look or taste like their parents. Second, rootstocks are produced so that they have disease and fungus resistance. So, orchard owners graft a really strong rootstock to a desirable fruit-bearing shoot.

Pruning

Pruning is one of the most important activities in an orchard. Each species of fruit tree has an optimal pruning size and shape that must be maintained in order to produce high yields and high-quality fruits. Pruning allows sunlight to penetrate foliage and helps ripen the fruits. Pruning also removes dead and diseased limbs and branches.

STEM Connection Cloned Fruit

When you eat a golden delicious apple, you are eating a direct, cloned descendant of the original golden delicious apple tree. The original golden delicious apple tree was an apple tree growing on a family farm in Clay County, West Virginia, in 1832. Apples are somewhat unique in the plant world because when you plant the seeds from an apple, the resulting plant will not be genetically identical to the parent. Each apple tree produced from seeds is unique. Just like no two humans are alike, so is the case with apples. No two apple trees produced from seed are alike. Cornell University's Geneva Experiment Station houses the USDA apple collection.

©iStock/valentinab

Tree Nuts

U.S. tree nut production continues to increase, with almonds, walnuts, pistachios, and pecans making up most of the sales receipts. U.S. farm receipts for all tree nuts is approximately $18 billion annually. Exports of tree nuts continue to rise. The increased demand for tree nuts can be attributed to the healthful effects of tree nut consumption. Other common tree nuts grown in the United States are hazelnuts and Macadamia nuts.

Most tree nuts are grown in orchards. Farmers work hard to keep the floor of the orchard clean and weed-free, as shown in the almond grove in **Figure 13-72**. Weeds compete with the trees for water and other nutrients. Weeds also harbor insect pests. Harvesting tree nuts usually involves shaking the tree with a large mechanical *shaker* when the nuts are mature. The machine also has a basket to catch the nuts as they fall from the tree.

Did You Know?

You cannot purchase cashews in the shell. Cashews belong to the same plant family as poison ivy and poison sumac. The shells of cashews contain an itchy oil similar to poison ivy and poison sumac.

Fruit and Nut Crop Production

Producing fruit and nut crops requires a lot of hard work. Planting, cultural practices, pest management, harvest, processing, and marketing require a lot of hands-on work and management knowledge. Fruits and nuts are perishable products, so there is no rest for the grower until the crop has left the farm and is on its way to the fresh market or processor.

Fruit and nut growers need to have a comprehensive knowledge of all areas of production. The grower must also be able to lead farm employees and communicate with customers. Daily supervision is required to manage the crop, lead the employees, and conduct effective marketing for farm products, **Figure 13-73**.

Olga Glinsky/Shutterstock.com

Figure 13-72. The orchard floor in this almond grove is free of litter and weeds, making management of the grove easier for the grower. ***Are there disadvantages to having bare ground as an orchard floor?***

As mentioned earlier, most fruit and nut production goes into the fresh market or into processing. Small farms will typically sell at farmers markets and to restaurants and specialty grocery stores that emphasize locally grown products. Larger farms sell to the fresh market through larger grocery chains and to processors. Many of these larger farms have contracts with grocery chains and processors to secure a favorable and consistent price for products.

Vlad Teodor/Shutterstock.com

Figure 13-73. Growers need to be able to lead employees and communicate effectively with customers.

Resource Management

Successful production requires the effective management of land, water resources, labor, and machinery. Soil conditions must be carefully monitored, with the right amount of moisture and nutrients for optimum plant growth. Good tillage practices protect soil moisture and soil structure. Cover crops are used to further increase the availability of soil moisture and increase the amount of organic fertilizer in the soil. Plastic mulches may be used to warm the soil and cut down on weed competition.

Weed, Insect, and Disease Control

Weed, insect, and disease control in fruit production are similar to those used in vegetable production. It is important to note that pesticides should be applied only when needed, by the recommended methods, and only in the amount needed. Consumers are concerned about the presence of pesticides in their fresh fruit.

Weather Hazards

Fruit and nut crops are highly susceptible to bad weather. Citrus can be damaged by frost and freezing weather, **Figure 13-74**. Fruit trees can be damaged by high wind. As with vegetable crops, fruits and nuts need to be grown in environmental conditions for which they are best suited. They require a lot of water for optimum production.

To provide plants with the best possible chance at survival and optimum growth, growers plant disease-resistant varieties. They cultivate carefully to avoid damage to the crop plants and use only the amount of fertilizers and pesticides needed for optimum production.

Pollination in Fruit and Nut Crops

Fruit and nut crops need to have their flowers pollinated in order to produce. Pollination occurs by either wind or insects that carry the pollen from flower to flower, accomplishing the task of fertilization. Honeybees work to pollinate most of the fruit trees. Often you will see honeybee hives along the edges of orchards, especially during flowering. Many ***apiarists***, or honeybee farmers, rent their hives to orchards during pollination.

mikhail/Shutterstock.com

Figure 13-74. Cold weather can seriously damage citrus crops. *How do growers supply heat to a citrus grove when the area experiences a cold front?*

Harvesting

Special machinery is required for proper harvesting, handling, grading, and packaging of fruit and nuts in order to ensure optimum produce quality at the marketplace. It makes little difference what the quality is at harvest if it is reduced by poor handling, packaging, or storage conditions. Because of the perishable nature of fruit, harvest should proceed carefully to avoid bruising or otherwise damaging the fruit, **Figure 13-75**.

Fruit should reach the consumer as quickly as possible. Unfortunately, growers have little control over fruit produce once it

Paolo Bona/Shutterstock.com

Figure 13-75. Quality at harvest is reduced by poor handling, packaging, or storage conditions.

leaves the farm. However, fruit produce can last much longer on the shelf if it is handled with care on the farm. Bruised fruit has a shorter shelf life.

Brambles

Brambles are any rough, tangled, prickly shrub that produces a berry. The most common brambles native to North America are raspberries and blackberries. Boysenberries, loganberries, and marionberries are hybrid

STEM Connection Comparing Nutritional Content

Many tree nuts provide health benefits for humans. Identify as many tree nuts as you can. Once you have this list, select 10 tree nuts and research their nutritional information. Prepare a graph or graphs of the nutritional value of tree nuts. Which nut provides the greatest health benefit? Why?

Once you have completed your study of the nutritional value of various nuts, expand your research by answering the following questions:

- Which tree nuts are grown in the United States?
- Which harvested tree nuts are the most expensive to purchase? Why?
- What pests or diseases are most common in tree nut production?

©iStock/Zerbor

Modfos/Shutterstock.com

Figure 13-76. Tame brambles are grown on trellis systems. The trellis systems must be designed to accommodate the harvesting machinery.

Did You Know?

Because of their dark blue color, blackberries have one of the highest antioxidant levels of any fruit. And, did you know that blackberry leaves calm stomach irritation in rabbits?

crosses of blackberries and raspberries. They are grown predominantly in the Pacific Northwest.

While wild blackberries and raspberries are delicious, harvesting them often results in scraped arms and pricked fingers due to the prickly thorns. Thankfully, agricultural scientists have developed tame bramble varieties of blackberries and raspberries that do not produce thorns. Tame brambles are trained to grow on a trellis or fence system. As the ***canes*** (stalks) grow, they are tied to the structure to keep the fruits off the ground and make harvest easier. Some growers use specialized machinery to harvest the berries, **Figure 13-76**.

Career Connection Food Safety Inspector

Job description: A food safety inspector may work with growers to inspect their production fields, review their production records, and certify fresh fruit and vegetable processing facilities.

Education required: Certification programs in food safety and good agricultural practices. Bachelor's degree in food safety, agronomy, crop science, post-harvest physiology, or agricultural education.

Job fit: This job may be a fit for you if you enjoy collecting and analyzing grain data and working both indoors and outside.

©iStock/LiudmylaSupynska

Lesson 13.5 Review and Assessment

Words to Know

Match the key terms from the lesson to the correct definition.

1. A rough, tangled, prickly shrub that produces a berry.
2. A fertilization process that occurs when pollen from the male portions of the flower is transferred to the female portions of the flower.
3. The technique of splicing the shoot of one plant onto the rootstock of another plant.
4. A plant disease that occurs in excessively humid conditions; causes young seedlings to collapse and die.
5. A plant's ability to grow in cold or warm environments.
6. The ripened ovary surrounding the seed of the plant.
7. A root and growing stem that is cut off just a few inches aboveground.
8. The process in which pesticide application records are kept and transferred to each entity handling produce, enabling each handler to certify the pesticide records.
9. A vegetable grown for its edible leaves.
10. Any organism that transmits disease, viruses, or fungi from one host organism to another.
11. Fresh crop products that have been prepared and packaged for easy consumer use.
12. A vegetable plant grown for its edible root.
13. Method of planting a different crop each growing season to help control pests and diseases that may become established in the soil over time.
14. Pesticides that are applied to the seed to provide pest control to the growing seedling.
15. A person who raises honeybees.
16. A vegetable grown for its edible underground stem.
17. Term describing produce that is consumed almost immediately without much processing, other than washing.
18. A vegetable grown for its edible stalk.
19. The amount of money earned by the farmer or grower for his or her crop when he or she sells it to the marketplace.
20. The branch of horticulture that studies the production of fruit and tree nut crops.
21. A vegetable that produces a fruit, such as bell peppers, squash, and cucumbers.
22. The nodes on potatoes from which new growth forms.

A. apiarist
B. bramble
C. bulb vegetable
D. canes
E. cash farm receipt
F. crop rotation
G. damping off
H. eyes
I. fresh-market consumption
J. fruit
K. fruit vegetable
L. grafting
M. leafy vegetable
N. plant hardiness
O. pollination
P. pomology
Q. rootstock
R. root vegetable
S. seed treatment
T. stalk vegetable
U. traceability
V. tuber vegetable
W. value-added
X. vector

23. A vegetable, such as the onion, grown for its edible below-ground plant material.
24. The stalks of brambles that are often tied to a fence or other support.

Know and Understand

Answer the following questions using the information provided in this lesson.

1. How many pounds of fruits and nuts does the average American consume each year?
2. What are pesticides called that are applied to seeds before planting?
3. What is traceability?
4. What is one example of a root vegetable?
5. Tuber vegetables are high in starches and _____.
6. What is one example of a leafy vegetable?
7. Stalk vegetables are high in _____.
8. After harvest, what tends to reduce the shelf life of fruits and vegetables?
9. How do many processers cool vegetables?
10. Fruits generally grow on _____.
11. What proportion of all fruits and vegetables are consumed fresh?
12. What is the term given to the process of splicing a scion to a rootstock?
13. What are three examples of tree nuts?
14. What is the machine that harvests tree nuts called?
15. Why are honeybees important to fruit production?

Analyze and Apply

1. What is the plant hardiness zone where you live? Which kinds of vegetables would be ideally grown in your area? Why?
2. Why are insects such a threat to a vegetable crop?
3. Why is orchard management of weeds and other pests important for the orchard owner?

Thinking Critically

1. What are the production challenges faced by fruit and vegetable farmers today? What modifications do you think scientists might make to fruits and vegetables to enhance production? Why?
2. Why would you want to thoroughly wash any fruits and vegetables before you consume them?
3. What would happen to apple varieties if a genetic disorder were discovered in one of the varieties?
4. If you were fighting a cold and wanted to consume fruits and vegetables that would help improve your immune system, which fruits and vegetables would you eat?

Lesson 13.6
Ornamental Horticulture

Lesson Outcomes

By the end of this lesson, you should be able to:

- Realize the size and scope of the ornamental horticulture industry in the United States.
- Identify the components of the ornamental horticulture industry.
- Examine and explain the floriculture industry in the United States.
- Analyze the components of the landscape horticulture industry.
- List common plants in ornamental horticulture.
- List common diseases, pests, and physiological disorders of ornamental plants.

Words to Know

bedding plant
cut flowers
floral design
floriculture
growing zones
houseplants
landscape architect
landscape horticulture
mother plants
nursery
ornamental horticulture
plugs
sod
stock plants
turf

Before You Read

Look through the lesson and write down each heading. With a partner, go through the list of headings and write down something that you think will be covered in each section. After you read the lesson, review your list with your partner to see how correct your thoughts about the section were.

Have you ever sent someone flowers? Have you ever received flowers? Flowers have come to be seen as a cultural symbol for wishing people well. We send flowers to cheer someone up, to congratulate someone, to show our sympathy, and even to apologize. Humans have long enjoyed a relationship with plants. Most of our plant interaction is based on the usefulness of plants for food and fiber; however, there are some plants that humans enjoy just because they add natural beauty to their environment.

Ornamental horticulture is the growing and marketing of plants used by humans for their aesthetic appeal. That means that this section of agriculture involves plants that exist to create beauty in the world around them. Can you think of some applications where plants like this are used? The first and most obvious component of ornamental horticulture is floriculture. ***Floriculture*** is the cultivation and management of flowering and ornamental plants, **Figure 13-77**. The other component of ornamental horticulture is ***landscape horticulture***, which involves the cultivation and management of living plants grown for aesthetic purposes in the human environment, both indoor and outdoor.

In the lesson, we will examine the ornamental horticulture industry. We will first look at floriculture and the implications for both living and harvested flowering plants. Then, we will look at the aspects of growing, maintaining, and designing plants to be used in landscape applications.

Did You Know?

In Victorian times, sending someone flowers was similar to sending someone a text message today. Each different type of flower meant a specific emotion or sentiment, and "flower books" were published to help the recipients decode their floral message.

Floriculture

The floriculture industry includes three basic types of ornamental plants:

- *Cut flowers*—flowers, including stems and leaves, which are cut off the plant and used in indoor decorative arrangements.
- *Bedding plants*—plants grown for outdoor ornamentation; may be annuals or perennials.
- *Houseplants*—plants grown for the purpose of indoor ornamentation.

Understanding a little bit about the production of each of these types of flowers will help us understand the floriculture industry as a whole.

Neirfy/Shutterstock.com

Figure 13-77. Floriculture is the science of growing flowering plants and the art of designing them. ***What is your favorite variety of flowering plant?***

Cut Flowers

There is urgency to the cut flower market as cut flowers have the shortest shelf life of any of the ornamental horticulture products. Imagine having a valuable product that is harvested, sold to wholesalers, processed, sold to retailers, sold to consumers, and completely used up all within seven days. Planning for cut flowers takes careful coordination by growers and the wholesale and retail market.

Typically, cut flowers begin their lives in a greenhouse or field in an area where they are natively grown. For example, most of the roses used in the floral designs in the United States are grown in Ecuador or Columbia, where growing conditions are optimal year-round, and most of the tulips in the world are grown in the optimal conditions in the Netherlands, **Figure 13-78**. American consumers import nearly 65% of their fresh cut flowers from overseas. Colombia, the largest exporter, provides nearly 80% of all fresh cut flowers in the United States.

In the United States, more than 75% of the cut flower farms are in California. Unfortunately, the cut flower industry in the United States has been in decline for the last twenty years. Foreign competitors can produce flowers at a lower cost than growers in the United States. In addition, abundant air travel and refrigerated shipping enables foreign growers to harvest and deliver cut flowers to florists or directly to your home in a matter of hours.

Once they have reached maturity, the flowers are cut and either shipped to a retailer who has contracted with the grower, or, more commonly, sent to a wholesaler.

West Coast Scapes/Shutterstock.com

Figure 13-78. Most cut flowers are grown in the environment that best suits them and then shipped to flower markets. These tulips are growing near the world's largest flower market in Aalsmeer, Holland.

The wholesaler has connections to flower shops, floral designers, and businesses which regularly use cut flowers for floral designs. These retailers then create floral designs that are sold to consumers.

Because cut flowers deteriorate rapidly, producers constantly monitor historic supply and demand to ensure that there is an adequate supply of the flower for the type and timing of consumer demand. For example, growers gear their planting and harvesting schedule to make sure that there are an abundance of the most common wedding flowers in early June, when there are more weddings in the United States than any other time of the year. Can you think which time of year rose producers make sure to hit peak production?

Luchi_a/Shutterstock.com

Figure 13-79. Floral design is more than just putting flowers together. Successful floral designs use artistic design principles to make sure they are pleasing to the eye. *How did the floral designer create balance in this floral bouquet?*

Floral Design

Floral design is the art and science of using cut flowers and other materials to create balanced and composed arrangements. Floral design is truly an art form, relying on the artistic principles of rhythm, balance, form, harmony, and unity, **Figure 13-79**. Creating a successful floral design is something that takes a lot of practice.

Many high school agriculture programs offer a floriculture class that includes learning the knowledge and practicing the skills required to become proficient at floral design. Ask your agricultural science teacher if this is a class that is an option for you.

Career Connection Floral Designer

Job description: Floral designers work with flowers and plants to create bouquets and floral displays. They are able to create arrangements that have the ability to set different tones, moods, and expressions.

Education required: Although some floral designers learn on the job, it is becoming much more common for floral designers to attend a one- or two-year floral design specific training school, or even obtain a four-year degree in design.

Job fit: This job may be a fit for you if you enjoy working with people and using creativity, working with your hands, and working under deadlines.

monkeybusinessimages/iStock/Thinkstock

Bedding Plants

Outdoor landscapes rely heavily on the use of bedding plants to increase their overall eye-appealing looks. In contrast to cut flowers, which can be grown and shipped thousands of miles away to a different continent, bedding plants are typically produced in the same general area where they are sold. Can you think of some reasons this is the best production plan? Hopefully, you thought about the fact that bedding plants are grown outside, and need to be able to survive in a fairly specific climate. That means that bedding plants can be grown in areas close to where they will be planted by consumers.

Growing zones are regions shown on the USDA Plant Hardiness Zone map, **Figure 13-80**. These regions show the geographical areas of the United States based on climate. Plants are given a zone rating to let growers know what will grow in their area.

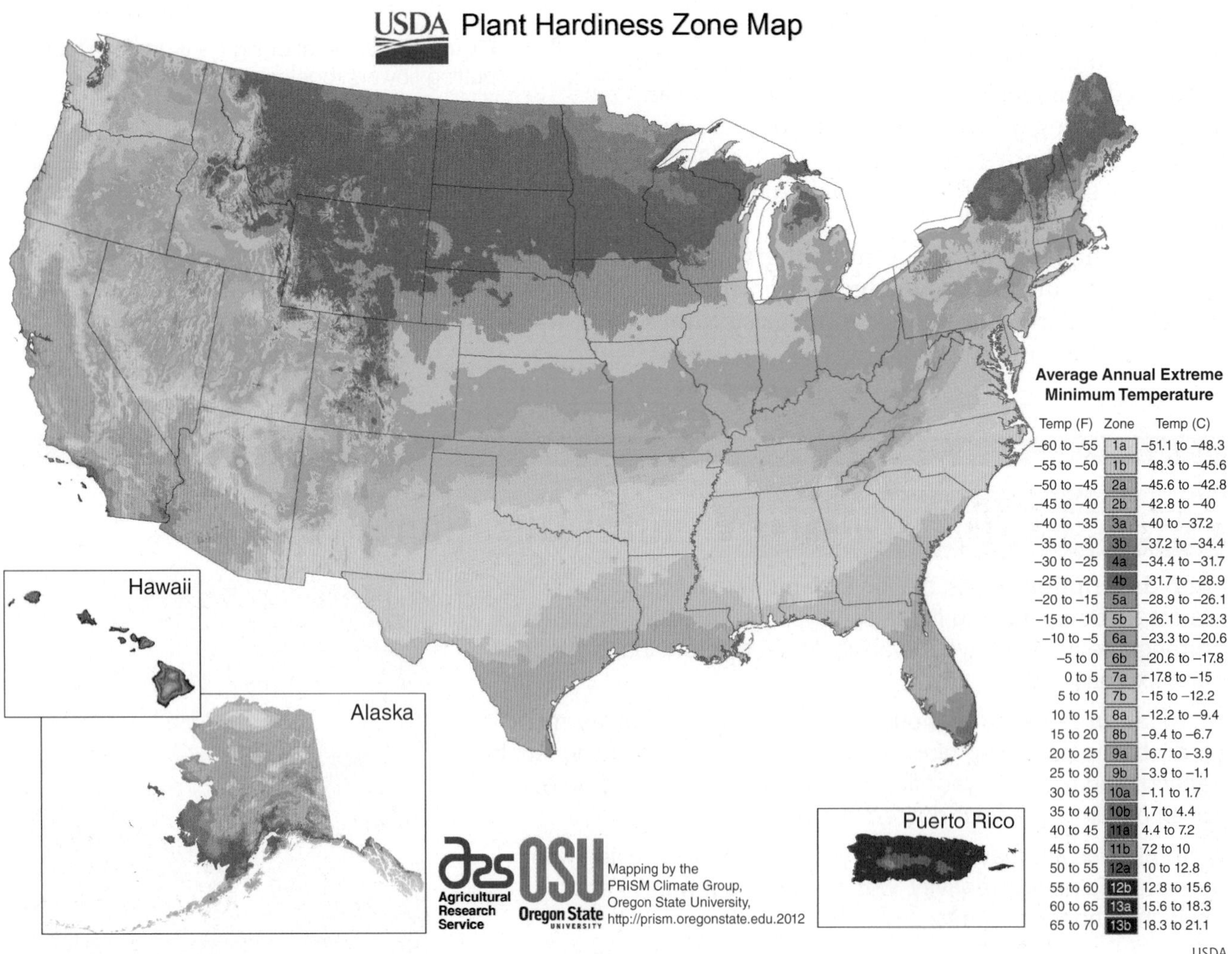

USDA

Figure 13-80. The USDA has divided the country into different climate zones. Checking this map will allow you to determine which bedding plants will grow best at your home. ***What growing zone do you live in?***

There are many different types of bedding plants. Most are categorized by the length of their growing season. If you remember from *Lesson 13.1*, annual plants live and complete their growing cycle in a single year, and perennial plants live multiple years, completing their life cycle continually throughout their lives.

The production of annual bedding plants typically begins in the winter for bedding plants that are ready to be sold to the consumer in the spring. Annual bedding plants are started in greenhouses from either seeds or plugs. ***Plugs*** are small seedlings germinated and sold to bedding plant producers, **Figure 13-81**. Plugs are a good option for seeds that are hard to germinate or for very valuable plants, as the risk of failed germination has been eliminated. Seeds are started in germination trays, then transplanted into larger containers, **Figure 13-82**.

tortoon/iStock/Thinkstock

Figure 13-81. Plugs are small seedlings that have been germinated by the grower prior to being sold to the greenhouse operator. *Do you think they are more expensive or less expensive than purchasing seeds?*

Perennial bedding plants are sometimes started much earlier than the winter before they are sold, depending on the desired size and rate of growth. Producers will generally keep perennial plants in the greenhouse that they can use to create new plants through asexual propagation. These plants used for cuttings are called ***mother plants*** or ***stock plants***. Typically, new plants are produced through cuttings and are genetically identical to the mother plant.

nakorn/Shutterstock.com

Figure 13-82. Seedlings can be seen in the germination tray in the background of this photo and are being transplanted into the larger container. *Why do you think the worker is using a stick to push the roots of the seedling into the growing media?*

danielo/Shutterstock.com Milosz_M/Shutterstock.com David Dea/Shutterstock.com Phonlaphat/Shutterstock.com

Figure 13-83. There are many common annual bedding plants used across the country. Some of the most common include A—Pansies, B—Petunias, C—Impatiens, and D—Marigolds. ***Do you have any of these plants growing in the flower areas of your landscape?***

Common Bedding Plants

There are literally thousands of varieties of bedding plants grown and sold in the United States. Some common bedding plants are shown in **Figure 13-83.**

Houseplants

Houseplants are grown and marketed in much the same way as perennial bedding plants. These plants are typically perennial and will last for years with proper care. The largest difference in houseplants and other plants in the floriculture sector is that houseplants come from temperate climates, and are well suited to the 65°F–80°F (18°C–27°C) temperature found indoors.

The amount of care a houseplant requires is dependent on the type of plant. There are houseplants that require minimal care and maintenance, and those that require a great deal of time and energy to keep in good health. Some basic tips for managing houseplants include:

- Learn about your plant. Some houseplants require more water, light, or different temperatures than others. Understanding your plant will allow you to tailor your care to its needs.
- If possible, make sure that the plant is in an area where natural sunlight can reach it. Plants will perform better if they can conduct photosynthesis with sunlight.
- Ensure good root health. A common problem in houseplants is that they become root bound, meaning the roots have no more room to expand, **Figure 13-84**. Monitor your plant and transplant it to a bigger pot if needed.
- Make sure your plant is getting nutrients. In the ground, plants get their nutrients from the soil, which is continually changing. A houseplant is in the same soil essentially its entire life. Research the nutrient requirements for your plant, and use a houseplant fertilizer if needed.

Scott Latham/Shutterstock.com

Figure 13-84. A common problem in houseplants is that they become rootbound. Transplanting this plant to a larger pot with new media will help the roots be able to expand and the plant to continue to grow. ***Can you see how much of the previous container was taken up by roots?***

FFA Connection Floriculture CDE

Are you intrigued by flowers and floral design? The National FFA Organization allows students who are interested in floriculture to participate in the Floriculture CDE. This CDE includes sections that hone student skills in:

- Creating floral designs.
- Identifying floriculture plants, equipment, and plant disorders.
- Propagating floriculture plants.
- Marketing and selling floriculture plants and designs.

If you are interested in joining or starting a Floriculture CDE Team, talk to your agricultural science teacher/FFA advisor.

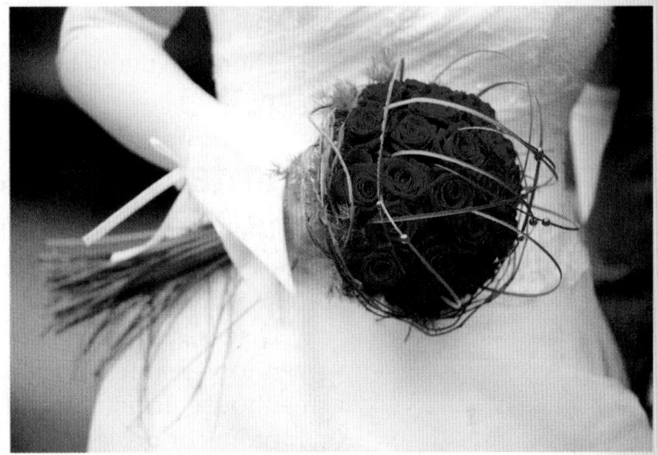

©iStock/jdwild

Figure 13-85. Houseplants come in many different varieties. Careful attention to the specific needs of each individual plant will help you develop your green thumb. A—African Violet. B—Sansevieria. C—Phiodendron. D—Dieffenbachia. E—Polka Dot Plant. F—Pepormia.

Common Houseplants

There are many houseplants, several of which are illustrated in **Figure 13-85**.

Maintaining Health of Floriculture Plants

The floriculture industry, like all plant production environments, has diseases, pests, and nutritional deficiencies that threaten the health and profitability of plants.

Diseases in floriculture plants cannot always be prevented. Good greenhouse sanitation is an important factor in making sure that your greenhouse floriculture plants stay healthy. Learning the common symptoms of disease and keeping a watchful eye on your plants will help you separate infected plants and prevent spread of diseases before all plants become infected. Some of the common floriculture plant diseases include: powdery mildew, leaf spot, verticillium wilt, rust, and bacterial wilt, **Figure 13-86**.

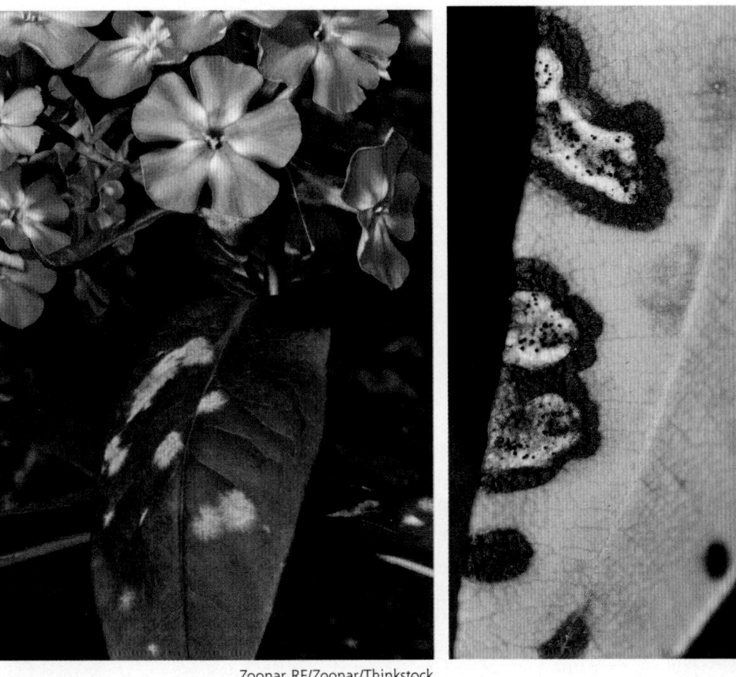

Zoonar RF/Zoonar/Thinkstock Floki/Shutterstock.com Damian Herde/Shutterstock.com

Figure 13-86. Common problems in floriculture plants include mildew, leaf spot due to fungus, and rust. ***What remedies are used for each of these problems? Should diseased plants be composted?***

Insects and other pests can also be a concern in a greenhouse. Even in a closed and carefully monitored greenhouse system, these pests can become a problem when they enter through unsterilized soil, plants being brought in, and even through an open door or ventilation system. Pests can be controlled through chemical or biological control. Having a good integrated pest management plan (IPM) can give you a step-by-step action plan for managing pests before they cut into the profitability of your floriculture operation. Some common pests that greenhouse managers deal with in the floriculture industry are: whiteflies, slugs, aphids, spider mites, thrips, mealybugs, and leafhoppers, **Figure 13-87**.

Another category of problems that floriculture plants encounter are nutritional and environmental disorders. Proper training will ensure that a greenhouse manager can tell the difference between plants who have failed due to the following environmental or nutritional conditions: overwatering; underwatering; too much or too little light; nitrogen, phosphorus, or iron deficiency; and improper temperature.

Did You Know?

Some of the most intensive management of agricultural crops occurs not in traditional operations, but on golf courses. The turf on golf courses often requires up to three times as much labor to maintain in a single season than traditional agricultural plants.

Landscape Horticulture

Another section of ornamental horticulture to consider is landscape horticulture. This includes the ornamental plants, trees, and turf involved in landscape settings. ***Turf*** is the upper layer of grass including the root structure. Some researchers estimate that there are more than 50 million

Figure 13-87. The whitefly is a common pest in the floriculture world. They are not actually flies, but are more closely related to aphids. Left unchecked, these insects can spread and destroy entire crops of greenhouse plants. ***Is there more than one species of aphids? If so, do they attack specific plants?***

acres of managed turf in the United States, **Figure 13-88**. To put that in perspective, that is an area larger than the state of South Dakota. Managing both the turf and other growing ornamental plants is a multibillion dollar industry employing millions of Americans. Landscape horticulture is based on the design, growing, and maintenance of landscape plants.

Mike Flippo/Shutterstock.com

Figure 13-88. There are more than 50 million acres of managed turf in the United States. Turf is found in residential yards, golf courses, baseball fields, soccer fields, and—if they use natural turf—on the field of your favorite football team.

Landscape Design

A huge component of the landscape horticulture industry is the planning of landscapes, **Figure 13-89**. A ***landscape architect*** is someone who develops the plan for the space and the specific plants that will be used. Designing a landscape requires several specific skills:

- Understanding of design principles to ensure that the landscape will be suited to the aesthetic needs of the client.
- Knowledge of the growth patterns, nutrient requirements, and space needs for a wide variety of landscape plants.
- Ability to design the layout of plants, water features, irrigation systems, walkways, and outdoor living spaces to customer specifications.

All of the parts of landscape design come together to create a cohesive outdoor environment. Often, the homeowner or business owner has specific needs for the landscape design. For example, the homeowners may live in an area with little rainfall. Thus, they may want to design and install a landscape that includes plants with low water requirements. Businesses may want to use the landscape design to provide cooling to the building or even hide some parts of the business from customers. So the landscape design can include plants that shade and shield.

Whatever the purpose, a well-designed and maintained landscape adds value to the home or business. A pretty and functional landscape provides curb appeal

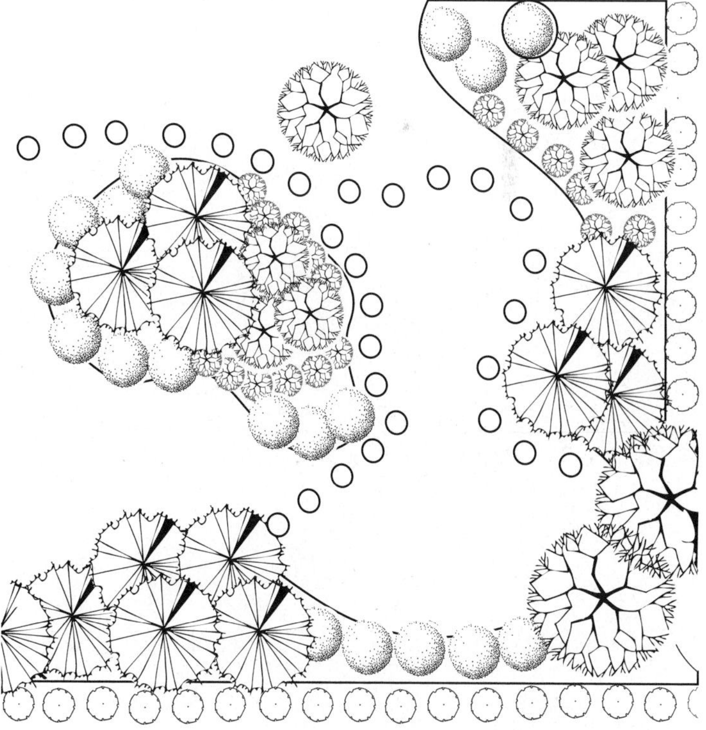

charobnica/Shutterstock.com

Figure 13-89. Landscape designers use computer programs to work up a sample layout, like this, for clients to approve before beginning their work on the project. The landscape includes all aspects of the outside space, including turf, trees, bushes, and bedding plants.

to the home. People like to look at a pretty home and relax in a comfortable yard. Also, most landscapes help reduce energy needs and are easy to maintain. No homeowner wants to spend all his or her time worrying about weeding, pruning, and maintaining a landscape.

Growing

Growing landscape plants is a little bit different than growing floriculture plants. While most floriculture plants are grown in indoor greenhouses, many landscape plants are grown outdoors. The trees and shrubs used in landscape designs are grown in a production facility called a ***nursery***. Nursery growers are responsible for the growing of both woody and herbaceous plants, and often need to cultivate plants for several growing seasons to produce plants that are ready for the consumer, **Figure 13-90**.

Growing turf can occur on location or on a turf farm. Growing turf at the site is often chosen because it is less expensive than purchasing pre-grown turf. A drawback to this type of turf is that the entire landscape is not accessible during the time the grass is germinating and rooting. Those who choose pre-grown turf, called ***sod***, generally make their decision based on the fact that the turf can be walked on and used much more quickly with sod. Like bedding plants, sod is grown according to climate regions and typically grown near the place where it is going to be used. To grow sod at a sod farm, the specific variety of grass seed is planted in large fields and grown until it has mature roots and adequate height. Then, the entire turf structure, including the roots and uppermost layer of soil, is removed and transported for installation, **Figure 13-91**. One of the drawbacks to installing

patboon/Shutterstock.com

Figure 13-90. Nurseries are where trees and shrubs are grown and marketed to consumers. This nursery in Oregon specializes in the production of ornamental juniper trees.

sod is that it is much more expensive than seeding grass.

Common Landscape Plants

Landscape plants vary almost as much as the climates that they go into. Can you guess the most common landscape plant in the country? If you guessed roses, you would be correct, **Figure 13-92**.

Maintenance

Have you ever watched golf on television? Even if you only caught a passing glance at the golf channel as you were clicking through the channels, it is likely you have seen this: the camera begins with a wide angle shot of the golf course and then slowly zooms

Ingrid Balabanova/Shutterstock.com

Figure 13-91. Sod is the term for turf that has been cut and removed for installation. ***What are some of the considerations you think you would need to think about when laying sod?***

Sandra Voogt/Shutterstock.com

Darla Hallmark/Shutterstock.com

ER_09/Shutterstock.com

V.J. Matthew/Shutterstock.com

Figure 13-92. Roses are available in many colors and sizes. ***Did you know that there are thousands of different varieties of roses?***

Hank Shiffman/Shutterstock.com

Figure 13-93. The care required for landscape maintenance is easily seen if you have ever been on a golf course. Mowing, fertilizing, controlling pests, and eradicating weeds are all important maintenance tasks that are accomplished for successful landscape horticulture maintenance.

in on a perfectly manicured golf green. It is a pretty impressive thing to see. Golf courses are one of the greatest examples of the care and effort required to maintain landscapes, **Figure 13-93**.

Maintenance in landscape includes cutting and trimming turf; pruning trees, bushes, and shrubs; and fertilizing; and managing pests that could threaten the landscape beauty. In addition, maintenance also includes the care and upkeep of irrigation (sprinkler) systems. Do you know someone who started mowing lawns in high school and now owns a landscape maintenance company? In the United States, landscape maintenance is a multibillion dollar industry, and has been ranked as one of the easiest jobs for high school students to translate into an adult occupation.

Diseases and Pests

Landscape plants require careful monitoring to ensure that they are kept healthy. Because they are exposed to any pathogen or insect that is in the area, diligence is important to controlling a situation before it gets out of hand. There are four categories of potential disorders in landscape plants: insects, diseases, weeds, and physiological problems.

Insect pest problems can cause issues in landscape plants because, if not treated, they are constantly vulnerable to the insects in the area, **Figure 13-94**. Some of the most common insects that landscape plants deal with are aphids, leafhoppers, slugs, and white grubs. When controlling insects in landscape areas, it is important to consider that many times pets and young children are in contact with the plants. Care should be taken when determining which pesticide chemicals to use.

FFA Connection Nursery/Landscape CDE

The Nursery/Landscape CDE is a perfect fit for those who like to work in the designing and maintenance of landscape horticulture plants. This CDE allows FFA members to complete practicums in:

- Designing landscapes.
- Completing landscape maintenance tasks.
- Identifying plants, pests, disorders, equipment, and supplies.
- Propagating nursery plants.
- Assisting customers, sales, and marketing.

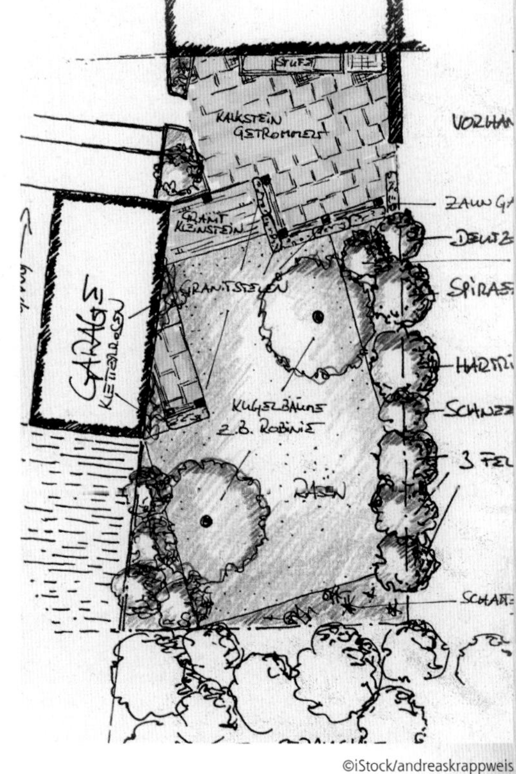

©iStock/andreaskrappweis

Christian Delbert/Shutterstock.com

Figure 13-94. Improper care of landscape plants can result in unsightly and costly landscape repairs. ***What do you think happened to this landscape?***

jarih/iStock/Thinkstock

Figure 13-95. Weeds can be managed through several different methods. Physical removal, like pulling weeds, is likely the *least* fun.

Landscape plants are subject to many different diseases which can attack the roots, stems, leaves, and fruiting bodies of the plants. One of the most important things to do to protect plants against diseases is to ensure that plants are being properly watered and have adequate drainage. Standing water on landscape plants can amplify the spread of diseases. Some of the more common landscape plant diseases are: black spot, canker, powdery mildew, and root rot.

Any plant growing in a place it is undesired is considered a weed. Landscape managers try to create optimal growing conditions for plants, and so it would make sense that weeds would find landscapes an ideal place to make their home, **Figure 13-95**. The most common weeds include chickweed, crabgrass, dandelion, and purslane. Controlling weeds is one of the most challenging tasks in maintaining any landscape. Weed control can be managed manually (by pulling the weeds out), chemically, or physically by putting down weed barrier in areas like flower beds.

The last category of problems is physiological problems. These disorders are due to the plant environment and include issues like frost injury, nitrogen deficiency, and overapplication of chemical herbicides. While some of these issues are due to climate factors beyond your control, many of these problems can be avoided with proper management.

Lesson 13.6 Review and Assessment

Words to Know

Match the key terms from the lesson to the correct definition.

1. A production facility in which the trees and shrubs used in landscape designs are grown.
2. Small seedlings germinated and sold to bedding plant producers.
3. Flowers which are cut off the plant and used in indoor decorative arrangements.
4. Plants grown for the purpose of indoor ornamentation.
5. Someone who develops the plan for a space and the specific plants that will be used in that space.
6. The art and science of using cut flowers and other materials to create balanced and composed arrangements.
7. Turf grass that has been cut in uniform pieces from the earth with the roots intact to be transplanted to create an "instant" lawn.
8. The growing and marketing of plants used by humans for their aesthetic appeal.
9. Plants kept by producers from which cuttings are taken to produce new plants through asexual propagation.
10. The cultivation and management of flowering and ornamental plants.
11. The cultivation and management of living plants grown for aesthetic purposes in the human environment (indoor and outdoor).
12. A plant grown for outdoor ornamentation; may be an annual or a perennial.
13. The geographical regions shown on the USDA Plant Hardiness Zone map to show consumers what plants will thrive in any given area.
14. The upper layer of grass including the root structure.

A. bedding plant
B. cut flowers
C. floral design
D. floriculture
E. growing zones
F. houseplants
G. landscape architect
H. landscape horticulture
I. mother plants
J. nursery
K. ornamental horticulture
L. plugs
M. sod
N. turf

Know and Understand

Answer the following questions using the information provided in this lesson.

1. What are the three types of floriculture plants?
2. Why are bedding plants grown close to where they are marketed?

3. How are bedding plants started in a greenhouse?
4. What is the name given to the original plant from which perennials are propagated?
5. When growing a houseplant, the plant will sometimes outgrow its _____.
6. What are common diseases in greenhouse plants?
7. What is another name for the grass growing in your yard?
8. Where are trees and shrubs used in landscape designs grown?
9. What is involved in maintaining a landscape?
10. When applying chemicals in a landscape, what is one precaution that *must* be undertaken?
11. What are common landscape diseases?
12. What are common landscape weeds?

Analyze and Apply

1. In which states are most cut flowers grown in the United States? Why?
2. Which times of the year are popular cut flower markets?
3. How can a well-designed landscape add value to a home?

Thinking Critically

1. Use the Internet to research some examples of landscape designs. Then, make your own landscape design. Be sure that you draw where you want features like flower beds, patios, and water features. Label all of the different plants you would use in the design and where you would put them.
2. After you have created the landscape design, make a materials list of all the plants that you would need to complete the plan. Research the cost for each plant online and create a complete plant materials list for your landscape.
3. Some people, especially those living in newer homes, find that their houseplants do not receive enough sunlight, even when placed directly in front of a window. Can you think of a possible reason?
4. Think about your house. What would be the requirements you would have for a redesigned landscape around your home? Which plants would you select to turn this plan into a reality?
5. How could you plan and work a Supervised Agricultural Experience in ornamental horticulture in your area? Where would you start? What would you grow?

Chapter 13

Review and Assessment

Lesson 13.1

Plant Anatomy and Physiology

Key Points

- Although plants look very different, all plants have the same basic structural components.
- The main plant parts include roots, stems, leaves, and flowers.
- Plants can be classified by their growth cycle or structure.
- Plants use the process of photosynthesis to capture sunlight and convert it into energy.
- Plants reproduce either through asexual means or sexually by producing seeds.

Words to Know

Use the following list and the textbook glossary to review and study the *Words to Know* from *Lesson 13.1*.

annual
anther
asexual reproduction
autotroph
biennial
blade
bulb
cambium
cellular respiration
chlorophyll
chloroplast
composite flower
corm
cotyledon
crop
cuticle
dark reaction
dicot
dicotyledon
disk flower
double fertilization
endosperm
epidermal tissue
fertilization
fibrous roots
filament
ground tissue
growth cycle
guard cell
herbaceous stem
imperfect flower
internode
lateral root
leaf
legume
lenticel
light reaction
meiosis
meristem
mesophyll
micropropagation
midrib
mitosis
monocot
monocotyledon
node
nodule
ornamental plant
ovary
ovule
palisade layer
peduncle
perennial
perfect flower
petal
petiole
phloem

photosynthesis
pistil
pistillate flower
pollen
pollen tube
pollination
primary root
ray flower
receptacle
rhizomes
root
root cap
root hair
runner
sepal
sexual reproduction
spongy mesophyll
stamen
staminate flower
stem
stigma
stolon
stomata
style
symbiotic relationship
taproot
tuber
vascular bundle
vascular tissue
vein
weed
woody stem
xylem
zone of elongation

Check Your Understanding

Answer the following questions using the information provided in *Lesson 13.1*.

1. What are the three main types of plant tissues?
2. What is the difference between herbaceous and woody stems?
3. The _____ distance can be important in determining plant height and shape. Explain why you think this is so.
4. What are the layers of a leaf?
5. Explain the difference between a perfect flower and an imperfect flower.
6. List four major differences between monocots and dicots.
7. *True or False?* Some plants may grow as an annual in certain climates and as a perennial in others.
8. What is the difference between light and dark reactions?
9. What are the two methods of plant reproduction?
10. *True or False?* Micropropagation requires the combination of a plant cell from two separate plants.

Lesson 13.2

Cereal Grain Production

Key Points

- Cereal grains belong to the grass family and provide an excellent source of vitamins, minerals, protein, carbohydrates, and fats in an unprocessed form.
- Cereal grains include barley, oats, corn, wheat, rye, millet, rice, and grain sorghum.

- Corn is the most widely produced feed grain in the United States, with almost 80 million acres in production.
- Oats taste mildly sweet, which makes them useful for human breakfast cereals.
- Rice is one of the most important grains in the human diet and is widely grown around the world.
- More than 2.1 billion bushels of winter wheat, durum, and spring wheat are grown in the United States.

Words to Know

Use the following list and the textbook glossary to review and study the *Words to Know* from *Lesson 13.2*.

barley yellow dwarf
boot stage
bran
creep feeding
crown rust
dent corn
embryo
endosperm
ergot
flint corn
flour corn
germ
Helminthosporium leaf blotch
hull-less barley
integrated pest management (IPM)
jointing
kernel
lodge
loose smut
moisture content
panicle
popcorn
pseudo-cereals
seedhead
six-row barley
smut
sweet corn
tillering
tillers
two-row barley
windrow

Check Your Understanding

Answer the following questions using the information provided in *Lesson 13.2*.

1. *True or False?* Pseudo-cereals produce a grain-type fruit.
2. What are the six types of corn?
3. Why does a popcorn kernel pop when heated?
4. *True or False?* Wheat adapts to many climates and is grown more than any other crop.
5. Where is rice grown in the United States? Why do you think it is grown in that area?
6. Which parts of oats are included in whole-grain products?
7. *True or False?* Corn contains less protein than oats.
8. Where does the seedhead grow on a cereal grain plant?
9. What are three uses for rye?
10. What are three uses for grain sorghum?
11. What are three factors growers consider when selecting species and varieties of crops to plant?
12. Seedbed preparation controls weeds while conserving soil _____.
13. At what point in a wheat plant's life does vegetative growth transform into reproductive growth?

14. How is rice planted?
15. When are oats planted? When are they harvested?
16. *True or False?* Barley is a cool season crop.
17. When is millet planted?
18. When is rye planted?
19. Why would growers plant grain sorghum in rows only 10″ apart?
20. _____ should preserve the natural structure of the soil and reduce wind and water erosion.

Lesson 13.3

Oil Crop Production

Key Points

- Oil crops are essential for human and livestock nutrition and industrial products.
- Oil crops include corn, soybeans, sunflowers, canola (rapeseed), and peanuts.
- Soybean oil is used in cooking oils, margarine, soy candles, soy ink, soy crayons, many lubricants, and even biodiesel.
- Sunflowers are grown for three major reasons: for human and livestock nutrition, bird feed, and habitat restoration.
- Canola production has increased in the past 40 years, making canola oil the second largest oil crop in the world.
- Peanuts are produced underground, unlike any other major nut crop plant.
- Like soybeans, peanut plants collect nitrogen from the air and convert it into a form suitable as fertilizer for the growing plant.

Words to Know

Use the following list and the textbook glossary to review and study the *Words to Know* from *Lesson 13.3.*

cash farm receipt
inoculation
legume
pegs
rust

Check Your Understanding

Answer the following questions using the information provided in *Lesson 13.3.*

1. What are three uses for corn oil?
2. Soybeans belong to the special class of plants called _____.

3. *True or False?* Sunflower varieties typically mature in approximately three months.
4. What are three factors farmers consider when selecting varieties of canola to plant?
5. What are the four major market types of peanuts?
6. *True or False?* The germination rate is more efficient when the soil is cooler.
7. Soybeans are planted in narrow row widths to reduce competition from _____.
8. Why should sunflowers grown for seed not be spaced too close together?
9. At what moisture content are soybeans harvested?
10. Why are sunflower seeds harvested at 18% to 20% moisture content?
11. Why must the farmer time canola harvesting with the plant's natural adaptive traits?
12. What type of machine is used to prepare the peanuts for drying in the field?

Lesson 13.4

Fiber Crop Production

Key Points

- Farmers grow a number of fiber crops that are used to make fabric, paper, rope, and other specialized products used in industry.
- Cotton has many uses, including: a livestock feed additive; oil for soap, cosmetics, paint, and detergents.
- Cotton fiber is used to manufacture fabrics for clothing, automotive tire cord, cardboard, paper, and as a fuel oil ingredient.
- Flax is both a food and a fiber crop.

Words to Know

Use the following list and the textbook glossary to review and study the *Words to Know* from *Lesson 13.4*.

Bacillus thuringiensis (Bt)	doffers	square stage
boll	linen	windrow
bract	linseed oil	
defoliant	lint	

Check Your Understanding

Answer the following questions using the information provided in *Lesson 13.4*.

1. When does the square stage of the cotton plant occur?
2. At what critical stage does a cotton plant need fertilizer?
3. How can insects and disease in cotton production be controlled?

4. Why must the leaves be removed before cotton is harvested?
5. Why should cotton be harvested after the morning dew has dried?
6. What is the purpose of ginning cotton?
7. What are three uses for flax?
8. *True or False?* Flax requires extensive weed and pest control through fertilizers and pesticides.
9. Why is flax straw laid out in windrows before harvesting?
10. Which other crops are grown for fiber production?

Lesson 13.5

Fruit and Vegetable Production

Key Points

- Vegetables are versatile crops. Almost anyone can grow vegetables for personal consumption.
- Fruit and vegetable production requires careful management at all levels in order to get the most returns.
- Good harvesting, packing, and storage methods are very important to ensuring the quality of the fruit and vegetable crop.

Words to Know

Use the following list and the textbook glossary to review and study the *Words to Know* from *Lesson 13.5*.

apiarist
bramble
bulb vegetable
canes
cash farm receipt
crop rotation
damping off
eyes
fresh-market consumption
fruit
fruit vegetable
grafting
leafy vegetable
plant hardiness
pollination
pomology
rootstock
root vegetable
scion
seed treatment
stalk vegetable
traceability
tuber vegetable
value-added
vector

Check Your Understanding

Answer the following questions using the information provided in *Lesson 13.5*.

1. What are value-added products? Does the added value add cost?
2. Why has American production of fruits and vegetables slowed?

3. Where are most of the fresh commercially produced vegetables in the United States grown?
4. Please explain why soil conditions are important to the grower.
5. What is a disadvantage of plastic mulch?
6. Why do growers need to keep accurate records of the exact timing and amounts of pesticide applications?
7. How does crop rotation help with pest control?
8. What can growers do to help control weeds?
9. Explain why bees are vital to crop production.
10. Why is careful handling, packaging, and storage vital in vegetable harvesting?
11. Why do fruit growers graft trees?
12. Why is pruning an important factor in fruit production?

Lesson 13.6

Ornamental Horticulture

Key Points

- Ornamental horticulture includes floriculture and landscape horticulture.
- Floriculture is the management of flowering and ornamental plants. It includes cut flowers, bedding plants, and houseplants.
- Floral design combines the knowledge of plants with artistic principles and requires the industry to get the cut flowers to the consumer quickly.
- Landscape horticulture is the management of turf, trees, and shrubs.
- The maintenance of landscape is one of the largest sectors of ornamental horticulture.
- Careful attention to all plants in ornamental horticulture can help to identify and treat common diseases and pests.

Words to Know

Use the following list and the textbook glossary to review and study the *Words to Know* from *Lesson 13.6*.

bedding plant
cut flowers
floral design
floriculture
growing zones
houseplants
landscape architect
landscape horticulture
mother plants
nursery
ornamental horticulture
plugs
sod
stock plants
turf

Check Your Understanding

Answer the following questions using the information provided in *Lesson 13.6*.

1. What is the difference between an ornamental plant and a crop?
2. Please list and define the three main areas of floriculture.
3. Why are most cut flowers produced in other countries?
4. Why do cut flower producers constantly monitor historic supply and demand?
5. How does the floral design industry relate to the cut flower industry?
6. If you were looking at bedding plants online and wanted to know if a particular type of plant would be able to handle the climate in your area, which resource could you use to help you?
7. Why are plugs a good option for some plants?
8. Why is there a difference in the time that perennial and annual bedding plants are started?
9. What are two main differences between houseplants and bedding plants?
10. What happens to houseplants which become rootbound?
11. What is the first step to preventing disease and insect infestations with floricultural plants in the greenhouse?
12. What are common insect pests in greenhouse plants?
13. Compare a landscape architect to a traditional architect who designs buildings.
14. How does a nursery differ from a traditional greenhouse?
15. What are the four categories of potential disorders in landscape plants?

Chapter 13 Skill Development

STEM and Academic Activities

1. **Science.** How does narrow row spacing in oats, barley, and millet reduce weed competition? Think of a hypothesis that would test the row spacing with regard to weed competition. Design and conduct an experiment to determine how narrow rows reduce weed competition.
2. **Math.** What are the average yields of cereal grains in your state? What is the current market price for these grains? Based on that information, which grains would you grow on your farm? What other economic factors would you need to consider before making a final decision on which grains to grow?
3. **Social Science.** Who was Norman Borlaug? What did he accomplish in agriculture? Which cereal crops were the focus of his research?

4. **Technology.** When growers attempt to control insect pests in their crops, one of the first lines of defense is beneficial insects. Research beneficial insects for different varieties of vegetables. How do growers introduce beneficial insects into their fields? How could you develop an SAE to produce beneficial insects for fruit and vegetable production in your area?
5. **Math.** Hypothesize the trend for acreage planted in fresh-market vegetables in the United States since 2010. Research the USDA agriculture statistics for fresh-market vegetable production in the United States. Plot the acreages of major vegetables. What are the trends? Which vegetables are increasing in acreage? Which are decreasing? Why do you think these trends exist? What is the percentage increase/decrease for each vegetable?
6. **Social Science.** Research the origin of common apple varieties. What is the story behind the development of that particular apple variety? Develop a travel guide pamphlet to highlight the origin of the apple tree.

Communicating about Agriculture

1. **Reading and Speaking.** What are the purposes of genetically modified seeds in cereal gain production? Research how and why scientists have genetically modified cereal grain seeds. Prepare a 3–5-minute speech on one of these advancements in agriculture. Argue either for or against genetically modified seeds, citing evidence from your research.
2. **Reading and Writing.** Which states produce the most of each type of cereal grain? Research this information from the U.S. Department of Agriculture. Prepare a poster about the importance of these grains to the United States and your own state.

Extending Your Knowledge

1. Using seeds provided by your instructor, germinate several species of cereal grains. Examine the seeds as they sprout to see if you can identify emerging plant parts.
2. Purchase dried peanuts from the grocery store. Grind the peanuts up and make your own peanut butter in the lab using the materials supplied by your agriculture teacher. How much oil do the peanuts produce as you grind them?
3. Using the materials supplied by your agriculture teacher, germinate tomato seeds, grow them to transplant size, and then transplant them into a garden or field plot. Take notes on the growth and development of the plants at each growth stage. Note any problems or concerns that arise. Share your results with your instructor.
4. Which fruits and vegetables would you consume if you were a vegetarian and wanted to increase the level of protein in your diet? Why?
5. How does salad dressing change the nutritional value of a salad?
6. Monitor your diet for a week. How many different fruits and vegetables do you consume in a week's time? What is the nutritional value of those fruits and vegetables?

Chapter 14

Environmental Systems Impacting Agriculture

TTphoto/Shutterstock.com

©iStock/elenavolkova

©iStock/ian600f

G-WLEARNING.com

While studying, look for the activity icon to:

- **Practice** vocabulary terms with e-flash cards and matching activities.
- **Expand** learning with video clips, animations, and interactive activities.
- **Reinforce** what you learn by completing the end-of-lesson activities.
- **Test your knowledge** by completing the end-of-chapter questions.

Lesson 14.1 Ecosystems

Words to Know

abiotic factor
aquatic ecosystem
biotic factor
brackish water
community
coral reef
deciduous tree
desert ecosystem
dissolved oxygen (DO)
ecology
ecosystem
estuary
evergreen tree
fauna
flora
food chain
food cycle
food web
forest ecosystem
freshwater ecosystem
grassland ecosystem
habitat
jungle
lentic ecosystem
lotic ecosystem
marine ecosystem
mountain ecosystem
prairie ecosystem
riparian zone
saline
salt marsh
taiga ecosystem
temperate deciduous forest

(*Continued*)

Lesson Outcomes

By the end of this lesson, you should be able to:

- Define ecology and ecosystem.
- Explain the differences between biotic and abiotic factors in an ecosystem.
- Identify and describe ecosystems found in the United States.
- Compare and contrast aquatic and terrestrial ecosystems.
- Explain how food production has impacted ecosystems.

Before You Read

As you read the lesson, put sticky notes next to the sections where you have questions. Write your questions on the sticky notes. Once you have finished reading, discuss the questions with your classmates or teacher.

When you come to school, you enter a place that is different from your home, the mall, or a restaurant. School has a structure and system with its own characteristics, including its living organisms (classmates and teachers) and its surroundings or environment (bells, books, lockers, gym uniforms, etc.). You might think of school as a special kind of ecosystem for learning. An ***ecosystem*** is a community of organisms living and interacting with one another and their environment. ***Ecology*** is the study of the relations or interaction of living things with one another and their environment. See **Figure 14-1**.

In a similar way, ecosystems make up our planet. All living things interact with one another and their environment. Some living organisms have adapted to, and prefer, certain types of living conditions, just as you probably like school more than your peers. The ecosystems found throughout the United States and in other parts of the world are discussed in this lesson.

Ecosystems

As stated previously, an ecosystem is a community of organisms living and interacting with one another and their environment. A ***community*** is a group of living organisms that interact with one another in an environment, or ecosystem. For example, in a ***prairie ecosystem***, the types of organisms that make up the community include grasses, antelope, prairie dogs, and butterflies, plus many other plants, animals, insects, and microorganisms.

Monkey Business Images/Shutterstock.com

Figure 14-1. Your school is actually an ecosystem. Many organisms (people) interact with one another on a daily basis. ***What are the biotic and abiotic factors in your school's ecosystem?***

Words to Know

(*Continued*)

temperate evergreen forest
terrestrial ecosystem
transpiration
tropical deciduous forest
tropical evergreen forest
wetland

Each species of plant or animal also has a preferred living area, or ***habitat***, within the ecosystem, **Figure 14-2**.

Biotic and Abiotic Factors

We often think of an ecosystem containing living (biotic) and nonliving (abiotic) things. The ***biotic factors*** of an ecosystem include things that are living, moving, growing, and reproducing (plants and animals). The biotic factors are divided into two categories: flora and fauna. ***Flora*** refers to the plant life in an ecosystem, and ***fauna*** refers to the animal life, **Figure 14-3**. Scientists also include microorganisms as biotic factors, but these do not have a distinctive name like flora and fauna. Again, these living things interact with one another in the living community of the ecosystem.

The ***abiotic factors*** in an ecosystem are nonliving things. Abiotic factors include rocks, minerals, water, and nutrients. These abiotic factors are important for supporting life in the ecosystem. For example, what would happen if an ecosystem had absolutely no water? Could life exist?

Did You Know?

Scientists from Microsoft Research and the United Nations Environment Programme World Conservation Monitoring Center (UNEP-WCMC) are building a giant computer model of the global ecosystem. This model will help our understanding of the biosphere as a whole, and perhaps help us better manage our natural resources.

Interdependent Communities

Communities of living organisms are interdependent. *Interdependent* means that organisms rely on one another. One of the most common examples of interdependency is the ***food chain***, or ***food cycle***, in which most living

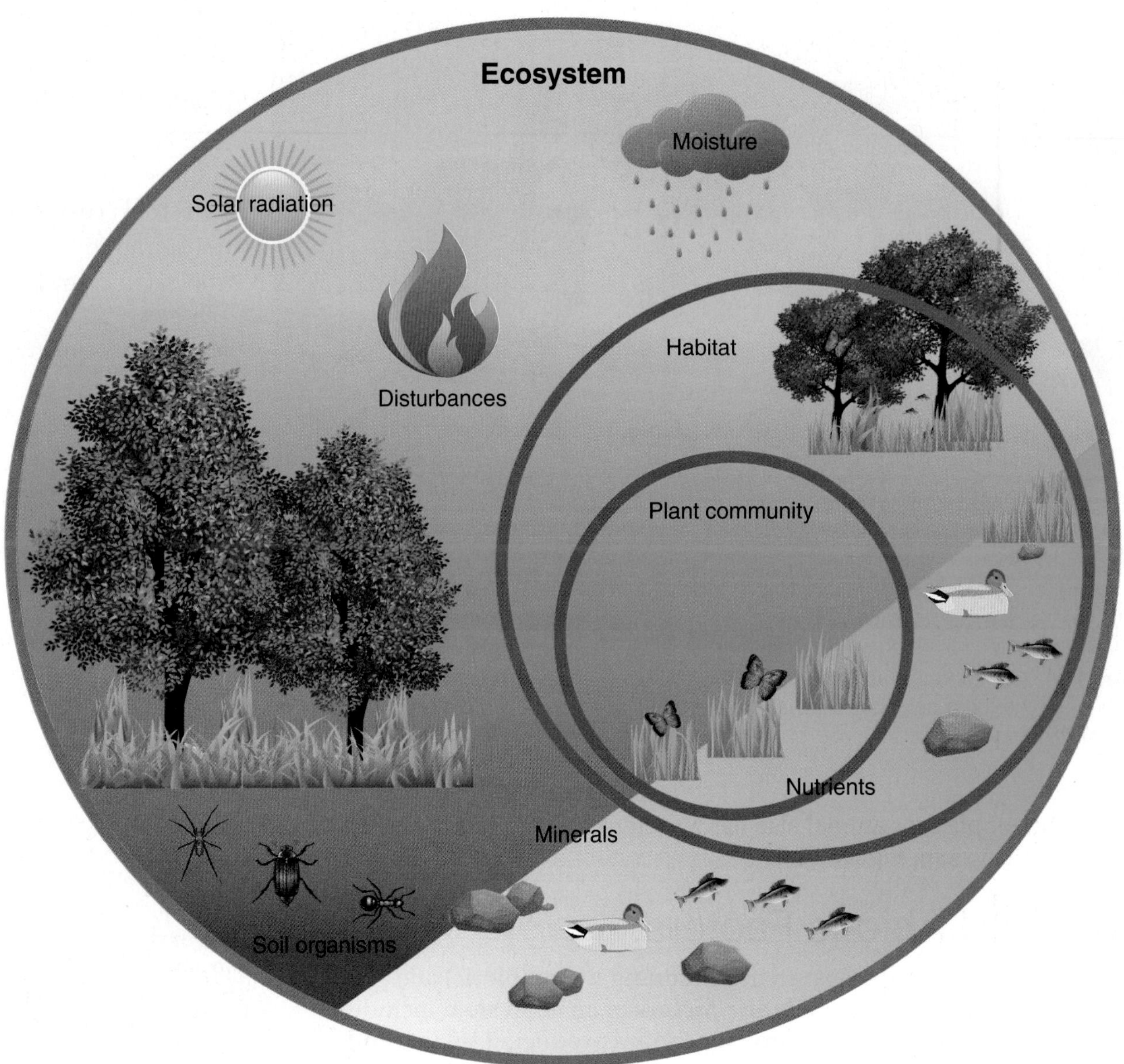

mmar/Shutterstock.com; La Gorda/Shutterstock.com; Helga Pataki/Shutterstock.com; Rendix Alextian/Shutterstock.com; Svinkin/Shutterstock.com; Svinkin/Shutterstock.com; Matthew Cole/Shutterstock.com; NataliyaF/Shutterstock.com; goGOgo/Shutterstock.com; iadams/Shutterstock.com

Figure 14-2. Ecosystems occur naturally when organisms interact with one another regularly.

organisms are food for other living organisms. This cycle is a model of how some animals eat plants and, in turn, are eaten by other animals. For example, a field mouse might eat grass seed and then be eaten by a snake. That snake might then be eaten by a hawk. When the hawk dies and decomposes, it is no longer biotic. However, the abiotic minerals, like calcium and phosphorus, return to the soil to be used by grasses to grow and produce seeds. A ***food web*** is a system of interlocking and interdependent food chains.

Ecosystems in the United States

How do scientists classify ecosystems? Scientists classify ecosystems based upon the predominant flora and fauna (plants and animals) present in the ecosystem. The flora and fauna are determined by various factors,

Aleksander Bolbot/Shutterstock.com

Erik Mandre/Shutterstock.com

Figure 14-3. Flora includes trees, grasses, ferns, mosses, algae, and broadleaf plants in an ecosystem. Fauna includes all animals, from rodents to bears, living in an ecosystem. ***How many types of flora can you identify on the school grounds? How about in your front yard or backyard?***

including weather, soils, and water availability. The predominant flora and fauna are the common plants and animals that have adapted to the special weather, soils, and water in the ecosystem.

What are the different types of ecosystems? There are essentially two types of ecosystems: aquatic and terrestrial. ***Aquatic ecosystems*** are those found in a body of water, like a stream, river, wetland, or ocean. ***Terrestrial ecosystems*** are those found on land, like prairies, woodlands, and deserts. Ecosystems are present in every part of the United States. The continental United States lacks tropical and arctic ecosystems, but has nearly every other kind of ecosystem. What types of ecosystems surround your home and school?

Did You Know?

According to the United Nations Educational, Scientific, and Cultural Organization (UNESCO), an estimated 50%–80% of all life on Earth is found under the ocean surface.

Aquatic Ecosystems

Aquatic ecosystems include communities of water animals, like fish and amphibians, and water plants, like algae and water lilies. There are two types of aquatic ecosystems: marine and freshwater.

Marine Ecosystems

Marine ecosystems are the largest ecosystems on the planet. Marine ecosystems cover more than 70% of Earth's surface. They also contain 97% of our planet's water. The unique feature of marine ecosystems is that the water contains high amounts of salt and other minerals. This high concentration of salt makes the water ***saline***.

Marine ecosystems are divided into smaller categories because of the diversity of life they contain. Marine ecosystems contain everything from whales, the largest mammals on the planet, to plankton, some of the smallest organisms. Marine ecosystems include oceans, estuaries, coral reefs, salt marshes, and several really unique ecosystems, such as *profundal* or *hydrothermal vents* (deep-sea volcanic vents). Sharks, seaweed, and sea turtles are other examples of well-known fauna found in oceans.

JuneJ/Shutterstock.com

Figure 14-4. Salt marshes occur on the coasts of oceans. At high tide, twice each day, the land is covered with salty ocean water. As the tide recedes, the land is exposed to sunlight and terrestrial creatures. ***Are there salt marshes near your home? If so, take pictures of the area at high tide and when the tide recedes. Compare and contrast the images.***

Estuaries are partially enclosed bodies of water on the coasts of oceans and larger seas. Although estuaries have freshwater flowing into them from rivers and streams, the water is still brackish. ***Brackish water*** has a higher salt content than freshwater, but a lower salt content than seawater. Estuaries are a transitional zone between the freshwater ecosystem of a river or stream and the marine ecosystem of the ocean or sea.

Salt marshes are the areas of land and water between the high and low tides of the ocean. You can see a salt marsh in **Figure 14-4**. For part of the day, these areas are covered by salty tidal waters. At other times, the salt marshes are not covered with water, but remain wet. The plant life in salt marshes tolerates the high salt content of the seawater and also helps prevent tidal erosion by anchoring the marshy soil in place. Salt marshes support a wide diversity and abundance of life by providing food and shelter for both fish and terrestrial animals.

Freshwater Ecosystems

Freshwater ecosystems are ecosystems where the water is fresh, not salty. Many freshwater ecosystems eventually flow into marine ecosystems. Freshwater ecosystems cover less than 1% of Earth's surface, but they contain *all* of the world's liquid freshwater. So, they are important for maintaining life on Earth, including human life. The three main types of freshwater ecosystems are lakes and ponds, streams and rivers, and wetlands.

Lakes and ponds, or ***lentic ecosystems***, are freshwater ecosystems that are slow-moving or still bodies of water, **Figure 14-5**. Lakes and ponds can be relatively small, like a farm pond, or enormous, like the Great Lakes in the Upper Midwest. These freshwater ecosystems may be a few or hundreds of feet deep. The depth of the water affects plant life in a lake or pond. Since sunlight can reach the bottom of a shallow pond, aquatic plants often grow there, especially around the edges. On the other hand, if very little sunlight reaches the bottom of a deep lake, few plants (if any) will grow there.

Did You Know?

More than 40% of the world's fish species live in freshwater ecosystems.

Streams and rivers, or ***lotic ecosystems***, are freshwater ecosystems that rise from lakes, ponds, or freshwater springs coming out of the ground. Streams and rivers are generally fast-moving, turbulent bodies of water. Because of this turbulence, rivers and streams usually contain high levels of dissolved oxygen. ***Dissolved oxygen (DO)*** is the oxygen that becomes dissolved in water when the water is turbulent with lots of ripples and rapids. Aquatic organisms need dissolved oxygen to breathe because they cannot split oxygen from water or other oxygen compounds. For this reason, we find abundant animal life, including many types of fish, plants, and invertebrates where the water has a high content of dissolved oxygen. Dissolved oxygen is also needed for the efficient decomposition of organic materials in aquatic ecosystems.

Vadim Petrakov/Shutterstock.com

Zadiraka Evgenii/Shutterstock.com

Czesnak Zsolt/Shutterstock.com

Figure 14-5. Lentic ecosystems may include farm ponds, seasonal puddles, and both large and small lakes, where the water does not flow at all or quickly. Lake or pond ecosystems include fish, birds, amphibians, reptiles, insects, and mammals. They also include abundant plant life. *How does drought affect lentic ecosystems?*

Wetlands are ecosystems where the soil is saturated with water for most of the time, but may dry out periodically. Wetlands are also referred to as bogs or swamps. Wetlands serve as a natural water purification system. The combination of soil, fauna, and stagnant or slowly moving water, allows the soil and plant life to filter impurities from the water. The "filtered" water recharges groundwater supplies.

Terrestrial Ecosystems

Several types of terrestrial ecosystems exist around the world and in the United States. Terrestrial ecosystems include forest ecosystems, desert ecosystems, grassland ecosystems, and mountain ecosystems.

Forest Ecosystems

Forest ecosystems are the most diverse of the terrestrial ecosystems. ***Forest ecosystems*** are characterized by a large number of trees and other plant life, as well as abundant fauna. The trees and shrubs in a forest make up layers of canopy, as shown in **Figure 14-6**. The upper canopy trees absorb the most sunlight, and the lower canopy absorbs the least sunlight. In some forests, the upper and middle canopy filter out nearly all sunlight so that very little direct sunlight reaches the forest floor.

Depending on the type of trees, evergreen or deciduous, and the climate of the area, forest ecosystems are classified into sub-ecosystems. ***Evergreen trees*** are trees whose leaves are needles or scales. Evergreen trees remain green throughout the year, even in winter. ***Deciduous trees*** are trees that drop their leaves in the fall.

Temperate evergreen forests are found where the climate is cooler and drier than in the tropics. Because of the drier climate, many trees in the temperate evergreen forest have adapted leaves in the form of needles. Tree needles are covered with a thick, waxy epidermis and have less surface area than a traditional tree leaf. These adaptations slow the rate of ***transpiration***, or loss of water from the leaf surface. Examples of temperate evergreen forests are found

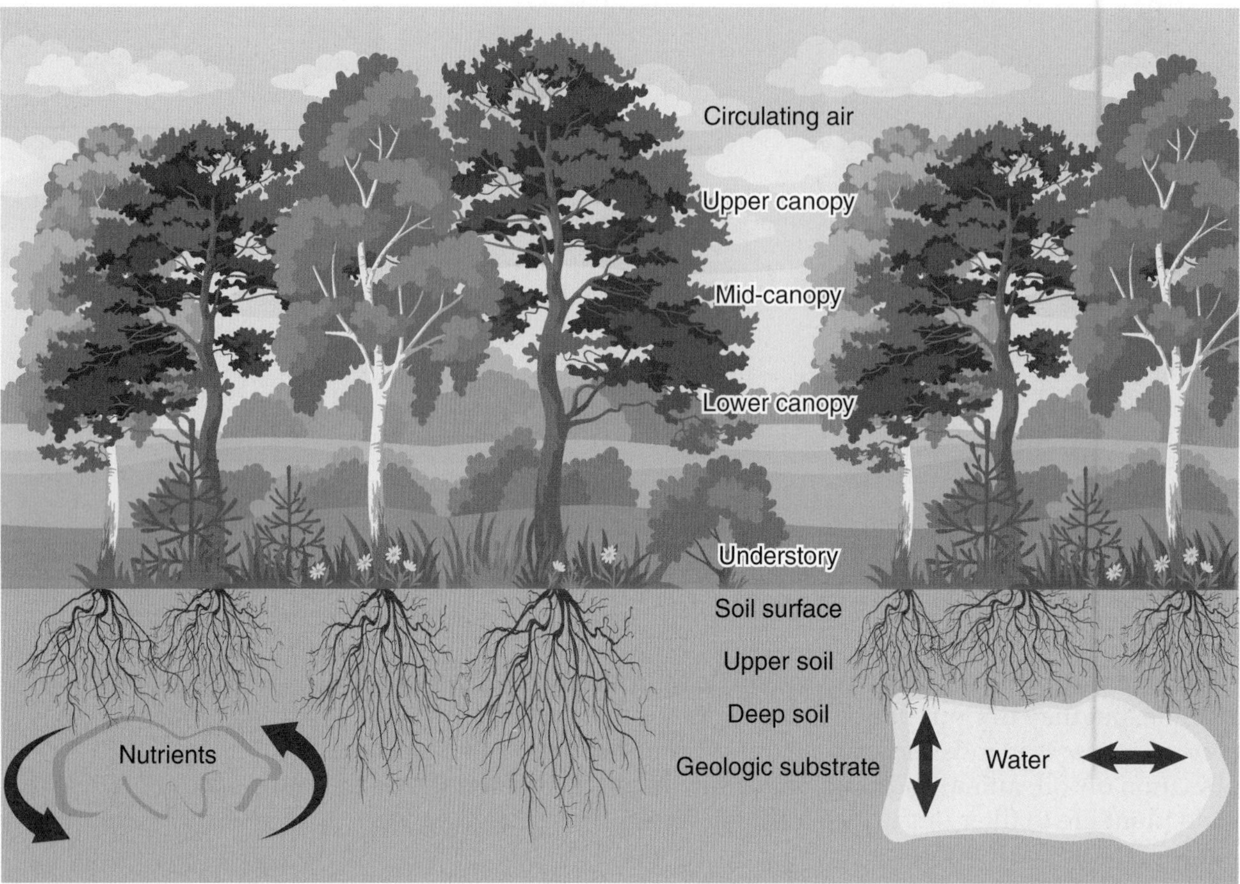

alexcoolok/Shutterstock.com; Glasscage/Shutterstock.com

Figure 14-6. Forest ecosystems include trees and shrubs of various heights. The upper canopy receives the most sunlight, and the lower canopy receives the least sunlight. ***How have plants adapted to these differences in light conditions?***

in the Sierra Nevada Mountains in California and in the upper Appalachian Mountains in the northeastern part of the United States. Because of the cooler climate, temperate evergreen forests support the growth of ferns and mosses. Evergreen trees do not seasonally shed their leaves, but remain green yearlong.

Temperate deciduous forests are the most common forest ecosystem in the United States. ***Temperate deciduous forests*** are found where sufficient rainfall supports the growth of trees. The climate includes summers and winters that are clearly defined by differences in temperature. The differences in temperature between summer and winter cause the trees to shed their leaves in the fall.

The taiga is a special kind of forest ecosystem found in the northern latitudes of Alaska and Canada, just before the arctic regions, **Figure 14-7**. ***Taiga ecosystems*** are characterized by plant life that is mainly evergreen conifers. Temperatures in the taiga fall below freezing many months out of the year. Because of these low temperatures, animal life is often limited to migratory birds and insects.

Desert Ecosystems

Desert ecosystems are characterized by low annual rainfall, often less than 10 inches per year. Because of the low rainfall, desert conditions are hostile to plant and animal life. The soil is sand or rock. Temperatures are

Dmitry Strizhakov/Shutterstock.com

Figure 14-7. The taiga forest is found in northern latitudes and consists of broad expanses of various tree types. *Is there a lot of undergrowth in a taiga forest?*

often extremely high during the day and cold at night. Cacti and reptiles are among the few forms of life that can live in a desert. In the United States, deserts are found in the Southwest.

Grassland Ecosystems

Grassland ecosystems are composed mainly of grasses with few trees and shrubs. **Figure 14-8** shows the extent of the original grasslands in the United States. Grasslands are also referred to as *prairies* in temperate climates like those found in the United States. Since grasslands contain a lot of grass, they provide food for many grazing animals, such as bison, antelope, and deer. The grasses also provide food and cover for insects, birds, and nongrazing animals. In the United States, we find both tall grass and short grass prairies. The Great Plains of the United States and Canada is one of the largest prairie ecosystems in the world.

Grassland ecosystems have enough rainfall to support grasses, but not enough rainfall to support the growth of trees. In a grassland ecosystem, trees are usually only found near small streams, springs, or ponds.

Did You Know?

Grasslands cover one-fifth of the land on Earth.

Mountain Ecosystems

Mountain ecosystems are characterized by a diverse array of habitats that are influenced by dramatic changes in altitude. In the highest mountain ecosystems, as in the peaks of the Rocky Mountains, the soil is too thin

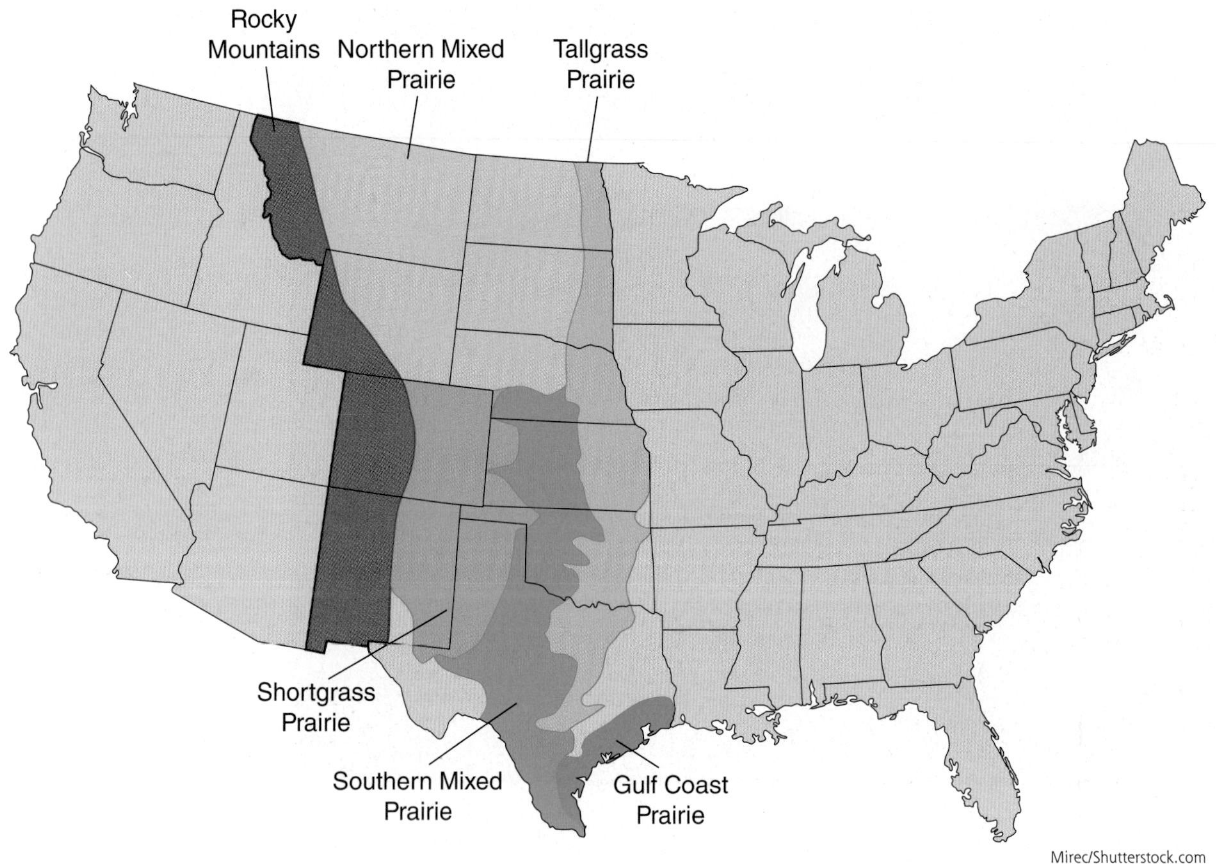

Figure 14-8. Grasslands cover much of the Midwestern United States. The grassland prairies are marked with deep, fertile soils that produce abundant crops of corn, soybeans, and wheat.

and the temperatures too low to support tree growth. Below the tree line, mountain forests are mainly temperate evergreen forests. Animals living in mountain ecosystems often have thick coats of fur and hibernate throughout the cold winter months.

In the western half of the United States, the Rocky Mountains play a major role in the environment. **Figure 14-9** shows how the Rocky Mountains can affect the environment. Warm, moist air blowing off the Pacific Ocean must rise in altitude to cross the mountain range. This means that the air cools. Cooler air cannot hold as much moisture, or precipitation, as warm air. So, rain or snow falls on the western slopes of the Rocky Mountains. Once the air crosses the peaks of the mountains, it has lost much of its moisture, so relatively little rain and snow fall on the eastern slopes of the mountains.

Ecosystems Outside the United States

Although the continental United States has a large variety of natural ecosystems, it does not have tropical forests, or what people commonly refer to as jungles. Ecosystems not found in the United States include tropical evergreen forests, tropical deciduous forests, and coral reefs.

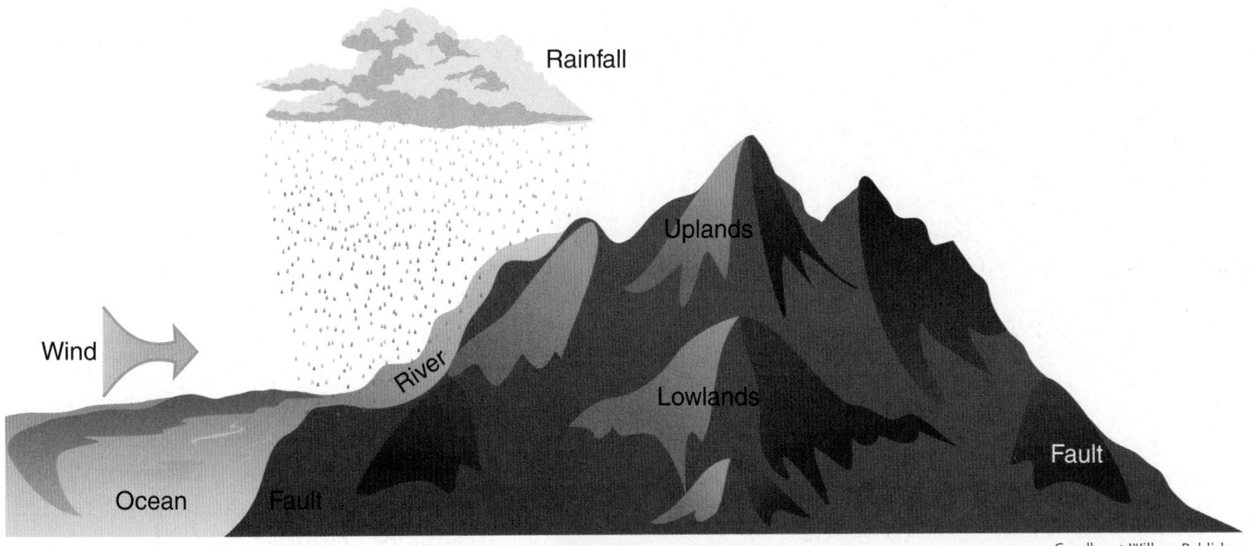

Figure 14-9. Warm, moist air flowing from the Pacific Ocean must rise to cross the Rocky Mountains. When the air rises, it cools and releases its moisture as rain or snow on the western slopes of the Rocky Mountains.

Tropical Evergreen Forests

Tropical evergreen forests are ecosystems that receive abundant rainfall, include dense tree vegetation at various heights, and are found in the tropics near the equator. Trees in tropical evergreen forests do not lose their leaves. We often think of these tropical evergreen forests as the ***jungles*** in Central and South America, as well as in Africa.

Did You Know?

Studies indicate that there are possibly more than 30 million species of insects dwelling in the canopies of tropical forests.

Tropical Deciduous Forests

Tropical deciduous forests are ecosystems that also receive abundant rainfall and include dense flora. However, unlike the evergreens, the deciduous trees lose their leaves once throughout the year. When the trees lose their leaves, more sunlight reaches the ground, allowing the increased growth of grasses and shrubs. These grasses and shrubs provide cover for small animals and insects, so tropical deciduous forests also support high populations of animal and insect life in addition to an abundance of plants.

Coral Reefs

Coral reefs are underwater ecosystems with structures made up of colonies of tiny animals called corals. The basic component of coral reefs is calcium carbonate, which comes from the secretions of corals. Coral reefs are vibrant, diverse ecosystems. In fact, they are often called the "rainforests of the sea" because of the abundance and diversity of life they support. Clownfish, **Figure 14-10**, are examples of the colorful, vibrant fish found living in or near coral reefs. The Great Barrier Reef off the coast of Australia is one of the most well-known coral reef ecosystems. Other coral reefs are located in the warm ocean waters off the coasts of Florida and Hawaii.

Willyam Bradberry/Shutterstock.com

Figure 14-10. This clownfish and its coral reef surroundings is an example of an aquatic ecosystem. More specifically, it is a marine ecosystem.

Brian Prechtel/USDA ARS

Figure 14-11. Riparian zones slow the runoff from agricultural cropland so that the water can be absorbed into the ground and the fertilizers used by the grass for nourishment.

Agriculture's Impact on Ecosystems

A continuously growing need for food production has greatly impacted the natural ecosystems in our country. Grasslands once dominated the Great Plains. However, due to the population growth of the late 1800s and early 1900s, these grasslands were plowed and sown to raise corn, wheat, and soybeans.

Farmers also drained natural wetlands to gain even more cropland. They dug drainage ditches and buried drainage tile under the wetlands in order to drain the water. Due to the accumulation of nutrients from years of plant decomposition, these drained wetlands became some of the most fertile cropland in America. Little regard was given to the negative impacts these practices could have on our natural ecosystems.

Today, agricultural practices are designed for maximum production and the preservation of valuable cropland and water resources. To reduce soil erosion, farmers have begun to plant wooded and grass strips along stream and river banks. These grassed or wooded strips, called ***riparian zones***, slow the runoff from agricultural cropland so that the water can be absorbed into the ground and the fertilizers used by the grass for nourishment, **Figure 14-11**. These practices have increased the health of rivers and streams throughout the United States. In many areas, wetland drainage systems have also been removed, allowing wetlands to return to their natural state. These renewed wetlands now attract migratory waterfowl and other wildlife.

Career Connection Ecologist

Job description: Ecologists are scientists who work to study, conserve, and rebuild ecosystems. Ecologists generally specialize in one type of ecosystem. Because of the complexity of interactions in the ecosystems, ecologists have a lot of information to understand. For example, some ecologists focus on forests, and others focus on marine ecosystems. Ecologists will study ecosystems and provide information to other professionals, like foresters and wildlife biologists, about potential changes in the ecosystem. They conduct field surveys of wildlife and prepare reports that maximize the potential of an ecosystem. Where land development is proposed, ecologists conduct impact studies. These impact studies determine the potential impact of the development on species, habitats, and the ecosystem of the area.

Mila Supinskaya/Shutterstock.com

An ecologist studies plants, animals, and their interactions with their environment.

Education required: A bachelor's degree in ecology or natural resources is required.

Job fit: This job may be a fit for you if you are interested in the environment, like studying wildlife, and can make recommendations about human impacts on those ecosystems.

Lesson 14.1 Review and Assessment

Words to Know

Match the key terms from the lesson to the correct definition.

1. The study of the relations or interaction of living things with one another and their environment.
2. A forest ecosystem that receives abundant rainfall, has dense flora, and in which the trees shed their leaves annually.
3. A forest ecosystem found in the tropics in which the trees do not shed their leaves, and there is abundant rainfall and dense tree vegetation at various heights.
4. An ecosystem found in oceans and seas in which the water contains high amounts of salt and other minerals.
5. An ecosystem characterized by low annual rainfall, soil composed of sand or rock, and temperatures that are often extremely high during the day and cold at night.
6. The areas of land and water between the high and low tides of the ocean.
7. A freshwater ecosystem in still or slow-moving water.
8. A wooded or grass strip planted along a stream or river bank to slow the runoff from agricultural cropland.
9. An underwater ecosystem with structures made up of colonies of tiny animals.
10. A system of interlocking and interdependent food chains.
11. An ecosystem characterized by a diverse array of habitats that are influenced by dramatic changes in altitude.
12. An ecosystem found on land, like prairies, woodlands, and deserts.
13. An ecosystem found in a body of water, like a stream, river, wetland, or ocean.
14. An ecosystem composed mainly of grasses with few trees and shrubs.
15. The preferred living area (within an ecosystem) of a plant or animal.

A. aquatic ecosystem
B. coral reef
C. desert ecosystem
D. ecology
E. estuary
F. fauna
G. food web
H. grassland ecosystem
I. habitat
J. jungle
K. lentic ecosystem
L. lotic ecosystem
M. marine ecosystem
N. mountain ecosystem
O. riparian zone
P. salt marsh
Q. terrestrial ecosystem
R. tropical deciduous forest

16. A partially enclosed body of water on the coasts of oceans and larger seas.
17. A freshwater ecosystem found in streams or rivers that rise from lakes, ponds, or freshwater springs.
18. The animal life in an ecosystem.

Know and Understand

Answer the following questions using the information provided in this lesson.

1. A(n) _____ is a group of living organisms that interact with one another in an environment.
2. Biotic factors are those factors in the ecosystem that are _____, whereas abiotic factors are those factors that are _____.
3. Flora refers to _____ life in an ecosystem.
4. The animal life in an ecosystem is referred to as _____.
5. Briefly explain how organisms in a food chain are interdependent. Give an example.
6. Which ecosystems are prevalent in the United States?
7. Land-based ecosystems are called _____ ecosystems.
8. *True or False?* Marine ecosystems are the second largest ecosystems on the planet.
9. Saline means that the water has a high quantity of _____.
   A. salt
   B. dissolved oxygen
   C. wildlife
   D. freshwater
10. Explain why estuaries have brackish water.
11. *True or False?* Salt marshes prevent tidal erosion by anchoring marshy soil in place.
12. Aquatic ecosystems include ______.
    A. wetlands
    B. forests
    C. deserts
    D. mountains
13. Briefly explain why dissolved oxygen is important for aquatic life in freshwater ecosystems.
14. What are three biotic factors that characterize a forest ecosystem?
15. Ecosystems where there is sufficient rainfall to support trees, and where summers and winters are clearly defined by differences in temperatures, are called _____.
    A. grasslands
    B. deserts
    C. temperate deciduous forests
    D. tropical coniferous forests

16. Which ecosystems are not prevalent in the United States?
17. Which ecosystems in the United States have been most impacted by agriculture and food production?
18. Explain how the use of riparian zones is helpful to the environment.

Analyze and Apply

1. Where is each type of ecosystem found in the United States? Download an outline map of the United States and map the location of each type of ecosystem.
2. In a forest that has a dense upper canopy, what impact does filtering out nearly all of the direct sunlight have on the forest floor?
3. The lesson highlighted a few ways that agriculture and food production have impacted ecosystems. Can you find other ways? If so, what are they?

Thinking Critically

1. What characteristics of grasslands made them ideal for conversion to farmland? Why? What characteristics of grasslands made or make farming a challenge? Why?
2. Why are wetlands vital to maintaining healthy surface water?
3. How have agricultural practices and food production impacted ecosystems?

Lesson 14.2 Ecological Cycles

Lesson Outcomes

By the end of this lesson, you should be able to:

- Define an ecological cycle.
- Explain how water cycles in the biosphere provide moisture for crops and livestock.
- Assess the impact of agriculture on the water cycle and the water cycle's impact on agriculture.
- Diagram the nitrogen cycle.
- Explain how nitrogen becomes usable to agronomic plants.
- Assess the impact of agriculture on the nitrogen cycle and the nitrogen cycle's impact on agriculture.
- Compare and contrast carbon sinks and carbon sources.
- Assess the impacts of agriculture on the carbon cycle.

Words to Know

acidification
acid rain
algae bloom
ammonification
aquifer
biogeochemical cycle
biosphere
blue baby syndrome
carbon credit system
carbon cycle
carbon dioxide (CO_2)
carbon pool
carbon sink
carbon source
cycle
denitrification
desalinate
ecological cycle
fixation
freshwater
groundwater
hydrological cycle
legume
methane (CH_4)
methemoglobinemia
nitrate
nitrification
nitrogen cycle
runoff water
sediment
surface water
water cycle
water rights
water-soluble
zone of life

Before You Read

After reading each section (separated by main headings), stop and write a three- to four-sentence summary of what you just read. Be sure to paraphrase and use your own words.

The concept of recycling is not new. Nature has been recycling for eons. Cycles exist to recycle water, carbon, nitrogen, and nutrients in our environment. It is almost as if Earth has a built-in mechanism for reusing these precious resources. In this lesson, we will look at how elements, nutrients, and water are recycled. We will explore and understand how water, and other substances, are used by plants, animals, and humans on Earth, and then recycled to support the life of future generations, **Figure 14-12**. Let us get started by exploring *bio-geo-chem-i-cal* cycles!

What Are Biogeochemical Cycles?

Scientists have found the natural interactions of plants, animals, and abiotic factors in ecosystems to be important within an ecosystem as well as between ecosystems. For an ecosystem to be functioning in a healthy way, energy, water, and nutrients must cycle between plants, animals, and even the soil. This is a system of ***ecological cycles***. The broad term that scientists use to define the cycling of biological and chemical substances in any of Earth's ecosystems is ***biogeochemical cycles***.

The prefix *bio* indicates we are dealing with biological factors. Biological factors include plants and animals. The *geo* part of the word

Marina Lohrbach/Shutterstock.com

Figure 14-12. Composting provides rich, nutritious material to the soil to help plants grow. *Does your family compost? Does your school cafeteria?*

indicates that we are concerned with geological, or Earth-based, factors. Geological factors might include the soil, rocks, and mountains. Finally, the *chemical* part of the word refers to naturally occurring chemicals that are vital to life on our planet. These primary chemicals are carbon, oxygen, nitrogen, and water.

A ***cycle*** is the pathway that a substance moves in the biosphere. The term ***biosphere*** is used to describe Earth's ecosystems as a whole, or the accumulation of all of the ecosystems on Earth. You might also hear the biosphere referred to as the ***zone of life***. A simplified example of a cycle would be the pathway a water molecule might take. A water molecule may begin in the ocean, move through various stages (atmosphere, stream, animal's body), and finally return to the ocean.

We are concerned with biogeochemical cycles in agriculture and natural resources because these cycles keep Earth functioning as it should. The total amount of nitrogen, carbon, oxygen, and water on our planet is relatively stable. But, the usable and clean quantities of these valuable nutrients is always in flux. These cycles provide agricultural crops and livestock with clean water and valuable nutrients. Important biogeochemical cycles include the water cycle, the nitrogen cycle, and the carbon cycle.

Thinking Green

According to the Can Manufacturers Institute, 105,800 aluminum cans are recycled every minute and 20 recycled cans can be made with the energy needed to produce one can using virgin ore!

Did You Know?

Water is the most important nutrient on the planet, and we have a limited supply. Less than 4% of all water on Earth is fresh or available for use by plants, animals, and humans.

What Is the Water Cycle?

The ***water cycle***, shown in **Figure 14-13**, is the pathway or cycle that water follows *underneath* the surface, *at* the surface, and *above* the surface of Earth. Another name for the water cycle is the ***hydrological cycle***. When you observe water falling to Earth as precipitation (rain, snow, hail, etc.), and collecting in rivers and streams which run into larger bodies of water, you are observing part of the water cycle. But, there is much more to the water cycle than meets the eye, since water exists in three phases: solid, liquid, and gas.

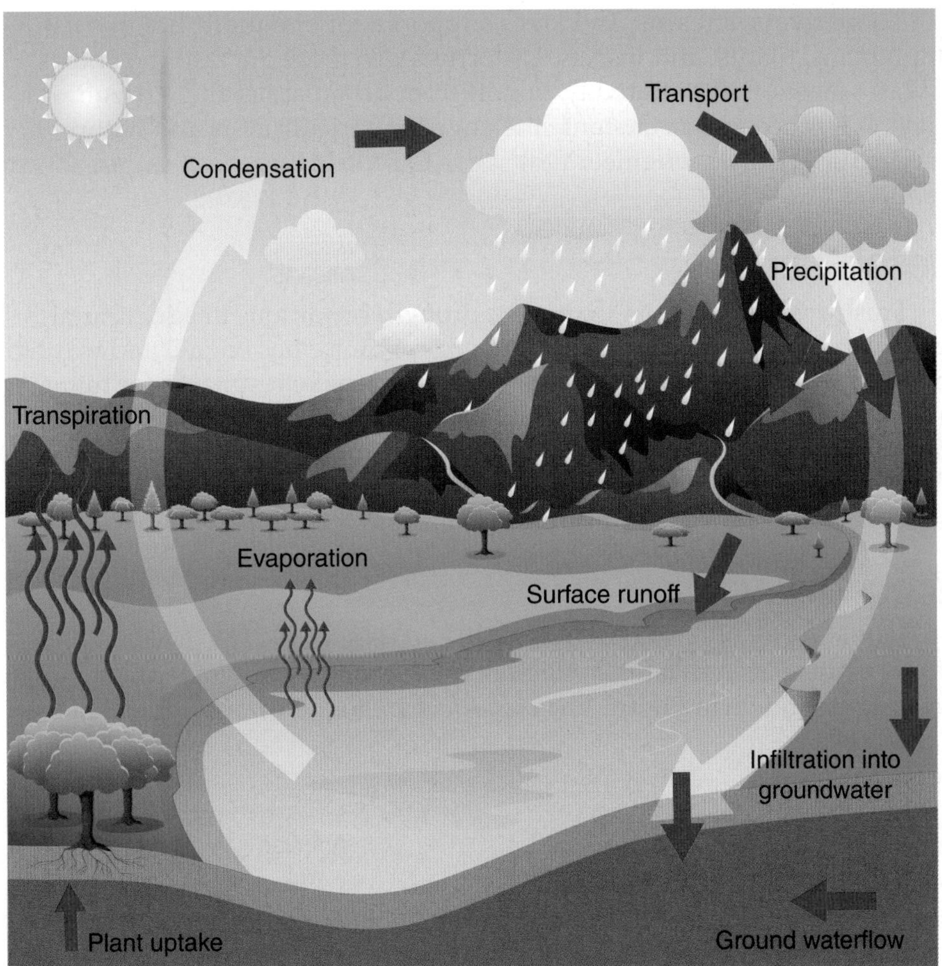

stockshoppe/Shutterstock.com

Figure 14-13. The water cycle refreshes and renews water on Earth. Water moves between the atmosphere, the land, the oceans, and even ice and snow. This process provides clean water that can be used by agricultural crops and livestock. ***Do you live in an area that has experienced severe drought conditions? What types of restrictions have been placed on water use?***

Solid Water

What is solid water and where can you find it on Earth? Solid water is simply snow and ice. The snow that falls in winter and remains for much of the year on mountaintops is solid water. The glaciers and polar ice caps in both the Arctic and Antarctic regions of the world are composed of solid water. As a matter of fact, half of the freshwater on the planet is locked up in ice. This means 2% of the world's usable water is frozen!

Liquid Water

Where can you find the majority of Earth's water in its liquid state? Oceans and seas are the primary places to find liquid water on Earth—about 96% of Earth's water. So, you are probably wondering, where do we find the other 4%? Where is the freshwater that supports most life? Most of it can be found in rivers, streams, and lakes. ***Freshwater*** is the water that is not salty and supports all life, both plant and animal.

Did You Know?

Freshwater makes up less than 1% of all the water on Earth.

These rivers, streams, and lakes support a lot of aquatic life like fish, amphibians, plants, and insects. Unfortunately, these are also the bodies of water most often polluted by runoff from urban areas and agricultural lands. It has become a constant challenge for agriculturists and natural resource scientists to engineer new ways to protect and clean our freshwater resources.

Desalination

Freshwater is essential to prevent drought conditions in agricultural fields and enable agricultural crops to grow. Science has engineered ways to remove (***desalinate***) the salt from seawater and make it suitable for human, animal, and plant consumption. Industrial desalination plants remove salt from seawater. This is a time-consuming and expensive process. Aside from the cost and energy required to build the facilities, the amount of energy required to move the water through the process is extremely high.

Groundwater

Liquid water also resides *underneath* Earth's surface. This water is called ***groundwater***. Groundwater is stored in underground reservoirs called ***aquifers***. As shown in **Figure 14-14**, aquifers are found under most all of the

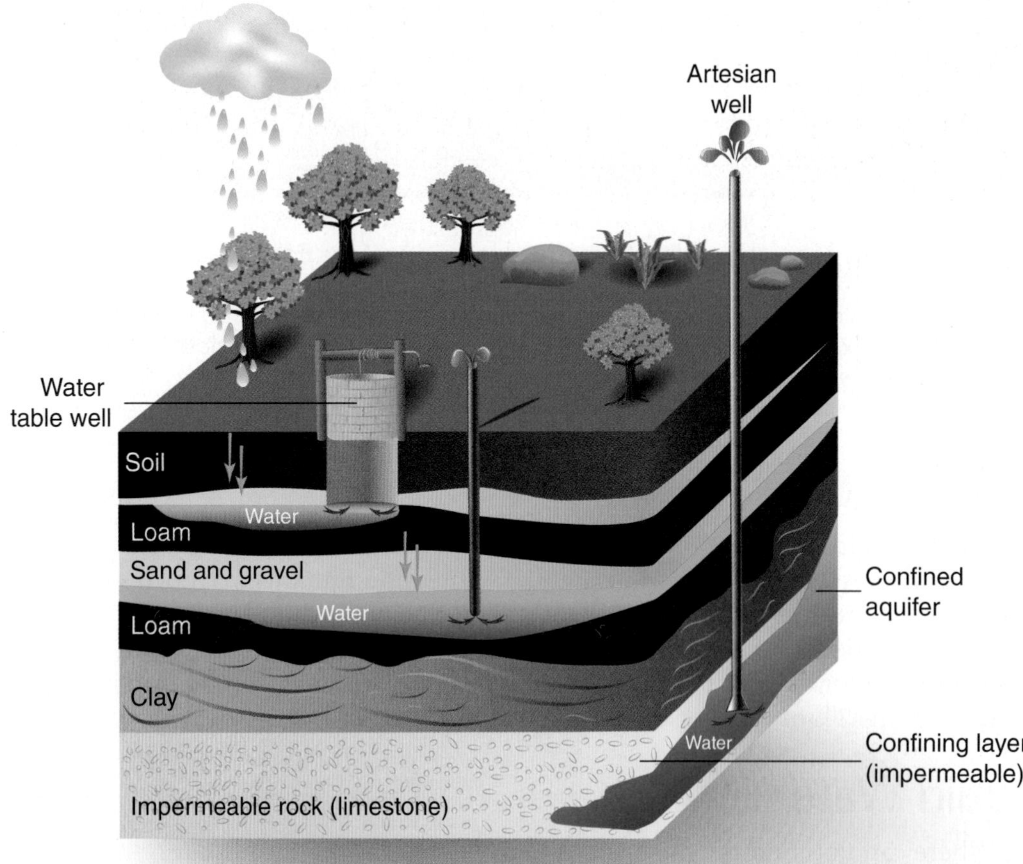

Designua/Shutterstock.com

Figure 14-14. Most freshwater used for drinking and irrigating agricultural cropland comes from underground aquifers. ***Where do you get your drinking water supply?***

surface of Earth. Aquifers store most of the freshwater found on Earth. One of the most prominent aquifers in the United States resides under the Great Plains, below the states of Nebraska, Kansas, Colorado, Oklahoma, and Texas. It is called the Ogallala Aquifer and was created by glaciers melting thousands of years ago.

Gaseous Water

A small fraction of water exists as vapor in the atmosphere. The stickiness you feel on hot and humid summer days is water vapor in the air. Also, think about the feeling of the air just before a summer storm. Does the air feel heavy and close? That heaviness and closeness is the water vapor in the air.

Cycling of Water on Earth

How does water cycle on Earth? We can start at any point in the water cycle; refer back to **Figure 14-13**, and trace a path that a water molecule might follow through the biosphere.

Consider a raindrop as it falls from the sky. As the raindrop hits the ground, it can go one of two places. It can pool with other raindrops and flow into a stream, or it might soak into the ground. If the drop flows into a stream, the stream may then flow into a larger river, which may eventually flow into the ocean. From the ocean, the drop of water might evaporate and form part of a cloud over the ocean. That cloud could become part of a storm and rain over the ocean or move over land. If the droplet became part of a cloud and rained down to Earth, it has completed a turn in the water cycle.

If the raindrop soaks into the ground, it will likely be used as water to support plant life, or to recharge one of those underground reservoirs, or aquifers. Let us suppose that the raindrop soaked into the soil and was absorbed by the root of a corn plant. The corn plant would use the water for cellular functions and to grow. Eventually, the corn plant would release the "raindrop" through transpiration. At this point, the molecules of the raindrop would be released as water vapor and float in the atmosphere. Most of this water vapor would condense into clouds and form precipitation. Once again, the droplet has completed a turn in the water cycle.

So, where does this water vapor originate? Water vapor in the atmosphere comes from several places. Most water vapor originates from evaporation of water from the oceans and other large bodies of water. Water vapor also enters the atmosphere when plants and animals respire. Plants release water through transpiration, and each time an animal (yourself included) takes a breath, it exhales a small volume of water vapor. So, when you take your next breath, think about your place in the water cycle.

The Water Cycle and Agriculture

The sturdiest plants sown on the healthiest land will not thrive without water. And, with such a limited amount of Earth's water being usable, there is always the threat of water shortage and the depletion of our freshwater resources.

Depletion of Aquifers

In the United States, cropland irrigation threatens to disrupt the water cycle by using water from the Ogallala Aquifer faster than it can be recharged by rainwater. As stated earlier, water in the Ogallala Aquifer was deposited by glaciers thousands of years ago, and it cannot be easily replenished. Much irrigated cropland is found in the land just above the Ogallala Aquifer. Without irrigation, the land is too dry to produce high yields of corn and soybeans. Agricultural scientists are trying to create crop varieties that can thrive in low moisture conditions, so farmers do not have to irrigate as much or as often. Developing crop varieties that require less water is one of the grand challenges facing agriculture today.

Surface Water Pollution

Although farmers and growers are constantly experimenting with more targeted applications of fertilizers and chemicals to reduce pollution runoff, agricultural lands are still one of the main sources of surface water pollution in the United States. ***Runoff water*** is water that pools on the surface of the soil and then flows into nearby streams, rivers, and lakes. It is often contaminated with chemical fertilizers, manure, and chemicals.

Actually, soil is one of the primary pollutants of rivers and streams in the United States. Soil enters rivers and streams during heavy rains when unprotected soil is splashed into pools of water that are then washed into bodies of surface water. ***Surface water*** is the water found in streams, rivers, and lakes. Soil erosion results in the loss of valuable topsoil. Erosion also leads to sedimentation of rivers and streams. This kills fish and other aquatic life.

To reduce soil erosion and runoff, farmers have begun to plant grass strips along stream and river banks. These grass strips, called *riparian zones*, slow the runoff from agricultural cropland so that the water can be absorbed into the ground and the fertilizers used by the grass for nourishment.

Did You Know?

Acid rain is more problematic in some environments. For example, areas lacking natural alkalinity are unable to neutralize acid and the acid rain creates more damage.

Acid Rain

Acid rain is rain that contains higher than normal amounts of pollutants, especially nitric and sulfuric acids. Actually, acid rain is a natural phenomenon when produced at a moderate level. However, human and industrial activities can elevate the levels of nitric and sulfuric acids in rain. Burning fossil fuels, like gasoline, diesel fuel, and especially coal, contribute to excessive levels of acid rain. Electrical power plants that burn coal are major contributors to acid rain because burning coal releases sulfur and nitrogen into the atmosphere, and these form the acidic compounds that form acid rain.

Acid rain causes damage to, or even death of, trees in higher elevations, as shown in **Figure 14-15**. The damage is often in the form of slowed tree growth. Acid rain also lowers the pH of surface water. Most aquatic life thrives in a pH range between six and eight. Acid rain can lower the pH of lakes to below five. This lowering of water acidity is called ***acidification***. Fish like bass, other aquatic species like clams, and snails cannot survive in acidified waters. As these animals are reduced in number in the ecosystems, so are the populations of animals that live in balance with them.

Mary Terriberry/Shutterstock.com

Figure 14-15. Acid rain causes tree damage or death in high elevations. Acid rain is caused by burning fossil fuels, including coal, gasoline, and diesel fuel. ***Have you seen damage that is due to acid rain near your home?***

Water Rights

In "water law," ***water rights*** refer to the legal right of a user to use water from a particular source. In places where water is plentiful, the rules and regulations are not usually complicated. However, in areas where water is not plentiful, the rules and regulations may become complicated and also become the source of conflict, both legal and physical, **Figure 14-16**.

In several places in our country, people argue over who has the right to freshwater resources. For example, in Florida, most of the people live in the southern part of the state. However, most of the freshwater is found in the northern part of the state. Landowners in northern Florida understand the value of their water and do not want to provide it free of charge to people in the southern part of the state. They also do not want to create a water shortage for their own crops and orchards. Government officials at the state and local levels have long debated the merits of piping freshwater from northern Florida to the thirsty people (and plants) living in South Florida.

Richard Thornton/Shutterstock.com

Figure 14-16. One of the many irrigation canals in California. Since California's interconnected water system serves more than 30 million people and irrigates over 5 million acres of farmland, it is not surprising that water rights and the distribution of water is continuously under debate.

Thinking Green

What can you do to conserve water? Consider this short list of water-saving tips as only the beginning to your own water conservation program. Encourage your family, friends, and coworkers to employ water-saving habits, too.

- Store a pitcher of drinking water in the refrigerator.
- Check faucets, pipes, and toilets for leaks.
- Insulate your water pipes for hot water faster.
- Install water-saving showerheads.
- Install low-flush toilets and avoid flushing unnecessarily.
- Take shorter showers.
- Do not let the water run when brushing your teeth.
- Use mulch around plants and trees to slow evaporation.
- Install and use a rainwater catch system to water your plants.
- Do not let the hose run while washing your car.
- Use a broom instead of the hose and water to clean driveways and sidewalks.

Denise Lett/Shutterstock.com

Rainwater recycling does not require an elaborate system. A wooden or plastic barrel can be used to collect runoff from the rain gutters.

Did You Know?

According to studies at the University of Twente in the Netherlands, the world is using more than nine billion cubic meters of water each year. China, India, and the United States are the top consumers.

Water rights are also a major issue for agriculture in parts of the West, such as California, New Mexico, and Texas. Ranchers rely on surface water in rivers and streams to provide water for their cattle, and farmers rely on this same water to irrigate their crops. In addition, both new and established housing developments also need access to the same water resources. It is a constant struggle to fairly distribute limited water resources amongst those who need them, and to not exhaust these resources in the process.

What Is the Nitrogen Cycle?

Nitrogen makes up approximately 78% of Earth's atmosphere. Most of the nitrogen in the atmosphere is *inert*, meaning that it cannot be used by many plants or animals, nor is it harmful to plants and animals. The ***nitrogen cycle*** is a natural process through which nitrogen cycles through various chemical forms in the atmosphere, in plants, in animals, and in the soil, **Figure 14-17**.

Nitrogen in the soil:

- Affects plant growth.
- Is a primary source of food for many animals.
- Affects the rate of decomposition of dead plants and animals.

Four processes, or stages, make up the nitrogen cycle. They are fixation, ammonification, nitrification, and denitrification.

Fixation

In order to make nitrogen usable by plants and animals, atmospheric nitrogen must be converted into usable forms, such as ammonium or nitrates.

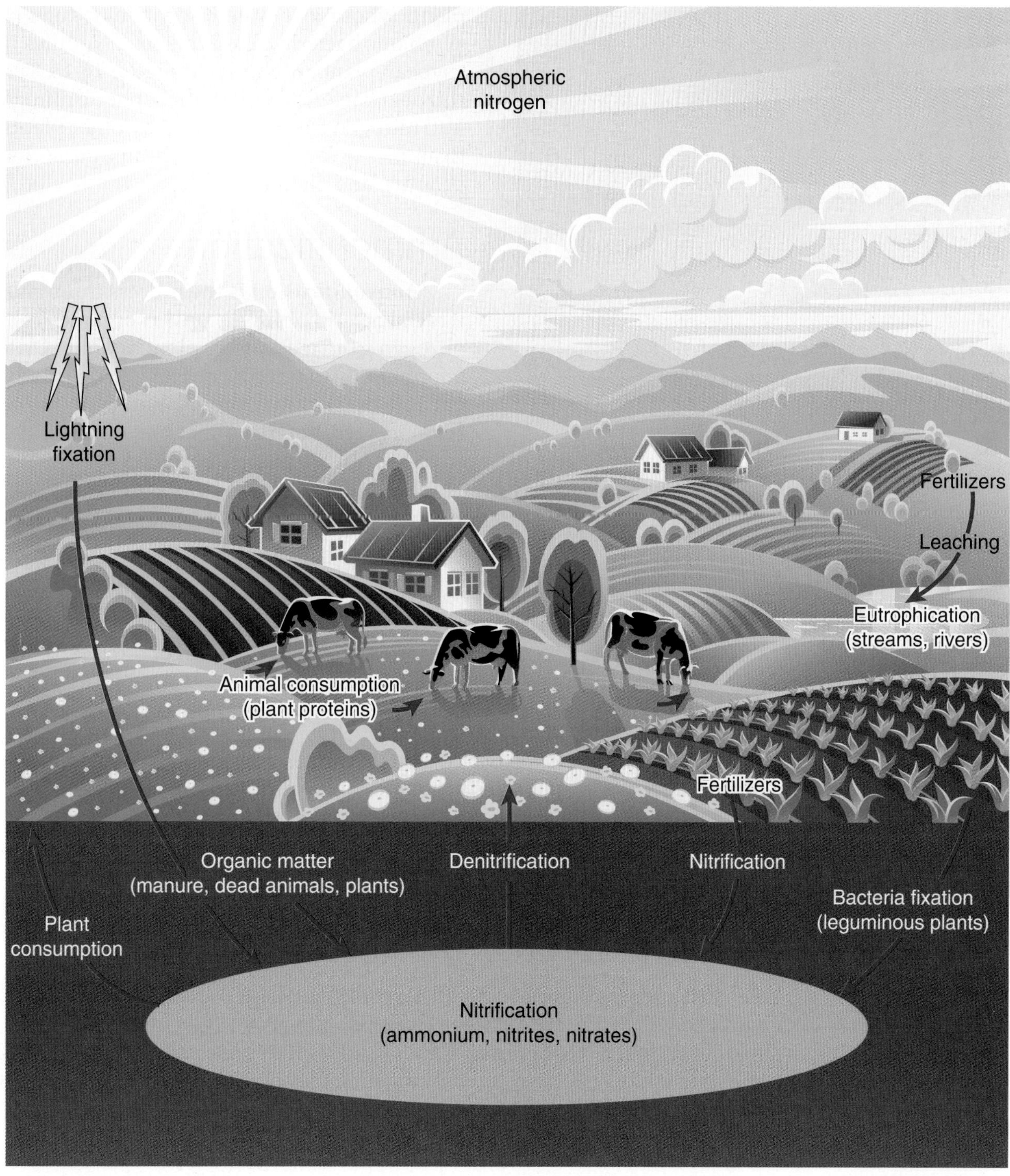

Crop/Shutterstock.com

Figure 14-17. The nitrogen cycle converts atmospheric nitrogen into forms of nitrogen that are usable by plants.

Fixation is the process through which atmospheric nitrogen is transformed into ammonium. As you can see, nitrogen fixation supports a great deal of life on our planet.

Not only does nitrogen fixation occur every time lightning strikes, it also occurs through biological means. For example, a special group of plants called legumes convert atmospheric nitrogen into ammonium. ***Legumes*** are

Travis Park/Goodheart-Willcox Publisher

Figure 14-18. Legumes are plants whose roots contain nodules of symbiotic bacteria called rhizobia in their root systems.

plants whose root systems contain nodules of symbiotic bacteria called rhizobia. Nodules in a soybean plant are shown in **Figure 14-18**. These rhizobia bacteria convert atmospheric nitrogen into nitrogen that can be used by other plants. Examples of legumes are soybeans, peas, clover, peanuts, and alfalfa.

Ammonification

Ammonification is the process through which bacteria or fungi convert organic nitrogen from a decaying plant or animal into ammonium. Ammonification also occurs when bacteria break down animal manure in the soil. Farmers incorporate animal manure into their fields primarily to add nitrogen to the soil through ammonification. Again, ammonium is a form of nitrogen that can be used by plants for growth.

Nitrification

Nitrifying bacteria in the soil convert ammonia into nitrates in a process called ***nitrification***. ***Nitrates*** are the plant-usable forms of nitrogen. The production of plant proteins requires nitrogen in the form of nitrates.

Denitrification

Denitrification is the conversion of nitrates into inert atmospheric nitrogen. Denitrifying bacteria in the soil perform this conversion process. These are a different species of bacteria from the bacteria that perform nitrification. Denitrification, the return of nitrogen to the atmosphere, completes the nitrogen cycle.

Did You Know?

Approximately 78% of the atmosphere is made up of a nonusable form of nitrogen gas. About 21% is made of oxygen.

The Nitrogen Cycle and Agriculture

Just as crops denied water will not thrive, plants denied sufficient nitrogen will not fare well. Some crops, like soybeans and clover, are able to remove nitrogen from the atmosphere and fix it into the soil. Other plants, like corn and cotton, use this stored nitrogen in the soil to grow strong and healthy.

Nitrogen for Corn

Corn is an agronomic crop that requires a lot of nitrogen. Nitrogen for corn production comes from the atmosphere. You might think that since nitrogen makes up approximately 78% of the atmosphere, it would be easy to find enough nitrogen for corn crops. Actually, finding the nitrogen is easy, but transforming it into a plant-usable form is neither easy, nor inexpensive. Converting atmospheric nitrogen to either *urea* or *anhydrous ammonia* requires a lot of energy and is quite expensive. Developing varieties of corn that use less nitrogen, or use the nitrogen that is available in the soil more efficiently, has been a standing challenge for the agricultural industry.

Nitrification of Water Supplies

Nitrification is the part of the nitrogen cycle that can actually introduce pollutants into surface and/or groundwater. Nitrates are ***water-soluble***, meaning that they are easily dissolved in water. Aquatic algae are sensitive to high concentrations of nitrates in the water, so nitrates actually create ***algae blooms***, or excessive algae growth. Excessive algae growth depletes the water of oxygen, which, in turn, kills fish and other aquatic animals. Algae blooms can be extremely devastating, **Figure 14-19**.

Nitrates in our drinking water pose a serious health threat because they interfere with the blood's ability to carry oxygen. If people or animals drink water high in nitrate, it may cause ***methemoglobinemia***, an illness found especially in infants. The disease is also called ***blue baby syndrome*** because the infant's skin turns a bluish color due to the blood's inability to carry oxygen.

Heike Kampe/iStock/Thinkstock

Figure 14-19. Algae blooms deplete the water of oxygen, which, in turn, kills fish and other aquatic animals. ***Are there bodies of water near your home with algae blooms? Have you noticed a change in the flora or fauna since the algae blooms covered the water surface?***

What Is the Carbon Cycle?

The ***carbon cycle*** is the natural process whereby carbon cycles through the oceans, atmosphere, plants, animals, and soil. Like the water cycle and the nitrogen cycle, the carbon cycle enables life to flourish on Earth. An example of the carbon cycle is shown in **Figure 14-20**.

Carbon Pool

When discussing the carbon cycle, scientists use terms such as carbon pools, sinks, and sources. A ***carbon pool*** is an area of Earth where carbon accumulates and is also released. There are four main carbon pools:

- The atmosphere
- Oceans
- The biosphere
- Sediments

Actually, there is one other large carbon pool—Earth's core. But the core holds its carbon tightly and does not exchange carbon to any great extent with the other pools.

Did You Know?

Coastal systems, such as mangroves, salt marshes, and seagrass meadows, can sequester carbon at rates up to 50 times those of the same area of tropical forests. Since 1980, almost 14,000 square miles of mangroves have been removed globally.

Carbon Sink

A ***carbon sink*** is a place that absorbs and stores carbon for long periods of time. Carbon sinks generally absorb more carbon than they release. The ocean is the largest, natural carbon sink on Earth. Forests and their plants, especially tropical rainforests, are also natural carbon sinks. An example of an

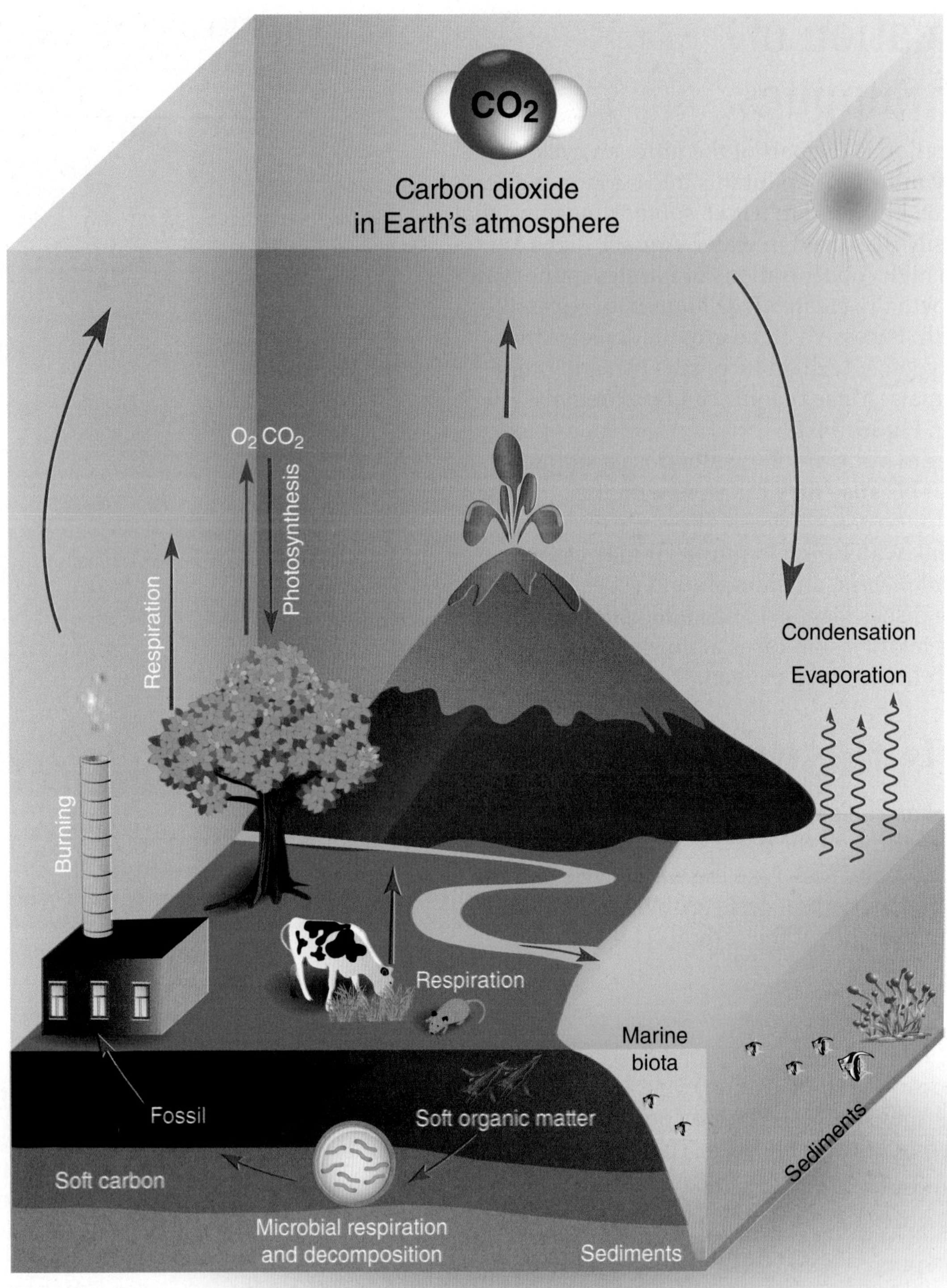

Designua/Shutterstock.com

Figure 14-20. Like the water cycle and the nitrogen cycle, the carbon cycle enables life to flourish on Earth.

artificial carbon sink would be a landfill. Artificial carbon sinks are typically not as large or efficient as natural carbon sinks.

One way that agricultural activities can enhance soil's ability to act as a carbon sink is to use reduced tillage systems. By reducing the amount of tillage used for row crops, farmers burn fewer fossil fuels to operate tractors. Reduced tillage preserves more organic matter from dead and decaying plants in the soil. An example of a farmer planting with a no-till planter is seen in **Figure 14-21**. Carbon stored in soil can remain in the soil for thousands of years, thus reducing the amount released into the atmosphere.

Travis Park/Goodheart-Willcox Publisher

Figure 14-21. Reduced tillage not only leaves plant residue on the soil surface to prevent erosion, but the plant residue also serves as a carbon sink, thus reducing the amount of carbon in the atmosphere. ***How does the crop residue prevent erosion?***

Carbon Source

A ***carbon source*** is a place that releases more carbon than it stores. So, anything that burns fossil fuels is a carbon source. Your car, the appliances in your house, the heating and cooling system in your school, and electric power plants are all examples of carbon sources. Actually, you are a carbon source. Through respiration, you breathe in oxygen and exhale carbon dioxide, which contributes to the store of atmospheric carbon.

Carbon in the Atmosphere

Scientists believe that the amount of carbon dioxide in our current atmosphere has been elevated by human activities since the Industrial Revolution in the late 1800s. During the Industrial Revolution, factories grew in size and scale, requiring more energy from the burning of fossil fuels, like coal, petroleum, and natural gas. Inventions like the automobile also burned fossil fuels such as gasoline and diesel fuel. This increased combustion of fossil fuels contributed large volumes of carbon dioxide to our atmosphere, **Figure 14-22**.

Bob Stilwell/Shutterstock.com

Figure 14-22. Although the Industrial Revolution spawned many great inventions, it also increased our combustion of fossil fuels and contributed large volumes of carbon dioxide to the atmosphere. ***How are our thoughts about fossil fuels and carbon dioxide in the atmosphere shaping food and environmental policy today?***

Further, the human population on Earth has grown tremendously since the mid-1800s. More people breathe more, thus expelling greater volumes of carbon dioxide into the atmosphere. A growing population also requires a growing food source. Much of this food comes from livestock, animals who also breathe and expel carbon dioxide.

Of all the carbon pools, carbon in the atmosphere causes the most concerns for human life. Carbon in the atmosphere exists as carbon dioxide (CO_2)

and methane (CH_4). Carbon dioxide and methane are two of the more dangerous greenhouse gases. ***Carbon dioxide (CO_2)*** is a colorless, odorless gas produced by burning carbon and organic compounds and by respiration. ***Methane (CH_4)*** is also a colorless and odorless gas. It is the main constituent of natural gas and highly flammable.

Both gases absorb and hold heat from the sun's rays, and together, they are responsible for much of the warming of our climate.

Did You Know?

According to the EPA, methane is more than 20 times more effective at trapping heat than carbon dioxide, and it accounts for 16% of the greenhouse gas emissions produced by human activities.

Carbon Dioxide

How does carbon dioxide get into the atmosphere? When animals respire, they breathe in oxygen and exhale carbon dioxide. And, although a significant source of carbon dioxide, this is not the primary source. The primary source of carbon dioxide in the atmosphere is from burning fossil fuels for electricity and for moving our cars from one place to another. The burning of tropical rainforests also releases great volumes of carbon into the atmosphere.

Photosynthesis

How does carbon dioxide leave the atmosphere? The primary means for the atmosphere to rid itself of carbon dioxide is through photosynthesis in plants. Plants use carbon dioxide in photosynthesis to create carbohydrates and simple sugars. Carbon dioxide also dissolves into surface water and even raindrops as they fall to Earth.

Methane

How does methane get into the atmosphere? Methane is the primary component of natural gas and is emitted by natural sources such as wetlands, as well as through industry activities such as leakage from natural gas systems, and the raising of livestock, **Figure 14-23**.

Warren Rice Photography/Shutterstock.com

Leonid Ikan/Shutterstock.com

Figure 14-23. Methane enters the atmosphere through the decomposition of organic materials in natural wetlands and during the production and processing of crude oil.

Natural Sources

Although methane is emitted into the atmosphere through many natural sources, wetlands are the largest natural source. Methane is emitted from bacteria that decomposes organic materials in wetlands through anaerobic (without oxygen) decomposition.

Industrial Sources

Methane entering the atmosphere due to industry occurs primarily during the production, processing, and distribution of natural gas. Since natural gas is often found alongside petroleum, methane is also emitted during the production and processing of crude oil. Methane emissions from industrial resources can be reduced by upgrading and maintaining basic production, storage, and transportation equipment.

Agricultural Sources

Methane entering the atmosphere through agricultural means occurs primarily through the digestive process of domestic livestock, and through the management of livestock manure (lagoons and holding tanks). Methane emissions from agricultural sources can be reduced and captured by changing manure management strategies. Refer to **Figure 14-24**. Scientists believe that cattle are major contributors to methane in the atmosphere. As the diets of many people around the world increasingly include meat-based proteins, more beef cattle will be necessary to meet this demand. That means more methane will be produced by feeding and growing cattle.

pinthong nakon/Shutterstock.com

Figure 14-24. Methane is emitted into the atmosphere through the digestive process of domestic livestock. With proper management, methane emissions can be captured from manure decomposition and used as biodiesel energy.

Carbon in the Oceans

A great deal of carbon is stored in the oceans. Actually, our oceans store most of the carbon that can be found at or near the surface of Earth. Most carbon enters the ocean as it dissolves out of the atmosphere. In a lesser fashion, carbon enters the oceans from rivers and streams that carry dissolved carbon in freshwater.

Carbon in the Biosphere

Remember that the biosphere is the accumulation of all of the ecosystems on Earth. Since all living organisms are made up of carbon, a lot of carbon is found in the biosphere. Plants and animals make up about one-quarter of the carbon in the biosphere, while the living organisms in the soil make up the rest.

Carbon in Sediments

Sediments include fossil fuels, freshwater, and the nonliving parts of the soil. As living organisms die and decompose, carbon is stored in sediments. Some of the carbon that is dissolved in rivers and streams settles to the bottom of the river or stream. Over hundreds and thousands of years, this sediment stores volumes of carbon. Some of the sediment eventually turns into sedimentary rock, which locks up the carbon for eons, **Figure 14-25**.

dominique landau/Shutterstock.com

Figure 14-25. Some of the carbon found in sediments is locked into sedimentary rock. Erosion releases this carbon into the atmosphere.

Carbon Credits

The ***carbon credit system*** was created to stop the increase of carbon dioxide emissions. The idea being that companies can trade or sell certificates or permits to produce carbon with other companies that have reduced their carbon production or act as carbon sinks. A carbon credit is a permit that allows the holder to emit one ton of carbon dioxide.

For instance, an auto manufacturer requires a lot of energy and must burn a lot of fossil fuels to acquire that energy. The business is a carbon source. In contrast, a landowner who plants enough trees to reduce carbon dioxide emissions by one ton has created a carbon sink. The auto manufacturer may purchase the landowner's carbon credit, continue to do business, and not have its total activities contribute to carbon in the atmosphere.

Career Connection Reservoir Manager

Job description: A reservoir manager is responsible for monitoring water pollution, game fish populations, and microbial activity in water reservoirs. The reservoir manager's duties will vary, depending on whether the reservoir serves as a source of water for human consumption or for recreational activities. A reservoir manager works with other scientists to monitor and maintain the water quality in the reservoir. This often means that the reservoir manager must have good knowledge of the surrounding watershed and activities within that watershed.

A reservoir manager also works with law enforcement officials to maintain water quality and monitor aquatic sporting activities and pollution in or near the reservoir's watershed.

Education required: A bachelor's degree in ecology, water resources, or natural resources is required. A specialization in chemistry or water chemistry is helpful.

Job fit: This job may be a fit for you if you have an interest in water and fish. You should also enjoy the outdoors in all seasons.

Goodluz/Shutterstock.com

A water reservoir scientist manages the water reservoir for human water consumption and/or recreational water activities.

Lesson 14.2 Review and Assessment

Words to Know

Match the key terms from the lesson to the correct definition.

1. A place that releases more carbon than it stores.
2. The process through which atmospheric nitrogen is transformed into ammonium.
3. The process through which bacteria or fungi convert organic nitrogen from a decaying plant or animal into ammonium.
4. The natural process whereby carbon moves through the oceans, atmosphere, plants, animals, and soil.
5. A process through which nitrifying bacteria in the soil converts ammonia into nitrates.
6. The lowering of water acidity due especially to acid rain.
7. The pathway that water follows underneath the surface, at the surface, and above the surface of Earth.
8. An area of Earth where carbon accumulates and is also released.
9. A natural process through which nitrogen moves through various chemical forms in the atmosphere, in plants, in animals, and in the soil.
10. A place that absorbs and stores carbon for long periods of time.
11. The accumulation of all of the ecosystems on Earth.
12. The conversion of nitrates into inert atmospheric nitrogen.
13. The plant-usable form of nitrogen.
14. The process of removing salt from seawater to make it suitable for human, plant, and animal consumption.
15. Excessive algae growth that depletes the water of oxygen.

A. acidification
B. algae bloom
C. ammonification
D. biosphere
E. carbon cycle
F. carbon pool
G. carbon sink
H. carbon source
I. denitrification
J. desalinate
K. fixation
L. hydrological cycle
M. nitrate
N. nitrification
O. nitrogen cycle

Know and Understand

Answer the following questions using the information provided in this lesson.

1. A(n) _____ is a pathway that a substance moves in the biosphere.
2. List the three phases of water and give examples of each.
3. Liquid water may be stored underground in reservoirs called _____.
4. *True or False?* More than one-third of Earth's water exists as vapor in the atmosphere.
5. How is cropland irrigation threatening to disrupt the water cycle in the United States?
6. The primary pollutants of rivers and streams in the United States are _____.
   A. waste water and acid rain
   B. invasive fish species
   C. runoff water and soil
   D. All of the above.

7. Why are farmers planting riparian zones? How does this help the environment?
8. *True or False?* Acid rain is a natural phenomenon,
9. Explain how human activity contributes to acid rain.
10. Explain why water rights are a major issue for agriculture.
11. *True or False?* Nitrogen fixation occurs every time lightning strikes.
12. The name given to the special class of plants that converts atmospheric nitrogen into ammonium is _____.
    A. Rhizobia
    B. fixaters
    C. nodules
    D. legumes
13. Why would the agricultural industry be interested in a corn variety that requires little nitrogen?
14. List and explain two reasons nitrification of water supplies is detrimental to the environment and its inhabitants.
15. Explain why carbon in the atmosphere causes the most concerns for human life.
16. How does carbon dioxide get into the atmosphere? What is the primary means for ridding the atmosphere of carbon dioxide?
17. What is methane and how does it get into the atmosphere?
18. *True or False?* Plants and animals make up about one-quarter of the carbon in the biosphere.

Analyze and Apply

1. If more than 96% of all the water on the planet is located in the oceans, then approximately what percentage of time would a droplet of water exist as saltwater in the ocean?
2. Identify areas in the United States that are challenged with freshwater issues.
3. What are some of the arguments against the practice of corporations purchasing and trading carbon credits? Do you agree? Why?

Thinking Critically

1. If a water drop or water molecules fall as snow on a glacier or one of the polar ice caps, what are the next steps for those water molecules in the water cycle? How long might the water molecules "live" as solid water in the glacier or on the polar ice? Why?
2. How has precision agriculture technology impacted the carbon, nitrogen, and water cycles on Earth? Why?
3. How might the nitrogen cycle and the water cycle interact with one another in agricultural areas?

Lesson 14.3

The Influence of Climate on Agriculture

Words to Know

air masses
altitude
arctic zone
climate
convection current
equator
front
greenhouse effect
greenhouse gases
jet stream
lake effect snow
latitude
polar zone
temperate zone
terrain
topography
tropical zone
tropics
uneven solar heating
weather
weather pattern

Lesson Outcomes

By the end of this lesson, you should be able to:

- Define climate.
- Differentiate between weather and climate.
- Describe the factors that influence climate.
- Explain USDA planting zones.
- Describe the greenhouse effect and its influence on our global climate.
- Evaluate how agriculture influences our climate.
- Explain how changes in climate impact agriculture.

Before You Read

Before you read the lesson, read all of the table and photo captions. What do you know about the material covered in this lesson just from reading the captions?

Since agricultural activities occur mainly outdoors, a region's climate plays a major role in the area's agriculture. And, with many scientists agreeing that human activities, including agricultural activities, have an impact on climate change, there is a reciprocal effect between agriculture and the climate.

What Is Climate?

Climate is the measure of the average temperature, precipitation, wind, and humidity of an area over a long period of time. ***Weather*** is the measure of the current temperature, precipitation, wind, and humidity of an area. The accumulation of all the weather in a place, over an extended period of time, makes up the climate of that place, **Figure 14-26**. This means an area's climate is actually its long-term weather pattern. ***Weather patterns*** are repeated incidences of similar temperatures and precipitation. For example, a summer weather pattern in much of the United States would be warmer temperatures and fairly frequent storms producing rainfall. The weather and weather patterns of an area combined over several decades, or even hundreds of years, are its climate.

Earth's climate has changed periodically from one extreme to another. For example, Earth was warm when dinosaurs roamed the planet and inland seas covered much of what is now the United States. At other times, like in the ice ages, the planet was quite cool and glaciers covered much of the planet. There are many theories and speculations as to what

SUSAN LEGGETT/Shutterstock.com

Songquan Deng/Shutterstock.com

Figure 14-26. What is the climate of the area in which you live? *Is it warm and sunny like Florida? Do you enjoy a change of seasons like they do in the Midwest?*

natural phenomena caused these drastic climate changes. Today, however, scientists believe that human activity, such as burning fossil fuels and deforestation, is having a significant effect on our climate. Scientists believe that effects of climate changes include overall warming of the atmosphere and increased frequencies of dramatic weather events, like severe storms, hurricanes, and blizzards.

Factors that Influence Climate

Many factors influence the climate of an area, including a location's latitude, altitude, and terrain, plus its proximity to larger bodies of water.

Latitude

Latitude is a measure of the distance from the *equator* to either of Earth's poles. The ***equator*** is an imaginary line drawn around the center of Earth, equally dividing the globe horizontally. This division places the equator an equal

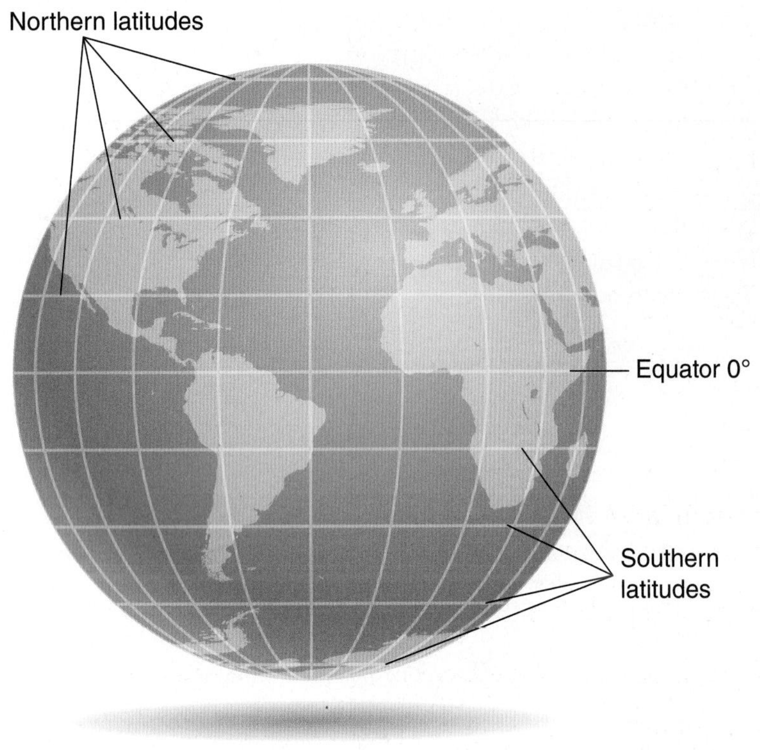

Figure 14-27. The equator is an equal distance from both the North Pole and the South Pole.

distance from both the North Pole and the South Pole, **Figure 14-27**. Between the equator and either pole, there are imaginary, parallel lines circling Earth. These lines indicate the degrees of latitude. The equator is given a value of 0° latitude and each of the poles is given a value of 90° latitude.

An area's latitude is a good indication of its climate because parts of Earth receive more direct sunlight than others, depending on their latitude, or distance from the equator. This occurs because Earth is not positioned fully north-to-south, but tilted on its axis. Therefore, different temperature zones exist between certain latitude ranges. For example, the sun's rays strike Earth at more or less 90° angles, or most directly overhead, between the latitude of 23.5°N and 23.5°S, **Figure 14-28**. This latitude range is the area immediately north and south of the equator. Since the sun's rays strike the surface of this area at a mostly direct angle, the atmosphere is consistently heated and temperatures are predominantly warm. This zone, or latitudinal range, is referred to as the ***tropics*** or ***tropical zone***. In contrast, the sun's rays never strike Earth at a 90° angle at latitudes higher than 66.5°N and, in the northern hemisphere, we call this region the ***arctic zone*** or ***polar zone***. Although there is some seasonal variation in temperature, every month in a polar climate has an average temperature of less than 50°F (10°C).

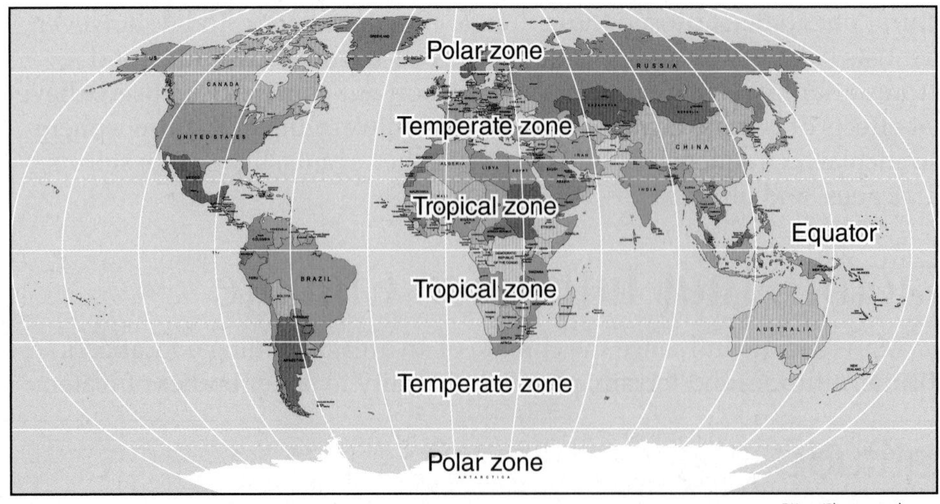

Figure 14-28. Latitude affects climate because the sun's rays strike Earth most directly between 23.5° north and 23.5° south.

The area in the latitudinal range between 25.5° and 66.5°, both north and south, is referred to as the ***temperate zone***. Areas in the temperate zone experience weather patterns that produce four distinct seasons. For example, during summer in the Northern Hemisphere, the sun's rays strike Earth at 90° angles between 25.5°N and 66.5°N, so the atmosphere warms. In the winter, when the Northern Hemisphere of Earth is tilted away from the sun, the sun's rays strike the Northern Hemisphere at a lower angle, thus creating much less heat energy in the atmosphere. The seasons in the Southern Hemisphere are somewhat opposite those in the Northern Hemisphere and generally much milder.

So latitude, or really, the angle of the sun's rays striking Earth, plays a significant role in determining climate.

Uneven Solar Heating

As evident by the differences in climate of each latitudinal zone, some parts of Earth receive more direct sunlight than others. Also, as Earth spins, the part of Earth that is in the sun's light is receiving more heat than the part that is in Earth's shadow. This ***uneven solar heating*** of Earth's surface causes global weather patterns.

As the air in the atmosphere at the equator is heated, it rises (heated air is lighter than cool air), carrying water vapor in the form of humidity. Some of the water vapor condenses into clouds. This hot, humid, cloudy air mass floats up into the atmosphere and then flows either north or south toward the North or South Pole. This flow of hot, humid air is called a ***convection current***.

Uneven heating of Earth's surface causes the convection currents. Convection currents then cause air masses. ***Air masses*** are bodies of either cool or warm air that flow together. If an air mass forms over warm ocean water at or near the equator, then a warm air mass is formed. If the air mass forms over cooler, drier land, then a cool air mass is formed. The coldest, driest air masses are formed over mountains near the poles.

The place where two air masses meet is called a ***front***, **Figure 14-29**. Weather fronts usually bring a change in temperature and humidity. Weather fronts also trigger rain or snowstorms.

Did You Know?

The Solar Energy Generating Systems (SEGS) in California's Mojave Desert is one of the world's largest solar thermal energy generating facilities. At peak power, the solar plants power more than 200,000 homes and displace almost 4,000 tons of pollution that would have been produced if fossil fuels had been used to generate the electricity.

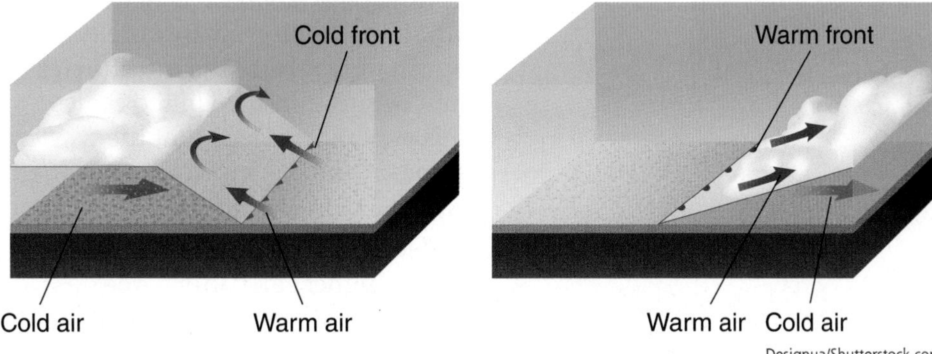

Designua/Shutterstock.com

Figure 14-29. Weather fronts are places where two air masses meet. Weather fronts usually bring a change in temperature and humidity. *Can you remember a time when you saw a weather front approaching? What did it look like? How did the air feel before the front moved through? How did it feel afterward?*

Did You Know?

Global winds are named by the direction from which they blow.

Global Winds

Warm and cold air masses move on their own accord, albeit slowly. The real movers and shakers in the atmosphere are the global winds. Four rivers of fast-moving wind called ***jet streams*** are constantly circling the globe. There are two polar jet streams and two subtropical jet streams, one for each of the Northern and Southern Hemispheres. The polar jet stream influences most of the weather patterns in the United States.

In the jet streams, air generally moves at about 140 miles per hour, but can reach speeds of 275 miles per hour. Thankfully, these jet streams remain high in the atmosphere and pose no threat to those of us living on Earth's surface.

Jet streams move air masses and fronts around the globe. In essence, they are the mixers in our atmosphere. Through this mixing, jet streams transfer heat and precipitation around the planet. They do not follow straight lines around the planet, but are wavy. Sometimes, the jet stream dips far south or far north, depending on the strength of warm and cool air masses.

Best works/Shutterstock.com

Figure 14-30. Altitude is the height of the land above sea level and affects the climate of the region. *Have you climbed a mountain? How did climbing in elevation affect your breathing?*

Altitude

Altitude is the height measurement of the land above sea level, **Figure 14-30.** The highest elevations above sea level are, of course, mountaintops. What happens to the climate as the altitude increases up the side of a mountain? Several significant changes occur:

- For each 1,000-foot rise in elevation, the average temperature drops 5.4°F. This affects the kinds of plants and animals that can grow and thrive in these elevations.
- Air blowing across the mountain range cools as it is forced to rise over the mountain. Cooler air cannot hold as much water vapor as warmer air, so the cooling air forms clouds, which then release the moisture as rain or snow. In the United States, this generally means that the western slopes of the Rocky Mountains and the Appalachian Mountains receive more precipitation than the eastern slopes.
- At the highest elevations, the temperature may never rise above freezing for much of the year, so snow accumulates and either does not melt or does so only for a few months out of the year. This creates a different climate at the higher elevations of the mountain range than in the valley.

Terrain

Terrain is the physical feature or general ***topography*** of the land. Mountains, swamps, lowland, and highland all make up an area's terrain. The terrain of an area greatly affects its climate because, depending on its structure and components (mountains, plains, lakes, etc.), the terrain will influence the movement of weather.

In the United States, most of the weather moves from the west coast to the east coast. Many storms begin in the Pacific Ocean, cross the Rockies, and then mix with cold air from Canada and dry, humid air from the Gulf of Mexico. Depending on the time of year, this mixing of cold and warm, humid air may result in violent thunderstorms or even tornados in the Great Plains. These storms usually lose intensity as they move east across the Appalachian Mountains.

North America is a fairly unique land mass in that, because of its terrain, it has thunder, lightning, hail, tornadoes, flash floods, and even blizzards unlike anywhere in the world, **Figure 14-31**. The Rockies and the Appalachian mountain ranges force the mixing of air masses across the Great Plains to create the unique weather experienced across the United States. In most other places on Earth, mountain ranges running east to west help keep the cold, arctic air confined to the polar regions and prevent it from mixing with warmer air masses.

Bodies of Water

An area's climate is also influenced by its proximity to larger bodies of water. When a land mass is near a body of water, the water tends to moderate the temperature and climate of the region. This moderating effect lasts only a few miles inland.

deepspacedave/Shutterstock.com

Figure 14-31. The United States is one of the few places in the world where conditions are right for spawning tornados. *Do you live in an area that is prone to tornados? Do you and your family have an emergency plan in place?*

Henryk Sadura/Shutterstock.com

Figure 14-32. Lake effect snow may bury one area of a state in several feet of snow, while other nearby areas will receive minimal snowfall. *Which U.S. city receives the most "lake effect" snow each year?*

Large bodies of water absorb and store much of the sun's solar heat energy in warm periods and release it slowly in cold periods. So, during the summer, the land near the water may not heat up as much as land as little as five or ten miles inland (because the water is absorbing much of the solar heat). This is why people can swim comfortably in the ocean along the Atlantic Coast, as far north as New York's Long Island and New Jersey's Atlantic City, during the summer months. The water retains the heat and maintains a more comfortable "swimming" temperature. Coastal areas may also be less cold than inland areas during winter months because the water would be releasing the stored solar heat.

Lake Effect Snow

Areas along the eastern coasts of large lakes, such as the Great Lakes, experience moderate temperatures for much of the year. But when the temperatures fall in the winter, these areas can also experience a lot of snow. Why so much snow on the eastern shores of the Great Lakes? Remember, most of the weather in the United States moves from west to east. As the wind blows across the lakes, it picks up moisture. That air cools as it encounters land, and since cool air cannot hold as much moisture as warm air, precipitation occurs. In the winter, this precipitation falls as snow, or as in this case, ***lake effect snow***, **Figure 14-32**.

Climate Changes

Unlike weather, the climate of an area does not change on a daily basis. A week's worth of rainy days will not change a temperate climate to a tropical one. Whereas weather describes the temperature and precipitation of an area over a few days, climate describes the temperatures and precipitation over thousands of years.

Greenhouse Effect

The ***greenhouse effect*** is the process by which the sun's solar energy is trapped in the atmosphere and reflected back to Earth. Think of the greenhouse in your agriculture program. The greenhouse glass or plastic allows the sun's light and heat energy to pass through, but it also traps the sun's light and heat energy in the greenhouse. The trapped solar energy heats the air inside the greenhouse and raises the temperature.

Unlike the greenhouses in your agriculture program, Earth is not surrounded by a layer of glass or plastic. However, it does have a naturally occurring combination of gases, water vapor, and clouds that

Hands-On Agriculture

If you want to experience the greenhouse effect, think about the temperature on the next cloudy evening and night, especially during the summer. After the sun sets, the temperature should begin to fall. If the air is not humid and the sky is free of clouds, then the temperature falls much more quickly than it does if the air is humid and the sky is full of clouds.

allow solar energy to pass through to Earth's surface and also prevent it from leaving the atmosphere. These substances essentially act like the greenhouse glass or plastic.

Greenhouse Gases

The naturally occurring combination of gases that acts like the glass or plastic covering on a greenhouse is referred to as ***greenhouse gases***. There are several greenhouse gases, but water vapor, carbon dioxide, methane, and ozone are the primary greenhouse gases.

The amount of water vapor in the atmosphere is relatively constant and contributes the most to the greenhouse effect. Carbon dioxide contributes the second most impact on the greenhouse effect. Scientists believe that carbon dioxide contributes 9%–26% of the greenhouse effect. Scientists also know that the amount of carbon dioxide in the atmosphere has been increasing for the last 150 years or so. This is one indication that humans are having an effect on our climate.

Did You Know?

The amount of carbon dioxide in the atmosphere has been measured from an observatory on Mauna Loa (a volcano in Hawaii) since 1958.

Agriculture and Climate

Over time, agriculture has had significant effects on the global climate, and these effects have clearly impacted agriculture. Early farmers were unaware of, and probably couldn't even imagine, the effect poor farming and logging practices would have on the local climate, let alone the global climate.

Early Agriculture

To understand how agriculture has impacted the climate, first consider that the American farmlands where crops like corn and soybeans are grown were either forest or prairie ecosystems prior to being converted into crop fields. Early pioneers landing on the east coast of the United States found lush forests reaching nearly to the beaches of the Atlantic Ocean. While forests initially provided enough food from game animals like deer, bear, and other animals, the population soon needed more food than the forests could provide. So, pioneer farmers cut down trees to build homes and convert forest land into fields for growing crops. They also used burns to clear areas of stumps and other undergrowth. Burning trees and other plant materials released carbon dioxide into the atmosphere. Remember that carbon dioxide is one of the primary greenhouse gases. These actions, on such a large scale, caused the atmosphere to begin warming. As the atmosphere warmed, the climate began to change.

As settlers moved across America toward the Pacific Ocean, they cut down more forests and plowed under prairie grasses to create more farm fields. All plants that use photosynthesis consume carbon dioxide and produce oxygen. So, by removing vast quantities of trees and prairie grasses, the pioneers were greatly reducing the number of plants and trees that would naturally remove carbon dioxide from, and provide oxygen to, the atmosphere. So, once again, agriculture contributed to climate change.

It is easy to lay blame on early farmers, but what choice did they really have? The population of the United States was growing and needed food. Farming provided that food. In order to farm, farmers needed fields, which meant that they had to remove forests and prairies to create those fields. Unfortunately, people were unaware of the long-term effects these practices would have. They also lacked the agricultural knowledge and technology we use today to reduce negative effects and preserve our natural resources.

Today's Agriculture

Even today, agriculture has an impact on climate and the environment. Unfortunately, some of it is still negative. Although advances in forestry help reduce the negative effects of our need to produce wood products, many forests, especially in third world countries, are still carelessly cleared to create farm fields. This is especially true in South America where acres of the Amazon rainforest are continuously cut and burned, **Figure 14-33**. These actions, much like those of the early American pioneers, have a significant impact on altering the carbon dioxide in our atmosphere.

Fortunately, today's agriculturalists are much more environmentally conscious and apply new technologies and practices to reduce the negative

guentermanaus/Shuttertstock.com

Figure 14-33. Cutting down and burning trees in the Amazon rainforest has a significant impact on altering the carbon dioxide in our atmosphere.

Meg Wallace Photography/Shutterstock.com

Figure 14-34. By applying pesticides instead of cultivating the soil to remove weeds, farmers actually contribute less greenhouse gases to the atmosphere. *What is the tradeoff of using pesticides on farmers' fields?*

impact agriculture may have on our environment. Today's farmers use less energy to raise crops like corn, soybeans, cotton, wheat, and vegetables. How? Aside from the use of newer, fuel-efficient machines, today's farmers use new cultivation technologies like no-till and minimum tillage practices. With these tillage practices, farmers drive across the field fewer times, reducing the amount of fuel needed to produce the crop. Farmers also use precision technologies to apply pesticides and fertilizers to eliminate weeds and insect pests and provide precise nutrients to growing crops, **Figure 14-34**. Again, this reduces the amount of energy needed to produce the crop in the field.

Climate Changes and Agriculture

In recent years, many of the climate changes that are attributed to the greenhouse effect have begun to have a significant impact on agriculture. Climate change accounts for increased frequency and severity of drought in the Midwestern Corn Belt. Climate change also means that dairy farmers may consider air conditioning to cool their cow barns so that the cows produce high volumes of milk. Less severe winter temperatures also mean that many insect pests can overwinter without being killed, which means larger infestations of insect pests in crops during the growing season.

Extremes in Temperature

Today's farmers and ranchers experience greater extremes in temperatures during the growing season than any previous generation.

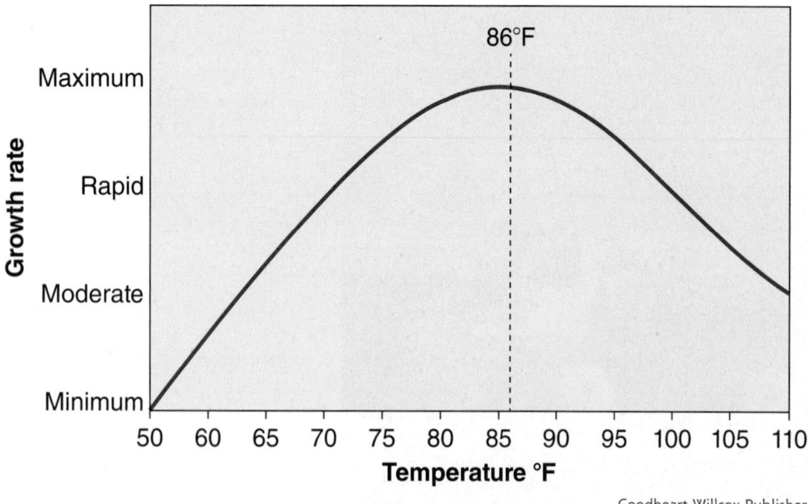

Figure 14-35. Most agronomic crops grow best at temperatures between 74°F (23°C) and 92°F (33°C).

More areas of our country now have days where the temperature rises above 100°F (37.7°C) than ever before, and days above 90°F (32°C) are now more common in areas of the United States where they were once rare. These higher temperatures affect agriculture in various ways, including:

- Plant growth
- Plant dehydration
- Pests and disease
- Livestock stress

Plant Growth

What happens to the ability of corn and soybeans to engage in photosynthesis and grow in extreme temperatures? Most agricultural crops generally thrive at temperatures ranging between 74°F (23°C) and 92°F (33°C), **Figure 14-35**. When the temperature rises much above this temperature range, plants do not grow as much. On a positive note, as the climate warms, more northern areas of the United States are able to grow new crops and increase the yields on crops like corn and soybeans because of higher temperatures and increased sunlight.

Plant Dehydration

As the temperature rises, what happens to the rate of water evaporation from fields? Water evaporates more quickly as the temperature rises. High rates of evaporation from both the soil and plant leaf surfaces increases the stress on growing plants, **Figure 14-36**. This is yet another indication that the warming climate has a detrimental effect on agriculture. Agronomists constantly work to develop crop varieties that are capable of withstanding drought and dry growing conditions.

Pests and Disease

As the temperature rises, new pests become problems for farmers. Diseases and fungi generally thrive in warm temperatures and high humidity. In an area where a disease was never seen a few decades ago, it may now occur with regularity because of changes in the climate.

Further, cold winters traditionally limit the growth of many pests and diseases. When winters are mild, more insects, bacteria, fungi, and other pests may survive the winter. Thus, more individual pests are alive to reproduce when the spring thaw arrives. These increased populations of pests damage crops and livestock during the growing season.

Livestock Stress

As the temperature rises, livestock becomes more stressed. Higher temperatures make the animals uncomfortable and thirsty. To thrive, livestock need adequate water. Consistently warmer temperatures also cause

Condor 36/Shutterstock.com

Figure 14-36. With changes to our climate, scientific data indicates that drought conditions will happen more frequently. ***Can you remember a drought in your area? What do you remember about the drought? How concerned were your parents, people in the news, and local farmers?***

water resources to evaporate at a higher rate, limiting water accessibility on many ranches.

Dairy cattle are especially prone to stress from hot temperatures. Producing milk requires a lot of metabolic energy, which causes the cow's body temperature to rise. Dairy cattle typically produce more milk in cooler climates where they are comfortable grazing, drinking, and moving around. In high temperatures, cattle are less likely to feed, drink, and move, and thus produce less milk.

Extremes in Precipitation

One of the real challenges to the warming climate is the wide ranges of precipitation that may fall in an area during the growing season. Meteorologists expect that the warming climate will produce more frequent and more severe droughts that last longer than in previous years. And, they also believe that when it rains, the rains will be higher in volume and will produce flooding of rivers and fields.

As shown in **Figure 14-37**, major droughts of 1983, 1988, 1993, and 2012 in the Corn Belt of the United States had significant impacts on corn production. Advances in biotechnology in the area of seed production lessened the impact of the 2012 drought. Corn varieties were developed so that corn could withstand lower amounts of precipitation and still produce a viable crop.

Agricultural scientists continue to experiment with new corn, soybean, rice, cotton, wheat, and other crop varieties each day. They are searching for varieties that will grow well in extremely high temperatures and with less

Market Year	Production (million metric tons)	Growth Rate	Period of Drought	Market Year	Production (million metric tons)	Growth Rate	Period of Drought
1980	168,648	−16.26%		1998	247,882	5.99%	
1981	206,223	22.28%		1999	239,549	−3.36%	
1982	209,181	1.43%		2000	251,854	5.14%	
1983	106,031	−49.31%	drought	2001	241,377	−4.16%	
1984	194,881	83.80%		2002	227,767	−5.64%	
1985	225,447	15.68%		2003	256,229	12.50%	
1986	208,944	−7.32%		2004	299,876	17.03%	
1987	181,143	−13.31%		2005	282,263	−5.87%	
1988	125,194	−30.89%	drought	2006	267,503	−5.23%	
1989	191,320	52.82%		2007	331,177	23.80%	
1990	201,534	5.34%		2008	307,142	−7.26%	
1991	189,868	−5.79%		2009	332,549	8.27%	
1992	240,719	26.78%		2010	316,165	−4.93%	
1993	160,986	−33.12%	drought	2011	313,949	−0.70%	
1994	255,295	58.58%		2012	273,832	−12.78%	drought
1995	187,970	−26.37%		2013	353,715	29.17%	
1996	234,518	24.76%		2014	353,965	0.07%	
1997	233,864	−0.28%					

USDA

Figure 14-37. Major droughts of 1983, 1988, 1993, and 2012 in the Corn Belt of the United States have had significant impacts on corn production. *Can you draw a chain reaction for the effects of severe drought?*

water than previous generations of crops. The challenge to find crops that will perform well in drought conditions is one of the toughest challenges facing agriculture.

Irrigation

The extremes in precipitation from the warming climate cause farmers to consider irrigation in areas where it was not previously needed, **Figure 14-38**. Irrigation is expensive to install and costly to use. Plus, irrigation draws on underground water supplies that are already threatened in many parts of our country.

alexmisu/Shutterstock.com

Figure 14-38. With increased incidents of drought, farmers turn to irrigation to supply water to their growing crops. Irrigation uses groundwater supplies that require hundreds or thousands of years to be replenished. So agricultural scientists are continually seeking new crop varieties that require less water to produce a significant yield.

Career Connection Meteorologist

Job description: Meteorologists study weather and climate. You probably have seen a meteorologist on your local television news programming. The meteorologist gives the weather report and weather forecast for a given area. Meteorologists also work in areas outside of television stations. Most large agricultural seed, chemical, and even fertilizer companies hire meteorologists. In these capacities, meteorologists help predict weather patterns over the growing season. Weather impacts the use of seed, fertilizers, and chemicals to control pests and diseases in important agricultural crops. Meteorologists provide weather information so that companies can predict how to produce their products so that farmers can maximize yields.

Fineart1/Shutterstock.com

This meteorologist may be working with agricultural seed, chemical, or insurance companies to provide weather-related information to predict insect infestations, drought, or crop yields.

Education required: A bachelor's degree in meteorology or atmospheric sciences is required, and a master's degree is often preferred.

Job fit: This job may be a fit for you if you enjoy and understand weather patterns and like studying scientific models to make predictions. If you want to be a television meteorologist, then your ability to speak in public is important.

Lesson 14.3 Review and Assessment

Words to Know

Match the key terms from the lesson to the correct definition.

1. The measure of the current temperature, precipitation, wind, and humidity of an area.
2. A measure of the distance from the equator to either of Earth's poles.
3. The flow of hot, humid air from the equator to either the North or South Pole.
4. The latitudinal range above 66.5°N in which the average temperature is less than 50°F (10°C).
5. The snow falling on the lee (sheltered) side of a lake, generated by cold, dry air passing over warmer water.
6. The measure of the average temperature, precipitation, wind, and humidity of an area over a long period of time.
7. An imaginary line drawn around the center of Earth, equally dividing the globe.
8. The general terrain of the land, including features such as mountains, swamps, lowland, and highland.
9. The place where two air masses meet.
10. The natural phenomena in which parts of Earth receive more sunlight and heat than others.
11. The process by which the sun's solar energy is trapped in the atmosphere and reflected back to Earth.
12. The latitudinal range between the latitude of 25.5° and 66.5°, both north and south, in which weather patterns produce four distinct seasons.
13. The height measurement of the land above sea level.
14. The latitudinal range between the latitude of 23.5°N and 23.5°S, in which the atmosphere is consistently heated and temperatures are predominantly warm.

A. altitude
B. arctic zone
C. climate
D. convection current
E. equator
F. front
G. greenhouse effect
H. lake effect snow
I. latitude
J. temperate zone
K. topography
L. tropical zone
M. uneven solar heating
N. weather

Know and Understand

Answer the following questions using the information provided in this lesson.

1. Explain the difference between climate and weather.
2. List four factors that influence the climate of an area.
3. Explain why some areas of Earth receive more sunlight than others.
4. *True or False?* The seasons in the Southern Hemisphere are somewhat opposite those in the Northern Hemisphere and are generally much milder.
5. Under what conditions does a warm air mass form? Under what conditions does a cold air mass form?

6. Explain how the jet streams affect the weather patterns on Earth.
7. How does the cooling of air blowing across a mountain range affect the precipitation on the mountain?
8. Why does the terrain of an area greatly affect its climate?
9. Explain the moderating effect of larger bodies of water on nearby land.
10. How does the greenhouse effect help Earth? How does it harm Earth?
11. Identify the four primary greenhouse gases.
12. Explain how the agricultural practices of early pioneers greatly affected the atmosphere.
13. *True or False?* Today's agricultural practices no longer have a negative effect on Earth's climate.
14. Identify three environmentally friendly agriculture methods used by today's agriculturists.
15. What are four ways today's temperature extremes affect agriculture?
16. How will the anticipated precipitation extremes affect agriculture?

Analyze and Apply

1. What are the major factors that influence the climate in your school district?
2. How does latitude affect the kinds of crops grown in the United States?
3. Where do bodies of water affect the kinds of crops grown in certain regions of the United States?
4. How do bodies of water affect the kinds of crops grown in certain regions of the United States?

Thinking Critically

1. How have new seed technologies, like stacked traits and Roundup Ready© seeds, affected agriculture's impact on the atmosphere and climate? Why?
2. How does agriculture play a role in mitigating the effects of climate change on the production of crops and livestock?
3. How do extremes in temperature affect farming and ranching? Why?
4. What are other ways that agriculture is being impacted by climate change, other than those illustrated in this lesson?

Lesson **14.4**

Animals, Plants, and Geography

Words to Know

apex consumer
apex predator
autotroph
carnivore
carrying capacity
consumer
decompose
decomposer
detritivore
food chain
food cycle
food web
herbivore
heterotroph
host
introduced species
invasive species
native species
omnivore
parasite
parasitism
predator
prey
primary consumer
producer
scavenger
secondary consumer
tertiary consumer

Lesson Outcomes

By the end of this lesson, you should be able to:

- Explain food webs and the interdependency of animals and plants in a food system.
- Analyze species that are producers, consumers, and/or decomposers.
- Differentiate the feeding relationships among herbivores, carnivores, scavengers, and parasites.
- Estimate carrying capacity for various species in an ecosystem.
- Differentiate between native and invasive species.

Before You Read

Write the lesson's *Words to Know* on a sheet of paper. Highlight the words that you do not know. Before you begin reading, look up the highlighted words in the glossary and write the definitions.

All living organisms interact with one another and their environment. Some species live in harmony with one another while others fight over food, water, or territory. And, of course, some species feed on others in order to gain energy and live. Regardless of how they interact, each species is part of an ecosystem and fits somewhere in the flow of energy, or food, in that ecosystem.

In this lesson, we will explore the hierarchy that exists where, at some point in its life, an organism is either the diner or the dinner. By studying this hierarchy, you will come to understand that our environment needs organisms that produce and consume energy.

Food Chains and Food Webs

A ***food chain*** is a simple model of how energy flows from one living organism to another, **Figure 14-39**. The interaction of many food chains is described as a ***food web*** or ***food cycle***. So, food chains and food webs are really energy cycles similar to the water, nitrogen, and carbon cycles we learned about earlier in this chapter.

In a food chain, all life starts with the sun. The sun provides solar energy that enables plants to make food through photosynthesis. For example, a corn plant grows tall in the summer sun and produces abundant ears of corn filled with rows of golden kernels. A field mouse eats some of those corn kernels in the fall and, at some point, becomes

LSkywalker/Shutterstock.com

Figure 14-39. A food chain describes the flow of energy (food) in a community. All food originates from the sun through the process of photosynthesis.

food for a hawk. The hawk that ate the mouse eventually dies and decomposes. Bacteria and other microorganisms break down the hawk's body. The decomposition of the hawk's body returns nutrients to the soil so that new plants, like corn, can grow in the following spring. This straight line "linkage" is a *food chain*. As organisms eat their way higher in the food chain, energy is primarily lost through heat. Animals need energy to heat their bodies to allow them to grow, move, and reproduce. So, the flow of energy in an ecosystem is a one-way model.

A food web depicts the interactions of many food chains in a community. Food webs contain many species of plants, animals, insects, and microorganisms. There are three basic categories of organisms in a food web: producers, consumers, and decomposers, **Figure 14-40**.

Did You Know?

No organism ever collects 100% of the energy stored in the plant or animal it eats.

Producers

Producers, or ***autotrophs***, are plants. Plants use photosynthesis, the foundation of all life on Earth, to live, grow, and reproduce. Plants are called *autotrophs* because they produce their own food. The prefix *auto* means "referring to oneself," and *troph* is Greek for "food" or "feeding." So, autotrophs are living organisms that produce their own food.

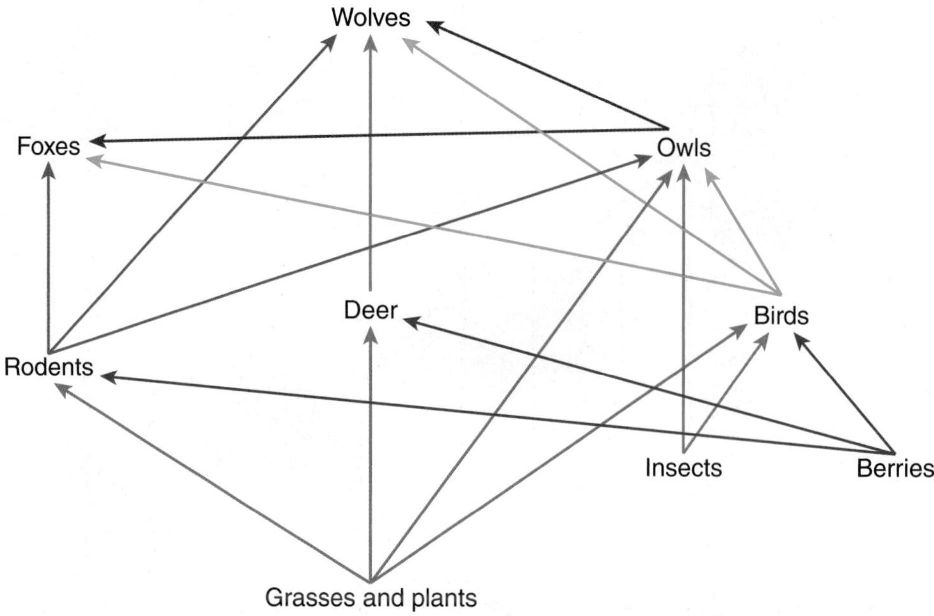

Producers	Consumers	Decomposers
Autotrophs	Heterotrophs	Heterotrophs
Plants, algae, many bacteria	Animals, insects	Fungi, many bacteria
Consume CO_2	Produce CO_2	Produce CO_2
Produce O_2	Consume O_2	Consume O_2

Goodheart-Willcox Publisher

Figure 14-40. A food web is a complex model of many species and many food chains interacting with one another in a community. One species may feed upon several other species, and not just one as depicted in a food chain. ***Where would you place humans in a food chain? Why?*** The table compares some of the differences and similarities among producers, consumers, and decomposers.

The primary gas used in photosynthesis is carbon dioxide. So, in addition to being part of the food cycle, producers are also part of the carbon cycle.

Consumers

Consumers, or ***heterotrophs***, are organisms that consume, or eat, other organisms. The prefix *hetero* means "other," and *troph* means "food" or "feeding." So heterotrophs get their food from other living organisms.

Almost all consumers are either insects or animals. Consumers cannot make their own food through any process. Animals and insects cannot photosynthesize (use the sun's energy to produce food). They may require sunlight to warm their bodies or produce some vitamins like vitamin D, but they cannot produce food with sunlight.

Since heterotrophs cannot use photosynthesis to produce their own food, they must feed on autotrophs or other heterotrophs. For example, deer only eat autotrophs, like grasses and other plants, whereas owls feed on small rodents and some insects (heterotrophs). Animals like opossums may

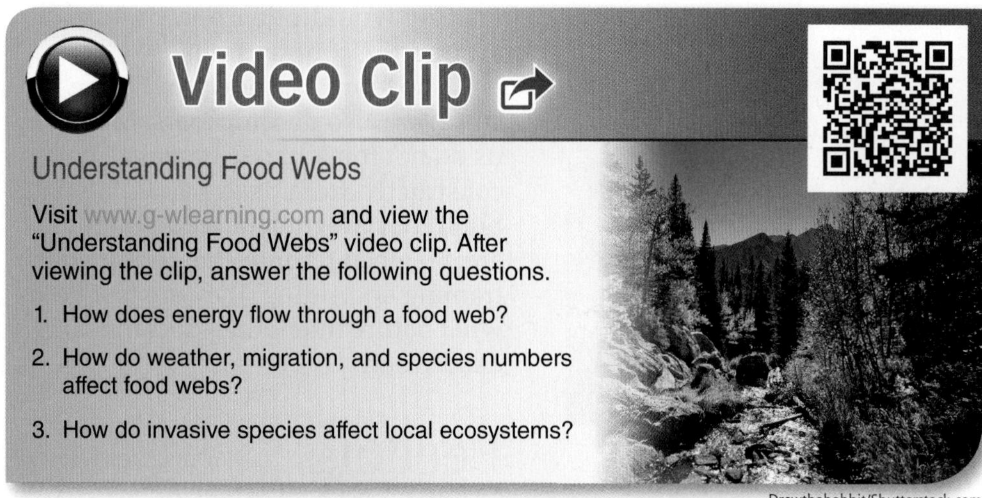

Drewthehobbit/Shutterstock.com

eat nuts, berries, or small animals, so they feed on both autotrophs and heterotrophs. See **Figure 14-41.**

Consumer Classifications

As scientists classify consumers, they make considerations for how high or low in the food chain a species normally feeds. Species that feed only on plants are called ***primary consumers***. Deer, bison, and rabbits are primary consumers. Species that feed on primary consumers are called ***secondary consumers***. A fox, raccoon, or opossum might be a secondary consumer. Species that feed on secondary consumers are called ***tertiary consumers***. Examples of tertiary consumers might be owls or coyotes.

Apex predators, or ***apex consumers***, are those species that are at the top of the food chain. Apex predators occupy a special place in food webs because nothing feeds on them. Their only concern is that they have sufficient food, meaning enough primary, secondary, and tertiary consumers, to eat. Cougars, crocodiles, sharks, and bald eagles are examples of apex predators.

Decomposers

Decomposers are organisms that break down, or ***decompose***, dead plant and animal material. Decomposers also break down manure and other animal waste. Decomposers are often fungi or bacteria. Some insects, like dung beetles and worms, are also decomposers. Without decomposers, dead plant and animal waste would pile up and slowly rot. For example, without dung beetles, mountains of cow

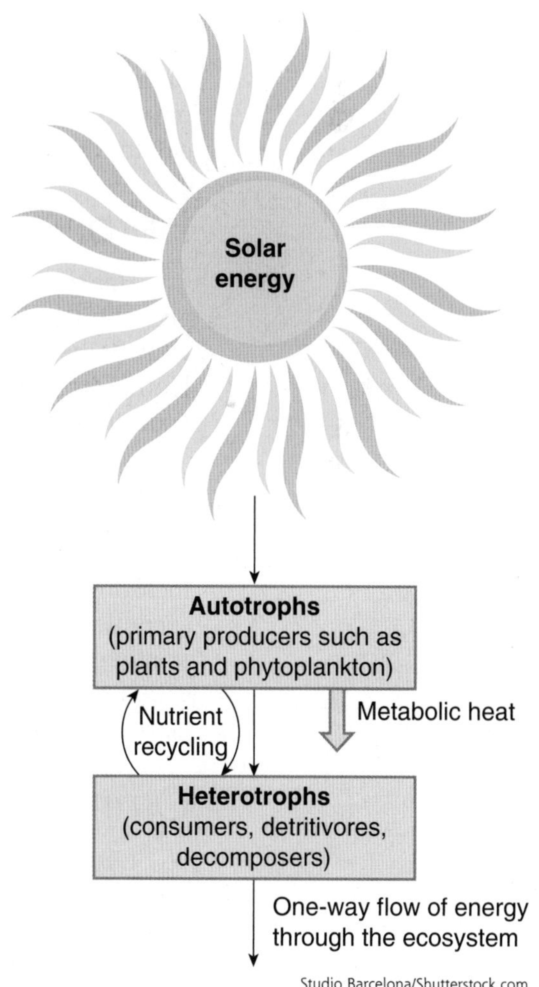

Studio Barcelona/Shutterstock.com

Figure 14-41. Energy flows from the sun to the environment in a typical ecosystem. It passes through autotrophs first, and then through heterotrophs.

Peter Kai/Shutterstock.com

Figure 14-42. Termites feeding on dead wood are an example of decomposers at work. *Why do termites feed on the wood in homes?*

manure would pile up in pastures around the countryside.

The termites, shown in **Figure 14-42**, are also decomposers. Many of us do not think of our homes as being constructed from dead plant material, but if they are built of wood, they are. So when termites are eating your framing and siding, they are just doing what nature intended—decomposing dead wood.

Feeding Relationships

As we just discussed, all living organisms rely on one another for food and energy. Most animals have a preference for their feeding habits. Some prefer to eat plants, while others eat only other animals. And then there are some animals, like humans, who will eat anything that they can catch. As the old adage goes, "Eat or be eaten!"

Herbivores

Herbivores are animals that eat only plants. Examples of herbivores are rabbits, deer, and bison. Herbivores are primary consumers in a food web. Some herbivores are small, like mice, while others are large, like bison. Herbivores have adapted unique anatomical features, like hooves and quick running speeds, as means of escaping from secondary consumers.

Carnivores

Carnivores are animals that eat other animals. Examples of carnivores include snakes, owls, and wolves, as shown in **Figure 14-43**. Carnivores are secondary, tertiary, and apex consumers in a food web. An animal that eats another animal is called the ***predator***. The animal that is the food animal is called the ***prey***. Ecologists often study the predator-prey interactions in an ecosystem to determine the health of animal populations.

Omnivores

Omnivores are animals that eat both plants and animals. Opossums are one of the most common species of omnivores. Omnivores can be primary, secondary, or even tertiary consumers in the food chain. Bears and humans might be among the few examples of omnivores that are apex consumers.

Detritivores

Detritivores are organisms that feed on dead matter. All detritivores are also decomposers. Examples of detritivores include millipedes, earthworms, dung beetles, starfish, fungi, and bacteria.

Ihar Byshniou/iStock/Thinkstock

Figure 14-43. Wolves are examples of carnivores. They are also apex consumers since they occupy the top of the food chain. The wolves in this picture are predators, and the bison is the prey.

Scavengers

Scavengers are organisms that feed on dead plants and animals. An example of a scavenger is shown in **Figure 14-44**. Scavengers may be either herbivores or carnivores, depending on whether they feed on plants or animals. While eating dead, decaying, leftover plants or animals may seem disgusting to us, scavengers are important to the food web and to the environment. Vultures, blowflies, raccoons, and wolves are examples of scavengers. Some of these species will also eat live animals.

Vinay Sinha/Shutterstock.com

Figure 14-44. Scavengers feed on dead plants and animals. *In addition to feeding on dead plants and animals, what other roles do scavengers play in an ecosystem?*

Parasitism

Parasitism is a unique feeding relationship where one species, the *parasite*, lives in or on, and feeds on another species, the ***host***. So, a ***parasite*** is an organism that lives in or on another species and consumes it for food. Parasites are classified as decomposers. An example of a parasite that is commonly found in mammals is a tapeworm, **Figure 14-45A**. An example of a plant parasite is mistletoe, **Figure 14-45B**. Other examples of parasites include many viruses and bacteria.

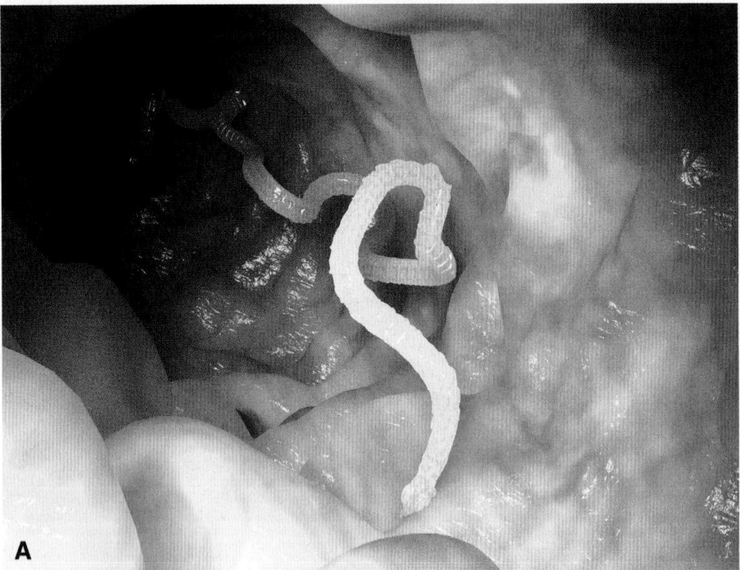

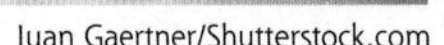

Juan Gaertner/Shutterstock.com

LianeM/Shutterstock.com

Figure 14-45. A—A tapeworm living inside an intestine is an example of a parasite. B—Mistletoe is another example of a parasite. Mistletoe is a parasitic plant that grows on trees.

Food Webs and Land

Another important component of food webs is the land on which all levels of consumers live. ***Carrying capacity*** is the maximum population of a species that a parcel of land can sustain indefinitely. Carrying capacity depends on available habitat, range, food, shelter, and water. Most communities of plants and animals coexist in a relatively delicate and dynamic balance. The balance is delicate in that the introduction of new species can oftentimes throw off the balance of both higher and lower food chain species. The balance is dynamic because the numbers of species are constantly changing, depending on weather patterns, migrations, and changes in individual species numbers.

Carrying capacity may also change if the food supply increases or decreases substantially. The quality and quantity of water in a habitat can also influence the carrying capacity. For example, in times of drought, the carrying capacity in a drought-stricken area will be reduced. A species that can migrate toward abundant water will leave the drought-stricken area since the area can no longer support it. See **Figure 14-46**.

Species Diversity

As stated previously, most communities of plants and animals coexist in a relatively delicate and dynamic balance. The introduction of a new plant or animal into an ecosystem may be extremely destructive to the native plant and animal species, and even to the health of the environment.

Native Species

Native species are the species of plants and animals that are naturally occurring in an area. When the first Europeans arrived in America, they

Johan Swanepoel/Shutterstock.com

Figure 14-46. In times of drought, the carrying capacity in a drought-stricken area will be reduced.

encountered native people who used hunting and gathering as their primary means of sustenance. To supplement their diets, these people also cultivated native plant species and grew gardens of squash, beans, and corn. These three plants were known as the *Three Sisters* and provided the foundation of the diet for many Native Americans.

Invasive Species

Invasive species are species that are not naturally occurring in a particular ecosystem or locale, but have been introduced primarily by humans. All invasive species start out as ***introduced species***, meaning that they were introduced into an area where they were not naturally occurring. The introduction may have been intentional or accidental. For example, Europeans intentionally introduced the house sparrow to the Americas because they thought it would rid the cities of some insect infestations, **Figure 14-47**. On the other hand, common and Norway rats were introduced to the Americas as stowaways on ships traveling across the Atlantic Ocean from Europe.

Invasive species, both plants and animals, share traits that enable them to outcompete native species. Invasive species generally have:

- Rapid development
- Short reproduction cycles
- High dispersal ability
- Environmental tolerance
- Flexible food or nutrition tolerance

Rob Christiaans/Shutterstock.com

Figure 14-47. Unfortunately, through repeated introductions, the house sparrow finally became established as an invasive species. *Is it possible to introduce a new species to an area for a specific purpose and prevent it from edging out native species?*

Although many species may be introduced to an area, relatively few become invasive enough to compete with native species for food, water, and space.

Rapid Development

In order for introduced or invasive species to outcompete native species, their young must develop and reach sexual maturity rapidly. If a species is slow to develop defenses and slow to reproduce, then native species can either outcompete or even consume the introduced species.

Short Reproduction Cycles

To become truly invasive, it is beneficial for an introduced species to have a short reproduction cycle. If a species can reproduce early in its development and also go through several reproduction cycles quickly, it can overwhelm native species with sheer numbers.

High Dispersal Ability

High dispersal ability refers to the introduced species ability to produce many offspring. The introduced species benefits from producing many offspring and many generations of offspring quickly because it can easily compete with native species for food, water, and habitat. When a species produces many offspring, they can spread across a wide area, reach sexual maturity quickly, and outcompete native species.

STEM Connection Invasive Species

What other invasive species have entered the United States? Research invasive species, such as African killer bees and fire ants, in the United States. Find out where they came from, how they arrived, the type of damage they are causing, and what is being done to control or eliminate them.

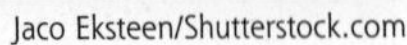

Jaco Eksteen/Shutterstock.com

Elliotte Rusty Harold

What types of problems are African bees and fire ants causing in the United States?

Environmental Tolerance

All species are native to some part of the world. If a new environment is not similar to its native environment, an introduced species must quickly learn about and adapt to new foods, water sources, and predators. It must also develop tolerance to weather patterns, precipitation, and temperatures in the new habitat.

Flexible Food or Nutrition Tolerance

An introduced species must often change its diet to meet the offerings of a new habitat. An introduced species that is unable to change its diet and find a substantial food source is not likely to survive, let alone become invasive. It appears that gypsy moths are a good example of an insect with a varied diet. Gypsy moths feed on many trees, including oaks, apples, sweetgum, birch, poplar, willow, and cherry trees.

Invasive Species in the United States

Unfortunately, several examples of invasive species exist in the United States. Some were introduced with good intentions, like serving as livestock feed or erosion control, and others were introduced unintentionally as stowaways on cargo ships or as released, unwanted pets. However, all invasive species share the common traits of being difficult to eradicate and harmful to native species and ecosystems.

Burmese Python

The Burmese python is one of the more recent and most destructive invasive species in the United States, **Figure 14-48**. Burmese pythons have become invasive in the Florida Everglades located primarily in the southern half of the state. These pythons were not introduced intentionally. They were introduced unintentionally when their owners carelessly released the unwanted pets into the wilds of the Florida Everglades.

The Everglades provide a habitat and food sources similar to the snake's natural habitat in Southeast Asia. Most of the pythons were fully developed adults when they were released into the Everglades, and since they are apex predators, almost nothing feeds on them. They are larger and more agile hunters than the common and relatively small herbivores and carnivores, like birds, foxes, rabbits, and other rodents, in the Everglades. Attempts to eradicate the species have proven unsuccessful.

Heiko Kiera/Shutterstock.com

Figure 14-48. Burmese pythons are invasive reptiles now threatening the native animal species in the Florida Everglades. As apex consumers, these snakes have no natural predators and can hunt almost all other animals in the Everglades. *How difficult is it to rid an area of an invasive species?*

Vladyslav Danilin/Shutterstock.com

Figure 14-49. Asian carp are invasive fish species that are threatening the Great Lakes. They grow rapidly and eat nearly all of the plankton in a body of water. *How successful have government efforts to control the spread of Asian carp been to date? Has any eradication method proved more successful than another?*

Asian Carp

Asian carp are a fast-growing, aggressive group of fish species that have invaded the Mississippi River and its tributaries, **Figure 14-49**. Asian carp include silver, bighead, and grass carp. Asian carp are native to Eastern Asia, where they have been domesticated for hundreds of years and are an important part of the diets of many Asian people.

In the 1970s, Asian carp were introduced to Arkansas fish farms to naturally filter pond water. They escaped the contained ponds with floodwaters in the 1980s and migrated to the Mississippi River. Asian carp have been found in more than 23 rivers in the United States.

Asian carp have no natural predators in North America, and the females can lay nearly a half million eggs each time they spawn. So, with no predators and the ability to reproduce rapidly, it is easy to see how Asian carp became invasive.

Currently, they have not infiltrated the Great Lakes, and ecologists hope to prevent them from gaining a foothold there. If Asian carp gain a foothold in the Great Lakes, scientists believe it will be almost impossible to eradicate them. Scientists also believe that Asian carp will eat nearly all of the plankton in the Great Lakes, thus starving many of the native species that thrive on the plankton.

Lynn Bunting/istock/Thinkstock;

Figure 14-50. Brown marmorated stink bugs are invasive insects from Asia that feed on vegetables, tree fruits, and vineyards. They cause damage to tree fruits, such as apples. This fruit is not marketable.

Brown Marmorated Stink Bugs

Brown marmorated stink bugs are an invasive insect that feeds on many fruits and vegetables over a wide range of territory in the United States. They were not intentionally introduced and probably arrived on fruit shipped from Asia. Annually, these invasive insects damage millions of dollars of vegetable crops, tree fruits, ornamental plants, and even vineyards, **Figure 14-50**. Just a few stink bugs can ruin hundreds of gallons of wine.

In their native habitat, parasitic wasps control the populations of stink bugs. Lacking this natural predator, the stink bugs have reproduced quickly and become invasive. Stink bugs reproduce quickly because they typically produce two generations per year.

Feral Pigs

As explained in *Lesson 12.1*, feral animals are animals living in the wild but descended from domesticated animals. In most instances, such as with feral pigs, feral animals are considered an invasive species. Feral pigs plague ranchers and farmers across the United States by eating crops, tearing down fencing, damaging and killing tree seedlings, and causing extensive damage to wetlands and other natural areas. Feral pigs may also be referred to as wild boars or wild pigs.

Feral pigs are aggressive animals, and their prolific breeding and natural rooting and wallowing behaviors make them a serious ecological, economic, and health threat. Although they prefer habitats with dense brush, marsh vegetation, and an abundant water supply, feral pigs are highly adaptable to most areas and are currently found in at least 45 states. In addition to eating crops, feral pigs will eat eggs, nuts, berries, roots, small mammals, and even reptiles. As their numbers increase, they outcompete the native fauna by consuming the natural mast and also destroy habitats for many ground-dwelling animals. In hot weather, feral pigs seek out ponds, springs, and streams in which to wallow. Their wallowing muddies the water and often destroys aquatic vegetation, which, in turn, contributes to erosion and water contamination.

Although there are some natural predators, their numbers are not sufficient to control the increasing feral swine population. In the attempt to curb the population growth, some states allow boar hunting with few restrictions.

Kudzu

Kudzu is an invasive plant species found in the southeastern part of the United States. See **Figure 14-51**. More than two million acres of land are covered with kudzu. Kudzu is one of our older invasive species. It was introduced in Philadelphia in 1876 as a forage crop for livestock and an ornamental plant for gardeners. After the Dust Bowl in the 1930s, the United States Department of Agriculture encouraged farmers to plant kudzu as a means of controlling soil erosion.

Kudzu competes with native trees and shrubs for space and sunlight. It forms a dense tangle of growth with large leaves that covers established plants and prevents sunlight from reaching them. Kudzu grows quickly, up to one foot per day in mature plants. It also reproduces via two highly productive means: vegetatively via stolons and sexually via seeds.

One of the means of controlling kudzu is to graze livestock on land infested with kudzu. Kudzu is nutritious and especially tasty to goats.

Rob Hainer/Shutterstock.com

Figure 14-51. Kudzu is an invasive terrestrial plant in the southeastern part of the United States. Kudzu grows quickly and overruns everything in its path. *Is there any Kudzu growing near your home?*

Water Hyacinth

Water hyacinth is a warm-water plant that is native to the Amazon River basin in South America. It is a free-floating aquatic plant, **Figure 14-52**. Water hyacinth is a perennial, so it grows back each year.

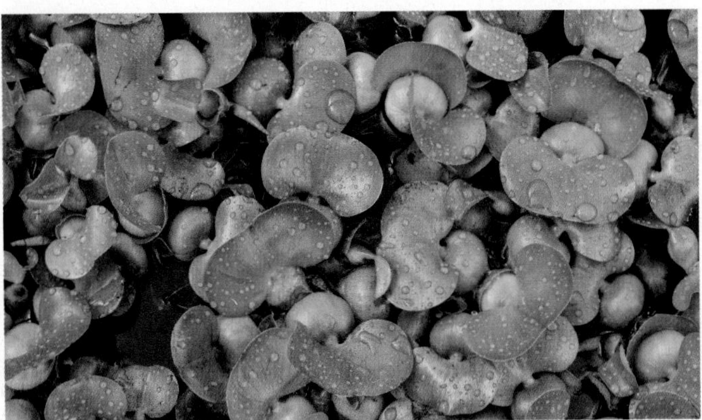

Zigzag Mountain Art/Shutterstock.com

Figure 14-52. Water hyacinth is an invasive aquatic plant that chokes the oxygen out of lakes, blocks sunlight, and provides an ideal breeding ground for mosquitoes.

Water hyacinth was introduced to the United States in 1884 in New Orleans. It was given freely to visitors to the World's Fair. Once it became established and invasive in Louisiana, several attempts were made to eradicate it. Policymakers even went so far as to envision introducing hippopotamuses to eat the plant. Luckily, the legislation to introduce hippos failed.

Like kudzu, water hyacinth reproduces via stolons. Water hyacinth also produces thousands of seeds each year. These multiple modes of reproduction, plus the plant's ability to grow quickly, translate into its ability to double its mass in as little as two weeks in the heat of the summer.

Water hyacinth can quickly cover an entire lake or pond. When it does so, it blocks sunlight, starves the water of oxygen, and may choke the flow of water. Blocking sunlight also prevents light from reaching native plants. When the water hyacinth deprives the water of oxygen, fish, frogs, turtles, and other aquatic species cannot live. The dense mat of vegetation produced by water hyacinth also provides an ideal breeding space for mosquitoes.

Additional examples of invasive species running rampant in the United States and along its coasts include Zebra Mussels, Multiflora Rose, and Johnsongrass. You can likely find other examples, as well.

Career Connection Global Positioning Systems Technician

Job description: A global positioning systems technician works with GPS technologies and even drones to map agricultural and natural areas. They use GPS and photo technologies to map soils, plant life, animal life, and even water usage by crops and forests. A GPS technician may map the same area several times over the growing season or year to determine changes in the available water, plant life, or animal life. They gather information and present it to other scientists and agriculturists to make recommendations about agricultural and natural resources.

Education required: A two- or four-year degree in agriculture or related sciences is required.

Job fit: This job may be a fit for you if you like working with GPS technology and drones for gathering information that is useful for agriculture and natural resources.

Henryk Sadura/Shutterstock.com

A GPS technician may also be trained as a surveyor to determine slopes and mapping of land for agricultural and natural resources purposes.

Lesson 14.4 Review and Assessment

Words to Know

Match the key terms from the lesson to the correct definition.

1. Species that feed on secondary consumers.
2. The maximum population of a species that a parcel of land can sustain indefinitely.
3. An organism that breaks down dead plant and animal material.
4. An organism that lives in or on another species and consumes it for food.
5. An organism that feeds on dead matter.
6. Species that feed only on plants.
7. The plants and animals that are naturally occurring in an area.
8. A plant that produces its own food.
9. An organism that eats other organisms.
10. An animal that eats only plants.
11. Animals or plants that are not naturally occurring in a particular ecosystem or locale, but have been introduced primarily by humans.
12. A simple model of how energy flows from one living organism to another.
13. The species at the top of the food chain.
14. An organism that eats both plants and animals.
15. An animal that is the food animal for predators.
16. An organism that feeds on dead plants and animals; may be an herbivore or carnivore.
17. Species that feed on primary consumers.
18. An organism that eats other animals.

A. apex consumer
B. autotroph
C. carnivore
D. carrying capacity
E. consumer
F. decomposer
G. detritivore
H. food chain
I. herbivore
J. invasive species
K. native species
L. omnivore
M. parasite
N. prey
O. primary consumer
P. scavenger
Q. secondary consumer
R. tertiary consumer

Know and Understand

Answer the following questions using the information provided in this lesson.

1. The interaction of many food chains is described as a(n) _____.
2. Briefly explain how organisms in a food web are interdependent.
3. Producers use a process called _____ to use the sun's solar energy to manufacture simple sugars.

4. Consumers gain energy by _____.
   A. using the sun's energy
   B. eating other animals or plants
   C. decomposing dead animal and plant materials
   D. photosynthesis
5. The group of consumers found at the top of the food chain is called _____.
6. Explain how producers, consumers, and decomposers are different from one another.
7. A parasite lives in or on another organism, called a(n) _____.
   A. omnivore
   B. detritivore
   C. host
   D. decomposer
8. Carrying capacity depends on all of the following, except _____.
   A. food and habitat
   B. shelter and water
   C. range or space
   D. reproductive ability
9. Species not naturally found in a particular ecosystem are called _____.
10. What are the factors that enable a species to become invasive to a new region?

Analyze and Apply

1. Hunters are interested in carrying capacity. What might a landowner do to increase the carrying capacity of his or her land? Which factors can he or she influence? How would these increase the carrying capacity?
2. What are the native species of plants and animals in your area? What has happened to the populations of these species over time? What factors have contributed to these trends?

Thinking Critically

1. There are several factors that limit the number of species that can live in an ecosystem. Why does each of the factors limit the carrying capacity in an ecosystem?
2. In addition to the characteristics of invasive species that were described in this lesson, what are other factors that enable an introduced species to become invasive?

Chapter 14

Review and Assessment

Lesson 14.1

Ecosystems

Key Points

- An ecosystem is a community of organisms living with and interacting with one another and their environment.
- There are two types of ecosystems: aquatic and terrestrial.
- Marine ecosystems, classified as aquatic, cover more than 70% of Earth's surface.
- Ecosystems are made up of biotic factors (living things) and abiotic factors (nonliving things).
- Ecosystems are vital to the health and well-being of wildlife of all kinds.
- Ecosystems and the flora and fauna found within them clean the air we breathe and the water we drink.
- Ecosystems refresh Earth's supply of oxygen and cycle nutrients for future generations of living organisms.
- The United States contains a diversity of ecosystems including deserts, mountains, wetlands, and forests.
- America's Midwest is one of the most fertile areas on the planet, and it was once classified as a grassland ecosystem.

Words to Know

Use the following list and the textbook glossary to review and study the *Words to Know* from *Lesson 14.1*.

abiotic factor
aquatic ecosystem
biotic factor
brackish water
community
coral reef
deciduous tree
desert ecosystem
dissolved oxygen (DO)
ecology
ecosystem
estuary
evergreen tree
fauna
flora
food chain
food cycle
food web
forest ecosystem
freshwater ecosystem
grassland ecosystem
habitat
jungle
lentic ecosystem
lotic ecosystem
marine ecosystem
mountain ecosystem

prairie ecosystem
riparian zone
saline
salt marsh
taiga ecosystem
temperate deciduous forest
temperate evergreen forest
terrestrial ecosystem
transpiration
tropical deciduous forest
tropical evergreen forest
wetland

Check Your Understanding

Answer the following questions using the information provided in *Lesson 14.1*.

1. Explain how the depth of water in lentic ecosystems affects their plant life.
2. Why do lotic ecosystems usually have more dissolved oxygen than lentic ecosystems?
3. *True or False?* Forest ecosystems are the most diverse of the terrestrial ecosystems.
4. How does a mountain range affect the surrounding environment?
5. *True or False?* Coral reefs support a diverse and abundant variety of life.

Lesson 14.2

Ecological Cycles

Key Points

- A biogeochemical cycle is the cycling of biological and chemical substances in any of Earth's ecosystems. Biogeochemical cycles include the water cycle, the nitrogen cycle, and the carbon cycle.
- Agricultural plants and animals require water, carbon, and nitrogen to grow and produce abundant harvests.
- Carbon is one of the more prevalent greenhouse gases, so agriculturists and naturalists try to find carbon sinks that will sequester carbon for many years, thus reducing the impact of greenhouse gases.
- Ecological cycles are important to sustain life on Earth.
- Farmers and ranchers in the American Midwest are concerned with the depletion of aquifers, especially the Ogallala Aquifer.
- Legumes fix nitrogen from the atmosphere into ammonium, which can be used by plants.
- The total amounts of water, carbon, and nitrogen on the planet are relatively stable.
- The water, carbon, and nitrogen cycles are among the more important ecological cycles.
- Water cycles among solid (ice), liquid, and gaseous (vapor) phases on Earth. Most water on Earth is contained in the oceans.

Words to Know

Use the following list and the textbook glossary to review and study the *Words to Know* from *Lesson 14.2*.

acidification
acid rain
algae bloom
ammonification
aquifer
biogeochemical cycle
biosphere
blue baby syndrome
carbon credit system
carbon cycle
carbon dioxide (CO_2)
carbon pool
carbon sink
carbon source
cycle
denitrification
desalinate
ecological cycle
fixation
freshwater
groundwater
hydrological cycle
legume
methane (CH_4)
methemoglobinemia
nitrate
nitrification
nitrogen cycle
runoff water
sediment
surface water
water cycle
water rights
water-soluble
zone of life

Check Your Understanding

Answer the following questions using the information provided in *Lesson 14.2*.

1. *True or False?* For an ecosystem to be functioning in a healthy way, energy, water, and nutrients must cycle between plants, animals, and soil.
2. List three important biogeochemical cycles.
3. What is solid water and where can it be found on Earth?
4. *True or False?* More than 60% of the water on Earth is freshwater.
5. Explain why the use of desalination is not more commonly used to make seawater suitable for human, animal, and plant consumption.
6. Explain how the depletion of aquifers would affect production agriculture.
7. What are the primary sources of river and stream pollutants in the United States?
8. Briefly explain how acid rain negatively affects waterways and aquatic life.
9. Nitrogen in the soil _____.
   A. is a primary source of food for many animals
   B. affects the rate of decomposition
   C. affects plant growth
   D. All of the above.
10. _____ is the process through which atmospheric nitrogen is transformed into ammonium.
11. *True or False?* Denitrification, the return of nitrogen to the atmosphere, completes the nitrogen cycle.
12. How do high concentrations of nitrates affect waterways?

13. Explain the difference between carbon sinks and carbon pools.
14. *True or False*? A place that releases more carbon than it stores is a carbon source.
15. The primary means for the atmosphere to rid itself of carbon dioxide is through _____ in plants.
16. *True or False*? Methane is emitted by natural sources as well as by industrial activities.
17. What is the primary agricultural source of methane?
18. *True or False*? Carbon enters the oceans as it dissolves out of the atmosphere.
19. Why are manufacturers issued carbon credits?

Lesson 14.3

The Influence of Climate on Agriculture

Key Points

- Climate is the long-term weather patterns of a region.
- Several factors influence climate, including latitude, altitude, and proximity to large bodies of water.
- Chemicals in the atmosphere, such as carbon dioxide and methane, trap the sun's solar energy near Earth's surface, thus elevating Earth's surface temperatures. This greenhouse effect contributes to global climate change.
- Agricultural practices have influenced the climate in the past and will be influenced by the climate in the future.

Words to Know

Use the following list and the textbook glossary to review and study the *Words to Know* from *Lesson 14.3*.

air masses
altitude
arctic zone
climate
convection current
equator
front
greenhouse effect
greenhouse gases
jet stream
lake effect snow
latitude
polar zone
temperate zone
terrain
topography
tropical zone
tropics
uneven solar heating
weather
weather pattern

Check Your Understanding

Answer the following questions using the information provided in *Lesson 14.3*.

1. Explain the difference between climate and weather.
2. List four factors that influence the climate of an area.
3. Explain why some areas of Earth receive more sunlight than others.
4. *True or False?* The seasons in the Southern Hemisphere are somewhat opposite those in the Northern Hemisphere and are generally much milder.
5. Under what conditions does a warm air mass form? Under what conditions does a cold air mass form?
6. Explain how the jet streams affect the weather patterns on Earth.
7. How does the cooling of air blowing across a mountain range affect the precipitation on the mountain?
8. Why does the terrain of an area greatly affect its climate?
9. Explain the moderating effect of larger bodies of water on nearby land.
10. How does the greenhouse effect help Earth? How does it harm Earth?
11. Identify the four primary greenhouse gases.
12. Explain how the agricultural practices of early pioneers greatly affected the atmosphere.
13. *True or False?* Today's agricultural practices no longer have a negative affect on Earth's climate.
14. Identify three environmentally friendly agriculture methods used by today's agriculturists.
15. What are four ways today's temperature extremes affect agriculture?
16. How will the anticipated precipitation extremes affect agriculture?

Lesson 14.4

Animals, Plants, and Geography

Key Points

- The cycling of energy is one of the most important cycles in any ecosystem.
- Plants, animals, insects, and even microorganisms interact with one another, often in predatory relationships to cycle energy in the ecosystem.
- Plants produce energy via photosynthesis and solar energy.
- Consumers eat plants and other animals, and when they die, decomposers move the living organism's energy back into the soil and atmosphere.

- Food webs are diagrams that explain the interdependency of plants and animals on one another for food.
- Producers, consumers, and decomposers interact to make up food webs.
- Consumers are classified by what they eat. They are classified as herbivores, carnivores, omnivores, scavengers, and parasites.
- All ecosystems have a definite carrying capacity based on the available food, water, shelter, and space in the ecosystem.
- Not all species in an ecosystem are native to the region; some have been introduced via human interactions in the ecosystem.
- Plants produce energy via photosynthesis and solar energy.

Words to Know

Use the following list and the textbook glossary to review and study the *Words to Know* from *Lesson 14.4*.

apex consumer
apex predator
autotroph
carnivore
carrying capacity
consumer
decompose
decomposer
detritivore
food chain
food cycle
food web
herbivore
heterotroph
host
introduced species
invasive species
native species
omnivore
parasite
parasitism
predator
prey
primary consumer
producer
scavenger
secondary consumer
tertiary consumer

Check Your Understanding

Answer the following questions using the information provided in *Lesson 14.4*.

1. *True or False?* Each species is part of an ecosystem and fits somewhere in the flow of energy in that ecosystem.
2. The interaction of many food _____ is described as a food web or food cycle.
3. What are the three basic categories of organisms in a food web?
4. Producers, or autotrophs, _____.
   A. produce their own food
   B. use photosynthesis to live, grow, and reproduce
   C. are part of the carbon cycle
   D. All of the above.
5. *True or False?* Consumers, or heterotrophs, get their food from other living organisms.
6. *True or False?* Owls and coyotes are examples of primary consumers.

7. Secondary consumers include animals such as _____.
   A. raccoons and opossums
   B. cougars, sharks, and eagles
   C. deer, bison, and rabbits
   D. All of the above.
8. Omnivores are animals that eat _____.
   A. only plants
   B. only other animals
   C. both plants and animals
   D. only fish
9. *True or False?* All detritivores are also decomposers.
10. Why are scavengers important components of a food web?
11. Why would the carrying capacity of a particular area be affected by the quality and quantity of water available?
12. *True or False?* The species of plants and animals that are naturally occurring in an area are native species.
13. Invasive species generally have _____.
    A. flexible food or nutrition tolerance
    B. high dispersal ability
    C. environmental tolerance
    D. All of the above.

Chapter 14 Skill Development

STEM and Academic Activities

1. **Science.** Find a river or stream in your community. What is the quality of water in that stream? How does it vary over the course of a year? Ask your agriculture teacher if he or she has scientific tools to analyze the water quality. Then, monitor the health of the water on a weekly basis over several months, or even an entire year. How does water quality change with time? What are the causes of these changes?
2. **Technology.** Research the predominant ecosystems in your school district or county or state. Use satellite technology to examine changes in the local ecosystems. Have the ecosystems changed? If so, when did the ecosystems begin to change? What caused those changes? What other ecological information can you obtain from satellite technology?
3. **Technology.** Research ways of desalinating water. Use two or three different methods to desalinate a gallon of saltwater. Which method was most efficient? Which was least efficient? Why?

4. **Engineering.** How does field tile work to drain a wetland? Create a model of a wetland and draining that wetland with common materials found in your home.
5. **Math.** Create a pie chart showing the percentages of Earth's water that are contained in aquatic and freshwater ecosystems.
6. **Social Science.** Research invasive species other than the ones described in this chapter. How did the species get introduced in America? What makes the species invasive? Where is the species a problem? What is being done to combat the invasive species? How can we control the introduction of new species into our natural habitats? Should people be held responsible for releasing pets that may become, or contribute to, an invasive species?
7. **Social Science.** Create a water law for your community. Research where your community gets its water and how residents affect the water supply. Using the information, create a water law that you believe to be fair. Present your water law to the class. Be prepared to defend your law.
8. **Language Arts.** What impact is agriculture having on the Ogallala Aquifer? Research the effects of irrigation water use on the Ogallala Aquifer. Use the information to write a short essay on the subject.

Communicating about Agriculture

1. **Reading and Writing.** Wetlands are an important aquatic ecosystem. Find a local wetland area, research it, and then prepare an online blog entry about preserving and/or restoring the wetland. Explain why wetlands in general, and this one specifically, are important.
2. **Reading and Speaking.** Water rights trigger heated debates among people from different parts of the country. Identify a water-based agricultural issue either in your state or your region of the country. Research that issue and follow the National FFA Organization's career development event guidelines for the Agricultural Issues Forum to prepare an analysis and presentation of the issue for local civic groups.
3. **Reading and Writing.** Create a food web with common plant, insect, animal, and fungi species in your school area. Design a poster that accurately depicts the trophic levels each of the species occupies.

Extending Your Knowledge

1. Design an experiment that analyzes the decomposition of common materials, such as vegetables or even manure, in different environments. How much time is required for decomposition in a "sterile" environment versus the schoolyard? What factors increase the speed of decomposition?
2. Examine the impact of saline water on plant growth. Design an experiment to evaluate the effect of water with various levels of salt on the growth of common agricultural crops.

3. Evaluate your school district. Where are the carbon sinks and sources? How might the sources be changed in order to reduce their production of carbon?
4. As we think about energy needed for providing us with food, let us think about the foods that we eat on a weekly basis. Where are the foods that you eat produced? Are they produced within walking distance of your home or school? If not, how are they transported from their fields to where you eat them? Are some of those foods produced overseas? How much energy is required just to transport the food from the field to your dinner plate? How does that fuel consumption affect the climate?
5. Analyze **Figure 14-37**. Which additional data could we analyze in order to see the effects of drought on agricultural crops? Research that data to see if they agree with the data in **Figure 14-37**.
6. Select a species of animal that is a consumer in your area. Determine the carrying capacity for that species. What are the limiting factors that keep the population in check?

Chapter 15

Soil and Water Management

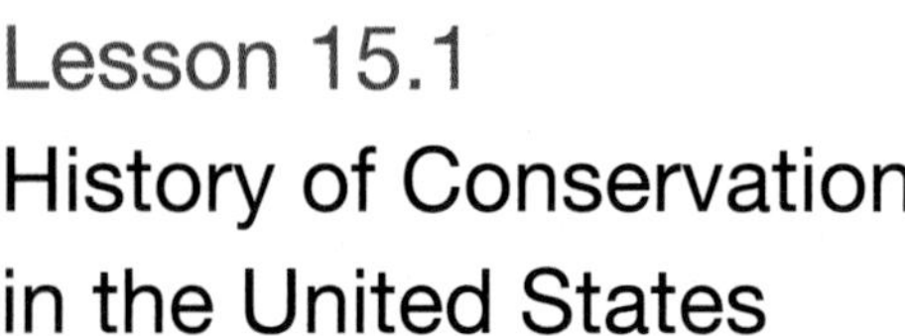

Lesson 15.1
History of Conservation in the United States

Lesson 15.2
Soil Formation and Properties

Lesson 15.3
Hydrological Cycles

Lesson 15.4
Water Quality

Lesson 15.5
Conservation Practices in Agriculture

©iStock/vovan13

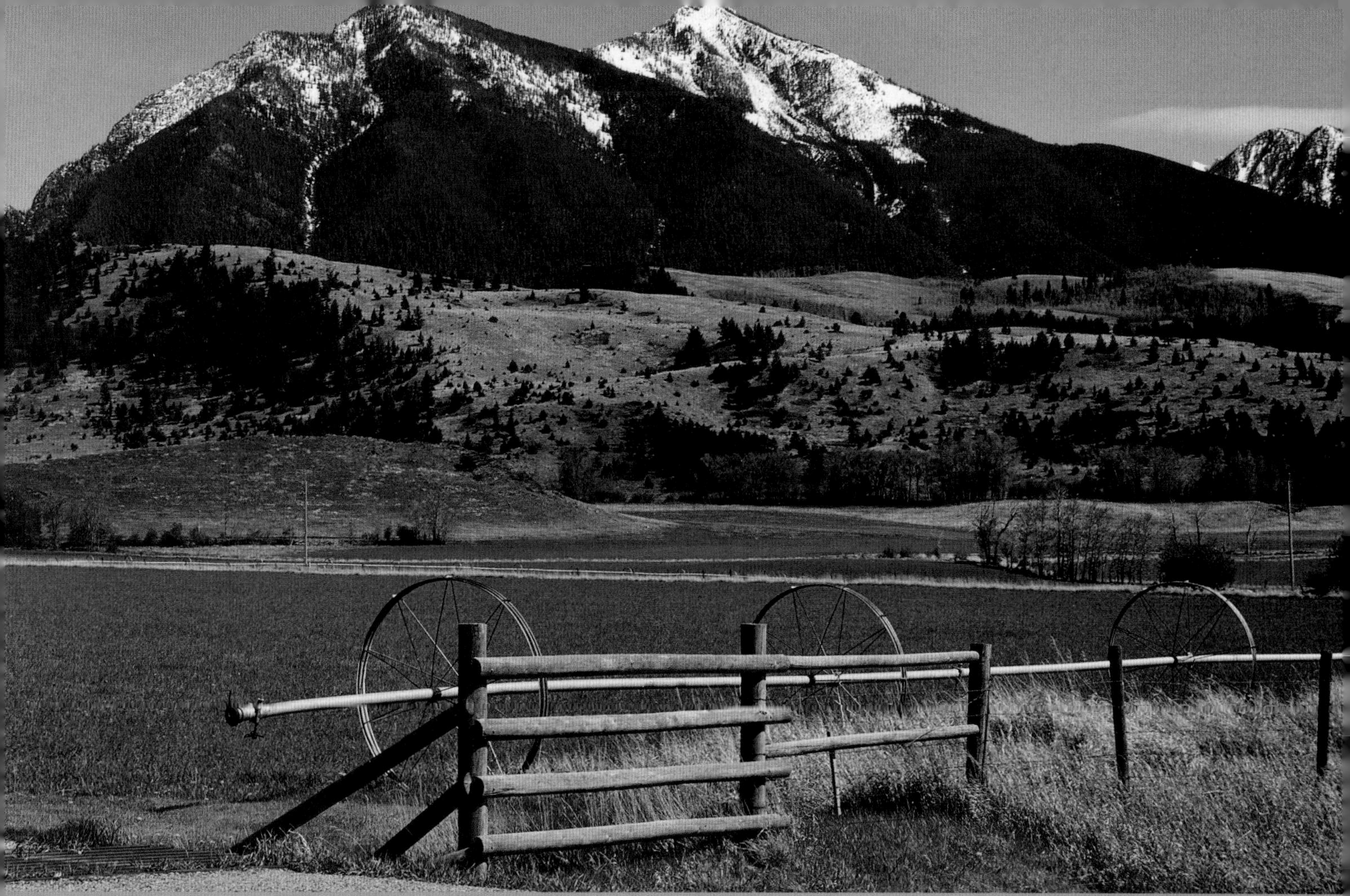

Sue Smith/Shutterstock.com

©iStock/steverts

©iStock/oticki

While studying, look for the activity icon to:

- **Practice** vocabulary terms with e-flash cards and matching activities.
- **Expand** learning with video clips, animations, and interactive activities.
- **Reinforce** what you learn by completing the end-of-lesson activities.
- **Test your knowledge** by completing the end-of-chapter questions.

Lesson 15.1

History of Conservation in the United States

Words to Know

aquifer
Civilian Conservation Corps (CCC)
Clean Air Act (CAA)
Clean Water Act (CWA)
conservation
conservation reserve program
DDT
Environmental Protection Agency (EPA)
National Association of Conservation Districts (NACD)
National Oceanic and Atmospheric Administration (NOAA)
Natural Resources Conservation Service (NRCS)
preservation
Soil and Water Conservation District (SWCD)
soil conservation practice
United States Forest Service (USFS)

Lesson Outcomes

By the end of this lesson, you should be able to:

- Describe early conservationists and explain their contributions to the conservation movement.
- Discuss the Dust Bowl as a major agricultural and environmental event that influenced both agricultural and environmental policies and practices.
- Compare and contrast the ideas of conservation and preservation.

Before You Read

Read the lesson title and tell a classmate what you have experienced or already know about the topic. Write a paragraph describing what you would like to learn about the topic. After reading the lesson, share two things you have learned with the classmate.

The conservation movement in the United States has its roots in the late 1800s. Curiously, this is approximately the same time frame as the Industrial Revolution. During the Industrial Revolution, large factories sprang up in cities around the nation. Because of jobs and the promise of a better life, millions of people moved out of the countryside and into cities. Thus, cities became crowded, air quality in cities declined, and people began to cherish wide-open, pure spaces of natural beauty.

Parks were incorporated in greater focus in city planning. And, influential writers and policymakers began to think about setting aside natural areas to preserve them for future generations of Americans. In this lesson, we will discuss the impact of some of these influential leaders on conservation, and learn about the history of the conservation movement in the United States.

Early Conservationists in America

During the late 1800s, three ideas came to the forefront of Americans' minds with regard to natural resources. First, since wood was a primary building material, and because we were cutting down thousands of acres of forests for lumber and farmland, people began to consider the idea that the country may run out of wood and other natural resources. Secondly, people began to ponder the future of wild spaces. They wondered if any wild spaces would be left for future generations to visit and appreciate. Thirdly, because of the rapidly growing cities, people began to experience pollution, especially of the air and water.

Henry David Thoreau

One of the initiators of the conservation movement in the United States was Henry David Thoreau. You have probably read about Henry David Thoreau in your English or literature class. He was a noted writer, poet, and naturalist who lived in the United States in the 1800s. In 1860, Thoreau presented a speech to a livestock show of the Middlesex Agricultural Society in Massachusetts. In his presentation, Thoreau explained how seeds grew from trees in the forest. In the end, he encouraged farmers to plant trees in their own woodlots.

John Muir

John Muir also contributed to conservationist thought in the 1800s. He was a naturalist and author, as well. John Muir wanted to see the American West, so he traveled to the Yosemite Valley and the Sierra Nevada Mountains. As a result of these travels and seeing the natural wonders, Muir cofounded the *Sierra Club* and eventually helped establish Yosemite National Park, the first national park, in 1872. The Sierra Club's mission was, and continues to be, protection of the wild spaces of America and advancing the conservation movement.

Teddy Roosevelt and the Conservation Movement

President Theodore (Teddy) Roosevelt is generally considered the "Father of the Conservation Movement" in the United States, **Figure 15-1**. President Roosevelt was an avid outdoorsman and hunter. He believed in the preservation of wild spaces and of the notion that our natural resources were not inexhaustible. Because of his ideas that we could potentially use up our natural resources, Roosevelt set aside millions of acres of forests and other wild areas to conserve them for future generations of Americans. In 1905, President Roosevelt created the United States Forest Service (USFS). The ***United States Forest Service (USFS)*** is an agency of the U.S. Department of Agriculture that administers the nation's national forests and national grasslands. During his presidency, Roosevelt greatly increased the amount of protected forests from approximately 43 million acres to nearly 200 million acres.

Underwood and Underwood/Library of Congress

Figure 15-1. Naturalist John Muir and President Theodore Roosevelt were influential leaders in the Conservation Movement in the United States. They are pictured here at Glacier Point in Yosemite Valley, California. ***How different would our country be if Roosevelt had not advocated for land and forest conservation?***

Conservation versus Preservation

While Roosevelt enacted laws and dedicated much land to National Parks and National Forests, his ideas on conservation were not accepted by everyone. Roosevelt believed in the idea of ***conservation***, meaning that natural resources were to be protected, but also used for public good, both now and with future generations. On the other side of the natural resources debate were people like John Muir who believed in preservation. ***Preservation*** means that natural resources are to be protected and not used.

For instance, a conservationist would set aside forested land and allow timber companies to harvest parts of that land. They would expect the timber company to use methods like selective harvesting and replanting tree seedlings. A preservationist would set aside forested land, but never allow a timber company to harvest trees. They would believe that the forest should be protected for future generations to enjoy in its natural state.

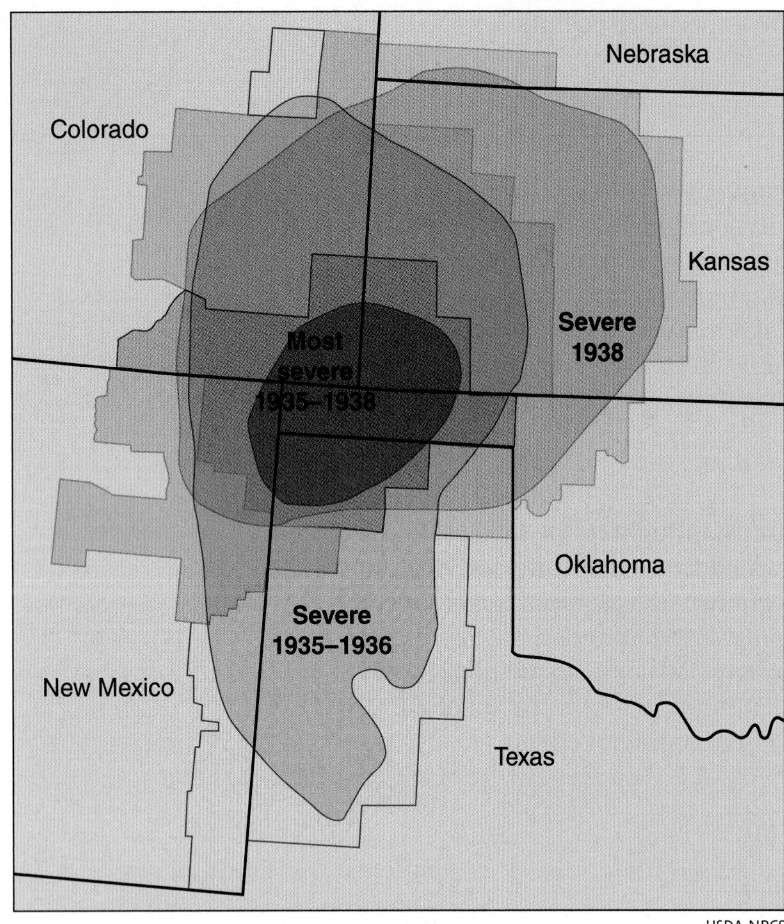

Figure 15-2. The Dust Bowl of the 1930s covered the Oklahoma Panhandle, the Texas Panhandle, southeastern Colorado, and western Kansas. The most severe areas impacted were in the Oklahoma Panhandle. *Have you ever seen images of farmers or families living during the Dust Bowl? What were your impressions of the images?*

Dust Bowl and the Great Depression

For agriculturalists, the conservation movement gained momentum in the years following the Dust Bowl. The Dust Bowl occurred during the 1930s, during the same era as the Great Depression. The Dust Bowl centered in the Oklahoma and Texas Panhandle area, but also covered a large expanse of the Great Plains, as shown in **Figure 15-2**. This area experienced repeated years of extreme drought. The extreme drought combined with the lack of soil conservation practices by farmers led to severe wind erosion of the soil and the drifting piles of dust seen in **Figure 15-3**. This erosion also produced immense dust storms.

Causes of the Dust Bowl

Beginning in the 1920s, the southern Great Plains experienced significant rainfall. Farmers then plowed under the native shortgrass prairies in order to plant corn, cotton, and wheat in the deep, fertile soil of the plains. When the rains were abundant in the 1920s, the land produced bumper (high yield) crops. During prosperous times, farmers plowed every acre they could. The relatively moist soils and covering of crops prevented the strong prairie winds from eroding the soil. As long as ample rains fell periodically, this region produced high yields and allowed farmers to use larger and larger machinery to plow the prairie soil.

Unfortunately, with the lack of advanced climatic data available, farmers were unaware that the abundant rains of the 1920s were not normal. In reality, this part of the United States was normally quite dry with annual rainfalls of less than 20″ per year and relatively frequent drought seasons.

What caused the Dust Bowl was the unusual pattern of drought year followed by drought year for most of the 1930s. Normally, the region might experience one or two years of drought, but those years were normally followed by abundant rainfall. Between 1930 and 1940, each of the 11 years in that span of time experienced at least 15%–25% lower rainfall than normal, and some years as much as one-half the normal rainfall.

Farming practices and the lack of soil conservation also contributed to the Dust Bowl. Farmers plowed thousands of acres of prairie grasses. In normal drought years, these prairie grasses held soil in place and prevented the prairie winds from loosening and eroding the soil. In years with abundant rainfall, crops held the soil in place. But, during drought years, crops failed and were unable to hold the soil in place.

Arthur Rothstein/Library of Congress

Figure 15-3. Winds dislodged soil particles and blew them into massive drifts of dust during the drought years of the Dust Bowl. *What would you imagine life would be like during a major dust storm?*

Impacts of the Dust Bowl

During the Dust Bowl, winds picked up immense volumes of topsoil, producing tremendous dust storms that sometimes lasted days. Instead of producing rain, these clouds simply blew dust everywhere, **Figure 15-4**.

USDA NRCS

Figure 15-4. Dust storms like this one were common during the 1930s in the area known as the Dust Bowl. *Are there areas in the world today experiencing dust storms? If so, what is the cause?*

John Hugh O'Neill/NRCS

Figure 15-5. Dust storms from the Great Plains blew across the country and produced massive dust storms in Washington, D.C. This is an image of a dust storm over the Lincoln Memorial on the Mall in Washington, D.C.

People working outside had to wear dampened handkerchiefs over their mouths to prevent breathing dust. People hung dampened towels over their windows to attempt to keep the dust out of their houses. The air was so dry and electrically charged that people tied chains to the undersides of their automobiles so that they would operate and to prevent severe static electric shock to the riders.

Still, these measures were not sufficient. Many people died of lung infections, such as dust pneumonia, caused by breathing the dust. Crops failed repeatedly. Crop failures forced farmers and their families to move to other areas of the country in order to reestablish their lives. You have likely read about the joys and, especially, discomforts of agricultural life of farmers during the Dust Bowl in books like *The Grapes of Wrath*.

The severe dust storms were of such immensity that they moved all the way across the country and deposited Texas and Oklahoma dust in Boston, New York City, and even Washington, D.C.! An image of a dust storm over the Lincoln Memorial in March 1935 is shown in **Figure 15-5**. The winter of 1933–34 produced red snowstorms in New England caused by the reddish topsoil dust from Texas and Oklahoma in the atmosphere. Some scientists estimate that as much as 75% of the topsoil was blown away from some areas of the Dust Bowl.

Conservation Outcomes of the Dust Bowl

As a result of most natural or human created disasters, we often search for ways to prevent a recurrence, or at least minimize the damage to our homes and environment should something similar occur. The Dust Bowl and the damage it caused to our environment spurred the government and agriculturists to implement conservation measures and practices to help prevent a similar catastrophic event.

Soil Conservation Service

When President Franklin D. Roosevelt was elected to office, one of his first acts was to establish the *Soil Erosion Service* in 1933. Within a few years, that office was transferred to the U.S. Department of Agriculture and renamed the *Soil Conservation Service*. More recently, it was renamed the ***Natural Resources Conservation Service (NRCS)***. It still exists today.

The purpose of the Soil Conservation Service was to educate farmers throughout the nation about soil conservation practices. A ***soil conservation practice*** is a method of farming that prevents the erosion of soil due to wind or water. Soil conservation practices include crop rotation, contour plowing, buffer strips, strip farming, grassed waterways, riparian zones, and reduced tillage. For examples of several soil conservation practices, including strip cropping, buffers, and contour farming, **Figure 15-6**. Soil conservation practices will be discussed in depth in *Lesson 15.4*. The NRCS also helps landowners enhance water supplies, improve water quality, increase wildlife habitat, and reduce damages caused by floods and other natural disasters.

Lynn Betts/NRCS

Figure 15-6. This image of farmland in Iowa provides examples of several soil conservation practices, including strip cropping, buffers, and contour farming. *What conservation or agricultural practices have you seen from the air?*

Civilian Conservation Corps

President Roosevelt also established the ***Civilian Conservation Corps (CCC)*** in 1933. It operated until 1942. The CCC was comprised of unmarried, unemployed men of 18–25 years of age who were employed by the federal government to install conservation practices throughout the nation.

One of their first actions was a project called the *Shelterbelt Project*. Starting in 1934 and concluding in 1942, men of the CCC planted nearly 220 million trees to help prevent wind erosion. The Shelterbelt Project was a 100-mile wide zone stretching from the Canadian border in the north to the Brazos River of Texas in the south, **Figure 15-7**. The idea was that this immense belt of trees would prevent the wind from blowing so hard across the plains and eroding the soil. The idea seemed to work as the trees grew and provided a natural buffer against the winds.

The CCC also worked to establish pathways, bridges, roads, and buildings in the nation's national parks. Members of the CCC educated farmers about soil conservation and methods to control erosion.

Farmlands Returned to Native Prairies

Beginning in 1937, the federal government started purchasing farmland in the Dust Bowl region. During the 1930s, the government purchased nearly

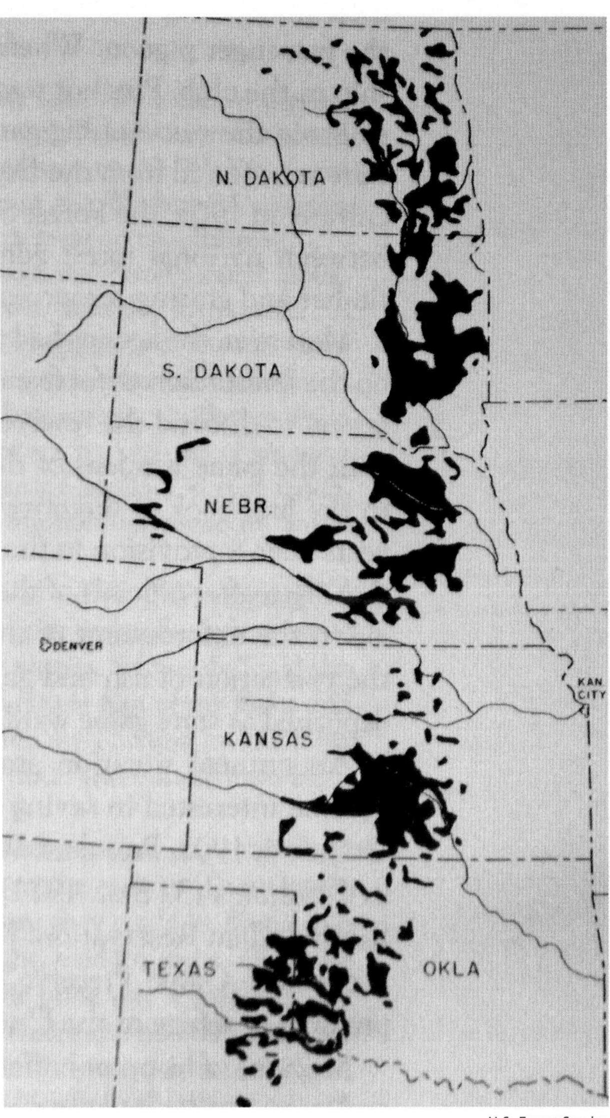

U.S. Forest Service

Figure 15-7. The Civilian Conservation Corps planted more than 220 million trees in the *Shelterbelt Project* that stretched from the Canadian border to the Brazos River in Texas. *Do you believe the Shelterbelt Project had any impact on wind erosion? Why?*

four million acres of farmland and replanted native prairie grasses. Once established, these prairie lands were less susceptible to the eroding forces of the punishing prairie winds. An example of prairie reclamation is the Pawnee Grasslands in Colorado, **Figure 15-8**.

Irrigation

Farmers also realized that they could not rely on rainfall for all of their cropping needs. So, they began to experiment with irrigation using well water. Within a few years of the Dust Bowl, farmers installed irrigation systems that tapped the vast Ogallala Aquifer. An ***aquifer*** is a vast underground reservoir of water. The Ogallala Aquifer is one of the most expansive aquifers in the world.

Modern Natural Resources Conservation

Following the Dust Bowl and World War II, the conservation movement gained momentum. Congress passed the *Federal Water Pollution Act* in 1948 and the *Air Pollution Control Act* in 1955. These would become the precursors for the ***Clean Water Act (CWA)*** and the ***Clean Air Act (CAA)***, respectively.

In agriculture, the post-World War II era was one of unprecedented increases in agricultural productivity. Hybrid corn, chemical fertilizers, and growing foreign markets made it a good time to be a farmer. Chemical fertilizers and pesticides were a boon for farmers. They were more effective than traditional approaches and boosted productivity. During the 1950s and 1960s, farmers relied more and more heavily on chemical approaches to soil fertility and pesticide management.

Rachel Carson and *Silent Spring*

In 1962, Rachel Carson published *Silent Spring*. The book highlighted the dangers of chemical pesticides to the environment, especially DDT. ***DDT*** is a synthetic insecticide that was used primarily to control mosquitos. In agricultural applications, DDT is used to control potato beetles, coddling moths that attack apples, corn earworm, cotton bollworm, and tobacco budworms. In *Silent Spring*, Carson highlighted the impact of DDT on wildlife, especially birds. The chemicals in DDT appeared to soften the shells of birds' eggs, especially those of the bald eagle, which was becoming endangered. With soft shells, the chicks cannot develop properly and die. So, she advocated the discontinued use and ban of DDT and other synthetic agricultural pesticides.

Carson's book spawned a renewed environmental revolution. It was one of the first environmental texts that relied on science as the basis for its arguments. *Silent Spring* also predicted pesticide resistance, weakened ecosystems, and even human cancers and illnesses attributed to chemical pesticides. In 1972, DDT was banned in the United States.

Federal Conservation and Environmental Agencies

The ***Environmental Protection Agency (EPA)*** was established in 1970. This agency is tasked with oversight and protection of the environment of the United States. It is a regulatory agency, so it proposes laws and regulations that protect our environment. The EPA is especially interested in protecting human health from the overuse and misuse of all forms of chemicals.

The ***National Oceanic and Atmospheric Administration (NOAA)*** was also established in 1970. Its mission is to focus on the conditions of the oceans and atmosphere. Today, NOAA warns of impending severe weather, monitors air and water quality, and conducts research to protect the environment.

Doug Andrews Photography

Figure 15-8. Prairie areas like the Pawnee Grasslands were purchased by the federal government during the Dust Bowl and returned to their native prairie state.

Natural Resources Conservation Service

Today, agriculturists rely upon the Natural Resources Conservation Service (NRCS) to provide programs, policies, and research about conserving our precious natural resources. Since its reorganization and renaming in 1994, the NRCS has been given financial resources to assist farmers and landowners in establishing conservation programs and projects on their land.

One of the primary goals of NRCS is to have every farmer and landowner establish and follow a conservation plan for every acre of their land. Because of this initiative, farmers have reporting obligations to NRCS each year. Another aspect of the work of the NRCS is the establishment of conservation reserve programs. A ***conservation reserve program*** is a decision by a farmer or landowner to set aside some of his or her land for forests, grasslands, or wetlands, **Figure 15-9**. In essence, they remove farmland from growing

Lynn Betts/NRCS

Figure 15-9. Conservation reserve program acreage must be planted in some kind of cover crop that controls erosion and provides habitat for wildlife, as shown in this planting of switchgrass in Iowa. ***Why are farmers known as America's frontline conservationists?***

Jeff Vanuga/NRCS

Figure 15-10. Grassed waterways are a part of conservation programs. The roots of grasses in grassed waterways hold the soil in place, thereby reducing soil erosion by water.

crops and allocate it to being conserved wild spaces.

One of the organizations supported by the NRCS is the National Association of Conservation Districts (NACD). The ***National Association of Conservation Districts (NACD)*** is a local conservation organization that focuses on agricultural conservation of water and soil resources. Local ***Soil and Water Conservation Districts (SWCDs)*** educate farmers and landowners about soil erosion practices, such as the grassed waterway pictured in **Figure 15-10**. They also teach about water quality and watershed protection. Most counties in the United States have a local Soil and Water Conservation District office.

Career Connection Natural Resources Conservation Service Civil Engineer

Job description: Civil engineers with the Natural Resources Conservation Service (NRCS) provide information about use of land and water resources to landowners, builders, and farmers. Engineers study information about natural areas and then design solutions to problems of building roads, constructing home developments, managing waste, and growing crops and raising livestock on land. Often, a civil engineer will work to conserve water and land resources for maximum benefits to humans.

Education required: A bachelor's degree in civil engineering is required, and a master's degree is often preferred.

Job fit: This job may be a fit for you if you enjoy working both outdoors and indoors with computer technology. You may also like making recommendations about land use for homes, development, and agriculture.

Shestakoff/Shuterstock.com

A civil engineer works to develop solutions to human activities with minimal impact on the environment.

Lesson 15.1 Review and Assessment

Words to Know

Match the key terms from the lesson to the correct definition.

1. A government sector of the U.S. Department of Agriculture that helps landowners reduce soil erosion, enhance water supplies, improve water quality, increase wildlife habitat, and reduce damages caused by floods and other natural disasters.
2. A United States federal law designed to control air pollution on a national level.
3. A government organization established in 1933 to build pathways, bridges, roads, and buildings in the nation's national parks and to educate farmers about soil conservation and methods to control erosion.
4. The protection of natural resources that allows use of the land by the public.
5. A vast underground reservoir of water.
6. A synthetic insecticide once used to control mosquitos and to control potato beetles, coddling moths that attack apples, corn earworm, cotton bollworm, and tobacco budworms.
7. A decision by a farmer or landowner to set aside some of his or her land for forests, grasslands, or wetlands.
8. A government action designed to regulate the discharges of pollutants into waters of the United States and regulate quality standards for surface waters.
9. Local organizations established to educate farmers and landowners about soil erosion practices.
10. A government agency tasked with oversight and protection of the environment of the United States. It is a regulatory agency, so it proposes laws and regulations that protect our environment.
11. An agency of the U.S. Department of Agriculture that administers the nation's national forests and national grasslands.
12. A local conservation organization that focuses on agricultural conservation of water and soil resources.
13. A government agency whose mission is to focus on the conditions of the oceans and atmosphere; the agency warns of impending severe weather, monitors air and water quality, and conducts research to protect the environment.
14. The protection of natural resources in which the resources are to be protected, but not used.
15. A method of farming that prevents the erosion of soil due to wind or water.

A. aquifer
B. Civilian Conservation Corps (CCC)
C. Clean Air Act (CAA)
D. Clean Water Act (CWA)
E. conservation
F. conservation reserve program
G. DDT
H. Environmental Protection Agency (EPA)
I. National Association of Conservation Districts (NACD)
J. National Oceanic and Atmospheric Administration (NOAA)
K. Natural Resources Conservation Service (NRCS)
L. preservation
M. Soil and Water Conservation District (SWCD)
N. soil conservation practice
O. United States Forest Service (USFS)

Know and Understand

Answer the following questions using the information provided in this lesson.

1. Explain why people in the United States became interested in conserving wildlife and wild spaces in the late 1800s.
2. Explain why Theodore Roosevelt is considered the "Father of the Conservation Movement."
3. Explain the difference between conservation and preservation.
4. How did agricultural practices in the 1920s contribute to the Dust Bowl in the 1930s?
5. Identify five soil conservation practices.
6. Factors that contributed to the unprecedented increases in agricultural productivity after World War II include _____.
   A. chemical fertilizers
   B. chemical pesticides
   C. hybrid corn and foreign markets
   D. All of the above.
7. Explain why DDT was banned in the United States.
8. *True or False?* One of the primary goals of NRCS is to have every farmer and landowner establish and follow a conservation plan for every acre of their land.
9. The _____ is a local conservation organization that focuses on agricultural conservation of water and soil resources.
10. Local _____ educate farmers and landowners about soil erosion practices.

Analyze and Apply

1. Develop a timeline of major federal legislation that impacted the conservation movement in the United States. What were the important dates and pieces of legislation that helped advance the conservation movement?
2. Create a Venn diagram comparing the ideas of conservation and preservation. Which characteristics do the two movements have in common? Which characteristics are different?
3. Create a timeline of important events in the history of natural resources conservation in the United States. Consider events and milestones beyond the ones explained in this lesson.

Thinking Critically

1. Which event in the history of natural resources conservation would you consider to be the most important? Why?
2. Was there a connection between the Dust Bowl and the Great Depression? Explain your answer.
3. What does your local county Soil and Water Conservation District do to educate farmers and the public about soil and water conservation? What opportunities exist for youth to be involved in these programs?

Lesson 15.2

Soil Formation and Properties

Lesson Outcomes

By the end of this lesson, you should be able to:

- Explain the formation of soil.
- Differentiate between soil particles.
- Explain the impact of pore spaces on soil drainage.
- Describe threats to soils, including erosion, salinization, and compaction.

Words to Know

"A" layer
"B" layer
carbonates
clay
"C" layer
cover crop
gully erosion
humus
land capability classes
leaching
loamy soil
no-till program
"O" layer
organic layer
organic matter
parent material
peds
pore spaces
reduced tillage
rill erosion
salinization
sand
saturated
silt
soil compaction
soil erosion
soil particles
soil profile
soil structure
soil texture
subsoil
tillage system
tilth
topsoil
weathering
windbreak

Before You Read

Read the lesson title and tell a classmate what you have experienced or already know about the topic. Write a paragraph describing what you would like to learn about the topic. After reading the lesson, share two things you have learned with the classmate.

Soil is one of our most valuable natural resources. A thin layer of soil, often only a few feet deep, covers only parts of Earth's land mass. If Earth were an apple, its soil would be like the skin of that apple, only thinner—a lot thinner. And, instead of covering the entire Earth, as the skin covers the entire apple, the productive, agricultural soil on our planet covers less than 10% of Earth's surface. It is on this soil that farmers produce enough food to feed the global population.

In this lesson, we will:

- Explore soil composition and the process of soil formation.
- Discuss the importance of soil.
- Determine which soil components allow farmers to grow crops and raise food for livestock.
- Discuss soil surveys and land capability classes.

Soil Horizons and Layers

Soil is the layer of material, both living and nonliving, on Earth's surface that is capable of supporting plant life. In most places, it takes almost 500 years to form an inch of new topsoil. On the other hand, unprotected, bare land may lose tons of topsoil during a thunderstorm.

"A" Layer

Most agriculturists are primarily concerned with the top two layers of soil. These are called the topsoil and the subsoil. When soil scientists dig a hole in the ground, they look at the ***soil profile***, a vertical slice down into the

STEM Connection Arable Land

Cut an apple into quarters. Hold 3/4 in one hand. This is the amount of Earth that is water. The remaining 1/4 is land. Cut that 1/4 in half, lengthwise. The 1/8 of Earth is not inhabitable by humans (ice, mountains, deserts, etc.). The remaining 1/8 is habitable land. Slice the 1/8 crosswise into four equal pieces. Hold 3/32 in one hand. This is the amount of land where people can live, but food cannot grow. The remaining 1/32 is arable land. Peel the remaining 1/32 as thinly as possible. This skin represents the topsoil, or the portion of the entire planet that is capable of producing crops.

enzodebernardo/Shutterstock.com

ground. A soil profile is shown in **Figure 15-11**. ***Topsoil*** is the top layer of soil. It is usually the most fertile and darkest in color, indicating high organic matter content. Topsoil is also called the ***"A" layer*** of the soil horizon.

"B" Layer

The layer right below the topsoil is the ***subsoil***. Subsoil is usually lighter in color, contains less organic matter, and is generally more responsible for drainage and aeration than the topsoil. Subsoil is also called the ***"B" layer*** of the soil horizon. The subsoil in **Figure 15-12** is clearly different than the topsoil. It is also gravelly, which helps improve soil drainage. If you look closely, you can see roots in the topsoil, but fewer roots in the subsoil, and no roots below this layer.

The roots of plants penetrate both the topsoil and the subsoil. However, most of the crops' nutrients are absorbed from the topsoil. This is also the cultivation layer of soil. Roots absorb water and some nutrients from the subsoil.

"O" Layer

Sometimes there is a layer of soil on top of the topsoil. This happens in grasslands and woodlots that have not been tilled. This layer is called the ***"O" layer***. It is also known as the ***organic layer***, meaning that the material found here has the highest organic matter content of any soil layer. This layer is often made up of partially decomposed leaves, stems, sticks, and other plant matter. The "O" layer is nature's mulch. Leaves and other plant material that decay are similar to compost in a compost bin or mulch that slowly decays in your landscape under shrubs and trees. The "O" layer, mulch, and compost all provide plants with needed nutrients.

On cropland, the "O" layer is absent. By plowing and tilling the soil, farmers incorporate the "O" layer into the "A" layer. The mixing of organic

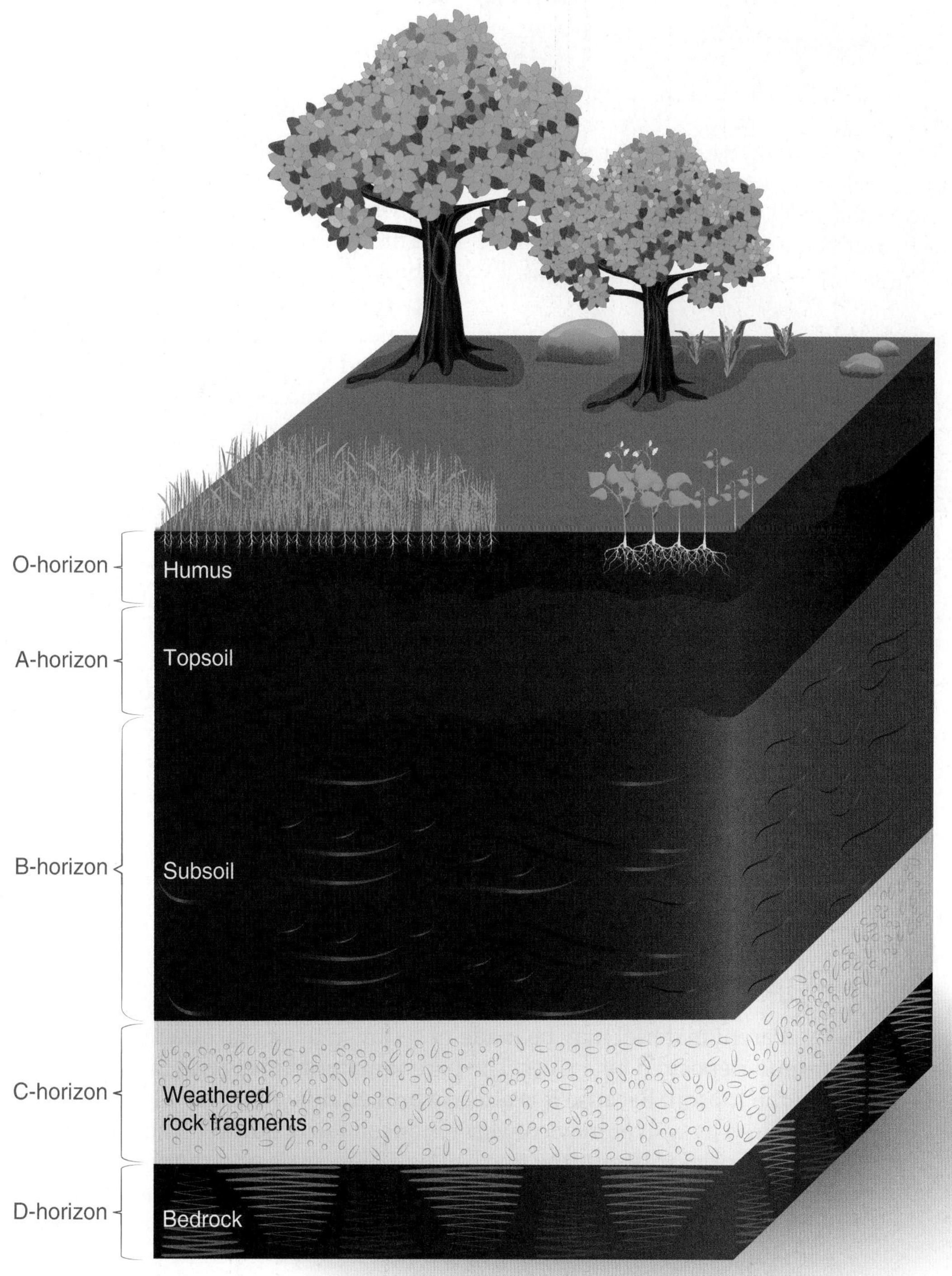

Designua/Shutterstock.com

Figure 15-11. A soil profile shows a slice of Earth where you can see the layers, or horizons, of soil, including the "O," "A," "B," and "C" layers. It is unlikely you will be digging deep enough to see the bedrock, or "D" layer of the soil horizon. *Do you know how deep each of these layers typically is?*

strixcode/iStock/Thinkstock

Figure 15-12. Roots do not penetrate the subsoil in this image where the subsoil is quite gravelly. ***Does the gravel in the subsoil eventually break down?***

matter with the "A" layer provides additional nutrients for the growing crop.

"C" Layer

Below the topsoil and the subsoil is the parent material. The ***parent material*** is the underlying soil material that gives rise to the subsoil and topsoil. Through weathering and the mechanical breakdown of rocks and heavier soil particles by plant roots, freezing, and water action, the topsoil and subsoil are created from the parent material. Parent material makes up the ***"C" layer*** of the soil horizon.

Soil Formation

So, how does soil form? Soil begins to form through the weathering of rocks and other materials. ***Weathering*** is the process of eroding rock over time, primarily through the action of rainfall and snowfall. The amount and rate of weathering is determined by the following five factors:

- Parent material
- Organisms
- Climate
- Topography
- Time

Parent Material

Parent material may be made of sedimentary rock, other types of rock, or mud deposited by a river. The parent material gives rise to the subsoil and the topsoil. Because it is already broken down into small particles, mud deposited by a river may convert to topsoil fairly quickly. The particles in the mud are primarily ***silt*** (the fine soil particles carried by running water and deposited as sediment). Mud is formed when those particles are saturated with water. Once the mud dries out, it begins to develop some structure and, with time, soil may be formed.

Soil formed through the weathering of rock requires much more time than mud and silt to develop into topsoil. The rock must slowly erode into its constituent minerals. This does not happen quickly. The parent material has a great deal of influence on a soil's physical and chemical properties.

Organisms

Organisms, like insects, animals, bacteria, and plants, play an important role in the development of soil. Together, these organisms help break down organic matter and mix the soil. As they move throughout the soil, microorganisms also create channels for water and air infiltration. Some of these are large channels, as with earthworm tubes, and others are microscopic. These actions also contribute to soil structure.

- Insects—soil-dwelling insects like dung beetles, termites, and ants move soil around and contribute to soil aeration and water filtration.
- Animals—burrowing animals like gophers, ground squirrels, and moles create spaces for air and water flow.
- Bacteria—microorganisms that break down plant and animal material and may also break down some pesticides and herbicides.
- Plants—roots penetrate soil and cracks in rocks or rotting organic materials.

Did You Know?

Earthworms ingest soil particles and organic residues and enhance the stability of the soil.

Plant Roots

Plant roots are one of the more important organisms affecting soil formation. The image in **Figure 15-13** shows how deeply turfgrass roots can penetrate soil. The roots of plants are slow-growing and can find minute cracks and crevices in rocks. Once roots grow into these cracks, they force the cracks to expand, allowing water to infiltrate. In areas that experience freezing temperatures, the water in the cracks freezes and expands, thus increasing the size of the crack. Over time, this crack will eventually break the rock.

Decaying Organisms

Organisms that live and then die in the forming soil also contribute to the soil's formation. These dead organisms increase the soil's fertility by contributing organic matter to its composition. The increased fertility encourages the growth and reproduction of new organisms that will eventually die and decay in the developing soil. This cyclical process creates the soil.

Climate

The climate of an area also affects soil formation. In colder climates, temperature extremes (freezing and thawing) greatly impact the breakdown of parent material. The decay of organic material is also slower in colder temperatures. In warmer climates, chemical reactions take place more quickly and speed up the decaying process.

The climate also includes factors such as wind and sun exposure. Regular exposure to high winds and the heat of the sun will speed up the breakdown of rocks and organic materials.

SCOTTCHAN/Shutterstock.com

Figure 15-13. Many organisms live in the topsoil, including the roots of plants, insects, moles, bacteria, and other microorganisms. *Have you ever dug up worms for fishing? In what type of soil do you find the most worms?*

Temperature

Temperatures have an impact on soil formation. Warm temperatures in summer foster the growth and development of microorganisms that build soil. But, cold temperatures in winter also contribute to soil formation. As water penetrates cracks in rocks and then freezes, this action causes the water to expand and make the crack in the rock larger. Over time, more and more water can enter the crack and freeze. This makes the crack bigger and bigger. Eventually the rock breaks into two parts. This process may take decades. Over hundreds of years, the rock eventually breaks down into smaller and smaller pieces that may eventually become soil.

Precipitation

The level of precipitation and the pH of rainfall also affect soil formation. Higher levels of precipitation may cause moist conditions that hasten the decay of organic materials. In many parts of the country, rainfall may be acidic. Acidic rainfall speeds the weathering process of rocks and organic matter. The acid in rainfall breaks down materials more quickly than neutral pH rainfall.

The image in **Figure 15-14** shows how stream banks of sandstone can weather to form soil. Rainfall and snowfall introduce water to the developing soil. Moving water has tremendous energy to cut into rock and dislodge both large and small pieces of it over time. Thus, with time, a large rock will eventually weather into smaller rocks, then sandy material, and then soil.

Leaching

Water is the universal solvent. So, water dissolves minerals and moves them deeper within the soil profile. This process is called ***leaching***, **Figure 15-15**. Leaching occurs primarily with carbonates. ***Carbonates*** are minerals that affect the acidity of the soil. Carbonates in the soil raise the pH of soil and keep it near a neutral state. As carbonates are leached from

Larry Knupp/Shutterstock.com

Figure 15-14. Weathering is one of the factors that break down rock into smaller particles, including soil. The sandstone cliffs along this stream bank have been weathered by eons of rain and stream erosion. *Have you ever hiked in areas like the one pictured above? Is the sandstone gritty or smooth?*

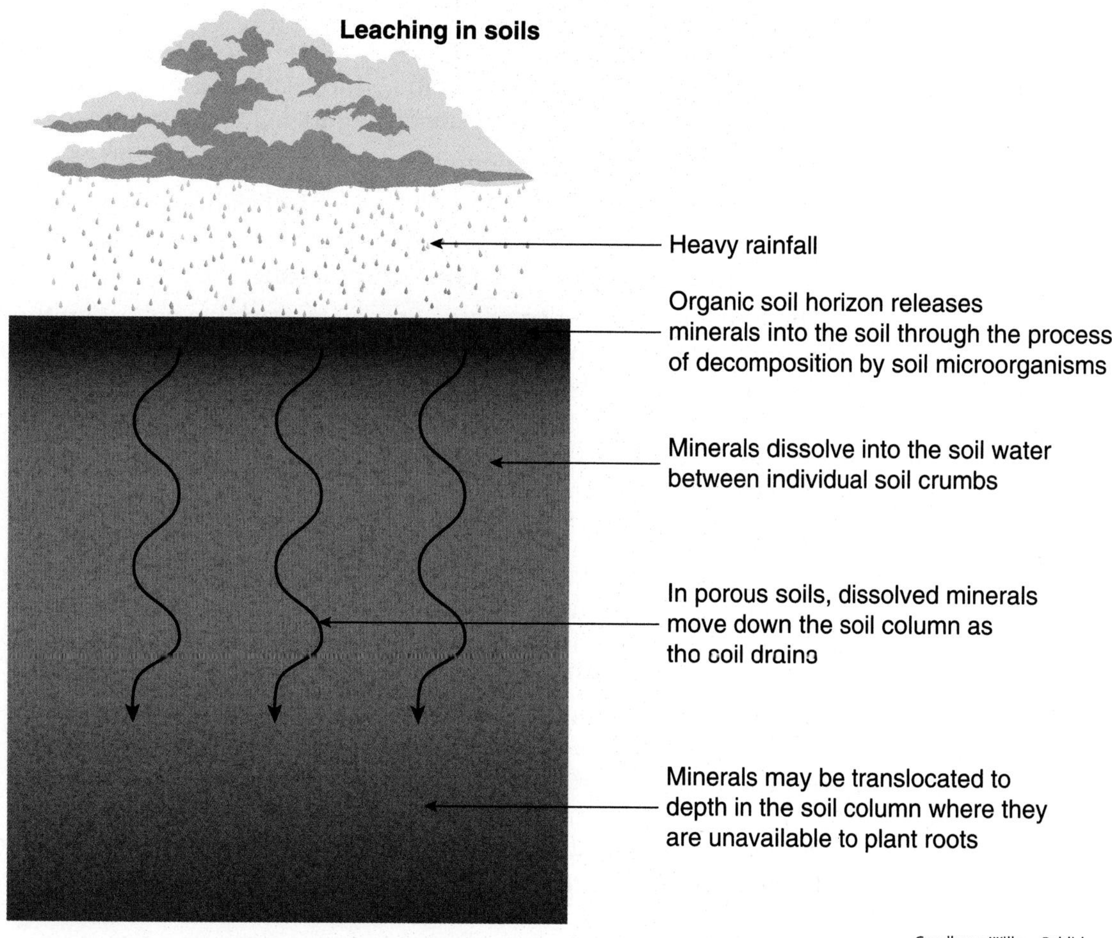

Figure 15-15. Leaching occurs when rainfall dissolves minerals and moves them deeper within the soil profile. *Can leaching occur with harmful chemicals?*

the soil, the soil becomes more acidic. Most soils in the eastern half of the United States are acidic soils and susceptible to leaching. Soils in the eastern half of the United States have pH that are generally acidic, or lower than 7.0. Applying lime in the form of crushed limestone to soil can raise the pH of soil and make nutrients in the soil more available for plant growth.

Topography

The topography, or terrain, of a region impacts soil formation by influencing soil temperature, wetness, and erosion rate. For example, completely exposed material on the ridge of a mountain may deteriorate more quickly than material in an area that is protected from the elements.

Higher elevations are generally cooler and dryer than lower elevations. Wind and erosion have a greater effect on material in higher regions than heat and moisture. However, this weathered material is washed down mountain slopes to lower areas during rainstorms and snowmelt. Lower areas on the mountain experience faster rates of soil formation due to higher temperatures and humidity. Organisms like bacteria and plants tend to thrive in wetter conditions. Their presence and actions increase the amount of organic matter available and its rate of decay in the developing soil.

Time

Time is also a significant factor in soil formation. Elements may be washed or blown away before contributing to soil formation in a given area.

- Younger soils are characterized by having only an "A" horizon.
- Older soils develop "B" and "O" horizons with time. Over time, water washes minerals from the "A" horizon into lower areas of the soil profile. This allows the formation of the "B" horizon.
- The passing of time also allows organic material to collect on the soil surface, primarily in the form of dead and decaying plant material. As this material collects, the "O" horizon is formed.

Soil Composition and Texture

Soil may look like a brown mass of stuff to the novice, but to an agriculturist or soil scientist, soil is a world unto its own. Soil is actually not just one mass, but a complex living system. Soil is so complex that some scientists treat it as a living organism. Soil is comprised of minerals, organic matter, water, air, and even microorganisms. A soil's composition will determine its texture and its structure.

Minerals

Minerals make up the largest portion of most soils. These are collectively referred to as ***soil particles***. The three soil particles are *sand*, *silt*, and *clay*. Soil particles make up 40%–50% of the total volume of soil. It is this proportion of different-sized particles that determines the soil texture. ***Soil texture*** refers to the amount of sand, silt, and clay present in a given soil, whereas ***soil structure*** refers to the binding of these soil particles into aggregates called ***peds***. The combination of particles also influences the soil's water-holding and drainage capacities.

Healthy soil suitable for growing crops has a mixture of these three particles in a ratio that allows the soil to:

- Hold water.
- Drain relatively quickly.
- Enable soil fertility.
- Allow the roots to anchor the plants.

A soil with good texture is able to breathe and support plants with enough nutrients. To some growers, soil texture is the singularly most important physical property to consider when managing soils for cultivation. Soil scientists determine soil texture with a soil texture triangle, **Figure 15-16**.

Sand

Sand is the largest soil particle. Individual grains of sand can be seen with the naked eye. A soil that is mostly sand feels gritty, almost like the sand in a sandbox or at the beach. If we were to magnify soil particles, the sand particles would look like basketballs, in relative proportion to the other soil particles, **Figure 15-17**.

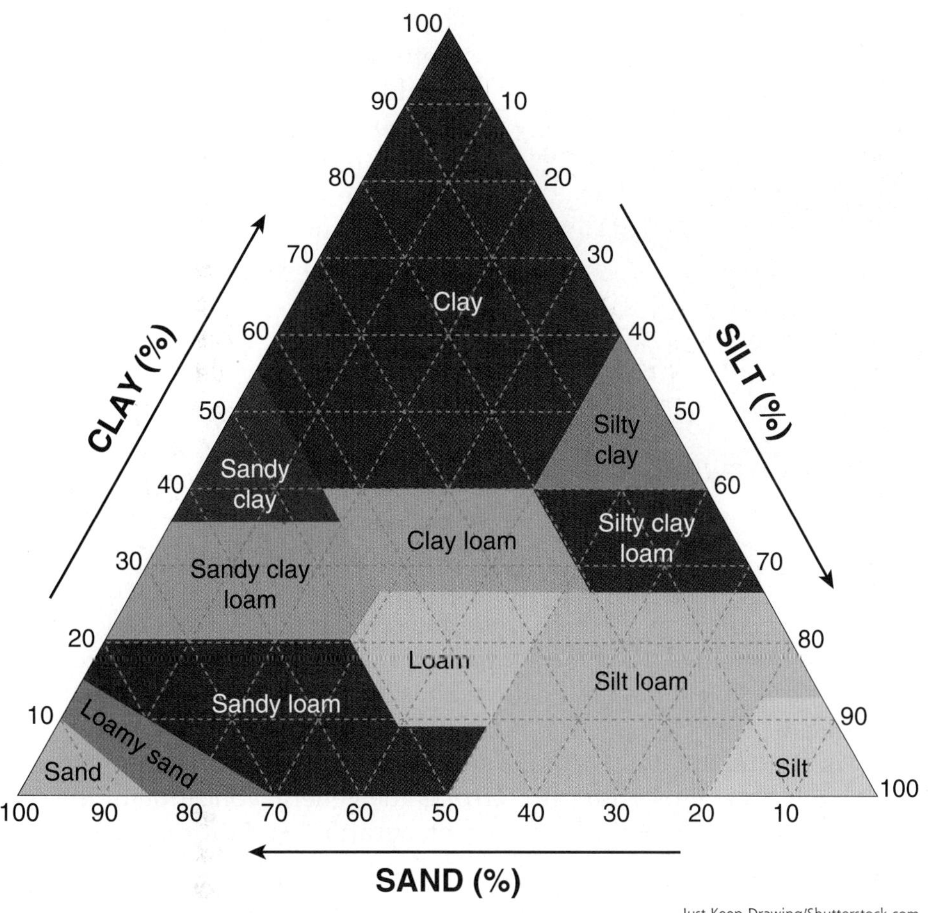

Just Keep Drawing/Shutterstock.com

Figure 15-16. The soil texture triangle uses the percentages of sand, silt, and clay to establish a soil's texture. Texture is critical to allowing the soil to drain water, hold oxygen in pore spaces, and support plant life with crucial nutrients. ***Where does the soil in your yard fall on the triangle?***

Soil Particle	Size	Feels...	Feels Like...
Sand	.05–2.0 mm	Gritty	Sugar
Silt	.002–.05 mm	Smooth	Flour
Clay	Less than .002 mm	Sticky	Caramel

Aaron Amat/Shutterstock.com; cloki/Shutterstock.com; Goodheart-Willcox Publisher

Figure 15-17. When we think of soil particles, it is sometimes easiest to think of them in relative sizes. If sand were the size of a basketball, then silt would be the size of a golf ball, and clay would be the size of a period at the end of a sentence. This table presents the actual sizes of the three soil particles, explains their feel, and compares them to other substances with which you might be familiar. ***What is the soil texture in your family's vegetable or flower garden?***

Sand is relatively inert, meaning that it does not affect soil fertility. Sand does not hold soil nutrients and make them available to the plant for growth and development.

Sand is important for soil drainage and aeration. This is because the pore spaces between particles of sand are relatively large. Imagine a barrel full of

VadiCo/Shutterstock.com

Figure 15-18. Soil with good tilth is light, fluffy, absorbs moisture well, and is easy to work. *What type of tillage system was used in this field?*

basketballs. It is easy to see the spaces between the basketballs. These spaces could hold lots of water. But, if you poked a hole in the bottom of the barrel, the water would drain quickly. Since the spaces between the sand particles are large, sandy soils drain quickly and are prone to drought conditions.

Silt

Silt is the second largest soil particle. You cannot see individual grains of silt without the aid of a microscope. Silt feels a lot like baking powder in your hands. It is smooth and cool to the touch. If we magnified the soil particles so that sand was the size of a basketball, then the silt particles would be the size of a golf ball under the same level of magnification. So, silt particles are much smaller than sand particles.

Silt is also relatively inert. Like sand, silt helps with soil drainage and aeration. Silt also helps with tilth. ***Tilth*** is the general condition of the soil with respect to texture, moisture content, aeration, and drainage. Soil that has good tilth looks light and fluffy, like the field in **Figure 15-18.**

Clay

Clay is the smallest soil particle. Individual grains of clay cannot be seen with the naked eye. Clay soils feel slippery, especially when wet. Many people think of modeling clay from art class when they think of clay soils.

If we were to magnify sand to the size of a basketball, and silt to the size of a golf ball, clay would be about the same size as the period at the end of this sentence. Imagine a barrel filled with particles the size of a period. If we were to fill the barrel with water and cut a hole in the bottom, the water would drain very slowly. Clayey soils have tremendous water-holding capacity.

Clayey soils are also chemically active. Most clayey soils have a negative electrical charge. They are not inert. Clay particles in the soil are what give the soil its fertility characteristics. Clay particles in the soil attract and hold valuable plant nutrients until the plant's roots are ready to use them.

STEM Connection Clay in Kaopectate™

Kaopectate™ is a relatively common medicine to remedy upset stomachs. At one time, Kaopectate™ was made with kaolinite, a type of clay particle in the soil. Kaolinite has many uses, including toothpaste, cosmetics, and glossy paints. In Kaopectate™, kaolinite adsorbed the acids and irritants in upset stomachs. It also soothed irritated linings of a person's bowels. Today, kaolinite is not used in Kaopectate™ because improved synthetic chemicals have been engineered as adsorbents.

Organic Matter

From a productivity standpoint, organic matter is one of the most critical components of soil. ***Organic matter*** is material composed of decomposing and decomposed living organisms, including their waste products. Another name for organic matter is ***humus***. Organic matter is usually about 1%–7% of the total volume of soil. Organic matter gives fertile soil its dark color and makes it smell "Earthy," **Figure 15-19**.

Even though organic matter is a small portion of the soil, farmers rely on organic matter to provide crops with essential nutrients like nitrogen, potassium, and phosphorus.

Julija Sapic/Shutterstock.com

Figure 15-19. Organic matter gives fertile soil its dark color and makes it smell "Earthy." *Does healthy topsoil come in other colors besides black or brown?*

Water, Air, and Pore Spaces

From a plant's perspective, the most important component of soil may be what is found, or not found, in the spaces between soil particles. ***Pore spaces*** are the spaces between soil particles, **Figure 15-20**. Pore spaces are filled with either water or air. After a heavy rainfall, nearly all of the pore spaces are filled with water. We call this soil ***saturated***. Even when the soil is not saturated, microscopic amounts of water still fill the smallest pore spaces. This water provides a consistent source of water needed by plants to grow and develop. Microscopic root hairs seek this water and absorb it into the plant.

When the soil dries, some of the pore spaces, especially the larger ones, allow water to evaporate. Once the water evaporates, the spaces fill with air. The growing roots need this oxygen source for respiration. The microscopic root hairs absorb oxygen from these pore spaces.

Pore spaces account for about 50% of the volume of soil. When the soil is saturated, water occupies these spaces and makes up about 50% of the volume of soil. As the soil dries, air begins to comprise a larger portion of the volume of soil.

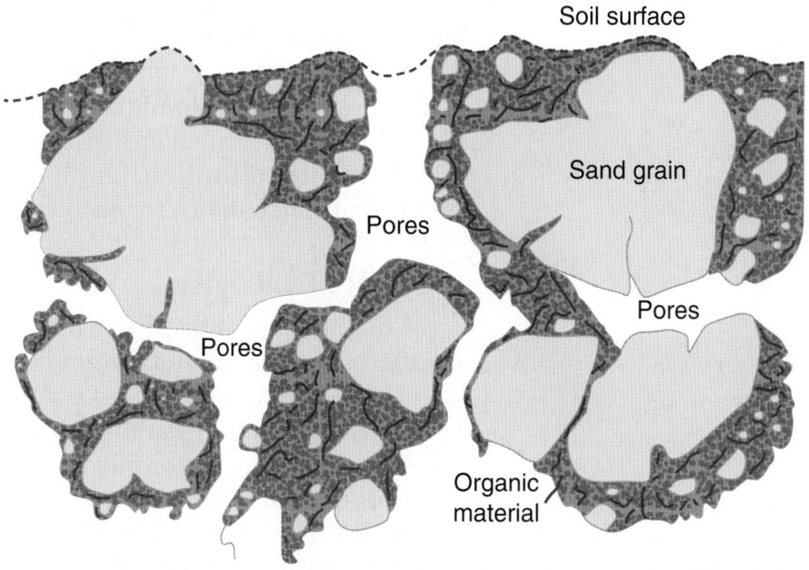

Goodheart-Willcox Publisher

Figure 15-20. Pore spaces can be filled with either air or water, but not both at the same time. Sandy soils have larger pore spaces, which drain more quickly than clayey soils with smaller pore spaces.

Microorganisms

Microorganisms make up a small, but important, portion of soil. Overall, they are about 1% of the total soil volume. Scientists estimate that one thimble of soil may contain more than 20,000 microorganisms! You can even see some of these organisms. Earthworms and other insects are part of the microorganism community in soil. Microorganisms such as bacteria, algae, and fungi are also found in soil. Most soil microorganisms are decomposers, so they live off the dead and decaying organisms in the soil. Soil microorganisms contribute to the nitrogen and carbon cycles by eating dead and decaying organic matter.

Loamy Soils

Soils that have good tilth are loamy. A ***loamy soil*** is one that contains approximately 40% sand, 40% silt, and 20% clay. Loamy soils hold more nutrients and moisture than sandy soils. They also generally contain more organic matter than sandy soils. Loamy soils drain more quickly and have greater aeration than clay soils. Because of this, they are easier to till than clayey soils.

Soil Enrichment and Preservation

With each passing year of use, growers face challenges with the soil on their land. Maintaining healthy, productive soil that does not erode or present problems to the water supply is a constant battle. Thankfully, we have the past to learn from and the future to develop methods of soil enrichment and preservation. Some of the soil challenges farmers face include:

- Maintaining biodiversity
- Limiting erosion
- Preventing compaction
- Counteracting salinization

Maintaining Biodiversity

As stated earlier, soil is a world unto its own. It is one of the most biologically diverse systems of Earth, **Figure 15-21**. The organisms that live and die in soil play an important role in both its development and health. And, in order to remain healthy, soil must maintain a certain level of biodiversity. Farmers ensure a good biodiversity in their soils by rotating crops, limiting cultivation, limiting the use of pesticides, and maintaining the proper nutrient balance for each type of crop.

Rotating Crops

Rotating crops introduces new microorganisms to the soil, as different microbes live in and on different crops. Different types of crops also contribute different kinds and volumes of crop residue after harvest. Legumes contain bacteria that grow in root nodules and fix nitrogen in the soil. These bacteria grow only on soybeans and other legumes, but not on the roots of corn, wheat, or rice. Many farmers will plant corn one year, and then soybeans the next year. The soybeans (which are legumes) replenish the nitrogen supply depleted from the soil by the first crop (the corn).

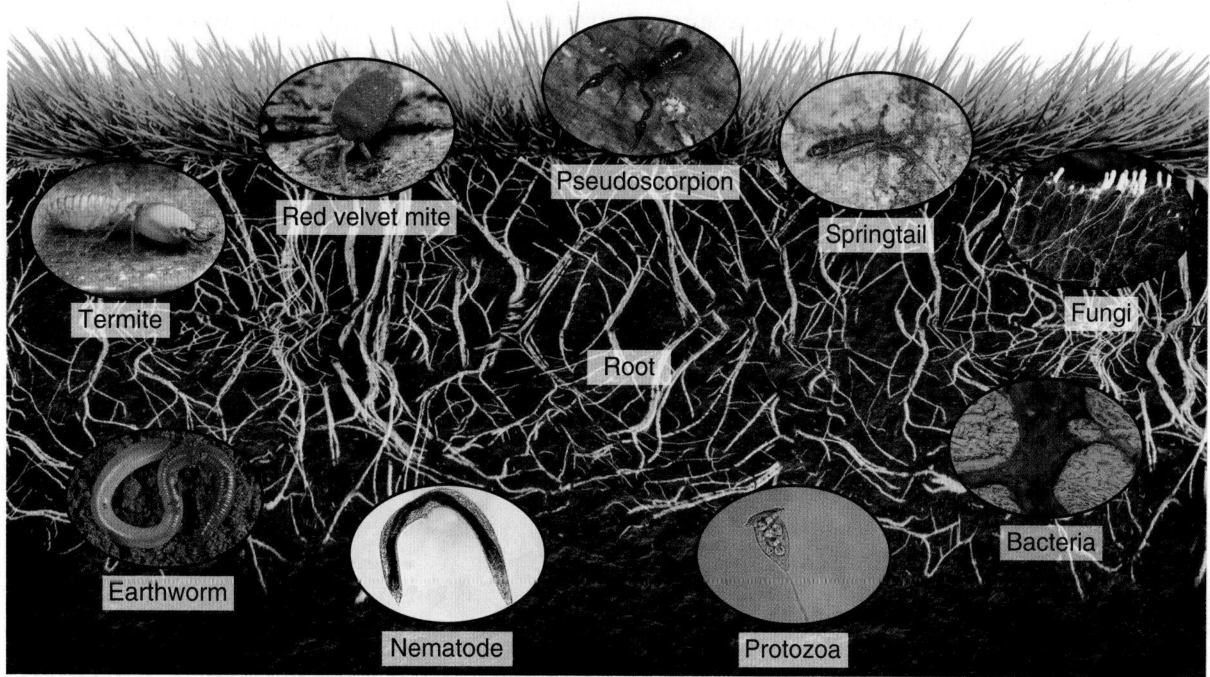

(clockwise from top left) Pan Xunbin/Shutterstock.com; D. Kucharski K. Kucharska/Shutterstock.com; Cosmin Manci/Shutterstock.com; Henrik Larsson/Shutterstock.com; Kichigin/Shutterstock.com; Poul Riishede/Shutterstock.com; Jubal Harshaw/Shutterstock.com; D. Kucharski K. Kuscharska/Shutterstock.com; D. Kucharski K. Kuscharska/Shutterstock.com; (background) varuna/Shutterstock.com

Figure 15-21. If you look closely at a handful of soil, you are sure to see much microbial life, including termites, worms, and burrowing insects. With the help of a microscope, you will see even more microbial life in the soil, including nematodes, protozoa, fungi, and bacteria. ***In which layer of soil do most organisms live?***

Limiting Cultivation

By limiting cultivation, the farmer limits compaction. Appropriate cultivation also ensures good soil tilth and pore spaces. Cultivation also disrupts the ecosystem of the soil. Undisturbed soil allows colonies of microorganisms to develop.

Limiting the Use of Pesticides

The application of both pesticides and fertilizers ensures that beneficial insects and microorganisms are encouraged over the ones that are truly pests. Chemical manufacturers strive to produce herbicides and insecticides that do no harm to beneficial insects.

STEM Connection Soil Erosion

Conduct a soil erosion experiment. In aluminum casserole pans, prepare three field models. First, prepare bare soil. Second, prepare bare soil covered with plant residue. Third, prepare grassed sod. Once prepared, cut a V-shaped notch in the end of each of the pans. Then, tilt them at a slight angle. Just below the V-shaped notch, place a clear container to catch the runoff water. With a garden watering can, measure and pour the same amount of water onto each pan. Which "field" produced the cleanest water? Why was the water cleaner? Which "field" allowed the least amount of runoff water? Why did it have less runoff water?

Soil Erosion

Soil erosion from agricultural fields is one of the major concerns in agriculture and natural resources. ***Soil erosion*** is the wearing away of soil by wind or water. Farmers want to prevent all soil erosion. Once soil erodes from the field into a stream or the atmosphere, the field is barren and cannot support crop production.

Wind Erosion

Wind can cause soil erosion. Dry fields that lack crop cover are susceptible to wind erosion, especially if those fields are newly tilled. Winds blowing across the surface of the soil have tremendous energy to pick up loose particles of soil and make them airborne. The Dust Bowl of the 1930s was created by wind erosion.

To help prevent or reduce wind erosion, farmers plant windbreaks, maintain cover crops, and practice various tillage systems.

- ***Windbreaks***—rows of trees planted on the western edge of a field that create a natural barrier to slow the flow of wind and minimize *wind erosion*. Windbreaks also encourage biodiversity by providing food and homes to various forms of wildlife.
- ***Cover crops***—crops planted primarily to manage soil erosion. Cover crops prevent wind erosion because the plants' stems provide a buffer to the wind at the soil surface and the plants' roots hold the soil tightly in place. Cover crops remain in place throughout winter to protect fields from winter winds. Cover crops also contribute to soil fertility and biodiversity.
- ***Tillage systems***—methods of turning the soil in crop fields.

Did You Know?

For years, especially in rural areas, people have planted evergreen trees on the north and west sides of their homes as windbreaks. Deciduous trees are planted on the south and east sides to provide shade.

No-Till and Reduced Tillage

Since the mid-1980s, farmers in the United States have adopted no-till and reduced tillage practices to help combat soil erosion. A ***no-till program*** is a method of planting where the farmer does not plow or disk the soil, but plants a new crop in the remnants of the previous year's crop, **Figure 15-22**. So, there is no tillage or disturbance of the soil surface.

Travis Park/Goodheart-Willcox Publisher

Figure 15-22. The soybeans planted into this wheat stubble are an example of no-till planting. The wheat stubble is the residue from the previous crop, which holds the soil in place until the soybeans can grow larger and their roots can hold the soil in place during rains.

A ***reduced tillage program***, often called a *minimum tillage program*, is a method in which the farmer minimally tills the soil, often with a machine that minimizes the disturbance of the soil surface. The farmer likely makes only one pass with a tillage machine and then plants the new crop. With reduced tillage systems, more of the crop residue from last year's crop remains on the surface of the soil to prevent wind and water erosion. With reduced tillage programs, the farmer will disturb the soil only once before planting. They will use a combination tillage tool that breaks up the roots of weeds and creates pore spaces for water penetration.

Water Erosion

On most agricultural lands, water causes the most erosion of the soil. When soils become waterlogged (oversaturated), the particles are susceptible to movement in the water. Think of a bucket of mud. If the mud is soupy enough, most of it would flow out of the bucket when poured. Water tends to move soil in a similar manner. Running water has tremendous energy to dislodge and move large volumes of soil.

In heavy rainstorms, water can pool and then form streams that run across the surface of the soil. ***Rill erosion*** occurs when many small streams of water form and are running across the surface of the soil. When many streams come together and form a larger stream, often in the valleys or low areas of a field, then we call this larger body of moving water ***gully erosion***, **Figure 15-23**. Both rill and gully erosion are detrimental. Waters in a rill or gully can move tons of soil in a single thunderstorm.

Meryll/Shutterstock.com

Figure 15-23. Gully erosion happens where water accumulates in a low area of a field and moves quickly across the surface of the soil. If unchecked, gully erosion will continue to the point where the gully is too deep for the farmer to drive a tractor across. *How could this type of erosion be remedied?*

Farmers install grassed waterways in low areas of their fields to combat the forces of water erosion. The stalks of the grass plants slow the force of the water across the surface of the land. The roots of the grass perform two erosion control functions:

- The roots absorb a lot of the water to help the grass grow.
- The roots anchor the soil tightly in place.

Compaction

Soil compaction occurs in areas of fields where large machinery or livestock frequently travel. ***Soil compaction*** is the compression or elimination of pore spaces in soil due to frequent heavy pressures, as with the heavy farm machinery, **Figure 15-24**. In essence, the soil becomes denser when it is compacted.

You have probably seen compaction in high traffic areas where people frequently walk across a lawn. In these areas, the heavy volume of traffic kills the grass first. With time, the foot traffic presses the soil so that it actually becomes lower than the surrounding soil. The foot traffic creates a dent in the soil.

Compacted soil is a problem for plant growth because it eliminates pore space and makes it impossible for the roots to penetrate the soil, **Figure 15-25**. Without pore spaces, there is little water and air

oticki/Shutterstock.com

Figure 15-24. The tire tracks in this image show how heavy farm machinery compacts the soil. *Is there more of an impact if you drive on the soil when it is really wet or extremely dry?*

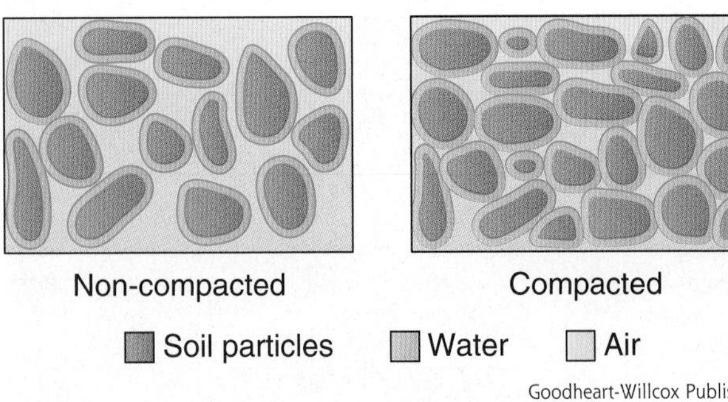

Goodheart-Willcox Publisher

Figure 15-25. When heavy farm machinery repeatedly runs over the same part of the field, the soil is compacted. The compaction reduces the volume of pore spaces in the soil.

available in the soil. And, although microscopic root hairs are strong, they cannot penetrate compacted soil because the cracks and crevices are too small.

To combat compaction, farmers may occasionally use a deep tillage tool. This is an implement that looks kind of like a large knife with wings that is pulled behind a high horsepower tractor. As the implement is dragged through the ground, it creates cracks and fissures in the soil. It basically recreates some pore spaces to allow water, air, and plant roots to create even more pore spaces.

Salinization

On cropland where farmers use frequent irrigation, salts build up in the soil. This is a process called ***salinization***. Farmers increasingly use irrigation to help their crops grow. This poses a long-term problem for soil. Almost all water contains some dissolved salts. As the irrigation machinery adds more water to the soil, these salts are filtered out of the irrigation water by the soil. Thus the soil acts like a filter. In this natural soil filter, salts begin to accumulate and build up in the soil.

Although salt, or sodium, is an essential nutrient to human life, it is actually one of the most poisonous materials on the planet. Plants cannot tolerate high concentrations of salt. Salt inhibits the plant's ability to absorb soil moisture. A prime example of how high salt concentrations affect the surrounding ecology is the Great Salt Lake in Utah, **Figure 15-26**.

Jason Patrick Ross/Shutterstock.com

Figure 15-26. The effect of high salinity is apparent when you view the land around the Great Salt Lake in Utah. The vegetation is sparse, but the marshes surrounding the lake support a great number of birds who feed on its brine flies and shrimp. *Do you know why the lake contains so much salt?*

Land Capability Classes

Depending on the slope of the land and the amount of erosion that has occurred or tends to occur on the land, it is classified into different classes. ***Land capability classes*** are the determination of the suitability of land for different cropping systems. Classifications range from the best farmland that is suitable for all types of cropping systems to land that is not suitable for any row crops or disturbance of the topsoil. The most ideal land is classified as Class 1 soils, **Figure 15-27A**. Classes 1, 2, 3, and 4 are the types of landscapes and soils that are generally suitable for crop production. Class 1 land is the flattest, most fertile classification of land. When you think of Class 1 land, think of the flatlands

S-F (/gallery-480532p1.html)

Sinelev (/gallery-639536p1.html)

Figure 15-27. A—These crops are growing on Class 1 land, which is suitable for all cropping systems and is generally not susceptible to severe erosion. B—Class 8 land is not suitable for any kind of cultivation, pasturing, or cropping of any kind. It is suitable for wildlife habitat, watersheds, and natural areas.

of the Midwest. Classes 5, 6, and 7 may be used for pasture, wildlife habitat, or forests. Soils that are classified as Class 8 soils are restricted to wildlife habitat, natural areas, watersheds, and other similar undisturbed uses, **Figure 15-27B**.

Career Connection Soil Scientist

Job description: A soil scientist is a person who studies soils and makes recommendations about cropping and building practices on that soil. Soil scientists make recommendations to farmers about cropping systems and fertilizers on their soils. They also work with urban planners and home construction builders to recommend soil conservation and usage practices. For instance, a home builder might contact a soil scientist when planning a septic system or constructing a basement for a home. They want to ensure proper drainage and percolation for the soil, and a soil scientist can provide accurate information on these phenomena.

Education required: A bachelor's degree in soil science.

Job fit: This job may be a fit for you if you like working outdoors and are interested in soil chemistry.

Roger Assmus, NRCS, Pierre, SD

NRCS Soil Scientist Kent Cooley explains the properties of a soil profile at the Rangeland and Soils Days event in Phillip, SD.

Lesson 15.2 Review and Assessment

Words to Know

Match the key terms from the lesson to the correct definition.

1. The spaces between soil particles that are filled with either water or air.
2. The soil particle that holds valuable plant nutrients.
3. Soil erosion that occurs when many streams of water come together and form a larger stream.
4. Material composed of decomposing and decomposed living organisms, including their waste products.
5. The buildup of salts in soil that occurs with the frequent use of irrigation.
6. The process in which water dissolves minerals and moves them deeper within the soil profile.
7. A view of the different soil layers made possible with a vertical slice down into the ground.
8. A method of planting where the farmer does not plow or disk the soil, but plants a new crop in the remnants of the previous year's crop.
9. The general condition of the soil with respect to texture, moisture content, aeration, and drainage.
10. The fine, relatively inert soil particles carried by running water and deposited as sediment.
11. The layer of soil found on top of the topsoil.
12. When all pore spaces of soil are filled with water.
13. The top layer of the soil horizon.
14. The underlying soil material that gives rise to the subsoil and topsoil.
15. Soil erosion that occurs when many small streams of water form and are running across the surface of the soil.
16. The process of eroding rock over time, primarily through the action of rainfall and snowfall.
17. The minerals that make up the largest portion of most soils.
18. The layer of soil right below the topsoil.
19. The largest soil particle that does not affect soil fertility, but is important for drainage and aeration.
20. The compression or elimination of pore spaces in soil due to frequent heavy pressures.
21. Rows of trees planted along the edge of a field to slow the speed of the wind.

A. "A" layer
B. clay
C. gully erosion
D. humus
E. leaching
F. no-till program
G. "O" layer
H. parent material
I. pore spaces
J. rill erosion
K. salinization
L. sand
M. saturated
N. silt
O. soil compaction
P. soil particles
Q. soil profile
R. subsoil
S. tilth
T. weathering
U. windbreak

Know and Understand

Answer the following questions using the information provided in this lesson.

1. How many years are required to form one inch of topsoil?
2. Identify the four layers of the soil horizon. Explain how each one affects crop production.
3. List five factors that affect weathering of rock.
4. *True or False?* The particles in mud are primarily sand.
5. Identify three types of organisms that affect the development of soil.
6. Explain why plant roots are important to soil formation.
7. Is organic decay faster or slower in warmer climates? Explain your answer.
8. Explain why acid rain speeds the weathering process of rocks and organic matter.
9. What are the three main soil particles? Which is the largest? Which is the smallest?
10. Which soil component has the most impact on soil fertility? Explain your answer.
11. *True or False?* Humus is usually about 40%–50% of the total volume of soil.
12. Explain why pore spaces are important to soil composition.
13. *True or False?* Loamy soils contain approximately 40% sand, 40% silt, and 20% clay.
14. Explain why it is important to maintain biodiversity in soil.
15. By which natural forces may soil be eroded?

Analyze and Apply

1. What is the connection between pore spaces and a soil's water-holding capacity?
2. How have farmers attempted to reduce compaction in their fields?
3. How do reduced tillage methods reduce soil erosion?

Thinking Critically

1. With climate change and global warming, do you think the rate of soil formation will increase or decrease? Why?
2. If you mow your yard in the same pattern each week, what happens to the soil beneath the lawn mower's tire tracks? Is this healthy for your lawn's soil? Why? What can you do to change this condition?
3. Water erosion of soil also has negative effects on other parts of the ecosystem. What are these other parts of the ecosystem? What are the impacts of water erosion in these areas?

Lesson 15.3

Hydrological Cycles

Words to Know

bog
constructed wetland
fen
hydric soil
marsh
nontidal marsh
septic system
swamp
tidal marsh
watershed
wetland creation
wetland enhancement
wetland restoration

Lesson Outcomes

By the end of this lesson, you should be able to:

- Define and identify watersheds in your state or region.
- Differentiate between groundwater and surface water.
- Explain methods of preventing groundwater contamination.
- Describe and evaluate the function of wetlands.
- Explain the challenges associated with wetland restoration.

Before You Read

As you read the lesson, put sticky notes next to the sections where you have questions. Write your questions on the sticky notes. Discuss the questions with your classmates or teacher.

Water is one of the most abundant substances on the planet. Nearly 70% of Earth is covered with water, mostly in the form of saltwater. Only about 2.5% of the water is fresh, meaning that it is suitable for drinking, watering livestock, and irrigating plants. Even most of this is trapped in ice and glaciers. Freshwater available for agricultural purposes is preciously limited.

The hydrological cycle is the pathway, or cycle, that water follows underneath the surface, at the surface, and above the surface of Earth, **Figure 15-28**. Hydrological cycles, also known as water cycles, impact agriculture in many ways. The parts of the hydrological cycles that impact agriculture the most include: watersheds, groundwater, surface water, and wetlands. These components of the water cycle provide irrigation and drainage, help water filtration, and increase biodiversity that benefits agriculture and the health of the environment in general.

Watersheds

Much of the rain that falls on farmland, rangeland, or even forests is absorbed into the ground. What is not absorbed, pools and flows across the land as *runoff water*. Water always flows along the path of least resistance, which usually means it flows downhill to lower ground. This flow of water begins as small trickling streams. These streams run together across the land and eventually flow into roadside ditches. This water empties into streams, which flow into rivers, which eventually flow into an ocean. This is called a watershed. A ***watershed*** is an area of land that is drained by a

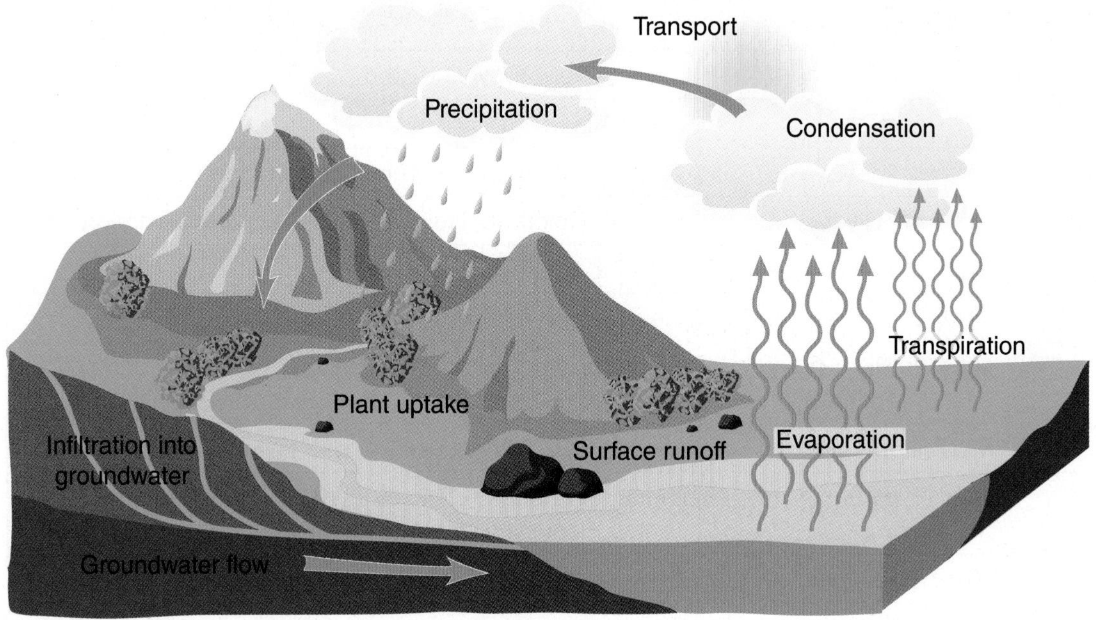

Merkushev Vasiliy/Shutterstock.com

Figure 15-28. The hydrological cycle, or water cycle, includes water above, in, and under the surface of Earth. Every living thing relies on the water cycle.

major body of water. Think of the watershed as a half-funnel of land that drains to one point.

We have several major watersheds in the continental United States, **Figure 15-29**. The largest of these watersheds is the Missouri River watershed, which also happens to drain some of the most productive cropland in the world. Combined, the Missouri River, Upper Mississippi River, and Ohio River watersheds, drain the Corn Belt and Soybean Belt of the United States. This area of our country is highly productive cropland.

Groundwater

The two primary sources of water for agricultural purposes are groundwater and surface water. *Groundwater* is water that resides underneath or in the soil. *Surface water* is the water found in streams, rivers, and lakes.

Aquifers and Wells

Groundwater is stored in underground reservoirs called aquifers, **Figure 15-30**. Aquifers store water in the large spaces formed by cracks and crevices between deep underground rocks. Some of this water has been stored underground for thousands of years. Because it has filtered slowly through many layers of soil and rock, groundwater has been cleansed of most contaminants and is generally suitable for drinking. Farmers and ranchers use wells to tap groundwater and pump it to the surface for irrigation and livestock. Groundwater wells may be less than 100′ deep to over 2,000′ deep. Drilling these wells is expensive.

Watershed Name	Major Waterways	States
Pacific Northwest	Columbia River	Idaho, Oregon, Washington
California	—	California
Great Basin	—	Nevada; parts of California, Oregon, and Utah
Upper Colorado River	Colorado River	Parts of Colorado, New Mexico, Utah, and Wyoming
Lower Colorado River	Colorado River	Arizona
Rio Grande	—	New Mexico and parts of Texas
Missouri	Missouri River	Montana, Nebraska, South Dakota, Wyoming; parts of Colorado, Kansas, Iowa, Missouri, and North Dakota
Upper Mississippi	Mississippi River	Illinois, Iowa, Minnesota; parts of Missouri and Wisconsin
Ohio	Ohio River	Indiana, Kentucky, Ohio, and West Virginia; and parts of Illinois, Pennsylvania, and Tennessee
Lower Mississippi	Mississippi River	Louisiana; parts of Arkansas, Mississippi, and Tennessee
Texas-Gulf Coast	—	Texas
Great Lakes and St. Lawrence River	St. Lawrence Seaway	Michigan; parts of New York, Ohio, and Wisconsin
Mid Atlantic	—	Delaware, Maryland, New Jersey; parts of New York, Pennsylvania, Vermont, Virginia, and West Virginia
New England	Chesapeake Bay	Connecticut, Maine, Massachusetts, New Hampshire, and Rhode Island
South Atlantic Gulf	—	Alabama, Florida, Georgia, North Carolina, and South Carolina; parts of Virginia
Tennessee	Tennessee River	Tennessee; parts of Alabama and North Carolina
Arkansas-White-Red	—	Oklahoma; parts of Arkansas, Colorado, Kansas, Missouri, and Texas

Goodheart-Willcox Publisher

Figure 15-29. The United States has several watersheds that drain our land mass. They come in all shapes and sizes.

Groundwater Contamination

While most groundwater is clean after being filtered by the layers of soil and rock in the earth, it may be contaminated by human activity. Once an aquifer's water is contaminated, it is expensive and nearly impossible to return the water to a pure form.

Groundwater can be contaminated in several ways, including:

- Improper chemical disposal.
- Poorly constructed or maintained septic systems.
- Abandoned wells.

Figure 15-30. The Ogallala Aquifer is one of the largest aquifers in the United States. It also happens to provide water for some of the most productive farmland in America's Breadbasket.

Chemicals

Groundwater is commonly contaminated when people improperly dispose of chemicals, including gasoline, oil, paint, and pesticides. By dumping chemical substances into the soil, or even in a driveway, we introduce the risk of contaminating a groundwater supply. As rains fall on the contaminated soil, it pushes the chemicals deeper and deeper into the soil, where they will eventually encounter the aquifer. Unwanted chemicals should be disposed of properly, **Figure 15-31**. Directions for disposal are available through local municipal waste districts.

Septic Systems

Groundwater contamination can also be caused by poor construction or improper maintenance of septic systems. ***Septic systems*** are wastewater systems used for homes and businesses that do not have access to municipal sewer systems, **Figure 15-32**. All septic systems should be constructed according to state and local regulations. Guidelines for maintaining healthy septic systems must also be followed. If the homeowner or business owner believes the septic system is malfunctioning, a septic professional should be consulted immediately. Neglecting a septic system often leads to expensive

Safety Note

To prevent groundwater contamination from toxic substances in landfills, it is illegal to dispose of some materials with your regular garbage. This includes items such as batteries, medications, paint, and electronic devices. Contact your waste disposal company or municipal waste district for a list of toxic or prohibited materials and the methods or places where they may be disposed of properly or reclaimed for recycling.

s-ts/Shutterstock.com

Figure 15-31. Chemical contamination of soil can occur when pesticides or other chemicals are disposed of improperly. *Have you ever failed to dispose of chemicals properly on your home property?*

repairs that far exceed the cost of proper maintenance.

Abandoned Wells

Unused, abandoned wells may also create opportunities for groundwater contamination. Abandoned and uncapped wells serve as a direct pipeline for contaminants to flow into the aquifer. If livestock are allowed to graze near an open well in the ground, their manure may run into the well during rainstorms. This exposes the aquifer to manure, bacteria, and other microbes. All unused and abandoned wells should be properly capped to prevent contamination, **Figure 15-33**.

Did You Know?

Bacteria living in septic systems break down solid waste and help keep the system functioning properly.

Safety Note

Unused and abandoned wells pose a danger to people and animals. They should be properly capped to prevent anyone or anything from falling in and being trapped, injured, or killed.

Surface Water

Farmers also pump surface water from rivers and streams to irrigate cropland and water livestock. Sometimes the livestock are even allowed to graze near or in the stream if it is slow-moving and not deep. However, the practice of allowing cattle and sheep to graze directly in streams is discouraged for several reasons. Livestock may urinate or defecate in the stream while crossing it, thus contaminating the water supply for other organisms downstream. Also, livestock tend to degrade the stream bank, contributing to bank erosion.

Jo Ann Sover/Shutterstock.com

Figure 15-32. Septic system leaking is one way that groundwater can become contaminated with bacteria.

Dja65/Shutterstock.com

trainman111/Shutterstock.com

Figure 15-33. Abandoned or uncovered wells are common sources of groundwater pollution. With such an open well, contaminants have a direct line to pollute groundwater. ***Have you ever run across an abandoned well?***

Wetlands

As described in *Lesson 14.1*, wetlands are defined as areas of land that are saturated with water throughout the year, either seasonally or permanently, **Figure 15-34**. To meet the definition, they must also support plants adapted for living in saturated soil conditions. A few decades ago, wetlands were thought of as wastelands, and many have been cleared, filled in, and built on. Today, scientists and agriculturists understand the value and vital contributions that wetlands make to our environment. Wetland contributions include:

- The filtering, cleaning, and storing of water as well as the recharging of groundwater supplies.
- Removal of pollution by absorbing chemicals and other pollutants from water.
- Collection and holding floodwaters, keeping rivers at normal levels.
- Absorption of wind and tidal forces to help prevent erosion.
- Feed downstream waters and provide fish and wildlife habitats.
- Release vegetative matter into rivers, which helps feed aquatic life.

As you can see, wetlands are hardworking land areas. Some environmentalists compare them to the human kidneys, as they filter waste and clean water.

Wetlands are commonly found along lakes, rivers, and oceans, and in flood plains. Some wetlands are not connected to any of these types of waterways, but have groundwater

Did You Know?

Wetlands are found on every continent except Antarctica.

Did You Know?

Wetlands make up 5% of the land surface area in the United States but are home to more than 30% of our plant species.

Betty LaRue/Shutterstock.com

Figure 15-34. Wetlands, such as this one in New England, provide vital and valuable environmental services, such as purifying water supplies, controlling flood waters, and providing habitat for wildlife.

proklovvalera/Shutterstock.com

Figure 15-35. Wetlands purify water in the environment. Slow-moving water allows the soil to absorb harmful contaminants from the water. ***After learning about wetlands such as this swamp, how has your impression of them changed? Why?***

connections instead. They may be fed by underground streams or rivers or result from land with high water tables.

Plants

Because the soil is saturated for long periods of time, the roots of wetland plants have adapted to growing in soils that are relatively devoid of oxygen. These plants become firmly rooted in the muddy ground and help control water erosion by retaining the soil against winds and the tide, and slowing the flow of water. Because water moves slowly through a wetland, it is filtered more thoroughly as it flows through the plants and soil, **Figure 15-35**.

Wetland plants themselves also provide some of the filtering function in wetlands. Plants like cattail, algae, and microorganisms use contaminants from runoff water, like excess nitrogen, phosphorus, and other fertilizers, for nutrients to grow. Microbes in the soil also absorb and use contaminants from the water to grow and live. They prevent these contaminants from spilling into waterways where they may cause harm to aquatic life.

Floodwaters

Wetlands also act like sponges during floods. Because the plant life and soils in wetlands are accustomed to being saturated, they can store an excessive amount of water during floods. They are able to hold large volumes of water until the floodwaters recede.

Groundwater Recharge

Wetlands are one area where groundwater supplies are recharged. Water supplies are recharged when rains are allowed to soak slowly into the ground, as with a wetland. Because the water moves so slowly in a wetland, it can seep down through layers of soil and rock beneath the wetland to recharge aquifers. Because aquifers are recharged through wetlands, agriculturists must prevent them from being contaminated.

Wetland Ecosystems

Wetlands are some of the most productive ecosystems in the world. If you recall, an ecosystem is a community of organisms living and interacting with one another and their environment. Wetlands include this living and interaction; however, they also perform the actions listed in the preceding section. And, each wetland is a unique ecosystem because of differences in soils, topography, climate, hydrology, water chemistry, vegetation, animal life, and even the amount of human disturbance it experiences.

Did You Know?

Two-thirds of wetlands in the United States are in Alaska.

Wetland Classification

Wetlands can be freshwater, saltwater, or even brackish water. The kind of water, duration of the water saturation, and the geographic location of the wetland help determine its type. The four main types of wetlands are:

- Marshes.
- Bogs.
- Swamps.
- Fens.

Marshes

A ***marsh*** is an area that is saturated, either occasionally or continually, with water and that fosters soft-stemmed plant growth (mostly grassy plants). Common marshes are prairie potholes and the Florida Everglades, **Figure 15-36**.

There are two types of marshes: tidal and nontidal. In the United States, ***tidal marshes*** are found near the coastlines of the Atlantic Ocean and Gulf of Mexico, **Figure 15-37**. As the ocean tide rises, water flows inland and saturates the soil in low-lying areas. As the tide recedes, or lowers, the water drains out of the marshes and flows back into the ocean. These areas are tidal marshes. Tidal marshes often have areas of open water interspersed with clumps of grasses and a few trees. Tidal marshes are either saltwater or brackish water wetlands. Tidal marshes protect coastlines from wave erosion and absorb excess nutrients in rivers and streams before they reach the ocean. Migratory waterfowl frequently use tidal marshes for food and rest on their migratory flights during the spring and fall. Clams, crabs, and fish are found in tidal marshes.

Nontidal marshes are the most common wetland in the United States and North America. These are inland wetlands, so they are not influenced by the rising tides of the ocean. Most of these are freshwater wetlands. They are depression areas near rivers,

pzig98/Shutterstock.com

Galyna Andrushko/Shutterstock.com

Figure 15-36. A—Canadian geese take off from this South Dakota prairie pothole. Prairie potholes are examples of nontidal marshes. B—The Florida Everglades are another example of a nontidal marsh. Actually, the Everglades are a very wide, slow-moving inland river. The Everglades are home to many wildlife, including the American alligator.

Jorge Moro/Shutterstock.com

Figure 15-37. At low tide, only a small amount of water remains and much mud is exposed in this Maine tidal marsh. Shore birds will quickly scour the area for meals of insects and other aquatic life.

Wildnerdpix/Shutterstock.com

Figure 15-38. Cypress trees dominate this swamp in southern Illinois. *How have trees adapted to survive in swamp environments?*

streams, and even lakes, ranging from a few inches to a few feet deep. The level of water varies throughout the year, and in drought conditions, nontidal marshes may dry out completely.

Common plants found in marshes are cattail, algae, lily pads, reeds, and bulrush. Common animal wildlife include waterfowl, red-winged blackbirds, frogs, turtles, and muskrats.

Swamps

A ***swamp*** is a wetland that is dominated by woody plants, including trees, like these cypress trees pictured in **Figure 15-38**. Swamps, like marshes, receive their water supply through streams or rivers. Swamps protect neighboring lands during flooding by absorbing excess water. Similar to marshes, swamps also remove pollutants from water. Swamp plants also use nitrogen and phosphorus from runoff water that would otherwise kill fish and other aquatic life.

Swamp tree species include maples, oaks, tupelo, and cypress. Swamps also contain shrubs like buttonbush and smooth alder. Animal life in a swamp may include shrimp, crayfish, clams, snakes, otters, fish, and crocodile.

stocksolutions/Shutterstock.com

Figure 15-39. A close-up view of sphagnum moss shows the intricate detail of a prevalent bog plant. *What other types of plants are prevalent in bogs?*

Bogs

A ***bog*** is a wetland that contains peat deposits, acidic water, and carpets of sphagnum moss, **Figure 15-39**. Bogs are critical in mitigating climate change. Peat is rich in carbon, so bogs act as carbon sinks and remove carbon from the atmosphere. Historically, this peat was harvested as a fuel source and soil conditioner. Recently, scientists and agriculturists recognized bogs for their unique contribution to the biosphere and encourage the responsible harvesting and restoration of bogs.

Unlike marshes and swamps, most of the water in a bog comes from precipitation. Because the water in a bog does not come from other surface water sources or runoff, the water does not contain many of the nutrients normally needed to support plant growth. Plant

pls/Shutterstock.com

Figure 15-40. Grasses and sedges have adapted to the surface water sources and runoff water that make up this bog in the northeastern United States. ***Are there any bogs near your home? What types of wildlife inhabit these bogs?***

and animal life in a bog has adapted to these special conditions. Bogs, like the one shown in **Figure 15-40**, are commonly found in the northeastern part of the United States and around the Great Lakes. Carnivorous plants are common in a bog, **Figure 15-41**.

Fens

A *fen* is a wetland that is fed by groundwater and contains peat. Fens are also fed by runoff from surrounding hills. Fens are less acidic than bogs and have higher nutrient levels. Fens are similar to bogs in their functions of aiding in flood control and improving water quality. Fens are commonly found in the northeastern United States and around the Great Lakes. Common plants in a fen are grasses, sedges, rushes, and wildflowers, **Figure 15-42**.

francesco de marco/Shutterstock.com

Figure 15-41. Many types of carnivorous plants may be found in bogs. ***What attracts these types of plants to bogs?***

Wetland Species

Freshwater wetlands, especially nontidal marshes, are some of the most productive ecosystems on Earth. Because

Andrea Wilhelm/Shutterstock.com

Figure 15-42. Plants like these in a fen slow the flow of water so that it can recharge groundwater supplies.

of the abundant water and nutrients, wildlife are also abundant in nontidal marshes. The soils are usually very rich in organic matter from the dead and decaying plant and animal life. Thus, plants adapted to saturated soils grow vigorously in marshes. Marshes also support abundant insect, microbial, and other wildlife.

Did You Know?

An acre of wetland can store 1–1.5 million gallons of floodwater.

Nurseries and Feeding Grounds

Many species use wetlands as nurseries for raising young, and feeding grounds for supplying food and water, even if they do not live there. Different species of fish, amphibians, and reptiles use wetlands as places to reproduce and grow their young. Some mammals and many bird species use wetlands as feeding grounds because of the abundance and diversity of animal life.

Endangered Species

Wetlands are important for the survival of America's threatened and endangered species. More than one-third of threatened and endangered wildlife species live in wetlands, and many more use wetlands at some point in their lives.

Did You Know?

Many of the endangered animals and plants in the United States depend on wetlands in some way. Destroying wetlands may cause their extinction.

Menhaden, striped bass, and sea trout, pictured in **Figure 15-43**, are all commercial and game fish that rely on coastal marshes to breed and raise their young. Shrimp, oysters, and clams spend nearly all of their lives eating, growing, and reproducing in coastal marshes, as well. Migratory birds, such as ducks, geese, and other waterfowl, use wetlands for resting, feeding, and nesting. Without wetlands as places of rest for migratory birds, they would soon become extinct. For some mammals, such as muskrats, beavers, and otters, wetlands are the only places they can live. Peregrine falcons, black bears, raccoons, and deer use wetlands for shelter, food, and especially water.

Wetland Creation and Restoration

In the early and middle 1900s, many wetlands across the Midwestern United States were drained for farmland. Farmers used perforated field tile and drainage ditches to remove the water and prevent it from flooding again. After World War II, more land was needed to grow crops for an ever-increasing population, **Figure 15-44**. Some of these drained wetland areas had fertile topsoil up to two feet deep! This was

Dan Bach Kristensen/Shutterstock.com

Figure 15-43. Sea trout are one of the commercial species of fish that rely on wetlands for feeding, breeding, and nursing young. *What other types of fish return to wetlands for breeding and nursing young?*

Stu49/Shutterstock.com

Figure 15-44. Farmers drained wetlands after World War II using perforated field tile and drainage ditches. *Are there any field tile and drainage ditches near your home? Are there plans to remove them?*

John Panella/Shutterstock.com

Figure 15-45. Constructing a wetland requires much engineering and effort to restore land to a naturally wet, productive state. *Are there any wetland construction or restoration sites near your home?*

incredibly productive for corn, soybeans, and wheat. Today, we know that draining these wetlands had detrimental impacts on migratory birds, flood control, and surface water quality. In an attempt to reestablish these lands, scientists, agriculturists, and environmentalists have begun to create new and restore previously existing wetlands.

Constructed Wetlands

Wetland creation is the process of developing a wetland where one did not previously exist. Another name for a newly created wetland is a ***constructed wetland***. Construction of a wetland may require as much effort as restoring one, **Figure 15-45**. An example of a constructed wetland is found in a housing subdivision or near a large parking lot. Small areas are set aside to catch and control floodwaters running from the subdivision streets and off the parking lots.

Restored Wetlands

Wetland restoration is the process of reconstructing and rehabilitating a previously existing wetland. Another name for this process is ***wetland enhancement***. This often means removing the installed tile drainage and dikes to allow the wetland to flood once again.

One of the challenges for wetland restoration is that once drained, the land that was once a wetland continues to evolve and change. Wetland soils adapted to wet conditions over many centuries. These are called ***hydric soils***, or soils that develop as a result of being saturated long enough to develop anaerobic conditions. When these soils dry out, they require many years of flooded conditions to return to their natural state. The size of the wetland and slope of the sides of the basin, as shown in **Figure 15-46**, also affect the types of plant and animal species that will reinhabit the wetland.

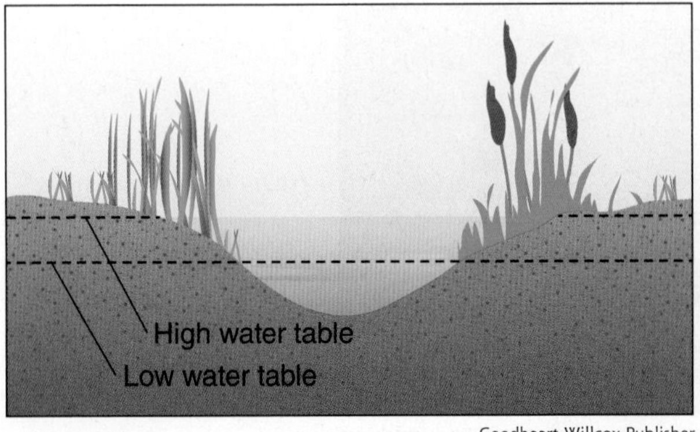

Goodheart-Willcox Publisher

Goodheart-Willcox Publisher

Figure 15-46. Environmental engineers must consider the depth of the water table and the slope of the sides of the wetland basin when restoring wetlands. *What other types of details must environmental engineers consider when constructing wetlands?*

The water quality in a restored wetland also plays a role in the revitalization of wildlife. Many times the land surrounding a wetland has changed since the wetland was last active. If roads are nearby in areas that receive snow, wintertime runoff water may contain deicing salts and other contaminants. These are not suitable for plant and animal life.

Career Connection Hydrologist

Job description: A hydrologist studies how water moves across the top of and within Earth's surface. They have a good understanding of surface and groundwater. Hydrologists work in the field studying water and its movement, and they prepare detailed reports for both public and private use. Hydrologists may help landowners determine where to drill wells. They may also work with hydrofracking operators to determine the impact of natural gas extraction on aquifers. Hydrologists may work anywhere on the globe, including in developing countries.

Education required: A master's degree in water science.

Job fit: This job may be a fit for you if you like working outdoors and are interested in earth and water sciences. A good understanding of physics is also helpful.

africa924/Shutterstock.com

Hydrologists are especially valuable in developing countries where pure, clean water for human and agricultural purposes may be scarce.

Lesson 15.3 Review and Assessment

Words to Know

Match the key terms from the lesson to the correct definition.

1. A wetland that is fed primarily by groundwater and some runoff.
2. An area of land that is drained by a major body of water.
3. An area that is saturated, either occasionally or continually, with water and that fosters soft-stemmed plant growth.
4. The process of developing a wetland where one did not previously exist.
5. Soil that has developed anaerobic conditions due to long-term saturation.
6. A wastewater system used for a home or business that does not have access to a municipal sewer system.
7. The process of reconstructing and rehabilitating a previously existing wetland.
8. A wetland that contains peat deposits, acidic water, and carpets of sphagnum moss and receives its water primarily through precipitation.
9. A wetland dominated by woody plants that receives its water supply through streams or rivers.
10. A wetland found near the coastline of an ocean, gulf, or sea that is subjected to the ebb of the tide.
11. An inland wetland commonly found along rivers, streams, and even lakes, ranging from a few inches to a few feet deep.

A. bog
B. constructed wetland
C. fen
D. hydric soil
E. marsh
F. nontidal marsh
G. septic system
H. swamp
I. tidal marsh
J. watershed
K. wetland restoration

Know and Understand

Answer the following questions using the information provided in this lesson.

1. Explain why freshwater is preciously limited.
2. Identify the parts of the hydrological cycle that impact agriculture the most. What do these components of the water cycle provide?
3. Explain what occurs with precipitation that does not get absorbed into the ground.
4. List the ways in which groundwater is often contaminated.
5. Explain why livestock should not be allowed to graze directly in streams.
6. List three wetland contributions.
7. How do wetland plants help control water erosion?
8. Explain why each wetland is a unique ecosystem.
9. List the four main types of wetlands.

10. Which wetland ecosystems are prevalent in the United States?
11. *True or False?* Swamps receive most of their water supply through precipitation.
12. *True or False?* Bogs act as carbon sinks and are critical in mitigating climate change.
13. *True or False?* Fens receive most of their water from underground rivers and streams.
14. Explain why wetlands are important for the survival of threatened and endangered species.
15. Identify at least two challenges faced by those attempting to recreate wetlands.

Analyze and Apply

1. Where is each ecosystem found in the United States? Download an outline map of the United States and map the location of each type of ecosystem.
2. Identify three species of mammals, fish, invertebrates, reptiles, and birds that live in or use wetlands. Describe how each of these uses the wetlands.
3. Use the U.S. Geological Survey rainfall calculator to determine how many gallons of rain fall in a one-inch rainstorm over your entire school district. How many acres of wetlands would be needed to contain 20% of that rain as runoff?
4. What are your local waste management district's requirements for managing household and agricultural waste? What can be recycled? Why are these policies in place?

Thinking Critically

1. What does an environmental engineer need to consider when planning to restore a wetland?
2. Why would migratory birds become extinct without wetlands?
3. If a wetland were going to be drained for construction of a shopping mall, how would you argue that the wetland provides more value than the shopping mall?

Lesson 15.4
Water Quality

Words to Know

aquaculture
biological availability
chemical pollution
dissolved oxygen (DO)
fecal coliform bacteria
irrigation
macroinvertebrates
microbial pollution
nonpoint source pollution
oxygen-depleting pollution
point source pollution
sediment pollution
seechi disk
solubility
thermal pollution
total suspended solids
turbidity

Lesson Outcomes

By the end of this lesson, you should be able to:

- Identify and explain agricultural uses of water.
- Analyze water quality based on biological and chemical factors.
- Compare and contrast point and nonpoint source water pollution.

Before You Read

Review the lesson headings and use them to create an outline for taking notes during reading and class discussion. Under each heading, list any term in ***bold italics***. Write two questions that you expect the lesson to answer.

Water is the most important nutrient for both livestock and crop production in the United States. Our supply of freshwater is in constant jeopardy from pollutants, drought, and global climate change. Currently, irrigation of agricultural crops accounts for as much as 70% of all freshwater for human use, which includes agriculture and food production. This water must be clean, abundant, and available where crops and livestock are produced.

Maria Uspenskaya/Shutterstock.com

Figure 15-47. A clean, quality water supply is necessary for both crop and livestock production in the United States. *Will livestock drink water from any source?*

Agricultural Use of Water

In agriculture, water is used primarily to water crops and to provide livestock with drinking water, as demonstrated by our friendly sheep in **Figure 15-47**. All crops, whether they are grown in a field or in a greenhouse, rely on freshwater for growth. Many growers use irrigation systems to water their crops. ***Irrigation*** systems use water pumped from a well or river and apply it to a growing crop with sprinklers, flooding, or subsurface piping. Water is the nutrient that most directly affects crop yields.

Livestock

All livestock drink lots of water each day in order to grow, feed their young, and move around. Dairy cows drink as much as 35 to 50 gallons of water each day. That is enough water to completely fill a bathtub! Comparatively speaking, your daily allowance of water would only fill a gallon jug about two-thirds full, **Figure 15-48**.

Farmers and ranchers also use water to clean and cool animals and facilities in livestock and dairy production. They may use less water for these purposes than for livestock drinking water, but it is still significant.

Coprid/iStock/Thinkstock; lontur-vid/Shutterstock.com

Figure 15-48. The small gallon jug represents the daily water intake that you need, and the large 50-gallon barrel represents the daily water intake of a dairy cow. *How many cows are there on a large dairy operation? How much water is needed for their total daily intake?*

Produce

Agricultural uses of water do not stop with raising crops and livestock. Even after harvest, water is used in many ways to prepare and process crops of all types. Growers use water to cool, wash, and move fruits and vegetables throughout processing, **Figure 15-49**.

Aquaculture

Aquaculture industries use freshwater to grow hybrid striped bass, tilapia, and other aquatic livestock. ***Aquaculture*** is the industry that raises fish and other aquatic species for human consumption. These aquatic species rely on a large quantity of freshwater in order to live and grow. Much of this water comes from underground sources, but some of it also originates in surface bodies of water.

Water Health

What factors influence the health of a stream or river? Scientists, and even many classes of agriculture students across the nation, monitor streams regularly. They measure several factors to determine the overall health of the stream. Those factors include:

- Water temperature
- Dissolved oxygen
- Total suspended solids and turbidity
- Acidity and alkalinity (pH)
- Nutrients
- Fecal coliform bacteria
- Macroinvertebrates

Paula Cobleigh/Shutterstock.com

Figure 15-49. Processing fruits and vegetables requires volumes of water for cleaning, cooling, and moving the produce through the processing plant. *How much, if any, water is recycled in this type of processing plant?*

Water Temperature

Water temperature is important because different kinds of fish, reptiles, and amphibians live best in different water temperatures. Each animal species has a preferred temperature range for optimum growth. If the stream's temperature moves too far outside of these preferred ranges, the aquatic life will move on to other more suitable environments or die off in that particular area.

Oxygen Levels

Water temperature also affects how much dissolved oxygen can be carried by the water. Chemical reactions generally increase in speed at higher temperatures and slow down with lower temperatures. The primary chemical reactions involve oxygen. Warmer waters carry less dissolved oxygen than cooler waters, so streams that are generally cooler in temperature are healthier.

Seasonal Changes

Fluctuations in water temperature occur naturally with seasonal changes, shading, rate of flow, and precipitation. Aquatic life have adapted over centuries to these natural changes. But, agriculture and other human activities can also influence the temperature of a stream or river.

Industrial Discharge

The water temperature of rivers or streams is raised primarily through heated industrial discharge water. Hot water coming from discharge pipes, like the one shown in **Figure 15-50**, contribute to elevating a stream's or river's water temperature. The unnaturally high water temperature fosters algae blooms, which eventually deplete oxygen content.

Pavement Runoff

Heated water also flows into streams and rivers off paved roads and parking lots after rainstorms. On hot summer days, pavement can reach temperatures higher than 140°F. This is much warmer than the surface temperatures of either exposed or shaded ground or turf. The rainwater running off pavement, especially blacktop, will be much hotter than water coming off nearly any other surface.

WvdM/Shutterstock.com

Figure 15-50. Hot, contaminated water from a discharge pipe like this one can pollute an entire stream or river in a short amount of time. *Are there criminal consequences for someone who intentionally pollutes public waters?*

Shade

In the past, farmers tried to farm as close to streams or river banks as possible. They cut down trees and removed vegetation on land near streams and rivers as well as along the banks. Removing the trees also removed shading for the stream and the surrounding land. Rain falling on the once shaded land and waterways was quickly warmed by the warmer land temperatures. This made any runoff water flowing into the waterway much warmer than it would have been if it had been falling first on shaded land.

STEM Connection Water Temperature

Measure the surface temperature of pavement in your school's parking lot, the surface ground temperature in grass, and the surface temperature in a shaded area over the course of a couple of weeks. Measure the temperature at the same time each day. Also, measure the temperature of any runoff water from the pavement, sod, and shaded areas. What can you conclude from your monitoring?

J5M/Shutterstock.com

Dissolved Oxygen

All aquatic life require small amounts of oxygen to be dissolved in the water in order to live. ***Dissolved oxygen (DO)*** is the microscopic molecules of oxygen that are dissolved in water. These animals do not breathe air as we do, but their gills and other body organs are specially adapted to harvest dissolved oxygen from the water.

All water carries some level of dissolved oxygen. Healthy waters contain high levels of dissolved oxygen. In unhealthy waters, these levels are too low for some forms of aquatic life to live. Dissolved oxygen is measured in milligrams of oxygen per liter of water (mg/L).

Turbulence

How can farmers and natural resource professionals increase the levels of dissolved oxygen? The amount of dissolved oxygen in a body of water increases when the water is turbulent. Turbulence increases when water flows quickly, when it flows across rocks, and when it falls. Waves also increase levels of dissolved oxygen in water. You can see the bubbles and foam in the image in **Figure 15-51**. Many pond owners install small fountain pumps to circulate pond water and increase the levels of dissolved oxygen.

Sharon Day/Shutterstock.com

Figure 15-51. As water flows over rocks, bubbles of air infiltrate the water. This allows oxygen to become dissolved in water, creating dissolved oxygen.

Temperature

Cooler water carries higher levels of dissolved oxygen than warmer waters. So, shading streams and increasing flow also increase the levels of dissolved oxygen in the water. Also, during summer months, as the air and water temperatures increase, the water carries less dissolved oxygen than in the early spring and fall. Planting trees and other plants along banks will help maintain cooler water temperatures.

Fertilizer Runoff

Agriculture's impact on dissolved oxygen comes from increased levels of chemical fertilizers in runoff water from fields. This nutrient-rich runoff effectively fertilizes the algae and other biological life in a stream, thus depleting the stream of oxygen. When farmers plant buffer strips and riparian zones along streams, they lessen the impact of fertilizer runoff.

Total Suspended Solids and Turbidity

Water scientists are concerned with both the amount of suspended organic material (i.e. algae) and minerals (i.e. soil) found in water because they also affect the health of a waterway. ***Total suspended solids*** is the measure of the weight of these materials in water. ***Turbidity*** is the measure of light penetration and scatter in water. When there are higher concentrations of suspended particles in water, light scatters more and penetrates less. When there are lower concentrations of suspended particles, light penetrates more and scatters less. An image of water with high levels of total suspended solids and high turbidity is shown in **Figure 15-52**.

Agricultural activities can have a dramatic impact on total suspended solids and turbidity. Erosion runoff from agricultural fields carries muddy water with high levels of total suspended solids into rivers and streams. This increases the weight of total suspended solids and turbidity in the water. Total suspended solids is measured in milligrams of suspended solids per liter of water (mg/L). Turbidity is measured by the depth to which you can see a black-and-white circular disk called a ***seechi disk***.

Francesco Scatena/Shutterstock.com

Figure 15-52. Floodwaters may also increase turbidity of streams, rivers, and ponds. *What else might cause turbidity in a stream, pond, or river?*

Acidity and Alkalinity (pH)

Aquatic life generally thrive in pH neutral water. The acidity level (pH) of the stream matters because it determines solubility and biological availability of nutrients and other chemicals in water. ***Solubility*** is the amount of a substance that can be dissolved in water. ***Biological availability*** is the amount of the substance that can be used by aquatic life. The water's pH affects the biological availability of nutrients, such as nitrogen, phosphorus, and potassium, as well as heavy metals, like lead and copper. Aquatic life may need a lot of nitrogen, but not much lead. Yet, pH affects the availability of both.

STEM Connection Turbidity

Think about muddy water. Does it have high or low turbidity? If you stick your hand into muddy water, you soon lose sight of your fingers. When you stick your hand into clearer water, you can see your fingers clearly. Which type of water has more turbidity and a higher volume of suspended solids?

Create an experiment that uses a bucket or buckets of water, measured amounts of topsoil, and stirring action to determine how much time is required for mud to settle in still water. Measure the turbidity at spaced time intervals.

A turbidity curtain was installed at this construction site (right) along a waterway to trap silt and sediment to prevent it from polluting the water during excavations.

Steven Frame/Shutterstock.com

Pollution impacts pH through acid rain and through the discharge of municipal and industrial waters into streams. Acid rain lowers the pH, making water more acidic. Municipalities and industries must monitor their discharge water for certain pH ranges, **Figure 15-53**.

Perfect Gui/Shutterstock.com

Figure 15-53. Municipalities monitor pH and other water quality indicators on a regular basis. Farmers who irrigate fresh fruits and vegetables must also monitor and record water quality regularly.

Nutrients in the Water

Phosphorus and nitrogen are the nutrients that are of major concern in streams. Excess nitrogen and phosphorus may come from runoff from cropland. These nutrients also originate from home yards and gardening. Homeowners may not test their soil and/or they may not apply fertilizers accurately. This often leads to overfertilization. Excess nutrients are not used by the turf, but run off or leach into streams.

Nitrogen and phosphorus work much the same in streams as they do in crops. They both promote the growth of plants. Excess plant and algae growth in streams depletes the water of dissolved oxygen, slows the flow of water, and increases the temperature of the stream. The stream in **Figure 15-54** is choked with algae caused by excessive nutrients in the water.

Chin Kit Sen/Shutterstock.com

Figure 15-54. The algae bloom in this image was likely caused by stagnant water and elevated levels of nitrogen and phosphorus in the water. *Have you ever seen an algae bloom? What did it look like? How did it smell? What impact did it have on wildlife and fish?*

Fecal Coliform Bacteria

Fecal coliform bacteria are microscopic bacteria that originate in the intestine of warm-blooded animals and are passed through the intestines via feces. When livestock manure or human waste enter the stream, fecal coliform bacteria may grow in the water. Fecal coliform bacteria are indicators of disease-carrying organisms that the manure may have introduced into the water.

Fecal coliform bacteria *may* come from livestock manure runoff, **Figure 15-55**, but other nonagricultural sources as well. Failing septic systems and municipal discharge are often contributors to fecal coliform bacteria in streams. Some fecal coliform bacteria introduction occurs naturally when wildlife defecate in streams. This introduction is usually limited in nature. Sunlight and low water temperatures limit the growth of fecal coliform bacteria in streams. The ultraviolet rays of sunlight act as a disinfectant to kill the bacteria.

rthoma/Shutterstock.com

Figure 15-55. When confined animal feeding operations fail to manage manure and feedlot runoff effectively, fecal coliform bacteria may enter local waterways.

Macroinvertebrates

Many organisms in a stream or river provide clues about the health of the water. We often think of fish as the primary life forms found in a river or stream. But, there are many other animals living in the water, as well. ***Macroinvertebrates*** are small animals without a backbone. These include many insects in their larval forms, crayfish, worms, snails, and shellfish, like clams and mollusks. Crayfish, as shown in **Figure 15-56**, are one common macroinvertebrate. How many of these have you seen in a stream?

Kathy Clark/Shutterstock.com

Figure 15-56. Macroinvertebrates are good indicators of stream water quality, even though this one appears to prefer muddy water.

Macroinvertebrates are good indicators of water quality in a stream for several reasons. They are not tolerant of pollution. When the water quality is good, they are found in large numbers. When water quality is poor, they disappear. They are also easy to see and count in a stream. Macroinvertebrates are important parts of the food web in a stream. Thus, other organisms rely on them for food. Macroinvertebrates are sensitive to chemical, biological, and physical changes in the stream. Some of these are undetectable by humans, but no less important to aquatic life.

Water Pollution

Water pollution of any kind can have detrimental effects. Most types of water pollution fall under one of the following types:

- Thermal pollution
- Chemical pollution
- Sediment pollution
- Microbial pollution

Thermal Pollution

Thermal pollution is pollution that elevates the temperature of the water. This type of pollution comes from the discharge of heated water from factories and industries into a stream or lake. Production agriculture seldom introduces this type of pollution into bodies of water.

Chemical Pollution

Chemical pollution is pollution that comes from introducing chemicals that are not naturally occurring into the water. Chemical pollution can arise from agricultural practices. When heavy rains fall just after a farmer has sprayed chemical pesticides or applied surface fertilizers on crops, there is the potential for chemical pollution. The rain washes the chemicals and fertilizers off the crops, and this runoff may flow into a nearby ditch or stream.

pisaphotography/Shutterstock.com

Figure 15-57. Muddy Mississippi River waters carry tons of fertile topsoil from the Midwest and transplant it along the Mississippi River and even the Gulf of Mexico.

Sediment Pollution

Sediment pollution is soil runoff pollution in the form of suspended soil particles from fields. Sediment pollution may be one of the most pressing agricultural issues. Erosion of topsoil, or sediment, is a problem because rains wash valuable topsoil off farmers' fields and because that sediment chokes rivers and streams. In the 1960s and 1970s, sedimentation was a real issue for farmers. Since then, with the implementation of erosion-control practices, farmers have reduced the amount of erosion and sedimentation from their fields. Still, rivers like the Ohio, Missouri, and Mississippi turn muddy from spring rains because of the excessive amounts of topsoil that are washing off agricultural cropland, **Figure 15-57**.

Microbial Pollution

Microbial pollution is pollution that comes from the introduction of detrimental microorganisms into the body of water. As mentioned before, fecal coliform bacteria would be one example of microbial pollution. But, microbial pollution may be the introduction of any microorganisms that do not naturally occur in the stream.

Oxygen-Depleting Pollution

Some types of microbial pollution may also be ***oxygen-depleting pollution***. These types of microorganisms take up oxygen and deplete the water of dissolved oxygen. Many of the microorganisms that might be introduced to a body of water would consume oxygen. Since dissolved oxygen is relatively limited in most streams and lakes, any introduction of microorganisms that consume oxygen reduces the oxygen that is available for desirable species.

Pollution Sources

Pollution entering the stream, river, pond, or lake, regardless of the kind, can come from many sources. Sometimes the pollution source is easy to identify and more difficult to identify at other times.

Point Source Pollution

Pollution that comes from a single source that is easily identifiable is called ***point source pollution***. For example, thermal pollution from a discharge pipe from a factory is an example of point source pollution. Another example of point source pollution is microbial pollution from manure running off a beef cattle or dairy feedlot.

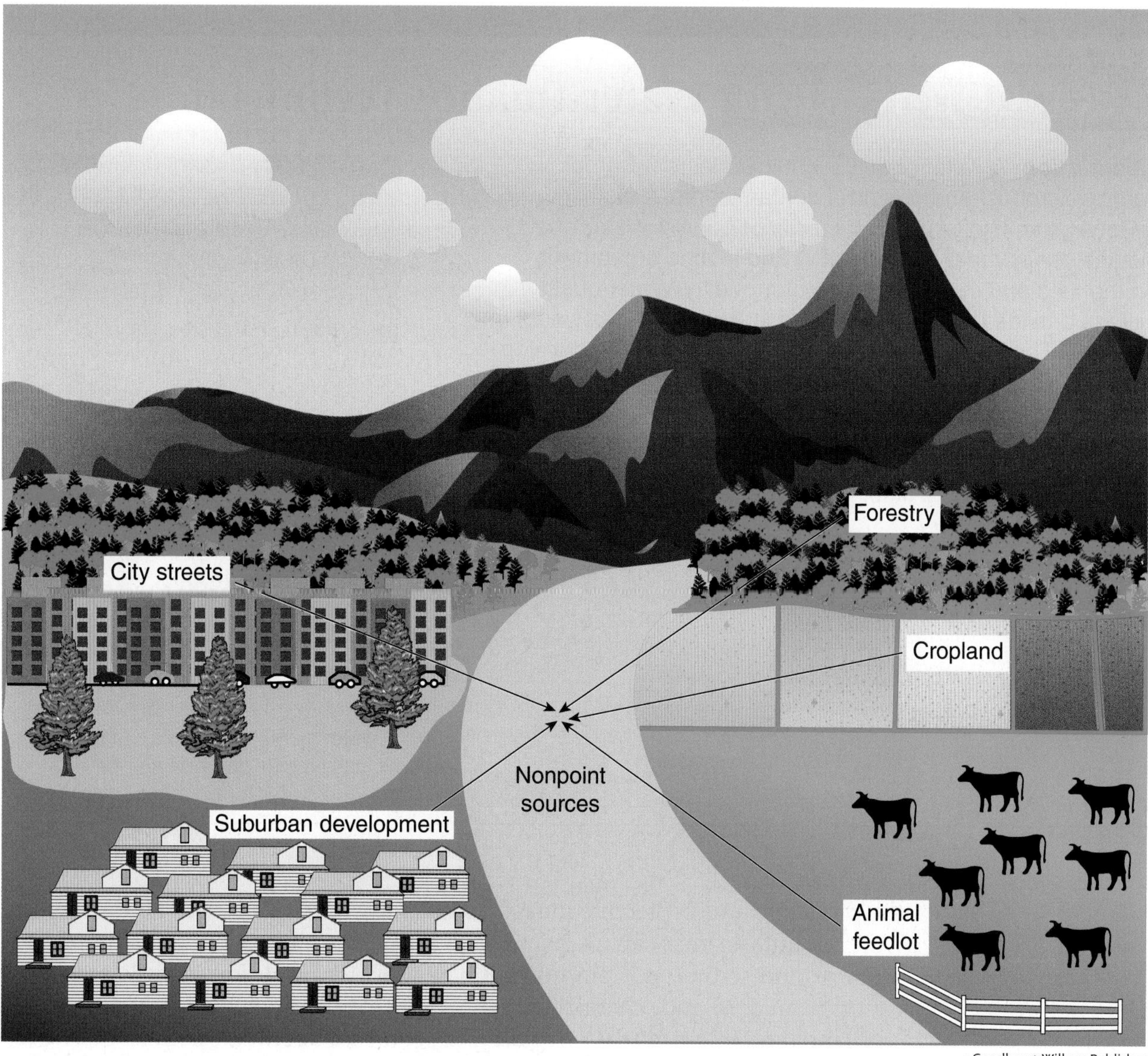

Goodheart-Willcox Publisher

Figure 15-58. Nonpoint source pollution is pollution whose source cannot be identified, and it may come from cities, homes, farms, or industry.

Nonpoint Source Pollution

Pollution whose source cannot be identified is called ***nonpoint source pollution***, as shown in **Figure 15-58**. Acid rain is an example of chemical pollution that is also nonpoint source pollution. It would be impossible to identify the source of the contaminants that caused the acid rain. Another example of nonpoint source pollution is the sediment runoff from cropland. In a muddy stream, scientists and farmers cannot identify which fields contributed to the sedimentation.

Agricultural Pollution

Agriculturists are good stewards of our soil, water, and natural resources. Because of this, they protect our natural resources with many strategies and production methods. Some agricultural pollutants enter waterways in spite of

Career Connection Water Quality Technician

Job description: A water quality technician tests water for pollutants, microbes, and chemical contaminants. A water quality technician is often hired by a city or county. Water quality technicians use a variety of collecting and testing equipment, both in the field and in labs, to test water quality. Water quality technicians provide data and reports to individual homeowners and agriculturists. They work with other ecology scientists to develop solutions for conserving water resources.

Education required: A two- or four-year degree in agriculture or water sciences is required.

Job fit: This job may be a fit for you if you like working outdoors and understand chemistry.

Goodluz/Shutterstock.com

A water quality technician may monitor either groundwater or surface water for contamination.

the proactive and protective measures taken by agriculturists. The four main water pollutants that can originate with farmers are nitrogen, phosphorus, manure, and sediment. Common sources of water pollution include erosion runoff from cropland, chemical runoff from cropland, fertilizer leaching from cropland, and manure waste runoff from confined feeding operations. The intensity of some irrigation systems may also cause chemicals and fertilizers to leach from the soil if they are improperly applied.

Lesson 15.4 Review and Assessment

Words to Know

Match the key terms from the lesson to the correct definition.

1. Water pollution that elevates the temperature of the water.
2. Water pollution that comes from introducing chemicals that are not naturally occurring into the water.
3. The amount of a soluble substance that can be used by aquatic life.
4. Water pollution caused by soil runoff in the form of suspended soil particles from fields.
5. Small, but visible, animals without a backbone.
6. Pollution that comes from a single source that is easily identifiable.
7. The microscopic molecules of oxygen that are dissolved in water.
8. The measure of weight of suspended organic material and minerals in water.
9. System in which water is pumped from a well or river and applied to a growing crop with sprinklers, flooding, or subsurface piping.
10. Water pollution that comes from the introduction of detrimental microorganisms into a body of water.
11. Pollution whose source cannot be identified.
12. Microscopic bacteria that originate in the intestine of warm-blooded animals and are passed through the intestines via feces.
13. The measure of light penetration and scatter in water.
14. Water pollution from microorganisms that deplete the water of oxygen.
15. The amount of a substance that can be dissolved in water.

A. biological availability
B. chemical pollution
C. dissolved oxygen (DO)
D. fecal coliform bacteria
E. irrigation
F. macroinvertebrates
G. microbial pollution
H. nonpoint source pollution
I. oxygen-depleting pollution
J. point source pollution
K. sediment pollution
L. solubility
M. thermal pollution
N. total suspended solids
O. turbidity

Know and Understand

Answer the following questions using the information provided in this lesson.

1. Which nutrient most directly affects crop yields?
2. Identify four factors measured to determine the health of a stream.
3. Explain what occurs with aquatic life if the water temperature moves outside the preferred temperature range.
4. *True or False?* Warmer waters carry more oxygen than cooler waters.
5. Explain why pavement runoff affects water temperatures of streams and rivers.

6. Describe how oxygen becomes dissolved in water.
7. Explain the relationship between water temperature and dissolved oxygen.
8. How can farmers lessen the impact of fertilizer runoff?
9. Which suspended solids affect water quality?
10. How does the water's pH affect biological availability?
11. What are the two most concerning nutrient pollutants in water?
12. Explain how animal and human fecal matter may enter a waterway. What effect might this have on the water?
13. Explain why macroinvertebrates are good indicators of water quality.
14. Describe the difference between point source and nonpoint source pollution.
15. List the four main kinds of agricultural pollution in water.

Analyze and Apply

1. Research 20 vegetables and determine the amount of water used to produce each. Rank order the vegetables by the amount of water used in their production.
2. Create a T-chart with the types of water pollution listed on the left, and provide examples of each kind of pollution on the right.
3. Read the definition of point source pollution in the Clean Water Act. Translate the complex definition into one that you can readily understand.
4. Why would a farmer want to eliminate water pollution on his or her farm?

Thinking Critically

1. Explain how total suspended solids and turbidity affect the health of a waterway.
2. Which conservation practices can farmers use on their farms to improve water quality in nearby rivers and streams? Please explain how two of these practices maintain or improve water quality.
3. Which is more dangerous, point source or nonpoint source pollution? Why?
4. From a land capability classification perspective, why is some land more valuable than others for agricultural purposes?
5. What type of water pollution causes the most damage? What are three ways to reduce this type of pollution?

Lesson 15.5 Conservation Practices in Agriculture

Lesson Outcomes

By the end of this lesson, you should be able to:

- Identify common soil erosion control practices and structures employed by farmers to reduce the impact of soil erosion.
- Identify means of controlling water pollution from agricultural lands.
- Discuss how farmers and ranchers manage the manure waste from livestock operations.
- Explain water conservation practices that are used on irrigated lands.
- Explain the roles of major agricultural conservation organizations in protecting soil and water resources.

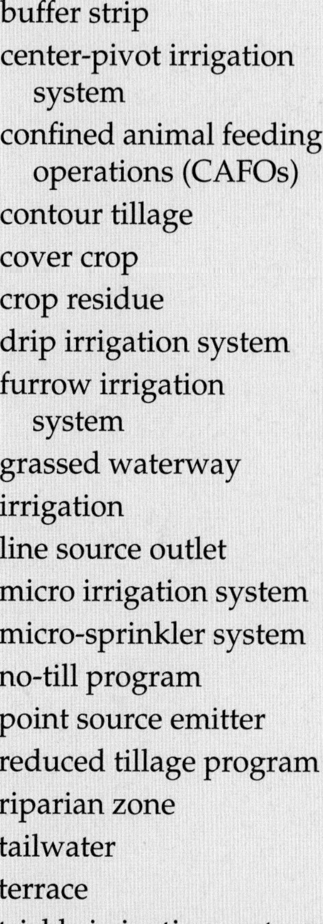

Words to Know

barren
buffer strip
center-pivot irrigation system
confined animal feeding operations (CAFOs)
contour tillage
cover crop
crop residue
drip irrigation system
furrow irrigation system
grassed waterway
irrigation
line source outlet
micro irrigation system
micro-sprinkler system
no-till program
point source emitter
reduced tillage program
riparian zone
tailwater
terrace
trickle irrigation system
water and sediment control basins (WASCoBs)
wheel line irrigation system
wind erosion

Before You Read

Take two-column notes as you read the lesson. Fold a piece of notebook paper in half lengthwise. On the left side of the column, write main ideas. On the right side, write subtopics and detailed information. After reading the lesson, use the notes as a study guide. Fold the paper in half so you only see the main ideas. Quiz yourself on the details and subtopics.

Agriculturists know that their most precious resources are clean water and fertile soil. Water enables crops and livestock to grow and produce abundant harvests. Our soil is the life-giving skin on the surface of Earth and produces food for both humans and livestock. In the 1700s and 1800s, colonists and early farmers did not know how to protect their soil and water resources. Growing corn, cotton, and tobacco year after year on the same acreage depleted the soil of necessary nutrients. With no means of replacing these, colonists became pioneers and simply moved west, cut down forests, and grew crops on new land.

Today, we cannot grow crops on "new land." Instead, agriculturists must conserve their soil and water resources for future generations. Actually, farmers and ranchers are the first line of defense in protecting our soil and water resources. These resources allow Americans to spend relatively little of their income on food, less than any other country in the world. So, what are these soil and water conservation resources that agriculturists use to ensure sufficient soil and water for continual food production?

Soil Erosion Control

Although water is the most important nutrient for both livestock and crop production, it is also the main threat to soil erosion. Rains that fall on barren soil dislodge soil particles. Once dislodged, these soil particles flow

KAMONRAT/Shutterstock.com

Figure 15-59. Eroded gullies like this one can wash tons of topsoil into nearby rivers and streams over the course of a growing season. *What do you think this farmer should do to remedy the gully erosion in this image?*

easily in muddy streams across the surface of the land. A muddy stream can soon wash tons of topsoil off fields and into nearby streams, **Figure 15-59**. Once the topsoil is washed from the field, the land will never regain its original level of productivity.

Topsoil may be recovered as sediment in streams and rivers by dredging, but this is costly and time-consuming. Spreading the sediment on fields is also time-consuming. And, sediment that is returned to a field is never as productive as the original topsoil. Additionally, sediment from fields reduces the quality of water in waterways and can even kill some aquatic life. So, farmers' best interests are served by keeping soil in the fields.

Farmers use several methods to diminish erosion caused by rain and other forms of precipitation, including:

- Leaving crop residue on fields
- Planting cover crops
- Using reduced tillage programs
- Terracing land
- Planting grassed waterways
- Installing water and sediment control basins

Crop Residue

A field is most vulnerable when it is ***barren***, meaning that no crops are growing. Crops in many parts of the United States are harvested in late fall, and the new crop is not planted until spring. So, from October through April, much farmland is barren and highly susceptible to erosion. Barren fields have no plant leaves and stems to break the fall of raindrops on the surface of the soil.

Plants need not be green and growing to break the fall of raindrops. ***Crop residue*** is the remnants of the previous year's crops left on the field after the harvest, like the dried cornstalks in **Figure 15-60**. In a corn field that has been harvested, the harvester leaves behind stalks, leaves, and cobs. These dry plants break the fall of raindrops on the soil and help keep the soil in place. The slow decomposition of these materials will also add nutrients to the soil.

Dan Thornberg/Shutterstock.com

Figure 15-60. Crop residue, like these dried cornstalks and leaves, helps protect barren soil from the force of raindrops. The dry stalks also absorb water and help provide a natural mulch for the topsoil.

Slow the Pace of Water

Crop residue creates a lot of small, natural obstacles over which runoff water must flow. The dry plants also absorb quite a lot of rainwater. All of this slows the pace of the runoff water. Slow water has less energy to dislodge soil particles than fast-moving water does. The roots of the dry plants also hold soil particles in place. When a field has plant roots and/or harvested plant residue on or near the surface, soil erosion is reduced.

Mind the Hillsides and Slopes

A perfectly flat field will not produce a lot of erosion in regular rainstorms. But, fields are not typically flat. Soil erosion is a greater threat where fields are sloped or on hillsides. Since fast-moving water can more easily dislodge soil particles, the most erosion occurs where the slopes are the steepest. Farmers can leave crop residue on these fields to slow water flow and minimize soil erosion.

Cover Crops

A ***cover crop*** is a crop that is planted in the fall after the previous growing season's crop has been harvested. The cover crop grows a little bit to provide a blanket of vegetation over the soil to control erosion and reduce weeds and pests. Unlike other agricultural crops, the cover crop is not harvested and sold. Generally, the cover crop is plowed under in the spring to provide a green manure to the soil. Cover crops protect the soil and add fertility to the land. Typical cover crops include hairy vetch, red clover, white clover, birdsfoot trefoil, and ryegrass, as shown in **Figure 15-61**.

Richard Thornton/Shutterstock.com

Figure 15-61. A cover crop of triticale protects the soil from wind and water erosion when rowcrops are not growing on the field.

Lurin/Shutterstock.com

Taina Sohlman/Shutterstock.com

Figure 15-62. A—Using a moldboard plow to plow a field turns all of the crop residue under the surface of the soil and exposes bare soil to wind and water erosion. B—The tines of this tillage tool loosen the soil but do not turn under all of the crop residue. *Think about all the farm fields you have observed lately. How many were moldboard plowed? Probably very few. How can you explain this?*

Reduced Tillage Programs

Reduced tillage programs are tillage systems where farmers leave some or all of a previous crop's residue on the surface of the soil. In conventional tillage systems, a farmer might plow the field with a moldboard plow, then disk the field a couple of times, and finally plant the crop. This process turns all of the crop residue underneath the soil surface. With reduced tillage programs, farmers will typically till the soil with a combination tillage implement only once before planting the crop, **Figure 15-62**. This leaves more crop residue on the surface of the soil and does not chop the soil into quite so fine particles as conventional tillage systems.

No-Till Programs

Progressive farmers have adopted ***no-till programs***, where they never till the soil, but plant directly into last year's crop residue. Undisturbed soil is much less susceptible to soil erosion. So no-till programs reduce soil erosion more than any other tillage system. However, no-till programs rely on the application of pesticides, especially herbicides, more than other tillage systems.

With both reduced tillage and no-till programs, farmers do not make as many passes across the field. This reduces erosion, but has other benefits as well. Fewer passes across the field mean that less diesel fuel is used; therefore, costs are reduced. With fewer passes, there is also less soil compaction, so the soil can absorb more water and the crop's roots penetrate the soil with greater ease.

Terraces

Where slopes are especially steep and rainfall is frequent, farmers may install terraces in their fields. ***Terraces*** are soil-formed structures in the field that break a steep slope into a series of steps. Sometimes the steps are quite dramatic and look like the steps in your house. Here in the United States, most of the terraces are less dramatic. Terraces are generally a mound of soil that follows the contour of the slope at fairly regular intervals to slow the flow of runoff water across the surface of the soil.

American Spirit/Shutterstock.com

Figure 15-63. Fields on steep slopes like these rely on contour tillage and small mounds of soil to slow the speed of surface water rushing down the slope. *Are there specific challenges contour tillage presents to producers?*

Contour Tillage

Farmers generally till and plant in lines that are perpendicular to the slope. ***Contour tillage*** is the tillage that follows the contour of the slope in this fashion. The rows of a crop are generally raised just a bit, so contour tillage creates a series of small obstacles to slow the flow of water down the hillside, **Figure 15.63**.

Grassed Waterways

Grassed waterways are planted as barriers to prevent water erosion. ***Grassed waterways*** are grassy areas planted to channel water down hillsides, **Figure 15-64**. These waterways are always planted and mowed as grass, so no crops grow in them. They provide a protected channel for water to run over the surface of the soil. The grass and its roots protect the soil from forming gullies when water flows.

Travis Park/Goodheart-Willcox Publisher

Figure 15-64. Grassed waterways provide protected channels for water to flow off a field.

Water and Sediment Control Basins (WASCoBs)

Water and sediment control basins (WASCoBs) are relatively low earthen embankments built perpendicularly across a watercourse. A sophisticated basin is shown in **Figure 15-65**. They typically have a downspout in the center of the basin. The ridge stops the flow of water, and then the downspout channels surface flowing water to underground tile drainage.

Travis Park/Goodheart-Willcox Publisher

Figure 15-65. This erosion control structure collects rainwater flowing off a large field into grassed waterways, which empty into the rip-rap stone and collection basin. *Are there places near your home or school that would benefit from an erosion control structure?*

KPG_Payless/Shutterstock.com

Figure 15-66. Rows of trees planted on the edge of a field create a windbreak to reduce soil erosion by wind, and also provide refuge for wildlife. ***For what other purposes are windbreaks used?***

WASCoBs reduce the erosive potential of running water by stopping the flow of that water. They also channel the water off the surface of the soil to underground drainage systems. This prevents surface water from creating gullies of erosion in drainage paths of fields. On the other hand, WASCoBs are fairly expensive to install and permanently remove part of the field from production.

Windbreaks

Windbreaks are rows of trees planted on the western edge of a field that create a natural barrier to slow the flow of wind, **Figure 15-66**. In many parts of the United States, winds create almost as much erosion as water. Dry soil is quite susceptible to wind erosion. ***Wind erosion*** is erosion of the soil caused by wind.

Water creates bonds that make soil sticky. When soil dries out, these bonds disappear. When these bonds disappear, the soil is susceptible to wind erosion. Think about the Dust Bowl during the Great Depression. After years of drought conditions in the Great Plains, winds blew and dislodged soil particles, creating major dust storms. Some of the dust storms were so massive that they deposited dust from Oklahoma on the steps of the United States Capitol in Washington, D.C.

Water Pollution Control

In addition to minimizing water waste, water conservation includes the maintenance of both surface water and groundwater health. Ways in which agriculturists can minimize the pollution of waterways include the employment of soil erosion controls and the controlled application of fertilizers and pesticides.

Agriculturists attempt to control water pollution for a variety of reasons:

- Soil particles that erode and wash into streams and ditches can never be totally reclaimed as productive soil.
- Fertilizers and chemicals that leach or wash into waterways cannot nourish a growing crop or control pests.
- Manure that washes into a stream often kills fish and other aquatic life.
- The cleanup and fines from the Environmental Protection Agency (EPA) are often expensive enough to drive the farmer out of business.

So, farmers understand the need to prevent water pollution from agricultural sources.

Chemical Application

Growers implement several measures to prevent water pollution. The least expensive method of pollution abatement is the wise application of chemicals and fertilizers. Farmers watch weather conditions and apply chemicals and fertilizers when conditions are optimal. Generally, this means that the farmer applies chemicals on sunny days with low wind and low probability of heavy rains within 24 hours. They spread fertilizer when the fields are dry and no rains are forecast for a few days.

Water and Sediment Control Basins (WASCoBs)

Farmers also install more expensive and invasive structures in their fields to control erosion and fertilizer and chemical runoff. These structures include grassed waterways, terraces, and water and sediment control basins (WASCoBs). These structures slow rainwater runoff and trap sediment before it can infiltrate waterways.

Riparian Zones and Buffer Strips

The land surrounding bodies of water is especially susceptible to erosion, pollution, and other forms of environmental degradation. Sometimes, farmers must plant either buffer strips or riparian zones to protect streams, rivers, and lakes from contamination.

Buffer Strips

A ***buffer strip*** is a parcel of land that is planted with permanent vegetation to prevent soil erosion, pesticide pollution, or fertilizer contamination. Buffer strips can buffer wind, water, or chemical contaminants. A grassed waterway and a riparian zone are examples of buffer strips. Windbreaks are buffer strips used to control the flow of wind across a field or farm.

Riparian Zones

A ***riparian zone*** is a buffer strip that buffers between land and a body of water. Plant vegetation slows the flow of surface water into the stream or river, and its roots hold soil particles in place. A drainage ditch with a small

Nagel Photography/Shutterstock.com

Figure 15-67. The drainage ditch in this image should have a wider riparian zone between the growing crop and the open body of water. *What other benefits would the riparian zone provide?*

riparian zone is shown in **Figure 15-67**. When the runoff contains fertilizers or pesticides, the plants in the buffer strips can absorb and use these nutrients and chemicals in a productive manner.

Wildlife Habitats

Both buffer strips and riparian zones provide habitat for an abundance of wildlife. These areas are generally grasslands with a few trees, so they provide good food sources for game birds and mammals. Reptiles, amphibians, and fish also thrive in or near riparian zones. Some wildlife even use these areas as connecting transition zones between larger bodies of habitat. Farmers typically manage these areas minimally, so they attain a natural habitat fairly quickly.

Livestock Waste Management

Livestock waste management is a special concern for agriculture. ***Confined animal feeding operations (CAFOs)*** are generally large-scale livestock operations that feed many animals in a relatively small space. This space may be indoors, as with swine and poultry, or outdoors, as with beef and dairy cattle. One of the primary concerns with CAFOs, from an environmental standpoint, is the management of livestock manure and waste water. As you can imagine, when a large number of livestock are raised in a small space, they produce a large quantity of waste, **Figure 15-68**. This waste must be managed effectively so that it does not pollute any part of the environment, especially waterways.

rthoma/Shutterstock.com

Figure 15-68. Large-scale livestock operations produce a large amount of animal waste. Responsible management of animal waste is essential to prevent pollution to water and land resources.

Livestock Waste as Fertilizer

Livestock manure is an excellent source of nitrogen and other plant nutrients, so farmers generally use the manure as a crop fertilizer. When applying livestock waste as fertilizer, farmers use equipment that injects the manure into the soil instead of just spreading it on top. This prevents the waste product from being washed away as runoff during rainstorms. Manure may also be spread on the soil surface and quickly tilled into the ground for the same effect. Regardless of the waste management practices they employ, farmers keep strict records of how livestock waste is managed in their facilities.

Irrigation and Water Conservation

In the simplest definition, ***irrigation*** is the artificial application of water to the land or soil. It is used in the production of agricultural crops as well as for turfgrass and in the maintenance of landscapes. For crop production, irrigation systems commonly apply water on soils that are not adapted to heavy water use. In order to prevent waste and both surface water and groundwater contamination, the farmer must closely monitor soil conditions and tightly regulate water application.

Irrigation Scheduling

Scheduling irrigation is important. Farmers monitor the soil moisture and weather patterns to ensure that the soil receives only the amount of water the growing crop needs. Farmers also monitor the crop to match the stage of growth and water requirements for that stage and the amount of water applied to the field. Farmers schedule irrigation to prevent water pollution and reduce their production costs.

Agricultural scientists work to develop new technologies to identify and match these water requirements. Some have even developed computer chips that can be inserted into target plants. These computer chips send information about the plants' water conditions to computers that regulate the amount of irrigation water applied to the field.

Irrigation Systems

There are various types of irrigation systems available, including:

- Surface (furrow, flood, level basin)
- Localized (drip, spray, micro sprinkler, bubbler)
- Sprinkler (center pivot, wheel line)

Irrigation systems vary in price, fuel use, labor requirements, and efficiency. The best type to use depends on the crop(s) being grown, the soil type, available water source, and the sizes and shapes of the fields.

Surface Irrigation

Different kinds of irrigation systems introduce varying degrees of opportunity for water pollution. Surface irrigation systems, such as ***furrow irrigation systems***, flood the field with water, allowing it to flow slowly across the field in furrows, **Figure 15-69**. The slow-moving water filters into the soil as it flows across the field. Because the high end of the field receives water first, this type of irrigation requires more water and some inherent runoff to adequately soak crops on the lower end of the field. Surface irrigation systems may be the least expensive to use, but they require more care to prevent excess runoff and water contamination.

Localized Irrigation

Drip irrigation systems are watering systems in which water drips slowly to the roots of plants through a network of tubing, valves, pipes, and emitters. They are also referred to as ***trickle irrigation*** or ***micro irrigation systems***. Drip irrigation systems are either point source or line source systems.

sunsetman/Shutterstock.com

Figure 15-69. Furrow irrigation is often used in vegetable fields. *What concern does this tailwater (runoff water) present to the farmer?*

They may be installed above- or belowground, often depending on the permanency of the application. Point source systems use ***point source emitters*** (output devices used for individual plants), and line source systems use ***line source outlets*** (tape or hose with outlets spaced between 4″ and 24″ apart). Compared to other irrigation systems, drip irrigation systems use less water, are energy efficient, require less labor, and can be easily automated. Drip irrigation systems are commonly used with permanent crops such as fruit and nut trees, **Figure 15-70.**

vallefrias/Shutterstock.com

Figure 15-70. Some drip irrigation systems are designed so the grower may add more water to a particular plant or tree and less to another simply by adding another ring of tubing around the base of the tree.

Micro-sprinkler systems are similar to drip irrigation systems except they use small sprinkler devices that disperse water over a larger surface area. The main issue with drip irrigation systems and micro-sprinkler systems is clogging due to algae or bacteria growth, the buildup of mineral deposits, or other organic matter. These types of systems are not practical for use with large-scale annual crops such as soybeans, cotton, wheat, corn, or even most vegetables.

Sprinkler Irrigation

Sprinkler irrigation systems are more precise in their water use than flood or furrow irrigation systems. Center-pivot irrigation and wheel line irrigation are the two most common sprinkler irrigation systems. They are shown in **Figures 15-71, 15-72,** and **15-73**. ***Center-pivot irrigation*** is a tall sprinkler that pivots in a circle across the field. At the pivot point is the water source, usually a well or water main line. ***Wheel line irrigation systems*** use an irrigation pipe that runs through the center of a series of wheels, which moves across the field in a line parallel to the sides of the field. Wheel line irrigation is closer to the top of the crop. Both of these systems provide a uniform application of water to the field that can be strictly regulated with a computer, depending on the crop needs and soil dryness.

B. Brown/Shutterstock.com

Figure 15-71. This is the center pivot of a center-pivot irrigation system, where water is pumped from a well out to the sprinklers along the expanse of the system.

Tailwater Collection and Return

Tailwater is the runoff water from irrigation systems. Again, farmers manage their irrigation systems to reduce and hopefully eliminate any tailwater from their fields. Tailwater is most commonly associated with flood irrigation systems. Farmers collect this water from one or a few drainage points at the lower edge of a field. They must either return this water to the field to allow it to infiltrate the soil or treat the water before releasing it to nonagricultural land.

Tim Roberts Photography/Shutterstock.com

Figure 15-72. Center-pivot irrigation systems rotate in a circle around the field with the well in the center of the field. ***Are there any irrigated fields near your home or school? What type(s) of irrigation is the producer using?***

Denton Rumsey/Shutterstock.com

Figure 15-73. Wheel line irrigation systems move across the field in a line that is parallel to one of the sides of the field.

Water Management Organizations

Several government agencies work with farmers, ranchers, and growers to conserve water on agricultural lands. The frontline for this work is the *Natural Resources Conservation Service (NRCS)*. The NRCS works to preserve watershed water quality throughout the United States. NRCS also provides programs, incentives, and advice for farmers who want to prevent water pollution on their lands.

The Environmental Protection Agency (EPA) monitors water quality in lakes and streams throughout the United States. From a water perspective, the EPA works to monitor water quality and clean up polluted waterways when and where they happen. EPA scientists pay attention to rivers, lakes, watersheds, wetlands, and even our oceans and coastlines. They are often the first people to detect where pollution has occurred.

Most states have an association of the *Soil and Water Conservation District (SWCD)* that works to provide advice and education about soil and water conservation to landowners and the general public. SWCDs are organizations made up of agriculturists who are committed to soil and water conservation and who implement local conservation programs. The SWCD implements conservation programs on both public and private lands.

Career Connection Heavy Equipment Operator

Job description: A heavy equipment operator may work with ecologists, civil engineers, and farmers to change the landscape in order to accomplish agricultural or developmental objectives. Some heavy equipment operators install and maintain tile drainage under row crop farmland. Others work with civil engineers to construct water retention ponds and runoff diversion systems to control erosion and water pollution. They work with GPS technologies to ensure that the constructed landscape performs as desired to protect and preserve as much of the environment as possible.

Education required: A two-year degree in heavy equipment or diesel mechanics.

Job fit: This job may be a fit for you if you like operating bulldozers, excavators, tilling machines, and other heavy equipment.

michaeljung/Shutterstock.com

A heavy equipment operator works with environmental scientists to manage use of the environment and conservation of natural resources.

Lesson 15.5 Review and Assessment

Words to Know

Match the key terms from the lesson to the correct definition.

1. A large-scale livestock operation that feeds many animals in a relatively small space.
2. The runoff water from irrigation systems.
3. A field with no crop growing.
4. When farmers till and plant in lines that are perpendicular to the slope of the field.
5. The remnants of the previous year's crops left on the field after the harvest.
6. Soil-formed structure that breaks a steep slope into a series of steps.
7. Crop watering system in which the field is flooded with water, allowing it to flow slowly across the field in furrows.
8. Grassy area planted by farmers to channel water down hillsides.
9. A parcel of land that is planted with permanent vegetation to prevent soil erosion, pesticide pollution, or fertilizer contamination.
10. Crop watering system in which an irrigation pipe runs through the center of a series of wheels that moves across the field in a line parallel to the sides of the field.
11. A buffer strip that buffers between land and a body of water.
12. A planting method in which the farmer does not till the soil, but plants directly into last year's crop residue.
13. A crop that is planted in the fall after the previous growing season's crop has been harvested.
14. Tillage systems where farmers leave some or all of a previous crop's residue on the surface of the soil.
15. The erosion of the soil caused by wind.
16. Crop watering system in which a tall sprinkler pivots in a circle across the field; the water source is typically at the pivot point.
17. Relatively low earthen embankments built perpendicularly across a watercourse; designed to stop the flow of water across crop fields, trap sediment, and channel water to underground drainage systems.

A. barren
B. buffer strip
C. center-pivot irrigation system
D. confined animal feeding operations (CAFOs)
E. contour tillage
F. cover crop
G. crop residue
H. furrow irrigation system
I. grassed waterway
J. no-till program
K. reduced tillage program
L. riparian zone
M. tailwater
N. terrace
O. water and sediment control basins (WASCoBs)
P. wheel line irrigation system
Q. wind erosion

Know and Understand

1. *True or False?* Water is the most important nutrient for both livestock and crop production.
2. Identify five methods used by farmers to diminish erosion caused by rain and other forms of precipitation.
3. Explain reasons why crop residue is left on a field after harvest.
4. How does slowing the pace of water across a field help prevent soil erosion?
5. Which tillage methods are used to prevent soil erosion? Explain how each method helps prevent soil erosion of cropland.
6. How do farmers create windbreaks? What purpose do windbreaks serve?
7. Identify and explain four ways in which agriculturists can minimize the pollution of waterways.
8. Why is waste management important for confined animal feeding operations (CAFOs)?
9. *True or False?* Farmers can prevent waste product from being washed away as runoff by injecting manure into the soil instead of just spreading it on top.
10. What factors are used to determine the best type of irrigation systems for a specific application?

Analyze and Apply

1. Using the principles of controlling soil erosion, how does a cover crop reduce soil erosion?
2. How does soil erosion present two losses, one to the farmer and the other to the environment?
3. Why does soil that is dredged out of rivers and streams and returned to a field never regain its previous level of productivity?
4. There are many different organizations that work with soil and water conservation. Identify and research each organization. Create a graphic organizer to connect the similarities of organizations and to show their differences. You may present your findings to the class.

Thinking Critically

1. If you were a farmer and you had a gently sloping field of about 4%–6% slope, what types of soil and water conservation practices would you use while producing a bumper crop?
2. How would you argue to a farmer that he or she remove a portion of their productive farmland and convert it into a riparian zone? What are the benefits for wildlife, the environment, and even the farmer?
3. Irrigation presents a trade-off between producing food and artificially changing the soil and water in an area. What are the benefits and drawbacks of irrigating cropland?
4. What are the potential benefits of returning tailwater to cropland?

Chapter 15

Review and Assessment

Lesson 15.1

History of Conservation in the United States

Key Points

- The conservation movement has been around since the middle 1800s in the United States.
- Early conservationists included Henry David Thoreau, John Muir, and Theodore Roosevelt.
- Conservation means that natural resources are protected, but also used for human purposes.
- Preservation means that natural resources are to be protected, but not used for human purposes.
- The Dust Bowl of the 1930s initiated many conservation programs and practices that are still in place today.
- Rachel Carson's book *Silent Spring* challenged our notions about the use of chemical pesticides in the environment.
- Several federal agencies have conservation missions.

Words to Know

Use the following list and the textbook glossary to review and study the *Words to Know* from *Lesson 15.1*.

aquifer
Civilian Conservation Corps (CCC)
Clean Air Act (CAA)
Clean Water Act (CWA)
conservation
conservation reserve program
DDT
Environmental Protection Agency (EPA)
National Association of Conservation Districts (NACD)
National Oceanic and Atmospheric Administration (NOAA)
Natural Resources Conservation Service (NRCS)
preservation
Soil and Water Conservation District (SWCD)
soil conservation practice
United States Forest Service (USFS)

Check Your Understanding

Answer the following questions using the information provided in *Lesson 15.1*.

1. Who were two prominent conservationists before Theodore Roosevelt?
2. Which current federal agency related to natural resources conservation was created by Theodore Roosevelt?
3. A person who would recommend that we cease all timber harvests and cease all hunting activities would be classified as a(n) _____.
4. A person who would recommend that we harvest timber if it is managed so that new trees are planted and that we allow hunting to control game populations would be considered a(n) _____.
5. What natural resources held the soil of the Great Plains in place prior to the Dust Bowl?
6. How far away was dust from the Great Plains deposited during the Dust Bowl?
7. What was the name of the major Civilian Conservation Corps project to reduce wind erosion?
8. Where was the previously mentioned project located?
9. Farmers in the Midwest draw water from which aquifer?
10. Which author and what book highlighted the need to manage agricultural chemicals in the environment?
11. If people wanted to work in their local area to promote soil and water conservation, which local organization would they contact for assistance?
12. What does a farmer do with his or her land when he or she enters it into a conservation reserve program?

Lesson 15.2

Soil Formation and Properties

Key Points

- Soils provide the necessary nutrition and support for the crops that feed the world.
- Healthy soils contribute to the carbon, nitrogen, water, and energy cycles.
- The soil particles, pore spaces, and living organisms in soils create a living ecosystem that is unparalleled in its value to life on Earth.
- The thin layer of soils covering our planet must be protected against erosion, salinization, and compaction in order for it to remain fertile and productive.

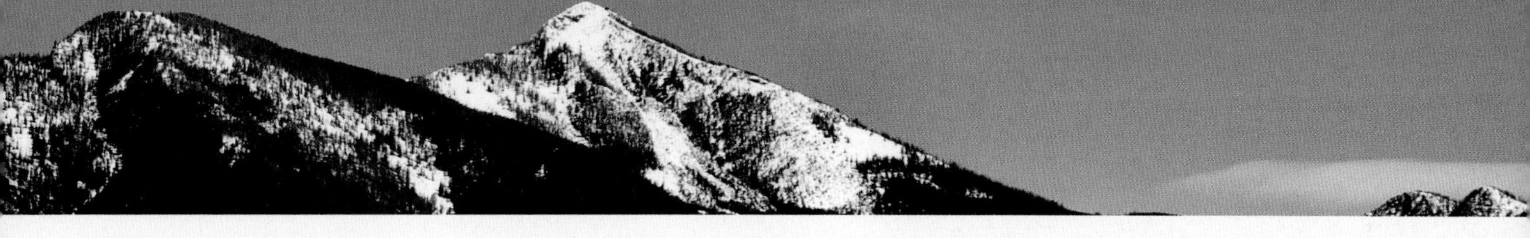

Words to Know

Use the following list and the textbook glossary to review and study the *Words to Know* from *Lesson 15.2.*

"A" layer
"B" layer
carbonates
clay
"C" layer
cover crop
gully erosion
humus
land capability classes
leaching
loamy soil
no-till program
"O" layer
organic layer
organic matter
parent material
peds
pore spaces
reduced tillage
rill erosion
salinization
sand
saturated
silt
soil compaction
soil erosion
soil particles
soil profile
soil structure
soil texture
subsoil
tillage system
tilth
topsoil
weathering
windbreak

Check Your Understanding

Answer the following questions using the information provided in *Lesson 15.2.*

1. Which layer of soil contains the most nutrients for the growing crop?
2. *True or False?* The "B" layer of soil is important to growing crops because it helps with soil drainage and anchoring plants into the soil.
3. Explain how freezing and thawing contribute to soil formation.
4. How can a farmer raise the pH of his or her soil?
5. In which part of the United States are soils most acidic?
6. Which type of soil has the largest pore spaces?
7. What materials can occupy pore spaces in soil?
8. What does a loamy or silty soil feel like in your hand?
9. What does a sandy soil feel like in your hand?
10. Which kinds of microorganisms live in soil?
11. How can a farmer increase biodiversity in soil?
12. Which tillage system disrupts soil the least?
13. How does rill erosion differ from gully erosion?
14. Which soil class is most usable by all cropping systems?

Lesson 15.3

Hydrological Cycles

Key Points

- A watershed is an area of land that is drained by a major body of water.
- Groundwater is water that resides underneath or in the soil.
- Surface water is the water found in streams, rivers, and lakes.
- Preventing groundwater contamination may be accomplished by properly disposing of chemicals, properly constructing and maintaining septic systems, and capping abandoned water wells.
- Wetlands filter, clean, and store water; recharge groundwater supplies; remove pollution by absorbing chemicals and other pollutants; collect and hold floodwaters; absorb wind and tidal forces to help prevent erosion; feed downstream waters and provide fish and wildlife habitats; and release vegetative matter into rivers, which helps feed aquatic life.
- Restoring a wetland involves much ecosystem engineering, including determining the size of the wetland, sloping the sides of the wetland, restoring plant and animal life, and restoring soils to saturated states.

Words to Know

Use the following list and the textbook glossary to review and study the *Words to Know* from *Lesson 15.3.*

bog
constructed wetland
fen
hydric soil
marsh
nontidal marsh
septic system
swamp
tidal marsh
watershed
wetland creation
wetland enhancement
wetland restoration

Check Your Understanding

Answer the following questions using the information provided in *Lesson 15.3.*

1. What is a watershed?
2. Which is the largest watershed in the United States?
3. Where is groundwater stored?
4. How do chemicals migrate into the groundwater?
5. How might livestock contaminate surface water?
6. How do wetlands recharge groundwater?

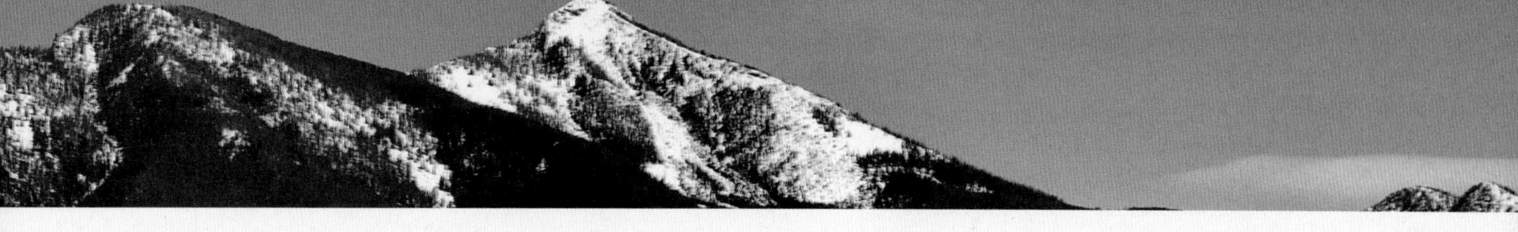

7. What are the differences between marshes, bogs, swamps, and fens?
8. Why were wetlands drained in the early and middle 1900s?
9. Is it easier and cheaper to create or restore a wetland?
10. Which kind of soils develop after they have been saturated for long periods of time?

Lesson 15.4

Water Quality

Key Points

- Water is used primarily to water crops and to provide livestock with drinking water.
- Water is also used to clean and cool animals and facilities in livestock and dairy production.
- Growers use water to cool, wash, and move fruits and vegetables throughout processing.
- Water quality is affected by the following factors: water temperature, dissolved oxygen, total suspended solids and turbidity, acidity and alkalinity (pH), nutrients, fecal coliform bacteria, and microorganisms.
- Pollution that comes from a single source that is easily identifiable is called point source pollution. Pollution whose source cannot be identified is called nonpoint source pollution.

Words to Know

Use the following list and the textbook glossary to review and study the *Words to Know* from *Lesson 15.4*.

aquaculture
biological availability
chemical pollution
dissolved oxygen (DO)
fecal coliform bacteria
irrigation
macroinvertebrates
microbial pollution
nonpoint source pollution
oxygen-depleting pollution
point source pollution
sediment pollution
seechi disk
solubility
thermal pollution
total suspended solids
turbidity

Check Your Understanding

Answer the following questions using the information provided in *Lesson 15.4*.

1. What is aquaculture?
2. How does water temperature affect the amount of oxygen that the water is able to carry?
3. What can increase water temperatures?
4. What can increase the amount of dissolved oxygen in water?

5. Water that is muddy after a rainstorm has high levels of total _____, leading to high turbidity.
6. What pH is preferred by most aquatic life?
7. What are sources of fecal coliform bacteria in water?
8. What are the four main classifications of water pollutants?
9. If a hog farm's manure lagoon breaks and spills manure into a stream, would this be point source or nonpoint source pollution?
10. What are common sources of agricultural pollution?

Lesson 15.5

Conservation Practices in Agriculture

Key Points

- Farmers use several methods to diminish erosion caused by rain and other forms of precipitation, including: leaving crop residue on fields, planting cover crops, using reduced tillage programs, terracing land, planting grassed waterways, and installing water and sediment control basins.
- Farmers use reduced tillage, terraces, grassed waterways, and buffer strips to reduce water pollution from agricultural lands.
- When applying livestock waste as fertilizer, farmers use equipment that injects the manure into the soil instead of just spreading it on top. This prevents the waste product from being washed away as runoff during rainstorms.
- Water conservation practices on irrigated lands include appropriate scheduling of irrigation, selection of irrigation systems, and tailwater collection and return.
- The NRCS works to preserve watershed water quality.
- The Environmental Protection Agency (EPA) monitors water quality in lakes and streams.
- SWCDs are organizations made up of agriculturists who are committed to soil and water conservation and who implement local conservation programs.

Words to Know

Use the following list and the textbook glossary to review and study the *Words to Know* from *Lesson 15.5*.

barren
buffer strip
center-pivot irrigation system
confined animal feeding operations (CAFOs)
contour tillage
cover crop
crop residue
drip irrigation system
furrow irrigation system
grassed waterway
irrigation
line source outlet
micro irrigation system
micro-sprinkler system

no-till program
point source emitter
reduced tillage program
riparian zone
tailwater
terrace
trickle irrigation system
water and sediment control basins (WASCoBs)
wheel line irrigation system
wind erosion

Check Your Understanding

Answer the following questions using the information provided in *Lesson 15.5*.

1. When is cropland most susceptible to erosion?
2. List two examples of crop residue.
3. What is a cover crop?
4. Which tillage method leaves the most crop residue on the soil surface?
5. Explain how contour tillage reduces soil erosion.
6. How are dry soils more susceptible to erosion?
7. Which organization regulates the cleanup of contaminated surface water?
8. Why do farmers inject animal manure into the soil?
9. What is tailwater?
10. Which governmental organization monitors water quality in lakes and streams?

Chapter 15 Skill Development

STEM and Academic Activities

1. **Science.** Conduct simple percolation tests on various areas of your schoolyard. Percolation is one way to measure the compaction of soil. Where water percolates into the soil quickly, there is usually little compaction. Where water fails to percolate into the soil quickly, there is high compaction. Where do you think soils would be most and least compacted around your school? Identify five areas of your school where you think there would be differences in compaction.

   Cut the top and bottom out of a small soup can so that you essentially have a metal tube. At each site, gently push the can about 1/4″ to 1/2″ into the ground. Then, measure a specific amount of water, for example, 100 mL. Pour the water into the can that is pressed into the soil and measure the amount of time required for all of the water to percolate into the soil. Repeat this process for each site you selected. Graph the results. What are your conclusions? For bonus, conduct this test on the football field by selecting different areas on the field where you think the soil would have differences in levels of compaction. Where is the soil most compact? Why?

2. **Engineering.** Many agriculture departments operate a greenhouse. What happens to the tailwater from irrigation in the greenhouse? What is the quality of that tailwater? What happens to the tailwater in your greenhouse? How could the tailwater be reused in the greenhouse? Design a system that could reclaim tailwater in your school's greenhouse.
3. **Engineering.** Design a demonstration that shows the effects of tillage and crop practices on soil erosion on agricultural lands. Compare tillage techniques, conservation structures, and other factors that affect erosion.
4. **Math.** Find a table of historic corn production in the United States. Graph historic corn production for the past 100 years. Hypothesize what has caused the major spikes and troughs in corn production. Gather data and evidence to support or refute your hypothesis.
5. **Social Science.** During the Dust Bowl, life for small farmers in Oklahoma, Kansas, New Mexico, Texas, and parts of Colorado was pretty tough. Read and research about what it was like to live on the Great Plains during the Dust Bowl. Create a poster with images and text that help others understand the harshness of life during these years.
6. **Language Arts.** Research some of Henry David Thoreau's writings and quotes about nature. Select three that might have meaning for you. What do they mean? How would you interpret these in light of our natural environment today? Do you agree or disagree with Thoreau? Why?

Communicating about Agriculture

1. **Speaking and Writing.** Identify a local farmer or grower who is an advocate for conservation. Interview him or her about his or her thoughts on conservation agriculture. Write an agricultural interest story for your local newspaper or school website about the farmer or grower.
2. **Speaking.** Contact the local SWCD representative and arrange an interview. Prepare a list of questions, including topics such as the major conservation practices being used in the area. Ask the representative if he or she could give a presentation to your class.
3. **Reading and Speaking.** What are the soil and water conservation issues in your local area? Select one issue, and then work with a small group of students to examine the issue. Prepare an agricultural issues forum career development event presentation on the issue.
4. **Writing.** If you were a modern-day Henry David Thoreau, what would you write about natural resources conservation and agriculture? Write a poem or short essay about the topic.
5. **Reading and Speaking.** Watersheds are important for many reasons. Prepare a short presentation that explains watersheds, identifies your watershed(s), and communicates the value of watersheds. Prepare the presentation for an audience of elementary school students. Then, once prepared, present it to those students.

Extending Your Knowledge

1. Conduct a walking survey of your school grounds. Diagram the layout of the buildings, sports fields, spectator stands, and parking lots on your school's property. Identify any existing soil and/or water conservation practices. Based on this analysis, where would you recommend additional soil and/or water conservation practices be implemented?

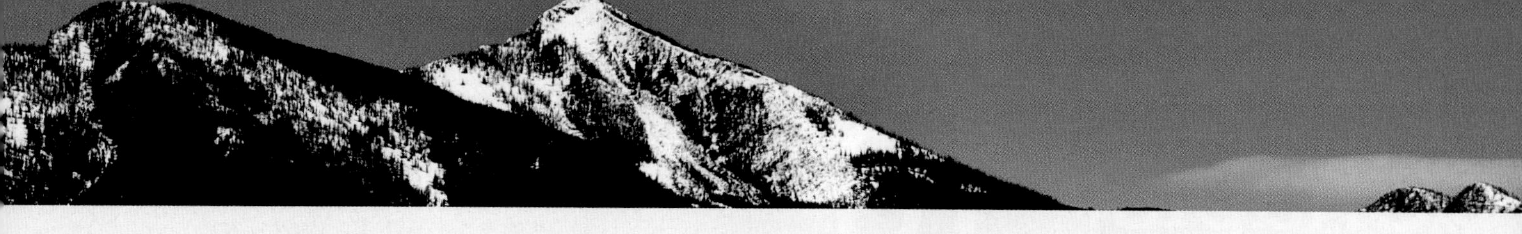

2. Hypothesize why soil dredged out of rivers and streams and returned to a field never regains its previous level of productivity. Use the Internet and other resources to determine reasons why this soil is never as productive as soil that is native to a field.
3. Over many years, much soil from the Midwest has been deposited at the mouth of the Mississippi River. What trouble do these deposits cause?
4. Search for the U.S. Drought Monitor website. Is your state experiencing drought? How do the current drought conditions in the Dust Bowl area compare to the drought conditions in the same area in the 1930s?
5. Research local state parks in your area. Did any parks experience the good works of the CCC? If so, which projects were CCC projects?
6. Conservation of our natural resources is not without controversy. Identify and explain an example of a natural resources controversy in your state or region.
7. Climatologists believe that severe droughts will become more prevalent as global climate change continues to develop. Research how scientists are working to develop crops that are more tolerant of drought.
8. Find a river or stream in your community. Where does it begin? Where does it end? Does it supply or run into any wetlands? Work with your teacher to monitor water quality over the course of a semester or year.

Chapter 16

Natural Resources and Recreation

©iStock/iofoto

Arina P Habich/Shutterstock.com

©iStock/htrnr

©iStock/Wirepec

G-WLEARNING.com

While studying, look for the activity icon  **to:**

- **Practice** vocabulary terms with e-flash cards and matching activities.
- **Expand** learning with video clips, animations, and interactive activities.
- **Reinforce** what you learn by completing the end-of-lesson activities.
- **Test your knowledge** by completing the end-of-chapter questions.

Lesson 16.1

Outdoor Recreation

Words to Know

conservation
fair chase rules
hunting season
managed hunt
parks
preservation
preserved land
private land
public land

Lesson Outcomes

By the end of this lesson, you should be able to:

- Explain public uses of natural resources for recreation.
- Identify and discuss factors involved with property ownership rights.
- Compare and contrast private and public property.
- Describe responsibilities and obligations of using natural lands for various purposes.

Before You Read

Write all of the lesson terms on a sheet of paper. Highlight the words that you do not know. Before you begin reading, look up the highlighted words in the glossary and write the definitions.

The fertile croplands, beautiful and unique national parks, vast numbers of lakes, and myriad open spaces in our country are a large part of our national identity. Although millions of acres are dedicated to agriculture, farmers and ranchers are not the only people who use our natural resources. American citizens and the millions who visit our country use both public and private lands for everything from hiking, hunting, and camping to sightseeing and animal observation. The local parks, walking trails, bike trails, horseback riding trails, ATV or snowmobile trails, and golf courses surrounding the places we live are available for our recreational enjoyment due to the foresight of people who saw the benefit of preserving lands for public use, **Figure 16-1**.

Even before the days of the great conservationist president Theodore Roosevelt, Americans understood the need to preserve and conserve outdoor resources for the enjoyment of all Americans. Our national park system is an example of the movement to preserve wild spaces. When we use the outdoors for recreation, we need to be mindful that our actions do not diminish the resources for future generations.

Land Ownership

There are two broad classifications of land ownership: private lands and public lands.

Private Lands

Private lands are lands owned by an individual or a corporation. Your parents may own the land where your house sits, and they may also

James Marvin Phelps/Shutterstock.com

Figure 16-1. National parks are preserved lands set aside for all people to enjoy, both now and in the future. ***Is there a national park near your home?***

own lands that are cropland or woodlots. Many companies own extensive acreages of land as real estate investment, or for harvesting resources. These are examples of privately owned lands.

Private landowners can use their lands pretty much as they wish. They can hunt, hike, and harvest off their land, and allow others to use their lands as well, **Figure 16-2**. In northern states, many private landowners allow snowmobilers to sled across their land in winter. Private landowners are also allowed to sell their land.

Public Lands

Public lands are lands owned by the government. This means the public, or the citizens of United States, own the land collectively. State and national parks, army bases, and many rivers and streams are examples of publicly owned lands. Although we may own these lands, many have access or use restrictions, and we cannot just walk onto them or use their resources as we please. For example, military bases are used to train and house military personnel. The bases require restricted access to ensure the safety of the soldiers and to protect the military equipment kept on the base.

TFoxFoto/Shutterstock.com

Figure 16-2. Private lands can be used for almost any purpose the landowner wishes, including logging. ***Are there instances when landowners can be prohibited from using their land for a particular purpose?***

Brandon Bourdages/Shutterstock.com

Figure 16-3. Public parks often have restrictions regarding the amount or type of traffic allowed, as well as hours of operation. They may also have restricted areas from which visitors are prohibited from entering. *Should public parks be allowed to close at a specific time or restrict access to certain areas of the park?*

Public lands such as local, state, and national parks are usually open for public use, but may have specific hours of operation, and/or limit the amount or type of traffic allowed, **Figure 16-3**. For example, some parks close at dusk due to safety and budget restraints. Some parks prohibit or restrict the use of horses to certain areas to ensure the safety of hikers and the riders as well as the horses. Public parks may also have trails dedicated to ATV use because the machines are loud, fast, and may disturb or cause irreparable damage to more delicate flora and fauna. To help fund their upkeep, some parks charge admission or camping fees.

Public land may also be leased for mining, timber harvest, and ranching rights to private citizens. These private citizens pay the government to use the land for a specific period of time. The citizen or company leasing the land is expected to maintain the land and return it to the government at the end of the lease. While under lease, public access may be limited or restricted.

State and National Parks

Parks are natural areas that are set aside as preserved land for public use. ***Preservation***, or ***preserved land***, means the land will always be maintained in its natural state. It cannot be used for farming, forestry, or mining. Hunting is generally prohibited, as well. State and national parks are examples of preserved lands. ***Conservation*** means that the land and its natural resources can be used, but not misused or used up. Public and private lands may be used for conservation. For example, timber corporations may harvest only a small portion of the land for timber each year. After the timber is harvested, the corporation replants trees that will be harvested decades later. In this way, the forest is used, but never used up. Government land that is leased may also be used in this manner.

The first national park was Yellowstone National Park. It was created in 1872. Millions of people visit national parks each year. They camp, hike, bike, and enjoy the natural beauty of these areas. National parks house some of the most spectacular sights on the planet. Old Faithful in Yellowstone National Park, Yosemite Valley in Yosemite National Park, and the hills of the Cumberland Gap National Historical Park, are just a few examples. Almost every state contains a national park of some kind, **Figure 16-4**.

Landowner Rights

As a private property owner, a person is entitled to certain rights. A landowner has the right to use his or her land as she sees fit. A landowner

Zack Frank/Shutterstock.com

Krzysztof Wiktor/Shutterstock.com

Figure 16-4. Yellowstone National Park is a national treasure for its wild spaces and unique geology, like this geyser pool. *Which natural landmarks are located in Yellowstone National Park?*

can construct different types of buildings on his or her land if they receive the correct permits for such construction. In most places, a landowner can even construct something that neighbors might find objectionable. For instance, a landowner could construct a windmill on his or her property to generate electricity even if the neighbors really did not like the looks of the windmill.

Some landowners discover minerals, such as oil, gold, or diamonds, on their land. It is their right to mine and extract and sell those minerals. The landowners may have to reclaim the land after the mining is complete, and return the land to as natural a state as possible. The landowners must also ensure the mining or other production on their land does not have adverse effects on surrounding land and water resources.

Landowner Responsibilities

Private landowners have responsibilities for the use of their lands. Landowners must do no harm to the land. This means they cannot pollute a stream on land they own by dumping household waste or chemicals into the water. Further, they cannot create a garbage dump on the property unless it is zoned for such use. Some lands are zoned for specific purposes, such as production agriculture, commercial business, or residential building construction. If land is zoned a certain way, then it may only be used for the intended purpose. Landowners cannot violate zoning laws, but they may petition for changes in the zoning restrictions.

Impact of Recreational Land Use

Any use of natural lands by humans will have some impact. Some activities have little or no impact, while others may create serious problems if they are not regulated. Most landowners (the government included) work to limit those impacts, so the resources will not be diminished beyond the point of recovery. For example, if someone hunts on their land or charges others to hunt on their land, they will not want the game to be overhunted and eliminated. To ensure this does not happen, hunting seasons and limits are established.

Steve Bower/Shutterstock.com

Figure 16-5. The American bison roams free in various national parks, including Yellowstone National Park.

Let us look at some of the more common outdoor recreational land uses, and the impacts these activities might have on the land.

Hunting

Many people enjoy hunting for sport and for food, and regulated hunting ensures game will be available for years to come. As a nation, we have learned that uncontrolled hunting can destroy wildlife populations. The American bison almost met extinction in the late 1800s due to commercial hunting and careless slaughter for sport, **Figure 16-5**. The gray wolf also came close to extinction due to commercial hunting. Today, to prevent the extinction of game animals, hunting is highly regulated in the United States. Regulations vary by state, but all include:

- When certain animals may be hunted.
- Where certain animals may be hunted.
- What types of weapons may be used.
- The number of animals each hunter may harvest per season.

Did You Know?

Due to overhunting, the American bison went from an estimated 40 million in the 1830s to less than 1,000 in 1890. Today, there are approximately 500,000 American bison in the United States but only a fraction are pure, wild animals. Because of effective management of these bison herds, they are growing and thriving today.

Hunting Seasons

Hunting seasons are specific times of the year in which certain game animals may be hunted. Hunting seasons are mostly determined based on the biology of each species. For example, the hunting season for wild turkeys is generally November through early January. This allows time for breeding adults to breed and hatch young, and for the young to grow before hunting season begins. Established hunting seasons allow time for the animals to reproduce, and ensure there will be ample game for the hunters to pursue.

Did You Know?

To control wildlife populations, hunters are needed to help with game management.

Managed Hunts

Sometimes the department of wildlife or natural resources in a state offers a managed hunt. A ***managed hunt*** is a hunt where a limited number of permits are issued, or a certain number of males and females of the species may be killed. During a managed hunt, the department of wildlife gathers data on the animals harvested, and uses that information to manage wildlife populations in the area. An example of a managed hunt is the tundra swan managed hunt in North Carolina.

Hunting on Public and Private Lands

Although the government has established public hunting areas, many hunters often gain permission from private landowners to hunt on their property. Some hunters or hunting clubs may even pay private landowners for access to their property. *It is illegal in all states to hunt on someone's property without permission.* Whether hunting on public or private land, hunters must adhere to legal regulations and obey guidelines designed for both safety and courtesy.

Nate Allred/Shutterstock.com

Figure 16-6. It is important to learn how to handle high-powered hunting weapons before you attempt to hunt any type of game. *Do any local organizations offer weapons training and hunter safety courses?*

Hunter Safety

Hunters have responsibility and accountability to themselves, other hunters, landowners, and the general public. High-powered hunting weapons can cause injury or death to the hunter as well as innocent bystanders when mishandled, so it is imperative for hunters to have the proper training, **Figure 16-6**. Although the following safety precautions are not an exhaustive list, keep these points in mind when hunting or handling a firearm or bow:

- Always keep the safety of your firearm "on" until you are ready to shoot.
- Make sure you are wearing the appropriate hunter safety attire at all times to protect yourself.
- Never consume alcohol or drugs prior to or while using a firearm or bow.
- Always unload your firearm or bow before climbing into a tree stand, entering a blind, or crossing a stream or fence.
- Always unload your firearm or bow when entering a vehicle or building. The action should be open or the gun broken down when transporting a gun.
- Never point your firearm or bow at anything unless you intend to shoot it.
- Never fire your firearm or bow into the air, unless it is at game. Rifles should never be fired into the air.
- Never fire a firearm or bow across a road or into a river, stream, or lake.
- Always positively identify the animal you are shooting *before* you discharge your firearm or bow.
- Keep all firearms and bows locked away from children. Store ammunition separately from firearms.
- Always unload your firearm before cleaning it.

Hunter's Code of Ethics and Courtesy

Following a hunter's code of ethics and using hunter courtesy will help you be a safer and more courteous hunter. A hunter's code of ethics includes the following:

- Leave the land and property you are hunting on better than you found it. This may mean that you pick up trash, even if it is not yours.
- Follow the ***fair chase rules***, which mean not hunting with automatic weapons, not using electronic calling, not spot lighting, and not hunting from all-terrain vehicles, among other rules.
- Shoot for a quick, clean kill and track any game that you injured until you find it.
- Do not waste usable parts of any game animal that you kill.
- Work with local conservation officers and other game management professionals.
- Respect landowners by asking permission to hunt on property, avoiding damage to livestock and crops, closing gates, and notifying the landowner immediately if anything seems out of place.
- Respect nonhunters by keeping firearms and harvested game out of sight and not overtly displaying harvested game.
- Respect other hunters and landowners by abiding by safe hunting practices.

Safety Note

ATV tires have deep grooves and treads designed to handle muddy and rocky land and do not perform well on paved surfaces. They are not designed for use on paved surfaces and, in most areas, cannot be legally driven on roads.

All-Terrain Vehicles (ATVs)

All-terrain vehicles, or ATVs, are motorized vehicles designed for off-road use. Four-wheelers and dirt bikes are the most common types of ATVs, **Figure 16-7**. Although they may be used for work, many people purchase ATVs for off-road recreation activities. They may be driven through creeks and swamps and across dunes and mountain trails. Because ATVs are rather large machines, they have the potential of wreaking havoc on farmed fields and natural areas.

homydesign/Shutterstock.com

Figure 16-7. Riding ATVs, like this four-wheeler, is a lot of fun for many Americans and a great way to see many wild areas in a short period of time. ***Why is it unwise to carry a loaded gun or crossbow while four-wheeling?***

Responsible ATV Use

ATV riders should always seek permission of property owners before riding across any land, **Figure 16-8**. There are several reasons for this, and chief among them is that this is the ethical and legal thing to do. Trespassing is illegal and may result in criminal charges and fines. It may also be dangerous to ride across land without permission because of unknown hazards, such as electric fences made of thin wire that is nearly invisible at high speeds or low-light conditions.

Cropland and Livestock Pastures

Landowners may have special structures, fences, and even crops or livestock on their property that would be harmed if someone rode an ATV across the land. The ruts and soil compaction caused to cropland by ATV tires may ruin a crop or prevent a farmer from planting in time for the coming season. Riding ATVs across pastures may spook livestock or damage fences, allowing livestock to wander into unsafe areas.

David Kocherhans/Shutterstock.com

Figure 16-8. Always respect the landowner's rights. As tempting as it may be to ride your ATV across private property, trespassing is illegal and may result in criminal charges and fines.

Creeks, Streams, and Swamps

Driving through or across creeks, streams, and swamps is a lot of fun for ATV riders. Although it may be fun to drive through these areas, extensive traffic through these waterways may damage habitat and endanger food sources for critical wildlife. Fish and aquatic life are disturbed when an ATV is driven through a creek. Usually this is just temporary, but sustained driving in the creek or stream can alter the flow of the stream permanently. Riding an ATV through a waterway also disturbs the sediment in the stream and creates temporary water pollution. This may have detrimental effects on stream life.

Protected Areas

Some natural areas are fragile and especially susceptible to damage from ATVs. They have species of wildlife that require certain conditions in order to grow and thrive. These areas are off-limits to ATV traffic.

ATV Safety

Each year, many people are severely injured or killed on ATVs. About one-third of all ATV-related fatalities were of riders under the age of 16. Many of these injuries and deaths could be prevented if the operators were properly trained and employed certain safety precautions. The following list is not all-inclusive, but it does include key points you should keep in mind when purchasing and using an ATV—for fun or for work.

- Anyone who will be using an ATV should receive training through an accredited program. There are programs available through 4-H, the All-Terrain Vehicle Safety Institute, and state-run programs.
- Use the right size ATV for your age. Children should never drive an adult-sized ATV. ATVs with an engine size of 70cc to 90cc should only be operated by people 12 years or older. ATVs with an engine size greater than 90cc should only be operated by people 16 years or older.
- Appropriate riding gear should always be worn. Approved riding gear is designed to help protect you from head and spinal injuries, as well as cuts and abrasions. Appropriate riding gear includes a DOT- or Snell

Did You Know?

Government studies have shown that the use of helmets reduces deaths on ATVs by up to 40%.

ANSI-approved helmet, goggles, gloves, over-the-ankle boots, long-sleeve shirt, and long pants.

- Do not allow additional riders. Most ATVs are designed for single riders. There are no handholds or footholds for additional riders, and they are easily thrown from the vehicle.
- Do not drive on paved roads. ATVs are not designed for on-road use and do not handle well on smooth pavement. When you must cross a paved road, drive slowly and watch for vehicles.
- Never consume alcohol or drugs before driving or while riding on an ATV.
- Never hunt from an ATV and make sure all firearms are unloaded and broken down when transporting them on an ATV.
- Make sure your ATV trailer is properly maintained and hitched securely when transporting ATVs. ATVs must be securely attached to the trailer during transport.
- Include a GPS unit with your gear to help prevent getting lost and to make it easier to find your way if you do get lost.

As with any machinery, you should regularly inspect and perform standard maintenance procedures to keep ATVs in safe operating condition. Use the Internet to research ATV safety and locate training programs available in your area.

Snowmobiles

Snowmobiles are a type of ATV popular in northern states when snows pile up in winter, **Figure 16-9**. Many states, like Michigan, Wisconsin, and New York, have local snowmobile clubs that build and maintain snowmobile trails across both public and private lands. People in these clubs check the trails in

bikeriderlondon/Shutterstock.com

Figure 16-9. Snowmobiling allows people to experience remote natural spaces in winter.

late fall for downed trees or other debris that might impact snowmobile travel. They also mow the trails to reduce the height of the grass on the trail. Club members maintain signs that regulate speeds and inform snowmobilers about dangers on the trail. The partnerships of many snowmobile clubs produce trail maps for people to use. The snowmobile club also assumes liability for accidents that occur on the land and ensures that the property will not be damaged by its members.

Snowmobile Trail Etiquette

When riding snowmobiles on approved and maintained trails, snowmobilers are expected to follow general rules of etiquette. Everyone who uses the trails, including other riders and even wildlife, has as much right to use the trail as you do. These trail etiquette guidelines include the following:

- Drivers are to stay to the right at all times, **Figure 16-10**.
- When you meet oncoming snowmobilers, people on foot, skiers, wildlife, or other traffic, slow down and approach them cautiously. As you pass, give each other as much room as possible to avoid collisions.
- Never block the snowmobile trail. This means that you should never stop in the middle of a trail or side by side with another snowmobile.
- When encountering other riders or riding with a group, use the universal hand signals for turning, slowing, and stopping.
- Always ride respectfully.

©iStock/VV-pics

Figure 16-10. Most areas across the country follow the same traffic rules for snowmobiling. For example, snowmobilers should always stay to the right when riding on designated trails.

Snowmobile Safety

To ensure your own safety, as well as that of other riders, keep the following in mind when riding snowmobiles:

- Complete an approved snowmobile safety and training course before operating a snowmobile.
- Be prepared for your ride. Dress properly: use layers, eye protection, a helmet, and insulated boots, **Figure 16-11**. Carry safety equipment, tools, and emergency kits. Repair and recovery kits often include basic tools, wire, duct and electrical tapes, spark plugs, spare drive belt, rags,

eye-for-photos/Shutterstock.com

Figure 16-11. Dressed for snowmobiling.

Lester Balajadia/Shutterstock.com

Figure 16-12. Follow signs and markers on trails to ensure safe riding.

siphon hose, and a tow rope. A first-aid and survival kit often includes bandages, antiseptic, waterproof matches, heating packs, a knife, nonperishable food, a whistle, space blanket, and perhaps flares.

- Always gain permission of landowners before riding snowmobiles across private land. Riding on private property without permission is considered trespassing.
- Follow all snowmobile trail markers and signs, **Figure 16-12.**
- If riding on public snowmobile trails, make sure you are familiar with the trail maps and local snowmobile club. When crossing roadways, stop, look, and listen for oncoming traffic. Always cross roadways at a 90-degree angle to the road.
- Because snowmobiling occurs in cold weather, snowmobilers should be prepared for cold-weather emergencies. This may mean carrying emergency blankets, water, food, and first-aid kits. This also means knowing the weather forecast and preparing for emergency contingencies if severe weather strikes while you are out on the trail.
- When following another snowmobiler, especially at night, make sure you have a good field of view in front of you. Allow enough distance to stop suddenly or swerve safely to avoid collisions.
- When riding snowmobiles at night, make sure that all lights on the snowmobile are working and visible. This means that they should be free of ice or caked snow.
- Never consume alcohol or drugs when operating a snowmobile.
- Make sure your snowmobile trailer is properly maintained and hitched securely when transporting snowmobiles. Snowmobiles must be securely attached to the trailer during transport.
- Carrying a GPS unit is recommended. You should also let someone know where you are going and when you expect to return.

Hiking and Trail Running

Hiking is one of the most popular outdoor recreation activities, **Figure 16-13**. It is a low-impact, relatively safe activity. When properly prepared, most people will not be harmed by hiking, save the occasional encounter with poison ivy or a pesky mosquito. Trail running is a higher-impact exercise that ensures a challenging run while allowing the runner to enjoy the beauty of the terrain. Both activities provide a sense of exploration, and often the opportunity of seeing something that few others have seen.

Jon Bilous/Shutterstock.com

Figure 16-13. Hiking trails provide good cardiovascular exercise, amazing scenery, and a sense of exploration. This trail in the Shenandoah National Park just seems to invite you to explore.

Hiking Safety

As with any outdoor activity, you must be properly prepared to ensure your safety while hiking. To be a safe hiker, keep the following in mind:

- Wear appropriate attire and shoes for the kind of hiking and weather that you will encounter on your hike, **Figure 16-14**. Your shoes are perhaps your most important piece of equipment. Shoes should be broken in, sturdy, and equipped for the kind of hiking that you intend to do.
- Prepare for inclement weather by packing an additional layer of clothing and/or rain gear. On the other hand, extra gear and clothing add weight to your hike, which can affect balance and exertion. So, pack as lightly as possible.
- Know your physical limits when hiking. Everyone has a different level of physical fitness.
- Know the weather forecast for the area where you are hiking. Be prepared for extreme weather if hiking for long durations and/or if you are changing elevations. Weather and temperatures can change quickly.
- Always seek permission from landowners if hiking across private land.

baranq/Shutterstock.com

Figure 16-14. Dress properly for the weather and the area in which you will be hiking.

- Leave all land better than you found it. Again, this may mean picking up trash even if it is not yours and clearing trails of debris and limbs.
- Carry water and stay hydrated on your hike. Dehydration contributes to fatigue and slowed reflexes.
- Carry nonperishable foods, such as granola bars and beef jerky.
- Carrying a GPS unit is recommended, **Figure 16-15**.

Occasionally, hikers get lost. You can take steps to prepare yourself before your hike. First, make a plan that includes knowing the terrain over which you will be hiking. Also, develop a plan and timing for where you will be at critical points on the hike. Let people know where you are going and what time you will arrive at your destination on the hike. If you are hiking on an extensive trip, prepare extra food, water, and clothing in the event that you become lost or disoriented.

As you are hiking, be aware of your surroundings. Look for landmarks by which you can identify a trail back to your starting point if needed. Stop and look over your back trail frequently. The landscape looks markedly different when you turn and look 180° behind you. If you feel like you are becoming lost, stop. Stopping allows you to take control of your situation and develop a plan for getting "unlost." If you are lost and uncertain as to which direction you should proceed, stay put and make a call for help if possible. You will be easier to find if you remain in one location.

Sander van der Werf/Shutterstock.com

Figure 16-15. Carrying a GPS is advisable for all hikers, no matter how experienced.

Career Connection Conservation Officer

Job description: Similar to a police officer, a conservation officer is a law enforcement officer. However, the conservation officer serves and protects wildlife and natural areas. Conservation officers ensure that hunting laws, snowmobiling laws, ATV laws, and park guidelines are followed. The job of a conservation officer is interesting, with no two days being identical. The conservation officer patrols parks and other natural areas. His or her time on the job may be in greater demand during peak hunting and outdoor recreation seasons. They may also work with other law enforcement officials to solve cases and protect the environment. Conservation officers may have jobs similar to park rangers.

A park ranger is a type of conservation officer who serves and protects wildlife and the environment. Part of the job of protecting the environment is also educating people about natural areas. Conservation officers may also be called game wardens, park rangers, forest rangers, or wildlife officers.

Education required: A bachelor's degree in criminal justice, natural resources, or a related area is required. The training for a conservation officer is similar to that of a police officer.

William Silver/Shutterstock.com

Job fit: This job may be a fit for you if you are interested in protecting the environment and ensuring that natural resources rules and guidelines are followed by hunters and people engaging in outdoor recreation.

Lesson 16.1 Review and Assessment

Words to Know

Match the key terms from the lesson to the correct definition.

1. Lands that are owned by an individual or a corporation.
2. Lands that are owned by the government.
3. The land will always be maintained in its natural state.
4. A certain time of the year in which certain game may be hunted.
5. Not hunting with automatic weapons, electronic calling, or spot lighting, and not hunting from all-terrain vehicles.
6. When a limited number of permits are issued, or a certain number of males and females of the species may be killed.
7. The land and its natural resources can be used, but not misused or used up.

A. conservation
B. fair chase rules
C. hunting season
D. managed hunt
E. preservation
F. private lands
G. public lands

Know and Understand

Answer the following questions using the information provided in this lesson.

1. What is a common right associated with private land ownership?
2. What is the difference between publicly owned land and privately owned land?
3. Why is hunting regulated?
4. What are four factors covered by hunting regulations?
5. What are three rules for hunter safety that relate to the use of firearms or bows?
6. When should you point your firearm at an object?
7. Which areas should be avoided when riding ATVs?
8. What functions do snowmobile clubs serve in the local area?
9. What is the hand signal for approaching snowmobilers?
10. To avoid fatigue, what should hikers carry with them?

Analyze and Apply

1. What are the benefits that people receive from outdoor recreation?
2. What are other outdoor recreation activities not mentioned in this lesson that are common in your area? Why are these common?
3. What are opportunities for outdoor recreation in your area?
4. When reviewing the hunter, ATV, snowmobile, and hiker safety and etiquette guidelines, what are the common guidelines? Why do you think these are common?
5. Why should a person never fire a gun into a body of water? Why should a person never fire a gun into the air?
6. Why is carrying beef jerky a good idea when hiking?

Thinking Critically

1. Look up hunting seasons for game species in your state. Analyze the species of game. What are the specific biological reasons for limiting the hunting season for these species?
2. What is the history of fair chase rules? Why are fair chase rules enacted? What would be an example of an *unfair* chase?
3. What is one other hunter, ATV, snowmobile, or hiker rule that the lesson did not mention? Why is this rule a good safety rule to follow?
4. List some infractions you, or someone you know, has made in regards to hunting, ATV/ snowmobile use, and/or hiking. What consequences resulted from these infractions?

Lesson 16.2

Wildlife Management

Words to Know

blue tongue
chronic wasting disease (CWD)
Colony Collapse Disorder (CCD)
corridors of habitat
endangered species
extinct species
fencerow
game
game species
genetic reproductive potential
midge
nongame species
population dynamics
threatened species
wildlife inventory

Lesson Outcomes

By the end of this lesson, you should be able to:

- Explain the importance of wildlife, fisheries, and ecology.
- Compare endangered, threatened, and extinct species.
- Explain how game management improves ecosystems.
- Compare game species to nongame species.
- Identify and describe diseases that affect game species.
- Describe organizations that work with wildlife management.

Before You Read

Arrange a study session to read the lesson aloud with a classmate. Take turns reading each section. Stop at the end of each section to discuss what you think its main points are. Take notes of your study session to share with the class.

When early settlers arrived in North America, many species of animals were abundant. As populations grew, so did the need to feed these people. More and more wildlife was hunted for food and clothing. Farmers cut down forests to make way for cropland, reducing natural habitats and, in turn, the numbers of many wildlife species. As wildlife species dwindled, North Americans realized the need to manage wildlife populations to ensure their survival for future generations.

So what is wildlife management and who is in charge of this monumental task? Let us look at the organizations that manage wildlife in the United States, and the ways in which they work to conserve and enhance both wildlife and their habitats.

Wildlife Management Organizations

Managing wildlife populations is a complicated and huge undertaking. In the United States, the federal government has several agencies dedicated to the task, primarily the Fish and Wildlife Service (FWS), **Figure 16-16.**

Fish and Wildlife Service (FWS)

In the United States, the Fish and Wildlife Service (FWS) is the primary agency in charge of wildlife management. The Fish and Wildlife Service manages hunting, monitors wildlife populations, and conducts research to help preserve and maintain species. They are also charged with enforcing federal wildlife laws. The Fish and Wildlife Service is composed of a number of divisions, each responsible for specific areas of wildlife management. These areas include wildlife refuges, migratory birds, fish hatcheries, and endangered species.

The Fish and Wildlife Service works with many nongovernmental organizations to manage wildlife. This is due in part because a large number of fish and wildlife habitats are not located on federal lands.

USFWS/Becky Skiba

Figure 16-16. Through the dedicated efforts of FWS employees and thousands of volunteers, wildlife such as this young Loggerhead may survive for many more generations. ***Have you ever volunteered to assist with an organization such as the FWS?***

Nongovernmental Organizations

Nongovernmental organizations that share the mission to protect and preserve wildlife and wildlife habitats include Ducks Unlimited, the National Wild Turkey Federation, and the Nature Conservancy.

Ducks Unlimited focuses its efforts on wetland preservation and restoration and waterfowl conservation. The National Wild Turkey Federation is a nonprofit organization that works to protect wildlife habitat, especially habitat that conserves wild turkey populations. The Nature Conservancy works to protect wild areas and wild populations. Because protection and preservation of wild areas and wildlife is such a huge undertaking, these organizations and many others work with citizens, businesses, and even the government. The ultimate goal is to protect wildlife for future generations of Americans. These organizations and the Fish and Wildlife Service accomplish this goal through a variety of processes including:

- Wildlife inventories
- Population dynamics
- Disease treatment
- Game species

Mark Cantrell/USFWS

Figure 16-17. Wildlife inventories are tedious and time-consuming work. However, the statistics gathered with wildlife inventories are extremely valuable to the scientists working with various organizations to preserve and conserve wildlife. Here, biologists are searching for mussels in Denson's Creek in North Carolina prior to a dam rebuilding.

Wildlife Inventory

In order to assess the status and trends of wildlife, ecologists regularly count the wildlife in an area. This is called a ***wildlife inventory***. The wildlife inventory may include a count of all mammals, birds, reptiles, fish, and amphibians in a given area, **Figure 16-17**.

Counting all the wildlife in an area is a time-consuming and difficult task. You might wonder how ecologists can even count some species, like owls or salamanders that are small, move quickly, and may be hidden from sight, especially in daylight hours. Ecologists use a combination of their senses to see, hear, and even smell species. They look for the actual species and for their tracks and excrement. They listen for calls and other sounds to indicate species. And, sometimes, they can even smell certain species.

Backyard Bird Count

Sometimes ecologists enlist the help of citizens to count some species. Many communities host a backyard bird count periodically throughout the year. With a backyard bird count, citizens are encouraged to look and listen in their backyard and then identify bird species. People then count the number of each species of bird and report them to a common website. Ecologists use this data to estimate the number and kinds of birds in the area.

Banding, GPS, and other Methods

Other methods used to take wildlife inventories and track wildlife habits include banding of birds, GPS tracking (collars/tags), and mark and recapture, **Figure 16-18**. Most of the wildlife counts taken are estimates, as it

Char Binstock/USFWS

Figure 16-18. Color bands are placed on birds because they are much easier to "resight" than the traditional metal bands. The bands provide information such as foraging locations and distribution of the species. ***Have you ever noticed banding on wildlife near your home?***

would be extremely difficult and very costly to count each and every animal of any given species.

Wildlife Records

Nearly all states also maintain historical records of wildlife. These include records of the size and number of game species harvested. In addition to records of game species, the forestry service in many states may maintain a record of the largest individual trees of each species in the state. These big trees are mapped so that the trees can be monitored for growth and disease.

Classification of Species That Are Low in Number

A wildlife inventory may include only certain species that are economically important or those that are low in number. Most people do not want to see any species, no matter how small, disappear forever from our planet. So, ecologists and wildlife biologists often track and monitor closely the populations of rare, threatened, and endangered species, so that they do not become extinct.

Extinct Species

An ***extinct species*** is one that is no longer living anywhere on Earth. Extinction may be caused by many factors. Some species have been hunted to the point of extinction, such as the Eastern elk, which became extinct in 1913 in the United States. The Eastern cougar is a large carnivore that became extinct in 2011 in the United States because of the loss of habitat and food sources. The ivory-billed woodpecker was extinct in 1944 in Louisiana due to harvesting by bird collectors and the loss of habitat, **Figure 16-19**.

Marzolino/Shutterstock.com

Figure 16-19. The ivory-billed woodpecker is one of the most famous extinct species from North America. Efforts to find a single ivory-billed woodpecker proved unsuccessful in 2005.

cdelacy/Shutterstock.com

Figure 16-20. The hawksbill sea turtle is an example of an endangered species. Even some of the coral in this image may be endangered. ***What is the biggest threat to the hawksbill sea turtle species survival?***

Endangered Species

An ***endangered species*** is one that is low in total population and likely to become extinct. Laws and policies are enacted to protect endangered species so the population can rebound. Hunting or harvesting of endangered species is forbidden. Also, ecosystems where endangered species are found are protected from timber harvest, land development, or other ecosystem degradation. Examples of endangered species include the American black bear, the Indiana bat, key deer, ocelot, and the Florida panther. The immediate goal of protecting endangered species is that their populations grow to the point that they are no longer classified as endangered, **Figure 16-20**.

Jason Mintzer/Shutterstock.com

Figure 16-21. The burrowing owl of California is one of the species that is of special concern to scientists because of its dwindling population. ***Why is the population dwinding? What efforts are being made to help the burrowing owl survive?***

Threatened Species

A ***threatened species*** is one that is low in total population and likely to become endangered in the near future. The burrowing owl of California is one such animal, **Figure 16-21**. Species on this list are monitored to detect changes in populations. Ecosystems where threatened species live are carefully monitored so that the species has a good opportunity of survival. Examples of threatened species include the Florida manatee, Texas kangaroo rat, red wolf, American crocodile, and the loggerhead sea turtle.

Like endangered species, the immediate goal of protecting threatened species is that their populations grow, so that they are no longer classified as threatened. It is a major accomplishment to have a species removed from the threatened species list.

Game Species

People have hunted birds and animals for centuries. Hunters number into the tens of millions in the United States. You may be a hunter and eagerly await the opening day of deer season in the fall. Groups of animals hunted for food and/or sport are called ***game species***, or simply ***game***. However, most animals are not hunted. ***Nongame species*** are species that are not hunted and/or are protected from hunting because they are threatened or endangered.

Hunters are regulated by seasons and daily hunting limits for each game species. This means they may only hunt the animal in certain months of the year, and they may only harvest a limited number of animals during the season, or each day during the season. Examples of game species hunted for food include deer, pheasant, turkey, duck, and elk, **Figures 16-22**, **16-23**, and **16-24**. Game species hunted because they are a nuisance include coyote, crow, raven, and opossum. Some states even allow hunting of these animals all year because they are such a pervasive nuisance.

How Is Game Management Good for Ecosystems?

All species in an ecosystem depend on one another for survival. Ecologists and agriculturists manage game in order to limit the damage to crops by game species, such as deer foraging on corn or vegetable fields. Other game species are hunted because they kill and eat livestock. Coyotes are often hunted because they kill sheep, goats, and even cattle.

Game management and hunting provide many benefits for an ecosystem, **Figure 16-25**. Game management controls the population so that consumers at the top of the food chain do not expand in population to the point that they eat all the animals lower in the food web. Likewise, game management provides a mechanism to harvest weak and old animals from the ecosystem. This enables stronger animals to access food, water, and shelter without competition.

With game management, ecologists may also monitor the health of species that are critical to the ecosystem. For example, if fewer deer are harvested in one season, this may be an indicator of diseases in the deer herd. Once identified, these diseases may be controlled in the wild population.

Tom Reichner/Shutterstock.com

Figure 16-22. Whitetail deer are one of the most common large game species in the United States. They are found in nearly all of the eastern two-thirds of the country.

Adam Fichna/Shutterstock.com

Figure 16-23. Pheasant are an example of game birds that are hunted in many parts of the Midwestern United States. *Have you ever hunted or eaten pheasant? Are the feathers valuable?*

kzww/Shutterstock.com

Figure 16-24. Wild turkeys are an American success story for game animals. By the early 1900s, wild turkey populations were gone from 18 states. Since that time, the populations have staged a dramatic comeback. *Is there a population of wild turkeys in your state? Where do they reside?*

Steve Oehlenschlager/Shutterstock.com

Figure 16-25. Wildlife management and hunting help to control populations of game animals so that the surviving animals are stronger and have plenty of space, water, and food to survive.

Eduard Kyslynskyy/Shutterstock.com

Figure 16-26. Brown bears need access to food and large expanses of space in order to have sufficient habitat for survival. All species have specific requirements for food, water, and space to survive.

Ecological Principles of Game Management

In attempting to manage populations of game animals, landowners must track the populations of important species on their property. They must manage several aspects of the game animals' life cycle in order to attract and maintain healthy populations of game animals. First, the game animals must have access to the right kinds of habitat and food. Within this habitat, landowners generally want to encourage a diversity of wildlife. Landowners also monitor and manage the predator-prey relationships with game animals.

Feeding

In order to foster large numbers of healthy game animals, like deer, quail, and pheasants, landowners must provide adequate supplies of the correct food source for the species. Deer like to feed on grains and high-quality grasses and forages. They will also eat acorns and nuts from trees. So, feeding a deer herd on your property might mean planting corn and leaving it standing in the field to provide food for deer in the fall. It also might mean planting oak and apple trees to provide acorns and fruit for deer to eat.

Habitat

Healthy habitat with sufficient cover is also important for game management. Quail and pheasants need dense underbrush to provide nesting habitat. The dense underbrush also provide food and a place for quail and pheasant to hide from predators like fox and coyotes. Species like bears require lots of space in order to thrive, **Figure 16-26**.

Another means of creating habitat for wildlife is for landowners to leave, maintain, or create corridors of habitat. ***Corridors of habitat*** are generally narrow expanses of trees and shrubs that connect larger, wider, more diverse areas of habitat. Fencerows are an example of corridors. A ***fencerow*** is an uncultivated strip of land on each side and below the fence on a landowner's property. A fencerow along a field may connect two large areas of forest. These corridors allow deer and other game animals to move from one location to another with some degree of protection.

Biodiversity

Farmers and landowners fostering game animals are most interested in the population of the game species. However, they also know that the health and vitality of the game population is partially determined by the health of the overall food web on their land. Food webs are generally enhanced when there is rich biodiversity. This means that landowners also do what they can to ensure that animals lower in the food chain have access to food and habitat. The landowner may establish food plants on their land to attract rabbits, raccoons, and other smaller species on which fox, coyotes, and wolves feed. By managing this biodiversity, they will attract and retain game species on their lands.

Population Dynamics

Population dynamics is the study of the changes in size and ages of populations over both short and long periods of time. Scientists who study population dynamics also study the environmental and biological factors that influence these changes in population.

Factors that may affect the population of a species include weather, diseases, human interactions, and land development. Harsh winters may contribute to the deaths of older or weaker individual animals, thus decreasing the size of the population. Diseases, as will be explored further in this chapter, may also contribute to reductions in population size. Human interactions and land development generally degrade or destroy game habitat, thus reducing the population. Predator-prey relationships and genetic reproductive potential of the species are two additional factors that control the populations of game species.

Predator-Prey Relationships

As the population of game species grows, so too might the population of predator species on their land. For example, providing feed and habitat for rabbits or deer will foster a healthy and growing population of these game species. Where there are more rabbits and deer, you will also find more foxes, coyotes, and other apex consumers. These apex consumers generally feed on the weak and older deer and rabbits, but they may also be fortunate enough to kill and eat healthy, reproducing adults. So landowners and wildlife managers must also be aware of and attempt to manage the predator-prey relationships among game species.

Genetic Reproductive Potential

Each species is genetically predisposed to give birth to a certain number of offspring each year and over its lifetime. This is called ***genetic reproductive potential***. For example, a deer doe typically gives birth to only one or two fawns each year, **Figure 16-27**. On the other hand, rabbits may have a litter containing up to nine or 10 kits. The rabbit doe may also give birth to a range of three to five litters per year. The deer doe will have

Tony Campbell/Shutterstock.com

Figure 16-27. Deer have relatively low genetic potential since a single doe usually gives birth to one or two fawns each year.

AndreyEzhov/Shutterstock.com

Ideas_supermarket/Shutterstock.com

Figure 16-28. Colony Collapse Disorder (CCD) affects both domestic and wild hives of honeybees. *Are there other diseases or nonnative species that affect honeybee colonies?*

one or two young to feed and care for over the course of a year, while the rabbit doe may have as many as 50 young.

Wildlife managers must understand the reproductive potential of species and maximize the opportunities for game species to reproduce. Most management of this sort involves protecting or limiting females from harvest. In deer species, the doe hunting season is generally shorter in duration than the buck season, and hunters can generally harvest fewer does than bucks.

Diseases That Affect Wildlife Species

Wildlife ecologists and biologists monitor and try to cure diseases in wildlife populations. In recent years, several diseases have ravaged wild populations. We will look at three such diseases here: Colony Collapse Disorder, chronic wasting disease, and blue tongue.

Colony Collapse Disorder

Colony Collapse Disorder (CCD) is a disease that affects both wild and domesticated hives of honeybees, **Figure 16-28**. With Colony Collapse Disorder, the worker bees in a hive suddenly disappear. No scientific cause for Colony Collapse Disorder has been established. Possible causes of the disorder include pesticides, varroa mites, viruses, fungi, malnutrition, and other environmental factors.

Colony Collapse Disorder is a major concern for agriculturists because both wild and domesticated bees pollinate millions of acres of crops in the United States each year. And, honey and other bee-based products are important sources of income for many beekeepers. Beekeepers must maintain close observations of their hives to attempt to manage Colony Collapse Disorder.

Chronic Wasting Disease

Chronic wasting disease (CWD) is a neurological disease in deer and elk. CWD can be transmitted from infected animals to noninfected animals. CWD

Career Connection Wildlife Biologist

Job description: Wildlife biologists are scientists who study wildlife. They are concerned with population dynamics and how the wildlife interact with their environment, humans, and other factors. Many wildlife biologists become specialists with a certain species of animal. Wildlife biologists work outdoors to study wildlife. They may count individuals, track animals, and study environmental factors affecting populations in a habitat. After they have gathered data from the field, wildlife biologists often use computers and software to analyze data and make predictions about the species. A wildlife biologist may also help band and track certain wildlife species to monitor population health and dynamics.

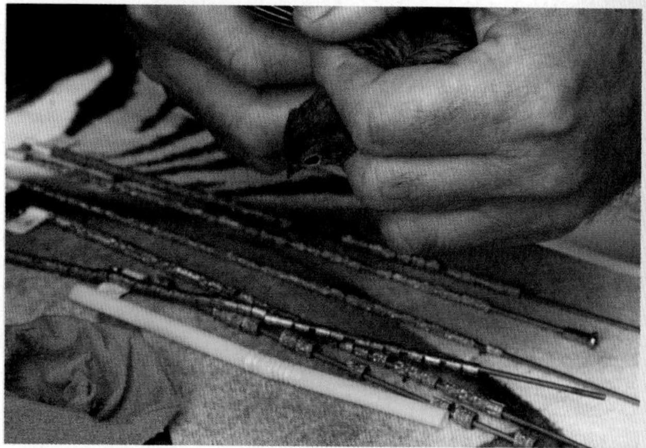

Melinda Fawver/Shutterstock.com

Education required: A bachelor's degree in biology is required. Often, a master's or even doctoral degree is required to advance in the field.

Job fit: This job may be a fit for you if you have an interest in wildlife and enjoy working both outdoors and with computers and data.

causes small brain lesions, or areas that are injured or diseased. The outward symptoms of CWD are a loss of body condition and abnormal behavior resulting from the brain lesions. Animals with CWD eventually die. CWD is similar to Mad Cow Disease in beef and dairy cattle. Hunters are advised to avoid eating the meat of elk or deer that are suspected of carrying CWD.

Blue Tongue

Blue tongue is an insect-borne viral disease that affects ruminants, such as deer, elk, antelope, and bison. The characteristic symptoms are swelling of the face and tongue, which gives the tongue a bluish hue. Blue tongue disease generally kills its victims quickly, sometimes within a week. Because the disease kills individual deer and elk so quickly, the risk of eating infected meat is low.

The insect that generally carries the disease is the small biting fly called a *midge*. Midges reproduce in stagnant, muddy water. So, they increase in population rapidly during periods of drought. As streams and other water supplies dry up, water collects in muddy, stagnant pools. These are perfect breeding grounds for midges. And, when deer come to drink water from these sources, the midges bite the deer, thus infecting them.

Lesson 16.2 Review and Assessment

Words to Know

Match the key terms from the lesson to the correct definition.

1. A count of all mammals, birds, reptiles, fish, and amphibians in a given area.
2. Animals that are no longer living anywhere on Earth.
3. Animals that are low in total population and likely to become extinct.
4. Animals that are low in total population and likely to become endangered in the near future.
5. Animals that are not hunted and/or are protected from hunting because they are threatened or endangered.
6. Animals that are hunted for food and/or sport.
7. Narrow expanses of trees and shrubs that connect larger, wider, more diverse areas of habitat.
8. The study of the changes in size and ages of populations over both short and long periods of time.
9. A genetic predisposition to give birth to a certain number of offspring each year and over its lifetime.
10. A neurological disease in deer and elk.
11. An insect-borne viral disease that affects ruminants.
12. An insect that carries the blue tongue disease.
13. A disease that affects both wild and domesticated hives of honeybees.

A. blue tongue
B. chronic wasting disease (CWD)
C. Colony Collapse Disorder (CCD)
D. corridors of habitat
E. endangered species
F. extinct species
G. game species
H. genetic reproductive potential
I. midge
J. nongame species
K. population dynamics
L. threatened species
M. wildlife inventory

Know and Understand

Answer the following questions using the information provided in this lesson.

1. What is a wildlife inventory?
2. What happens to the population of a species when it becomes extinct?
3. How are endangered species protected?
4. What is an example of a threatened species?
5. What is the difference between a game species and a nongame species?
6. What is an example of a game species?
7. Which game species are hunted because they are nuisances?
8. What factors do landowners consider with ecological game management?
9. Which insect spreads blue tongue disease?
10. What are possible causes of Colony Collapse Disorder?

Analyze and Apply

1. What is an example of wildlife aesthetics?
2. Why are wildlife, fisheries, and ecosystems important in the United States?
3. Which factors may control the population of a game species?
4. Why is Colony Collapse Disorder a major concern for agriculturalists?

Thinking Critically

1. Which endangered species are found in your state? What is being done to protect those species?
2. What is the major game species in your state? How many individual animals have been harvested in each of the last 20 years? Create a graph of the harvest. Is there a fluctuation in the numbers? Why?

Lesson 16.3

Stewardship of Natural Resources

Words to Know

carbon credit
carbon footprint
carbon sequestration
compost
ecological footprint
nonrenewable natural resource
renewable natural resource
stewardship
sustainability

Lesson Outcomes

By the end of this lesson, you should be able to:

- Define stewardship in the context of natural resources.
- Identify ways that individuals consume natural resources.
- Explain how individual consumers can reduce their ecological impact.
- Explain how agricultural companies can reduce their ecological impact.

Before you Read

Before you read the lesson, interview someone in the workforce (your supervisor, a parent, relative, or friend). Ask the person why it is important to know about the lesson topic and how this topic affects the workplace. Take notes during the interview. As you read the lesson, highlight the items from your notes that are discussed in the lesson.

Over time, we have learned that only through conservation and preservation can we ensure that our natural resources will be available for future generations. Although some of our natural resources are currently plentiful, careless actions and poor practices could quickly change things. For example, air can be contaminated from unregulated industrial exhaust and become harmful to breathe. Water sources may also be polluted through industrial discharge and excess runoff from fertilized lawns and agricultural lands. Wildlife species may be decimated by overhunting and habitat loss. So how can we ensure that our natural resources will be available to future generations? We can work toward this goal by practicing good stewardship of the land, water, and other natural resources.

What Is Stewardship?

Stewardship is the ethic that people care for and conserve natural resources. Stewardship does not mean that a person or group does not use natural resources, rather, stewardship is the wise use of natural resources. As individuals, we can demonstrate good stewardship of natural resources by recycling materials and composting biodegradable waste. We can also properly dispose of garbage, and practice leaving areas such as public parks "better than we found them."

Companies, like those in the timber industry, can also practice good stewardship. A timber company cuts down trees for lumber and wood

Janelle Lugge/Shutterstock.com

Figure 16-29. Timber is a renewable resource because foresters can replant tree seedlings, which will mature for harvest in 15–20 years. *Which species of trees are harvested most commonly in your state?*

products, but it also plants many more acres of trees after the timber has been harvested, **Figure 16-29**. In this way, it provides for its timber needs as well as for future generations. Farmers and livestock producers also practice good stewardship of natural resources through such efforts as using conservative tillage practices, planting buffer zones, and implementing precision agriculture methods.

Types of Natural Resources

There are two main types of natural resources: renewable and nonrenewable. ***Renewable natural resources*** are those that may be regenerated in a short period of time. The trees harvested and replanted in the timber industry are an example of a renewable resource. Most trees can reach maturity in about 20 years, allowing a consistent supply of timber. Water is also considered a renewable resource, as is the air that we breathe. Both water and air can be cleaned and replenished through natural processes. The water cycle cleans water so that it can be used over and over, and various atmospheric nutrient cycles replenish air so that it is clean enough to breathe.

Nonrenewable natural resources are those resources that cannot be renewed in a relatively short period of time. Soil is an example of a nonrenewable natural resource, **Figure 16-30**. Soil can be formed through natural processes like erosion, freezing, and thawing. However, nearly 500 years are required to create even an inch of topsoil. So

Simon Bratt/Shutterstock.com

Figure 16-30. Soil is a nonrenewable resource because hundreds of years are required to create an inch of rich, fertile topsoil.

for practical purposes, when rains wash topsoil into rivers and streams, that natural resource is gone for many generations.

Nonrenewable natural sources are also those that took millions of years to develop and cannot be replenished in our lifetime or even a few generations. Nonrenewable resources include fossil fuels such as coal, natural gas, and crude oil.

Consumption of Natural Resources

Individuals consume natural resources in many ways. Some of these ways are more obvious than others. For example, when you warm your home in the winter or cool it in the summer, you and your family are likely using nonrenewable fossil fuels. Other ways we use natural resources are often hidden. For example, when you eat a banana, you consume the natural resource of the banana, as well as the fossil fuels, minerals, and other natural resources used to grow, harvest, and transport the banana. The machinery used to plant, harvest, and transport the banana is powered by fossil fuels, **Figure 16-31**.

Even the very act of your living and breathing consumes natural resources. When you breathe, your body removes oxygen from the atmospheric air and expels carbon dioxide, which contributes to air pollution and climate change—in minute proportions.

guentermanaus/Shutterstock.com

Figure 16-31. Growing and transporting bananas requires machinery powered by fossil fuels. ***What is the average distance a banana travels to reach consumers in the United States?***

Ecological Footprints

All human activities require the use or consumption of natural resources, so each individual's very existence on Earth leaves a mark on our planet, **Figure 16-32**. The measure of each person's mark on the planet is called an ***ecological footprint***. Ecological footprints include the amount of land needed to produce food, generate water, handle waste, produce energy, and replenish the atmosphere. More developed countries, like the United States and Australia, have relatively large ecological footprints. Scientists use the sum total of individual footprints involved in an activity to determine the activity's overall impact.

Mopic/Shutterstock.com

Figure 16-32. One of the footprints that we leave with the planet is the ecological footprint. All of your human activities will leave a mark on our planet. *How big is your ecological footprint? What can you do to reduce it?*

Carbon Footprints

Another measure of how humans consume natural resources and leave a mark on the planet is our carbon footprint. A ***carbon footprint*** is the measure of the greenhouse gases that are emitted as a result of human activity. That human activity could arise from an individual, the manufacture of a product, or a set of activities, **Figure 16-33**. Carbon is a

ssuaphotos/Shutterstock.com

Figure 16-33. Our carbon footprint in the United States is heavily influenced by the number of automobiles in this country. Combustion of fossil fuels, such as gasoline and coal, is a major contributor to carbon in the atmosphere. *Do regular vehicle tune-ups affect an engine's exhaust contents or level?*

powerful greenhouse gas, and with the changing global climate, scientists have become increasingly concerned with the amount of carbon human activities are emitting into the atmosphere.

A great deal of our carbon footprints come from indirect carbon sources. These indirect sources come from everyday things we use such as soap, food, clothing, and furniture. The construction of these products emits carbon into the atmosphere via carbon dioxide and methane. By purchasing these products, you enlarge your carbon footprint.

Did You Know?

According to The Water Project, more than 2 million tons of discarded water bottles are in U.S. landfills.

Reducing Our Carbon and Ecological Footprints

One of the keys to reducing our ecological and carbon footprints is to recycle containers, reduce consumption, and reuse products and their containers. For example, by washing plastic food containers and using them as containers for other foods or to store other products in our homes, we reduce the need to purchase more containers. Another example would be using glass jelly jars as portable and reusable drinking glasses instead of drinking bottled water. Reusing these types of containers also prevents the containers from ending up in a landfill. Using reusable containers to transport drinking water instead of drinking bottled water will also reduce your carbon footprint. You can also reduce your carbon footprint by becoming an informed consumer and purchasing products from companies dedicated to reducing their own carbon footprint.

Carbon Sequestration

One of the most potent greenhouse gases is carbon dioxide. Many scientists are researching ways to reduce the amount of carbon dioxide in the atmosphere. Carbon sequestration is one such method. ***Carbon sequestration*** is the process of capturing and storing carbon for long periods of time. Agriculture is an effective industry for carbon sequestration. The crop residue that is tilled into farmers' fields is mainly carbon. Thus, this crop residue is a place where carbon can be stored in the soil and not released into the atmosphere.

Carbon Credits

All industries are currently seeking ways to reduce the amount of carbon dioxide they introduce into the atmosphere. Companies that produce a lot of carbon dioxide may purchase carbon credits from industries that sequester carbon. A ***carbon credit*** is a fee paid by a company to allow it to emit a certain amount of carbon into the atmosphere. It is essentially a tax to produce carbon.

General Motors (GM) is one of the leading American companies in using carbon credits. GM produces a large amount of carbon dioxide in the production of automobiles. The company purchases carbon credits, but also engages in activities to reduce or offset the atmospheric carbon it produces. GM invests in wind farms, methane capturing processes at landfills, and planting forests to capture carbon dioxide.

Agriculture's Impact on the Environment

Just as individuals can reduce their impact on the environment, so can agriculturists. Agriculture uses a great deal of land and natural resources to feed the world's growing population. Agriculturists are concerned with maintaining and prolonging Earth's productivity because the livelihoods of farmers and ranchers depend on a productive planet. Individual farmers and ranchers can reduce their footprint on the land by using more environmentally conscious practices and reducing or eliminating negative-impact activities.

Scruggelgreen/Shutterstock.com

Figure 16-34. Open burning of trash is one of the high-risk environmental activities on the farm. This practice was common in rural areas because there was no regular garbage pickup service.

Negative-Impact Activities on the Farm

Open burning is one of the most detrimental activities for the atmosphere. When people burn trash, including items such as old pesticide containers, and other materials, the burning process releases toxic chemicals and pollutants into the atmosphere. Burning materials also releases carbon into the atmosphere, contributing to the greenhouse gas effect, **Figure 16-34**. Instead of burning trash and other refuse, farmers and ranchers try to recycle as many containers as possible.

Today's modern agriculture relies on the judicious use of chemical fertilizers and pesticides. If disposed of improperly, these chemicals can leach into the soil, run off into surface water, or evaporate into the atmosphere. Once the chemicals have escaped containment, they may pollute the soil, water, and air for long periods of time.

Dumping waste is another example of a negative-impact activity that was historically conducted on farms. Many decades ago, farmers dumped waste into ravines and gullies to get rid of it. The only problem was that this form of dumping did not get rid of the waste, it just moved it to another site. The table in **Figure 16-35** shows how long various materials take to decompose. Further, waste dumped into ravines and gullies eventually polluted surface water.

Reducing

Agriculturists try to find ways to reduce their use of resources on the farm. The development of new production technologies have enabled farmers

How long does it take to decompose?	
Paper towel	2–4 weeks
Newspaper	6 weeks
Apple core	2 months
Waxed milk carton	3 months
Plywood	1–3 years
Cigarette butt	1–5 years
Plastic bag	10–20 years
Styrofoam™ cup	50 years
Tin can	50 years
Rubber boot	50–80 years
Aluminum can	80–200 years
Plastic bottle	450 years
Disposable diaper	450 years
Glass bottle	1 million years

EPA, 2013

Figure 16-35. How long does it take to decompose? *Do any of these numbers surprise you?*

and ranchers to reduce the amount of fossil fuels needed to produce crops and livestock. Agricultural businesses produce products that require less packaging or reusable packaging. Also, new technologies like global positioning systems and drones reduce the amount of chemicals that farmers apply to their fields.

Agricultural Energy

Today's farmers who grow corn, cotton, soybeans, and other row crops rely on integrated pest management systems and the use of agricultural pesticides to produce abundant crops, **Figure 16-36**. The use of chemical pesticides requires fewer passes across the field, so farmers use much less diesel fuel for tractors than in the past. Also, no-till and reduced tillage production systems eliminate three or more passes across the field to produce the same amount of crop yield.

If you pass a farmer using a tractor to till a field today, you will not see any contaminating black smoke emitting from the tractor's smokestack. This

Balefire/Shutterstock.com

Figure 16-36. Growers, like this cotton farmer in California, manage pesticide applications to reduce the need for spraying and the need to burn fossil fuels by making too many passes over the field.

is because today's farm equipment is more fuel efficient and less polluting than the equipment used just a few decades ago. Farmers also use fewer tillage tools today. This means that the production of these tools is less, thus factories producing them have a lower impact on the environment.

Some farmers even "grow" energy through the implementation of wind farms. In many parts of our country, such as Texas, northwestern Indiana, and the Tug Hill Plateau area of New York, the wind conditions are right for constructing arrays of windmills. In Indiana, for example, the winds at about 150′ in the air almost never drop below seven miles per hour. This is ideal for a windmill to produce electricity. Hundreds of windmills have been constructed in the area.

Packaging Agricultural Seed, Feed, Fertilizer, and Chemicals

As the size of farms has increased, today's farmers can take advantage of larger and reusable containers for seed, feed, fertilizers, and chemicals. Most livestock feed is delivered in bulk trucks instead of trucks with bags of feed. Eliminating the feed bag eliminates the need for trees to produce the paper and water and fossil fuels to produce the bags. The same process of delivering bulk products to the farm has been used with seed, fertilizer, and pesticides.

Global Positioning Systems and Pesticide Application

With the increased use of global positioning systems and crop monitoring systems in the 1990s, farmers were able to more precisely target the application of pesticides and fertilizers to their fields. They could monitor the crop and determine the specific location of weed or insect infestations. Farmers more accurately mapped the soil types and fertilizer needs of their fields. This resulted in lower quantities of pesticides and fertilizer applications. So fewer chemicals were introduced into the ecosystem of the field.

Global positioning systems and now drones are changing the ways that agricultural chemical manufacturers think about producing and applying chemicals. They realize that farmers can pinpoint the application of fertilizers and pesticides. Formulations may be more potent and directly target the pest, while not exposing surrounding plants, soil, or water to the chemical. In this way, global positioning systems and drones have assisted farmers in accurately monitoring their crops and fields. This has reduced the amount of pesticide and fertilizers that farmers apply to their cropland.

Reusing

Agriculturists try to reuse as much as possible. For example, after careful cleaning, containers in the greenhouse and horticulture industry can be reused with future plantings.

Plastic Greenhouse Pots

Many greenhouse owners help reduce their ecological footprint by reusing plastic plant pots whenever possible. Plastic pots can be washed and/or sterilized and used with new plantings. Washing pots prior to reuse is important to prevent the spread of plant diseases. Reusing plastic pots saves greenhouse owners money because they do not need to purchase as many new pots each season, **Figure 16-37**. It also reduces landfill waste. Reusing greenhouse pots also conserves natural resources because energy is not used to create new pots.

Rigucci/Shutterstock.com

Figure 16-37. Reusing plastic greenhouse pots, such as these, can help reduce agriculture's impact on the environment. Growers may also use biodegradable seed pots to contain new seedlings and add nutrients to the soil as they decompose. *What other agricultural materials can you think of that can be reused?*

Recycling

Recycling on the farm takes on many forms, including composting organic matter and recycling scrap metal and plastic wrap.

Composting

Compost is organic matter that has been decomposed by microbial activity over time, **Figure 16-38**. Compost is a nutrient-rich soil amendment. In the soil, compost provides needed nutrients and even helpful microorganisms. Compost improves soil aeration and tilth.

Many agricultural by-products are composted. Composting is common in livestock industries, as well as greenhouse operations. In livestock industries, the manure and used bedding from cattle and sheep are composted. This compost is often sold to gardeners and homeowners to improve the soil around their homes. Compost is used commonly in agriculture, especially in small-scale organic farming.

Svend77/Shutterstock.com

Figure 16-38. By composting manure, farmers recycle nutrients from livestock back into a usable plant food. Dairy and beef producers often benefit financially from the sale of manure to be used for composting and enriching garden soil.

Scrap Metal

Farmers recycle steel, aluminum, copper, and other metals. The parts of implements like cultivator shovels and disk blades that contact the soil become dull with use over time. When the farmer replaces these, he or she recycles the steel through a local recycling outlet. Often several hundred pounds of steel and other metals are recycled.

Plastic Bale Wrap

Dairy farmers in many parts of the country use plastic wrap to protect large round hay bales and silage (or haylage) from spoiling, **Figure 16-39**. In the past, agricultural plastics were either burned or buried on the farm.

Burning exposed the atmosphere to harmful pollutants, and burying the plastics created contaminants in the ground, namely the plastics. Today this plastic wrap is recycled. It cannot be used again, so the farmers recycle the plastic bale wrap to be used as other consumer plastic products. Recycling is the most environmentally friendly and economical method of disposing of agricultural plastics. In order to recycle agricultural plastics, the materials must be relatively clean, dry, and bundled with like kinds of plastics.

mrcmos/Shutterstock.com

Figure 16-39. Plastic bale wrap is a useful tool for livestock farmers but can be difficult to dispose of. Farmers can actually recycle agricultural plastics, thus demonstrating good stewardship. *Are there plastics that cannot be recycled? If so, why can they not be recycled?*

Sustainability

Sustainability is the idea that we use Earth's resources in a way that they are not used up, but are available for future generations. We know that agriculture and food production require the use of natural resources. All farmers and ranchers, and even agricultural businesses, want to ensure that food, environmental resources, and clean water are available in the future. To make this possible, all agriculturists work together to conserve our natural resources through stewardship.

Career Connection Recycler

Job description: A recycler is a person who coordinates recycling and operates a recycling center. These people may also provide services, such as paper shredding, composting, and recycling materials pickup. A recycler interacts with people. The secret to a successful recycling business is two-fold. First, you need to encourage people to bring recyclables to your center. And then, you need to constantly seek the highest market prices for your recycled products.

bikeriderlondon/Shutterstock.com

Education required: A two-year degree in business or a related area.

Job fit: This job may be a fit for you if you enjoy interacting with people and can find creative ways to engage people with recycling.

Lesson 16.3 Review and Assessment

Words to Know

Match the key terms from the lesson to the correct definition.

1. The idea that we use Earth's resources in a way that they are not used up, but are available for future generations.
2. The ethic that people care for and conserve natural resources.
3. The resources that may be regenerated in a short period of time so that they do not run out.
4. The resources that cannot be renewed in a relatively short period of time.
5. The measure of an individual's impact on Earth.
6. The measure of the greenhouse gases that are emitted as a result of human activity.
7. The process of capturing and storing carbon for long periods of time.
8. A fee paid by a company to allow it to produce excess carbon into the atmosphere.
9. Organic matter that has been decomposed by microbial activity over time.

A. carbon credit
B. carbon footprint
C. carbon sequestration
D. compost
E. ecological footprint
F. nonrenewable natural resource
G. renewable natural resource
H. stewardship
I. sustainability

Know and Understand

Answer the following questions using the information provided in this lesson.

1. *True or False.* Stewardship means that people do not use natural resources.
2. What are two examples of a renewable resource?
3. What is the difference between an ecological footprint and a carbon footprint?
4. Give two examples of how you can reduce your carbon/ecological footprints.
5. A carbon credit is essentially a(n) _____ to produce carbon.
6. How do farmers and agriculturists reduce their use of natural resources?
7. Which on-farm substances can be recycled?
8. How does composting recycle materials on the farm?
9. Where are plastic films used on the farm?
10. What is the notion that we protect our natural resources for future generations called?

Analyze and Apply

1. Why do developed countries like the United States and Australia have large ecological footprints?
2. Why are we most concerned with reducing, reusing, and recycling nonrenewable natural resources?
3. How are carbon credits useful to saving the environment?

Thinking Critically

1. How do you show good stewardship of natural resources in your home?
2. What are ways that your agriculture program can be better stewards of natural resources?
3. How can you manage your SAE to be more sustainable? What type of SAE could you start in your area that would be built on a sustainability model?

Chapter 16

Review and Assessment

Lesson 16.1

Outdoor Recreation

Key Points

- Public use of land is a benefit of our natural resources.
- Even before the first national parks in the late 1800s, people set aside land for city parks so that everyone could enjoy the peace, serenity, and beauty of natural landscapes and wildlife.
- Public lands are those lands that are owned by the government.
- Private lands are those lands that are owned by an individual or a corporation.
- As a private property owner, a person is entitled to the right to use his or her land as he or she sees fit.
- Outdoor recreation incudes hunting, ATV and snowmobile riding, and hiking.

Words to Know

Use the following list and the textbook glossary to review and learn the *Words to Know* from *Lesson 16.1*.

conservation
fair chase rules
hunting season
managed hunt
parks
preservation
preserved land
private land
public land

Check Your Understanding

Answer the following questions using the information provided in *Lesson 16.1*.

1. What is an example of a publicly owned land?
2. When was the first national park in the United States created?
3. What are common responsibilities associated with private land ownership?
4. What is a hunting season?
5. What should not be consumed when hunting or operating an ATV or snowmobile?
6. When should firearms be unloaded?
7. What is the rule of fair chase?

8. What protective equipment should be worn when riding an ATV?
9. Why is it advisable to carry a GPS device when riding a snowmobile or hiking?
10. What types of supplies should you carry when hiking?

Lesson 16.2

Wildlife Management

Key Points

- Wildlife, fisheries, and ecosystems are important because many people like seeing wildlife and fish, visiting state and national parks, enjoying economic benefits, and enjoying wildlife and fish for food.
- An extinct species is one that is no longer living anywhere on Earth.
- An endangered species is one that is likely to become extinct.
- A threatened species is one that is low in total population and likely to become endangered in the near future.
- Game species are those species that are hunted for food and/or sport.
- Nongame species are those species that are not hunted and/or are protected from hunting because they are threatened or endangered.
- Game management controls the population, provides a mechanism to harvest weak and old animals from the ecosystem, and enables stronger animals to access food, water, and shelter without competition.
- Chronic wasting disease, blue tongue, and Colony Collapse Disorder are diseases that impact important species of animals.
- Organizations that work with wildlife management include the Fish and Wildlife Service, Ducks Unlimited, the National Wild Turkey Federation, and the Nature Conservancy.

Words to Know

Use the following list and the textbook glossary to review and learn the *Words to Know* from *Lesson 16.2.*

blue tongue
chronic wasting disease (CWD)
Colony Collapse Disorder (CCD)
corridors of habitat
endangered species
extinct species
fencerow
game

game species
genetic reproductive potential
midge
nongame species
population dynamics
threatened species
wildlife inventory

Check Your Understanding

Answer the following questions using the information provided in *Lesson 16.2*.

1. Which organization manages hunting, monitors wildlife populations, and conducts research to help preserve and maintain species?
2. What do scientists look for when conducting a wildlife inventory?
3. What are included in historical records of wildlife?
4. What happens to the population of a species when it becomes extinct? Give an example of an extinct species.
5. How are endangered species protected?
6. What is the difference between a game species and a nongame species?
7. Which game species are hunted because they are nuisances?
8. Why is biodiversity important to farmers and landowners?
9. Which species has a higher genetic reproductive potential: deer or rabbits? Explain your answer.
10. Which species does chronic wasting disease infect?
11. Which insect spreads blue tongue disease?
12. What are possible causes of hive collapse disorder?

Lesson 16.3

Stewardship of Natural Resources

Key Points

- Stewardship is the ethic that people care for and conserve natural resources.
- Renewable natural resources are those resources that may be regenerated in a short period of time so that they do not run out.
- Individuals consume natural resources by heating their homes, driving cars, eating food, wearing clothes, and even breathing.
- Individuals can reduce their impact on the environment by avoiding activities that negatively impact the environment.
- Agriculturists reduce their impact on the environment by reducing the use of energy to produce crops and livestock, reducing the packaging of agricultural products, and using global positioning systems.

- Agriculturists reduce their impact on the environment by reusing plastic greenhouse pots.
- Agriculturists reduce their impact on the environment by recycling manure as compost and recycling agricultural plastics and scrap metals.

Words to Know

Use the following list and the textbook glossary to review and learn the *Words to Know* from *Lesson 16.3*.

carbon credit
carbon footprint
carbon sequestration
compost
ecological footprint
nonrenewable natural resource
renewable natural resource
stewardship
sustainability

Check Your Understanding

Answer the following questions using the information provided in *Lesson 16.3*.

1. What is the difference between a renewable and a nonrenewable resource?
2. What are two examples of a nonrenewable resource?
3. How much time is required to replenish one inch of topsoil?
4. Carbon is a powerful _____ gas.
5. How is agriculture an effective industry for carbon sequestration?
6. What are two negative-impact activities on the farm?
7. How do today's farmers use less energy than farmers from a few decades ago?
8. How can agriculturists reuse natural resources on the farm and in agricultural businesses?
9. How can agriculturists recycle natural resources on the farm and in agricultural businesses?
10. *True or False?* Farmers and other agriculturists are interested in sustainability of our natural resources.

Chapter 16 Skill Development

STEM and Academic Activities

1. **Science.** Design a scientific research project to measure the amount of trash produced by your family in one week. Institute changes in reducing, reusing, and recycling. What is the impact of those changes on your family's ecological footprint? How much impact would there be if all the families in your school implemented similar changes? Why?

2. **Technology.** Research major wind farms in the United States. Where is each located? Map these wind farms. What can you conclude about the location of wind farms? How much electricity is produced in each? For how many years could one of these wind farms power your home?
3. **Engineering.** Many models of home composters are available on the market through hardware stores. Research these designs and see if students in an agricultural mechanics shop could engineer similar or better designs for home composters. Prepare a blueprint and model of your redesigned composter.
4. **Math.** Where do the big trees grow in your state? When were the record game species of wildlife and fish harvested? Where were they harvested? Think of these questions as you develop hypotheses about big trees and record fish and wildlife in your state. Research the big trees in your state and/or the game species records. Were your hypotheses supported? Why? What do you notice about where and when the records were set? How does this relate to what you have learned about outdoor recreation?
5. **Math.** Using an online carbon footprint calculator, determine the total amount of carbon emissions produced in one school year by your agriculture class in driving to school. Which years, makes, and models produce the most carbon? How would your class carbon footprint change if everyone rode the school bus? What is the magnitude of change caused by using this public transportation?
6. **Social Science.** Select one natural resource that is used in agriculture. Trace that resource's history of use in agriculture. Are we using more or less of the resource today? Why? What has been done to conserve the resource?

Communicating about Agriculture

1. **Speaking and Listening.** Identify an issue related to ecological footprints and what we can do to reduce our footprints. Using the National FFA Organization's prepared public speaking career development event format, prepare a speech to present to local community organizations about reducing, reusing, and recycling.
2. **Writing.** Think about stewardship of natural resources. What can your school do to better steward natural resources at school? Think of and research a plan to improve stewardship of natural resources at your school. Develop an implementation plan for your ideas. Write it and present it to the student council or school administration.
3. **Reading and Speaking.** Using the National FFA Organization's agricultural issues forum career development event format, identify an issue related to ecological or carbon footprints in your area or state. Work with fellow students to research information and perspectives about the issue. Then, prepare a 15-minute presentation for public audiences.
4. **Technical Writing.** Research home composting. Prepare a one-page flyer about how to start home composting.

Extending Your Knowledge

1. Inventory the wildlife on your school property. How do you know which species of animals are present on the grounds of your school? Where should you look for signs of wildlife?
2. Find online examples of population dynamics. For example, what happens to the fox population when snowshoe hare populations increase? Why?
3. Interview farmers and agriculturists in your school district about their stewardship of natural resources. What changes have they seen over their lifetimes in the ability of agriculture to reduce, reuse, and recycle?
4. Use the Internet to find the percentages of land in your state that are owned by private individuals, corporations, and the public or government. Compare these data to other states. Which states have the highest percentage of public-owned land? Why do you think this is the case?

Chapter 17

Forestry

Lesson 17.1
Forest Production

Lesson 17.2
Technology in Forest Management

©iStock/nzgmw

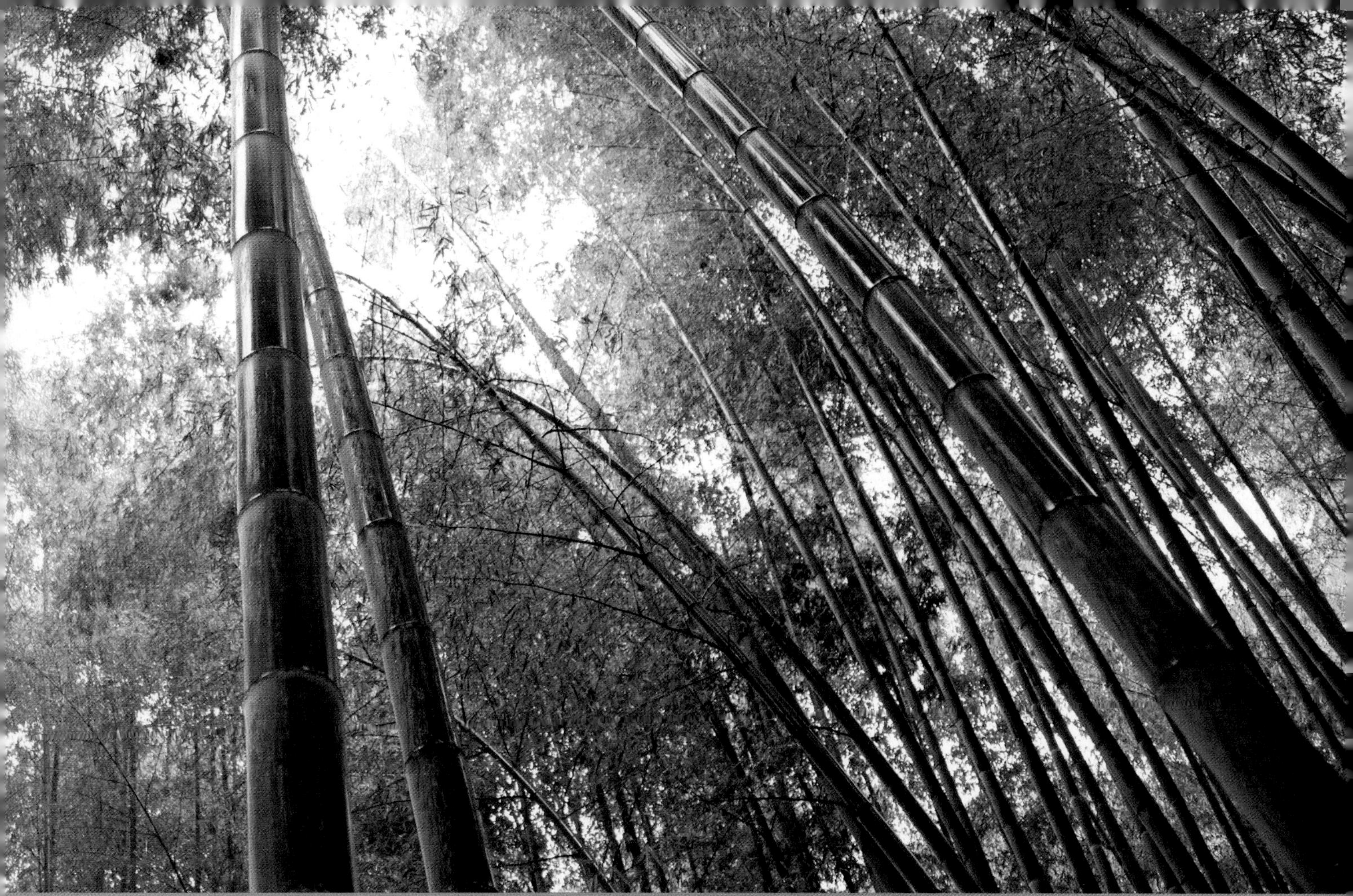

TAGSTOCK1/Shutterstock.com

©iStock/Darinburt

Gene Alexander/USDA Natural Resources Conservation Service

While studying, look for the activity icon  **to:**

- **Practice** vocabulary terms with e-flash cards and matching activities.
- **Expand** learning with video clips, animations, and interactive activities.
- **Reinforce** what you learn by completing the end-of-lesson activities.
- **Test your knowledge** by completing the end-of-chapter questions.

Lesson 17.1

Forest Production

Words to Know

angiosperm
broadleaf tree
cambium
coniferous tree
crown
deciduous coniferous tree
deciduous tree
evergreen broadleaf tree
evergreen tree
forestry
gymnosperm
hardwood species
heartwood
inner bark
knot
lumber
mast
outer bark
phloem
plywood
pulpwood
sap
sap flow
sapwood
sawlogs
silviculture
softwood species
spile
sugar bush
veneer
veneer logs

Lesson Outcomes

By the end of this lesson, you should be able to:

- Discuss the economic significance of forestry in the United States.
- Identify economically important tree species.
- Discuss forest products.
- Explain the parts of a tree and their functions.
- Differentiate between coniferous and deciduous trees.

Before You Read

As you read the lesson, record any questions that come to mind. Indicate where the answer to each question can be found: within the text, by asking your teacher, in another book, on the Internet, or by reflecting on your own knowledge and experiences. Pursue the answers to your questions.

In the United States, we are fortunate to have many types of trees and forests growing within our borders. In fact, the United States owns a disproportionate amount of the overall forests in the world, **Figure 17-1**. When the first European settlers arrived in what would become the United States, almost 50% of the land was covered in forests. Today, about 35% of our land is covered in forests. What may surprise you is that the amount of forested land has been growing since the year 2000, and there are more trees in the United States now than when you were born.

Silviculture is the practice of growing trees for production. More commonly, this is called ***forestry***. The forestry and timber industries are a significant portion of our overall economy. Let us look at how forests impact our lives in America.

The Economics of Forestry

In the United States, forests cover 35% of our land and account for about 4% of the total gross domestic product (GDP), according to the American Forest and Paper Association. The forestry and timber industries directly employ nearly one million people and support the jobs of another 1.5 million people in the United States.

Sawlogs, Pulpwood, and Veneer Logs

Trees grown for timber production are grown for three main purposes: sawlogs, pulpwood, and veneer logs. ***Sawlogs*** are logs that are sawn into lumber used for building houses, furniture, or other wood products. ***Lumber*** is timber that has been sawn into long boards for planks. Lumber is typically used in construction and making furniture. ***Pulpwood*** are the smaller coniferous trees grown to produce paper and paper products like cardboard. Pulpwood ready for processing is shown in **Figure 17-2**. ***Veneer logs*** are harvested from trees like walnut, cherry, and oak. ***Veneer*** is a thin layer of decorative wood that can be glued over the surface of less expensive or less decorative wood.

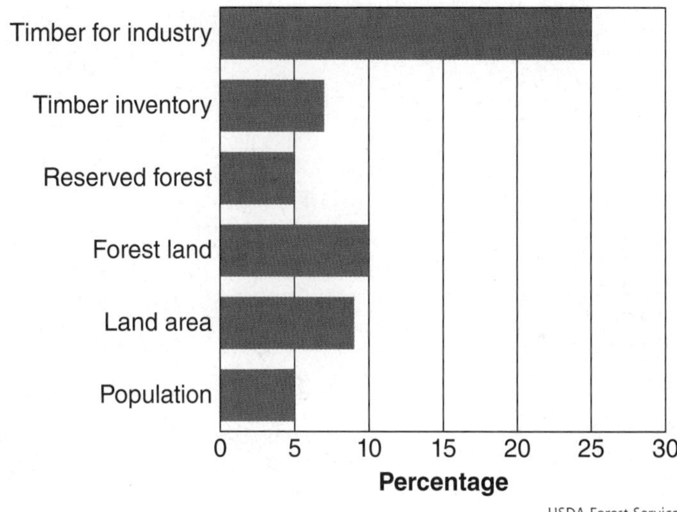

Figure 17-1. We are fortunate to have substantial forest resources in the United States.

Tree Species

Not all trees are created the same. Tree species grow differently from one another, and each is usually best suited for a specific purpose. The two main kinds of tree species are deciduous and coniferous. ***Deciduous trees*** are trees that grow new leaves in the spring of each year, and then lose those leaves in the fall. Deciduous trees are also known as ***hardwood species***. ***Coniferous trees*** are trees whose leaves are needles or scales. Coniferous trees remain green throughout the year, even in winter. They may shed a few needles or scales in the fall, but will not become bare at any time throughout the year. Coniferous trees are also called ***softwood species***. Economically important trees may be either hardwood species or softwood species.

Did You Know?

More than half of the forests in the United States are privately owned.

Dainis Derics/Shutterstock.com

Figure 17-2. Pulpwood are smaller coniferous trees grown to produce paper and paper products. Here pulpwood is ready for processing. ***Have you ever made paper or visited a paper processing plant?***

3523studio/Shutterstock.com

Figure 17-3. Oaks are some of the most diverse and common of the tree species. Acorns not only produce new trees, but they provide an important source of nutrition to many animals. *How many types of oak trees can you identify? Which of these grow in your state?*

Hardwood Species

Hardwood species are used for nearly all veneer logs and for most furniture construction. As a general rule of thumb, hardwood species grow much more slowly than softwood species. Because they grow more slowly, the growth rings of these trees are closer together and more dense. Wood that is denser is also stronger.

Wood from hardwood species tends to have more diverse colors and patterns than wood from softwood species. This diversity of colors and patterns makes certain hardwood species more desirable as well as more expensive. To reduce the cost of the end product, but still maintain a more desirable finish, the manufacturer will construct the product with a less expensive wood and cover the exposed surfaces with a hardwood veneer.

Let us look at some hardwood species that are economically important.

Oak

Oaks are one of the most diverse and common of the tree species, **Figure 17-3**. Oak trees are grown throughout the United States. The two major families of oak trees are the red oaks and the white oaks. Oaks can be either deciduous or coniferous. Oak lumber is used to construct cabinets and furniture, wood flooring, and as veneer. Other common uses for oak lumber include the oak barrels used in wine, liquor, and beer production, as well as the wooden railroad ties that span our nation. Depending on the species, oak trees may reach heights of 100′ (30 m) with trunk diameters ranging from 2′–5′ (0.6 m−1.5 m).

STEM Connection Density of Wood Species

The density of wood is one indicator of the hardness of the wood. Think about the terms *softwood* and *hardwood*. What hypotheses can you make about the average density of tree species in each of these categories? Create a hypothesis. Then, categorize 30 tree species as either softwood or hardwood. Locate one or two wood density charts online. Using the information from these charts, calculate the average density of each category. Does that support or refute your hypothesis? Which tree species are outliers to your hypothesis? Can you think of reasons why this might be the case?

Density also affects other properties, like nail-holding ability, ease of working, and even the firewood quality of the wood. You can find measures of these wood features online. Create hypotheses about these features and species of wood. Then, find data and evidence online to support or refute your hypotheses.

Also, look at the common names of some of these species. Can you see how they earned their names? Why? For instance, why is silver maple also commonly referred to as *soft maple*? And why is sugar maple referred to as *hard maple*?

Figure 17-4. White ash leaves have an opposite compound leaf structure. A compound leaf structure has more than one leaflet and a bud at its stem base. Ash leaves tend to have five to nine leaflets per leaf. Ash seeds are commonly known to most of us as helicopter seeds. ***How many species of ash are there in the United States?***

Steve Cordory/Shutterstock.com

vvoe/Shutterstock.com

Ash

White ash is famously used to make most wooden baseball bats. White ash wood is quite resistant to shock, which is why it is an ideal wood for baseball bats. This characteristic also makes ash wood ideal for tool handles. Ash is used to make handles for high-impact hand tools such as shovels and hammers. Ash is also used to make wooden boxes and crates. Unique ash tree characteristics include their branching and leaf structure, as well as their seed design, **Figure 17-4**. With more than 40 species available, heights and trunk diameters cover a wide range.

Black Cherry

Black cherry is one of the more expensive hardwood species. Black cherry is expensive, in part, because of the desirability of its slightly reddish wood. Black cherry is used for cabinetry, high-end furniture, flooring, and veneer, **Figure 17-5**. The trees range from 50′–100′ (15 m−30 m) tall and have trunk diameters between 3′–5′ (1 m−1.5 m). They are commonly found throughout the eastern part of the United States.

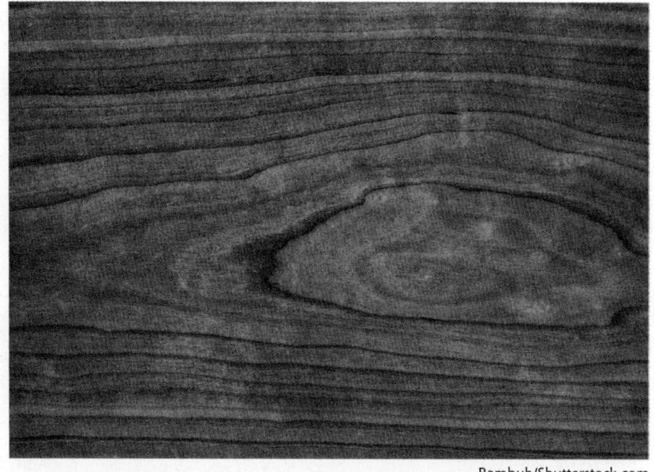

Bambuh/Shutterstock.com

Figure 17-5. Cherry is expensive because of the desirability of its slightly pinkish or reddish wood when freshly cut. ***Are cherry trees grown specifically for lumber?***

Black Walnut

Black walnut is a hardwood species that is dark brown or even purplish in color. Walnuts from the black walnut tree are edible and

yoshi0511/Shutterstock.com

Figure 17-6. Black walnut has a rich, dark color, and it is popular because of its ease of working, preferred grain patterns, and strength.

quite commonly used in salads, cookies, and other foods. Black walnut is a prized hardwood that is highly resistant to decay. It is an expensive hardwood lumber used to construct products such as furniture, gunstocks, and veneer. Among woodworkers, black walnut is perhaps the most popular of the hardwoods because of its ease of working, preferred grain patterns, and strength. Black walnut has a rich, dark color, **Figure 17-6**. Black walnut trees range from 100′–120′ (30 m–37 m) tall with trunk diameters between 2′–3′ (0.6 m–1 m).

Tulip or Yellow Poplar

Tulip or yellow poplar is a common tree in the eastern United States. In the forest, its trunk grows tall and straight with few branches. The leaf is shaped like a tulip, **Figure 17-7**. Poplar is a low-density hardwood, which makes it easy to work. The sapwood is pale yellow or even white, while the heartwood is dark gray or green. Because of the low density and light color of the wood, it can be stained to almost any color. Poplar is commonly used to construct wooden crates, shipping pallets, and upholstered furniture frames. Depending on the species, poplar trees range in height from 49′–164′ (15 m–50 m) with trunk diameters up to 8′ (2.4 m).

Did You Know?

Poplar tree roots are excellent at seeking water sources and may cause problems with underground water pipes.

Hickory

There are between 17 and 20 species of hickory trees, several of which are native to the eastern and midwestern United States. A shagbark hickory is one of the easiest trees to identify because of its shaggy bark, **Figure 17-8**. Hickory trees all produce hickory nuts, which are a valuable

lorenzobovi/Shutterstock.com

StevenRussellSmithPhotos/Shutterstock.com

Figure 17-7. The straight trunks of poplar trees make them ideal for lumber and make an impressive forest. The leaves have a tulip shape. ***Do poplar trees have other unique characteristics?***

source of food for wildlife. Hickory has high shock resistance. Because of this characteristic, hickory is often used to make wooden handles in shovels and hammers. Hickory wood is among the hardest and strongest woods, which make it ideal for wooden flooring and even cabinetry. Depending on the species, hickory trees range in height from 10′–90′ (3 m–27.4 m) with trunk diameters ranging from 8″ (20.3 cm) to 4′ (1.2 m).

Hard Maple

Maple is one of the more common hardwood species in the United States. The most common type of maple for lumber is the sugar or hard maple, **Figure 17-9**. Maple wood has a fine texture and is very dense. Maple is also very durable, so it is used in flooring, especially in basketball courts, dance floors, and bowling alleys. Maple is the preferred wood for cutting boards, work benches, and butcher blocks. Because maple wood has a light, open texture, it is also used in cabinet construction. Maple is also used in baseball bats, but it has a greater tendency to shatter compared to ash. Maple tree species range in size from small 20′ (6 m) trees to giants reaching over 100′ (30.5 m).

Jeffrey M. Frank/Shutterstock.com

Figure 17-8. A shagbark hickory is one of the easiest trees to identify because of its shaggy bark. ***Have you seen a shagbark hickory tree? Which other tree species are relatively easy to identify because of their bark?***

Nancy Kennedy/Shutterstock.com

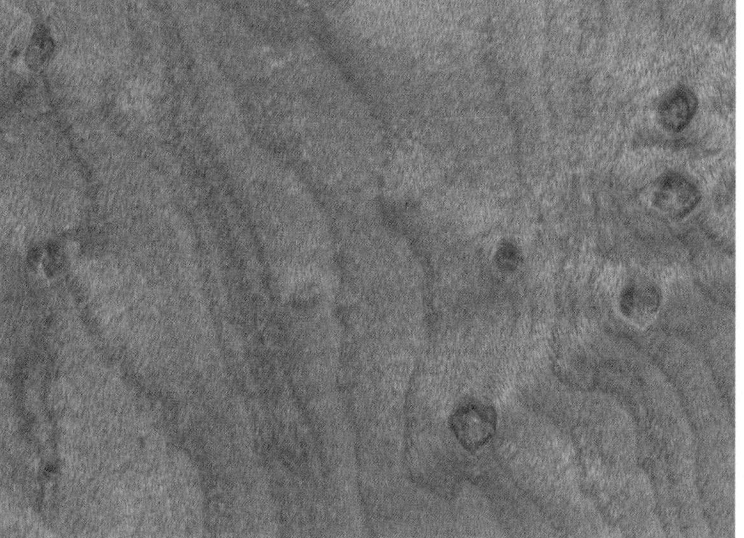

Lakeview Images/Shutterstock.com

Figure 17-9. Sugar maple trees have some of the most colorful leaves in autumn. They also have a beautiful clean shape that makes a wonderful addition to any landscape. The maple wood illustrated above, known as bird's eye maple, is most often found in wood from the sugar maple.

TFoxFoto/Shutterstock.com

srekap/Shutterstock.com

Figure 17-10. Douglas fir is used in making plywood and structural lumber.

Softwood Species

Nearly all pulpwood comes from softwood species. Also, nearly all sawlogs used for home or building construction are softwood species. The wood of softwood tree species has high strength-to-weight ratios, which allows them to be flexible and strong. Softwood species are able to bear heavy loads without breaking. They may bend, but are less prone to breaking under stress than hardwoods.

Douglas Fir

Douglas firs are evergreen trees used in making plywood and structural lumber, **Figure 17-10**. Douglas fir is also commonly used to make window frames. Douglas firs are popular Christmas trees in many parts of the United States.

White Pine

White pine lumber is the mainstay of construction. White pine trees grow tall, straight, and without many branches. ***Knots*** in the wood are areas that make lumber weak, **Figure 17-11**. The knots occur where the branches are attached to the trunk. So, the lack of branches and knots helps preserve the strength of white pine lumber. Because white pine is highly resistant to splitting, it is used to construct furniture, houses, and various moldings. White pine holds paints well and can be machined easily.

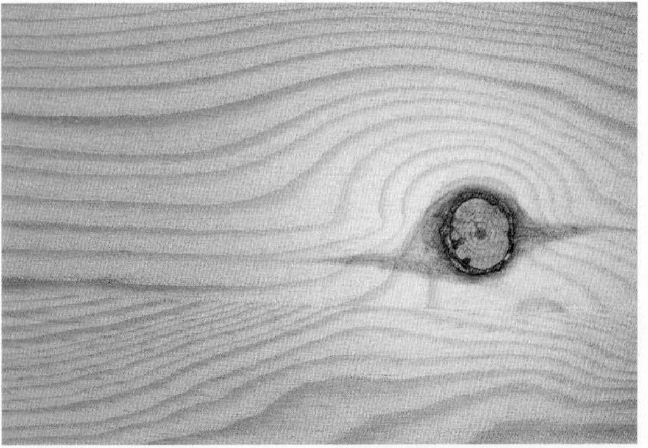
Patryk Michalski/Shutterstock.com

Figure 17-11. Knots in the wood are areas that make lumber weak. Therefore, the lack of branches and knots helps to preserve the strength of lumber.

Western Red Cedar

Western red cedar is a fragrant, highly decay-resistant wood that is commonly used for shingles and exterior coverings of houses and other buildings. Some expensive outdoor decks are constructed of cedar. Natural fence posts are commonly red cedar because of the wood's resistance to decay. Someone in your family may own a cedar-lined chest or closet. Cedar maintains its fragrance. The fragrance of cedar inhibits many insects, so clothes stored in a cedar chest or closet are somewhat protected from insect damage.

Other Softwood Species

Other common softwood species include southern yellow pine and western hemlock. These species are used in timber for construction and various plywood and engineered timbers. Southern yellow pine is the most common tree species used in constructing utility poles.

Fraser fir and balsam fir are rapidly becoming a popular Christmas tree type in many parts of the country. Fraser fir is currently the most common. Colorado blue spruce is another favorite type of Christmas tree.

What Are Forest Products?

Forests certainly provide lumber, and they also provide the raw materials for the manufacturing of many other products that we use in our everyday lives. Forest products include lumber, paper, veneer, energy, and food.

Lumber and Plywood

Lumber is wood that has been cut into boards of various sizes. Two common sizes of lumber are 2 × 4 and 2 × 6. A 2 × 4 is roughly two inches by four inches when viewed from the end of the board. It may be any length. A 2 × 6 is roughly two inches by six inches when viewed from the end. These dimensions of boards are common for framing houses and other buildings.

Trees cut for lumber are generally cut into 16′ lengths. Boards of this length are common in the building industries. If you measure most rooms in your house, you will find that the rooms are roughly dimensions in increments of four feet. So, a 16′ board can be used in many places in your home. How a log is cut into lumber is shown in **Figure 17-12**.

Plywood is an engineered lumber product that includes three or more layers of veneer glued together. Plywood is generally sold in sheets that measure four feet wide by eight feet long. Most plywood varies in thickness between about one-quarter inch to one inch in depth. Plywood is shown in **Figure 17-13**.

Slavoljub Pantelic/Shutterstock.com

Figure 17-12. Boards are cut from the trunks of trees in very precise patterns to obtain the maximum yield of lumber from the tree.

noppharat/Shutterstock.com

Figure 17-13. Plywood is an engineered lumber product that includes three or more layers of veneer glued together. It can vary in thickness.

hxdyl/Shutterstock.com

Figure 17-14. Paper may be rolled onto huge rolls that are used on large printing presses. *A piece of paper is lightweight, but a ream of paper in a package is quite heavy. How much do you think one of these paper rolls weighs?*

MARGRIT HIRSCH/Shutterstock.com

Figure 17-15. Wood veneer is available to woodworkers in various formats. The rolls shown here (cherry, oak, and pine) may be used to create elaborate inlays. *In what other formats may wood veneers be purchased?*

Paper

Pulpwood is processed into paper and paper products, such as cardboard. To make paper from wood (paper can be made from many other materials), the trees must first be shred into very fine fragments. These fragments are then soaked in a pulp slurry. The slurry contains chemicals that help break down the lignin in the wood.

Once the slurry is sufficiently mixed, it is poured through a fine screen that captures the cellulose fibers from the wood in a crisscross pattern. These fibers are pressed and dried into the thin film of wood cellulose that we call paper. Once pressed and dried, the paper may be bleached or further processed for different applications. Some paper is rolled onto huge rolls that are used for large print runs of products such as this textbook, **Figure 17-14**.

Veneer

As stated earlier, *veneer* is a thin slice of decorative wood that can be glued over less decorative wood, **Figure 17-15**. Many veneer factories in the United States are located in Indiana because of the vast hardwood forests in the southern half of the state. It is a highly specialized segment of the wood industry.

To manufacture veneer, a log is turned against a rotary lathe with a very sharp blade. The veneer is peeled away from the trunk in one continuous sheet. Think of unrolling a roll of wrapping paper. This is what creating veneer is like, except that an extremely sharp blade is required to create the sheet of wood.

Energy

Trees also provide energy. People in many parts of the United States cut and burn firewood for heating their homes in winter. In the papermaking processes, excess heat is produced. This heat is used to provide energy for nearby factories and homes.

Food

Trees and forests provide food for both wildlife and humans. Various nuts, like hickory nuts, beechnuts, and acorns are important food for wildlife. We call this natural wildlife food ***mast***. Mast sustains deer, squirrels, raccoons, and other forest critters throughout the winter.

Maple Syrup

Have you ever eaten pancakes or waffles with maple syrup? If so, then you have eaten a forest food product. Maple syrup comes from tapping maple trees in the late winter and early spring. The two species of maple trees that produce the most maple syrup are the sugar and black maple. Maple syrup production is especially important in northern states like Michigan, Pennsylvania, New York, New Hampshire, and Vermont.

As winter days grow longer and more sunshine hits the forest, trees begin to awaken from their winter dormancy. They start to move stored sugars from their roots to their shoots and leaf buds. The sugary watery substance in trees is called ***sap***. The nighttime freezing and daytime thawing facilitates this translocation of stored sugars. ***Sap flow*** is the term used to describe the translocation of sugar from roots to shoots of the tree.

Maple producers drill small holes in trees and then place a small metal spout, called a ***spile***, in the tree, **Figure 17-16**. This spigot captures sap and allows it to drain into either a bucket or network of sap tubes running through the forest. A collection of maple trees that are tapped is called a ***sugar bush***, **Figure 17-17**.

Once the sap is collected, it is boiled until most of the water is removed, **Figure 17-18**. At this point, we have maple syrup. Approximately 40 gallons of sap are needed to produce one gallon of maple syrup. Pure maple syrup is expensive. One gallon of maple syrup may cost up to nearly $50 dollars, depending on the grade of the syrup. Maple syrup is graded based on the color and clarity of the syrup. Generally, the darker syrups have a stronger maple flavor.

What Are the Parts of a Tree?

The parts of a tree are similar to the parts of most other plants. All trees have roots, a trunk, branches, leaves, and flowers. Let us explore the parts of a tree in greater detail.

O Driscoll Imaging/Shutterstock.com

Figure 17-16. Maple producers drill small holes in trees, place spiles in the holes, and then collect sap in either buckets or a network of plastic tubing. ***Have you ever tasted pure maple syrup? How do you like it?***

Alexandralaw1977/Shutterstock.com

Figure 17-17. These tapped maple trees are called a sugar bush. ***How much sap does a single tree produce?***

Lurin/Shutterstock.com

Figure 17-18. Once the sap is collected, it is boiled until most of the water is removed. ***Is maple syrup harvested near your home?***

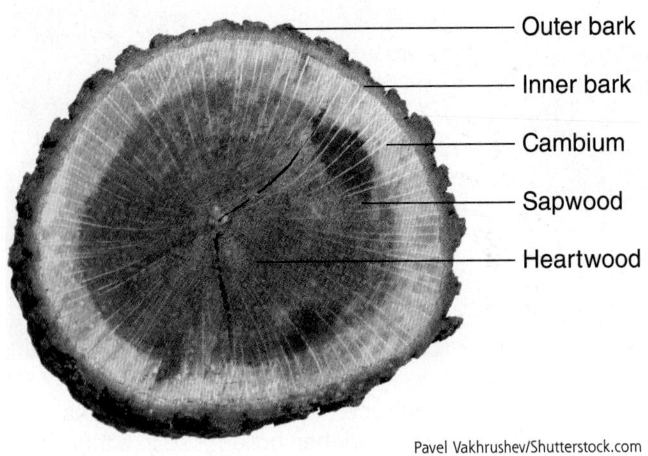

Pavel Vakhrushev/Shutterstock.com

Figure 17-19. The layers of a tree trunk. *What might the difference in width or color of each age ring signify?*

Roots

The roots of trees absorb water and nutrients from the soil. They also store sugar in winter so that the tree can produce new leaves in the spring. Roots anchor the tree deep in the soil. Trees can have either taproots or fibrous roots. Taproots may extend up to 15′ into the ground. With fibrous root systems, almost all of the roots are found within the upper 12″ to 18″ of soil because this is where oxygen is most prevalent in the soil.

Trunk

The trunk of the tree supports the branches and leaves. There are four layers of tissues in the trunk that move water and nutrients throughout the tree. We will explore these layers moving out from the center of the tree trunk, as shown in **Figure 17-19**.

Heartwood

Heartwood is the innermost part of the trunk of the tree. Heartwood consists of old, inactive xylem cells. Heartwood is where sugars and oils are stored in the tree. These stored materials give heartwood a darker color than the other wood in the trunk. Heartwood also gives the tree trunk its strength.

Xylem or Sapwood

Sapwood is the xylem of the tree trunk, which carries water and nutrients from the roots to the shoots of the tree. Xylem cells are the lighter colored wood in the trunk.

Cambium

The ***cambium*** is the thin layer of actively growing tissue in the tree trunk. The cambium is what produces the growth rings seen when the tree is cut down. The trunk of a tree does not actually grow taller. Rather, it grows larger in girth. This is caused by new layers of cambium being grown each growing season. Cambium will become xylem, phloem, or more cambium as it ages.

Phloem or Inner Bark

The ***phloem***, or ***inner bark***, is the thin layer of phloem cells between the cambium and outer bark. The inner bark provides the food supply to the entire tree. Sugars produced in the leaves are transported throughout the tree by the inner bark.

Did You Know?

Due to the subtle difference between wet and dry seasons, relatively few species of trees in the tropical rain forest have visible annual rings.

Outer Bark

Outer bark is the outer covering of the trunk, branches, and twigs. These are dead phloem cells that protect the tree from insects, disease, and even fire. Over time, bark may become quite thick and creviced. Tree bark is different from one tree species to another. So, trees can be identified in the winter by their bark.

Crown

The ***crown*** of the tree consists of the leaves and branches at the top of the tree. The crown of the tree is the site of photosynthesis. Because of the dense leaf foliage, the crown is also the part of the tree that provides the most shade and cooling.

How Do Trees Grow?

Trees are generally classified by how they grow, or rather whether or not they lose their leaves in the winter. As stated earlier, the two main types of trees are deciduous and coniferous. Two minor types of trees are deciduous coniferous trees and evergreen broadleaf trees.

Deciduous Trees

Deciduous trees are trees that drop their leaves in the fall. Because they drop their leaves in the fall, deciduous trees are dormant in winter. They grow new leaves in the spring when the days begin getting longer and temperatures warm. Most deciduous trees are also ***broadleaf trees***. This means that their leaves have a broad, flat blade. Maples, cherries, and many oak trees are examples of deciduous trees. Deciduous trees are generally referred to as hardwoods. Deciduous trees are classified as angiosperms. ***Angiosperms*** are plants that bear seeds in vessels. These vessels could be fruits, nuts, or capsules.

Coniferous Trees

Coniferous trees do not lose their leaves in the fall. They remain green throughout the year, hence their other name, ***evergreen tree***. Most coniferous trees have long, needle-like leaves. Pines, spruces, cedars, and firs are common coniferous trees. Most coniferous trees also bear cones which contain the seeds of the tree. Coniferous trees are generally referred to as softwoods. Coniferous trees are classified as ***gymnosperms***. Gymnosperms are evolutionarily older than angiosperms. The seeds of gymnosperms are bare. But, they may be protected by a sturdy cone.

Deciduous Coniferous Trees

Some coniferous trees do lose their leaves in the fall. Larch and tamarack are common examples. A small grove of tamaracks changing colors in the fall is shown in **Figure 17-20**. ***Deciduous coniferous trees*** are trees that have needle-like leaves, but lose those leaves in the fall.

Mark Herreid/Shutterstock.com

Figure 17-20. Tamarack trees are examples of deciduous coniferous trees. These are changing colors in preparation to lose their needles in the fall.

Serge Skiba/Shutterstock.com

Figure 17-21. Live oak trees are evergreen broadleaf trees. They will lose some but not all of their leaves in winter. *In what parts of our country do live oak trees grow? Why do you think this is the case?*

Evergreen Broadleaf Trees

Some trees have broad, flat leaves, but remain green throughout the year. These are called ***evergreen broadleaf trees***. Rhododendron, holly, and live oak are examples of evergreen broadleaf trees. A live oak tree from South Carolina is pictured in **Figure 17-21**.

Career Connection Arborist/Urban Forester

Job description: An arborist, or urban forester, is a specialist who prunes and removes trees in an urban setting. Trees in an urban setting are generally high-dollar trees. They are expensive because of their beauty and landscape value. So arborists must take care to manage the health of these trees, some of which may be hundreds of years old. Arborists use specialized equipment, such as sophisticated block-and-tackle systems, special chain saws, and climbing equipment. If a tree needs to be removed, they work to make sure the tree does not fall and damage homes or other buildings. Some arborists work in city parks to maintain trees there.

Education required: A two- or four-year degree in forestry is required.

Job fit: This job may be a fit for you if you like working with trees and interacting in an urban setting.

Ian Tragen/Shutterstock.com

An arborist or urban forester cuts and manages trees in urban settings. They may prune or remove trees, taking care not to damage houses and other urban structures.

Lesson 17.1 Review and Assessment

Words to Know

Match the key terms from the lesson to the correct definition.

1. The innermost part of the trunk of the tree.
2. A tree whose leaves are needles or scales that remain green throughout the year, even in winter.
3. The timber that has been sawn into long boards for planks.
4. A tree that has needle-like leaves, but loses those leaves in the fall.
5. A tree with broad, flat leaves that remain green throughout the year.
6. The smaller coniferous trees grown to produce paper and paper products, like cardboard.
7. A tree that grows new leaves in the spring of each year, and then loses those leaves in the fall.
8. Logs that are sawn into lumber used for building houses, furniture, or other wood products.
9. A collection of maple trees that are tapped for collecting sap.
10. The outer covering of the trunk, branches, and twigs.
11. The leaves and branches at the top of the tree.
12. The practice of growing trees for production.
13. The xylem of the tree trunk which carries water and nutrients from the roots to the shoots of the tree.
14. Plants that bear seeds in vessels that could be fruits, nuts, or capsules.
15. The thin layer of actively growing tissue in the tree trunk.
16. A small metal spout placed in trees to drain the sap.
17. The thin layer of phloem cells between the cambium and outer bark.
18. Natural wildlife food like hickory nuts, beechnuts, and acorns.
19. Hardwood trees, like walnut, cherry, and oak, that are grown to provide thin layers of decorative wood that are glued over a base wood.
20. An engineered lumber product that includes three or more layers of veneer glued together.

A. angiosperm
B. cambium
C. coniferous tree
D. crown
E. deciduous coniferous tree
F. deciduous tree
G. evergreen broadleaf tree
H. heartwood
I. inner bark
J. lumber
K. mast
L. outer bark
M. plywood
N. pulpwood
O. sapwood
P. sawlogs
Q. silviculture
R. spile
S. sugar bush
T. veneer logs

Know and Understand

Answer the following questions using the information provided in this lesson.

1. What are veneer logs?
2. What is the difference between sawlogs and pulpwood?
3. What is the difference between a deciduous tree and a coniferous tree?
4. What characteristic makes white ash a good choice for baseball bats and wooden handles?
5. What makes hickory an ideal wood for flooring?
6. Which tree species are common Christmas trees?
7. Why are knot-free boards preferred in construction?
8. Why is western red cedar used to construct fence posts?
9. Why are cedar chests and closets popular?
10. Besides lumber, what other products come from a forest?
11. Why are boards cut into 16′ lengths?
12. How much sap is required to produce one gallon of maple syrup?
13. What are the main layers of a tree trunk?
14. What function does outer bark provide the tree?
15. What is the difference between an angiosperm and a gymnosperm?
16. What are examples of evergreen broadleaf trees?

Analyze and Apply

1. Why would a forester prefer to use a tree for veneer rather than for sawlogs?

Thinking Critically

1. Identify other tree species aside from the ones mentioned in this lesson that are common in your part of the country. What are the uses of the wood from those trees?
2. Identify other products aside from the ones mentioned in this lesson that come from forests. Where do you find them in your home?
3. Why is pure maple syrup expensive to purchase?

Lesson 17.2

Technology in Forest Management

Lesson Outcomes

By the end of this lesson, you should be able to:

- Discuss forest diseases and pests.
- Explain timber stand improvement.
- Demonstrate timber cruising.
- Explain harvesting practices used in forest production.

Words to Know

ash decline
Biltmore stick
clearcut harvest
dendrochronology
diameter at breast height
Dutch elm disease
increment borer
merchantable height
prescribed burn
seed tree harvest
seed trees
selective cutting
shelterwood harvesting
timber cruising
timber harvesting
timber stand improvement
timber thinning
tree volume

Before You Read

Take two-column notes as you read the lesson. Fold a piece of notebook paper in half lengthwise. On the left side of the column, write main ideas. On the right side, write subtopics and detailed information. After reading the lesson, use the notes as a study guide. Fold the paper in half so you only see the main ideas. Quiz yourself on the details and subtopics.

A significant part of forest management is controlling pests that can devastate individual trees or even an entire forest. Foresters must understand which trees are most desirable as either sawlogs or veneer trees in a hardwood forest. They must monitor the growth of trees regardless of the purpose for the timber. And, when harvest time arrives, foresters must decide on which of several methods of cutting will be most advantageous for the trees.

What Diseases and Pests Are Common in Forests?

Unfortunately, several diseases and pests have become more common in forests. Eradication of these pests is nearly impossible; therefore, foresters try to limit the spread of diseases and pests from one area of forest to another. Common diseases and pests in forests include Dutch elm disease, emerald ash borer, ash decline, mountain pine beetle, and gypsy moths.

Dutch Elm Disease

Dutch elm disease is a fungus that affects elm trees and is spread by the elm bark beetle, **Figure 17-22**. Dutch elm disease has destroyed nearly

dragon_fang/Shutterstock.com

Figure 17-22. This sticky trap is used to capture elm bark beetles, which carry the fungus Dutch elm disease. Once the fungus has taken hold, it will eventually destroy the elm tree.

all native elm trees in the United States. Elm trees were once highly popular trees in yards and urban areas. They had a nice shape and provided a lot of shade, but between 1930 and 1990, nearly 75% of the elm trees in the United States succumbed to Dutch elm disease. Today, scientists in many countries are trying to find species and varieties of elm trees that are resistant to this devastating disease.

Emerald Ash Borer

The emerald ash borer, shown in **Figure 17-23**, is a green beetle that feeds on ash trees. It is native to Asia and Eastern Russia. The larvae of the beetle are the stage of the insect's life that causes the most damage. Larvae feed on the phloem, or inner bark, of the tree. This type of feeding cuts off the flow of nutrients throughout the tree; thus, the tree dies from a lack of nutrients and water.

The first emerald ash borer found in the United States was discovered in Michigan in 2002. By 2014, it had spread to 24 other states, including states as far west as Colorado and as far south as Georgia, **Figure 17-24**. To try to slow the spread of the emerald ash borer, moving firewood across state lines is prohibited from states where it is found.

USDA Forest Service

Figure 17-23. The emerald ash borer is a green beetle that feeds on the phloem of ash trees. *Which kinds of insect pests that attack trees have you seen in your area?*

Scientists are currently trying to find some biological control for emerald ash borer. Several parasitic wasps from Asia feed on emerald ash borers in their native ranges. Consequently, these wasps have been carefully introduced in the United States and have had some level of success in reducing the infestation.

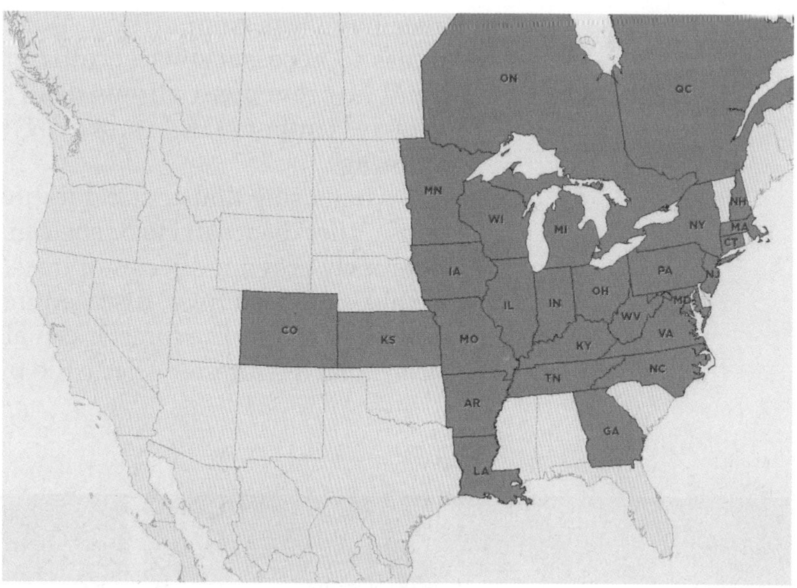

USDA Forest Service/Michigan State University

Figure 17-24. The maps above show the infestation of the emerald ash borer in 2002 and in 2015.

Ash Decline

Ash trees are also under attack from other diseases. The term ***ash decline*** is used to identify any decline of ash trees from either disease or wilt. Oftentimes, scientists are at a loss as to why individual sections of foliage die within ash trees. Once the decline starts, the ash tree will die within a couple of growing seasons.

Mountain Pine Beetle

Mountain pine beetles are one of the most destructive pests in western pine forests. The mountain pine beetle tunnels through the outer bark of pine trees to the inner bark where mating pairs lay their eggs. Once the eggs hatch, the larvae tunnel away from the main egg gallery, again feeding on the inner bark. Thus, these beetles effectively girdle the tree and cut off the phloem and nutrients.

Ponderosa, lodgepole, Scotch, and limber pines are most commonly infested, **Figure 17-25**. Older trees are generally most susceptible to mountain pine beetle infestations. A large enough beetle population can kill a tree in one growing season, and then spread their offspring to other trees the next growing season.

Gypsy Moth

The gypsy moth is one of the most invasive forest pests. Unlike the previously mentioned insect pests, gypsy moths feed on multiple species of hardwoods, including oaks, aspens, apple, sweetgum, poplar, and willow.

Don Becker, USGS

Figure 17-25. The reddish pine trees in this image have been killed by the mountain pine beetle.

Gypsy moth caterpillars build large cocoons in the branches of trees, **Figure 17-26**. Large infestations can cause the complete defoliation of trees. Most healthy trees can withstand one year of defoliation, but the defoliation still slows the growth of the tree and may enable easier infestation of other pests.

Timber Stand Improvement

Foresters work with landowners to improve the timber stand. ***Timber stand improvement*** is a general term for effectively weeding a forest of diseased or unwanted trees. By removing diseased or unwanted trees from the forest, desirable trees have less competition. With less competition, they have greater access to necessary water, nutrients, space, and sunlight. So, timber stand improvement allows desirable trees to grow more vigorously and quickly. Timber stand improvement may also be referred to as ***timber thinning***.

When using timber stand improvement, the forester has three main tools at his or her disposal. He or she can thin the stand of trees by removing less vigorously growing species of trees, as shown in **Figure 17-27**. The forester can also cull dead trees, diseased trees, or undesirable species of trees from the timber stand. The forester can also manage trees for crop tree release. This means that the forester selects the trees that are growing best and removes

USDA

4736101690/Shutterstock.com

Figure 17-26. A—The gypsy moth is one of the most invasive forest pests. B—Gypsy moth caterpillars build large cocoons in the branches of trees. *Have you seen this type of infestation in trees near your home or school?*

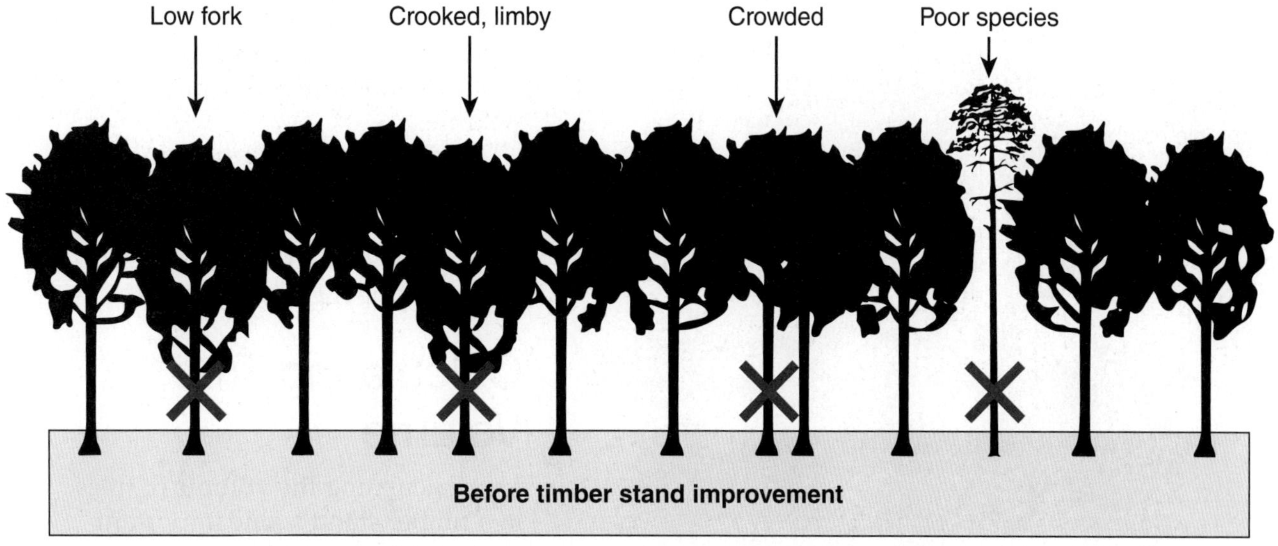

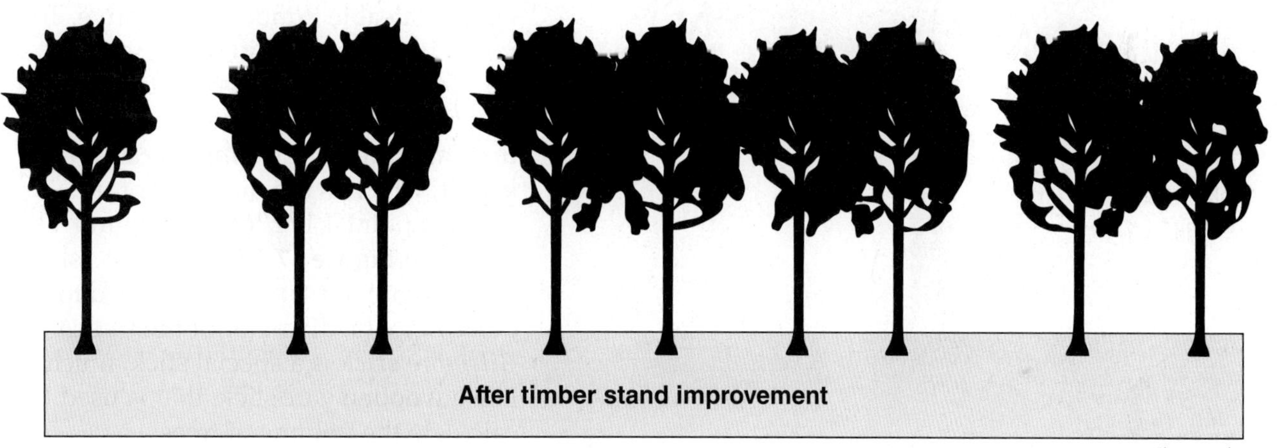

ntnt/Shutterstock.com

Figure 17-27. When using the methods of timber stand improvement, the forester can thin the stand of trees by removing less vigorously growing species.

competing trees nearby. Basically, timber stand improvement involves many different management practices that provide for optimal tree growth leading to harvest.

Timber Cruising

Timber cruising is the process of walking the forest to provide needed management information for timber stand improvement. When timber cruising, a forester may create an inventory of the tree species in the timber stand. The inventory will typically contain the names of species, numbers of trees of each species, and size of trees.

The forester may mark trees that are ready for harvest or in need of removal. They may identify and monitor pest infestations. When cruising the timber, the forester may note the wildlife and ecology of the forest. The

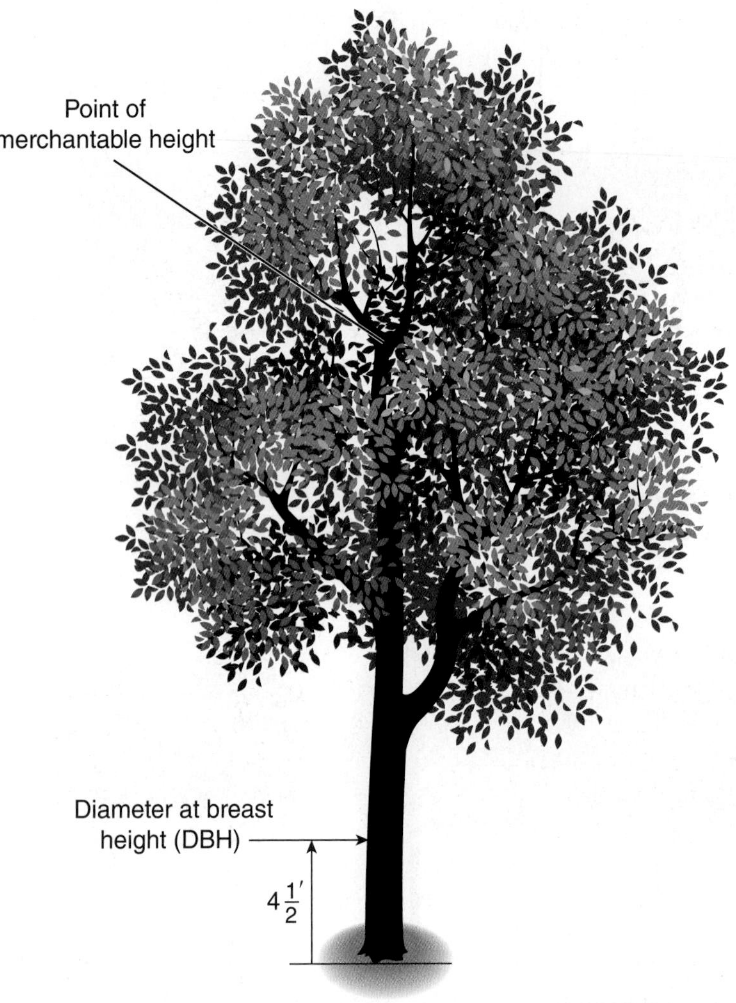

Svinkin/Shutterstock.com

Figure 17-28. Determining tree volume requires two measurements: diameter at breast height and merchantable height of the tree. *What is the largest tree that you have ever seen? How many 16′ logs could be cut from that tree?*

forester can use his or her observations to maximize the positive effects of timber stand improvement on forest wildlife.

Not all trees have the same value to a forester. For instance, pulpwood is generally the lowest value wood. Sawlogs have high value, and veneer logs have the highest value.

Calculating Tree Volume

Calculating the volume of trees in a timber stand is one important aspect of timber cruising. ***Tree volume*** calculates the size of individual trees from a saleable lumber perspective. Tree volume is measured in board feet.

Two measurements of a tree are necessary to calculate tree volume. The first measurement is ***diameter at breast height***, which is the diameter of the tree at a point 4 1/2′ off the ground, as shown in **Figure 17-28**. A Biltmore stick or tree diameter tape is often used to determine the diameter of the tree. A ***Biltmore stick*** is a special stick, much like a wooden yardstick, that is used to calculate the volume of trees.

The second measurement needed to calculate the amount of saleable timber is the merchantable height of the tree. The ***merchantable height*** is the height of the tree measured in the number of 16′ logs. The main trunk of the tree must be at least 16′ tall to have a merchantable log in it. Some trees have two or three merchantable logs in them. Once a tree produces a major branch, this is the topmost stopping point for calculating merchantable logs. Again, a specialized scale is marked on one side of the Biltmore stick to help calculate the merchantable height of a tree. A hypsometer may also be used to determine merchantable height of a tree.

Once the diameter at breast height and merchantable height of the tree are known, the forester can use a table to calculate the approximate number of board feet in the tree.

Estimating Timber Growth and Yield

Foresters try to predict the overall growth of saleable timber in a forest. All trees do not grow the same every year. Trees grow at different rates depending on the growing season, age of trees, competition, and species of the tree. Once a forester has a tree inventory and has calculated the tree volume of a stand of trees, he or she can estimate the growth trajectory of the stand. This gives the forester accurate information with which to predict the optimal timing for tree harvest.

Determining the Age of Trees

Foresters are also interested in the age of trees in a forest. As trees age, they certainly grow taller, and they also grow larger in diameter of the main trunk. The most certain way to determine a tree's age is to count the growth rings in the trunk. The study of using a tree's growth rings to determine its age is called ***dendrochronology***. A forester can determine the age of a tree by counting the rings once it is harvested, but he or she may also use a tool called an increment borer. The ***increment borer*** cuts a small hole in the trunk of the tree. With this tool, the forester can remove a core sample of the trunk to count the rings.

Timber Harvesting Practices

Timber harvesting is the cutting of trees from a forest and making them available for sale as sawlogs, veneer logs, or pulpwood. The goal of most timber stand improvement is to maximize the profit from a timber harvest, while minimizing the impact of that harvest on the forest ecology.

Forests are not typically harvested annually. Depending on the harvesting technique, harvest may occur every eight to 10 years or once every 20 years. Some harvests can yield $600 per acre and others can yield $4,500 per acre, depending on the ages and species of trees. Tall, straight veneer trees can fetch several thousands of dollars per tree, so using the most appropriate harvesting technique for the timber stand is important.

Clearcut Harvest

Clearcut harvest is the removal of all trees in a given area, **Figure 17-29.** This is seldom practiced in hardwood forests. Clearcut harvest is practiced with pulpwood plantations because pulpwood trees are generally grown much like row crops. Clearcut harvest is also conducted with pine forests because young pine trees require full sunlight in order to grow. Clearcut harvests also work well with pine trees because they grow relatively quickly compared to other tree species.

Although clearcutting removes all the trees and may be unsightly for a period of time, it does have benefits for wildlife. Mature forests typically do not support high volumes of wildlife compared to grasslands or immature forests. Clearcutting removes all the trees from the land. This allows sunlight to reach the soil once again and

chris kolaczan/Shutterstock.com

Figure 17-29. Clearcut harvest is the removal of all trees in a given area. Here new pine trees have been planted to start a new crop of trees.

rainwater to nurture new plant life. Most small animals, insects, and birds fare better with low growing plants and grasses. Shrubs and grasses provide food and habitat for many wildlife species. Within a few months of a clearcut, new wildlife will have moved into the area.

Kamee14u/Shutterstock.com

Figure 17-30. With seed tree harvest, a few desirable trees of each species are left to grow on each acre of land. The forester marks which trees to cut down.

Shelterwood Harvest

Shelterwood harvesting is the process of removing mature trees periodically over 10 to 15 years. Young trees of certain species like oak, cherry, and hickory can grow effectively in the shade of older trees. The slower growth at early stages of development helps the tree to become stronger and establish deep roots. Over time, the older trees that provide shelter for the younger trees are removed to allow more sunlight, water, and other nutrients to reach the maturing young trees.

Seed Tree Harvest

With ***seed tree harvest,*** a few desirable trees of each species are left to grow on each acre of land, **Figure 17-30**. These mature ***seed trees*** provide seeds for new growth and mast for wildlife, **Figure 17-31**. The seed trees left

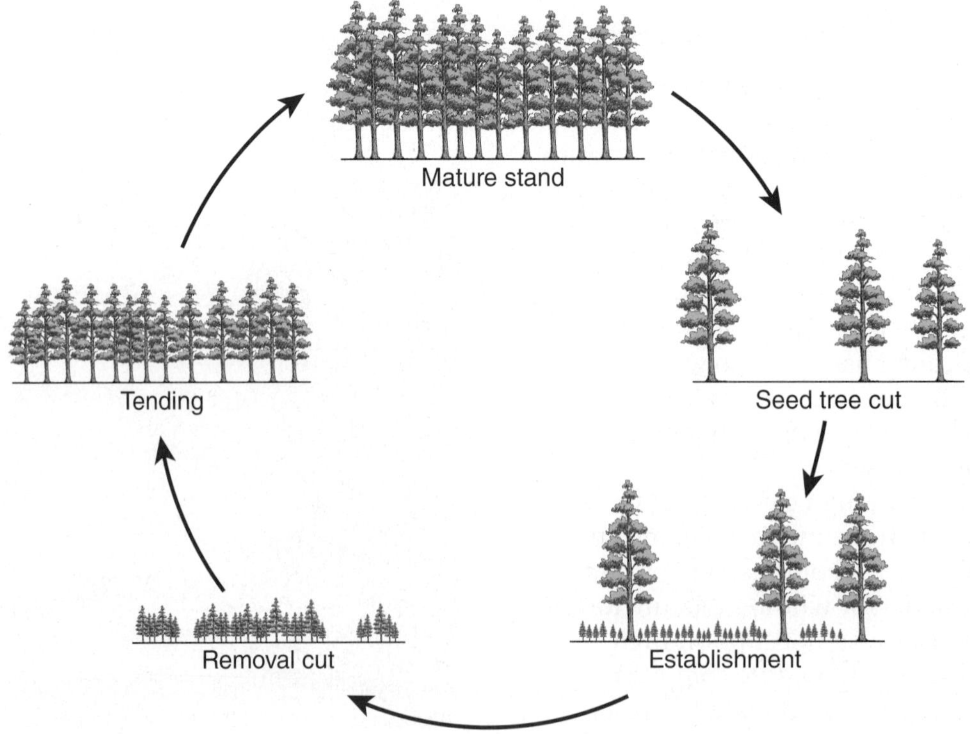

Vectomart/Shutterstock.com

Figure 17-31. Seed tree harvest presents benefits to wildlife similar to a clearcut harvest, except the wildlife also have the added benefit of seeds being produced by the trees.

standing on the now "open" land are more susceptible to wind damage. In a forest, the effects of strong winds are diminished because trees buffer one another. When most of the trees in a stand are removed, this buffering ability is diminished. These single trees are more susceptible to lightning and insect damage as well.

Selective Cutting

Selective cutting is the removal of only certain trees in a stand. Trees may be removed if they are mature and ready for harvest. Also, low-value, diseased, or damaged trees may be removed and cut into firewood. Selective cutting is also used for veneer logs.

There are several disadvantages to selective cutting. First, when any tree in a forest is cut down, it often damages nearby trees as it falls. Second, both selective cutting and shelterwood cuttings require that logging roads be developed in the forest. Logging roads disrupt the forest soils and leave them prone to erosion. The road and ensuing traffic also disrupt the wildlife.

Prescribed Burning

Sometimes foresters need to reduce the amount of dead leaves, branches, or undergrowth in a forest. This may be accomplished with a prescribed burn. A ***prescribed burn*** is the process of intentionally burning small areas of forest to remove excess buildup of organic material that would feed a wildfire if left unchecked, **Figure 17-32**. Prescribed burns are employed in the western United States to help control forest fires.

Did You Know?

Certain tree species, including the longleaf, shortleaf, and loblolly pine, are naturally resistant to fire.

Dmytro Gilitukha/Shutterstock.com

Figure 17-32. A prescribed burn is the process of intentionally burning small areas of forest to remove excess buildup of material that could feed a wildfire if unchecked. *When are forest fires dangerous? When are they beneficial?*

Josephsohm/Shutterstock.com

Figure 17-33. Forest fires have destroyed many homes and forests in the western portion of the United States. *How do foresters begin rebuilding a forest that has been destroyed by fire?*

Prescribed burns are an inexpensive and effective control method and do not damage the healthy, mature trees.

Natural Burns

Fire is part of the natural ecology of a forest. All forests would naturally burn periodically if we did not intervene. As with prescribed burns, natural fires would clear the undergrowth and the decaying organic materials, but not the healthy mature trees. The fire would also eliminate the weak and diseased flora and fauna, ensuring a healthier forest ecology.

As more homes are built in or near forests, more effort is made to prevent forest fires. Although preventing fires will protect the homes, it also enables the accumulation of excessive amounts of decaying organic materials and the establishment of more undergrowth. When lightning or another cause triggers a forest fire, the fire is more likely to spread wildly and burn uncontrollably. The large amount of combustible materials also creates more intense fires that are more likely to completely destroy the forest, **Figure 17-33.**

Career Connection Forester

Job description: A forester is a person who manages forests. The forester surveys the forest to monitor for pests and disease. Through timber cruising, he or she also determines which trees to keep and which to remove to improve the overall health of the forest. The forester may use any of the timber stand improvement and harvesting practices at one time or another. The forester has extensive knowledge of tree species and forest ecology.

A forester is different from a logger. A logger may be employed by a forester to harvest timber if the forest is mature and ready for harvest. A forester may cut a few trees but primarily manages the forest and works with other professionals to improve forest health.

Education required: A bachelor's degree in forestry is required, and a master's degree is often preferred.

Job fit: This job may be a fit for you if you enjoy working both outdoors in the forest and managing forest ecosystems.

Robert Crum/Shutterstock.com

The forester is using an increment borer to determine the age of this Douglas fir tree.

Lesson 17.2 Review and Assessment

Words to Know

Match the key terms from the lesson to the correct definition.

1. The process of walking the forest to provide needed management information for timber stand improvement.
2. The removal of all trees in a given area.
3. The first measurement of a tree at a point 4 1/2′ off the ground.
4. A special stick, much like a wooden yardstick, that is used to calculate the volume of trees.
5. A fungus that affects elm trees and is spread by the elm bark beetle.
6. The decline of ash trees from either disease or wilt.
7. When a few desirable trees of each species are left to grow on each acre of land.
8. The process of removing mature trees periodically over 10 to 15 years.
9. The study of using a tree's growth rings to determine its age.
10. A tool used to cut a small hole in the trunk of a tree to remove a core sample of the trunk to count the rings.
11. The effective weeding of a forest of diseased or unwanted trees.
12. The height of the tree measured in the number of 16′ logs that can be cut from the tree.
13. The calculation of the size of individual trees from a saleable lumber perspective.
14. The removal of only certain trees in a stand.
15. The process of intentionally burning small areas of forest to remove excess buildup of material that could feed a wildfire if unchecked.
16. Trees left standing from a seed tree harvest to provide seeds for new trees.
17. The cutting of trees from a forest and making them available for sale as sawlogs, veneer logs, or pulpwood.

A. ash decline
B. Biltmore stick
C. clearcut harvest
D. dendrochronology
E. diameter at breast height
F. Dutch elm disease
G. increment borer
H. merchantable height
I. prescribed burn
J. seed tree harvest
K. seed trees
L. selective cutting
M. shelterwood harvesting
N. timber cruising
O. timber harvesting
P. timber stand improvement
Q. tree volume

Know and Understand

Answer the following questions using the information provided in this lesson.

1. How do emerald ash borers damage ash trees?
2. How do gypsy moths damage trees?
3. How does timber stand improvement improve forests?
4. What is one practice used in timber stand improvement?
5. Which two measurements are needed to determine the lumber volume of a tree?
6. How often are trees harvested in a woodlot or forest?
7. Clearcutting is typically used for trees that are destined to become _____.
8. What is one of the threats to seed trees that are left standing after harvest?
9. Why might a forester initiate a prescribed burn?

Analyze and Apply

1. In which types of trees would you look for gypsy moths? What signs would you look for?
2. As trees in a forest grow older, what happens to their need for water, space, sunlight, and nutrients? Why?
3. How can timber cruising help slow the spread of forest pests?

Thinking Critically

1. Why do you think that the wood from hardwood tree species is generally more expensive than the wood from softwood species?
2. For several of the forest pests, the larval stage of the insect's life seems to be the most damaging phase of life. Why do you think this is the case? Why do you think the larval stage may be the most difficult stage in which to kill these insect pests?
3. How can clearcutting actually help wildlife?

Chapter 17

Review and Assessment

Lesson 17.1

Forest Production

Key Points

- In the United States, forests cover 35% of our land and account for about 4% of the total gross domestic product.
- The forestry and timber industries directly employ nearly one million people and support the jobs of another 1.5 million people in the United States.
- More than half of the forests in the United States are privately owned.
- Deciduous trees are trees that grow new leaves in the spring of each year, and then lose those leaves in the fall.
- Coniferous trees are trees whose leaves are needles or scales. Coniferous trees remain green throughout the year, even in winter.
- Economically important tree species include pine, oak, ash, black cherry, and black walnut.
- In addition to lumber and timber products, forests also produce food, paper, veneer, and even energy.
- The roots of trees absorb water and nutrients from the soil.
- The trunk of the tree supports the branches and leaves.
- The crown of the tree consists of the leaves and branches at the top of the tree.

Words to Know

Use the following list and the textbook glossary to review and study the *Words to Know* from *Lesson 17.1*.

angiosperm
broadleaf tree
cambium
coniferous tree
crown
deciduous coniferous tree
deciduous tree
evergreen broadleaf tree
evergreen tree
forestry
gymnosperm
hardwood species
heartwood
inner bark
knot
lumber
mast
outer bark
phloem
plywood
pulpwood
sap
sap flow
sapwood

sawlogs
silviculture
softwood species
spile
sugar bush
veneer
veneer logs

Check Your Understanding

Answer the following questions using the information provided in *Lesson 17.1*.

1. What is the practice of growing trees for production called?
2. What is an example of a coniferous tree?
3. Which type of trees comprise most of the pulpwood industry?
4. Which type of trees grow more slowly?
5. Which tree species is used to make baseball bats?
6. Which tree species are used in cabinetry?
7. Which tree species is the primary wood used in construction?
8. What is considered mast?
9. What is a common length of boards used in construction?
10. Which layer of the tree trunk is the actively growing part?

Lesson 17.2

Technology in Forest Management

Key Points

- Common diseases and pests in forests include Dutch elm disease, emerald ash borer, ash decline, mountain pine beetle, and gypsy moth.
- Timber stand improvement is a general term for effectively weeding a forest of diseased or unwanted trees.
- By removing problem trees from the forest, desirable trees have less competition and greater access to water, nutrients, space, and sunlight.
- Timber cruising is the process of walking the forest to provide needed management information for timber stand improvement.
- Clearcut harvest is the removal of all trees in a given area.
- Shelterwood harvesting is the process of removing mature trees periodically over 10 to 15 years.
- With seed tree harvest, a few desirable trees of each species are left to grow on each acre of land.
- Selective cutting is the removal of only certain trees in a stand.

Words to Know

Use the following list and the textbook glossary to review and study the *Words to Know* from *Lesson 17.2*.

- ash decline
- Biltmore stick
- clearcut harvest
- dendrochronology
- diameter at breast height
- Dutch elm disease
- increment borer
- merchantable height
- prescribed burn
- seed tree harvest
- seed trees
- selective cutting
- shelterwood harvesting
- timber cruising
- timber harvesting
- timber stand improvement
- timber thinning
- tree volume

Check Your Understanding

Answer the following questions using the information provided in *Lesson 17.2*.

1. How is Dutch elm disease spread?
2. Which of the forest diseases causes scientists to be at a loss for explanation?
3. Which of the forest insects are least harmful to trees?
4. What does timber stand improvement strive to provide to healthy, vigorous trees?
5. What do foresters do with dead or diseased trees?
6. What does a forester look for when timber cruising?
7. What is a tree's merchantable height?
8. What tools are used to calculate diameter at breast height and merchantable height?
9. How is a tree's age determined?
10. Which type of harvesting method would be used for veneer logs?
11. What is a prescribed burn?

Chapter 17 Skill Development

STEM and Academic Activities

1. **Science.** Which insect pests threaten forests and trees in your area? What is being done to monitor and control those insect pests?
2. **Technology.** With your teacher's permission, use an increment borer to determine and catalog the age of trees on your school's property.
3. **Engineering.** Design an experiment to test the hardness of different species of wood. Which species is the hardest? Which species is the softest? How much difference is there between species of wood?

4. **Math.** Using Biltmore sticks, measure trees on your school property, a nearby park, or a local woodlot. Which tree species are present? Calculate the volume of lumber in the trees and sort the volumes by species. Contact a local forester to determine approximate value for board feet of these species of trees.
5. **Social Science.** Some trees have become quite popular in our nation's history. Research famous trees and their unique contributions to our nation's history. Why do you think these trees contributed so much to our nation's foundations?
6. **Language Arts.** Identify species of trees that are "famous" in literature. Why are trees so prevalent in poems, quotes, and other literature? Write your own poem about trees.

Communicating about Agriculture

1. **Speaking and Listening.** Visit a nearby arboretum. Ask to interview the botanical expert. Prepare a list of questions before your interview. Here are some questions you might ask: What is your work environment like? What are your job duties? What type of research are you currently doing? What type of facilities do you use for your research? What impact will your research have on the floriculture industry? Ask if you can have a tour of the facilities. Report your findings to the class, giving reasons why you would or would not want to pursue a career similar to that of the person you interviewed.
2. **Speaking.** Using the National FFA Organization's agricultural issues forum career development event format, identify an issue related to forests and forest management in your area or state. Work with fellow students to research information and perspectives about the issue. Then, prepare a 15-minute presentation for public audiences on the issue.
3. **Writing.** After identifying insect pests that may be a problem in your area or state, prepare one-page information sheets about those pests. What do they look like? Which tree species do they attack? Which parts of the trees do they attack? During the year, when are these insect pests most likely to be observed? What can be done to control or eliminate the pest? Who should a landowner notify if they find the pest? To make your information sheets engaging and informative, use images from the Internet. You may consider posting these information sheets in your school, in public libraries, or on your school's website.
4. **Writing.** Research different kinds of building materials to use for decking when constructing an outdoor wooden deck. Compare and contrast untreated lumber, pressure-treated lumber, naturally resistant woods (cedar, redwood, etc.), and plastic decking. What are the factors to consider? How do each of these materials rate on your criteria? Prepare a report of your findings.

Extending Your Knowledge

1. Use the Internet to research how to make paper in a school classroom. Then, create paper in your high school agriculture lab. You might consider using various types of materials with which to create your paper. How can you recycle paper products to process new paper? What are the differences in quality of your paper?

2. Which types of trees are growing in your area? Create a tree inventory of your school, home, or town. You might even consider using global positioning systems and a tablet computer to register your town as a Tree Town, USA.
3. How have forests been used in your state? What is the condition of forests in your state?
4. The state-level forestry service in many states maintains an inventory of "big" trees for each species. Determine if your state maintains such an inventory. If so, find it. Are any of the "big" trees near your school? If so, visit the tree and the landowner. Consider interviewing the landowner about the tree and its history.
5. Write and conduct a survey comparing the relative strength of plywood compared to straight lumber. Which type is most flexible? Which type is prone to splitting? What happens to plywood as more plies of wood are added to the material?
6. Design an experiment to compare people's preferences for corn syrup-based pancake syrup versus pure maple syrup. Which do people in your school prefer? Why?

Glossary

A

abiotic factors. Nonliving things (i.e. rocks, minerals, water) in an ecosystem. (14.1)

abomasum. The compartment of the ruminant digestive system that is considered the true stomach; location where digestive chemicals are added for the chemical breakdown of food. (9.4)

accounts payable. Charges typically due in full when received; they have a short time frame for payment. (8.1)

acidification. The lowering of water acidity due especially to *acid rain*. (14.2)

acidosis. A condition caused when the pH level of a ruminant animal remains acidic for an extended period of time. This condition is usually due to a rapid intake of highly digestible concentrates. (9.4)

acid rain. Rain that contains higher than normal amounts of pollutants, especially nitric and sulfuric acids. (14.2)

acre. A special unit of measurement for area in agriculture (equivalent to 43,560 ft^2). (8.2)

active listening. The act of fully participating as you process what other people say. (2.2)

adjustable wrench. An open-ended wrench with one stationary jaw and one adjustable jaw. May also be referred to as a *Crescent wrench*. (7.3)

aeration. The addition of oxygen to the water of an aquaculture system to ensure proper oxygen levels for all species of aquatic life. (12.2)

agrarian civilization. Organized groups whose society is based in agricultural production. (1.2)

agribusiness. The agricultural industry area concerned with managing the profitability of agricultural products and companies. (3.2)

agricultural communications. The agricultural industry area concerned with informing consumers and industry professionals about topics related to agriculture and agricultural products. (3.2)

agricultural processing. The agricultural industry area concerned with the transformation of raw goods to consumer-ready products. (3.2)

agricultural support service. The agricultural industry area concerned with providing the logistical, technological, and maintenance needs of the agricultural industry. (3.2)

agricultural systems. The agricultural industry area concerned with designing, manufacturing, and maintaining the mechanical equipment and structures that agriculturalists need to prepare, produce, and process agricultural commodities. (3.2)

agriculture. The art and science of cultivating plants, animals, and other life forms for use by humans to sustain life. (1.1)

agriscience. The agricultural industry area concerned with the scientific principles behind agriculture and the research and development of emerging agricultural technologies. The application of science to agriculture. (3.2, 5.1)

air masses. Bodies of either cool or warm air that flow together. (14.3)

air-purifying respirator. A respirator with an air-purifying filter, cartridge, or canister that removes specific air contaminants by passing air through the air-purifying element. Air-purifying respirators can be used only in an environment that has enough oxygen to sustain life. (4.2)

air turbine. A modern version of the windmill used to convert kinetic wind energy into mechanical power or electricity. Also referred to as a *wind turbine*. (7.1)

"A" layer. The top layer of the soil horizon. It is usually the most fertile and darkest in color. Also referred to as *topsoil*. (15.2)

algae bloom. Excessive algae growth that depletes the water of oxygen, which, in turn, kills fish and other aquatic animals. (14.2)

algebra. The field of mathematics in which symbols are used to represent unknown values;

involves using specific mathematically derived rules to find the unknown values. (8.2)

alkyl esters. The chemical name for *biodiesel.* (7.2)

allelomimetic behavior. A range of activities in which the performance of a behavior increases the probability of that behavior being performed by other nearby animals. (10.3)

Allen wrench. An L-shaped or T-shaped wrench with a hexagonal-shaped shaft. (7.3)

allergen. Specific proteins in food that cause some people to have an immune reaction. (5.3)

all-terrain vehicle (ATV). A motorized vehicle with four low-pressure tires designed to travel over various types of rough terrain. (4.3)

altitude. The height measurement of the land above sea level. (14.3)

amendable motion. A motion that may be modified per parliamentary rules. (2.3)

American Berkshire Association (ABA). A swine breed organization committed to the promotion and success of the purebred swine industry. (11.2)

American Society for Testing Materials (ASTM). An organization that sets the industry standard on the quality of construction materials and their manufacture. (7.4)

ammonia (NH_3). A colorless gas with a strong, pungent smell that is released during decomposition of manure. Also used in agriculture as a part of a source of nitrogen fertilizer. (4.3)

ammonification. The process through which bacteria or fungi convert organic nitrogen from a decaying plant or animal into ammonium. It also occurs when bacteria break down animal manure in the soil. (14.2)

amortization. The process of splitting a loan repayment into equal installments. (8.1)

anadromous. An aquatic animal born in freshwater that migrates to sea and returns to freshwater to reproduce or *spawn*. (12.2)

anaerobic digestion. A process in which microorganisms break down organic materials such as food scraps, manure, and plant residue. The process creates biogas, such as methane, that can be harnessed and used to create energy, and solids that may be used to amend soil. (7.1)

angiosperm. Plants that bear seeds in vessels that could be fruits, nuts, or capsules. (17.1)

anhydrous ammonia. Caustic liquid used as a source of nitrogen fertilizer. (4.3)

Animal and Plant Health Inspection Service (APHIS). A government agency responsible for protecting consumers against plant and animal pests and diseases in our food supply. (9.3)

annual. A plant that completes its growth cycle in one growing season, and all roots, stems, and leaves die at the end of each growing season. (13.1)

anther. The part of a *stamen* that contains the *pollen*. (13.1)

antler. Bone structure that grows as an extension of the skull in members of the deer family. (12.3)

apex consumer. The species at the top of the food chain that consumes primary, secondary, and tertiary consumers. (14.4)

apex predator. The species at the top of the food chain that consumes primary, secondary, and tertiary consumers. (14.4)

apiarist. A person who raises honeybees. (13.5)

apiculture. The process of keeping bees for commercial production. The study and keeping of bees. (12.3)

appetite. How much an animal desires food of any type. (9.4)

applied mathematics. The field of mathematics involved in the study of physical, biological, or sociological areas. (8.2)

apprentice. A worker bound by legal agreement to work for an employer for the purpose of learning a craft. (4.1)

aptitude test. A personal questionnaire designed to help you determine what types of activities suit you and your talents. (3.2)

aquaculture. Fish production or farming that may be done outdoors in natural lakes, ponds, or the sea/ocean or indoors using various types of containers. The breeding, rearing, and harvesting of aquatic plants and animals in all types of water environments including ponds, rivers, lakes, oceans, and land-based facilities. May also be referred to as *concentrated aquatic animal production (CAAP)* or *fish farming*. The industry that raises fish and other aquatic species for human consumption. (6.1, 12.2, 15.4)

aquatic ecosystem. An ecosystem found in a body of water, like a stream, river, wetland, or ocean. (14.1)

aquifer. An underground water reservoir. (14.2, 15.1)

arable land. Land that is suitable for growing crops. (1.1, 9.2)

arctic zone. The latitudinal range above 66.5°N in which the average temperature is less than 50°F (10°C). Also referred to as the *polar zone*. (14.3)

area. The size of a two-dimensional space. (8.2)

arithmetic. The mathematics of counting quantities and manipulating numbers. (8.2)

artificial insemination (AI). A fertilization process in which semen is collected from the male animal and placed through artificial means in the reproductive tract of the female. The semen may be frozen and stored for an extended period of time. (5.3, 10.1, 10.2)

artificial selection. The process of humans selectively breeding plants and animals to choose traits/characteristics which will be more beneficial/desirable to humans. (1.2, 5.3)

asexual reproduction. A type of plant reproduction in which one cell is made into two identical cells using *mitosis*. The plants produced are genetically identical and have the exact same DNA. (13.1)

ash decline. The decline of ash trees from either disease or wilt. (17.2)

assembly drawing. A drawing created to show how the components of a project fit together. Commonly used for small projects to show how the structure or piece of equipment is to be constructed. (7.4)

asset. An item of value owned by a business. (8.1)

atmosphere-supplying respirator. A respirator that supplies the user with breathing air from a source other than the air in the surrounding area (i.e. *supplied-air respirators (SARs)* and *self-contained breathing apparatus (SCBA)* units). (4.2)

autocratic leadership. A type of leadership style in which the leader makes most of the decisions and sets the expectations for the followers. There is a clear division between the leader and the followers, and the followers have little input related to the decisions being made. (2.1)

autotroph. An organism that can perform photosynthesis to provide its own nutrients. A plant that produces its own food. Also referred to as a *producer*. (13.1, 14.4)

avian. Of or relating to birds. (11.1)

avian digestive system. The digestive system used by poultry and all birds. (9.4)

aviation snips. A hand tool designed to cut sheet metal and wire. (7.3)

B

Bacillus thuringiensis (Bt). A bacterium that kills certain insect species that prey upon cotton and other crops. (13.4)

backgrounding operation. A beef cattle production operation in which young calves are allowed to graze pastures for moderate weight gain, before moving to *feedlots*. Also referred to as a *stocker operation*. (10.1)

back saw. A push stroke saw with fine teeth used to make smooth cuts. It has a stiffening rib on the edge opposite the cutting edge, allowing for better control and more precise cutting. (7.3)

balanced ration. A *ration* that contains all of the necessary nutrients for an animal's intended growth and development. (9.4)

ball-peen hammer. A hammer with a rounded peen that is used in metalworking and mechanics. It is used for striking chisels and punches, and rounding the heads of metal fasteners such as rivets. (7.3)

band saw. A stationary power tool that uses a circular, metal band with serrated teeth to cut through everything from bone to logs. The circular band travels at high speeds around two or more large wheels. (7.3)

barley yellow dwarf. A worldwide virus that affects barley and other grasses such as wheat and rice. The virus causes plants to yellow and dwarf in size. (13.2)

barren. A field with no crop growing. (15.5)

barrow. A castrated male swine. (11.2)

bars. Depressions in the sole of a horse's hoof which provide stability for the walls, and allow the *frog* to expand as the hoof impacts the ground. (10.3)

bedding plant. A plant grown for outdoor ornamentation; may be an annual or perennial. (13.6)

beef. The meat harvested from cattle. (10.1)

beef cattle. Cattle bred and raised to produce beef. (10.1)

belt-and-disk sander. A combination, stationary sander that is useful for sanding square or curved ends on narrow boards. It may also have a tilting worktable that allows precise sanding of angled workpieces. (7.3)

belt sander. A strong power tool best used for the initial smoothing of large, flat surfaces. Belt sanders use a continuous loop or belt of sandpaper that is stretched across two pulleys. (7.3)

biennial. A plant that lives for two years to complete its growing cycle. (13.1)

billy. A male goat of any age. Also referred to as a *buck*. (11.4)

Biltmore stick. A special stick, much like a wooden yardstick, that is used to calculate the volume of trees. (17.2)

biodegradable. Something that can be broken down by bacteria or other living organisms. (7.2)

biodiesel. A *biofuel* derived from vegetable oils, used cooking oils, and animal fats. (6.2, 7.2)

biodiversity. The biological diversity in an environment as indicated by the numbers of different plant and animal species present. (5.3)

biofuel. A liquid fuel made from organic materials that is intended to one day replace nonrenewable fossil fuels. See also *ethanol*, *biodiesel*, and *methanol*. (6.2, 7.2)

biogas. A gaseous biofuel produced by the biological breakdown or fermentation of organic matter. (7.2)

biogeochemical cycle. The cycling of biological and chemical substances in any of Earth's ecosystems. (14.2)

biological availability. The amount of a soluble substance that can be used by aquatic life. (15.4)

biology. The study of life. (5.2)

biomass. Any organic matter, besides fossil fuels, that can be used as an energy source; includes wood (firewood, charcoal, chips, sheets, pellets, sawdust, pulp sludge), animal waste, and selected agricultural crops. (7.1)

biomass energy. The energy harnessed from organic matter as it decomposes, when it is burned, or when it is treated with chemicals. (7.1)

biosphere. Earth's ecosystems as a whole, or the accumulation of all of the ecosystems on Earth. Also referred to as the *zone of life*. (14.2)

biotechnology. The use of scientific modification to the genetic material of living cells to produce new substances or functions; includes the process of genetic engineering. A collection of scientific techniques used to modify plants, animals, and microorganisms to improve their longevity, disease resistance, drought resistance, flavor, and adaptability. The use of living systems and organisms to develop or make products useful to society. (1.2, 5.2)

biotic factors. Things in an ecosystem that are living, moving, growing, and reproducing, like plants and animals. (14.1)

black-out house. A poultry production facility in which the lighting is controlled completely by the producer. (11.1)

blacksmith hammer. A hammer used for striking unhardened metals such as that used to make horseshoes. (7.3)

blade. The main part and shape of a *leaf*. (13.1)

"B" layer. The layer of soil right below the *topsoil*. It is usually lighter in color, contains less organic matter, and is more responsible for drainage and aeration than the *topsoil*. Also referred to as *subsoil*. (15.2)

bloat. A condition in livestock animals where excessive gases build up in the rumen. (9.4)

block plane. A handheld woodworking tool used for smaller jobs such as planing the ends of wood and molding and trim. (7.3)

blue baby syndrome. An illness caused by the consumption of water high in nitrates in which the skin turns a bluish color due to the blood's inability to carry oxygen. Also referred to as *methemoglobinemia*. (14.2)

blue tongue. An insect-borne viral disease that affects ruminants. (16.2)

boar. A noncastrated young or mature male swine. (11.2)

bog. A wetland that contains peat deposits, acidic water, and carpets of sphagnum moss. It receives its water primarily through precipitation. (15.3)

boll. The rounded seed capsule of the cotton or flax plant. (13.4)

boneless/skinless breast. Poultry meat from a split breast which has been deboned and skinned. (11.1)

boot stage. When the seedhead forms and is protected by the flag leaf at the top of the stem of a wheat plant. (13.2)

border line. A heavy, bold line used to define the edges of the drawing surface. (7.4)

Bos Indicus. The cattle breeds that originated in tropical countries such as Asia, Africa, and India. (10.1)

Bos Taurus. The cattle breeds that originated in Europe (England, Scotland, France, Germany, and Italy). (10.1)

Boston butt. A tougher primal cut of pork meat from the shoulder of swine that contains muscles and sinew. It may also be referred to as *pork butt* or *Boston shoulder*. (11.2)

Boston shoulder. A tougher primal cut of pork meat from the shoulder of swine that contains muscles and sinew. It may also be referred to as *pork butt* or *Boston butt*. (11.2)

botany. The plant life section of biology; includes the study of plant structure, physiology, genetics, and classification. (5.2)

bovine. The species name for cattle. (10.1)

box end wrench. A nonadjustable wrench that is closed at each end. A box end wrench uses points around the inside diameter of its jaws to grip hexagonal bolt heads and nuts. (7.3)

brackish water. The water found in *estuaries*; it has a higher salt content than freshwater, but a lower salt content than seawater. (14.1)

bract. A small cluster of flowers found in the axil of a stem. (13.4)

bramble. A rough, tangled, prickly shrub that produces a berry. (13.5)

bran. The hard outer coating of cereal grains. (13.2)

breast quarter. A bone-in poultry cut (with skin attached) that includes a portion of the back, the breast, and the wing. (11.1)

breed. A specific group of animals that has similar appearance, characteristics, and behaviors that distinguish it from other animals in the same species. For example, a breed may be a specific group of cattle that has similar appearance, characteristics, and behaviors that distinguish it from other cattle. (10.1)

breed association. An organization that maintains documented pedigrees and performance data for purebred animals such as cattle, horses, sheep, and swine. (10.1)

breeder farm. A poultry production facility in which pullets are placed with roosters to produce fertilized eggs. (11.1)

brisket. A fairly tough, boneless primal cut that lies over the sternum, ribs, and connecting cartilage of beef cattle. (10.1)

broadleaf tree. A deciduous tree with leaves with broad, flat blades. (17.1)

broiler. A chicken raised strictly for meat. (11.1)

broiler production. A production operation in which poultry are raised as a meat source. (6.1)

broodmare. A mature female horse kept for breeding. (10.3)

browse. The preferred eating method of goats in which they eat leaves, twigs, and shoots instead of grazing on grasses. (11.4)

buck. A male goat of any age. Also referred to as a *billy*. A male rabbit. A mature male deer. (11.4, 12.1, 12.3)

budget. A list of the money a business plans to deal with over the course of the upcoming year. (8.1)

buffer strip. A parcel of land that is planted with permanent vegetation to prevent soil erosion, pesticide pollution, or fertilizer contamination. (15.5)

bulb. The thickened leaves of a plant that grow underground attached to a shortened stem and roots. (13.1)

bulb vegetable. A vegetable, such as onion, grown for its edible belowground plant material (bulb). (13.5)

bull. A term to describe young and mature male beef and dairy cattle. (10.1)

bull tongs. A larger version of a *strap wrench* used with large diameter pipe. (7.3)

bushel. A unit of dry measurement equivalent in volume to 8 gallons. (8.2)

bush hammer. A masonry, tool used to texturize stone or concrete. (7.3)

business ethics. The practice of holding businesses accountable for making decisions that are in the best interest of their consumers, even if they are not what the numbers say is most profitable. (8.1)

business plan. A statement of business goals, along with the strategies and actions for achieving those goals. (8.1)

butter. A soft yellow or white food churned from milk or cream and used to spread on food, in cooking, and in baking. (10.2)

buttermilk. The slightly sour liquid left over after butter has been churned. It is used in baking and consumed as a drink. (10.2)

by-product. Something that is produced incidentally in the production of a *primary product*. (12.4)

C

cabrito. Spanish word for young, milk-fed goat. (11.4)

caeca. Small pouches located at the end of the large intestine of a bird's digestive system. An opening in a bird's digestive system through which waste products from the kidneys enter the digestive system. (9.4, 11.1)

calf. A term to describe young beef/dairy cattle. (10.1)

calving. The term used to describe a cow who is in the process of giving birth. (10.1)

cambium. A layer of cells that make up the *xylem* and *phloem* of plants. The thin layer of actively growing tissue in a tree trunk that produces the growth rings seen when the tree is cut down. (13.1, 17.1)

candling. The act of using light to check embryo development through the shell. (11.1)

canes. The stalks of *brambles*. (13.5)

cape. The red or brown coloring extending down from the head to the shoulder region of a Boer goat. (11.4)

capital. The financial and material resources a business has available for use. (8.1)

capital loan. A business loan for the costs necessary to run a business. (8.1)

capital purchase. A purchase involving the spending of cash to purchase a noncurrent inventory item. (8.1)

capital sale. The sale of a company's noncurrent inventory item, like a piece of equipment or a breeding animal. (8.1)

capon. A castrated male chicken. (11.1)

caprine. The species name for goats. (11.4)

carbohydrate. A nutrient that gives the organism readily available energy. (9.4)

carbonates. The minerals that affect the acidity of the soil. (15.2)

carbon credit. A fee paid by a company to allow it to release excess carbon into the atmosphere. (16.3)

carbon credit system. A carbon credit is a permit that allows the holder to emit one ton of carbon dioxide. The system allows companies to trade or sell certificates or permits to produce carbon with other companies. (14.2)

carbon cycle. The natural process whereby carbon cycles through the oceans, atmosphere, plants, animals, and soil. (14.2)

carbon dioxide (CO_2). An odorless and colorless and highly dangerous gas that is commonly found in silos and manure storage areas. It is produced by burning carbon and organic compounds and by respiration. It is also naturally present in air (about 0.03%) and is absorbed by plants in photosynthesis. (4.3, 14.2)

carbon footprint. The measure of the greenhouse gases that are emitted as a result of human activity. (16.3)

carbon pool. An area of Earth where carbon accumulates and is also released. The atmosphere, oceans, biosphere, and sediments are all carbon pools. (14.2)

carbon sequestration. The process of capturing and storing carbon for long periods of time. (16.3)

carbon sink. A place that absorbs and stores carbon for long periods of time. The ocean is the largest carbon sink on Earth. (14.2)

carbon source. A place that releases more carbon than it stores. Anything that burns fossil fuels is a carbon source. (14.2)

carnivore. An organism that eats other animals. (14.4)

carrying capacity. The maximum population of a species that a parcel of land can sustain indefinitely. (14.4)

cash farm receipt. The amount of money earned by the farmer or grower for his or her crop when he or she sells it to the marketplace. (13.3, 13.5)

cash flow statement. A financial report that shows how changes in income and expenses affect the amount of available cash a business has to work with at a given point in time. (8.1)

cash purchase. A purchase in which a business owner exchanges money for a current inventory item. (8.1)

cash sale. A sale in which a business is receiving money for either services provided, or for the sale of a current inventory item. (8.1)

castrate. To remove the testicles from a male animal. (11.2)

cecum. A pouch-like area found at the beginning of the large intestine (in a *modified monogastric digestive system*) where digestive microbes are housed to break down roughages. (9.4, 10.3)

cell tissue culture. The growth of new plants or animals from a cell or small group of cells taken from the parent tissue. (5.3)

cellular respiration. The breaking down of glucose molecules in all living organisms to release energy. (13.1)

cellulose. The main substance that makes up the cell walls and fibers of plants. It may be used to produce *biofuels*. (7.2)

centerline. A thin line with a crosshair at the center of the round object that is used to represent the diameter of a round object. (7.4)

center-pivot irrigation system. Crop watering system in which a tall sprinkler pivots in a circle across the field (the water source is typically at the pivot point). (15.5)

Centers for Disease Control and Prevention (CDC). A government agency responsible for identifying and investigating foodborne illnesses. The CDC plays a large role in local and state health departments and monitors surveillance of foodborne disease and illness outbreaks. (9.3)

certified pedigreed swine (CPS). A swine breed organization committed to the promotion and success of the purebred swine industry. (11.2)

chair. The person responsible for conducting the meeting. (2.3)

chairman. The person responsible for conducting the meeting. (2.3)

character. A group of personal traits that define moral or ethical quality. (2.1)

chart. Information shown in a diagram form rather than in a written narrative. (5.1)

chattel mortgage. A loan in which movable personal property is used to secure the loan. (8.1)

cheese. A solid food made from milk curd that is produced in a range of flavors, textures, and forms. (10.2)

chemical digestion. The addition of digestive chemicals that break chemical bonds in foods at a molecular level. (9.4)

chemical pollution. Water pollution that comes from introducing nonnaturally occurring chemicals into the water. (15.4)

chemistry. The branch of science that studies the composition, structure, properties, and change of matter; the interaction of molecules to form different substances. (5.2)

chevon. The meat from a mature goat. (11.4)

chick. A baby chicken (remains a chick until its body is covered with feathers). (11.1)

chicken. A domestic fowl kept for its meat or eggs; the meat from the bird. (11.1)

chisel. A long-bladed hand tool with a beveled cutting edge that is struck with a hammer or mallet to cut or shape wood, stone, or metal. (7.3)

chlorophyll. A green substance in plants which absorbs sunlight to help complete the process of *photosynthesis*. (13.1)

chloroplast. The structure in plant cells that collects sun energy and changes it into nutrients for the plant. (13.1)

Chronic Wasting Disease (CWD). A disease similar to mad cow disease found in wild populations of deer and elk that can be transmitted to domestic populations. (12.3, 16.2)

chuck. A device that clamps down and holds the bit in place on a power drill. *Primal cut* from the shoulder and neck of beef cattle; flavorful but may be tough and fatty and contain excess bone and gristle. (7.3, 10.1)

circuit breaker. An automatic device for stopping the flow of current in an electric circuit. (4.2)

circular saw. A portable, handheld power saw that uses a circular blade to cut through materials. (7.3)

Civilian Conservation Corps (CCC). A government organization established in 1933 to build pathways, bridges, roads, and buildings in the nation's national parks and to educate farmers about soil conservation and methods to control erosion. (15.1)

claw hammer. The most commonly used hammer, available in a number of sizes and weights with handles made from wood, fiberglass, steel, or composite material. It has a face for driving nails and a curved claw for pulling nails. Also referred to as a *nail hammer*. (7.3)

clay. The smallest soil particle. It is chemically active, has a negative electrical charge, and holds valuable plant nutrients. (15.2)

"C" layer. The underlying soil material that gives rise to the *subsoil* and *topsoil*; may be made of sedimentary rock, other types of rock, or mud deposited by a river. Also referred to as the *parent material* of the soil horizon. (15.2)

Clean Air Act (CAA). A United States federal law designed to control air pollution on a national level. (15.1)

Clean Water Act (CWA). A government action designed to regulate the discharges of pollutants into waters of the United States and regulating quality standards for surface waters. (15.1)

clearcut harvest. The removal of all trees in a given area. (17.2)

climate. The measure of the average temperature, precipitation, wind, and humidity of an area over a long period of time. (14.3)

cloaca. The opening through which waste leaves the *avian* body and through which eggs are laid. (9.4, 11.1)

cloning. The complicated and time-consuming process of creating a genetically identical copy of an organism. (5.3)

closed ecological system. An enclosed community that does not allow for the exchange of any material inputs outside of the system; an attempt to create an ecosystem that is self-sustaining. (6.2)

club hammer. A small sledge hammer used to drive stakes, cold chisels, or demolish masonry. (7.3)

coal. A combustible, dark rock consisting mainly of carbonized or fossilized plant matter. (7.1)

cockerel. A male chicken under one year old. (11.1)

codominance. Situation in which some traits show as a mixture when the *genotype* is *heterozygous*. May be referred to as *incomplete dominance*, depending on the nature of the specific combination. (5.3)

coldblood. Term used to describe larger draft horses. See *warmblood*. (10.3)

cold chisel. A long-bladed hand tool with a beveled cutting edge that is struck with a hammer or mallet to cut metal. (7.3)

colic. A digestive disorder that is the number one cause of death in horses. (9.4)

collateral. The property used to secure a loan. (8.1)

colony. A large group of bees that lives and works together in a beehive. (12.3)

Colony Collapse Disorder (CCD). A disease that affects both wild and domesticated hives of honeybees. (12.3, 16.2)

colt. A male horse two years or younger. (10.3)

combination square. A measuring tool with a grooved blade and head that can be adjusted to different locations along its 12″ blade. It may be used for scribing, crosscutting, mitering, checking for level and plumb, and for setting angles along the 180° range. (7.3)

combination wrench. A wrench with one open end and one closed end. Both ends generally fit the same size fasteners. (7.3)

commercial cattle. Usually crossbred beef cattle produced for sale to stocker operations and feedlots. (10.1)

commodity. A raw material that can be bought and sold. (1.1)

communication. The process of sending and receiving information using verbal and nonverbal cues. (2.2)

communication systems. The interaction of people during the act of communication. (2.2)

community. A group of living organisms that interact with one another in an environment or ecosystem. (14.1)

community garden. Plots of land that are gardened collectively by community members. (9.1)

companion animal. The animals that live, play, and work with people on a daily basis; animals kept as pets. (12.1)

comparative advantage. A statement looking at potential competitors for a certain business, and how the business fits into the same market. (8.1)

compass saw. A push stroke handsaw with a narrow blade made for cutting holes and gentle curves in wood. It is useful for cutting holes in walls or floors for pipes, electrical outlets, or other fixtures. May be referred to as a *keyhole saw* or *drywall saw*. (7.3)

composite flower. A flower made up of many small flowers in one large structure. (13.1)

compost. Organic matter that has been decomposed by microbial activity over time. (16.3)

computational mathematics. The area of math involved in developing computer programs. (8.2)

concentrate. A feed that is high in energy and *total digestible nutrients (TDN)* and low in *fiber* content. (9.4)

concentrated aquatic animal production (CAAP). The breeding, rearing, and harvesting of plants and animals in all types of water environments, including ponds, rivers, lakes, oceans, and land-based facilities. May also be referred to as *aquaculture* or *fish farming*. (12.2)

condensed milk. A heavily sweetened milk product made by removing about 60% of the water from ordinary milk. (10.2)

confined animal feeding operations (CAFOs). Large-scale livestock operations that feed many animals in a relatively small space. (15.5)

confinement operation. A swine production facility in which pigs are not allowed outdoors. (6.1, 11.2)

conformation. The correctness of an animal's bone structure, musculature, and body proportions in relation to each other. (10.3)

coniferous tree. A tree whose leaves are needles or scales that remain green throughout the year, even in winter. They are also referred to *softwood species* or *evergreen trees*. (17.1)

conservation. The protection of natural resources that allows use of the land by the public. The land and its natural resources can be used, but not misused or used up. (15.1, 16.1)

conservation reserve program. A decision by a farmer or landowner to set aside some of his or her land for forests, grasslands, or wetlands. (15.1)

constant. Something that is kept the same between experimental groups (in a scientific experiment) to ensure that only one thing is being changed at a time. (5.1)

constructed wetland. The process of developing a wetland where one did not previously exist. Also referred to as *wetland creation*. (15.3)

consumer. Person who purchases and uses agricultural products. An organism that consumes (eats) other organisms. Also referred to as a *heterotroph*. (1.1, 14.4)

consumption. The actual use of the food product by the consumer. (9.1)

continuously operating reference station (CORS). Telemetric system used to locate pest infestations, measure soil moisture and temperature, and monitor weather conditions. It may also be used to control the amount of irrigation a field will receive based on measurements of soil moisture. (6.2)

contour tillage. When farmers till and plant in lines that are perpendicular to the slope of the field. (15.5)

control. A section of a scientific experiment that does not receive treatment. (5.1)

convection current. The flow of hot, humid air from the equator to either the North or South Pole. (14.3)

conversation. Two-way oral communication that occurs between two people. (2.2)

conversion. The act of changing one unit of measurement into another. (8.2)

conversion factor. A number that allows an understanding of how many of one unit of measurement are in another unit of measurement. (8.2)

cooperative. A farm or business that is owned and run jointly by its members. (9.1)

coping saw. A pull stroke saw that uses a very thin blade and fine teeth to make smooth-edged, curved cuts. It may also be used to make cuts inside a piece of wood. (7.3)

coral reef. An underwater ecosystem with structures made up of colonies of tiny animals called corals. (14.1)

corm. A thick, modified underground plant stem with scaly leaves and buds from which new plants grow. (13.1)

corn stover. The stalks and leaves of the corn plant used in *ethanol* production in which plant *cellulose* is used to produce the *biofuel*. (7.2)

correspondence. Two-way written communication between two or more people. (2.2)

corridors of habitat. Narrow expanses of trees and shrubs that connect larger, wider, more diverse areas of habitat. (16.2)

cotyledon. The very first seed leaf produced by a plant. (13.1)

cover crop. A crop planted primarily to manage soil erosion. A crop that is planted in the fall after the previous growing season's crop has been harvested. (15.2, 15.5)

cover letter. A document that is sent with a résumé to give the employer information about your interest in the position and additional information about your qualifications. (3.2)

cow. A term to describe mature female beef/dairy cattle. (10.1)

cow-calf operation. An operation used to breed and birth beef cattle for sale. (10.1)

cream. The thick white or pale yellow fatty liquid that rises to the top of *milk* that is left to stand (unless it is homogenized). (10.2)

creed speaking. When first-year FFA members memorize the FFA Creed and present it to judges, who score them on their presentation skills, along with their answers to questions based on the creed. (2.2)

creep feeding. A method of supplementing the diet of young livestock by offering feed to animals who are still nursing. (13.2)

Crescent wrench. An open-ended wrench with one stationary jaw and one adjustable jaw. May also be referred to as an *adjustable wrench*. (7.3)

cria. The offspring of a llama or alpaca. (12.3)

crimpers. A hand tool designed to connect (crimp) a connector to the end of a cable. (7.3)

crop. A storage area at the base of the esophagus in an *avian digestive system*. Any plant grown in the care of humans as a food source. (9.4, 11.1, 13.1)

crop biomass. The plant material or agricultural waste used as a fuel or energy source. (7.2)

crop residue. The remnants of the previous year's crops left on the field after the harvest. (15.5)

crop rotation. Method of planting a different crop each growing season to help control pests and diseases that may become established in the soil over time. (13.5)

crossbred-wool breed. Sheep that are a cross between *long-wool breeds* and *fine-wool breeds*. (11.3)

cross-contamination. When harmful bacteria are transferred to other foods through utensils, cutting surfaces, or workers' hands or gloves that have not been properly cleaned. (9.3)

crosscut saw. A handsaw used to cut wood across the grain. (7.3)

crowbar. A prying tool made of solid metal with a hooked end and a claw designed to open or break apart materials. May also be used to pull out fasteners. (7.3)

crown. The leaves and branches at the top of the tree; the site of photosynthesis. (17.1)

crown rust. A fungus that commonly affects both cultivated and wild oat. The fungus appears as orange (rust-colored) pustules of spores on the plant's leaves. (13.2)

crude oil. A nonrenewable fossil fuel found in underground areas called reservoirs. It is composed mainly of hydrocarbons and organic compounds. (7.1)

crustaceans. A large group of arthropods that has exoskeletons that they molt. The group includes crabs, lobsters, shrimp, crawfish, and prawns. (12.2)

cud. Partially digested food that is regurgitated into the mouth and rechewed so that it can pass to the rest of the digestive system. A regurgitation of the roughage and feed a ruminant has eaten to aid in further digestion. (9.4, 10.1)

current asset. Item that has a short-term usefulness for a business. (8.1)

cutability. The proportion of lean, salable meat yielded by a carcass. (10.1)

cut flowers. Flowers, including stems and leaves, which are cut off the plant and used in indoor decorative arrangements. (13.6)

cuticle. The waxy covering on a plant's *leaf*. (13.1)

cutting-plane line. A solid, black line (same weight as *object lines*) used to indicate where the theoretical cut is taken for a sectional view. It may also be a series of dashes or long dashes with intermittent short dashes. (7.4)

cycle. The pathway that a substance moves in the *biosphere*. (14.2)

D

dam. A female llama or alpaca. (12.3)

damping off. A plant disease that occurs in excessively damp conditions and causes young seedlings to collapse and die. (13.5)

dark reaction. A part of *photosynthesis* in which the energy molecules produced by *light reactions* are converted into sugar molecules for energy storage. (13.1)

DDT. A synthetic insecticide once used to control mosquitoes and to control potato beetles, coddling moths that attack apples, corn earworms, cotton bollworms, and tobacco budworms. (15.1)

debatable motion. A motion that may be put up for debate. All main motions and amendments to main motions are debatable. (2.3)

deciduous coniferous tree. A tree that has needle-like leaves, but loses those leaves in the fall. (17.1)

deciduous tree. A tree that grows new leaves in the spring of each year, and then loses those leaves in the fall. Also known as *hardwood species*. (14.1, 17.1)

decompose. The breaking down of dead plant and animal material. (14.4)

decomposer. An organism that breaks down dead plant and animal material. (14.4)

defoliant. A chemical herbicide used to kill the cotton plant (or other plants) and make the leaves fall off to ensure a cleaner harvest. (13.4)

delegation. The process of giving another person the control and responsibility for a given task. (2.1)

delegative leadership. A type of leadership style in which the leader provides little or no guidance and allows the group members to make the decisions. The leader is present merely to provide tools and resources. Also referred to as *laissez-faire leadership*. (2.1)

demand. The measure of consumers' desire for a particular product. (9.2)

democratic leadership. A type of leadership style in which the leader provides guidance while encouraging group members to share their ideas and opinions. There is less division between the leader and followers and a greater contribution of ideas and creativity from the group. Also referred to as *participative leadership*. (2.1)

dendrochronology. The study of using a tree's growth rings to determine its age. (17.2)

denitrification. The conversion of nitrates into inert atmospheric nitrogen which is performed by denitrifying bacteria in the soil. (14.2)

dent corn. A variety of field corn having yellow or white kernels that become indented as they ripen. It has a high starch content and is used for animal feed, manufacturing plastics, and fructose. (13.2)

dependent variable (DV). Something that changes because of the outcome of a scientific experiment. (5.1)

depreciation. The reduction in the value of an *asset* with the passage of time. (8.1)

derrick. A steel tower on a drilling platform used to support the drilling bit. (7.1)

desalinate. The process of removing salt from seawater to make it suitable for human, plant, and animal consumption. (14.2)

desert ecosystem. An ecosystem characterized by low annual rainfall, soil composed of sand or rock, and temperatures that are often extremely high during the day and cold at night. (14.1)

detail drawing. A drawing that includes view(s) of the product with dimensions and other important information (name of the object, quantity, drawing number, material used, *scale*, revisions). (7.4)

detritivore. An organism that feeds on dead matter. All detritivores are *decomposers*. (14.4)

diameter at breast height. The first measurement of a tree at a point 4 1/2′ off the ground. (17.2)

dicot. A plant that has two seed leaves when it firsts emerges. Also referred to as a *dicotyledon*. (13.1)

dicotyledon. A plant that has two seed leaves when it firsts emerges. Also referred to as a *dicot*. (13.1)

digestion. The process of breaking down food through mechanical and chemical means and absorbing nutrients into the bloodstream. (9.4)

dimension line. A thin line (usually terminated with arrowheads at each end) used to indicate measurements. (7.4)

diode. A semiconductor device that allows the flow of electrical current in one direction only. (7.1)

disk flower. Wider, flat flowers with large *stigmas*. (13.1)

disposal. The management of the unused portion of the product purchased for consumption. (9.1)

dissolved oxygen (DO). The amount of gaseous oxygen (O_2) dissolved in the water. The oxygen that is dissolved in water by diffusion from the surrounding air, through aeration of turbulent water, and as a waste product of photosynthesis. The microscopic molecules of oxygen that are dissolved in water. (12.2, 14.1, 15.4)

distillation. A liquid purifying process that involves heating, cooling, and recovery. (7.2)

distribution. The transportation of products from their processing point to the place where they are available for sale. (9.1)

doe. A female goat of any age. Also referred to as a *nanny*. A female rabbit. A mature female deer. (11.4, 12.1, 12.3)

doffers. Rubber pads with ridges or fingers that remove the fibers from the spindle of a cotton picker machine. (13.4)

domesticated animal. Any animal that has been tamed from its wild form to a domesticated form for the benefit of humans. (12.1)

domestication. The process of humans changing a plant or animal from its wild form to a domesticated form for the benefit of humans. (1.2)

dominant genes. Genes that will be expressed if they are in the organism's genetic makeup. (5.3)

double fertilization. A complex fertilization process in which a grain of pollen splits into two pollen grains that travel into the ovary to fertilize the *ovule* and the *endosperm*. (13.1)

draft horse. The classification given to horses measuring 16 to 18 hands tall and weighing upward of 2,000 lb. (10.3)

drag. The force that acts opposite to the direction of motion. (7.1)

drake. A male duck. (11.1)

dressing percentage. The ratio of live animal weight to carcass. (12.4)

drill. A versatile, handheld power tool that can be used to drill holes in wood, metal, and other materials; drive screws; turn nuts and bolts; and sand or grind material. (7.3)

drilling fluid. A mixture of fluids, solids, and chemicals that lubricate the bit used to drill for oil. Also referred to as *drilling mud*. (7.1)

drilling mud. A mixture of fluids, solids, and chemicals that lubricate the bit used to drill for oil. Also referred to as *drilling fluid*. (7.1)

drill press. A stationary, electrically powered drive mounted vertically on a stand that is used to drill holes in wood, metal, and other materials. It is much more powerful than handheld power drills. (7.3)

drip irrigation system. A watering system in which water drips slowly to the roots of plants through a network of tubing, valves, pipes, and emitters. Also referred to as *trickle* or *micro irrigation*. (15.5)

drone. A worker bee. A male bee in the hive that fertilizes the queen bee. (12.3)

drumstick. The lower portion of the leg quarter (between the knee joint and the hock) of poultry. (11.1)

dry mill process. An *ethanol* production method in which the dry *feedstock* is ground first, mixed with liquid, the sugars are extracted and fermented to produce ethanol, and the ethanol is purified through *distillation*. (7.2)

drywall hammer. A hammer with a serrated face for a better grip on the nail when installing drywall. It has a hatchet-shaped claw that is used to remove old material and/or mark cutouts for outlets and switches. (7.3)

drywall saw. A push stroke handsaw with a narrow blade made for cutting holes and gentle curves in wood. It is useful for cutting holes in walls or floors for pipes, electrical outlets, or other fixtures. May be referred to as a *keyhole saw* or *compass saw*. (7.3)

duck. A female duck. (11.1)

duckling. A baby duck of either sex. (11.1)

Dust Bowl. The time period (1930–1936) in which mismanagement of American cropland led, in part, to massive dust storms throughout the Midwest region of the United States. The term may also be used to describe the region most severely affected. (1.2)

Dutch elm disease. A fungus that affects elm trees and is spread by the elm bark beetle. (17.2)

dyadic communication. Communication that occurs between two people. Includes phone calls, emails, and text messaging between two people. (2.2)

E

Earth science. The examination of the physical Earth and its atmosphere; includes geology, oceanography, meteorology, and ecology. (5.2)

ecological cycles. The various self-regulating processes that recycle Earth's resources that are essential to sustain life. (14.2)

ecological footprint. The measure of your impact on Earth. (16.3)

ecology. The study of the relations or interaction of living things with one another and their environment. (14.1)

economics. The study of the production, distribution, and consumption of commodities. (5.2)

ecosystem. A *community* of organisms living and interacting with one another and their environment. (14.1)

elevation plan. A drawing used to portray the building from one side, showing the vertical features of the structure. (7.4)

embryo. The portion of a cereal grain's seed that forms the living plant. Also referred to as the *germ*. (13.2)

embryo transfer (ET). An artificial fertilization process in which fertilized embryos are collected from a female in order to store them or place them into other females for growth and development. (5.3)

endangered species. Animals that are low in total population and likely to become extinct. (16.2)

endosperm. A tissue produced inside the seeds of a flowering plant around the time of fertilization. It surrounds the embryo and provides nutrition in the form of starch. (13.1, 13.2)

engineering. The branch of science and technology concerned with the design, building, and use of engines, machines, and structures. (1.3)

English riding. A riding discipline in which the rider uses an English saddle; includes dressage, hunter under saddle, hunt seat equitation, and working hunter. (10.3)

enterprise. One specific section of your total SAE program. (3.1)

entrepreneurship SAE. A *Supervised Agricultural Experience* in which students own and operate an agricultural business. (3.1)

Environmental Protection Agency (EPA). A government agency tasked with regulating the environment in which our food is produced. The EPA works to protect public health by monitoring the use of pesticides on crops and promotes safe means of pest management. A government agency tasked with oversight and protection of the environment of the United States. It is a regulatory agency that proposes laws and regulations that protect our environment. (9.3, 15.1)

enzyme. A chemical that helps break down food. (9.4)

epidermal tissue. The outside covering of a plant designed to protect the plants internal parts and retain water. (13.1)

equator. An imaginary line drawn around the center of Earth, equally dividing the globe. (14.3)

equine. The animals in the family *Equidae*, which includes single-toed hooved animals like horses, asses, and zebras. The species name given to all horses, donkeys, and asses. (6.1, 10.3)

equine assisted therapy. Riding programs that use equine activities to increase the physical, occupational, or psychological health of the riders. (10.3)

***Equus asinus*.** The scientific name for donkeys and asses; often called the *long-eared breeds*. (10.3)

***Equus caballus*.** The scientific name for wild and domesticated horses; includes subcategories of *pony*, *light horse*, and *draft horse*. (10.3)

ergot. A type of fungus to which rye is susceptible and that causes severe illness in humans. (13.2)

estuary. A partially enclosed body of water on the coasts of oceans and larger seas. (14.1)

ethanol. A *biofuel* that is a natural by-product of the *fermentation* of sugars derived from plants. Also referred to as *ethyl alcohol*. (6.2, 7.2)

ethanol soluble. Somewhat dissolvable remnants from the *feedstock* used to make *ethanol*. These solids are concentrated to make a syrup that is mixed and dried with the coarse grain to make livestock feed. (7.2)

ethyl alcohol. A biofuel that is a natural by-product of the fermentation of sugars derived from plants. A clear, colorless, volatile liquid that is also the intoxicating component of alcoholic beverages. Also referred to as *ethanol*. (7.2)

etiquette. Behavior that is considered polite by society. (2.1)

euthanasia. The act of humanely putting an animal to death through chemical means, or allowing it to die by withholding extreme medical measures. (12.1)

evaporated milk. A milk product made by removing about 60% of the water from ordinary milk. It may be used as a substitute for milk or cream. (10.2)

evergreen broadleaf tree. A tree with broad, flat leaves that remain green throughout the year. (17.1)

evergreen tree. A tree whose leaves are needles or scales that remain green throughout the year. They are also referred to *softwood species* or *coniferous trees* and classified as *gymnosperms*. (14.1, 17.1)

ewe. A mature female sheep. (11.3)

ewe lamb. A young female sheep. (11.3)

exotic foods. Foods that are not able to be produced in the area where they are consumed. (9.1)

expense. Any money a business pays out. (8.1)

experiential learning. The process of creating understanding from doing something. (3.1)

experiment. A controlled scientific test to determine the validity of a hypothesis. (5.1)

exploratory SAE. A *Supervised Agricultural Experience* designed to increase student agricultural career awareness through exploration activities. Exploration activities may include observing, interviewing, or assisting an individual in an agriculture-related profession, participating in a field day or event, or giving a classroom demonstration. (3.1)

export. A commodity that is shipped to another country (abroad) for sale and/or consumption. (9.2, 11.2)

extemporaneous speaking. A speech format in which the speaker has a very short period of time between receiving the topic and presenting the speech. (2.2)

extension line. A thin line that extends beyond the outline of a view to make it easier to read the dimensions. (7.4)

extinct species. Animals that are no longer living anywhere on Earth. (16.2)

eyes. The nodes on potatoes from which new growth forms. (13.5)

F

fair chase rules. Hunting rules which mean not hunting with automatic weapons, not using electronic calling, not spot lighting, and not hunting from all-terrain vehicles, among other rules. (16.1)

Fair Labor Standards Act (FLSA). A law that defines the standards for paying wages. It also establishes the maximum number of hours that youth can work, and the types of work that they can do. (4.1)

farmers market. A place where groups of agricultural producers gather together to market their products directly to consumers. (9.1)

farm flock. A smaller sheep operation that typically has less than 100 ewes and is a secondary enterprise for the producer. (11.3)

farrier. A craftsperson who trims hooves and shoes horses. (10.3)

farrow. When a sow goes into labor and gives birth. (11.2)

farrowing. The act of a sow giving birth. (11.2)

farrowing crate. A special birthing/feeding stall designed to prevent the sow from laying on and killing the piglets. (11.2)

farrowing facility. An agricultural structure designed to provide an area for the sow to birth and nurse pigs until weaning. (6.1)

farrow-to-finish operation. Swine production system in which the swine herd is maintained with sows farrowing year-round. (11.2)

farrow-to-wean operation. Swine production system in which weaned piglets (5–8 weeks) are sold to *finishing operations* to be fed and marketed. Also referred to as a *feeder pig operation*. (11.2)

fat. A nutrient that provides long-term storage of energy for living organisms. Also referred to as a *lipid*. (9.4)

fauna. The animal life in an ecosystem. (14.1)

fawn. The offspring of a mature female deer. (12.3)

fecal coliform bacteria. Microscopic bacteria that originate in the intestine of warm-blooded animals and are passed through the intestines via feces. (15.4)

feeder pig operation. A swine production system in which weaned piglets (5–8 weeks) are sold to *finishing operations* to be fed and marketed. Also referred to as a *farrow-to-wean operation*. (11.2)

feedlot. A cattle feeding operation in which large numbers of cattle are grouped based on size, genetics, and consistency in order to maximize profit for the feedlot; cattle are fed high-grain diets and are not allowed to graze. (10.1)

feedstock. The raw material supplied for processing into another product. (7.2)

feet. A poultry cut used to make stock and that is also harvested and sold for a premium to export markets. May also be referred to as paws. (11.1)

fen. A wetland that is fed primarily by groundwater and some runoff. It contains peat and a variety of vegetation. (15.3)

fencerow. An uncultivated strip of land on each side and below the fence on a landowner's property. (16.2)

feral. Animals that have historically been domesticated and now live in the wild with no human assistance. (10.3)

feral animal. An animal that has historically been domesticated and now lives in the wild with no human assistance. An animal living in the wild, but descended from domesticated individuals. (12.1)

fermentation. A chemical process in which an agent, such as bacteria, yeasts, or other microorganisms, causes an organic substance to break down into simpler substances; in ethanol production, the process breaks down sugar or starch into *ethyl alcohol*. (7.2)

Fertile Crescent. An area that extends from the eastern part of the Mediterranean to the lower Zagros Mountains in Iraq and Iran. This area is referred to as the birthplace of agriculture. (1.2)

fertilization. When a male and female sex cell combine to create a new organism. (13.1)

feudal system. An agricultural management system in which the land was owned by those who completed military service. These landowners would lease use of the land, along with supplies, to a farmer in exchange for the payment of taxes and goods as rent. (1.2)

fiber. A product with long, thin components often used to create woven or composite materials. Very complex carbohydrates, which require a lot of digestive effort to break down into absorbable nutrients. (1.1, 9.4)

fibrous roots. Plant roots that are heavily branched and spread out underground from the plant's *stem*. (13.1)

filament. The stalk that holds the *anther* of the male portion of a flower. (13.1)

filly. A female horse two years or younger. (10.3)

filtering facepiece (dust mask). A respirator with a filter as an integral part of the facepiece or with the entire facepiece composed of the filtering medium. (4.2)

fine-wool breed. Sheep that are primarily used for finer wool production and have less value as a meat animal. (11.3)

fingerling. Small, young fish under one year old. (12.2)

finishing facility. An agricultural structure designed to provide space and feed for pigs to grow from weaning weight to market weight. (6.1)

finishing operation. A swine production system in which feeder pigs are purchased weighing approximately 25 lb and are fed to desirable market weight (230–280 lb). (11.2)

fishery. An *aquaculture* operation or the practice of aquaculture. (12.2)

fish farming. The breeding, rearing, and harvesting of plants and animals in all types of water environments including ponds, rivers, lakes, oceans, and land-based facilities. May also be referred to as *aquaculture* or *concentrated aquatic animal production (CAAP)*. (12.2)

fixation. The process through which atmospheric nitrogen is transformed into ammonium. (14.2)

flank. A *primal cut* from the abdominal muscles of beef cattle. (10.1)

flat-head screwdriver. A screwdriver with a flat, wedge-shaped tip that fits into the slot in the head of a screw. May also be referred to as a *slotted-head screwdriver*. (7.3)

fleece. The woolly coat of sheep. (11.3)

flint corn. A variety of corn that has a hard outer layer on each kernel. It has a lower starch content than dent corn and a very low water content. Commonly used in animal feed. (13.2)

flock. A number of chickens or ducks feeding and living together. A group of goats that may also be referred to as a *herd*, *tribe*, or *trip*. A group of sheep that keep or feed together or are herded together. (11.1, 11.3, 11.4)

floor plan. A common construction drawing used to represent the layout of rooms, doors, and other features within a building; used to indicate where walls will be placed, and provide a top view of rooms within the building. (7.4)

flora. The plant life in an ecosystem. (14.1)

floral design. The art and science of using cut flowers and other materials to create balanced and composed arrangements. (13.6)

floriculture. The cultivation and management of flowering and ornamental plants. (13.6)

flour corn. A type of corn with a high, soft starch content that is primarily used for making corn flour. (13.2)

foal. A young equine of either gender. (10.3)

foaling. The term for the process of giving birth in an equine animal. (10.3)

folding rule. A measuring tool composed of strips of wood joined by rivets so as to be foldable. All of the opening and closing parts are in parallel planes. Also referred to as a *zigzag rule*. (7.3)

followership. The ability to actively follow a leader. (2.1)

food. All materials and substances used for human consumption. (1.1)

Food and Drug Administration (FDA). A government agency responsible for ensuring that products other than meat, poultry, and eggs are safe, pure, and labeled correctly for consumers. (9.3)

Food and Nutrition Services (FNS). A government agency responsible for nutrition assistance programs administered through the United States Department of Agriculture (USDA). (9.3)

foodborne illness. An ailment caused by improper handling, storage, or processing of food products. (9.2, 9.3)

food chain. A cycle in which some animals eat plants and, in turn, are eaten by other animals. Also referred to as a *food cycle*. A simple model of how energy flows from one living organism to another. (14.1, 14.4)

food cycle. A cycle in which some animals eat plants and, in turn, are eaten by other animals. Also referred to as a *food chain*. The interaction of many *food chains*. Also referred to as a *food web*. (14.1, 14.4)

food desert. An area with low income levels, where more than 500 people live more than one mile from the nearest grocery store. (9.2)

food insecurity. A situation in which consistent food is unavailable year-round due to a lack of money or resources. (9.2)

food miles. The total distance that the ingredients of a food product travel before they reach the consumer. (9.1)

food safety. The handling, preparation, and storage of food to prevent foodborne illnesses. (9.3)

Food Safety and Inspection Service (FSIS). An agency administered through the United States Department of Agriculture (USDA) that is responsible for food safety related to meat, poultry, and egg products. (9.3)

food security. The ability to access and secure adequate food. (9.3)

food system. The collaboration between producers, processors, distributors, consumers, and waste managers related to agriculture in a particular area. (9.1)

food web. A system of interlocking and interdependent food chains. The interaction of many *food chains*. Also referred to as a *food cycle*. (14.1, 14.4)

forest ecosystem. An ecosystem characterized by a large number of trees and other plant life, as well as abundant fauna. (14.1)

forestry. The practice of growing trees for production. Also referred to as *silviculture*. (17.1)

formal drawing. A more detailed drawing of a project that is drawn to *scale*. (7.4)

formal science. Disciplines concerned with abstract scientific concepts; includes logic, mathematics, theoretical computer science, and statistics. (5.2)

formula. Predetermined mathematical patterns used to solve for unknown values. (8.2)

fossil fuel. An organic compound used to generate energy; includes coal, natural gas, and crude oil, and metals such as uranium. (7.1)

four-cycle engine. The type of petroleum-powered engine commonly used for lawn mowers and other large landscaping tools. (7.3)

4-H. A youth organization administered by the National Institute of Food and Agriculture of the United States Department of Agriculture (USDA), with the mission of "engaging youth to reach their fullest potential while advancing the field of youth development." (2.1)

fraction. Numerical expression that represents one number divided by another to explain the part of a whole that is being measured. (8.2)

framing hammer. A slightly heavier hammer with a straighter claw for use as a prying tool as well as a nail remover. Typically used in construction for framing. May also be referred to as a *ripping hammer*. (7.3)

framing square. An L-shaped measuring tool made from one piece of metal that is used for laying out rafters and marking stair stringers. (7.3)

free-range system. A poultry production system in which the birds are allowed to roam and forage for insects, seeds, and herb material during the day. (11.1)

fresh-market consumption. Term describing produce that is consumed almost immediately without much processing, other than washing. (13.5)

freshwater. The water that is not salty and supports all life, both plant and animal. (14.2)

freshwater aquaculture. The production of aquatic life that is native to rivers, lakes, and streams, or spends the majority of its life in freshwater. (12.2)

freshwater ecosystem. An ecosystem in which the water is fresh, not salty. (14.1)

frog. A triangular area in the middle of a horse's sole (of the hoof) that helps circulate blood in the foot and from the bottom of the leg to the heart. (10.3)

front. The place where two air masses meet. (14.3)

frostbite. Body tissue injury caused by overexposure to extreme cold that typically affects exposed skin and extremities like the nose, fingers, and toes. (4.3)

frozen yogurt. A frozen dessert that is lower in fat than ice cream and contains yogurt cultures that may or may not be active. (10.2)

fruit. The ripened ovary surrounding the seed of the plant. (13.5)

fruit vegetable. A vegetable that produces a fruit, such as bell peppers, squash, and cucumbers. (13.5)

fry. Young fish ready to begin eating on their own. (12.2)

fryer. A young rabbit raised for meat. (12.1)

furrow irrigation system. Crop watering system in which the field is flooded with water, allowing it to flow slowly across the field in furrows. (15.5)

G

gaggle. A group of geese. (11.1)

game. Animals that are hunted for food and/or sport. (16.2)

game bird. Any type of fowl species and subspecies that is legally hunted. (11.1)

game species. Animals that are hunted for food and/or sport. (16.2)

gander. A male goose. (11.1)

gang. A group of turkeys. It may also be called a *rafter*. (11.1)

gas mask. A respirator that is more effective than chemical cartridge respirators against high concentrations of toxic gases. (4.2)

gavel. The hammer-shaped tool that the chairperson uses to send specific cues to the assembly. (2.3)

gelatin. An odorless, colorless, sterile protein powder that is used in many food products. (12.4)

gelding. A castrated male equine, llama, or alpaca. (10.3, 12.3)

genetically modified organism (GMO). A plant or animal that has been created using *genetic engineering*. May also be referred to as genetically modified product (GMP), biotech food, biotechnology-enhanced food, or a *transgenic organism*. (5.3)

genetic engineering. The technique of removing, modifying, or adding genes to a DNA molecule in order to change the information it contains; changing this information changes the type or amount of proteins an organism is capable of producing, thus enabling it to make new substances or perform new functions. (5.3)

genetic reproductive potential. A genetic predisposition to give birth to a certain number of offspring each year and over a lifetime. (16.2)

genotype. The actual genes an organism has for a specific trait. (5.3)

geometry. The field of mathematics related to understanding spatial relationships and geometric shapes. (8.2)

geothermal energy. Energy harnessed from the natural supply of heat beneath Earth's surface. (7.1)

geothermal heat pump. A heating and/or cooling system that takes advantage of the relatively constant year-round ground temperature to pump heat to or from the ground. Also referred to as a *ground-source heat pump*. (7.1)

germ. The portion of a cereal grain's seed that forms the living plant. Also referred to as the *embryo*. (13.2)

germane. It means that the amendment made to a main motion must have something to do with the main motion. (2.3)

gestation. The process of carrying young in the womb between conception and birth (pregnancy). (10.1)

giblets. The edible offal of the chicken, including the gizzard, liver, and heart. (11.1)

gilt. A young female swine. (11.2)

gizzard. An area where muscular contractions, combined with pebbles or rocks consumed by the bird, work to mechanically break down the food. (9.4, 11.1)

global food system. The systematic production of food for distribution to consumers around the globe. (9.2)

global navigation satellite system (GNSS). A satellite system that uses receivers and specialized computer programs to pinpoint the geographic location of a user's receiver. (6.2)

global positioning system (GPS). A system of navigation that uses space satellites to provide information on the location of places and objects. (6.2)

glycerin. A valuable co- or by-product of *biodiesel* production used in soaps and other products; originally formed the chemical bonds between the fatty acids of the oil or fats being converted. (7.2)

goose. A female goose. (11.1)

gooseneck bar. A prying tool made of solid metal with a hooked end and a claw designed to open or break apart materials. Commonly referred to as a *crowbar*. (7.3)

gosling. A baby goose of either sex. (11.1)

grafting. The technique of splicing the shoot of one plant onto the rootstock of another plant. (13.5)

grain bin. Highly specialized structures designed to store grain until it is transported to market. (6.1)

graph. A type of *chart* that shows the relation between two or more variables. (5.1)

grassed waterway. Grassy area planted by farmers to channel water down hillsides. (15.5)

grassland ecosystem. An ecosystem composed mainly of grasses with few trees and shrubs. Also referred to as a *prairie ecosystem*. (14.1)

greasy wool. Wool in its natural state. (11.3)

greenhouse. A structure used to recreate ideal growing conditions for plants to extend the growing range and season. (6.1)

greenhouse effect. The process by which the sun's solar energy is trapped in the atmosphere and reflected back to Earth. (14.3)

greenhouse gases. The naturally occurring combination of gases surrounding Earth that behave as the glass or plastic covering on a greenhouse. The primary gases are water vapor, carbon dioxide, methane, and ozone. (14.3)

Green Revolution. A time period between 1940 and the late 1960s in which scientists focused on developing new technology that would increase agricultural production on a worldwide scale. (1.2)

grinder. A power tool that uses an abrasive material (on a wheel) to smooth and shape metal, polish metal surfaces, and to remove small amounts of metal from surfaces and edges. Grinders may be stationary or handheld. (7.3)

grit. The pebbles and other hard objects birds consume to help grind up food in the *gizzard*. (11.1)

gross domestic product (GDP). A number that represents the total dollar value of all goods and services produced over a specific time period (usually a year). Also used to gauge the health of a country's economy. (7.2)

ground. On an electrical plug, it is the third prong that serves as a return path for electrical current to return safely to "ground" without danger to anyone. (4.2)

ground fault circuit interrupter (GFCI). A fast-acting circuit breaker designed to cut off the current within as little as 1/40 of a second if the outlet becomes ungrounded. (4.2)

ground-source heat pump. A heating and/or cooling system that takes advantage of the relatively constant year-round ground temperature to pump heat to or from the ground. Also referred to as a *geothermal heat pump*. (7.1)

ground tissue. The tissue that makes up the largest part of most plants, and includes all the parts of the plant that perform photosynthesis. (13.1)

groundwater. Liquid water that resides underneath Earth's surface. (14.2)

group communication. Communication that involves a selected group of people. It may be as few as three people or a large crowd. (2.2)

growing zones. The geographical regions based on climate shown on the USDA Plant Hardiness map that are used to show consumers what plants will thrive in any given area. (13.6)

growth cycle. The amount of time it takes for a plant to germinate, grow to maturity, and reproduce. (13.1)

guard cell. Kidney-shaped cells in the lower epidermis of a leaf that create openings that allow or prevent gases and water from leaving the leaf. (13.1)

gully erosion. Soil erosion that occurs when many streams of water come together and form a larger stream, often in the valleys or low areas of a field. (15.2)

gymnosperm. A woody, vascular seed plant that produces seeds that are bare, but may be protected by a sturdy cone. (17.1)

H

habitat. The preferred living area (within an ecosystem) of a plant or animal. (14.1)

hacksaw. A push stroke saw that uses disposable blades held in place with front and back pins. Hacksaws are used primarily to cut metals ranging from thin metal tubing to cables and wire ropes. (7.3)

hair sheep. Breeds of sheep that have a higher ratio of hair-to-wool fibers than wool sheep. (11.3)

ham. A primal cut of pork meat from the upper hind leg of swine. (11.2)

hammer. A basic hand tool used for pounding in and pulling out nails. (7.3)

hand. A four-inch increment used to measure the height of equines. (10.3)

hands-on learning. Activities provided for students to gain a better understanding of concepts by applying knowledge in a way that allows them to interact with material objects. (3.1)

hardwood species. Trees that grow new leaves in the spring of each year, and then lose those leaves in the fall. Also known as *deciduous trees*. (17.1)

harvesting. The act of gathering raw agricultural products for consumer use. (9.1)

hatchery. A carefully designed and specially constructed poultry facility in which fertilized eggs are stored, incubated, and hatched. An aquaculture production facility focused on breeding, nursing, and rearing fish, invertebrates, and aquatic plants to fry, fingerling, or juvenile stages. (11.1, 12.2)

Hazard Analysis and Critical Control Points (HACCP). Food safety guidelines developed by the FDA to serve as a management system for food safety involving control of biological, chemical, and physical hazards that may occur during agricultural product production, handling, manufacturing, distribution, and consumption. (9.3)

heartwood. The innermost part of the trunk of the tree. (17.1)

heat cramp. Muscular spasms that occur when a person is sweating a lot and not replacing lost fluids. (4.3)

heat exhaustion. A condition that occurs when you do not drink enough fluids to replace what is being lost through heavy perspiration, and you have a slightly higher than normal body temperature. Symptoms include severe thirst, headache, nausea, diarrhea, profuse sweating, dizziness, clammy skin, and a rapid pulse. Heat exhaustion may lead to heatstroke if not treated. (4.3)

heatstroke. A condition that occurs when your body is unable to release enough heat, and your temperature rises to life-threatening levels. Symptoms include skin that feels feverish, confusion or disorientation, seizures, and ultimately, the victim may fall into a coma. (4.3)

heifer. A term to describe young, female beef/dairy cattle that have not been bred. (10.1)

Helminthosporium leaf blotch/spot. A fungal disease that affects many types of grasses, including crop (especially wheat) and turf grasses. (13.2)

hen. An adult female chicken over a year old. An adult female turkey. (11.1)

herbaceous stem. A type of flexible plant *stem* that continually grows upward, and has its *vascular* system in a bundle. (13.1)

herbivore. An animal that eats only plants. (14.4)

herd. A group of cattle of a single kind that is kept together for a specific purpose. A group of goats. May also be referred to as a *flock*, *tribe*, or *trip*. (10.1, 11.4)

heterotroph. An organism that consumes (eats) other organisms. Also referred to as a *consumer*. (14.4)

heterozygous. The genes an organism has from the parents are different for a specific trait. (5.3)

hidden line. A thin, dashed, black line used to outline edges and intersections that are hidden from view. (7.4)

hide. The skin of an animal. (12.4)

hinny. The equine offspring of a female donkey and a *stallion*. (10.3)

homogenization. A process in which the milk's fat droplets are emulsified and the cream does not separate. (10.2)

homozygous. The genes an organism has from both parents are the same for a specific trait. (5.3)

honeycomb. A chamber of the ruminant digestive system that serves as an area to sort feed coming from the mouth and remove any foreign objects from the feed prior to passage into the rumen. Officially referred to as the *reticulum*. (9.4)

horsepower. A unit for measuring the work of a motor. It was derived from the power required from a horse to lift 150 lb a distance of 220′ in one minute. (4.1)

host. The species on which a *parasite* lives and feeds. (14.4)

hotblood. Term used to describe more spirited light breeds like the Arabian and Thoroughbred. See *warmblood*. (10.3)

houseplants. Plants grown for the purpose of indoor ornamentation. (13.6)

hull-less barley. A cereal grain with an easy-to-remove hull. (13.2)

humus. Material composed of decomposing and decomposed living organisms, including their waste products. Also referred to as *organic matter*. (15.2)

hunger. The scarcity of food in a particular region. (9.2)

hunting season. Certain time of the year in which a particular type of game may be hunted. (16.1)

hutch. A type of cage typically designed for keeping rabbits. (12.1)

hybrid. A cross between two poultry breeds. (11.1)

hydraulic system. A power system that uses fluids to do work. (7.5)

hydric soil. Soil that has developed anaerobic conditions due to long-term saturation. (15.3)

hydroelectric power plant. A power plant in which water passing through gates in a dam is directed through turbine generators that create electricity on a commercial scale. (7.1)

hydrogen sulfide (H_2S). A colorless, flammable, and poisonous gas with a smell of rotten eggs. It is emitted during the decomposition of organic matter such as manure. (4.3)

hydrological cycle. The pathway or cycle that water follows underneath the surface, at the surface, and above the surface of Earth. Also referred to as the *water cycle*. (14.2)

hydroponic system. A crop growing system in which plants are grown in nutrient-rich water, without the use of soil. (6.2)

hydropower. The power derived from the kinetic energy of moving water. (7.1)

hypothermia. A condition that occurs when your body loses heat faster that it can produce it, causing a dangerously low body temperature. (4.3)

hypothesis. An educated prediction of the outcome of a scientific experiment. (5.1)

I

ice cream. A sweet, frozen food made from no less than 10% butterfat and made in a myriad of flavors. (10.2)

imperfect flower. A flower that is missing one or more of the basic flower parts (*petals, sepals, pistils,* and *stamen*). (13.1)

import. A commodity brought into a country for consumption. (9.2)

imprinting. Training given to newborn equines to ensure they are accustomed to human interaction and will be safer to handle when they are older. (10.3)

impromptu speaking. Extemporaneous speaking in which the speaker is expected to give a speech immediately following the receipt of the topic. (2.2)

incidental motion. A motion related to parliamentary rules and procedures, not to other motions. May be a motion that asks questions, corrects mistakes, ensures the accuracy of a vote, or changes or modifies rules. (2.3)

income. The money a business earns. (8.1)

incomplete dominance. Situation in which some traits show as a mixture when the *genotype* is *heterozygous.* May be referred to as *codominance,* depending on the nature of the specific combination. (5.3)

increment borer. A tool used to cut a small hole in the trunk of a tree to remove a core sample of the trunk to count the rings. (17.2)

indentured servant. In colonial times, a person paid to work for an employer for a certain number of years in return for paying their passage from Europe to the American colonies. (4.1)

indentured servitude. A system of agricultural labor in which a person is indebted to someone and required to work for that person until their debt is paid. (1.2)

independent variable (IV). Something that the researcher changes in order to test the hypothesis. (5.1)

informational written correspondence. Written communication in which information is being transferred one way. (2.2)

information system. The equipment and software used to collect, filter, process, and distribute information, or to process data. (1.3)

informed consumer. A person who is knowledgeable about the agricultural processes and origins of agricultural commodities who can make good decisions based on this knowledge. (1.3)

inner bark. The thin layer of phloem cells between the cambium and outer bark that provides the food supply to the entire tree. Also referred to as *phloem.* (17.1)

inoculation. The process of introducing beneficial bacteria to the soybean seeds prior to planting. (13.3)

in-situ leach (ISL) mining. Uranium mining process that involves drilling wells and pumping water from prospective uranium deposits. Also referred to as *in-situ recover (ISR) mining.* (7.1)

in-situ recover (ISR) mining. Please refer to *in-situ leach (ISL) mining.* (7.1)

integrated pest management (IPM). An ecosystem-based method of pest control. It focuses on long-term prevention of pests and damage through a combination of techniques including modification of cultural practices, biological control, and the use of resistant varieties. (13.2)

intensive housing system. A total confinement poultry production system in which the birds stay inside a house or a cage their entire life. (11.1)

intensive recirculating aquaculture system (IRAS). An aquaculture system that filters water from the tanks so it can be reused within the tanks. (12.2)

interest. The fee charged by the lender in exchange for loaning the money. (8.1)

internal combustion engine. An engine that generates power by burning fuel combined with air inside the engine. (7.5)

internode. A section of a plant stem between *nodes.* This distance is often essential in determining plant height and shape. (13.1)

intrapersonal communication. Communication in which the thought or language use is internal to the communicator. It is often called self-reflection. (2.2)

introduced species. Animal or plant species that are not naturally occurring in a particular ecosystem or locale, but have been introduced by humans. (14.4)

invasive species. Animal or plant species that are not naturally occurring in a particular ecosystem or locale, but have been introduced primarily by humans. (14.4)

inventory. A complete list of a company's *assets*. (8.1)

irrigation. Crop watering system in which water is pumped from a well or river and applied to a growing crop via sprinklers, flooding, or subsurface piping. The artificial application of water to the land or soil. (15.4, 15.5)

J

jack. A male ass or donkey. (10.3)

jack plane. A handheld woodworking tool ranging in length from 12″ to 18″ that is used to smooth and square rough lumber. (7.3)

jennet. A female ass or donkey. Also referred to as a *jenny*. (10.3)

jenny. A female ass or donkey. Also referred to as a *jennet*. (10.3)

jet stream. A narrow band of fast-moving wind that is constantly circling the globe. These winds transfer heat and precipitation around the planet. (14.3)

jig saw. A handheld power tool that uses a thin blade that moves back and forth to make a cut. Jig saws are useful for making curved cuts to wood or metal. (7.3)

job application. An electronic/paper formal request for employment. (3.2)

jointer. A woodworking tool that is 12″ or longer and used to trim, square, and straighten the edges of doors and long boards. (7.3)

jointing. The point in a wheat plant's life when vegetative growth transforms into reproductive growth. (13.2)

jungle. A forest ecosystem that receives abundant rainfall, includes dense tree vegetation at various heights, and is found in the tropics near the equator; the trees do not shed their leaves. Also referred to as a *tropical evergreen forest*. (14.1)

K

kernel. The fruit of a cereal grain. (13.2)

keyhole saw. A push stroke handsaw with a narrow blade made for cutting holes and gentle curves in wood. It is useful for cutting holes in walls or floors for pipes, electrical outlets, or other fixtures. May be referred to as a *compass saw* or *drywall saw*. (7.3)

kid. A young goat of either sex. (11.4)

kidding. The act of a goat giving birth. (11.4)

kindling. The act of a rabbit giving birth. (12.1)

kinetic energy. The energy possessed by an object or system as a result of its motion. (7.1)

kit. A baby rabbit. (12.1)

knot. An imperfection in lumber that occurs where a branch is attached to the trunk; weakens the strength of the wood. (17.1)

L

laissez-faire leadership. A type of leadership style in which the leader provides little or no guidance and allows the group members to make the decisions; the leader is present merely to provide tools and resources. Also referred to as *delegative leadership*. (2.1)

lake effect snow. The snow falling on the lee (sheltered) side of a lake, generated by cold dry air passing over warmer water. (14.3)

lamb. The term used to describe young sheep. The meat from sheep less than one year old. (11.3)

lambing. The term used to describe a ewe in the process of giving birth. (11.3)

land capability classes. The determination of the suitability of land for different cropping systems. (15.2)

land grant institutions. Colleges with land set aside by the Morrill Act for agricultural education. (1.2)

landscape architect. Someone who develops the plan for a space and the specific plants that will be used in that space. (13.6)

landscape horticulture. The cultivation and management of living plants grown for aesthetic purposes in the human environment (indoor and outdoor). (13.6)

lanolin. The natural grease found on sheep's wool. (12.3)

lapin. A castrated male rabbit. (12.1)

lateral root. A large root growing out from the *primary root* of a plant. (13.1)

latitude. A measure of the distance from the equator to either of Earth's poles. (14.3)

layer. A chicken raised for laying eggs. (11.1)

leaching. The process in which water dissolves minerals and moves them deeper within the *soil profile*. (15.2)

leader. A person who guides or directs a group or organization. (2.1)

leader line. A line with an arrow at one end that is used to call out and explain or describe components of the drawing. (7.4)

leadership. A personal quality related to being able to guide or direct others. (2.1)

leaf. A part of a plant whose primary purpose is to collect sunlight for the plant to perform *photosynthesis*. (13.1)

leafy vegetable. A vegetable grown for its edible leaves. (13.5)

leather. Tanned animal hide that has had the hair removed. (12.4)

leg. A tender primal cut of lamb/mutton that may be cooked as a roast or cut into chops. (11.3)

leg quarter. A poultry cut of meat composed of both the *drumstick* and *thigh* when still connected. (11.1)

legume. A plant that has a symbiotic relationship with bacteria (rhizobia) housed in nodules on its roots, allowing it to take nitrogen from the soil and convert/uptake it as a protein source. The rhizobia convert atmospheric nitrogen into nitrogen that can be used by other plants. (9.4, 13.1, 13.3, 14.2)

lentic ecosystem. A freshwater ecosystem in still or slow-moving water. (14.1)

lenticel. A small opening on a plant stem that controls air and water leaving the plant. (13.1)

level. A tool used to measure the degree to which an object is horizontal and the degree to which it is vertical. (7.4)

liability. A debt or financial obligation of a business. (8.1)

lift. The force created by differences in air pressure that acts at a right angle to the direction of motion through the air. (7.1)

light horse. The classification typically given to horses around 14 to 16 hands tall and weighing between 900 and 1,400 lb at maturity. (10.3)

light reaction. A part of *photosynthesis* in which light energy is converted to chemical energy. Oxygen is also released during this stage. (13.1)

lineman's pliers. Handheld tool that combines the functions of *side-cutting pliers* and *slip-joint pliers*. (7.3)

linen. A natural material made from flax used in bed sheets, napkins, tablecloths, and clothing. (13.4)

line source outlet. The point where water is emitted on a drip irrigation system (typically spaced between 4″ and 24″ apart throughout the system). (15.5)

lining bar. A prying tool made of solid metal with a hooked end and a claw designed to open or break apart materials. Commonly referred to as a *crowbar*. (7.3)

linseed oil. Oil from flaxseed used in both food and other products. (13.4)

lint. The white or yellow fibers produced by the cotton plant. (13.4)

lipid. A nutrient that provides long-term storage of energy for living organisms. Also referred to as *fat*. (9.4)

listening. The act of bringing in information through hearing verbal communication. (2.2)

litter. The bedding materials (including manure and molted feathers) from a poultry house. The collective group of piglets born to a *sow*. A group of newborn rabbits. (11.1, 11.2, 12.1)

loamy soil. Soil that contains approximately 40% sand, 40% silt, and 20% clay; holds more nutrients and moisture and contains more organic matter than sandy soils. (15.2)

loan. A sum of money that is paid back to the lender over a certain length of time. (8.1)

local food system. The systematic production of food that is consumed in close proximity to its origin. (9.1)

locking pliers. A hand tool similar to slip-joint pliers, but with the added benefit of being able to be locked in place (clamped down on materials), allowing the user free use of his or her hand. (7.3)

lodge. Condition when wheat stalks fall over. (13.2)

logbook. A notebook containing detailed notes of a scientific experiment. (5.1)

loin. A *primal cut* from the section between the ribs and the round, and above the flank of beef cattle. A tender primal cut of pork meat located on the upper back and side of swine. A tender primal cut of lamb/mutton. (10.1, 11.2, 11.3)

long-eared breed. A common name for donkeys and asses. (10.3)

long tape. A measuring tool consisting of a steel or cloth measuring tape on a reel. Long tapes are available in lengths from 25′ to 300′ and are useful for measuring long distances for such jobs as fencing, pipes, and foundations. (7.3)

long-term goal. An aim or desired result that will be accomplished over an extended period of time. (3.1)

long-wool breed. Large sheep that are primarily raised for their long, coarse, wool fiber. (11.3)

loose smut. A fungal disease that replaces barley grain heads with masses of spores that infect the open flowers and grow into the seed. (13.2)

lotic ecosystem. A freshwater ecosystem found in streams or rivers that rise from lakes, ponds, or freshwater springs. (14.1)

lumber. The timber that has been sawn into long boards for planks. Wood that has been cut into boards of various sizes. (17.1)

M

macroinvertebrates. Small, but visible, animals without a backbone. (15.4)

main motion. A motion that brings up a new topic of discussion before the assembly. (2.3)

malnourishment. When a person does not receive the proper nutrients to sustain life. (9.2)

managed hunt. A hunt where a limited number of permits are issued, or a certain number of males and females of the species may be killed. (16.1)

manure pit. Manure storage area that may be an open lagoon, an earthen basin, sit below grade, aboveground, or contained in a closed building. (4.3)

mare. A mature female horse. (10.3)

mariculture. The production of species of aquatic plants and animals that live in the ocean (saltwater). Also referred to as *marine aquaculture*. (12.2)

marine aquaculture. The production of species of aquatic plants and animals that live in the ocean (saltwater). May also be referred to as *mariculture*. (12.2)

marine cage. An aquaculture enclosure with a rigid frame on all sides that is suspended in a body of water. (12.2)

marine ecosystem. An ecosystem found in oceans and seas in which the water contains high amounts of salt and other minerals. (14.1)

market. The group of consumers that are likely to purchase a specific product. (8.1)

market hog. Swine that is bound for market for meat production. (11.2)

marketing. Everything that is done to convince consumers to purchase a product. (9.1)

marsh. An area that is saturated, either occasionally or continually, with water and that fosters soft-stemmed plant growth (mostly grassy plants); receives its water supply through streams or rivers. (15.3)

mason's hammer. A hammer designed specifically to cut and set bricks and stone. The hammer head is moderately heavy with a square face for striking and a chisel edge for cutting and dressing bricks and stone. (7.3)

mass media. A communication system in which information is transferred to a broad and diverse public audience, mediated by technology. (2.2)

mast. Natural wildlife food like hickory nuts, beechnuts, and acorns. (17.1)

mastitis. The inflammation of the mammary glands, which is usually caused by bacteria. (10.2)

material safety data sheet (MSDS). A document that contains information on the potential hazards (fire, health, reactivity, and environmental) of a material, and instructions for how to work safely with the product. It also contains essential information for emergency crews in case of an emergency. Also referred to as a *safety data sheet (SDS)*. (4.2)

maternal breed. A swine breed known for their mothering ability, litter size, and longevity. (11.2)

mathematics. The abstract science of numbers, quantities, and space or shapes and the relationships between them. The study of numbers, equations, and geometric shapes and their relationships. (5.2, 8.2)

mean. The arithmetic average of the numbers in a set of data. (8.2)

measuring square. A triangular-shaped measuring tool that is technically a try square. It is used for drawing 90° or 45° angles and for laying out rafters and stairs. May also be referred to as a *speed square*. (7.3)

mechanical digestion. The physical separation of the foodstuff into smaller and smaller pieces. (9.4)

median. The number in the center of a list when numbers are ordered from least to greatest. (8.2)

medium-wool breed. Sheep that are primarily used for lamb and mutton and produce a less valuable fleece than fine-wool breeds. (11.3)

meiosis. A type of cell division that results in four daughter cells, each with half the number of chromosomes of the parent cell. (13.1)

mentor. A person who advises and gives guidance to another person to help develop his or her potential. (2.1)

mentorship. The act of advising and giving guidance to another person to help develop his or her potential. (2.1)

merchantable height. The height of the tree measured in the number of 16′ logs that can be cut from the tree. (17.2)

meristem. The very bottom of a plant's root tip. (13.1)

mesophyll. A layer of loosely arranged cells beneath the *palisade layer* in plants that allows gases to pass through for use in *photosynthesis*. Also referred to as the *spongy mesophyll*. (13.1)

metabolism. The sum of all the chemical processes conducted by the body to sustain life. (9.4)

methane (CH_4). A colorless, odorless, and flammable gas that is the main constituent of natural gas. It is emitted during the decomposition of organic matter such as manure. (4.3, 14.2)

methanol. A biofuel derived from natural gas, coal, plant vegetation, and agricultural crop biomass. Also referred to as *wood alcohol*. (6.2, 7.2)

methemoglobinemia. An illness caused by the consumption of water high in nitrates. The skin turns a bluish color due to the blood's inability to carry oxygen. Also referred to as *blue baby syndrome*. (14.2)

microbial pollution. Water pollution that comes from the introduction of detrimental microorganisms into the body of water. (15.4)

micro irrigation system. A watering system in which water drips slowly to the roots of plants through a network of tubing, valves, pipes, and emitters. Also referred to as *trickle* or *drip irrigation*. (15.5)

micropropagation. The process of using tissue culture to create new plants. An asexual plant reproduction method in which a single plant cell is placed in a special nutrient growing medium, and grown into an entire plant. (5.3, 13.1)

micro-sprinkler system. A crop watering system in which water is dispersed through a series of small sprinklers using a network of tubing, valves, and pipes. (15.5)

midge. An insect that carries the blue tongue disease. (16.2)

midrib. The central vein of a leaf. (13.1)

milk. A fluid rich in protein and fat, secreted by female mammals for the nourishment of their young. (10.2)

milking parlor. The section of a dairy operation where the cow is moved to in order for milking to occur. (6.1, 10.2)

mineral. An inorganic compound that is required by animals to sustain their life functions. (9.4)

minutes. A complete, written record of the proceedings of a formal meeting. (2.3)

mission statement. A summary of the aims and values of a group. (2.1)

miter box. A three-sided box made of plastic, metal, or wood that is used to guide the saw to make precise angled cuts. (7.3)

miter square. Hand tool used for measuring and marking corner miter joints. (7.3)

mitosis. A type of cell division that results in two daughter cells each having the same number and kind of chromosomes as the parent nucleus. (13.1)

mode. The number that occurs most often in a data set. (8.2)

modified monogastric. A type of digestive system found in *equines* (and other animals), with a stomach that has a single compartment and a special compartment in their intestine called the *cecum*, which helps them digest roughages. May also refer to an animal with this type of digestive system. (10.3)

modified monogastric digestive system. A digestive system that has a single compartment stomach and an enlarged *cecum* for microbial digestion. (9.4)

mohair. The *fleece* from Angora goats. (11.4)

moisture content. The percentage of moisture by weight found in a crop. (13.2)

molted. The forced shedding of poultry feathers to induce egg laying. (11.1)

monocot. A plant that has one seed leaf when it first emerges. Also referred to as a *monocotyledon*. (13.1)

monocotyledon. A plant that has one seed leaf when it first emerges. Also referred to as a *monocot*. (13.1)

monogastric. An animal (including humans) with a single-chambered stomach. (11.2)

monogastric digestive system. A digestive system that has a single compartment to the stomach and is unable to process cellulose. (9.4)

mother plant. A plant kept by producers from which cuttings are taken to produce new plants through asexual propagation. Also referred to as a *stock plant*. (13.6)

motion. A formal proposal by a member which, if agreed upon by the members of the assembly, will result in a certain action. (2.3)

mountain ecosystem. An ecosystem characterized by a diverse array of habitats that are influenced by dramatic changes in altitude. (14.1)

mule. The equine offspring of a *mare* bred to a *jack*. (10.3)

mutton. The meat from sheep older than one year. (11.3)

myotonic goats. A breed of meat goat that appears to faint when startled. (11.4)

N

nacelle. The fixture on a wind turbine that houses the operating components. (7.1)

nail hammer. See *claw hammer*. (7.3)

nail puller. A straight bar with a curved end and a V-shaped notch used to grip nail heads. (7.3)

nail set. A type of punch used to set the heads of nails below the surface of wood. (7.3)

nanny. A female goat of any age. Also referred to as a *doe*. (11.4)

National Agricultural Library (NAL). An organization that gathers and shares U.S. and international agricultural information. (9.3)

National Association of Conservation Districts (NACD). A local conservation organization that focuses on agricultural conservation of water and soil resources. (15.1)

National FFA Organization. The middle school and high school organization most closely tied to agriculture, food systems, and natural resources management. The FFA is an intracurricular program that works to make a positive difference in the lives of students by developing their potential for premier leadership, personal growth, and career success. (2.1)

National Fire Protection Association (NFPA) A global, nonprofit organization that develops electrical wiring codes. (7.2)

National Institute of Food and Agriculture (NIFA). An organization that promotes knowledge of agriculture and the environment through teaching, research, and extension at Land-Grant Universities (Colleges of Agriculture) in the United States. (9.3)

National Oceanic and Atmospheric Administration (NOAA). A government agency whose mission is to focus on the conditions of the oceans and atmosphere. NOAA warns of impending severe weather, monitors air and water quality, and conducts research to protect the environment. (15.1)

National Swine Registry (NSR). A swine association formed to consolidate many of the old purebred swine associations. (11.2)

native species. The species of plants and animals that are naturally occurring in an area. (14.4)

natural gas. An odorless, colorless, flammable gas consisting mainly of methane and other hydrocarbons. It occurs underground and is often found with coal or crude oil deposits, but may also be found in natural gas fields. (7.1)

natural resource. Material or substance that occurs in nature, such as minerals, wood, water, and wildlife. (1.1)

Natural Resources Conservation Service (NRCS). A government sector of the United States Department of Agriculture (USDA) that helps landowners reduce soil erosion, enhance water supplies, improve water quality, increase wildlife habitat, and reduce damages caused by floods and other natural disasters. The agency works with farmers, ranchers, and growers to preserve watershed water quality and to conserve water on agricultural lands. (15.1)

natural resources management. The processing, marketing, and distribution of natural resources to consumers. The agricultural industry area concerned with the conservation and use of cultivated and uncultivated lands in our country and around the world. (1.1, 3.2)

natural science. The field of science that deals with the study of the physical world. The process of genetic selection in which only the strongest organisms will live to reproduce; through this process, plant and animal species retain or develop traits that are more favorable for survival. (5.2)

natural selection. The process of genetic selection in which only the strongest organisms will live to reproduce. (5.3)

needle-nose pliers. Handheld tool with narrow jaws that allow the user to grasp and hold small objects. They are commonly used to bend and twist electrical wire. (7.3)

net pen. A mesh enclosure suspended in a body of water from a rigid structure on the water surface. (12.2)

net worth. The actual value of a business determined by total liabilities compared to the total assets. (8.1)

neuter. The castration of a male animal to prevent breeding and reproduction. (12.1)

niche market. A specific group of consumers likely to purchase a product sold by specific producers. (8.1)

nitrate. The plant-usable form of nitrogen. (14.2)

nitrification. A process through which nitrifying bacteria in the soil converts ammonia into *nitrates*. (14.2)

nitrogen cycle. A natural process through which nitrogen cycles through various chemical forms in the atmosphere, in plants, in animals, and in the soil. (14.2)

nitrogen dioxide (NO_2). A highly toxic gas with a strong bleach-type odor, low-lying yellow, red, or dark brown fumes; formed in silos during filling. (4.3)

node. A section of a plant stem where a leaf, flower, or main branch of the stem will emerge. (13.1)

nodule. A small bump on the roots of *legumes* which houses a special form of bacteria that converts nitrogen into a form the plant can use. (13.1)

nomadic tribe. Groups of people who traveled from place to place, hunting available food sources. (1.2)

noncash expense. When a business receives a gift or exchanges labor or products for the inventory the business needs. (8.1)

noncurrent asset. An item that has a longer useful life for a business. (8.1)

nongame species. Animals that are not hunted and/or are protected from hunting because they are threatened or endangered. (16.2)

nonpoint source pollution. Pollution whose source cannot be identified. (15.4)

nonrenewable energy. Energy that comes from natural sources that took millions of years to develop and cannot be replenished. (7.1)

nonrenewable natural resource. A resource that cannot be renewed in a relatively short period of time. (16.3)

nontidal marsh. An inland wetland commonly found along rivers, streams, and even lakes, ranging from a few inches to a few feet deep; level of water varies throughout the year. (15.3)

nontraditional animal industry. An industry that produces products from animals not generally considered livestock or companion animals. (12.3)

nonverbal communication. The exchange of knowledge through the senses. (2.2)

Norman Borlaug (1914–2009). The "father" of the Green Revolution whose scientific exploration of the genetics and production of wheat is credited with saving more than a billion people from starvation. (1.2)

no-till program. A method of planting where the farmer does not plow or disk the soil, but plants a new crop in the remnants of the previous year's crop. (15.2, 15.5)

nursery. Confined area in which weaned piglets are placed for approximately six weeks before being moved to a grower-finisher facility. A production facility in which the trees and shrubs used in landscape designs are grown. (11.2, 13.6)

nutrient. A substance that provides nourishment essential for growth and the maintenance of life. (9.4)

nutrition. The process by which an organism obtains food which is used to provide energy and sustain life. (9.4)

O

object line. A solid, black line that shows the outline of the object being drawn. (7.4)

Occupational Safety and Health Administration (OSHA). A division of the United States Department of Labor that enforces the laws regarding worker safety, and provides training and education that promotes safe working conditions. (4.1)

offal. The entrails and internal organs of animals processed for food. (10.1, 11.1)

"O" layer. The layer of soil found on top of the *topsoil*. This layer has the highest organic matter content of any soil layer. Also referred to as the *organic layer*. (15.2)

OLED lighting. A lightweight, cool, lighting system that uses organic light-emitting diodes intended for use in indoor crop growing facilities. (6.2)

omasum. The compartment of the ruminant digestive system that works to take feed that has already been broken down in the rumen and remove water as it passes into the final compartment. (9.4)

omnivore. An organism that eats both plants and animals. (14.4)

open end wrench. A nonadjustable wrench that is open at each end. Open end wrenches are usually double-ended with different sized openings. (7.3)

open-pit mining. Mining performed from the surface of Earth. Also referred to as *strip mining*. (7.1)

oral communication. The act of communicating using spoken words and language. (2.2)

orbital sander. A lightweight power tool that uses square pad sandpaper to finish sanding and removing old finish from smaller workpieces. (7.3)

order of business. The order in which items should be presented in a meeting. (2.3)

organic farming. A system of agricultural production that attempts to reduce or eliminate the use of synthetic agricultural inputs such as synthetic antibiotics, fertilizers, and pesticides. (6.2)

organic layer. The layer of soil found on top of the *topsoil*. This layer has the highest organic matter content of any soil layer. Also referred to as the *"O" layer*. (15.2)

organic matter. Material composed of decomposing and decomposed living organisms, including their waste products. Also referred to as *humus*. (15.2)

ornamental horticulture. The growing and marketing of plants used by humans for their aesthetic appeal. (13.6)

ornamental plant. A plant grown and cared for by humans to increase the beauty of their surroundings. (13.1)

orthographic drawing. A drawing that shows all surfaces (top, front, sides, bottom) of an object projected onto flat planes, and set at 90° angles to one another. It includes detailed dimensions essential to constructing the project. (7.4)

orthographic projection. When an object is drawn as a two-dimensional drawing. (7.4)

oscillating spindle sander. A stationary power sander that uses a spindle that rotates and oscillates (moves up and down) at the same time. Most models have interchangeable abrasive drums or spindles in varying diameters. (7.3)

outer bark. The outer covering of the trunk, branches, and twigs; dead phloem cells that protect the tree from insects, disease, and even fire. (17.1)

ovary. The site of fertilization in plants. (13.1)

ovine. The species name for sheep. (11.3)

ovule. A female sex cell of plants produced in the *ovary* portion of the *pistil.* (13.1)

oxygen-depleting pollution. Water pollution from microorganisms that take up oxygen and deplete the water of dissolved oxygen. (15.4)

P

palatability. How desirable an animal finds a foodstuff. (9.4)

palisade layer. A layer of very tightly organized cells underneath the upper *epidermal tissue* of a plant. (13.1)

panicle. The part of the plant made up of clusters of seeds. Also referred to as the *seedhead.* (13.2)

parasite. An organism that lives in or on another species and consumes it for food. (14.4)

parasitism. A feeding relationship in which one species lives in or on, and feeds on another species. (14.4)

parent material. The underlying soil material that gives rise to the *subsoil* and *topsoil.* This material may be made of sedimentary rock, other types of rock, or mud deposited by a river. Also referred to as the *"C" layer* of the soil horizon. (15.2)

parks. Natural areas, set aside as preserved land for public use. (16.1)

parliamentary procedure. A set of rules and regulations for properly conducting meetings. (2.3)

participative leadership. A type of leadership style in which the leader provides guidance while encouraging group members to share their ideas and opinions. There is less division between the leader and followers and a greater contribution of ideas and creativity from the group. Also referred to as *democratic leadership.* (2.1)

parturition. The act of giving birth. (10.2)

passive listening. The act of casually listening to the speaker; you may not *hear* everything that is said. (2.2)

pathogen. An infectious or biological agent that causes disease or illness to its host. (9.3)

paws. Chicken feet which are often used to make stock. (11.1)

peds. The aggregates formed by the binding of soil particles. (15.2)

peduncle. The *stem* which supports a flower. (13.1)

pegs. The flower on a peanut plant produces this type of stem that goes into the soil where the seedpod is formed. (13.3)

pelt. A tanned animal hide that has been processed with the hair on. (12.4)

percentage. The proportion of a whole, expressed as a part out of 100. (8.2)

perennial. A plant that grows for many growing seasons. (13.1)

perfect flower. A flower that contains all four of the basic flower parts (*petals, sepals, pistils,* and *stamen*). (13.1)

personal leadership. The ability of a person to embody the characteristics of a good leader and work to becoming a better leader. (2.1)

personal leadership plan. A strategy for how you will accomplish your goals. (2.1)

personal protective equipment (PPE). Equipment and clothing worn by a person to protect themselves from harm in potentially hazardous situations (includes eyewear, hearing protection, respiration protection, and clothing or footwear). (4.2)

petal. A part of a flower that serves the plant by protecting the reproductive structures and, for many plants, helps to attract insects or birds to pollinate the flower. (13.1)

petiole. The portion of a plant's structure that connects the *leaf* to the *stem.* (13.1)

petroleum-powered tool. Power tool that uses gasoline or diesel fuel to operate. They are commonly used in the turf and landscape industry where access to electricity is not practical. (7.3)

phantom line. A thin line used to indicate alternate positions of moving parts, repeated details (like threads), and motion. (7.4)

phenotype. The outward expression of the genes an organism has for a specific trait. (5.3)

Phillips-head screwdriver. A screwdriver with a cross-shape tip. (7.3)

phloem. *Vascular tissue* responsible for carrying nutrients produced in the plant down toward the roots. The thin layer of *phloem* cells between the *cambium* and *outer bark* that provides the food supply to the entire tree. Also referred to as *inner bark*. (13.1, 17.1)

photosynthesis. The process plants use to convert light energy to chemical energy and store it in the form of sugar. (13.1)

photovoltaic cell (PV cell). A specialized semiconductor *diode* that converts visible light into direct current (DC) electricity. (7.1)

physical science. The field of science related to the study of inanimate objects; includes areas like physics, chemistry, and Earth science. (5.2)

physics. The study of the properties of matter and energy and the forces and interactions they exert on one another. (5.2)

phytoremediation. The use of living plants to remove organic and inorganic contaminants from soil. (5.3)

picnic shoulder. A tough, primal pork cut which is frequently cured or smoked. Often used to make ground pork or sausage meat. (11.2)

pictorial drawing. A drawing designed to show a likeness of the object or structure as viewed by the human eye. (7.4)

pig. Term used to describe young swine. May also be referred to as *piglet*. (11.2)

piglet. Term used to describe young swine. May also be referred to as *pig*. (11.2)

pincers. A hand tool made of two pieces of metal with blunt concave jaws that are arranged like the blades of scissors. They are used for gripping and pulling things and sometimes for cutting thin metal or wire. (7.3)

pinch bar. A prying tool made of solid metal with a hooked end and a claw designed to open or break apart materials. Commonly referred to as a *crowbar*. (7.3)

pin punch. A hardened steel tool designed to drive pins out of fasteners. Pin punches are designed to be used with a hammer or wooden mallet. (7.3)

pipe wrench. An adjustable, self-tightening wrench used (especially in plumbing) to tighten and loosen threaded, round pipe. (7.3)

pistil. The female part of the flower made up of the *stigma*, *style*, and *ovary*. (13.1)

pistillate flower. A flower that has only female reproductive parts (*stigma*, *style*, *ovary*). (13.1)

placement SAE. A *Supervised Agricultural Experience* that requires the student to work for someone else in an agriculture-related job. The work may or may not be paid. (3.1)

planes. Woodworking tools used for trimming, squaring, and smoothing the edges of lumber. (7.3)

plant hardiness. A plant's ability to grow in cold or warm environments. (13.5)

plate. A tough, fatty *primal cut* of meat from the front belly of beef cattle, just below the *rib* cut. (10.1)

pliers. Hand tools designed to hold objects firmly when pressure is applied to the handles. The clamping or squeezing action brings the jaws together to hold an object. (7.3)

plugs. Small seedlings germinated and sold to bedding plant producers. (13.6)

plywood. An engineered lumber product that includes three or more layers of veneer glued together. (17.1)

pneumatic system. A power system that uses gas to transmit and control energy. (7.5)

pneumatic tool. A power tool that uses a *pneumatic system* to perform. (7.3)

point source emitter. The point where water is emitted on a drip irrigation system (typically customized per plant throughout the system). (15.5)

point source pollution. Pollution that comes from a single source that is easily identifiable. (15.4)

polar zone. The latitudinal range above 66.5°N and 66.5°S in which the average temperature is less than 50°F (10°C). Also referred to as the *arctic zone*. (14.3)

political science. The study of how public policy is regulated and developed. (5.2)

polled. Cattle that lack horns by natural or artificial means. (10.1)

pollen. The male sex cell in a plant. (13.1)

pollen tube. A path to a flower's *ovary* created by pollen grains. (13.1)

pollination. The process of fertilization in which *pollen* is transferred from the *anther* in a flower's *stamen* to the *style* of a flower's *pistil*. A fertilization process that occurs when pollen from the male portions of the flower is transferred to the female portions of the flower. (13.1, 13.5)

pomology. The branch of horticulture that studies the production of fruit and tree nut crops. (13.5)

pond. A body of standing water that is usually smaller than a lake. A pond may be a natural or constructed structure. Large, aboveground aquaculture tanks may also be referred to as ponds. (12.2)

pony. The classification typically given to horses under 14.2 hands tall, and weighing up to 900 lb. (10.3)

popcorn. A corn variety with small roundish seeds (kernels) that have a hard outer coating that seals in the corn's starch, oil, and water. When heated, the kernels pop into puffy popcorn. (13.2)

population dynamics. The study of the changes in size and ages of populations over both short and longer periods of time. (16.2)

porcine. The species name for swine. (11.2)

pore spaces. The spaces between soil particles that are filled with either water or air. The size and amount of these spaces are essential to healthy soil. (15.2)

pork. Meat from swine. (11.2)

pork butt. A tougher primal cut of pork meat from the shoulder of swine that contains muscles and sinew. It may also be referred to as *Boston butt* or *Boston shoulder*. (11.2)

position announcement. A full job description published by companies when they are looking to hire for a specific position. (3.2)

poult. A newly hatched turkey not fully covered with feathers. (11.1)

power. The rate at which work is performed or energy is applied. (7.5)

powered air-purifying respirator (PAPR). A respirator that uses a blower to force the air through air-purifying elements. It may use a mechanical filter, chemical cartridge, or a combination of both. They are not designed for use in oxygen-limited environments. (4.2)

power system. A system that harnesses, converts, transmits, and controls energy to perform work. (7.5)

power take off (PTO) shaft. A cylindrical metal rod that is attached to a power source (tractor) at one end and an attachment at the other. When the engine is running, power flows along the shaft and transfers energy from the engine to the attachment. (4.3)

prairie ecosystem. An ecosystem composed mainly of grasses with few trees and shrubs. Also referred to as a *grassland ecosystem*. (14.1)

precedence. The list of motions in the order that they must be handled. (2.3)

precision agriculture. The method of managing agricultural land with the assistance of computer or satellite information. A set of technologies that has helped agriculture advance into the digital information-based world. (1.2, 6.2)

predator. An animal that eats another animal. (14.4)

prepared public speaking. When the speaker has had ample time to prepare the content and delivery of the information. (2.2)

prescribed burn. The process of intentionally burning small areas of forest to remove excess buildup of material that could feed a wildfire if unchecked. (17.2)

presentation. A type of oral communication where the audience has the opportunity to ask for clarification after the information has been presented. The specific set of skills required to effectively transfer information through oral communication. (2.2)

preservation. The protection of natural resources in which the resources are to be protected but not used. The land will always be maintained in its natural state. It cannot be used for farming, forestry, or mining. (15.1, 16.1)

preserved land. Land that will always be maintained in its natural state. It cannot be used for farming, forestry, or mining. Hunting is generally prohibited, as well. (16.1)

prey. An animal that is the food animal for predators. (14.4)

primal cut. One of the large primary pieces (chuck, rib, loin, round, brisket, plate, flank) into which a beef carcass is divided. (10.1)

primary consumer. Species that feed only on plants. (14.4)

primary product. The main purpose for which an animal is produced. (12.4)

primary root. The main portion of a plant's root system. (13.1)

private land. Land that is owned by an individual or a corporation. (16.1)

privileged motion. A motion with the highest precedence of all motions, that deals with the rights or needs of the organization. Privileged motions include those concerned with starting, stopping, pausing, or focusing the purpose of the meeting. (2.3)

processing. The conversion of a raw product into a product ready for human consumption. (9.1)

producer. Someone actively involved in the production of raw animal and plant goods for human use. A plant that produces its own food. Also referred to as an *autotroph*. (1.1, 14.4)

production. The act of growing a crop or animal for food. (9.1)

production agriculture. The agricultural industry area directly involved in the management and production of agricultural commodities for sale to the consumer. (3.2)

proficiency awards. Awards given by the National FFA Organization on the local, regional, state, and national levels for student SAE projects that have shown growth and development of skills related to success in the SAE area. (3.1)

profit. The amount of money that a business earns over its expenses. (8.1)

protein. The nutrient responsible for building and repairing cells in an animal's body. (9.4)

proventriculus. A small area in an *avian digestive system* where digestive enzymes for chemical breakdown are added to the food. (9.4, 11.1)

pry bar. A solid metal tool with a hooked end and a claw designed to open or break apart materials. Commonly referred to as a *crowbar*. (7.3)

pseudo-cereals. Crops with the same properties as cereal grains, but that do not produce a grain-type fruit. (13.2)

public land. Land that is owned by the government. (16.1)

pullet. A female chicken under one year old. (11.1)

pulpwood. The smaller coniferous trees grown to produce paper and paper products, like cardboard. (17.1)

punches. Hardened steel tools designed to be struck by a hammer or wooden mallet for the purpose of driving pins out of fasteners. (7.3)

Punnett Square. A diagram used to predict the outcome of a crossbreeding or breeding experiment. The diagram is used to determine the probability of an offspring having a particular *genotype*. Named for Reginald C. Punnett. (5.3)

purebred. A cultivated variety of animal species achieved through the process of selective breeding. Breed associations keep detailed records of the lineage of registered animals to establish the animals' pedigrees. (10.1)

purebred breeder. A rancher or cattleman who mainly raises purebred cattle for breed associations. May also be referred to as a *seedstock cattle producer*. (10.1)

purebred producers. Ranchers or sheep producers who mainly raise purebred replacement breeding sheep to market to other purebred and commercial sheep producers. (11.3)

purebred swine. Swine with documented pedigree that is usually registered with breed clubs and national registries. (11.2)

push stick. A narrow strip of wood with a notch in one end. Push sticks are used to push short or narrow pieces of material through table saws. (7.3)

Q

quorum. The number of members that must be in attendance at a meeting for business to be conducted. It is typically set at one more than half of the total members in the group. (2.3)

R

raceway. Straight-sided channels typically made of poured concrete or concrete blocks used to house aquatic life. (12.2)

rack. The rib primal cut of lamb/mutton. (11.3)

radial arm saw. A power tool consisting of a circular saw mounted on a sliding horizontal arm. It is an excellent tool for fast and convenient crosscutting of lumber. (7.3)

radio frequency identification (RFID). Small electronic tag that is attached to livestock in some way and used to identify and track animal movement. (6.2)

rafter. A group of turkeys. It may also be called a *gang*. (11.1)

ram. A mature male sheep. (11.3)

ram lamb. A young male sheep. (11.3)

random orbital sander. A lightweight power tool that uses two simultaneous motions. The circular pad rotates in small circles while simultaneously moving in a random elliptical loop; may be used for both finish sanding and slower stock removal. (7.3)

range production. Typically large, commercial (non-registered) flocks of sheep that have more than 1,000 head of ewes. (11.3)

ratio. The comparative value of two or more quantities. (8.2)

ration. The amount of feed that an animal consumes in a day. (9.4)

ray flower. Small, narrow flowers that usually have no *stamens*, and one long *petal*. (13.1)

receptacle. The wide bottom portion of a flower. (13.1)

recessive genes. Genes that are only expressed when no dominant gene is found in the genetic makeup. (5.3)

recipient. A female animal who carries the embryo collected from the donor animal. (5.3)

reciprocating saw. A handheld power tool that uses a thin blade that moves back and forth to make a cut. Reciprocating saws are useful for making curved cuts and may be used to cut wood or metal, depending on the blade type. (7.3)

reduced tillage. A planting method in which the farmer minimally tills the soil, often with a machine that minimizes the disturbance of the soil surface. (15.2)

reduced tillage program. A planting method in which the farmer minimally tills the soil, often with a machine that minimizes the disturbance of the soil surface. Farmers may leave some or all of a previous crop's residue on the surface of the soil. (15.5)

rendering. The taking of *by-products* and converting them into useful goods. (12.4)

renewable energy. Energy that comes from sources continually replenished by nature. (7.1)

renewable natural resources. The resources that may be regenerated in a short period of time so that they do not run out. (16.3)

research SAE. A *Supervised Agricultural Experience* designed for students to use the scientific method to analyze a research question or test a hypothesis. Research SAEs may be experimental projects (comparing plant growth using different growing methods), as well as nonexperimental ones (developing a marketing plan or ad campaign for an agriculture commodity). (3.1)

résumé. A brief overview of your education, qualifications, and skills. (3.2)

reticulum. A chamber of the ruminant digestive system that serves as an area to sort feed coming from the mouth and remove any foreign objects from the feed prior to passage into the rumen. Also referred to as the *honeycomb*. (9.4)

return on investment (ROI). A mathematical calculation that allows the owners to see how much money they are making compared to the money that they are spending. (8.1)

rhizome. A modified plant *stem* that grows underground horizontally and sends plant shoots upward. (13.1)

rib. A tender primal cut made from the center section of rib (specifically the sixth through twelfth ribs). (10.1)

rill erosion. Soil erosion that occurs when many small streams of water form and are running across the surface of the soil. (15.2)

riparian zone. A wooded or grass strip planted along a stream or river bank to slow the runoff from agricultural cropland. A strip of land that buffers between the cropland and a body of water. (14.1, 15.5)

ripping hammer. A slightly heavier hammer with a straighter claw for use as a prying tool as well as a nail remover. Typically used in construction for framing. May also be referred to as a *framing hammer*. (7.3)

rip saw. A handheld saw designed to make cuts parallel to the direction of the wood grain. (7.3)

rodeo. A sporting event comprised of events based on ranch skills that cowboys and horses needed to have in order to work cattle. (10.3)

rollover protection system (ROPS). The operator compartment structures (usually cabs, frames, or bars) intended to protect operators from injuries caused by vehicle overturns or rollovers. May also be referred to as a rollover protection structure. (4.3)

rooster. An adult male chicken over a year old (may also be called a cock). (11.1)

root. The part of a plant that functions as a means of water absorption, food storage, and/or as a means of anchorage and support. (13.1)

root cap. A hardened portion of a plant's root tip (*meristem*) designed to protect the tip from harm. (13.1)

root hair. Small outshoots on plant roots that increase the area of the roots touching the soil, increasing the amount of water and nutrients that can be brought into the plant. (13.1)

rootstock. A root and growing stem that is cut off just a few inches aboveground. Also referred to as a *scion*. (13.5)

root vegetable. A plant grown for its edible root. (13.5)

roughage. A feed that is low in energy and *total digestible nutrients (TDN)* and high in *fiber* content. (9.4)

roughstock. The bucking animals that cowboys attempt to ride and are scored on their ability during rodeos. (10.3)

round. A primal cut of meat from the hindquarters of beef cattle. (10.1)

router. A handheld power tool typically used in cabinetmaking and millwork. A rotating router bit uses blades to shape the surfaces and edges of wood stock. (7.3)

rubber mallet. A special type of driving tool used for hammering on a finished metal surface without making marks and dents. (7.3)

rumen. The largest section of the ruminant stomach, which contains microbes that process feed and help break down fiber. (9.4)

ruminant. An animal whose digestive system has four compartments to its stomach. A mammal that chews cud regurgitated from its rumen. These mammals are able to acquire nutrients from plant-based food by fermenting it in a specialized stomach prior to digestion, principally through bacterial actions. (9.4, 10.1)

ruminant digestive system. An animal digestive system with four compartments to the stomach. (9.4)

runner. A plant *stem* that grows above the ground horizontally and connects plants or roots in a new place. Also referred to as a *stolon*. (13.1)

runoff water. The water that pools on the surface of the soil and then flows into nearby streams, rivers, and lakes. It is often contaminated with chemical fertilizers, manure, and chemicals. (14.2)

rust. A plant disease caused by fungus. (13.3)

S

SAE improvement project. An enterprise undertaken to make changes and improve an existing SAE. (3.1)

safety data sheet (SDS). A document that contains information on the potential hazards (fire, health, reactivity, and environmental) of a material, and instructions for how to work safely with the product. It also contains essential information for emergency crews in case of an emergency. Also referred to as a *material safety data sheet (MSDS)*. (4.2)

safety glasses. Impact-resistant glasses designed to protect eyes when working with hand and power tools or when working around harmful liquids. Safety glasses have side shields to provide all-around protection. (7.3)

saline. The condition of water containing high concentrations of salt. (14.1)

salinity. The dissolved salt content of the water. (12.2)

salinization. The buildup of salts in soil that occurs with the frequent use of irrigation. (15.2)

salt marsh. The areas of land and water between the high and low tides of the ocean. (14.1)

salvage value. The amount a depreciable asset is worth when all of the usefulness to the business is gone. (8.1)

sand. The largest soil particle. It is inert and does not affect soil fertility, but is important for drainage and aeration. (15.2)

sander. A handheld power tool that uses an abrasive material to smooth surfaces or remove surface material. (7.3)

sap. The sugary watery substance in trees. (17.1)

sap flow. The term used to describe the translocation of sugar from roots to shoots of the tree. (17.1)

sapwood. The *xylem* of the tree trunk which carries water and nutrients from the roots to the shoots of the tree. Xylem cells are the lighter-colored wood in the trunk. (17.1)

saturated. When all pore spaces of soil are filled with water. (15.2)

sausage. A meat product commonly made from ground pork. (11.2)

sawlogs. Logs that are sawn into lumber used for building houses, furniture, or other wood products. (17.1)

scale. A specific ratio relative to the actual size of a place or object. It is used in *formal drawings*. (7.4)

scavenger. An organism that feeds on dead plants and animals; may be an herbivore or carnivore. (14.4)

scientific method. A means of research that uses systematic rules and procedures to investigate a problem. (5.1)

scion. A root and growing stem that is cut off just a few inches aboveground. Also referred to as a *rootstock*. (13.5)

scours. A digestive disorder characterized by extreme diarrhea in young animals. (9.4)

screwdriver. A driving tool that uses torque to drive or remove fasteners. (7.3)

seasonal-use production. Seasonal-use systems make use of farm-produced roughage such as winter pasture or crop residue to serve as a source of nutrition for bred or open ewes. (11.3)

second. An indication that there is at least one person besides the mover that is interested in seeing a motion come before a meeting. (2.3)

secondary consumer. Species that feed on primary consumers. (14.4)

sectional drawing. A drawing created to show how an object would look if a cut were made through the object. (7.4)

section line. A parallel, inclined line used to indicate the cut surface of an object (in sectional view). (7.4)

sediment. Pieces of material from fossil fuels, freshwater, and the nonliving parts of the soil that accumulate and form layers on Earth's surface. (14.2)

sediment pollution. Water pollution caused by soil runoff in the form of suspended soil particles from fields. (15.4)

seechi disk. A black-and-white circular instrument used to measure the *turbidity* of water. (15.4)

seedhead. The part of the plant made up of clusters of seeds. Also referred to as the *panicle*. (13.2)

seedstock. Bulls, heifers, and cows that are used in registered breeding herds as well as commercial cattle operations. Boars, gilts, and sows used in breeding operations. (10.1, 11.2)

seedstock cattle producer. A rancher or cattleman who mainly raises purebred cattle for breed associations. May also be referred to as a *purebred breeder*. (10.1)

seed treatment. Pesticides that are applied to the seed to provide pest control to the growing seedling. (13.5)

seed trees. Trees left standing from a seed tree harvest to provide seeds for new trees. (17.2)

seed tree harvest. When a few desirable trees of each species are left to grow on each acre of land. (17.2)

selective cutting. The removal of only certain trees in a stand. (17.2)

self-contained breathing apparatus (SCBA). A respirator for which the breathing air source is carried by the user (i.e., a tank). (4.2)

semi-intensive housing system. A poultry production system in which the birds are allowed some access to outdoor areas for fresh air and foraging. Free space in this system is limited, but outdoor runs may be attached to the poultry house. (11.1)

sepal. A small leaf at the base of a flower that helps protect the flower when it is in bud form. (13.1)

septic system. A wastewater system used for a home or business that does not have access to a municipal sewer system. (15.3)

servant leadership. A leadership method in which the leader approaches their role as a way to serve those who follow them. (2.1)

sexed semen. Semen which has been separated to have a higher concentration of either X or Y chromosome sperm cells. It is used in applications where offspring of a particular gender are desired. (5.3)

sexually dimorphic. When the male and female of a species are different in color and size. (11.1)

sexual reproduction. A method of plant reproduction involving the creation of sex cells through *meiosis*. (13.1)

sharecropping. A method of agricultural labor in which the landowner provided the use of arable land to a farmer in exchange for a share of the crop produced. (1.2)

shelf-stable. A means of packaging food in which all pathogens are removed, allowing the packaged food to remain unrefrigerated until it is opened. (9.3)

shelterwood harvesting. The process of removing mature trees periodically over 10 to 15 years. (17.2)

shingler hammer. A special type of driving tool used for splitting wood shakes/shingles. The striking face is designed to set roofing nails, and it has a blade instead of a claw used for cutting roofing materials. (7.3)

short circuit. The interruption of the proper flow of electricity along its intended path in an electrical system. (4.2)

short-term goal. An aim or desired result that will be accomplished in the immediate future. (3.1)

shoulder. A tender primal cut of lamb or mutton that is usually deboned and cooked as a roast. (11.3)

shutoff switch. An emergency device designed to immediately turn off all power to the machines in the shop. (4.2)

side-cutting pliers. A special type of pliers that has a set of jaws for cutting wire instead of jaws that grip and hold material. (7.3)

silage. Grass, fermented feed, or other fodder that is harvested green, stored in a silo, and used to feed cattle. (6.1)

silo. A tall metal or concrete storage container that holds *silage*. A deep round aquaculture tank that is usually belowground. (6.1, 12.2)

silo gas. Toxic gas formed by the natural fermentation of the silage shortly after it is placed in the silo. (4.3)

silt. The fine, relatively inert soil particles carried by running water and deposited as a sediment. (15.2)

silviculture. The practice of growing trees for production. Also referred to as *forestry*. (17.1)

site plan. A drawing of the land where a project will be constructed. These plans show property boundaries and the location of buildings on the site. (7.4)

six-row barley. Cereal grain with six rows of seeds along the main shaft of the seedhead. (13.2)

sketch. A rough drawing that captures the basic likeness of a project. It is used to create a detailed and accurate drawing. (7.4)

sledge hammer. A long-handled hammer with a steel head weighing 8 to 20 lb. Used for demolition and other heavy construction tasks. It is designed to be swung with both hands to provide a forceful impact. (7.3)

sliding T-bevel. A measuring tool used to lay out angles. (7.4)

slip-joint pliers. An adjustable hand tool with a slot in one jaw through which the other jaw slides. (7.3)

slotted-head screwdriver. A screwdriver with a flat, wedge-shaped tip that fits into the slot in the head of a screw. May also be referred to as a *flat-head screwdriver*. (7.3)

SMART goals. Personal goals that are specific, measurable, attainable, realistic, and timely. (2.1)

smut. A fungal disease of grains in which parts of the ear change to black powder. (13.2)

social media. Communication through publically available websites and the use of applications that enable users to share content and engage in social networking. (2.2)

social science. The study of human society and social relationships. (5.2)

Society of Automotive Engineers (SAE). A globally active professional association and standards organization for engineering professionals in various industries. Now referred to more commonly as SAE International. (7.3)

socket wrench. A wrench that uses a ratcheting mechanism to turn a socket, which turns a nut or bolt. The ratcheting mechanism eliminates the need to remove the socket for repositioning after each turn. (7.3)

sod. Turf grass that has been cut in uniform pieces from the earth with the roots intact to be transplanted to create an "instant" lawn. (13.6)

soft-face hammer. A driving tool with a face made of plastic or rubber that is used to drive a nail without marring the material's surface. (7.3)

softwood species. Trees whose leaves are needles or scales that remain green throughout the year. They are also called *coniferous trees* or *evergreen trees*. (17.1)

Soil and Water Conservation District (SWCD). A state association that works to provide advice and education about soil and water conservation to landowners and the general public. (15.1)

soil compaction. The compression or elimination of pore spaces in soil due to frequent heavy pressures, as with the heavy farm machinery. (15.2)

soil conservation practice. A method of farming that prevents the erosion of soil due to wind or water. (15.1)

soil erosion. The wearing away of soil by wind or water. (15.2)

soil particles. The minerals (sand, silt, and clay) that make up the largest portion (40%–50%) of most soils. (15.2)

soil profile. A view of the different soil layers made possible with a vertical slice down into the ground. (15.2)

soil structure. The binding together of soil particles into aggregates called *peds*. (15.2)

soil texture. The amount of sand, silt, and clay present in a given soil. (15.2)

solar energy. Radiant energy emitted by the sun. (7.1)

solar power. Power obtained by harnessing the radiant energy emitted by the sun. (7.1)

sole. The bottom of a horse's hoof. (10.3)

solubility. The amount of a substance that can be dissolved in water. (15.4)

somatic cell. A white blood cell. (10.2)

sour cream. Cream that has been fermented with certain kinds of bacteria. The bacterial culture sours and thickens the cream. (10.2)

sow. A mature female swine. (11.2)

spareribs. A common cut of pork cut from below the loin and above the stomach of swine. (11.2)

spawn. The release or deposit of eggs by a fish, frog, mollusk, crustacean, or other such organism. (12.2)

spay. The removal of the ovaries and/or uterus (ovariohysterectomy) of a female animal to prevent breeding and reproduction. (12.1)

speech. A type of formal, oral communication that typically involves one-way transfer of information. (2.2)

speed square. A triangular-shaped measuring tool that is technically a try square. It is used for drawing 90° or 45° angles and for laying out rafters and stairs. May also be referred to as a *measuring square*. (7.3)

spile. A small metal spout placed in trees to drain the sap. (17.1)

split breast. A poultry *breast quarter* with the wing removed. It may or may not come with the back portion. (11.1)

spoilage. The amount of a food product that cannot be consumed because of a decrease in quality or safety during storage. (9.2)

spongy mesophyll. A layer of loosely arranged cells beneath the *palisade layer* in plants that allow gases to pass through for use in *photosynthesis*. Also referred to as the *mesophyll*. (13.1)

spotter. A person who helps guide the driver of farm equipment and work vehicles to prevent accidents. (4.3)

square stage. A growing stage in cotton plants in which three bracts form a triangle to protect the growing flower bud. (13.4)

stalk vegetable. A vegetable grown for its edible stalk. (13.5)

stallion. A mature male horse. (10.3)

stamen. The male portion of a flower which is comprised of an *anther*, where pollen is produced, and a *filament*, the stalk that holds the anther. (13.1)

staminate flower. A flower that has only male reproductive parts (*anther*, *filament*). (13.1)

standard deviation. The average distance from the center point. (8.2)

staple crop. Any one of the plants which make up the majority of the calories consumed by people in a region. (9.2)

statistics. The formal science related to collecting and analyzing large quantities of numerical data, to relate findings to unknown data. (5.2)

steepwater. The liquid in which the *feedstock* was soaked before being processed into *ethanol* using the *wet mill process*. (7.2)

steer. A castrated beef/dairy male bovine. (10.1)

stem. The main trunk of a plant that gives the plant structure and shape. (13.1)

STEM. An acronym standing for science, technology, engineering, and mathematics. The acronym describes the areas of education that are vital to the development of students and new technologies. (1.3)

sterile. When an animal cannot produce offspring (as is the case with *mules*). (10.3)

stewardship. The ethic that people care for and conserve natural resources. (16.3)

stigma. A part of a flower's *pistil* (female part) that collects *pollen*. (13.1)

stillage. The solids remaining once the liquid *ethanol* is separated from the solid remnants of the *feedstock*. (7.2)

stocker operation. A beef cattle production operation in which young calves are allowed to graze pastures for moderate weight gain, before moving to *feedlots*. Also referred to as a *backgrounding operation*. (10.1)

stockhorse. Horses that have been selected for ranch working traits such as a specific level of athleticism, speed, and instinctive ability to work with cattle. (10.3)

stocking rate. The number of animals on a given amount of land over a certain period of time. It is generally expressed as animal units per unit of land area; the number of fish per the amount of space in an outdoor system, or the number of gallons of water per fish in an indoor system. (11.3, 12.2)

stock plant. A plant kept by producers from which cuttings are taken to produce new plants through asexual propagation. Also referred to as *mother plant*. (13.6)

stolon. A plant stem that grows above the ground horizontally and connects plants or roots in a new place. Also referred to as a *runner*. (13.1)

stomata. Openings in the lower epidermis of a leaf that allow or prevent gases and water from leaving the leaf. (13.1)

strap wrench. A self-tightening wrench with a strap or chain that pulls tight around cylindrical objects. The friction created by the strap keeps the object from slipping. (7.3)

strip mining. Mining performed from the surface of Earth. Also referred to as *open-pit mining*. (7.1)

stud. The place a stallion is kept for breeding. A male llama or alpaca. (10.3, 12.3)

style. A part of a flower's *pistil* (female part) that allows *pollen* to travel to the *ovary*. (13.1)

subprimal cut. A secondary cut made from a *primal cut* of beef cattle. (10.1)

subsidiary motion. A motion that deals with managing other motions. This type of motion works to set aside a motion, modifies the amount of debate, or changes the motion in some way. (2.3)

subsoil. The layer of soil right below the *topsoil*. It is usually lighter in color, contains less organic matter, and is more responsible for drainage and aeration than the *topsoil*. Also referred to as the *"B" layer*. (15.2)

sugar bush. A collection of maple trees that are tapped for collecting sap. (17.1)

Supervised Agricultural Experience (SAE). The independent study of agricultural subjects, supervised by a responsible and knowledgeable adult that may be outside of regular classroom instruction and is a manifestation of the personal career interests and needs of the student. (3.1)

Supplied-Air Respirator (SAR). A respirator for which the source of breathing air is not designed to be carried by the user. An SAR will have a hose mask with a blower and an emergency air supply. SARs are for use in oxygen-deficient areas such as manure pits, silos containing silo gas, airtight silos, or bins containing high moisture grain. (4.2)

supply. The amount of a specific product that is available at a particular price point. (9.2)

surface water. The water found in streams, rivers, and lakes. (14.2)

sustainability. The concept that those in the agricultural fields need to use our natural resources wisely and practice sustainable agriculture. The idea that we use Earth's resources in a way that they are not used up, but are available for future generations. (1.3, 16.3)

sustainable. When something, especially resources, is able to be used without being used up or permanently damaged. (1.3)

sustainable agriculture. The process of producing agricultural products using techniques that protect the environment and all living beings, while allowing agricultural land to maintain production for many years. (1.3)

sustainable energy. Energy that can be used to meet current needs without compromising the ability to meet future energy needs. (1.3)

swamp. A wetland dominated by woody plants, including trees. It receives its water supply through streams or rivers. (15.3)

sweet corn. A variety of corn with a high sugar content. Most commonly used for human consumption. (13.2)

swine. Pigs and hogs at various stages of growth and development. (11.2)

symbiotic relationship. A mutually beneficial relationship between two organisms. (13.1)

syngas. A mixture of hydrogen and carbon monoxide produced from biomass. *Syngas* is an abbreviation for synthesis gas. (7.1)

T

table saw. A stationary power tool that includes a work "table" that houses a circular blade and provides a platform for resting the workpiece. (7.3)

tack hammer. A hammer with a small face designed for driving tacks in upholstery work. This type of hammer should never be used for masonry, nails, cold chisels, or other metals. May also be referred to as an *upholsterer's hammer*. (7.3)

taiga ecosystem. A forest ecosystem characterized by plant life that is mainly evergreen conifers, temperatures that fall below freezing many months out of the year, and fauna that is limited to migratory birds and insects. (14.1)

tailwater. The runoff water from irrigation systems. (15.5)

tanning. A chemical or mechanical process whereby the hide of an animal is preserved. (12.4)

tape measure. A measuring tool that consists of a steel tape housed in a handheld case. The flexible steel tape extends and retracts from a handheld case. (7.3)

taproot. A plant root system with one large primary root, and few (if any) lateral roots growing from it. (13.1)

tariff. A tax charged by an importing country on imported goods. (9.2)

team. A group of people who come together to achieve a common goal. (2.1)

teats. The nipples of the mammary gland from which the milk is extracted. (10.2)

technology. The application of scientific knowledge, tools, or processes for practical purposes. (1.3)

telematics. The sending and receiving of data using wireless telecommunications. This is typically used in the transportation industry to track vehicles, and is used in agriculture for the same purpose. (6.2)

telemetry. An automated communication system that collects data from a remote location and transmits it to receiving equipment in another location. (6.2)

temperate deciduous forest. A forest ecosystem where sufficient rainfall supports the growth of trees. The trees and other flora shed their leaves annually due to large temperature differences. (14.1)

temperate evergreen forest. A forest ecosystem in which the trees do not seasonally shed their leaves. The climate is cool and dry, and many trees have adapted leaves in the form of needles. (14.1)

temperate zone. The latitudinal range between the latitude of 25.5° and 66.5°, both north and south, in which weather patterns produce four distinct seasons. (14.3)

tenderloin. A poultry cut of white meat from a small muscle that sits adjacent to the breast. (11.1)

terminal breed. Swine breeds known for muscling, carcass quality, and cutability. (11.2)

terrace. Soil-formed structure that breaks a steep slope into a series of steps. (15.5)

terrain. The general *topography* of the land; mountains, swamps, lowland, and highlands all make up an area's terrain. (14.3)

terrestrial ecosystem. An ecosystem found on land, like prairies, woodlands, and deserts. (14.1)

tertiary consumer. Species that feed on secondary consumers. (14.4)

theoretical computer science. The science behind creating computer programs. (5.2)

thermal pollution. Water pollution that elevates the temperature of the water. (15.4)

thigh. A poultry cut of dark meat that tends to be firmer than the breast meat and may contain more fat. The thigh is the portion of the chicken leg cut above the joint of the knee. (11.1)

threatened species. Animals that are low in total population and likely to become endangered in the near future. (16.2)

tidal marsh. A wetland found near the coastline of an ocean, gulf, or sea that is subjected to the ebb of the tide. (15.3)

tillage system. A method of turning the soil in crop fields. (15.2)

tillers. The extra stems rising out of the root system of a rice plant at ground level. (13.2)

tillering. The production of *tillers* in rice plants. (13.2)

tilth. The general condition of the soil with respect to texture, moisture content, aeration, and drainage. (15.2)

timber cruising. The process of walking the forest to provide needed management information for timber stand improvement. (17.2)

timber harvesting. The cutting of trees from a forest and making them available for sale as sawlogs, veneer logs, or pulpwood. (17.2)

timber stand improvement. The effective weeding of a forest of diseased or unwanted trees. Also referred to as *timber thinning*. (17.2)

timber thinning. The effective weeding of a forest of diseased or unwanted trees. Also referred to as *timber stand improvement*. (17.2)

tom. A male turkey. (11.1)

topline. The muscles going over the haunches, back, and neck of a horse. (10.3)

topography. The general *terrain* of the land; mountains, swamps, lowland, and highlands all make up an area's topography. Also referred to as *terrain*. (14.3)

topsoil. The top layer of soil. It is usually the most fertile and darkest in color. Also referred to as the *"A" layer* of the soil horizon. (15.2)

torque. A force that produces or tends to produce rotation. (7.4)

total digestible nutrients (TDN). The sum of the digestible fiber, protein, lipid, and carbohydrates in a feed. (9.4)

total suspended solids. The measure of weight of suspended organic material and minerals in water. (15.4)

traceability. The process in which pesticide application records are kept and transferred to each entity handling produce, enabling each handler to certify the pesticide records. (13.5)

trade surplus. The state in which a country exports more than they import. (9.2)

transesterification. A chemical reaction process used to convert the fat or vegetable oil to *biodiesel*. The chemical reaction between the fats or oils is created with an alcohol such as *ethanol*, along with a catalyst such as lye. (7.2)

transgenic animal. An animal whose genome has been changed to carry genes from other species. (5.3)

transgenic organism. A plant or animal that has been created using genetic engineering. Also referred to as a *genetically modified organism (GMO)*. (5.3)

transpiration. The loss of water through a leaf's surface. (14.1)

treatment. One of the different levels of *independent variables* in a scientific experiment. (5.1)

tree volume. The calculation of the size of individual trees from a saleable lumber perspective. (17.2)

trend. The general direction that something is developing or changing. (1.3)

trial. Each repetition of an experiment, or each of the organisms to which the treatments are applied. (5.1)

tribe. A group of goats. May also be referred to as a *flock*, *herd*, or *trip*. (11.4)

trickle irrigation system. A watering system in which water drips slowly to the roots of plants through a network of tubing, valves, pipes, and emitters. Also referred to as *drip* or *micro irrigation*. (15.5)

trip. A group of goats. May also be referred to as a *flock*, *herd*, or *tribe*. (11.4)

tropical deciduous forest. A forest ecosystem that receives abundant rainfall and includes dense flora. The trees shed their leaves annually, allowing more sunlight to reach the ground and enabling the increased growth of grasses and shrubs on the forest floor. (14.1)

tropical evergreen forest. A forest ecosystem that receives abundant rainfall, includes dense tree vegetation at various heights, and is found in the tropics near the equator. The trees do not shed their leaves. Also referred to as a *jungle*. (14.1)

tropical zone. The latitudinal range between the latitude of 23.5°N and 23.5°S, in which the atmosphere is consistently heated and temperatures are predominantly warm. Also referred to as the *tropics*. (14.3)

tropics. The zone, or latitudinal range, between the latitude of 23.5°N and 23.5°S, in which the atmosphere is consistently heated and temperatures are predominantly warm. (14.3)

try square. An L-shaped tool used as a guide for marking 90° angles and checking the edges and ends of boards for squareness. It may also be used to determine whether a board is the same thickness for its entire length. (7.3)

T-square. A T-shaped measuring tool used primarily to draw horizontal lines. (7.3)

tuber. The thickened part of an underground stem bearing buds from which new plant shoots arise. (13.1)

tuber vegetable. A vegetable grown for its edible underground stem. (13.5)

turbidity. The measure of light penetration and scatter in water. (15.4)

turf. The upper layer of grass including the root structure. (13.6)

two-cycle engine. A smaller engine that requires blended fuel. Commonly used on smaller landscaping tools such as string trimmers and chainsaws. (7.3)

two-row barley. Cereal grain with two rows of seeds opposite from each other on the seedhead. (13.2)

U

udder. The baglike mammary gland of cattle (and other livestock) that has two or more *teats* hanging near the hind legs. (10.2)

uneven solar heating. The natural phenomena in which parts of Earth receive more sunlight and heat than others. (14.3)

United States Forest Service (USFS). An agency of the United States Department of Agriculture (USDA) that administers the nation's national forests and national grasslands. (15.1)

upholsterer's hammer. A hammer with a small face designed for driving tacks in upholstery work. This type of hammer should never be used for masonry, nails, cold chisels, or other metals. May also be referred to as a *tack hammer*. (7.3)

uranium. A naturally occurring radioactive metal used as fuel in nuclear reactors. (7.1)

urbanization. When the areas surrounding an urban area take on more of the urban area's attributes and lose more of their rural attributes. (9.2)

V

value-added. Fresh crop products that have been prepared and packaged for easy consumer use (cut, peeled, sliced or diced, and packaged in convenient quantities). (13.5)

variable. Something that will change over the course of the experiment. (5.1)

variety meat. Meat that comes from a nonskeletal muscle. (12.4)

vascular bundle. The transport system of plants composed of *vascular tissues* like the *xylem* and *phloem*. (13.1)

vascular tissue. Plant tissue that transports materials around the plant through a series of tubes. (13.1)

vector. Any organism that transmits disease, viruses, or fungi from one host organism to another. (13.5)

vein. The *vascular bundles* contained in plant leaves. (13.1)

velvet. The vascular covering of deer or elk antlers. (12.3)

veneer. A thin slice of decorative wood that is glued over the top of a foundation of other less decorative wood. (17.1)

veneer logs. *Hardwood species* trees, like walnut, cherry, and oak, that are grown to provide thin layers of decorative wood that is glued over a base wood. (17.1)

venison. The meat from deer. (12.3)

verbal communication. The transfer of information using words and language. (2.2)

vertical farming. An emerging technology in urban areas in which buildings taller than one story may be used to cultivate raw agricultural products. (6.2)

vertical integration. A method of poultry production in which every step in the production cycle, from birth to harvested poultry products, is managed by a poultry company. When a major company controls the production process from birth to a finished pork product. (11.1, 11.2)

viscosity. The property of resistance to flow in a fluid or semifluid. (7.2)

vitamin. An organic compound that is essential for life. (9.4)

volume. The amount of cubic area that a three-dimensional space contains. (8.2)

W

warmblood. Horses that are a combination of larger draft horses, which are considered *coldbloods*, and more spirited light breeds like the Arabian and Thoroughbred, that are considered *hotbloods*. (10.3)

waste rock. The material removed from the surface of a mining pit and materials from the pit that are not used. (7.1)

water and sediment control basins (WASCoBs). Relatively low earthen embankments built perpendicularly across a watercourse. They are designed to stop the flow of water across crop fields, trap sediment, and channel water to underground drainage systems. (15.5)

water cycle. The pathway or cycle that water follows underneath the surface, at the surface, and above the surface of Earth. Also referred to as the *hydrological cycle*. (14.2)

water rights. The legal right of a user to use water from a particular source. (14.2)

watershed. An area of land that is drained by a major body of water. (15.3)

water-soluble. The state of something that is easily dissolved in water. (14.2)

weaned. When a young animal is no longer nursing. (11.2)

weather. The measure of the current temperature, precipitation, wind, and humidity of an area. (14.3)

weathering. The process of eroding rock over time, primarily through the action of rainfall and snowfall. (15.2)

weather pattern. Repeated incidences of similar temperatures and precipitation. (14.3)

weed. Any plant that is not cultivated by humans and growing in competition with desired or cultivated plants. (13.1)

western riding. The discipline of horse showing in which horses are shown in a western saddle. Includes trail, western pleasure, reining, and cutting. (10.3)

wether. A castrated male sheep or goat. (11.3, 11.4)

wetland. An ecosystem or area of land that is saturated with water throughout the year, either seasonally or permanently (it may dry out periodically). (14.1)

wetland creation. The process of developing a wetland where one did not previously exist. Also referred to as a *constructed wetland*. (15.3)

wetland enhancement. The process of reconstructing and rehabilitating a previously existing wetland. Also referred to as *wetland restoration*. (15.3)

wetland restoration. The process of reconstructing and rehabilitating a previously existing wetland. Also referred to as *wetland enhancement*. (15.3)

wet mill process. An *ethanol* production method in which the dry *feedstock* is soaked in liquid to more readily separate its components before being ground. The ground feedstock is processed to separate the oil, fiber, gluten, and starch. The starch and remaining liquid is processed through fermentation into ethanol, dried and sold as modified corn starch, or processed into corn syrup. (7.2)

wheel line irrigation system. Crop watering system in which an irrigation pipe runs through the center of a series of wheels that moves across the field in a line parallel to the sides of the field. (15.5)

whelping. The act of dogs giving birth. (12.1)

white line. The point at which the outer layer of a horse's hoof attaches to the outer wall of the hoof. It is used as a guide for trimming and setting nails. (10.3)

whole bird. Poultry sold as an uncut bird, with or without the giblets. (11.1)

whole fryer. The entire chicken with packaged giblets. (11.1)

wild animal. An animal that lives in a natural setting, must get food and water on its own, and is not cared for by humans. (12.1)

wildlife inventory. A count of all mammals, birds, reptiles, fish, and amphibians in a given area. (16.2)

windbreak. Row of trees planted on the western edge of a field that creates a natural barrier to slow the flow of wind and minimize *wind erosion*. (15.2)

wind energy. The *kinetic energy* of air in motion. Also referred to as *wind power*. (7.1)

wind erosion. The erosion of the soil caused by wind. (15.5)

wind farm. A group of *wind turbines* in the same location used for production of electricity. (7.1)

wind power. The *kinetic energy* of air in motion. Also referred to as *wind energy*. (7.1)

windrow. A long line of raked hay or sheaves of grain laid out to dry in the wind. (13.2, 13.4)

wind turbine. A modern version of the windmill used to convert kinetic wind energy into mechanical power or electricity. Also referred to as an *air turbine*. (7.1)

wing. The least expensive poultry cut with little meat. (11.1)

wire strippers. A specialized set of pliers used to strip insulation from various sizes of electrical wire. (7.3)

WOG. A whole fryer (chicken) packaged without the *giblets*. (11.1)

wood alcohol. A *biofuel* derived from natural gas, coal, plant vegetation, and agricultural crop *biomass*. Also referred to as *methanol*. (6.2)

wood chisel. A long-bladed hand tool with a beveled cutting edge that is struck with a hammer or mallet to cut or shape wood (7.3)

wooden mallet. A handheld, wooden driving tool designed for use with wood chisels. (7.3)

woody stem. A rigorous plant *stem* that is continually growing upward and outward. The vascular system grows in large rings around the plant stem. (13.1)

work. The application of force over distance. (7.5)

world hunger. The combined scarcity of food for all countries in the world. (9.2)

wrecking bar. A prying tool made of solid metal with a hooked end and a claw designed to open or break apart materials. May also be used to pull out fasteners. Commonly referred to as a *crowbar*. (7.3)

wrench. A hand tool designed to apply torque and either tighten or loosen nuts and bolts. (7.3)

written communication. The transfer of information through writing. (2.2)

X

xenotransplantation. The process of grafting or transplanting tissues and organs from one species to another. (12.4)

xylem. Vascular plant tissue responsible for carrying water and nutrients up the plant. (13.1)

Y

yogurt. A semisolid food made of milk and milk solids fermented by two added cultures of bacteria (*Lactobacillus bulgaricus* and *Streptococcus thermophilus*). (10.2)

YouthRules!. Safety initiative launched by the Department of Labor to promote positive and safe work experiences for young workers. (4.1)

Z

zigzag rule. A measuring tool composed of strips of wood joined by rivets so as to be foldable. All of the opening and closing parts are in parallel planes. Also referred to as a *folding rule*. (7.3)

zone of elongation. The rapidly growing section of a plant's root where cells are rapidly dividing through *mitosis*. (13.1)

zone of life. Earth's ecosystems as a whole, or the accumulation of all of the ecosystems on the planet. Also referred to as the *biosphere*. (14.2)

zoology. The scientific study of the behavior, structure, and physiology of animals. (5.2)

zoonoses. Diseases and infections that can be transmitted from animals to humans. May also be referred to as zoonotic diseases. (12.1)

Index

B

C

D

E

F

G

H

I

J

K

L

M

N

O

P

Q

R

S

T

U

V

W

X

Y

Z